W9-BAT-868

FOURTH EDITION

SYSTEMS ANALYSIS AND DESIGN METHODS

Instructor's Edition

JEFFREY L. WHITTEN

Professor

LONNIE D. BENTLEY

Professor

Both at Purdue University

West Lafayette, IN

Special contributions by

Kevin C. Dittman

Assistant Professor, Purdue University

Boston, Massachusetts Burr Ridge, Illinois Dubuque, Iowa
Madison, Wisconsin New York, New York San Francisco, California St. Louis, Missouri

PREFACE TO INSTRUCTOR'S EDITION

PURPOSE

Once again, we are pleased to offer this Instructor's Edition of *Systems Analysis and Design Methods.* New to the Instructor's Edition are course planning guidelines. We'll give you some ideas about how to select and organize chapters and modules for different types of systems development courses and audiences. Then, for each chapter and module, we provide the following:

- An instructor's overview.
- Course sequencing options for that chapter.
- A summary of changes in the fourth edition and why we made them.
- A topical outline that further explains changes and rationale.

In this edition we omitted the key terms and common synonyms due to mixed feelings of the third edition adopters.

SUMMARY OF CHANGES FOR FOURTH EDITION

We believe that we have preserved all of the features you liked in the previous editions. To help you with your adoption decision, we summarize the changes as follows:

- The fourth edition returns to the legacy of its namesake, *systems analysis and design methods*. The coverage of systems planning, systems implementation, and systems support is downsized, and primarily serves to place those peripheral subjects into the context of systems analysis and design. The basic philosophical change was that, "if we want to teach anything worthwhile, we must not try to teach everything." In order to maintain currency in analysis and design, we elected to summarize all other phases.
- The *pyramid* model for conceptual foundations has served us well for three editions, but now gives way to a *matrix* model that more clearly reinforces the concepts of information systems and their development. The Zachman *Framework for Information Systems Architecture* continues to organize the subject's conceptual foundations. The "network" column of the third edition and Zachman's framework has been renamed "geography" to avoid any confusion with "computer networking and technology." Finally, an "interface" column has been added to reflect both the emergence of the graphical user interface and the importance of systems integration.
- This edition further emphasizes systems analysis and design techniques for developing *client/server* or *distributed computing applications*. The mainframe is considered just another server.
- The fundamentals unit (Part One) has been downsized from five chapters to three as a direct response to adopters' requests to accelerate the path into the skill units and chapters.
- The opening chapter (1) relates the overall subject matter to current business and information strategies such as *information strategy planning, total quality management, business process redesign,* and *continuous process improvement.*
- The *information systems building blocks* chapter (2) has been downsized at the request of third edition adopters. Its sole intent in this edition is to introduce the Zachman framework that will organize the concepts, tools, and techniques taught in the course.
- The *systems development life cycle* and *methodology* chapters have been integrated—yet simplified. Life cycle/methodology coverage has been modernized; however, it returns to a derivative of the second edition life cycle that offered a simpler and cleaner "problem-solving approach" to the subject. Details about the phases are deferred to the analysis, design, and implementation chapters that introduce (or conclude) their respective units.
- The *CASE* chapter has been deleted from the fundamentals unit; however, CASE is introduced early and reinforced often throughout the text.
- This edition replaces DOS-based CASE tools with *Windows*-based equivalents. Popkin's *System Architect* provides an inexpensive upper-CASE environment that supports *both* structured and

object-oriented methods. Microsoft's *Visual Basic* provides a popular lower-CASE, object-based, rapid application development (RAD) environment that supports design and prototyping. Very similar functionality can be achieved with *Delphi*, *Powerbuilder*, and other visual programming environments.

- The *data modeling* chapter (5) now uses the more popular Martin-style entity relationship diagrams that are common to information engineering.
- The *process modeling* chapter (6) has undergone significant revisions for the first time since the first edition. The top-down approach to DFDs is replaced by the more modern event-driven approach as recommended by McMenamin, Palmer, Yourdon, and many information engineering-based methodologies.
- The *network modeling* (as in data and process distribution) chapter (7) is improved by the incorporation of association matrices from information engineering.
- Object-oriented analysis and design are truly emerging. In the previous edition, we retreated from extensive coverage because we did not sense that any one OO paradigm had developed the requisite industry following. *Object modeling* (chapter 8) is formally taught in this edition as an *alternative* to data and process modeling. After studying competing methods and paradigms, we have elected to synthesize concepts from several methodologies, but present them within the context of Rumbaugh's *object modeling technique (OMT)* that had a wider industry acceptance than either the Booch or Coad/Yourdon techniques. We also incorporated elements from the very popular use case techniques as developed by Jacobson.

  Like many others, we greatly anticipate the arrival of the so-called unified *object modeling language* that is being developed through the collaboration of Rumbaugh, Jacobson, and Booch. Even Ed Yourdon (of the Coad/Yourdon technique fame) has conceded in his newsletters that the unified language will probably prevail.
- The *project repository* chapter from third edition has been effectively merged into the data, process, network, and object modeling chapters.
- The design unit (Part Three) pervasively applies a *rapid application development (RAD)* technique using Microsoft's *Visual Basic* for design prototyping. The conceptual framework for this approach is also compatible with other RAD technologies such as Powersoft's *Powerbuilder* and Borland's *Delphi*.
- The *analysis-to-design transition* coverage from third edition has been modified to better support distributed computing client/server environments. The chapter title has been changed to *Application Architecture and Process Design* (Chapter 10).
- The *database design* chapter (11) more adequately addresses modern *distributed database technology* (using *Oracle* as an example), and *data distribution* techniques that are becoming commonplace in client/server computing environments.
- *Input and output design* have been split into separate chapters (12 and 13). Each chapter has been significantly modernized to reflect newer, pervasive technologies and paperless design techniques.
- The *user interface design* chapter (14) has been completely modernized to focus on GUI design strategies and techniques using the popular *Windows* client as an example. Examples of such GUI designs are provided through *Visual Basic*, but they apply to almost any GUI-based development environment.
- The *structured design* chapter (15) has been both simplified and modernized to better teach transition from data flow models to structure charts using a relatively simple and elegant algorithm.
- An *object-oriented design* chapter (16), using the aforementioned OMT technique, has been added to complement the object modeling chapter (8) in the analysis unit (Part Two). It is reasonable to expect that object methods will eventually replace structured methods, but that may not happen until the fifth or sixth edition.
- The *project management* module (A) has received its first-ever complete revision. The revision is based on computer-based project management using Microsoft *Project* tools, and techniques.
- An entire module (D) has been added to teach *joint requirements planning/joint application development* (JRP/JAD) techniques as they apply to systems analysis and design.

PEDAGOGICAL USE OF COLOR

The fourth edition continues the use of full color applied to an adaptation of Zachman's *Framework for Information Systems Architecture.* Based on adopter feedback, we have reduced the number of base colors to four (with a fifth color used only in the two *object-oriented* chapters). The color mappings are shown in the student edition preface. Throughout the textbook, we use the colors to map concepts from the conceptual framework to the various systems development models.

TEXTBOOK ORGANIZATION

Systems Analysis and Design Methods, fourth edition, is divided into five parts. Past experience indicates that instructors can omit and resequence chapters as they feel is important to their audience. Every effort has been made to decouple chapters from one another as much as possible to assist in resequencing the material—even to the extent of reintroducing selected concepts and terminology. The Instructor's Edition and the Instructor's Guide should clearly indicate any exceptions to this general rule.

Part One, "The Context of Systems Analysis and Design Methods," presents the information systems development situation and environment. The chapters introduce the student to systems analysts and the systems development team, information systems building blocks, and a contemporary systems development methodology. Part One has been redesigned so that it can be covered more quickly.

Part Two, "Systems Analysis Methods," covers the front-end life cycle activities, tools, and techniques for analyzing business requirements for an improved system. Coverage can be restricted to either structured or object-oriented techniques, but both are recommended since industry is caught in what should prove to be a lengthy transition from structured methods to object-oriented methods. The network modeling chapter (7) is optional, but recommended in light of client/server application trends that will distribute data, processes, and objects throughout a distributed computing network.

Part Three, "Systems Design and Construction Methods," covers the middle life cycle activities, tools, and techniques. It includes coverage of both general and detailed design with a particular emphasis on design decision making, human factors, and rapid application development.

Part Four, "Beyond Systems Analysis and Design," is a capstone unit that places systems analysis and design into perspective by surveying the back-end life cycle activities. The systems implementation and support chapters (17 and 18) are intended primarily to bridge this textbook's systems analysis and design coverage with the students' prior programming experiences.

Part Five, "Cross Life Cycle Activities and Skills," introduces material that spans multiple phases of systems development. Chapters are replaced by modules that can be interwoven into the course at various points in time (or assigned as supplemental reading). The modules include project and process management (A), fact-finding and information gathering (B), feasibility analysis and cost-benefit analysis (C), joint application development (D), and interpersonal skills and communications (E). The Instructor's Edition provides threads in appropriate chapters that might be enhanced by introducing material from these modules.

See the Instructor's Annotated Edition or the Instructor's Guide for more information on how to use this book in a variety of course formats and sequences.

OTHER INSTRUCTOR-RELATED SUPPLEMENTS AVAILABLE FOR THIS BOOK

It has always been our intent to provide our adopters with a complete course, not just a textbook. We are especially excited about this edition's comprehensive support package. The supplements for this edition include the following components.

Instructor's Guide

For those instructors more comfortable with a separate instructor supplement, an Instructor's Guide is available. It provides the same front- and back-matter as the Instructor's Edition, but also provides instructional notes embedded within a "notes pages" printout of complete Microsoft *PowerPoint* presentations. The electronic versions of these presentations will be either available through your McGraw-Hill sales representative, or downloadable from the textbook's World Wide Web site.

Test Bank

A written test bank covering all the chapters and modules includes more than 3,000 questions. Each chapter and module has questions in the following formats: true/false, multiple choice, sentence completion, and matching. The test bank and answers are cross-referenced to the page numbers in the textbook. An electronic version of the test bank is available for the creation of customized exams.

Laboratory CASE Tools

Popkin Software & Systems, Inc. is making available a *University Edition* program for their CASE product, *System Architect. System Architect* provides *Windows*-based CASE tools for all of the techniques taught in this edition, inclusive of object-oriented analysis and design. The *University Edition* software provides identical functionality to the commercial product to support large student projects, at a nominal cost. This software requires a network. To obtain the software, contact:

University Edition Program Manager
Popkin Software
11 Park Place
New York, NY 10007
(212) 571-3434

Academic grant programs are available for other CASE tools including Visible Systems *Visible Analyst.* Your McGraw-Hill representative can provide further details.

RAD Tools

Academic grants or educational discounts are widely available for RAD software development environments such as *Visual Basic, Powerbuilder,* and *Delphi.* All of these development environments support visual construction of graphical user interfaces and various degrees of database connectivity and rapid application development.

Internet and Electronic Mail Support

Soon after the publication of this book, we will establish an Internet support site for adopters. We plan to use this site to share adopter materials and insights, post corrections and new or revised supplements, post news of interest for students, solicit instructor recommendations for fifth edition changes, and facilitate email communication between adopters. The email and Internet addresses follow:

Email:

jlwhitten@tech.purdue.edu

ldbentley@tech.purdue.edu

Internet:

www.tech.purdue.edu/textbooks/sadm

www.mhhe.com

Jeffrey L. Whitten
Lonnie D. Bentley

C O N T E N T S

Instructor's Introduction I–viii

PART ONE

The Context of Systems Analysis and Design Methods I–1

1 The Modern Systems Analyst I–2
2 Information Systems Building Blocks I–6
3 Information System Development I–10

PART TWO

Systems Analysis Methods I–15

4 Systems Analysis I–17
5 Data Modeling I–22
6 Process Modeling I–25
7 Network Modeling I–29
8 Object Modeling I–31

PART THREE

Systems Design and Construction Methods I–33

9 Systems Design and Construction I–34
10 Application Architecture and Process Design I–38
11 Database Design I–42
12 Input Design and Prototyping I–45
13 Output Design and Prototyping I–48
14 User Interface Design and Prototyping I–50
15 Software Design I–52
16 Object-Oriented Design I–54

PART FOUR

Beyond Systems Analysis and Design I–57

17 Systems Implementation I–58
18 Systems Support I–61

PART FIVE

Cross Life Cycle Activities and Skills I–63

A Project and Process Management I–65
B Fact-Finding and Information Gathering I–68
C Feasibility and Cost-Benefit Analysis I–70
D Joint Application Development I–72
E Interpersonal Skills and Communications I–74

PLANNING A SYSTEMS ANALYSIS AND DESIGN COURSE

COURSE PLANNING PARAMETERS

Course planning is a very personal task. And we don't intend to impose on your academic freedom. Therefore, these guidelines should be considered just that—guidelines! They are offered to you in the interest of sharing ideas. And we especially offer them to first-time instructors of the course, part-time instructors whose course preparation time may be limited, and graduate teaching assistants whose teaching experience may be limited. Therefore, we ask your indulgence as we may cover items that, to the experienced teacher, may seem obvious.

At many schools, most of the questions that will be addressed are not left to the discretion of the individual instructor. However, we will cover all the questions and leave it to you to deal with those which are pertinent to your situation. The questions are as follows:

- What is the duration of your course? Our book can be used in any of the following: one quarter, one semester, two quarters, or two semesters. Several guidelines will be presented for each of these options.
- To what degree is your course balanced toward systems analysis or systems design? We'll call this the course *emphasis*. Our market research and reviews have revealed three general approaches:
  - Courses that emphasize **systems analysis** over systems design. These courses introduce and cover the entire systems development life cycle; however, they tend to emphasize the systems analysis and problem solving aspects of the life cycle. Frequently, they are followed by a sequel course that emphasizes systems design and implementation. The course frequently includes many nonmajors since systems analysis addresses business issues of systems development.
  - Courses that emphasize **systems design** over systems analysis. This is usually a sequel to a first course on *systems analysis.* Because the issues become more technical, there may be fewer business majors enrolled (unless they are pursuing a minor in information systems).
  - Courses that offer a fairly balanced treatment of systems analysis and design.
- What level of skill are you trying to teach to the students? This is often a function of the following:
  - **Enrollment**—the larger the class, the less likely the instructor is to require extensive application of tools and techniques to a sizable project. The grading workload could become unbearable.
  - **Audience**—the greater the number of noncomputing majors, the less likely the instructor will focus on the application of tools and techniques. Nonmajors typically need an understanding of the *process* of analysis and design. They tend, therefore, to be less interested in mastering the *tools and techniques* of analysis and design.

    With these factors in mind, our research has noted two general skill levels for the course:
  - **Application**—the student is expected to *apply* specific tools and techniques to an analysis and/or design project.

- **Understanding**—the student is expected to understand the process, tools, and techniques of analysis and design, but not necessarily apply them to projects.

Our textbook was designed to meet the needs of all of the above course options. That, in part, explains the large size of the textbook (even though it was downsized for this edition). It is our feeling that the instructor chooses the level of coverage, but the students should also be left with a suitable post-course reference.

COURSE TOPICS AND COVERAGE

We feel very strongly that we have designed the book to be useful in different formats. Here's how! Given the three course planning questions, use the *Course Planning Template* in Figure A. It is intended as a point of departure for your own planning. Before you study the template, you need to understand the assumptions and notations used.

Assumptions

1 quarter = 10 weeks of instruction (not including final exams)

1 semester = 15 weeks of instruction (not including final exams)

1 week = 3 class meetings of 50–60 minutes' duration each

You may need to adjust the template if your calendar and schedule differ from these assumptions.

Every class instructional hour was not scheduled. A buffer equal to one class for every five one-hour lessons was added. The buffer can be used for catch-up, reviews, exams, duplicate coverage when students encounter problems, help sessions, sick days, conference leaves, and so on.

Space is provided in the template for you to do your own course planning. Also, notice that the template allows you to include topics not covered in our textbook. Before using the template, you need to answer the following questions:

Number of class meetings in term	______	
– 1/5 number of class meetings in term	(______)	* see Note 1 below
– Number of exams during class meetings	(______)	
– Number of meetings for local or instructor needs	(______)	
= Number of class meetings available for lessons	______	

*NOTE 1: As a rule of thumb, it is recommended that you reserve 1 overhead period (e.g., for catch-up, reviews, questions, supplementary discussions, etc.) for every five regular class meetings. As the course matures, a schedule becomes more firm and you can schedule most regular class meetings.

Given the above parameters, Figure A shows different combinations of suggested topics based on audience and desired student skill levels. This should serve as a point of departure for your own planning based on the number of available class meetings as described above and any unique requirements (e.g., topics not included in our textbook).

A COURSE SYLLABUS TEMPLATE

This may be the first time you've taught this particular course. Or perhaps you are one of the growing group of part-time or adjunct faculty. Then again, you may be a graduate teaching assistant. Or maybe you just want some new ideas. In this section, a skeleton course syllabus is presented to show some of the possibilities. Once again, the skeleton syllabus doesn't dictate how the course should be structured—it only provides some guidelines for your own course planning.

The discussion that follows includes a number of suggestions intended to help part-time and less experienced instructors. They are based on questions we're frequently asked and hints we wish we had known when we were in that same position. It is not intended to be condescending to those of you with significant teaching experience!

FIGURE A

CHAPTER and TOPIC	Course Duration	Quarter	Quarter	Quarter	Semester	Semester	Semester	YOUR COURSE
	Course Emphasis	Analysis	Design	Both	Analysis	Design	Both	
	Targeted Student Skill Level	Application	Application	Knowledge	Application	Application	Knowledge	
		24 lessons	24 lessons	24 lessons	36 lessons	36 lessons	36 lessons	Lesson numbers
1 The Modern Systems Analyst		1	prerequisite	1	1	prerequisite	1	
2 Information System Building Blocks		2, 3	1 (review)	2, 3	2, 3	1 (review)	2, 3, 4	
3 Information System Development		4, 5	2 (review)	4, 5	4, 5, 6	2 (review)	5, 6, 7	
4 Systems Analysis		8, 9	3 (review)	8, 9	9, 10, 11	3 (review)	10, 11	
5 Data Modeling		14, 15, 16	prerequisite	10	16, 17, 18, 19	prerequisite	14	
6 Process Modeling		17, 18, 19	prerequisite	11	20, 21, 22, 23	prerequisite	15, 16	
7 Network Modeling		20, 21	prerequisite	12	24, 25	prerequisite	17	
8 Object Modeling		omit	omit	omit	26, 27, 28, 29	prerequisite	18	
9 Systems Design and Construction		22	4, 5	15, 16	30, 31	4, 5, 6	20, 21	
10 Application Architecture and Process Design		omit	7, 8	17	omit	7, 8, 9	24	

FIGURE A *Continued*

CHAPTER and TOPIC	Course Duration	Quarter	Quarter	Quarter	Semester	Semester	Semester	YOUR COURSE
	Course Emphasis	Analysis	Design	Both	Analysis	Design	Both	
	Targeted Student Skill Level	Application	Application	Knowledge	Application	Application	Knowledge	
		24 lessons	24 lessons	24 lessons	36 lessons	36 lessons	36 lessons	Lesson numbers
11 Database Design		omit	11, 12, 13	18	omit	14, 15, 16, 17	25	
12 Input Design and Prototyping		omit	14, 15	19	omit	18, 19, 20	26	
13 Output Design and Prototyping		omit	16, 17	20	omit	21, 22, 23	27	
14 User Interface Design and Prototyping		omit	18, 19, 20	21	omit	24, 25, 26, 27	28	
15 Software Design		omit	21, 22	omit	omit	28, 29	29	
16 Object-Oriented Design		omit	omit	omit	omit	30, 31, 32, 33	30	
17 Systems Implementation		23	23	22	32	34, 35	31, 32	
18 Systems Support		24	24	23	33	36	33, 34	
A Project and Process Management		6, 7	omit	6, 7	7, 8	10 (review)	8, 9	
B Fact-Finding and Information Gathering		10, 11	omit	13	12, 13	prerequisite	12	

FIGURE A *Concluded*

CHAPTER and TOPIC	Course Duration	Quarter	Quarter	Quarter	Semester	Semester	Semester	YOUR COURSE
	Course Emphasis	Analysis	Design	Both	Analysis	Design	Both	
	Targeted Student Skill Level	Application	Application	Knowledge	Application	Application	Knowledge	
		24 lessons	24 lessons	24 lessons	36 lessons	36 lessons	36 lessons	Lesson numbers
C Feasibility and Cost-Benefit Analysis		omit	9, 10	14	omit	11, 12, 13	22, 23	
D Joint Application Development		12, 13	6	omit	14, 15	prerequisite	13	
E Interpersonal Skills and Communications		assigned reading	assigned reading	24	34, 35, 36	omit	19	
ASSESSMENT		quizzes, exams, homework	quizzes, exams, homework	quizzes, exams	quizzes, exams, homework, project	quizzes, exams, homework, project	quizzes, exams, homework	

Before you use our template, be aware that many institutions and departments have well-defined standards and policies for course syllabi and outlines. Regardless, almost all educators agree that it is a good idea to have as complete a course syllabus prepared as possible, *before the first class meets.* Think of the course syllabus as a contract between you and your students.

The following skeleton outlines some of the sections that might be included in the syllabus.

Course Number and Title

Self explanatory.

Term and Year

In our experience, it is important to date each syllabus so that the policies for any given term and year can be remembered. This is especially useful when clearing incomplete grades given to students.

As a course progresses, you will undoubtedly conceive ideas to improve the next offering. Unfortunately, you may forget those great ideas if you don't record them promptly. Try this! Make a separate copy of your final syllabus as a working copy of the next syllabus. As the course progresses, you can record notes for improved requirements, exercises, policies, etc., directly in the copy and delay formatting until the next semester approaches.

Instructor

Some faculty like to publish the home phone numbers; most do not. If you do so, we suggest that you be specific in telling students whether you'll accept late evening calls or calls at your primary business office.

Office Hours

Your school may have policies or guidelines on minimum office hours.

Prerequisites (and Corequisites)

Many registration systems, including our own, don't automatically enforce prerequisites. And many course catalogs aren't current. State the latest prerequisites and exceptions. We've found that our students find it helpful to know what specific concepts, knowledge, and skills are expected—as opposed to mere course numbers.

Course Materials

State both required and recommended textbooks and supplements. Indicate the availability of any materials that you have placed on reserve in your local library. Also state any incidental materials that must or should be acquired (e.g., forms, templates, binders, etc.). Finally, warn your students about any required or recommended duplication expenses that may be incurred in the course.

Systems Analysis and Design Methods, fourth edition, is available with several student software bundles that may be useful for the course.

Course Description

School catalogs are notorious for concise, but cryptic, course descriptions. If permitted by school policy, consider publishing a more thorough and up-to-date course description in the syllabus. By the way, employers find such descriptions very useful.

Course Goals and Objectives

As in project management, we like to state a single, high-level goal for the course. The course goal is then expanded upon by a set of student-oriented behavioral objectives that describe what the student *should* know and/or what the student should be able to do after successfully completing your course. Behavioral objectives are a time-honored and respected means of learning assessment in the education industry. If you really want to develop an educationally sound approach to objectives, research Bloom's *Taxonomy of Learning Objectives* (documented in many education books and papers).

Course Requirements

In this section, you describe the learning activities and assessment mechanisms to be used in the course. These are the graded elements of the course. The selection of course requirements is usually a function of teaching philosophy, and level of skill you want to teach the students. Also, the number of students enrolled in the course can have a significant impact on your grading workload and influence your choice of requirements. We generally include some combination of the basic options in our majors-oriented courses:

- **Reading**—usually assessed in rote memory quizzes over reading material; primarily to encourage and reward keeping up with the reading assignments.
- **Concepts**—usually assessed in more comprehensive examinations; to determine whether the students "understand" the subject matter, concepts, and applications beyond the rote memory level.
- **Application**—usually assessed in individual and/or team exercises; to determine if students can repeat the application of concepts and techniques in a fairly mechanical fashion to small and controlled problem domains. This may be sufficient for survey courses and non-IS majors courses.
- **Integration and adaptation**—usually assessed in application-oriented semester projects that require students to integrate concepts and techniques, and possibly adapt them to larger, less controlled (or uncontrolled) cases or live projects. Live projects, even if small and constrained, tend to result in greater "learning."
- **Expansion**—usually assessed in a research-oriented project that requires students to leap beyond the course and textbook to study some course topic in greater detail; usually implemented as extra credit in an undergraduate course, or expectation in a graduate course.

Grading Policies

Arguably, this may be the most important section of any syllabus. Students will want to know how you will award grades, and a strong "management-oriented" argument can be made for their point of view. Well-defined grading policies, if enforced to the letter, are your best deterrent to the unpleasantness of formal grade appeals.

If you haven't already done so, you should research local grading customs and policies. How would your superiors feel if you gave a large percentage of higher than "normal" grades? There is this rather curious custom in American education that says most students are average and should receive Cs (the infamous normal distribution curve). On the other hand, some educators argue that if most of your students earn Cs, then you've earned a C as a teacher, and they also know where you stand. Other educators argue that grade inflation has become a serious issue, and that employers no longer know how to interpret grades.

However you decide to assign final grades, the following items should be specified.

- What weight will the various items be assigned (quizzes, tests, exercises, projects, etc.)? We find (and are disturbed that) many students struggle with percentages. They prefer course weights we expressed in points.
- What is your policy for missed quizzes and exams? Can they be made up?
- What penalties will be assessed on late submission of exercises and projects? A word of advice: Follow your own policies. Don't extend deadlines or waive late penalties if you are not willing to compensate those who get their work done on time.
- Under what circumstances will you (or can you) give an incomplete grade? Our colleagues from other institutions confirm our perceptions that incompletes are increasingly requested (and in some cases awarded) to cover up failure in a course.

Teamwork Policies

If your course requirements include team projects, you will need to specify policies and procedures for managing team problems. You will probably discover that some team members pull less than their fair share of the workload. You will need to develop some methods, both informal and formal, for mediating such problems. And in some cases, you may have to implement methods for adjusting grades accordingly.

We have tried numerous approaches to the latter problem including: (1) instructor-assessed milestone penalties, (2) peer evaluations with project penalties, (3) allowing teams to "terminate" a problem student from a team, (4) reassigning problem students to other teams with a penalty to discourage repeat offenses, (5) assigning the terminated team member to work alone on the project (with penalty), and (6) awarding a project score of zero for getting terminated from a team (which almost always results in a failing grade for the course). Each approach has advantages and disadvantages—so much so that we aren't sure we will ever find an ideal solution. But we remain committed to the educational value of the teamwork experience, good or bad, to the student's understanding of group behavior and dynamics.

Other Course Policies

Describe any policies such as cheating issues and consequences, withdrawals, attendance, etc.

Course Outline and Schedule

A course outline helps the students to appreciate the structure and organization of course topics as part of the whole course. We find it useful to place the outline on a transparency and revisit it frequently as a reminder of where we came from and where we are going in the course. Until your outline becomes stable from term to term, we suggest you annotate it as "subject to change."

A course schedule of reading assignments, lessons, exercise due dates, and project milestone due dates is the mark of a mature syllabus. It usually takes one or two semesters to get a new course stable enough to justify a schedule. But once established, it is a useful student supplement. If you do publish a schedule, we highly recommend that it be annotated as "subject to change" since even the most mature courses must adapt to events such as instructor sick days, snow recesses, and other events beyond the instructor's control.

Conclusion

Once again, we apologize if the aforementioned guidelines seem condescending. They were intended for those who have had less teaching experience. Remember your first time?

PART ONE

THE CONTEXT OF SYSTEMS ANALYSIS AND DESIGN METHODS

Part One of the textbook is about **the context of systems analysis and design methods**. That context is defined in terms of the three Ps: *participants, product,* and *process*. The participants are presented in Chapter 1 from the context of the **systems analyst** who plays a crucial role as the facilitator of information systems development. (User roles are also discussed in detail for non-information systems majors.) The product, **information systems,** is presented in Chapter 2 from the perspective of architectural building blocks. The process of building information systems, **information system development,** is presented in Chapter 3.

COURSE ORGANIZATION AND SEQUENCING

Part One is intended as an essential unit for a first course or a review unit for a second course in systems development, systems analysis, or systems design. Although we like to introduce the systems analyst and participants first (Chapter 1), many adopters have reported that they prefer to begin with Chapter 2. Both approaches work equally well. Since Chapter 2 introduces a graphical framework used throughout the book, it should probably precede Chapter 3 (which immediately reinforces the framework).

WHAT'S DIFFERENT HERE AND WHY?

Downsizing! That was the message we received from third edition adopters. In other words, adopters requested a downsized Part One to provide a faster path into the tools and techniques chapters. In response, we have reduced Part One from five chapters to three! Specifically, here's what we did:

- Chapter 1 was basically modernized. For the benefit of non-IS majors, we have tried to slightly expand coverage of system owner and user roles in systems development.
- Chapter 2 was downsized by (1) condensing the information systems review, and (2) eliminating all the preview figures of the various models that will be taught in the book. Many adopters indicated their students were somewhat overwhelmed by those models, which, in fact, they are not expected to learn until later in the book.
- Chapter 2 introduces a new version of our conceptual framework. The pyramid model has been replaced by a matrix model. The matrix model will allow students to see the entire framework—not just two aspects at a time, as in the pyramid model. This provides a better pedagogical tool throughout the book. An additional focus, INTERFACES, is provided because of the increased emphasis on user and system interfaces as system components become increasingly distributed.
- Chapter 3 integrates material from the previous edition's second, third, and fourth chapters. The emphasis is a methodology-based life cycle. We eliminated specific chapters on methodologies and CASE because adopters reported that the material, while interesting, did not really sink in until the later chapters anyway. (For example, students had trouble appreciating upper-CASE tools because they had not yet studied modeling tools and techniques.)

1

THE MODERN SYSTEMS ANALYST

OVERVIEW

Chapter 1 introduces the *systems analyst* and other participants and partners in the information system development game. It has long been our perception that students want to begin every course with an assessment of how this course will impact their own career. For that reason, we begin with an audience-, people-, and education-oriented chapter instead of the traditional product (information system) or process (systems development life cycle) chapter. Through three previous editions, this approach has served us well by immediately providing career context:

- Am I an aspiring analyst who will have to become an expert in this subject?
- Am I an aspiring information technology specialist (such as a database professional or network specialist) who will need to understand these techniques in the context of my interactions with systems analysts?
- Am I a future end-user of information systems and computer applications who will have to participate in analyst-facilitated teams to deliver information services and solutions that will directly affect my chosen career path and job responsibilities?

The focus in this chapter is placed on the systems analyst, not as the most important player, but instead as the facilitator whose responsibilities cross the entire systems problem-solving process. The most significant fourth edition improvement has been to improve the coverage and interests of the analyst's customers and partners.

CHAPTER-TO-COURSE SEQUENCING

Reviews of the third edition were inconclusive as to whether this chapter should be the first or second in the book. Some reviewers agreed with our premise that audience (analysts and users) should precede product (information system). Others felt that product should precede audience. While we elected to retain the current organization, it should be noted that Chapter 1 can either precede or follow Chapters 2, *and* even Chapter 3. For adopters who elect to cover Chapter 1 first, but still feel the need for some product context, we rewrote the chapter introduction to provide that context in the form of some basic contextual definitions for product and process.

How do we handle sequencing at Purdue? Our first course is survey oriented. Currently, we assign and quiz Chapter 1, but do not extensively lecture on the topic beyond the first class meeting (for which the majority of time is spent covering requisite topics such as course requirements and policies). Thus, the first true lecture in our first course actually covers Chapter 2.

THE SOUNDSTAGE RUNNING CASE STUDY

Through three editions, there has existed some confusion as to whether an episode of the running case study was part of a chapter, a preface to one or more chapters (correct!), or a sequel to one or more chapters. In this fourth edition, we have eliminated the confusion by placing the SoundStage episodes at the beginning of each chapter to introduce the subject matter.

This chapter's episode is intended only to introduce the case study. Subsequent episodes will include discussion questions.

Adopters of our previous editions should note that most of the chapter opening minicases and vignettes (along with their discussion questions) have been retained as end-of-chapter Minicases.

WHAT'S DIFFERENT HERE AND WHY?

1. We modified the chapter introduction to quickly introduce key terms based on the book and chapter titles. These definitions for **systems analyst, information system, computer application, systems analysis,** and **system design** provide immediate focus and context for the reader.
2. We have replaced most references to *computers* and *computer technology* with the more contemporary term **information technology.** We have embraced the popular definition of information technology as the "combination of computer technology (inclusive of hardware, software, and peripherals) and telecommunications technology (inclusive of local, wide, metropolitan, and global networks for data, voice, and images)."
3. We have modernized the chapter by recognizing the new roles of **business analyst** (from the user community) and **application/systems analyst** (from the information services community). Since writing the previous edition, we have had the happy privilege of teaching systems analysis and design to newly assigned business analysts at our university. Business analysts bring a unique and overdue perspective to the study and application of this subject.
4. The impacts of downsizing, **decentralization of information services,** and **outsourcing** forced us to slightly expand our coverage about where systems analysts work. Because different organizations have matured on different timelines, we show traditional, modern, and outsourced organizations.
5. There was some adopter and reviewer concern that we shortchanged *users* in the third edition. For this chapter, we expanded coverage of the stakeholders in systems development. In particular, we expanded the scope of the term *user* to include external businesses and consumers to reflect the modern trend of extending the reach of information systems and computer applications beyond the walls of the business itself. We also briefly described their roles in the system problem-solving process.
6. We have added coverage of major business trends and influences on the systems analyst. Examples include **total quality management, business process redesign, continuous process improvement, globalization of the economy,** and **empowerment.** In each case, we describe the impact on systems analysts and systems development.
7. We modernized the career preparation section with special emphasis on **object technology, visual programming,** and the aforementioned business trends.

CHAPTER OUTLINE

Section headings and subheadings are indicated in boldface. Key terms are italicized.

1. **Who should read this book?** We introduced the chapter with a series of contextual definitions.
   a. *Information system.*
   b. *Computer application.*
   c. *Systems analyst.*
   d. *Systems analysis.*
   e. *Systems design.*
2. **The systems analyst as a modern business problem solver.** Because the tools and techniques taught in this book are practiced by, or with, a systems analyst, the role of the analyst is explored in some depth.

a. **Why do businesses need systems analysts?**
b. **What is a systems analyst?**
   i. *Information technology* is introduced as the enabling technology that an analyst tries to exploit when solving business problems.
   ii. *Business analysts* are introduced as a new breed of analysts from the user community.
   iii. *Application analysts* are introduced as application design specialists.
c. **What does a systems analyst do?** This section provides readers with their first, brief glimpse of the *systems development life cycle*—presented in the context of *systems problem solving.*
d. **Where do systems analysts work?**
   i. **The systems analyst in the traditional business.** This section presents the traditional function-centered organization of information services and the analyst's position in that organization.
   ii. **Modern information services in a business.** This section presents a contemporary, decentralized organization for information services and describe's various analyst positions in that organization.
   iii. **Outsourcing in the modern business.** This new section examines the impact of outsourcing on the systems analyst, as well as opportunities created by outsourcing.
   iv. **Consulting.** Consulting firms present young, aggressive college graduates with immediate opportunities to apply systems analysis and design.
   v. **Application software solution providers.** This is our new term for software houses, another growing opportunity for young analysts.

3. **Customers, partners, and expectations.** This is our expanded coverage for business and management majors who will become customers of the systems analyst.
   a. **Customers—users and management.**
      i. *System owners.*
      ii. *System users.*
         (a) *Internal users.*
         (b) *External users.* New to this edition!
      iii. **The roles of management and users in systems problem solving.** The systems problem-solving life cycle presented earlier in the chapter is revisited—this time from the perspective of *user roles.*
      iv. **Modern business trends and implications for the systems analyst.**
         (a) **Total quality management (TQM).**
         (b) **Business process redesign (BPR).**
         (c) **Continuous process improvement (CPI).**
         (d) **Globalization of the economy.**
         (e) **Empowerment.**
      v. **Partners for the systems analyst.**

4. **Preparing for a career as a systems analyst.** This section focuses on skill sets and how they might be acquired.
   a. **Working knowledge of information technology.**
   b. **Computer programming experience and expertise.**
      i. *Visual programming.*
      ii. *Object-oriented programming.*
      iii. *Client/server* and *distributed computing.*
   c. **General business knowledge.**
   d. **Problem-solving skills.**
   e. **Interpersonal communication skills.**
   f. **Interpersonal relations skills.**

   g. **Flexibility and adaptability.**
   h. **Character and ethics.**
   i. **Systems analysis and design skills.**
5. **The next generation.**
   a. **Career prospects.**
   b. **Predictions.**
6. **Where do you go from here?**

2

INFORMATION SYSTEM BUILDING BLOCKS

OVERVIEW

Chapter 2 introduces *information system building blocks* that will be reinforced throughout the book. These building blocks are adapted from John Zachman's internationally acclaimed *Framework for Information Systems Architecture* (see recommended readings at the end of the chapter). Essentially, our adaptation of Mr. Zachman's framework is compatible with a currently popular Gartner Group application infrastructure that suggests a contemporary application or system consists of four distinct layers: *data,* logic (which we call *processes*), presentation (which we call *interfaces*), and networks (which we call *geography*). These layers become the basis for contemporary distributed computing, which partitions or distributes various aspects of each of the first three layers (data, process, and interface) across the fourth layer (geography).

Because some time may have passed since the student completed the introduction to information systems course, a quick review of key definitions and applications is provided as an introduction to the information systems framework and building blocks.

A specific color scheme is used throughout the chapter and textbook to associate the symbols in various systems analysis and design tools and techniques with the appropriate layer (which both we and Zachman call a *focus* in the actual chapter—the term *layer* did not translate properly into the matrix-format of the Zachman framework).

CHAPTER-TO-COURSE SEQUENCING

Reviews of the third edition were inconclusive as to whether this chapter should be the first or second in the book. Adopters should feel free to assign and teach Chapter 2 first. We have intentionally repeated any key definitions that might be affected by swapping Chapters 1 and 2. It may even be possible to assign and teach Chapter 3 before returning to Chapter 1.

At Purdue we spend about one week examining this framework from top to bottom, bottom to top, left to right, and so forth. The framework has become an integral conceptual homebase for several of our courses, both in the systems analysis and design as well as the data management course sequences. We are forever grateful for having the good fortune to have heard a keynote presentation on the original framework by John Zachman. He has changed our philosophic approach to the study and teaching of information systems.

WHAT'S DIFFERENT HERE AND WHY?

The following changes have been made to the fourth edition of the information system building blocks chapter:

1. In response to adopter requests for a downsized chapter, we have simplified the information systems review in recognition that most students have already taken an introduction to information systems course.
   a. We have reduced the number of figures by half. Adopters felt that the third edition's previews of modeling tools were premature and somewhat intimidating to the readers who felt they should try to understand those models.

The models have been reduced to icons so as to deemphasize them in the context of this chapter.

b. We eliminated the formal discussion of the TECHNOLOGY building block.

2. The most noticeable change is the replacement of the pyramid model with a matrix model. This was a difficult decision since, through three editions, the pyramid model had become almost synonymous with Whitten/Bentley/Ho/Barlow. But the pyramid model's inability to depict more than two information system dimensions at one time became a liability to the model's purpose—to provide a simplifying and unifying framework for organizing systems analysis and design concepts. In our own classrooms, the matrix model has proven superior for quickly providing students with context and synchronization of important concepts, tools, and techniques. Other framework improvements include:
   a. INTERFACE building blocks were added to recognize the contemporary dimension of system INTERFACES, both user and system, in the information system landscape. This addition provides the conceptual foundation for the much improved user interface design chapters in the design unit.
   b. While no longer presented as formal building blocks, the PEOPLE dimension of an information system is retained in its original Zachman form as "stakeholders" in systems development.
   c. In response to reviewer and adopter requests to simplify the framework, we eliminated TECHNOLOGY as a formal building block.
   d. Some reviewers suggested that we rename the NETWORK building blocks to technology. After careful consideration, we concluded that information technology is pertinent to all the other focuses: data, process, interface, *and* networks.
   e. In response to the above concern, we renamed the NETWORK building blocks to GEOGRAPHY to eliminate some confusion about business versus computer networks.

CHAPTER OUTLINE

Section headings and subheadings are boldfaced.

1. **A review of fundamentals of information systems.** This section is provided to establish the authors' terminology and context, which may differ from the students' prior coursework and experiences.
   a. Definitions provided for context.
      i. Data versus information.
      ii. Information system. The authors view information systems as a collective term that includes the various application classes described below.
      iii. Information technology.
   b. Categories of information systems to which systems analysis and design methods are applied.
      i. **Transaction processing systems.** These are also called *data processing systems* and they support business processes. Students should be made aware of the trend toward business process redesign, which is triggering the redesign of many transaction processing systems.
      ii. **Management information systems.** Some books call these *management reporting systems* or *operations information systems.*
      iii. **Decision support systems.** DSS is enjoying something of a renaissance with the emergence of *data warehouses* and operational data stores. The distinction between DSS and *executive information systems* continues to be somewhat fuzzy since both types of systems support decision information requirements.
      iv. **Expert systems.** The interest in expert systems seems to ebb and flow.

v. **Office information systems.** Also called *office automation,* interest in this class of application remains high because of the focus on business process reengineering.

vi. **Personal and work group information systems.** Interest in the latter class of application is being fueled by the popularity of work group technologies such as Lotus *Notes,* Microsoft *Exchange,* and Delrina *Forms,* the likes of which are being integrated into contemporary business applications.

c. **Putting it all together.** A walkthrough of Figure 2.1 may reinforce the reality that the above application classes don't work in isolation. Such a walkthrough might also reinforce the importance of the new INTERFACES building blocks in our information system model.

2. **A framework for information systems architecture.** The majority of the chapter presents information systems in the context of basic building blocks that will be reinforced throughout the textbook. The framework is illustrated using a matrix.
   a. *Perspectives* are the *rows* of the matrix. They represent the perspectives of various stakeholders in systems development.
   b. Focuses are the *columns* of the matrix. They represent different architectural dimensions of the system.
   c. Each row/column intersection is a building block that represents a stakeholder's view of a focus.
3. **Perspectives—the people side of systems.** This section introduces the rows of the framework and each stakeholder's general perspective on the system. It is important to emphasize that these are *roles,* not *titles.* Any one individual may play many roles depending on the corporate development culture.
   a. **System owners.**
   b. **System users.** The following classes of users are briefly described.
      i. **Internal users.**
      ii. **Remote and mobile users.** The issue of telecommuting is introduced.
      iii. **External users.**
   c. **System designers.**
   d. **System builders.**
   e. **The role of the systems analyst.** Students need to realize that while the analyst may play the role of system designer and system builder, he or she plays a more important role as *facilitator* of the activities performed by the other roles.
4. **Building blocks—expanding the information systems framework.** This section introduces the individual cells (or building blocks) of the framework.
   a. **Building blocks of DATA.** These building blocks culminate in the DATA layer or database components of an application's architecture.
      i. **System owners' view of DATA.**
      ii. **System users' view of DATA.**
      iii. **System designers' view of DATA.**
      iv. **System builders' view of DATA.**
   b. **Building blocks of PROCESSES.** These building blocks culminate in the logic layer or application programs of an application's architecture.
      i. **System owners' view of PROCESSES.**
      ii. **System users' view of PROCESSES.**
      iii. **System designers' view of PROCESSES.**
      iv. **System builders' view of PROCESSES.**
   c. **Building blocks of INTERFACES.** These building blocks culminate in the presentation layer or user and system interfaces of an application's architecture.

i. **System owners' view of INTERFACES.**
ii. **System users' view of INTERFACES.**
iii. **System designers' view of INTERFACES.**
iv. **System builders' view of INTERFACES.**

d. **Building blocks of GEOGRAPHY.** These building blocks culminate in the network layer of an application's architecture.

i. **System owners' view of GEOGRAPHY.**
ii. **System users' view of GEOGRAPHY.**
iii. **System designers' view of GEOGRAPHY.**
iv. **System builders' view of GEOGRAPHY.**

5. **Where do you go from here?**

3

INFORMATION SYSTEM DEVELOPMENT

OVERVIEW

Chapter 3 provides a comprehensive introduction to **information system development.** The chapter's intent is to introduce *principles* and *processes* used to develop information systems. The information system framework is used to reinforce the information system building blocks introduced in Chapter 2. A hypothetical methodology called *FAST* is used to teach a representative, physical implementation of the system development life cycle. This chapter provides a "phase-level" overview. Later chapters provide "activity-level" descriptions and coverage of tools and techniques.

The methodology as presented supports all contemporary paradigms including structured methods, information engineering, and rapid application development, as well as emerging object-oriented methods. As the readers progress through the book, they will learn how to "plug and play" various tools and techniques into the common methodological framework provided by *FAST*.

CHAPTER-TO-COURSE SEQUENCING

This chapter should follow Chapter 2 in all but exceptional scenarios. Chapter 2 provides the fundamental building blocks of information systems and Chapter 3 immediately reinforces those building blocks with its system development methodology.

This chapter concludes the dramatically downsized fundamentals unit, thus providing a much-accelerated path into the meat of the course, systems analysis and design methods.

WHAT'S DIFFERENT HERE AND WHY?

The following changes have been made to the fourth edition of the system development chapter:

1. This chapter consolidates Chapters 3, 4, and 5 from the third edition. The bulk of the material comes from the previous edition's third chapter which was titled "A Systems Development Life Cycle."
2. In the second and third editions, we separated the life cycle and methodology chapters. A third edition adopter suggested we consolidate the chapters by merely characterizing a methodology as one *physical* implementation of the *logical* system development life cycle. This seemed quite elegant to us and this edition adopted that strategy.
   a. We had originally planned to incorporate a real-world, commercial methodology in this edition, but we quickly concluded such a methodology would overwhelm the inexperienced reader.
   b. We elected to create a hypothetical methodology called *FAST* to demonstrate a representative methodology's structure and strategy.
3. The biggest complaint about the third edition's life cycle chapter is that it was overwhelming to students. For the fourth edition, we have produced a kinder, gentler life cycle. We have simplified the life cycle back to a one-page phase-level diagram (as in the first and second editions), while modernizing the coverage to emphasize contemporary trends such as rapid application development.

4. Each phase is described with easily recognized third-level headings that describe the purpose, participants, prerequisites (= inputs), activities, deliverables (= outputs), postrequisite phases, and feasibility checkpoints. This provides for better organization and review.
   a. This edition doesn't go into as much depth (in this chapter) for each phase.
   b. The third edition's use of underlining was eliminated because it was deemed hard on the eyes.
5. We have changed a couple of phase names in an attempt to clarify them.
   a. The third edition's *evaluation* phase is now called the *configuration* phase because it establishes the target physical system.
   b. The third edition's *acquisition* phase is now called the *procurement* phase to better emphasize the buy option in make-versus-buy.
6. We eliminated most of the paradigm discussions (e.g., structured programming, structured analysis, information engineering, etc.) that were in Chapter 4 of the third edition. Although reviewers found them relevant and informative, adopters reported that most of the material did not add value until later chapters when the actual tools and techniques of the paradigm were taught. We have shifted any relevant introductions to those chapters.
7. The CASE chapter was considerably downsized as an introduction for this chapter. Again, adopters reported that students could not truly comprehend CASE outside of the context of the tools and techniques chapters. Since we extensively use CASE throughout the textbook, the detailed material was shifted to those chapters where the impact of CASE could be immediately comprehended by the reader.
   a. With the death of *AD/Cycle,* the interest in CASE data integration has dramatically subsided, and so has the coverage.
   b. The extensive coverage of specific CASE tools for specific phases has been deferred until the system analysis, system design, system implementation, and system support chapters of the book.

CHAPTER OUTLINE

Section headings and subheadings are boldfaced.

1. **System development life cycles and methodologies.** This section differentiates the life cycle and methodologies using the classical logical/physical distinction that will be reinforced throughout the book.
2. **Underlying principles of system development.** It is reassuring that these principles have held up through 10 years of writing this textbook.
   a. **Principle 1: Get the owners and users involved.** System owners have been added to the mix to reinforce the first two chapters.
   b. **Principle 2: Use a problem-solving approach.**
   c. **Principle 3: Establish phases and activities.** Phases are introduced in Chapter 3. Activities are described in Chapters 4, 10, 18, 19, and 20.
   d. **Principle 4: Establish standards for consistent development and documentation.**
   e. **Principle 5: Justify systems as capital investments.**
   f. **Principle 6: Don't be afraid to cancel the project or revise scope.**
   g. **Principle 7: Divide and conquer . . .** even in an object-oriented age!
   h. **Principle 8: Design systems for growth and change . . .** and reuse.
3. ***FAST*—a system development methodology.** *FAST* will be used throughout the running case study and provide the context for the application of the book's structured, object, and rapid development tools and techniques.
   a. **How a *FAST* project gets started.**
      i. Unplanned system requests.
      ii. Planned system initiatives.

(a) From an *information strategy plan.*
(b) From a *business process redesign.*
iii. For both planned and unplanned projects.
(a) *Problems.*
(b) *Opportunities.*
(c) *Directives.*
(d) The PIECES framework.

b. **An overview of the *FAST* life cycle and methodology.**
i. The *repository.*
ii. Relationship to a *database* and a *program library.*
iii. The phases.

c. **The survey phase.**
i. **Purpose.**
ii. **Participants and roles.**
iii. **Prerequisites.**
iv. **Activities.**
v. **Deliverables.**
vi. **Postrequisites and feasibility checkpoints.**
vii. **Impact analysis.**

d. **The study phase.**
i. **Purpose.**
ii. **Participants and roles.**
iii. **Prerequisites.**
iv. **Activities.**
v. **Deliverables.**
vi. **Postrequisites and feasibility checkpoints.**
vii. **Impact analysis.**

e. **The definition phase.**
i. **Purpose.**
ii. **Participants and roles.**
iii. **Prerequisites.**
iv. **Activities.**
(a) *Modeling.*
(b) *Prototyping.*
v. **Deliverables.**
vi. **Postrequisites and feasibility checkpoints.**
(a) *Time boxing.*
vii. **Impact analysis.**

f. **The configuration phase.**
i. **Purpose.**
ii. **Participants and roles.**
iii. **Prerequisites.**
iv. **Activities.**
(a) *Technical feasibility.*
(b) *Operational feasibility.*
(c) *Economic feasibility.*
(d) *Schedule feasibility.*
v. **Deliverables.**
vi. **Postrequisites and feasibility checkpoints.**
vii. **Impact analysis.**
(a) *Application architecture.*

g. **The procurement phase.**
i. **Purpose.**
ii. **Participants and roles.**
iii. **Prerequisites.**
iv. **Activities.**

v. **Deliverables.**
vi. **Postrequisites and feasibility checkpoints.**
vii. **Impact analysis.**

h. **The design phase.** Because *FAST* is a rapid application development methodology, the design and construction phases are integrated to create a rapid prototyping loop.
   i. **Purpose.**
   ii. **Participants and roles.**
   iii. **Prerequisites.**
   iv. **Activities.** Includes a description of the *rapid application development* strategy in *FAST*.
   v. **Deliverables.**
   vi. **Postrequisites and feasibility checkpoints.**
   vii. **Impact analysis.**

i. **The construction phase.** Because *FAST* is a rapid application development methodology, the design and construction phases are integrated to create a rapid prototyping loop.
   i. **Purpose.**
   ii. **Participants and roles.**
   iii. **Prerequisites.**
   iv. **Activities.**
      (a) *Unit testing.*
      (b) *System testing.*
   v. **Deliverables.**
   vi. **Postrequisites and feasibility checkpoints.**
   vii. **Impact analysis.**

j. **The delivery phase.**
   i. **Purpose.**
   ii. **Participants and roles.**
   iii. **Prerequisites.**
   iv. **Activities.**
   v. **Deliverables.**
   vi. **Postrequisites and feasibility checkpoints.**
   vii. **Impact analysis.**

k. **Beyond systems development—system support.**

4. **Cross life cycle activities.**
   a. **Fact-finding.**
   b. **Documentation and presentations.**
   c. **Estimation and measurement.**
   d. **Feasibility analysis.**
   e. **Project management.**
   f. **Process management.**

5. **Computer-aided systems engineering (CASE).**
   a. **The history and evolution of CASE technology.**
   b. **A CASE tool framework.**
      i. *Upper-CASE.*
      ii. *Lower-CASE.*
   c. **CASE tool architecture.**
      i. **Repositories.**
      ii. **Facilities and functions.**
   d. **The benefits of CASE.**
      i. **The development center.**

6. **Where do you go from here?**

PART TWO

SYSTEMS ANALYSIS METHODS

Part Two of the textbook is about **systems analysis methods.** Chapter 4 expands on the systems analysis phases that were introduced in Chapter 3. Each phase is expanded to activity descriptions that describe purpose, roles, inputs, outputs, and steps. This gives the student a simple methodology perspective on systems analysis. Chapters 5 to 8 teach specific modeling techniques applicable to systems analysis: **data modeling, process modeling, distribution modeling,** and **object modeling,** respectively.

COURSE ORGANIZATION AND SEQUENCING

The choice and sequencing of the modeling chapters is left to the discretion of the instructor. The textbook's sequencing reflects the authors' sequencing; however, we teach data, process, and distribution modeling in our first course and object modeling in our second course. The sequencing of the data and process modeling chapters has always been the subject of debate. Information engineers prefer data modeling first. Structured analysts usually prefer process modeling first. To accommodate both audiences, the brief introduction to system modeling was duplicated in each chapter (slightly customized for each chapter). The new object modeling chapter depends somewhat on the data and process modeling chapters since we adopted the market-leading object modeling technique (OMT) that attempts to leverage some of the concepts and techniques of data and process modeling.

Through three editions, many adopters have integrated several of the Part Five modules into the analysis unit. We do that ourselves!

WHAT'S DIFFERENT HERE AND WHY?

We have significantly improved the systems analysis unit as follows:

- Consistent with this edition's overriding goal to return to its classical strengths of systems analysis and design, we deleted the former systems planning chapter. The only activity-level chapter focuses entirely on the systems analysis phases.
- The former data dictionary/repository chapter has been integrated into the appropriate modeling chapters to make room for the new object modeling chapter.
- Chapter 5, **Data Modeling,** now uses the popular Martin or information engineering data modeling notation.
- Chapter 6, **Process Modeling,** has been rewritten to reflect the more practiced event-driven approach to data flow diagramming (as first proposed by McMenamin and Palmer, and subsequently popularized by Yourdon). This approach replaces the top-down strategy with an event-based, bottom-up strategy.
- Chapter 7, **Distribution Modeling,** was renamed from *network modeling* to eliminate confusion that the chapter was teaching computer network design. We had inadvertently contributed to that misperception by introducing networking computer concepts in the chapter. The new chapter clearly focuses on the business system's geography, independent of computer networking alternatives (which are more appropriately deferred to the system design unit).

- We have formally introduced **object modeling** in Chapter 8. The collaboration of three of the leading OOA methodologists (Grady Booch, Ivar Jacobson, and James Rumbaugh) to unify three of the four most popular OOA methodologies gives instant credibility to object modeling as a mainstream topic for systems analysis courses. It can no longer be relegated to a box feature or appendix. We teach the object modeling technique (OMT) as supported by our CASE tool, *System Architect.*

4

SYSTEMS ANALYSIS

OVERVIEW

Chapter 4 provides a comprehensive look at the *process* of **systems analysis.** This process builds on the information systems framework introduced in Chapter 2 and the system development methodology introduced in Chapter 3. We continue to use our representative methodology called *FAST* to teach a realistic simulation of the systems development life cycle. Whereas Chapter 3 provided a phase-level overview, Chapter 4 explores the activities of the three systems analysis phases—*survey, study,* and *definition.*

Instructors are encouraged to remind their students that although Chapter 4 demonstrates certain models and techniques of systems analysis, the chapter does not intend to teach those techniques; some readers become too preoccupied with understanding the data flow diagrams and entity relationship diagrams that are provided merely to place the systems analysis activities into perspective and make them seem real. Chapters 5 to 8 will teach the reader how to build the system models that are pervasive in systems analysis.

CHAPTER-TO-COURSE SEQUENCING

This chapter should follow Chapter 3 in all but exceptional scenarios. Chapter 3 introduced the life cycle at the phase level. This chapter, like most in the book, reinforces the fundamental building blocks of information systems that were introduced in Chapter 2. Consequently, those chapters are essential prerequisites to Chapter 4.

Chapter 4 can be used in various course scenarios:

- In a single course that must cover systems analysis *and* design, Chapter 4 introduces the systems analysis unit. Instructors can then pick which subsequent systems analysis chapters will follow.
- In the first course of a two-course sequence, Chapter 4 can be used to set up a semester project that will emphasize systems analysis over system design. That course can subsequently draw on Chapters 5 to 8 and Modules A to E to complete that emphasis.
- In the second course of a two-course sequence, Chapter 4 can serve as either a review of systems analysis before a focus on systems design and construction or as the reference for an entrance exam covering the systems analysis prerequisite.

WHAT'S DIFFERENT HERE AND WHY?

The following changes have been made to the fourth edition of the systems analysis chapter:

1. The prerequisite chapter on systems planning (see the third edition, Chapter 6) was deleted. This was done for several reasons:
   a. To increase book capacity for the object modeling chapters without further increasing the size of the overall textbook.
   b. To return to our historical roots—emphasizing systems analysis and design over other phases.
   c. To respond to increasing concerns that the information systems strategy planning era had ended—and with more failures than successes. It's not that information systems planning is wrong. Organizations are still planning

systems and developing standard information technology architectures. The problem was that the process (as exemplified by pure information "engineering") is very costly, very time consuming, and wrought with politics—all this for a deliverable (*the plan*) that does not return direct value to the business, at least not until the plan is implemented. And that implementation frequently proves difficult because today's businesses are changing very rapidly, rendering the plan suboptimal or obsolete. And then, when expectations have been increased due to the visibility of the plan, those expectations are dashed by a slow and compromised implementation.[1]

2. The value of the matrix framework for information systems becomes more apparent in this chapter. As each phase is introduced, the framework identifies both the principal audience (e.g., system owners or system users) as well as the building blocks that are developed for those audiences. This integration of the life cycle and methodology into the conceptual framework will continue in the design and implementation chapters.
3. The methodology overviews that disappeared from Part One of the book resurface here. Those methods that are applicable to systems analysis are surveyed under the heading of "strategies." We included several contemporary additions to these strategies including business process analysis and object-oriented analysis. These themes are carried forward to the design and implementation chapters as appropriate.
4. We updated our systems analysis activities. We gratefully acknowledge the many suggestions we've received from students, faculty, and adopters alike. We will continue to refine our methodology in subsequent editions.
5. One complaint about the third edition systems analysis coverage was that the activity descriptions were not consistent in their organization and level of coverage. We adopted a consistent methodological set of third-level headings that enabled us to consistently present each activity. The headings include "Purpose," "Participants and Roles," "Inputs and Prerequisites," "Outputs and Deliverables," "Techniques" (conveniently cross-referenced to the chapters and modules that teach them), and "Steps."
6. The above organization permitted us to eliminate the annoying underlining of inputs and outputs from the third edition.

CHAPTER OUTLINE

Section headings and subheadings are boldfaced.

1. **What is systems analysis?**
   a. Classical definitions.
      i. *System analysis.*
      ii. *System synthesis.*
   b. Contemporary definition.
      i. Phases.
      ii. Role of the *repository.*
2. **Strategies for systems analysis and problem solving.** Strategies are public methodologies (e.g., *structured analysis*) that are typically published in the professional book market. Commercial methodologies usually build on one or

[1]For one interesting quick take on the decline and fall of information systems strategy planning, see Dr. Tom Davenport's *Management Agenda* segment, "Maybe It's Time to Move On" in *Information Week,* September 26, 1994, p. 68.

more public methodologies (e.g., Ernst & Young's *Navigator* is based on the public methodology called information engineering).

a. **Modern structured analysis.** We have elected in this edition to cover only the modern, event-driven version of structured analysis (Chapter 6, Process Modeling) since the pure top-down approaches of Demarco and of Gane and Sarson are no longer widely practiced.
   i. This section also introduces *model-driven* development.
b. **Information engineering.**
c. **Prototyping.**
   i. *Feasibility prototyping.*
   ii. *Discovery prototyping.*
   iii. *Rapid application development (RAD).*
d. **Joint application development (JAD).**
e. **Business process analysis.** This is introduced as a component of systems analysis for business process redesign projects.
f. **Object-oriented analysis.** We provide a very brief introduction to the concepts, but defer the real coverage to Chapter 8, Object Modeling.
g. ***FAST* systems analysis strategies.** This section explains how *FAST* implements the above strategies, and how the *FAST* activities will be presented in the remainder of the chapter.

3. **The survey phase of systems analysis.**
   a. **Activity: Survey problems, opportunities, and directives.**
      i. **Purpose.**
      ii. **Roles.**
      iii. **Prerequisites (inputs).**
      iv. **Deliverables (outputs).**
      v. **Applicable techniques.**
      vi. **Steps.**
   b. **Activity: Negotiate project scope.**
      i. **Purpose.**
      ii. **Roles.**
      iii. **Prerequisites (inputs).**
      iv. **Deliverables (outputs).**
      v. **Applicable techniques.**
      vi. **Steps.**
   c. **Activity: Plan the project.**
      i. **Purpose.**
      ii. **Roles.**
      iii. **Prerequisites (inputs).**
      iv. **Deliverables (outputs).**
      v. **Applicable techniques.**
      vi. **Steps.**
   d. **Activity: Present the project.**
      i. **Purpose.**
      ii. **Roles.**
      iii. **Prerequisites (inputs).**
      iv. **Deliverables (outputs).**
      v. **Applicable techniques.**
      vi. **Steps.**

4. **The study phase of systems analysis.**
   a. **Activity: Model the current system.** This activity has been *downsized* consistent with modern structured analysis thinking.
      i. **Purpose.**
      ii. **Roles.**
      iii. **Prerequisites (inputs).**
      iv. **Deliverables (outputs).**
      v. **Applicable techniques.**
      vi. **Steps.**
   b. **(OPT) Activity: Analyze business processes.** This activity was added to cover emerging BPR projects.
      i. **Purpose.**
      ii. **Roles.**
      iii. **Prerequisites (inputs).**
      iv. **Deliverables (outputs).**
      v. **Applicable techniques.**
      vi. **Steps.**
   c. **Activity: Analyze problems and opportunities.**
      i. **Purpose.**
      ii. **Roles.**
      iii. **Prerequisites (inputs).**
      iv. **Deliverables (outputs).**
      v. **Applicable techniques.**
      vi. **Steps.**
   d. **Activity: Establish system improvement objectives and constraints.**
      i. **Purpose.**
      ii. **Roles.**
      iii. **Prerequisites (inputs).**
      iv. **Deliverables (outputs).**
      v. **Applicable techniques.**
      vi. **Steps.**
   e. **Activity: Modify project scope and plan.**
      i. **Purpose.**
      ii. **Roles.**
      iii. **Prerequisites (inputs).**
      iv. **Deliverables (outputs).**
      v. **Applicable techniques.**
      vi. **Steps.**
   f. **(REC) Activity: Present findings and recommendations.**
      i. **Purpose.**
      ii. **Roles.**
      iii. **Prerequisites (inputs).**
      iv. **Deliverables (outputs).**
      v. **Applicable techniques.**
      vi. **Steps.**
5. **The definition phase of systems analysis.**
   a. **Activity: Outline business requirements.**
      i. **Purpose.**

ii. **Roles.**

iii. **Prerequisites (inputs).**

iv. **Deliverables (outputs).**

v. **Applicable techniques.**

vi. **Steps.**

b. **Activity: Model business system requirements.** The introduction to this activity includes definitions for *logical models* and *logical design.*

i. **Purpose.**

ii. **Roles.**

iii. **Prerequisites (inputs).**

iv. **Deliverables (outputs).**

(1) *Data models.*

(2) *Process models.*

(3) *Interface models.*

(4) *Distribution models.*

(5) *Object models* as an emerging trend.

v. **Applicable techniques.**

vi. **Steps.**

c. **(OPT) Activity: Build discovery prototypes.**

i. **Purpose.**

ii. **Roles.**

iii. **Prerequisites (inputs).**

iv. **Deliverables (outputs).**

v. **Applicable techniques.**

vi. **Steps.**

d. **Activity: Prioritize business requirements.** The notion of *time boxing* is introduced to emphasize the importance of requirements prioritization.

i. **Purpose.**

ii. **Roles.**

iii. **Prerequisites (inputs).**

iv. **Deliverables (outputs).**

v. **Applicable techniques.**

vi. **Steps.**

e. **Activity: Modify the project plan and scope.**

i. **Purpose.**

ii. **Roles.**

iii. **Prerequisites (inputs).**

iv. **Deliverables (outputs).**

v. **Applicable techniques.**

vi. **Steps.**

f. **Some final words about system requirements.**

i. The *requirements statement.*

ii. A caution about "freezing" the requirements.

6. **The next generation of requirements analysis** provides a brief prediction of some forthcoming trends.

7. **Where do you go from here?** For most students, they begin to study the modeling techniques in Chapters 5 to 8.

5

DATA MODELING

OVERVIEW

Chapter 5 is the first technique chapter. It teaches students the important skill of **data modeling.** Students learn the underlying system concepts that apply to data models, and then they learn how to construct them.

CHAPTER-TO-COURSE SEQUENCING

Students are encouraged to read Chapter 4 to provide perspective for any of the four modeling chapters. There has always been disagreement concerning the sequencing of the modeling chapters. Classical structured techniques taught process modeling first. Contemporary information engineering techniques suggest that data models should be taught first. We prefer to teach data modeling first because

- More of the industry now practices information engineering than structured analysis.
- Synchronization of data and process models is simpler if data models are constructed first.
- Data models can usually be constructed faster than process models.
- We sincerely believe that analysts gain a deeper and quicker understanding of business terminology and business rules through data modeling.

Recognizing that this sequencing preference is subject to honest debate, we have designed the process modeling chapter such that it can be used as a preface to the data modeling chapter.

WHAT'S DIFFERENT HERE AND WHY?

The fourth edition's coverage of data modeling has been substantially improved. The following changes have been made to the chapter:

1. We have replaced the Chen notation with the much more popular Martin notation that is supported by a greater number of CASE tools.
   a. We eliminated the coverage of competing notations. Our students found it confusing. Some of our adopters (and our colleagues) suggested the alternative notations might be better covered in a database course.
   b. We have adapted the notation for compatibility with our new CASE tool, *System Architect* (from Popkin Software & Systems, Inc.). We changed to this new CASE tool because of academic cost, Popkin's willingness to produce a student edition, the *Windows* interface, and Popkin's inclusion of structured, information engineering, and object-oriented facilities.
2. We continue to improve the SoundStage case study. In this edition, we included a supertype/subtype hierarchy and a ternary relationship.
3. Third edition adopters did not seem to like our deviation from the traditional terms, *logical* and *physical* (using *essential* and *implementation*). We have switched back to the traditional terminology.
4. We have also replaced the term *identifier* with *key* at the request of some adopters. This also better matches the terminology in most CASE tools.
5. In previous editions, we avoided the term *foreign key.* Based in part on adopter feedback, and in part on our most recent experiences with users and other

analysts, we recognize the need to introduce foreign keys as the mechanism by which our users identify the parent entity in a one-to-many relationship. We also introduce a popular term for many-to-many relationships—nonspecific relationships.

6. We have improved the coverage of supertypes and subtypes in this edition.
7. Through our experiences in a database modeling and design course at Purdue, and its use of the feature-rich IDEF1X data modeling technique, we have significantly improved our own data modeling steps.
8. This edition does not have a separate dictionary chapter; therefore, we have integrated the attribute description material into this chapter.

CHAPTER OUTLINE

Section headings and subheadings are boldfaced.

1. **An introduction to systems modeling.** This section will be repeated for the process modeling chapter to enable instructors to choose their own sequence.
   a. *Model.*
   b. *Logical model.*
      i. *Data modeling.*
   c. *Physical model.*
2. **System concepts for data modeling.** The section begins with an introduction to systems thinking and the Entity Relationship Diagram as a systems thinking tool.
   a. **Entities.**
      i. *Entity.*
      ii. *Entity instance.*
   b. **Attributes.** The concepts of an *attribute* and *compound attribute* are initially introduced and affiliated with the concept of an *entity.*
      i. **Domains.** In the absence of a fourth edition data dictionary chapter, we have integrated the material into the modeling chapters.
         (a) *Data type.*
         (b) *Data range.*
         (c) *Data default.*
      ii. **Identification.**
         (a) *Keys* and *concatenated keys.*
         (b) *Candidate, primary, and alternate keys.*
         (c) *Subsetting criteria.*
3. **Relationships.**
   a. **Cardinality.** We have consolidated ordinality into the term *cardinality* at the request of some adopters.
   b. **Degree.**
      i. *Recursive relationship.*
      ii. *Binary relationship.*
      iii. *Ternary and N-ary relationships.*
         (a) *Associative entity.*
   c. **Foreign keys.**
   d. **Generalization.**
      i. *Supertype.*
      ii. *Subtype.*
4. **The process of logical data modeling.**
   a. **Strategic data modeling.**
   b. **Data modeling during systems analysis.**
      i. *Context data model.*
      ii. *Key-based data model.*
      iii. *Fully attributed data model.*
      iv. *Fully described data model.*

   c. **Looking ahead to system design and construction.**
   d. **Fact-finding and information gathering for data modeling.**
   e. **Computer-aided systems engineering (CASE) in data modeling.**
5. **How to construct data models.**
   a. **Entity discovery.**
   b. **The context data model.**
   c. **The key-based data model.**
      i. How to identify or construct keys.
      ii. Business codes.
         (a) *Serial codes.*
         (b) *Block codes.*
         (c) *Significant digit codes.*
         (d) *Hierarchical codes.*
   d. **The fully attributed model.**
   e. **The fully described model.**
6. **The next generation.**
7. **Where do you go from here?**

6

PROCESS MODELING

OVERVIEW

Chapter 6 is the second technique chapter. It teaches students the *still* important skill of **process modeling.** Students learn the underlying system concepts that apply to process models, and then they learn how to construct them.

CHAPTER-TO-COURSE SEQUENCING

Students are encouraged to read Chapter 4 to provide perspective for any of the four modeling chapters. There has always been disagreement concerning the sequencing of the modeling chapters. Classical structured techniques taught process modeling first. Contemporary information engineering techniques suggest that data models should be taught first. Although we prefer the latter, we have designed the process modeling chapter such that it can also be used as a preface to the data modeling chapter.

Given the emergence of data flow diagrams as a component in popular object-oriented analysis techniques (such as Rumbaugh's OMT), we recommend this chapter precede the object modeling chapter.

WHAT'S DIFFERENT HERE AND WHY?

This is the first major upgrade to this chapter since the first edition. We believe this chapter, at least for the moment, provides significant competitive advantage to your students. Changes for the fourth edition include the following.

1. Virtually all academic textbooks, including the most popular ones, have been teaching process modeling based on Tom Demarco's classic approach as first presented in 1978. This approach is easily recognized by (1) production of four complete sets of data flow diagrams—*current physical, current logical, target logical, target physical*—and (2) the pure *top-down* approach from context diagram to system diagram to subsystem diagrams and so forth until detailed diagrams are completed.

   While the academic world has steadfastly published textbooks based on that process modeling strategy, professional market books and methodologies have shifted to an *event-driven strategy* based on the works of McMenamin and Palmer (1984), Yourdon (1991), the Robertsons (1994), and a host of methodology vendors. This approach abandons the four-set DFD strategy in the interest of more rapid modeling and development and abandons the pure top-down approach, replacing it with what might best be described as an inside-out approach that builds many of the middle diagrams (in context format) before building those above and below them. The result is simpler, faster, more elegant, and less cumbersome.

   We believe that this edition of our textbook is the first academic-level textbook to embrace the event-driven process modeling approach that has become the foundation for methodologies such as Ernst & Young's *Navigator* and Structured Solutions' *AD/Method*.
2. To make room for new object-oriented chapters in this edition, several chapters from the third edition were consolidated. For this process modeling chapter, we

integrated the associated coverage of data structures and procedural logic specification from the former "repository" chapter. To avoid doubling the size of the resulting chapter, we made the following concessions:

a. Data structure coverage was reduced to cover only the algebraic notation as applied to data flows (equally applicable to data stores). That notation was chosen as the most often encountered notation in structured analysis books and methods. The coverage itself was downsized through tighter editing and use of tables.
b. Structured English coverage was similarly reduced to cover the basic constructs. This material was also downsized through tighter editing and use of tables.
c. With some regret, decision table coverage was dramatically reduced. We still think decision tables are valuable tools, but few CASE vendors seem to provide support for this tool. You still find the popular "poker chip problem" in the minicases at the end of the chapter.
d. We have attempted to fully integrate the above material into the process modeling coverage.
e. We could not provide a complete set of examples for the SoundStage case study. Even without the new material, the event-driven strategy for data flow diagramming would have at least doubled the number of examples. Empirical evidence suggested that a complete set of diagrams was overkill. Most students admitted they studied only one or two diagrams of each type before moving on. In this edition, we provide only one or two diagrams of each type.

3. All SoundStage process models were drawn with our new CASE tool, *System Architect* from Popkin Software & Systems. Other than support for "and" and "or" junctions associated with diverging and converging data flows, the diagrammatic changes should not be discernible from the *Excelerator* equivalents in prior editions.
4. We no longer provide examples of the Demarco/Yourdon notation, but we do acknowledge it. Neither students nor instructors seemed to feel it was significant.
5. We no longer use the classical process numbers of structured analysis. They are quite inconvenient in practice, especially when the need to restructure a system's decomposition presents itself.
6. Third edition adopters did not seem to like our deviation from the traditional terms *logical* and *physical* (using *essential* and *implementation*). We have switched back to the traditional terminology.

CHAPTER OUTLINE

Section headings and subheadings are boldfaced.

1. **An introduction to systems modeling**. Most of this section, revised from Chapter 5, can be skipped by those students who read it in Chapter 5. The definitions for process modeling and data flow diagrams and the sample DFD should not be skipped.
   a. *Model.*
   b. *Logical model.*
   c. *Physical model.*
   d. *Process modeling.*
   e. *Data flow diagram.*
2. **System concepts for process modeling**. The section begins with a review of the importance of *systems thinking*.
   a. Process concepts.
      i. A system *is* a process. This section introduces some classic system concepts that impact process modeling.

ii. **Process decomposition**. The concept of factoring a large system into pieces is introduced, along with a tool for doing it—decomposition diagrams.

iii. **Logical processes and conventions.** This includes the classical notational and naming conventions. Three types of processes are introduced:
(a) *Functions.*
(b) *Event (handlers).*
(c) *Elementary processes* (also called primitives).

iv. **Process logic.** This section introduces Structured English and decision tables as specification tools for elementary processes on a model.
(a) *Structured English.*
(b) *Policies.*
(c) *Decision tables.*

b. Data flows.

i. **Data in motion.**
(a) *Data flows.*
(b) *Composite data flows.*
(c) *Control flows.* Control flows are introduced in this edition to help with some of the temporal event concepts in modern structured analysis.

ii. **Logical data flows and conventions.**

iii. **Data flow conservation.** This concept of data flow diagrams is introduced to provide transition to the coverage of data structures and data attributes for data flows.

iv. **Data structures.**
(a) *Data attributes.* An appropriate back-reference to the data modeling coverage is provided.
(b) *Data structures.* The Boolean algebraic notation for describing data structures was integrated here.

v. **Domains.** This material was covered more extensively in Chapter 5. Here those concepts that are applicable to process modeling are reviewed, and the instructor is referenced back to Chapter 5 for details and figures.

vi. **Divergent and convergent flows.** In prior editions, we discouraged such flows on *logical* models. At the encouragement of adopters, and with the addition of AND and OR junctions, we have deleted that opinion.

c. External agents. We still like that term better than *external entity* which was frequently confused with data entity.

d. Data stores. We continue to emphasize the similarity between data stores and data entities.

3. **The process of logical process modeling**. Given the concepts behind process models, the chapter next teaches students how process models are built.

a. Strategic systems planning. A brief discussion of the role of process models in planning projects is offered.

b. Process modeling for business process redesign.

c. Process modeling during systems analysis. This section introduces the event-driven approach to process modeling.

i. *Context diagram.*
ii. *Functional decomposition diagram.*
iii. *Event-response list.*
iv. *Event handler.*
v. *Event diagram.*
vi. *System diagram.*

vii. *Primitive diagram.*

d. **Looking ahead to systems organization and design.**

e. **Fact-finding and information gathering for process modeling.**

f. **Computer-aided systems engineering (CASE) for process modeling.**

4. **How to construct process models**. Using pieces of the SoundStage process model, students are taught how to apply the event-driven approach.

a. **Context diagram.**

b. **Functional decomposition diagram.**

c. **Event-response list.**

i. *External events.*

ii. *Temporal events.*

iii. *State events.*

d. **The event decomposition diagram.**

e. **Event diagram.**

f. **System diagram.**

g. **Primitive diagrams.**

5. **The next generation.**

6. **Where do you go from here?**

7

NETWORK MODELING

OVERVIEW

Chapter 7 introduces business network modeling as an important skill for tomorrow's systems analysts and end-users who must ultimately deal with the modern partitioning decisions in a distributed computing environment. Unique to our textbook, the authors argue that the business case and requirements for distributed solutions are too often overlooked. Just as we have learned to appreciate the need for logical and physical data modeling, we must develop an appreciation for logical (or business) network modeling. Only then can we truly make optimal business data, process, and interface distribution decisions in a distributed computing architecture.

CHAPTER-TO-COURSE SEQUENCING

Students are encouraged to read Chapter 4 to provide a complete systems analysis perspective for any of the four modeling chapters. In particular, this chapter also assumes a working knowledge of data and process modeling (Chapters 5 and 6). At the instructor's discretion, object modeling (Chapter 8) can precede this network modeling chapter. (The current industry preoccupation with distributed object architectures may eventually force that sequence upon us.)

WHAT'S DIFFERENT HERE AND WHY?

This is the second edition in which this unique subject is featured. We have made the following changes for this edition:

1. We have deleted most of the distributed computing coverage. We relocated that coverage to the system design unit based on third edition adopters' concerns that the coverage was entirely too physical (absolutely correct!) and premature for a *systems analysis* and *business requirements* focused chapters. We now provide just enough coverage to justify the need for *logical* network modeling.
2. We have changed our logical network modeling notation. In the prior edition, we were overly concerned with using shapes that differed from those used in data and process modeling. Unfortunately, the shapes chosen for the prior edition were unavailable in any CASE tool. For this edition, we adapted the shapes on generic system flowcharts for our network models. Adopters should find that most CASE tools support system flowchart symbology.
3. We added coverage that links the data models and process models to the network models through *synchronization matrixes*. This should help the student to better appreciate the need for network modeling. (We added data-to-process CRUD matrices to this discussion since it seemed reasonable to introduce that level of synchronization at the same time.)
4. This chapter continues to break new ground. If you have ideas on how to further develop this subject, please contact us.

CHAPTER OUTLINE

Section headings and subheadings are boldfaced.

1. **Network modeling—not just for computer networks.** This section uses computer networking and distributed computing to justify the need for logical network modeling.
   a. *Network and distribution modeling.*

b. *Distributed and client/server computing.*

2. **System concepts for network modeling.** This section introduces *network models* using applicable system concepts.
   a. **Business geography.**
      i. *Location connectivity diagrams.*
   b. **Locations.**
      i. *Location (defined).*
      ii. *Logical versus physical locations.*
      iii. Clusters.
      iv. Mobile or moving locations.
      v. External locations.
   c. **Decomposition.**
      i. *Decomposition (defined).*
      ii. *Location decomposition diagrams.* We have improved on our treatment by eliminating the vague concept of "locations that consist of other locations" and used the decomposition diagram to more elegantly illustrate the same concept.
   d. **Connectivity.**
      i. *Connectivity (defined).*
   e. **Miscellaneous constructs.** This section describes how relevant constructs from data flow diagrams might be added to location connectivity diagrams.
   f. **Synchronization of models.** This section was added to (*a*) tie together the various models learned in Chapters 5–7, and (*b*) establish the missing link (from third edition) between network models and their data and process counterparts. This section begins to deal with the partitioning issues of application partitioning, albeit at a logical, business level of interest.
      i. **Data and process model synchronization.**
         (a) The relationship between data stores and data entities is reinforced.
         (b) *Data-to-process-CRUD matrices* are introduced as a completeness and consistency check that should precede any distribution analysis.
      ii. **Data and network model synchronization.**
         (a) *Data-to-location-CRUD matrices* are introduced for two purposes:
            (1) To determine *business distribution and replication* requirements.
            (2) To determine *business data access* requirements for each location.
      iii. **Process and interface model synchronization.**
      iv. **Process and network model synchronization.**
         (a) *Process-to-location-association matrices* are introduced to describe which processes from the process model are needed at each location.
3. **The process of network modeling.**
   a. **Network modeling during strategic system planning projects.**
   b. **Network modeling during systems analysis.**
   c. **Fact-finding and information gathering for network modeling.**
   d. **Computer-aided systems engineering (CASE) for network modeling.**
4. **How to construct logical network models.** This section reviews the SoundStage case study models.
   a. **Location decomposition diagram.**
   b. **Location connectivity diagram.**
5. **The next generation.**
6. **Where do you go from here?**

8

OBJECT MODELING

OVERVIEW

Chapter 8 is a new chapter on object modeling. It teaches students the important skill of object modeling during systems analysis. Students learn the underlying object-oriented systems development concepts that apply to data models, and then they learn how to construct them.

CHAPTER-TO-COURSE SEQUENCING

Students are encouraged to read Chapter 4 to provide perspective for object modeling. Although it is not necessary, it is suggested that the chapters covering process and data modeling be covered before this object modeling chapter. Object modeling is intended to address the difficulty of working with two separate models—process and data models.

After reading this chapter, it is suggested Chapters 9 and 16 be covered in sequence. Chapter 9 will provide a perspective for continuing study of the object-oriented approach to systems development. Chapter 16 refines the object models developed in Chapter 8 to include design decisions.

WHAT'S DIFFERENT HERE AND WHY?

This new chapter introduces object modeling. The following points regarding this coverage should be noted:

1. The authors have chosen to use the OMT object modeling notation. OMT currently holds a substantial majority of the object modeling market. Most of the competition has chosen to use the Coad/Yourdon object modeling notation. However, Coad/Yourdon has only a small niche of users in the industry. We did consider the newer unified method. However, the unified method was still in the development stage when this chapter was written. We also are concerned that the unified method notation, in its attempt to address all aspects of object modeling, may be too complex and overwhelming to use in a basic introduction to object modeling.
2. Consistent with our coverage of other modeling tools and techniques, the authors provide numerous examples pertaining to our running case study, SoundStage Member Services System. This offers the students the benefit of seeing the three techniques—process modeling, data modeling, and object modeling—applied to a common systems project.
3. This chapter includes coverage of the *use case* technique. Use case is a very important tool and technique that is used in many popular object modeling approaches.

CHAPTER OUTLINE

Section headings and subheadings are boldfaced.

1. **An introduction to object modeling.** This section will introduce the student to object-oriented analysis and object modeling.
2. **System concepts for object modeling.** This section introduces several concepts that object-oriented analysis is based on, including:
   a. **Objects, attributes, methods, and encapsulation.**
   b. **Classes, generalization, and specification.**

c. **Object/class relationships.**
   i. *Inheritance.*
   ii. *Supertype.*
   iii. *Subtype.*
   iv. *Multiplicity.*
   v. *Aggregation.*

d. **Messages and message sending.**

e. **Polymorphism.**

3. **The process of object modeling.** This section provides a step-by-step explanation of how object modeling is performed during systems analysis.
   a. **Finding and identifying the business objects.** This discusses how use case modeling is used to identify use cases and actor for the purpose of identifying potential system objects for object modeling.
      i. *Step 1: Identifying actors and use cases.*
      ii. *Step 2: Constructing a use case model.*
      iii. *Step 3: Documenting the use case course of events.*
      iv. *Step 4: Identifying use case dependencies.*
      v. *Step 5: Documenting the use case alternate course of events.*
      vi. *Step 6: Finding the potential objects.*
      vii. *Step 7: Selecting the proposed objects.*
   b. **Organizing the objects and identifying their relationships.** This section presents steps involved in organizing system objects and documenting their relationships
      i. *Step 1: Identifying associations and multiplicity.*
      ii. *Step 2: Identifying generalization/specialization relationships.*
      iii. *Step 3: Identifying aggregation relationships.*
      iv. *Step 4: Preparing the object association model.*
4. **Where do you go from here?**

PART THREE

SYSTEMS DESIGN AND CONSTRUCTION METHODS

Part Three is about **systems design and construction.** Chapter 9 expands on the systems design phases that were introduced in Chapter 3. Each phase is expanded to activity descriptions that describe purpose, roles, inputs, outputs, and steps. This gives the student a simple methodology perspective on systems design. Chapters 10 to 16 teach specific concepts, tools, and techniques applicable to systems design.

COURSE ORGANIZATION AND SEQUENCING

Where systems analysis ends and systems design begins has always been subject to different opinions. We include Chapter 10's coverage of application architecture and design in the design unit, but some may prefer to cover the chapter's material in the context of systems analysis. Regardless, we recommend Chapter 9 be covered first to provide a foundation for subsequent design chapters.

Given today's emphasis on prototyping and rapid application development (RAD) as approaches to systems design, we highly recommend covering database design (Chapter 11) before subsequent design chapters. Prototyping begins with the design and development of a database that can be used to prototype interface components that use that data. Chapters 12 (input design and prototyping) and 13 (output design and prototyping) could then be covered in either order, and we recommend covering both chapters before Chapter 14 on interface design and prototyping which necessitates first designing inputs and outputs.

WHAT'S DIFFERENT HERE AND WHY?

We have significantly improved the systems design unit as follows:

- Data analysis coverage has been moved into Chapter 11, "Database Design." This lets students immediately see the impact that data analysis will have on the resulting database design.
- Chapter 12, "Input Design and Prototyping," now includes coverage of newer, increasingly popular technologies to capture data. We discuss six automatic data collection (ADC) technologies. Updates also emphasize the design of on-line inputs using a graphical user interface (GUI). We provide the students with guidelines on selecting proper screen-based controls. Also, we introduce the new repository-based programming strategy for prototyping GUI input screens.
- Chapter 13, "Output Design and Prototyping," now includes more emphasis on the design of graphical outputs. We introduce several of the more commonly used graph types. Our examples and how-to coverage are geared to display outputs for a GUI-based application.
- Chapter 14, "User Interface Design and Prototyping," has been significantly updated to reflect today's practice of developing graphical user interfaces for applications.
- Chapter 15, "Software Design," has been revised to include detailed guidelines for deriving program designs. In this edition, we provide specific instructions on how to analyze a data flow diagram to derive a good structure chart.
- Chapter 16, "Object-Oriented Design," is a new chapter. It provides an overview of the increasingly popular object-oriented approach, including tools and techniques. The coverage is a continuation of Chapter 8, which introduced object-oriented analysis using the OMT approach.

9

SYSTEMS DESIGN AND CONSTRUCTION

OVERVIEW

Chapter 9 provides a comprehensive look at the *process* of **systems design.** This process builds on the information systems framework introduced in Chapter 2 and the systems development methodology introduced in Chapter 3. We continue to use our representative methodology called *FAST* to teach a realistic simulation of the systems development life cycle. Whereas Chapter 3 provided a phase-level overview, Chapter 9 explores the activities of the three systems design phases—configuration, procurement, and design and integration.

Instructors are encouraged to remind their students that although Chapter 9 demonstrates certain models and techniques of systems analysis, it is not the intention of the chapter to teach those techniques. Some readers become too preoccupied with understanding the tools that are merely provided to place the systems design activities into perspective and make them seem real. Chapters 10 to 16 will teach the reader how to develop the design specifications used in modern systems design.

CHAPTER-TO-COURSE SEQUENCING

This chapter should normally follow Part Two. However, in those cases where the students have already been exposed to systems analysis, this chapter could immediately follow Chapter 3. Chapter 3 introduced the life cycle at the phase level. It is recommended that Chapter 3, at a minimum, be covered first to establish a common vocabulary and provide an overview of where systems design fits into the overall systems development life cycle. This chapter, like most in the book, reinforces the fundamental building blocks of information systems that were introduced in Chapter 2. Consequently, those chapters are essential prerequisites to Chapter 9.

Chapter 9 can be used in various course scenarios:

- In a single course that must cover systems analysis *and* design, Part Two addresses systems analysis concepts, tools, and techniques. Chapter 9 begins Part Three, which introduces the systems design unit. Instructors can then pick which subsequent systems design chapters they wish to follow.
- In a second course of a two-course sequence, Chapter 9 can be used to follow up on a semester project that will emphasize system design. That course can subsequently draw on Chapters 10 to 16 and Modules A to E to complete that emphasis.

WHAT'S DIFFERENT HERE AND WHY?

The following changes have been made to the fourth edition of the systems design chapter:

1. The selection and acquisition phases were renamed to configuration and procurement, respectively.
2. The value of the matrix framework for information systems is again exploited in this chapter. As each phase is introduced, the framework identifies both the principal audience (e.g., system owners or system users) as well as the building blocks that are developed for those audiences.
3. The paradigm overviews (e.g., structured design) that disappeared from Part One of the book are included here. Those methods that are applicable to

systems design are surveyed under the heading of "Strategies for Systems Design." We address several contemporary strategies including object-oriented design. These themes are carried forward to the design and implementation chapters as appropriate.

4. We updated our systems design activities. We gratefully acknowledge the many suggestions we've received from students, faculty, and adopters alike. We will continue to refine our *FAST* methodology in subsequent editions.
5. One complaint about the third edition systems design coverage was that the activity descriptions were not consistent in their organization and level of coverage. To improve on this situation, we adopted a consistent methodological set of third-level headings that enabled us to consistently present each activity. The headings include PURPOSE, PARTICIPANTS AND ROLES, INPUTS AND PREREQUISITES, OUTPUTS AND DELIVERABLES, TECHNIQUES (conveniently cross-referenced to the chapters and modules that teach them), and STEPS.
6. The above organization permitted us to eliminate the annoying underlining of inputs and outputs from the third edition.

CHAPTER OUTLINE

Section headings and subheadings are boldfaced.

1. **What is systems design?**
   a. *Systems design,* alias *physical design.*
   b. *Configure a feasible solution*—a transition phase considered by some to be part of systems analysis.
2. **Strategies for systems design.**
   a. **Modern structured design.**
   b. **Information engineering (IE).**
   c. **Prototyping.**
   d. **Joint application development (JAD).**
   e. **Rapid application development (RAD).**
   f. **Object-oriented design (OOD).** We provide a very brief introduction to the concepts, but defer the real coverage to Chapter 16.
3. ***FAST* systems design methods.** This section explains how *FAST* implements the above strategies and how the *FAST* activities will be presented in the remainder of the chapter.
4. **The configuration phase of systems design.**
   a. **Activity: Define candidate solutions.**
      i. **Purpose.**
      ii. **Roles.**
      iii. **Prerequisites (inputs).**
      iv. **Deliverables (outputs).**
      v. **Applicable techniques.**
      vi. **Steps.**
   b. **Activity: Analyze feasibility of alternative solutions.**
      - *Technical feasibility.*
      - *Operational feasibility.*
      - *Economic feasibility.*
      - *Schedule feasibility.*
      i. **Purpose.**
      ii. **Roles.**
      iii. **Prerequisites (inputs).**
      iv. **Deliverables (outputs).**
      v. **Applicable techniques.**
      vi. **Steps.**
   c. **Activity: Recommend a system solution.**
      i. **Purpose.**

- ii. **Roles.**
- iii. **Prerequisites (inputs).**
- iv. **Deliverables (outputs).**
- v. **Applicable techniques.**
- vi. **Steps.**

5. **The procurement phase of systems design.**
   - a. **Activity: Research technical criteria and options.**
     - i. **Purpose.**
     - ii. **Roles.**
     - iii. **Prerequisites (inputs).**
     - iv. **Deliverables (outputs).**
     - v. **Applicable techniques.**
     - vi. **Steps.**
   - b. **Activity: Solicit proposals (or quotes) from vendors.**
     - • *Request for quotations (RFQ).*
     - • *Request for proposals (RFP).*
     - i. **Purpose.**
     - ii. **Roles.**
     - iii. **Prerequisites (inputs).**
     - iv. **Deliverables (outputs).**
     - v. **Applicable techniques.**
     - vi. **Steps.**
   - c. **Activity: Validate vendor claims and performance.**
     - i. **Purpose.**
     - ii. **Roles.**
     - iii. **Prerequisites (inputs).**
     - iv. **Deliverables (outputs).**
     - v. **Applicable techniques.**
     - vi. **Steps.**
   - d. **Activity: Evaluate and rank vendor proposals.**
     - i. **Purpose.**
     - ii. **Roles.**
     - iii. **Prerequisites (inputs).**
     - iv. **Deliverables (outputs).**
     - v. **Applicable techniques.**
     - vi. **Steps.**
   - e. **Activity: Award (or let) contract and debrief vendors.**
     - i. **Purpose.**
     - ii. **Roles.**
     - iii. **Prerequisites (inputs).**
     - iv. **Deliverables (outputs).**
     - v. **Applicable techniques.**
     - vi. **Steps.**
   - f. **Activity: Establish integration requirements.**
     - i. **Purpose.**
     - ii. **Roles.**
     - iii. **Prerequisites (inputs).**
     - iv. **Deliverables (outputs).**
     - v. **Applicable techniques.**
     - vi. **Steps.**
6. **The design and integration phase of systems design.**
   - a. **Activity: Analyze and distribute data.**
     - i. **Purpose.**
       - *Data analysis.*
       - *Normalization.*
       - *Event analysis.*

   ii. **Roles.**
   iii. **Prerequisites (inputs).**
   iv. **Deliverables (outputs).**
   v. **Applicable techniques.**
   vi. **Steps.**
b. **Activity: Analyze and distribute processes.**
   i. **Purpose.**
   ii. **Roles.**
   iii. **Prerequisites (inputs).**
   iv. **Deliverables (outputs).**
   v. **Applicable techniques.**
   vi. **Steps.**
c. **Activity: Design databases.**
   i. **Purpose.**
   ii. **Roles.**
   iii. **Prerequisites (inputs).**
   iv. **Deliverables (outputs).**
   v. **Applicable techniques.**
   vi. **Steps.**
d. **Activity: Design computer outputs and inputs.**
   i. **Purpose.**
   ii. **Roles.**
   iii. **Prerequisites (inputs).**
   iv. **Deliverables (outputs).**
   v. **Applicable techniques.**
   vi. **Steps.**
e. **Activity: Design on-line user interface.**
   i. **Purpose.**
   ii. **Roles.**
   iii. **Prerequisites (inputs).**
   iv. **Deliverables (outputs).**
   v. **Applicable techniques.**
   vi. **Steps.**
f. **Activity: Present and review design.**
   i. **Purpose.**
   ii. **Roles.**
   iii. **Prerequisites (inputs).**
   iv. **Deliverables (outputs).**
   v. **Applicable techniques.**
   vi. **Steps.**

7. **Where do you go from here?** For most students, they begin to study the design techniques in Chapters 10 to 16.

10

APPLICATION ARCHITECTURE AND PROCESS DESIGN

OVERVIEW

Chapter 10 teaches the student how to make the up-front architectural decisions that we call general systems design and how to model those decisions using physical data flow diagrams. System flowcharts are briefly discussed as a legacy modeling technique. This chapter also stimulates the reader's imagination and creativity with respect to the many technological choices that can and should be made during general systems design inclusive of network, database, interface, and software development technologies.

CHAPTER-TO-COURSE SEQUENCING

Students are encouraged to first read both Chapter 9 and Chapter 6 before this chapter. Chapter 9 places general design into the perspective of overall information system design. It is also assumed that the reader is familiar with *logical* data flow diagrams (Chapter 6). The geographic modeling techniques taught in Chapter 7 are recommended, but not essential.

WHAT'S DIFFERENT HERE AND WHY?

This chapter significantly updates the third edition's Chapter 14. We decided to remove the title's reference to "analysis" for clarity—this is the "design" unit of the book. The fourth edition's coverage of process design has been substantially improved. The following changes have been made to the chapter:

1. We use the industry concept of *application architecture* as a conceptual umbrella for the developers' requirement to assign various technologies to the implementation of an information system.
   a. The coverage of *client/server* architecture is pervasive. Various flavors of client/server architecture are presented, including the emergence of *intranets*.
   b. We use the information systems framework once again, this time to classify the technologies that must be specified—network, database, user and system interfaces, and software processes.
   c. We have updated the network technologies with the aforementioned emphasis on client/server architectures.
   d. We have updated the database technology with a focus on distributed relational database systems and differentiation between distribution and replication.
   e. We have updated the input and output technology coverage to include electronic messaging and the emergence of Internet and intranet interfaces.
   f. We have added process technology coverage with a focus on the software development environments (SDEs) used to implement computerized processes.
2. We have slightly updated our coverage of physical data flow diagrams and system flowcharts. We expect interest in both tools to rapidly decline as object-oriented tools become more pervasive. As in previous chapters, we now refer to DFDs as *logical* or *physical*—not *essential* and *implementation* (in keeping with the mainstream terms).
3. We have abandoned, for now, the elaborate "derived design" technique for identifying and documenting *design units*. We did this for two reasons. First, the

technique is more suited to non-client/server applications of the past. We are still hopeful that, with some additional research and development, we can resurrect the technique in a form better suited to today's partitioning requirements for n-tiered client/server applications. Second, the fourth edition's new event-driven approach for logical DFDs (from Chapter 6) results in semi*natural* design units without the overhead of the derived design technique.

CHAPTER OUTLINE

Section headings and subheadings are boldfaced.

1. **General system design**. This section reviews the difference between general and detailed design and establishes *general system design* as the focus of this chapter. The concept of *application architecture* is also defined.
2. **Information technology architecture**. An application's architecture is defined in terms of the information technologies to be used. This section quickly reuses the book's information systems framework (building blocks) to conceptualize the role of architecture in general system design.
   a. **Network architectures for client/server computing**. The client/server paradigm is used to classify network technology options.
      i. Client.
      ii. Server.
      iii. Client/server computing.
      iv. Centralized computing. A new strategy was employed here. We no longer characterize centralized computing as an alternative to distributed computing. Instead, we characterize it as the simplest flavor of client/server! We owe the strategy to the Gartner Group models.
         (a) **Centralized computing**.
         (b) **Distributed presentation**. This flavor of client/server computing distributes the user interface from the centralized computer to a client PC (usually in a GUI format).
         (c) **Distributed data**. This *two-tiered* flavor of client/server computing partitions the user interface and business logic to the client, and data and database processing to the server.
            (1) Local area network.
            (2) Wide area network.
            (3) File servers.
            (4) Database servers.
         (d) **Distributed data and logic**. This *three-* or *n-tiered* flavor of client/server computing partitions the user interface, the user logic, and the user's data and database processing *each* to separate servers.
            (1) Partitioning.
         (e) **The Internet and intranets**. This section introduces how the Internet and corporate intranets are changing the fabric of client/server computing.
            (1) Internet.
            (2) Intranets.
         (f) **The role of network technologies** in implementing client/server architecture.
            (1) *Connectivity* as a goal.
            (2) *Interoperability* as a goal.
            (3) *Network topology*
               (a) Bus.
               (b) Ring.
               (c) Star.
               (d) Hierachical.

b. **DATA architectures for distributed relational databases**.
   i. Relational database.
   ii. Distributed relational database.
   iii. Distributed relational database management system.
   iv. Database engine.
   v. Data distribution.
   vi. Data replication.
c. **INTERFACE architectures—inputs, outputs, and middleware**.
   i. **Batch input/output**.
   ii. **On-line input/output**.
   iii. **Remote batch**.
   iv. **Keyless data entry**.
      (a) Optical character reading.
      (b) Optical mark reading.
      (c) Auto-identification and bar-coding.
   v. **Pen input**
      (a) Handheld PCs.
   vi. **Graphical user interfaces**.
      (a) Internet/intranet *browsers*.
   vii. **Electronic messaging and work group technology**.
   viii. **Electronic data interchange**.
   ix. **Imaging and document interchange**.
   x. **Middleware**.
      (a) System integration.
      (b) *Middleware* defined.
   xi. **Selecting user and system interface technologies**.
d. **PROCESS architecture—the software development environment and system management**.
   i. *Software development environments* that are built around programming languages and that provide the tools to automate software processes.
   ii. **SDEs for centralized computing and distributed presentation**.
   iii. **SDEs for two-tiered client/server**.
   iv. **SDEs for multitiered client/server**.
   v. **SDEs for Internet and intranet client/server**.
      (a) HTML.
      (b) CGI.
      (c) Java.
   vi. **System management**.
      (a) Transaction processing monitors.
      (b) Version control and configuration management.

3. **Application architecture strategies and design implications**. This brief section sets the stage for process design by describing two strategies for selecting an application architecture.
   a. **The enterprise application architecture strategy**. This is the strategic approach that is applied to all subsequent system development projects.
   b. **The tactical application architecture strategy**. This is the project-by-project approach.
   c. **Build versus buy implications**.

4. **Modeling application architecture and information system processes.** This section is the modeling constructs for the chapter.
   a. **Physical data flow diagrams**. *Physical data flow diagrams* defined and differentiated from their logical counterparts.
      i. **Physical processes**. This section describes how to label them.
         (a) Network topology data flow diagram.
      ii. **Physical data flows**.

iii. **Physical external agents**.
iv. **Physical data stores**.

b. **System flowcharts**. System flowcharts are only introduced from a legacy perspective. This is likely the last edition in which they will appear.
   i. **System flowchart symbols**.
   ii. **Reading system flowcharts**.

c. **Computer-aided systems engineering (CASE) for physical DFDs and flowcharts**.

5. **Designing the application architecture and the information system processes**. This section applies the modeling constructs to a portion of the SoundStage case study.
   a. **Drawing Physical DFDs**.
      i. *Design unit.*
      ii. *Fully attributed data model.*
      iii. *Fully described data model.*
   b. **Prerequisites**.
   c. **The network topology DFD**. This physical DFD defines the processors to which processes and data stores will ultimately be assigned.
   d. **Data distribution and technology assignments**. Physical data stores are assigned to the network topology DFD.
   e. **Process distribution and technology assignments**. Events and their primitive processes are partitioned between clients and servers.
   f. **Person machine boundaries**. Manual processes and subprocesses are factored out into their own design units.
6. **Where do you go from here?**

11

DATABASE DESIGN

OVERVIEW

Chapter 11 teaches the student how to transform the logical data models that were constructed in Chapter 5 into physical, relational database schemas. There is some coverage of file design concepts.

CHAPTER-TO-COURSE SEQUENCING

Students are encouraged to first read both Chapter 9, "System Design and Construction," and Chapter 10, "Application Architecture and Process Design." The former places database design into the perspective of overall information system design. The latter, in part, provides the perspective for database distribution and replication—important prerequisites to database design.

In the not too distant past, conventional file design was preceded by output and input design because files were designed for specific applications. Today, the opposite is true. Because databases are intended to be a shared resource for scalable or multiple applications, we strongly recommend that the database design precede the output and input design chapters. Doing otherwise would reinforce dependency of the data on the application.

WHAT'S DIFFERENT HERE AND WHY?

This chapter integrates Chapter 13, "Output Design and Prototyping," and Chapter 15, "Structured Design," from the third edition. The following changes have been made to the chapter:

1. We have significantly deemphasized classical file design in favor of modern database design. Our empirical discussions with employers and IS administrators suggest that an overwhelming majority of new systems development and reengineering is being supported by database technology, specifically relational database technology. It has been suggested that most of what students need to know about classical file organizations is adequately taught in the legacy COBOL courses.
2. We have updated the database concepts to include modern concepts such as triggers and stored procedures.
3. Consistent with Chapter 5, we use the Martin logical data modeling notation.
4. We have further improved our normalization coverage. Because normalization techniques were implicitly covered in Chapter 5 (through the introduction of associative entities and attribution guidelines), this chapter presents normalization as a quality management technique that further refines the logical data model before physical design.

   Because the final Chapter 5 logical data model was so clean, we had to step back in time and introduce an earlier rendition of the model to include anomalies that could be corrected through normalization. The end result is a slightly more refined version of the analysis phase data model.
5. We have provided improved guidelines for transforming the logical data model into a physical, relational database schema. We also acknowledge the capability of modern CASE tools to automatically generate a first-cut physical database schema.

6. We have introduced the topics of key, domain, and referential integrity, as well as role names and database capacity planning.
7. We demonstrate the modern capability of CASE tools to automatically generate the SQL/DDL code to construct the database.

CHAPTER OUTLINE

Section headings and subheadings are boldfaced.

1. **Conventional files versus the database.**
   a. *Files.*
   b. *Databases.*
   c. **The pros and cons of conventional files.**
   d. **The pros and cons of databases**
   e. **Database design in perspective.** This brief subsection places database design into the context of the ongoing information system building blocks framework.
2. **Database concepts for the systems analysts.** The section is intended for those who have little or no prior database background.
   a. **Fields.**
      i. *Field* (defined).
      ii. *Primary key.*
      iii. *Secondary key.*
      iv. *Foreign key.*
      v. *Descriptive field.*
   b. **Records.** (We are building up from familiar file terminology to relational terminology.)
      i. **Record** (defined).
         (a) Fixed-length record structure.
         (b) Variable-length record structure.
         (c) Blocking factor.
      ii. **Files and tables.** (It is at this point in the chapter when we begin to make the transition from transition file terminology to relational database terminology.)
         (a) *File* (defined).
         (b) *Table* (defined).
            (1) *Master files and tables.*
            (2) *Transaction files and tables.*
            (3) *Document files and tables.*
            (4) *Document files and tables.*
            (5) *Archival files and tables.*
            (6) *Table look-up files.*
            (7) *Audit files.*
         (c) *File organization.*
         (d) *File access.*
   c. **Databases.**
      i. **Data architecture.**
         (a) *Data architecture* (defined).
         (b) *Operational databases* (previously called *production databases*).
         (c) *Data warehouses.*
         (d) *Personal and work group databases.*
         (e) *Data administrator.*
         (f) *Database administrator.*
      ii. **Database architecture.**
         (a) *Database architecture* (defined).
         (b) *Database management system* (and *database engine*).
         (c) *Data definition language (DDL).*
         (d) *Data manipulation language (DML).*

(e) *Transaction processing monitor.*
(f) **Relational database management systems.**
(1) *SQL.*
(2) *Triggers.*
(3) *Stored procedures.*

3. **Data analysis for database design.**
   a. **What is a *good* data model?**
   b. **Data analysis.**
      i. *Data analysis* (defined).
      ii. *Normalization.*
         (a) *First normal form.*
         (b) *Second normal form.*
         (c) *Third normal form.*
      iii. **Normalization example.**
         (a) First normal form.
         (b) Second normal form.
         (c) Third normal form.
         (d) Simplification by inspection.
      iv. **CASE support for normalization.**
4. **File design.** This is a very brief introduction (farewell?) to the topic.
5. **Database design.**
   a. *Computer-aided systems engineering (CASE).*
   b. **Goals and prerequisites to database design.**
      i. *Data distribution.*
      ii. *Data replication.*
   c. **The database schema.**
      i. *Database schema* (defined).
      ii. Rules and guidelines for transforming the logical data model into a physical, relational database schema.
   d. **Data and referential integrity.**
      i. *Key integrity.*
      ii. *Domain integrity.*
      iii. *Referential integrity.* In the interest of space, we elected to cover only the deletion rules for referential integrity.
         (a) Delete: No restrictions.
         (b) Delete: Cascade.
         (c) Delete: Restrict.
         (d) Delete: Set null.
   e. **Roles.** Roles are introduced at the request of those instructors who also teach a database course.
      i. *Role name.*
   f. **Database capacity planning.** This section presents an (over)simplified, but conceptually sound technique for estimating disk space requirements for a typical relational database.
   g. **Database generation.**
6. **The next generation. Provides a peek into object-oriented database possibilities.**
7. **Where do you go from here?**

12

INPUT DESIGN AND PROTOTYPING

OVERVIEW

Chapter 12 is a technique chapter. It teaches students the important skill of **input design and prototyping.** Students learn the underlying system concepts that apply to input design, and then they learn how to design on-line inputs. The chapter focuses on the design of the increasingly more common graphical user interface screen designs. Focus is placed on a new trend called repository-driven programming, which is explained in the chapter.The chapter stresses the importance of focusing on finalizing the content of imputs and the screen-based controls used to input data. Unlike approaches used to design traditional text-based screens, GUI input screens are not designed in isolation. Thus, the final input screen designs are appropriately deferred to Chapter 14, "User Interface Design and Prototyping," where the overall functionality look and feel of the application is addressed.

CHAPTER-TO-COURSE SEQUENCING

Students are encouraged to read Chapter 9 to provide perspective for where input design fits into systems design. It is also recommended that this chapter follow Chapters 10 and 11. Chapter 10 provides the process design and application frameworks that determine what inputs need to be designed. Chapter 11 covers database design. This chapter emphasizes prototyping, and prototyping a working system assumes that a database has already been designed and constructed. If desired, Chapter 13 could be covered before Chapter 12. However, it is not recommended that Chapter 14 precede Chapter 13. Chapter 14 is concerned with designing the overall user interface for an application—with the assumption that users' input and output design requirements have been initially addressed.

WHAT'S DIFFERENT HERE AND WHY?

The fourth edition's coverage of data modeling has been substantially improved. The following changes have been made to the chapter:

1. We split the third edition's Chapter 12, "Input and Output Design," into separate chapters. Reviewers suggested we return to this first and second edition approach to using separate chapters.
2. We updated our coverage of input design to deal most heavily with the more common types of applications being built today—on-line inputs having a graphical user interface (GUI).
3. Our more expanded coverage of GUI input screen designs is based on a new strategy called respository-driven programming. This approach is demonstrated in the later portion of the chapter using SoundStage examples.
4. Significant advancements have been made in recent years in the area of automatic data collection technology. These new technologies are rapidly finding their way into a wide variety of business applications, often giving companies a significant competitive advantage. It is no longer adequate for students to simply be knowledgeable of error-prone keyboard and mouse input methods for capturing data. Thus, we included coverage of several types of automatic data collection technologies.

CHAPTER OUTLINE

Section headings and subheadings are boldfaced.

1. **Methods and issues for data capture and input.** This section addresses traditional input concepts and introduces new coverage on automatic data collection technology.
   a. **Data capture, data entry, and data input.**
      i. Transaction.
      ii. Data capture.
      iii. Source document.
      iv. Data entry.
      v. Data input.
   b. **Modern input methods: batch versus on-line inputs.**
      i. Batch input.
      ii. On-line input.
      iii. Remote batch.
   c. **Trends in automatic data collection technology.**
      i. Biometric.
      ii. Electromagnetic.
      iii. Magnetic.
      iv. Optical.
      v. Smart cards.
      vi. Touch.
   d. **System user issues for input design.**
   e. **Internal controls for inputs.**
      i. Completeness checks.
      ii. Limit and range checks.
      iii. Combination checks.
      iv. Self-checking digits.
      v. Picture checks.
2. **GUI controls for input design.** This section introduces the new trend in programming called repository-driven programming. It also discusses the proper usage of the most common screen-based controls for inputting data.
   a. **Text box.**
      i. When to use text boxes for input.
      ii. Suggested guidelines for using text boxes.
   b. **Radio button.**
      i. When to use radio buttons for input.
      ii. Suggested guidelines for using radio buttons.
   c. **Check box.**
      i. When to use check boxes for input.
      ii. Suggested guidelines for using check boxes.
   d. **List box.**
      i. When to use list boxes for input.
      ii. Suggested guidelines for using list boxes.
   e. **Drop-down list.**
      i. When to use drop-down lists for input.
      ii. Suggested guidelines for using drop-down lists.
   f. **Combination box.**
      i. When to use combo boxes for input.
      ii. Suggested guidelines for using combo boxes.
   g. **Spin box.**
      i. When to use spin boxes for input.
      ii. Suggested guidelines for using spin boxes.
3. **How to prototype and design computer inputs.** Consistent with the previous three editions, this section provides numerous examples.

a. Step 1: Review input requirements.
b. Step 2: Select the GUI controls.
c. Step 3: Prototype the input screen.
d. Step 4: If necessary, design or prototype the source document.

4. Where do you go from here?

13

OUTPUT DESIGN AND PROTOTYPING

OVERVIEW

Chapter 13 is a technique chapter. It teaches students the important skill of **output design and prototyping.** Students learn the underlying system concepts that apply to output design, and then they learn how to design and prototype computer outputs. The chapter focuses most heavily on the design of video outputs—the fastest growing medium for computer outputs. The chapter distinguishes between the most common types of charts used in graphic outputs. Consistent with Chapter 12, this chapter stresses the importance of user involvement during system development.

CHAPTER-TO-COURSE SEQUENCING

Students are encouraged to read Chapter 9 to provide perspective for where output design fits into systems design. It is also recommended that this chapter follow Chapters 10 and 11. Chapter 10 provides the process design and application frameworks that determine what outputs need to be designed. Chapter 11 covers database design. This chapter emphasizes prototyping a working system, assuming that a database has already been designed and constructed. If desired, Chapter 13 could be covered before Chapter 12. However, it is not recommended that Chapter 14 precede Chapter 13. Chapter 14 is concerned with designing the overall user interface for an application—with the assumption that users' input and output design requirements have been initially addressed.

WHAT'S DIFFERENT HERE AND WHY?

The fourth edition's coverage of output design has been substantially improved. The following changes have been made to the chapter:

1. We split Chapter 13 from the third edition into separate chapters. Reviewers suggested that we return to this first and second edition approach of using separate chapters.
2. We updated our coverage of output design to deal most heavily with the more common types of applications being built today—on-line outputs having a graphical user interface (GUI).
3. Our coverage of graphic outputs was expanded with a discussion of the most common types of graphs/charts and their different uses.
4. Significant advancements have been made in recent years in the area of output technology. These new technologies are rapidly finding their way into a wide variety of business applications—often giving companies a significant competitive advantage. It is no longer adequate for students to simply be knowledgeable about paper, video, microfiche, and microfilm media for outputs. Thus, we included discussion of several newer, increasingly popular output technologies: Web pages, EDI files, e-mail messages, to name a few.

CHAPTER OUTLINE

Section headings and subheadings are boldfaced.

1. **Principles and guidelines for output design.** This section addresses traditional output concepts for designing effective outputs for system users.
   a. **Types of outputs.**

- i. External.
  - Turnaround.
- ii. Internal.
  - Detailed.
  - Summary.
  - Exception.

b. **Output media and formats.**

- i. **Alternative media for presenting information.**
  - Paper.
  - Microfilm.
  - Microfiche.
  - Video.
  - Trends in output media.
- ii. **Alternative formats for presenting information.**
  - *Tabular.*
  - *Zoned.*
  - *Graphic—most common types are discussed.*
  - *Narrative.*

c. **System user issues for output design.**

- i. Computer outputs should be simple to read and interpret.
- ii. The timing of computer outputs is important.
- iii. The distribution of computer outputs must be sufficient to assist all relevant system users.
- iv. The computer outputs must be acceptable to the system users who will receive them.

2. **How to design and prototype computer outputs.** Consistent with the previous three editions, this section provides examples.
   a. **Step 1: Identify system outputs.**
   b. **Step 2: Select output medium and format.**
   c. **Step 3: Prototype the output for system users.**
3. **Where do you go from here?**

14

USER INTERFACE DESIGN AND PROTOTYPING

OVERVIEW

Chapter 14 is a technique chapter. It teaches students the important skill of **user interface design and prototyping.** The sample application that is demonstrated represents a GUI application. Students learn about four user interface styles. This chapter also focuses on human factors that should be considered when developing an interface. In addition, technology considerations are also examined. Finally, the chapter discusses steps involved in developing an application's interface. Consistent with earlier chapters, this chapter stresses the importance of user involvement during this system development activity.

CHAPTER-TO-COURSE SEQUENCING

Students are encouraged to read Chapter 9 to gain a perspective for where user interface design fits into systems design. It is also recommended that this chapter follow Chapters 11 and 12. Chapter 11 covers the design and prototyping of inputs. Chapter 12 covers the design of outputs. This chapter assumes that both inputs and outputs for the application have been designed and prototyped. It focuses on developing those screens that will provide the user with a mechanism for gaining access to those desired inputs and outputs.

WHAT'S DIFFERENT HERE AND WHY?

The fourth edition's coverage of user interface design and prototyping has been substantially improved. The following changes have been made to the chapter:

1. We updated our coverage of user interface design and prototyping to deal most heavily with the more common types of applications being built—on-line outputs having a graphical user interface (GUI).
2. The fourth edition expands the coverage of the four interface styles: menu selection, question-answer dialogue, instruction sets, and direct manipulation. The styles are no longer viewed as mutually exclusive choices. Rather, this edition demonstrates how all four interface styles have evolved and are being used in today's graphical-based development environments.
   a. The coverage of menu selection was expanded beyond mere discussion of the classical hierarchical approach to include discussion of several menu features found in graphical systems, including menu bars, pull-down menus, cascading menus, pop-up menus, and iconic menus.
   b. The coverage of instruction set user interface style was expanded to demonstrate how graphical-based systems feature this approach (for example, Microsoft's *Access Query Facility* that generates SQL syntax).
   c. The coverage of question-answer dialogue style was also expanded to demonstrate how graphical-based systems feature this approach (for example, Microsoft's *Access Answer Wizard* that features a question-and-answer approach).
   d. The coverage of direct manipulation stresses the need to be knowledgeable of the various features available in today's graphical systems. Some excellent references are made available in the Suggested Readings section.

CHAPTER OUTLINE

Section headings and subheadings are boldfaced.

1. **Styles of user interface.** This section addresses four types of strategies for designing user interfaces.
   a. **Menu selection** strategy of dialogue design presents a list of alternatives or options to the user. The system user selects the desired alternative or option by keying in the number or letter that is associated with that option.
      i. **Menu bars.**
      ii. **Pull-down menus.**
      iii. **Cascading menus.**
      iv. **Pop-up menus.**
      v. **Iconic menus.**
   b. **Instruction sets** is an approach that uses a dialogue around an instruction set (also called a command language interface). Because the user must learn the syntax of the instruction set, this approach is suitable only for dedicated users. There are three types of syntax that can be defined. Determining which type should be used depends on the available technology.
      i. **Structured English.**
      ii. **Mnemonic syntax.**
      iii. **Natural language syntax.**
   c. **Question-answer dialogue strategy** is a style that was primarily used to supplement either menu-driven or syntax-driven dialogues. The simplest questions involve yes or no answers.
   d. **Direct manipulation** focuses on using icons, small graphic images, to suggest functions to the user.
2. **Human factors for user interface design.**
   a. **General human engineering guidelines.**
   b. **Dialogue tone and terminology.**
3. **Display features that affect user interface design.**
   a. **Display area.**
   b. **Character sets and graphics.**
   c. **Paging and scrolling.**
   d. **Display properties.**
   e. **Split-screen and windowing capabilities.**
   f. **Keyboards and function keys.**
   g. **Pointer options.**
4. **How to design and prototype a user interface.**
   a. **Step 1: Chart the dialogue.**
   b. **Step 2: Prototype the dialogue and user interface.**
   c. **Step 3: Obtain user feedback.**
5. **Where do you go from here?**

15

SOFTWARE DESIGN

OVERVIEW

Chapter 15 is a technique chapter. It teaches students the important skill of **software design.** Software design consists of two parts: modular design and packaging. Students learn about the challenges of maintaining legacy mainframe applications. They learn about modular design and how it is being applied to improve legacy programs' structure and code to enhance their maintainability. Specifically, the students learn about the most popular technique for modular design, structured design. They learn both techniques of structured design—transform analysis and transaction analysis. In addition, students learn how to evaluate the resulting modular design. Specifically, they learn about types coupling and cohesion measures. Finally, the students learn about packaging. They learn about the importance of including design specifications in a technical design statement that will guide systems implementation.

CHAPTER-TO-COURSE SEQUENCING

Students are encouraged to read Chapter 9 to provide perspective for where software design fits into systems design. Since this chapter addresses packaging design specifications that were developed earlier in systems design, it is also highly recommended that the students first cover the design technique chapters, 10 through 14.

WHAT'S DIFFERENT HERE AND WHY?

The fourth edition's coverage of data modeling has been substantially improved. The following changes have been made to the chapter:

1. We chose to address modular design from the standpoint that the students will likely apply this activity to the maintenance or reengineering of existing, older applications, rather than during the development of new applications, which are dominated by client/server-based applications that negate the need for modular design.
2. We emphasize the most popular modular design strategy, structured design. We provide extensive coverage and examples of structured design's transform analysis and transaction analysis approaches to deriving structure charts.
3. Previous editions provided a "recipe" approach to packaging—stressing that packaging of design specifications be done around the program's inputs, processes, outputs, and data. However, the prototyping approach to systems development has proven this cookbook approach to no longer work. Thus, our coverage is from the standpoint that the only sure thing is the content of the package. The actual organization of the package (technical design statement) depends on the tools and techniques used during systems design, company standards, and designer preferences.

CHAPTER OUTLINE

Section headings and subheadings are boldfaced.

1. **What is software design?** This section introduces software design and establishes motivation for learning modular design—a design activity normally associated with yesterday's applications.

2. **Structured design.** This section provides an overview of the most popular strategy for deriving program module designs. Tools and techniques of structured design are presented, as well as criteria for measuring the quality of those designs. We provide both conceptual and realistic examples of all.
   a. **Structure charts.** This section introduces the primary tool of structured design.
   b. **Data flow diagrams of programs.** This section stresses the need to revise existing DFDs to reflect detailed functions that are often omitted from DFDs used in systems analysis.
   c. **Transform analysis.** This section explains a strategy on studying a program DFD to identify afferent, transform, and efferent processes. Identification of these processes provides a basis for constructing a structure chart.
   d. **Transaction analysis.** This section explains a strategy on studying a program DFD to identify its transaction centers. The transaction center is used as a basis for developing the structure chart.
3. **Structure chart quality assurance checks.** This section provides an in-depth look at two structured design measures for evaluating the quality of a program's modular design.
   a. **Coupling**—The many levels or degrees of coupling exhibited by program modules are examined.
      i. Data.
      ii. Stamp.
      iii. Control.
      iv. Common.
      v. Content.
   b. **Cohesion**—The many levels or degrees of module cohesion are discussed.
      i. Functional.
      ii. Sequential.
      iii. Communicational.
      iv. Procedural.
      v. Temporal.
      vi. Logical.
      vii. Coincidental.
4. **Packaging program specifications.**
5. **Where do you go from here?**

16

OBJECT-ORIENTED DESIGN

OVERVIEW

Chapter 16 is a new chapter on object-oriented design. It teaches students the important skill of using object modeling during systems design. Students learn how to build on the underlying object-oriented concepts, tools, and techniques that were first presented in Chapter 8. Recall that Chapter 8 presented an introduction to object modeling during systems analysis. A smooth transition into systems design is presented. Students learn additional object-oriented concepts, tools, and techniques for dealing with design decisions that will ultimately drive the implementation process using object-oriented programming technology.

CHAPTER-TO-COURSE SEQUENCING

Students are encouraged to read Chapters 4 and 9 to provide context for object analysis and object-oriented design. At a minimum, it is important that the students first read Chapter 8, "Object Modeling," to gain an understanding of object modeling during systems analysis.

After reading this chapter, it is suggested that Chapters 12, 13, and 14 be covered. These chapters cover input, output, and user interface prototyping. Prototyping is an important strategy used in most object-oriented development technologies.

Finally, students should be encouraged to further their exposure to object-oriented development by taking a course, reading a book, or conducting research on object-oriented programming.

WHAT'S DIFFERENT HERE AND WHY?

This is an entirely new chapter that introduces object-oriented design. The following points regarding this coverage should be noted:

1. Consistent with Chapter 8, the authors have continued to use the OMT object modeling notation. OMT currently holds a substantial share of the object modeling market. Most of the academic competition has chosen to use the Coad/Yourdon object modeling notation. However, Coad/Yourdon has only a small niche of users in industry. We did consider the newer Unified Method. However, the Unified Method was still in the development stage when this chapter was written. Based on research, we also became concerned that the Unified Method notation, in its attempt to address all aspects of object modeling, may be too complex and overwhelming to use in providing a basic introduction to object modeling.
2. Consistent with our coverage of other modeling tools and techniques, the authors provide numerous examples pertaining to our running case study, SoundStage Entertainment Club. This offers the students the benefit of seeing the three techniques—process modeling, data modeling, and object modeling—applied to a common systems project.

CHAPTER OUTLINE

Section headings and subheadings are boldfaced.

1. **An introduction to object-oriented design.** This section will introduce the student to object-oriented design concepts, including

a. **Design objects**—Objects are classified during systems design into one of the following categories of object types:
   i. *Interface objects* through which the users communicate with the system.
   ii. *Entity objects* of interest to the users that usually correspond to items in real life. Entity objects contain information, known as attributes, that describes the different instances of the entity. They also encapsulate those behaviors that maintain its information or attributes.
   iii. *Control objects* that manage the interactions of other objects to support the functionality of a use case. They serve as a "traffic cop" containing the application logic or business rules of the event for managing or directing the interaction between the various objects involved.
   iv. *Model-view controller.*
   v. **Interface objects.**
   vi. **Entity objects.**
   vii. **Control objects.**
b. **Object responsibilities.**
   i. *Persistent.*
   ii. *Object responsibility.*
   iii. *Object framework.*

2. **The process of object design.** This section introduces the activities involved in completing object-oriented design, including
   a. **Refining the use case model to reflect the implementation environment.** This activity involves the following steps:
      i. **Step 1—Transforming the "analysis" use cases to "design" use cases.**
      ii. **Step 2—Updating the use case model diagram and other documentation to reflect any new use cases.**
   b. **Modeling object interactions and behavior that support the use case scenario.** This activity involves the following steps:
      i. **Step 1—Identify and classify use case design objects.**
      ii. **Step 2—Identify object attributes.**
      iii. **Step 3—Model high-level object interactions for a use case.**
      iv. **Step 4—Identify object behaviors and responsibilities.**
      v. **Step 5—Model detailed object interactions for a use case.**
   c. **Updating the object model to reflect the implementation environment.**
3. **Where do you go from here?**

PART FOUR

BEYOND SYSTEMS ANALYSIS AND DESIGN

Part Four of the textbook is about **systems implementation and support**. This unit consists of two chapters. Chapter 17 expands on the systems implementation phases introduced in Chapter 3. Chapter 18 expands on the systems support phase that was also introduced in Chapter 3.

COURSE ORGANIZATION AND SEQUENCING

Chapters 17 and 18 should normally follow Part Three and be covered in sequence. However, when students have already been exposed to systems analysis and design, these chapters could immediately follow Chapter 3. Chapter 3 introduced the life cycle at the phase level. It is recommended that Chapter 3, at a minimum, be covered first to establish a common vocabulary and provide an overview of where systems implementation and systems support fit into the overall systems development life cycle. This chapter, like most in the book, reinforces the fundamental building blocks of information systems that were introduced in Chapter 2. Consequently, those chapters are essential prerequisites to Chapters 17 and 18.

WHAT'S DIFFERENT HERE AND WHY?

The following changes have been made to the fourth edition:

1. The systems implementation chapter was simplified from four phases to two—construction and delivery. Once again, the value of the matrix framework for information systems is exploited in this chapter. As each phase is introduced, the framework identifies both the principal audience (e.g., system owners or system users) as well as the building blocks that are developed for those audiences. One complaint about the third edition systems design coverage was that the activity descriptions were not consistent in their organization and level of coverage. To improve on this, we adopted a consistent methodological set of third-level headings that enabled us to consistently present each activity. The headings include "Purpose," "Roles," "Prerequisites (Inputs)," "Deliverables (Outputs)," "Applicable Techniques" (conveniently cross-referenced to the chapters and modules that teach them), and "Steps."
2. The systems support chapter did not necessitate many changes from the third edition. The changes primarily pertain to reorganization and terminology. The most significant content change is inclusion of a section that discusses the impact of the year 2000 on systems support.

17

SYSTEMS IMPLEMENTATION

OVERVIEW

Chapter 17 provides a comprehensive look at the *process* of **systems implementation.** This process builds on the information systems framework introduced in Chapter 2 and the systems development methodology introduced in Chapter 3. We continue to use our representative methodology called *FAST* to teach a realistic simulation of the systems development process. Whereas Chapter 3 provided a phase-level overview, Chapter 17 explores the activities of the two systems implementation phases—construction and delivery.

CHAPTER-TO-COURSE SEQUENCING

This chapter should normally follow Part Three. However, in those cases where the students have already been exposed to systems analysis and design, this chapter could immediately follow Chapter 3. Chapter 3 introduced the life cycle at the phase level. It is recommended that Chapter 3, at a minimum, be covered first to establish a common vocabulary and provide an overview of where systems implementation fits into the overall systems development life cycle. This chapter, like most in the book, reinforces the fundamental building blocks of information systems that were introduced in Chapter 2. Consequently, those chapters are essential prerequisites to Chapter 17.

WHAT'S DIFFERENT HERE AND WHY?

The following changes have been made to the fourth edition of the systems implementation chapter:

1. The systems implementation process was simplified from four phases to two phases—construction and delivery.
2. The value of the matrix framework for information systems is exploited in this chapter. As each phase is introduced, the framework identifies both the principal audience (e.g., system builders) as well as the building blocks that are developed for those audiences.
3. One complaint about the third edition systems design coverage was that the activity descriptions were not consistent in their organization and level of coverage. To improve on this situation, we adopted a consistent methodological set of third-level headings that enabled us to consistently present each activity. The headings include "Purpose," "Roles," "Prerequisites (Inputs)," "Deliverables (Outputs)," "Applicable Techniques" (conveniently cross-referenced to the chapters and modules that teach them), and "Steps."

CHAPTER OUTLINE

Section headings and subheadings are boldfaced.

1. **What is systems implementation?**
   a. *Systems implementation,* alias *systems development.*
   b. *Construction and delivery.*
2. ***FAST* systems design strategies.** This section explains how *FAST* implements the above strategies, and how the *FAST* activities will be presented in the remainder of the chapter.

3. The configuration phase of systems design.
  a. Activity: build and test networks.
    i. Purpose.
    ii. Roles.
    iii. Prerequisites (inputs).
    iv. Deliverables (outputs).
    v. Applicable techniques.
    vi. Steps.
  b. Activity: Build and test databases.
    i. Purpose.
    ii. Roles.
    iii. Prerequisites (inputs).
    iv. Deliverables (outputs).
    v. Applicable techniques.
    vi. Steps.
  c. Activity: install and test new software packages.
    i. Purpose.
    ii. Roles.
    iii. Prerequisites (inputs).
    iv. Deliverables (outputs).
    v. Applicable techniques.
    vi. Steps.
  d. Activity: write and test new programs.
    i. Purpose.
    ii. Roles.
    iii. Prerequisites (inputs).
    iv. Deliverables (outputs).
    v. Applicable techniques.
      - Stub testing.
      - Unit or program testing.
      - Systems testing.
    v. Steps.

4. The delivery phase of systems implementation.
  a. Activity: conduct systems test.
    i. Purpose.
    ii. Roles.
    iii. Prerequisites (inputs).
    iv. Deliverables (outputs).
    v. Applicable techniques.
    vi. Steps.
  b. Activity: prepare conversion plan.
    i. Purpose.
    ii. Roles.
    iii. Prerequisites (inputs).
    iv. Deliverables (outputs).
    v. Applicable techniques.
    vi. Steps.
      - Conversion.
      - Abrupt cut-over.
      - Parallel conversion.
      - Location conversion.
      - Staged conversion.
      - Systems acceptance test.
      - Verification testing.
      - Validation testing.
      - Audit testing.

- **c. Activity: install databases.**
  - **i. Purpose.**
  - **ii. Roles.**
  - **iii. Prerequisites (inputs).**
  - **iv. Deliverables (outputs).**
  - **v. Applicable techniques.**
  - **vi. Steps.**
- **d. Activity: train system users.**
  - **i. Purpose.**
  - **ii. Roles.**
  - **iii. Prerequisites (inputs).**
  - **iv. Deliverables (outputs).**
  - **v. Applicable techniques.**
  - **vi. Steps.**
- **e. Activity: convert to new system.**
  - **i. Purpose.**
  - **ii. Roles.**
  - **iii. Prerequisites (inputs).**
  - **iv. Deliverables (outputs).**
  - **v. Applicable techniques.**
  - **vi. Steps.**

5. Where do you go from here? For most students, they begin to study systems support in Chapter 18.

18

SYSTEMS SUPPORT

Chapter 18 provides a comprehensive look at the *process* of **systems support.** This chapter builds on the systems development methodology introduced in Chapter 3. Whereas Chapter 3 provided a phase-level overview, Chapter 18 explores the four types of systems support—maintenance, enhancement, reengineering, and design recovery.

CHAPTER-TO-COURSE SEQUENCING

This chapter should normally follow Chapter 17. However, when students have already been exposed to systems analysis and design, this chapter could immediately follow Chapter 3. Chapter 3 introduced the life cycle at the phase level. It is recommended that Chapter 3, at a minimum, be covered first to establish a common vocabulary and provide an overview of where systems support fits into the overall systems development life cycle.

WHAT'S DIFFERENT HERE AND WHY?

This chapter did not necessitate many changes from the third edition. The changes primarily pertain to reorganization and terminology. The most significant content change is inclusion of a section that discusses the impact of the year 2000 on systems support.

CHAPTER OUTLINE

Section headings and subheadings are boldfaced.

1. **What is systems support?**
2. **Systems maintenance—correcting errors.**
   a. **Define and validate the problems.**
   b. **Benchmark the programs and application.**
   c. **Understand the application and its programs.**
   d. **Edit and test the programs.**
      i. *Unit testing.*
      ii. *System testing.*
      iii. *Regression testing.*
      iv. *Version control.*
   e. **Update documentation**
3. **System recovery—overcoming the "crash."**
4. **End-user assistance.**
5. **Systems enhancement and reengineering.**
   a. **Analyze enhancement request.**
   b. **Write simple, new programs.**
   c. **Restructure files or databases.**
   d. **Analyze program library and maintenance costs.**
      i. *Software metrics.*
      ii. *Control flow knots.*
      iii. *Cycle complexity.*

e. Reengineer and test programs.

i. *Code reorganization.*

ii. *Code conversion.*

iii. *Code slicing.*

6. The year 2000 and systems support.

7. Where do you go from here?

PART FIVE

CROSS LIFE CYCLE ACTIVITIES AND SKILLS

Part Five of the textbook is about cross life cycle skills, those skills and techniques that are important during most or all phases of any systems development methodology. These modules can and should be integrated into the course and reading assignments according to the preferences of the instructor.

Module A provides an overview of **project and process management** concepts and techniques. Module B introduces the important skill of **fact-finding and information gathering.** Module C introduces **feasibility and cost-benefit analysis** techniques. Module D introduces **joint application development.** Finally, Module E introduces **interpersonal and communications skills** as they apply to systems developers.

COURSE ORGANIZATION AND SEQUENCING

The modules are not necessarily presented in the sequence that they should be introduced. Each instructor is left to his or her preferences as to which modules should be covered, in what sequence, and in what level of detail. Similarly, the modules are not necessarily intended to be covered in a single unit of the course. Different modules can be allocated to different course units, and reviewed in different course units. This flexibility nonwithstanding, we are often asked how we sequence the material. We do it as follows:

- Module A (Project and Process Management) is typically covered after Chapter 3, the introduction to systems development and methodologies.
- Module B (Fact-Finding and Information Gathering) is typically covered just prior to the first modeling chapter (usually Chapter 5, Data Modeling, or Chapter 6, Process Modeling).
- Module C (Feasibility and Cost-Benefit Analysis) is usually covered just after the last systems analysis chapter, and just before Chapter 9, the overview of systems design.
- Module D (Joint Application Development) is introduced just after Chapter 9, the system design overview, or Chapter 10, the general design and application architecture chapter.
- Module E (Interpersonal Skills and Communications) is introduced just about any time, frequently as early as after Chapter 1.

WHAT'S DIFFERENT HERE AND WHY?

We have significantly improved the cross life cycle skills unit as follows:

- We rewrote the project management module for the first time since the first edition, with a strong focus on computer-aided project management-compatible principles and techniques. We also added process management issues to that module. Process management focuses on the reproducible process (methodology) of systems development.
- We improved the currency of fact finding by recognizing the value of electronic research via the Internet.
- We improved the feasibility analysis matrices and economic analysis spreadsheets in the cost-benefit module.

- We added an entirely new module on joint application development (sometimes called joint application design or joint application requirements planning). This JAD coverage reinforces this edition's focus on *rapid development* methods, in this case by teaching a facilitated workshop approach to requirements definition and prototyping.

PROJECT AND PROCESS MANAGEMENT TECHNIQUES

OVERVIEW

Module A introduces project and process management guidelines, tools, and techniques. **Project management** refers to the planning, staffing, scheduling, directing, and controlling of an individual project. The unit covers Gantt and PERT charts, as well as expectations management techniques. **Process management,** new to this edition, refers to the management of the system development process and technology across multiple projects. The latter is usually driven by the chosen methodology of an organization.

MODULE-TO-COURSE SEQUENCING

Students are expected to have completed Chapters 1 through 3 before reading this module. Chapter 3, the methodology chapter, is the most important of these prerequisites. We believe the module is best introduced immediately after Chapter 3, or after Chapter 4, the systems analysis chapter. We feel it provides a nice change of pace when sandwiched between Chapters 3 and 4. The module can also be introduced or reviewed before Chapter 9, the systems design chapter.

WHAT'S DIFFERENT HERE AND WHY?

This is the first major revision of the module since the first edition. Specifically, the following changes have been made.

1. We revised the module introduction to the notion and dimensions of project management.
2. We introduced an entirely new section on process management. Process management is discussed in terms of methodology management, development technology management, total quality management, and measurement and metrics. These aspects of process management were driven by the SEI *Capability Maturity Model.*
3. Because today virtually no one does Gantt and PERT charts without project management software, we reorganized the coverage of Gantt and PERT to be compatible with the use of such software. (It is, however, *not* our intent to teach the software.) And because most project management software tends to default in the Gantt chart view, we switched the coverage to introduce Gantt charts first. Finally, because Microsoft *Project* is so widely available via academic pricing, we used that product to demonstrate the project modeling techniques.
4. The Gantt and PERT charts included are based on the *FAST* methodology that is covered extensively in Chapters 3, 4, and 9. We plan to post those models as *Project* templates on our Internet site.
5. With respect to Gantt charts, we introduced *Project*-compatible concepts such as forward and reverse scheduling, calendars, work breakdown structures, durations, predecessors, constraints, resources, and resource-driven scheduling.
6. We have introduced critical path analysis to Gantt charts (as well as retaining the coverage in PERT charts).
7. We downsized the PERT chart coverage since that tool seems to play a peripheral role in most of the project management software tools.

MODULE OUTLINE

Section headings and subheadings are indicated in boldface.

1. **What is project management?**
   a. *Project.* We define the term in the context of those properties that make a project worth managing.
   b. *Project management.*
   c. **Why projects fail** with a focus on project mismanagement.
      i. Shortcuts.
      ii. *Scope creep.*
      iii. *Feature creep.*
      iv. Premature estimates.
      v. Poor estimating.
      vi. Schedule delays.
      vii. The *mythical man-month.*
      viii. Poor people management.
      ix. Inevitable business change (external factors).
   d. **The basic functions of the project manager.**
      i. **Scoping the project.** Note the new emphasis on project definition.
      ii. **Planning project tasks and staffing the project team.**
      iii. **Organizing and scheduling the project effort.**
      iv. **Directing and controlling the project.**
   e. **Project management software.**
2. **Process management.**
   a. *Process management* defined.
   b. **Management of methodology.**
   c. **Management of system development technology** including CASE tools.
   d. **Total quality management.**
   e. **Metrics and measurement.**
   f. **The development center.** This coverage came from Chapter 5 in the third edition.
3. **Project management tools and techniques.**
   a. **Gantt charts.**
      i. *Gantt chart* defined.
      ii. **Forward and reverse scheduling.**
         (a) Forward scheduling from a start date.
         (b) Reverse scheduling from a deadline.
      iii. **Calendars.**
      iv. **Work breakdown structures.**
         (a) Work breakdown structures are defined as a phase/activity/task/step outline of the work to be performed in a project.
            (1) Summary task.
            (2) Detailed or primitive task.
            (3) Milestone.
      v. **Effort and duration.**
      vi. **Predecessors and constraints.**
      vii. **Critical path and slack resources.**
         (a) Critical path defined.
         (b) Slack time defined.
      viii. **Resource assignments and management.**
         (a) Resource.
         (b) Resource driven scheduling.
      ix. **Using Gantt charts to evaluate progress.**
   b. **PERT charts.**
      i. **PERT definition and symbols** updated for compatibility with MS *Project.*

ii. **The critical path in a PERT network.**
(a) Critical path redefined.
iii. **Using PERT for planning and control.**
iv. **PERT versus Gantt charting.**

5. **Expectations management.**
   a. **The expectations management matrix.**
6. **People management.**
   a. **The** *One-Minute Manager.*
   b. **The subtle art of delegation and accountability.**

B

FACT-FINDING AND INFORMATION GATHERING

OVERVIEW

Module B presents fact-finding and information gathering guidelines, tools, and techniques that are crucial to the application development process. Seven popular techniques are introduced, including sampling, research and site visits, observation, questionnaires, interviews, rapid application development (RAD), and joint application development (JAD). Because of the rising popularity of JAD, we have dedicated a separate module for its discussion. This module also presents the ethics involved with fact-finding and information gathering, as well as a fact-finding strategy that is intended to decrease the amount of user-release time required during fact-finding.

MODULE-TO-COURSE SEQUENCING

Students are expected to have already read Chapters 1 through 3. This module may be introduced at any point thereafter. We feel the material is best introduced after Chapter 4.

WHAT'S DIFFERENT HERE AND WHY?

The following changes have been made to the fourth edition of this module:

1. We have expanded our coverage of selecting a technique to do sampling and have included an example using Microsoft *Excel* to assist in selecting random samples.
2. We have expanded the module to include references to the vastly popular Internet and the World Wide Web as tools for fact-finding. We explain what they are, how they can be used, and what types of information can be found by using them.
3. We have included a section on computer and fact-finding ethics. Companies are stressing employee awareness of computer and business ethics as a requirement for employment. We provide a discussion of different ethical scenarios and the possible impacts. Also included is a list of "The Ten Commandments of Computer Ethics" as published by the Computer Ethics Institute.
4. The material on joint application development (JAD) was removed and placed in a separate module dedicated to its discussion. Because of its popularity with industry today and its widespread use, we felt it important to expand the coverage and provide a better understanding of its concepts.

MODULE OUTLINE

Section headings and subheadings are boldfaced.

1. **What is fact-finding?**
2. **What facts does the systems analyst need to collect and when?**
   a. *Fact finding*
3. **What fact-finding methods are available?**
4. **Sampling of existing documentation, forms, and files.**
   a. **Collecting facts from existing documentation.**
   b. **Document and file sampling techniques.**

i. *Sampling*
ii. **How to determine the sample size.**
iii. **Selecting the sample.**
(a) *Randomization.*
(b) *Stratification.*

4. **Research and site visits.**
   a. *Internet.*
   b. *World Wide Web.*
   c. *Intranets*
5. **Observation of the work environment.**
   a. *Observation*
   b. **Collecting facts by observing people at work.**
      i. **The railroad paradox.**
      ii. **Observation advantages and disadvantages.**
   c. **Guidelines for observation.**
      i. *Work Sampling.*
6. **Questionnaires.**
   a. *Questionnaires.*
   b. **Collecting facts by using questionnaires.**
      i. **Advantages and disadvantages.**
   c. **Types of questionnaires.**
      i. *Free-format questionnaires.*
      ii. *Fixed-format questionnaires.*
         (a) *Multiple choice questions.*
         (b) *Rating questions.*
         (c) *Ranking questions.*
   d. **Developing a questionnaire.**
7. **Interviews.**
   a. *Interviews.*
   b. *Interviewer.*
   c. *Interviewee.*
   d. **Collecting facts by interviewing people.**
      i. **Advantages.**
      ii. **Disadvantages.**
   e. **Interview types and techniques.**
      i. *Unstructured interviews.*
      ii. *Structured interviews.*
      iii. *Open-ended questions.*
      iv. *Closed-ended questions.*
   f. **How to conduct an interview.**
      i. **Select interviewees.**
      ii. **Prepare for the interview**
         (a) *Interview guide.*
      iii. **Conduct the interview.**
         (a) *Interview body.*
         (b) *Interview conclusion.*
      iv. **Follow up on the interview.**
8. **Rapid application prototyping.**
   a. *User centered approach.*
9. **Fact-finding ethics.**
   a. *Code of ethics.*
10. **A fact-finding strategy.**

C

FEASIBILITY AND COST-BENEFIT ANALYSIS

OVERVIEW

This module introduces feasibility analysis guidelines, tools, and techniques. We do not view feasibility analysis as an early phase of the life cycle; instead, we view feasibility analysis as an ongoing activity of the life cycle. Also, we do not view feasibility strictly in terms of economics and cost-benefit analysis; instead, we prefer to think of feasibility as achieving an acceptable balance among four categories of feasibility: operational (or political), technical, schedule, and economic. All are covered in this module, although all are not important during every phase of the life cycle.

MODULE-TO-COURSE SEQUENCING

Students are expected to have already read Chapters 1 through 3. This module may be introduced at any point thereafter. We feel the material is best introduced after Chapter 4.

WHAT'S DIFFERENT HERE AND WHY?

The following changes have been made to the fourth edition of the systems analysis module:

1. This module corresponds to Module C in the third edition.
2. We have introduced the concept of usability analysis. When determining operational feasibility in the later stages of the development life cycle, usability analysis is often performed with a working prototype of the proposed system. This is a test of the system's user interfaces and is measured in how easy they are to learn and to use and how they support the desired productivity levels of the users.
3. We have refined and expanded the examples in this module to include screen captures of Microsoft *Excel* and *Word* to demonstrate the techniques of economic feasibility analysis and system candidate feasibility analysis.

MODULE OUTLINE

Section headings and subheadings are boldfaced.

1. **Feasibility analysis—a creeping commitment approach.**
   a. **Feasibility checkpoints in the life cycle.**
      i. *Feasibility.*
      ii. *Feasibility analysis.*
      iii. *Creeping commitment.*
      iv. **Systems analysis—a survey phase checkpoint.**
      v. **Systems analysis—a study phase checkpoint.**
      vi. **Systems analysis—a definition phase checkpoint.**
      vii. **Systems design—a selection phase checkpoint.**
      viii. **Systems design—a procurement phase checkpoint.**
      ix. **Systems design—a design phase checkpoint.**
2. **Four tests for feasibility.**
   a. **Operational feasibility.**
      i. **Is the problem worth solving, or will the solution to the problem work?**

ii. **How do the end-users and managers feel about the problem (solution)?**
iii. **Usability analysis.**
b. **Technical feasibility.**
i. **Is the proposed technology or solution practical?**
ii. **Do we currently possess the necessary technology?**
iii. **Do we possess the necessary technical expertise, and is the schedule reasonable?**
c. **Economic feasibility.**
d. **The bottom line.**

3. **Cost-benefit analysis techniques.**
a. **How much will the system cost?**
b. **What benefits will the system provide?**
i. *Tangible benefits*
ii. *Intangible benefits*
c. **Is the proposed system cost-effective?**
i. **The time value of money.**
ii. **Payback analysis.**
(a) *Payback period.*
(b) *Discount rate.*
(c) *Opportunity cost.*
(d) *Present value.*
iii. **Return-on-investment analysis.**
iv. **Net present value.**

4. **Feasibility analysis of candidate systems.**
a. **Candidate systems matrix.**
b. **Feasibility analysis matrix.**

D

JOINT APPLICATION DEVELOPMENT

OVERVIEW

Module D is a new module on joint application development (JAD). It teaches students the important skill of planning and conducting a JAD session during systems development. Students learn about JAD participants and their roles during a typical session. They also learn about the many benefits that can be realized through successfully applying JAD as a fact-finding and confirmation technique during development.

MODULE-TO-COURSE SEQUENCING

Since joint application development (JAD) is a cross-life-cycle activity, students should be encouraged to, at a minimum, first read Chapter 3 to gain a basic understanding of the systems development process for which JAD is incorporated. It is also suggested that Chapters 4 and 9 be covered to provide more in-depth coverage to the two primary life cycle phases in which JAD plays a substantial role—systems analysis and systems design.

WHAT'S DIFFERENT HERE AND WHY?

This is an entirely new module! The third edition included some coverage of JAD, but this module provides much more extensive coverage including:

1. The module provides a detailed discussion of the various types of individuals that typically participate in a JAD session. The module describes each participant and the participant's role in a JAD session.
2. This module provides extensive coverage of the steps involved in planning and conducting a JAD session.
3. Finally, this module identifies numerous benefits that can be realized through effectively applying JAD to a systems development project.

MODULE OUTLINE

Section headings and subheadings are boldfaced.

1. **Joint application development.** This section will introduce the student to joint application development as an alternative fact-finding technique to separate interviews of individual system users and managers.
2. **JAD participants.** This section introduces the different types of participants in a typical JAD session, including:
   a. **Sponsor.**
   b. **JAD leader.**
   c. **Users and managers.**
   d. **Scribe(s).**
   e. **IS staff.**
3. **How to plan and conduct JAD sessions.** The section provides a step-by-step explanation of how JAD sessions are planned and conducted for systems development projects.

a. **Planning the JAD session.**
   i. **Selecting a location for JAD sessions.**
   ii. **Selecting JAD participants.**
   iii. **Preparing a JAD session agenda.**
b. **Conducting a JAD session.**

4. **Benefits of JAD.**

E

INTERPERSONAL SKILLS AND COMMUNICATIONS

OVERVIEW

Module E surveys interpersonal skills, a cornerstone of successful systems development. It covers these skills independent of specific discussions on how they are applied during systems planning, systems analysis, systems design, systems implementation, and systems support. Today's students don't seem to understand that their technical skills may get them a job but their interpersonal skills will determine their management potential. The tools and techniques of systems development will fail if the analyst can't communicate the findings, proposals, and designs to the nontechnical audience.

MODULE-TO-COURSE SEQUENCING

Students are expected to have already read Chapters 1 through 3. This module may be introduced at any point thereafter. We feel the material is best introduced after Chapter 4.

WHAT'S DIFFERENT HERE AND WHY?

The following changes have been made to the fourth edition of this module:

1. This module corresponds to Module D in the third edition.
2. We have expanded the module to include a section on the skill of listening. We have been conditioned throughout our lives to learn how not to listen. Now, we must learn how to listen effectively to be able to understand what people want so we can be successful in our profession. We have provided six guidelines on how to improve your listening skills.
3. We have included a section on speaking. Learning to speak interestingly and effectively is one of the most admired skills in our business and is an attribute that separates the leaders from the followers in our business.
4. We have updated our material on preparing formal presentations to include the use of Microsoft *PowerPoint* and the use of wizards to assist in the creation of such presentations.

MODULE OUTLINE

Section headings and subheadings are boldfaced.

1. **Communicating with people.**
   a. **Four audiences for interpersonal communication during systems projects.**
   b. **Listening.**
   c. **Guidelines in effective listening.**
      i. *Approach the session with a positive attitude.*
      ii. *Set the other person at ease.*
      iii. *Let them know you are listening.*
      iv. *Ask questions.*
      v. *Don't assume anything.*
      vi. *Take notes.*
   d. **Speaking.**
   e. **Use of words: turn-ons and turnoffs.**

   f. **Electronic mail.**
   g. **Body language and proxemics.**
2. **Meetings.**
   a. **Preparing for a meeting.**
      i. *Step 1: Determine the need for and purpose of the meeting.*
      ii. *Step 2: Schedule the meeting and arrange for facilities.*
      iii. *Step 3: Prepare an agenda.*
   b. **Conducting a meeting.**
   c. **Following up on a meeting.**
3. **Formal presentations.**
   a. **Preparing for the formal presentation.**
   b. **Conducting the formal presentatio.n**
      i. *Ways to keep the audience listening.*
      ii. *Answering questions.*
   c. **Following up the formal presentation.**
4. **Project walkthroughs.**
   a. **Who should participate in the walkthrough?**
   b. **Conducting a walkthrough.**
   c. **Following up on the walkthrough.**
5. **Written reports.**
   a. **Business and technical reports.**
      i. *Systems planning reports.*
      ii. *Systems analysis reports.*
      iii. *Systems design reports.*
      iv. *Systems implementation reports.*
   b. **Length of a written report.**
   c. **Organizing the written report.**
      i. *Primary elements.*
      ii. *Secondary elements.*
6. **Writing the business or technical report.**

NOTES

NOTES

NOTES

NOTES

NOTES

SYSTEMS ANALYSIS AND DESIGN METHODS

FOURTH EDITION

SYSTEMS ANALYSIS AND DESIGN METHODS

JEFFREY L. WHITTEN
Professor

LONNIE D. BENTLEY
Professor

Both at Purdue University
West Lafayette, IN

Special contributions by
Kevin C. Dittman
Assistant Professor, Purdue University

Boston, Massachusetts Burr Ridge, Illinois Dubuque, Iowa
Madison, Wisconsin New York, New York San Francisco, California St. Louis, Missouri

Irwin/McGraw-Hill

A Division of The **McGraw-Hill** *Companies*

SYSTEMS ANALYSIS AND DESIGN METHODS

This book is printed on acid-free paper.

1 2 3 4 5 6 7 8 9 0 VNH/VNH 90987

ISBN 0-256-19906-X (student edition); ISBN 0-256-23826-X (instructor's edition)

Publisher: *Tom Casson*
Senior sponsoring editor: *Rick Williamson*
Developmental editor: *Christine Wright*
Marketing manager: *Jim Rogers*
Project manager: *Susan Trentacosti*
Production supervisor: *Bob Lange*
Designer: *Heidi J. Baughman/Maureen McCutcheon*
Cover illustration: *NewBorn Group*
Photo research coordinator: *Keri Johnson*
Prepress buyer: *Charlene R. Perez*
Compositor: *GTS Graphics, Inc.*
Typeface: *10/12 Garamond Light*
Printer: *Von Hoffmann Press, Inc.*

Library of Congress Cataloging-in-Publication Data

Whittin, Jeffrey L.
Systems analysis and design methods / Jeffrey L. Whitten, Lonnie D. Bentley; special contributions by Kevin C. Dittman.—4th ed.
p. cm.
Includes bibliographical references and index.
ISBN 0-256-19906-X (acid-free paper)
0-256-23826-X (instructor's ed.: acid-free paper)
1. System design. 2. System analysis. I. Bentley, Lonnie D. II. Dittman, Kevin C. III. Title.
QA76.9.S88W48 1998
005.1—dc21 97-8260

http://www.mhcollege.com

To my wife, Deb. Your love and encouragement have given new significance to my every accomplishment. Also, to my mother and father. You have always been a source of inspiration.—*Jeff*

To my wife and best friend, Cheryl. And to my children, Coty, Robert, and Heath. God blessed me with a wonderful family.—*Lonnie*

To the students and alumni of the Computer Technology Department at Purdue University, West Lafayette, and Stateside Technology sites. May all your experiences be successful.—*Jeff and Lonnie*

PREFACE

INTENDED AUDIENCE

Systems Analysis and Design Methods, fourth edition, is intended to support a practical course in information systems development. This course is normally taught at the sophomore, junior, senior, or graduate level. It is taught in vocational trade schools, junior colleges, colleges, and universities. The course is taught to both information systems and business majors. Previous editions have been used successfully in one-, two-, and three-semester and quarter-based systems.

The textbook can support a superset of the following courses from the *IS'95 Information Systems Model Curriculum* as jointly developed by the Association of Computing Machinery (ACM), the Association for Information Systems (AIS), and the Data Processing Management Association (DPMA):

IS 95.7	Analysis and Logical Design
IS 95.8	Physical Design and Implementation with DBMS
IS 95.9	Physical Design and Implementation with a Programming Environment

We recommend that students should have taken a computer and information systems literacy course. While **not** required or assumed, a personal computing literacy course, and a programming course can enhance the learning experience provided by this textbook.

WHY WE WROTE THIS BOOK

More than ever, today's students are "consumer-oriented," due in part to the changing world economy that promotes quality, competition, and professional currency. They expect to walk away from a course with more than a grade and a promise that they'll someday appreciate what they've learned. They want to "practice" the application of concepts, not just study applications of concepts. We wrote the previous editions of this book to deal with the following perceived problems:

- Some books were (and still are) *too* conceptual. Examples were too few, too late, or too loosely integrated to generate student interest.
- Some books were (and still are) *too mechanical*. They ignored concepts in favor of techniques. We continue to strive toward a balance of concepts, techniques, and applications.
- To our surprise, many books *still* perpetuate the myth that classical, structured, and modern techniques are mutually exclusive. Students need to develop an appreciation of the value added by continuously improving tools and techniques.
- Most books *still* lack sufficient examples and integrated cases to adequately demonstrate concepts and techniques.

Our goal was to write a textbook that overcomes these problems. Additionally, we wanted to write a textbook that would serve the reader as a post-course, professional reference for best current practices.

Consistent with the first three editions, we have written the book using a lively, conversational tone. Our experience suggests that the more traditional, academic tone detracts from student interest. Our conversational approach admittedly results in a somewhat longer text; however, the "talk with you—not at you" style seems to work well with a wider variety of students. We hope that our style does not offend or patronize any specific audience. We apologize if it does.

CHANGES FOR FOURTH EDITION

We believe that we have preserved all of the features you liked in the previous editions. And in the spirit of continuous improvement we have made the following changes:

- The fourth edition returns to the legacy of its namesake, *systems analysis and design methods*. The coverage of systems planning, systems implementation, and systems support is downsized, and primarily serves to place those peripheral subjects into the context of systems analysis and design. The basic philosophical change was that, "if we want to teach anything worthwhile, we must not try to teach everything." In order to maintain currency in analysis and design, we elected to summarize all other phases.

- The *pyramid* model for conceptual foundations has served us well for three editions, but now gives way to a *matrix* model that more clearly reinforces the concepts of information systems and their development. The Zachman *Framework for Information Systems Architecture* continues to organize the subject's conceptual foundations. The "network" column of the third edition and Zachman's framework has been renamed "geography" to avoid any confusion with "computer networking and technology." Finally, an "interface" column has been added to reflect both the emergence of the graphical user interface and the importance of systems integration.
- This edition further emphasizes systems analysis and design techniques for developing *client/server* or *distributed computing applications*. The mainframe is considered just another server.
- The fundamentals unit (Part One) has been downsized from five chapters to three as a direct response to adopters' requests to accelerate the path into the skill units and chapters.
- The opening chapter (1) relates the overall subject matter to current business and information strategies such as *information strategy planning, total quality management, business process redesign,* and *continuous process improvement.*
- The *information systems building blocks* chapter (2) has been downsized at the request of third edition adopters. Its sole intent in this edition is to introduce the Zachman framework that will organize the concepts, tools, and techniques taught in the course.
- The *systems development life cycle* and *methodology* chapters have been integrated—yet simplified. Life cycle/methodology coverage has been modernized; however, it returns to a derivative of the second edition life cycle that offered a simpler and cleaner "problem-solving approach" to the subject. Details about the phases are deferred to the analysis, design, and implementation chapters that introduce (or conclude) their respective units.
- The *CASE* chapter has been deleted from the fundamentals unit; however, CASE is introduced early and reinforced often throughout the text.
- This edition replaces DOS-based CASE tools with *Windows*-based equivalents. Popkin's *System Architect* provides an inexpensive upper-CASE environment that supports *both* structured and object-oriented methods. Microsoft's *Visual Basic* provides a popular lower-CASE, object-based, rapid application development (RAD) environment that supports design and prototyping. Very similar functionality can be achieved with *Delphi*, *Powerbuilder*, and other visual programming environments.
- The *data modeling* chapter (5) now uses the more popular Martin-style entity relationship diagrams that are common to information engineering.
- The *process modeling* chapter (6) has undergone significant revisions for the first time since the first edition. The top-down approach to DFDs is replaced by the more modern event-driven approach as recommended by McMenamin, Palmer, Yourdon, and many information engineering-based methodologies.
- The *network modeling* (as in data and process distribution) chapter (7) is improved by the incorporation of association matrices from information engineering.
- Object-oriented analysis and design are truly emerging. In the previous edition, we retreated from extensive coverage because we did not sense that any one OO paradigm had developed the requisite industry following. *Object modeling* (Chapter 8) is formally taught in this edition as an *alternative* to data and process modeling. After studying competing methods and paradigms, we have elected to synthesize concepts from several methodologies, but present them within the context of Rumbaugh's *object modeling technique (OMT)* that had a wider industry acceptance than either the Booch or Coad/Yourdon techniques. We also incorporated elements from the very popular use case techniques as developed by Jacobson.

  Like many others, we greatly anticipate the arrival of the so-called unified *Object Modeling Language* that is being developed through the collaboration of Rumbaugh, Jacobson, and Booch. Even Ed Yourdon (of the Coad/Yourdon technique fame) has conceded in his newsletters that the unified language will probably prevail.
- The *project repository* chapter from the third edition has been effectively merged into the data, process, network, and object modeling chapters.
- The design unit (Part Three) pervasively applies a *rapid application development (RAD)* technique using Microsoft's *Visual Basic* for design prototyping. The conceptual framework for this approach is also compatible with other RAD technologies such as Powersoft's *Powerbuilder* and Borland's *Delphi*.
- The *analysis-to-design transition* coverage from third edition has been modified to better support distributed computing client/server environments. The chapter title has been changed to *Application Architecture and Process Design* (Chapter 10).
- The *database design* chapter (11) more adequately addresses modern *distributed database technology* (using *Oracle* as an example) and *data distribution* techniques that are becoming commonplace in client/server computing environments.

- *Input and output design* has been split into separate chapters (12 and 13). Each chapter has been significantly modernized to reflect newer, pervasive technologies and paperless design techniques.
- The *user interface design* chapter (14) has been completely modernized to focus on GUI design strategies and techniques using the popular *Windows* client as an example. Examples of such GUI designs are provided through *Visual Basic*, but they apply to almost any GUI-based development environment.
- The *structured design* chapter (15) has been both simplified and modernized to better teach transition from data flow models to structure charts using a relatively simple and elegant algorithm.
- An *object-oriented design* chapter (16), using the aforementioned OMT technique, has been added to complement the object modeling chapter (8) in the analysis unit (Part Two). It is reasonable to expect that object methods will eventually replace structured methods, but that may not happen until the fifth or sixth edition.
- The *project management* module (A) has received its first-ever complete revision. The revision is based on computer-based project management using Microsoft *Project* tools and techniques.
- An entire module (D) has been added to teach *joint requirements planning/joint application development* (JRP/JAD) techniques as they apply to systems analysis and design.

PEDAGOGICAL USE OF COLOR

The fourth edition continues the use of full color applied to an adaptation of Zachman's *Framework for Information Systems Architecture.* Based on adopter feedback, we have reduced the number of base colors to four (with a fifth color used only in the two object-oriented chapters). The color mappings are as follows:

- Indicates something to do with "data"
- Indicates something to do with "processes"
- Indicates something to do with "geography"
- Indicates something to do with "interface"
- Indicates something to do with "objects" (only applicable to Chapters 8 and 16)

The Information Systems Architecture matrix uses these colors to associate concepts with the model.

ORGANIZATION

Systems Analysis and Design Methods, fourth edition, is divided into five parts. Past experience indicates that instructors can omit and resequence chapters as they feel is important to their audience. Every effort has been made to decouple chapters from one another as much as possible to assist in resequencing the material—even to the extent of reintroducing selected concepts and terminology. The Instructor's Edition and the Instructor's Guide should clearly indicate any exceptions to this general rule.

Part One, "The Context of Systems Analysis and Design Methods," presents the information systems development situation and environment. The chapters introduce the student to systems analysts and the systems development team, information systems building blocks, and a contemporary systems development methodology. Part One has been redesigned so that it can be covered more quickly.

Part Two, "Systems Analysis Methods," covers the front-end life cycle activities, tools, and techniques for analyzing business requirements for an improved system. Coverage can be restricted to either structured or object-oriented techniques, but both are recommended since industry is caught in what should prove to be a lengthy transition from structured methods to object-oriented methods. The network modeling chapter (7) is optional, but recommended in light of client/server application trends that will distribute data, processes, and objects throughout a distributed computing network.

Part Three, "Systems Design and Construction Methods," covers the middle life cycle activities, tools, and techniques. It includes coverage of both general and detailed design with a particular emphasis on design decision making, human factors, and rapid application development.

Part Four, "Beyond Systems Analysis and Design," is a capstone unit that places systems analysis and design into perspective by surveying the back-end life cycle activities. The systems implementation and support chapters (17 and 18) are intended primarily to bridge this textbook's systems analysis and design coverage with the students' prior programming experiences.

Part Five, "Cross Life Cycle Activities and Skills," introduces material that spans multiple phases of systems development. Chapters are replaced by modules that can be interwoven into the course at various points in time (or assigned as supplemental reading). The modules include project and process management (A), fact-finding and information gathering (B), feasibility analysis and cost-benefit analysis (C), joint application development (D), and interpersonal skills and communications (E). The Instructor's Edition provides threads in appropriate chapters that might be enhanced by introducing material from these modules.

See the Instructor's Annotated Edition or the Instructor's Guide for more information on how to use this book in a variety of course formats and sequences.

SUPPLEMENTS

It has always been our intent to provide our adopters with a complete course, not just a textbook. We are especially excited about this edition's comprehensive support package. It includes software and instructional bundles for both the student and instructor. The supplements for this edition include the following components.

For the Instructor

Instructor's Edition. Once again, an Instructor's Edition of the textbook is available. For each chapter, it provides an instructor's overview, chapter-to-course sequencing (new!), summary of changes for the fourth edition, and a topical/sentence outline (new!).

Instructor's Guide. For those instructors more comfortable with a separate instructor supplement, an Instructor's Guide is available. It provides the same front- and back-matter as the Instructor's Edition, but also provides instructional notes embedded within a "notes pages" printout of complete Microsoft *PowerPoint* presentations. The electronic versions of these presentations will either be available through your McGraw-Hill sales representative, or downloadable from the textbook's World Wide Web site.

Test Bank. A written Test Bank covering all the chapters and modules includes more than 3,000 questions. Each chapter and module has questions in the following formats: true/false, multiple choice, sentence completion, and matching. The test bank and answers are cross referenced to the page numbers in the textbook. An electronic version of the test bank is available for the creation of customized exams.

Laboratory CASE Tools. Popkin Software & Systems, Inc. is making available a *University Edition* program for their CASE product, *System Architect*. *System Architect* provides *Windows*-based CASE tools for all of the techniques taught in this edition, inclusive of object-oriented analysis and design. The *University Edition* software provides identical functionality to the commercial product to support large student projects, at a nominal cost. This software requires a network. To obtain the software, contact:

University Edition Program Manager
Popkin Software
11 Park Place
New York, NY 10007
(212) 571-3434

Academic grant programs are available for other CASE tools including Visible Systems' *Visible Analyst Workbench*. Your McGraw-Hill representative can provide further details.

RAD Tools. Academic grants or educational discounts are widely available for RAD software development environments such as *Visual Basic*, *Powerbuilder*, and *Delphi*. All of these development environments support visual construction of graphical user interfaces and various degrees of database connectivity and rapid application development.

For the Student

Application Cases Workbook. The application cases workbook has been rewritten by Professor Kevin C. Dittman under our direct review. The new format provides for multiple minicases that are cross-referenced for use in specific chapters and modules. Additionally, the workbook provides three challenging case studies suitable for a term project with milestones. Also, the *Build Your Own Case* option has been further refined for this edition.

CASE Tools. We recognize that greater numbers of students own their own PCs. Now they can install CASE technology on those PCs. Popkin Software is providing a *Student Edition* of *System Architect (Book Mode)*. It supports all of the diagrams covered in the textbook, and it is restricted to 10 diagrams, 300 symbols, and 400 definitions per encyclopedia. That should be more than sufficient to support homework and subsets of project work. (If the adopting school purchases the networked *University Edition,* students can directly upgrade to the *Student Edition (Lab Mode),* which triples these limits.) The diagrams developed in both *Student Editions* can be exported to, and imported from, the full-function *University Edition* of *System Architect.* The software is packaged with a tutorial written by our colleague, Professor Kevin C. Dittman. It is available only through McGraw-Hill and your local bookstore.

In addition, Visible Systems' *Visible Analyst Workbench (Student Edition)* can be packaged with the text. It has similar functionality to *System Architect's Student Edition.*

ACKNOWLEDGMENTS

We are indebted to many individuals who have contributed to the development of four editions of this textbook. First, we wish to thank Susan Solomon, Larry Alexander, and Rick Williamson who in succession managed the editorial development of the first three editions. These individuals each contributed support to their management for what, at the time, must have seemed to be unusual author demands for subject matter content, format, and strategy. We owe a debt of gratitude for all those times they supported our ideas to top management! Thanks for believing! We also thank their colleagues and staffs who contributed to the success of the prior editions.

We would also like thank our colleague Kevin Dittman who wrote several modules, collaborated with us on the object-oriented chapters, and wrote many of the supplements. We look forward to Kevin's increased role in the future editions of this book.

We would also like to thank our colleague Carlin R. Smith who developed database and screen prototypes for this edition. His detailed understanding of the *Visual Basic* and *SQL Server* client/server environments added value in the form of technical currency.

We wish to thank the reviewers and critics of this and prior editions:

Jeanne M. Alm, *Moorhead State University*
Charles P. Bilbrey, *James Madison University*
Ned Chapin, *California State University—Hayward*
Carol Clark, *Middle Tennessee State University*
Gail Corbitt, *California State University—Chico*
Barbara B. Denison, *Wright State University*
Linda Duxbury, *Carleton University*
Craig W. Fisher, *Marist College*
Dennis D. Gagnon, *Santa Barbara City College*
Abhijit Gopal, *University of Calgary*
Patricia J. Guinan, *Boston University*
Constance Knapp, *Pace University*
Riki S. Kuchek, *Orange Coast College*
Thom Luce, *Ohio University*
Charles M. Lutz, *Utah State University*
William H. Moates, *Indiana State University*
Ronald J. Norman, *San Diego State University*
Charles E. Paddock, *University of Nevada—Las Vegas*
June A Parsons, *Northern Michigan University*
Gail L. Rein, *SUNY—Buffalo*
Rebecca H. Rutherfoord, *Southern College of Technology*
Jerry Sitek, *Southern Illinois University—Edwardsville*
Craig W. Slinkman, *University of Texas—Arlington*
Mary Thurber, *Northern Alberta Institute of Technology*
Jerry Tillman, *Appalachian State University*
Margaret S. Wu, *University of Iowa*
Jacqueline E. Wyatt, *Middle Tennessee State University*
Ahmed S. Zaki, *College of William and Mary*

Your patience and constructive criticism were essential and appreciated. As you might guess, opinions from different individuals were varied. We have incorporated as many as possible into this edition. And as usual, we have already started a file for fifth edition—after we catch our breath, of course.

Special thanks are offered to Dorothy Jane Miller and JoAnne Anderson, our secretaries, who provided special assistance throughout the project. Not a single edition would have ever seen the light of day without their support and service.

We also offer thanks to our colleagues, J. P. Lisack (for his regular manpower reports that demonstrate the continued demand for qualified systems analysts) and Duane Dunlap (for his insightful contributions to the world of automatic identification).

Dean Don K. Gentry of the School of Technology is singled out for sustaining an academic environment that encourages and rewards this type of creative endeavor and scholarship.

We also thank our students and alumni. Your future is our incentive. You make teaching a worthwhile experience. We truly grow through your successes!

Finally, we acknowledge the contributions, encouragement, and patience of the staff of Irwin/McGraw-Hill. For aspiring authors, you can't do any better than these folks! For this edition, special thanks to Rick Williamson for never losing his cool in the wake of deadlines we missed; to Christine Wright, our developmental editor; to Susan Trentacosti, who as production editor for two editions has established the benchmark of excellence in that role. We also thank Bob Lange, Charlene Perez, Heidi Baughman, Keith McPherson, Keri Johnson, and Jim Rogers.

We hope we haven't forgotten anyone. And we assume full responsibility for any inadequacies or errors in this text. We eagerly await your reactions and willingly provide the following Internet forums for your comments and suggestions:

Email:

jlwhitten@tech.purdue.edu

ldbentley@tech.purdue.edu

Internet:

www.tech.purdue.edu/textbooks/sadm

www.mhhe.com

To those who used our previous three editions, thank you for your continued support. And for new adopters, we hope you'll see a difference in this text. We hope you enjoy using the textbook as much as we have enjoyed writing it. And until the *fifth edition*, go forth and teach . . . there is no nobler profession!

Jeffrey L. Whitten
Lonnie D. Bentley

BRIEF CONTENTS

PART ONE

The Context of Systems Analysis and Design Methods 1

1 The Modern Systems Analyst 2

2 Information System Building Blocks 32

3 Information System Development 70

PART TWO

Systems Analysis Methods 117

4 Systems Analysis 118

5 Data Modeling 170

6 Process Modeling 208

7 Network Modeling 262

8 Object Modeling 284

PART THREE

Systems Design and Construction Methods 309

9 Systems Design and Construction 310

10 Application Architecture and Process Design 350

11 Database Design 392

12 Input Design and Prototyping 436

13 Output Design and Prototyping 462

14 User Interface Design and Prototyping 482

15 Software Design 508

16 Object-Oriented Design 532

PART FOUR

Beyond Systems Analysis and Design 553

17 Systems Implementation 554

18 Systems Support 576

PART FIVE

Cross Life Cycle Activities and Skills 593

A Project and Process Management 594

B Fact-Finding and Information Gathering 622

C Feasibility and Cost-Benefit Analysis 642

D Joint Application Development 662

E Interpersonal Skills and Communications 670

Glossary/Index 692

CONTENTS

PART ONE THE CONTEXT OF SYSTEMS ANALYSIS AND DESIGN METHODS 1

CHAPTER 1

The Modern Systems Analyst 2

SoundStage Entertainment Club 3

Who Should Read This Book? 7

The Systems Analyst as a Modern Business Problem Solver 7

- Why Do Businesses Need Systems Analysts? 8
- What Is a Systems Analyst? 8
- What Does a Systems Analyst Do? 9
- Where Do Systems Analysts Work? 11

Customers, Partners, and Expectations 16

- Customers—Users and Management 16
- The Roles of Management and Users in Systems Problem Solving 17
- Modern Business Trends and Implications for the Systems Analyst 18
- Partners for the Systems Analyst—Information Technologists and Vendors 20

Preparing for a Career as a Systems Analyst 20

- Working Knowledge of Information Technology 21
- Computer Programming Experience and Expertise 22
- General Business Knowledge 22
- Problem-Solving Skills 23
- Interpersonal Communications Skills 23
- Interpersonal Relations Skills 23
- Flexibility and Adaptability 24
- Character and Ethics 24
- Systems Analysis and Design Skills 25

The Next Generation 25

- Career Prospects 25
- Predictions 26

Where Do You Go from Here? 26

CHAPTER 2

Information System Building Blocks 32

SoundStage Entertainment Club 33

A Review of Fundamentals of Information Systems 37

- Transaction Processing Systems 38
- Management Information Systems 39
- Decision Support Systems 39
- Expert Systems 40
- Office Information Systems 40
- Personal and Work Group Information Systems 41
- Putting It All Together 42

A Framework for Information Systems Architecture 42

Perspectives—The People Side of Information Systems 42

- System Owners 44
- System Users 45
- System Designers 47
- System Builders 47
- The Role of the Systems Analyst 47

Building Blocks—Expanding the Information System Framework 48

- Building Blocks of DATA 50
- Building Blocks of PROCESSES 54
- Building Blocks of INTERFACES 57
- Building Blocks of GEOGRAPHY 60

Where Do You Go from Here? 65

CHAPTER 3

Information System Development 70

SoundStage Entertainment Club 71

System Development Life

Cycles and Methodologies 72
Underlying Principles of Systems Development 73
Principle 1: Get the Owners and Users Involved 73
Principle 2: Use a Problem-Solving Approach 74
Principle 3: Establish Phases and Activities 74
Principle 4: Establish Standards for Consistent Development and Documentation 75
Principle 5: Justify Systems as Capital Investments 76
Principle 6: Don't Be Afraid to Cancel or Revise Scope 76
Principle 7: Divide and Conquer 76
Principle 8: Design Systems for Growth and Change 77
FAST—A System Development Methodology 78
How a *FAST* Project Gets Started 78
An Overview of the *FAST* Life Cycle and Methodology 79
The Survey Phase 84
The Study Phase 87
The Definition Phase 88
The Configuration Phase 91
The Procurement Phase 93
The Design Phase 94
The Construction Phase 96
The Delivery Phase 98
Beyond Systems Development—Systems Support 99
Cross Life Cycle Activities 99
Fact-Finding 99
Documentation and Presentations 100
Estimation and Measurement 101
Feasibility Analysis 102
Project and Process Management 102
Computer-Aided Systems Engineering (CASE) 103
The History and Evolution of CASE Technology 103
A CASE Tool Framework 104
CASE Tool Architecture 105
The Benefits of CASE 107
Where Do You Go from Here? 108

PART TWO SYSTEMS ANALYSIS METHODS 117

CHAPTER 4

Systems Analysis 118
SoundStage Entertainment Club 119
What Is Systems Analysis? 120
Strategies for Systems Analysis and Problem Solving 122
Modern Structured Analysis 122
Information Engineering (IE) 123
Prototyping 124
Joint Application Development (JAD) 125
Business Process Redesign (BPR) 126
Object-Oriented Analysis (OOA) 126
FAST Systems Analysis Strategies 127
The Survey Phase of Systems Analysis 128
Activity: Survey Problems, Opportunities, and Directives 129
Activity: Negotiate Project Scope 132
Activity: Plan the Project 134
Activity: Present the Project 135
The Study Phase of Systems Analysis 137
Activity: Model the Current System 139
(OPT) Activity: Analyze Business Processes 142
Activity: Analyze Problems and Opportunities 143
Activity: Establish System Improvement Objectives and Constraints 145
Activity: Modify Project Scope and Plan 147
(REC) Activity: Present Findings and Recommendations 149
The Definition Phase of Systems Analysis 151
Activity: Outline Business Requirements 151
Activity: Model Business System Requirements 154
(OPT) Activity: Build Discovery Prototypes 157
Activity: Prioritize Business Requirements 159
Activity: Modify the Project Plan and Scope 161
Some Final Words about System Requirements 163
The Next Generation of Requirements Analysis 163
Where Do You Go from Here? 164

CHAPTER 5

Data Modeling 170
SoundStage Entertainment Club 171
An Introduction to Systems Modeling 172
System Concepts for Data Modeling 175
Entities 175
Attributes 176
Relationships 179
The Process of Logical Data Modeling 186

Strategic Data Modeling 186
Data Modeling during Systems Analysis 187
Looking Ahead to Systems Configuration and Design 188
Fact-Finding and Information Gathering for Data Modeling 189
Computer-Aided Systems Engineering (CASE) for Data Modeling 189
How to Construct Data Models 190
Entity Discovery 190
The Context Data Model 193
The Key-Based Data Model 194
Generalized Hierarchies 197
The Fully Attributed Data Model 197
The Fully Described Model 199
The Next Generation 199
Where Do You Go from Here? 202

CHAPTER 6

Process Modeling 208
SoundStage Entertainment Club 209
An Introduction to Systems Modeling 210
System Concepts for Process Modeling 214
Process Concepts 215
Data Flows 225
External Agents 233
Data Stores 235
The Process of Logical Process Modeling 235
Strategic Systems Planning 235
Process Modeling for Business Process Redesign 236
Process Modeling during Systems Analysis 236
Looking Ahead to Systems Configuration and Design 237
Fact-Finding and Information Gathering for Process Modeling 239
Computer-Aided Systems Engineering (CASE) for Process Modeling 239
How to Construct Process Models 239
The Context Diagram 239
The Functional Decomposition Diagram 241
The Event-Response List 243
The Event Decomposition Diagram 244
Event Diagram 245
The System Diagram 247
Primitive Diagrams 250
The Next Generation 253
Where Do You Go from Here? 254

CHAPTER 7

Network Modeling 262
SoundStage Entertainment Club 263
Network Modeling—Not Just for Computer Networks 264
System Concepts for Network Modeling 266
Business Geography 267
Miscellaneous Constructs 272
Synchronizing of System Models 272
The Process of Logical Network Modeling 275
Network Modeling during Strategic Systems Planning Projects 276
Network Modeling during Systems Analysis 276
Looking Ahead to Systems Design 276
Fact-Finding and Information Gathering for Network Modeling 276
Computer-Aided Systems Engineering (CASE) for Network Modeling 276
How to Construct Logical Network Models 277
Location Decomposition Diagram 277
Location Connectivity Diagram 277
The Next Generation 279
Where Do You Go from Here? 280

CHAPTER 8

Object Modeling 284
SoundStage Entertainment Club 285
An Introduction to Object Modeling 286
System Concepts for Object Modeling 286
Objects, Attributes, Methods, and Encapsulation 286
Classes, Generalization, and Specialization 288
Object/Class Relationships 290
Messages and Message Sending 293
Polymorphism 293
The Process of Object Modeling 294
Finding and Identifying the Business Objects 294
Organizing the Objects and Identifying Their Relationships 303
Where Do You Go from Here? 305

PART THREE SYSTEMS DESIGN AND CONSTRUCTION METHODS 309

CHAPTER 9

Systems Design and Construction 310

SoundStage Entertainment Club 311

What Is Systems Design? 312

Strategies for Systems Design 312

- Modern Structured Design 312
- Information Engineering (IE) 314
- Prototyping 314
- Joint Application Development (JAD) 316
- Rapid Application Development (RAD) 316
- Object-Oriented Design (OOD) 317

FAST Systems Design Methods 317

The Configuration Phase of Systems Design 319

- Activity: Define Candidate Solutions 319
- Activity: Analyze Feasibility of Alternative Solutions 321
- Activity: Recommend a System Solution 324

The Procurement Phase of Systems Design 326

- Activity: Research Technical Criteria and Options 326
- Activity: Solicit Proposals (or Quotes) from Vendors 329
- Activity: Validate Vendor Claims and Performances 330
- Activity: Evaluate and Rank Vendor Proposals 332
- Activity: Award (or Let) Contract and Debrief Vendors 333
- Activity: Establish Integration Requirements 334

The Design and Integration Phase of Systems Design 335

- Activity: Analyze and Distribute Data 336
- Activity: Analyze and Distribute Processes 339
- Activity: Design Databases 340
- Activity: Design Computer Outputs and Inputs 341
- Activity: Design On-Line User Interface 342
- Activity: Present and Review Design 343

Where Do You Go from Here? 344

CHAPTER 10

Application Architecture and Process Design 350

SoundStage Entertainment Club 351

General System Design 352

Information Technology Architecture 353

- NETWORK Architectures for Client/Server Computing 353
- DATA Architectures for Distributed Relational Databases 362
- INTERFACE Architectures—Inputs, Outputs, and Middleware 364
- PROCESS Architecture—The Software Development Environment and System Management 368

Application Architecture Strategies and Design Implications 370

- The Enterprise Application Architecture Strategy 371
- The Tactical Application Architecture Strategy 371
- Build versus Buy Implications 371

Modeling Application Architecture and Information System Processes 372

- Physical Data Flow Diagrams 372
- System Flowcharts 376
- Computer-Aided Systems Engineering (CASE) for Physical DFDs and Flowcharts 380

Designing the Application Architecture and the Information System Processes 380

- Drawing Physical DFDs 380
- Prerequisites 380
- The Network Topology DFD 380
- Data Distribution and Technology Assignments 383
- Process Distribution and Technology Assignments 383
- The Person/Machine Boundaries 384

Where Do You Go from Here? 387

CHAPTER 11

Database Design 392

SoundStage Entertainment Club 393

Conventional Files versus the Database 395

- The Pros and Cons of Conventional Files 395
- The Pros and Cons of Databases 397
- Database Design in Perspective 397

Database Concepts for the Systems Analyst 397

- Fields 398
- Records 399
- Files and Tables 400
- Databases 401

Data Analysis for Database Design 407

- What Is a Good Data Model? 407
- Data Analysis 408
- Normalization Example 409

File Design 418

Database Design 418

- Goals and Prerequisites to Database Design 419
- The Database Schema 419
- Data and Referential Integrity 423

Roles 424
Database Prototypes 424
Database Capacity Planning 426
Database Structure Generation 426
The Next Generation of Database Design 426
Where Do You Go from Here? 427

CHAPTER 12

Input Design and Prototyping 436
SoundStage Entertainment Club 437
Methods and Issues for Data Capture and Input 438
Data Capture, Data Entry, and Data Input 438
Modern Input Methods: Batch versus On-Line Inputs 439
Trends in Automatic Data Collection Technology 440
System User Issues for Input Design 442
Internal Controls for Inputs 443
GUI Controls for Input Design 445
Text Box 446
Radio Button 447
Check Box 447
List Box 448
Drop-Down List 449
Combination (Combo) Box 450
Spin (Spinner) Box 450
How to Prototype and Design Computer Inputs 450
Step 1: Review Input Requirements 451
Step 2: Select the GUI Controls 452
Step 3: Prototype the Input Screen 453
Step 4: If Necessary, Design or Prototype the Source Document 455
Where Do You Go from Here? 457

CHAPTER 13

Output Design and Prototyping 462
SoundStage Entertainment Club 463
Principles and Guidelines for Output Design 464
Types of Outputs 464
Output Media and Formats 466
System User Issues for Output Design 471
How to Prototype and Design Computer Outputs 471
Step 1: Identify System Outputs 471
Step 2: Select Output Medium and Format 472
Step 3: Prototype the Output for System Users 473
Where Do You Go from Here? 477

CHAPTER 14

User Interface Design and Prototyping 482
SoundStage Entertainment Club 483
Styles of User Interfaces 484
Menu Selection 484
Instruction Sets 489
Question-Answer Dialogues 491
Direct Manipulation 492
Human Factors for User Interface Design 493
General Human Engineering Guidelines 494
Dialogue Tone and Terminology 495
Display Features That Affect User Interface Design 495
Display Area 496
Character Sets and Graphics 496
Paging and Scrolling 496
Display Properties 496
Split-Screen and Windowing Capabilities 496
Keyboards and Function Keys 496
Pointer Options 497
How to Design and Prototype a User Interface 497
Step 1: Chart the Dialogue 497
Step 2: Prototype the Dialogue and User Interface 499
Step 3: Obtain User Feedback 500
Where Do You Go from Here? 503

CHAPTER 15

Software Design 508
SoundStage Entertainment Club 509
What Is Software Design? 510
Structured Design 511
Structure Charts 511
Data Flow Diagrams of Programs 513
Transform Analysis 515
Transaction Analysis 521
Structure Chart Quality Assurance Checks 523
Packaging Program Specifications 527
Where Do You Go from Here? 527

CHAPTER 16

Object-Oriented Design 532
SoundStage Entertainment Club 533
An Introduction to Object-Oriented Design 534
- Design Objects 534
- Object Responsibilities 535
- Object Reusability 535

The Process of Object Design 536
- Refining the Use Case Model to Reflect the Implementation Environment 536
- Modeling Object Interactions and Behaviors that Support the Use Case Scenario 539
- Updating the Object Model to Reflect the Implementation Environment 547

Where Do You Go from Here? 547

PART FOUR BEYOND SYSTEMS ANALYSIS AND DESIGN 553

CHAPTER 17

Systems Implementation 554
SoundStage Entertainment Club 555
What Is Systems Implementation? 556
FAST Systems Implementation Methods 556
The Construction Phase of Systems Implementation 557
- Activity: Build and Test Networks (if Necessary) 557
- Activity: Build and Test Databases 560
- Activity: Install and Test New Software Package (if Necessary) 561
- Activity: Write and Test New Programs 562

The Delivery Phase of Systems Implementation 564
- Activity: Conduct System Test 564
- Activity: Prepare Conversion Plan 566
- Activity: Install Databases 569
- Activity: Train System Users 569
- Activity: Convert to New System 571

Where Do You Go from Here? 572

CHAPTER 18

Systems Support 576
SoundStage Entertainment Club 577
What Is Systems Support? 578
System Maintenance—Correcting Errors 580
- Define and Validate the Problems 580
- Benchmark the Programs and Applications 581
- Understand the Application and Its Programs 582
- Edit and Test the Programs 583
- Update Documentation 583

System Recovery—Overcoming the "Crash" 584
End-User Assistance 584
Systems Enhancement and Reengineering 584
- Analyze Enhancement Request 585
- Write Simple, New Programs 586
- Restructure Files or Databases 587
- Analyze Program Library and Maintenance Costs 588
- Reengineer and Test Programs 588

The Year 2000 and Systems Support 589
Where Do You Go from Here? 589

PART FIVE CROSS LIFE CYCLE ACTIVITIES AND SKILLS 593

MODULE A

Project and Process Management Techniques 594
What Is Project Management? 595
- Project Management Causes of Failed Projects 596
- The Basic Functions of the Project Manager 597
- Project Management Software 599

Process Management 599
- Management of the Methodology 601
- Management of System Development Technology 601
- Total Quality Management 601
- Metrics and Measurement 602
- The Development Center 602

Project Management Tools and Techniques 602
- Gantt Charts 602
- PERT Charts 609

Expectations Management 613
The Expectations Management Matrix 613
Using the Expectations Management Matrix 615
People Management 617
The One Minute Manager 617
The Subtle Art of Delegation and Accountability 617

MODULE B

Fact-Finding and Information Gathering 622
What Is Fact-Finding? 623
What Facts Does the Systems Analyst Need to Collect and When? 623
What Fact-Finding Methods Are Available? 623
Sampling of Existing Documentation, Forms, and Files 624
Collecting Facts from Existing Documentation 624
Document and File Sampling Techniques 624
Research and Site Visits 626
Observation of the Work Environment 628
Collecting Facts by Observing People at Work 628
Guidelines for Observation 629
Questionnaires 630
Collecting Facts by Using Questionnaires 630
Types of Questionnaires 631
Developing a Questionnaire 632
Interviews 632
Collecting Facts by Interviewing People 632
Interview Types and Techniques 633
How to Conduct an Interview 633
Rapid Application Development (RAD) 636
Fact-Finding Ethics 637
A Fact-Finding Strategy 638

MODULE C

Feasibility and Cost-Benefit Analysis 642
Feasibility Analysis—A Creeping Commitment Approach 643
Feasibility Checkpoints in the Life Cycle 643
Four Tests for Feasibility 646
Operational Feasibility 646
Technical Feasibility 647
Schedule Feasibility 648
Economic Feasibility 648
The Bottom Line 648
Cost-Benefit Analysis Techniques 649
How Much Will the System Cost? 649
What Benefits Will the System Provide? 650
Is the Proposed System Cost-Effective? 652
Feasibility Analysis of Candidate Systems 656
Candidate Systems Matrix 656
Feasibility Analysis Matrix 657

MODULE D

Joint Application Development 662
Joint Application Development 663
JAD Participants 663
Sponsor 663
JAD Leader (or Facilitator) 663
Users and Managers 664
Scribe(s) 664
IS Staff 664
How to Plan and Conduct JAD Sessions 664
Planning the JAD Session 664
Conducting a JAD Session 667
Benefits of JAD 667

MODULE E

Interpersonal Skills and Communications 670
Communicating with People 671
Four Audiences for Interpersonal Communication during Systems Projects 672
Listening 672
Guidelines in Effective Listening 672
Speaking 673
Use of Words: Turn-ons and Turnoffs 674
Electronic Mail 675
Body Language and Proxemics 675
Meetings 677
Preparing for a Meeting 677
Conducting a Meeting 678
Following Up on a Meeting 679
Formal Presentations 679
Preparing for the Formal Presentation 679
Conducting the Formal Presentation 681
Following Up the Formal Presentation 683
Project Walkthroughs 683
Who Should Participate in the Walkthrough? 683
Conducting a Walkthrough 683

Following Up on the Walkthrough 684

Written Reports 685

Business and Technical Reports 685

Length of a Written Report 687

Organizing the Written Report 687

Writing the Business or Technical Report 688

Glossary/Index 692

PART ONE

THE CONTEXT OF SYSTEMS ANALYSIS AND DESIGN METHODS

This is a practical book about information systems development methods. All businesses and organizations develop information systems. You can be assured that you will play some role in the systems analysis and design for those systems—either as a customer or user of those systems or as a developer of those systems. Systems analysis and design is about business problem solving and computer applications. The methods you will learn in this book can be applied to a wide variety of problem domains, not just those involving the computer.

Before we begin, we assume you've completed an introductory course in computer-based information systems. Many of you have also completed one or more application development or programming courses (using technologies such as *Access, dBASE, COBOL, C/C++, Visual BASIC,* or *Delphi*). That will prove helpful, since systems analysis and design are activities that precede and/or integrate with those activities. But don't worry—we'll review all the necessary principles on which the subject of systems analysis and design is based.

Part One focuses on the big picture. Before you learn about specific activities, tools, techniques, methods, and technology, you need to understand this big picture. As you explore the context of systems analysis and design, we will be introducing many ideas, tools, and techniques that are not explored in great detail until later in the book. Try to keep that in mind as you explore the big picture.

Systems development isn't a mechanical activity. There are no magic secrets for success, no perfect tools, techniques, or methods. To be sure, there are skills that can be mastered. But the complete and consistent application of those skills is still somewhat of an art.

We start in Part One with fundamental concepts, philosophies, and trends that provide the context of systems analysis and design methods—in other words, the basics! If you understand these basics, you will be better able to apply, with confidence, the practical tools and techniques you will learn in Parts Two through Five. Furthermore, you will find yourself able to adapt to new situations and methods.

Three chapters make up this part. Chapter 1, The Modern Systems Analyst, introduces you to the *PARTICIPANTS* in systems analysis and design with special emphasis on the modern systems analyst as the facilitator of systems work. You'll also learn about the relationships among systems analysts, end-users, managers, and other information systems professionals. Finally, you'll learn to prepare yourself for a career as an analyst (if that is your goal). Regardless, you will understand how you will interact with this important professional.

Chapter 2, Information System Building Blocks, introduces the *PRODUCT* we will teach you how to build—*information systems.* Specifically, you will learn to examine information system in terms of common information system building blocks: DATA, PROCESSES, INTERFACES, and GEOGRAPHY—each from the perspective of different participants or stakeholders. A visual matrix framework will help you organize these building blocks so that you can see them applied in the subsequent chapters.

Chapter 3, Information System Development, introduces a high-level (meaning general) process for information system development. This is called a *system development life cycle.* We will present the life cycle in a form that most of you will experience it—a *system development methodology.* The methodology presented consists of eight logical phases: *survey, study, definition, configuring, purchasing, design, construction,* and *implementation.* This methodology will be the context in which you will learn to use and apply the systems analysis and design methods taught in the remainder of the book.

1

THE MODERN SYSTEMS ANALYST

CHAPTER PREVIEW AND OBJECTIVES

No matter what your occupation or position in any business, you will likely encounter *systems analysts.* Some of you will become systems analysts. This chapter introduces you to systems analysts, the people who usually facilitate computer applications development through the methods described in this book. You will also learn about the systems analysts' customers and partners. The chapter will address the following questions:

- What are the systems analyst's role and responsibilities in the modern business?
- Why are organizations recruiting computer end-users to partner with the traditional systems analyst?
- What are the roles of systems analysts and business users in systems problem solving?
- Where are the career opportunities for systems analysts?
- Who are the systems analyst's customers and partners in systems development?
- What business trends and drivers are influencing the careers of systems analysts?
- How can you prepare yourself for a career as a systems or business analyst?
- What does the future hold for systems analysts?

CASE STUDY

This is the continuing story of Sandra Shepherd and Bob Martinez, systems analysts for SoundStage Entertainment Club. SoundStage Entertainment Club is one of the fastest-growing music and video clubs in America. The company headquarters, central region warehouse, and sales office are located in Indianapolis, Indiana. Other sales offices are located in Baltimore and Seattle.

Systems analysis and design is more than concepts, tools, techniques, and methods. It is about people working with people. Although experience is the best teacher, you can learn a great deal by observing other systems analysts in action. Nancy Picard, vice president of Information Systems Services (ISS) for SoundStage Entertainment Club, has kindly consented to let you watch two of her analysts on a typical project.

Sandra Shepherd, a senior systems analyst and project manager, has volunteered for this demonstration. She has successfully implemented several information systems for SoundStage and should be able to provide you with a valuable learning experience. Bob Martinez, Sandra's partner, is a new programmer/analyst at SoundStage. In fact, today is his first day!

Bob has to go through orientation today, and Nancy has invited you to observe the orientation. It'll be a good way for you and Bob to get acquainted with SoundStage.

SCENE

In Nancy's office where Bob has been directed after completing his employment paperwork.

NANCY

Hi, Bob! It's great to have you aboard. My name's Nancy Picard, and I'm the vice president of Information Services. I didn't get a chance to meet you when you interviewed last month. Why don't you tell me a little about yourself?

BOB

Well, I just received my bachelor's degree in information systems from State University. I was president of my local student chapter of the Association for Information Technology Professionals and hope to get involved in the Indianapolis professional chapter as well. My career goals are oriented toward applying the systems analysis and design skills I learned in college. That's the main reason I accepted this job. It looked like I'd get a chance to do some analysis and design here—not just programming.

NANCY

That's why we hired you, Bob. You may be interested to know that we were especially impressed by your classroom experience with computer tools for systems analysis and design. We just bought a package called *System Architect* from Popkin Software & Systems, Inc. We want you to learn and use that package on your first project and report your experiences back to the management staff. You'll learn more about this technology from your partner, Sandra Shepherd, a senior systems analyst who is very familiar with the tools and techniques you learned in college.

BOB

I've heard of *System Architect.* We used Intersolv's *Excelerator* in my systems analysis courses at State, but our instructors really impressed on us the importance of learning the underlying tools, not just the CASE technology. I'm confident that I can quickly teach myself any new tool that supports standard analysis and design techniques.

NANCY

Sandra will show you the ropes and help you learn about SoundStage and our way of doing things. You'll meet with Sandra soon. OK, let's take a tour of the building.

[Walking up to the second floor . . .]

NANCY

Did you hear the big news . . . ? *[pause]* . . . We just acquired controlling interest in Private Screenings Video Club. Along with our recent acquisition of GameScreen, the electronic games club, we can now offer comprehensive entertainment merchandise to our members. Of course, consolidating our operations and information services will be a real challenge. Your first assignment will directly address that challenge.

Here we are. The second floor houses several departments including Personnel, Building Services, Accounting, Marketing, and Finance. These outside wall offices belong to executive managers, staff, and assistants.

I don't know how much you know or remember about SoundStage, so I'll give you a quick overview. We used to be called SoundStage Record and Tape Club–a long-play record and cassette tape only subscription service. As you know, LP records are obsolete, and audio entertainment alternatives have been expanded to include compact discs and minidiscs. The GameScreen merger will add video games to our inventory. And the Private Screenings merger will add videotapes and discs to the product mix. Our management believes the new CD-sized video discs will become a big market. This new product mix is why we changed our name to SoundStage *Entertainment* Club.

Customers join the club through advertisements and member referrals. We want to start looking at electronic commerce options on the Internet and commercial nets. Our advertisements typically dangle a carrot such as, "Choose any 10 CDs for a penny and agree to buy 10 more within two years at regular club prices." I'm sure you've seen such offers.

BOB

Yes, I've been a SoundStage member for three years.

NANCY

Club members receive monthly promotions and catalogs that offer a selection of the month. They must respond to the offer within a few weeks or that selection will automatically be shipped and billed to their account. Customers can also order alternative selections and special merchandise from the catalogs. After members fulfill their original subscription agreement, they are eligible for bonus coupons that may be redeemed for free merchandise from our catalogs.

SOUNDSTAGE

SOUNDSTAGE ENTERTAINMENT CLUB

BOB

How many members are there?

NANCY

The last number I heard was 340,750. Of that total, about 180,000 accounts are active, having purchased merchandise in the last 12 months. But that doesn't include the GameScreen and Private Screenings members that we will soon inherit.

[Walking along the hallway of executive offices.]

NANCY

The office on your left belongs to our president and chief executive officer, Steven Short. This is our organization chart, Bob (see Figure A). As you can see, SoundStage is divided into three main divisions including the Information Services division. The offices you walked through on your way to this office belong to the Administrative Services division.

Down on the first floor we have the Operations division, which handles day-to-day operations including customer services, purchasing, inventory control, warehousing, and shipping and receiving. I will soon take you through those facilities.

In the basement, you'll find the Information Services division, where you'll have your office. Our computer operations are located down there, along with my staff, which totals about 35 people.

[Down to the first floor; another maze of offices.]

NANCY

These are the offices for the Operations division, including Purchasing and Inventory Control. They buy the merchandise we will resell to the customer.

And this office area we are approaching is the Customer Services division. Those clerks are processing orders, back orders, follow-ups, and other customer transactions. That's where your first project is going to be. Sandra will be taking you there to meet your customers.

BOB

Customers?

NANCY

We refer to the users of our services as our customers. The customer focus is a cornerstone of our total quality culture

FIGURE A *SoundStage Entertainment Club Organization Chart*

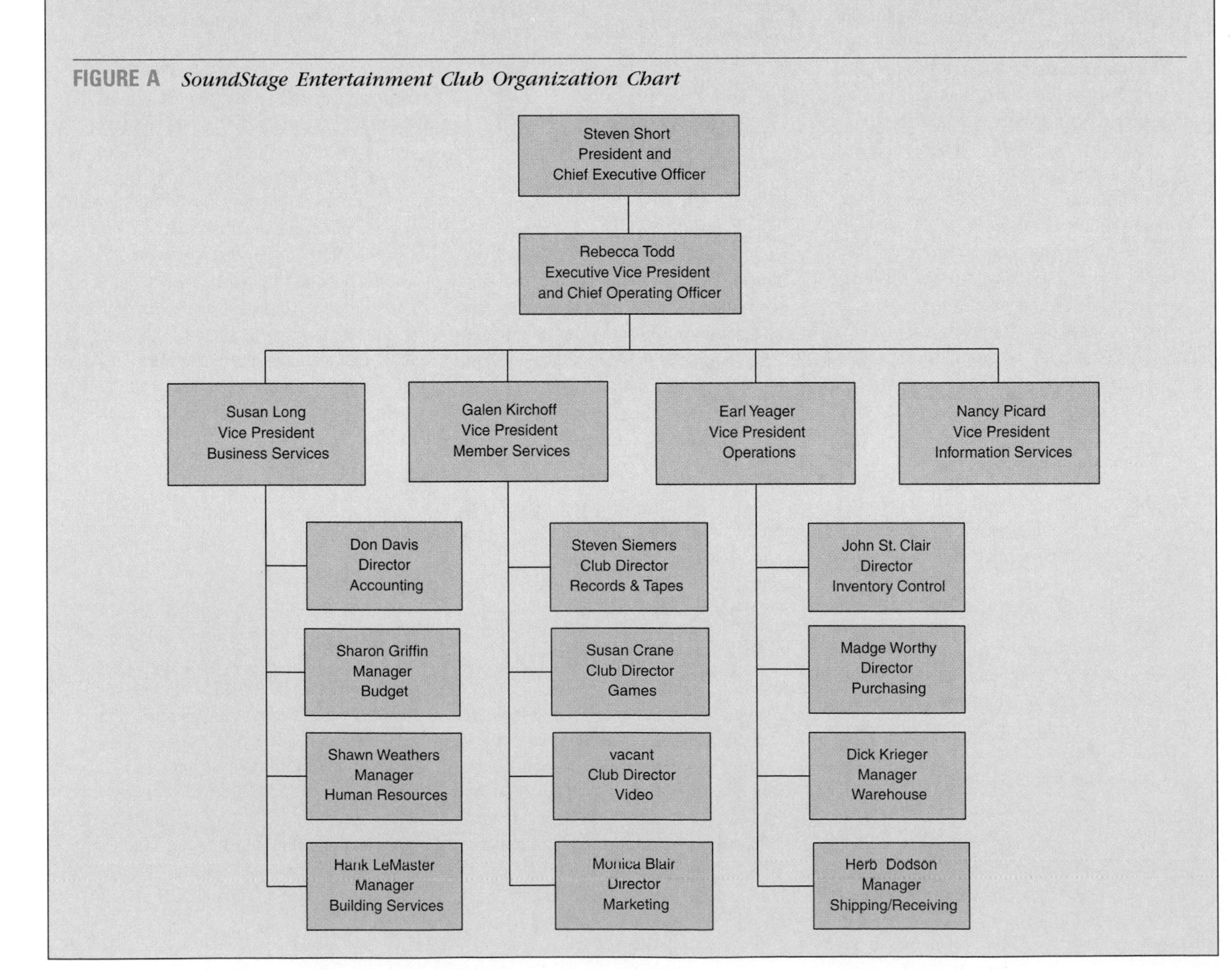

SOUNDSTAGE

SOUNDSTAGE ENTERTAINMENT CLUB

(TQC). You'll be indoctrinated to the TQC in your corporate orientation next week. It is taken very seriously here. Promotions and pay increases are driven entirely by your contribution to the company's bottom line, which is a direct function of our overall quality.

BOB

Doesn't bother me a bit. Where to next?

[Through a set of double doors into a huge warehouse.]

NANCY

This is the warehouse. This is where the action is!

BOB

I saw this before. I'm still taken back by the size and activity. This has to be a very large operation to coordinate.

NANCY

I wouldn't want to do it. And it will get larger with the mergers. There is already talk of expanding the facility by 50 percent. We haven't done much in the way of information services for the warehouse. We do support Purchasing and Inventory Control, but that is as close as we come to supporting the warehouse. Our information strategy plan, however, has placed great importance on information services for optimizing this operation. The first project is slated to begin next quarter.

Well, let's go downstairs and tour our own Information Services division, and then I'll take you to Sandra's office.

[Downstairs, in the basement facilities, beginning with the Computer Operations Center.]

NANCY

This is an organization chart for Information Services, Bob (see Figure B).

FIGURE B *SoundStage Information Services Organization Chart*

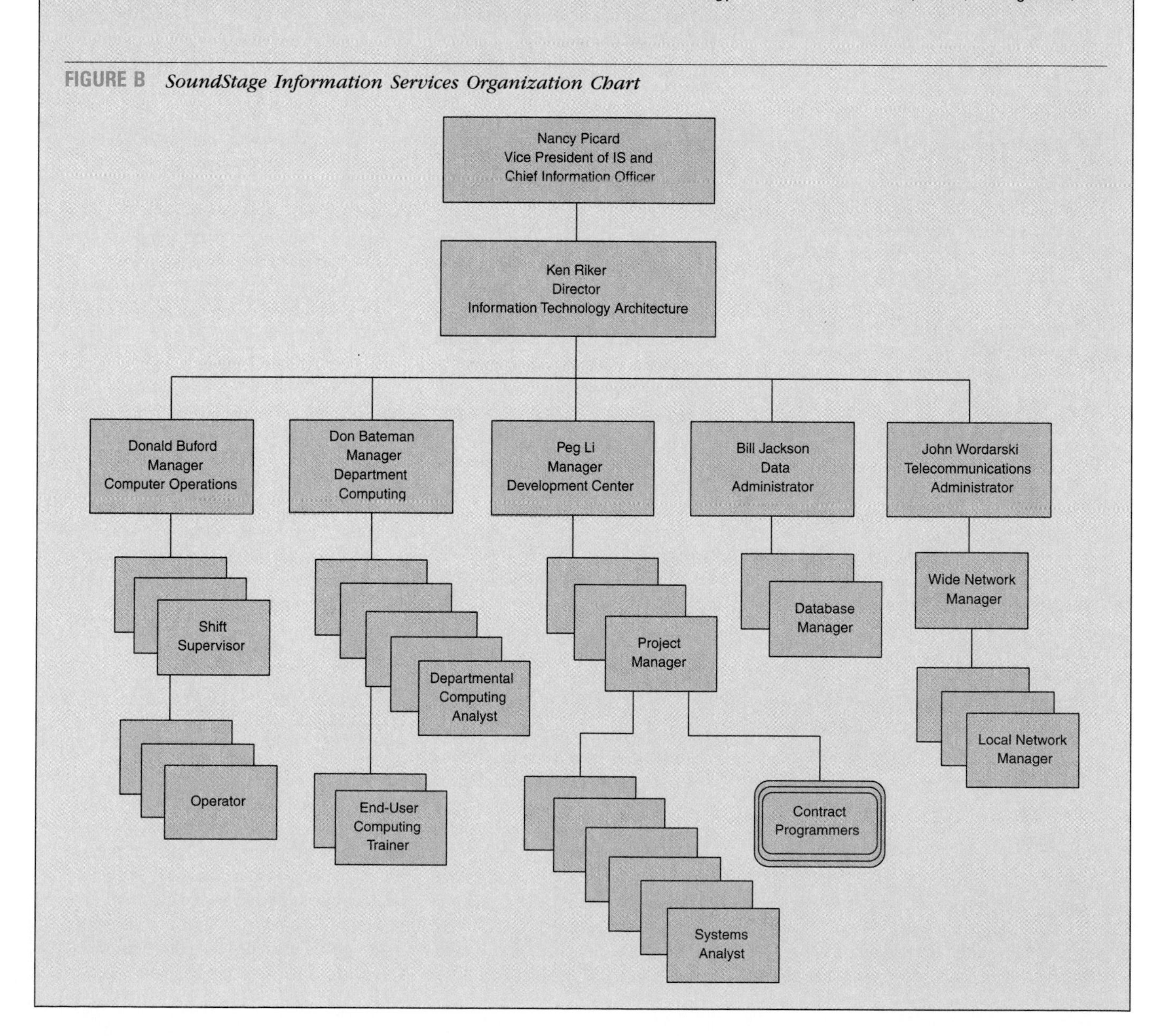

SOUNDSTAGE

SOUNDSTAGE ENTERTAINMENT CLUB

You might want to refer to it as we complete this part of the tour. I now report directly to Steve Short. That reflects the recent reorganization. I used to report to the vice president of the Administrative Services division, but it caused some problems with prioritizing requests from the other divisions.

This is the Computer Operations Center. You are looking at our IBM AS/400 computer. Historically, this midrange computer has supported most of our computer-processing needs. But our strategic information technology architecture will migrate us from the centralized AS/400 solution to a distributed computing solution based on servers and workstations.

BOB

Is that one of them? *[pointing to a small rack unit]*

NANCY

Yes, that is our new Compaq communications server. It runs Windows NT Server and Microsoft *Exchange Server,* our electronic mail, and fax system. Our distributed systems will use a combination of Windows NT and UNIX servers, plus Windows NT and Windows 95 client workstations. That is the architecture you'll be designing to for your first project.

BOB

This is pretty impressive stuff. You've definitely piqued my interest.

[Through the double-doors (using Nancy's security badge), and into an adjacent open-office area.]

NANCY

And this is where you'll be working. We call it the Development Center. All of our systems analysts, contract programmers, and technical specialists work here.

BOB

Contract programmers?

NANCY

Yes, as we've moved away from COBOL and RPG programming to *Powerbuilder* and *C++* programming, we have retooled our existing programmers to become either analysts or other technical specialists. To meet our needs of programming, we contract for programmers on a project-by-project basis. Of course, we keep a few good programmers around for management, technical support, and maintenance.

BOB

Interesting approach! Does that mean I won't program?

NANCY

Actually, you will program in the rapid application development and prototyping sense of the word. But you will have contract programmers on your team to help you, and they will do the heavy-duty programming that transforms prototypes into working applications.

BOB

What's this area with the gigantic diagram on the wall?

NANCY

These offices are assigned to our data and database administrators. They manage our corporate data model, on the wall, and our database management systems. You'll work with the data administrator's staff on any information systems projects that use the database. We are mandated by policy to implement all new systems using that database; therefore, I'm sure you'll soon be visiting his staff.

BOB

What methodology do you use to develop your databases and applications?

NANCY

We use a methodology called *FAST.* We purchased and customized it with the help of a couple of professors at one of our state universities. It is part methodology and part framework. For example, it provides frameworks for flavors of development such as "structured," "rapid development," and soon, "object-oriented." Your partner, Sandra, is an experienced *FAST* developer.

Our methodology analyst, Susan Clark, coordinates *FAST* training and helps our analysts with their *FAST* questions. She has two tools analysts who help our analysts apply tools like *System Architect* and *Powerbuilder* within the *FAST* methodology.

BOB

Obviously, I've got a lot to learn.

[Into an empty office.]

NANCY

And this is your office—I wonder where . . . Here she is! Sandra, I want you to meet Bob Martinez, your new partner.

SANDRA

Bob and I met during his interview, Nancy. Welcome, Bob.

NANCY

Terrific! I didn't know you had already met. I don't know what you've told Bob about yourself, Sandra, but I'd like to do a little biographical sketch.

Sandra has been with us for seven years. She was recently promoted to senior systems analyst because she has proven herself to be one of our most competent, progressive, and personable analysts. Sandra has a bachelor's degree in business. She had little formal computing education, but she has done well because she always seeks out opportunities to learn more through reading, seminars, and company training courses. Sandra's credentials also include recognition as a Certified Systems Professional or CSP.

Bob, you'll be learning from one of our best people! Once again, Bob, welcome! I'll leave you with Sandra now. She'll help you get organized and start teaching you all the things you'll need to start learning. We'll sit down in a week or so to set some goals for your first six months. Bye!

HOW TO USE THE DEMONSTRATION CASE

You've just been introduced to a case study that will be continued throughout this book. The purpose of the continuing case is to show you that tools and techniques alone do not make a systems analyst. Systems analysis and design involves a commitment to work for and with a number of people.

When we started writing this book, we wanted to make sure the chapters would

SOUNDSTAGE

SOUNDSTAGE ENTERTAINMENT CLUB

teach you the important concepts, tools, and techniques. But we were also afraid that you might begin to believe that, if you knew those tools and techniques, you'd have all the knowledge necessary to be a systems analyst.

Each chapter of this book will begin with a SoundStage episode that introduces new ideas, tools, and/or techniques that will be examined in that chapter. In almost all cases, you will see Sandra and Bob working closely with their team, which includes management, system users, and technical specialists. This self-managed team will build a quality- and improvement-driven system solution for SoundStage, consistent with the corporate total quality culture initiative described in this episode.

WHO SHOULD READ THIS BOOK?

Today, it is hard to imagine any industry or business that has not been affected by computer information systems and computer applications. Many businesses consider management of their information resource to be equal in importance to managing their other key resources: property, facilities, employees, and capital. For purposes of this book, these terms are defined as follows:

An **information system** is an arrangement of people, data, processes, interfaces, networks, and technology that interact to support and improve both day-to-day operations in a business (sometimes called **data processing**), as well as support the problem-solving and decision-making needs of management (sometimes called **information services**).

A **computer application** is a computer-based solution to one or more business problems and needs. One or more computer applications are typically contained within an information system.

Many organizations consider information systems and computer applications as essential to their ability to compete or gain competitive advantage. Information has become a management resource equal in importance to property, facilities, employees, and capital. And many businesses have come to realize that all their workers need to participate in the development of these systems and applications—not just the computer and information specialists. But one specialist plays a special role in systems and applications development, the *systems analyst.*

A **systems analyst** *facilitates* the development of information systems and computer applications.

As part of this facilitation, the systems analyst performs *systems analysis* and *design.*

Systems analysis is the study of a business problem domain to recommend improvements and specify the business requirements for the solution.

Systems design is the specification or construction of a technical, computer-based solution for the business requirements identified in a systems analysis. (Note: Increasingly, the design takes the form of a working prototype.)

Some of you will routinely work with systems analysts. The rest of you will be *customers* of systems analysts who will try to *help you* solve your business and industrial problems by creating and improving your access to the data and information needed to do your job. This chapter is about systems analysts and their customers.

THE SYSTEMS ANALYST AS A MODERN BUSINESS PROBLEM SOLVER

In this section, we will examine the origins of the systems analyst. Then we will look at a more formal definition of the systems analyst and classify the roles and activities performed by the systems analyst. Finally, we will describe where systems analysts work.

Why Do Businesses Need Systems Analysts?

The first systems analysts were born out of the industrial revolution. They were industrial engineers whose responsibilities centered around the design of efficient and effective manufacturing systems. Information systems analysts evolved from the need to improve the use of computer resources for the information-processing needs of business applications. In other words, they designed computer-based systems that manufacture information.

Despite all its current and future technological capabilities, the computer still owes its power and usefulness to people. Businesspeople define the applications and problems to be solved by the computer. Computer programmers and technicians apply the technology to well-defined applications and problems.

Information technology offers the opportunity to collect and store enormous volumes of data, process business transactions with great speed and accuracy, and provide timely and relevant information for management. Unfortunately, this potential has not been fully or even adequately realized in most businesses. Why? Business users may not fully understand the capabilities and limitations of modern information technology. Similarly, computer programmers and information technologists frequently do not understand the business applications they are trying to computerize or support. Worse still, some computer professionals become overly preoccupied with technology.

A communications gap has always existed between those who need the computer and those who understand the technology. The systems analyst bridges that gap. You can (and probably will) play a role as either a systems analyst or someone who works with systems analysts.

What Is a Systems Analyst?

In simple terms, systems analysts are people who understand both business and computing. Systems analysts study business problems and opportunities and then transform business and information requirements of the business into the computer-based information systems and computer applications that are implemented by various technical specialists including computer programmers. A more formal definition follows.

> A **systems analyst** facilitates the study of the problems and needs of a business to determine how the business system and information technology can best solve the problems and accomplish improvements for the business. The *product* of this activity may be improved business processes, improved information systems, or new or improved computer applications—frequently all three.

When *information technology* is used, the analyst is responsible for the efficient capture of data from its business source, the flow of that data to the computer, the processing and storage of that data by the computer, and the flow of useful and timely information back to the business and its people.

> **Information technology** is a contemporary term that describes the combination of computer technology (hardware and software) with telecommunications technology (data, image, and voice networks).

Essentially, a systems analyst is a business problem solver. Computers and information systems are of value to a business only if they help solve problems or effect improvements. Accordingly, a systems analyst helps the business by solving its problems using system concepts and information technology. Note in the definition that modern systems analysts develop, or help to develop, *both* an organization's business processes *and* its information systems.

Systems analysts sell business management and computer users the services of information technology. More importantly, they sell *change* (which doesn't always make systems analysts popular). Every new system changes the business. Increasingly, the very best systems analysts literally change their organizations—providing information that can be used for competitive advantage, finding new markets and services, and even dramatically changing and improving the way the organization does business.

The role of systems analyst is changing rather dramatically. Many organizations are splitting the role into two distinct positions or roles, *business analyst* and *application analyst.*

A **business analyst** is a systems analyst that specializes in business problem analysis and technology-independent requirements analysis.

Typically recruited from the user community, business analysts focus on business and nontechnical aspects of systems problem solving. As experts in their business area, they help define system requirements for business problems and coordinate interactions between business users and technical staff. Business analysts are usually appointed to a specific project, or for a fixed duration. Business analysts will typically be paired with one or more application analysts.

An **application analyst** is a systems analyst that specializes in application design and technology-dependent aspects of development. A synonym is *system* or *application architect.*

Application analysts usually come from computer or information systems backgrounds and education. While they are frequently capable of business problem analysis and requirements analysis, they are experts in translating business requirements into technical designs. Both business and application analysts share many skills, particularly systems analysis skills.

There are also several legitimate, but often confusing, variations on the position we are calling systems analyst. A *programmer/analyst* (or analyst/programmer) includes the responsibilities of both the computer programmer and the systems analyst. (In reality, most systems analysts do some programming and most programmers do some systems analysis.)

Other synonyms for systems analyst include *consultant, systems architect, systems engineer, information engineer, information analyst,* and *systems integrator.*

What Does a Systems Analyst Do?

As previously stated, the systems analyst is basically a system-oriented problem solver.

System problem solving is the act of studying a problem environment in order to implement corrective solutions that take the form of new or improved systems.

Throughout this book, the term *problem* will be used to describe many situations including: (1) situations, either real or anticipated, that require corrective action; (2) opportunities to improve a situation despite the absence of complaints; and (3) directives to change a situation regardless of whether anyone has complained about the current situation.

Most systems analysts use some variation of a system problem-solving approach called a systems development life cycle.

A **systems development life cycle** is a systematic and orderly approach to solving system problems.

While this problem-solving approach comes in many flavors, it usually incorporates the following general problem-solving steps (see Figure 1.1):

1. *Planning*—identify the scope and boundary of the problem, and plan the development strategy and goals.
2. *Analysis*—study and analyze the problems, causes, and effects. Then, identify and analyze the requirements that must be fulfilled by *any* successful solution.
3. *Design*—if necessary, design the solution—not all solutions require design.

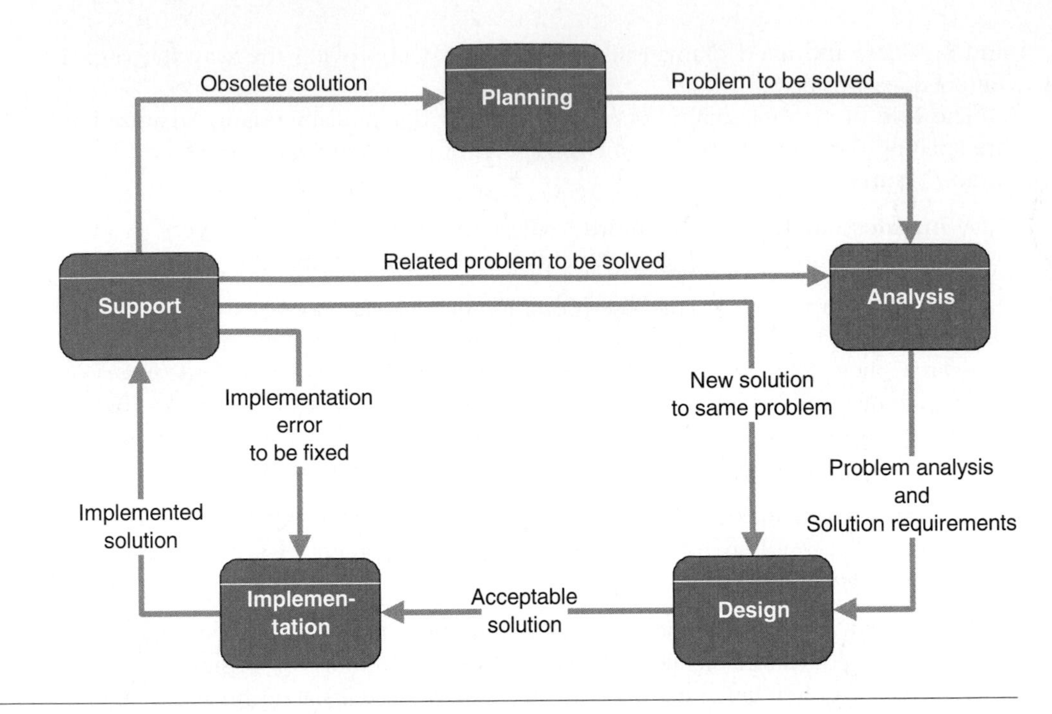

FIGURE 1.1 *A Systems Development Life Cycle*

4. *Implementation*—implement the solution.
5. *Support*—analyze the implemented solution, refine the design, and implement improvements to the solution. Different support situations can thread back into the previous steps.

The term *cycle* in systems development life cycle refers to the natural tendency for systems to cycle through these activities, as was shown in Figure 1.1.[1]

How do the activities of the computer programmer compare with those of the systems analyst? First, systems analysts are typically involved in all the aforementioned problem-solving steps. On the other hand, programmers are typically involved only in the last three or four steps. Second, analysts typically communicate business requirements and design specifications to the programmer. Finally, programmers tend to be only concerned with information technology. Systems analysts, on the other hand, are responsible for other aspects of a system or application, including:

- PEOPLE, including managers, users, and other developers—and including the organizational behaviors and politics that occur when people interact with one another.
- DATA, including capture, validation, organization, storage, and usage.
- PROCESSES, both automated and manual, that combine to process data and produce information.
- INTERFACES, both to other systems and applications, as well to the actual users (e.g., reports and display screens).
- GEOGRAPHY, which effectively distribute data, processes, and information to the people.

Consequently, the systems analyst's job presents a fascinating and exciting challenge to many individuals. It offers high management visibility and opportunities

[1]Although these activities appear to be sequential, in practice they tend to overlap.

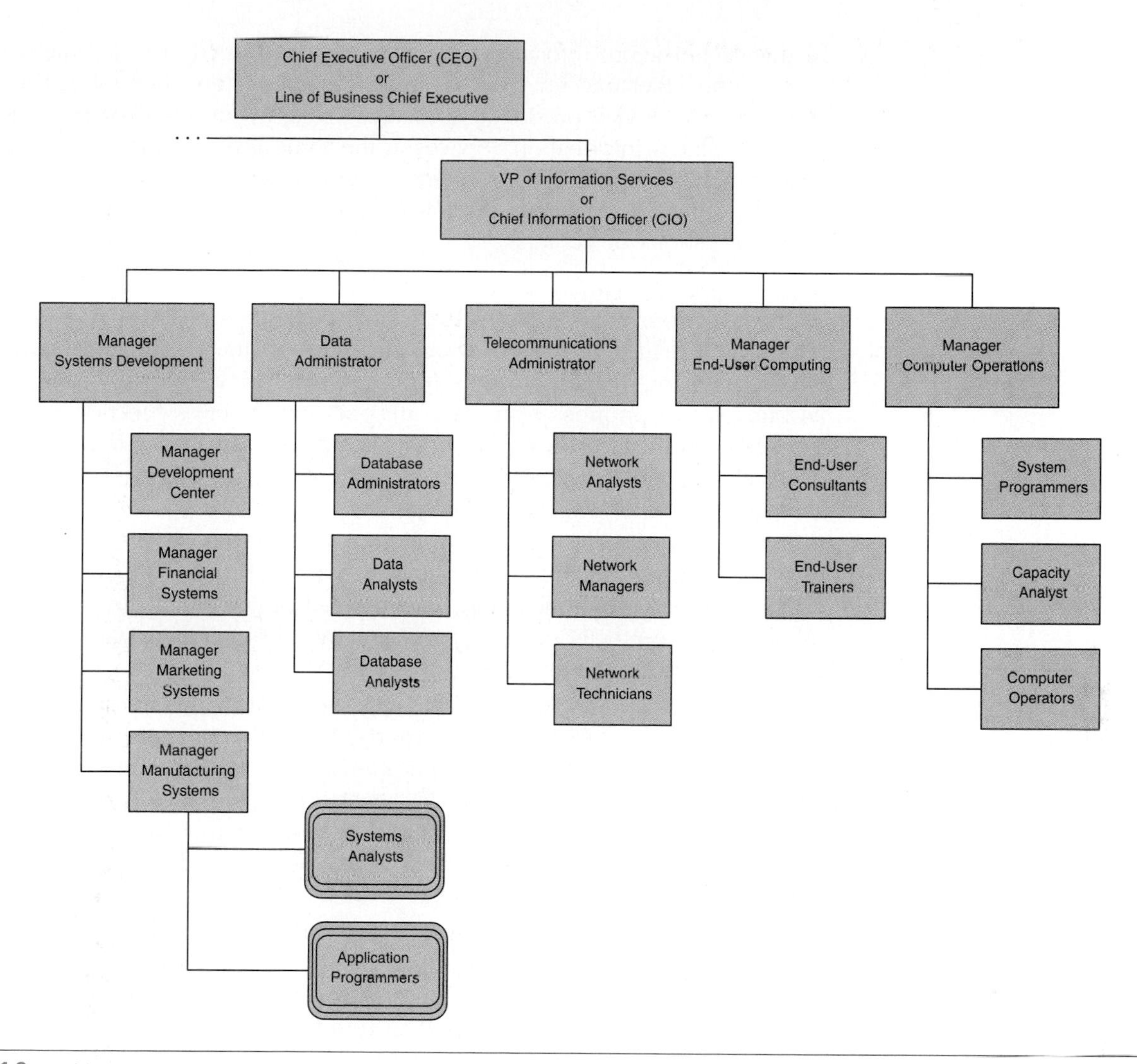

FIGURE 1.2 *A Traditional Information Services Organization*

for important decision making and creativity that may affect an entire organization. Furthermore, this job can offer these benefits relatively early in your career (compared to other jobs and careers at the same level).

Where Do Systems Analysts Work?

Systems analysts can be found in most businesses; however, the organization of information services in many businesses is in turmoil as those businesses reorganize to improve service, quality, and value. In this section, we'll examine where analysts work in both traditional and modern organizations. Then we'll examine an important trend called *outsourcing* and discuss implications for systems analysts. Finally, we'll examine alternatives to working in either a traditional or modern business information services organization.

The Systems Analyst in the Traditional Business. Every business organizes itself uniquely. But certain patterns of organization seem to reoccur. In the traditional business, *information services* are centralized for the entire organization or a specific line of business.[2] Figure 1.2 is an *organization chart* for our traditional Information Services unit.

[2]The term *line of business* refers to a complete collection of business functions (e.g., human resources, financial services, facilities, manufacturing, etc.) that support a single product line or family of services. For example, laundry detergents might be a line of business for Procter & Gamble Corporation. That product line may include many similar products such as Tide and Cheer.

In this organization, Information Services reports directly to the chief executive officer, or the chief executive for a line of business. The highest ranking information officer is a vice president, sometimes called a **chief information officer** (CIO). This places Information Services at the same level of importance as the traditional business functions (e.g., Finance, Human Resources, Sales and Marketing, Production, etc.). The rest of Information Services is organized according to the following functions or centers:

- *Systems and applications development.* Most systems analysts work here, along with most programmers. Notice that the systems analysts and programmers are organized into permanent teams that support the information systems and applications for specific business functions. Each development team derives its budget and priorities from its corresponding business function unit. Typically, the systems analysts (and programmers) in the systems development unit have various levels of assignments ranging from *project manager* to *senior analyst* to *analyst.*

  Notice that our systems and applications development unit includes a **development center.** A development center establishes and enforces the methods, tools, techniques, and quality of all development projects. The development center is staffed by assignments systems analysts who provide consulting services to all other projects.
- *Data administration.* This center manages the data and information resource in the organization. This includes the databases that are (will be) used by systems developers to support applications. Data administration usually employs several systems analyst-like specialists called *data analysts* who analyze database requirements and design and construct the corresponding databases. To learn more about data administration and data analysts, refer to your local data management course or textbook.
- *Telecommunications.* This center designs, constructs, and manages the computer networks that have become integral to most businesses. They do not employ traditional systems analysts; however, *network analysts* perform many of the same tasks as applied to designing local and wide area networks that will ultimately be used by systems and applications.
- *End-user computing.* This center supports the growing base of personal computers and local area networks in the end-user community. They provide installation services, training, and help-desk services (call-in help for various PC-related problems). In mature businesses, they also provide standards and consulting to end-users who develop their own systems with PC power tools such as spreadsheets and PC database management systems. In this latter role, they employ analyst-like *end-user computing consultants.*
- *Computer operations.* This centers runs all of the *shared* computers including mainframes, minicomputers, and nondepartmental servers. This unit rarely employs systems analysts.

Modern Information Services in a Business. There is a dramatic reorganization trend in medium-to-large information services units for the modern business. This reorganization is highly *decentralized* with a focus on *empowerment* and *dynamic teams.*

As shown in Figure 1.3, the resulting organization might best be described as a *federation* of information systems centers that report directly to functional business units (or groups of business units). Each of these centers is empowered to set priorities and make decisions on behalf of its constituent management and users. In other words, Information Services is trying to get closer to users in an effort to improve services and value. The net impact on systems analysts is that many more systems analysts (and programmers) are working directly for the busi-

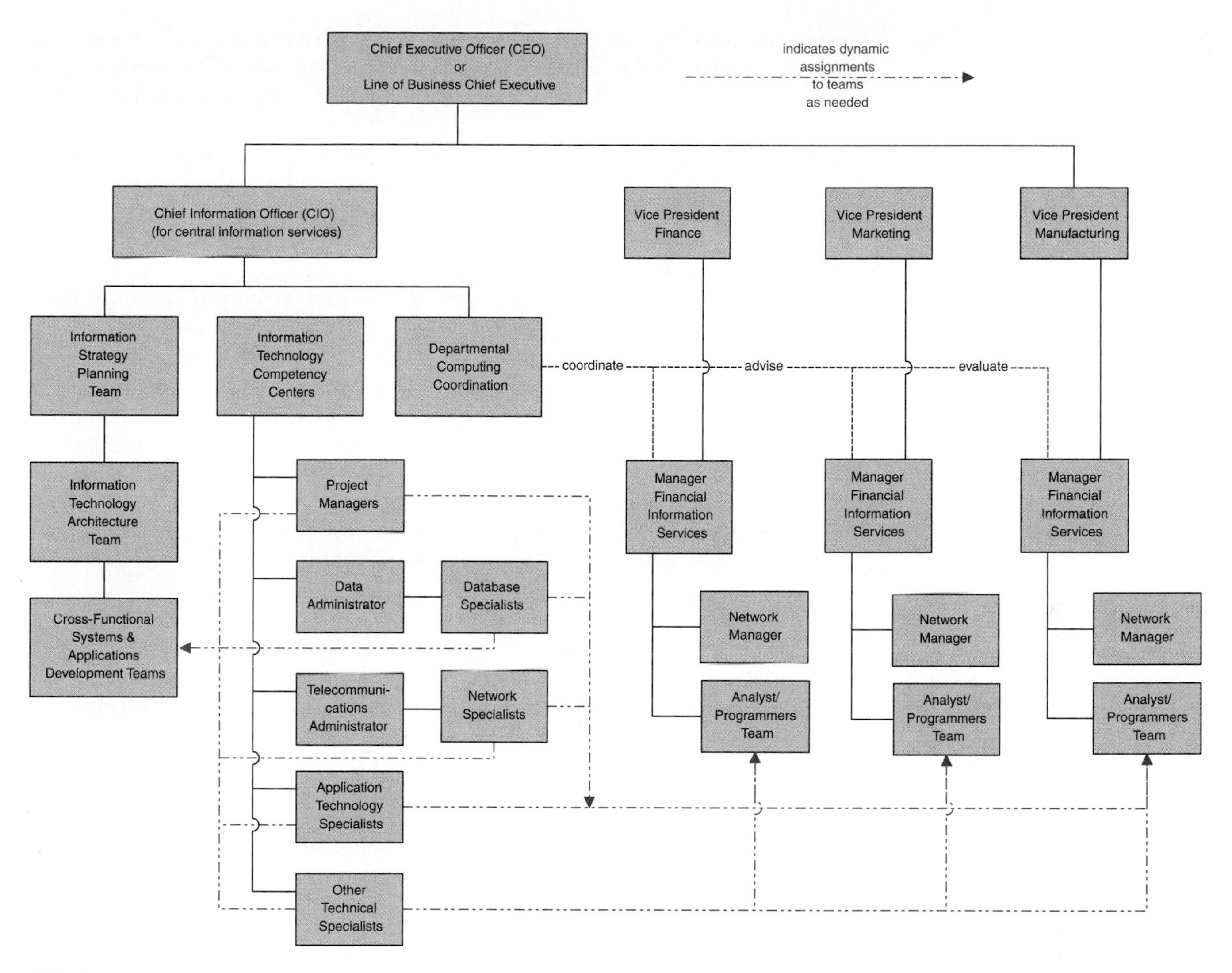

FIGURE 1.3 *A Modern Information Services Organization*

ness units. This often requires a different mind-set in terms of accountability to the bottom line of that business unit.

Decentralized information services can, however, lead to information anarchy and systems that do not interoperate to the benefit of the business as a whole. Also, there will always be systems and applications that support more than one business function—perhaps the entire enterprise. We call these **cross-functional applications.** Examples include financial and human resources systems.

Consequently, there still exists a need for a central information services unit in our modern organization (as shown in Figure 1.3). The central information services unit is responsible for

- *Information strategy planning.* The information strategy planning team establishes direction and priorities for aligning information services *for the entire business* with the corporate mission, vision, and goals. The plan identifies projects and priorities for building shared databases, wide area networks, and cross-functional applications. Experienced systems analysts often play key roles in development.
- *Information technology architecture.* An information technology architecture team establishes and maintains a blueprint for which technologies will be

approved for the entire business, including the decentralized information services centers. These standards ensure compatibility and interoperability can be achieved between centralized and decentralized information systems and computer applications.

- *Information technology competency centers.* The centers provide a pool of technology-specific specialists who are provided to both centralized and decentralized units for project work. Examples include specialists for project management, data management and database design, telecommunications management and network design, applications technologies (such as graphical user interfaces), and so forth. Each specialist contributes his or her expertise to any project to which he or she is assigned, for both centralized and decentralized projects (as shown in Figure 1.3 by the red lines).
- *Cross-functional systems and applications development.* This center develops and supports the shared information systems and cross-functional applications for the business. Additionally, and when necessary, they help to integrate these centralized systems and applications with the decentralized systems and applications developed by the business units.

  Obviously, this center employs experienced systems analysts. It also draws heavily on the technology competency specialists described above. As projects are started and completed, both systems analysts and technical specialists are assigned to and released from project teams. This allows the organization to use its human resources in a more efficient manner.
- *Departmental computing coordination.* This unit provides both consulting services and quality management services to the decentralized information and computing centers. Experienced systems analysts may be employed here to help establish standards and guidelines and to provide training and consultation to departmental projects. The black dashed lines indicate a staff or support relationship exists between departmental computing coordination and the decentralized departmental information services managers and teams.

In this decentralized information services organization, central information services serves much the same role as the federal government of the United States. It develops systems for which there is a businesswide need and provides regulation and standards for the decentralized systems. The decentralized information systems units exercise autonomy over their own systems, but ideally within the parameters established by the central information services coordination. This so-called federated organization is intended to balance the advantages of agility (through the decentralized units) with the advantages of interoperability (through the centralized unit). The current projection is that as many as 70 percent of all organizations of all sizes will reorganize information services according to this model by 1998.[3]

Outsourcing in the Modern Business Another very significant trend in businesses that will affect many systems analysts is outsourcing.

> **Outsourcing** is the act of contracting a service or function to an external third party.

Typically, the business must retain enough of a central information services unit to monitor and manage the outsourcing agreement. Also, some mission critical information services functions and projects may not be outsourced. Representative outsourcing vendors include Electronic Data Systems (EDS), ICCS, Computer Sciences Corporation, Cap/Gemini, IBM, and Andersen Consulting.

As many as 50 percent of medium-to-large businesses have already outsourced some or all of the their information services. The initial business driver was *cost*

[3]Gartner Group 1995 Annual Symposium on the Future of the Information Technology Industry.

reduction. Outsourcers were able to sell and/or demonstrate that they could provide the information technology and services cheaper than the current in-house information services unit. Despite some well-publicized horror stories, the data suggest that most businesses are satisfied with their outsourcing deals. The Gartner Group predicts that 70 percent or more of businesses will be outsourcing information services by the year 2000. However, the evidence also suggests most outsourcing contracts will be written or restructured to outsource only certain functions (such as end-user computing support or network management). Also, future outsourcing will be based more on *value added* to the business than on cost reduction. In other words, the outsourcer will be contracted for technology or expertise that will return real value to the business (and for which the business perceives it cannot provide through its current workforce).

How will outsourcing impact the systems analyst? First, many systems analysts (and other computer professionals) can expect to be absorbed by outsourcing. When an information services unit or function is outsourced, its employees typically become new employees of the outsourcing vendor. Over time, the outsourcer may choose to retain the employee (which is common), terminate the employee (if he or she cannot contribute to the contracted mission), or reassign the analyst to a different customer (meaning a different company). While you can understand the anxiety this may cause outsourced employees, the Gartner Group has compiled data that suggest the experience for most information technology professionals has been positive.

A second implication is the employment opportunity afforded by outsourcing vendors. These outsourcers must retain a high-quality, technically competent workforce of information technology managers and specialists, including systems analysts. They must also invest in that workforce to keep them current and maintain their business. Otherwise, their customers will not renew their outsourcing agreements. Thus, outsourcers offer systems analysts many of the advantages of *consulting* (see below), but with one major difference—less travel. Why? Because outsourcing agreements tend to be long term (usually 5 to 10 years), job assignments last longer than one project. There are, however, no guarantees that outsourcer employees won't be transferred to different companies (called *accounts*) since outsourcers also depend on a mobile work force that can quickly respond to any account's project needs.

Consulting Management and systems consulting firms build information systems and applications for other organizations. Why wouldn't an organization build all systems through its own information systems unit? Perhaps the information systems unit is understaffed. Perhaps the information systems unit's management is looking for technical expertise that its own staff doesn't (yet) possess. Perhaps the information systems unit's management is looking for an unbiased opinion and fresh ideas. The list of reasons is endless. Examples of well-known management consulting firms include Ernst & Young, Andersen Consulting, Price Waterhouse, American Management Systems, and IBM AD Consulting. Many of these firms also provide accounting and auditing services.

The systems analysts employed by management consulting firms are usually called **management consultants** or **systems consultants.** They are loaned (for a fee) to the *client* for *engagements* (a consulting term that means "project") that result in a new system for the client. Once the engagement is completed, they are reassigned to a new engagement, frequently for a new business client. Management consulting firms represent an attractive employment option for aspiring systems analysts. The engagements tend to be very challenging and provide a wide variety of exposure and experiences. Also, management consulting firms keep their consultants on the cutting edge of technology and techniques to compete for business. For college graduates who are particularly well schooled in the latest systems analysis and design methods and programming, management consulting firms represent an interesting and challenging alternative to employment in a traditional information systems unit.

A variation on consulting firms is **systems integration.** This involves helping organizations integrate systems and applications that don't work together properly, or that run on very different technical platforms from different computer manufacturers. Systems analysts that specialize in systems integration are frequently called **systems engineers** or **systems integrators.**

Application Software Solution Providers One systems analysis opportunity often overlooked by college graduates is the software market. Application software solution providers are in the business of building information systems and application software packages for resale to other businesses. Examples include *Peoplesoft, SAP,* and *Oracle (Applications).*

Many businesses have a policy of not building any system they can purchase. Software packages are typically written to the greatest common denominator of their intended market—that is, they are designed to meet general requirements and offer limited customizability. The software packages may be geared either to specific business functions (e.g., accounting, payroll, purchasing) or a specific industry/business (health care, retail, education, government, etc.). The development of packaged software follows a problem-solving approach similar to that of developing custom information systems within an organization. In some ways, their development is more challenging since the vendor wants its package to appeal to as large a market as possible. Software and solutions vendors usually hire two types of systems analyst. The first, called a **software engineer,** is responsible for designing (and programming) the package itself. The second, sometimes called a **sales engineer,** is responsible for helping customers that purchase the package to integrate it into their business operations.

CUSTOMERS, PARTNERS, AND EXPECTATIONS

In this section, we'll briefly examine the systems analyst's customers and partners. We'll also briefly look at some business trends that will influence future systems analysts.

Customers—Users and Management

Let's begin with a definition.

> A **user** is a person, or group of persons, for whom the systems analyst builds and maintains business information systems and computer applications. A common synonym is the **client.**

Users don't always refer to themselves as users; some take offense at the term. According to Ed Yourdon, a noted systems author and consultant, "the user is the 'customer' in two important respects: (1) as in many other professions, the customer is always right,' regardless of how demanding, unpleasant, or irrational he or she may seem; and (2) the customer is ultimately the person paying for the system and usually has the right or ability to refuse to pay if he or she is unhappy with the product received."[4]

Based on Yourdon's definition, we can identify at least two specific user/customer groups: system users and system owners.

> **System users** are those individuals who either have direct contact with an information system or application (e.g., they use a terminal or PC to enter, store, or retrieve data) or use information (reports) generated by a system.

[4]Edward Yourdon, *Modern Structured Analysis* (Englewood Cliffs, NJ: Prentice Hall, Yourdon Press Computing Series, 1989), p. 41.

System owners provide sponsorship of information systems and computer applications. In other words, they pay to have the systems and applications developed and maintained. They may also approve technology, and most certainly approve significant business changes caused by using technology.

Clearly, a manager can also be one of the end-users of a system.

Traditionally, most system users were **internal users,** that is employees of the business for which a system or application is designed. That has changed significantly! Today's user community includes **external users** as businesses seek to make their information systems and applications interoperate with other businesses and the consumer. For example, it is not unusual to extend the reach of information systems and computer applications directly to your customers, suppliers, business partners, and contractors. This has been enabled by a whole range of telecommunications technologies such as electronic data interchange and the Internet.

Until recently, systems analysts and their internal customers have had an unhealthy relationship. This has manifested itself in projects and systems that exceeded budgets, were almost always late, and frequently did not meet expectations. Analysts were insulated from the business problems they were trying to solve. To correct that problem, information technology managers are making a demonstrated attempt to get closer to their customers. This is especially true of systems analysts. The new buzzword is partnership. Successful partnerships can only occur when management and users are viewed as true customers and participants in the information systems development process. That will be a theme throughout this book.

Clearly, tomorrow's managers and users need an understanding of the tools and methods employed by systems analysts. This book will provide the necessary survey for that audience. Equally important, analysts must recognize the growing breadth of internal and external users, and apply appropriate technologies to their increasingly complex problems. The running case study, SoundStage Entertainment Club, will demonstrate this breadth of users as it attempts to directly connect with new and current customers on the Internet.

The Roles of Management and Users in Systems Problem Solving

Earlier we described the problem-solving process employed by systems analysts. The analyst facilitates that process, but management and users are stakeholders in the result. For that reason, management and users should be active participants (as opposed to passive observers) of the process. In fact, best practice also suggests that project teams include full- and part-time management and user membership. The roles of management and users are:

1. *Planning.* Management must sponsor and fund all projects. Users must define the domain and boundaries of the problem. Ideally, management and users should establish the measures of the project's success.
2. *Analysis.* As subject matter experts, management and users must analyze the problem domain for causes, effects, and opportunities, as well as communicate the requirements to be fulfilled by any successful solution, regardless of technology chosen.
3. *Design.* At the least, users must react to high-level solution designs. Increasingly, they participate in the solution's design.
4. *Implementation.* Users participate in system construction and testing. They are the recipients of training necessary to enable the full user community to work with the solution.
5. *Support.* Users and management should routinely evaluate the working solution and suggest improvements.

As suggested earlier, one or more users may participate in all these roles, working side by side with a systems or applications analyst as a business analyst.

Modern Business Trends and Implications for the Systems Analyst

This book is about systems analysis and design. Clearly, systems analysts must keep up with rapidly changing technologies, but today's priorities are rapidly shifting from technology-driven solutions to business-driven solutions. What business trends and management expectations will most impact the systems analyst in the coming decade? Many trends quickly become fads, but here are some nontechnical trends we believe will influence systems analysts in the coming years. Many of these trends are related and integrated such that they form a new business philosophy that will impact the way every systems analyst works in the next century.

Total Quality Management (TQM) One major business trend of the 1990s is total quality management (TQM).

> **Total quality management** or TQM is a comprehensive approach to facilitating quality improvements and management within a business.

The key word is *comprehensive.* Businesses have learned that quality has become a critical success factor to compete. They have also learned that quality management does not begin and end with the products and services sold by the business. Instead, it begins with a culture that recognizes that everyone in the business is responsible for quality. TQM commitments require every business function, including information services, to identify quality indicators, measure quality, and make appropriate changes to improve quality.

TQM impacts systems analysts on at least two fronts. First, the very nature of systems analysis encourages analysts to look for business quality problems. The two most important questions in the analyst's repertoire are "why?" and "why not?" The answers often uncover quality problems or contributors. Second, systems analysis and design provides the specifications for the number one quality problem in modern information systems—buggy software. Incomplete and inconsistent specifications from analysts are a significant contributor to poor software quality.

Our discussions with recruiters suggest that an almost obsessive attitude toward quality will become an essential characteristic of successful systems analysts (and all information technology professionals). Throughout this book, quality management will be a theme.

Business Process Redesign (BPR) Total quality management has forced many businesses to radically rethink and redesign their fundamental business processes.

> **Business process redesign** is the study, analysis, and redesign of fundamental business processes to reduce costs and improve value added to the business.

The BPR phenomenon was initiated by economic hardship and increased competition. Most businesses are learning that their fundamental business processes have not changed in decades, and that those business processes are grossly inefficient and/or costly. Many processes are overly bureaucratic and do not truly contribute value to the business. While we have automated many of these business processes with computers, in many cases we have not truly addressed any manual inefficiencies in the system. The tired excuse for these problems was always "we've always done it that way."

Enter business process redesign. A BPR project begins with identification of a value chain, a combination of processes that should result in some value added to the business. For example, a value chain might be all those processes that respond to a customer order and result in a satisfied customer. The key here is that these business processes cut across department boundaries. The business processes are documented and analyzed in excruciating detail. Every facet of every process is analyzed for timeliness, bottlenecks, costs, and whether or not it truly adds value to the organization (or conversely, only adds bureaucracy). Business processes are subsequently streamlined for maximum efficiency. Finally, new business processes are analyzed for opportunities for further improvement through information technology.

Systems analysts figure prominently in BPR. First, systems analysts are often included in BPR projects because their "system" perspective is valued. Also, the skill competencies for BPR and systems analysis and design are somewhat similar. Chapter 6 will present some introductory business process analysis techniques. Second, a typical BPR project identifies several opportunities for new and revised computer applications. You've already learned that systems analysts facilitate such projects. Some studies suggest that a majority of new applications development in the next decade will be initiated by business process redesign.

Continuous Process Improvement (CPI) Another TQM-related trend is continuous process improvement.

> **Continuous process improvement** is the continuous monitoring of business processes to effect small but measurable improvements to cost reduction and value added.

In a sense, CPI is the opposite of BPR. Whereas BPR is intended to implement dramatic change, CPI implements a continuous series of smaller changes. Continuous improvement contributes to cost reductions, improved efficiencies, and increased value and profit. Systems analysts may be called on to participate in continuous process improvement initiatives for any business process, including the design and implementation of improvements to associated computer applications.

Interestingly, and as described earlier, most systems analysts follow a problem-solving methodology to facilitate systems and applications development. This methodology is itself a business process. As such, it should not be blindly followed. Instead, a systems methodology should be monitored, measured, and improved—just like any other process.

Globalization of the Economy The 1980s will be remembered as the era of economic globalization. Competition became global with emerging industrial nations offering lower-cost or higher-quality alternatives to many products. American businesses suddenly found themselves with international competitors. On the other hand, many American businesses have also discovered new and expanded international markets for their own goods and services. The bottom line is that most businesses have been forced to reorganize to compete globally.

A related phenomenon has been the trend toward industrial consolidation. Business headlines have been dominated by news of corporate mergers, acquisitions, and partnerships. In many cases, these acquisitions were international or intended to stimulate internationalization of products and services.

How does economic globalization affect systems analysts? First, information systems and computer applications must be internationalized. They must support multiple languages, currency exchange rates, international trade regulations, accepted business practices (which differ in different countries), and so forth. Second, most information systems ultimately require information consolidation for performance analysis and decision making. Such consolidation is complicated by the aforementioned language barriers, currency exchange rates, transborder information regulations, and the like. Finally, systems development itself is complicated by the need for systems analysts who can communicate, orally and in writing, with management and users that speak different languages, dialects, and slang. Opportunities for international employment of systems analysts should continue to expand.

Another international business trend will have an increasingly significant impact on systems analysts—the *outsourcing* of the programming effort. Traditionally, systems analysts wrote business specifications and technical designs to be implemented by local applications programmers. But many businesses have discovered that other countries offer competent contract programming services at a fraction of the cost of traditional programmers. Hence, they outsource the programming and testing work overseas. But this outsourcing option creates new pressures on the systems analyst who must provide extraordinarily complete, consistent, and precise specifications to the contract programmers—to a degree never before expected.

Empowerment Many organizations have responded to the above-described business trends by downsizing their workforce. In particular, most businesses targeted middle management as a source of bureaucratic inefficiencies and high costs. As a result, there are far fewer managers than in any recent era of business. Given fewer managers and the same amount of work (or more), remaining managers have been forced to rethink decades of management style based on authoritative chains of command. In its place, the new watchword is *empowerment.*

> **Empowerment** is the business trend of driving the authority to make decisions downstream to nonmanagers and teams.

Through empowerment, individuals and teams are allowed to make decisions that would have normally required considerable bureaucratic approval. To succeed, the remaining managers must make every effort to encourage this decision making, and also stand behind the decisions. The team must also be held accountable for the decisions made. This will increasingly be accomplished through a compensation approach that puts some percentage of everyone's salary at risk, meaning dependent on the team or organization's performance.[5]

Systems analysts can expect greater degrees of empowerment in their assignments. More importantly, empowerment will stimulate the user community to play more active and participative roles in systems and application development projects. (Note: The traditional role of the user has frequently been characterized as passive.)

Partners for the Systems Analyst—Information Technologists and Vendors

It takes more than systems analysts and users to build effective systems. In addition to clients, the systems analyst works with a number of technical peers. These **information technologists** include programmers, database designers, networking specialists, computer operators, and hardware and software vendors (sales and technical support representatives). As illustrated in Figure 1.4, the systems analyst's role in the typical project is to act as a facilitator. The analyst may be the only individual who sees the system or application as a whole. Thus, systems analysts must possess a unique set of skills and abilities to accomplish the complex task of facilitating systems.

PREPARING FOR A CAREER AS A SYSTEMS ANALYST

What does it take to become a successful systems analyst? One writer suggests the following timeless answer:

> I submit that systems analysts are people who communicate with management and users at the management/user level; document their experience; understand problems before proposing solutions; think before they speak; facilitate systems development, not originate it; are supportive of the organization in question and understand its goals and objectives; use good tools and approaches to help solve systems problems; and enjoy working with people.[6]

[5]Gartner Group 1995 Annual Symposium on the Future of the Information Technology Industry.

[6]Michael Wood, "Systems Analyst Title Most Abused in Industry: Redefinition Imperative," *Computer World,* April 30, 1979, pp. 24, 26.

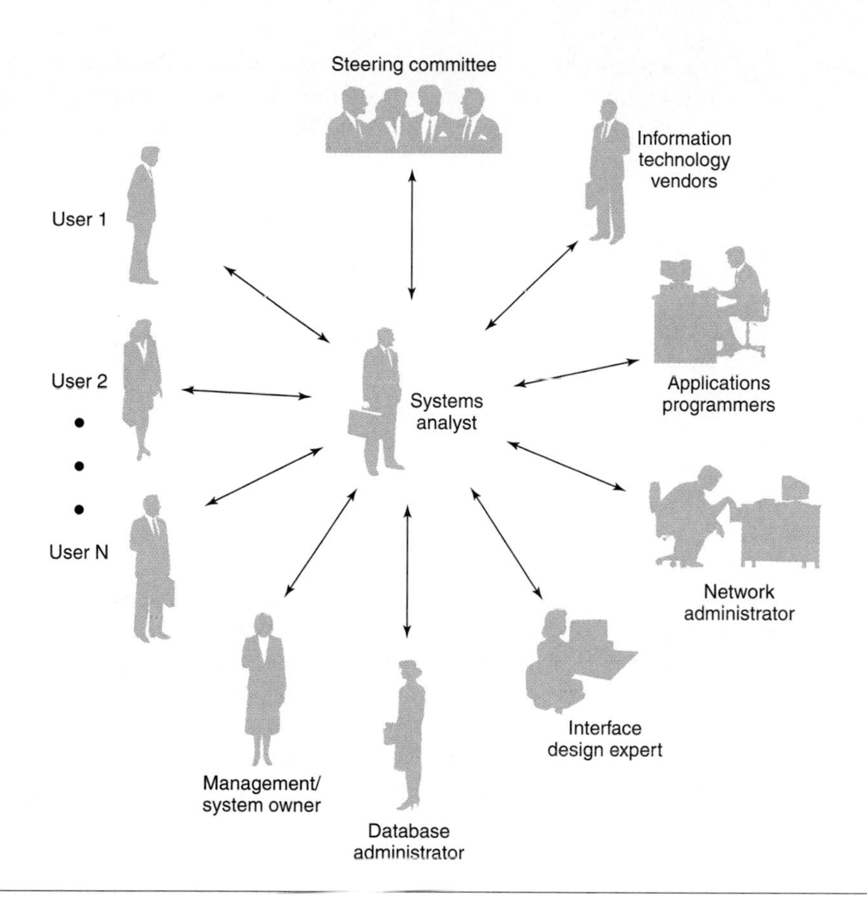

FIGURE 1.4 *The Systems Analyst as a Facilitator*

This seems like a tall order, doesn't it? It is often difficult to pinpoint those skills and attributes necessary to succeed. However, the following subsections describe those skills most frequently cited by practicing systems analysts.

Working Knowledge of Information Technology

The systems analyst is an *agent of change*. He or she is responsible for showing end-users and management how new technologies can benefit their business and its operations. To that end, the analyst must be aware of both existing and emerging information technologies and techniques. Such knowledge can be acquired in college courses, professional development seminars/courses, and in-house corporate training programs. Some technologies and topics that you should be studying today include:

- Distributed, relational database management systems.
- Telecommunications and networking.
- Client/server and distributed computing architecture.
- Object technology.
- Rapid application development technology.
- Graphical user interfaces.
- The Internet.

One good way to keep up on what's happening is to develop a disciplined and organized habit of skimming and reading various trade periodicals about computers and information systems. Examples of helpful trade publications include *Information Week* and *Computer World*. We call special attention to *Information Week* as an outstanding periodical that summarizes the week's news, includes numerous "trend" and "applications" articles, and surveys other periodicals and articles of interest to the busy IS professional.

Another way to keep current is through professional association. Consider joining a professional association such as the Association for Information Technology Professionals (AITP), the Association for Computing Machinery (ACM), or the Association for Systems Management (ASM).

Computer Programming Experience and Expertise

Whether or not systems analysts write programs, they must know how to program because they are the principal link between business users and computer programmers. Consequently, many, but by no means all, organizations consider experience in computer programming to be a prerequisite to systems analysis and design.

You should not, however, assume that a good programmer will become a good analyst or that a bad programmer could not become a good analyst. There is no such correlation. Unfortunately, many organizations insist on promoting good programmers who become poor or mediocre systems analysts. Worse still, mediocre programmers are often passed over in the belief that they cannot become good analysts.

Regardless of opinions concerning the need for programming experience, it is difficult to imagine how systems analysts could adequately prepare business and technical specifications for a programmer if they didn't have some programming experience. This experience can be obtained at virtually any college or vocational school.

Most systems analysts need to be proficient in one or more high-level programming languages. Historically, the language of choice has been COBOL for business applications. But a revolution is occurring in the languages of choice. Many organizations are shifting to **visual programming languages** such as *Visual BASIC* or *Powerbuilder* or to **object-oriented programming languages** such as *Visual Smalltalk, Delphi,* or *Visual C++*. This revolution is being fueled by the following factors:

- The transition to graphical user interfaces (such as Windows). Organizations want their custom business applications to have the same user interface as their office productivity applications (e.g., word processors and spreadsheets).
- The desire to downsize applications from the mainframe to networks of PCs—called **client/server** or **distributed computing.**
- The pressures to improve productivity in applications development through rapid, iterative prototyping and the reuse of programming modules called *objects* and *components.*

Visual and object-oriented programming require a completely different style of program design, construction, coding, and testing than did COBOL. Because most organizations have an abundance of experienced COBOL programmers, there is sufficient anecdotal evidence to suggest that these organizations are counting on colleges and universities to educate (and where appropriate, retool) the next generation of visual programmers. The visual languages are very similar in style and approach. Your career as an analyst will be significantly enhanced if you take early advantage of any visual programming courses offered at your school.

General Business Knowledge

Increasingly, systems analysts are expected to immerse themselves in the business. They are expected to be able to specify and defend technical solutions that address the bottom-line value returned to the business. Obviously, this ability comes with experience, but aspiring analysts can and should take advantage of all opportunities to improve their business literacy (just as aspiring business and management students must improve their computer literacy).

Systems analysts should be able to communicate with business experts to gain knowledge of problems and needs. We strongly suggest you include courses in as many of the following subjects as possible: accounting, finance, marketing, human resources, operations, manufacturing, and organizational behavior. Specializations such as accounting or manufacturing can be very valuable in some instances. These subjects are taught in many colleges.

Some general business subjects should be sought out by prospective systems analysts. Earlier in the chapter, we mentioned a few general business trends for which many colleges and universities are developing courses: business process redesign, continuous process improvement, and total quality management. These courses may be offered in either the business/management/organizational leadership schools or the information systems/computer science schools. Look for them!

It should be noted that by working with business experts, systems analysts gradually acquire business expertise. It is not uncommon for analysts to develop so much expertise over time they move out of information systems and into the user community.

Problem-Solving Skills

The systems analyst must be able to take a large business problem, break down that problem into its component parts, analyze the various aspects of the problem, and then assemble an improved system to solve the problem. Engineers call this problem-solving process analysis and synthesis. The analyst must learn to analyze problems in terms of causes and effects rather than in terms of simple remedies. Being well organized is also part of developing good problem-solving skills.

Analysts must be able to creatively define alternative solutions to problems and needs. Creativity and insight are more likely to be gifts than skills, although they can certainly be developed to some degree. Perhaps the best inspiration for students and young analysts comes from the late USN Rear Admiral Grace Hopper, mother of the COBOL language. She suggests, "The most damaging phrase in the English language is 'We've always done it that way.'" Always be willing to look beyond your first idea for other solutions.

Most of this book is about systems problem solving. But if you're looking for a general education or liberal arts elective, research the offerings of your philosophy department. They often teach courses on general problem solving and logic.

Interpersonal Communications Skills

Without exception, an analyst must be able to communicate effectively, both orally and in writing. The analyst should actively seek help or training in business writing, technical writing, interviewing, presentations, and listening. A good command of the English language is considered essential. These skills are learnable, but most of us must force ourselves to seek help and work hard to improve them. College recruiters and business managers will emphatically tell you that communications skills are the single most important ingredient for success. Many college seniors lose job offers because of their inability to write and speak at the expected level, and others face a slow career development due to inadequate communications skills. Almost without exception, your communications skills, not your technical skills, will prove to be the single biggest factor in your career success or failure.

This book surveys the importance of communications skills for the systems analyst; however, there is no substitute for the wealth of English and speech/communications courses offered by the typical college and university. Take as many courses as possible; you will never have as many opportunities to improve your ability to speak and write as you do today!

Interpersonal Relations Skills

It has been suggested that analysts "need to exercise the boldness of Lady Godiva, the introspection of Sherlock Holmes, the methodology of Andrew Carnegie, and the down-home common sense of Will Rogers."[7] In other words, systems work is

---

[7]Kenniston W. Lord, Jr., and James B. Steiner, *CDP Review Manual: A Data Processing Handbook*, 2nd ed. (New York: Van Nostrand Reinhold, 1978), p. 349.

people-oriented and systems analysts must be extroverted or people-oriented. Interpersonal skills help us work effectively with people. Although these skills can be developed, some people simply do not possess the necessary outgoing personality. The interpersonal nature of systems work is demonstrated in the running SoundStage case study that appears throughout this book.

Interpersonal skills are also important because of the political nature of the systems analyst's job. The analyst's first responsibility is to the business, its management, and its workers. Individuals frequently have conflicting goals and needs. They have personality clashes. They fight turf battles over who should be responsible for what and who should have decision authority over what. The analyst must mediate such problems and achieve benefits for the business as a whole.

Another aspect of interpersonal relations is recognition of the analyst's role as an agent of change. The systems analyst is frequently as welcome as an IRS auditor! Many individuals feel comfortable with the status quo and resent the change the systems analyst brings. An analyst should study the theory and techniques of effecting change. Persuasion is an art that can be learned. Begin by studying sales techniques—after all, systems analysts sell change.

Finally, systems analysts work in teams composed of IS professionals, end-users, and management. Being able to cooperate, to compromise, and to function as part of a team is critical for success in most projects. Because development teams include people with dramatically different levels of education and experience, group dynamics is an important skill to develop.

Many business or organizational leadership schools offer valuable courses on topics such as change management, team dynamics, leadership, and conflict resolution. If available, these courses can contribute significantly to the career growth of the modern systems analyst.

Flexibility and Adaptability

No two systems development projects encountered by a systems analyst are identical. Each project offers its own unique challenges. Thus, there is no single, magical approach or solution applicable to systems development. Successful systems analysts recognize this and learn to be flexible and adapt to special challenges or situations presented by specific systems development projects.

Many organizations have standards that dictate specific approaches, tools, and techniques that must be adhered to when developing a system. Although these standards should be followed as closely as possible, the systems analyst must be able to recognize when variations on (or single-instance exceptions to) those standards are necessary and beneficial to a particular project. At the same time, the analyst must be aware of the implications of not following the standards. It's a balancing act that usually improves with experience.

Character and Ethics

The nature of the systems analyst's job requires a strong character and sense of ethics.

Ethics is a personal character trait in which an individual understands the difference between "right" and "wrong" and acts accordingly.

Because systems analysts require information about the organization to develop systems that properly support the organization, those analysts are often privy to sensitive plans and secrets. Consequently, the analyst must be very careful not to share that information with others, either within or outside the organization.

Systems analysts also frequently uncover dissent in the ranks of employees and gain access to sensitive and private data and information (through sampling of memos, files, and forms) about customers, suppliers, employees, and the like. The analyst must be very careful not to share such feelings or information with the wrong people. Trust is sacred! Confidence is earned!

Systems analysts also design systems and write programs. But who owns such intellectual property? In most cases, the design and programs are the property of the organization because it paid for the services of the analyst and programmers. It would be unethical to take (or sell) such designs and programs to another company.

Finally, systems analysts are a key interface between the computing industry and end-users and management. They have a moral obligation to set a good example for end-users, especially in the area of software copyrights. Systems analysts should help end-users and management appreciate the importance of honoring the terms of software licensing agreements.

Systems Analysis and Design Skills

All systems analysts need thorough and ongoing training in systems analysis and design. Systems analysis and design skills can be conveniently factored into three subsets—concepts and principles, tools, and techniques. This book begins that training.

When all else fails, the systems analyst who remembers the basic concepts and principles of systems work will still succeed. No tool, technique, process, or methodology is perfect in all situations! Concepts and principles will help you adapt to new and different situations and methods as they become available. We have purposefully emphasized applied concepts and principles in this book. This is not a mechanical, "monkey see, monkey do" book! We believe that if you carefully study the concepts presented in Part One, you will be better able to communicate with potential employers, business users, and computer programmers alike. Also note the references at the end of this chapter for books that emphasize problem solving.

Not too long ago, it was thought that the systems analyst's only tools were paper, pencil, and flowchart template. Over the years, several tools and techniques have been developed to help the analyst. Today, a new generation of computer-based tools is emerging. Tools and their associated techniques help the analyst build systems faster and with greater reliability. This book comprehensively covers the modern tools and techniques of the trade.

Techniques are specific approaches for applying tools in a disciplined manner to successfully develop systems. There are numerous popular techniques; each has its own supporters. We will present the most popular techniques throughout this book. You will learn to use and integrate these techniques and avoid the pitfall of blind devotion to any one technique.

Systems analysis and design techniques are constantly evolving. Sensible systems analysts avail themselves of any opportunity to improve their skills. Books provide the easiest source of self-improvement. Forward-thinking organizations are willing to invest in courses and seminars to keep their analysts current.

THE NEXT GENERATION

The life of a systems analyst is both challenging and rewarding. But what are the prospects for the future? Do organizations need systems analysts? Will they need them into the foreseeable future? Is the job changing for the future, and if so, how? These questions are addressed in this section.

Career Prospects

Is it worth preparing yourself for a career as a systems analyst? Absolutely! The job outlook is bright. According to the Bureau of Labor Statistics, opportunities for systems analysts are expected to increase much faster than the average for all professions, even more than for programmers. Depending on the economy, businesses will need between 173,000 and 264,000 new systems analysts by the year 2000, an increase of 24 to 37 percent since 1988. During this same period, the overall job force is expected to increase only 8 to 22 percent. Systems analyst is ranked as the 12th fastest-growing occupation between now and the year 2000. In terms of total demand, it is ranked fourth. The demand is increasing because industry needs systems analysts to meet the seemingly endless demand for more computer-based systems.[8]

What happens to the successful systems analyst? Does a position as a systems analyst lead to any other careers? Indeed, there are many career paths. Some analysts leave the information systems field and join the user community. Their experience with developing business applications, combined with their total systems perspective, can make experienced analysts unique business specialists. Alternatively, analysts could become project managers, information systems managers, or technical specialists (for database, telecommunications, microcomputers, and so forth). The opportunities are virtually limitless.

Predictions

As with any profession, systems analysts can expect change. While it is always dangerous to predict changes, we'll take a shot at it.

We believe that a greater percentage of tomorrow's systems analysts will not work in the information systems department. Instead, they will work directly for their end-users. This will enable them to better serve their users. It will also give users more power over what systems are built and supported.

We also believe that a greater percentage of systems analysts will come from noncomputing backgrounds. At one time most analysts were computer specialists. Today's computer graduates are becoming more business literate. Similarly, today's business and noncomputing graduates are becoming more computer literate. Their full-time help and insight will be needed to meet demand and to provide the business background necessary for tomorrow's more complex applications.

WHERE DO YOU GO FROM HERE?

Each chapter and module will provide guidance for self-paced instruction under the heading, "Where Do You Go from Here?" Recognizing that different students and readers have different backgrounds and interests, we will propose appropriate learning paths—most within this book, but some beyond the scope of this book.

Most readers should proceed directly to Chapter 2. The first three chapters provide much of the context for the remainder of the book. Several recurring themes, models, and terms are introduced in those chapters to allow you to define your own learning path from that point forward. This chapter focused on the *people* who develop information systems. Chapter 2 will take a closer look at the *product* itself—**information systems**—from an architectural perspective appropriate for systems development. Chapter 3 completes the foundation with an introduction to the *process* of systems development.

This would be a good time to call your attention to Part Five in the book. This unit expands on many of the soft skills mentioned in this chapter as crucial to the systems analyst's success. For example, we have provided modules on interpersonal communications and interpersonal relations skills. These modules are intended to be woven into whatever sequence of chapters you will study. We'll use this section of each chapter to direct you to appropriate modules.

[8] *Monthly Labor Review* 112, no. 11 (November 1989), pp. 42–47.

SUMMARY

1. Systems analysis and design methods are applied by systems analysts to facilitate the development of information systems and computer applications.
2. A systems analyst facilitates the study of business problems and needs to determine how the business system and information technology can best solve the problem and accomplish improvements for the business.
3. Some businesses have separated the role of the traditional systems analyst into two roles: a business analyst to focus on business requirements, and an application analyst to focus on technical design.
4. A systems analyst uses a systems problem-solving approach called a systems development life cycle. It includes steps for project planning, problem analysis, requirements analysis, solutions analysis, solution design, solution construction, solution implementation, and solution refinement.
5. The systems perspective of the systems analyst extends beyond technology to include people, data, processes, interfaces, and networks.
6. Most systems analysts work in central information services units of businesses, but they are increasingly being relocated to decentralized business units to get them closer to their customers. Variations of the systems analyst role can be found in third-party businesses such as outsourcers, consultants, and software solutions providers.
7. The systems analyst's customers are the owners and users of information systems and computer applications.
8. Systems analysts are being significantly influenced and affected by several business trends including total quality management, business process redesign, continuous process improvement, globalization of the economy, and empowerment of individuals and teams.
9. Systems analysts act as facilitators who coordinate systems and application development with users, management, and other information technology specialists.
10. Systems analysts require a broad knowledge and skill set including a working knowledge of systems and technology, computer programming, general business, problem solving, interpersonal communications, interpersonal relations, flexibility and adaptability, character and ethics, and formal systems analysis and design.
11. Career prospects for systems analysts will remain strong through the end of the decade.

KEY TERMS

application analyst, p. 9
business analyst, p. 9
business process redesign, p. 18
chief information officer, p. 12
client, p. 16
client/server computing, p. 22
computer application, p. 7
continuous process improvement, p. 19
cross-functional applications, p. 13
development center, p. 12
distributed computing, p. 22
empowerment, p. 20
ethics, p. 24
external user, p. 17
information system, p. 7
information technologists, p. 20
information technology, p. 8
internal user, p. 17
management consultant, p. 15
object-oriented programming language, p. 22
outsourcing, p. 14
sales engineer, p. 16
software engineer, p. 16
system owner, p. 17
system problem solving, p. 9
systems analysis, p. 7
systems analyst, p. 7
systems consultant, p. 15
systems design, p. 7
systems development life cycle, p. 9
systems engineer, p. 16
systems integration, p. 16
systems integrator, p. 16
system user, p. 16
total quality management, p. 18
user, p. 16
visual programming languages, p. 22

REVIEW QUESTIONS

1. What is the difference between an information system and a computer application?
2. Explain why a noncomputer professional (for instance, engineer, business manager, accountant, and the like) needs to understand systems development.
3. Differentiate between a systems analyst, business analyst, and application analyst.
4. What is the difference between computer technology and information technology?
5. What is the role of the systems analyst when developing a computer application? To whom is the systems analyst responsible?
6. Describe three different types of problems.
7. What is the relationship between systems problem solving and the systems development life cycle?
8. What are the centers of activity in the traditional information services unit of a business? Where do most of the systems analysts work?

9. With respect to the systems analyst, what is the difference between the traditional and modern information services organizations.
10. What is outsourcing and how does it affect systems analysts?
11. What is total quality management and how does it affect systems analysts?
12. What is business process redesign and how does it affect systems analysts?
13. Differentiate between business process redesign and continuous process improvement.

PROBLEMS AND EXERCISES

1. Using the systems problem-solving methodology described in this chapter, write a letter to your instructor that proposes development of an improved personal financial management system (to plan and control your own finances). Tell your instructor what has to be done. Assume your instructor knows nothing about computers or systems analysts. In other words, be careful with your use of new terms.
2. Kathy Thomas has been asked to reclassify her systems analysts. Virtually all her analysts perform planning, analysis, design, implementation, and support. However, depending on their experience, the percentage of time in the five phases varies. Younger analysts do 80 percent programming, whereas the most experienced analysts do 80 percent systems planning and analysis—largely because of their greater understanding of the business and its users and management. How should Kathy reclassify her personnel?
3. Construct a matrix that maps the roles of the systems analyst versus the programmers in terms of the systems problem-solving methodology described in this chapter. Next, add a column for the roles of system owners and system users.
4. Diversified Plastics, Inc., has adopted an unusual data processing organization. The Management Systems group consists of systems analysts who perform only planning, analysis, and very general design. The Technical Systems group consists of programmers who perform only detailed design, implementation, and support. All analysts must come from a business, engineering, or management background, with no computer experience requirement. Programmers must come from a computing background. Transfers between the groups are discouraged and in most cases not allowed. What are the advantages and disadvantages of such an organization and its policies?
5. Federated Mortgages' corporate information officer (CIO) is facing a budget dilemma. End-users have been buying microcomputers at an alarming rate. In a sense, the business users of these microcomputers, who have little or no DP background, are developing their own application systems. Because of this, the CIO has been asked to justify the continued growth of his budget, especially the growth of his programming and systems analysis staff. There is even some feeling that the number of programmers and analysts should be reduced. How can the DP manager justify his staff? Will the roles of his programmers and analysts change? If so, how? Can the users completely replace the programmers and analysts?
6. Which of the following environments would you prefer to work in? An information services unit of a business? An outsourcer? A consulting firm? A software vendor? Why?
7. Within your college or university environment, describe one cross-functional application and explain why you believe it to be cross-functional.
8. Within your college or university environment, who are the system owners and system users of your course registration and scheduling system?
9. The students in an introductory programming course would like to know how systems analysis and systems design differ from computer programming. Specifically, they want to know how to choose between the two careers. Help them out by explaining the differences between the two and pointing out factors that might influence their decision.
10. You need to hire two systems analysts. Explain to a Personnel department recruiter the characteristics and background you seek in an experienced systems analyst.

PROJECTS AND RESEARCH

1. Make an appointment to visit with a systems analyst or programmer/analyst in a local business. Try to obtain a job description from the analyst. Compare that job description with the job description provided in this chapter.
2. Visit a local data processing department in your business community. Compare its organization with the generic organization described in this chapter. Are the five centers of activity present? What are they called? How is its orga-

nization different? How is it better? Do you see any disadvantages? Where do the systems analysts fit in?

3. Visit a local data processing department in your business community. How are its project teams formed? How are those teams organized? How does this compare with the generic structure described in this chapter?
4. Based on the systems analyst's job characteristics and requirements described in this chapter, evaluate your own skills and personality traits. In what areas would you need to improve?
5. Your library probably subscribes to at least one big-city newspaper. Additionally, your library, academic department, or instructor may subscribe to a data processing newspaper such as *Computer World* or *MIS Week*. Study the job advertisements for systems analysts and programmer/analysts. What skills are being sought? What experience is being required? How are those skills and experiences important to the role of the analyst as described in this chapter?
6. Prepare a curriculum plan for your education as a systems analyst. If you are already working, prepare a statement that expresses your personal need for continuing education to become a systems analyst.
7. A systems analyst applies new technologies to business and industrial problems. As a prerequisite to this "technology transfer," the analyst must keep abreast of the latest trends and techniques. The best way to accomplish this is to develop a disciplined reading program. This extended project will help you develop this program.
   a. Visit your local school, community, or business library. Make a list of all computing, data processing, and systems-oriented publications.
   b. Skim two or three issues of each publication to get a feel for their contents and orientation. Select the five periodicals that you find most interesting and helpful. We recommend that you select five publications that address the areas of microcomputers, mainframes, data communications, applications, and management issues.
   c. Set up a browsing schedule. This should consist of one or two hours a week that you will spend browsing the list of journals. You should try to maintain this schedule for 10 to 15 weeks. If you miss a day, make it up within one week.
   d. Set up a journal to track your progress. Record the date, the journals browsed, the title or subject of the cover story or headlines, and the title of one other article that caught your eye.
   e. Learn to browse. You won't have time to read. If you try to read everything, you will get discouraged and quit the program. Instead, study the table of contents. Read only the first paragraph or two of each article along with any highlighted text in the article. Read the conclusion or last paragraph of the article. Then move on to the next article no matter how interesting the present one. Note any article that you want to fully read after browsing your reading list.
   f. After browsing each of your selected publications, select at least two articles to read thoroughly. The number you read is limited only by your interest and available time. Record these articles in your journal.

   This project will show you how to keep up with a rapidly changing technological world without consuming excessive time and effort.

MINICASES

1. For whom should the systems analyst work? Rolland Industries is facing a data processing reorganization dilemma. Non-DP management is pressing for a new structure whereby most systems analysts would directly report to their application user group (such as Accounting, Finance, Manufacturing, Personnel) as opposed to reporting to Data Processing management. Non-DP management believes that, in the existing structure, systems analysts are too influenced to "change everything" for the sake of computing because they report to Data Processing. To ensure that systems meet Data Processing standards, a small contingent of analysts would remain in Data Processing as a quality assurance group that has final sign-off on all systems projects.

   Data Processing managers are resisting this change. They believe systems analysts will become technologically "out of tune" if removed from DP. They also feel that separating the systems analysts from one another will result in less sharing of ideas and, subsequently, reduce innovation. They also think data files and programs will be unnecessarily duplicated. Conflicts between analysts and programmers, who will remain in DP, will likely increase. DP also believes users will hire new systems analysts without regard to programming and technical experience or familiarity with DP's technical environment.

   Systems analysts are split on the issues. They see the benefits of users being more directly in control of their own systems' destinies; however, they are concerned that users and user management will be less forgiving when faced with budget overruns and schedule delays that have historically plagued DP. Analysts are also concerned that they will become more prone to technological obsolescence if they are physically relocated outside of Data Processing and its more technically oriented staff.

   The decision will likely be made at a higher level than DP. What do you think should be done?
2. The following fable, which appears in a book by Jerry Weinberg called *Rethinking Systems Analysis and Design,* is an entertaining yet effective minicase. Read and enjoy the fable, but pay attention to its moral. There are many systems analysts who have discovered the moral of this fable the hard way!

Three ostriches had a running argument over the best way for an ostrich to defend himself. Although they were brothers, their mother always said she couldn't understand how three eggs from the same nest could be so different. The youngest brother practiced biting and kicking incessantly and held the black belt. He asserted, "The best defense is a good offense." The middle brother, however, lived by the maxim that "he who fights and runs away, lives to fight another day." Through arduous practice, he had become the fastest ostrich in the desert, which you must admit is rather fast. The eldest brother, being wiser and more worldly, adopted the typical attitude of mature ostriches: "What you don't know can't hurt you." He was far and away the best head-burier that any ostrich could recall.

One day a feather hunter came to the desert and started robbing ostriches of their precious tail feathers. Now, an ostrich without his tail feather is an ostrich without pride, so most ostriches came to the three brothers for advice on how best to defend their family honor. "You three have practiced self-defense for years," said their spokesman. "You have the know-how to save us, if you will teach it to us." And so each of the three brothers took on a group of followers for instruction in the proper method of self-defense—according to each one's separate gospel.

Eventually, the feather hunter turned up outside the camp of the youngest brother, where he heard the grunts and snorts of all the disciples who were busily practicing kicking and biting. The hunter was on foot but armed with an enormous club, which he brandished menacingly as the youngest brother went out undaunted to engage him in combat. Yet fearless as he was, the ostrich was no match for the hunter, because the club was much longer than an ostrich's leg or neck. After taking many lumps and bumps, and not getting in a single kick or bite, the ostrich fell exhausted to the ground. The hunter casually plucked his precious tail feather, after which all his disciples gave up without a fight.

When the youngest ostrich told his brothers how his feather had been lost, they both scoffed at him. "Why didn't you run?" demanded the middle one. "A man cannot catch an ostrich."

"If you had put your head in the sand and ruffled your feathers properly," chimed the eldest, "he would have thought you were a yucca and passed you by."

The next day the hunter left his club at home and went out hunting on a motorcycle. When he discovered the middle brother's training camp, all the ostriches began to run, the brother in the lead. But the motorcycle was much faster, and the hunter simply sped up alongside each ostrich and plucked his tail feather on the run.

That night the other two brothers had the last word. "Why didn't you turn on him and give him a good kick?" asked the youngest. "One solid kick and he would have fallen off that bike and broken his neck."

"No need to be so violent," added the eldest. "With your head buried and your body held low, he would have gone past you so fast he would have thought you were a sand dune."

A few days later, the hunter was out walking without his club when he came upon the eldest brother's camp. "Eyes under!" the leader ordered and was instantly obeyed. The hunter was unable to believe his luck, for all he had to do was walk slowly among the ostriches and pluck an enormous supply of tail feathers.

When the younger brothers heard this story, they felt impelled to remind their supposedly more mature sibling of their advice. "He was unarmed," said the youngest. "One good bite on the neck and you'd never have seen him again."

"And he didn't even have that infernal motorcycle," added the middle brother. "Why, you could have outdistanced him at half a trot."[9]

But the brothers' arguments had no more effect on the eldest than his had on them, so they all kept practicing their own methods while they patiently grew new tail feathers.

a. What is the moral of the fable?
b. This book will teach you a variety of tools, techniques, and methodologies for developing information systems and computer applications. How might the fable and its moral relate to your study of systems analysis and design?

[9]Gerald M. Weinberg, *Rethinking Systems Analysis and Design,* pp. 23–24. Copyright ©1988, 1982 by Gerald M. Weinberg. Reprinted by permission of Dorset House Publishing, 353 W. 12th St., New York, NY 10014 (212-620-4053/1-800-DH-BOOKS/www.dorsethouse.com).

SUGGESTED READINGS

Gartner Group 1995 Annual Symposium on the Future of the Information Technology Industry. Our institution's management information unit has long subscribed to the Gartner Group's service that reports on industry trends, the probabilities for success of trends and technologies, and suggested strategies for information technology transfer. This edition, and its predecessors, has been influenced by Gartner Group reports and symposiums.

MacDonald, Robert D. *Intuition to Implementation: Communicating about Systems Toward a Language of Structure in Data Processing System Development.* Englewood Cliffs, NJ: Prentice Hall, Yourdon Press Computing Series, 1987. This is a good conceptual systems problem-solving textbook. Although copyrighted in 1987, its timeless wisdom demonstrates that system concepts have withstood the test of time.

Martin, James. *An Information Systems Manifesto.* Englewood Cliffs, NJ: Prentice Hall, 1982. While somewhat dated in some chapters, this book still describes several important trends in information systems, technology, management, and systems development; it was published ahead of its time! We particularly like the end-of-book manifestos for educators, students, and various computer professionals. They can easily be modernized to reflect some of the issues and trends covered in our book.

Yourdon, Edward. *Modern Structured Analysis.* Englewood Cliffs, NJ: Prentice Hall, Yourdon Press Computing Series, 1989. This book was the initial source of our classification scheme for end-users. Yourdon's coverage is in Chapter 3, "Players in the Systems Game."

2

INFORMATION SYSTEM BUILDING BLOCKS

CHAPTER PREVIEW AND OBJECTIVES

Systems analysis and design methods are used to develop information systems and computer applications for organizations. Before learning the *process* of building systems, you need a clear understanding of the *product* you are trying to build. This chapter takes an architectural look at information systems and applications. We will build a framework for information systems architecture that will subsequently be used to organize and relate all of the chapters in this book. The chapter will address the following areas:

- Describe the difference between data and information.
- Define the product called an *information system.*
- Describe six classes of information system applications and how they interoperate.
- Describe the role of information systems architecture in systems development.
- Describe four groups of stakeholders in information systems development and the unique role of the systems analyst in relation to the four groups.
- Recognize categories of systems users and managers who become stakeholders in systems development.
- Differentiate between a perspective and a view as it relates to information systems architecture.
- Describe four perspectives of the DATA focus for an information system.
- Describe four perspectives of the PROCESS focus for an information system.
- Describe four perspectives of the INTERFACE focus for an information system.
- Describe four perspectives of the GEOGRAPHY focus for an information system.

SCENE

A conference room where Sandra Shepherd has called a project launch meeting. Sitting around the table are Bob Martinez (a teammate), Galen Kirchoff (vice president of Member Services), Steven Siemers (director of the Record and Tape Club), Susan Crane (director of the Game Club), Debbie Lopez (interim director of the Video Club), Monica Blair (director of Marketing), and Dick Krieger (director of Warehouse Operations).

SANDRA

Good morning! I see that you all found the coffee, juice, and rolls. Thank you for coming. As you know, we are launching a new systems project this morning. This is the launch meeting, and our goal is merely to orient everyone so that we can assemble the team and establish some vision.

First, I'd like to introduce you to two new faces. On my left is Bob Martinez. Bob has just joined Information Services as a systems analyst and will be assigned to the project. On my right is Debbie Lopez who has joined us from our new Video Club acquisition. Her title is interim director of the Video Club. Welcome to both of you. I know everybody introduced one another before the meeting, so let's get started.

First, it is required in our methodology that every project have an executive sponsor. I'd like to turn the floor over to our executive sponsor, Mr. Galen Kirchoff.

GALEN

Good morning. I'll make my remarks brief. This morning, it is my happy privilege to empower this steering body to begin a long-anticipated project, the reengineering of our member services information system.

[Galen distributes copies of an administrative memorandum, Figure A.]

This is a directive from Rebecca Todd, our executive vice president and chief operating officer, in her capacity as chairperson of the Strategic Planning group. That group hired the IBM Business System Planning [BSP] consultants to help us develop a strategic plan for business process redesign and information services. As part of that plan, they documented management's business plan and then developed an overall architecture for our future databases, networks, and applications.

They also developed a prioritized list of information systems development projects based on perceived value to the business plan. Well, the Order Entry and Member Services system is first on the list. That's why you are all here.

I have personally committed this group up to one-quarter time to direct this project. This group will assemble a project team.

From Information Services, Sandra will serve as project manager. Bob and various other IS staff will provide technical services. The three club directors have agreed to appoint one experienced manager full time to this project as a business analyst. That manager, Sarah Hartman, will serve a two-year appointment to Information Services as a business analyst. Additionally, I am asking that each of you designate one individual to work with the team one-quarter time as necessary to complete various aspects of the project.

Folks, this project is important. The business plan suggested a major expansion of marketing and member services. The Order Entry and Member Services system must be completely overhauled to enable the business plan. I know that I can count on each of you for your full cooperation. Thank you, Sandra.

SANDRA

Thank you, Galen. Everybody except Bob should be somewhat familiar with our *FAST* methodology for continuous systems improvement. We want a well-controlled *FAST* project on this one—it'll be a great learning experience for Bob. The strategic planning repository includes several high-level system models that should provide us with existing documentation.

GALEN

In a nutshell, we want to see what you can do to improve our Order Entry and Follow-Up system. As you know, our product mix is rapidly changing, especially with the addition of games and videos. As part of the business process redesign, we want to disband the current club membership structure that ties members to a particular medium such as compact discs, audiocassettes, or videotapes. In its place, we want a flexible membership club that is not dependent on type of merchandise. Marketing has the details.

SANDRA

My team has not historically supported Marketing.

GALEN

The business plan suggested the development of cross-functional information systems. We're ignoring organizational boundaries and territories. The goal is to design systems across multiple organization boundaries according to common data needs and functional efficiency. Clearly, Marketing and Order Services functions need to be integrated, regardless of where we place them in the organization chart. Monica, why don't you describe the marketing dimension?

MONICA

Galen has only touched the tip of the iceberg in describing the Marketing/Order Services business plan to you. We're looking to new markets, new marketing strategies, new membership and sales goals, even new order technologies. We want you to prototype a new phone-based member response technology as part of this project. We want to explore the Internet as both a marketing and order entry framework. And we want to solve some of the existing system's problems while we're at it.

SANDRA

I take it this project doesn't go through the project steering committee?

GALEN

The steering committee only evaluates user-initiated system requests to assess feasibility and priority. This project comes from the strategic planning committee, which has already assessed its importance and feasibility—they assigned it top priority for new systems development.

SOUNDSTAGE

SOUNDSTAGE ENTERTAINMENT CLUB

Interoffice Memo

Information System Services
Phone: 494-0666 Fax: 494-0999

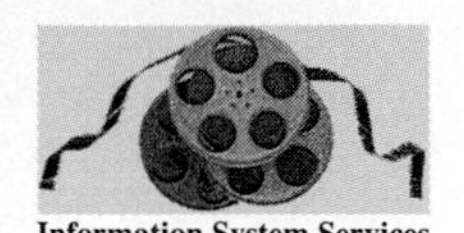

Information System Services

To: Galen Kirchoff, VP for Member Services and Executive Sponsor for MSIS project
Information Services Steering Committee

From: Sandra Shepherd, Project Manager, MSIS Project
Robert Martinez, Systems Analyst, MSIS Project

Date: January 23, 1997

Subject: Project Charter for a new Member Services Information System (MSIS)

We have just completed our initial assessment for the proposed Member Services Information System in accordance with the FAST methodology "survey phase" (which was briefly summarized in our last meeting). This project charter is the final deliverable of that phase. By way of this report, we reaffirm the strategic plan's assertion that "A redesigned member services information system will substantially improve support for not only Member Services, but all peripheral business units, and especially our customers."

Problem Statement

Member Services handles membership subscriptions and member orders. Subscription and order processing is, for the most part, based on manual and computerized processes that have remain largely unchanged for twenty years. Existing computer processes are based on dated batch processing that does not keep pace with the contemporary economy and industry in which we compete. Existing computer processes have been supplemented by rudimentary PC database and spreadsheet applications that are not always fully compatible or consistent with their IS counterparts. Finally, the team conceded that most computerization was merely automating what appear to be outdated business processes. The following specific problems were discussed in a full-day meeting of the survey phase project team:

1. The constantly changing product mix has led to incompatible, and often jury-rigged systems and procedures that have created numerous internal inefficiencies and customer relation problems.
2. The changing product mix creates new opportunities to create new clubs and membership options that would appeal to prospective customers; however, the current system will not support such changes.
3. Directives to increase membership and sales through aggressive advertising will soon overload the current system's ability to process transactions on a timely basis. Customer shipment delays and cash flow problems are anticipated.
4. Response times to orders have already doubled during peak periods from those measures just one year past.
5. Management has suggested a "Preferred Member Program" that cannot be implemented with current data.
6. Unpaid orders have increased from 2%, only two years ago, to 4%. The current credit checking process has contributed significantly to the problem.
7. Member defaults on contracts have increased 7% in three years. It is believed that the current system inadequately enforces contracts.
8. Members have begun to complain about automatic cancellation of memberships after too brief periods of inactivity. This problem has been traced to a data integrity problem in current files.
9. Competition from other companies has led management to proposed dynamic contract adjustments to retain members. The current system cannot handle this requirement.
10. Backorders are not receiving proper priority. Some backorders go for as long as three months, with many cancellations and refused deliveries. New orders frequently deplete inventory before backorders can be processed.

Scope of the Project

The strategic IS plan (and the Member Services management team) are requesting a system that will:

1. Expedite the processing of subscriptions and orders through improved data capture technology, methods, and decision support. If possible, management would like a system that (at least eventually) extends to the Internet World Wide Web.
2. Interface to the new bar-coding automatic identification system being implemented in the warehouse.
3. Reduce unpaid orders to 2% by the end of fiscal year 1997.

FIGURE A *Request for Information Services*

4. Reduce contract defaults to 5% by the end of fiscal year 1997, and 3% by the end of fiscal year 1998.
5. Support constantly changing club and agreement structures, including dynamic agreement changes during the term of an agreement.
6. Triple the order processing capacity of the unit by the end of fiscal year 1997.
7. Reduce order response time by 50% by the end of fiscal year 1997. Management has changed the definition of order response from 'order-receipt-to-warehouse' to 'order-receipt-to-member-delivery'.
8. Rethink any and all underlying business processes, procedures, and policies that have any visible impact on member satisfaction and complaints.
9. Provide improved marketing analysis of subscription and promotion programs.
10. Provide improved follow-up mechanisms for orders and backorders.

The primary users of the new system would be the management and staff of the Member Services Division. The improved system will affect or interface with Purchasing, the Warehouse, Distribution Centers (in two other cities), Accounts Receivable, Marketing, as well as members and prospective members.

Constraints

1. The initial version of the system must be operational in nine months. Subsequent versions should be released in six-month increments.
2. The system must be developed in accordance with the FAST development methodology and CASE tools. A special exception is permitted for the consultant team that will prototype new object technology and techniques.
3. The system cannot alter any existing file or database structures in the Accounts Receivable Information System.
4. The system must conform to the technology architecture approved as part of the IS strategic plan. Exceptions must be pre-approved by both the Technology Architecture Committee and the Information Services Steering Committee. The system should harness the recent plan to invest in state-of-the-art desktop computing and client/server network technology.

Recommendations

We highly recommend that this project be expedited for complete systems analysis. The basis for our recommendation is that this system is essential to the strategic mission and plan of SoundStage Entertainment Clubs. Unless the member services system is overhauled and dramatically improved, it is unlikely that business will be able realize any of its strategic vision.

Since the strategic planning project had already completed a detailed study of this system, we request a waiver on the FAST study phase and request permission to proceed to the definition (of business requirements) phase. (Note: The findings of that detailed study phase can be found at the following intranet address – http://soundstage.com/secure/iss/reports/member services analysis – you will need to provide your network id and password).

Resource Requirements for Systems Analysis

The complete project plan can be found on the intranet at the following address:

http://soundstage.com/secure/iss/projects/msis/baseline plan.html

The next phase is to complete our systems analysis by defining business requirements for the new system. This FAST definition phase has been budgeted as follows:

2 FTE systems analysts for 2 months	
1 FTE business analyst for 2 months	$20,000
2.5 FTE, various end users and management release time	8,000
.5 FTE technical support	18,000
Supplies and expenses	3,000
Overhead	1,500
TOTAL	1,600
	$52,100

Please contact us in person or by email if you have any questions.

CC: Sarah Hartman
Ken Riker
Peg Li
MSIS project team

FIGURE A *Concluded*

SOUNDSTAGE

SOUNDSTAGE ENTERTAINMENT CLUB

SANDRA

Let's establish some vision for this project.

STEVEN

The way I see it, the system should provide several essential functions. First, the system should generate and process three kinds of orders: dated orders, priority orders, and merchandise orders for each club.

BOB

Could you differentiate those types of orders?

STEVEN

Sure. Dated orders are those that are automatically filled if a member doesn't return a dated order card to cancel or change the monthly title of the month's offer. Priority orders are for those cards that are returned. Merchandise orders are for any other type of merchandise we sell, including shirts, posters, computer software, videotapes, and so on.

SANDRA

What kind of problems exist in the current system, Steven?

STEVEN

I've got a big problem coming! I had a meeting with Rebecca Todd yesterday. I was told to expect an aggressive new marketing program over the next three years . . . TV, radio, newspapers, and magazines. Also, a single integrated agreement structure will replace existing clubs' current agreement structures. Instead of tying members to an agreement based on a certain number of purchases over some time period, they will be tied to agreements based on a certain number of credits over a period of time. This will allow members to purchase different types of merchandise with various levels of credit toward fulfilling the purchase agreement.

SUSAN

I'm not sure I see the difference.

STEVEN

All right. Let's say you join the compact disc club today. You must buy six discs in the next two years. You can also purchase cassettes and videotapes, but those don't count against your membership agreement. Under the new approach, you will still join the club, and each compact disc you buy will establish a certain number of credits toward your membership agreement. But so will cassettes, videotapes, videodiscs, computer games, and any other merchandise. And when you fulfill your membership agreement, you'll receive SoundStage Dollars with subsequent purchases. Those dollars can be credited toward the purchase of any of our products.

BOB

Wow! Where do I sign up?

STEVEN

That's the problem. The current system cannot handle any of this. To give customers this new level of service, and to give Marketing the go-ahead to start advertising the service, we have to totally redesign the supporting information systems, in both Marketing and Order Services.

The second essential function is to fully integrate marketing into order entry. The third is to extend the integration into the warehouse. We have to recognize that the members are not satisfied when their order is processed. They are only satisfied when the ordered products are delivered to their home. The last essential function is to provide management with faster and more reliable information to support marketing and member services decision making.

SANDRA

Dick, what's the warehouse operations angle?

DICK

Yeah, we're not used to working so closely with you Marketing and Order Entry types. Basically, I think I'm here because of the new auto-ID system.

SUSAN

Auto-ID system?

DICK

Bar coding! We are in the middle of converting all inventory over to a bar coding scheme. It should eliminate order filling errors when we finish. For some period of time, most products will have their existing product numbers and a new bar code number. Also, the new video title numbering scheme will have to be overhauled to match SoundStage standards.

SANDRA

Whew! That's news to me! We'll need to gather some more facts about this bar code system. We should talk with you and your staff very soon.

DEBBIE

That's why I'm here. A couple of years ago I was working for Private Screenings Video Club. I'm at least somewhat familiar with their current products and inventory schemes. We now have a good grasp of which Private Screenings employees will transfer to SoundStage as part of the merger. I intend to assign one of their staff directly to the project team to represent our interests.

SUSAN

Can I change the subject? While I concur with the basic functions that Steven has outlined, from my perspective, the biggest problem we currently have is with the data—it's out of control. My top priority would be to get control of the data.

SANDRA

Please explain.

SUSAN

Order management requires us to bring together data from various sources. Marketing provides us with promotion and product data. The warehouse provides inventory data. The clubs provide data about agreements and members, and now we have three clubs with different management approaches. I'm probably missing someone, but the problem is coordinating all this data in an organized fashion. If we get control of that data, all of Steven's functions could be built around that data.

BOB

We can do that. Sandra and I can work with the DA to design and implement an Oracle database that consolidates all this data into a highly organized database. We'll write SQL programs to

SOUNDSTAGE

SOUNDSTAGE ENTERTAINMENT CLUB

properly maintain that data and provide users with 4GLs to . . .

SUSAN
Hold on! You're speaking a foreign language to me!

DEBBIE
Can I interrupt? I think we might be missing the big picture here. We now have three warehouses in different cities. We have regional membership offices in five cities. I'm sure that we all do things somewhat differently, but we all do a lot of the same things. Shouldn't our system communicate with each site, or duplicate itself at each site? Why reinvent the wheel at each operating location?

GALEN
I think Debbie has a good point, maybe even better than she realizes. I'm a novice at this network stuff, but the Internet presumably gives us the potential of taking the store directly to members and prospective members. I'd like to think that we should try to creatively exploit network technology, both at and between our current operating locations and direct to our members and suppliers.

BOB
We can do that too! We can LAN each operating site and create a WAN to other sites and use the Web to reach our members.

SUSAN
More technology terms! Ugh! But I do like the idea. While we're on the technology, why can't you guys make our homegrown systems look and act like my PC applications? I use *Word, Excel,* and *PowerPoint,* and I really like the way they all put the commands and buttons in one place. If companies like Microsoft and Lotus can do it, why not us?

SANDRA
We can, and we will. And we'll try to control our use of technical jargon and focus on business issues early in this project. Bob and I both have a tendency to speak the jargon because it is part of what we do everyday. Everybody here should feel free to stop us, just like Susan did.

It seems like we have a lot of opinions. Believe it or not, I think you have helped start to establish that vision. To sum up, you want a system that (1) provides various functional capabilities, (2) integrates and coordinates a wide variety of data and information, (3) takes into consideration multiple operating locations including the member's home address, and (4) is as easy to learn and use as your Windows and Mac applications.

Let's take a 10-minute break. When we return, we'll try to formalize some of these ideas and establish a project plan.

DISCUSSION QUESTIONS

1. Why did the different participants in this meeting have entirely different views of the same basic system?
2. Each of the different participants was concerned with different aspects of the system. Briefly organize their concerns into four or five categories.
3. Why did Bob's view of the system cause communications problems with Susan? How could Bob have better communicated with Susan and the group?
4. How do the different views of the system affect Sandra's job? How should she deal with such diverse perspectives?

A REVIEW OF FUNDAMENTALS OF INFORMATION SYSTEMS

"Wherever people are active and working together in an organization, they work out some sort of system, without the help of an analyst. Through these systems the people get their day-to-day work done—buying, selling, making, growing, sending, shipping, transporting, paying money, collecting money, and so on."[1] In other words, information systems are the natural by-product of people working with people.

Most of you have had at least one information systems course before coming to this course. For that reason, we'll keep this review to a minimum. The basic terminology of information systems varies slightly from author to author and course to course. It is, therefore, important that we establish our basic concepts and terminology to provide context for this book.

Most experts agree on the fundamental difference between data and information.

> **Data** are raw facts about the organization and its business transactions. Most data items have little meaning and use by themselves.

[1]Keith London, *The Management System: Systems Are for People* (New York: Wiley-Interscience, 1976), p. 30.

Information is data that has been refined and organized by processing and purposeful intelligence. The latter, purposeful intelligence, is crucial to the definition—People provide the purpose and the intelligence that produces true information.

In other words, data are a by-product of doing business. Information is a resource created from the data to serve the management and decision-making needs of the business. Technology has created a data and information explosion in virtually all businesses. The ability of businesses to harness and manage this data and information has become a critical success factor in most businesses.

Stated simply, information systems transform data into useful information. To serve the analysis and design focus of this book, we offer a more formal definition.

An **information system** is an arrangement of people, data, processes, interfaces, and geography that are integrated for the purposes of supporting and improving the day-to-day operations in a business, as well as fulfilling the problem-solving and decision-making information needs of business managers.

The role of the computer was intentionally left out of this definition. An information system exists with or without a computer. But information technology has significantly expanded the power and potential of most information systems.

Information technology is a contemporary term that describes the combination of computer technology (hardware and software) with telecommunications technology (data, image, and voice networks).

As we shall see throughout this chapter, various information technologies significantly influence or drive the fundamental building blocks of our information system definition: people, data, processes, interfaces, and geography.

In practice, there are several types of information systems, each serving the needs of different users. The tools and techniques taught in this book are intended to help you analyze, design, and build the following types of systems.[2]

Transaction Processing Systems

Business transactions are events that serve the mission of the business. They are the primary means by which the business interacts with its suppliers, customers, partners, employees, and government. Transactions are significant to us because they capture and or create data about and for the business. Examples of transactions include purchases, orders, sales, reservations, registrations, vouchers, shipments, invoices, and payments.

Transaction processing systems are information system applications that capture and process data about (or for) business transactions. They are sometimes called **data processing systems.**

Transaction processing systems can either (1) respond to business transactions (such as orders, time cards, or payments) or (2) initiate transactions (such as invoices, paychecks, or receipts), possibly both. Also, transaction processing systems can respond to both external events (such as processing orders from customers) or internal events (such as generating production orders for the shop floor).

A third type of transaction processing system, **data maintenance,** provides for routine updates to stored data. For example, a system must provide for the ability to add and delete CUSTOMERS and PRODUCTS, as well as to change specific facts such as CUSTOMER ADDRESS and PRODUCT PRICE.

[2]For a more detailed explanation of each of these applications, see any Introduction to or Fundamentals of Information Systems textbook.

The analysis and design of transaction processing systems tends to focus on factors such as response time, throughput (volume of transactions), accuracy, consistency, and service. Although most transaction processing systems have long been computerized, new opportunities for analysis and design are being driven by the trend toward **business process redesign.** In many cases, existing systems automate the inefficiencies and bureaucracy of obsolete business practices and policies. When the business redesigns those processes and practices, the underlying transaction processing systems must also be redesigned. Thus, there are many new opportunities to practice systems analysis and design for transaction processing systems.

Management Information Systems

Management information systems supplement transaction processing systems with management reports required to plan, monitor, and control business operations.

> A **management information system** (MIS) is an information system application that provides for management-oriented reporting, usually in a predetermined, fixed format.

Management information systems produce information based on accepted management or mathematical/statistical models. For instance, *materials requirements planning (or MRP)* is a formal model for building production and material procurement schedules based on sales projections. An MIS for MRP would generate these schedules based on the model.

Management information is normally produced from a shared database that stores data from many sources, including transaction processing systems. Thus, data analysis and database design become critical to MIS design. Both skills are taught in this book.

Management information systems can present detailed information, summary information, and exception information. Detailed information is used for operations management as well as regulatory requirements (as imposed by the government). Summary information consolidates raw data to quickly indicate trends and possible problems. Exception information filters data to report exceptions to some rule or criteria (such as reporting only those products that are low in inventory).

So long as organizations continue to recognize information as an important management resource, opportunities for MIS analysis and design will expand. This book will teach you numerous tools and techniques for analyzing MIS requirements and designing MIS solutions.

Decision Support Systems

Decision support systems further extend the power of information systems.

> A **decision support system** (DSS) is an information system application that provides its users with decision-oriented information whenever a decision-making situation arises. When applied to executive managers, these systems are sometimes called **executive information systems.**

A DSS does not typically make decisions or solve problems—people do. Decision support systems are concerned with providing useful information to support the decision process. In particular, decision support systems are usually designed to support **unstructured decisions,** that is, those decision-making situations that cannot be predicted in advance.

Does this seem impossible? Not really! DSS is based on the reality that the information needed to support unstructured decisions has already been captured by transaction processing and management information systems. The DSS provides the decision maker with tools to access that information and analyze it to make a decision. In general, a DSS provides one or more of the following types of support to the decision maker:

- Identification of problems or decision-making opportunities (similar to exception reporting).
- Identification of possible solutions or decisions.
- Access to information needed to solve a problem or make a decision.
- Analysis of possible decisions, or of variables that will impact a decision. Sometimes this is called "what if" analysis.
- Simulation of possible solutions and their likely results.

Interest in decision support systems is at an all-time high. This interest is being driven by projects to develop data warehouses.

A **data warehouse** is a read-only, informational database that is populated with detailed, summary, and exception information that can be accessed by end-users and managers with DSS tools that generate a virtually limitless variety of information in support of unstructured decisions.

The DSS tools include spreadsheets (such as *Excel*), PC-database management systems (such as *Access*), custom reporting tools (such as *Focus*), and statistical analysis programs (such as *SAS*).

The analysis and design of data warehouses and decision support systems use many of the tools and techniques taught in this book.

Expert Systems

Expert systems are an extension of the decision support system.

An **expert system** is an information system application that captures the knowledge and expertise of a problem solver or decision maker and then simulates the "thinking" of that expert for those who have less expertise.

Expert systems address the critical need to duplicate the expertise of experienced problem solvers, managers, professionals, and technicians. These experts often possess knowledge and expertise that cannot easily be duplicated or replaced.

Expert systems imitate the logic and reasoning of the experts within their respective fields. The following examples are real:

- A food manufacturer uses an expert system to preserve the production expertise of experienced engineers who are nearing retirement.
- A major credit card broker uses an expert system to accelerate credit screening that requires data from multiple sites and databases.
- A plastics manufacturer uses an expert system to determine the cause of quality control problems associated with shop floor machines.

Expert systems are implemented with **artificial intelligence** technology, often called **expert system shells,** that captures and simulates the reasoning of the experts. Expert systems require data and information like any other information system, but they are unique in their requirement of rules that simulate the reasoning of the experts who use the data and information. We'll explore a few tools and techniques in this book for expressing such rules.

Office Information Systems

Office automation is more than word processing and form processing.

Office information systems support the wide range of business office activities that provide for improved work flow and communications between workers, regardless of whether or not those workers are physically located in an office.

Office information systems are concerned with getting all relevant information to all those who need it.

Office automation functions include word processing, electronic messages (or electronic mail), work group computing, work group scheduling, facsimile (fax) processing, imaging and electronic documents, and work flow management.

What are the implications for systems analysis and design? As a direct result of business process redesign, office information systems and transaction processing systems are slowly being integrated using the following technologies:

- **Electronic forms** technology (such as Delrina's *PerForms*) that supports the creation of electronic forms with programmable routing and integration into transaction processing databases and applications.
- **Work group** technology (such as Lotus *Notes*) that provides a way of coordinating work group access to and updates to both structured and unstructured transaction information.
- **Electronic messaging** technology that enables individuals to electronically communicate with one another directly within the context of the transaction processing application.
- **Office automation suite** technology that integrates the functionality of office PC tools, such as word processors and spreadsheets, directly into transaction processing applications.
- **Imaging** technology that combines imaging with electronic forms to simulate the familiar paper- and form-based metaphor that preceded computerized transaction processing.

We'll explore the design implications of some of these technologies in this book.

Personal and Work Group Information Systems

Personal and work group information systems represent another opportunity for systems analysis and design to build applications for individuals.

> **Personal information systems** are those designed to meet the needs of a single user. They are designed to boost an individual's productivity.

> **Work group information systems** are those designed to meet the needs of a work group. They are designed to boost the group's productivity.

Typically, both personal and work group information systems are built using personal computer technology and software. Work group information systems usually interconnect the PCs with local area network technology.

Most personal information systems are developed directly by the end-user using common PC tools such as word processors, spreadsheets, and databases (often in an integrated fashion). While developed by and for the individual user, such systems might be deployed to others, making them work group information systems. Also, clever end-users can often integrate their personal information systems into the other types of information systems we've discussed.

Work group applications are typically developed by departmental computing specialists using the same PC tools in conjunction with work group technology such as electronic mail (such as Microsoft *Exchange*) and work group data sharing (such as Lotus *Notes*). Through this work group technology, users in the group can collaborate on projects. For example, the partners in a legal firm could collaboratively develop a new contract using a combination of electronic research, word processing, spreadsheets, and databases—all while working from their own offices on their own time schedules.

Although this book is about systems analysis and design tools and techniques for mainstream information systems and applications, the future end-user can also benefit significantly from their applicability to personal and work group information systems. For example, the database design techniques taught in this book are equally applicable to PC database applications!

Putting It All Together

We've discussed several classes of information systems that can be built with systems analysis and design methods. In practice, these classes overlap such that you can't always differentiate one from the other. Even if you could, the various applications should ideally interoperate to complement and supplement one another. Take a few moments to study the flow diagram in Figure 2.1. In the average business, there will be many instances of each of these different applications.

A FRAMEWORK FOR INFORMATION SYSTEMS ARCHITECTURE

It has become fashionable to deal with the complexity of modern information systems using *architecture*. Information technology professionals speak of data architectures, application architectures, network architectures, technology architectures, and so forth. For purposes of this book, we'll define an *information systems architecture* as follows.

> An **information systems architecture** provides a unifying framework into which various people with different perspectives can organize and view the fundamental building blocks of information systems.

Essentially, information systems architecture provides a foundation for organizing the various components of any information system you care to develop.

In this chapter, we'll build a framework for information systems architecture that is inspired by the work of John Zachman.[3] The Zachman *Framework for Information Systems Architecture* has achieved international recognition and use. We have adapted and extended the framework to teach (and practice) systems analysis and design.

Let's assume you want to build an information system. Different people have different views of the system. Managers, users, and technical specialists each view the system in different ways, and in different levels of detail. We call these people **stakeholders** in the system. They can be broadly classified into four groups:

- *System owners* pay for the system to be built and maintained. They own the system, set priorities for the system, and determine policies for its use. In some cases, system owners may also be system users.
- *System users* are the people who actually use the system to perform or support the work to be completed. In today's team-oriented business world, system users frequently work side by side with system designers.
- *System designers* are the technical specialists who design the system to meet the users' requirements. In many cases, system designers may also be system builders.
- *Systems builders* are the technical specialists who construct, test, and deliver the system into operation.

As shown in Figure 2.2, each group of stakeholders is afforded one row in our information systems framework. Furthermore, each row has its own **perspective** or view of the information system.

PERSPECTIVES—THE PEOPLE SIDE OF INFORMATION SYSTEMS

As you just learned, there are four general perspectives on information systems: owners, users, designers, and builders. They comprise the systems development team for any given project. All these participants in the information systems game share one thing—they are what the U.S. Department of Labor now calls information workers.

> The term **information worker** (also called knowledge worker) was coined to describe those people whose jobs involve the creation, collection, processing, distribution, and use of information.

[3]John A. Zachman, "A Framework for Information Systems Architecture," *IBM Systems Journal* 26, no. 3 (1987), pp. 276–92.

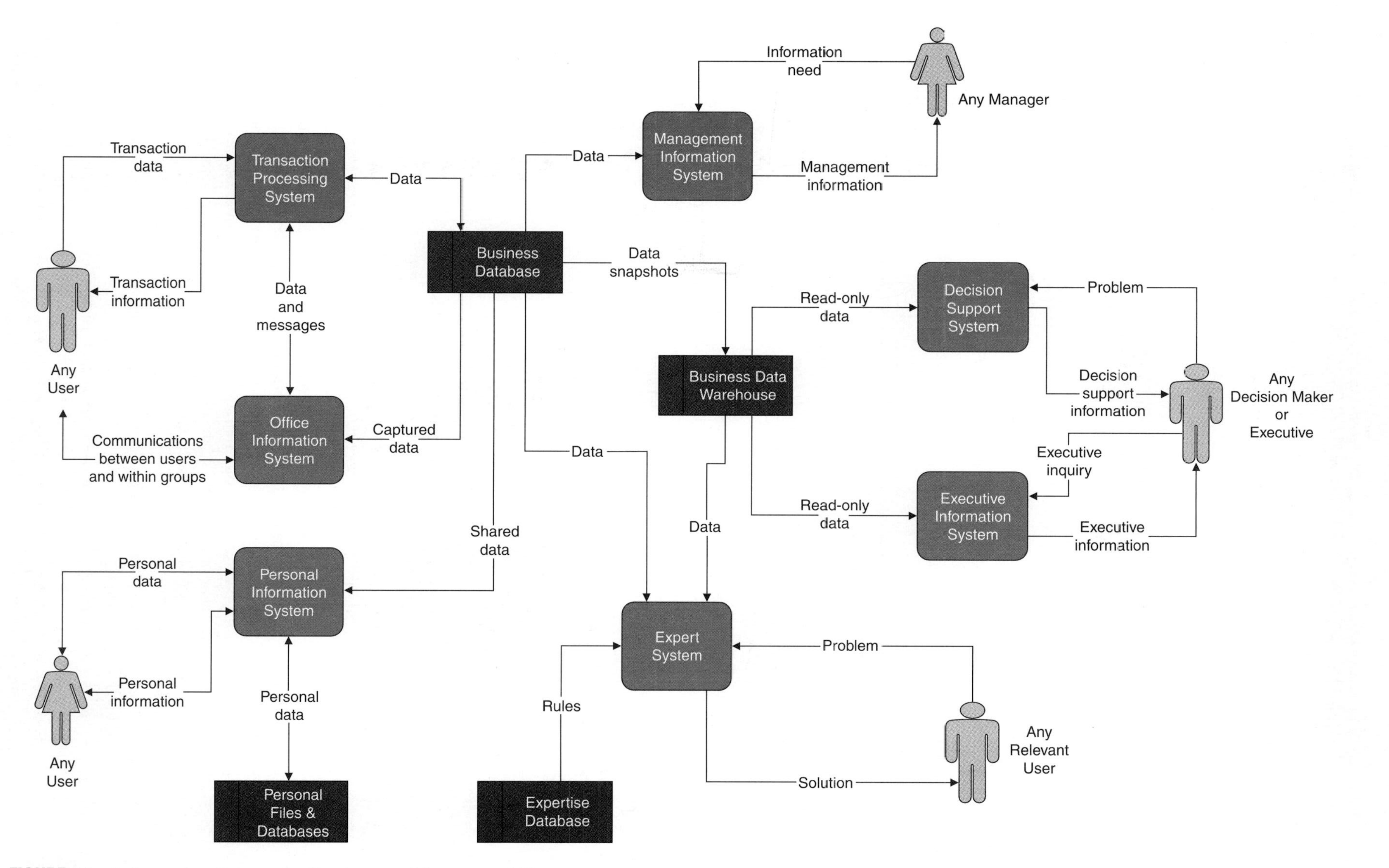

FIGURE 2.1 *Information System Applications and Interoperability*

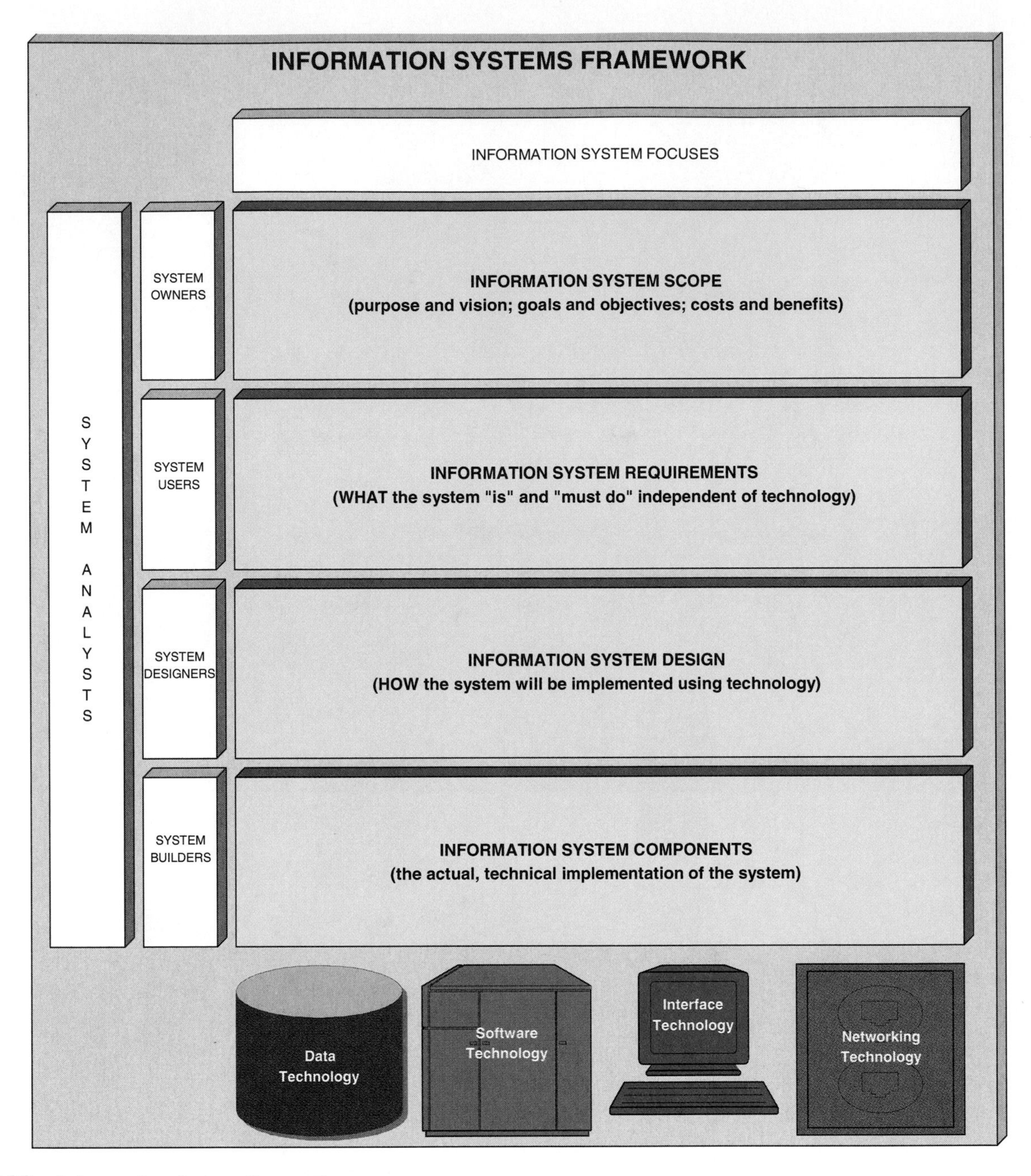

FIGURE 2.2 *Information System Perspectives*

The livelihoods of information workers depend on information and the decisions made from information. Some information workers (such as systems analysts and programmers) create systems that process and distribute information. Others (such as clerks, secretaries, and managers) primarily capture, distribute, and use data and information. Today, more than 60 percent of the U.S. labor force is involved in the production, distribution, and use of information. Not surprisingly, an information services industry (which includes the computer, software, networks, and information-consulting industries) has developed to support the growing information needs of businesses.

Let's examine the four groups of information workers in greater detail.

System Owners

For any system, large or small, there will be one or more system owners. Owners usually come from the ranks of management. For medium-to-large information systems, the owners are usually middle or executive managers. For smaller systems, the owners may be middle managers or supervisors. For personal information systems, the owner and user are the same person.

> **System owners** are an information system's sponsors and chief advocates. They are usually responsible for budgeting the money and time to develop, operate, and maintain the information system. They are also ultimately responsible for the system's justification and acceptance.

System owners tend to think in very general terms, not in details. You'll see this when we examine the system owners' view of the DATA, PROCESSES, INTERFACES, and GEOGRAPHY.

System owners tend to be the least interested (or impressed) with the technology used in any information system. As was shown in Figure 2.2, they are concerned with the value returned by the system. Value is measured in different ways. What is the *purpose* of the system? What is the vision of the system—goals and objectives? How much will the system cost to build? How much will the system cost to operate? Will those costs be offset by measurable benefits? What about intangible benefits?

System Users

System users make up the vast majority of the information workers in any information system.

> **System users** are the people who use (and directly benefit from) the information system on a regular basis—capturing, validating, entering, responding to, storing, and exchanging data and information.

System users define (1) the problems to be solved, (2) the opportunities to be exploited, (3) the requirements to be fulfilled, and (4) the business constraints to be imposed by (or for) the information systems. They also tend to be concerned with how easy (or difficult) the system is to learn and use.

Unlike system owners, system users tend to be less concerned with costs and benefits of the system. Instead, they are concerned with business requirements of the system (again, see Figure 2.2). Although users have become more technology-literate over the years, their primary concern is to get the job done. Consequently, discussions with users need to be kept at the business detail level as opposed to the technical detail level. Much of this book is dedicated to teaching you how to effectively communicate business requirements for an information system.

There are many classes of system users. Each class should be directly involved in any information system development project that affects them. Let's briefly examine these classes.

Internal Users Internal users are employees of the business for which an information system is built. Internal users make up the largest percentage of system users in most businesses. Examples include clerical and service staff, technical and professional staff, supervisors, middle management, and executive management.

Clerical and service workers perform most of the day-to-day data processing in the average business. They process orders, invoices, and payments. They type and file correspondence. They fill orders in the warehouse. And they manufacture goods on the shop floor. Most of the fundamental data in any business is captured or created by these workers, many of whom perform manual labors in addition to processing of data. The volume of data captured or created by these workers in the average organization is staggering. Information systems that target these workers tend to focus on transaction processing speed and accuracy.

Technical and professional staff consists largely of business and industrial specialists who perform highly skilled and specialized work. Examples include lawyers, accountants, engineers, scientists, market analysts, advertising designers, and statisticians. Their work is based on well-defined bodies of knowledge; hence, they are sometimes called knowledge workers.

Knowledge workers are a subset of information workers whose responsibilities are based on a specialized body of knowledge.

Most knowledge workers are college educated. Their jobs are dependent not only on information, but also on their ability to properly use and react to that information. By way of their education and professional activities, technical and professional staff tend to be demanding users. Information systems that target these knowledge works tend to focus on data analysis as well as generating timely information for problem solving.

Supervisors, middle managers, and **executive managers** are all decision makers. Supervisors tend to focus on day-to-day management issues. Middle managers are more concerned with tactical or short-term management plans and problems. Executive managers are concerned with overall business performance, any strategic or long-term planning, and problem solving. Information systems for management tend to focus entirely on information access. Managers need the right information at the right time to solve problems and make good decisions.

Remote and Mobile Users A relatively new class of internal users deserves special mention—remote and mobile users. Like traditional internal users, they are employees of the business for which you are building information systems. Unlike traditional internal users, they are geographically separated from the business.

The classic example is sales and service representatives. These **mobile users** live their professional lives on the road—customer to customer, client to client, buyer to buyer, and so forth. Historically, they have been second-class information system users, receiving little value from those systems. But with the advent of global telecommunications technology, and global competition, these road warriors must be included in the new mix of information system users. Information systems that support mobile users are certainly more complex, but they also offer significant potential value to the modern business. Sales representatives who can directly tap databases and order filling systems have a competitive advantage over those who cannot.

The new kid in the system user community is the **remote user.** Many businesses are looking to **telecommuting** to reduce costs and improve worker productivity. Telecommuting, stated simply, is working from home. There is considerable evidence to suggest that many employees can be just as productive working at home if (and that's a big if) they can be connected to the company's information system through modern telecommunications technology. This is especially true of knowledge workers (described earlier).

External Users Modern information systems are now reaching beyond the boundaries of the traditional business to include other businesses and individuals as system users. Again, driven by global competition, businesses are redesigning their information systems to directly connect to and interoperate with their business and trading partners, suppliers, customers, and even the end consumer.

Consider Sears, the retailing giant. For many years, Sears has insisted that its suppliers directly connect to Sears to eliminate paper-based inventory purchases in favor of *electronic data interchange.* If you want to sell Sears your product, you must be able to electronically exchange orders with it. Your information system must directly interact with its information system—they have become external users for one another. Again, this complicates information system development, but the benefits are substantial—you often cannot compete unless you are willing to consider system-to system interfaces.

And it gets even more interesting! If you can connect with other businesses, why not with your own customers. For example, why should you complete a paper admissions application to get into your college or university? Why not just electronically apply, directly through the college's admission information system? If you can do this, you have become an external user of the college's information system. Today, that represents competitive advantage. Tomorrow, it may become essential just to compete.

Extending the metaphor, why not market directly to the consumer through your information system? The explosive growth of the Internet for electronic commerce is making the consumer an external user of information systems. Currently, World Wide Web pages on the Internet are mostly used to market information to the end consumer of products. Eventually, as the information superhighway evolves, those consumers will be able to place orders and pay bills directly to the information systems of businesses. Some of this is already here. As the security issues get solved, information systems builders will have an almost limitless base of new external users to consider when building the next generation systems.

System Designers

System designers are the technology specialists for information systems.

> **System designers** translate users' business requirements and constraints into technical solutions. They design the computer files, databases, inputs, outputs, screens, networks, and programs that will meet the system users' requirements. They also integrate the technical solution back into the day-to-day business environment.

Again referring to Figure 2.2, system designers are clearly located closer to the information technology base than are system owners and users. Frequently, they must make technology choices and/or design systems within the constraints of the chosen technology. Today's system designers tend to focus on technical specialties.

- Database designers have a DATA focus.
- Software engineers and programmers have a PROCESS (or program) focus.
- Personal computing specialists and systems integrators usually have an INTERFACE focus.
- Network and telecommunications specialists have a GEOGRAPHY focus.

System Builders

System builders represent our final category of system development roles.

> **System builders** construct the information system components based on the design specifications from the system designers. In many cases, the system designer and builder for a component are one and the same.

The applications programmer is the classic example of a system builder. However, other technical specialists may also be involved, such as systems programmers, database programmers, network administrators, and microcomputer software specialists.

And one last time in Figure 2.2, system builders are located directly above the information technology base; they directly use the technology to build the information system. Although this book is not directly intended to educate or train the system builder, it is intended to teach system designers how to better communicate with system builders.

The Role of the Systems Analyst

The systems analyst was introduced in Chapter 1 as a facilitator of information systems development. In this capacity, and in Figure 2.2, the systems analyst must be comfortable with each of the stakeholders' views of the system. For the system owners and users, the analyst typically constructs and validates their views.

For the system designers and builders, the analyst (at the very least) ensures that the technical views are consistent and compatible with the business views.

BUILDING BLOCKS—EXPANDING THE INFORMATION SYSTEM FRAMEWORK

The perspectives shown in Figure 2.2 are oversimplified. Within a single row, different stakeholders may focus on different aspects of the system. For example, one designer may be assigned to design a database while a different designer is assigned to design the programs. Today, we can identify at least four distinct **focuses** in a system. They are:

- DATA—the raw material used to create useful information.
- PROCESSES—the activities (including management) that carry out the mission of the business.
- INTERFACES—how the system interacts with people and other systems
- GEOGRAPHY—where the data is captured and stored; where the processes happen; where the interfaces happen.

As shown in Figure 2.3, each focus is represented by its own column. The intersection of each perspective (row) and each focus (column) defines a fundamental building block (cell) of the information system. Depending on who you are (owner, user, designer, builder) and on what you want to focus (data, processes, interfaces, or networks), you tend to *view* the system's architecture differently. A database designer sees a database schema. A programmer sees application programs.

NOTE Throughout this book, we have used a consistent color scheme for both the matrix model and the various tools that relate to, or document, the building blocks. The color scheme is based on the building blocks as follows:

■ represents something to do with DATA.
■ represents something to do with PROCESSES.
■ represents something to do with INTERFACES.
■ represents something to do with GEOGRAPHY.

An example might help you to better understand the framework. Most of you are at least somewhat familiar with programming, so let's look at the PROCESS focus by itself (Figure 2.4). Let's assume a technology with which at least some of you are familiar—the *COBOL* programming environment (we could have just as easily chosen *BASIC, C,* or *PASCAL*). Let's work our way up the column.

- *Programmers* are the system builder for the process column. They view the processes of the system as COBOL/**program code.**
- The system designers will typically be *programmers, software engineers,* or *systems analysts.* They view the processes of the system as **program design** documentation such as structure charts, flowcharts, and/or state models.
- The system users could care less about the above programs, as long as those programs support their **business processes.** They might be more interested in viewing how data flows through those business processes and what work is performed by each process—a simple process map might be something they could understand.
- Finally, the system owners are rarely interested in even that level of detail. They care only about high-level **business functions,** their costs, and what they contribute to the bottom line of the business.

Each of these stakeholders' perspectives is legitimate and valuable. One of the challenges in building systems is to communicate with each stakeholder at his or her level of interest and, at the same time, to ensure a complete and compatible transition from one view to the next.

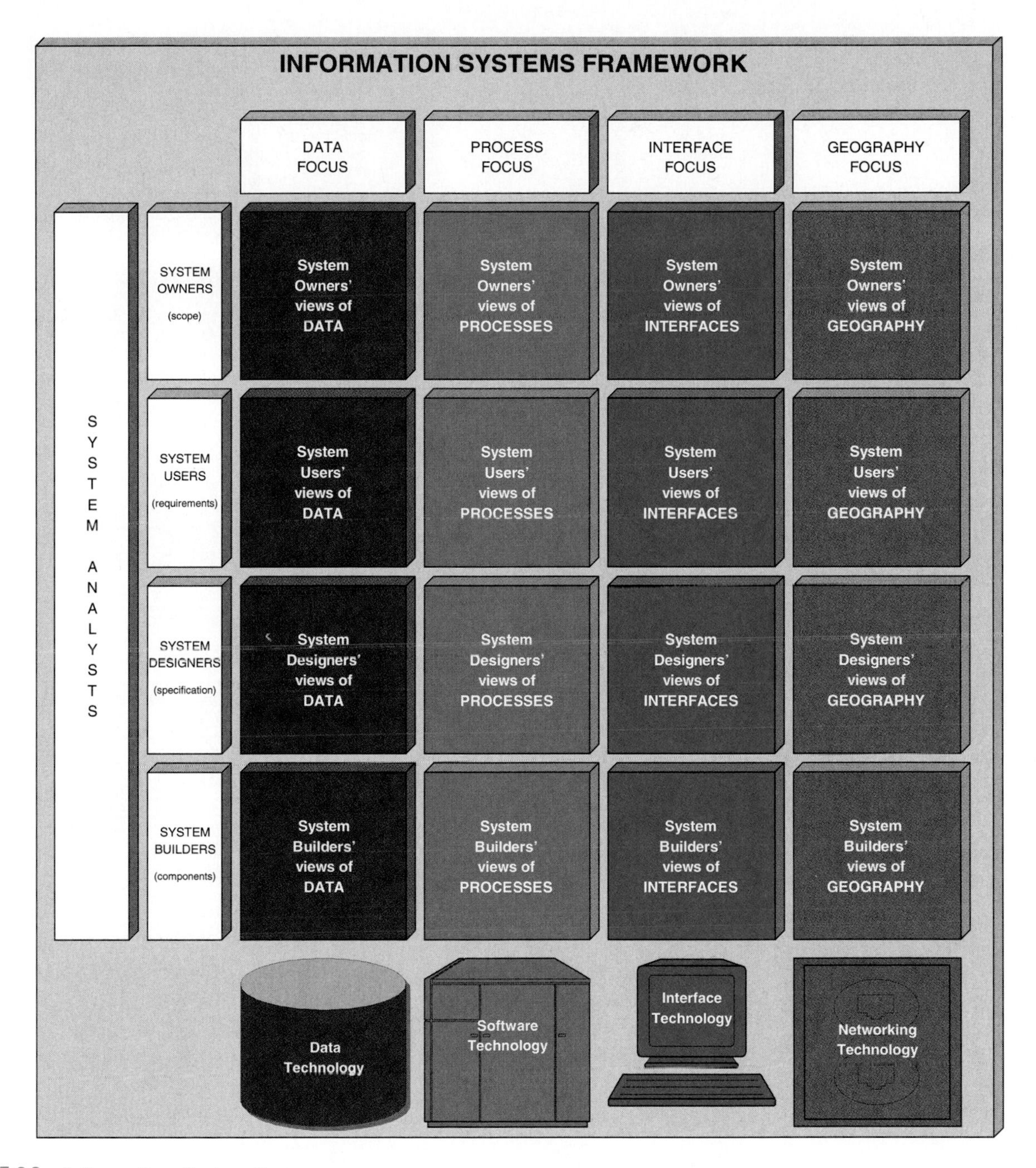

FIGURE 2.3 *Information System Focuses*

Not surprisingly, the information system building blocks do not exist in isolation. They must be carefully synchronized to avoid inconsistencies and incompatibilities within the system. For example, while our database designer and programmer have their own architectural view of the system, these views must be coordinated and compatible if the system is going to work properly. Synchronization occurs both horizontally (across any given row) and vertically (down any given column). We'll study the synchronization requirements as we apply the framework throughout this book.

The Zachman framework is not unique to information systems. Zachman discovered the framework by studying how architects design buildings and how engineers design new products. Consider the latter. The new product can be viewed from different perspectives: management, intended customers, product designers,

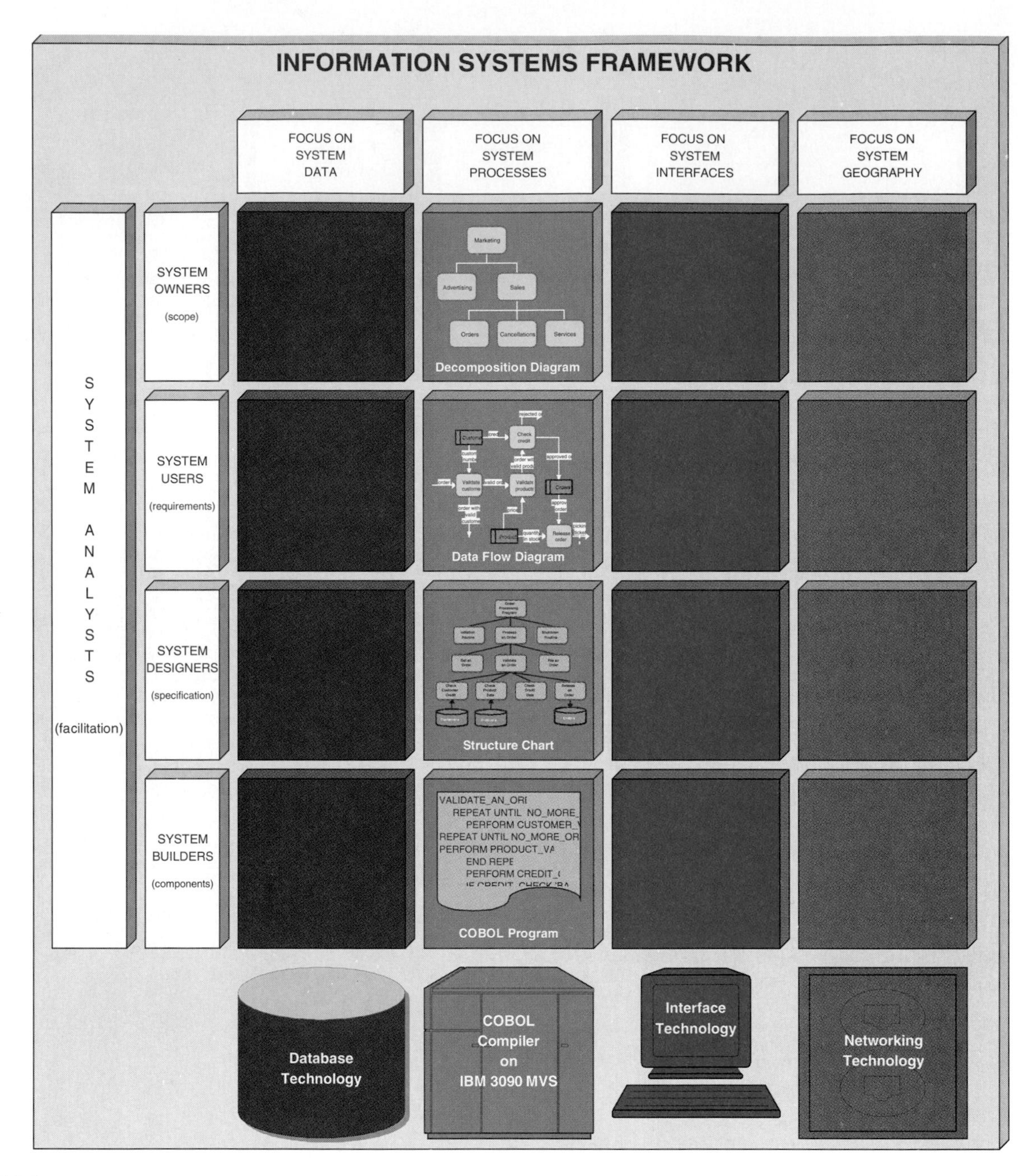

FIGURE 2.4 *Using the Information Systems Framework—a Sample*

and workers on the manufacturing shop floor. The new product can also be viewed from different focuses: the bill of materials needed to produce the product; the functionality and tolerances required of the product; and spatial geometry for the product. Did you note the similarity of architectures? All products—be they cars, buildings, or information systems—can be developed architecturally.

In the remainder of this chapter, we'll briefly examine the perspectives, focuses, and building blocks in the framework.

Building Blocks of DATA

When engineers design a new product, they must create a bill of materials for that product. A bill of materials says nothing about what the product is intended to do. It states only that certain raw materials and subassemblies will make up the finished product. The same analogy can be used for information systems. Data

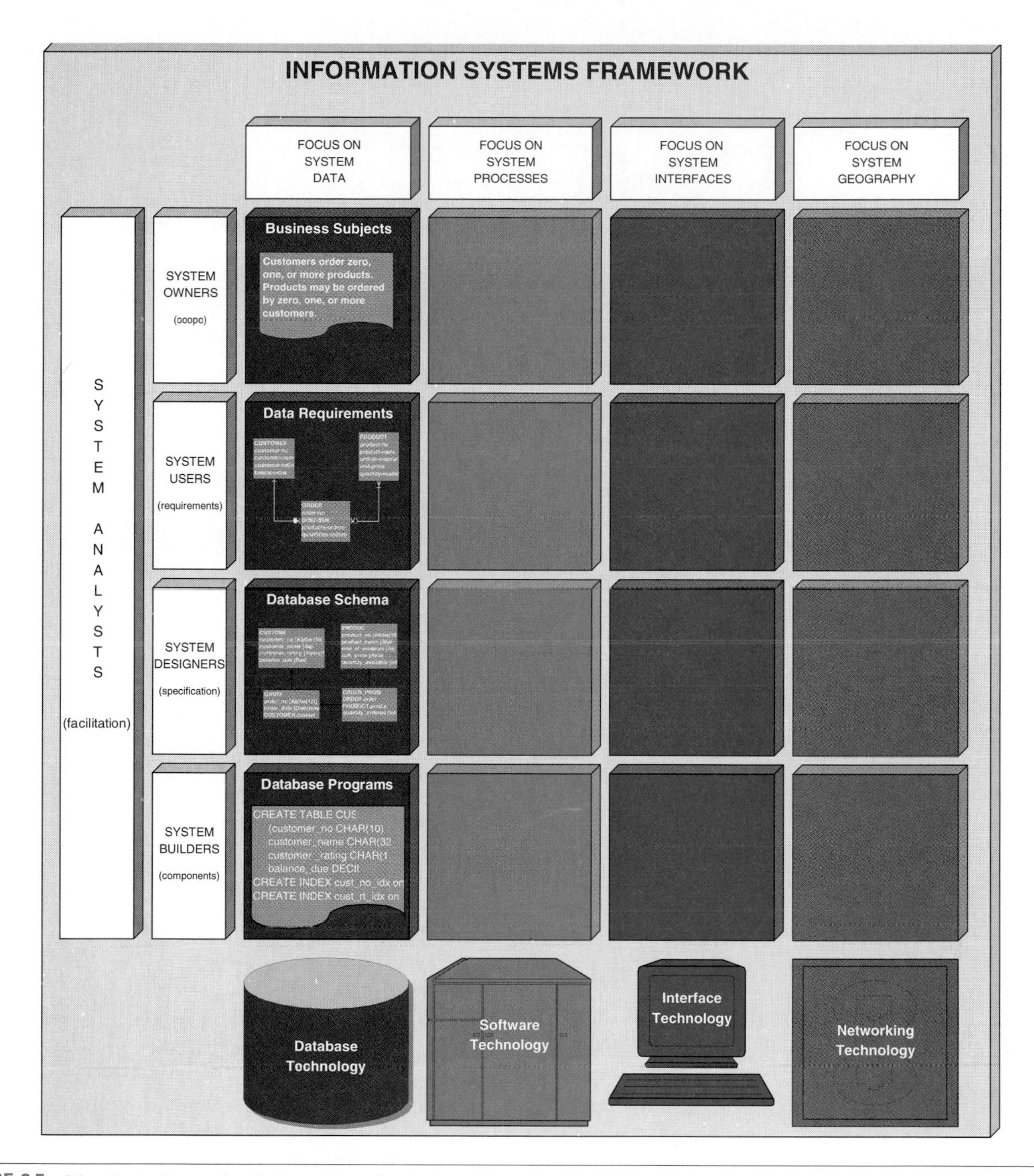

FIGURE 2.5 *The Data Focus for Information Systems*

can and should be thought of as the raw material used to produce information. Consequently, we consider DATA to be one of the fundamental building blocks of an information system.

The DATA column of your framework is illustrated in Figure 2.5. Notice at the bottom of the DATA column that our goal is to capture and store business data using *database* (or file) *technology*. Database technology will be used to facilitate the data storage. Also, as you look down the DATA column, each of our different stakeholders has different views of the system's data. Let's examine those views and discuss their relevance to systems analysis and design.

System Owners' View of Data Strictly speaking, the average system owner is not interested in raw data. The owner is interested in business resources.

Business resources are (1) things that are essential to the system's purpose or mission; or (2) things that must be managed or controlled to achieve business goals and objectives.

Examples of business resources might include CUSTOMERS, PRODUCTS, EQUIPMENT, BUILDINGS, ORDERS, and PAYMENTS.

So what do business resources have to do with data? Resources must be managed. Effective resource management requires information about those resources. And we already know that information is produced from raw data; in this case, data about the business resources and how they interact to operate the business.

In Figure 2.5, we see that system owners view data in terms of "things" about which we need to capture and store data and simple business associations and rules that describe how those entities interact. The icon shown in the matrix is merely one representation of these entities and associations. Other representations, including simple pictures, could also be used.

Examples of entities for a sales system might include CUSTOMERS, PRODUCTS, SALES FORECASTS, SALES REGIONS, ORDERS, and SALES REPRESENTATIVES. These entities are things about which we might need to capture and store data. Similarly, relationships might be expressed with simple, declarative statements such as

- CUSTOMERS place ORDERS.
- ORDERS sell PRODUCTS.
- CUSTOMERS are located in SALES REGIONS.

Intuitively, our system needs to track these relationships (for example, "Does CUSTOMER 2846 have any unfilled ORDERS?").

When dealing with system owners, each entity should be defined in business terms to ensure that everybody understands the system context. Relationships should be expressed either as simple statements (such as those expressed above) or in some intuitive picture or model. Finally, for each entity and relationship, system owners should be asked to identify any perceived problems, opportunities, goals, objectives, and constraints. Together, all these items establish the data context for the information system.

System owners are rarely interested in details about entities and relationships (unless they also happen to be system users).

System Users' View of DATA The users of an information system are the experts about the data that describe the business system. As information workers, they capture, store, process, edit, and use that data every day. Unfortunately, they frequently see the data only in terms of how data are currently implemented or how they think data should be implemented. To them, the data are recorded on forms, stored in file cabinets, recorded in books and binders, organized into spreadsheets, or stored in computer files and databases. The challenge in systems analysis is to identify and verify users' business *data requirements* exclusively in business terms (so alternative implementations might be considered).

Data requirements are a representation of users' data in terms of entities, attributes, relationships, and rules. Data requirements should be expressed in a format that is independent of the technology that can or will be used to implement the data.

Data requirements are an extension of the entities and relationships identified with the system owners. System users may identify additional entities and relationships because of the greater familiarity with the data. More importantly, system users must specify the exact data to be stored and the business rules for maintaining that data. Consider the following example.

A system owner may have identified the need to store data about an entity called CUSTOMER. System users might tell us that we need to differentiate between

PROSPECTIVE CUSTOMERS, ACTIVE CUSTOMERS, and INACTIVE CUSTOMERS because they know that slightly different types of data describe each type of customer. System users can also tell us precisely what data must be stored about each type of customer. For example, an ACTIVE CUSTOMER might require such data attributes as CUSTOMER NUMBER, NAME, BILLING ADDRESS, CREDIT RATING, and CURRENT BALANCE. Finally, system users are also knowledgeable about the rules that govern entities and relationships. For example, they might tell us the credit rating for an ACTIVE CUSTOMER must be PREFERRED, NORMAL, or PROBATIONARY, and the default for a new customer is NORMAL. They might also specify that only an ACTIVE CUSTOMER can place an ORDER, but an ACTIVE CUSTOMER might not necessarily have any current ORDERS at any given time. All the examples in this paragraph represent fundamental data requirements that must be implemented regardless of our choice of technology!

Figure 2.5 illustrates the data requirements cell using the format of a graphical picture called a *data model.* Systems analysts use such a tool to document and confirm the system users' view of data requirements (meaning entities, relationships, attributes, and rules). You'll learn how to draw data models in this book.

System Designers' View of DATA System users define the data requirements for an information system. System designers translate those requirements into computer files and databases that will be made available via the information system. Obviously, the system designers' view of data is constrained by the limitations of whatever data technology is chosen. Often, the choice has already been made and the developers must use that technology. For example, many businesses have standardized on an enterprise database engine (such as *Oracle* or *DB2*) and a personal computer database tool (such as *Access* or *dBASE*).

In any case, the system designer's view of data consists of data structures, database schemas, file organizations, fields, indexes, and other technology-dependent components. Some of these, if presented properly, can be interpreted by system users; most cannot!

The systems analyst and/or database specialists design and document these technical views of the data. As shown in Figure 2.5, the system designers' view of data is a **database schema.** A database schema is the transformation of the data model (system users' view) into a set of data structures that can be implemented using the chosen database technology. The designers' intent is to represent the design such that: (1) it fulfills the data requirements of the users; and (2) it provides sufficient detail and consistency for communicating the database design to the system builders. Once again, this book will teach tools and techniques for transforming user data requirements into a technical database schema.

System Builders' View of DATA The final view of data is relevant to the system builders. In the DATA column of Figure 2.5, system builders are closest to the database technology foundation. Not surprisingly, they are forced to represent data in very precise and unforgiving languages. The most commonly encountered database construction language is *SQL (Structured Query Language)*. Alternatively, many database management systems, such as *Access* and *Paradox,* include proprietary facilities for constructing a new database.

Not all information systems use database technology to store their business data. Older legacy systems were built with *flat-file* technologies such as ISAM and VSAM. These flat-file data structures were constructed directly within the programming language used to write the programs that use those files. For example, in a COBOL program the flat-file data structures are expressed as PICTURE clauses in a DATA DIVISION. It is not the intent of this book to teach either database or flat-file construction languages, only to place them into the context of the DATA building block of information systems.

Building Blocks of PROCESSES

Let's start the same way we did with the DATA building block. When engineers design a new product, that product should provide some level of functionality or service. It must do something useful. Prospective customers define the desired functionality of the product and the engineer creates a design to provide that functionality. PROCESSES deliver the functionality of an information system. Processes perform the *work* in a system. Some processes are performed by people. Others are performed by machines, including computers. Some processes are performed repetitively. Others occur less frequently or rarely.

The PROCESS building blocks of information systems are added to our matrix in Figure 2.6. Notice at the bottom of the PROCESS column that our goal is to automate appropriate processes with *software technology*. Also, as you look down the PROCESS column, each of our different stakeholders has different views of the

FIGURE 2.6 *The Process Focus for Information Systems*

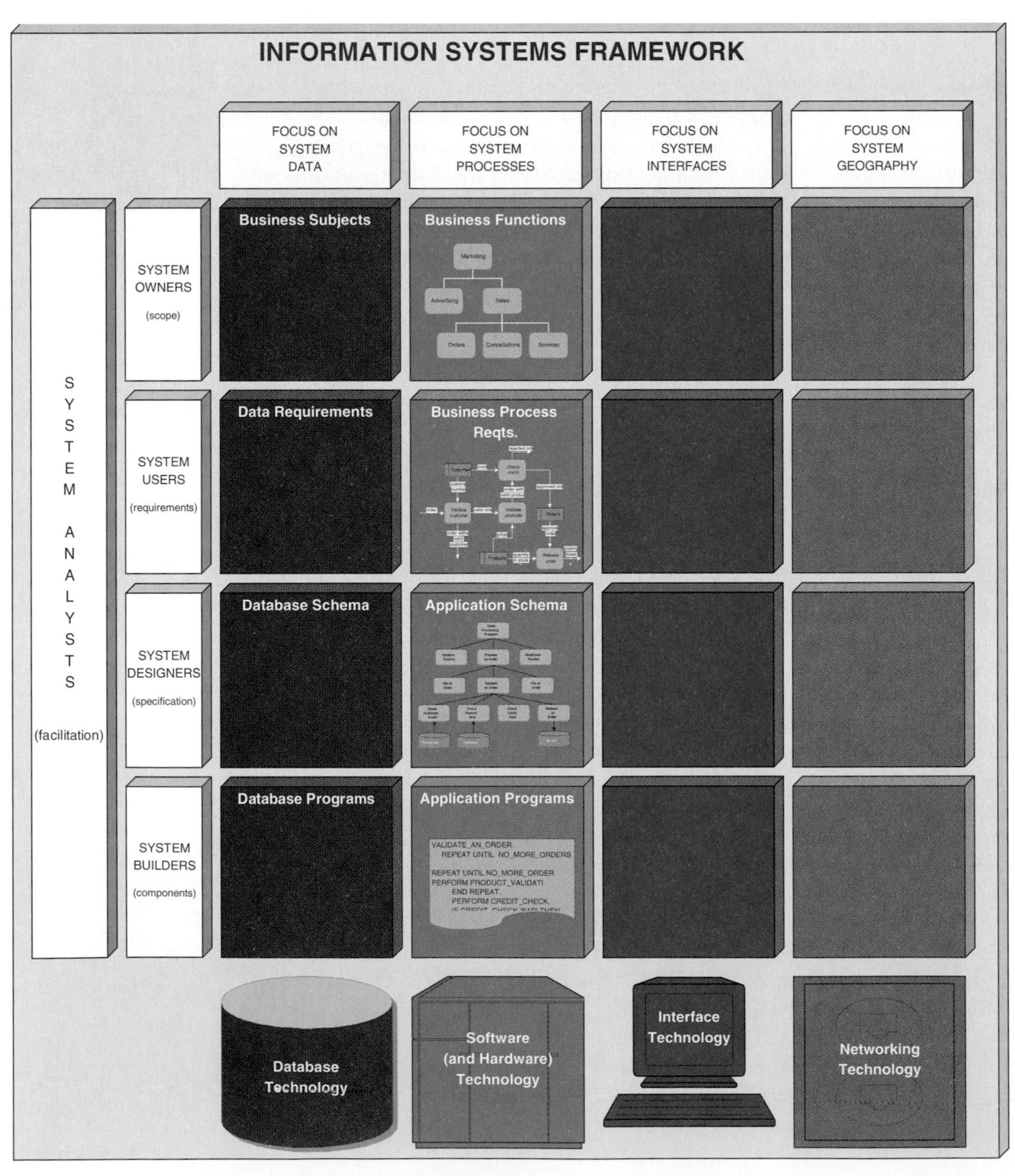

system's processes. Let's examine those views and discuss their relevance to systems analysis and design.

System Owners' View of PROCESSES As usual, system owners are usually interested in the big picture—in this case, groups of high-level processes called *business functions*.

> **Business functions** are ongoing activities that support the business. Functions can be decomposed into other functions and eventually into discrete processes that do specific tasks.

Think of functions as groups of processes that have something in common. Typical business functions include sales, service, manufacturing, shipping, receiving, accounting, and so forth. Each function is ongoing; that is; it has no starting time or stopping time.

Historically, most information systems were (or are) *function-centered*. That means the system supports one business function or functional area. An example would be a SALES INFORMATION SYSTEM. Today, many of these single-function information systems are being redesigned as *cross-functional* systems.

> A **cross-functional information system** supports relevant business processes from several business functions without regard to traditional organizational boundaries such as divisions, departments, centers, and offices.

For example, reconsider the traditional SALES INFORMATION SYSTEM that processes only customer orders. Alternatively, a cross-functional ORDER FULFILLMENT INFORMATION SYSTEM would support all relevant business processes from sales, warehousing, shipping, billing, and service—in other words, all processes required to ensure a complete response to a customer order, regardless of which departments are involved. This trend is being driven by **total quality management** and **business process redesign** initiatives that are intended to reinvent and streamline the way organizations do business.[4]

Development teams frequently express these functions or cross-functions in terms of some sort of simple, hierarchical decomposition diagram as shown in the cell (Figure 2.6). With respect to each subfunction identified, system owners would be queried about perceived problems, opportunities, goals, objectives, and constraints. Obviously, the costs and benefits of developing information systems to support functions would also be discussed. You'll learn how to draw decomposition models (and their equivalent) in this book.

As was the case with DATA, system owners are not concerned with detailed processes, tasks, or procedures (unless those owners also happen to be users). That level of detail is identified and documented as part of the system users' view of PROCESSES.

System Users' View of PROCESSES Returning to Figure 2.6, we are ready to examine the system users' view of PROCESSES. Users see processes in terms of discrete business processes.

> **Business processes** are discrete activities that have inputs and outputs, as well as starting times and stopping times. Some business processes happen repetitively, while others happen occasionally or even rarely. Business processes may be implemented by people, machines, computers, or a combination of all three.

Specific policies and procedures underlie these business processes.

---

[4]See Chapter 1 for more about total quality management and business process redesign.

Policies are rules that apply to a business process.

Procedures are step-by-step instructions and logic for accomplishing a business process.

An example of a policy is CREDIT APPROVAL, a set of rules for determining whether to extend credit to a customer. That credit approval policy is usually applied within the context of a specific CREDIT CHECK procedure that established the correct steps for checking credit against the credit policy. Many of you have implemented policies and procedures in computer programming assignments.

Unfortunately, users tend to see their processes only in terms of how they are currently implemented or how they think they should be implemented. Most businesses need to rethink and redesign business processes to eliminate redundancy, increase efficiency, and streamline the entire business—all independently of any information technology to be used! The fashionable term, introduced earlier, is *business process redesign*.

Once again, the challenge in systems analysis is to identify, express, and analyze business process requirements exclusively in business terms (such that alternative approaches and technologies might then be considered). To this end, most systems analysis methods include process models and work flow models (illustrated in Figure 2.6) that express business process requirements in nontechnical ways. Supplemental documentation is needed to document the underlying policies and procedures. Tools and techniques for process modeling and documentation are taught extensively in this book.

System Designers' View of PROCESSES As was the case with the DATA building block, the system designer's view of activities is constrained by the limitations of specific technology. Sometimes the analyst is able to choose that technology. But often the choices are limited by a standardized **application architecture** that specifies which software (and hardware) technologies must be used. In either case, the designers' view of processes is technical.

Given the business processes from the system users' view, the designer must first determine which processes to automate and how to best automate those processes. In other words, the designer tends to focus on an application schema.

An **application schema** is a model that communicates how selected business processes are, or will be, implemented using the computer and programs.

These schemas usually take the form of computer program design specifications such as structure charts, logic flowcharts (as shown in Figure 2.6), and/or state models. You may have encountered some of these program design tools in a programming course. As was the case with DATA, some of these technical views of PROCESSES can be understood by users, but most cannot. The designers' intent is to prepare specifications that (1) fulfill the business process requirements of system users; and (2) provide sufficient detail and consistency for communicating the computer process design to system builders. The systems design chapters in this book teach tools and techniques for transforming business process requirements into a computer process design.

System Builders' View of PROCESSES System builders represent PROCESSES using precise computer programming languages that describe inputs, outputs, logic, and control. Examples include *COBOL, C++, Visual BASIC, Powerbuilder, Smalltalk,* and *Object Pascal.* Additionally, some database management systems provide their own embedded languages for programming. Examples include *Visual BASIC for Applications* (included in *Access*) and *PL-SQL* (included in *Oracle*). All these languages are used to write applications programs.

Applications programs are language-based, machine-readable representations of what a computer process is supposed to do, or how a computer process is supposed to accomplish its task.

It is not the intent of this book to teach application programming. We will, however, demonstrate how some of these languages provide an excellent environment for prototyping computer processes.

Prototyping is a technique for quickly building a functioning model of the information system using rapid application development tools (provided with most popular programming languages).

Prototyping has become the design technique of choice for many system designers. Prototypes frequently evolve into the final version of the system or application.

Building Blocks of INTERFACES

Let's begin the same way we did with the DATA and PROCESS building blocks. When engineers design a new product, that product should be easy to learn and use. Today's best product engineers are obsessed with *human engineering* and *ergonomics*. Consider, for example, the evolution of VCR programming. Engineers have greatly simplified VCR programming through innovations such as VCR+, a system whereby you merely punch in a number from your TV guide to automatically record a program—no need to select date, start time, stop time, channel, etc. Similarly, customers also expect their products to work with their other products. For example, today's VCRs must fully integrate into your stereo system to provide a home theater experience.

Information systems have similar integration and interoperability expectations. Our INTERFACES building blocks provide that dimension of the system. There are two critical components to information system INTERFACES.

- Information systems must provide effective and efficient interfaces to the system's users.
- Information systems must interface effectively and efficiently to other information systems, both within the business and increasingly with other businesses' information systems.

The INTERFACE building blocks of information systems are added to our matrix in Figure 2.7. Notice at the bottom of the INTERFACE column that technologies exist to implement interfaces. The classic example is graphical user interface technology used to implement *Windows 95* compliant user interfaces. Other technologies exist that can almost eliminate human error or intervention. And still other technologies exist for system integration.

And once again, as you look down the INTERFACE column, each of our different stakeholders has different views of the system's INTERFACES. Let's examine those views and discuss their relevance to systems analysis and design.

System Owners' View of INTERFACES As always, system owners are interested in the big picture—in this case, the overall context of the system as it relates to the business as a whole and other systems. They are virtually never concerned with details. And as always, system owners are concerned with costs and benefits of any interfacing solutions that will be developed.

System owners' view of INTERFACES is greatly simplified. When considering whether to sponsor a new information system, the system owners only want to know:

- With which business units, customers, and external businesses will the new system interface?

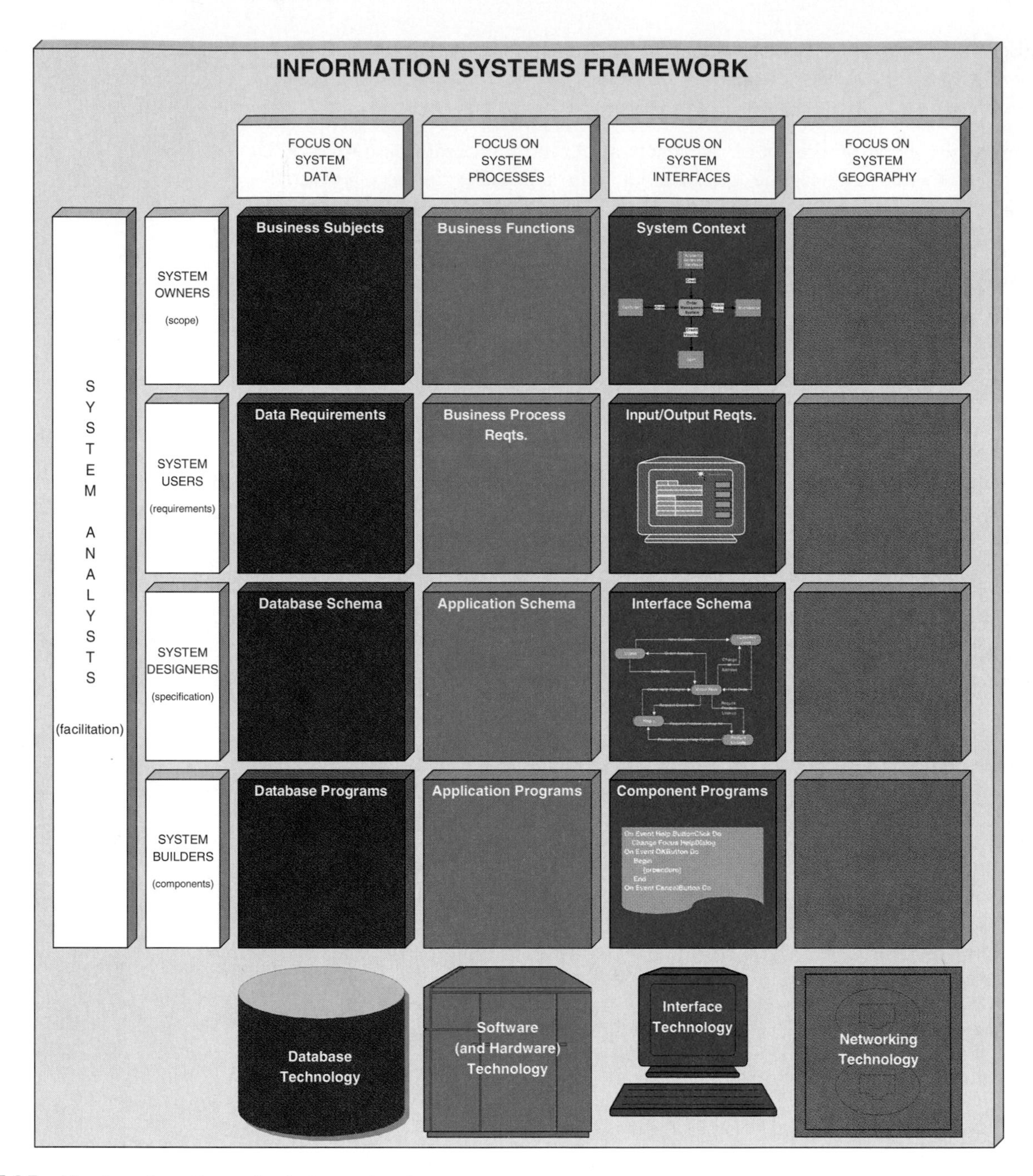

FIGURE 2.7 *The Interface Focus for Information Systems*

- What are the key inputs and outputs with respect to those business units, customers, and external businesses?
- Will the system have to interface with any other information systems or services?
- Are there any corporate or governmental regulations or policies that may constrain the system interfaces?

A suitable system owners' view of information system interfaces might be formatted as a simple **context model** as shown in Figure 2.7. Such a model represents the proposed system as a single process in the middle of the page (effectively hiding all the gory details). Inputs and outputs to that process provide a conceptual view of how the proposed system would interact with users, business units, customers, and other businesses. You'll learn to draw this relatively simple process model when you read the systems analysis chapters.

System Users' View of INTERFACES System users take a more active interest in interfaces, so much so that they often get closer to the technology very early in a modern information system development project. System users are most interested in what has come be called the user interface to the system.

> The **user interface** defines how the system users directly interact with the information system to provide inputs and queries and receive outputs and help.

The explosive growth of personal computers, combined with the popularity of graphical user environments such as Microsoft *Windows* (for Intel-based PCs) and Apple *Macintosh* (for Motorola-based PCs) has created a de facto standard—the **graphical user interface.** System users are increasingly demanding that their custom-built information system applications have the same "look and feel" as their favorite PC tools such as word processors and spreadsheets. This common graphical user interface makes each new application easier to learn and use.

Accordingly, system users' view of interfaces are increasingly presented as working prototypes of the user interface (Figure 2.7). The figure also notes that users are interested in printed reports. Again, mockups or prototypes can be used to represent this view. Although not depicted in this cell of the framework, system users also need to specify the business data requirements for interfaces to other systems and applications. For example, if a new payroll system must interface with an existing personnel information system, the inputs and outputs between those systems should be specified in terms of data content and flexibility (or inflexibility). Chances are that the existing system will impose certain constraints on the proposed system (to minimize the need to modify the existing system).

Systems analysis and design methods for user interface design and system interfaces will be addressed in various chapters of this book.

System Designers' View of INTERFACES System designers must be concerned with the details of both user and system interfaces. Let's begin with user interfaces.

In some ways, it is difficult to distinguish between the system users' and system designers' views of user interfaces. Both are involved in input, output, and screen design and construction. But whereas system users are interested in form and content, system designers have other interests such as consistency, completeness, and user dialogues.

> **User dialogues** describe how the user moves from screen to screen, interacting with the application programs to perform useful work.

The trend toward graphical user interfaces has simplified life for system users, but complicated the design process for system designers. Think about it! In a typical *Windows* application, there are many different things you can do at any given time—type something, click the left mouse button on a menu item or tool bar icon, press the F1 key for help, maximize the current window, minimize the current window, switch to a different program, and many others.

Accordingly, the system designer views the interface in terms of interface properties, system states, events that change the system states, and responses to events. Collectively, we'll call this the **interface schema** (Figure 2.7). The cell depicts an interface schema tool called a *state transition diagram*. You'll learn how to use this tool in the design unit of the book.

Although not depicted in Figure 2.7, modern system designers may also design *keyless interfaces* such as bar coding, optical character recognition, and pen-based input. These alternatives reduce errors by eliminating the keyboard as a source of human error. But these interfaces, like graphical user interfaces, must be carefully designed to both exploit the underlying technology and maximize the return on what can be a sizable investment.

Finally, and as suggested earlier, system designers are concerned with system-to-system interfaces. The existing information systems in most businesses were each built with the technologies and techniques that represented best practices at the time when they were developed. Some systems were built in-house. Others were purchased from software vendors or consultants. As a result, the integration of these heterogeneous systems can be difficult. But the need for different systems to interoperate is pervasive. Accordingly, system designers frequently spend as much or more time on system integration as they do on system development.

Finally, system designers have to design the system-to-system interfaces that allow a new information system to transparently interoperate with previously designed systems. This is not as easy as it sounds because systems developed at different times may use different technologies and different design philosophies (and quality). Some of the legacy systems may be very old and poorly maintained. The designer's mission is to find or build technical interfaces between these systems that (1) do not create maintenance projects for the legacy systems, (2) do not compromise the superior technologies and design of the new system, and (3) are ideally invisible to the end-users. The best system integrators are often freelance consultants or they work for consulting firms that "rent" their talents to companies that have complex integration problems.

System Builders' View of INTERFACES System builders construct, install, test, and implement both user and system interfaces. For user interfaces, the technology is usually embedded into the programming language environments used to construct the computer processes (described earlier in the chapter). For example, languages such as *Visual BASIC, Delphi,* and *Powerbuilder* include all the technology required to construct a *Windows*-compliant graphical user interface without the need for any actual programming. At the risk of oversimplifying the process, the graphical user interface is constructed by placing interface components on the screen, and those objects automatically generate a template computer program to be completed by the application programmer (described earlier under the heading of "System Builders' View of PROCESSES").

System interfaces are considerably more complex to construct. One system interfacing technology that is currently popular is middleware.

> **Middleware** is a layer of utility software that sits in between applications software and systems software to transparently integrate differing technologies so that they can operate.

One example of middleware is the *Open Database Connectivity* (*ODBC*) tools that allow application programs to work with different database management systems without having to be rewritten to take into consideration the nuances and differences of those database management systems. Programs written with ODBC commands can, for the most part, work with any ODBC-compliant database (which includes hundreds of different database management systems). Similar middleware products exist for each of the columns in our information system framework. System designers help to select and apply these products to integrate systems.

Once again, this book is not about system construction. But we present the builder's view because all the other INTERFACE views lead up to construction.

Building Blocks of GEOGRAPHY

For one last time, we'll begin with product engineering. When engineers design a new product (like an automobile engine), that product has spatial properties (or geometry) that describe the product in three dimensions. That's why engineering graphics or technical graphics courses are universal in engineering curricula. In a similar vein, information systems have spatial properties. We'll use the term GEOGRAPHY to describe those properties.

Information systems **geography** describes (1) the distribution of DATA, PROCESSES, and INTERFACES (the other building blocks) to appropriate business locations, and (2) the movement of data and information between those locations.

The inclusion of GEOGRAPHY in our framework is driven by the trend toward distributed computing.

Distributed computing is the decentralization of applications and databases to multiple computers across a computer network.

Currently, the most popular applications of distributed computing are called client/server computing.

In a **client/server computing** application, information system building blocks are distributed between client personal computers and server shared computers. The clients and servers effectively interoperate to share the overall workload.

There are always several or many client computers. There are fewer (but frequently more than one) server computers. Integrated across the computer network, a client/server information system effectively distributes the total workload across many computers. The evidence suggests that distributed, client/server computing is slowly replacing traditional centralized, mainframe computing. The mainframe will probably continue to exist, but its role may change to that of a superserver.

Distributed computing is being driven by several trends. First, the pace of business is increasing. Organizations that can gain faster access to critical information have a competitive advantage. This is sometimes accomplished by duplicating systems in multiple locations or distributing portions of systems to the locations that have the greatest need for those portions. Second, the reach of business is expanding. Organizations that can extend their information systems to include their customers and suppliers have a competitive advantage. For example, if a retailer can directly tie its purchasing information system into its supplier's order information system, it can bypass the delays caused by mailing, phoning, and keying orders. Finally, the complexity of business is increasing. Organizations operate in more locations, national and international, than ever before. Different laws, different markets, and different spoken languages make it difficult for organizations to coordinate their efforts and consolidate important corporate data.

Figure 2.8 summarizes different people's views of information system geography. Once again, the systems analyst who understands these views will be better equipped to deal with the tools and techniques taught in this book.

System Owners' View of GEOGRAPHY To the system owner, networking is not a technical issue. It is merely a business reality. In its simplest form, the system owner views the geography in terms of **operating locations.** This is illustrated in Figure 2.8, but try not to restrict your thinking to geography in the classic sense of the icon we depicted. True, geography includes such locations as cities, states, and countries. But it also includes locations such as campuses, sites, buildings, floors, and rooms. Also note, operating locations are not synonymous with computer centers because computer centers can network to many operating locations.

With respect to locations, systems analysts must learn which business functions are performed at which locations. Are some functions duplicated at some locations? Are some functions provided from a single (or few) location(s)? What locations are to be served or affected by the information system? These are concerns that systems analysts must negotiate with the owners. The system owners will ultimately decide the degree to which the system will be centralized, distributed, or duplicated. There are important cost and political considerations in these decisions.

Geography has taken on a new dimension in many businesses that strategically seek to integrate their information system into the businesses of their suppliers

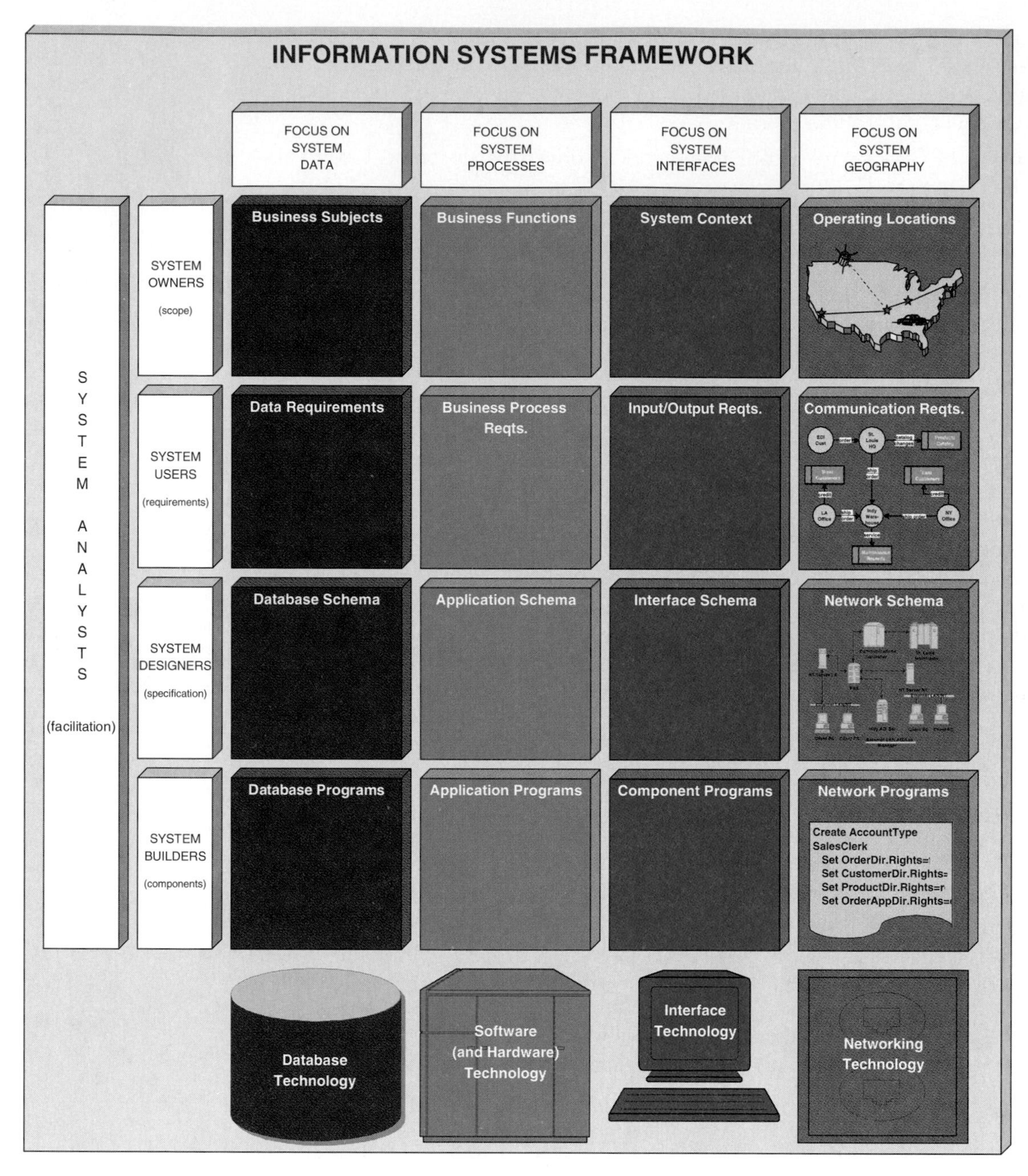

FIGURE 2.8 *The Geography Focus for Information Systems*

and customers. For example, Sears, the retailing giant, has directly linked its purchase order information system with the order processing information systems of many of its primary suppliers. Thus, Sears can avoid the delays associated with mailing or phoning orders to replenish stock. This system also gives the retailer a competitive advantage—keeping products on the shelves with minimum lead time—which, in turn, reduces inventory carrying costs and increases profits.

Another geographic location being tested by some executive managers is the cottage office. In this method, employees (permanent and temporary) are allowed to work out of a home office with complete access to company databases and communications tools. Some experts predict that cottage offices will become a permanent and important component of the workforce by the end of this decade. After all, many individuals (both students and workers) believe they are more productive in the quiet of their own home. So long as employees remain accountable for their productivity and quality, cottage offices may yield benefits over traditional business office architecture.

When dealing with system owners, an analyst might communicate an information system's geography with simple maps or floor plans. They might also use matrices to determine what business functions are performed at each location. As with the other building blocks, this book will provide you with effective techniques for defining and documenting system geography for system owners.

System Users' View of GEOGRAPHY Consistent with the user views of data and processing, system users are the experts about the requirements for any given location. Like system owners, system users are interested in operating locations. But because they are closer to the day-to-day locations, they might identify locations that are unknown or forgotten by system owners. Users also tend to have a more microscopic view of locations. For instance, owners might think of buildings, whereas users become concerned with individual offices or areas within buildings.

Users might also identify some truly unique definitions of location. For example, consider a traveling sales representative as part of an order entry information system. That sales representative's location varies. He or she might be in a car, an office, the office of a customer, or working from a home office. How do you define such a location? One possibility would be to define a single, generic location that represents "wherever the salesperson is at any given time." This forces the system users, designers, and builders to consider creative possibilities such as the use of portable computers and cellular modems.

More importantly, system users think in terms of communications requirements.

> **Communications requirements** define the information resource requirements for operating locations and how different operating locations need to communicate with one another. These communications requirements are expressed independently of any specific technology.

Each operating location's information resources can be expressed in terms of the building blocks you've already learned—PEOPLE, DATA, PROCESSES, and INTERFACES that are needed at the location. The communications requirements are further expressed in terms of location-to-location data flows. For example, the flow of a business transaction between different offices might be important.

How does this affect the modern systems analyst? Most analysts are at least somewhat familiar with computer networks. They must, however, initially divorce themselves from that technology to consider the business network. First, the analyst must identify all locations pertinent to the system. Whereas system owners tend to have a macro view of the system's locations (for instance, Building B, 3rd floor), system users frequently have a more detailed, micro view of those same locations. For instance:

> *Most purchasing agents are in Building B, 3rd floor, Offices 302 through 315. The purchasing managers are in Offices 320 and 325. Don't forget to include our special agents who share space with Engineering in Building A, 1st floor, Room 101. Oh yes, and what about the 10 procurement agents who are usually on the road—working with our many suppliers? When they are not on the road, they share space in Room 201 of Building B.*

It can get far more complicated for systems that must support different cities, states, or countries. It can also be complicated by systems that directly interface to other businesses (for example, suppliers and customers).

Even as systems analysts define system users' DATA and PROCESS requirements (or building blocks), they should also begin defining how those requirements should be distributed, duplicated, or shared with the different locations. In other words, the users' view of GEOGRAPHY depends on the other building blocks. A systems analyst might represent the business network for an information system using a flow-like diagram as shown in Figure 2.8. Although more detailed than a map, this picture is still not technical. You'll learn to use similar tools and techniques for user-oriented network representation in this book.

System Designers' View of GEOGRAPHY As was the case with the previous building blocks, system designers' view of GEOGRAPHY is influenced and/or constrained by the limitations of specific technology. The emphasis shifts to a network schema that can support the business network.

> A **network schema** (also called a **network configuration** or **topology**) is a technical model that identifies all the computing centers, computers, and networking hardware that will be involved in a computer application.

Sometimes the analyst may able to influence or choose this technology, but more often than not the choices have already been made and the analyst must use the standard networking technology. For example, an organization might standardize on Ethernet and Novell to implement all local area networks. In any case, the designer's view of networks is technical.

As with the system user's view of GEOGRAPHY, the system designer's view is also expressed in terms of locations, but in this case the locations are for information technology and computing centers. Given the network schematic, the designer's job is to determine the optimal distribution of DATA, PROCESSES, and INTERFACES across the network. This is sometimes called **application partitioning.** Working with telecommunications specialists, the systems analyst must also perform impact analysis to determine the volume of traffic that may be transported across the network.

The systems analyst and telecommunications specialists address and document these technical views of networks. Figure 2.8 illustrates a system designer's view of a network. As always, the designer's intent is to prepare specifications that (1) fulfill the business network requirements of the users; and (2) provide sufficient detail and consistency for communicating the network design to the system builders.

Much of the technical dimension of network design and performance is beyond the scope of this book. Because of the trend toward distributed computing, aspiring systems analysts should actively pursue a conversational level of understanding in data communications. If your school does not offer such a course, we have included an excellent introductory textbook in the recommended readings for this chapter.

System Builders' View of GEOGRAPHY The final view of GEOGRAPHY is relevant only to the system builders. They use telecommunications languages and standards to write network programs.

> **Network programs** are machine-readable specifications of computer communications parameters such as node addresses, protocols, line speeds, flow controls, security, privileges, and other complex, networking parameters.

The basic software technology for networks is purchased and installed. But it must be installed, configured, and tuned for performance. Examples of communications software include network operating systems such as *Netware, OS/2 LAN Manager, Windows/NT Server,* and *System Network Architecture* (SNA). Additionally, special-purpose communications software may need to be installed and configured. Examples include transaction monitors (e.g., *CICS*), terminal emulators (e.g., *ProComm* and *Kermit*), electronic mail post offices (e.g., *Exchange Server* and *cc:Mail Server*), and work group enablers (e.g., Lotus *Notes*).

It is not the intent of this book to teach these languages, only to place them into the context of the information system building block called GEOGRAPHY. Again, most information system programs offer courses that can expand your technical understanding of networking.

WHERE DO YOU GO FROM HERE?

That completes our framework for information system architecture. The completed framework, shown below, will be used throughout the book as a home base to organize the concepts and techniques that you will learn and apply as you study systems analysis and design methods. You will notice that we modified this framework to provide a graphical outline into the remaining chapters of the book. The framework clearly indicates which chapters focus on which building blocks. (A few chapters do not correspond to any of these building blocks, and a couple of chapters will address how the framework is evolving to accommodate something called *object technology.*)

So where are we now? If you have already read Chapter 1, you learned about systems analysts as the facilitators of information system development. In Chapter 2, you learned about the product itself—information systems—in terms of basic building blocks.

If you haven't already done so, you should at least skim Chapter 1 to learn about the *players in the information system development game.* But most readers should proceed directly to Chapter 3, which will introduce you to the *process* of information system development. You'll learn about information system problem solving, methodologies, and system development technology as you complete your education in the fundamentals for systems analysis and design. At that point, you will have the fundamental understanding required to truly begin to study and learn systems analysis and design tools, techniques, and their proper application to information system development.

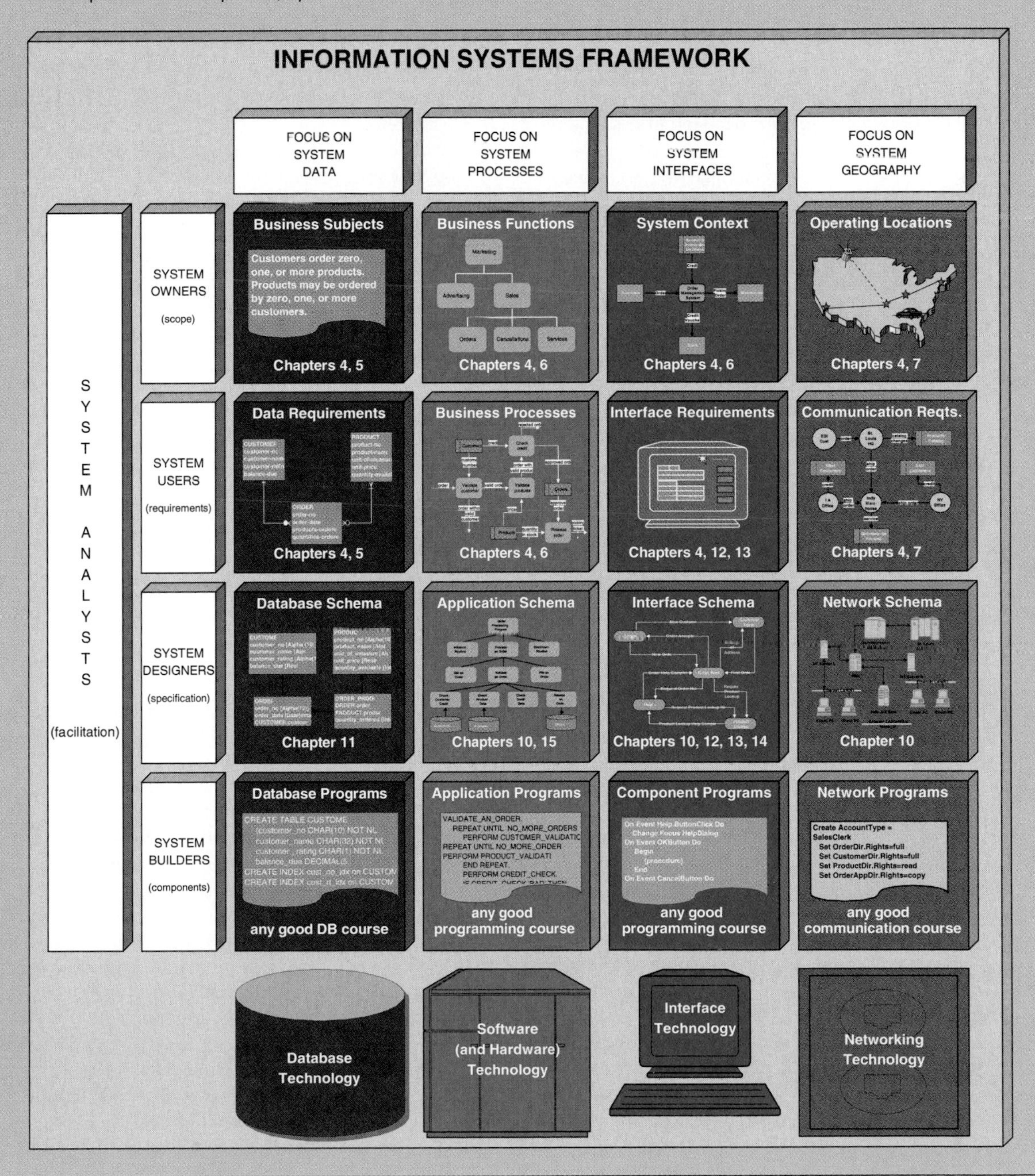

SUMMARY

1. Data are raw facts about a business and its business transactions. Information is data that has been refined and organized by processing and human intelligence.
2. An information system is an arrangement of people, data, processes, interfaces, and geography that are integrated to support and improve the day-to-day operations in a business, as well as fulfill the problem-solving and decision-making information needs of managers.
3. Information systems fulfill one or more of the following basic purposes:
   a. Transaction processing systems capture and process data about (or for) business transactions.
   b. Management information systems provide essential management reports required to plan, monitor, and control business operations.
   c. Decision support systems provide users, especially managers, with decision-oriented information in response to various unstructured decision and problem-solving opportunities.
   d. Expert systems capture and simulate the knowledge and expertise of subject matter experts so it may be applied by nonexperts.
   e. Office information systems automate a wide variety of routine and complex office activities through improved work flow and communications between workers.
   f. Personal information systems improve an individual worker's productivity through application of personal computing technology.
   g. Work group information systems enhance the ability of groups of workers to collaborate and communicate.
4. Information systems architecture provides a unifying framework for understanding the basic building blocks of information systems. The framework used throughout this book is based on, and adapted from, the Zachman *Framework for Information Systems Architecture*.
5. The information system framework is visually presented as a matrix. The rows correspond to the perspectives of different stakeholders in the information system development process. The columns correspond to a specific focus or dimension of the information system that must be analyzed, designed, and implemented. Each cell represents one perspective and one focus.
6. Although the cells can be studied and developed in isolation, they must be synchronized with the other cells (both across rows and down columns) to develop a successful information system.
7. The four perspectives are provided by:
   a. System owners who fund and sponsor the system's development.
   b. System users who define the business requirements for the information system.
   c. System designers who define the technical requirements for the information system.
   d. System builders who construct and test the information system.
8. Systems analysts may play some of the designer and builder roles, but their principal contribution is to facilitate the process of communicating between the various other stakeholders.
9. The four focuses represented in the model are:
   a. DATA—representations of business data that must be captured and stored in the information system's database.
   b. PROCESSES—representations of the business functions and processes that must be optimized and supported by the information system's computer programs.
   c. INTERFACES—representations of how the information system will interact with end-users and other information systems.
   d. GEOGRAPHY—representations that lead to the distribution of data, processes, and interfaces to multiple business operating locations.

KEY TERMS

application architecture, p. 56
application partitioning, p. 64
application program, p. 57
application schema, p. 56
artificial intelligence, p. 40
business function, p. 55
business process, p. 55
business process redesign, p. 55
business resource, p. 52
clerical and service worker, p. 45
client/server computing, p. 61
communications requirement, p. 63
context model, p. 58
cross-functional information system, p. 55
data, p. 37
data maintenance, p. 38
data processing system, p. 38
data requirement, p. 52
data warehouse, p. 40
database schema, p. 53
decision support system, p. 39
distributed computing, p. 61
electronic form, p. 41
electronic messaging, p. 41
executive information system, p. 39
expert system, p. 40
expert system shell, p. 40
focus (in the framework), p. 48
geography, p. 61
graphical user interface, p. 59
imaging, p. 51
information, p. 38
information system, p. 38
information systems architecture, p. 42
information technology, p. 38
information worker, p. 42
interface schema, p. 59
knowledge worker, p. 46
management information system, p. 39
middle manager, p. 46
middleware, p. 60
mobile user, p. 46
network configuration, p. 64
network program, p. 64
network schema, p. 64
network topology, p. 64
office automation suite, p. 41

office information system, p. 40
operating location, p. 61
personal information system, p. 41
perspective (in the framework), p. 42
policy, p. 56
procedure, p. 57
prototype, p. 46
remote user, p. 46
stakeholder, p. 42
system builder, p. 47
system designer, p. 47
system owner, p. 45
system user, p. 45
technical and professional staff, p. 45
telecommuting, p. 46
total quality management (TQM), p. 55
transaction processing system, p. 38
unstructured decision, p. 39
user dialogue, p. 59
user interface, p. 59
work group, p. 41
work group information system, p. 41

REVIEW QUESTIONS

1. Define data, information, information system, and information technology.
2. What is transaction processing? How is transaction processing being affected by business process redesign?
3. Explain the role of data maintenance in a transaction processing system.
4. Differentiate between management information systems and decision support systems.
5. What role does a data warehouse play in a decision support system?
6. Differentiate between a personal and work group information system.
7. What is an information systems architecture?
8. What is the difference between a perspective and focus in this chapter's information system framework?
9. Define information worker. Name three classes of users.
10. Briefly describe the four stakeholders in information system development. How does the systems analyst fit in?
11. Briefly describe the four focuses for an information system.
12. Briefly describe the four building blocks of data for an information system.
13. Explain how the system designer's and system builder's views of data differ.
14. Briefly describe the four building blocks of processes for an information system.
15. Differentiate between business functions and business processes.
16. Why are businesses interested in cross-functional systems?
17. Differentiate between business policies and business procedures.
18. Differentiate between user and system interfaces. With respect to user interfaces, what phenomenon is driving the trend toward graphical user interfaces?
19. Briefly define geography. Give several reasons why geography is becoming increasingly important in information system design.
20. Is distributed computing different from client/server computing? Why or why not?
21. Differentiate between the business network and a computer network.

PROBLEMS AND EXERCISES

1. Identify each of the following as data or information. Explain why you made the classification you did.
   a. A report that identifies, for the purchasing manager, parts that are low in stock.
   b. A customer's record in the customer master file.
   c. A report your boss must modify to be able to present statistics to her boss.
   d. Your monthly credit card invoice.
   e. A report that identifies, for the inventory manager, parts low in stock.
2. Information systems are all around you. From your last job, data processing or otherwise, give an example of a completely manual information system. Describe how a computer might improve that information system.
3. Explain three examples of transaction processing at your college or university.
4. Explain three examples of transaction processing that might occur for a retail store.
5. An office manager has described his company's office information system in terms of word processing and spreadsheets for his staff. Explain to him why his system is actually a personal information system. Explain to him how it might be extended to become a true office information system.
6. For each of the following classifications of system users, identify the individual as clerical staff, technical and professional staff, supervisory staff, middle management, or executive management. Defend your answer.
   a. Receptionist.
   b. Shop floor foreman.
   c. Financial manager.
   d. Assistant store manager.
   e. Chief operating officer.
   f. Manufacturing control manager.
   g. Terminal operator/data entry.
   h. Applications programmer.
   i. Programmer/analyst.
   j. Warehouse clerk who fills orders.
   k. Stockholder.
   l. Product engineer.
   m. Consultant.
   n. Broker.

7. Give three examples of system users from each of the following classifications: clerical workers, technical and professional staff, supervisory staff, middle management, and executive management. Explain the job responsibilities of each example you provided and state why your example represents that particular classification of system user. For each system user, describe a situation that would make that worker a client of an analyst.
8. Explain how the existence of remote users might complicate information system design. Describe a system in which *you* might be considered a remote user.
9. Consider an organization by which you were, or are, employed in any capacity. Identify the information workers at each level in the organization according to their classification of system user. (Alternative: Substitute your school for the organization. Students are part of the clerical staff classification. Do you see why?)
10. What is the relationship between data, information, input, processing, and output? Give an example of this relationship.
11. Systems analysts must identify system users for an information system development project. Who are system users? What is their role in system development projects? How do system users differ from system owners?
12. The system owner's view of data is primarily concerned with entities. If the owner of a course scheduling information system is the registrar, brainstorm the entities this registrar might identify.
13. Differentiate between entities and relationships. Using the course registration example in the previous exercise, give examples of each.
14. For a programming assignment that you wrote for a programming course, describe how each of the four process building blocks was either presented to you or developed by you.
15. Explain the difference between user interfaces and user dialogues. Why are user dialogues more difficult to design than user interfaces? How does each impact the end-user?
16. Identify the type of information system application described for each of the following:
    a. A customer presents a deposit slip and cash to a bank teller.
    b. A teller gets a report from the cash register that summarizes the total cash and checks that should be in the drawer.
    c. Before cashing a customer's check, the teller checks the customer's account balance.
    d. The bank manager gets an end-of-day report that shows all tellers whose cash drawers don't balance with the cash register summary report.
    e. The system prints a report of all deposits and withdrawals for a given day.
    f. A chief executive officer is using his computer to obtain access to the latest stock market trends.
    g. A doctor keys in data describing symptoms of a patient; she receives a report that suggests what illness the patient is likely suffering from and a detailed explanation concerning the rationale as to why the symptoms suggest that particular illness.
    h. An employee is electronically sending a memorandum concerning a scheduled committee meeting to all committee members who are to attend.
17. Describe the business and computing geography of a progressive college admissions information system.
18. Food for thought! Chapter 1 presented a simple information system development process used by systems analysts. Chapter 2 presented a system development framework. Extend the framework to show how the development phases in Chapter 1 might apply to the framework perspectives in Chapter 2.

PROJECTS AND RESEARCH

1. Obtain an organization chart from a local company or at your library. Classify each person and/or job position appearing on the chart according to the type of system user (such as clerical staff). (Alternative: Substitute the organization charts appearing in the SoundStage episode from Chapter 1.)
2. Make an appointment to visit a systems analyst at a local data processing installation. Discuss information system projects the analyst has worked on. What were some of the functions (e.g., transaction processing, decision support, office automation) being supported?
3. Do background research into two local data processing shops and some of their implemented computer-based systems. Now read Minicase 2. Were their clients' expectations similar to Mr. Oliver's before completion of the system? How did the analysts deal with the expectations, if they did so at all? How would you have dealt with Mr. Oliver's expectations before beginning the systems project? Were the outcomes of the implemented system similar? What factors might explain these outcomes?
4. Information systems are all around you. Consider one of your previous employers. Describe an information system that you used in terms of the four focuses, and from the perspective of a system owner, a system user, and a system designer.

MINICASES

1. Knowledgeable University plans to support course registration and scheduling on a computer. The following client community has been designated:
    a. Curriculum deputy—one per department, responsible for estimating demand by that department's own students for each course offered by the university. This

person may revise demand estimates from time to time.
b. Schedule deputy—one per department, responsible for deciding which courses from that department will be offered, at what times, by what teachers, and with what enrollment limits. These parameters may change during the registration period. This is the only person who can increase or decrease enrollment limits for a course (including adding or deleting a course to or from the schedule).
c. Schedule director—in charge of allocating classroom and lecture hall space and time to departments. Also prints the schedule of classes to show students what will be offered and when.
d. Students—submit course requests and revisions and receive schedules and fee statements.
e. Counselors—advise students and approve all course requests and revisions. They also help students resolve time conflicts (where student has registered for two courses that meet at the same time).

This is the cast of characters (which may be revised or supplemented by your instructor to more closely match your school). For each client, brainstorm and describe different types of functional support that might be provided.

2. James Oliver, president of Oliver Pest Control, was under the impression that computerization of his clerical functions would reduce the size of that staff. Since installing computer support eight years ago, he has seen his clerical staff increase 10 percent, whereas organization staffing has remained steady. His friends in the local chamber of commerce have experienced the same phenomenon. Mr. Oliver is particularly puzzled by another curious trend. Although the use of computers did not decrease his need for clerical staff, it has reduced his dependence on middle management and professional staff. This seems exactly the opposite of his expectations; however, it is typical.
3. Liz, an account collections manager for the bank card office of a large bank, has a problem. Each week, she receives a listing of accounts that are past due. This report has grown from a listing of 250 accounts (two years ago) to 1,250 accounts (today). Liz has to go through the report to identify those accounts that are seriously delinquent. A seriously delinquent account is identified by several different rules, each requiring Liz to examine one or more data fields for that customer. What used to be a half-day job has become a three-days-per-week job. Even after identifying seriously delinquent accounts, Liz cannot make a final credit decision (such as a stern phone call, cutting off credit, or turning the account over to a collections agency) without accessing a three-year history of the account. Additionally, Liz needs to report what percentage of all accounts are past due, delinquent, seriously delinquent, and uncollectible. The current report doesn't give her that information. What kind of report does Liz have—detail, summary, or exception? What kind of reports does Liz need? What kind of decision support aids would be useful?

SUGGESTED READINGS

Bruce, Thomas. *Designing Quality Databases Using IDEF1X Information Models.* New York: Dorset House Publishing, 1993. We mention this book here because it, like much of the database industry, has embraced the Zachman framework.

Database Programming and Design. This monthly periodical that focuses on the database industry and data management issues has strongly embraced the Zachman framework to unify its subject matter.

Davis, Gordon B. "Knowing the Knowledge Workers: A Look at the People Who Work with Knowledge and the Technology That Will Make Them Better." *ICP Software Review,* Spring 1982, pp. 70–75. This article provided our first exposure to the concept of information workers as the new majority.

Goldman, James E. *Applied Data Communications: A Business-Oriented Approach.* New York: John Wiley & Sons, 1995. For those students who are looking for a student-oriented introduction to data communications, we recommend Professor Goldman's book because it was written for management and information systems majors to provide both a business and technical introduction to data communications, networking, and telecommunications.

Laudon, Kenneth C., and Jane P. Laudon. *Management Information Systems: Organization and Technology.* New York: Macmillan, 1994. This is our favorite junior/senior-level information systems survey textbook for students.

O'Brien, James. *Information Systems: The Alternate Edition.* Burr Ridge, IL: Richard D. Irwin, 1994. This is our current favorite among the many information systems survey textbooks targeted at the freshman/sophomore audience.

Zachman, John A. "A Framework for Information System Architecture." *IBM Systems Journal* 26, no. 3 (1987). We adapted the matrix model for information system building blocks from Mr. Zachman's conceptual framework. We first encountered John Zachman on the lecture circuit where he delivers a remarkably informative and entertaining talk on the same subject as this article. Mr. Zachman's framework has drawn professional acclaim and inspired at least one conference on his model. His framework is based on the concept that architecture means different things to different people. His model suggests that information systems consist of: (1) three distinct "product-oriented" views—data, processes (which we call activities), and networks; and (2) six different audience-specific views for each of those product views—the ballpark and owner's views (which we renamed as owner's and user's views, respectively), the designer's and builder's views (which we combined into our designer's view), and an out-of-context view (which we called the builder's view). Our adaptations make the framework an ideal conceptual model for this entire textbook.

3

INFORMATION SYSTEM DEVELOPMENT

CHAPTER PREVIEW AND OBJECTIVES

This chapter introduces a system development life cycle-based methodology as the process used to develop information systems. System development is not a hit-or-miss process! As with any product, information systems must be carefully developed. Successful systems development is governed by some fundamental, underlying principles that we will introduce in this chapter. We also introduce a representative systems development methodology as a disciplined approach to developing information systems. Although such an approach will not guarantee success, it will improve the chances of success. You will know that you understand information systems development when you can:

- Differentiate between the system development life cycle and a methodology.
- Describe eight basic principles of systems development.
- Define problems, opportunities, and directives—the triggers for systems development projects.
- Describe a framework that can be used to categorize problems, opportunities, and directives.
- Describe a phased approach to systems development. For each phase or activity, describe its purpose, participants, prerequisites, deliverables, activities, postrequisites, and impact.
- Describe the cross life cycle activities that overlap the entire life cycle.
- Define computer-aided systems engineering (CASE) and describe the role of CASE tools in system development.

SCENE

A conference room where Sandra Shepherd has called a project planning meeting. Sitting around the table are Bob Martinez (a teammate), Terri Hitchcock (the business analyst assigned to the Member Services system project), Gary Goldstein (the Development Center manager), and Lonnie Bentley (a professor at State University).

SANDRA

Good afternoon. I thought the project launch meeting went very well. I called this meeting to start planning the Member Services project. I guess some introductions are in order. I am Sandra Shepherd, senior systems analyst and project manager for this project. The permanent members of our team will be Bob Martinez, a new systems analyst here at SoundStage, and Terri Hitchcock, who is on a two-year loan to Information Services from Member Services. She will be the business analyst for the project. I'd like to introduce you both to Gary Goldstein. He manages our Development Center. And Gary, I see you've brought a guest.

GARY

Yes. Bob, it is nice to meet you. Terri, it will be a pleasure working with you again. I'd like to introduce everybody to Professor Lonnie Bentley from State University. Lonnie has been hired by Nancy [Picard, vice president of Information Services] to perform a special role in this project. As you may be aware, I recently chaired a task force to study emerging object-oriented methodologies with the goal of determining how to best integrate object-oriented methods and technologies into future projects.

TERRI

I hate to interrupt, but I don't come from an information systems background. What is object-oriented?

LONNIE

Good question. For the last two decades, most of our methodologies have been based one way or the other on something called *structured* techniques—structured programming, structured design, structured analysis, information engineering, and so forth. These methodologies attempt to capture the essence of a business problem and its solution requirements in some sort of graphical picture called a model. The models become the basis for design. I believe that your methodology—What do you call it? *FAST?*—uses structured methods.

GARY

Yes, but in combination with prototyping methods to accelerate system development.

LONNIE

Yes. But all of these structured methods dealt with various aspects of the system differently and separately. You used one modeling approach to document data requirements and another to document process requirements. In some cases, you use different kinds of structured models to describe the same aspect of the system. It has all been very confusing, especially to users.

TERRI

Tell me about it!

LONNIE

Object methods are a dramatically different alternative to structured methods. They use the same models to describe all aspects of the system, and the chosen models don't change as you progress through the project?

TERRI

Sounds intriguing!

SANDRA

So, what's the angle on this project?

GARY

Well, Nancy has hired Lonnie and one of his colleagues to develop our system in parallel with our use of *FAST*'s structured methods. Lonnie will use object methods and technology. After the project is completed, Lonnie and his team will help me integrate object methods and technology into *FAST.* Initially, object methods will be an alternative strategy for us, but all indications are that object methods will eventually obsolete structured methods.

BOB

This sounds cool! Will we get to see this stuff?

SANDRA

Ditto on Bob's question. I'm fascinated!

GARY

Of course. Lonnie and his team will serve as observers in all of our facilitated sessions. They will develop the object models in parallel with our structured models and we will invite all interested staff to attend special comparison presentations. Of course, Lonnie's team will implement the system with object technology.

SANDRA

I thought our *Powerbuilder* language is object-oriented.

GARY

It is. But Lonnie has taught me the difference between object-oriented and object-based technology. *Powerbuilder* is object-based.

LONNIE

A detailed differentiation is not important now, but suffice it to say that any tool must meet three criteria to be object-oriented. First, it must support integration of data and methods that act upon that data into objects. *Powerbuilder* and languages like *Visual BASIC* meet that criterion. Second, it must support a concept called inheritance. Both *Powerbuilder* and *Visual BASIC* only partially meet that criterion in their current versions. Third, the tool must support something called polymorphism. Only true object languages do that.

SANDRA

I've heard the terms, and I assume you teach them to us at the appropriate time, but which language are you going to use?

BOB

I'd bet either *C++* or *Smalltalk.* I've read that they are pure object technologies.

LONNIE

No. Your current methodology is based on an important strategy called *rapid*

SOUNDSTAGE

SOUNDSTAGE ENTERTAINMENT CLUB

application development. My colleagues are not convinced that either *C++* or *Smalltalk* are especially conducive to rapid application development. Instead, we are going to apply Borland's *Delphi* or Microsoft's *Visual BASIC* to the project. It has the rapid application development capabilities of *Powerbuilder*, and it fulfills most of the three object-oriented criteria.

SANDRA
Too bad we can't use it for our structured methods. That would give us an even stronger basis for the final comparison. But neither *Delphi* nor *Visual BASIC* are in our approved application architecture.

GARY
Actually, you can. And Nancy is encouraging it. She has requested a waiver of the *Powerbuilder* standard for this project—if you're willing.

SANDRA
[*noting Bob's obvious interest*] I think I can support that!

[pause]

Welcome to the team, Lonnie. Let's get to the purpose of this meeting. Gary, Bob, and Terri are new to the *FAST* methodology.

GARY
I know. I've scheduled a two-day workshop for all new hires and business analysts starting Monday of next week. Since Lonnie is also new to the team, maybe I should provide a quick overview.

[noting no objections, Gary continues]

FAST is a rapid application development methodology. For SoundStage, it is our second-generation methodology.

BOB
Second generation?

GARY
Yes, our first methodology was based solely on structured analysis and design. It was shipped in the form of seven three-ring binders, which had to be purchased for every developer. It was so detailed, complex, and inflexible that many developers resisted its use and deployment. So we replaced it with *FAST.*

TERRI
Why is *FAST* better?

GARY
First, it is simpler and more flexible. We found that to be highly desirable in our development culture. It is customizable to our standards and extensible to emerging methods, like object-oriented. And it is on-line. The only documentation is a "Getting Started" booklet and a reference manual for the on-line service. There are also some training materials, but you receive those when you complete the various in-house courses I teach.

LONNIE
On-line?

GARY
Yes, the methodology database is stored on the local area network, and it's available in its most current version from any developer's workstation. Simple diagrams walk the developers through the various phases, activities, roles, deliverables, and so forth. It can even automatically invoke the correct CASE tools we use to design and build our systems at the correct time during the project.

BOB
This sounds like fun.

GARY
Let's introduce the phases briefly today. I'll project the on-line services from my laptop to the screen . . .

DISCUSSION QUESTIONS

1. Some people believe that an enforced methodology for building systems stifles creativity. Why would they think that is true? Why might SoundStage still implement a common methodology?
2. Why must a methodology be adaptable to emerging methods and technologies?
3. How do you perceive Gary's role in this meeting and project?

SYSTEM DEVELOPMENT LIFE CYCLES AND METHODOLOGIES

This chapter introduces the business process used to develop information systems. In practice, the process is called a methodology. In theory, all methodologies are derived from a logical problem-solving process that is sometimes called a system development life cycle.

> A **system development life cycle (SDLC)** is a logical process by which systems analysts, software engineers, programmers, and end-users build information systems and computer applications to solve business problems and needs. It is sometimes called an **application development life cycle.**

The life cycle is essentially a project management tool used to plan, execute, and control systems development projects.

System development methodologies are frequently confused with the system development life cycle. Is there a difference? Some experts claim that methodologies have replaced the life cycle. In reality, they are one and the same! The intent of the life cycle is to plan, execute, and control a systems development project. It defines the phases and tasks that are *essential* to systems development, no matter what type or size system you may try to build. Note the emphasis on the word *essential*. For instance, we should always study and analyze the current system (at some level of detail!) before defining and prioritizing user requirements. It's common sense!

So, what *is* a methodology? At the risk of oversimplification, we offer the following definition.

> A **methodology** is the physical implementation of the logical life cycle that incorporates (1) step-by-step activities for each phase, (2) individual and group roles to be played in each activity, (3) deliverables and quality standards for each activity, and (4) tools and techniques to be used for each activity.

Two points are important. First, a true methodology should encompass the entire systems development life cycle. Second, most modern methodologies incorporate the use of several development tools and techniques.

Why do companies use methodologies? Methodologies ensure that a consistent, reproducible approach is applied to all projects. Methodologies reduce the risk associated with shortcuts and mistakes. Finally, methodologies produce complete and consistent documentation from one project to the next. All of these advantages provide one overriding benefit—as development teams and staff constantly change, the results of prior work can be easily retrieved and understood by those who follow!

Methodologies can be home grown; however, many businesses purchase their development methodology. For example, SoundStage has purchased a methodology called *FAST*. Why purchase a methodology? For one thing, most information system organizations can't afford to dedicate staff to the continuous improvement of a homegrown methodology. Also, methodology vendors have a vested interest in keeping their methodologies current with the latest business and technology trends.

As you will discover in this book, *FAST*, like most modern, commercial methodologies, is based on some combination of techniques called best practices. In this chapter, we will explore *FAST* to learn about both the life cycle and a representative methodology.

UNDERLYING PRINCIPLES OF SYSTEMS DEVELOPMENT

Before we study the life cycle, let's introduce some general principles that should underlie all systems development methodologies.[1]

Principle 1: Get the Owners and Users Involved

Analysts, programmers, and other information technology specialists frequently refer to "my system." This attitude has, in part, created an "us-versus-them" attitude between technical staff and their users. Although analysts and programmers work hard to create technologically impressive solutions, those solutions often backfire because they don't address the real organization problems or they introduce new organization or technical problems. For this reason, owner and user involvement is an absolute necessity for successful systems development.[2] The individuals responsible for systems development must make time for owners and users, insist on their participation, and seek agreement from them on all decisions that may affect them.

[1] Adapted from R. I. Benjamin, *Control of the Information System Development Cycle* (New York: Wiley-Interscience, 1971).

[2] We are using the term *owner*, introduced in Chapters 1 and 2, to mean *management*.

Miscommunication and misunderstandings continue to be a significant problem in systems development. However, owner and user involvement and education minimize such problems and help to win acceptance of new ideas and technological change. Because people tend to resist change, the computer is often viewed as a threat. Through education, information systems and computers can be properly viewed by users as tools that will make their jobs less mundane and more enjoyable.

Principle 2: Use a Problem-Solving Approach

A methodology is, first and foremost, a problem-solving approach to building systems. The term *problem* is used throughout this book to include real problems, opportunities for improvement, and directives from management. The classical problem-solving approach is as follows:

1. Study and understand the problem (opportunity and/or directive) and its system context.
2. Define the requirements of a suitable solution.
3. Identify candidate solutions and select the "best" solution.
4. Design and/or implement the solution.
5. Observe and evaluate the solution's impact, and refine the solution accordingly.

The notion here is that systems analysts should approach all projects using some sort of problem-solving approach.

There is a tendency among inexperienced problem solvers to eliminate or abbreviate one or more of the above steps. The result can range from (1) solving the wrong problem, to (2) incorrectly solving the problem, to (3) picking the wrong solution. The problem-solving orientation of a methodology, when correctly applied, can reduce or eliminate the above risks.

Principle 3: Establish Phases and Activities

Most life cycles and methodologies consist of phases. In its simplest, classical form, the life cycle consists of four phases: systems planning, analysis, systems design, and systems implementation. (A fifth activity, systems support, refines the resulting system by iterating through the previous four phases on a smaller scale to refine and improve the system.)

Figure 3.1 illustrates the four key phases in the context of your information systems framework (from Chapter 2). In each phase, the analysts and participants define and/or develop the building blocks in that corresponding row of the framework. (Recall that the building blocks are DATA, PROCESSES, INTERFACES, and GEOGRAPHY. Also notice that as you progress top to bottom through the phases, you are moving closer to the technology base of the pyramid—suggesting that your early concerns are with the *business* and your later concerns become more *technology*-driven.

Because projects may be quite large and each phase usually represents considerable work and time, the phases are usually broken down into activities and tasks that can be more easily managed and accomplished. This chapter will focus only on phases, albeit somewhat more refined than shown in Figure 3.1. The underlying activities and tasks will be covered in later chapters.

Generally, the phases of a project should be completed top to bottom, in sequence. This may leave you with the impression that once you finish a phase, you are done with that phase. That is not true. At any given time, you may be performing tasks in more than one phase simultaneously. You may have to backtrack to previous phases and activities to make corrections or to respond to new requirements. Obviously, you can't get carried away with this type of backtracking or you might never implement the new system.

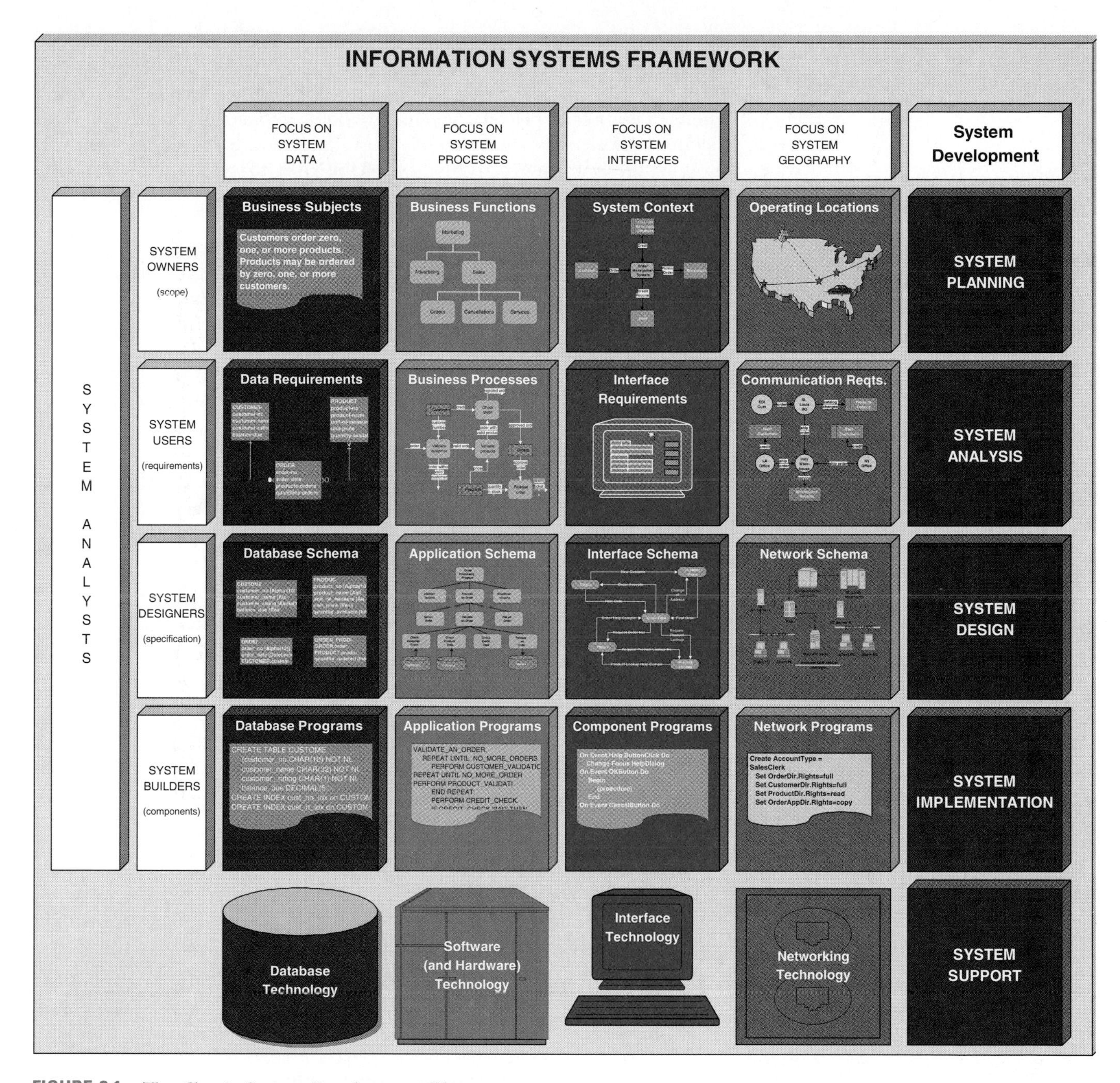

FIGURE 3.1 *The Classic System Development Phases*

Principle 4: Establish Standards for Consistent Development and Documentation

An organization has many information systems that may include thousands of programs and software packages. If all analysts and programmers were to adopt their own preferred methodology and use their own tools and techniques to develop and document systems, chaos would quickly result. In medium-to-large information systems shops, systems analysts, programmers, and technical specialists come and go (as do system owners and users!). Some will be promoted; some will quit the organization; and still others will be reassigned. To promote good communication between this constantly changing base of users and information systems professionals, you must develop standards to ensure consistent systems development.

Systems development standards usually describe (1) activities, (2) responsibilities, (3) documentation guidelines or requirements, and (4) quality checks. These four standards should be established for every phase in the methodology.

The need for documentation standards underscores a common failure of many analysts—the failure to document as an ongoing activity during the life cycle. Most students and practitioners talk about the importance of documentation, but talk is cheap! When do you really place comments in your computer programs? After you finish, of course! Therein lies the problem: Most of us tend to post-document software. Unfortunately, we often carry this bad habit over to systems development.

Documentation should be a working by-product of the entire systems development effort. Documentation reveals strengths and weaknesses of the system to others—before the system is built. It stimulates user involvement and reassures management about progress. This book teaches technique, but it also teaches documentation. Learn to use the tools and techniques to communicate with users during the life cycle, not after!

Principle 5: Justify Systems as Capital Investments

Information systems are capital investments, just as are a fleet of trucks or a new building. Even if management fails to recognize the system as an investment, you should not. When considering a capital investment, two issues must be addressed.

First, for any problem, there are likely to be several possible solutions. The analyst should not accept the first solution that comes to mind. The analyst who fails to look at several alternatives is an amateur. Second, after identifying alternative solutions, the systems analyst should evaluate each possible solution for feasibility, especially for cost-effectiveness.

> **Cost-effectiveness** is defined as the result obtained by striking a balance between the cost of developing and operating a system, and the benefits derived from that system.

Cost-benefit analysis is an important skill to be mastered.

Principle 6: Don't Be Afraid to Cancel or Revise Scope

A significant advantage of the phased approach to systems development is that it provides several opportunities to reevaluate feasibility. There is often a temptation to continue with a project only because of the investment already made. In the long run, canceled projects are less costly than implemented disasters! This is extremely important for young analysts to remember.

Similarly, many analysts allow project scope to increase during a project. Sometimes this is inevitable because the analyst learns more about the system as the project progresses. Unfortunately, most analysts fail to adjust estimated costs and schedules as scope increases. As a result, the analyst frequently and needlessly accepts responsibility for cost and schedule overruns.

The authors of this text advocate a creeping commitment approach to systems development.[3] Using the creeping commitment approach, multiple feasibility checkpoints are built into the systems development methodology. At any feasibility checkpoint, all costs are considered sunk (meaning irrecoverable). They are, therefore, irrelevant to the decision. Thus, the project should be reevaluated at each checkpoint to determine if it is still feasible.

At each checkpoint, the analyst should consider (1) cancellation of the project if it is no longer feasible, (2) reevaluation of costs and schedule if project scope is to be increased, or (3) reduction of scope if the project budget and schedule are frozen, but not sufficient to cover all project objectives.

The concept of sunk costs is more or less familiar to some financial analysts and managers, but it is frequently forgotten or not used by the majority of practicing analysts and users.

Principle 7: Divide and Conquer

All systems are part of larger systems (called supersystems). Similarly, virtually all systems contain smaller systems (called subsystems). These facts are significant for

[3] Thomas Gildersleeve, *Successful Data Processing Systems Analysis,* 2nd ed. (Englewood Cliffs, NJ: Prentice Hall, 1985), pp. 5–7.

two reasons. First, systems analysts must be mindful that any system they are working on interacts with its supersystem. If the supersystem is constantly changing, so might the scope of any given project. Most systems analysts can still be faulted for underestimating the size of projects. Most of the fault lies with not properly studying the implications of a given system relative to its larger whole—its supersystem.

Consider the old saying, "If you want to learn anything, you must not try to learn everything—at least not all at once." For this reason, we divide a system into its subsystems in order to more easily conquer the problem and build the larger system. By dividing a larger problem (system) into more easily managed pieces (subsystems), the analyst can simplify the problem-solving process. We'll be applying this principle throughout this book.

Principle 8: Design Systems for Growth and Change

There is a critical shortage of information systems professionals needed to develop systems. Combined with the ever-increasing demand for systems development, many systems analysts have fallen into the trap of developing systems to meet only today's user requirements. Although this may seem to be a necessary approach at first glance, it actually backfires in almost all cases.

Entropy is the term system scientists use to describe the natural and inevitable decay of all systems. Entropy is illustrated in the context of the system development life cycle in Figure 3.2. After a system is implemented, it enters the support phase of the life cycle. During the support phase, the analyst encounters the need for changes that range from correcting simple mistakes, to redesigning the system to accommodate changing technology, to making modifications to support changing user requirements. As indicated by the blue arrows, many of these changes direct the analyst and programmer to rework former phases of the life cycle. Eventually, the cost of maintenance exceeds the costs of starting over—the system has become obsolete. This is indicated by the red arrow in the figure.

Frequently, systems that are designed to meet only current requirements are difficult to modify in response to new requirements. The systems analyst is frequently forced to duplicate files and "patch" programs in ways that make the system very costly to support over the long run. As a result, many systems analysts become frustrated with how much time must be dedicated to supporting existing systems (often called **legacy systems**) and how little time is left to work on important, *new* systems development.

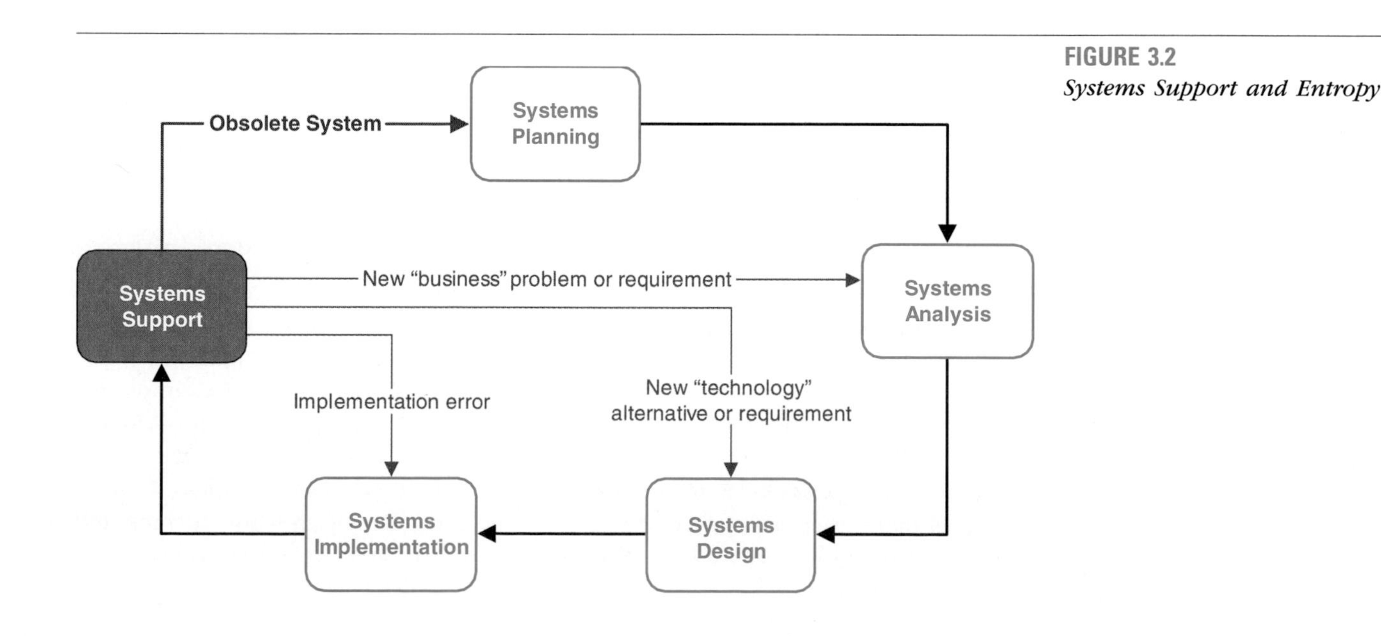

FIGURE 3.2
Systems Support and Entropy

FIGURE 3.3 *Principles of Systems Development*

- Get the owners and users involved.
- Use a problem-solving approach.
- Establish phases and activities.
- Establish standards for consistent development and documentation.
- Justify systems as capital investments.
- Don't be afraid to cancel.
- Divide and conquer.
- Design systems for growth and change.

Even if you design the system to easily adapt to change (our last principle), at some point, it will become too costly to simply support the existing system. Why? Perhaps the organization itself has changed too dramatically to be supported by the system. Or perhaps the requirements have become too complex to be integrated into the current system. In either case it is time to start over! This situation puts the term *cycle* into the term *systems development life cycle*. No system lasts forever (although many do last for a decade or longer).

But system entropy can be managed. Today's tools and techniques make it possible to design systems that can grow and change as requirements grow and change. This book will teach you many of those tools and techniques. For now, it's more important to simply recognize that flexibility and adaptability do not happen by accident—they must be built into a system.

We have presented eight principles that should underlie any methodology. These principles, summarized in Figure 3.3, can be used to evaluate any methodology, including *FAST*.

FAST—A SYSTEM DEVELOPMENT METHODOLOGY

In this section we'll examine a modern system development life cycle using a representative methodology we call *FAST* (used in the SoundStage case study). We'll begin by studying how a *FAST* project gets started. Then we'll examine *FAST*'s phases with a high-level picture that omits many details. Finally, we'll study a more complete picture of the *FAST* phases.

How a *FAST* Project Gets Started

Traditionally, system owners and system users have initiated most projects because they are closer to the organization's activities that need improvement. Alternatively, systems analysts are expected to survey the organization for possible improvements. When system owners, system users, or systems analysts initiate a project, *FAST* calls this an **unplanned system request**. These unplanned system requests can easily overwhelm the largest information services organization; therefore, they are frequently screened and prioritized by a **steering committee** of system owners to determine which requests get approved. Those requests that are not approved are often said to be **backlogged** until resources become available (which sometimes never happens).

The opposite of an unplanned system request is a **planned system initiative.** A planned system initiative is the result of one of the following earlier projects:

- An **information strategy plan** that has examined the business as a whole to identify those systems and application development projects that will return the greatest strategic (long-term) value to the business. An example of a methodology that incorporates an information strategy plan is *information engineering*.
- A **business process redesign** that has thoroughly analyzed a series of fundamental business processes to eliminate redundancy and bureaucracy and to improve efficiency and value added—now it is time to redesign the

supporting information systems for those business processes. Business process redesign (BPR) is a source of many current systems development projects.

Planned or unplanned, the impetus for most projects is some combination of problems, opportunities, or directives.

Problems are undesirable situations that prevent the organization from fully achieving its purpose, goals, and objectives.

For example, an unacceptable increase in the time required to fill an order can trigger a project to reduce that delay. Problems may either be current, suspected, or anticipated.

An **opportunity** is a chance to improve the organization even in the absence of specific problems. (Note: You could argue that any unexploited opportunity is, in fact, a problem.)

For instance, management is always receptive to cost-cutting ideas, even when costs are not currently considered a problem. Opportunistic improvement is expected to be the source of today's most important systems development projects.

A **directive** is a new requirement that's imposed by management, government, or some external influence. (Note: You could argue that until a directive is fully complied with, it is, in fact, a problem.)

For example, the Equal Employment Opportunity Commission, a government agency, may mandate that a new set of reports be produced each quarter. Similarly, company management may dictate support for a new product line or policy. Some directives may be technical. For instance, systems may be strategically directed to convert from batch to on-line processing or from conventional files to database. Such measures are appropriate if the current technology is obsolete, difficult to maintain, slow, or cumbersome to use.

There are far too many potential problems, opportunities, and directives to list them all in this book. However, James Wetherbe has developed a useful framework for classifying problems, opportunities, and directives.[4] He calls it **PIECES** because the letters of each of the six categories, when put together, spell the word *pieces*. The categories are:

P the need to improve *performance*.

I the need to improve *information* (and data).

E the need to improve *economics,* control costs, or increase profits.

C the need to improve *control* or security.

E the need to improve *efficiency* of people and processes.

S the need to improve *service* to customers, suppliers, partners, employees, etc.

Figure 3.4 expands on each of these categories.

The categories of the PIECES framework overlap. Any given project can be characterized by one or more categories, and any given problem or opportunity may have implications with respect to more than one category. PIECES is a practical framework (used in *FAST IV*), not just an academic exercise! We'll revisit PIECES later in the book.

An Overview of the *FAST* Life Cycle and Methodology

Figure 3.5 shows the context of a *FAST* system development project. You can clearly see the two aforementioned project triggers: **unplanned system request**

[4] James Wetherbe and Nicholas P. Vitalari, *Systems Analysis and Design: Traditional, Best Practices,* 4th ed. (St. Paul, MN: West Publishing, 1994), pp. 196–99.

FIGURE 3.4 *The PIECES Problem-Solving Framework*

The following checklist for problem, opportunity, and directive identification uses Wetherbe's PIECES framework. Note that the categories of PIECES are not mutually exclusive; some possible problems show up in multiple lists. Also, the list of possible problems is not exhaustive. The PIECES framework is equally suited to analyzing both manual and computerized systems and applications.

PERFORMANCE Problems, Opportunities, and Directives

A. Throughput—the amount of work performed over some period of time.
B. Response time—the average delay between a transaction or request and a response to that transaction or request.

INFORMATION (and Data) Problems, Opportunities, and Directives

A. Outputs
   1. Lack of any information
   2. Lack of necessary information
   3. Lack of relevant information
   4. Too much information—information overload
   5. Information that is not in a useful format
   6. Information that is not accurate
   7. Information that is difficult to produce
   8. Information is not timely to its subsequent use
B. Inputs
   1. Data are not captured
   2. Data are not captured in time to be useful
   3. Data are not accurately captured
   4. Data are difficult to capture
   5. Data are captured redundantly—same data captured more than once
   6. Too much data are captured
   7. Illegal data are captured
C. Stored data
   1. Data are stored redundantly in multiple files and/or databases
   2. Stored data are not accurate (may be related to 1)
   3. Data are not secure to accident or vandalism
   4. Data are not well organized
   5. Data are not flexible—not easy to meet new information needs from stored data
   6. Data are not accessible

ECONOMICS Problems, Opportunities, and Directives

A. Costs
   1. Costs are unknown
   2. Costs are untraceable to source
   3. Costs are too high
B. Profits
   1. New markets can be explored
   2. Current marketing can be improved
   3. Orders can be increased

CONTROL (and Security) Problems, Opportunities, and Directives

A. Too little security or control
   1. Input data are not adequately edited
   2. Crimes are (or can be) committed against data
      a. Fraud
      b. Embezzlement
   3. Ethics are breached on data or information—refers to data or information getting to unauthorized people
   4. Redundantly stored data are inconsistent in different files or databases
   5. Data privacy regulations or guidelines are being (or can be) violated
   6. Processing errors are occurring (either by people, machines, or software)
   7. Decision-making errors are occurring
B. Too much control or security
   1. Bureaucratic red tape slows the system
   2. Controls inconvenience customers or employees
   3. Excessive controls cause processing delays

EFFICIENCY Problems, Opportunities, and Directives

A. People, machines, or computers waste time
   1. Data are redundantly input or copied
   2. Data are redundantly processed

FIGURE 3.4 *Concluded*

3. Information is redundantly generated
B. People, machines, or computers waste materials and supplies
C. Effort required for tasks is excessive
D. Materials required for tasks are excessive

SERVICE Problems, Opportunities, and Directives

A. The system produces inaccurate results
B. The system produces inconsistent results
C. The system produces unreliable results
D. The system is not easy to learn
E. The system is not easy to use
F. The system is awkward to use
G. The system is inflexible to new or exceptional situations
H. The system is inflexible to change
I. The system is incompatible with other systems
J. The system is not coordinated with other systems

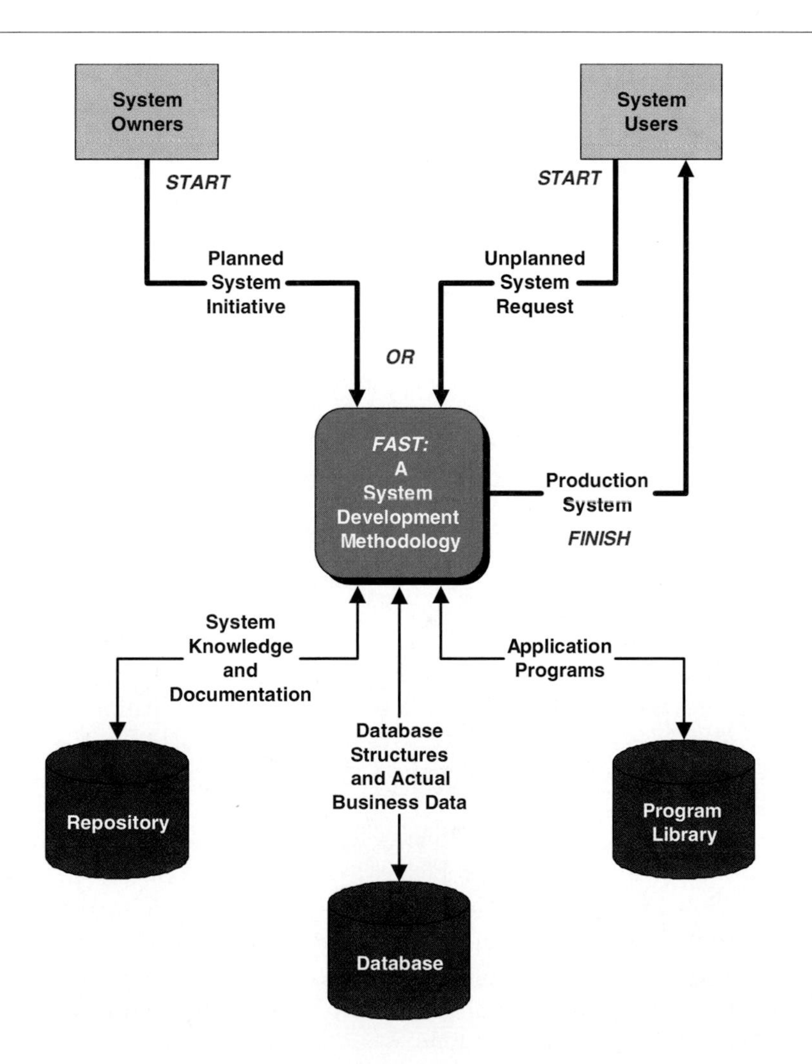

FIGURE 3.5
The Context of System Development

and **planned system initiative.** The final output of the methodology is the **production system** (so named because the system produces results). Also notice that, as you develop a system, you need a place to store various artifacts such as documentation, production data, and software. The three data stores are described as follows:

- The **repository** is a place where systems analysts and other developers store documentation about the system. Examples of such documentation might include *written memos, user requirements,* and *program flowcharts.*
- The **database** is built during the project to store actual business data about such things as CUSTOMERS, PRODUCTS, and ORDERS. This database will be maintained by the application programs written (or purchased) for the information system.
- The **program library** is where any application software and programs will be stored once they are written (or purchased).

We'll discuss these three data stores in more detail later in the chapter and book. Now we're ready to study the phases of a project in somewhat greater detail.

Let's look inside the *FAST* methodology process symbol. Figure 3.6 depicts a simplified *phase diagram* of the *FAST* methodology. The symbology, used throughout this chapter, is described as follows:

- The rounded rectangles represent *phases* in a *FAST* system development project.
- The thick black arrows represent the major deliverables (or outputs) of the phases. Each deliverable contains important documentation and/or specifications. Notice that the deliverable of one phase may serve as input to another phase.
- Although not used in Figure 3.6, we will eventually add thin black, double-ended arrows to represent other secondary information and communication flows. These flows can take the form of conversations, meetings, letters, memos, reports, and the like.
- The rectangles indicate people or organizations with whom the analyst may interact. Notice that the roles introduced in Chapter 2 have been carried over to this figure. For instance, we see system owners and system users participating in a project. Once again, we did not show all people interactions on this overview diagram of the methodology.
- Finally, consistent with our creeping commitment principle, the black circles indicate checkpoints at which time the project participants should reevaluate feasibility and/or project scope.

As you can see, the *FAST* methodology consists of eight phases. We'll get into details later, but each of these phases deserves a brief explanation:

1. The *survey phase* establishes the project context, scope, budget, staffing, and schedule.
2. The *study phase* identifies and analyzes both the business and technical problem domains for specific problems, causes, and effects.
3. The *definition phase* identifies and analyzes *business* requirements that should apply to any possible technical solution to the problems.
4. The *configuration phase* identifies and analyzes candidate technical solutions that might solve the problem and fulfill the business requirements. The result is a feasible application architecture.

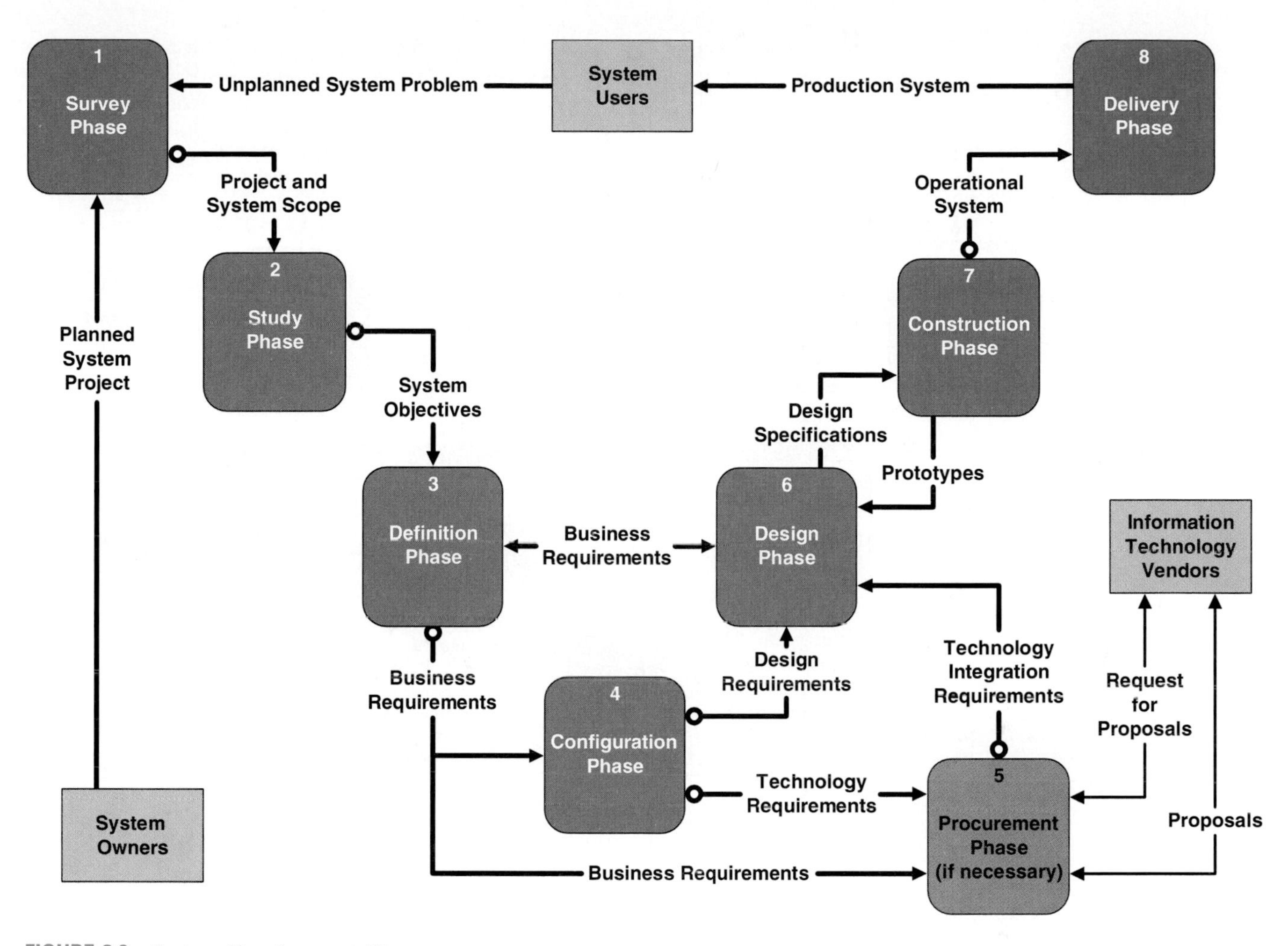

FIGURE 3.6 *System Development Phases*

5. The *procurement phase* (optional) identifies and analyzes hardware and software products that will be purchased as part of the target solution.
6. The *design phase* specifies the technical requirements for the target solution. Today, the design phase typically has significant overlap with the construction phase.
7. The *construction phase* builds and tests the actual solution (or interim prototypes of the solution).
8. The *delivery phase* puts the solution into daily production.

Some of these phases take little time to complete. The time required to complete other phases depends on strategy and risk. The phases are not necessarily sequential; they typically overlap.

Given this high-level overview of *FAST,* let's examine each of the phases with respect to (1) purpose, (2) participants and roles, (3) prerequisites, (4) activities, (5) deliverables, (6) postrequisites (based on continuous feasibility reassessment), and (7) impact analysis.

Figure 3.7 provides a single reference point for all this discussion. You might want to mark that page for easy access. Figure 3.8 illustrates the phases in the context of your information systems framework (from Chapter 2). You might also want to mark that figure for easy access.

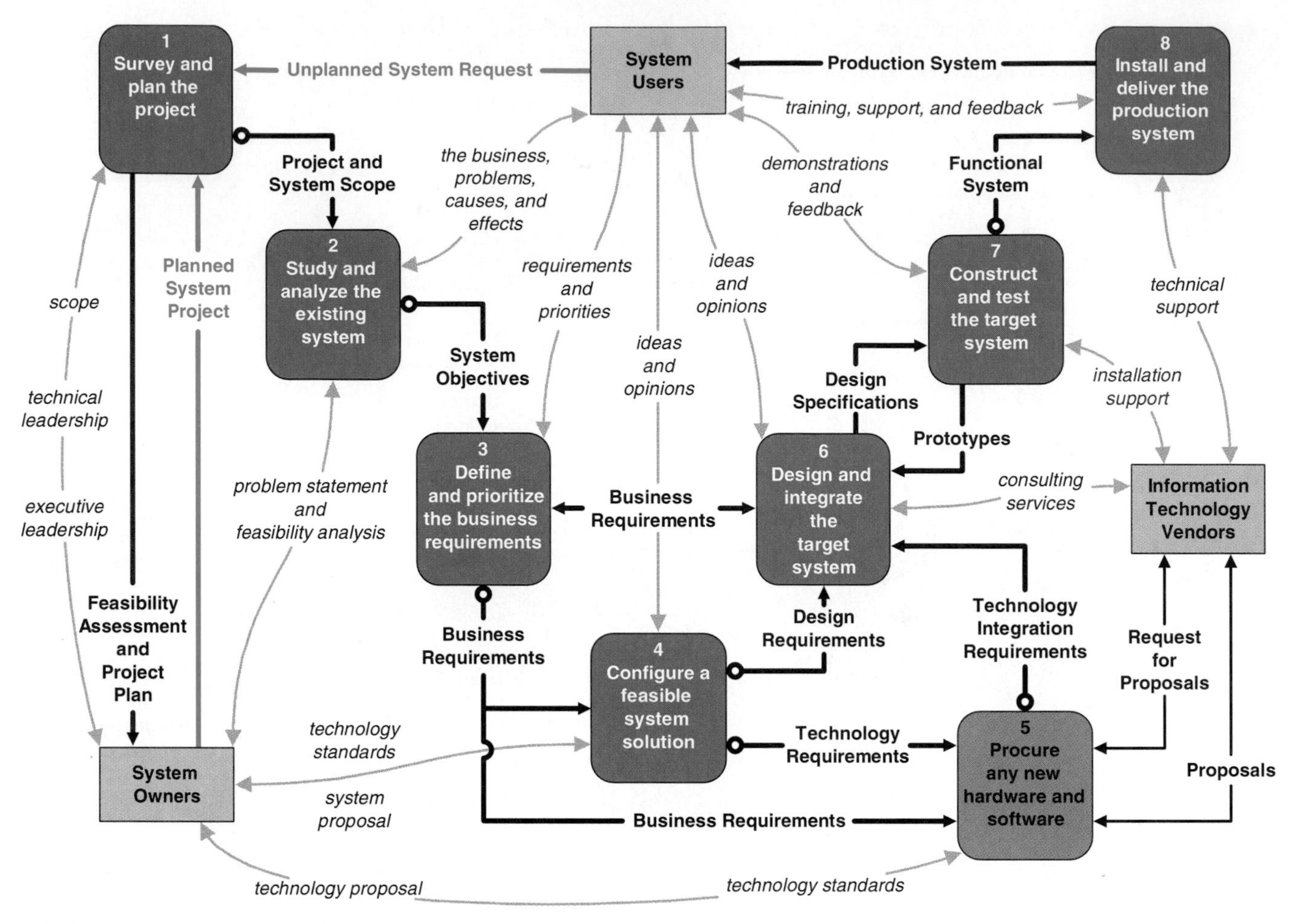

FIGURE 3.7 *The FAST Phase Diagram*

The Survey Phase

Systems development can be very expensive. Thus, it pays to answer the important question, "Is this project worth looking at?" Thus, the first phase of a project is to survey the project. The **survey phase** is also sometimes called a preliminary investigation or feasibility study. It amounts to a "quick-and-dirty" (usually two to three days) preliminary investigation of the problems, opportunities, and/or directives that triggered the project.

Purpose The purpose of the survey phase is threefold. First, the survey phase answers the question, "Is this project worth looking at?" To answer this question, the survey phase must define the scope of the project and the perceived problems, opportunities, and directives that triggered the project. Assuming the project *is* worth looking at, the survey phase must also establish the project team and participants, the project budget, and the project schedule.

Participants and Roles The facilitator of this phase is the systems analyst. Because this phase describes the system and project from the perspective of system owners, those owners are usually the only other participants. It is appropriate, however, to now expand our definition of system owner to include the following roles:

- *Executive sponsor*—the highest-level manager who will pay for the project. More importantly, this manager should provide visible support and leadership to the project participants. Cooperation and compromise among participants are often dependent on this leadership.

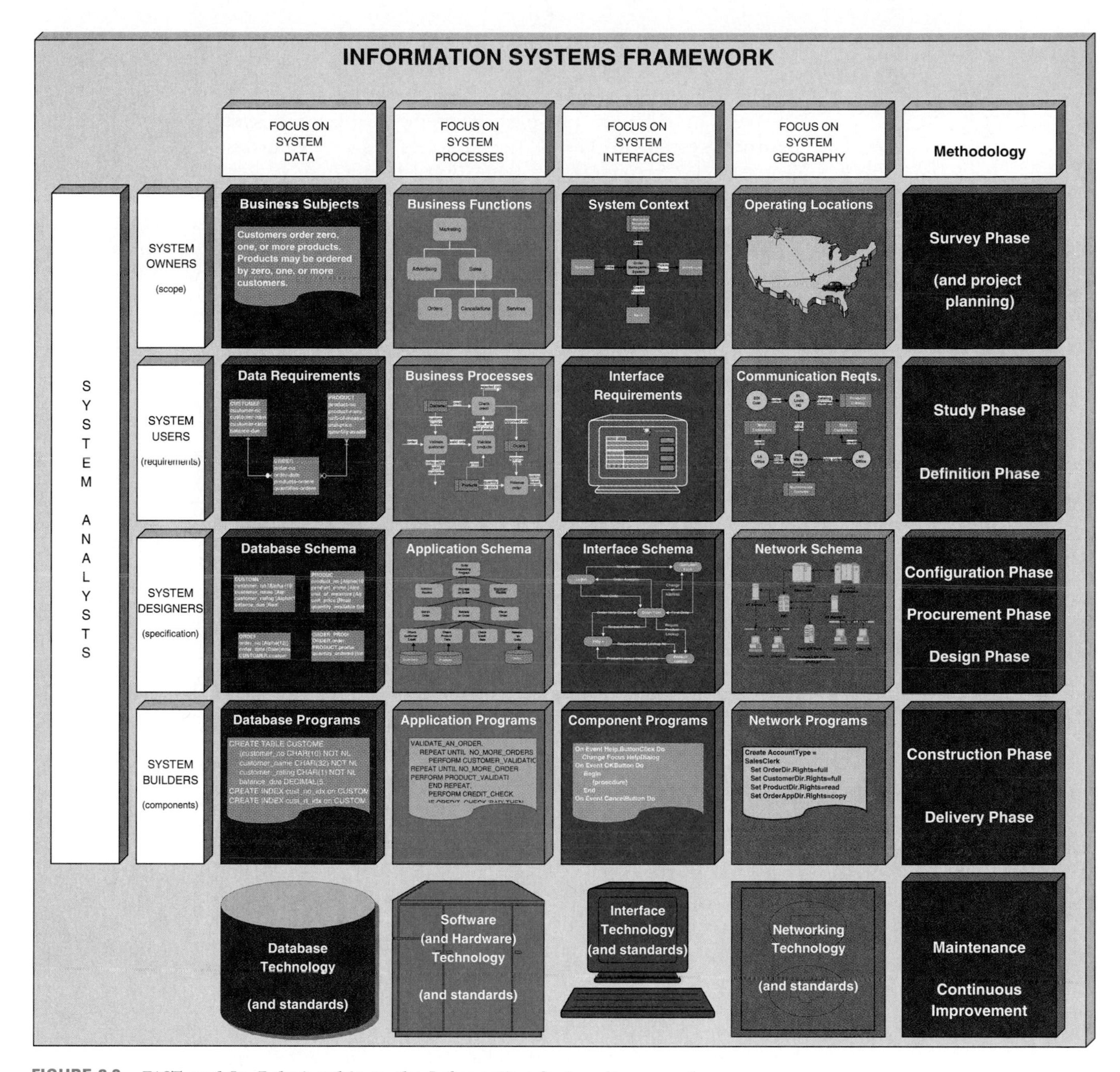

FIGURE 3.8 *FAST and Its Relationship to the Information System Framework*

- *Technical sponsor*—the highest-level manager from the information services organization who will pay for the project. This may be the chief information officer, the director of all systems development, or the manager of all information systems for a specific business function.
- *Project managers*—the managers of the project team. This person is responsible for the staffing, budget, and schedule. The project manager may be a senior systems analyst, a full-time project manager, or a significant staff member from the system user community. *FAST* encourages joint management of the project by an information services worker and a ranking manager from the user community.

Other system users are rarely involved until the next phase.

Prerequisites Prerequisites define the required and optional inputs for a phase. The key input to the phase is either the **unplanned system request** or the **planned system initiative.** These project triggers were described earlier in the chapter. *FAST* accommodates both formal triggers (such as a form) and informal triggers (such as a memo).

Figure 3.7 also illustrates the need for *executive leadership* and *technical leadership* from the system owners. (Note: This is our first example of secondary information and communication flows.) Also, *scope* is negotiated between the systems analyst and system owners.

Activities The most important activity in the survey phase is to define the scope or size of the project. Many projects fail because of poor scope management. If uncontrolled, scope tends to grow (along with costs and missed deadlines). You can't manage scope if you don't define it. Think of scope as a statement of the users' expectations of the project. Once defined, you can renegotiate for additional resources if users increase that scope. Using the information systems framework (Figure 3.8), project scope can be defined in terms of the system owner building blocks for DATA, PROCESSES, INTERFACES, and GEOGRAPHY. Project scope also includes the identification of other system owners and system users.

Once scope has been defined, we need to answer the question—"Is this project worth looking at?" To answer this question, the analyst should discuss the perceived problems, opportunities, and directives with the system owners. The PIECES framework provides an excellent outline for such a discussion. The goal here is not to solve the problems, just to catalog and categorize them. The analyst should also explore any constraints that may impact the project, such as budget, schedule, or technology standards. Finally, the analyst can discuss the question with the system owners. What impact would solving these problems have on the business? Generally, it is too early to assess the cost of developing the system.

Assuming the system is worth looking at, the project manager should formally plan the project. This includes establishing a preliminary budget and schedule and staffing the development team. (Note: *FAST* requires that the budget and schedule be reevaluated and refined at the end of each phase.)

The entire phase should not consume more than two to three days since the purpose of the survey phase is to decide whether or not more significant time and resources should be committed to the project.

Deliverables Deliverables are the outputs of a phase. As shown in Figure 3.7, a key deliverable for the survey phase is a *feasibility assessment and project plan*. This might be a report or verbal presentation, possibly both. The report version is sometimes called an *initial study report*. The analyst's recommendation may prescribe (1) a "quick fix," (2) an enhancement of the existing system and software, or (3) a completely new information system. For the latter possibility, a statement of **project and system scope** should be prepared as a deliverable to the next phase.

Postrequisites and Feasibility Checkpoints Recall that a circle at the beginning of any information flow in Figure 3.7 indicates the flow may or may not occur based on our creeping commitment principle. Thus, these circles define feasibility checkpoints in *FAST*. In other words, project and system scope will only occur if the project has been approved to continue to the next phase.

The **feasibility assessment and project plan** must usually be reviewed by the system owners (or a steering committee that includes system owners). This is especially common when many proposed projects are competing for the same resources. One of four decisions is possible: (1) approve the project to continue to the study phase, (2) change the scope and continue on to the study phase, (3) reject the project, or (4) delay the project in favor of some other project.

Impact Analysis Can the survey phase ever be skipped? Scope definition is critical to all projects, planned and unplanned, but it could be deferred until the study phase for those projects that have already been determined to be worth looking at. For example, some organizations plan all or most projects as part of an ongoing *strategic information systems plan*. For another example, some projects are driven by extremely strong administrative directives. Thus, if a vice president *insists* you look at an application, it would not be wise to question worthiness at this point in the project.

The Study Phase

There's an old saying that suggests, "Don't try to fix it unless you understand it." With those words of wisdom, the next phase of a *FAST* project is to study and analyze the existing system. There is always an existing business system, regardless of whether it currently uses a computer. The **study phase** provides the project team with a more thorough understanding of the problems, opportunities, and/or directives that triggered the project. The analyst frequently uncovers new problems and opportunities. The study phase may answer the questions, "Are the problems worth solving?" and "Is a new system worth building?"

Purpose The purpose of the study phase is threefold. First and foremost, the project team must gain an appropriate understanding of the business problem domain. Second, we need to answer the question, "Are these problems (opportunities and directives) worth solving?" Finally, we need to determine if the system is worth developing. The study phase provides the systems analyst and project team with a more thorough understanding of the problems, opportunities, and/or directives that triggered the project. In the process, they frequently uncover new problems and opportunities.

Participants and Roles Once again, the systems analyst facilitates this phase. But system users are actively involved in the study. Although some system users (or business analysts) may be permanently assigned to the project team, all system users should be involved to some degree (such as interviews, group meetings, formal presentations, progress reports, reports of findings, etc.)

System owners must visibly support the study to ensure that all system users actively participate. Also, the findings of the study phase should be presented to or reviewed with the system owners.

Prerequisites As shown in Figure 3.7, the key input is the statement of project and system scope from the survey phase. The project team studies the existing system by collecting factual information from the system users concerning *the business* and the perceived *problems, causes, and effects*. It is also appropriate to discuss system limitations with information services specialists who support the existing system. From all of this information, the project team gains a better understanding of the existing system's problems and limitations.

Activities Every existing system has its own terminology, history, culture, and nuances. Learning those aspects of the system is the principle activity in this phase. Depending on the complexity of the problem domain and the project schedule, the team may or may not choose to formally document the system. Your information system framework (Figure 3.8) provides an effective outline for studying the existing system from the perspective of system users as they see the DATA, PROCESS,

INTERFACE, and GEOGRAPHY building blocks. In the next unit of the book, you will discover and learn how to use specific tools and techniques for this purpose.

Given an understanding of the current system, the project team analyzes the perceived problems, opportunities, and directives. Do the problems and opportunities really exist? If so, how serious are they? Many times, the initial problems are mere symptoms, frequently of more serious or subtle problems. During the study phase, we need to address the causes and effects of the problems, opportunities, and directives. The PIECES framework presented earlier in this chapter can be used as an effective tool for organizing the analyses.

Deliverables The findings of the study phase are reviewed with the system owners as a business problem statement and feasibility analysis (sometimes called a detailed study report). *FAST* encourages that the findings also be shared with system users. This problem statement may take the form of a formal written report, an updated feasibility assessment, or a formal presentation to management and users. SoundStage has standardized on a formal written report distributed during a formal presentation to owners and users.

If the project continues to the next phase, the project team should include **system objectives** in the aforementioned report or presentation. These objectives do not define inputs, outputs, or processes. Instead, they define the business criteria on which any new system will be evaluated. For instance, we might define an objective that states the new system must REDUCE THE TIME BETWEEN ORDER PROCESSING AND SHIPPING BY THREE DAYS, OR REDUCE BAD CREDIT LOSSES BY 45 PERCENT. Think of objectives as the "grading curve" for evaluating any new system that you might eventually design and implement.

Postrequisites and Feasibility Checkpoints After reviewing the findings, the system owners will either agree or disagree with the recommendations of the study phase. Thus, the system objectives information flow is marked as optional (with the circle) based on a feasibility reassessment. The project can be (1) canceled if the problems prove not worth solving, or a new system is not worth building, (2) approved to continue to the definition phase, or (3) reduced in scope or increased in budget and schedule, and then approved to continue to the definition phase.

Impact Analysis Can you ever skip the study phase? Rarely! You almost always need some understanding of the existing system. But there may be reasons to complete the phase as quickly as possible. If the project was triggered by a planned system initiative, the worthiness of both the project and system was determined earlier. Accordingly, the study phase requirements are reduced to understanding the current system.

As another example, if the project was triggered by a management directive (e.g., "We must have this system by February 1 in order to comply with new federal regulations.") then worthiness is definitely not in question. In fact, we may want to get through the study phase rather quickly to increase the likelihood of meeting the deadline.

FAST is flexible insofar as determining how much time should be spent on this phase. Like many other businesses, SoundStage used to spend much time documenting the existing system using *structured analysis* techniques. As you'll learn by studying the SoundStage project, it now spends much less time on documenting the existing system and more time on cause/effect analysis. The goal is to move every project into a "new system focus" (the definition phase) as quickly as possible.

The Definition Phase

Given approval of the business problem statement, now you can design a new system, right? No, not yet! What capabilities should the new system provide for its users? What data must be captured and stored? What performance level is expected?

Careful! This requires decisions about what the system must do, not how it should do those things. The next phase of a *FAST* project is to define and prioritize the business requirements. It is sometimes called a requirements analysis phase or simply, the **definition phase.** Simply stated, the analyst approaches the users to find out what they need or want out of the new system. This is perhaps the most important phase of the life cycle. Errors and omissions in the definition phase result in user dissatisfaction with the final system and costly modifications to that system.

Purpose Essentially, the purpose of requirements analysis is to identify the DATA, PROCESS, INTERFACE, and GEOGRAPHY requirements for the users of a new system. Most importantly, the purpose is to specify these requirements without expressing computer alternatives and technology details; at this point, keep analysis at the business level!

Participants and Roles The systems analyst or business analyst facilitates the definition and prioritization of business requirements. System users assigned to the team play an essential role in specifying, clarifying, and documenting the business requirements. It is, however, extremely important to involve system users not on the team. These users may provide perspectives and requirements not known to the users who are assigned to the project team. *Joint requirements planning* sessions (Chapter 5) are highly recommending as a technique to fully involve the user community in the requirements identification process.

Prerequisites As shown in Figure 3.7, the definition phase is triggered by an approved statement of system objectives. From the system users, the team collects and discusses requirements and priorities. This information is collected by way of interviews, questionnaires, and facilitated meetings. The challenge to the team is to validate those requirements.

Notice the two-way information flow, **business requirements,** between the definition phase and the design phase. Through this information flow, *FAST* recognizes that business requirements are sometimes not discovered until some level of design or prototyping activity occurs (described below).

Activities Intuitively, the identification of business requirements is the principle activity in this phase. Your information systems framework (Figure 3.8) indicates that the definition phase explores the requirements from the system users' perspectives of the DATA, PROCESS, INTERFACE, and GEOGRAPHY building blocks. The real challenge is to organize and synchronize these requirements in a way that permits system users to validate and prioritize the business requirements.

Clearly, the new system objectives from the study phase provide some degree of validation. But detailed validation requires the analyst to translate the users' words into a more precise representation of the requirements. The system users can then validate the requirements and offer corrected requirements and priorities. The most popular approach to documenting and validating users' requirements is modeling.

> **Modeling** is the act of drawing one or more graphical (meaning *picture-oriented*) representations of a system. The resulting picture represents the users' DATA, PROCESSING, INTERFACE, or GEOGRAPHY requirements from a business point of view.

This is the "picture is worth a thousand words" approach to validation. Different models will describe data requirements, process requirements, interface requirements, and geographic requirements. (You may be familiar with computer program models such as structure charts, flowcharts, and pseudocode.) They model the modular structure and logic of a program. In this book you will learn several tools and techniques for modeling business requirements.

Another approach to documenting and validating requirements is prototyping.

> **Prototyping** is the act of building a small-scale, representative or working model of the users' requirements to discover or verify those requirements.

This is the "they'll know what they need when they see it" approach. The analyst uses powerful prototyping tools to quickly build computer-based prototypes. The users can then react to the prototype to help the analyst refine or add to the requirements. Prototypes can also be used to develop or refine the aforementioned system models.

Another activity in the definition phase is to prioritize requirements. This can be accomplished during or after the other activities. At the highest level, requirements can be classified as mandatory, desirable, or optional. Within these classifications, it may be necessary to rank the requirements to break the implementation into versions, the first of which could relatively quickly provide some value to the business.

Deliverables The final models and prototypes are usually organized into a business requirements statement. Some approaches call for great detail in this requirements statement, whereas others emphasize "the big picture." *FAST* supports both approaches based on the complexity and schedule of the project. We'll examine these approaches in Chapters 4 through 9. The requirements statement becomes the trigger for systems design.

Postrequisites and Feasibility Checkpoints Although it is rare, the project could still be canceled at the end of this phase—hence, the feasibility notation at the beginning of the business requirements information flow.

More realistically, the project scope (or schedule and budget) could be adjusted if it becomes apparent the new system's requirements are much more substantive than originally anticipated. Today, it is popular to time box a project based on the business requirements.

> **Time boxing** is a technique that divides the set of all business requirements for a system into subsets, each of which will be implemented as a version of the system. Essentially, the project team guarantees that new versions will be implemented on a regular and timely basis.

Time boxing is a modern response to historical system owner and user complaints that the Information Services unit of the business takes too long to develop systems. Usually, the time box is defined to be six to nine months in duration for each release. (Note: The packaged software industry has long used the time box approach to define and manage the release of new software versions.)

Regardless, if the project is not canceled, it proceeds to the targeting and design phases. Actually, the targeting phase can begin before we complete the definition phase because the targeting phase is not dependent on complete, detailed business requirements. Also, as previously noted, the design phase may have already started if we use prototyping techniques to define requirements.

Impact Analysis You can never skip the definition phase. One of the most common complaints about new systems and applications is that they don't really satisfy the users' needs. This is usually because information systems specialists found it difficult to separate "what" the user needed from "how" the new system would work. In other words, they became so preoccupied with the technical solution that they failed to consider the users' essential requirements. Requirements are a statement of what the system must do no matter how you design and implement it.

The definition phase formally separates "what" from "how" to properly define and prioritize those requirements.

FAST recognizes the need for more rapid system and application development, but the methodology also insists, "If you don't have time to do it right, how will you find (and justify) the time to do it over?"

The Configuration Phase

System design and construction are detailed, technical, and time-consuming phases. Given any statement of business requirements, there are usually numerous alternative ways to design the new system. Some of the pertinent questions include the following:

- How much of the system should be computerized?
- Should we purchase software or build it ourselves (called the make-versus-buy decision)?
- Should we design the system for centralized computing on a mainframe or minicomputer, or should we design a distributed computing solution utilizing PCs, server computers, and networks?
- What (emerging) information technologies might be useful for this application?

The questions are answered in the next phase of a *FAST* project, *configure a feasible system solution*.

Purpose There are almost always multiple candidate solutions to any set of business requirements. The purpose of the **configuration phase** is to identify candidate solutions, analyze those candidate solutions, and recommend a target system that will be designed and implemented.

Participants and Roles The systems analyst usually facilitates this phase. Alternatively, *FAST* allows information services to designate an *application architecture manager* to facilitate this phase for all projects.

All members of the project team including system owners, system users, and system designers must be involved in this key decision-making phase. Each class of participant brings its own ideas and perspectives to the table. Clearly, this is the first phase that involves system designers such as database administrators, software engineers or programmers, human interface specialists, and network administrators. Technology consultants may also become involved.

Prerequisites As shown in Figure 3.7, the configuration phase is triggered by a reasonably complete specification of business requirements. These business requirements are usually not extremely detailed; they identify inputs, outputs, databases, key functions or programs, and so forth. They do not typically specify precise details of those elements. In fact, the configuration phase begins before completion of the definition phase and may end after the design phase has already begun.

The project team also solicits ideas and opinions from a diverse audience. The project team also identifies or reviews any technology standards via the technology-oriented system owners.

Activities In your information system framework (Figure 3.8), the configuration phase is the first phase that begins to look at the perspectives of system designers. Also notice that the configuration phase is significantly influenced by the existing technology and standards that exist for the technology (shown at the bottom of the framework).

The first activity is to define the candidate solutions. Some candidate solutions will be proposed as design ideas and opinions by various sources (e.g., systems analysts and system designers, other information system managers and staff, technical consultants, system users, or information technology vendors). Some technical choices may be limited by a predefined approved technology architecture provided by system managers.

After defining candidates, each candidate is evaluated by the following criteria:

- *Technical feasibility.* Is the solution technically practical? Does our staff have the technical expertise to design and build this solution?
- *Operational feasibility.* Will the solution fulfill the user's requirements? To what degree? How will the solution change the user's work environment? How do users feel about such a solution?
- *Economic feasibility.* Is the solution cost-effective (as defined earlier in the chapter)?
- *Schedule feasibility.* Can the solution be designed and implemented within an acceptable time period?

The final activity is to recommend a feasible candidate as the target system. Infeasible candidates are immediately eliminated from further consideration; however, several alternatives usually prove to be feasible. The project team is usually looking for the *most* feasible solution—the solution that offers the best combination of technical, operational, economic, and schedule feasibility.

Deliverables The key deliverable of the configuration phase is a formal *systems proposal* to system owners. The double-ended arrows indicate the system proposal must be presented, and usually negotiated, with the system owners who will usually make the final business and financial decisions. This proposal may be written or verbal.

If it is decided to purchase some or all of the target system (hardware or application software), the technology requirements must be forwarded to the purchasing phase (as shown on Figure 3.7).

Regardless, the solution design requirements must be provided to the design phase (which may already be in progress). These design requirements are not specifications. They are high-level architectural decisions that will constrain the detailed design of the system.

Postrequisites and Feasibility Checkpoints Clearly, several outcomes are possible from the configuration phase. Notice the deliverables are all marked as optional (circles) based on the feasibility analyses described earlier. Specifically, system owners might choose any one of the following options:

- Approve and fund the system proposal (possibly including an increased budget and timetable if scope has significantly expanded).
- Approve or fund one of the alternative system proposals.
- Reject all the proposals and either cancel the project or send it back for new recommendations.
- Approve a reduced-scope version of the proposed system.

Based on the decision, a procurement phase may be triggered. Also, based on the decision, the design phase (possibly already in progress) may be canceled or modified in scope or direction.

Impact Analysis The configuration phase is not always required. Like many modern methodologies, *FAST* encourages businesses to develop an application architecture.

> An **application architecture** defines an approved set of technologies to be used when building any new information system.

Using *FAST*'s application architecture project template, SoundStage has already defined its approved technologies. For example, it has standardized on Microsoft *Access* as its PC database technology and Microsoft *SQL Server* as its enterprise database engine. It has also standardized its approved programming languages, network topologies, and user and system interface technologies. We'll see SoundStage apply these standards in its member services project.

Businesses that apply an application architecture must be willing to invest the time and effort to continuously improve that architecture based on business and technological change.

The Procurement Phase

This phase is missing from many methodologies. College graduates are often shocked to discover the high percentage of computer software that is purchased (or leased) rather than built. Also, any new system may present the need to acquire additional hardware, such as personal computers or printers. Recall that the make-versus-buy decision was made in the configuration phase. If the decision includes a "buy" component, then the next phase of systems design is to purchase any new hardware and software. It is sometimes called the acquisition or purchasing phase; *FAST* calls it the **procurement phase.** This phase is only required if new technology needs to be purchased.

Purpose The purpose of the procurement phase is to research the information technology marketplace, solicit vendor proposals, and recommend (to management) the proposal that best fulfills the business and technology requirements. Why include this phase in a methodology? The selection of hardware and software takes time. Much of that time can occur between order and delivery. This time lag must be figured into the methodology to schedule the subsequent life cycle phases!

Participants and Roles The key facilitator in the procurement phase is still the systems analyst; however, several other parties get involved. Clearly, the information technology vendors (who sell hardware and/or software) get involved. Also, users (both those on the project team and those in the user community) must be involved since they must ultimately live with the system. System owners must be involved because these purchases usually exceed the authorized spending limits of the average project team. In most businesses, purchasing agents and legal staff must be involved in negotiations for any contracts and service agreements.

Prerequisites As shown in Figure 3.7, the key inputs to the phase are business requirements from the definition phase and technology requirements from the configuration phase. The project team should also be aware of any technology standards imposed by systems management.

Activities In your information system framework (Figure 3.8), the procurement phase considers the perspectives of system designers. It should be noted, however, that the activities of the procurement phase ultimately result in the selection of one or more of the information technologies illustrated at the bottom of the information system framework. The most common purchased component is an application software package to implement the PROCESS column. But in some cases, the project team may only purchase the technologies required to design and construct the system (for example, a database management system, a new programming language, or a local area network operating system).

The project team's initial activity is to research the technology and marketplace. Although initial inquiries might be made to information technology vendors, the first source of information is usually gained by studying the many periodicals and services that survey the technology marketplace.

Subsequently, the project team organizes the business, technology, and relationship requirements and establishes the mechanisms that will be used to evaluate the technical alternatives. These requirements and mechanisms are communicated to the vendors as a **request for proposals.** The two-way arrow indicates that clarification of these requirements is normal.

The vendors usually respond with formal proposals that may also have to be clarified or negotiated. The project team must evaluate proposals and quotes to determine (1) which ones meet requirements and specifications and (2) which one is the *most* cost-effective. The analysts make a recommendation to the system owners (and usually the information system managers as well). This recommendation may also be negotiated. Finally, the authorized agents of the business execute the final orders, contracts, licenses, and service agreements.

Deliverables The key deliverable of the procurement phase (once again shown on Figure 3.7) is a technology proposal to system owners to acquire specific hardware and/or software. If that proposal is approved, then a technology integration requirements statement is passed on to the design phase. This statement describes, in detail, the approved hardware and/or software and requirements for integrating that technology into the overall system design. It is important to include any unfilled requirements in this statement since the design phase must then build in those missing features and capabilities.

Postrequisites and Feasibility Checkpoints The procurement phase is followed by the design phase unless the purchased software fully meets the business and technology requirements of the project, a very rare occurrence! The design phase will integrate the purchased system into the overall system, adding any necessary functionality and features that could not be purchased.

In the rare case where a purchased system fully meets requirements (sometimes called a *turn-key system* because you just turn the key to start the system), the project proceeds immediately to the *delivery phase*.

Sometimes the procurement phase results in a "no decision." Either the users were unsatisfied with the business solution of any package, or the technical staff was unsatisfied with the technology or support. In this case, the project proceeds directly to the design phase to be designed and constructed in-house as a custom solution. That explains the optional notation on the technology integration requirements information flow.

It is rare that the entire project would be canceled at this stage; however, if a purchased package was targeted as the *only* feasible solution, and no satisfactory package could be found, then the project could be canceled.

Impact Analysis This phase is entirely optional based on the make-versus-buy decision in the configuration phase. If the decision is to buy, shortcuts through *FAST*'s recommended procurement activities (covered in detail later in the book) are strongly discouraged. You tend to get what you pay for and only what you pay for in the software industry. Unfortunately, the software industry is somewhat notorious for unfulfilled promises; it pays to thoroughly research, evaluate, and negotiate all agreements.

The Design Phase

Given the approved, feasible solution from the configuration phase, you can finally design and integrate the target system. You understand what the business requirements are from the definition phase and how you plan to fulfill those requirements from the configuration phase. Thus, you can now justify the time and cost to fully design the new system. Ideally, any new system should work in harmony with other current information systems. Similarly, if we have purchased software packages, those packages must work in harmony with any components of the systems that are to be built in-house. For these reasons, we not only design the new system but we also integrate the new system as well.

Purpose The purpose of the **design phase** is to transform the business requirements from the definition phase into a set of technical design blueprints for construction. *FAST* does not support the traditional design-then-construct strategy wherein you complete a detailed design before starting construction. Instead, *FAST* encourages an iterative design-and-construct strategy. Some specification must precede the first pass through the design-and-construct approach; however, *FAST* encourages that most of the design be tested by developing a series of prototypes that evolve into the final system. This contemporary strategy is called rapid application development or RAD.

Participants and Roles The key facilitator of the design and integration phase is still the systems analyst. But various other design specialists play important roles. For instance, database specialists might design or approve the design of any new or modified databases. Network specialists might design or modify the structure of any computer networks. Microcomputer specialists may assist in the design of workstation-based software components. Human interface specialists may assist in the design of the user interface. And as always, the system users must be involved. They evaluate the new system's ease of learning, ease of use, and compatibility with the stated business requirements.

Prerequisites As shown in Figure 3.7, the design phase has two triggers: the business requirements (from the definition phase) and the design requirements (from the configuration phase). In those projects that will purchase hardware and/or software, the design phase also receives technology integration requirements (from the purchasing phase). System users provide various ideas and opinions into or about the system's design.

When appropriate and cost-effective, the design phase can employ consulting services from information technology vendors. For example, if a specific vendor's bar-coding technology has been chosen for the target system, the vendor of that technology might be engaged to help integrate the technology into the target system.

Activities In your information system framework (Figure 3.8), and as expected, the design phase builds views of the system from the perspective of the system designers. Various design specialists design databases, programs, user interfaces, and networks to support the new system.

As noted earlier, *FAST* is like many contemporary methodologies in that it has effectively merged the design and construction phases to form a **rapid application development** (or **RAD**) approach based on iterative prototyping. Shown more clearly in Figure 3.7, this strategy designs and constructs the system as a series of **prototypes** to which the system users react. (Notice the loop between the design and construction phases, and the two-way arrow between the design and definition phases.)

Prototyping was introduced in the discussion of the definition phase. Design by prototyping allows the analyst to quickly create scaled-down but working versions of a system or subsystem. In the interest of speed, certain features and capabilities, such as input editing, may be left out of prototypes. Prototypes will typically go through a series of iterations and user reviews until they evolve into an acceptable design. The entire process works something like this:

1. Define the base-level scope of the first (or next) version of the system.
2. Define, design, and construct the database (only!). Load that database with some test data and review it with the system users. Make any corrections they request (to the test database only!).
3. Define, design, and construct the inputs (only!). Demonstrate this prototype to the system users. Repeat step 3 until system users are satisfied. If necessary, return to step 1 to add new requirements to the database design.

4. Define, design, and construct the outputs (only!). Demonstrate this prototype to system users. Repeat step 4 until system users are satisfied. If necessary, return to step 1 to add new database requirements or step 2 to add new input requirements.
5. Define, design, and construct the interface. The interface is the glue that integrates all the above system components. Demonstrate this prototype to system users. Repeat step 5 until system users are satisfied. If necessary, return to step 1, 2, or 3 to add new database, input, or output requirements, respectively.
6. Design and construct any missing system controls such as security, backup, recovery, etc.
7. Implement this version of the system.
8. Go to step 1 to begin the RAD cycle for the next version of the system.

The basic idea of RAD is (1) to actively involve system users in the design process, (2) to accelerate the definition/design/construction process by catching errors and omissions earlier in the process, and (3) to reduce the amount of time that passes before the users begin to see a working system. RAD has become the strategy of choice in modern design methodologies.

Deliverables As described above, the design and construction phases are effectively merged in the *FAST* methodology. This section will describe only the design deliverables of the phase. The final deliverable is a technical set of design specifications. These specifications can take several forms, but the most common approach is modeling. (Modeling was introduced in the definition phase description.) Normally, general design models will depict:

- The structure of the database.
- The structure of the overall application.
- The overall look and feel of the user interface.
- The structure of the computer network.
- The design structures for any complex software to be written.

Postrequisites and Feasibility Checkpoints Once again, in a *FAST* project the design and construction phases are effectively merged. Thus, a project is rarely canceled in the design phase, unless it is hopelessly over budget or behind schedule. As shown in Figure 3.7, each constructed prototype is refined or expanded by another pass through system design until the final system is constructed.

Impact Analysis Not surprisingly, the design phase cannot be skipped. In the true spirit of classical engineering, there is no such thing as a prototype that was not preceded by some level of design.

The Construction Phase

We just learned that in the *FAST* methodology, the **construction phase** is actually part of a design/construction loop that implements rapid application development. Given some aspect of the system design, we construct and test the system components in that design. After several iterations of the design/construction loop, we will have built the functional system to be implemented.

Purpose The purpose of the construction phase is twofold: (1) to build and test a functional system that fulfills business and design requirements, and (2) to implement the interfaces between the new system and existing production systems.

Participants and Roles The construction phase is still facilitated by the systems analyst; however, the analysts and users often play a less visible role. The analyst serves as a general contractor for work done by technical specialists or subcontractors. Examples include database programmers, application programmers, and network administrators. (Keep in mind that *system builders* are roles, not job descriptions. In practice, it is common for a single person to play the role of both designer and builder.) System users' responsibilities are usually limited to reacting to the functional system's ease of learning and ease of use.

Prerequisites As shown in Figure 3.8, the design specifications (general or detailed) are the key input to the construction phase. Information technology vendors may provide installation support for any packaged software or software development tools.

Activities Your information system framework (Figure 3.8) identifies the relevant building blocks and activities for the construction phase. The project team must construct the database, application programs, user and system interfaces, and networks. Some of these elements may already exist (subject to enhancement).

Usually, the database and networks provide the system's infrastructure; therefore, they must be constructed first (unless they already exist). Next, any new software packages must be installed and tested. Finally, any new programs must be constructed and tested. You may already have some experience with the principal technique in this activity—application programming. Programs can be written in many different languages, but the current trend is toward the use of visual and object-oriented programming languages such as Powersoft's *Powerbuilder,* Microsoft's *Visual BASIC,* Borland's *Delphi,* Digitalk's *Visual Smalltalk,* Sun's *Java,* and Microsoft's *Visual C++*. (Note: Visual and object-oriented versions of our old standby, *COBOL,* are also beginning to emerge.)

One of the most important aspects of application programming is testing—both unit and system testing.

> **Unit tests** ensure that the applications programs work properly when tested in isolation from other applications programs.
>
> **System tests** ensure that applications programs written in isolation work properly when they are integrated into the total system.

It is common for programs that work perfectly by themselves to fail to work when combined with other related programs. If this happens, the programmer must often return to the build/test phase.

The install/test phase is also used to install purchased or leased software packages. Even those packages must be system tested to ensure they properly interact with other programs and packages. Furthermore, the programmer may have built special programs to connect the package with other programs, files, and databases. These integrated solutions must also be system tested.

Deliverables The final deliverable of the construction phase is the functional system; however, the rapid application development strategy of *FAST* results in several interim deliverables called prototypes. The prototypes are subject to demonstrations and feedback from the system users. Each prototype then cycles back through the design and definition phase until the final, functional system is acceptable.

Postrequisites, Feasibility Checkpoints, and Impact Analysis After the functional system has been completed, all that remains is the implementation phase. Although Figure 3.7 marks this phase as optional, the project is rarely canceled after the construction phase. It is possible that a prototype might be implemented as a first (next) version before the system has been fully constructed.

The Delivery Phase

What's left to do? New systems usually represent a departure from the way business is currently done; therefore, the analyst must provide for a smooth transition from the old system to the new system and help users cope with normal start-up problems. Analysts must also train users, write various manuals, and load files and databases. Thus, the last phase of implementation is to deliver the production system into operation.

Purpose The purpose of the **delivery phase** is to install, deploy, and place the new system into operation or production.

Participants and Roles The system analysts are still the facilitators but again become the most visible players as they communicate implementation problems and issues among system users, system designers, and system builders. The entire project team, inclusive of owners, users, designers, and builders, is active in this phase. Ideally, system owners and users step to the forefront as cheerleaders for the new system.

Prerequisites The functional information system is the key input to the delivery phase of the functional system (see Figure 3.7). System users provide continuous feedback as new problems and issues are common (note: no system has achieved the nirvana goal of perfection). For new information technology (hardware and software), the information technology vendors provide necessary technical support.

Activities In your information system framework (Figure 3.8) the delivery phase considers the system builders' perspective. During this phase, tests are conducted to ensure that the new system works properly. A conversion plan must then be prepared to provide a smooth transition to the new system. This plan may call for an abrupt cutover where the old system is terminated and replaced by the new system on a specific date. Alternatively, the plan may run the old and new systems in parallel until the new system has been monitored and deemed acceptable to replace the old system. Prior to converting to the new system, the system builders must install databases to be used by the new system. This involves loading production data to be used by the delivered system.

The delivery phase involves training individuals that will use the final system and developing documentation to aid the system users. The delivery phase normally concludes with a postaudit to gauge the success of the completed systems project. This activity represents a TQM effort that will contribute to the success of future systems projects.

Deliverables The final deliverable of the delivery phase (and the project) is the production system for the system users. Various follow-up reports or meetings (not depicted in Figure 3.7) are recommended to evaluate and improve both the final product (the information system), the process used to build it (the *FAST* methodology), and the participants (the project team).

Another output of the delivery phase is training and support. Training includes user documentation and formal classes. Support includes day-to-day assistance to system users when they encounter problems and undiscovered errors.

Postrequisites, Feasibility Checkpoints, and Impact Analysis The project is complete! There is no further feasibility analysis. But, as noted earlier, there may be a project postaudit to evaluate the system, methodology, and team. The system is continuously evaluated to log problems and new requirements that can be fixed in future versions. The system is also evaluated to determine whether it truly solved the intended problems and exploited the opportunities. The methodology is evaluated to reuse and improve the tools, techniques, and decision making of the systems development activities. The team is evaluated to assess education and training needs that can improve future performance.

Beyond Systems Development—Systems Support

Once the system is placed into production, the analyst's role changes to systems support. In fact, a significant portion of most systems analysts' time and effort is spent providing ongoing support for existing systems.

Systems support is the ongoing maintenance of a system after it has been placed into operation. This includes program maintenance and system improvements.

In your information system framework, systems support maintains all the building blocks (and their documentation) for a production information system. Systems analysts usually coordinate systems support, calling on the services of maintenance programmers and system designers as necessary. Systems support doesn't consist of phases so much as it does ongoing activities. These activities include:

- *Fixing software bugs.* Software bugs are errors that slipped through the testing phases during software construction.
- *Recovering the system.* From time to time, a system failure will result in an aborted program or loss of data. This may have been caused by human error or a hardware or software failure. The systems analyst may then be called on to recover the system—that is, to restore a system's files and databases, and to restart the system.
- *Assisting users.* Regardless of how well the users have been trained and how good the end-user documentation is, users will eventually require additional assistance—unanticipated problems arise, new users are added, and so forth.
- *Adapting the system to new requirements.* New requirements may include new business problems, new business requirements, new technical problems, or new technology requirements.

All these support activities continue through the lifetime of the production system. In *FAST,* the support activities are supported by a simplified version of the same phases used to build the system.

CROSS LIFE CYCLE ACTIVITIES

Systems development also involves a number of activities called cross life cycle activities.

Cross life cycle activities are activities that overlap many or all phases of the methodology. In fact, they are normally performed in conjunction with several phases of the methodology.

Cross life cycle activities include fact-finding, documentation and presentation, estimation and measurement, feasibility analysis, project management, and process management. Let's briefly examine each of these activities. The cross life cycle nature of these activities is illustrated in the time bar chart in Figure 3.9.

Fact-Finding

There are many occasions for fact-finding (or information gathering) during the project.

Fact-finding—also called information gathering or data collection—is the formal process of using research, interviews, meetings, questionnaires, sampling, and other techniques to collect information about systems, requirements, and preferences.

Fact-finding is most crucial to the survey, study, and definition phases of a *FAST* project. It is during these phases that the project team learns about a business's and system's vocabulary, problems, opportunities, constraints, requirements, and priorities.
Fact-finding is also used during the configuration, design, and construction phases but to a lesser extent. For instance, in systems design, fact-finding becomes technical as the project team attempts to learn more about the technology selected for the new system.

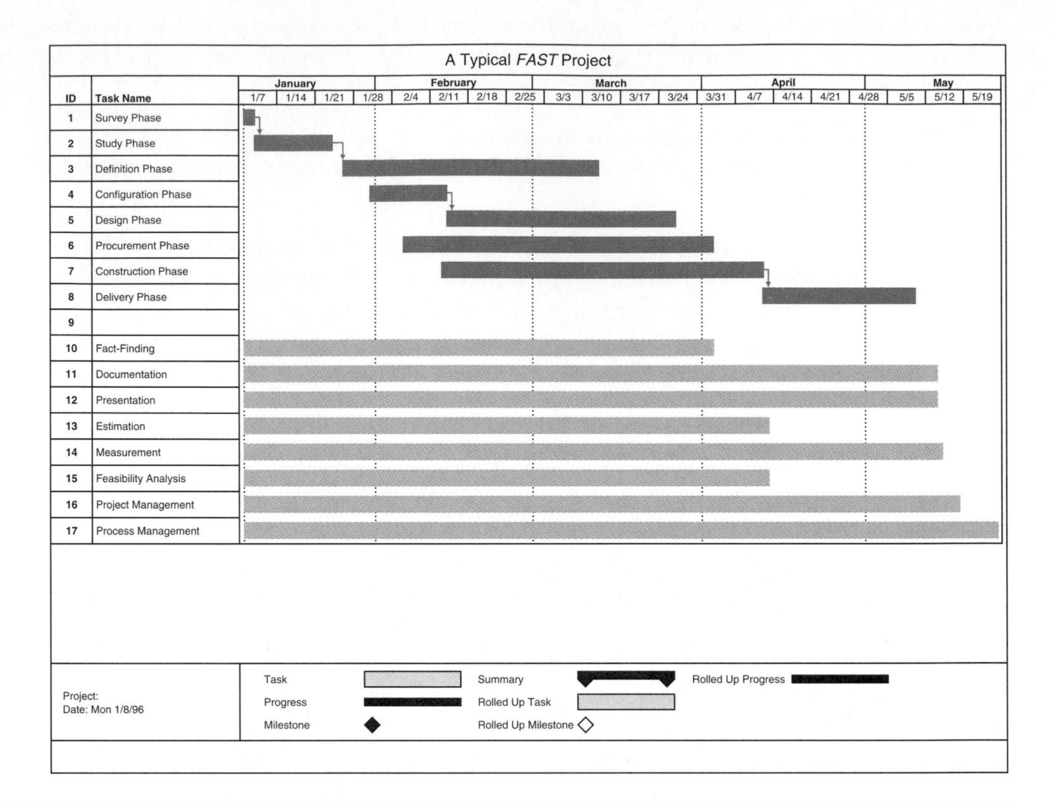

FIGURE 3.9
Cross Life Cycle Activities (and Overlap of the Other Development Phases)

Documentation and Presentations

Communication skills are essential to the successful completion of a project. Two forms of communication that are common to systems development projects are documentation and presentation.

> **Documentation** is the activity of recording facts and specifications for a system.
>
> **Presentation** is the related activity of formally packaging documentation for review by interested users and managers. Presentations may be either written or verbal.

Clearly, documentation and presentation opportunities span the entire *FAST* methodology and support all the phases.

As described throughout this chapter, documentation for a project is built during various phases and activities. Over time, a sizable base of documentation for various systems and applications is compiled. Many businesses try to get control over such documentation so it is kept and maintained for future use. Version control over documentation has become a critical success factor; it involves keeping and tracking multiple versions of a system's documentation. At a minimum and at any given time, most information systems shops want to keep documentation for all the following versions:

- One or more previous versions of the system.
- The current production version of the system.
- Any version of the system going through the build and test activity.
- Any version going through the life cycle to create a new version.

In Figure 3.7, the arrows on the flow-oriented models have been used to represent the typical use and updating of project documentation and the presentation of documentation to various audiences by systems analysts and programmers. In practice, these information flows don't go from phase to phase. Instead, interim results are stored in the repository, database, and program libraries introduced earlier in the chapter (refer back to Figure 3.6).

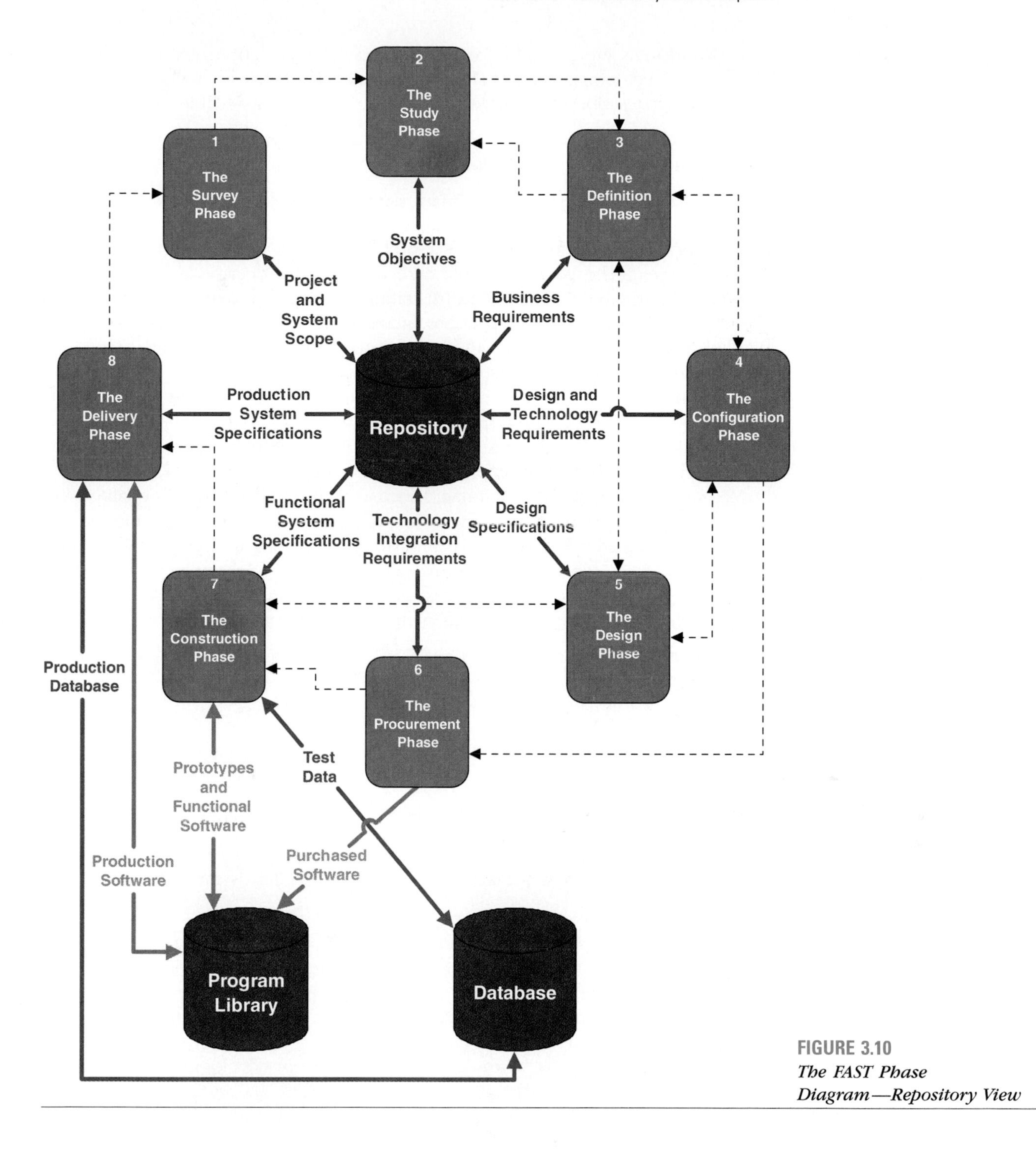

FIGURE 3.10
The FAST Phase Diagram—Repository View

To better illustrate the repository concept, the *FAST* methodology phases are redrawn in Figure 3.10 to illustrate the sharing of knowledge and system components. To simplify this diagram, we (1) used the phase nicknames and (2) eliminated the interpersonal communications information flows. The dashed lines show the flow of activities through the project. Because the phases communicate across the common data stores, they can overlap, thus accelerating the project.

Estimation and Measurement

Information systems are significant capital investments. For this reason, estimation and measurement activities are commonly performed to address the quality and productivity of systems.

Estimation is the activity of approximating the time, effort, costs, and benefits of developing systems. The term guesstimation (as in "make a guess") is used to describe the same activity in the absence of reliable data.

Measurement is the activity of measuring and analyzing developer productivity and quality (and sometimes costs).

Estimating is an extremely important activity because information systems, as noted earlier in the chapter, are capital investments. You don't want to spend $25,000 of time and effort to solve a problem that is costing your organization $2,000 per year. The payback would take more than a decade.

Estimating can be a difficult and frustrating activity. It is difficult to translate an abstract problem statement into a precise estimate of the time, effort, and costs needed to solve that problem. Multiple factors influence the estimate. There are two common approaches to estimation. First, some analysts avoid estimation out of fear, uncertainty, or lack of confidence. In this case, the analyst may resort to what are jokingly called "guesstimates." Alternatively, better analysts draw on experience and data (both their own and the collective experience of others) from previous projects to continually improve their estimates.

Measurement has become important because of the productivity and quality problems that plague systems development. In response to those problems, the industry has developed methods and tools to improve both quality and productivity. These methods and tools can be costly. Formal measurement of development productivity may be the only way to justify the cost of these expenses.

The field of software and systems metrics offers hope for the future.

Software and systems metrics provides an encyclopedia of techniques and tools that can both simplify the estimation process and provide a statistical database of estimates versus performance.

Feasibility Analysis

A system development life cycle that supports our creeping commitment approach to systems development recognizes feasibility analysis as a cross life cycle activity.

Feasibility is a measure of how beneficial the development of an information system would be to an organization.

Feasibility analysis is the activity by which feasibility is measured.

Too many projects call for premature solutions and estimates. This approach often results in an overcommitment to the project. If analysts are so accurate in feasibility estimates, why then are so many information system projects late and over budget? Systems analysts tend to be overly optimistic in the early stages of a project. They underestimate the size and scope of a project because they haven't yet completed a detailed study.

A project that is feasible at any given stage of system development may become less feasible or infeasible later. For this reason, we use the creeping commitment approach to reevaluate feasibility at appropriate checkpoints (indicated by small circles) in Figure 3.7.

Various measures of feasibility were introduced in the targeting phase. These measures included technical feasibility, operational feasibility, economic feasibility, and schedule feasibility.

Project and Process Management

Systems development projects may involve a team of analysts, programmers, users, and other IS professionals who work together.

Project management is the ongoing activity by which an analyst plans, delegates, directs, and controls progress to develop an acceptable system within the allotted time and budget.

Failures and limited successes of systems development projects far outnumber very successful information systems. Why is that? One reason is that many systems analysts are unfamiliar with or undisciplined in the tools and techniques of systems development. But most failures are attributed to poor leadership and management. This mismanagement results in unfulfilled or unidentified requirements, cost overruns, and late delivery.

The systems development life cycle provides the basic framework for the management of systems projects. Because projects may be quite large and complex, the life cycle's phased approach to the project results in smaller, more measurable milestones that are more easily managed.

Process management is an ongoing activity that establishes standards for activities, methods, tools, and deliverables of the life cycle.

Process management is a relatively new concept in systems development. The intent is to standardize both the way we approach projects and the deliverables we produce during projects.

COMPUTER-AIDED SYSTEMS ENGINEERING (CASE)

You may be familiar with the old story of the cobbler (shoemaker) whose own children had no shoes. That situation is not unlike the one faced by systems developers. For years we've been applying information technology to solve our users' business problems; however, we've been slow to apply that same technology to our own problem—developing information systems. In the not-too-distant past, the principal tools of the systems analyst were paper, pencil, and flowchart template.

Today, an entire technology has been developed, marketed, and installed to assist systems developers. Chances are that your future employer is using or will be using this technology to develop systems. That technology is called computer-aided systems engineering.

Computer-aided systems engineering (CASE) is the application of information technology to systems development activities, techniques, and methodologies. *CASE tools* are programs (software) that automate or support one or more phases of a systems development life cycle. The technology is intended to accelerate the process of developing systems and to improve the quality of the resulting systems.

Some people refer to this as computer-aided *software* engineering, but software is only one component of information systems. (Recall your information system building blocks.) Thus, we prefer the broader context of the term *systems*.

CASE is not a methodology (although some vendors sell it as such). Nor is CASE an alternative to methodologies. Instead, it is an enabling technology that supports a methodology's preferred strategies, techniques, and deliverables.

The term *systems engineering* is based on a vision for CASE technology—that systems development can and should be performed with engineering-like precision and rigor. In its broadest context, the use of CASE technology automates your entire systems development methodology. *FAST* recommends CASE tools to implement all its phases and activities. Let's take a look at the CASE concept and survey some representative product categories and tools.

The History and Evolution of CASE Technology

The concept of using computers to automate systems development is not new. In a sense, common language compilers and interpreters can be thought of as CASE tools. Let's briefly examine the history and evolution of this technology, a technology that will almost certainly affect your future.

The true history of CASE dates to the early- to mid-1970s. The ISDOS project, under the direction of Dr. Daniel Teichrowe at the University of Michigan, developed a language called *Problem Statement Language (PSL)* for describing user problems and solution requirements for an information system into a computerized dictionary. A companion product called *Problem Statement Analyzer (PSA)*

FIGURE 3.11
A Typical CASE Workbench

was created to analyze those problems and requirements for completeness and consistency. *PSL/PSA* ran on large mainframe computers and consumed precious and expensive machine resources. Few companies could afford to dedicate computer resources to *PSL/PSA*.

The real breakthrough came with the advent of the personal computer. Not long after, in 1984, an upstart company called Index Technology (now known as Intersolv) created a PC software tool called *Excelerator*. Its success established the CASE acronym and industry. Today, hundreds of CASE products are available to various systems developers.

Most CASE products run on personal computers or UNIX workstations. In some environments, these workstations are networked to shared CASE tools and a common repository. A typical CASE workbench is shown in Figure 3.11.

CASE technology bears a remarkable similarity to another engineering technology: computer-aided design/computer-aided manufacturing (CAD/CAM). Modern engineers use CAD tools to design and analyze new products. CAM tools then automatically generate the computer programs that will run the shop floor machinery needed to produce the design. CASE seeks to do for information system developers what CAD/CAM does for engineers: help them design better products (systems) and automatically generate the computer programs.

A CASE Tool Framework

Let's base our CASE tool framework on a concept that is already familiar to you, the systems development life cycle. In other words, we will classify tools according to which phases of the life cycle they support. Our CASE framework is based on the following popular terminology:

> The term **upper-CASE** describes tools that automate or support the upper or earliest phases of systems development—the survey, study, definition, and design phases.

> The term **lower-CASE** describes tools that automate or support the lower or later phases of systems development—detailed design, construction, and delivery (and also support).

There is some overlap between upper- and lower-CASE tools. This is because our profession has never reached agreement on when systems analysis ends and when systems design begins.

A typical business's complete CASE tool kit should include one or more products from each category. Although some CASE vendors offer an integrated CASE product family that rather comprehensively covers all categories, it is highly unlikely that any firm will find a single source for every CASE tool they need, or might want to use.

What about SoundStage? The *FAST* methodology recommends that CASE tools be used, but does not specify which CASE tools should be used. SoundStage has used various CASE tools, but it is currently using Popkin Software's *System Architect* as its upper-CASE tool, and Powersoft's *Powerbuilder* as its lower-CASE tool. As described in the chapter-opening minicase, they have decided to pilot (meaning "try") Borland's *Delphi* rapid application development environment as an object-oriented alternative to *Powerbuilder* for the new Member Services system.

CASE Tool Architecture

How is a CASE tool structured? At the center of any true CASE tool's architecture is a database called a repository (or a link into such a repository). Around that repository is a collection of tools or facilities to create documentation or other system components. Let's briefly examine this architecture.

Repositories The repository concept was introduced earlier as a data store for all the knowledge, specifications, and documentation by-products created during a systems development project.

Our experience suggests that most people are first attracted to CASE tools by their graphics capabilities (see Figure 3.12). "Wow! Here is a neat way to produce high-quality entity relation diagrams or system flowcharts!" But the real power of

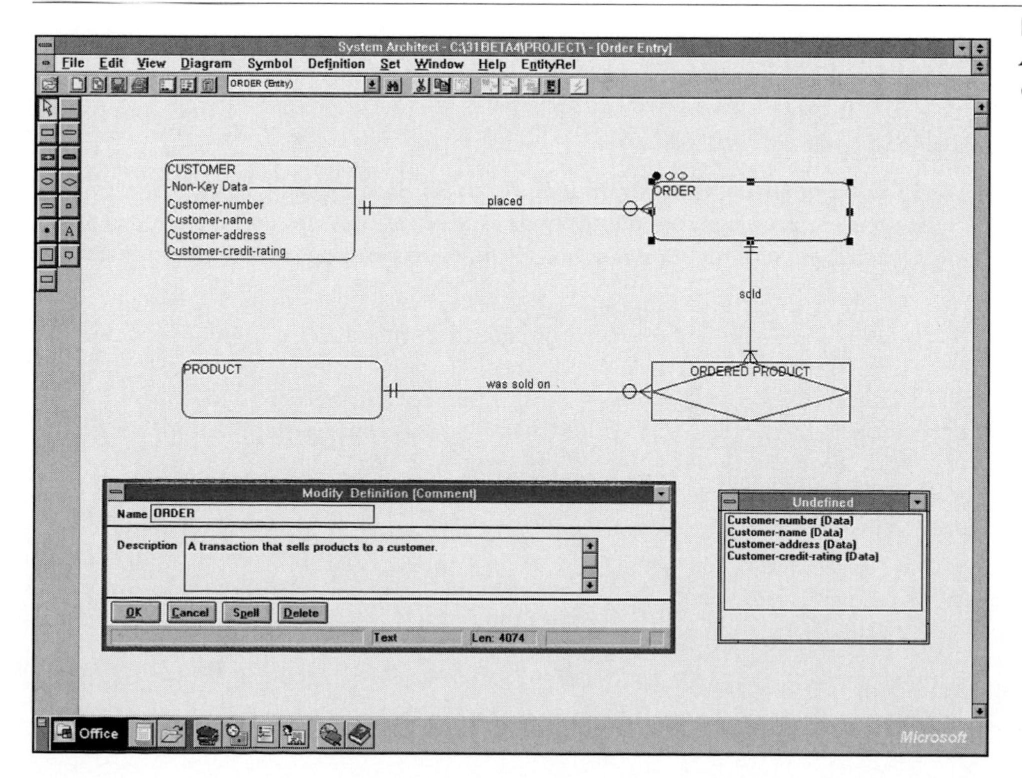

FIGURE 3.12
A CASE Tool Screen (Popkin System Architect*)*

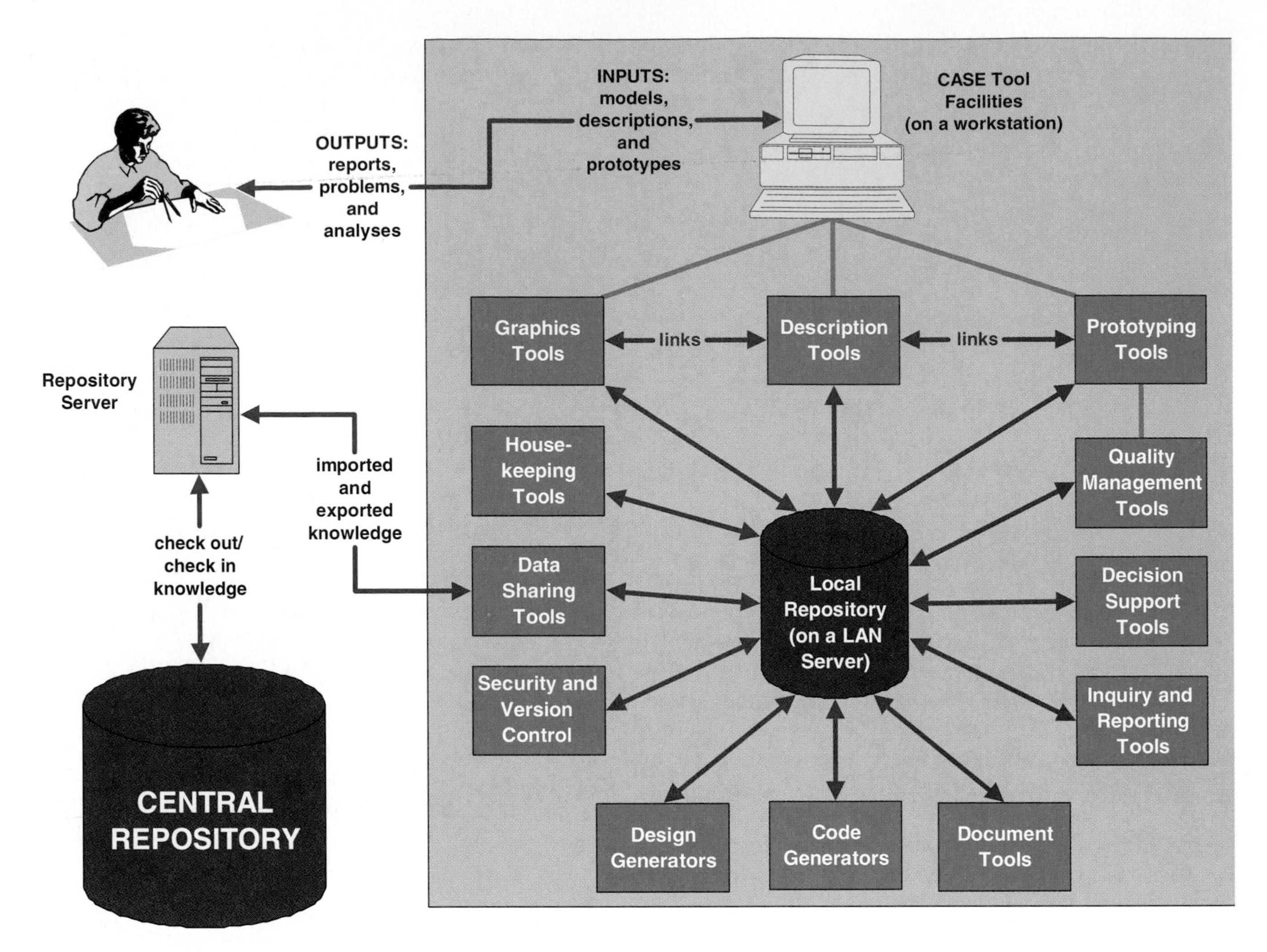

FIGURE 3.13 *CASE Architecture*

a true CASE tool is derived from its repository (or its ability to use and update some other tool's repository).

> A **CASE repository** is a developers' database. It is a place where the developers can store diagrams, descriptions, specifications, and other by-products of systems development. Synonyms include **dictionary** and **encyclopedia.**[5]

Many different CASE tools can share information across a single repository.

Typically, a mature development organization maintains a central repository of all system documentation developed since it started using CASE tools. This central repository (Figure 3.13) serves as a library for all projects. Existing repository information is then "checked out" to a work group repository for the project team. This work group repository might be stored on the team's local area network server. After various project milestones, or after the project is completed, the work group repository's information is "checked back into" the central repository.

Facilities and Functions To use the repository, the CASE tools provide input and output facilities and functions. Representative CASE tools provide some of the following facilities, illustrated in Figure 3.13:

[5] The *System Architect* CASE tool used by SoundStage calls its repository an encyclopedia.

- *Diagramming tools* are used to draw the system models required or recommended in most methodologies. Usually, the objects on one model can be linked to other graphical models and to descriptive details.
- *Description tools* are used to record, delete, edit, and output nongraphical documentation and specifications. As shown in Figure 3.13, the descriptions can be created and accessed directly by the description tools or directly from the objects on the aforementioned diagrams.
- *Prototyping tools* are used to construct system components including inputs, outputs, and programs. Today, most of these prototypes can evolve into the final, working system. As shown in the figure, the prototypes can directly use and maintain the models and descriptions in the repository.
- *Inquiry and reporting tools* are used to extract models, descriptions, and specifications from the repository.
- *Quality management tools* analyze models, descriptions, and prototypes for consistency, completeness, or conformance to accepted rules of the methodologies the CASE tools support. Some CASE tools can identify and report quality errors as they occur.
- *Decision support tools* provide information for various decisions that occur during systems development. For example, some CASE tools help systems analysts estimate and analyze feasibility.
- *Documentation organization tools* are used to assemble, organize, and report repository information that can be reviewed by system owners, users, designers, and builders. Based on customizable templates, they implement the deliverables of the chosen methodology.
- *Design generation tools* automatically generate first-draft designs for various system components based on the business requirements recorded in the repository and technology standards provided by the system designer. For example, many CASE tools can generate a database given from a business data model and the chosen database technology.
- *Code generator tools* automatically generate application programs or significant portions of those programs.
- *Testing tools* help the system designers and builders test databases and application programs.
- *Data sharing tools* provide for import and export of repository information to and from other software tools that cannot directly access the repository. (Increasingly, CASE tool vendors are providing direct access to their repositories, reducing the need for import and export capabilities.)
- *Version control tools* maintain the integrity of the repository by preventing unauthorized or inadvertent changes and saving prior versions of various information stored in the repository.
- *Housekeeping tools* establish user accounts, privileges, repository subsets, tool defaults, backup and recovery, and other essential facilities.

The Benefits of CASE

Most methodologies do not require CASE. But most methodologies do benefit from and recommend CASE technology be used. Some of the most commonly cited benefits include:

- Improved productivity (through automation of tasks and rapid application development).
- Improved quality (because CASE tools check for completeness, consistency, and contradictions).
- Better documentation (mostly because the tools make it easier to create and assemble consistent, high-quality documentation).

- Reduced lifetime maintenance (because of the aforementioned system quality improvements combined with better documentation).
- Methodologies that really work (through rule enforcement and built-in expertise).

These benefits do not come free. CASE can be very expensive, but some relatively inexpensive CASE tools now provide much of the functionality of CASE tools that used to cost $10,000 per workstation. Popkin's *System Architect* is one example of such a cost-effective CASE tool. But the real costs include more than the CASE software. Organizations must provide suitably configured workstations, training, and a support infrastructure to fully exploit CASE technology. Many information system organizations provide training and support through a development center.

A **development center** is a central group of information system professionals who plan, implement, and support a systems development environment for other developers. They provide training and support for both the methodology and CASE tools.

Thus, the development center might be thought of as consultants to the systems analysts, system designers, and system builders in the organization.

WHERE DO YOU GO FROM HERE?

This chapter, along with the first two, completes your context for systems analysis and design. We have described that context in terms of the three Ps—participants (with special emphasis on the systems analyst), product (the information system), and process (the system development methodology). Armed with this understanding, you are now ready to study systems analysis and/or design methods.

Depending on whether your course is an only course on the subject, a systems analysis course, or a systems design course, you can take different routes through this book. For a first systems development course or a systems analysis course, we recommend you continue your sequential path into the next unit, "Systems Analysis Methods." In Chapter 4, "Systems Analysis," you will look deeper into the survey, study, and definition phases of the *FAST* methodology. You will learn about alternative routes through those activities. Then, in Chapters 5 to 8, you will learn to apply specific tools and techniques to those systems analysis activities.

If you are in a system design course, you might want to quickly review Chapter 4 and then begin your detailed study in Chapter 9, "System Design." In Chapter 9, you will learn more about the configuration and design phases of the *FAST* methodology. Then in Chapters 10 to 16, you will learn to apply specific tools and techniques to those system design activities.

Regardless of the focus of your course, you and your instructor have a couple of additional options. You could begin by taking a detour to Part Five, "Cross Life Cycle Activities Skills." You already have the prerequisite background to study both modules in Part Five. Module A, "Project and Process Management," teaches you tools and techniques for managing a *FAST* project. Module D, "Joint Application Development," teaches you how to manage the methodology itself (and the CASE technology too).

SUMMARY

1. A **system development life cycle** is a logical process by which developers build information systems and computer applications. A **methodology** is the physical implementation of that logical life cycle including activities, roles, deliverables, and tools and techniques.
2. The following principles should underlie all system development methodologies:
   a. Get the owners and users involved.
   b. Use a problem-solving approach. The life cycle is such an approach.
   c. Establish phases and activities.
   d. Establish standards for consistent development and documentation.
   e. Justify systems as capital investments.
   f. Don't be afraid to cancel the project or revise scope.
   g. Divide and conquer.
   h. Design systems for growth and change.
3. There are two types of projects, unplanned and planned. Both types are responses to some combination of problems, opportunities, and directives.
   a. **Problems** are undesirable situations that prevent a business from achieving its purpose, goals, or objectives.
   b. **Opportunities** are chances to improve an organization even in the absence of problems.
   c. **Directives** are new requirements imposed by management or external influences.
4. Wetherbe's PIECES framework is useful for organizing problems, opportunities, and directives. The letters of PIECES correspond to *performance, information, economics, control, efficiency,* and *service.*
5. The book will teach systems analysis and design methods in the context of a typical, but simple methodology called *FAST. FAST* consists of eight phases:
   a. Survey and plan the project.
   b. Study and analyze the existing system.
   c. Define and prioritize the business requirements.
   d. Configure a feasible system solution.
   e. Procure any new hardware and software.
   f. Design and integrate the target system.
   g. Construct and test the target system.
   h. Deliver the production system.
6. During systems development, by-products are stored in the following data stores:
   a. A **repository** is a place where documentation about the system is stored.
   b. The **database** is where actual business data will be stored.
   c. The **program library** is where application software will be stored.
7. **Systems support** is the ongoing maintenance of a system after it has been placed into production.
8. **Cross life cycle activities** are activities that overlap many or all of the phases of system development. They include fact finding, documentation and presentation, estimation and measurement, feasibility analysis, project management, and process management.
9. **Computer-aided systems engineering (CASE)** is the application of information technology to systems development activities, techniques, and methodologies.
   a. **Upper-CASE** tools support the survey, study, definition, and design phases of the methodology.
   b. **Lower-CASE** tools support the design, construction, and delivery phases of the methodology.
   c. CASE tools are built around an automated **repository** (see above).
   d. A **development center** is a central group of information system professionals who plan, implement, and support a systems development environment for other developers.

KEY TERMS

application architecture, p. 92
application development life cycle, p. 72
backlog, p. 78
business process redesign, p. 78
CASE repository, p. 106
computer-aided systems engineering (CASE), p. 103
configuration phase, p. 91
construction phase, p. 96
cost-effectiveness, p. 76
cross life cycle activity, p. 99
database, p. 82
definition phase, p. 89
delivery phase, p. 98
design phase, p. 94
development center, p. 108
dictionary, p. 106
directive, p. 79
documentation, p. 100
encyclopedia, p. 106
entropy, p. 77
estimation, p. 102
fact-finding, p. 99
feasibility, p. 102
feasibility analysis, p. 102
feasibility, types of, p. 92
information strategy plan, p. 78
legacy system, p. 77
lower-CASE, p. 105
measurement, p. 102
methodology, p. 73
modeling, p. 89
opportunity, p. 79
PIECES, p. 79
planned system initiative, p. 78
presentation, p. 100
problem, p. 79
process management, p. 103
procurement phase, p. 93
production system, p. 82
program library, p. 82
project management, p. 102
prototyping, p. 90
rapid application development (RAD), p. 95

repository, p. 82
software and systems metrics, p. 102
steering committee, p. 78
study phase, p. 87
survey phase, p. 84
system development life cycle (SDLC), p. 72
systems support, p. 99
system test, p. 97
time boxing, p. 90
unit test, p. 97
unplanned system request, p. 78
upper-CASE, p. 104

REVIEW QUESTIONS

1. What is the difference between the *system development life cycle* and a *methodology*?
2. Why do companies use methodologies? Why do many choose to purchase a methodology?
3. What are the eight fundamental principles of systems development? Explain what you would do to incorporate those principles into a systems development process.
4. What is cost-effectiveness and how does its measurement impact system development?
5. What is the creeping commitment approach?
6. What is entropy and how does it impact system development?
7. What are the two triggers for a system development project?
8. Briefly describe the two types of planned system initiatives.
9. Differentiate among problems, opportunities, and directives.
10. Name the six problem classes of the PIECES framework.
11. What is a repository? How is it different from a database? How is it different from a program library?
12. Identify and briefly describe the eight high-level phases that are common to most modern system development life cycles.
13. Describe each phase of the *FAST* life cycle in terms of purpose, participants, inputs, outputs, and activities.
14. Differentiate between *modeling* and *prototyping* as requirements definition techniques.
15. What is time boxing? Why is it now popular?
16. Describe four types of feasibility.
17. What is an application architecture? How does it impact the targeting phase?
18. What's the difference between unit testing and system testing?
19. Name five cross life cycle activities.
20. What is CASE? Why is CASE not a methodology or an alternative to a methodology?
21. Differentiate between upper-CASE and lower-CASE.
22. What is the role of a repository in CASE?
23. List several common facilities that might be implemented by a CASE tool using a repository.
24. What is a development center?

PROBLEMS AND EXERCISES

1. Using the PIECES framework, evaluate your local course registration system. Do you see any problems or opportunities? (Alternative: Substitute any system with which you are familiar.)
2. Assume you are given a programming assignment that requires you to make some modifications to a computer program. Explain the problem-solving approach you would use. How is this approach similar to the phased approach of the methodology presented in this chapter?
3. Which phases of the *FAST* methodology presented in this chapter do the following tasks characterize?
   a. The analyst demonstrates a prototype of a new sequence of work order terminal screens.
   b. The analyst observes the order-entry clerks to determine how a work order is currently processed.
   c. The analyst develops the internal structure for a database to support work order processing.
   d. An analyst is teaching the plant supervisor how to inquire about work orders using the new microcomputer.
   e. A plant supervisor is describing the content of a new work order progress report that would simplify tracking.
   f. The analyst is reading an inquiry concerning whether or not a computer system might solve the current problems in work order tracking.
   g. The analyst is installing the microcomputer and database management system needed to run work order processing programs.
   h. The analyst is reviewing the company's organizational chart to identify who becomes involved in work order processing and fulfillment.
   i. The analyst is comparing the pros and cons of a software package versus writing the programs for a new work order system.
   j. An analyst is testing a computer program for entering work orders into the system.
   k. The analyst is correcting a program to more accurately summarize weekly progress.
   l. The analyst is buying a microcomputer and some available software.

m. Information systems management and top business executives are identifying and prioritizing business area applications that should be developed.

4. Management has approached you to develop a new system. The project will last seven months. Management wants a budget next week. You will not be allowed to deviate from that budget. Explain why you shouldn't overcommit to early estimates. Defend the creeping commitment approach as it applies to cost estimating. What would you do if management insisted on the up-front estimate with no adjustments?
5. You have a user who has a history of impatience—encouraging shortcuts through the systems development life cycle and then blaming the analyst for systems that fail to fulfill expectations. By now, you should understand the phased approach to systems development. For each activity, compile a list of possible consequences to use when the user suggests a shortcut through or around that activity.
6. Our company does not have a methodology, but realizes it needs one. The president has suggested that we start a project to build a methodology. Explain to her the advantages of purchasing a methodology instead.
7. You are in a meeting with two other systems analysts. One is arguing that the business should adopt a model-driven development strategy. The other is arguing that a prototyping strategy should be used instead. Explain to them how the two approaches might complement one another.
8. In a staff meeting, a middle-level manager says, "I suggest we dump our methodology and replace it with CASE technology." Respond to this manager.
9. Some people believe that CASE code generators will eventually replace programmers. What do you think? Why or why not? Now suppose such an evolution does occur. How will this impact programmers? How about systems analysts?
10. When describing the phased approach that you plan to follow in developing a new information system, your client asks why your approach is missing a feasibility analysis and project management phase. How would you respond?

PROJECTS AND RESEARCH

1. Make an appointment to visit a systems analyst at a local information system installation. Discuss the analyst's current project. What problems, opportunities, and directives triggered the project? How do they relate to the PIECES framework?
2. Visit a local information system installation. Compare the methodology with the one in this chapter. Evaluate the company's methodology with respect to the eight systems development principles. (Alternative: Substitute the system development life cycle or methodology used in another systems analysis and design book, possibly assigned by your instructor.)
3. Read the mini-book *The One Minute Methodology* (see Suggested Readings). Write a paper that describes what the author was trying to teach, and relate it to the subject presented in this chapter.
4. Visit a local information system installation. Pick a CASE tool that they are using and describe its capabilities using the CASE tool architecture described in this chapter. Did you discover any new capabilities?
5. Research the upper-CASE or lower-CASE marketplace. (Be careful. The industry's terminology is constantly changing.) What are some products and their vendors? On which operating systems do they run? How are they similar and different? What do they cost?
6. Research some of the current trade literature about CASE experiences. What do the writers like about their CASE tools? What are their major complaints? How would you rate their overall satisfaction? Why? What are some future directions of the CASE tool industry?

MINICASES

1. Jeannine Strothers, investments manager, has submitted numerous requests for a new investment tracking system. She needs to make quick decisions regarding possible investments and divestments. One hour can cost her thousands of dollars in profits for her company.

   She has finally given up on Information Systems for not giving her requests high enough priority to get service. Therefore, she goes to a computer store and buys a microcomputer along with spreadsheet, database, and word processing software. The computer store salesperson suggests she build a database of her investments and options, subscribe to a computer investment databank (accessed via a modem in the microcomputer), feed data from her database and the bulletin board into the spreadsheet, play "what if" investment games on the spreadsheet, and then update the database to reflect her final decisions. The word processor will draw data from the database for form letters and mailing lists.

After discussing her plans with Jeff, a systems analyst at another company, he suggests she take a systems analysis and design course before beginning to use the spreadsheet and database. The local computer store, on the other hand, says she doesn't need any systems analysis and design training to be able to develop systems using the spreadsheet and database programs. Their reasoning is that spreadsheets and database tools are not programming languages; therefore, she shouldn't need analysis and design to build systems with them. Is the computer store correct? Why or why not? Can you convince Jeannine to take the systems analysis and design course? What would your arguments be?

2. Jeannine Strothers, the impatient manager in the earlier minicase, did not take Jeff's advice. She built the new system, but she can't get top management to allow her to use it. And she's run into a number of other problems.

First, the financial comptroller has been reevaluating company investment strategies and policies. Jeannine wasn't aware of that. The new system does not account for many of the policies being considered.

Her own staff has rejected the investment and divestment orders generated by the system. She used Information Systems' existing file structure to design those orders, only to find out that her clerks had abandoned those files two years ago because they didn't include the data necessary to execute order transactions. Her staff is also critical of the design, saying that minor mistakes send them off into the "twilight zone" with no easy way to recover.

The computer link to the investment databank has been useless. The data received and its format are not compatible with systems requirements. Although other databanks are available, the current databank has been prepaid for two years. Additionally, Jeannine is now skeptical of such services.

Some of her subordinate managers are insisting on graphic reports. Unfortunately, neither her database management nor spreadsheet package supports graphics. She's not sure how to convert the data of either package to a graphic format (assuming it is possible).

To top off her problems, she isn't sure that her existing database structure can be modified to meet new requirements without having to rewrite all the programs, even those that appear to be working. And her boss is not sure he wants to invest the money in a consultant to fix the problems.

Jeannine's analyst friend Jeff is not very sympathetic to her problems: "Jeannine, I don't have any quick answers for you. You've taken too many shortcuts through the project life cycle. When we do a system, we go through a carefully thought-out procedure. We thoroughly study the problem, define needs, evaluate options, design the system and its interfaces, and only then do we begin programming."

Jeannine replies, "Wait a minute. I only bought a microcomputer. It's not the same as your mainframe computer. I didn't see the need for going through the ritual you guys use for mainframe applications. Besides, I didn't have the time to do all those steps."

Jeff's parting words are philosophical: "You didn't have the time to do it right. Where will you find the time to do it over?"

What principles did Jeannine violate? Why do so many people today fall prey to the belief that the life cycle for an application is somehow different when using microcomputers? What conclusions can you draw from Chapter 2 (the information systems chapter) that might help Jeannine learn from her mistakes? What would you recommend to Jeannine if she were to decide to approach her manager with a plan to salvage the system?

3. Evaluate the following scenarios using the PIECES framework. Do not be concerned that you are unfamiliar with the application. That situation isn't unusual for a systems analyst. Use the PIECES framework to brainstorm potential problems or opportunities you would ask the user about.

a. The staff benefits and payroll counselor is having some problems. Her job is to counsel employees on their benefit options. The company has just negotiated a new medical insurance package that requires employees to choose from among several health maintenance organizations (HMOs). The HMOs vary according to employee classifications, contributions, deductibles, beneficiaries, services covered, and service providers permitted. The intent was to provide the most flexible benefits possible for employees, to minimize costs to the company, and to control costs to the insurance agency (which would affect subsequent premiums charged back to the company).

The counselor will be called on to help employees select the best plan for themselves. She currently responds manually to such requests. But the current options are more straightforward than those under the new plan. She can explain the options, what they do and do not cover, what they cost and may cost, and the pros and cons. However, current employee distrust of the new plan suggests she will need to provide more specific suggestions and responses to employees.

She may have to work up scenarios—possibly worst-case scenarios—for many employees. The scenarios will have to be personalized for each employee's income, marital and family status, current health risks, and so on. In working up a few sample scenarios, she discovered first that it takes one full day to get salary and personnel data from the Information Systems department. Second, employee data are stored in many files that are not always properly updated. When conflicting data become apparent, she can't continue her projections until that conflict has been resolved. Third, the computations are complex. It often takes one full day or more to create investment and/or retirement scenarios for a single employee. Fourth, there are some concerns that projections are being provided to unauthorized individuals, such as former spouses or nonimmediate relatives. Finally, the complexity of the variations in the calculations (there are a lot of "If this, do that" calculations) results in frequent errors, many of which probably go undetected.

b. The manager of a tool and die shop needs help with job processing and control. Jobs are currently processed by hand. First, a job number is established. Next, the job supervisor estimates time and materials for the job. This is a time-consuming process, and delays are common. Then, the job is scheduled for a specific day and estimated time.

On the day the job is to be worked on, materials orders have to be issued. If materials aren't available, the order has to be rescheduled.

Time cards are completed in the shop when workers fulfill the work order. These time cards are used to charge back time to the customer. Time cards are processed by hand, and the final calculations are entered on the work order. The work orders are checked for accuracy and sent to CIS, where accounting records are updated and the customer is billed.

The problem is that the customer frequently calls to inquire about costs already incurred on a work order, but it's not possible to respond because CIS sends a report of all work orders only once a month. Also, management has no idea of how good initial estimates are or how much work is being done on any tool or machine, or by any worker.

c. State University's Development Office raises funds for improving instructional facilities and laboratories at the university. It has uncovered a sensitive problem: The data are out of control.

The Development Office keeps considerable redundant data on past gifts and givers, as well as prospective benefactors. This results in multiple contacts for the same donor—and people don't like to be asked to give to a single university over and over!

To further complicate matters, the faculty and administrators in most departments conduct their own fund-raising and development campaigns, again resulting in duplication of contact lists.

Contacts with possible benefactors are not well coordinated. Whereas some prospective givers are contacted too often, others are overlooked. It is currently impossible to generate lists of prospective givers based on specific criteria (e.g., prior history, socioeconomic level), despite the fact that data on hundreds of criteria have been collected and stored. Gift histories are nonexistent, which makes it impossible to establish contribution patterns that would help various fund-raising campaigns.

4. Century Tool and Die, Inc., is a major manufacturer of industrial tools and machines. It is located in Newark, New Jersey. Larry is the assistant A/R manager. Valerie is a systems analyst for the Information Systems department. Valerie and Larry have just sat down to discuss their current project—improving the recently implemented Accounts Receivable (A/R) information system. As they start to discuss the project, they are interrupted by Robert Washington, the executive vice president of finance, and Gene Burnett, the A/R manager. Larry suddenly looks very nervous. And for good reason—he had suggested the new system, and it has not turned out as promised. Gene's support had been lukewarm, at best. Robert initiates the conversation.

"We've got big problems," said Robert. "This new A/R system is a disaster. It has cost the company more than $625,000, not to mention lost customer goodwill and pending legal costs. I can't afford this when the board of directors is complaining about declining return on investment. I want some answers. What happened?"

Gene responds, "I was never really in favor of this project. Why did we need this new computer system?"

Larry gets defensive. "Look, we were experiencing cash flow problems on our accounts. The existing system was too slow to identify delinquencies and incapable of efficiently following up on those accounts. I was told to solve the problem. A manual system would be inefficient and error-prone. Therefore, I suggested an improved computer-based system."

Gene replies, "I'm not against the computer. I approved the original computer-based system. And I realized that a new system might be needed. It's just that you and Valerie decided to redesign the system without considering alternatives—just like that! In my opinion, you should always analyze options. And let's suppose that a new computerized system was our best option. Why did we have to build the system from scratch? There are good A/R software packages available for purchase. I . . ."

Sensing a confrontation, Robert interrupts, "Gene has a point, Larry. Still, the system you proposed was defended as feasible. And yet it failed! Valerie, as lead analyst, you proposed the new system, correct?"

Nervously, Valerie responds, "Yes, with Larry's help."

Robert continues, "And you wrote this feasibility report early in the project. Let's see. You proposed replacing the current batch A/R system with an on-line system using a database management package."

Valerie replies, "Strictly speaking, the database management package wasn't needed. We could have used the existing VSAM files."

With a troubling look, Robert says, "The report says you needed it. I paid $15,000 to get you that package!"

Valerie answers, "Bill, our database administrator, made that recommendation. The A/R system was to be the pilot database project."

Robert continues. "You also proposed using a network of microcomputers as a front end to the mainframe computer?"

"Yes," answers Valerie. "Larry felt a mainframe-based system would take too long to design and implement. With microcomputers, we could just start writing the necessary transaction programs and then transfer the data to the database on the mainframe computer."

Robert looked puzzled. "I'm a former engineer. It seems to me that some sort of design work should have been done no matter what size computer you used . . . *[brief pause]* In any case, the bottom line in this report is that your projected benefits outweighed the lifetime costs. You projected a 22 percent annual return on investment. Where did you get that number?"

Valerie answers, "I met with Larry four times—about six hours total, I'd say—and Larry explained the problems, described the requirements, made suggestions, and then projected the costs, benefits, and rate of return."

Robert replies, "But that return hasn't been realized, has it? Why not?" After a long, silent pause, Robert continues, "Valerie, what happened after this proposal was approved?"

"We spent the next nine months building the system."

Robert responds, "And how much did you have to do with that, Larry?"

"Not a lot, sir. Valerie occasionally popped into my office to clarify requirements. She showed me sample reports, files, and screens. Obviously, she was making progress, and I had no reason to believe that the project was off schedule."

Robert continues his investigation. "Were you on schedule, Valerie?"

"I don't believe so, Mr. Washington. My team and I were having some problems with certain business aspects of the system. Larry was unfamiliar with those aspects and he had to go to the account clerks and the accountants for answers."

Visibly irritated, Robert asks, "I don't get it! Why didn't you go to the clerks and accountants?"

Although the question was directed to Valerie, Gene replies, "I can answer that. I designated Larry as Valerie's contact. I didn't want her team wasting my people's time—they have jobs to do!"

Robert is silent for a moment, but then he responds, "Something about that bothers me, Gene. In any case, when this project got seriously behind, Larry, why didn't you consider canceling it, or at least reassessing the feasibility?"

"We did!" answers Larry. "Gene expressed concern about progress about seven months into the project. We called a meeting with Valerie. At that meeting, we learned that the new database system wasn't working properly. We also found that we needed more memory and storage on the microcomputers. And to top it off, Valerie and her staff seemed to have little understanding of the business nature of our problems and needs."

This time, Valerie gets defensive. "As I already pointed out, I wasn't permitted contact with the users during the first seven months. Besides, we were asked by Larry to start programming as quickly as possible so that we would be able to show evidence of progress."

Looking at Larry, Robert asks, "Larry, I'm no computer professional; however, my gut instinct suggests that some design or prototyping should have been done first."

Valerie responds for Larry, "Yes, but that would have required end-user participation, which Larry and Gene would not permit."

Larry interrupts. "As I was saying, we considered canceling the project. But I pointed out that $150,000 had already been spent. It would be stupid to cancel a project at that point. I did reassess the feasibility, and concluded that the project could be completed in four more months for another $50,000."

Robert responds, "Nobody asked me if I wanted to spend that extra money."

Larry answers, "We realize that the system hasn't worked out as well as we had hoped. We are trying to redesign . . ."

Robert interrupts, "As well as you hoped? That's an understatement! Let me read you some excerpts from Gene's last monthly report. Customer accounts have mysteriously disappeared, deleted without explanation. Later, we discovered that data-entry clerks didn't know that the F2 key deletes a record. Also, customers have been legally credited for payments that were never made! Customers have been double billed in some cases! Reports generated by the system are late, inaccurate, and inadequate. Cash flow has been decreased by 35 percent! My sales manager claims that some customers are taking their business elsewhere. And the Legal department says we may be sued by two customers and that it will be impossible to collect on those accounts where customers received credit for nonpayments. You tell me, what would you do if you were in my shoes?"

a. Evaluate this project against the eight principles of systems development.
b. What would you do if you were in Mr. Washington's shoes? How would you react to Larry's performance? Valerie's? Gene's?
c. What did Valerie do wrong? Was she in control of her own destiny on this project? Why or why not?
d. What did Larry or Gene do wrong? Can either be held responsible for the failure of a computer project when they have limited computer literacy or experience?
e. What was wrong with the feasibility report? Did Valerie and Larry meet often enough? Was the input to that report sufficient? Did the team commit to a solution too early? Did programming begin too soon? Why or why not?
f. Why were Valerie and her staff uncomfortable with the business problem and needs?
g. Should the project have been canceled? What about the $150,000 investment that had already been made?
h. If you were Valerie or Larry, what would you have done differently?

5. Assume that you are a newly hired CASE coordinator for a major company. You are to make a presentation to management proposing that large sums of money be spent to buy a methodology and CASE tools for your company. Research the market and prepare a spreadsheet that details projected costs and benefits. State all your assumptions. Don't forget training and support.

SUGGESTED READINGS

Application Development Strategies (monthly periodical). Arlington, MA: Cutter Information Corporation. This is our favorite theme-oriented periodical that follows system development strategies, methodologies, CASE, and other relevant trends. Each issue focuses on a single theme.

Application Development Trends (monthly periodical). Westborough, MA: Software Productivity Group, Inc. This is our favorite periodical for keeping up with the latest trends in methodology and CASE. Each month features several articles on different topics and products.

Benjamin, R. I. *Control of the Information System Development Cycle.* New York: Wiley-Interscience, 1971. Benjamin's 16 axioms for managing the systems development process inspired our adapted principles to guide successful systems development.

Bouldin, Barbara. *Agents of Change: Managing the Introduction of Automated Tools.* Englewood Cliffs, NJ: Prentice Hall, 1989. This book documents experiences and makes recommendations for introducing CASE tools into an organization.

Gane, Chris. *Rapid Systems Development.* Englewood Cliffs, NJ: Prentice Hall, 1989. This book presents a nice overview of RAD that combines model-driven development and prototyping in the correct balance.

Gildersleeve, Thomas. *Successful Data Processing Systems Analysis,* 2nd ed. Englewood Cliffs, NJ: Prentice Hall, 1985. We are indebted to Gildersleeve for the creeping commitment approach.

Hammer, Mike. "Reengineering Work: Don't Automate, Obliterate." *Harvard Business Review,* July–August 1990, pp. 104–11. Mike Hammer is a noted consultant and author (and entertaining speaker!) who talks of a new development paradigm that suggests systems analysts take a more "active" (as opposed to "passive") role in reshaping underlying business processes instead of merely automating existing and dated business approaches. This article documents several classic examples where the business process was successfully redefined (during systems analysis), resulting in a more competitive business and information system.

Hobus, James J. *Application Development Center: Implementation and Management.* New York: Van Nostrand Reinhold, 1991. This work describes the implementation of the development center concept described at the end of our chapter.

Martin, James. *Information Engineering: Volumes 1–3.* Englewood Cliffs, NJ: Prentice Hall, 1989 (Vol. 1), 1990 (Vols. 2, 3). These works present information engineering, the dominant methodology after which our *FAST* methodology is patterned.

Orr, Ken. *The One Minute Methodology.* New York: Dorsett House Publishing, 1990. Must reading for those interested in exploring the need for methodology. This very short book can be read in one sitting. It follows the story of an analyst's quest for the development silver bullet, "the one minute methodology."

Wetherbe, James. *Systems Analysis and Design: Best Practices,* 4th ed. St. Paul, MN: West, 1994. We are indebted to Wetherbe for the PIECES framework.

Zachman, John A. "A Framework for Information System Architecture." *IBM Systems Journal* 26, no. 3 (1987). This article presents a popular conceptual framework for information systems survey and the development of an information architecture.

PART TWO

SYSTEMS ANALYSIS METHODS

The five chapters in Part Two introduce you to systems analysis activities and methods. Chapter 4, "Systems Analysis," provides the context for all the subsequent chapters by introducing the activities of *systems analysis.* Systems analysis is the most critical phase of a project. It is during systems analysis that we learn about the existing business system, come to understand its problems, define objectives for improvement, and define the detailed business requirements that must be fulfilled by *any* subsequent technical solution. Clearly, any subsequent design and implementation of a new system depends on the quality of the preceding systems analysis. Systems analysis is often shortchanged in a project because (1) many analysts are not skilled in the concepts and logical modeling techniques to be used, and (2) many analysts do not understand the significant impact of those shortcuts. Chapter 4 introduces you to systems analysis and its overall importance in a project. Subsequent chapters teach you specific systems analysis skills with an emphasis on logical system modeling.

In Chapter 5, "Data Modeling," we teach you *data modeling,* a technique for organizing and documenting the stored data requirements for a system. You will learn to draw entity relationship diagrams as a tool for structuring business data that will eventually be designed as a database. These models will capture the business associations and rules that must govern the data.

Chapter 6, "Process Modeling," introduces *process modeling.* It explains how data flow diagrams can be used to depict the essential business processes in a system, the flow of data through a system, and policies and procedures to be implemented by processes. If you've done any programming, you recognize the importance of understanding the business processes for which you are trying to write the programs.

Chapter 7, "Distribution Modeling," explains the concept of *distribution modeling.* Today's computer networking technology allows us to extend the reach of information systems to a larger geography. But to intelligently use that technology, we need a better understanding of the business geography to be supported. You'll learn how to draw location diagrams and map data and process requirements to those locations.

Finally, Chapter 8, "Object Modeling," introduces the latest systems analysis trend—*object modeling.* Object technology was introduced in prior chapters as an emerging technology to accelerate systems development. But new technologies require new techniques to properly apply the technologies. Eventually, object modeling will become the preferred modeling technique, replacing (or to be more accurate, *integrating*) separate data, process, and distribution modeling. Your early understanding of the concepts and techniques should prove to be a career advantage.

4

SYSTEMS ANALYSIS

CHAPTER PREVIEW AND OBJECTIVES

In this chapter you will learn more about the systems analysis phases in a systems development project: survey, study, and definition. These three phases are collectively referred to as systems analysis. You will know that you understand the process of systems analysis when you can:

- Define systems analysis and relate the term to the survey, study, and definition phases of the *FAST* methodology.
- Describe a number of systems analysis strategies for solving business system problems.
- Describe the survey, study, and definition phases in terms of your information system building blocks.
- Describe the survey, study, and definition phases in terms of objectives, roles, inputs, outputs, techniques, and steps.
- Identify those chapters and modules in this textbook that can help you perform the activities of systems analysis.

Although some techniques of systems analysis are introduced in this chapter, it is *not* the intent of this chapter to teach the *techniques*. This chapter teaches only the *process* of systems analysis. The techniques will be taught in the subsequent four chapters.

SCENE

When we last visited SoundStage, a survey of the existing information system had been completed. (This chapter will more closely examine the activities that were part of that survey.) This episode begins shortly after the project's executive sponsor (the person who pays for the system to be built) has approved the feasibility assessment and scope and authorized a more detailed study of the current system. We join Sandra, Bob, and Sarah Hartman (the business analyst assigned to this project) as they plot strategy in Sandra's office.

SANDRA

Good morning! I asked you here so Bob and I could review our strategy for the next phase of the project. This won't take long.

SARAH

As you know, I was just appointed to this business analyst position, and I'm not that familiar with your methods and terminology yet. I'm kind of looking forward to working for the next year or so in Information System Services. I have to admit, I avoided most of the computer courses when I was in school. To be honest, I thought you were all nerds—no offense!

BOB

None taken.

SARAH

Anyway, Bob's e-mail message said we are starting the study phase?

SANDRA

That's right. This phase will be accomplished in five steps. First, we will draw some simple models of the current system. We want to get a basic grasp of how the current system works.

SANDRA

Models?

BOB

Actually pictures. We'll draw four of them. One will show how this system connects to the business and other systems. One will show the things about which the current system collects data. A third picture will show all of the operating locations that are affected by the current system. And the fourth will show the basic functions performed by the system.

SARAH

Sounds like a lot of trouble.

BOB

Not really. Our *FAST* methodology encourages rapid model development. There will be no multipage models. Four models—four pages! We'll do them in one morning or afternoon.

SARAH

Really? It seems it should take a lot longer than that just to collect the facts.

SANDRA

Ah, but that's where you come in, Sarah. As the business analyst, you are going to plan, organize, and schedule a facilitated group meeting to collect the facts as a team and construct the models.

SARAH

I'm not sure I can lead such a meeting. I mean, I don't have the skill or experience.

BOB

Relax! Sandra will serve as facilitator. She's gone through special training to lead such discussions. I'll be drawing the models on the white board, and you'll be recording any miscellaneous facts . . . as well as encouraging all the managers and users to actively participate.

SARAH

Cool! I can do that. What happened to interviews?

SANDRA

We use them when necessary, but these facilitated sessions are far more productive.

[pause]

The second step will be to analyze all problems and directives.

BOB

Can I assume that *FAST* has a preferred technique for that? I'm new here too

[nodding to Sarah].

SANDRA

Of course! *FAST* recommends that we use something called the PIECES framework to brainstorm our problems and opportunities. I'll get you both some information on that framework later. Then we'll analyze each problem and opportunity for causes and/or effects. Sometimes a cause might actually be another problem, but if we keep applying the problem analyses, we'll get to the root causes and effects.

SARAH

Don't tell me, another facilitated meeting?

SANDRA

Actually, a continuation of the first meeting; that is, if you can talk our participants into giving up a whole day. The productivity and informational benefits are enormous.

SARAH

I think we can do it, if I get the right managers to buy in.

SANDRA

The third step occurs in that same meeting. For each problem, we'll establish some business-oriented objectives to improve the system.

SARAH

Is that like a requirements statement?

BOB

No, the requirements statement will be developed in the next project phase. Think of objectives as business criteria for measuring how good the new system will have to be.

SARAH

Got it. And the fourth step?

SANDRA

We're starting to wind the phase down now. The three of us will reassess our project scope, to see if it has changed. Then we'll adjust our project plan. The fifth and final step will be to present our findings to management and seek approval to continue to the next phase.

SARAH

Which is . . . ?

SOUNDSTAGE

SOUNDSTAGE ENTERTAINMENT CLUB

BOB
Requirements definition. We're going to identify and document the business requirements for the new system, a system that will achieve the aforementioned system improvement objectives. We call it the definition phase.

SARAH
I'm impressed. So far, all you've talked about are business issues—business this and business that. I know that we'll eventually get technical, but it is refreshing to hear your business focus. Maybe you guys aren't all nerds after all.

SANDRA
Thank you. Bob, why don't you tell Sarah how we're going to document the *business* requirements.

BOB
My pleasure. We will identify and document four categories of business requirements—data, process, interfaces, and geography. In each case, we'll draw models.

SARAH
Pictures?

BOB
You got it! These pictures will be considerably more detailed than those we previously discussed. They will not only show more detail, but will also be described by even greater detail in a dictionary-like database called a repository.

SARAH
And that's when we start getting technical?

BOB
No! We'll still leave our taped eyeglasses and pocket protectors in Information Services. These models and their supporting details will still focus entirely on the business requirements—those requirements that must be fulfilled by *any* technical solution we might come up with later. It is this requirements statement that will bridge the gap to technical design.

SARAH
I'm really going to enjoy this a lot more than I thought!

SANDRA
Well, at the risk of destroying our non-nerd image, we are going to get slightly technical during one aspect of requirements definition.

BOB
We are?

SANDRA
Right! *FAST* does recommend that the user interface requirements be prototyped.

BOB
Oh yes! I forgot. The prototypes of inputs and outputs can help us expand and refine our understanding of the data and process requirements.

SARAH
User interface?

BOB
Forgive our techno-terminology. The user interface consists of the inputs, outputs, and screens.

SARAH
Well that's the only part of the technical solution that I'm opinionated about, so you haven't scared me off yet. So, the definition phase will deliver system models and prototypes of inputs and outputs. Don't tell me, more facilitated sessions to capture and document these requirements, right?

DISCUSSION QUESTIONS

1. How would you characterize the focus of the early stages of this project? Why are technical concerns being avoided?
2. What is the advantage of asking Sarah to organize the facilitated meetings instead of Sandra or Bob? How much time do you think the facilitated group meeting might take as compared to a series of one-on-one interviews? Do you see advantages other than speed? Do you see any disadvantages?
3. Prepare an agenda for the full-day facilitated meeting suggested by Sandra and Bob.
4. What kind of pictures would you draw to illustrate the system aspects that Bob described?
5. What kinds of questions would facilitate discussion of problems and their underlying causes and effects?
6. Looking ahead to the requirements definition phase, do you think facilitated group meetings accelerate that process? Why or why not?

WHAT IS SYSTEMS ANALYSIS?

Let's begin with the formal definition of systems analysis.

> **Systems analysis** is the dissection of a system into its component pieces to study how those component pieces interact and work.

We do a systems analysis to subsequently perform a systems synthesis.

> **Systems synthesis** is the re-assembly of a system's component pieces back into a whole system—it is hoped an improved system.

Through systems analysis and synthesis, we may add, delete, and modify system components toward our goal of improving the overall system.

Moving from the theoretical definition to something a bit more contemporary, *systems analysis* is a term that collectively describes the early phases of systems development. There has never been a universally accepted definition. And there has never been agreement on when analysis ends and design begins. To further confuse the issue, some methodologies refer to systems analysis as logical design. Typically, each organization's methodology of choice determines the defintion for that organization. In the *FAST* methodology, systems analysis is defined as those phases and activities that focus on the business problem, independent of technology (for the most part). Specifically, we refine our definition of systems analysis as follows.

> **Systems analysis** is (1) the survey and planning of the system and project, (2) the study and analysis of the existing business and information system, and (3) the definition of business requirements and priorities for a new or improved system. A popular synonym is **logical design.**

This definition corresponds to the first three phases of *FAST* (which were introduced in Chapter 3). The phase "configure a feasible solution" would be considered part of systems analysis by some experts. We prefer to think of it as an analysis-to-design transition phase.

Systems analysis is driven by business concerns, specifically, those of system users. Hence, it addresses the DATA, PROCESS, INTERFACE, and GEOGRAPHY building blocks from a system user perspective. Emphasis is placed on business issues, not technical or implementation concerns.

Figure 4.1 is a view of a *FAST* phase diagram that illustrates the three analysis phases only. Notice the repository.

> A **repository** is a collection of those places where we keep all documentation associated with the application and project.

Although we show only one project repository in the figure, the repository is normally implemented as some combination of the following:

- A disk or directory of word processing, spreadsheet, and other computer-generated files that contain project correspondence, reports, and data.

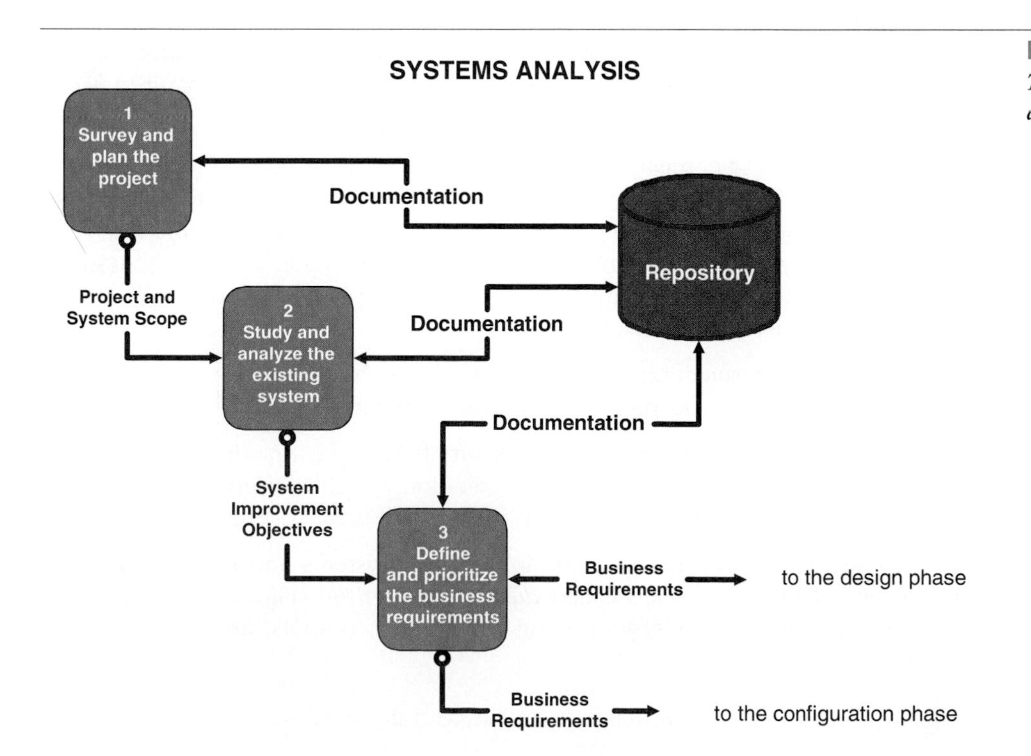

FIGURE 4.1
The Systems Analysis Phases of a Project

- One or more CASE local repositories (as discussed in Chapter 3).
- Hard-copy documentation (stored in notebooks, binders, and system libraries).

Hereafter, we will refer to these as making up a singular project repository.

FAST is a repository-based methodology. This means that phases (and activities included in phases) communicate across a shared repository. Thus, the phases and activities are not really sequential! Work in one phase can and should overlap work in another phase, so long as the necessary information is already in the repository. This accelerates development and allows *FAST* to live up to its name. Furthermore, this model permits the developer to backtrack when an error or omission is discovered.

This chapter examines each of the above phases in greater detail. But first, let's examine some overall strategies for systems analysis.

STRATEGIES FOR SYSTEMS ANALYSIS AND PROBLEM SOLVING

Traditionally, systems analysis is associated with application development projects, that is, projects that produce information systems and their associated computer applications. Your first experiences with systems analysis will likely fall into this category. But systems analysis methods can be applied to projects with different goals and scope. In addition to single information systems and computer applications, systems analysis techniques can be applied to strategic information systems planning and to the redesign of business processes.

There are also many strategies or techniques for performing systems analysis. They include modern structured analysis, information engineering, prototyping, and object-oriented analysis. These strategies are often viewed as competing alternatives. In reality, certain combinations complement one another. Let's briefly examine these strategies and the scope or goals of the projects to which they are suited. The intent is to develop a high-level understanding only. The subsequent chapters in this unit will actually teach you the techniques.

Modern Structured Analysis

Structured analysis was one of the first formal strategies developed for systems analysis of information systems and computer applications. Modern structured analysis[1] is still one of the most widely practiced techniques.

> **Modern structured analysis** is a process-centered technique that is used to model business requirements for a system. The models are structured pictures that illustrate the processes, inputs, outputs, and files required to respond to business events (such as ORDERS).

By process-centered, we mean the initial emphasis in this technique is on the PROCESS building blocks in our information system framework. The technique has evolved to also include the DATA building blocks as a secondary emphasis.

Structured analysis was not only the first popular systems analysis strategy; it also introduced an overall strategy that has been adopted by many of the other techniques—model-driven development.

> A **model** is a representation of reality. Just as "a picture is worth a thousand words," most models use pictures to represent reality.

> **Model-driven development** techniques emphasize the drawing of models to define business requirements and information system designs. The model becomes the design blueprint for constructing the final system.

Modern structured analysis is simple in concept. Systems and business analysts draw a series of process models called *data flow diagrams* (Figure 4.2) that depict the essential processes of a system along with inputs, outputs, and files. Because

---

[1] Edward Yourdon, *Modern Structured Analysis* (Englewood Cliffs, NJ: Yourdon Press, 1989.)

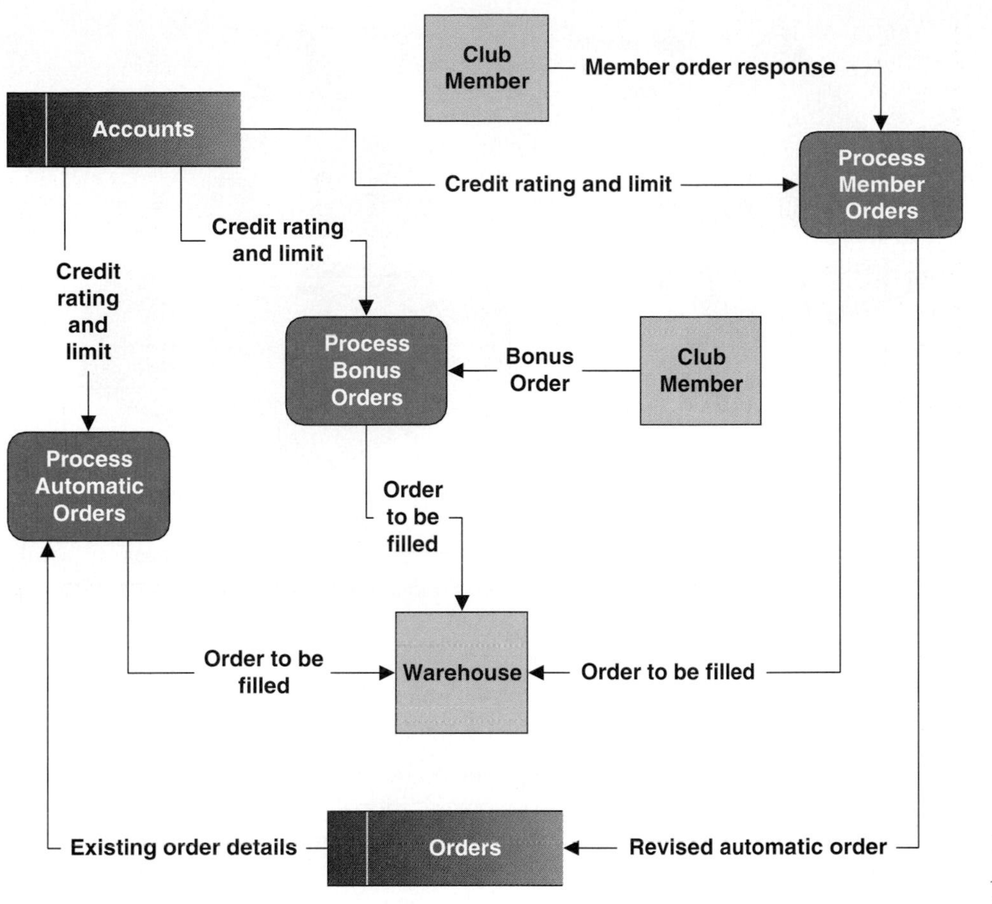

FIGURE 4.2
A Process Model (also Called a Data Flow Diagram)

these pictures represent the *logical* business requirements of the system independent of any *physical,* technical solution, the models are said to be a *logical design* for the system.

Information Engineering (IE)

Today, many organizations have evolved from a structured analysis approach to an information engineering[2] approach.

> **Information engineering** is a data-centered, but process-sensitive technique that is applied to the organization as a whole (or a significant part, such as a division), rather than on an ad-hoc, project-by-project basis (as in structured analysis).

The basic concept of information engineering is that information systems should be engineered like other products. Information engineering books typically use a pyramid framework to depict information systems building blocks and system development phases. As shown in Figure 4.3, the phases are:

1. **Information strategy planning (ISP)** applies systems analysis methods to examine the business as a whole to define an overall plan and architecture for subsequent information systems development. No actual information systems or computer applications are developed. Instead, the project team *studies* the business mission and goals and *defines* an information systems architecture and plan to optimally align information systems to help the organization achieve its business goals.

[2] James Martin, *Information Engineering: Volumes 1–3* (Englewood Cliffs, NJ: Prentice Hall, 1989).

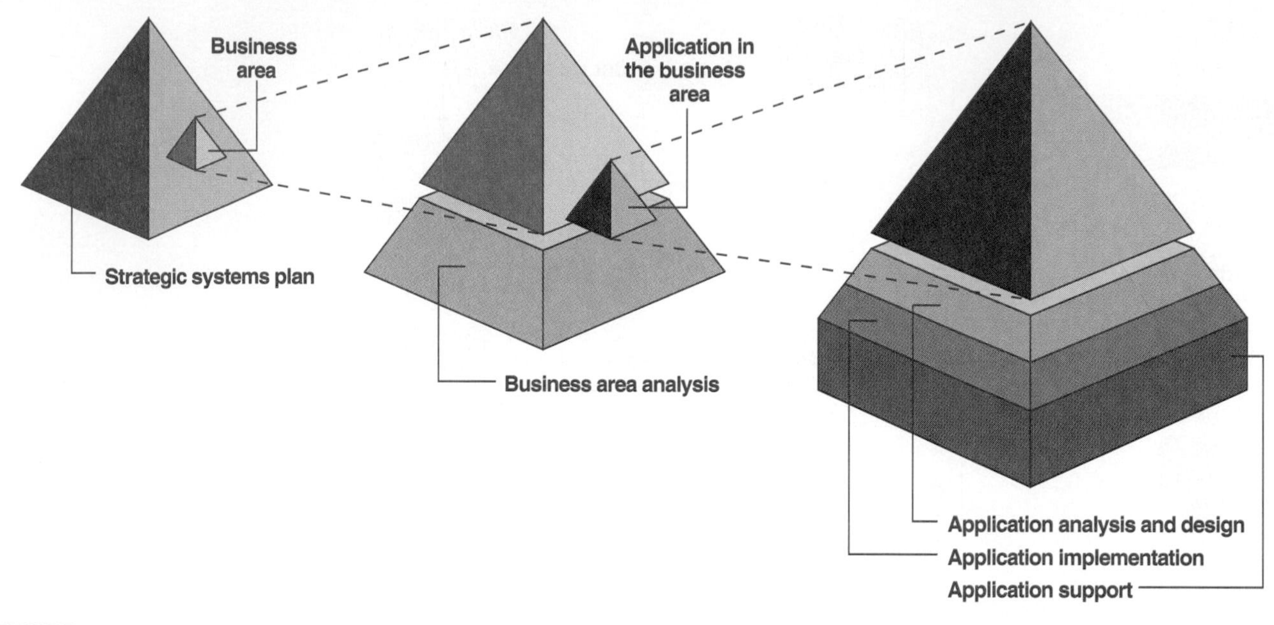

FIGURE 4.3 *Information Engineering Phases*

2. Based on the strategic plan, business areas are carved out and prioritized. A business area is a collection of cross-organizational business processes that should be highly integrated to achieve the information strategy plan (and business mission). A **business area analysis (BAA)** uses systems analysis methods to study the business area and define the business requirements for a highly streamlined and integrated set of information systems and computer applications to support that business area.
3. Based on the business area requirements analysis, information system applications are carved out and prioritized. These applications become projects to which other systems analysis *and* design methods are applied to develop production systems. These methods may include some combination of structured analysis and design, prototyping, and object-oriented analysis and design.

Information engineering is said to be a data-centered paradigm because it emphasizes the study and definition of DATA requirements before those of PROCESS, INTERFACE, or GEOGRAPHY requirements. This is consistent with the contemporary belief that information is a corporate resource that should be planned and managed. Since information is a product of data, data must be planned first! Data models, such as that shown in Figure 4.4, are drawn first. In addition to data models, information engineers also draw process models similar to those drawn in structured analysis.

Although information engineering has gradually replaced structured analysis and design as the most widely practiced strategy for systems analysis, information engineering actually integrates all the process models of structured analysis with its data models. That should make sense, since we know (from Chapter 2) that an information system must include both DATA and PROCESS building blocks. Information engineering was the first formal strategy for synchronizing those building blocks! Information engineering was also the first widely practiced strategy that considered GEOGRAPHY building blocks through application of tools that plan and document the distribution of data and processes to locations.

Prototyping

Another strategy for systems analysis is prototyping.

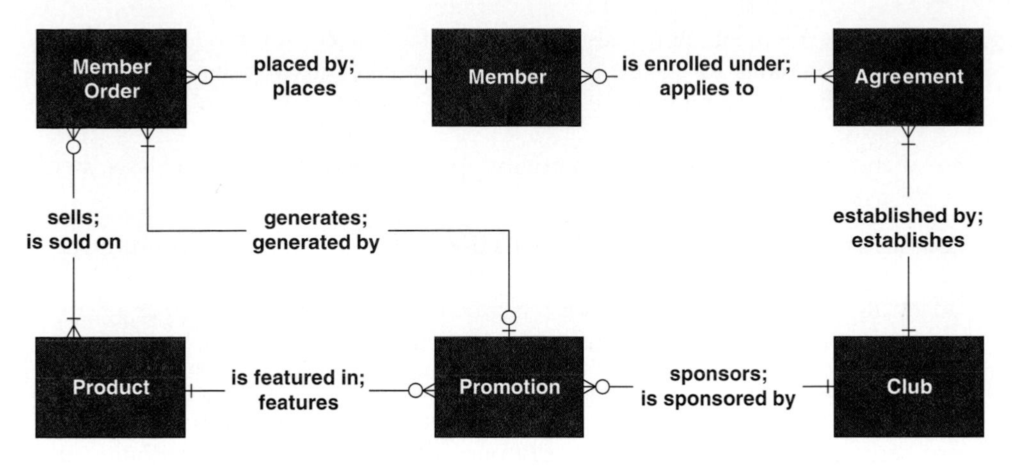

FIGURE 4.4 *A Data Model (also Called an Entity Relationship Diagram)*

Prototyping is an engineering technique used to develop partial but functional versions of a system or applications. When extended to system design and construction, a prototype can evolve into the final, implemented system.

Two flavors of prototyping are applicable to systems analysis:

- **Feasibility prototyping** is used to test the feasibility of a specific technology that might be applied to the business problem. For example, we might use Microsoft *Access* to build a quick-but-incomplete prototype of the feasibility of moving a mainframe application to a PC-based environment.
- **Discovery prototyping** (sometimes called *requirements prototyping*) is used to discover the users' business requirements by having them react to a quick-and-dirty implementation of those requirements. For example, we might again use Microsoft *Access* to create sample forms and reports to solicit user responses as to whether those forms and reports truly represent business requirements. (Note: In discovery prototyping, we try to discourage users from worrying about the style and format of the prototypes; that can be changed during system design!)

In response to the faster pace of the economy in general, prototyping has become a preferred technique for accelerating systems development. Many system developers extend the prototyping techniques to perform what they call **rapid application development.** Unfortunately, some developers are using prototyping to replace model-driven strategies, only to learn what true engineers have known for years—you cannot prototype without some degree of more formal design models.

Joint Application Development (JAD)

As previously described, modern structured analysis and information engineering both emphasize *model-driven development*. Prototyping places emphasis on construction of the working prototypes. Joint application development (JAD) complements both of these techniques by emphasizing *participative development* among system owners, users, designers, and builders.

Joint application development (JAD) uses highly organized and intensive workshops to bring together system owners, users, analysts, designers, and builders to jointly define and design systems. Synonyms include *joint application design* and *joint requirements planning*.

A JAD-trained systems analyst usually plays the role of facilitator for a workshop that will typically run from three to five full working days. This workshop may replace months of traditional interviews and follow-up meetings.

JAD provides a working environment in which to accelerate methodology activities and deliverables. It promotes enhanced system owner and user participation in system development. But it also requires a facilitator with superior mediation and negotiation skills to ensure that all parties receive appropriate opportunities to contribute to the system's development.

Business Process Redesign (BPR)

One of the most interesting contemporary applications of systems analysis methods is business process redesign.

> **Business process redesign** (also called *business process reengineering*) is the application of systems analysis (and design) methods to the goal of dramatically changing and improving the fundamental business processes of an organization, independent of information technology.

The interest in BPR was driven by the discovery that most current information systems and applications have merely automated existing and inefficient business processes. Automated bureaucracy is still bureaucracy; it does not contribute value to the business and may actually subtract value from the business. Introduced in Chapter 1, BPR is one of many types of projects triggered by the trend we call *total quality management* (TQM).

BPR projects focus almost entirely on noncomputer processes. Each process is studied and analyzed for bottlenecks, value returned, and opportunities for elimination or streamlining. Once the business processes have been redesigned, most BPR projects conclude by examining how information technology might best be applied to the improved business processes. This creates new application development projects to which the other techniques described in this section might be applied. Business process redesign is a subject that deserves its own course and book.

Object-Oriented Analysis (OOA)

Object-oriented analysis is the new kid on the block. The concepts behind this exciting new strategy (and technology) are covered extensively in Chapter 8, but a simplified introduction is appropriate here.

For the past 30 years, most systems development strategies have deliberately separated concerns of DATA from those of PROCESS. The COBOL language, which dominated business application programming for years, was representative of this separation—the DATA DIVISION was separated from the PROCEDURE DIVISION. Most systems analysis (and design) techniques similarly separated these concerns to maintain consistency with the programming technology. Although most systems analysis and design methods have made significant attempts to synchronize data and process models, the results have been less than fully successful.

Object technologies and techniques are an attempt to eliminate the separation of concerns about DATA and PROCESS. Instead, data and the processes that act on that data are combined or encapsulated into things called *objects*. The only way to create, delete, change, or use the data in an object (called *properties*) is through one of its encapsulated processes (called *methods*). The system and software development strategy is changed to focus on the "assembly" of the system from a library of reusable objects. Of course, those objects must be defined, designed, and constructed. Thus, in the early part of the systems development process, we need to use object-oriented analysis techniques.

> **Object-oriented analysis (OOA)** techniques are used to (1) study existing objects to see if they can be reused or adapted for new uses, and to (2) define new or modified objects that will be combined with existing objects into a useful business computing application.

Object-oriented analysis techniques are best suited to projects that will implement systems using emerging object technologies to construct, manage, and assemble those objects into useful computer applications. Examples include *Smalltalk, C++, Delphi,* and *Visual BASIC.*

Today, most computer operating systems use graphical user interfaces (GUIs) such as Microsoft *Windows* and IBM's *OS/2 Presentation Manager.* GUIs are built with object-oriented (or object-like) technologies. The development of GUI applications can be based on libraries of reusable objects (sometimes called *components*) that exhibit the same behaviors in all applications. For example, *Delphi* and *Visual BASIC* contain all the necessary objects (called components) to assemble the desired GUI screens for any new application (without programming!).

FAST Systems Analysis Strategies

The *FAST* methodology used by SoundStage Entertainment Club does not impose a single technique on system developers. Instead, it integrates all the popular techniques: structured analysis (via process modeling), information engineering (via data modeling), prototyping (via rapid application development), and joint application development (for all methods). Progressive *FAST* developers can use object-oriented analysis in conjunction with object technology for prototyping to fully exploit the object paradigm.[3]

Finally, the *FAST* methodology supports different types of projects including (1) application development, (2) information strategy planning, (3) business area analysis, (4) decision support system development, and (5) business process redesign. The SoundStage case study will demonstrate application development, a typical first assignment for a systems analyst.

FAST analysis techniques are applied within the framework of (1) your information system building blocks (which were introduced in Chapter 2), (2) the *FAST* phases (which were introduced in Chapter 3), and (3) *FAST* activities (described in this chapter). Given this overview of systems analysis scope and strategy, we can now explore the systems analysis activities. Each *FAST* phase will be described in terms of your information system building blocks and the activities that constitute that phase. For each activity, we will examine the following methodology elements:

- *Purpose* (self-explanatory).
- *Roles.* All *FAST* activities are completed by individuals who are assigned to **roles.** Roles are not the same as job titles. One person can play many roles in a project. Conversely, one role may require many people to adequately fulfill that role. For example, a system user role may require several users in order to adequately represent the interests of an entire system. *FAST* roles are assigned to the following role groups: system owner roles, system user roles, systems analyst roles, system designer roles, and system builder roles.

Notice that these groups correspond with the perspectives in your information system framework (from Chapters 2 and 3).

Every activity is described with respect to:

- *Prerequisites and inputs* (to the activity).
- *Deliverables and outputs* (produced by the activity).
- *Applicable techniques*—which techniques (from the previous section) are applicable to this phase.
- *Steps*—a brief description of the steps required to complete this activity. In the spirit of continuous improvement, all *FAST* steps are fully customizable for each organization.

Furthermore, all roles, inputs, outputs, techniques, and steps are presented with the following designations:

- (REQ) indicates that the role, input, output, technique, or step is REQuired.
- (REC) indicates that the role, input, output, technique, or step is RECommended but not required.

[3] Recall from the Chapter 3 opening case study that SoundStage has elected to test the use of object techniques and technology in the Member Services Information System project.

- (OPT) indicates that the role, input, output, technique, or step is OPTional but not required.

THE SURVEY PHASE OF SYSTEMS ANALYSIS

Recall from Chapter 3 that the first phase of a *FAST* project is to survey the project. The purpose of the survey phase is threefold. First, the survey phase answers the question, "Is this project worth looking at?" To answer this question, the survey phase must define the scope of the project and the perceived problems, opportunities, and directives that triggered the project. Assuming the project *is* worth looking at, the survey phase must also establish the project team and participants, the project budget, and the project schedule.

Your information system framework (Figure 4.5) provides the context for defining scope and understanding the basic problem domain. As shown in the model,

FIGURE 4.5 *Building Blocks for the Survey Phase*

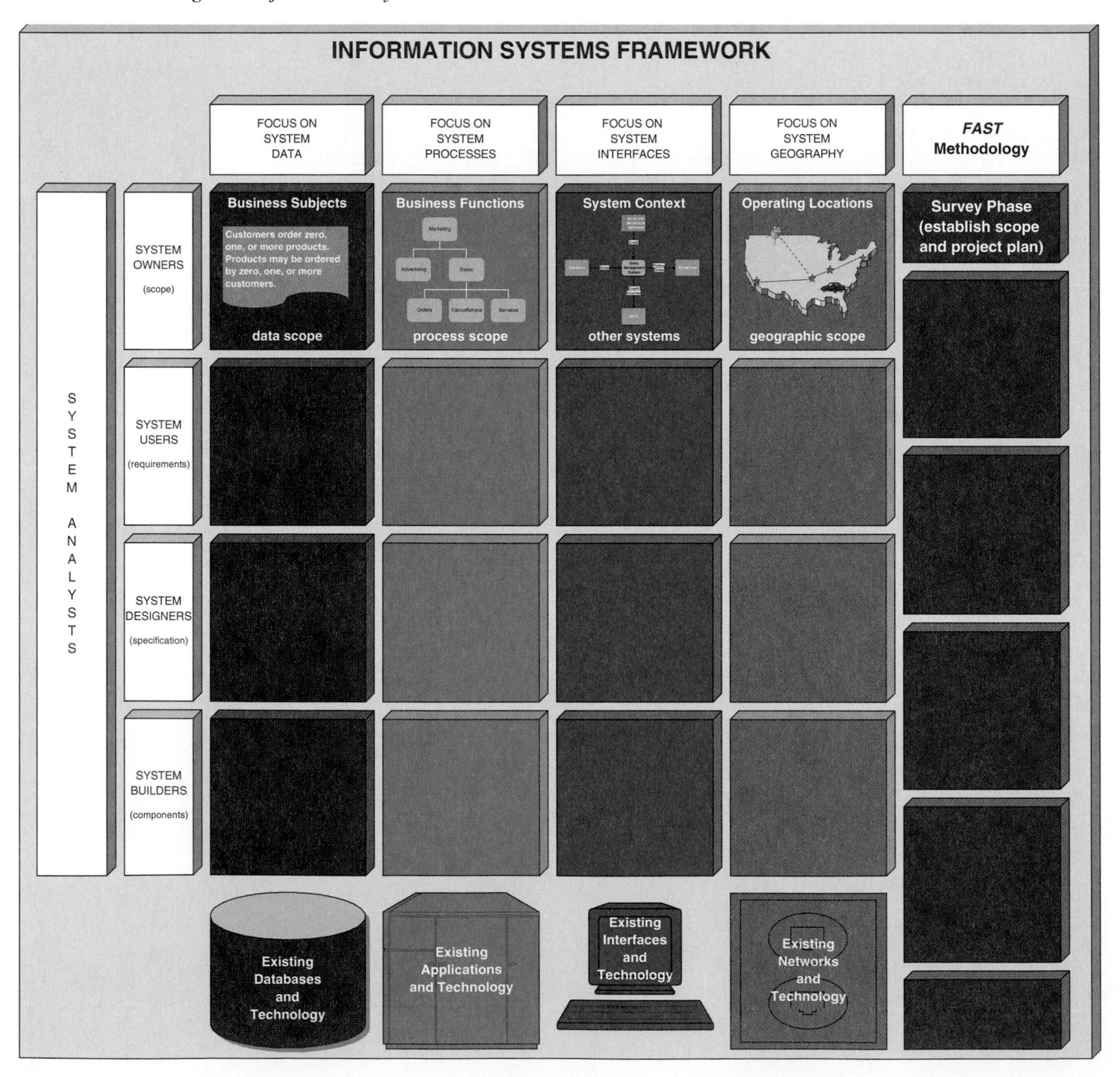

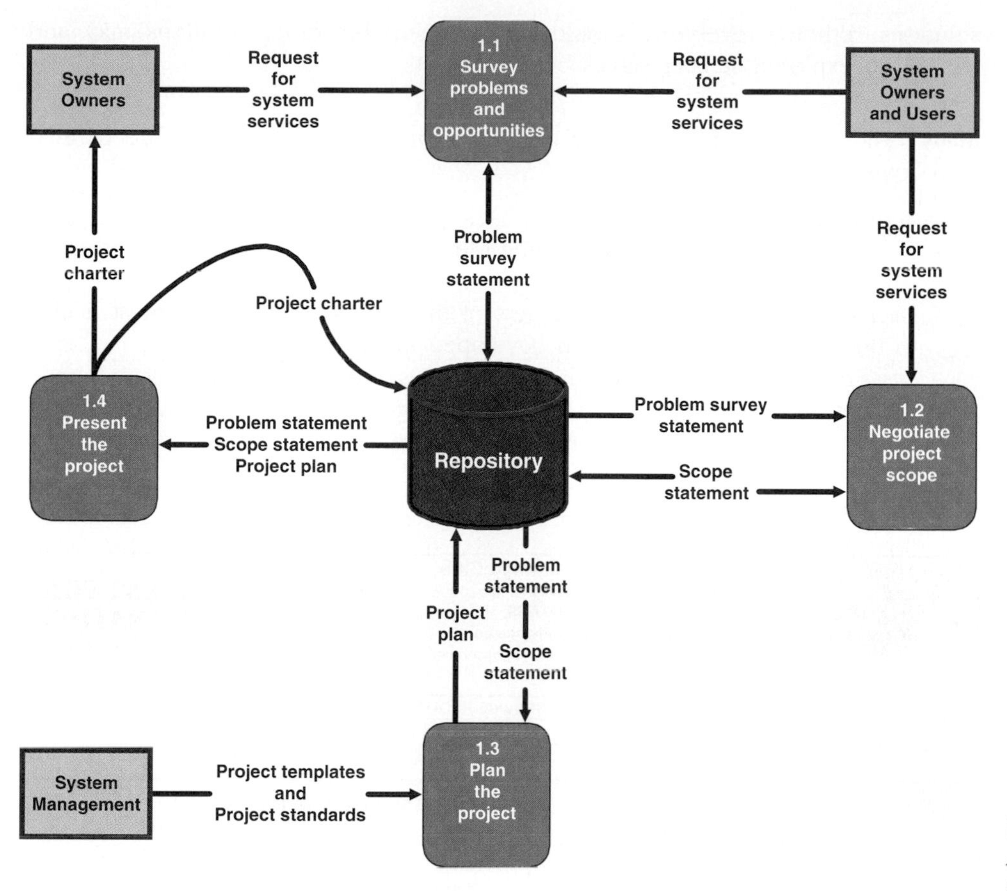

FIGURE 4.6
Activity Diagram for the Survey Phase

the survey phase is concerned with the system owner's view of the overall information system, which includes very few details. More detailed information will be collected if the project is approved to continue to the next phase.

Figure 4.6 is the first of three activity diagrams for systems analysis.

A *FAST* **activity diagram** shows the activities or work that must be completed to accomplish a *FAST* phase.

Solid lines indicate information and documentation flows. A small, shaded circle at the beginning of any input or output information flow indicates a feasibility checkpoint.[4]

Recall that the survey phase is intended to be quick. The entire phase should not exceed two or three days for most projects. Let's now examine each activity in greater depth.

Activity: Survey Problems, Opportunities, and Directives

One of the most important activities of the survey phase is to establish an initial reading of the problems, opportunities, and/or directives that triggered the project. Participants should keep in mind that a more thorough analysis of the same will occur during the study phase.

Purpose The purpose of this activity is to quickly survey and evaluate each identified problem, opportunity, and directive with respect to urgency, visibility, tangible benefits, and priority. Optionally, the participants can explore possible

[4] This is consistent with the *creeping commitment* principle of systems development that was introduced in Chapter 3.

solutions, although everyone should be informed that other solutions may and should be explored at later stages of the project.

Roles The activity is facilitated by the *project manager.* Other roles are defined as follows:

- System owner roles
  - Executive sponsor (REQ)—the highest-level manager who will pay for and support the project.
  - User managers (REQ)—the managers of the organizational units most likely to be supported by the system developed in this project.
  - System managers (OPT)—the information systems unit managers to whom the project manager reports (e.g., the manager of systems development).
  - Project manager (REQ)—the information systems unit manager who will

FIGURE 4.7
A Request for System Services

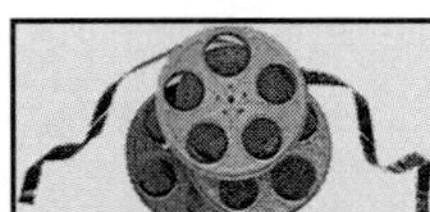

SoundStage Entertainment Club
Information System Services
Phone: 494-0666 Fax: 494-0999
Internet: http://www.soundstage.com
Intranet: http://www.soundstage.com/iss

REQUEST FOR INFORMATION SYSTEM SERVICES

DATE OF REQUEST	SERVICE REQUESTED FOR DEPARTMENT(S)
January 10, 1997	Member Services, Warehouse, Shipping

SUBMITTED BY (key user contact)
Name Sarah Hartman
Title Business Analyst, Member Services
Office B035
Phone 494-0867

EXECUTIVE SPONSOR (funding authority)
Name Galen Kirkhoff
Title Vice President, Member Services
Office G242
Phone 494-1242

TYPE OF SERVICE REQUESTED:
☐ Information Strategy Planning
☒ Business Process Analysis and Redesign
☒ New Application Development
☐ Other (please specify) ________
☐ Existing Application Enhancement
☐ Existing Application Maintenance (problem fix)
☐ Not Sure

BRIEF STATEMENT OF PROBLEM, OPPORTUNITY, OR DIRECTIVE (attach additional documentation as necessary)
The information strategy planning group has targeted member services, marketing, and order fulfillment (inclusive of shipping) for business process redesign and integrated application development. Currently serviced by separate information systems, these areas are not well integrated to maximize efficient order services to our members. The current systems are not adaptable to our rapidly changing products and services. In some cases, separate systems exist for similar products and services. Some of these systems were inherited through mergers that expanded our products and services. There also exist several marketing opportunities to increase our presence to our members. One example includes Internet commerce services. Finally, the automatic identification system being developed for the warehouse must fully interoperate with member services.

BRIEF STATEMENT OF EXPECTED SOLUTION
We envision completely new and streamlined business processes that minimize the response time to member orders for products and services. An order shall not be considered fulfilled until it has been received by the member. The new system should provide for expanded club and member flexibility and adaptability of basic business products and services.

We envision a system that extends to the desktop computers of both employees and members, with appropriate shared services provided across the network, consistent with the ISS distributed architecture. This is consistent with strategic plans to retire the AS/400 central computer and replace it with servers.

ACTION (ISS Office Use Only)

☐ **Feasibility assessment approved** **Assigned to** Sandra Shepherd

☒ **Feasibility assessment waived** **Approved Budget** $ 450,000

Start Date ASAP **Deadline** ASAP

☐ **Request delayed** **Backlogged until date:** ________

☐ **Request rejected** **Reason:** ________

Authorized Signatures:

Rebecca J. Todd
Chair, ISS Executive Steering Body

Galen Kirkhoff
Project Executive Sponsor

FORM ISS-100-RFSS (Last revised December, 1996)

directly manage the project team. This is usually a senior systems analyst (e.g., Sandra will play this role in the SoundStage project).

- System user roles.
  - Business analyst (OPT)—an analyst who is on loan from the user community to the information systems unit (applicable only to organizations that practice this concept).
  - Other users are typically not involved in this activity at this time.
- Systems analyst roles.
  - System modelers (REC)—systems analysts who are skilled with the system modeling techniques and CASE tools that will be used in the project.
  - System designer roles are not typically involved in this activity unless deemed appropriate by a system owner.
  - System builder roles are not typically involved in this activity unless deemed appropriate by a system owner.

Prerequisites (Inputs) This activity is triggered by a request for system services (REQ), shown in Figure 4.7. This input implements the two logical project triggers that were described in Chapter 3—a *planned system project directive* or an *unplanned system request*.

Deliverables (Outputs) The principle deliverable of this activity is a problem statement (REQ), which documents the problems, opportunities, and directives that were discussed. Figure 4.8 is a sample document that summarizes problems, opportunities, and directives in terms of:

- *Urgency*. In what time frame must/should the problem be solved or the opportunity or directive realized? A rating scale could be developed to consistently answer this question.
- *Visibility*. To what degree would a solution or new system be visible to customers and/or executive management? Again, a rating scale could be developed for the answers.
- *Benefits*. Approximately how much would a solution or new system increase annual revenues or reduce annual costs. This is often a guess, but if all participants are involved in that guess, it should prove sufficiently conservative.
- *Priority*. Based on the above answers, what are the consensus priorities for each problem, opportunity, or directive. If budget or schedule becomes a problem, these priorities will help to adjust project scope.
- *Possible solutions* (OPT). At this early stage of the project, possible solutions are best expressed in simple terms such as (1) leave well enough alone, (2) a quick fix, (3) a simple-to-moderate enhancement of the existing system, (4) redesign the existing system, or (5) design a new system. The participants listed for this activity are well suited to an appropriately high-level discussion of these options.

Applicable Techniques The following techniques are applicable to this activity. For each activity, we indicate which chapters or modules of the book teach that technique.

- *Fact-finding*. Fact-finding methods (REQ) are used to interact with people to identify problems, opportunities, and directives. Typically, scope is defined through interviews or a group meeting. Interviewing and meeting methods and frameworks are discussed in Part Five, Module B, "Fact-Finding and Information Gathering."

PROBLEM STATEMENTS

PROJECT:	Member Services Information System	**PROJECT MANAGER:**	Sandra Shepherd
CREATED BY:	Sandra Shepherd	**LAST UPDATED BY:**	Robert Martinez
DATE CREATED:	January 15, 1997	**DATE LAST UPDATED:**	January 17, 1997

Brief Statements of Problem, Opportunity, or Directive	Urgency	Visibility	Annual Benefits	Priority or Rank	Proposed Solution
1. Order response time as measured from time of order receipt to time of customer delivery has increased to an average of 15 days.	ASAP	High	$175,000	2	New development
2. The recent acquisitions of Private Screenings Video Club and GameScreen will further stress the throughput requirements for the current system.	6 months	Med	75,000	2	New development
3. Currently, three different order entry systems service the audio, video, and game divisions. Each system is designed to interface with a different warehousing system; therefore, the intent to merge inventory into a single warehouse has been delayed.	6 months	Med	515,000	2	New development
4. There is a general lack of access to management and decision-making information. This will become exasperated by the acquisition of two additional order processing systems (from Private Screenings and GameScreen)	12 months	Low	15,000	3	After new system is developed, provide users with easy-to-learn and use reporting tools.
5. There currently exists data inconsistencies in the member and order files.	3 months	High	35,000	1	Quick fix; then new development
6. The Private Screenings and GameScreen file systems are incompatible with the SoundStage equivalents. Business data problems include data inconsistencies and lack of input edit controls.	6 months	Med	unknown	2	New development. Additional quantification of benefit might increase urgency.
7. There is an opportunity to open order systems to the Internet, but security and control is an issue.	12 months	Low	unknown	4	Future version of newly developed system
8. The current order entry system is incompatible with the forthcoming automatic identification (bar coding) system being developed for the warehouse.	3 months	High	65,000	1	Quick fix; then new development

FIGURE 4.8 *A Representative Problem Statement*

- *Interpersonal skills.* Interpersonal skills (REQ) are related to fact-finding skills. They impact the way we communicate and negotiate with one another. Good interpersonal relations are essential to this activity. Interpersonal skills are taught in Part Five, Module E, "Interpersonal Skills and Communications."

Steps The following steps are suggested to complete this activity.

1. (REQ) Collect and review all documentation submitted to begin this project.
2. (REQ) Schedule and conduct a meeting of the people tentatively assigned to the aforementioned roles for this activity. (Alternative: Interview the people tentatively assigned to those roles.)
3. (REC) Document problems, opportunities, and constraints.

Activity: Negotiate Project Scope

Scope defines the boundary of the project—what aspects of the system will and will not be included in the project. Scope can change during the project; however, the initial project plan must be based on some agreement regarding scope. Then if the scope changes significantly, all parties involved will have a better appreciation for why the budget and schedule have also changed. This activity can occur in parallel with the prior activity.

Purpose The purpose of this activity is to define the boundary of the system and project. The boundary should be defined as precisely as possible to minimize the impact of creeping scope. Creeping scope is the subtle, but significant increase of scope that frequently occurs during system projects. Scope can increase for many legitimate reasons. By defining scope, we are not eliminating creeping scope. We are merely providing a mechanism to document and track that scope so the impact on budget and schedule can be continuously reassessed.

Roles The activity is facilitated by the *project manager*. Other roles are defined as follows:

- System owner roles (defined in the previous activity).
  - Executive sponsor (REC).
  - User managers (REQ).
  - System managers (OPT).
  - Project manager (REQ).
- System user roles.
  - Business analysts (OPT).
  - Other users are typically not involved in this activity at this time.
- Systems analyst roles.
  - System modelers (REC).
- System designer roles are not typically involved in this activity unless deemed appropriate by a system owner.
- System builder roles are not typically involved in this activity unless deemed appropriate by a system owner.

Prerequisites (Inputs) This activity is triggered by the same *request for system services* (REQ) described as an input to the previous activity. The *problem survey statement* (REQ) produced by the previous activity can be a useful input for defining scope.

Deliverables (Outputs) The principal deliverable of this activity is a *scope statement* (REQ) that corresponds to the four building blocks for this phase:

- *Business subjects* define the scope for DATA. This might be a simple list of things about which the system needs to know information.
- *Business functions* define the scope for PROCESSES. This might be a simple list of ongoing business functions that would be included in or affected by this system.
- *System context* defines the scope for INTERFACES. This might be a list of external people, organization units, organizations, or other systems with which the system might have to interact.
- *Operating locations* define the scope for GEOGRAPHY. This might be a simple list of specific business operating locations that will be included within the scope of the project.

Notice that each statement of scope was described as a simple list. We don't necessarily define the items in the list. Nor are we very concerned with precise definitions. And we definitely are not concerned with any time-consuming steps such as modeling or prototyping.

Applicable Techniques The following techniques are applicable to this activity. For each activity, we indicate which chapters or modules of the book teach that technique.

- *Fact-finding*. Fact-finding methods (REQ) are used to interact with people to define scope. Typically, scope is defined by interviews or a group meeting. Interviewing and meeting methods and frameworks are discussed in Part Five, Module B, "Fact-Finding and Information Gathering."

- *Interpersonal skills.* Once again, good interpersonal skills (REQ) are essential to this activity. Applicable interpersonal skills are taught in Part Five, Module E, "Interpersonal Skills and Communications."

Steps The following steps are suggested to complete this activity.

1. (REC) Collect and review all documentation submitted to begin this project.
2. (REQ) Schedule and plan a meeting of the people tentatively assigned to the aforementioned roles for this activity. (Alternative: Interview the people tentatively assigned to those roles.) The meeting or interviews should focus on negotiating the scope in terms of the four building blocks of information systems: DATA, PROCESSES, INTERFACES, and GEOGRAPHY. Figure 4.8 provides sample questions for the meeting or interviews.
3. (REQ) Document scope.

Activity: Plan the Project

Based on the prior activities, you can now answer the question, Is the project worth looking at? If the answer is yes, the group needs to formally plan the project. At this stage of the project, we know too little to accurately predict costs or schedules. Thus, our initial project plan should consist of the following:

- A first-draft master plan and schedule for completing the entire project. This schedule will be modified at the end of each phase of the project. This is sometimes called a *baseline plan*.
- A detailed plan and schedule for completing the next phase of the project (the study phase). In most cases this schedule will be more accurate but still subject to a lack of detailed knowledge about the current system and user requirements.

The project plan is derived from experience plus the problem survey and scope definitions produced in the prior activities.

Purpose The purpose of this activity is to develop the initial project schedule and resource assignments. This includes selecting the appropriate *FAST* project template, assigning *FAST* roles to personnel, estimating time commitments for both people and activities, and generating the initial schedule and resource requirements.

Roles The activity is facilitated by the *project manager*. Other roles are defined as follows:

- System owner roles (defined in the previous activity).
  - Executive sponsor (REQ).
  - User managers (REC).
  - System managers (REC).
  - Project manager (REQ).
  - Steering body (OPT)—many organizations require that all project plans be formally presented to a steering body (sometimes called a steering committee) for final approval.
- System user roles.
  - Business analysts (OPT).
- Systems analyst, system designer, and system builder roles are not typically involved in this activity unless deemed necessary by the project manager.

Prerequisites (Inputs) This activity is triggered by completion of the problem survey and scope definition activities. The *problem statement* (REC) and the *scope* *statement* (REC), if formally documented, are very helpful references for the project planning group.

Deliverables (Outputs) The principle deliverable of this activity is the *project plan* (REC). Many project managers use project management software to create and maintain their project plans. This initial project plan consists of two components:

- A *phase-level* plan that covers the entire project (REC).
- An *activity-level* plan that details the study phase of the project (REC).

Applicable Techniques The following techniques are applicable to this activity. For each activity, we indicate which chapters or modules of the book teach that technique.

- *Process management.* Process management (REC) (also called *methodology management*) defines the standards for applying the methodology to a project. It defines skill requirements and training for each role, CASE tool standards, documentation standards, quality management standards, and project management standards. Process management techniques are taught in Part Five, Module A, "Project and Process Management."
- *Project management.* Project management (REC) builds on process management by applying the methodology to specific projects in the form of schedule planning, staffing and supervision, progress reporting, management of expectations, budgeting, and schedule management. Project management techniques are taught in Part Five, Module A, "Project and Process Management."
- *Presentation skills.* The project charter and any verbal presentations of the project and plan obviously require presentation skills (REC). Presentation skills are taught in Part Five, Module E, "Interpersonal Skills and Communications."

Steps The following steps are suggested to complete this activity.

1. (REQ) Review system problems, opportunities, and directives, as well as project scope.
2. (REC) Select the appropriate *FAST* project template. *FAST* templates support different strategies and/or different system development goals (e.g., purchase a package versus object-oriented development).
3. (REQ) Assign specific people to each *FAST* role.
4. (REC) Estimate time required for each project activity, assign roles to activities, and construct a schedule.
5. (OPT) Negotiate expectations. (See Part Five, Module A, "Project and Process Management" for a simple expectations management instrument.)
6. (REC) Negotiate the schedule with system owners, adjusting resources, scope, and expectations as necessary.
7. (REC) Write the project charter.

Activity: Present the Project

In most organizations, there are more potential projects than resources to staff and fund those projects. If a project has not been predetermined to be of the highest priority (by some sort of prior tactical or strategic planning process), then it must be presented and defended to a steering body for approval.

> A **steering body** is a committee of executive business and system managers that studies and prioritizes competing project proposals to determine which projects will return the most value to the organization and thus should be approved for continued systems development.

The majority of any steering body should consist of noninformation systems professionals or managers. Many organizations designate vice presidents to serve on a steering body. Other organizations assign the direct reports of vice presidents to the steering body. And some organizations utilize two steering bodies, one for vice presidents and one for their direct reports. Information systems managers serve on the steering body only to answer questions and to communicate priorities back to developers and project managers.

Purpose The purpose of this activity is (1) to secure any required approvals to continue the project and (2) to communicate the project and goals to all staff.

Roles Ideally, the activity should be facilitated by the *executive sponsor.* Other roles are defined as follows:

- System owner roles (defined in the previous activity).
  - Executive sponsor (REQ).
  - User managers (REC).
  - System managers (REC).
  - Project manager (REQ).
  - Steering body (OPT or REQ—depends on the organization).
- System user roles.
  - Business analysts (REC).
  - All direct and indirect users (REC).
- System designers.
  - Any systems analysts assigned to the project (REC).
  - Any system designers and specialists likely to be assigned to the project (REC).
- System builders.
  - Any system builders likely to be assigned to the project (REC).
  - Representatives of any technology vendors whose products are likely to be involved in the project (OPT).

Prerequisites (Inputs) This activity is triggered by the completion of the project planning activity. The inputs include the *problem statement* (REC), the *scope statement* (REC), and the *project plan* (REC) generated by the prior activities. Additionally, *project templates* (OPT) and *project standards* (REC) may be provided by systems management.

Deliverables (Outputs) The key deliverable of this activity is the *project charter* (REQ). This charter is usually a formal consolidation of all the inputs to the activity. It might be thought of as an internal contract for the project, should the project continue to the next phase.

The final deliverables of the activity (as discussed in Chapter 3) are the *problem statement* and *scope statement* that become the triggers for various study phase activities. They may take the form of a verbal presentation, a written document (possibly the project charter or a summary thereof), a letter of authority from the executive sponsor, or some combination of these formats. Figure 4.9 provides a representative outline of a written report.

Applicable Techniques The following techniques are applicable to this activity. For each activity, we indicate which chapters or modules of the book teach that technique.

- *Interpersonal skills.* Once again, good interpersonal skills (REQ) are essential to this activity. These include persuasion, sales (of ideas), writing, and

FIGURE 4.9 *Outline for a Survey Phase Final Report*

Project Feasibility Assessment Report

I. Executive summary (1 page)
   A. Summary of recommendation
   B. Brief statement of anticipated benefits
   C. Brief explanation of report contents
II. Background information (1–2 pages)
   A. Brief description of project request
   B. Brief explanation of the summary phase activities
III. Findings (2–3 pages)
   A. Problems and analysis (optional: reference *problem statement matrix*)
   B. Opportunities and analysis (optional: reference *problem statement matrix*)
   C. Directives and implications
IV. Detailed recommendation
   A. Narrative recommendation (1 page)
      1. Immediate fixes
      2. Quick fixes
      3. Enhancements
      4. New systems development
   B. Project plan
      1. Initial project objectives
      2. Initial master project plan (phase level)
      3. Detailed plan for the study or definition phase
V. Appendixes
   A. Request for system services
   B. Problem statements matrix
   C. (other documents as appropriate)

speaking. Applicable interpersonal skills are taught in Part Five, Module E, "Interpersonal Skills and Communications."

Steps The following steps are suggested to complete this activity.

1. (REQ) Review the deliverables of all prior activities.
2. (OPT) Reformat the project charter for presentation to the steering body.
3. (REQ in some organizations) Present the project proposal (charter) to the steering body. Be prepared to defend recommendations, address issues and controversies, and answer questions as posed by the steering body.
4. (REC) Plan an event to communicate the approved project to any and all affected staff, or (OPT) distribute the project charter or summary over a cover letter of authority from the executive sponsor. This launch event (e.g., a pizza luncheon) presents the project and plan to both participants and all interested parties. The executive sponsor's visible support of the project can prevent many political problems from surfacing.

This concludes the survey phase. The participants in the survey phase might decide the project is not worth proposing. It is also possible the steering body may decide that other projects are *more* important. Or the executive sponsor might not endorse the project. In each of these instances, the project is terminated. Little time and effort have been expended.

On the other hand, with the blessing of all system owners, the project can now proceed to the study and/or definition phases.

THE STUDY PHASE OF SYSTEMS ANALYSIS

You may recall the old saying, "Don't try to fix it unless you understand it." That statement aptly describes the next phase of a systems analysis, *study and analyze the current system*. There is always a current system, regardless of whether or not

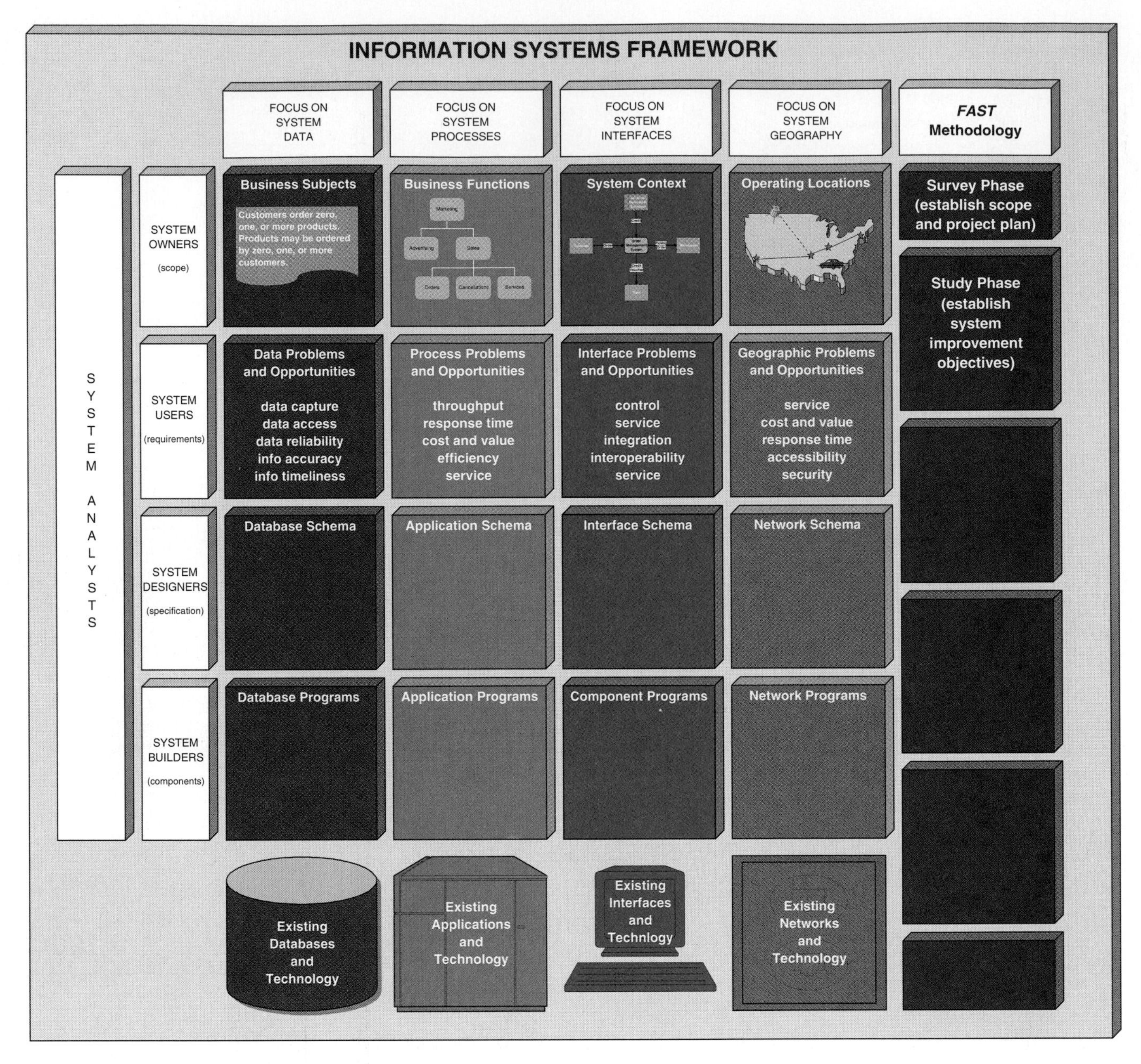

FIGURE 4.10 *Building Blocks for the Study Phase*

it currently uses computers. The study phase provides the analyst with a more thorough understanding of problems, opportunities, and/or directives. The study phase answers the questions, Are the problems really worth solving? and Is a new system really worth building? In different methodologies it may be called the detailed study or problem statement phase.

Can you ever skip the study phase? Rarely! You almost always need some understanding of the current system. But there may be some reasons to accelerate the study phase. First, if the project was triggered by systems planning, the worthiness of the project is not in doubt—the study phase is reduced to understanding the current system, not analyzing it. Second, if the project was initiated by a directive (such as, "Process financial aid applications according to new federal regulations that go into effect on July 1"), then worthiness is, once again, not in doubt.

The goal of the *FAST* study phase is to understand the problem domain well

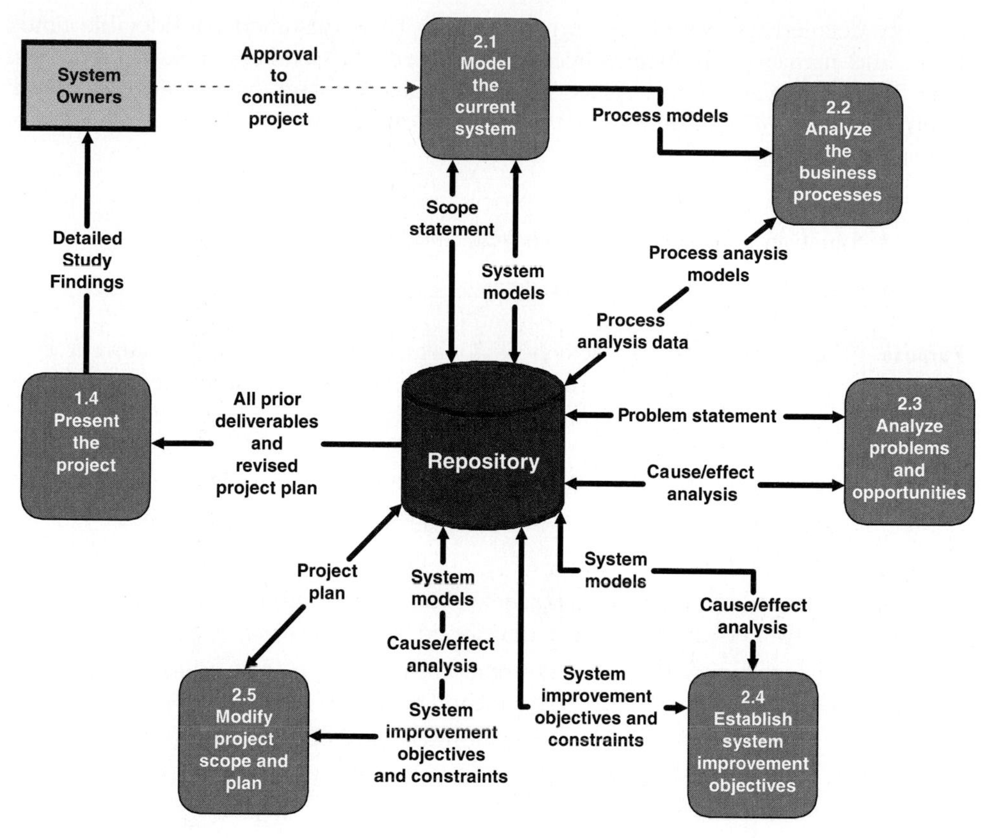

FIGURE 4.11
Activity Diagram for the Study Phase

enough to thoroughly analyze its problems, opportunities, and constraints. Like many other methodologies, *FAST* once encouraged a detailed understanding of the current system that was documented with painstaking precision in various models such as data flow diagrams. Historically, the value added by this documentation proved questionable. Thus, the current version of *FAST* encourages some modeling, but only enough to refine our understanding of project scope and to define a common vocabulary for the system.

Your information systems framework (Figure 4.10) illustrates the context of the study phase as it relates to the information systems building blocks. Notice that the study phase is concerned jointly with the system owners' and users' views of the overall information system in somewhat greater detail than the survey phase.

Figure 4.11 is the activity diagram for the study phase. Depending on the size of the system, its complexity, and the degree to which project worthiness is already known, these activities may consume one to six weeks. Most of these activities can be accelerated by JAD sessions. (Joint application development was introduced earlier in this chapter and is taught in depth in Module D.)

Let's explore the study phase activities in greater depth.

Activity: Model the Current System

During the study phase, the team is attempting to learn about the current system. Each team member brings a different level of understanding to the process—different scope, different detail, different vocabulary, different perceptions, and different opinions. A well-conducted study can prove revealing to all parties, including management and users. How do you verify the team's consensus understanding of the current system? One way is to *model the current system*. Recall that a model is a representation of reality, usually pictorial.

The value of modeling in the study phase has come under some debate in recent years. Many early techniques, such as De Marco's structured analysis, called for very detailed models of the current system. This consumed considerable time. Users and managers frequently became frustrated with endless reviews of a system that they wanted to replace anyway! This problem has sometimes been referred to as *analysis paralysis*. *FAST* suggests one of two modeling strategies for the study phase:

- A combination of high-level data, process, and geographic models.
- A combination of object and geographic models.

Let's examine the activity in greater detail.

Purpose The purpose of this activity is to learn enough about the current system's data, processes, interfaces, and geography to expand the understanding of scope and to establish a common working vocabulary for that scope.

Roles The activity may be facilitated by either the project manager or a systems analyst (sometimes one and the same person). Other roles are defined as follows:

- System owner roles.
  - User managers (REQ)—the managers of the organizational units most likely to be supported by the system developed in this project.
  - System managers (OPT)—the information systems unit managers to whom the project managers report (e.g., the manager of systems development).
  - Project manager (REQ)—the information systems unit manager who will directly manage the project team. This is usually a senior systems analyst (e.g., Sandra will play this role in the SoundStage project).
- System user roles.
  - Business analyst (REC)—an analyst who is on loan from the user community to the information systems unit (applicable only to organizations that practice this concept).
  - Other users (REC) as needed to fully represent the business scope of the project.
- System analyst roles.
  - System modeler (REQ)—a systems analyst who is skilled in the modeling techniques to be used in this activity.
- System designer roles are not typically involved in this activity unless deemed appropriate by a system owner.
- System builder roles are not typically involved in this activity unless deemed appropriate by a system owner.

Prerequisites (Inputs) This activity is triggered by completion of the survey phase activities and approval from the system owners to continue the project. The key informational input is the project and system *scope statement* (REC) that was completed as part of the survey phase. These information flows were covered in the previous section.

Deliverables (Outputs) The principal deliverable of this activity are *system models* (REQ) that serve two purposes: (1) to expand understanding of scope, and (2) to verify the team's consensus understanding of the business situation. The overriding modeling strategy is information hiding.

> The principle of **information hiding,** as applied to system models, suggests that models should hide inappropriate details in an effort to focus attention on what's really important. In other words, "If we want to learn *anything,* we must not try to learn *everything*—at least not all at once."

According to our information system framework, appropriate current system models may include:

- DATA—A one-page data model (REQ) is very useful for establishing business vocabulary, rules, and policies.
- PROCESSES—Today, it is widely accepted that a one- or two-page functional decomposition diagram (REC) should prove sufficient to get a feel for the current system processing.
- INTERFACES—A one-page context diagram (REQ) is useful for clarifying the system's inputs and outputs with other systems, organizations, and departments.
- GEOGRAPHY—A one-page geographic model (OPT) adequately identifies current operating locations that are relevant to the system.

Samples of some of these models were presented earlier in the chapter (as Figures 4.2 and 4.3).

Applicable Techniques The following techniques are applicable to this activity. For each activity, we indicate which chapters or modules of the book teach that technique.

- *Fact-finding.* By now, a common theme has emerged. Good fact-finding skills (REQ) are essential to most activities in the systems analysis phases. Fact-finding skills include interviewing, sampling, questionnaires, and research. These skills are covered in Part Five, Module B, "Fact-Finding and Information Gathering."
- *Joint application development.* The preferred technique for gathering information as rapidly as possible is joint application development (REC). The requisite system models can be developed in one or two facilitated group sessions with all the participants. JAD techniques are covered in Part Five, Module D, "Joint Application Development."
- *Data, process, and geographic modeling.* System modeling (REQ) is taught in the following chapters: Chapter 5, "Data Modeling"; Chapter 6, "Process Modeling"; and Chapter 7, "Distribution Modeling." These chapters cover modeling for both the study and definition phases.
- *Interpersonal skills.* And yet another common theme of systems analysis emerges—good interpersonal skills (REQ) are essential to most systems analysis activities. These skills are explored in Part Five, Module E, "Interpersonal Skills and Communications."

Steps The following steps are suggested to complete this activity.

1. (REC) Review the scope statement completed in the survey phase.
2. (REQ) Collect facts and gather information about the current system. The preferred technique is JAD, but JAD sessions may be preceded or followed by traditional fact-finding and information gathering activity.
3. (REQ) Draw system models. The recommended sequence of models is (1) INTERFACE, (2) DATA, (3) PROCESS, and (4) GEOGRAPHY. The interface model is first because it verifies and expands on the project scope. The data model is second because it helps establish basic business vocabulary and rules. The process model identifies high-level business functions. The geography model identifies the potential operating locations to which data, processes, and interfaces might eventually be distributed. Together, the models provide a solid foundation for problem and opportunity analysis.

4. (REQ) Verify the system models. The goal is to reach consensus on what the current system is all about. If JAD techniques are used, steps 2, 3, and 4 are consolidated into the group sessions.

(OPT) Activity: Analyze Business Processes

This activity is only applicable to *business process redesign* (BPR) projects. In such a process, the team is asked to examine business processes in much greater detail. The team models those business processes in greater detail and measures the value added or subtracted by each process as it relates to the total organization. Business process analysis can be politically charged. Managers and users alike can become very defensive about their existing business processes. The analysts involved must keep the focus on the processes, not the people, and constantly remind everyone that the goal is to identify opportunities for fundamental business change that will benefit the business and everyone in the business.

Entire books are written on the subject of business process analysis and redesign. This activity summary is intended only to skim the surface of the topic.

Purpose Applicable only to business process redesign projects, the purpose of this activity is to analyze each business process in a set of related business processes to determine if the process is necessary and what problems might exist in that business process.

Roles The activity is facilitated by a *business process analyst.* Other roles are defined as follows:

- System owner roles.
  - User managers (REQ) of the business processes that will be analyzed.
- System user roles.
  - Appropriate business users (REQ)—those who are involved or familiar with the intricacies of the business processes.
- Systems analyst roles.
  - Business process analyst (REQ)—an individual who is skilled in the BPR techniques being applied. This individual should also be skilled in keeping the team emphasis on the process, not the people who perform the process.
- System designers are never involved. The focus is on analyzing the business, not the technology.
- System builders are never involved in this activity.

Prerequisites (Inputs) Once again, this activity is applicable only to business process redesign projects. In those projects, this activity is triggered by completion of the *system models* (REQ) from the previous activity. This activity is only interested in the process models. These process models are much more detailed than in other types of projects. They show every possible work flow path through the system, including error processing.

Deliverables (Outputs) The deliverables of this activity are *process analysis models* (REQ) and *process analysis data* (REQ). The process analysis models look very much like data flow diagrams (Figure 4.2) except they are significantly annotated to show: (1) the volume of data flowing through the processes, (2) the response times of each process, and (3) any delays or bottlenecks that occur in the system. The process analysis data provides additional information such as: (1) the cost of each process, (2) the value added by each process, and (3) the consequences of eliminating or streamlining the process. The combination of the models and analysis will be used to redesign the business processes in the systems design phases.

Applicable Techniques The following techniques are applicable to this activity. For each activity, we indicate which chapters or modules of the book teach that technique.

- *Process modeling.* Basic process modeling (REQ) is taught in Chapter 6; however, this book defers any process model analysis material to books that can provide more extensive coverage.
- *Process analysis.* Process analysis techniques are not covered in this book. BPR books are a better source of such education. They also tend to include the use of CASE technology specifically geared to business process redesign.

Steps The following steps are suggested to complete this activity.

1. (REQ) If necessary, refine process models to include all possible work flows and data flows that can occur in the business area under examination.
2. (REQ) For each primitive business process, analyze throughput and response time, as well as any average delays that may occur.
3. (REQ) For each primitive business process, analyze cost and value added. Identify candidates for elimination, consolidation, and optimization.

Activity: Analyze Problems and Opportunities

In addition to learning about the current system, the project team must work with system owners and system users to *analyze problems and opportunities*. The team may use the current system model(s) to serve as a framework that identifies those aspects of the current system to be analyzed. (Note: Some problems and opportunities may be apparent on the models, but many will not.)

You might be asking, Weren't problems and opportunities identified earlier in the survey phase? Yes, they were. But initial problems may be only symptoms of other problems, perhaps not well known or understood by the users. Besides, we haven't yet really analyzed any problems in the classical sense.

True problem analysis is a difficult skill for inexperienced systems analysts. Experience suggests that most new analysts (and many users and managers) try to solve problems without truly analyzing them. They might state a problem like this: "We need to . . ." or "We want to . . ." They are stating the problem in terms of a solution. More effective problem solvers have learned to state and analyze the problem, not the solution. They analyze each perceived problem for causes and effects. That cause-effect analysis provides true understanding that can lead to not-so-obvious solutions. And that analysis can also uncover other, more significant problems to be analyzed. Let's study the activity in greater detail.

Purpose The purpose of this activity is to: (1) understand the underlying causes and effects of all perceived problems and opportunities, and (2) understand the effects and potential side effects of all perceived opportunities. The analysis should be kept at a business (nontechnical) level during the entire analysis.

Roles The activity is facilitated by either the *project manager* or *systems analyst*. Other roles are defined as follows:

- System owner roles.
  - User managers (REQ)—the managers of the organizational units most likely to be supported by the system developed in this project.
  - Project manager (REC)—the information systems unit manager who will directly manage the project team. This is usually a senior systems analyst (e.g., Sandra will play this role in the SoundStage project).
- System user roles.
  - Business analysts (OPT).

  - Other user experts (REC) as necessary to fully analyze the problems and opportunities.
- Systems analyst roles.
  - Systems analyst (REQ)—specifically, systems analysts who are skilled at cause-effect analysis.
- System designer roles are not typically involved in this activity unless deemed appropriate by a system owner.
- System builder roles are not typically involved in this activity unless deemed appropriate by a system owner.

Prerequisites (Inputs) Like the modeling activity, this activity is triggered by completion of the survey phase activities and approval from the system owners to continue the project. In fact, this activity begins in parallel with the modeling activity. One key informational input is the *problem statement* (REC) that was completed as part of the survey phase. Other key informational inputs are *problems and opportunities* and *causes and effects,* which are collected from the business analysts and other system users.

Deliverables (Outputs) The principal deliverable of this activity is the aforementioned *cause-effect analysis.* Figure 4.12 illustrates a representative way to document cause-and-effect analysis.

Applicable Techniques The techniques applicable to this activity are almost identical to those of the modeling activity. For each activity, we indicate which chapters or modules of the book teach that technique.

FIGURE 4.12 *Cause-Effect Analysis*

PROBLEMS, OPPORTUNITIES, OBJECTIVES, AND CONSTRAINTS MATRIX

Project:	Member Services Information System	Project Manager:	Sandra Shepherd
Created by:	Robert Martinez	Last Updated by:	Robert Martinez
Date Created:	January 21, 1997	Date Last Updated:	January 31, 1997

CAUSE-AND-EFFECT ANALYSIS		SYSTEM IMPROVEMENT OBJECTIVES	
Problem or Opportunity	**Causes and Effects**	**System Objective**	**System Constraint**
1. Order response time is unacceptable.	1. Throughput has increased while number of order clerks was downsized. Time to process a single order has remained relatively constant. 2. System is too keyboard dependent. Many of the same values are keyed for most orders. Net result is (with the current system) each order takes longer to process than is ideal. 3. Data editing is performed by the AS/400. As that computer has approached its capacity, order edit responses have slowed. Because order clerks are trying to work faster to keep up with the volume, the number of errors has increased. 4. Warehouse picking tickets for orders were never designed to maximize the efficiency of order fillers. As warehouse operations grew, order filling delays were inevitable.	1. Decrease the time required to process a single order by 30%. 2. Eliminate keyboard data entry for as much as 50% of all orders. 3. For remaining orders, reduce as many keystrokes as possible by replacing keystrokes with point-and-click objects on the computer display screen. 4. Move data editing from a shared computer to the desktop. 5. Replace existing picking tickets with a paperless communication system between member services and the warehouse.	1. There will be no increase in the order processing workforce. 2. Any system developed must be compatible with the existing *Windows 95* desktop standard. 3. New system must be compatible with the already approved automatic identification system (for bar coding).

Page 1 of 5

- *Fact-finding.* Fact-finding skills (REQ) are necessary to both identify and analyze the problems and opportunities. These skills are covered in Part Five, Module B, "Fact-Finding and Information Gathering."
- *Joint application development.* The preferred technique for rapid problem analysis is joint application development (JAD) (REC). The requisite analysis can usually be completed in one full-day session or less. The JAD facilitator must be especially skilled at conflict resolution because people tend to view problem analysis as personal criticism. JAD techniques and conflict resolution are explored in Part Five, Module D, "Joint Application Development."
- *Interpersonal skills.* This activity can easily generate controversy and conflict. Good interpersonal skills are necessary to maintain a focus on the problems and not the personalities. These skills are explored in Part Five, Module E, "Interpersonal Skills and Communications."
- *Cause-effect analysis.* Cause-effect analysis, when applied with discipline, can help the team avoid a premature concern with solutions. Cause-effect analysis is explored in Part Five, Module B, "Fact-Finding and Information Gathering."

Steps The following steps are suggested to complete this activity.

1. (REC) Review the problem statement completed in the survey phase.
2. (REQ) Collect facts and gather information about the perceived problems and opportunities in the current system. The preferred technique is JAD, but JAD sessions may be preceded or followed by traditional fact-finding and information gathering activity.
3. (REQ) Analyze and document each problem and opportunity. The PIECES framework (introduced in Chapter 3) is most useful for cause-effect analysis. As you collect facts, note problems and limitations according to the PIECES categories. Remember, a single problem may be recorded into more than one category of PIECES. Also, don't restrict yourself to only those problems and limitations noted by end-users. As the analyst, you may also identify potential problems! Next, for each problem, limitation, or opportunity, ask yourself the following questions and record answers to them.
   a. What is causing the problem? What situation has led to this problem? Understanding why is not as important. Many current systems were never designed; they simply evolved. It is usually pointless to dwell on history. In fact, you should be careful not to insult system owners and users who may have played a role in how things evolved.
   b. What are the negative effects of the problem or failure to exploit the opportunity? Learn to be specific. Don't just say, excessive costs. How excessive? You don't want to spend $20,000 to solve a $1,000 problem.
   c. The effect sometimes identifies another problem. If so, repeat steps 1 and 2.

Opportunities do not have causes. But they do have effects (and side effects) that will eventually have to be weighed against the cost of implementing the opportunity.

Activity: Establish System Improvement Objectives and Constraints

Given our understanding of the current system's scope, problems, and opportunities, we can now *establish system improvement objectives.* How can we determine the success of a systems development project? Success should be measured in terms of the degree to which objectives are met for the new system.

> An **objective** is a measure of success. It is something that you expect to achieve, if given sufficient resources.

Objectives represent the first attempt to establish expectations for any new system.

In addition to objectives, we must also identify any known constraints.

> A **constraint** is something that will limit your flexibility in defining a solution to your objectives. Essentially, constraints cannot be changed.

A deadline is an example of a constraint.

Given this overview, let's further study this important activity.

Purpose The purpose of this activity is to establish the criteria against which any improvements to the system will be measured and to identify any constraints that may limit flexibility in achieving those improvements.

Roles The activity is facilitated by the *project manager* or a *systems analyst*. Other roles are defined as follows:

- System owner roles.
  - User managers (REQ)—the managers of the organizational units most likely to be supported by the system developed in this project.
  - Project manager (REQ)—the information systems unit manager who will directly manage the project team. This is usually a senior systems analyst (e.g., Sandra will play this role in the SoundStage project).
- System user roles.
  - Business analysts (OPT).
  - Other user experts (REC) as necessary to fully analyze the problems and opportunities.
- Systems analyst roles.
  - Systems analysts (REC)—systems analysts who are skilled in the formulation and documentation of objectives and constraints.
- System designer roles are not typically involved in this activity unless deemed appropriate by a system owner.
- System builder roles are not typically involved in this activity unless deemed appropriate by a system owner.

Prerequisites (Inputs) This activity is triggered by the completion of the two previous activities. The inputs are the *system models* (REC) and the *cause-effect analysis* (REQ). Together, they define the context for establishing objectives and constraints.

Deliverables (Outputs) The deliverable of this activity is *system improvement objectives and constraints*. This deliverable also corresponds to the net deliverable of the study phase, *system objectives,* as described in Chapter 3. Objectives should be precise, measurable statements of business performance that define the expectations for the new system. Some examples might include the following:

- Reduce the number of uncollectible customer accounts by 50 percent within the next year.
- Increase by 25 percent the number of loan applications that can be processed during an eight-hour shift.
- Decrease by 50 percent the time required to reschedule a production lot when a workstation malfunctions.

The following is an example of a poor objective:

~~Create a delinquent accounts report.~~

This is a poor objective because it only states a requirement, not an actual objective. Now, let's reword that objective:

Reduce credit losses by 20 percent through earlier identification of delinquent accounts.

This gives us more flexibility. Yes, the delinquent accounts report would work. But a customer delinquency inquiry might provide an even better way to achieve the same objective.

Objectives must be tempered by identifiable constraints. Constraints fall into four categories as listed below (with examples):

- *Schedule:* The new system must be operational by April 15.
- *Cost:* The new system cannot cost more than $350,000.
- *Technology:* The new system must be on-line, or all new systems must use the DB2 database management system.
- *Policy:* The new system must use double-declining balance inventory techniques.

The last two columns of Figure 4.12 document typical system improvement objectives and constraints.

Applicable Techniques The following techniques are applicable to this activity. For each activity, we indicate which chapters or modules of the book teach that technique.

- *Joint application development.* The preferred technique for this activity is joint application development (REC). The requisite brainstorming can usually be completed in a half-day session or less. JAD techniques and conflict resolution are explored in Part Five, Module D, "Joint Application Development."
- *Benefit analysis.* Whenever possible, objectives should be stated in terms that can be measured. Tangible and intangible benefit estimation techniques are surveyed in Part Five, Module C, "Feasibility and Cost/Benefit Analysis."
- *Interpersonal skills.* Like the previous activity, this activity can easily generate controversy and conflict. Good interpersonal skills are necessary to maintain a focus on what's best for the organization. These skills are explored in Part Five, Module E, "Interpersonal Skills and Communications."

Steps The following steps are suggested to complete this activity.

1. (REC) Review scope and problem analyses from the prior activities.
2. (REQ) Negotiate business-oriented objectives to solve each problem and exploit each opportunity. Ideally, each objective should establish the way you will measure the improvement over the current situation. Measures should be as tangible (measurable) as you can possibly make them.
3. (REQ) Brainstorm any constraints that may limit your ability to fully achieve objectives. Use the four categories previously listed in this section (time, cost, technology, and policy) to organize your discussion.

Activity: Modify Project Scope and Plan

Recall that project scope is a moving target. Based on our initial understanding and estimates from the survey phase, it may have grown or diminished in size and complexity. (Growth is much more common!) Now that we're approaching the completion of the study phase, we should reevaluate project scope and revise the project plan accordingly. The systems analyst and system owner are the key individuals in this activity.

The analyst and system owner will consider the possibility that not all objectives may be met by the new system. Why? The new system may be larger than expected, and they may have to reduce the scope to meet a deadline. In this case the system owner will rank the objectives in order of importance. Then, if the scope must be reduced, the higher-priority objectives will tell the analyst what's most important.

Purpose The purpose of this activity is to reevaluate project scope, schedule, and expectations. The overall project plan is then adjusted as necessary, and a detailed plan is prepared for the next phase.

Roles The activity is facilitated by the *project manager*. Other roles are defined as follows:

- System owner roles (defined in the previous activity).
  - Executive sponsor (OPT)—It may be necessary to renegotiate scope, expectations, schedule, and budget with the executive sponsor.
  - User managers (OPT)—User managers should also be involved in any renegotiation of scope, expectations, and schedule.
  - System managers (OPT)—System managers commit information services resources to projects; therefore, they need to be made aware of any scope, schedule, or budget changes.
  - Project manager (REQ).
- System users are not typically involved in this activity unless deemed appropriate by the project manager.
- Systems analyst, system designer, and system builder roles are not typically involved in this activity unless deemed necessary by the project manager.

Prerequisites (Inputs) This activity is triggered by completion of the system modeling, problem analysis, and objective definition activities. The *system models* (REC), *cause-effect analysis* (REC), and *system improvement objectives and constraints* are inputs for the activity. The original *project plan* from the survey phase (REC, if available) is also an input.

Deliverables (Outputs) The principal deliverable of this activity is a revised *project plan* (REC). Many project managers use project management software to create and maintain their project plans. Additionally, a detailed *definition phase plan* (REC) may be produced.

Applicable Techniques The following techniques are applicable to this activity. For each activity, we indicate which chapters or modules of the book teach that technique.

- *Process management.* Process management (REC) (also called *methodology management*) defines the standards for applying the methodology to a project. Process management techniques are taught in Part Five, Module A, "Project and Process Management."
- *Project management.* Project management (REC) builds on process management by applying the methodology to specific projects in the form of schedule planning, staffing and supervision, progress reporting, management of expectations, budgeting, and schedule management. Project management techniques are taught in Part Five, Module A, "Project and Process Management."
- *Presentation skills.* The project charter and any verbal presentations of the project and plan obviously require presentation skills (REC). Presentation skills are taught in Part Five, Module E, "Interpersonal Skills and Communications."

Steps The following steps are suggested to complete this activity.

1. (REC) Review the original plan.

2. (REQ) Review the system models, problems and opportunities, cause-effect analyses, system improvement objectives, and scope. Ask yourself two questions:
   a. Has the scope of the project significantly expanded?
   b. Are the problems, opportunities, or objectives more difficult to solve than originally anticipated?
3. (REC) Estimate time required for each project activity in the next phase—the definition phase.
4. (REC) If necessary, refine baseline estimates for the overall project plan.
5. (OPT) If the answer is yes, renegotiate scope, schedule, and/or budget with the system owner group. (See Part Five, Module A, "Project and Process Management," for a simple expectations management instrument.)

(REC) Activity: Present Findings and Recommendations

As with the survey phase, the study phase findings and recommendations should be communicated to all affected personnel. The format may be a report, a verbal presentation, or an inspection by an auditor or peer group (called a walkthrough).

Purpose The purpose of this activity is to communicate the project and goals to all staff. The report or presentation, if developed, is a consolidation of the activities' documentation.

Roles Ideally, the activity should be facilitated by a *business analyst.* Alternatively, the activity could be facilitated by a systems analyst or the project manager. Other roles are defined as follows:

- System owner roles (defined in the previous activity).
  - Executive sponsor (OPT).
  - User managers (REC).
  - System managers (REC).
  - Project manager (REQ).
- System user roles.
  - Business analysts (REC).
  - All direct and indirect users (REC).
- Systems analysts.
  - Any systems analysts assigned to the project (REC).
- System designers are not typically involved in this activity.
- System builders are not typically involved in this activity.

Prerequisites (Inputs) This activity is triggered by the completion of the system objectives or project plan activity. The inputs include the *system models* (REC), the *cause-effect analysis* (REC), the *system improvement objectives and constraints,* and the *revised project plan* (REC) generated by the prior activities.

Deliverables (Outputs) The key deliverable of this activity is the *detailed study findings* (REC). It usually includes a feasibility update and the revised project plan. A representative outline for a written report is illustrated in Figure 4.13.

Applicable Techniques The following techniques are applicable to this activity. For each activity, we indicate which chapters or modules of the book teach that technique.

- *Interpersonal skills.* Once again, good interpersonal skills (REQ) are essential to this activity. These include persuasion, sales (of ideas), writing, and speaking. Applicable interpersonal skills are taught in Part Five, Module E, "Interpersonal Skills and Communications."

FIGURE 4.13 *Detailed Study Report Outline*

Analysis of the Current _____________ System

I. Executive summary (approximately 2 pages)
   A. Summary of recommendation
   B. Summary of problems, opportunities, and directives
   C. Brief statement of system improvement objectives
   D. Brief explanation of report contents
II. Background information (approximately 2 pages)
   A. List of interviews and facilitated group meetings conducted
   B. List of other sources of information that were exploited
   C. Description of analytical techniques used
III. Overview of the current system (approximately 5 pages)
   A. Strategic implications (if the project is part of or impacts an existing information systems strategic plan)
   B. Models of the current system
      1. Interface model (showing project scope)
      2. Data model (showing project scope)
      3. Geographic models (showing project scope)
      4. Process model (showing functional decomposition only)
IV. Analysis of the current system (approximately 5–10 pages)
   A. Performance problems, opportunities, and cause-effect analysis
   B. Information problems, opportunities, and cause-effect analysis
   C. Economic problems, opportunities, and cause-effect analysis
   D. Control problems, opportunities, and cause-effect analysis
   E. Efficiency problems, opportunities, and cause-effect analysis
   F. Service problems, opportunities, and cause-effect analysis
V. Detailed recommendations (approximately 5–10 pages)
   A. System improvement objectives and priorities
   B. Constraints
   C. Project plan
      1. Scope reassessment and refinement
      2. Revised master plan
      3. Detailed plan for the definition phase
VI. Appendixes
   A. Any detailed system models
   B. (other documents as appropriate)

Steps The following steps are suggested to complete this activity.

1. (REQ) Review the deliverables of all prior activities.
2. (OPT) Write the detailed study findings.
3. (REQ) Present the findings to the system owners. Be prepared to defend recommendations, address issues and controversies, and answer questions. One of the following decisions must be made:
   - Authorize the project to continue, as is, to the definition phase.
   - Adjust the scope, cost, and/or schedule for the project and then continue to the definition phase.
   - Cancel the project due to either (1) lack of resources to further develop the system, (2) realization that the problems and opportunities are simply not as important as anticipated, or (3) realization that the benefits of the new system are not likely to exceed the costs.
4. (REC) Present the findings to all affected staff.

This concludes the study phase. With some level of approval of the system owners, the project can now proceed to the definition phase.

THE DEFINITION PHASE OF SYSTEMS ANALYSIS

Many analysts make a critical mistake after completing the study of the current information system. The temptation at that point is to begin looking at alternative solutions, particularly technical solutions. The most frequently cited error in new information systems is illustrated in the statement, "Sure the system works, and it is technically impressive, but it just doesn't do what we wanted (or needed) it to do."

Did you catch the key word? The key word is *what*. Analysts are frequently so preoccupied with the *technical* solution that they inadequately define the *business* requirements for that solution. The definition phase answers the question, *What* does the user need and want from a new system? The definition phase is critical to the success of any new information system! In different methodologies the definition phase might be called the *requirements analysis* or *logical design* phase.

Can you ever skip the definition phase? Absolutely not! New systems will always be evaluated, first and foremost, on whether or not they fulfilled business objectives and requirements regardless of how impressive or complex the technological solution might be!

Here again, your information systems building blocks (Figure 4.14) can serve as a useful framework for identifying what information systems requirements need to be defined. Notice that we are still concerned with the system users' perspectives. Also notice that for most of the building blocks, it is possible to draw graphical models to document the requirements for a new and improved system. This information represents the system users' view of the new application's requirements.

Figure 4.15 illustrates the typical activities of the definition phase. Once again, let's examine the activities conducted during this phase, the individuals' roles in each activity, the inputs and outputs, and techniques that are commonly used to complete each activity.

Activity: Outline Business Requirements

The initial activity of the definition phase is to *outline business requirements*. Effective writers outline their publications. This activity serves the same purpose. The foundation for this activity was established in the study phase when we identified system improvement objectives. This activity translates those objectives into inputs, outputs, and processes that will be specified in greater detail during subsequent activities. Rarely will this activity identify *all* the business requirements—no more than an initial outline for a written publication will include all headings and subheadings for a final publication. But the outline will frame your thinking as you proceed. Neither completeness nor perfection is a goal. The entire activity can be completed in as little as an hour.

Purpose The purpose of this activity is to identify, in general terms, the business requirements for a new or improved information system. A classic *input-process-output* framework should prove sufficient to structure the activity.

Roles The activity is facilitated by a business analyst or systems analyst. Other roles are defined as follows:

- System owner roles (most were defined in the previous activity).
  - User managers (REQ)—User managers played a key role in defining system improvement objectives. In this activity, they are especially helpful in identifying the outputs (information) to be produced by a system.
  - Project manager (OPT).
- System user roles.
  - Business analysts (REC)—Because business analysts are actually users on loan from the business community, they are well suited to the task of identifying high-level requirements, especially the inputs and processes required in the system.

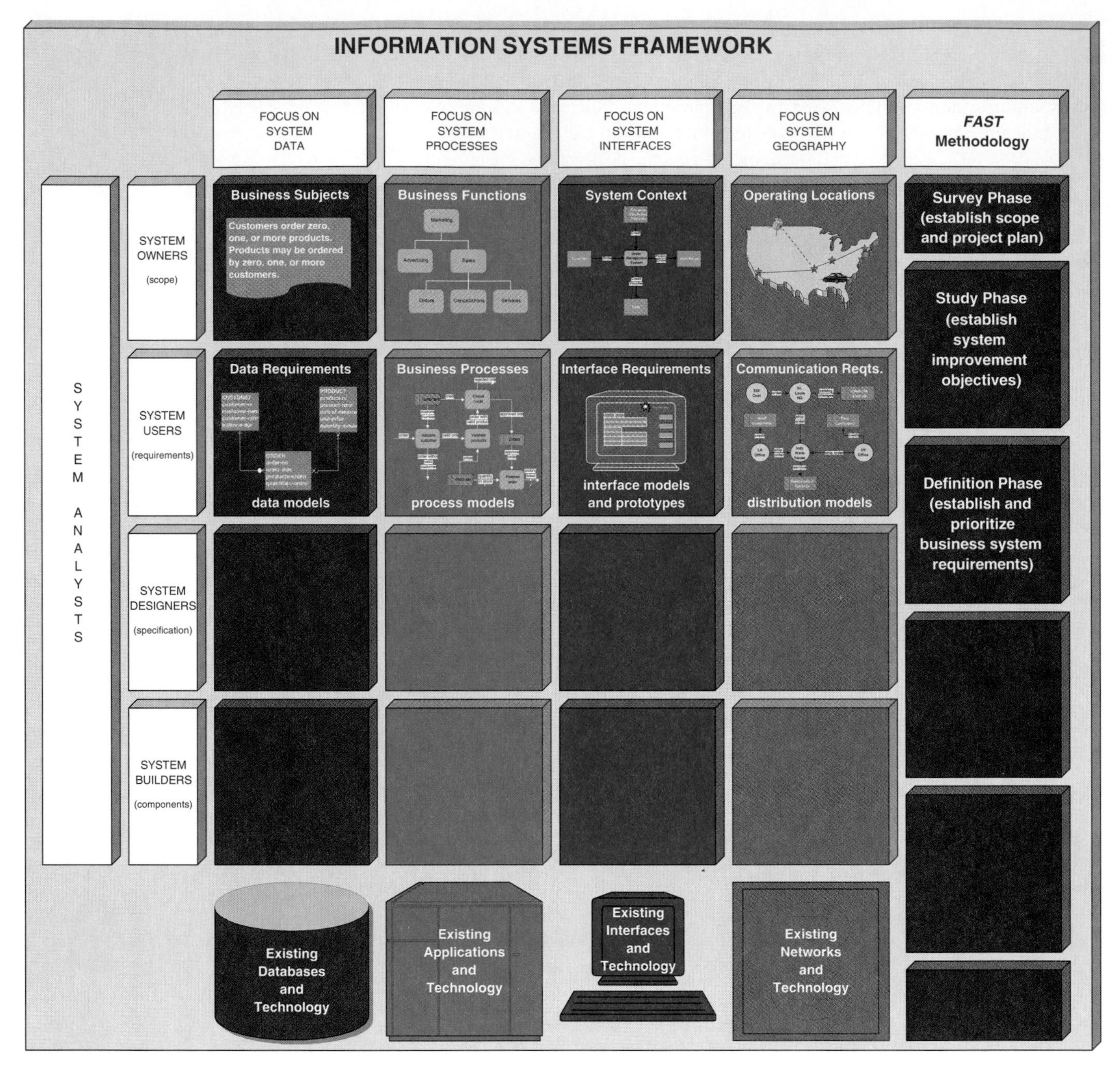

FIGURE 4.14 *Building Blocks for the Definition Phase*

- Appropriate direct and indirect users (OPT)—Be careful! If too many users are involved in this activity, the team can easily become overly preoccupied with details.
- Systems analysts.
  - System architects (REC)—We introduce this role here and define it to be systems analysts who are skilled in the modeling tools used to document and verify business system requirements. Although models are not a deliverable of this activity, this activity begins to capture the facts that will be used to construct models.
- System designers are not typically involved in this activity.
- System builders are not typically involved in this activity.

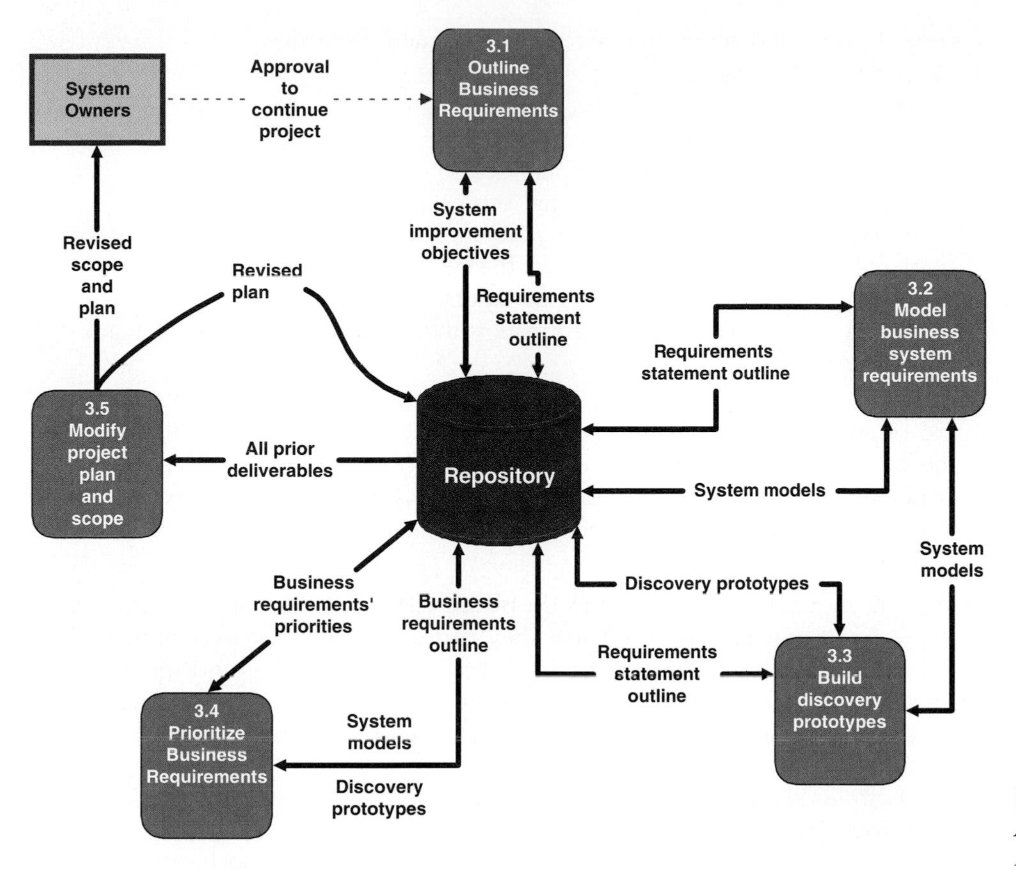

FIGURE 4.15
Activity Diagram for the Definition Phase

Prerequisites (Inputs) This activity is triggered by approval from the system owners to continue the project into the definition phase. The key input is the *system improvement objectives* (REQ) from the study phase. Of course, any and all relevant information from the study phase should be available for reference as needed.

Deliverables (Outputs) The only deliverable of this activity is a *requirements statement outline* (REQ). Various formats can work. In its simplest format, the outline could be divided into four logical sections: (1) the original list of objectives, (2) inputs, (3) processes, and (4) outputs. We prefer a slightly more complex format that associates inputs, processes, and outputs with each individual system improvement objective.

Applicable Techniques The following techniques are applicable to this activity. For each activity, we indicate which chapters or modules of the book teach that technique.

- *Joint application development.* The preferred technique for rapidly outlining business system requirements is joint application development (REC). The requisite analysis can usually be completed in less than one-half a working day. JAD techniques and conflict resolution are explored in Part Five, Module D, "Joint Application Development."
- *Interpersonal skills.* Good interpersonal skills are necessary to maintain a focus on the requirements. These skills are explored in Part Five, Module E, "Interpersonal Skills and Communications."

Steps The following steps are suggested to complete this activity.

1. (REQ) Review and refine the system improvement objectives.
2. (REQ) For each objective:
   a. Identify and document any business events or inputs to which the system must respond. Briefly define each event or input, but do not define the specific data content of any input.
   b. Identify and document any special business policies, processing, or decisions that must be made to adequately respond to each event or input.
   c. Identify and document the normal business outputs or responses to the aforementioned business events or inputs.
   d. Identify and document any information that must be produced or made available.
3. (REC) Compare the system improvement objectives and requirements against the original problem statements from the study phase. Are you still solving the original problems or is the scope of the project growing? Increased scope is not necessarily wrong; however, an appropriate adjustment of expectations (particularly schedule and budget) may eventually become necessary.

Activity: Model Business System Requirements

Identifying requirements is only the tip of the iceberg. We must also find a way to express the requirements such that they can be verified and communicated to both business and technical audiences. Technicians must understand requirements so they can transform them into appropriate technical solutions. Business users must understand requirements so they can prioritize the needs and justify the expenditures for any technical solution. Best current practice suggests that we *model business system requirements.* Recall that system models are pictures (similar to flowcharts) that express requirements or designs.

The best systems analysts can develop models that provide no hint of how the system will or might be implemented. This is called logical or essential system modeling.

> **Logical models** depict what a system is or what a system must do—*not* how the system will be implemented. Because logical models depict the *essence* of the system, they are sometimes called **essential models.**

Logical models express business requirements—sometimes referred to as the **logical design**—as opposed to the technical solution, which is called the *physical design.* In theory, by focusing on the logical design of the system, the project team will (1) appropriately separate business concerns from technical solutions, (2) be more likely to conceive and consider new and different ways to improve business processes, and (3) be more likely to consider different, alternative technical solutions (when the time comes for physical design).

Purpose The purpose of this activity is to model business system requirements such that they can be verified by system users and subsequently understood and transformed by system designers into a technical solution. In a sense, the system models bridge the communication gap that inevitably exists between business and technical staffs.

Roles The activity is facilitated by a *systems analyst.* Other roles are defined as follows:

- System owner roles.
  - User managers (REC).
  - Project manager (REC).
- System user roles.
  - Business analysts (REC).
  - Appropriate direct and indirect users (REQ).

- Systems analysts.
  - System architects (REQ).
- System designers are not recommended since they tend to talk in technical terms that intimidate and frustrate the users and user managers.
- System builders are not typically involved in this activity. On the other hand, programmers who are skilled in user interface construction might be invited to observe the activity as a preface to constructing rapid prototypes of user interfaces for later activities.

Prerequisites (Inputs) This activity is usually triggered by completion of the *requirements statement outline* (REC); however, it often begins as part of the same group meeting. Because system modeling is a mature technique, there may already exist many system models or templates in the corporate repository. These models may be very general or very detailed, depending on their source.

It is also possible to *reverse engineer* some system models directly from existing databases and program libraries. The resulting models tend to be very physical, meaning technology-oriented. The systems analyst would need to translate those physical system models into logical system models.

Deliverables (Outputs) The deliverables of this activity are the *system models* (REQ). Your information systems framework identifies the need for four system models:

- DATA—All systems capture and store data. **Data models** (such as the entity relationship diagram shown earlier in Figure 4.4) are used to model the data requirements for many new systems. These data models eventually serve as the starting point for designing databases.
- PROCESSES—All systems perform work. **Process models** (such as the data flow diagram shown earlier in Figure 4.2) are frequently used to model the work flow through business systems. These process models serve as a starting point for designing computer applications and programs.
- INTERFACES—No system exists in isolation from either people, other systems, or other organizations. Actually, **interface models** such as context diagrams (Figure 4.16) are drawn in the study phase, but they might be expanded to show more detail in this definition phase activity. Context diagrams depict net inputs to the system, their sources, net outputs from the system, their destinations, and shared databases. These interface models serve as the basis for designing user and system interfaces.
- GEOGRAPHY—Because today's business and information systems have greater geographic breadth, systems analysts are exploring ways to model geographical locations. These **distribution models** serve as a starting point for designing the communication systems for distributing the data, processes, and interfaces to the various geographical locations. We'll defer any samples of a distribution model to Chapter 7.

As suggested earlier in this chapter, many systems analysts are now experimenting with **object models** as an alternative to data and process models.

Most systems analysts use computer-aided systems engineering (CASE) software to construct and maintain system models and their underlying documentation. Most of the system models illustrated in this book were developed with a CASE tool called *System Architect*. An inexpensive, student edition of this CASE tool is available from Irwin along with a tutorial workbook. Most CASE tools support system modeling; therefore, gaining experience with a tool such as *System Architect* is easily transferable to organizations that use any other CASE tool.

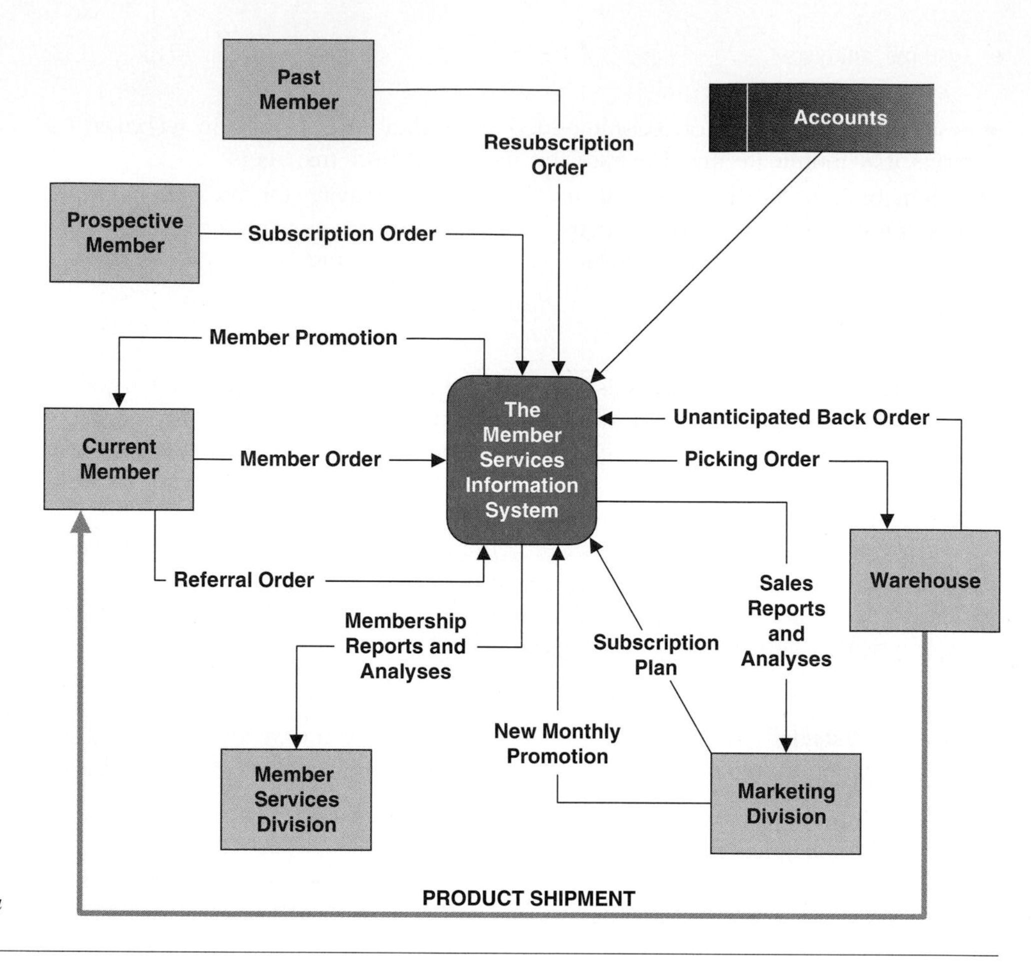

FIGURE 4.16
Interface Model (also Called a Context Diagram)

Applicable Techniques The following techniques are applicable to this activity. For each activity, we indicate which chapters or modules of the book teach that technique.

- *Data modeling.* Data modeling is the most popular technique for expressing the business requirements for data that will be stored in a system's database. This technique is extensively covered in Chapter 5, "Data Modeling."
- *Process modeling.* Arguably, process modeling is the oldest and most widely practiced technique for expressing business process requirements, work flow, inputs, and outputs. This technique is taught in Chapter 6, "Process Modeling." Chapter 6 also teaches you how to draw the interface model or context diagram.
- *Distribution modeling.* Distribution modeling is the least mature of the system modeling techniques covered in this textbook; however, its importance is growing as organizations seek ways to express the business geography to be supported by a system. Chapter 7 teaches distribution modeling.
- *Object modeling.* Object modeling has been in development for many years; however, it is only now emerging as a successor to traditional data and process modeling. Object modeling is being driven by the growing use of object technology and object-oriented analysis methods to exploit that technology. Chapter 8 teaches object modeling using the latest object modeling standards.
- *Fact-finding.* You can't build models without facts. These techniques are taught in Part Five, Module B, "Fact-Finding and Information Gathering."

- *Joint application development.* JAD has become the most popular technique for quickly constructing system models in direct cooperation with system owners and system users. JAD techniques merge the model construction and verification into the same meetings to accelerate the project. JAD is taught in Part Five, Module D, "Joint Application Development."

Steps The following steps are suggested to complete this activity.

1. (REC) Review the system improvement objectives and requirements statement outline.
2. (REQ) Collect or retrieve any system models that may have been developed in prior projects. High-level system models may have been created as part of an information strategy planning project or business process redesign project. Detailed models may have been created as part of prior application development projects. In either case, existing models are typically stored in the corporate repository. Many organizations have formal checkout/check-in procedures for using and updating existing system models.
3. (OPT) If the appropriate CASE technology is available, consider reverse engineering existing databases or applications into physical system models. Then translate those physical models into more business-friendly logical system models. The value of this step depends on the quality and value of the databases and applications to be reverse engineered. Many systems are so old or poorly designed that the value of reverse engineering is questionable.
4. (REQ) Draw the interface model. The interface model establishes the scope and boundary for the entire project.
5. (REQ) Depending on your modeling strategy of choice:
   a. If you practice *structured analysis:*
      i. (REQ) Construct and verify the process models.
      ii. (REC) Construct and verify data models.
      iii. (REC) Synchronize process and data models. This synchronization ensures that the models are consistent and compatible with one another.
      iv. (OPT) Construct and verify distribution models.
   b. If you practice *information engineering* (REC):
      i. (REQ) Construct and verify data models.
      ii. (REQ) Construct, verify, and synchronize the process models.
      iii. (REC) Construct and verify the distribution models.
   c. If you practice *object-oriented analysis* (the leading edge):
      i. (REC) Identify use cases. Use cases are an object method that connects objects to familiar business events. Use cases are taught in the object modeling chapter.
      ii. (REQ) Construct and verify object models. Several popular object model standards exist.

[OPT] Activity: Build Discovery Prototypes

An alternative or complementary approach to system modeling of business requirements is to *build discovery prototypes.* You may recall the following definitions:

> **Prototyping** is the act of building a small-scale, representative, or working model of the users' requirements to discover or verify the users' requirements.

Prototyping is typically used in the requirements definition phase to establish user interface requirements (as was shown in Figure 4.14, INTERFACE column, SYSTEM USER row). But these prototypes also help analysts and programmers identify detailed business requirements that show up on the screens. For this reason, user interface prototypes are often called discovery prototypes.

Discovery prototypes are simple mock-ups of screens and reports that are intended to help systems analysts discover requirements. The discovered requirements would normally be added to system models. A synonym is requirements prototypes.

Although discovery prototyping is optional, it is frequently applied to systems development projects, especially in those cases when the parties involved are having difficulty developing or completing system models. The philosophy is that the users will recognize their requirements when they see them.

Prototypes are developed using **fourth-generation languages** (4GLs), most of which include **rapid application development** (RAD) facilities for quickly "painting" screens, forms, and reports. Examples of such languages include *Powerbuilder* and *Focus;* however, most third-generation languages are evolving to include the same RAD facilities (e.g., *Visual BASIC* and *Delphi,* which is based on Pascal). Furthermore, most PC-based database management systems now include rapid application development facilities that can be used to prototype inputs and outputs (e.g., *Access, Visual dBASE,* and *Paradox*).

Purpose The purpose of this optional activity is to (1) establish user interface requirements, and (2) discover detailed data and processing requirements interactively with users through the rapid development of sample inputs and outputs.

Roles The activity is facilitated by the *business analyst* or *systems analyst.* Other roles are defined as follows:

- System owners usually do not elect to participate unless they are also system users.
- System user roles.
  - Business analyst (REC)—The first time, a phrase is added to define the role. After that, role name only unless some special explanation is required.
  - Direct system users.
- Systems analyst roles—Systems analysts facilitate, observe, and assist this activity. It should be recognized that many systems analysts have the skills necessary to play the system designer and builder roles described below.
- System designer roles.
  - User interface specialist (OPT)—If prototypes are to evolve into the final system, a user interface specialist is suggested. User interface specialists are system designers who have become skilled in the standards or guidelines of a particular graphical user interface (such as *MacOS* or *Windows 95*). They are also responsible for local interface standards that ensure that all internally developed applications exhibit the same look and feel.
- System builder roles.
  - Prototyper (REQ)—This is a systems analyst or programmer who is skilled in the use of rapid application development (RAD) technology for the purpose of developing discovery prototypes.
  - Programmer (OPT)—It is not the purpose to write complete application programs at this stage of the project; however, the current trend is toward using the same prototyping languages to rapidly complete the development of the applications. Consequently, programmers may get useful insights from observing this activity.

Prerequisites (Inputs) This activity is not triggered by any event. It occurs, as deemed appropriate or desirable, during the system modeling activity. Thus, it occurs in parallel with system modeling. It uses the *system requirements outline* (REC). It can also use any *system models* (REC) as they are developed.

Deliverables (Outputs) The deliverables of this activity are *discovery prototypes* (REQ) of selected inputs and outputs. These prototypes are shared with the system modeling activity since they can be very helpful in refining the system models. The models are also stored in the repository since they can eventually evolve into the final production system.

Applicable Techniques The following techniques are applicable to this activity. For each activity, we indicate which chapters or modules of the book teach that technique.

- *Prototyping.* Prototyping (REQ) is predominantly considered to be a design technique because it is based on design and construction of actual program components. For that reason, we teach it in the system design unit of this textbook (Chapters 12–14).
- *Technology.* It is not the intent of this book to teach specific technologies. There are many RAD technologies that are well suited to prototyping. Clearly, the actual use of prototyping will require an investment in learning the technology to be used. Fortunately, most of today's visual programming languages are easy to learn and use as prototyping tools. (It will take a much greater knowledge of these languages to complete the application's development beyond the prototype.)

Steps The following steps are suggested to complete this activity.

1. (REC) Review the system improvement objectives and requirements statement outline.
2. (REC) Study any system models that may have been developed.
3. (OPT) Working directly with system users, construct a simple, single-user prototype of the database and load it with some sample data. Do not become preoccupied with data editing and perfection.
4. (REQ) Working directly with the system users, construct input prototypes for each business event. Do not worry about input editing, system security, etc.; the focus is completely on business requirements. Do not spend too much time on any one input since this stage does not develop the final system.
5. (REQ) Working directly with system users, construct output prototypes for each business output. Do not worry about whether the data are real or whether or not they make sense. Focus on identifying the columns, totals, and graphs the users are seeking. If you built a sample database in step 3 and used step 4 to collect data for that database, you can probably use that database prototype to quickly generate sample reports.
6. (REC) Return to the system modeling activity to formalize the requirements that have been discovered through the above prototyping steps.

Activity: Prioritize Business Requirements

We stated earlier that the success of a systems development project can be measured in terms of the degree to which business requirements are met. But actually, all requirements are not equal. If a project gets behind schedule or over budget, it may be useful to recognize which requirements are more important than others. Systems can be built in versions. Early versions should deliver the most important requirements.

Thus, given the proposed system requirements, models, and prototypes defined earlier, the system owner and systems analyst should *prioritize business requirements.* The actual priorities should be specified jointly by the system owners and users. Prioritization of business requirements also enables a popular technique called timeboxing.

> **Timeboxing** is a technique that develops larger fully functional systems in versions. The development team selects the smallest subset of the system that, if fully implemented, will return immediate value to the system owners and users. That subset is developed, ideally with a time frame of six to nine months or less. Subsequently, value-added versions of the system are developed in similar time frames.

Timeboxing requires that the priorities be clearly understood.

Purpose The purpose of this activity is to prioritize business requirements for a new system. It is not the intent of this activity to eliminate any business requirement—only to recognize that some requirements can be implemented later than others.

Roles The activity is facilitated by the *business analyst* or *project manager*. Other roles are defined as follows:

- System owner roles.
  - Executive sponsor (OPT)—Most executive sponsors will empower their user managers (below) to make these decisions.
  - User managers (REQ)—The reason should be obvious. If management strongly disagrees with priorities, the project could lose support.
  - Project manager (REC).
- System user roles.
  - Business analyst (REC)—A great choice for facilitator of this activity since business analysts come from the user community and have the trust of that community.
  - Appropriate direct and indirect system users (REQ).
- Good systems analysts listen to discussion and answer questions during this activity. User buy-in to priorities is critical to the political feasibility of any new system if a systems analyst or project manager facilitates this activity.
- System designers are not typically involved in this activity because they tend to influence priorities for technical, nonbusiness reasons.
- System builders are not typically involved in this activity.

Prerequisites (Inputs) This activity can begin in parallel with the other definition phase activities. The only inputs are business requirements as expressed in the updated *business requirements outline* (REQ), *system models* (REC), and *discovery prototypes* (OPT).

Deliverables (Outputs) While some priorities can be set as requirements are established, the priorities can never be finalized until all the business requirements have been identified. The deliverables of this activity are *business requirements' priorities* (REQ) as recorded in the repository.

Applicable Techniques There are no special techniques for prioritizing requirements. A simple but effective technique is described as follows:

- Is the business requirement mandatory? In other words, is it something the system owner must have? Careful! There is a temptation to label too many requirements as mandatory. By definition, if a system does not include a mandatory requirement, that system cannot fulfill its purpose. Perform the following test on any suspected mandatory requirements: Rank them. If you can rank them, they are not mandatory. You should not be able to rank requirements that you absolutely must have. All mandatory requirements are essential to the first version of the system!

- Is the business requirement desirable but not essential? Desirable requirements are things the user eventually wants; however, the early versions of the system can provide value without them. Unlike mandatory requirements, desirable requirements can and should be ranked.
- Is the business requirement optional? This is a catchall category for those features and capabilities that you could live without indefinitely. Although these would be nice to have, they are not really requirements. These requirements can also be ranked.

The priorities resulting from this analysis will allow the system owner and systems analyst to make intelligent decisions if cost and schedule become constrained.

Steps The following steps are suggested to complete this activity.

1. (REQ) For each system input and output, categorize it as mandatory, optional, or desirable.
2. (REQ) For each desirable requirement above, rank it with respect to the other desirable requirements. Make note of any dependencies that exist between requirements (e.g., requirement 5 depends on the implementation of requirement 2).
3. (REQ) For each optional requirement, rank it with respect to the other optional requirements. Make note of any dependencies that exist between requirements.
4. (OPT) Define system versions. A recommended scheme follows:
   a. (REC) Version one consists of all mandatory requirements.
   b. (REC) Versions two through *x* consist of logical groupings of desirable requirements.
   c. (REC) Optional requirements are usually added to versions as time permits or deferred to maintenance releases of the system. Many such requirements are for new reports. Today, users can be given relatively simple technology (e.g., query tools and report generators) to fulfill many such requirements on their own.

Activity: Modify the Project Plan and Scope

Here again, recall that project scope is a moving target. Now that we've identified business system requirements, we should step back and redefine our understanding of the project scope and adapt our project plan accordingly. The team must consider the possibility that the new system may be larger than originally expected. If so, they must adjust the schedule, budget, or scope accordingly.

Purpose The purpose of this activity is to (1) modify the project plan to reflect changes in scope that have become apparent during requirements definition, and (2) secure approval to continue the project into the next phase. (Work may have already started on the configuration or design phases; however, the decisions still require review.)

Roles The activity is facilitated by the *project manager*. Other roles are defined as follows:

- System owner roles.
  - Executive sponsor (REQ)—As the final spending authority, the sponsor must approve project continuation.
  - User managers (REQ)—The system belongs to these managers; therefore, their input is crucial.
  - Project manager (REQ)—The project manager must make any changes to the schedule and budget.

- System user roles.
  - Business analysts (OPT).
- Other systems analysts are not usually involved in this activity.
- System designer roles.
  - Database administrator (REC)—Given the data model, the database administrator will reassess time to design and implement the database.
  - Network administrator (REC)—Given the geographic model, the network administrator will reassess the time to design, expand, or implement the network.
  - Application administrator (REC)—Given the process and data, or object models, the application administrator will reassess the time to design, construct, and implement the application programs.
- System builders are not involved in this activity.

Prerequisites (Inputs) This activity is triggered by initial completion of the *system models* (REQ), *discovery prototypes* (OPT), and the *business requirements priorities* (REC).

Deliverables (Outputs) The deliverable of this activity is a revised *project plan* (REC) that covers the remainder of the project. Additionally, a detailed *configuration plan* (OPT) and *design plan* (OPT) could be produced.

Applicable Techniques The following techniques are applicable to this activity. For each activity, we indicate which chapters or modules of the book teach that technique.

- *Process management.* Process management (REC) (also called methodology management) defines the standards for applying the methodology to a project. Process management techniques are taught in Part Five, Module A, "Project and Process Management."
- *Project management.* Project management (REC) builds on process management by applying the methodology to specific projects in the form of schedule planning, staffing and supervision, progress reporting, management of expectations, budgeting, and schedule management. Project management techniques are taught in Part Five, Module A, "Project and Process Management."
- *Presentation skills.* The project charter and any verbal presentations of the project and plan obviously require presentation skills (REC). Presentation skills are taught in Part Five, Module E, "Interpersonal Skills and Communications."

Steps The following steps are suggested to complete this activity.

1. (REC) Review the original plan.
2. (REQ) Review the up-to-date business requirements outline, system models, discovery prototypes, and business requirements' priorities. Ask yourself two questions:
   a. Has the scope of the project significantly expanded?
   b. Are the requirements more substantial than originally anticipated?
3. (REC) Estimate the time required for each project activity in the next phase—the design phase.
4. (REC) If necessary, refine baseline estimates for the overall project plan.
5. (OPT) If the answer is yes, then renegotiate scope, schedule, and/or budget with the system owner group. (See Part Five, Module A, "Project and Process Management," for a simple expectations management instrument.)

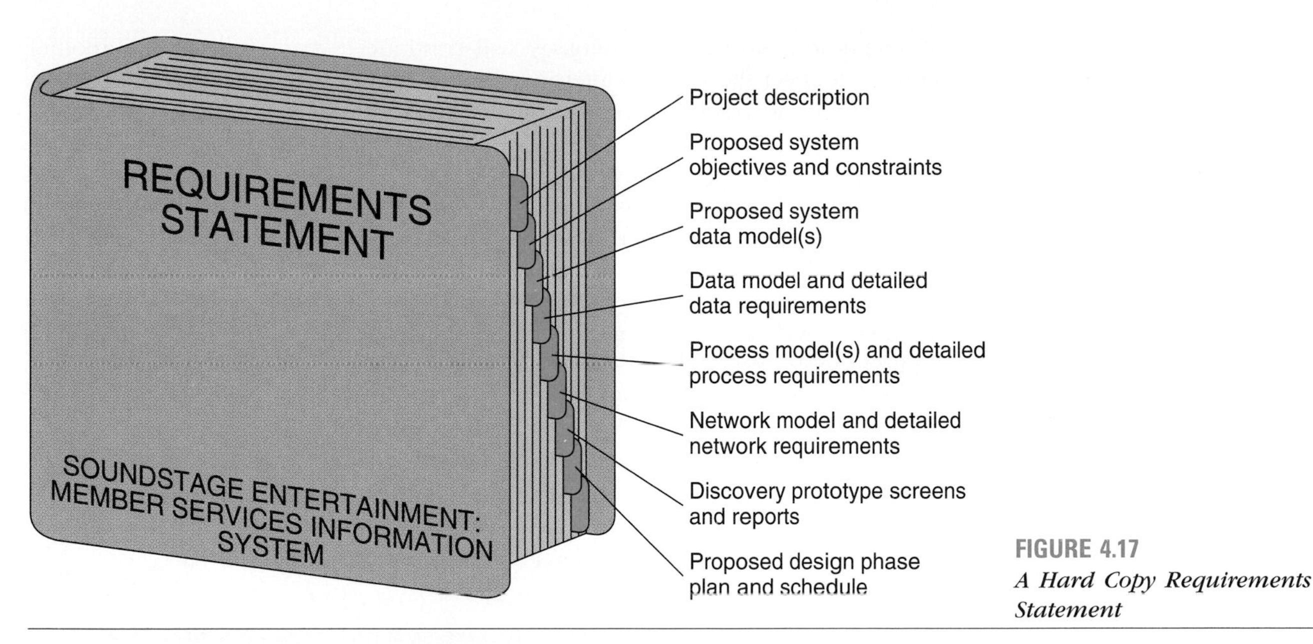

FIGURE 4.17
A Hard Copy Requirements Statement

Some Final Words about System Requirements

A consolidation of all system models, discovery prototypes, and supporting documentation is sometimes called a **requirements statement.** All elements of the requirements statement are stored in the repository, but most systems analysts find it useful to keep a printed copy of that documentation for reference and reporting. This hard-copy requirements statement might be organized as shown in Figure 4.17. It is not recommended that typical users be exposed to this rather imposing document. The contents of the requirements statement are best presented to users in subsets (for example, the data models and associated documentation).

Finally, the definition phase is now complete—or is it? It was once popular to freeze the business requirements before beginning the system design and construction phases. But today's economy has become increasingly fast paced. Businesses are measured on their ability to adapt to rapidly changing requirements and opportunities. Information systems can be no less responsive. *FAST* does not freeze the requirements statement. Nor does the methodology call for a gala presentation of requirements to the business community. Such an event might be construed to impart a false sense of completeness or perfection. The definition phase does not end with a bang. It quietly and efficiently transitions into the configuration and design phases. As new business requirements come up, they will be acknowledged and documented in the system models and repository. But they will also be carefully assessed to determine their impact on scope, schedule, and budget. System owners, when promptly supplied with new information, will ultimately decide whether the new requirements are mandatory, desirable, or optional and then make the decision on their inclusion.

THE NEXT GENERATION OF REQUIREMENTS ANALYSIS

Predicting the future of requirements analysis is not easy, but we'll make an attempt. CASE technology will continue to improve making it easier to model system requirements. Two CASE technologies will lead the charge. First, CASE tools will include object modeling to support emerging object-oriented analysis techniques. Some CASE tools will be purely object-oriented, but we believe that the demand for other types of modeling support (data modeling for databases and process modeling for business processes and work flow), and the need to integrate these models with object models, will place a premium on comprehensive CASE tools that can support many types of models. Second, the reverse engineering technology in CASE tools will improve our ability to more quickly generate first-draft system models from existing databases and application programs.

CASE technology and RAD technology will continue to complement one another. Indeed, we expect the tools to interoperate in both directions to simplify both system modeling and discovery prototyping.

Object-oriented analysis is poised to eventually replace structured analysis and information engineering as the method of choice among systems analysts. This change will not be as rapid as the object purists would like, but it will occur all too rapidly for a generation of systems analysts who were skilled in the older methods. There is a grand opportunity for young talent who are well educated and skilled in the use of object-oriented analysis (and design); however, career opportunities will remain strong for database specialists who know data modeling and information engineering. Also, process modeling is enjoying something of a renaissance thanks to the popularity of business process redesign projects.

One thing will *not* change! We will continue to need systems analysts and business analysts who understand how to fundamentally analyze business problems and define logical business requirements as a preface to system design. But we will all have to do that with increased speed and accuracy to meet the accelerated systems development schedules required in tomorrow's faster-paced economy.

WHERE DO YOU GO FROM HERE?

This chapter provided a detailed overview of the systems analysis phases of a project. You are now ready to learn some of the systems skills introduced in this chapter. Because system modeling is the most popular systems analysis technique, we recommend you complete Chapters 5 to 8. Each of these chapters teaches a different system modeling technique that you can immediately apply to projects.

The order of the system modeling chapters is flexible; however, we do recommend Chapter 5, "Data Modeling," first. All applications have databases, and data modeling is an essential skill for database development. Also, it is easier to synchronize early data models with later process models than vice versa.

This would also be an appropriate time to develop some of the soft skills of systems analysis. For example, models cannot be constructed without business facts and ideas. Part Five contains two modules that can help you learn about getting facts and ideas from system owners and users. Specifically, Module B, "Fact-Finding and Information Gathering," presents the traditional ways to collect facts and ideas. Module D, "Joint Application Development," presents a more contemporary approach based on facilitated group meetings. If your course requires a live project engagement with real managers and users, one or both modules will prove invaluable to your later system modeling efforts.

For those of you who have completed a systems analysis course already, this chapter was a review. For you, we suggest you merely review the system modeling chapters and proceed to Chapter 9, "Systems Design."

SUMMARY

1. Formally, systems analysis is the dissection of a system into its component pieces. As a problem-solving phase, it precedes systems synthesis or systems design. With respect to information systems development, systems analysis is the survey and planning of a project, the study and analysis of the existing system, and the definition of business requirements for the new system.
2. The results of systems analysis are stored in a repository for use in later phases and projects.
3. There are several popular or emerging strategies for systems analysis. These techniques can be used in combination with one another.
   a. Modern structured analysis, a technique that focuses on processes.
   b. Information engineering, a technique that focuses on data and strategic planning.
   c. Prototyping, a technique that focuses on building small-scale prototypes of solutions.
   d. Joint application development, a technique that focuses on facilitated group meetings with both technicians and users.
   e. Business process redesign, a technique that focuses on simplifying and streamlining fundamental business processes before applying information technology to those processes.
   f. Object-oriented analysis, a technique that integrates the concerns of data and processes to create objects that can be easily adapted and reused.
4. Each phase of systems analysis (survey, study, and definition) can be understood in the context of the information system building blocks: DATA, PROCESSES, INTERFACES, and GEOGRAPHY.
5. The purpose of the survey phase is to determine the worthiness of the project and to create a plan to complete those projects deemed worthy. To accomplish the survey phase objectives, the systems analyst will work with the system owner, system users, IS manager, and other IS staff to: *(a)* survey problems, opportunities, and solutions; *(b)* negotiate project scope; *(c)* plan the project; and *(d)* present the project. The deliverable for the survey phase is an oral or written project feasibility assessment that must go to a decision-making body, commonly referred to as the steering committee. A steering committee is a decision-making body that prioritizes potential information systems projects. Not all projects require this evaluation. For example, projects initiated by formal systems planning have already been justified by the planning team and management. In such cases the project skips directly to the study phase.
6. The purpose of the study phase is to answer the questions: Are the problems really worth solving? and Is a new system really worth building? To answer these questions, the study phase thoroughly analyzes the alleged problems and opportunities first identified in the survey phase. To complete the study phase, the analyst will continue to work with the system owner, system users, and other IS management and staff. The systems analyst and appropriate participants will: *(a)* optionally, model the current system; *(b)* optionally, analyze business processes; *(c)* analyze problems and opportunities; *(d)* establish system improvement objectives and constraints; *(e)* modify project scope and the plan; and *(f)* present the findings and recommendations.
7. The purpose of the definition phase is to identify what the new system is to do without the consideration of technology; in other words, to define the business requirements for a new system. As in the survey and study phases, the analyst actively works with system users and owners as well as other IS professionals to complete the definition phase. To complete the definition phase, the analyst and appropriate participants will: *(a)* outline business requirements; *(b)* model the business system requirements; *(c)* optionally, build discovery prototypes; *(d)* prioritize the business system requirements; and *(e)* modify the project plan and scope.
8. System models are a key deliverable in systems analysis. The project team typically builds:
   a. Data models for the business database requirements.
   b. Process models or object models for the business process requirements to be programmed.
   c. Interface models to demonstrate how the system should interact with both system users and other systems (including external organizations).
   d. Distribution models for the business geography to be supported by a network.

KEY TERMS

activity diagram, p. 129
business area analysis, p. 124
business process redesign, p. 126
constraint, p. 146
data model, p. 155
discovery prototypes, p. 158
distribution model, p. 155
essential model, p. 154
feasibility prototyping, p. 125
fourth-generation language, p. 158
information engineering, p. 123
information strategy planning, p. 123
interface model, p. 155
joint application development, p. 125
logical design, p. 121
logical model, p. 154
model, p. 122
model-driven development, p. 122
modern structured analysis, p. 122
object model, p. 155
object-oriented analysis, p. 126

objective, p. 145
process model, p. 155
prototyping, pp. 125, 157
rapid application development, p. 125
repository, p. 121
requirements prototyping, p. 125
requirements statement, p. 163
steering body, p. 135
systems analysis (classical), p. 120
systems analysis (contemporary), p. 121
systems synthesis, p. 120
timeboxing, p. 160

REVIEW QUESTIONS

1. What is the difference between systems analysis and systems synthesis?
2. What is the difference between systems analysis and logical design?
3. What role does a repository play in systems analysis?
4. What is the difference between modern structured analysis and information engineering? Which approach is most popular at the time of this book's publication?
5. What is model-driven development?
6. What is prototyping? Is it an alternative to model-driven development? Why or why not?
7. What is business process redesign?
8. What is object-oriented analsysis? How is it similar to, and different from, modern structured analysis and information engineering?
9. Identify and briefly describe the purpose of the three systems analysis phases.
10. What important question is addressed during the survey phase? What are the fundamental objectives of the survey phase? How might the information systems building blocks be used to identify the general level of understanding required for fulfilling these objectives?
11. What important question is addressed during the definition phase? What are the fundamental objectives of the definition phase? How might the information systems building blocks be used to identify what requirements must be specified to fulfill these objectives?
12. What is the end product of each systems analysis phase? Explain the purpose and content of each of these products.

PROBLEMS AND EXERCISES

1. How do the classical definitions of systems analysis and systems synthesis relate to contemporary systems analysis and design?
2. What triggers the survey phase? Do all systems projects go through a survey phase? Explain your answer.
3. Differentiate between the survey and study phases of the systems analysis process.
4. What are the fundamental objectives of the study phase? Explain how the information systems building blocks aid in identifying the required level of understanding to be obtained in the study phase.
5. What is the value of modeling during the systems analysis phases? What types of models might be developed during the survey, study, and definition phases?
6. Explain how the joint application development (JAD) technique can be used during the definition phase. Explain why prototypes and system models should complement one another.

PROJECTS AND RESEARCH

1. How might the PIECES framework introduced in Chapter 3 be used in the study phase of systems analysis? Use the PIECES framework to evaluate your local course registration system. Do you see problems or opportunities? (Alternative: Substitute any system with which you are familiar.)
2. Interview a systems analyst or systems manager in the information services unit of a local organization. How do they initiate projects? What type of report or document do they produce to justify a project in the early stages of development?
3. Interview a systems analyst in a local organization. What techniques do they use to ensure that the problems stated by their users are worthy of solution? Do they quantify that worth? If so, how? If not, why not?
4. Interview a systems manager in a local organization. What strategy do they use to verify and document business requirements for a system? Which of the strategies described in this book most closely correspond to their strategies? How do they teach these strategies to their analysts? How do their users react to these strategies? How do they enforce the use of their strategies? What new

techniques are they exploring? (Ask about object-oriented analysis, BPR, information engineering, and JAD.)

5. Research the subject of "getting the requirements correct." While strategies such as information engineering and object-oriented analysis provide mechanisms for documenting requirements, their value in discovering requirements has often been criticized. Find out why, and offer your perspectives on how the popular techniques might be improved.

MINICASES

1. Colleens Financial Services is a nationwide financial services company headquartered in Omaha, Nebraska. Senior systems analyst Fred McNamara is meeting with Ken Borelli, the MIS manager at Colleens Financial Services. Fred has just completed an evaluation of a new software package, a fourth-generation programming language (4GL). In addition to evaluating the 4GL, Fred has been asked to learn about a popular systems development approach called prototyping that uses 4GLs to build working models of systems. Fred is meeting with Ken to give him his assessment of the 4GL product and prototyping as an alternative approach to systems development.

   Ken starts the conversation. "So what do you think about that new software product? Is it worthy of being called a 4GL?"

   Fred replies, "Without a doubt! Third-generation programming languages like COBOL can't compare to it! I think this product can do wonders for the systems staff. It is very user friendly, and it provides a number of facilities that assist in developing a complete system."

   "What kind of facilities does it provide?" asks Ken.

   Fred answers, "To give you some idea, I used a facility to develop a database containing actual data, another facility to produce a relatively complex printed report against that database, and other facilities to develop menus and other input and output screens all in a fraction of the time that would have been required with COBOL."

   Ken says, "You said it was very user friendly. Does that mean you didn't have to spend a lot of time referencing manuals?"

   "I hardly used the manuals," answers Fred. "The facilities simply led me through a series of questions or prompts. All I had to do was answer the questions. The 4GL generated the program code, which I subsequently executed. The productivity implications are tremendous!"

   Ken responds enthusiastically, "Sounds like the product is a good investment. And what about prototyping? Do you think prototyping is something we should consider doing as an alternative to our current approach to developing systems?"

   "Well," answers Fred, "prototyping certainly takes advantage of tools such as 4GLs. The strategy is very simple. It emphasizes the development of a working model of the target system, instead of traditional paper specifications. You begin building the model by first defining the database requirements for the new or desired system. Identifying the database requirements is relatively simple, since the data for most systems already exist, either in computer files or manual forms. Afterward, you then build and load a database using the 4GL. Once you have the database built, the rest is easy. You can use various facilities to quickly generate the menus, reports, and input and output screens. The analyst does not have to worry about whether the working model is totally complete or accurate. The end-user is encouraged to review the model and provide the analyst with feedback. If the model, say a report, is not acceptable, the analyst simply makes requested changes and reviews it again with the end-user at some later time. This repetitive process and active end-user participation are considered essential and to be encouraged."

   "Now that's a new one!" exclaims Ken. "I certainly agree with the idea of encouraging end-user participation. But this attitude of encouraging or expecting to keep redoing work would be difficult to adjust to."

   Fred responds, "I'm sure some of us old-timers will have some difficulty adjusting to this type of thinking."

   Ken pauses momentarily and then says, "It sounds like this new 4GL and prototyping should be pursued further. Both 4GL and prototyping seem to offer some productivity gains. I've been looking for a way to get the end-users more involved in the systems development process, and I think this prototyping approach is the answer. There is one other thing: I suspect my staff's morale would be improved. I believe my staff would be motivated by this new 4GL product and by prototyping's emphasis on building a model as a basis for performing systems development."

   Fred nods, "I agree! I can't wait to start my next project. I've got just the project picked out. I received a request for a new employee benefits system from the Personnel Department. I thought I'd use the 4GL in conjunction with the prototyping approach to complete this project. I won't be needing any programmers since I'll be developing the system myself while I'm working with the end-users. I've already drafted a memo asking Personnel to provide me with some sample records, forms, reports, and other materials that will help me identify their data storage requirements. From those samples, I'll be able to build a database for the new system. Then I'll start meeting with the end-users to define and implement screens and reports."

   Ken suddenly looks concerned. "Hold your horses! I do have some concerns about this approach to systems development. Neither the tool nor the approach justifies a departure from the systems development life cycle concept we follow here at Colleens. And that's exactly what you're proposing. You're proposing to select a project, define some basic requirements, and jump right into the design and construction of a new system. That I won't have. I want you to reconsider things."

Ken reaches for a systems development standards manual on his bookshelf. He turns to a figure that depicts the systems development life cycle phases and continues, "Notice that our systems development life cycle includes several systems analysis phases, including a survey or preliminary investigation phase. Do you fully understand why we require this first phase?"

Fred answers, "Sure, that's where we perform a very quick study of the proposed project request. We try to gain a quick understanding of the size, scope, and complexity of the project."

"You're half right," replies Ken. "But why must we complete the phase? I'll tell you why! Because we receive numerous project requests from our end-users! We have a limited number of resources. We can't take on all the requests. It is the purpose of this phase to address the seriousness of the problems and to prioritize the project request against other requests."

Fred pauses, and then he replies, "I see what you're saying. I sort of jumped the gun by picking this Personnel project without considering other project requests that might be given higher priority."

"Good!" says Ken. "Now let's consider the second phase, the detailed study phase. If the project is to be pursued further, we then conduct a detailed study or investigation of the current system. We want to gain an understanding of the causes and effects of all problems, and to appreciate the benefits that might be derived from any existing opportunities. We don't want to bypass this phase. This phase ensures that any new system we propose solves the problems encountered in the current system."

Fred, after glancing at his notes, says, "That brings us to the requirements phase, right?"

"Right," answers Ken. "For all practical purposes, this is where you were proposing to begin your project. You were going to define data storage requirements for a new employee benefits system. I assume you would also attempt to identify other requirements. What particularly bothered me was that I didn't get an impression that you intended to study the database requirements to ensure data reliability and flexibility of form. And what about completeness checks? What were you going to do to ensure that processes were sufficient to ensure data will be properly maintained?"

Fred looks frustrated. "I'm beginning to lose my confidence in this prototyping approach. I was about to make some big mistakes by trying to bypass several important problem-solving phases and tasks."

Ken smiles and replies, "Listen, there's no reason to write off prototyping. So long as we base our prototypes on some sound design principles, I think we can still achieve all the benefits that you described earlier. I have another concern, though. We will eventually want to look at alternative solutions such as manual versus computer-based systems, on-line versus batch systems, and the like. These options should be evaluated for technical, operational, and economical feasibility. It seems to me that this activity should precede extensive prototyping. We don't want to prematurely commit to less feasible or infeasible solutions."

Fred says, "I understand, it looks like prototyping has potential, especially in the definition phase of analysis. And I suspect that prototyping will greatly accelerate our systems design and implementation phases."

a. Do you think Fred did a thorough evaluation of the fourth-generation software product? What benefits do you believe can be derived from using such a tool?
b. Did Fred view prototyping as an alternative to the traditional systems development life cycle? If so, how should he have viewed it?
c. What systems analysis phases would have been skipped by the prototyping approach Fred proposed to follow? What do you think would have been the results of the employee benefits project if Fred had approached the project in the manner he originally envisioned?

2. A company is considering awarding your consulting firm a contract to develop a new and improved system. But, at the beginning, the company only wants to commit to systems analysis. The firm is concerned about your ability to understand its problems and needs. And most of all, it wants to see what kind of computer-based solutions you propose before it contracts you to design and implement a new system. Write a letter of proposal that will address the company's concerns.
3. You have been a systems analyst with your current employer for the past five years. During this time, the projects that you were assigned came directly from your IS manager. All the projects were originally submitted to the IS manager on formal request forms by various users within the company. But now things are different! Several months ago, the company hired a consulting firm to help it develop a strategic plan for the business. How might the resulting strategic plan affect the IS manager in determining which future projects will receive commitment of IS resources? Explain how a systems plan might affect the phases and activities you perform during systems analysis.
4. You have recently completed the survey phase for an assigned systems project. You have become concerned with your ability to complete the study of the current system and the definition of new system requirements within a reasonable time frame. The project is particularly challenging given that it involves numerous users, having differing vocabularies and system perspectives, who are located across several departments. What strategy would you use to reduce the amount of time required to complete the study and definition phases? How would this strategy deal with the issues presented by the diverse user community?

SUGGESTED READINGS

Application Development Strategies (monthly periodical). Arlington, MA: Cutter Information Corporation. This is our favorite theme-oriented periodical that follows systems development strategies, methodologies, CASE, and other relevant trends. Each issue focuses on a single theme.

Coad, Peter, and Edward Yourdon. *Object-Oriented Analysis,* 2nd ed. Englewood Cliffs, NJ: Yourdon Press, 1991, chap. 1. This chapter is a great way to expose yourself to objects and the relationship of object methods to everything that preceded them.

Gane, Chris. *Rapid Systems Development.* Englewood Cliffs: NJ: Prentice Hall, 1989. This book presents a nice overview of RAD that combines model-driven development and prototyping in the correct balance.

Gause, Donald C., and Gerald M. Weinberg. *Are Your Lights On? How to Figure Out What the Problem REALLY Is.* New York: Dorsett House Publishing, 1990. Here's a title that should really get you thinking, and the entire book addresses one of the least published aspects of systems analysis; namely, problem solving.

Hammer, Mike. "Reengineering Work: Don't Automate, Obliterate." *Harvard Business Review,* July–August 1990, pp. 104–11. Dr. Hammer is a noted expert on business process redesign. This paper examines some classic cases where the technique dramatically added value to businesses.

Wetherbe, James. *Systems Analysis and Design: Best Practices,* 4th ed. St. Paul, MN: West Publishing, 1994. We are indebted to Wetherbe for the PIECES framework.

Wood, Jane, and Denise Silver. *Joint Application Design: How to Design Quality Systems in 40% Less Time.* New York: John Wiley & Sons, 1989. This book provides an excellent in-depth presentation of joint application development (JAD).

Yourdon, Edward. *Modern Structured Analysis.* Englewood Cliffs, NJ: Yourdon Press, 1989.

Zachman, John A. "A Framework for Information System Architecture." *IBM Systems Journal* 26, no. 3 (1987). This article presents a popular conceptual framework for information systems survey and the development of an information architecture.

5

DATA MODELING

CHAPTER PREVIEW AND OBJECTIVES

This is the first of three graphical systems modeling chapters. In this chapter you will learn how to use a popular data modeling tool, *entity-relationship diagrams,* to document the data that must be captured and stored by a system, independently of showing how that data is or will be used—that is, independently of specific inputs, outputs, and processing. You will know data modeling as a systems analysis tool when you can:

- Define systems modeling and differentiate between logical and physical system models.
- Define data modeling and explain its benefits.
- Recognize and understand the basic concepts and constructs of a data model.
- Read and interpret an entity relationship data model.
- Explain when in a project data models are constructed and where they are stored.
- Discover entities and relationships.
- Construct an entity-relationship context diagram.
- Discover or invent keys for entities.
- Construct a fully attributed entity relationship diagram and describe all data structures and attributes to the repository or encyclopedia.

SCENE

We begin this episode shortly after the project executive sponsor has approved the initial system improvement objectives (*FAST* definition phase). It's time to model the new system requirements, beginning with data. Sandra is facilitating a meeting with the following staff:

- David Hensley, representing Legal Services.
- Ann Martinelli, representing Member Services.
- Sally Hoover, representing Member Services.
- Joe Bosley, representing Marketing.
- Antonio Scarpachi, representing the warehouse.
- Bob Martinez, assigned to sketch the data model.
- Sarah Hartman, assigned to take detailed notes.

SANDRA

The purpose of this meeting is to discover the business data that your new system needs to capture and store. I'd like to keep this discussion as nontechnical as possible. Try not to think about files and databases. Instead, I'd like you to focus exclusively on the data and explain to us how that data describes your business.

[pause, noting consensus]

In an earlier meeting, you identified the following things or subjects about which we agreed that the system must capture and store data: agreements, clubs, members, member orders, products, and promotions. Have we missed anything?

ANTONIO

I'm not sure. What about merchandise?

SANDRA

Is that another name for product?

ANTONIO

Not exactly. I'm referring to general merchandise like posters and T-shirts. They do not fall into the same category as products like CDs, videos, and games.

ANN

I don't agree, Antonio. Merchandise is simply another type of product. At least that's the way we treat it in Marketing.

SANDRA

OK, are there other types of products? For example, should we treat audio, video, and game products as separate types?

ANN

I would! They are described by different pieces of data. For example, we need data such as category, media, screen aspect ratio, and motion picture rating for a video. On the other hand, we need a different category and media and a content advisory code for an audiotape or disc.

SANDRA

Different category and media?

ANN

Of course! The categories for a video include science fiction and adventure. But the categories for audio are entirely different, such as popular and country/western. Media for videos include VHS tapes and laserdiscs, while media for audio titles include cassettes and CDs.

ANTONIO

You're absolutely right, but not all the data are different.

SANDRA

What do you mean?

ANTONIO

For instance, every audio, video, and game title will have a universal product code in our bar coding scheme. Come to think of it, so will the general merchandise I spoke of earlier. Let's see, I'm sure that audio, video, and games share some unique data . . . Yes, they all have a title, catalog description, and copyright date.

SALLY

And let's not forget system objectives! We want every audio, video, and game title to count toward fulfilling any club agreement; therefore, they must each be assigned some sort of credit value that counts toward fulfilling an agreement. That value will be based on the price of the item . . .

JOE

or how badly we want to get rid of the inventory!

[laughter]

SANDRA

Bob, I think we have a generalization hierarchy here.

JOE

Say what?

SANDRA

Sorry, it's a technical term that Bob and I have to worry about. Let's get back to this discussion. Are there any data that describe all products, regardless of all these types of products?

ANTONIO

I already mentioned the universal product codes, but there are some other common data. Because the UPCs are not yet implemented, each product has an existing product number that we can't get rid of. For every product, we need to know the quantity in stock.

ANN

Almost every product has a manufacturer's suggested retail price and a club default retail price. At any given time, a product could have a special retail price. I'd also like to have some historical data about the sales of each product.

SANDRA

Such as?

ANN

Oh, number of units sold this month; this year; lifetime.

SANDRA

Anything else?

[pause]

Let's get back to this product type thing. So we now know that a product is either general merchandise, an audio title, a video title, or a game title—and that each of those types has its own unique data.

ANTONIO

No, I said that the three types of title have some common data just like all products have some common data.

SOUNDSTAGE

SOUNDSTAGE ENTERTAINMENT CLUB

SANDRA
You're right. I heard that. Let's try again. A product is either general merchandise or a title. Some data describe all products. Some data are unique to either merchandise or titles. Subsequently, a title is either an audio, video, or game title; each of which has some unique data requirements. Let's break one of those down. What attributes describe a game title?

SALLY
I can answer that one. A game title is described by all the general product and title data we've been discussing, but it is also described by its manufacturer, category, platform, media type, number of players, and a parent advisory code.

SANDRA
I assume this category is different from that for audio and video titles?

SALLY
Yes, it represents game categories such as sports and fantasy.

SANDRA
What is a platform?

SALLY
That is the type of computer the game runs on. Media type tells us whether the game comes on a cartridge or disk.

SANDRA
Now that brings up an interesting question. Many titles, including audio, video, and game, are distributed on several media. For example, I can get the *Beatles Anthology* on either CD or cassette . . .

DAVID
or DVD and possibly others.

SANDRA
That leads to my question. Is the same title on different media considered different products?

ANTONIO
Absolutely! They have different product numbers, and they will have different UPCs. That's how we avoid sending the wrong format to a member.

SANDRA
OK. Can a title be an audio, video, and game title?

DAVID
I'm not sure I understand your question.

SANDRA
I own the audio soundtrack to *Jurassic Park,* a copy of the videotape, and a game based on the movie. Are those three titles?

DAVID
Yes, they are three different titles. And I suspect that each title is offered on several different media; therefore, it could be said that we have a dozen or so *Jurassic Park* products.

SARAH
Are you sure you want to build a database? Sounds complicated!

[We can exit the meeting now.]

DISCUSSION QUESTIONS

1. Should the different types of products each have their own file (or database table) or should they be consolidated? What are the business implications of that decision?
2. How would you deal with the issue of category and media having different values for different types of products?
3. Why can't SoundStage simply replace the current product number scheme with the planned universal product code scheme for identifying its products?
4. The group spent a lot of time identifying characteristics of different subject areas. What else would you need to know about each characteristic (e.g., game category) to properly store that data?

AN INTRODUCTION TO SYSTEMS MODELING

In the last chapter you were introduced to activities that called for drawing system models. System models play an important role in systems development. As a systems analyst or user, you will constantly deal with unstructured problems. One way to structure such problems is to draw models.

> A **model** is a representation of reality. Just as a picture is worth a thousand words, most system models are pictorial representations of reality.

Models can be built for existing systems as a way to better understand those systems or for proposed systems as a way to document business requirements or technical designs. An important concept, in both this chapter and the next, is the distinction between logical and physical models.

> **Logical models** show *what* a system is or does. They are implementation *in*dependent; that is, they depict the system independent of any technical implementation. As such, logical models illustrate the *essence* of the system. Popular synonyms include **essential model,** *conceptual model,* and *business model.*

Physical models show not only *what* a system is or does, but also *how* the system is physically and technically implemented. They are implementation *de*pendent because they reflect technology choices and the limitations of those technology choices. Synonyms include *implementation model* and *technical model.*

An example might help to clarify the distinction. Let's say we need a system that can store students' GRADE POINT AVERAGES. For any given student, the GRADE POINT AVERAGE must be between 0.00 and 4.00 where 0.00 reflects an average grade of F and 4.00 reflects an average grade of A. We further know that a new student does not have a grade point average until he receives his first grade report. Every statement in this paragraph reflects a business fact. It does not matter which technology we use to implement these facts, the facts are constant—they are logical. Thus, GRADE POINT AVERAGE is a *logical* attribute with a *logical* domain of 0.00–4.00, and a *logical* default value of nothing.

Now, let's get physical. First we have to decide how we are going to store the above attribute. Let's say we will use a table column in a Microsoft *Access* database. First we need a column name. Let's assume our company's programming standards require the following name: COLGRADEPOINTAVG. That is the *physical* field name that will be used to implement the logical attribute GRADE POINT AVERAGE. (It is not important that you understand why.) Additionally, we could define the following *physical* properties for the database column COLGRADEPOINTAVG:

Data type	NUMBER	
Field size	SINGLE	(as in "single precision")
Decimal places	2	
Default value	NULL	(as in "none")
Validation rule	BETWEEN 0 AND 4	
Required?	NO	

Systems analysts have long recognized the importance of separating business and technical concerns. That is why they use logical system models to depict business requirements and physical system models to depict technical designs. Systems analysis activities tend to focus on the logical system models for the following reasons:

- Logical models remove biases that are the result of the way the current system is implemented or the way that any one person thinks the system might be implemented. Thus, we overcome the "we've always done it that way" syndrome. Consequently, logical models encourage creativity.
- Logical models reduce the risk of missing business requirements because we are too preoccupied with technical details. Such errors can be costly to correct after the system is implemented. By separating what the system must do from how the system will do it, we can better analyze the requirements for completeness, accuracy, and consistency.
- Logical models allow us to communicate with end-users in nontechnical or less technical languages. Thus, we don't lose requirements in the technical jargon of the computing discipline.

This chapter will present data modeling as a technique for defining business requirements for a database.

Data modeling is a technique for organizing and documenting a system's DATA. Data modeling is sometimes called database modeling because a data model is usually implemented as a database. It is sometimes called *information modeling*.

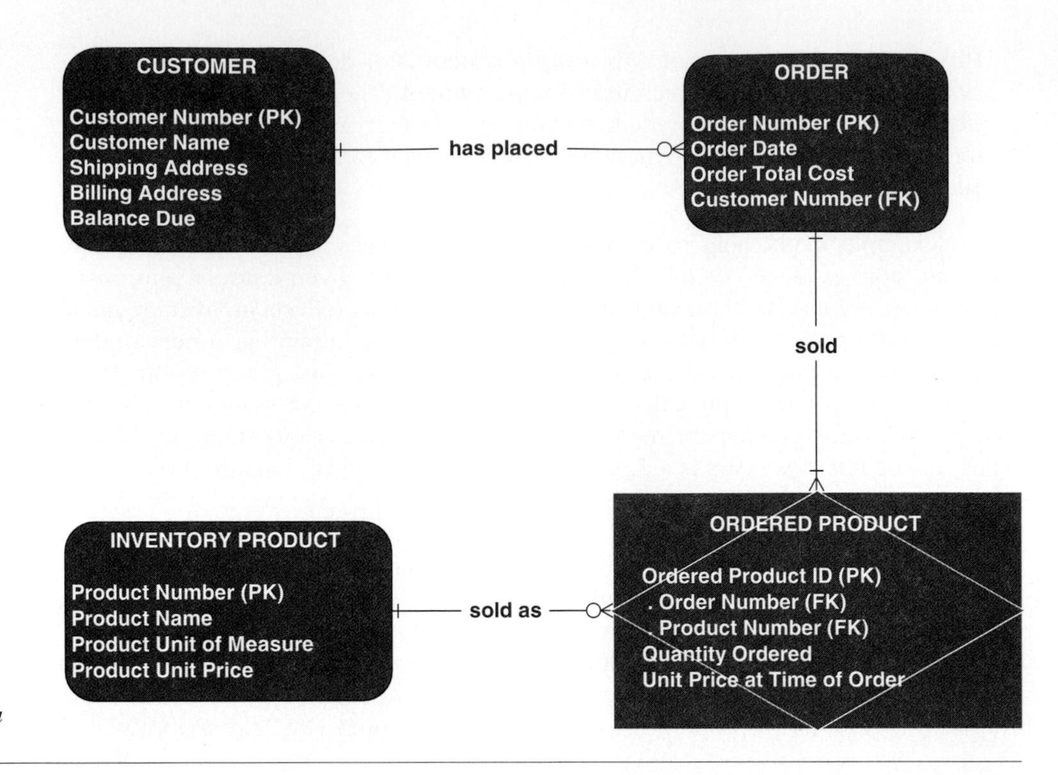

FIGURE 5.1
An Entity Relationship Data Model

Figure 5.1 is an example of a simple data model called an *entity relationship diagram* or ERD. This diagram makes the following business assertions:

- We need to store data about CUSTOMERS, ORDERS, and INVENTORY PRODUCTS.
- The value of CUSTOMER NUMBER uniquely identifies one and only one CUSTOMER. The value of ORDER NUMBER uniquely identifies one and only one ORDER. The value of PRODUCT NUMBER uniquely identifies one and only one INVENTORY PRODUCT.
- For a CUSTOMER we need to know the CUSTOMER NAME, SHIPPING ADDRESS, BILLING ADDRESS, and BALANCE DUE. For an ORDER we need to know ORDER DATE and ORDER TOTAL COST. For an INVENTORY PRODUCT we need to know PRODUCT NAME, PRODUCT UNIT OF MEASURE, and PRODUCT UNIT PRICE.
- A CUSTOMER has placed zero, one, or more ORDERS.
- An ORDER is placed by exactly one CUSTOMER. The value of CUSTOMER NUMBER (as recorded in ORDER) identifies that CUSTOMER.
- An ORDER sold one or more ORDERED PRODUCTS. Thus, an ORDER must contain at least one ORDERED PRODUCT.
- An INVENTORY PRODUCT may have been sold as zero, one, or more ORDERED PRODUCTS.
- An ORDERED PRODUCT identifies a single INVENTORY PRODUCT on a single ORDER. The ORDER NUMBER (for an ORDERED PRODUCT) identifies the ORDER, and the PRODUCT NUMBER (for an ORDERED PRODUCT) identifies the INVENTORY PRODUCT. Together, they identify one and only one ORDERED PRODUCT.
- For each ORDERED PRODUCT we need to know QUANTITY ORDERED and UNIT PRICE AT TIME OF ORDER.

After you study this chapter, you will be able read data models, and construct them.

Many experts consider data modeling to be the most important of the modeling techniques. To be sure, it is the most emphasized model in the information engineering-based methodologies that currently represent the most practiced strategy of systems analysis (Chapter 4). Why is data modeling considered crucial?

First, data is viewed as a resource to be shared by as many processes as possible. As a result, data must be organized in a way that is flexible and adaptable to unanticipated business requirements—and that is the purpose of data modeling. Second, data structures and properties are reasonably permanent, certainly a great deal more stable than the processes that use the data. Often the data model of a current system is nearly identical to that of the desired system. Third, data models are much smaller than process and object models (Chapters 6 and 8, respectively) and can be constructed more rapidly. Finally, the process of constructing data models helps analysts and users quickly reach consensus on business terminology and rules.

It should be acknowledged that data modeling is also taught in database courses, albeit from the perspective of data management, independent of applications and systems development. In database courses, you may encounter somewhat different terminology and graphical symbols, but for the most part, all data models have the same basic constructs.

SYSTEM CONCEPTS FOR DATA MODELING

This is the first chapter of the book that actually teaches a *technique* of systems analysis. Most systems analysis techniques are strongly rooted in systems thinking.

> **Systems thinking** is the application of formal systems theory and concepts to systems problem solving.

Theory? Concepts? Don't be intimidated. This is not a theory course or textbook. But some basic systems theory and concepts help us understand the way systems are organized and how they work. Techniques teach us how to apply the theory and concepts to build useful real-world systems.

Most people can learn the techniques of systems analysis. But if they understand the underlying concepts, they can adapt the techniques to ever-changing problems and conditions. They can also improve on the techniques, recognize the advantages of new techniques, and see opportunities to integrate different techniques. Therein lies your true opportunity for competitive advantage and security in today's business world. If you understand theory, concepts, *and* techniques, you will be able to do more than just solve textbook exercises. You will be able to solve real-world problems and command the premium salaries paid to today's best problem solvers!

There are several notations for data modeling. The actual model is frequently called an **entity relationship diagram** (ERD) because it depicts data in terms of the entities and relationships described by the data. There are several notations for ERDs. Most are named after their inventor (e.g., Chen, Martin, Bachman, Merise) or after a published standard (e.g., IDEF1X). These data modeling "languages" generally support the same fundamental concepts and constructs. We have adopted the Martin (information engineering) notation because of its widespread use and CASE tool support.

Let's explore some basic concepts that underlie all data models.

Entities

All systems contain data—usually lots of data! Data describe "things." Consider a school system. A school system includes data that describe things such as STUDENTS, TEACHERS, COURSES, CLASSROOMS. For any of these things, it is not difficult to imagine some of the data that describe any given instance of the thing. For example, the data that describe a particular student might include name, address, phone number, date of birth, gender, race, major, and grade point average, to name a few.

We need a concept to abstractly represent all instances of a group of similar things. We call this concept an entity.

> An **entity** is something about which we want to store data. Synonyms include *entity type* and *entity class*.

An entity

In system modeling, we find it useful to assign each abstract concept to a shape. An entity will be drawn as a rectangle with rounded corners (see margin). This shape represents all instances of the named entity. For example, the entity STUDENT represents all students in the system.

We can refine our definition of entity to identify specific classes of entities.

An **entity** is a class of persons, places, objects, events, or concepts about which we need to capture and store data.

Each entity represents a group of many instances of that entity, and each entity is distinguishable from the other entities. Examples of entities include:

Persons: AGENCY, CONTRACTOR, CUSTOMER, DEPARTMENT, DIVISION, EMPLOYEE, INSTRUCTOR, STUDENT, SUPPLIER. Notice that a person entity can represent either individuals, groups, or organizations.

Places: SALES REGION, BUILDING, ROOM, BRANCH OFFICE, CAMPUS.

Objects: BOOK, MACHINE, PART, PRODUCT, RAW MATERIAL, SOFTWARE LICENSE, SOFTWARE PACKAGE, TOOL, VEHICLE MODEL, VEHICLE. An object entity can represent actual objects (such as SOFTWARE LICENSE) or specifications for a type of object (such as SOFTWARE PACKAGE).

Events: APPLICATION, AWARD, CANCELLATION, CLASS, FLIGHT, INVOICE, ORDER, REGISTRATION, RENEWAL, REQUISITION, RESERVATION, SALE, TRIP.

Concepts: ACCOUNT, BLOCK OF TIME, BOND, COURSE, FUND, QUALIFICATION, STOCK.

It is important to distinguish between an entity and its instances.

An **entity instance** is a single occurrence of an entity.

For example, the entity STUDENT may have multiple instances: Mary, Joe, Mark, Susan, Deborah, and so forth. In data modeling, we do not concern ourselves with individual students because we recognize that each student is described by similar pieces of data.

Attributes

If an entity is something about which we want to store data, then intuitively, we need to identify what specific pieces of data we want to store about each instance of a given entity. We call these pieces of data attributes.

An **attribute** is a descriptive property or characteristic of an entity. Synonyms include *element, property,* and *field.*

As noted at the beginning of this section, each instance of the entity STUDENT might be described by the following attributes: NAME, ADDRESS, PHONE NUMBER, DATE OF BIRTH, GENDER, RACE, MAJOR, GRADE POINT AVERAGE, and others.

We can now extend our graphical abstraction of the entity to include attributes by recording those attributes inside the entity shape along with the name (see margin).

Some attributes can be logically grouped into superattributes called compound attributes.

A **compound attribute** is one that actually consists of more primitive attributes. Synonyms in different data modeling languages are numerous: *concatenated attribute, composite attribute,* and *data structure.*

For example, a student's NAME is actually a composite attribute that consists of LAST NAME, FIRST NAME, and MIDDLE INITIAL. In the margin, we demonstrate a notation for compound attributes.

STUDENT
Name
. Last Name
. First Name
. Middle Initial
Address
. Street Address
. City
. State or Province
. Country
. Postal Code
Phone Number
. Area Code
. Exchange Number
. Number Within Exchange
Date of Birth
Gender
Race
Major
Grade Point Average

Attributes and compound attributes

Domains An attribute is a piece of data. When analyzing a system, we should define those values for an attribute that are legitimate, or that make sense. The values for each attribute are defined in terms of three properties: data type, domain, and default.

> The **data type** for an attribute defines what class of data can be stored in that attribute.

Data typing should be familiar to those of you who have written computer programs; declaring types for variables is common to most programming languages. For purposes of systems analysis and business requirements definition, it is useful to declare logical (nontechnical) data types for our business attributes. For the sake of argument, we will use the logical data types shown in Table 5.1.

An attribute's data type determines its domain.

> The **domain** of an attribute defines what values an attribute can legitimately take on.

Eventually, system designers must use technology to enforce the business domains of all attributes. Table 5.2 demonstrates how logical domains might be expressed for each data type.

Finally, every attribute should have a logical default value.

> The **default** value for an attribute is the value that will be recorded if not specified by the user.

Table 5.3 shows possible default values for an attribute.

Identification An entity typically has many instances, perhaps thousands or millions. Conceptually, there exists a need to uniquely identify each instance based on the data value of one or more attributes. Thus, every entity must have an identifier or key.

> A **key** is an attribute, or a group of attributes, that assumes a unique value for each entity instance. It is sometimes called an *identifier.*

For example, each instance of the entity STUDENT might be uniquely identified by the key STUDENT NUMBER. No two students can have the same STUDENT NUMBER. Sometimes more than one attribute is required to uniquely identify an instance of an entity.

> A group of attributes that uniquely identifies an instance of an entity is called a **concatenated key.** Synonyms include *composite key* and *compound key.*

For example, each TAPE entity instance in a video store might be uniquely identified by the concatenation of TITLE NUMBER plus COPY NUMBER. TITLE NUMBER by itself

TABLE 5.1 *Representative Logical Data Types for Attributes*

Logical Data Type	Logical Business Meaning
NUMBER	Any number, real or integer
TEXT	A string of characters, inclusive of numbers. When numbers are included in a TEXT attribute, it means we do not expect to perform arithmetic or comparisons with those numbers.
MEMO	Same as TEXT but of an indeterminate size. Some business systems require the ability to attach a potentially lengthy note to a given database record.
DATE	Any date in any format.
TIME	Any time in any format.
YES/NO	An attribute that can only assume one of these two values.
VALUE SET	A finite set of values. In most cases, a coding scheme would be established (e.g., FR = freshman, SO = sophomore, JR = junior, SR = senior, etc.)
IMAGE	Any picture or image.

TABLE 5.2 *Representative Logical Domains for Logical Data Types*

Data Type	Domain	Examples
NUMBER	For integers, specify the range: {minimum–maximum} For real numbers, specify the range and precision: {minimum.precision–maximum.precision}	{10–99} {1.000–799.999}
TEXT	TEXT (maximum size of attribute) *Actual values are usually infinite; however, users may specify certain narrative restrictions.*	TEXT (30)
MEMO	*Not applicable. There are no restrictions on size or content.*	*Not applicable.*
DATE	Variation on the MMDDYYYY format. To accommodate the year 2000, do not abbreviate year to YY. Formatting characters are rarely stored; therefore, do not include hyphens or slashes.	MMDDYYYY MMYYYY YYYY
TIME	For AM/PM times: HHMMT –or– For military times: HHMM	HHMMT HHMM
YES/NO	{YES, NO}	{YES, NO}
VALUE SET	{value#1, value#2, . . . , value#n} –or– {table of codes and meanings}	{FRESHMAN, SOPHOMORE, JUNIOR, SENIOR} {FR = FRESHMAN, SO = SOPHOMORE, JR = JUNIOR SR = SENIOR}
IMAGE	*Not applicable; however, any known characteristics of the images will eventually prove useful to designers.*	*Not applicable.*

TABLE 5.3 *Permissible Default Values for Attributes*

Default Value	Interpretation	Examples
A legal value from the domain (as described above)	For an instance of the attribute, if the user does not specify a value, then use this value.	0 1.00 FR
NONE or NULL	For an instance of the attribute, if the user does not specify a value, then leave it blank.	NONE NULL
REQUIRED or NOT NULL	For an instance of the attribute, require the user to enter a legal value from the domain. (This is used when no value in the domain is common enough to be a default, but some value must be entered.)	REQUIRED NOT NULL

would be inadequate because we may own many copies of a single title. COPY NUMBER by itself would also be inadequate since we would have a *copy #1* for every title we own. We need both pieces of data to identify a specific tape (e.g., *copy #7* of *Jurassic Park*). In this book, we will give a name to the group as well as the individual attributes. For example, the concatenated key for TAPE would be recorded as follows:

TAPE ID (PRIMARY KEY)

. TITLE NUMBER

. COPY NUMBER

Frequently, an entity may have more than one key. For example, the entity EMPLOYEE may be uniquely identified by SOCIAL SECURITY NUMBER, or company-assigned EMPLOYEE NUMBER, or E-MAIL ADDRESS. Each of these attributes is called a candidate key.

> A **candidate key** is a "candidate to become the primary identifier" of instances of an entity. It is sometimes called a *candidate identifier.* (A candidate key may be a single attribute or a concatenated key.)

> A **primary key** is the candidate key that will most commonly be used to uniquely identify a single entity instance.

By the way, the default for a primary key is always NOT NULL. Why? Because if the key has no value, then it cannot identify an instance of an entity.

> Any candidate key that is not selected to become the primary key is called an **alternate key.**

In the margin, we demonstrate our notation for primary and alternate keys. All candidate keys must be either primary or alternate; therefore, they do not warrant a separate notation.

Sometimes, it is also necessary to identify a subset of entity instances as opposed to a single instance. For example, we may require a simple way to identify all male students and all female students.

> A **subsetting criteria** is an attribute (or concatenated attribute) whose finite values divide all entity instances into useful subsets. Some methods call this an *inversion entry.*

For example, in our STUDENT entity, the attribute GENDER divides the instances of STUDENT into two subsets: male students and female students. In general, subsetting criteria are useful only when an attribute has a finite (meaning limited) number of legitimate values. For example, GRADE POINT AVERAGE would not be a good subsetting criteria because there are 999 possible values of that attribute. The margin art demonstrates a notation for subsetting criteria.

STUDENT

Student Number (Primary Key 1)
Name (Alternate Key 1)
. Last Name
. First Name
. Middle Initial
Address
. Street Address
. City
. State or Province
. Country
. Postal Code
Phone Number
. Area Code
. Exchange Number
. Number Within Exchange
Date of Birth
Gender (Subsetting Criteria 1)
Race (Subsetting Criteria 2)
Major (Subsetting Criteria 3)
Grade Point Average

Keys and subsetting criteria

Relationships

Conceptually, entities and attributes do not exist in isolation. The things they represent interact with and impact one another to support the business mission. Thus, we introduce the concept of a relationship.

> A **relationship** is a natural business association that exists between one or more entities. The relationship may represent an event that links the entities or merely a logical affinity that exists between the entities.

Consider, for example, the entities STUDENT and CURRICULUM. We can make the following business assertions that link students and courses:

- A current STUDENT IS ENROLLED in one or more CURRICULA.
- A CURRICULUM IS BEING STUDIED BY zero, one, or more STUDENTS.

The underlined verb phrases define business relationships that exist between the two entities. Because a STUDENT can be enrolled in *many* CURRICULA, and a CURRIULUM can enroll *many* STUDENTS, this is often called a *many-to-many relationship.*

We can graphically illustrate this association between STUDENT and CURRICULUM as shown in Figure 5.2. The connecting line represents a relationship. The verb phrase describes the relationship. Notice that all relationships are implicitly bidirectional, meaning they can be interpreted in both directions (as suggested by the above business assertions).

FIGURE 5.2
A Relationship (Many-to-Many)

Cardinality Interpretation	Minimum Instances	Maximum Instances	Graphic Notation
Exactly one	1	1	
Zero or one	0	1	
One or more	1	many (> 1)	
Zero, one, or more	0	many (> 1)	
More than one	> 1	> 1	

FIGURE 5.3
Cardinality Notations

Cardinality Figure 5.2 also shows the complexity or *degree* of each relationship. For example, in the above business assertions, we must also answer the following questions:

- Must there exist an instance of STUDENT for each instance of CURRICULUM? No! Must there exist an instance of CURRICULUM for each instance of STUDENT? Yes!
- How many instances of CURRICULUM can exist for each instance of STUDENT? Many! How many instances of STUDENT can exist for each instance of CURRICULUM? Many!

We call this concept cardinality.

> **Cardinality** defines the minimum and maximum number of occurrences of one entity for a single occurrence of the related entity. Because all relationships are bidirectional, cardinality must be defined in both directions for every relationship.

A popular graphical notation for cardinality is shown in Figure 5.3. Sample cardinality symbols were demonstrated in Figure 5.2.

Conceptually, cardinality tells us the following rules about the data we want to store:

- When we insert a STUDENT instance in the database, we must link (associate) that STUDENT to at least one instance of CURRICULUM. In business terms, "a

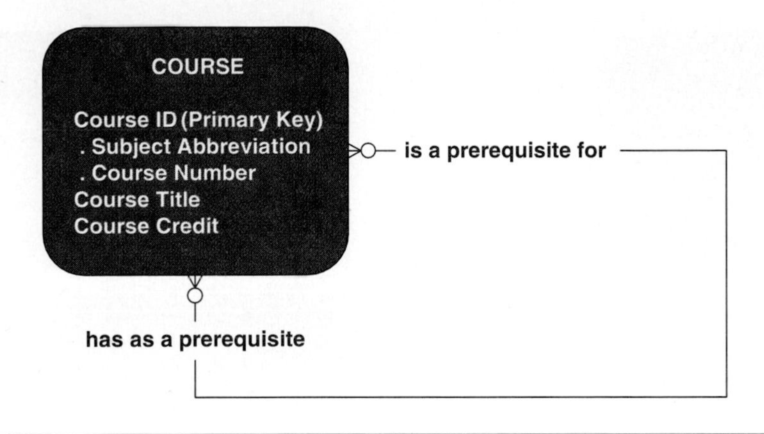

FIGURE 5.4
A Recursive Relationship

student cannot be admitted without declaring a major." (Most schools would include an instance of CURRICULUM called "undecided" or "undeclared.")

- A STUDENT <u>can</u> study more than one CURRICULUM, and we must be able to store data that indicates all CURRICULA for a given STUDENT.
- We must insert a CURRICULUM before we can link (associate) STUDENTS to that CURRICULUM. That is why a CURRICULUM can have zero students—no students have yet to be admitted to that CURRICULUM.
- Once a CURRICULUM has been inserted into the database, we can link (associate) many STUDENTS with that CURRICULUM.

Degree Another measure of the complexity of a data relationship is its degree.

> The **degree** of a relationship is the number of entities that participate in the relationship.

All the relationships we've explored so far are *binary* (degree = 2). In other words, two different entities participated in the relationship.

Relationships may also exist between different instances of the same entity. We call this a **recursive relationship** (sometimes called a unary relationship; degree = 1). For example, in your school a course may be a prerequisite for other courses. Similarly, a course may have several other courses as its prerequisite. Figure 5.4 demonstrates this many-to-many recursive relationship.

Relationships can also exist between more than two different entities. These are sometimes called N-ary relationships. An example of a *3-ary* or *ternary relationship* is shown in Figure 5.5. An N-ary relationship is illustrated with a new entity construct called an associative entity.

> An **associative entity** is an entity that inherits its primary key from more than one other entity (parents). Each part of that concatenated key points to one and only one instance of each of the connecting entities.

For example, in Figure 5.5 the associative entity SCHEDULED CLASS (notice the unique shape) matches a COURSE, a ROOM, and an INSTRUCTOR. For each instance of SCHEDULED CLASS, the key indicates which COURSE ID, which ROOM ID, and which INSTRUCTOR ID are combined to form that class.

Also as shown in Figure 5.5, an associative entity can be described by its own nonkey attributes. In addition to the primary key, a SCHEDULED CLASS is described by the attributes DIVISION NUMBER, DAYS OF WEEK, START TIME, and END TIME. If you think about it, none of these attributes describes a COURSE, ROOM, or INSTRUCTOR; they describe a single instance of the relationship between an instance of each of those three entities.

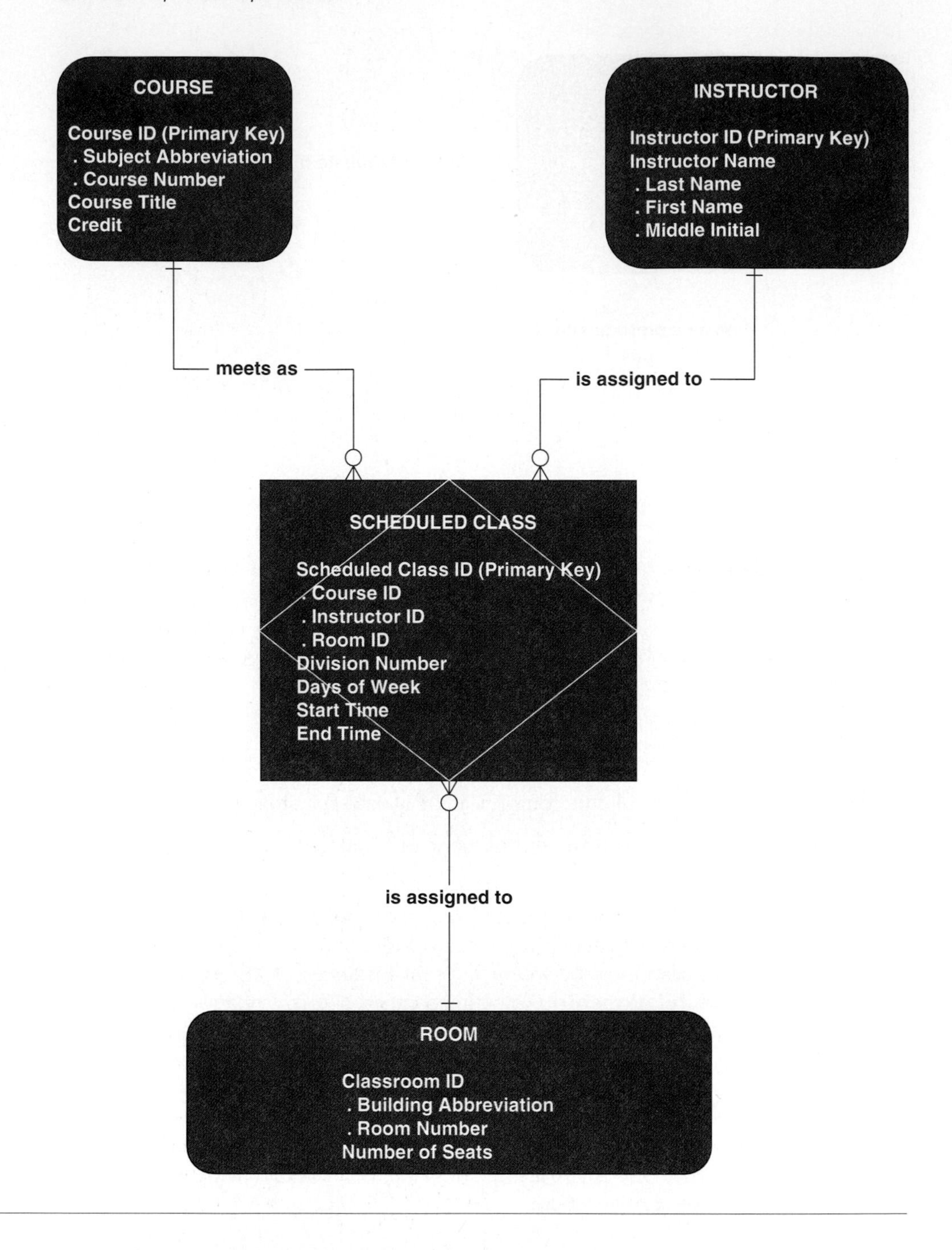

FIGURE 5.5
A Ternary Relationship

Foreign Keys A relationship implies that instances of one entity are related to instances of another entity. We should be able to identify those instances for any given entity. For example, consider a relationship between the entities MAJOR and DEPARTMENT.

- A CURRICULUM is taught by exactly one DEPARTMENT. For a CURRICULUM, which DEPARTMENT teaches it?
- A DEPARTMENT teaches one or more CURRICULA. For a CURRICULUM, which STUDENTS are enrolled in that CURRICULUM?

These are legitimate questions that must be addressed in a database. To answer these questions, we migrate the primary key of one entity into the other entity as a foreign key.

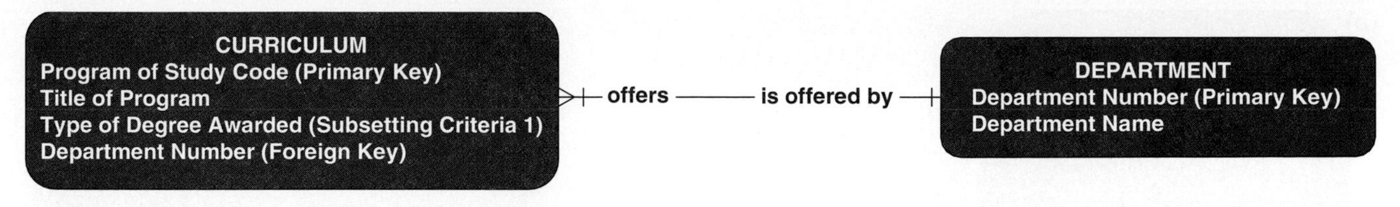

FIGURE 5.6 *How to Show Foreign Keys*

> A **foreign key** is a primary key of one entity that is contributed to (duplicated in) another entity to identify instances of a relationship. A foreign key (always in a child entity) always matches the primary key (in a parent entity).

In Figure 5.6, we demonstrate the concept of foreign keys with our simple data model. In this case, DEPARTMENT is called the *parent* entity and CURRICULUM is the *child* entity. The primary key is always contributed by the parent to the child as a foreign key. Thus, an instance of CURRICULUM now has a foreign key DEPARTMENT NAME whose value points to the correct instance of DEPARTMENT that offers that curriculum. (Foreign keys are never contributed from child to parent.)

What if you cannot differentiate between parent and child? For example, in Figure 5.7(a) we see that a CURRICULUM is being studied by zero, one, or more STUDENTS. At the same time, we see that a STUDENT is studying one or more CURRICULA. The maximum cardinality on both sides is "many." So, which is the parent and which is the child? You can't tell! This is called a nonspecific relationship.

> A **nonspecific relationship** (or **many-to-many relationship**) is one in which many instances of one entity are associated with many instances of another entity. Such relationships are suitable only for preliminary data models and should be resolved as quickly as possible.

All nonspecific relationships can be resolved into a pair of one-to-many relationships. As illustrated in Figure 5.7(b), each entity becomes a parent. A new, *associative entity* is introduced as the child of each parent. In Figure 5.7, each instance of MAJOR represents one STUDENT's enrollment in one CURRICULUM. If a student is pursuing two majors, that student will have two instances of the entity MAJOR.

Study Figure 5.7 carefully. For associative entities, the cardinality from child to parent is always exactly one. That makes sense since an instance of MAJOR must correspond to exactly one STUDENT and one CURRICULUM. The cardinality from parent to child depends on the business rule. In our example, a STUDENT must declare one or more MAJORS. Conversely, a CURRICULUM is being studied by zero, one, or more MAJORS—perhaps it is new and no one has been admitted to it yet. Finally, notice that an associative entity can also be described by its own nonkey attributes (such as DATE ENROLLED and CURRENT CANDIDATE FOR DEGREE?).

Generalization Most people associate the concept of generalization with modern object-oriented techniques. In reality, the concepts have been been applied by data modelers for many years. Generalization is an approach that seeks to discover and exploit the commonalties between entities.

> **Generalization** is a technique wherein the attributes that are common to several types of an entity are grouped into their own entity, called a *supertype.*

Consider, for example, an extension of the hypothetical academic scenario we've been using throughout this chapter. Our school enrolls STUDENTS and employs EMPLOYEES. There are several attributes that are common to both entities; for example, NAME, GENDER, RACE, MARITAL STATUS, and possibly even a key based on SOCIAL

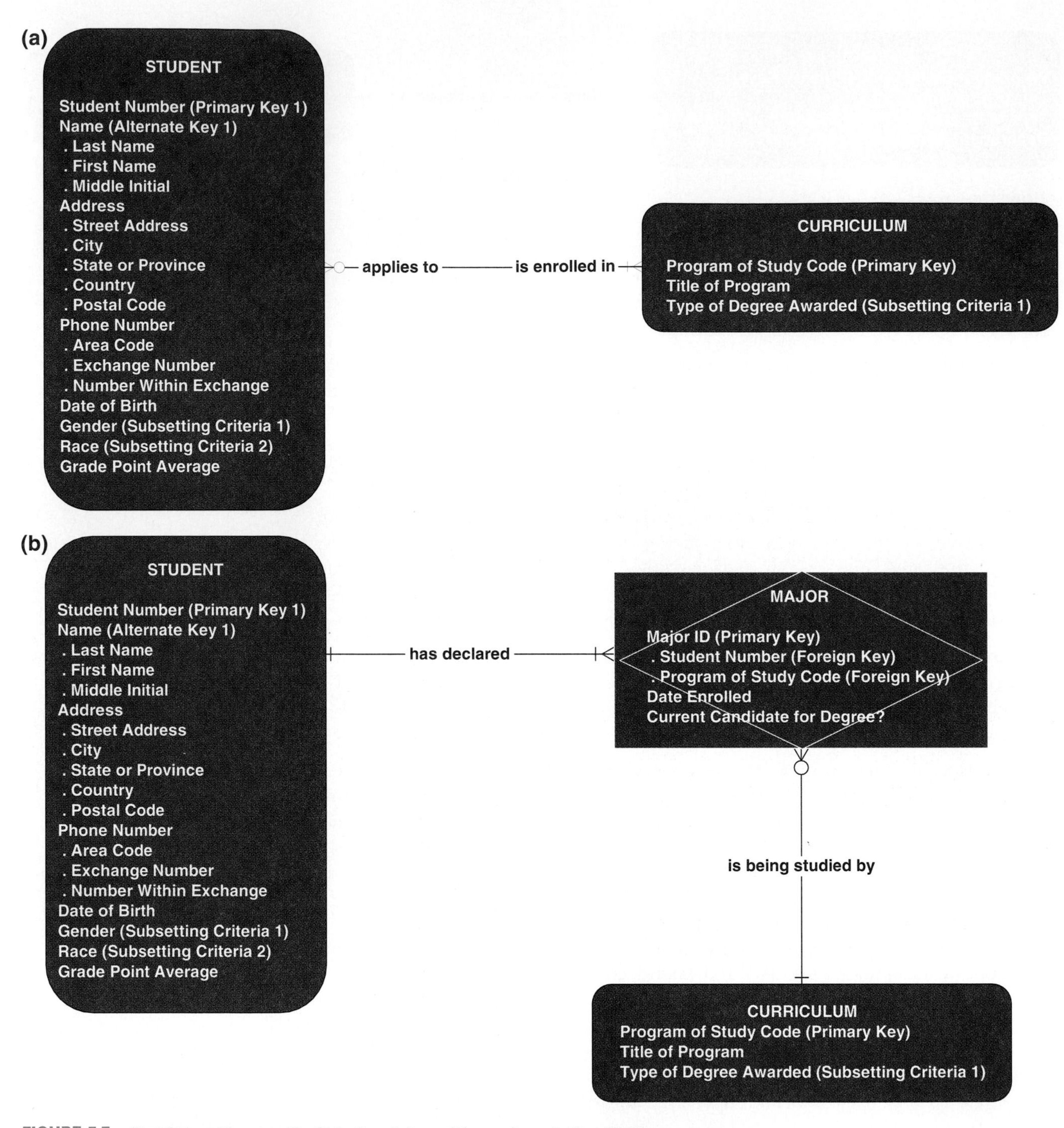

FIGURE 5.7 *Resolving Nonspecific Relationships with an Associative Entity*

SECURITY NUMBER. We could consolidate these common attributes into an entity supertype called PERSON.

> An entity **supertype** is an entity whose instances store attributes that are common to one or more entity subtypes.

The entity supertype will have one or more *one-to-one* relationships to entity *subtypes*. These relationships are sometimes called IS A relationships (or WAS A, or COULD BE A) because each instance of the supertype "is also an" instance of one or more subtypes.

An entity **subtype** is an entity whose instances inherit some common attributes from an entity supertype and then add other attributes that are unique to an instance of the subtype.

In our example, "a PERSON is an employee, or a student, or both." The top half of Figure 5.8 illustrates this generalization (1) as a hierarchy. Notice that the subtypes STUDENT and EMPLOYEE have inherited attributes from PERSON, as well as adding their own. (Unfortunately, most CASE tools do not actually migrate the inherited attributes.)

Extending the metaphor, an entity can be both a supertype and subtype. Returning to Figure 5.8, we see that a STUDENT (which was a subtype of PERSON) has its

FIGURE 5.8 *A Generalization Hierarchy*

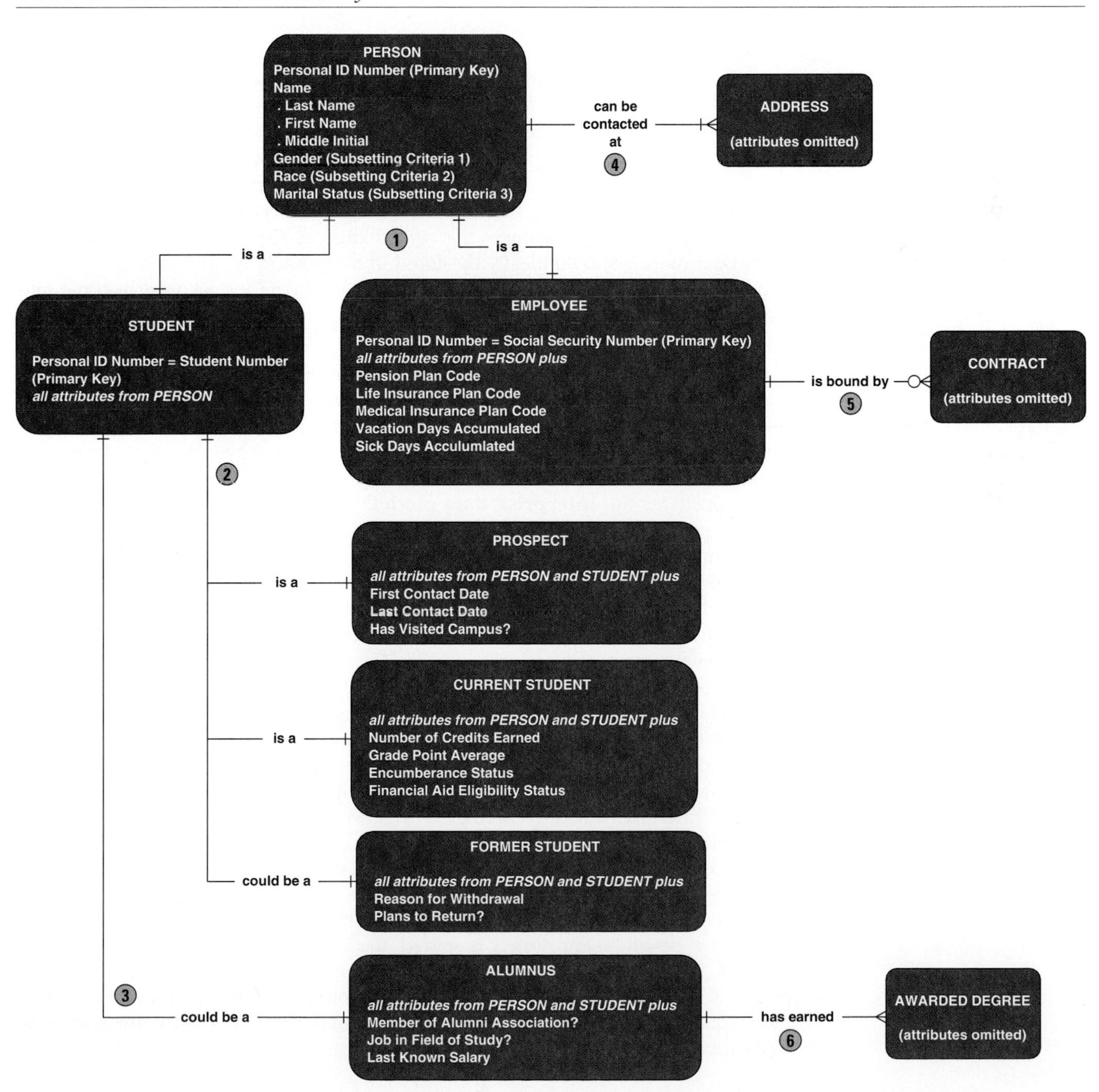

own subtypes. In the diagram, we see that a STUDENT is ② either a PROSPECT, or a CURRENT STUDENT, or a FORMER STUDENT (having left for any reason other than graduation), and ③ a STUDENT could be an ALUMNUS. These additional subtypes inherit all the attributes from STUDENT, as well as those from PERSON.

Through inheritance, the concept of generalization in data models permits us to reduce the number of attributes through the careful sharing of common attributes. The subtypes not only inherit the attributes, but also the data types, domains, and defaults of those attributes. This can greatly enhance the consistency with which we treat attributes that apply to many different entities (e.g., dates, names, addresses, currency, etc.).

In addition to inheriting attributes, subtypes also inherit relationships to other entities. For instance, all EMPLOYEES and STUDENTS inherit the relationship ④ between PERSON and PERSONAL ADDRESS. But only EMPLOYEES inherit the relationship ⑤ with EMPLOYMENT CONTRACTS. And only an ALUMNUS can be related to ⑥ a DEGREE.

THE PROCESS OF LOGICAL DATA MODELING

Now that you understand the basic concepts of data models, we can examine the process of data modeling. When do you do it? How many data models may be drawn? What technology exists to support the process?

Data modeling may be performed during various types of projects and in multiple phases of projects. Data models are progressive; there is no such thing as the "final" data model for a business or application. Instead, a data model should be considered a living document that will change in response to a changing business. Data models should ideally be stored in a repository so they can be retrieved, expanded, and edited over time. Let's examine how data modeling may come into play during systems planning and analysis.

Strategic Data Modeling

Many organizations select application development projects based on strategic information systems plans. This is especially true in organizations that practice information engineering-based methodologies. Strategic planning is a separate project. This project produces an information systems strategy plan that defines an overall vision and architecture for information systems. Almost always, this architecture includes an **enterprise data model.** (Note: There are other architectural components that are not important to this discussion.)

An enterprise data model typically identifies only the most fundamental of entities. The entities are typically defined (as in a dictionary), but they are not described in terms of keys or attributes. In many enterprise data models, some entities are composites of several entities that will not be discovered until application development projects are initiated. The enterprise data model may or may not include relationships (depending on the planning methodology's standards and the level of detail desired by executive management). If relationships are included, many of them will be nonspecific (a concept introduced earlier in the chapter).

How does an enterprise data model impact subsequent applications development? Part of the information strategy plan identifies applications development projects and prioritizes them according to whatever criteria that management deems appropriate. As those projects are started, the appropriate subsets of the information systems architecture, including a subset of the enterprise data model, are provided to the applications development team as a point of departure.

The enterprise data model is usually stored in a corporate repository. When the application development project is started, the subset of the enterprise data model (as well as the other models) is exported from the corporate repository into a project repository. Once the project team completes systems analysis and design, the expanded and refined data models are imported back into the corporate repository.

Data Modeling during Systems Analysis

In systems analysis and in this chapter, we will focus on *logical* data modeling as a part of systems analysis. The data model for a single system or application is usually called an **application data model.** In your information systems building blocks (see Figure 5.9), logical data models have a DATA focus and a SYSTEM USER perspective. Also as shown, they are typically constructed as deliverables of the study and definition phases of a project. Finally, notice that while logical data models are not concerned with implementation details or technology, they may be constructed (through reverse engineering) from existing databases.

Data models are rarely constructed during the survey phase of systems analysis. The short duration of that phase makes them impractical. On the other hand, if an enterprise data model exists, the subset of that model that is applicable to

FIGURE 5.9 *Data Modeling in the FAST Methodology*

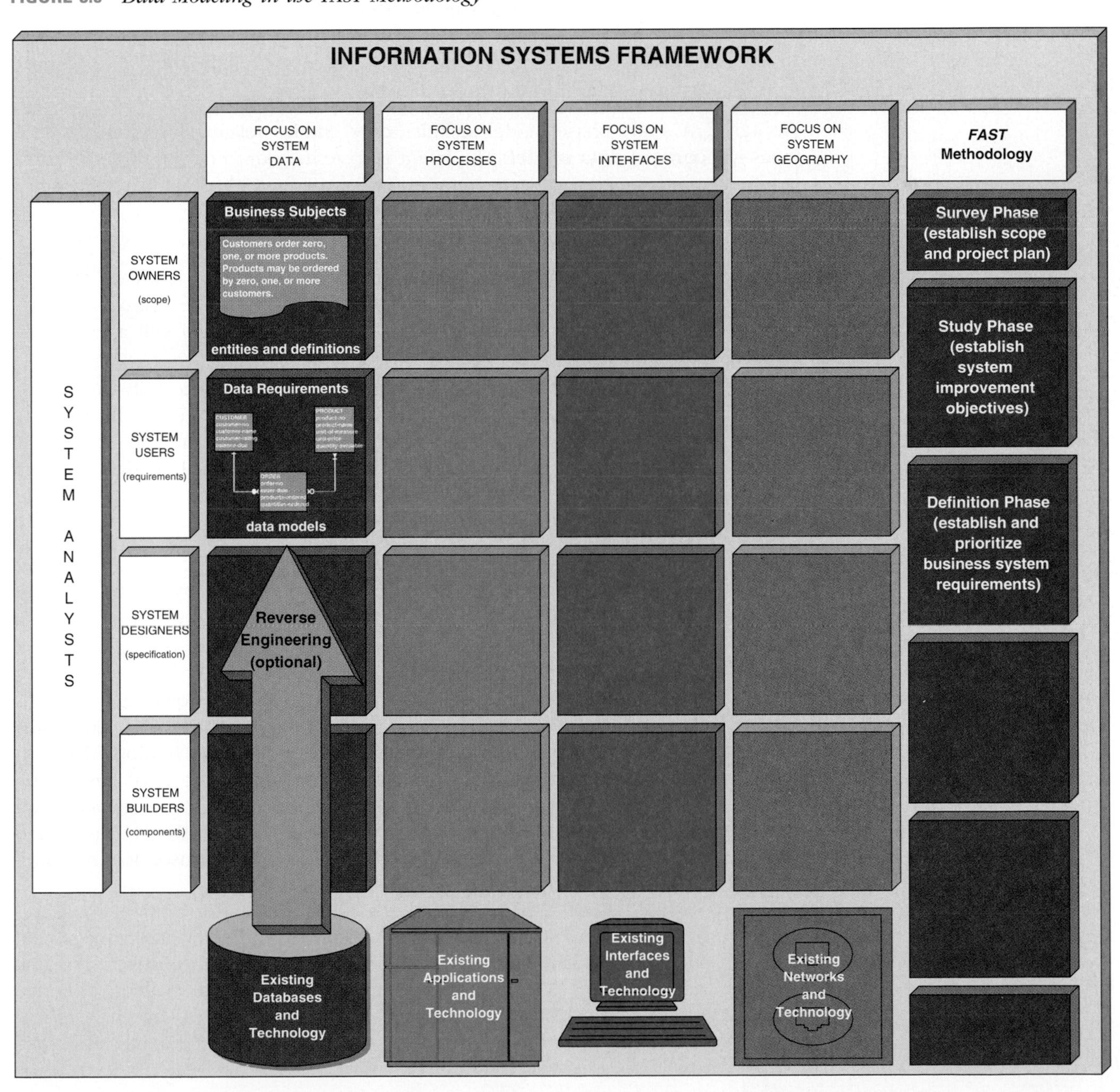

the project might be retrieved and reviewed as part of the survey phase requirement to establish context. Alternatively, the project team could identify a simple list of entities, the things about which they think the system will have to capture and store data.

Unfortunately, data modeling is rarely associated with the study phase of systems analysis. Most analysts prefer to draw process models (Chapter 6) to document the current system, but many analysts report that data models are far superior for the following reasons:

- Data models help analysts to quickly identify business vocabulary more completely than process models.
- Data models are almost always built more quickly than process models.
- A complete data model can fit on a single sheet of paper. Process models often require dozens of sheets of paper.
- Process modelers too easily get hung up on unnecessary detail.
- Data models for existing and proposed systems are far more similar than process models for existing and proposed systems. Consequently, there is less work to throw away as you move into later phases.

We agree! A study phase model includes only entities relationships, but no attributes—a **context data model.** The intent is to refine our understanding of scope, not to get into details about the entities and business rules. Many relationships may be nonspecific (meaning "many-to-many").

The definition phase data model will be constructed in at least two stages:

1. Initially, a **key-based data model** will be drawn. This model will eliminate nonspecific relationships, add associative entities, and include primary and alternate keys, as well as foreign keys. The key-based model will also include precise cardinalities and any generalization hierarchies.
2. Next, a **fully attributed data model** will be constructed. The fully attributed model includes all remaining descriptive attributes and subsetting criteria. Each attribute is defined in the repository with data types, domains, and defaults (in what is sometimes called a **fully described data model**). This data requirements model requires a team effort that includes systems analysts, users and managers, and data analysts. A data administrator often sets standards for and approves all data models.

The completed data model represents all the business requirements for a system's database.

Looking Ahead to Systems Configuration and Design

The logical data model from systems analysis describes business data requirements, not technical solutions. The purpose of the configuration phase is to determine the best way to implement those requirements with database technology. In practice, this decision may have already been standardized as part of a database architecture. For example, SoundStage has already standardized on two database management systems: Microsoft *Access* for personal and work-group databases, and Microsoft *SQL Server* for enterprise databases. The latter will be used for the Member Services Information System.

During system design, the logical data model will be transformed into a physical data model (called a *database schema*) for the chosen database management system. This model will reflect the technical capabilities and limitations of that database technology, as well as the performance tuning requirements suggested by the database administrator. The physical data model will also be analyzed for adaptability and flexibility through a process called *normalization*. Any further discussion of database design is deferred until Chapter 11.

TABLE 5.4 *JAD and Interview Questions for Data Modeling*

Purpose	Candidate Questions
Discover the system entities	What are the subjects of the business? In other words, what types of persons, organizations, organizational units, places, things, materials, or events are used in or interact with this system, about which data must be captured or maintained? How many instances of each subject exist?
Discover the entity keys	What unique characteristic (or characteristics) distinguishes an instance of each subject from other instances of the same subject? Are there any plans to change this identification scheme in the future?
Discover entity subsetting criteria	Are there any characteristics of a subject that divide all instances of the subject into useful subsets? Are there any subsets of the above subjects for which you have no convenient way to group instances?
Discover attributes and domains	What characteristics describe each subject? For each of these characteristics: (1) what type of data is stored? (2) who is responsible for defining legitimate values for the data? (3) what are the legitimate values for the data? (4) is a value required? and (5) is there any default value that should be assigned if you don't specify otherwise?
Discover security and control needs	Are there any restrictions on who can see or use the data? Who is allowed to create the data? Who is allowed to update the data? Who is allowed to delete the data?
Discover data timing needs	How often does the data change? Over what period of time is the data of value to the business? How long should we keep the data? Do you need historical data or trends? If a characteristic changes, must you know the former values?
Discover generalization hierarchies	Are all instances of each subject the same? That is, are there special types of each subject that are described or handled differently? Can any of the data be consolidated for sharing?
Discover relationships and degrees	What events occur that imply associations between subjects? What business activities or transactions require handling or changing data about several different subjects of the same or a different type?
Discover cardinalities	Is each business activity or event handled the same way or are there special circumstances? Can an event occur with only some of the associated subjects, or must all the subjects be involved?

Source: Adapted from Jeffrey A. Hoffer, Joey F. George, and Joseph S. Valacich, *Modern Systems Analysis and Design* (Menlo Park, CA: Benjamin/Cummings, 1996), p. 386.

Fact-Finding and Information Gathering for Data Modeling

Data models cannot be constructed without appropriate facts and information as supplied by the user community. These facts can be collected by a number of techniques such as sampling of existing forms and files, research of similar systems, surveys of users and management, and interviews of users and management. The fastest method of collecting facts and information and simultaneously constructing and verifying the data models is joint application sevelopment (JAD). JAD uses a carefully facilitated group meeting to collect the facts, build the models, and verify the models—usually in one or two full-day sessions.

Fact-finding and information gathering techniques are more fully explored in Part Five, Module B. JAD techniques are presented in Part Five, Module D. Table 5.4 summarizes some questions that may be useful for fact-finding and information gathering as it pertains to data modeling.

Computer-Aided Systems Engineering (CASE) for Data Modeling

Data models are stored in the repository. In a sense, the data model is **metadata**—that is, data about the business's data. Computer-aided systems engineering (CASE) technology, introduced in Chapter 3, provides the repository for storing the data model and its detailed descriptions. Most CASE products support computer-assisted data modeling and database design. Some CASE products (such as Logic Works *ERwin*) only support data modeling and database design. CASE takes the drudgery out of drawing and maintaining these models and their underlying details.

Using a CASE product, you can easily create professional, readable data models without the use of paper, pencil, erasers, and templates. The models can be easily modified to reflect corrections and changes suggested by end-users; you don't have to start over! Also, most CASE products provide powerful analytical tools that can check your models for mechanical errors, completeness, and consistency. Some CASE products can even help you analyze the data model for consistency, completeness, and flexibility. The potential time and quality savings are substantial.

CASE tools do have their limitations. Not all data model conventions are supported by all CASE products. Therefore, it is very likely that any given CASE product may force a company to adapt its methodology's data modeling symbols or approach so that it is workable within the limitations of the CASE tool.

All the SoundStage data models in the next section of this chapter were created with Popkin's CASE tool, *System Architect.* For the case study, we provide you the printouts exactly as they came off our printers. We did not add color. The only modifications by the artist were the bullets that call your attention to specific items of interest on the printouts. All of the entities, attributes, and relationships on the SoundStage data models were automatically cataloged into *System Architect*'s project repository (which it calls an encyclopedia). Figure 5.10 illustrates some of *System Architect*'s screens as used for data modeling.

HOW TO CONSTRUCT DATA MODELS

You now know enough about data models to read and interpret them. But as a systems analyst or knowledgeable end-user, you must learn how to construct them. We will use the SoundStage Entertainment Club project to teach you how to construct data models.

NOTE This example teaches you to draw the data model from scratch. In reality, you should always look for an existing data model. If such models exist, they are usually maintained by the data management or data administration group.

Entity Discovery

The first task in data modeling is relatively easy. You need to discover those fundamental entities in the system that are or might be described by data. You should not restrict your thinking to entities about which the end-users know they want to store data. There are several techniques that may be used to identify entities.

- During interviews or JAD sessions with system owners and users, pay attention to key words in their discussion. For example, during an interview with an individual discussing SoundStage's business environment and activities, a user may state, "We have to keep track of all our members and the many clubs in which they are enrolled." Notice that the key words in this statement are MEMBERS and CLUBS. Both are entities!
- During interviews or JAD sessions, specifically ask the system owners and users to identify things about which they would like to capture, store, and produce information. More often than not, those things represent entities that should be depicted on the data model.
- Another technique for identifying entities is to study existing forms and files. Some forms identify event entities. Examples include ORDERS, REQUISITIONS, PAYMENTS, DEPOSITS, and so forth. But most of these same forms also contain data that describe other entities. Consider a registration form used in your school's course registration system. A REGISTRATION is itself an event entity. But the average registration form also contains data that describe other

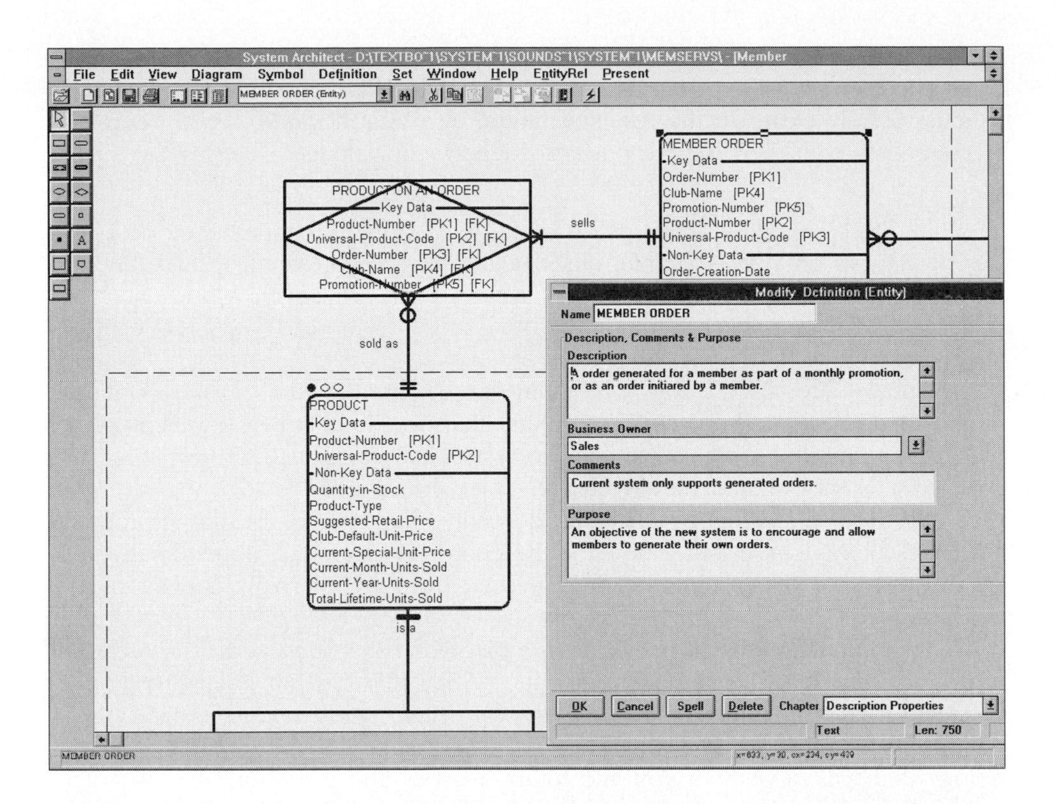

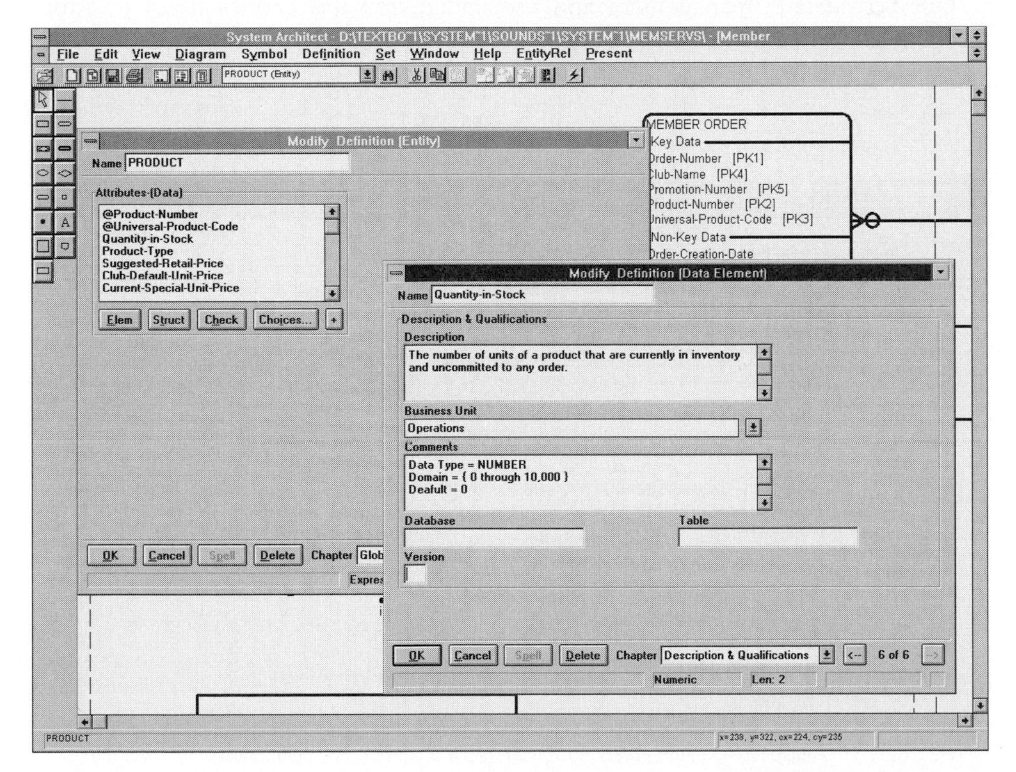

FIGURE 5.10
CASE Tools for Data Modeling

entities, such as STUDENT (a person), COURSES (which are concepts), INSTRUCTORS (other persons), ADVISOR (yet another person), DIVISIONS (another concept), and so forth. These same entities could also be discovered by studying the computerized registration system's computer files, databases, or outputs.

- Technology may also help you identify entities. Some CASE tools can reverse engineer existing files and databases into physical data models. The analyst must usually clean up the resulting model by physical names, codes, and comments with their logical, business-friendly equivalents.

While these techniques may prove useful in identifying entities, they occasionally play tricks on you. A simple, quick quality check can eliminate false entities. Ask your user to specify the number of instances of each entity. A true entity has multiple instances—dozens, hundreds, thousands, or more! If not, the entity is false.

As entities are discovered, give them simple, meaningful, business-oriented names. Entities should be named with nouns that describe the person, event, place, object, or thing about which we want to store data. Try not to abbreviate or use acronyms. Names should be singular so as to distinguish the logical concept of the entity from the actual instances of the entity. Names may include appropriate adjectives or clauses to better describe the entity—for instance, an externally generated CUSTOMER ORDER must be distinguished from an internally generated PURCHASE ORDER.

For each entity, define it in business terms. Don't define the entity in technical terms, and don't define it as "data about . . ." Try this! Use an English dictionary to create a draft definition, and then customize it for the business at hand. Your entity names and definitions should establish an initial glossary of business terminology that will serve both you and future analysts and users for years to come.

Our SoundStage management and users identified the entities listed in Table 5.5. Other entities may be discovered as the data model unfolds.

TABLE 5.5 *Fundamental Entities for the SoundStage Project*

Entity Name	Business Definition
AGREEMENT	A contract whereby a member agrees to purchase a certain number of products within a certain time. After fulfilling that agreement, the member becomes eligible for bonus credits that are redeemable for free or discounted products. Note: A major system improvement objective is to make agreements more flexible with respect to other clubs. Currently, only purchases within the club that issued an agreement count toward credits. Another system improvement objective would award bonus credits for each purchase leading up to fulfillment of the agreement, with accelerated bonuses after fulfillment of the agreeement.
CLUB	A SoundStage membership group to which members can belong. Clubs tend to be organized according to product interests such as music versus movies versus games; or specialized media interests such as Digital Video Disks (DVD) or Nintendo. Note: Cross-club interaction is a desired objective for the new system.
MEMBER	An active member of one or more clubs. Note: A target system objective is to re-enroll inactive members as opposed to deleting them.
MEMBER ORDER	An order generated for a member as part of a monthly promotion, or an order initiated by a member. Note: The current system only supports orders generated from promotions; however, customer-initiated orders have been given a high priority as an added option in the proposed system.
PRODUCT	An inventoried product available for promotion and sale to members. Note: System improvement objectives include (1) compatibility with new bar code system being developed for the warehouse, and (2) adaptability to a rapidly changing mix of products.
PROMOTION	A monthly or quarterly event whereby dated orders are generated for all members in a club. Members then have some period of time to cancel or accelerate fulfillment of that order, after which the order is automatically filled.

The Context Data Model

The next task in data modeling is to construct the context data model. The context data model includes the fundamental or independent entities that were previously discovered.

> An **independent entity** is one that exists regardless of the existence of any other entity. Its primary key contains no attributes that would make it dependent on the existence of another entity.

Independent entities are almost always the first entities discovered in your conversations with the users.

Relationships should be named with verb phrases that, when combined with the entity names, form simple business sentences or assertions. Some CASE tools, such as *System Architect,* let you name the relationships in both directions. Otherwise, always name the relationship from parent to child.

We have completed this task in Figure 5.11. Don't worry if you don't get it perfect the first time—that would be rare. Once we begin mapping attributes, new entities and relationships may surface. The numbers below reference those same numbers in the figure. The ERD communicates the following:

1. A CLUB establishes one or more AGREEMENTS. Members will learn about these agreements through advertisements and other marketing programs. An AGREEMENT is established by exactly one CLUB. The double hash marks mean one and only one.
2. An AGREEMENT binds zero, one, or more MEMBERS. Members join clubs via such an agreement. Why zero? Because a club may be new with no membership as yet. A MEMBER is bound by one or more AGREEMENTS. (Note: This is a nonspecific (or many-to-many) relationship.)
3. A MEMBER belongs to one or more CLUBS. A CLUB enrolls zero, one, or more MEMBERS. Again, the club may be new.
4. Each month or quarter, a CLUB sponsors zero, one, or more PROMOTIONS. Why zero? Again, a club may be just starting and not yet offering promotions. A PROMOTION is sponsored by exactly one CLUB.
5. Each PROMOTION features exactly one PRODUCT. A PRODUCT is featured in zero, one, or more PROMOTIONS. For example, a CD that appeals to both country/western and light rock audiences might be featured in the promotion for both. Since products greatly outnumber promotions, most products are never featured in a promotion.
6. A PROMOTION generates many MEMBER ORDERS. These are dated orders to which a member must reply by the specified date or else the order is filled. The promotion always generates more than one order; in fact, it generates one order per club member. A MEMBER ORDER is generated for zero or one PROMOTION. Why zero? In the desired system, a member can initiate his or her own order.
7. It is permissible for more than one relationship to exist between the same two entities if the separate relationships communicate different business events or associations. Thus, a MEMBER responds to zero, one, or more MEMBER ORDERS. This relationship supports the promotion-generated orders. A MEMBER places zero, one, or more MEMBER ORDERS. This relationship supports member-initiated orders. In both cases, a MEMBER ORDER is placed by (is responded to by) exactly one MEMBER.

   Although we didn't need it for this double relationship, some CASE tools (including *System Architect*) provide a symbol for recording Boolean relationships (such as AND, OR). Thus, for any two relationships, a Boolean symbol could be used to establish that instances of the relationships must be mutually exclusive (= OR) or mutually contingent (= AND).

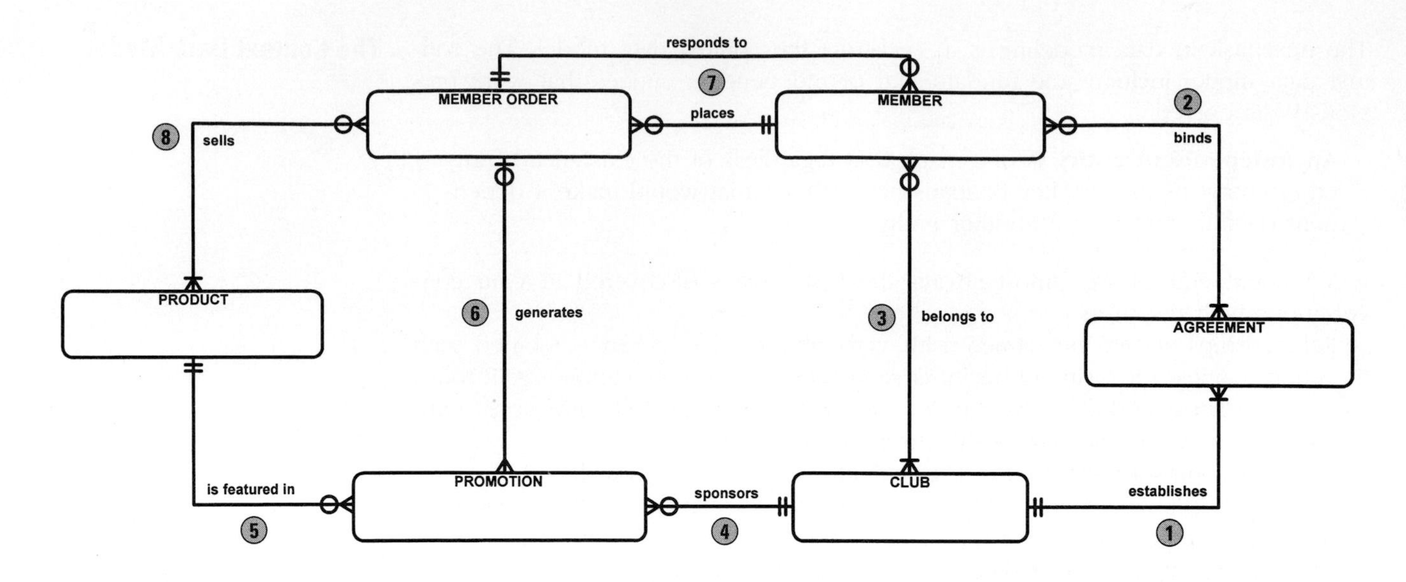

FIGURE 5.11 *The SoundStage Context Data Model*

(8) A MEMBER ORDER sells one or more PRODUCTS. A PRODUCT is sold on zero, one, or more MEMBER ORDERS.

If you read each of the preceding items carefully, you probably learned a great deal about the SoundStage system. Data models have become increasingly popular as a tool for describing the business context for system projects.

The Key-Based Data Model

The next task is to identify the keys of each entity. The following guidelines are suggested for keys:[1]

1. The value of a key should not change over the lifetime of each entity instance. For example, NAME would be a poor key since a person's last name could change by marriage or divorce.
2. The value of a key cannot be null.
3. Controls must be installed to ensure that the value of a key is valid. This can be accomplished by precisely defining the domain and using the database management system's validation controls to enforce that domain.
4. Some experts (Bruce) suggest you avoid **intelligent keys.** An intelligent key is a business code whose structure communicates data about an entity instance (such as its classification, size, or other properties). A code is a group of characters and/or digits that identifies and describes something in the business system. They argue that because those characteristics can change, it violates rule number 1 above.

   We respectfully disagree. Business codes can return value to the organization because they can be quickly processed by humans without the assistance of a computer.

[1] Thomas A. Bruce, *Designing Quality Databases with IDEFIX Information Models.* Copyright © 1992 by Thomas A. Bruce. Reprinted by permission of Dorset House Publishing, 353 W. 12th St., New York, NY 10014 (212-620-4053/1-800-DH-BOOKS/www.dorsethouse.com). All rights reserved.

a. There are several types of codes. They can be combined to form effective means for entity instance identification.
   i. **Serial codes** assign sequentially generated numbers to entity instances. Many database management systems can generate and constrain serial codes to a business's requirements.
   ii. **Block codes** are similar to serial codes except that block numbers are divided into groups that have some business meaning. For instance, a satellite television provider might assign 100–199 as PAY PER VIEW channels, 200–299 as CABLE channels, 300–399 to SPORT channels, 400–499 to ADULT PROGRAMMING channels, 500–599 to MUSIC-ONLY channels, 600–699 to INTERACTIVE GAMING channels, 700–799 to INTERNET channels, 800–899 to PREMIUM CABLE channels, and 900–999 to PREMIUM MOVIE AND EVENT channels.
   iii. **Alphabetic codes** use finite combinations of letters (and possibly numbers) to describe entity instances. For example, each STATE has a unique two-character alphabetic code. Alphabetic codes must usually be combined with serial or block codes to uniquely identify instances of most entities.
   iv. In **significant position codes,** each digit or group of digits describes a measurable or identifiable characteristic of the entity instance. Significant digit codes are frequently used to code inventory items. The codes you see on tires and lightbulbs are examples of significant position codes. They tell us about characteristics such as tire size and wattage, respectively.
   v. **Hierarchical codes** provide a top-down interpretation for an entity instance. Every item coded is factored into groups, subgroups, and so forth. For instance, we could code employee positions as follows:
   – First digit identifies classification (e.g., clerical, faculty, etc.).
   – Second and third digits indicate level within classification.
   – Fourth and fifth digits indicate calendar of employment.

b. The following guidelines are suggested when creating a business coding scheme:
   i. Codes should be expandable to accommodate growth.
   ii. The full code must result in a unique value for each entity instance.
   iii. Codes should be large enough to describe the distinguishing characteristics, but small enough to be interpreted by people *without a computer.*
   iv. Codes should be convenient. A new instance should be easy to create.

5. Consider inventing a surrogate key instead to substitute for large concatenated keys of independent entities. This suggestion is not practical for associative entities because each part of the concatenated key is a foreign key that must precisely match its parent entity's primary key.

Figure 5.12 is the key-based data model for the SoundStage project. We have eliminated all nonspecific relationships by resolving them into associative entities and one-to-many relationships (as described earlier in the chapter). Since all our relationships are now one-to-many, we have adopted the common practice of naming the relationship from parent to child. The inverse relationship, while not shown, is implicit. We call your attention to the following noteworthy items:

① Many entities have a simple, single-attribute primary key (PK1).

② In the PRODUCT entity, either one of two attributes could uniquely identify an instance of the entity. We designate them as separate primary keys (PK1 and PK2).

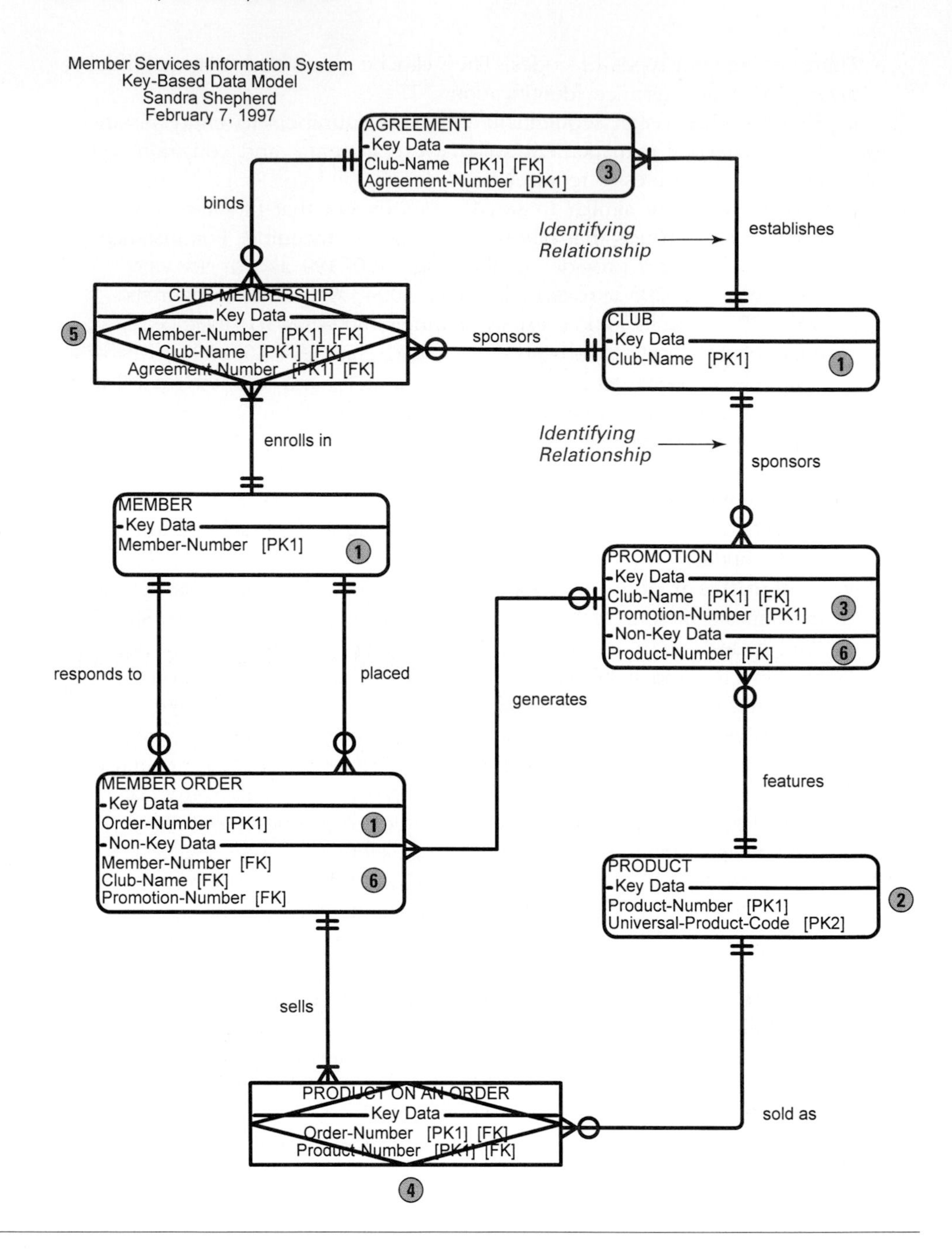

FIGURE 5.12
The SoundStage Key-Based Data Model

(3) Notice how the primary keys for AGREEMENT and PROMOTION were constructed. Each has a concatenated key. Part of that key is inherited from the parent entity CLUB. You can tell that because CLUB NAME is also a foreign key (FK). When one entity contributes its key to another entity across a relationship, the relationship is said to be **identifying**—because it helps to identify the child entity. Notice that all of the attributes that comprise the concatenated key have the same primary key number, PK1.

(4) We resolved the nonspecific relationship between ORDER and PRODUCT by introducing the associative entity ORDER ON A PRODUCT. Each associative entity instance represents one product on one order. The parent entities contributed their own primary keys to comprise the associative entity's concatenated key

(PK1). Also notice that each attribute in that concatenated key is a foreign key that points back to the correct parent instance.

5. CLUB MEMBERSHIP is a ternary relationship that simultaneously associates one MEMBER, CLUB, and AGREEMENT. Thus, the concatenated key consists of four attributes contributed by the three participating parent entities.

6. All relationships contribute foreign keys from parent to child. You just learned that if the contributed foreign key helps to uniquely identify instances of the child entity, the relationship is said to be identifying.

   On the other hand, if the foreign key plays no role in identifying instances of the child entity, then it is recorded as nonkey data in our model. Its only purpose is to point to a child entity's specific parent. For example, MEMBER NUMBER in the MEMBER ORDER entity serves only to point to the correct MEMBER entity instance for an order. In this case, the relationship is called **nonidentifying.**

One final comment is in order. If you cannot define keys for an entity, it may be that the entity doesn't really exist—that is, multiple occurrences of the so-called entity do not exist. Thus, assigning keys is a good quality check before fully attributing the data model.

Generalized Hierarchies

At this time, it would be useful to identify any generalization hierarchies in a business problem. The SoundStage project at the beginning of this chapter identified at least one supertype/subtype structure. Subsequent discussions uncovered the generalization hierarchy shown in Figure 5.13. We had to lay out the model somewhat differently because of the hierarchy; however, the relationships and keys that were previously defined have been retained. We call your attention to the following:

1. The SoundStage CASE tool automatically draws a dashed box around generalization hierarchy.
2. The subtypes inherited the keys of the supertypes.
3. We disconnected PROMOTION from PRODUCT as it was shown earlier and reconnected it to the subtype TITLE. This was done to accurately properly assert that MERCHANDISE is never featured on a PROMOTION.

The Fully Attributed Data Model

It may seem like a trivial task to identify the remaining data attributes; however, analysts not familiar with data modeling frequently encounter problems. To accomplish this task, you must have a thorough understanding of the data attributes for the system. These facts can be discovered using top-down approaches (such as brainstorming) or bottom-up approaches (such as form and file sampling). If an enterprise data model exists, some (perhaps many) of the attributes may have already been identified and recorded in a repository.

The following guidelines are offered for attribution.

- Many organizations have naming standards and approved abbreviations. The data or repository administrator usually maintains such standards.
- Choose attribute names carefully. Many attributes share common base names such as NAME, ADDRESS, DATE. Unless the attributes can be generalized into a supertype, it is best to give each variation a unique name such as:

CUSTOMER NAME	CUSTOMER ADDRESS	ORDER DATE
SUPPLIER NAME	SUPPLIER ADDRESS	INVOICE DATE
EMPLOYEE NAME	EMPLOYEE ADDRESS	FLIGHT DATE

  Also, remember that a project does not live in isolation from other projects, past or future. Names must be distinguishable across projects.

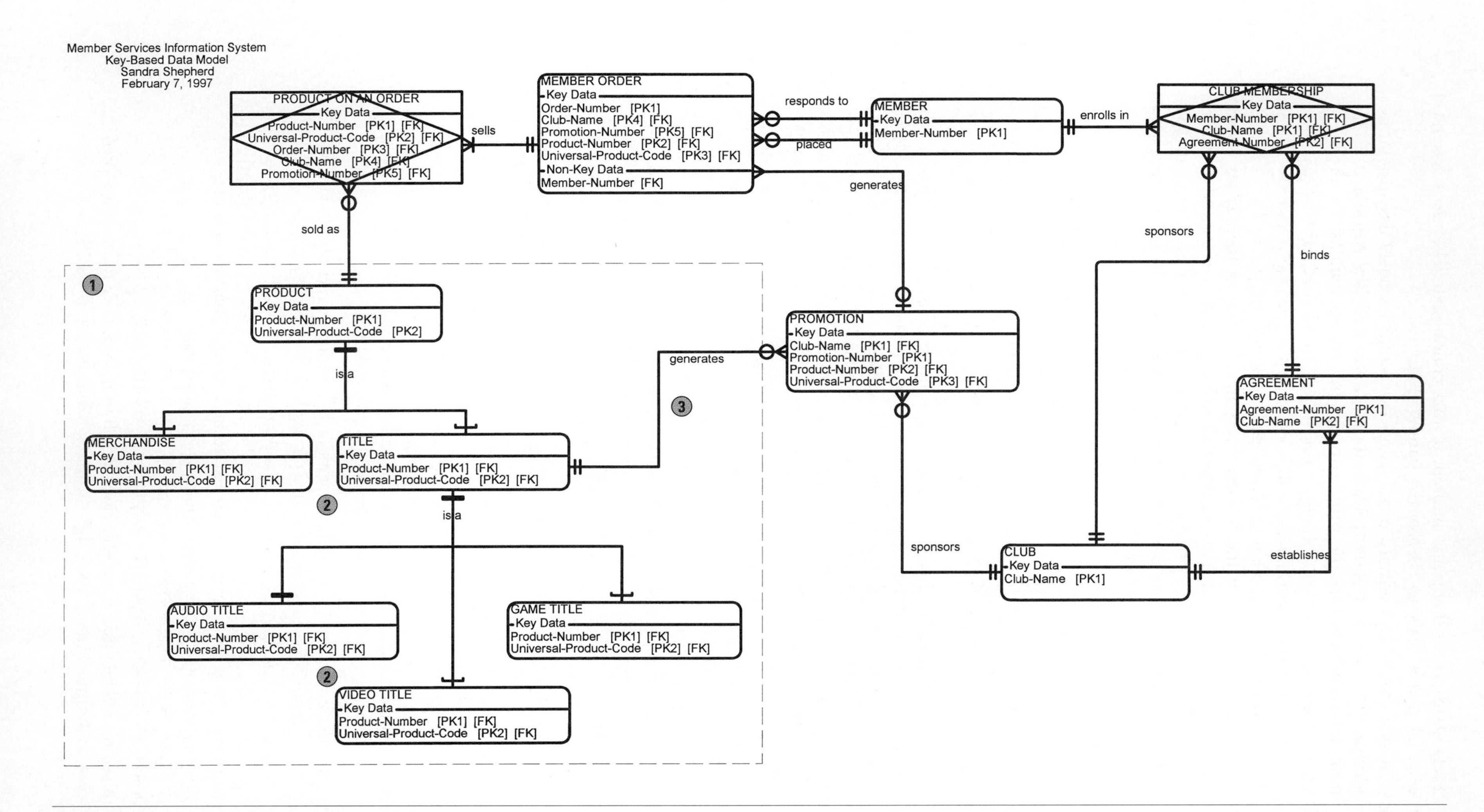

FIGURE 5.13 *The SoundStage Key-Based Data Model with a Generalization Hierarchy*

Some organizations maintain reusable, global templates for these common base attributes. This promotes consistent data types, domains, and defaults across all applications.

- Physical attribute names on existing forms and reports are frequently abbreviated to save space. Logical attribute names should be clearer—for example, translate the order form's attribute COD into its logical equivalent, AMOUNT TO COLLECT ON DELIVERY; translate QTY into QUANTITY ORDERED, and so forth.
- Many attributes take on only YES or NO values. Try naming these attributes as questions. For example, the attribute name CANDIDATE FOR A DEGREE? suggests the values are YES and NO.

  Each attribute should be mapped to only one entity. If an attribute truly describes different entities, it is probably several different attributes. Give each a unique name.
- Foreign keys are the exception to the nonredundancy rule—they identify associated instances of related entities.
- An attribute's domain should not be based on logic. For example, in the SoundStage case we learned the values of MEDIA were dependent on the type of product. If the product type is a video, the media could be VHS tape, 8mm tape, laserdisc, or DVD. If the product type is audio, the media could be cassette tape, CD, or MD. The best solution would be to assign separate attributes to each domain: AUDIO MEDIA and VIDEO MEDIA.

Figure 5.14 provides the mapping of data attributes to entities for the definition phase of our SoundStage systems project.

The Fully Described Model

The last task is the most time consuming. It can be started in parallel with the key-based model or fully attributed model, but it is usually the last data modeling task completed. The fully attributed model identifies all the attributes to be captured and stored in our future database. But the descriptions for those attributes are incomplete; they require domains.

Most CASE tools provide extensive facilities for describing the data types, domains, and defaults for all attributes to the repository. Additionally, each attribute should be defined for future reference.

Additional descriptive properties may be recorded for attributes at this time. For example, who should be able to create, delete, update, and access each attribute? How long should each attribute (or entity) be kept before the data is deleted or archived? The data administrator typically decides which of these attributes are important enough to be documented.

THE NEXT GENERATION

The demand for data modeling as a skill is dependent on two factors: (1) the need for databases and (2) the use of relational database management system technology to implement those databases. As for the former, databases will always be required in information systems. Will relational technology continue to dominate?

There is some belief that relational database technology will eventually be replaced by object technology. If that were to happen, data modeling would be replaced by object modeling techniques. We don't think this will happen anytime soon. Demand for relational database technology continues to grow, and that demand dwarfs the demand for any of the object database technologies that are beginning to emerge. Even as object database technology becomes available, we expect the relational database industry to add object features and technologies to its product lines. Data modeling should remain a value-added skill for many years.

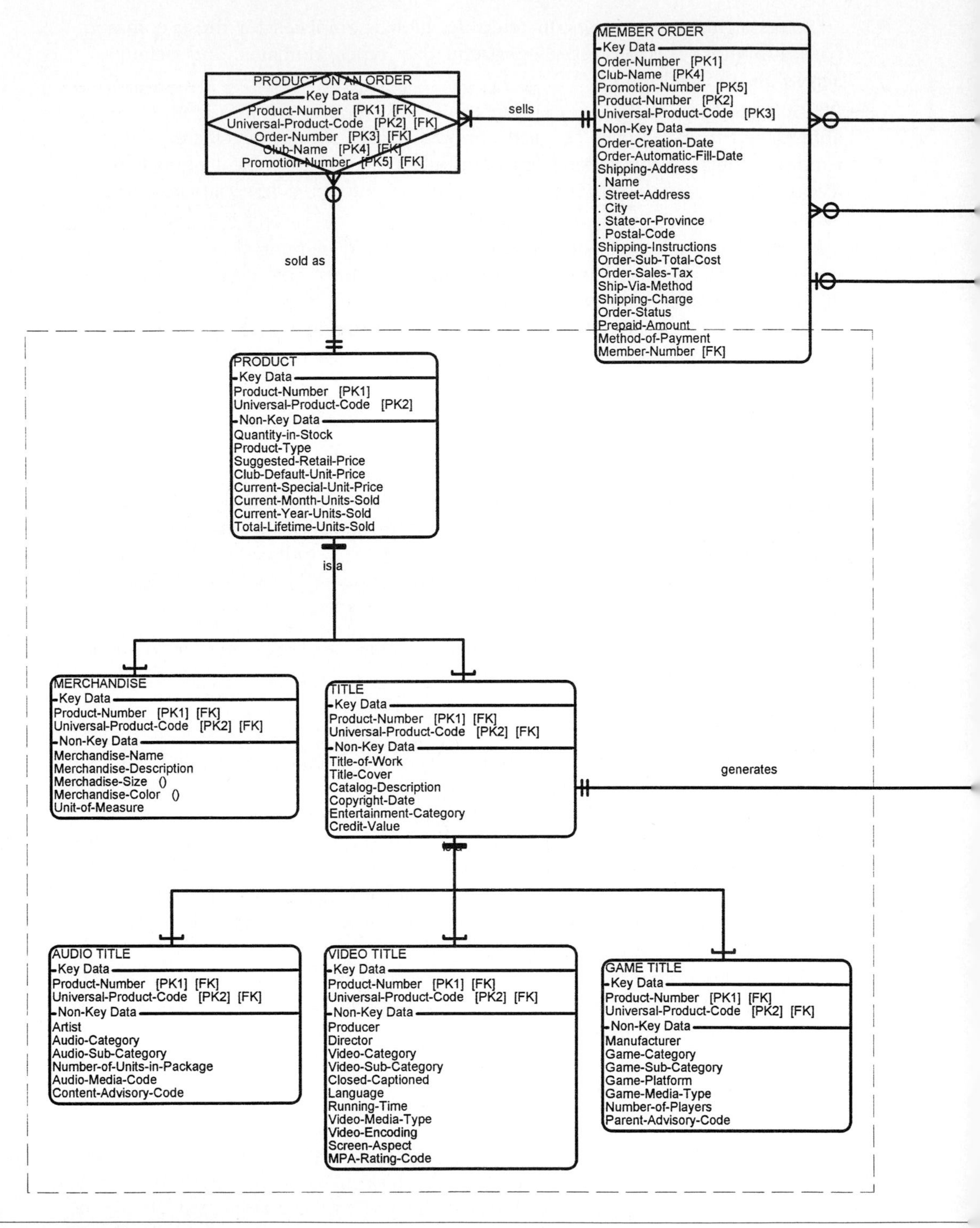

FIGURE 5.14 *The SoundStage Fully Attributed Data Model*

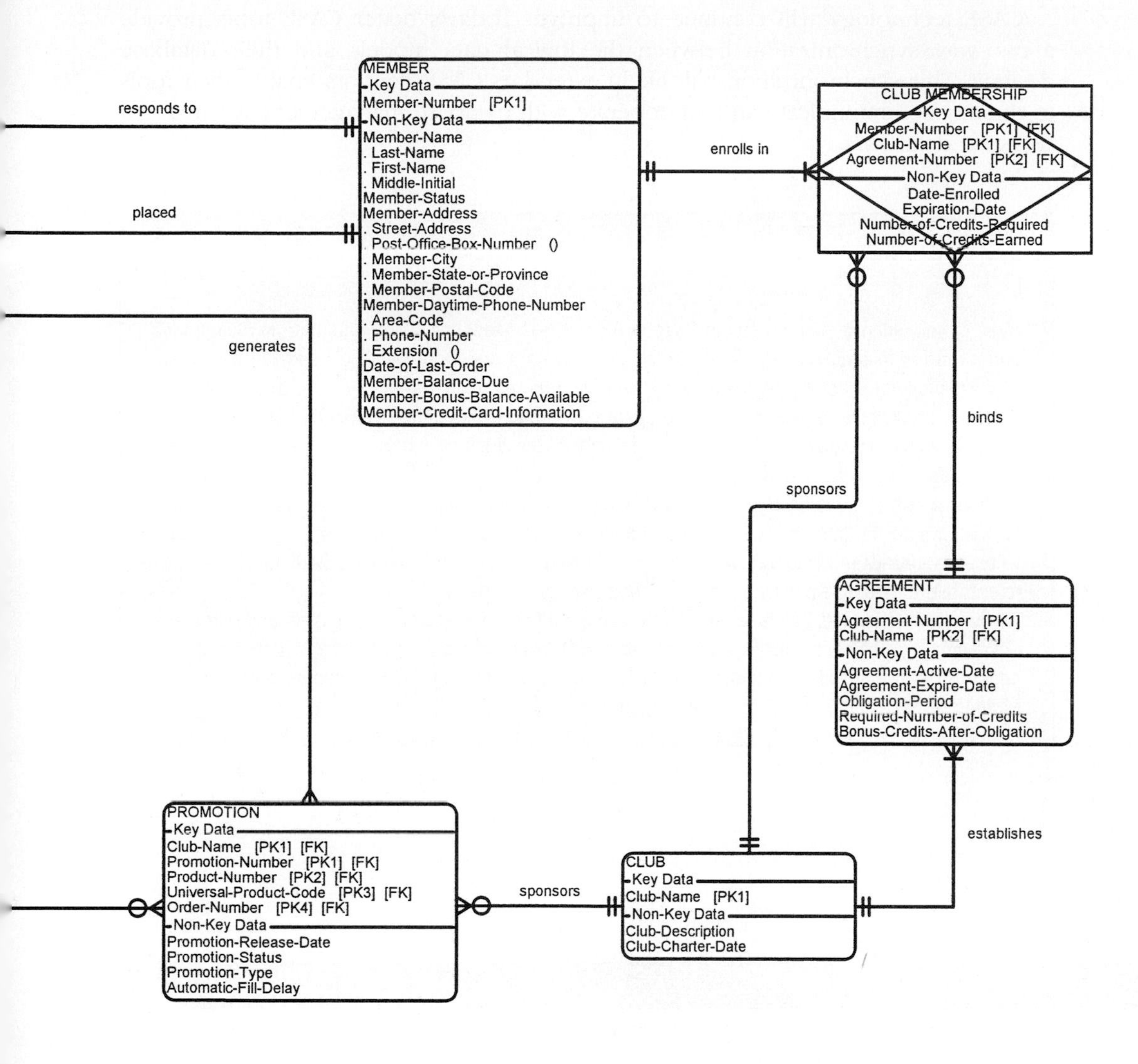

Member Services Information System
Fully Attributed Data Model
Sandra Shepherd
February 9, 1997

CASE technology will continue to improve. Today's better CASE tools provide a two-way synchronization between the logical data models and their database designs. This synchronization will likely extend as CASE vendors enable their tools to directly commmunicate and interoperate with database management systems and working databases.

WHERE DO YOU GO FROM HERE?

Most of you will proceed directly to Chapter 6, "Process Modeling." Whereas data modeling was concerned with data independently from how that data are captured and used (data at rest), process modeling shows how the data will be captured and used (data in motion). At your instructor's discretion, some of you may jump to Chapter 8, "Object Modeling." Object modeling has many parallels with data modeling. An object includes attributes, but it also includes all the processes that can act on and use those methods.

If you want to immediately learn how to implement data models as databases, you should skim or read Chapter 11. "Database Design." In that chapter, the logical data models are analyzed for stability and integrity and then transformed into physical database schemas. With CASE tools, the code to create the database can be generated automatically.

Finally, data models can be created only through effective interaction with users and managers who know the business. Part Five, Modules B and D, teach you how to collect the facts needed to construct real-world data models: Module B teaches fact-finding and information gathering and Module D teaches joint application development.

SUMMARY

1. A model is a representation of reality. Most models use pictures to represent reality.
   a. Logical models show *what* a system is or does. They are useful for business requirements documentation.
   b. Physical models show *how* a system will be technologically implemented.
2. Data modeling is a technique for organizing and documenting the data that must be stored in a database. The most popular logical data modeling techniques involve drawing entity relationship diagrams.
3. Data modeling is based on system thinking, the application of systems theory and concepts to problem solving.
4. The entity is the basic construct of data modeling. It is an abstraction of a collection of similar things about which we need to capture and store data. An entity instance is one occurrence of an entity.
5. Entities are described by attributes that hold data about entity instances.
6. Attributes have domains that define the legal set of data values for the attributes.
7. One or more attributes in every entity are used to uniquely identify each instance of the entity. We call this a key.
8. The instances of an entity may be divided into subsets based on the value of a subsetting criteria. A concatenated key is a combination of more than one attribute that uniquely identifies entity instances.
9. Relationships are natural business associations that exist between one or more entities. They may represent events or interactions between the entities. Regardless, they must be implemented in a database to integrate data about different entities.
10. Entities are described by their degree and cardinality. The degree indicates the number of different entities that participate in the relationship. Cardinality indicates the minimum and maximum number of associations between an entity and its related entity.
11. An associative entity is used to represent a logical association between two or more entities that have a many-to-many relationship. Each instance of an associative entity is the association of one (and only one) instance of each of the entities that participate in that relationship.
12. A foreign key is a nonprimary key in a child entity that matches a primary key in a parent entity in order to correctly identify a child entity's parent (instance).
13. A nonspecific relationship is a many-to-many relationship

that should be used only in the early stages of a data model. A nonspecific relationship can, and should, always be resolved into a pair of specific (one-to-many) relationships to an associative entity.

14. Generalization is a technique wherein entities share attributes through a supertype/subtype relationship. An entity supertype shares common attributes with one or more entity subtypes.
15. Data models are most commonly developed in strategic information systems planning projects (high-level data models) and systems development projects (detailed data models).
16. Many CASE tools support logical (and physical) data modeling. They provide the repository in which data models are stored for future use and reuse.
17. A logical data model is developed in the following stages:
    a. Entities are discovered and defined.
    b. A context data model is built. A context data model contains only fundamental entities and relationships. This is the only model in which nonspecific relationships should be shown.
    c. A key-based data model is built. The key-based model eliminates nonspecific relationships and adds associative entities. All entities in the model are given keys.
    d. A fully attributed model is built. This model shows all the attributes to be stored in the system.
    e. A fully described model is built. Each attribute is defined in the dictionary and described in terms of properties such as domain and security.

KEY TERMS

alphabetic code, p. 195
alternate key, p. 179
associative entity, p. 181
attribute, p.176
block code, p. 195
candidate key, p. 179
cardinality, p. 180
compound attribute, p. 176
concatenated key, p. 177
context data model, p. 188
data modeling, p. 173
data type, p. 177
default, p. 177
degree, p. 181
domain, p. 177
enterprise data model, p. 186
entity, pp. 175, 176
entity instance, p. 176
entity relationship diagram, p. 175
essential model, p. 172
foreign key, p. 183
fully attributed data model, p. 188
generalization, p. 183
hierarchical code, p. 195
independent entity, p. 193
intelligent key, p. 194
key, p. 177
key-based data model, p. 188
logical model, p. 172
many-to-many relationship, p. 183
metadata, p. 189
model, p. 172
nonidentifying (relationship), p. 197
nonspecific relationship, p. 183
physical model, p. 173
primary key, p. 179
recursive relationship, p. 181
relationship, p. 179
serial code, p. 195
significant position code, p. 195
subsetting criteria, p. 179
subtype, p. 185
supertype, p. 184
systems thinking, p. 175

REVIEW QUESTIONS

1. Differentiate between logical and physical models. Give three reasons why logical models are superior for structuring business requirements.
2. What is systems thinking, and how do data models represent systems thinking?
3. What is an entity? What are the five categories of entities?
4. Differentiate between entities and entity instances.
5. What are attributes? What are compound attributes? Give an example (not from the chapter) of each.
6. What are the three aspects of domain description for attributes?
7. Differentiate between candidate keys, primary keys, and alternate keys. Can each of these be a concatenated key?
8. What is a subsetting criteria?
9. What is a relationship? Why are relationships important to identify and describe? What is a nonspecific relationship?
10. Differentiate between cardinality and degree.
11. What is an associative entity? What role does it play in ternary relationships? What role does it play in resolving nonspecific relationships?
12. What role does a foreign key play in implementing a relationship?
13. What is generalization, and what is its value?
14. Differentiate between an enterprise and application data model.
15. During the survey and study phases, an analyst collected numerous samples, including documents, forms, and reports. Explain how these samples will prove useful for data modeling.
16. Explain the tasks used to construct an application data model.

PROBLEMS AND EXERCISES

1. A database designer has complained about plans to construct a logical data model. He believes we should just design the database with the database management system. Give three reasons why requirements should be specified in an implementation-independent fashion.
2. Using the final SoundStage data model, translate it into a set of business assertions (as we did with the explanation of Figure 5.1).
3. In Chapter 4, we provided classic definitions for analysis and synthesis. How does data modeling apply analysis and synthesis? How does this correspond to systems thinking?
4. Why do some systems analysts believe that data modeling is the most important aspect of business requirements modeling?
5. Most data entities correspond to persons, objects, events, or locations in the business environment. Give three examples of each data entity class.
6. Obtain three sample business forms from a business, your school, or your instructor. What entities are described by the fields on the forms?
7. Translate one of the above forms into a context data model.
8. Using Table 5.4 as a guide, develop a complete set of interview questions for the context data model you constructed in the previous problem.
9. Given the context data model from problem 7, attribute that model using a combination of the form and brainstorming.
10. What entities are described on your class schedule form or your school's course registration form? Draw a context data model to support academic scheduling at your school.
11. Give two examples of each of the following data relationship complexities: one-to-one (1:1), one-to-many (1:M or M:1), and many-to-many (M:M). Draw an ERD for each of your examples. Be sure to label data entities using nouns and label data relationships using verbs. Annotate the graph to communicate the relationship complexity.
12. Resolve the nonspecific relationship in the previous exercise.
13. During the definition phase, the actual data model is drawn, refined, and improved. Identify the data modeling issues that must be specified when alternative solutions are being identified and analyzed in the configuration phase.
14. Given the following narrative description of entities and their relationships, prepare a draft entity relationship diagram (ERD). Be sure to state any reasonable assumptions that you are making.

    Burger World Distribution Center serves as a supplier to 45 Burger World franchises. You are involved with a project to build a database system for distribution. Each franchise submits a day-by-day projection of sales for each of Burger World's menu products (the products listed on the menu at each restaurant) for the coming month. All menu products require ingredients and/or packaging items. Based on projected sales for the store, the system must generate a day-by-day ingredients need and then collapse those needs into one-per-week purchase requisitions and shipments.

15. Write a paragraph or two explaining how you would present and verify the ERD prepared in problem 14 to a group of end-users who are not familiar with computer concepts.
16. Explain why a concatenated key of NAME and ADDRESS would not be a good key.
17. All vehicles in the state of ______________ must be licensed. Some data are common to all vehicles, but certain types of vehicles require their own data. Construct a generalization hierarchy to represent this scenario.
18. Explain the origins of the primary key for an associative entity. Next, explain how the parts of the primary key serve as foreign keys.
19. For the final SoundStage data model, construct sample tables that contain enough instances of data to demonstrate every possible cardinality shown. Invent your own data.
20. Create an example of a ternary relationship other than those demonstrated in this book. Why is a ternary relationship different from three binary relationships? Do you think that ternary and binary relationships between the same entities are mutually exclusive, or can they coexist? Why?

PROJECTS AND RESEARCH

1. Using this chapter as a guide, draw a metadata model. A metadata model is a data model that stores data about the data model. (This is how CASE vendors design the repository for their tools.)
2. Obtain copies of all the forms used in your school's class scheduling and/or student scheduling system. (Your instructor should be able to secure these forms for the class.) Using the construction stages described in this chapter, build a complete data model including attribute descriptions. Feel free to suggest improvements in your data model.
3. As part of a semester project, identify a client with a

small application development need. Departmental applications are ideal! Construct a complete data model for that client.

4. Obtain a data model developed in an information systems shop. (You can also find complete data models in many database textbooks.) If the model uses a different graphical notation, convert it to the information engineering notation in this book. (That should prove to you that different notations are more alike than they are different.) Thoroughly critique the data model. Make improvements and explain why you feel they are improvements.

MINICASES

1. The MIS department our business wants to build a database to track all our hardware and software. We own workstations, network servers, and peripherals. The department wants to keep track of software packages, as well as the licenses for those packages. Some software licenses are for single machines. We can install them on network servers, but we can only permit as many network users as we own licenses. We also own network licenses. A single network license authorizes a specific number of users. Nonnetwork licenses may be installed on either workstations or servers. Network licenses may be installed only on servers. We want to keep track of where software licenses are installed. Some licenses may not be installed anywhere at any given time. We must also be able to prove the legality of any software we have installed. Each license must be traced to either a purchase order, a gift, or a loan. We may also have certain software on order. We order packages, but we receive licenses. Construct the data model and attribute it through brainstorming.
2. Most students have bank accounts. Construct a data model that shows the relationships that exist among customers, different types of accounts (e.g., checking, savings, loan, funds), and transactions (e.g., deposits, withdrawals, payments, ATMs). Attribute your model such that it could be used to produce a consolidated bank statement.
3. To schedule classes, your school needs to know about courses that can be offered, instructors and their availability, equipment requirements for courses, and rooms (and their equipment). From the courses that can be scheduled, they select the courses that will be scheduled. For each of those courses, they schedule one or more classes (sometimes called sections or divisions). The problem of the schedulers is to assign classes to instructors, rooms, and time slots. The schedulers are constrained by the reality that (1) some courses cannot conflict because many students take them during the same term, (2) instructors cannot be in two places at the same time, and (3) rooms cannot be double-booked. Construct a data model to help the schedulers. (Caution: This problem requires some thought. Depending on your instructor's course policies, it might help to work in groups.) Clearly state any reasonable assumptions.
4. Given the following data attributes and entities, indicate which attributes could be identifiers for each of the entities. You may have to combine attributes or even add some attributes that are not listed. Map all the attributes to their appropriate entity. Remember, each attribute should describe one and only one entity. Draw a rough draft entity relationship diagram.

Green Acres Real Estate System

Entities:

Seller	House	Closing
Buyer	Offer	Showing
Listing	Property	Room

Attributes:

Seller name	Offer amount
Square foot size	Listing date
Seller address	Property description
House style	Offer date
Closing location	Room type
Listing price	Property size
Number of bathrooms	Showing time
Garage size	Room size
Showing date	Elementary school zone
Garage location	Buyer phone number
Buyer name	Closing date
Basement size	Sales terms
House heating method	

SUGGESTED READINGS

Bruce, Thomas A. *Designing Quality Databases with IDEF1X Information Models.* New York: Dorsett House Publishing, 1992. We actually use this book as a textbook in our database analysis and design course. IDEF1X is a rich, standardized syntax for data modeling (which Bruce calls information modeling). The graphical language looks different, but it communicates the same system concepts presented in our book. The language is supported by at least two CASE tools: Logic Works' *ERwin* and Popkin's *System Architect*. The book includes two case studies.

Database Programming and Design. San Mateo, CA: Miller Freeman, Inc., published monthly. This periodical is our favorite source of current events and developments in the practice of data modeling.

Dittman, Kevin C., and Popkin Systems & Software, Inc. *SYSTEM ARCHITECT: A Guided Tour*. Burr Ridge, IL: Richard D. Irwin. Our colleague has adapted this Popkin document for academic use. It includes a student edition of the software that was used to build the SoundStage data models in our book. Professor Dittman is currently developing a more comprehensive tutorial, tentatively titled *Using System Architect for Systems Analysis and Design* (also to be an Irwin title). It should be available about the same time as this book or shortly thereafter.

Hay, David C. *Data Model Patterns: Conventions of Thought*. New York: Dorsett House Publishing, 1996. For those of you who seek additional data modeling knowledge, this is an excellent book. The book's hypothesis is intriguing. It suggests that all data models are extrapolations of a few basic templates. The book then proceeds to examine those templates. At the time of this writing, students in one of our database courses were extending one of the templates for a departmental data warehouse in our own academic unit. This book uses its own data modeling notations called *CASE*Method,* but they are conceptually equivalent to our own.

Martin, James, and Clive Finkelstein. *Information Engineering*. 3 volumes. New York: Savant Institute, 1981. Information engineering is a formal, database, and fourth-generation language-oriented methodology. The graphical data modeling language of information engineering is virtually identical to ours. Data modeling is covered in Volumes I and II.

Schlaer, Sally, and Stephen J. Mellor. *Object-Oriented Systems Analysis: Modeling the World in Data*. Englewood Cliffs, NJ: Yourdon Press, 1988. Forget the title! "Object-oriented" means something different than when this book was written, but it is still one of the easiest to read books on the subject of data modeling.

Teorey, Toby J. *Database Modeling & Design: The Fundamental Principles*. 2nd ed. San Francisco: Morgan Kaufman Publishers, Inc., 1994. This book is somewhat more conceptual than the others in the list, but it provides useful insights into the practice of data modeling.

6

PROCESS MODELING

CHAPTER PREVIEW AND OBJECTIVES

This is the second of three systems modeling chapters. In this chapter you will learn how to draw **data flow diagrams,** a popular process model that documents a system's processes and their data flows. You will know process modeling as a systems analysis tool when you can:

- Define systems modeling and differentiate between logical and physical system models.
- Define process modeling and explain its benefits.
- Recognize and understand the basic concepts and constructs of a process model.
- Read and interpret a data flow diagram.
- Explain when to construct process models and where to store them.
- Construct a context diagram to illustrate a system's interfaces with its environment.
- Identify external and temporal business events for a system.
- Perform event partitioning and organize events in a functional decomposition diagram.
- Draw event diagrams and then merge those event diagrams into a system diagram.
- Draw primitive data flow diagrams, and describe the elementary data flows and processes in terms of data structures and procedural logic (Structured English and decision tables), respectively.

SCENE

We begin this episode shortly after the project executive sponsor has approved the initial system improvement objectives (*FAST* definition phase). It's time to model the new system requirements, beginning with data. Sandra is facilitating a meeting with the following staff:

- David Hensley, representing Legal Services.
- Ann Martinelli, representing Member Services.
- Sally Hoover, representing Member Services.
- Joe Bosley, representing Marketing.
- Antonio Scarpachi, representing Warehouse.
- Bob Martinez, assigned to sketch the data model.
- Sarah Hartman, assigned to take detailed notes.

SANDRA

Well, I thought the data model session went well. You each have a copy of the model in your packet.

[You have it in Chapter 5.]

The purpose of this facilitated session is to discover all the business events to which your new system must provide a response. As usual, I'd like to keep this discussion as nontechnical as possible. Let's try not to think about computer programs, especially those of you who have written programs. Instead, I'd like you to focus exclusively on the business events, the inputs that trigger the events, and all possible responses or outputs from the events. Any questions?

DAVID

Just the obvious one! It may help to understand why we are interested in events today, as well as your definition of an event. Also, it would help me understand the relationship between event and and the data model that we constructed in the last session.

SANDRA

Good questions. I'd like to answer the last question first. The data model identified the things about which the system must capture and store data. We called them *entities*. But, as you may have noticed, I carefully steered our discussions away from when and how that data would be captured, stored, or even used. Today we want to identify the processes that will do that. Those processes will eventually become programs that our staff will have to write. But ultimately, those programs are business processes that respond to everyday events in your environment. So today, we will try to discover the events to make sure that we design a system that responds to each and every one of them.

ANTONIO

Makes sense to me. Now what do you consider an event to be?

SANDRA

An event is just something that happens.

ANTONIO

An input?

SANDRA

Not really. We recognize many, if not most, events because they show up as a transaction or input. But some events don't have an actual input. For example, the passage of time can trigger some events. For example, the last day of the month triggers various reporting events.

ANN

So far in this project, everything seems like common sense. I really appreciate the way we seem to keep focusing on business issues and requirements. I assume we will get technical sooner or later.

BOB

It's coming. But we don't want to write programs that do not fulfill your day-to-day needs or that fail to recognize less common events that are nonetheless important. This afternoon, we are going to form breakout groups to construct simple business pictures of each event. Tomorrow, we'll review those pictures as a group and make any changes. Then Sandra, Sarah, and I will combine those pictures into a picture of your overall system. That picture will show all of the *work* that must be performed as part of the system.

DAVID

Let's do it!

SANDRA

You'll notice that I have written the names of all the data entities from the data model on the blackboard. These are the things about which you decided the system must capture and store data. The stored attributes of each entity are documented in your packet for reference.

[Noting that everyone had found the data model, Sandra continued . . .]

One way to do this is to ask ourselves, "For each entity, what business events might cause us to create a new instance of the entity, delete or deactivate an instance of the entity, or update an instance of the entity?" Those are events. But the big question is, "Where do we start?"

SALLY

Do you mean with which entity?

SANDRA

Yes.

JOE

Well why don't we simply ask ourselves which entity has to exist first. We can't do business without *clubs*. Clubs have to create *agreements*. Members establish *memberships* in clubs using the agreements. Only after all this happens can we worry about things like *products, promotions,* and *orders.*

SANDRA

So you suggest we start with events that affect clubs?

JOE

Yes.

BOB

You know, this strategy does make some sense. Several methodologies advocate the study of an entity's life history to discover essential processes. Let's give it a try. Is there a format to these events?

SANDRA

I'd suggest simple sentences such as "customer places a new order" or "customer cancels an order" or "time

SOUNDSTAGE

SOUNDSTAGE ENTERTAINMENT CLUB

to invoice customers"—sentences that describe not only the event but also who or what triggers the event.

JOE
As the marketing representative, I should take the lead here. Try these on for size. One—Marketing establishes a new club. Two—Marketing deletes a club.

ANN
Do you like the word *delete*? A club doesn't just go away, does it?

JOE
Technically, you're right. First we decertify the club so that new members cannot join. Later, once members have been relocated or canceled, we delete the club.

DAVID
Do we ever make changes to a club between the time it is created and deleted?

JOE
We never have in the past, but I suppose it possible.

BOB
OK. Let me read the events back to you. One—Marketing establishes a new club. Two—Marketing decertifies a club. Three—Marketing relocates members to new clubs. Four—Marketing cancels members that don't relocate to a new club. Five—Marketing deletes a club.

ANN
Wait a minute. Marketing doesn't mess with memberships. We do that. Change "Marketing" to "Membership" in the sentences.

JOE
Sorry, Ann's correct . . . What's next, agreements?

SANDRA
Yes, but I'd like to spend a few more minutes with each of these events first. I'd like to identify the inputs and outputs for each of these events . . .

The meeting goes on . . .

DISCUSSION QUESTIONS

1. How can the study of data help to identify processes?
2. Given the information that Sandra and Bob are trying to collect, what type of picture are they trying to draw? Why?
3. How will event discovery benefit system design and programming?
4. What types of events might be missed if the group focuses exclusively on data entities and how they are created, updated, and deleted?

AN INTRODUCTION TO SYSTEMS MODELING

NOTE This section is repeated and adapted from Chapter 5, "Data Modeling". If you have already covered Chapter 5, you may skip or review this introduction to system models and the distinction between logical and physical models. You can begin reading with the definition of a process model.

In Chapter 4 you were introduced to systems analysis activities that called for drawing system models. System models play an important role in systems development. As a systems analyst or user, you will constantly deal with unstructured problems. One way to structure such problems is to draw models.

A **model** is a representation of reality. Just as a picture is worth a thousand words, most system models are pictorial representations of reality.

Models can be built for existing systems as a way to better understand those systems or for proposed systems as a way to document business requirements or technical designs. An important concept, in both this chapter and the next, is the distinction between logical and physical models.

Logical models show *what* a system is or does. They are implementation-*in*dependent; that is, they depict the system independent of any technical implementation. As such, logical models illustrate the *essence* of the system. Popular synonyms include **essential model,** *conceptual model,* and *business model.*

Physical models show not only *what* a system is or does, but also *how* the system is physically and technically implemented. They are implementation-*de*pendent because they reflect technology choices and the limitations of those technology choices. Synonyms include *implementation model* and *technical model.*

An example might help to clarify the distinction. A flowchart is a model of a business process or computer program. It communicates the design of the logic for the process or program. A *logical* flowchart would document the business logic regardless of the choice of programming language. It would not worry about declaring variables, using counters (e.g., FOR I = 1−N), or file processing concerns (e.g., UNTIL END OF FILE). It would include only those specifications that must be implemented by the programmer regardless of the choice of programming language or any other technology. Now, let's assume the programming language and technology have been selected. The flowchart can now be expanded and refined to reflect the requirements, capabilities, and limitations of the chosen language and technology. In other words, the model has become a *physical* flowchart.

Systems analysts have long recognized the value of separating business and technical concerns. That is why they use logical system models to depict business requirements and physical system models to depict technical designs. Systems analysis activities tend to focus on the logical system models for the following reasons:

- Logical models remove biases that are the result of the way the current system is implemented or the way that any one person thinks the system might be implemented. Thus, we overcome the "we've always done it that way" syndrome. Consequently, logical models encourage creativity.
- Logical models reduce the risk of missing business requirements because we are too preoccupied with technical details. Such errors can be costly to correct after the system is implemented. By separating what the system must do from how the system will do it, we can better analyze the requirements for completeness, accuracy, and consistency.
- Logical models allow us to communicate with end-users in nontechnical or less technical languages. Thus, we don't lose requirements in the technical jargon of the computing discipline.

In this chapter we will focus exclusively on *logical* process modeling.

Process modeling is a technique for organizing and documenting the structure and flow of data through a system's PROCESSES and/or the logic, policies, and procedures to be implemented by a system's PROCESSES.

In the context of your information system building blocks (Figure 6.1), *logical* process models are used to document an information system's PROCESS focus from the perspective of the system owners and system users (the intersection of the PROCESS column with the system owner and system user rows). Also notice that one special type of process model, called a *context diagram,* illustrates the INTERFACE focus from the perspective of the system owners. Theoretically, it is possible to recover data flow diagrams by reverse engineering existing application programs. In practice, the technology is not as mature as reverse engineering for data models. The resultant data flow diagrams are *too* physical and overly complicated by the poor (or absent) design practices that were used to develop the original software.

Process modeling originated in classical software engineering methods; therefore, you may have encountered various types of process models such as program structure charts, logic flowcharts, or decision tables in an application programming course. In this chapter, we'll focus on a systems analysis process model, *data flow diagrams* (DFDs).

A **data flow diagram (DFD)** is a tool that depicts the flow of data through a system and the work or processing performed by that system. Synonyms include *bubble chart, transformation graph,* and **process model.**

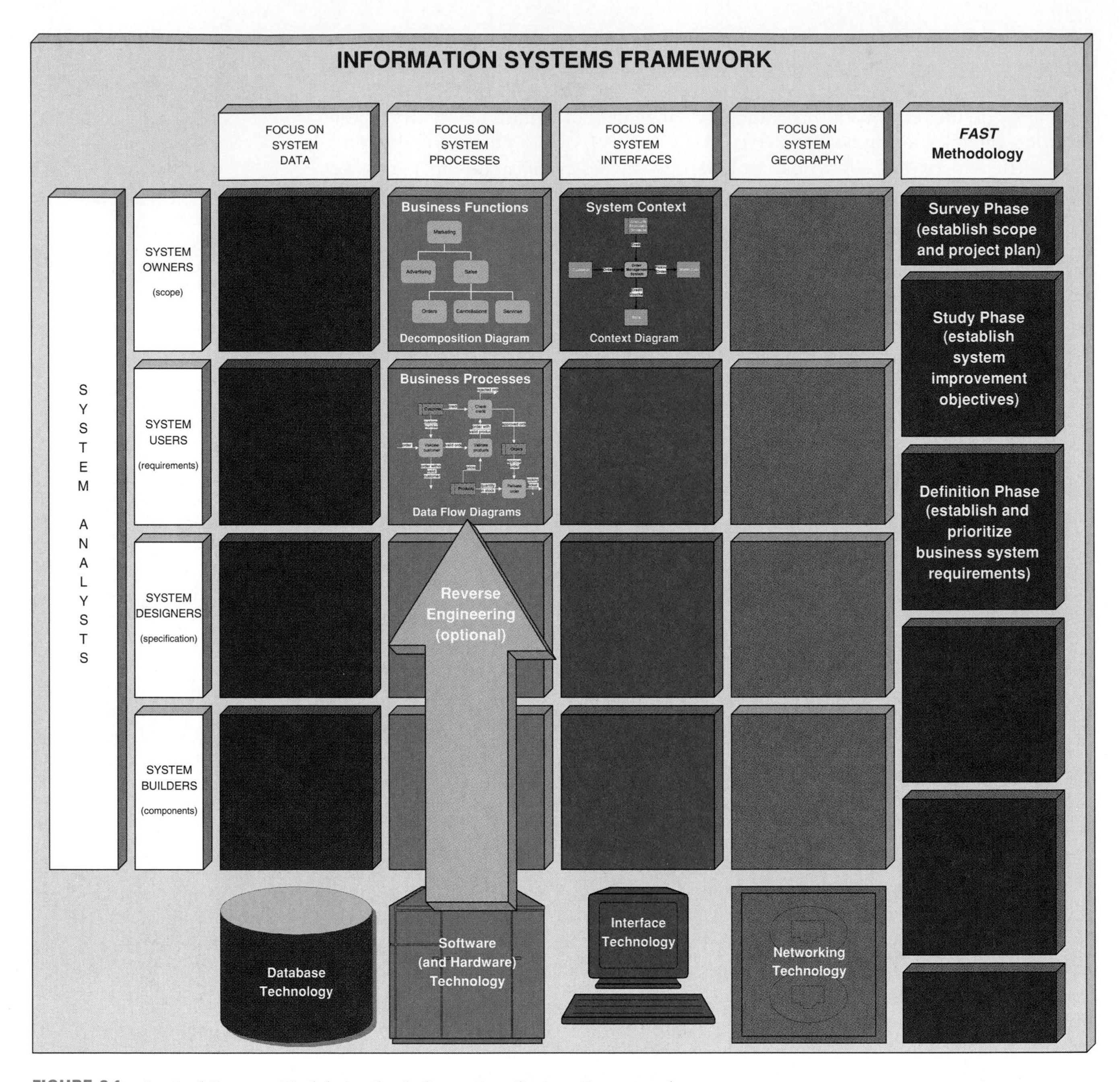

FIGURE 6.1 *Logical Process Models in the Information System Framework*

We'll also introduce a DFD planning tool called *decomposition diagrams*. Finally, we'll also study *context diagrams*, a process-like model that actually illustrates a system's interfaces to the business and outside world, including other information systems.

A simple data flow diagram is illustrated in Figure 6.2. In the design phase, some of these business processes might be implemented as computer software (either built in-house or purchased from a software provider). If you examine this data flow diagram, you should find it easy to read, even before you complete this chapter—that has always been the advantage of DFDs. There are only three symbols and one connection:

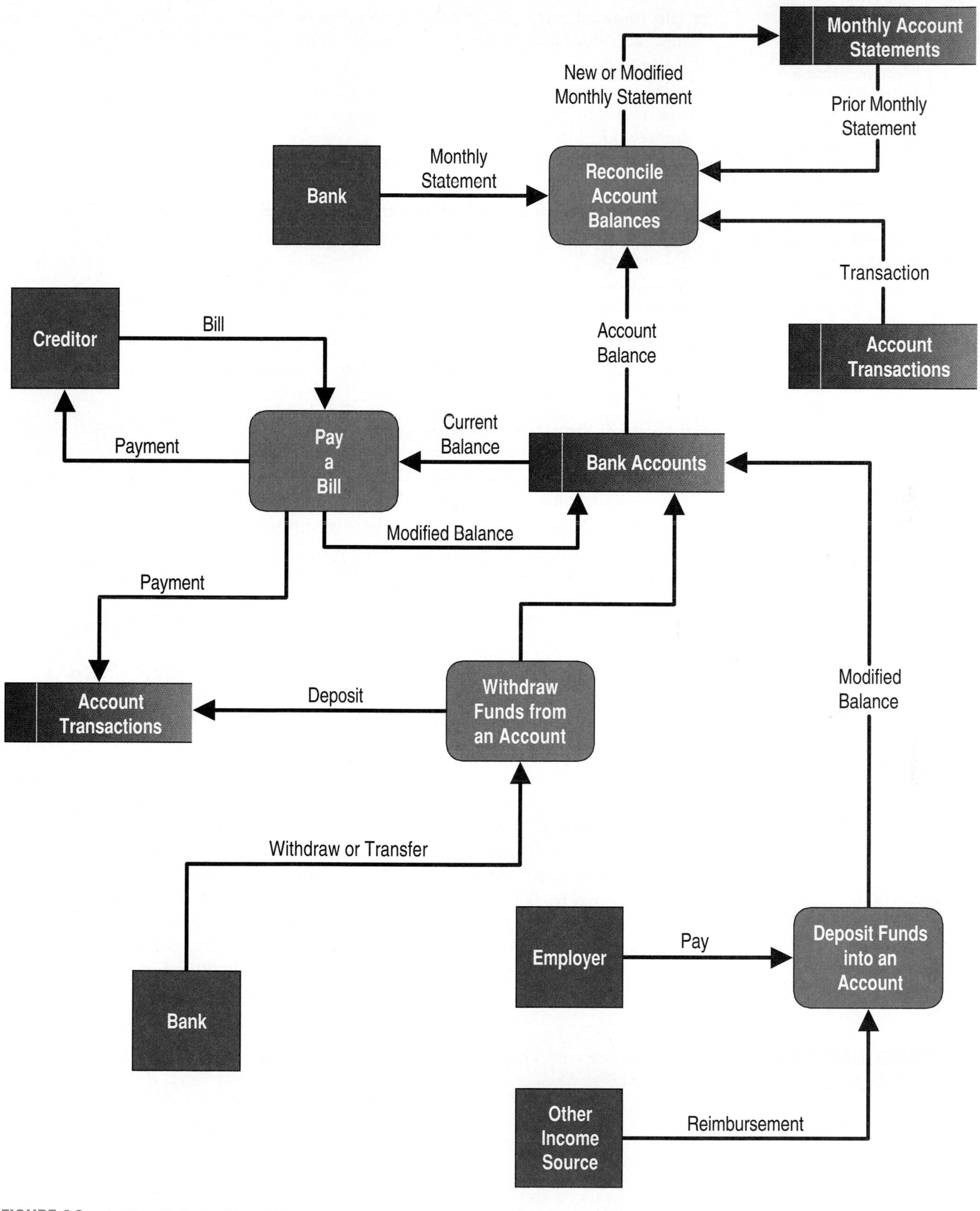

FIGURE 6.2 *A Simple Data Flow Diagram*

There are several competing symbol sets for DFDs. Most are named after their inventors (e.g., DeMarco/Yourdon, Gane/Sarson) or after a published standard (e.g., IDEFO, SSADM). Some analysts will argue semantics, but these data flow diagramming "languages" generally support the same fundamental concepts and constructs. We have adopted the Gane and Sarson (structured analysis) notation because of its popularity and CASE tool support.

- The rounded rectangles represent *processes* or work to be done. Notice that they are illustrated in the PROCESS color from your information system framework.
- The squares represent external agents—the *boundary* of the system. Notice that they are illustrated in the INTERFACE color from your information system framework.
- The open-ended boxes represent *data stores*, sometimes called files or databases. If you have already read Chapter 5, these data stores correspond to all instances of a single entity in a data model. Accordingly, they have been illustrated with the DATA color from your information systems framework.
- The arrows represent *data flows,* or inputs and outputs, to and from the processes.

Don't confuse data flow diagrams with flowcharts! Program design frequently involves the use of flowcharts. But data flow diagrams are very different! Let's summarize the differences.

- Processes on a data flow diagram can operate in parallel. Thus, several processes might be executing or working simultaneously. This is consistent with the way businesses work. On the other hand, processes on flowcharts can execute only one at a time.
- Data flow diagrams show the flow of data through the system. Their arrows represent paths down which data can flow. Looping and branching are not typically shown. On the other hand, flowcharts show the sequence of processes or operations in an algorithm or program. Their arrows represent pointers to the next process or operation. This may include looping and branching.
- Data flow diagrams can show processes that have dramatically different timing. For example, a single DFD might include processes that happen hourly, daily, weekly, yearly, and on-demand. This doesn't happen in flowcharts.

Data flow diagrams have been popular for nearly 20 years, but the interest in DFDs has been expanded recently because of their role in **business process redesign (BPR).** As businesses have come to realize that most data processing systems have merely automated outdated, inefficient, and bureaucratic business processes, there is renewed interest in streamlining those business processes. This is accomplished by first modeling those business processes for the purpose of analyzing, redesigning, and/or improving them. Subsequently, information technology can be applied to the improved business processes in creative ways that maximize the value returned to the business. We'll revisit this trend at the end of the chapter.

SYSTEM CONCEPTS FOR PROCESS MODELING

This is the second chapter of the book that actually teaches a *technique* of systems analysis. Most systems analysis techniques are strongly rooted in *systems thinking.* You may recall from Chapter 5,

> **Systems thinking** is the application of formal systems theory and concepts to systems problem solving.

Systems theory and concepts help us understand the way systems are organized and how they work. Techniques teach us how to apply the theory and concepts to build useful real-world systems. If you understand the underlying concepts, you can better adapt the techniques to ever-changing problems and conditions. Therein lies your true opportunity for competitive advantage and security in today's business world.

Let's explore some of the basic concepts that underlie all process models.

Process Concepts

Recall from Chapter 2 that a fundamental building block of information systems is PROCESSES. All information systems include processes—usually lots of them! Information system processes respond to business events and conditions and transform DATA (another building block) into useful information. We need a way to model processes and understand its interactions with its environment, other systems, and other processes.

A System *Is* a Process The word *system* is a common one that is used to describe almost any orderly arrangement of ideas or constructs. People speak of educational systems, computer systems, management systems, business systems, and, of course, information systems. In the oldest and simplest of all system models, a system *is* a process.

A common theme in systems analysis is the use of models to view or present a system. As shown in Figure 6.3, the simplest process model of a system is based on inputs, outputs, and the system itself—viewed as a process. The process symbol defines the boundary of the system. The system is inside the boundary; the environment is outside that boundary. The system exchanges inputs and outputs with its environment. Because the environment is always changing, well-designed systems have a feedback and control loop to allow the system to adapt itself to changing conditions.

Consider a business as a system. It operates within an environment that includes customers, suppliers, competitors, other industries, and the government. Its inputs include materials, services, new employees, new equipment, facilities, money, and orders (to name but a few). Its outputs include products and/or services, waste materials, retired equipment, former employees, and money (payments). It monitors its environment to make necessary changes to its product line, services, operating procedures, and the like.

A rounded rectangle (the Gane and Sarson notation) is used throughout this chapter to represent a process (see margin). Some other process modeling notations prefer a circle (the DeMarco/Yourdon notation) or a rectangle (the

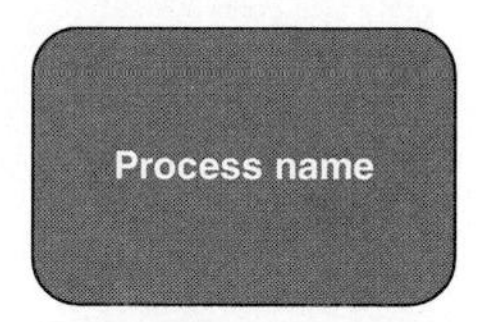

Gane and Sarson shape; used throughout this book

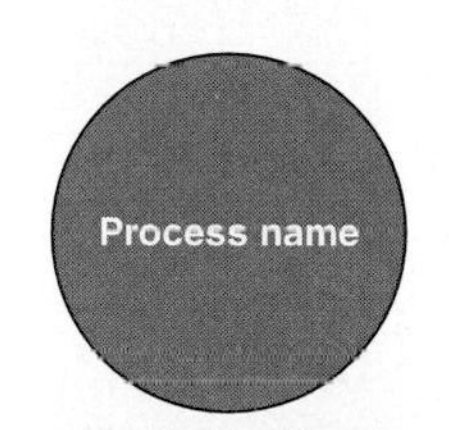

DeMarco/Yourdon shape

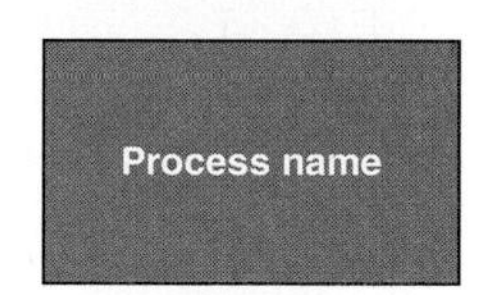

SSADM/IDEF0 shape

Process Symbols

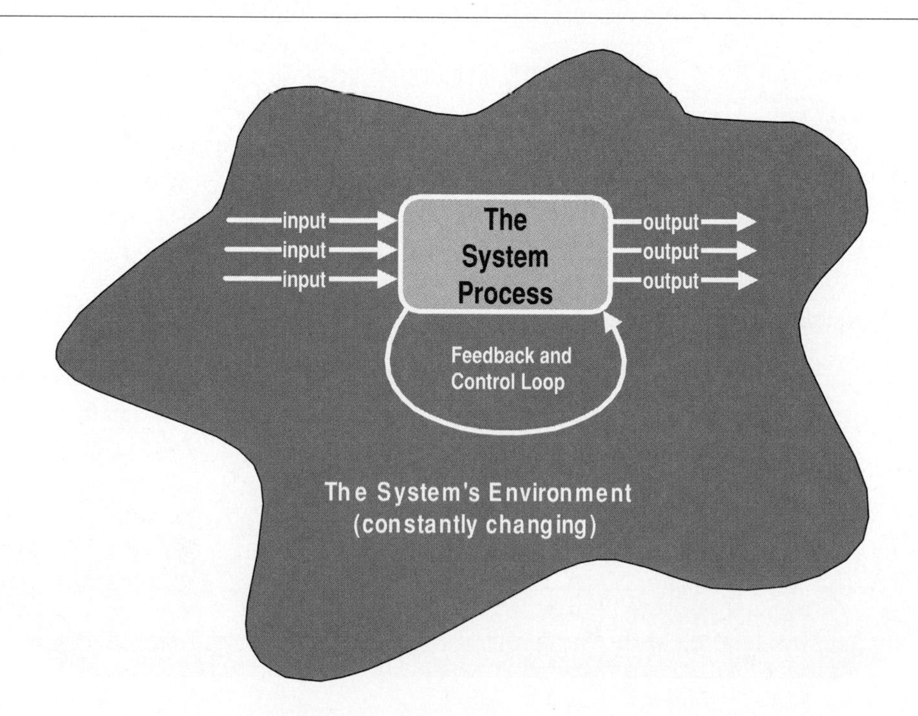

FIGURE 6.3 *The Classical Process Model of a System*

SSADM/IDEF0 notation). The choice is often dependent on your methodology and CASE tool features. But what is a process?

> A **process** is work performed on, or in response to, incoming data flows or conditions. A synonym is *transform.*

Although processes can be performed by people, departments, robots, machines, or computers, we once again want to focus on *what* work or action is being performed (the *logical* process), not on who or what is doing that work or activity (the *physical* process). For instance, in Figure 6.2 we included the logical process WITHDRAW FUNDS FROM AN ACCOUNT. We did not indicate how this would be done. Intuitively, we can think of several physical implementations such as using an ATM, a bank's drive-through service, or actually going inside the bank.

Process Decomposition A complex system is usually too difficult to fully understand when viewed as a whole (meaning, *as a single process*). Therefore, in systems analysis we separate a system into its component subsystems, which in turn are decomposed into smaller subsystems, until such a time as we have identified manageable subsets of the overall system (see Figure 6.4). We call this technique *decomposition.*

> **Decomposition** is the act of breaking a system into its component subsystems, processes, and subprocesses. Each level of *abstraction* reveals more or less detail (as desired) about the overall system or a subset of that system.

You have already applied decomposition in various ways. Most of you have *outlined* a term paper—this is a form of decomposition. Many of you have partitioned a medium-to-large-sized computer program into subprograms that could be developed and tested independently before they are integrated.

In systems analysis, decomposition allows you to partition a system into logical subsets of processes for improved communication, analysis, and design. A dia-

FIGURE 6.4 *A System Consists of Many Subsystems and Processes*

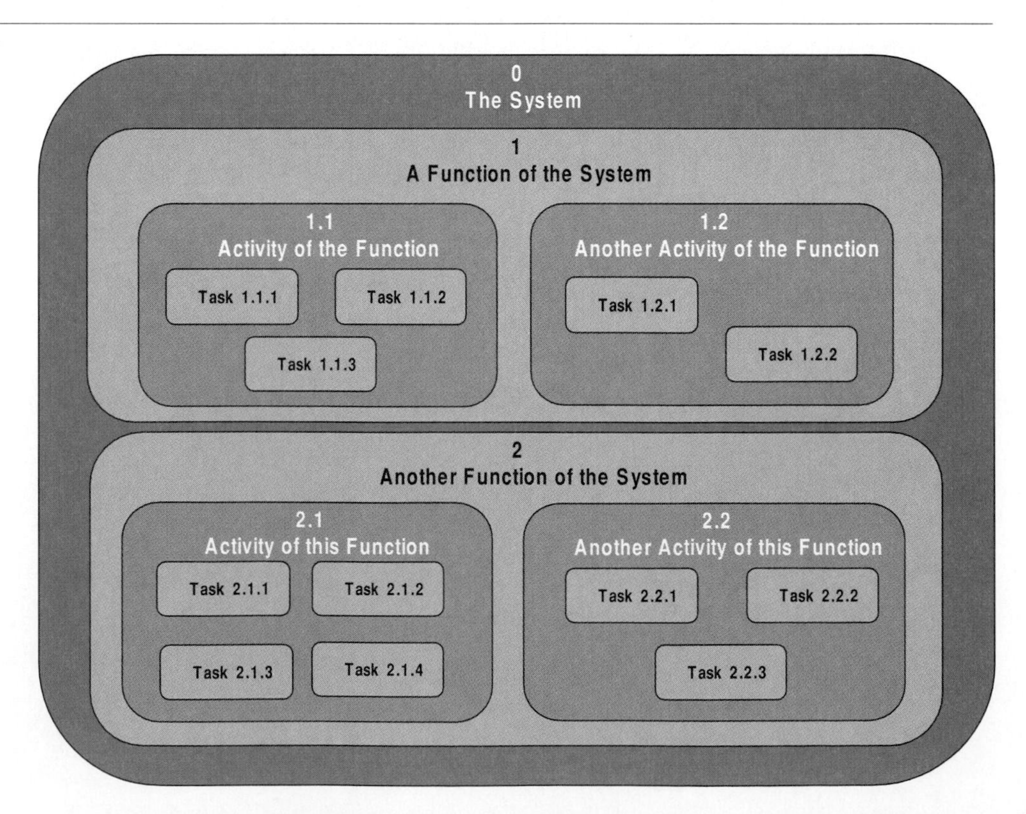

gram similar to Figure 6.4 can be a little difficult to construct when dealing with all but the smallest of systems. Figure 6.5 demonstrates an alternative layout that is supported by many CASE tools and development methodologies. It is called a *decomposition diagram.* We'll use it extensively in this chapter.

A **decomposition diagram,** also called a hierarchy chart, shows the top-down functional decomposition and structure of a system.

A decomposition diagram is essentially a planning tool for more detailed process models, namely, data flow diagrams. The following rules apply:

FIGURE 6.5 *A Decomposition Diagram (for Figure 6.4)*

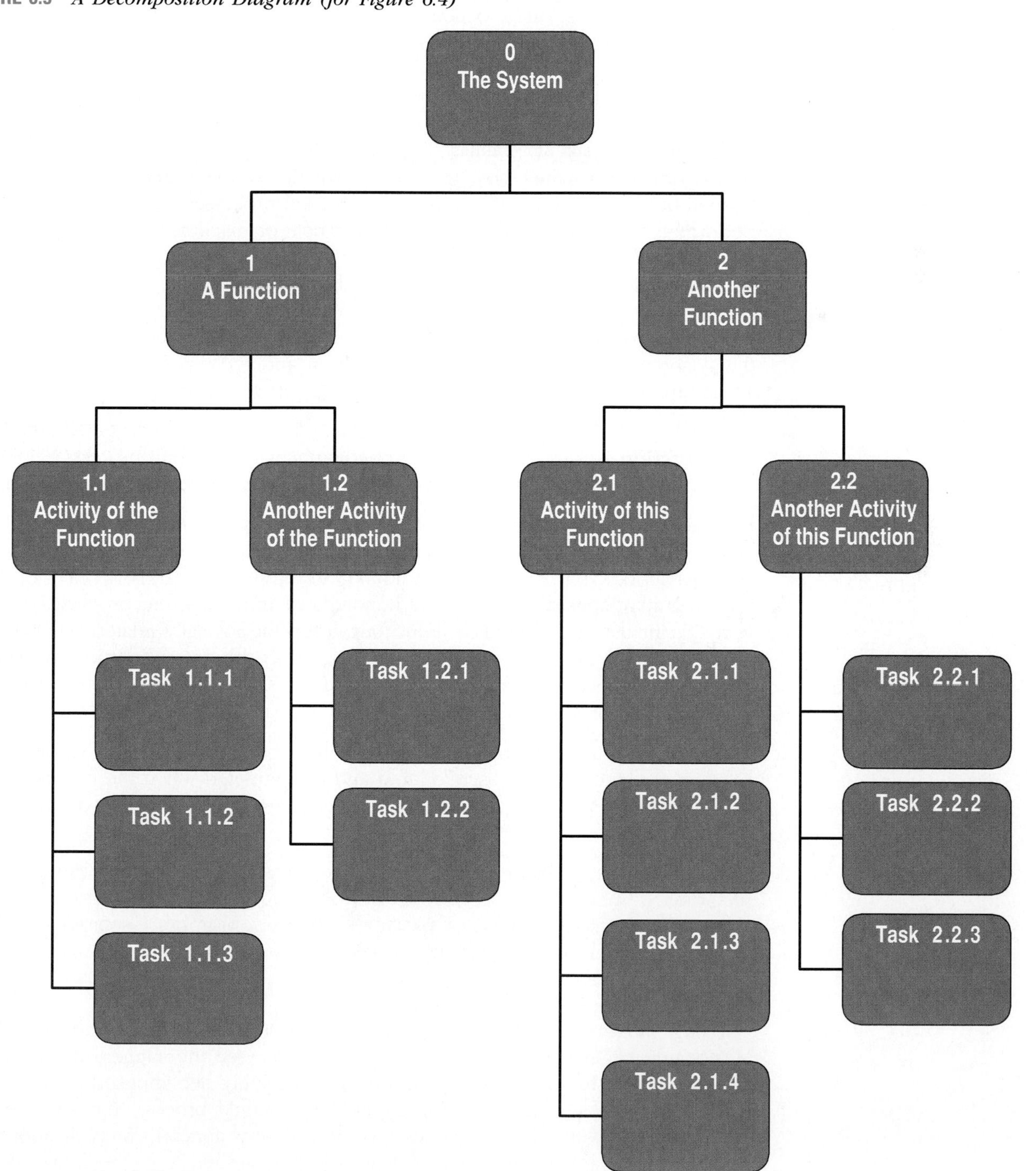

- Each process in a decomposition diagram is either a *parent process,* a *child process* (of a parent), or both.
- A parent *must* have two or more children—a single child does not make sense since that would not reveal any additional detail about the system.
- In most decomposition diagramming standards, a child may have only one parent.
- Finally, a child of one parent may be the parent of its own children.

The upper and lower halves of the decomposition diagram in Figure 6.5 demonstrate two styles for laying out the processes and connections. You may use either or both as necessary to present an uncluttered model. Some models may require multiple pages for maximum clarity.

The connections on a decomposition diagram do not contain arrowheads because the diagram is meant to show *structure,* not *flow.* Also, the connections are not named. Implicitly they all have the same name—CONSISTS OF—since the sum of the child processes for a parent process *equals* the parent process.

Logical Processes and Conventions Logical processes are work or actions that must be performed no matter how you implement the system. Each logical process is (or will be) implemented as one or more physical processes that may include work performed by people, work performed by robots or machines, or work performed by computer software. It doesn't matter which implementation is used, however, because logical processes should only indicate that there is work that must be done.

Naming conventions for logical processes depend on where the process is in the decomposition diagram/data flow diagram and the type of process depicted. There are three types of logical processes: *functions, events,* and *elementary processes.*

> A **function** is a set of related and ongoing activities of the business. A function has no start or end; it just continuously performs its work as needed.

For example, a manufacturing system may include the following functions (subsystems): PRODUCTION PLANNING, PRODUCTION SCHEDULING, MATERIALS MANAGEMENT, PRODUCTION CONTROL, QUALITY MANAGEMENT, and INVENTORY CONTROL. Each of these functions may consist of dozens or hundreds of more discrete processes to support specific activities and tasks. Functions group the logically related activities and tasks. Functions are named with nouns that reflect the entire function. Additional examples are: ORDER ENTRY, ORDER MANAGEMENT, SALES REPORTING, CUSTOMER RELATIONS, and RETURNS AND REFUNDS.

> An **event** is a logical unit of work that must be completed as a whole. An event is triggered by a discrete input and is completed when the process has responded with appropriate outputs. Events are sometimes called *transactions.*

Functions consist of processes that respond to events. For example, the MATERIALS MANAGEMENT function may consist of the following events: TEST MATERIAL QUALITY, STOCK NEW MATERIALS, DISPOSE OF DAMAGED MATERIALS, DISPOSE OF SPOILED MATERIALS, REQUISITION MATERIALS FOR PRODUCTION, RETURN UNUSED MATERIALS FROM PRODUCTION, ORDER NEW MATERIALS, and so on. Each of these events has a trigger and response that can be defined by its inputs and outputs.

Using *modern* structured analysis techniques such as those advocated by McMenamin, Palmer, Yourdon, and the Robertsons (see the suggested readings at the end of the chapter), system functions are ultimately decomposed into business events. Each business event is represented by a single process that will respond to that event. Event process names tend to be very general. We will adopt the convention of naming event processes as follows: PROCESS__________, RESPOND TO

___________, or GENERATE____________, where the blank would be the name of the event (or its corresponding input). Sample event process names are: PROCESS CUSTOMER ORDER, PROCESS CUSTOMER ORDER CHANGE, PROCESS CUSTOMER CHANGE OF ADDRESS, RESPOND TO CUSTOMER COMPLAINT, RESPOND TO ORDER INQUIRY, RESPOND TO PRODUCT PRICE CHECK, GENERATE BACKORDER REPORT, GENERATE CUSTOMER ACCOUNT STATEMENT, and GENERATE INVOICE.

An event process can be further decomposed into elementary processes that illustrate in detail how the system must respond to an event.

Finally, **elementary processes** are discrete, detailed activities or tasks required to complete the response to an event. In other words, they are the lowest level of detail depicted in a process model. A common synonym is **primitive process.**

Elementary processes should be named with a strong action verb followed by an object clause that describes what the work is performed on (or for). Examples of elementary process names are: VALIDATE CUSTOMER IDENTIFICATION, VALIDATE ORDERED PRODUCT NUMBER, CHECK PRODUCT AVAILABILITY, CALCULATE ORDER COST, CHECK CUSTOMER CREDIT, SORT BACKORDERS, GET CUSTOMER ADDRESS, UPDATE CUSTOMER ADDRESS, ADD NEW CUSTOMER, AND DELETE CUSTOMER.

Logical process models omit any processes that do nothing more than move or route data, thus leaving the data unchanged. Physical business systems frequently implement such processes, but they are not essential; in fact, they are increasingly considered unnecessary bureaucracy. Thus, you should omit any proccss that corresponds to a secretary or clerk receiving and simply forwarding a variety of documents to their next processing location. In the end, you should be left only with logical processes that:

- *Perform computations* (calculate grade point average).
- *Make decisions* (determine availability of ordered products).
- *Sort, filter, or otherwise summarize data* (identify overdue invoices).
- *Organize data into useful information* (generate a report or answer a question).
- *Trigger other processes* (turn on the furnace or instruct a robot).
- *Use stored data* (create, read, update, or delete a record).

Be careful to avoid three common mechanical errors with processes (illustrated in Figure 6.6):

- Process 3.1.2 has inputs but no outputs. We call this a **black hole** because data enter the process and then disappear. In most cases, the modeler simply forgot the output.
- Process 3.1.3 has outputs but no input. Unless you are David Copperfield, this is a **miracle**! In this case, the input flows were likely forgotten.
- In Process 3.1.1 the inputs are insufficient to produce the output. We call this a **gray hole.** There are several possible causes including: (1) a misnamed process, (2) misnamed inputs and/or outputs, or (3) incomplete facts. Gray holes are the most common errors—and the most embarrassing. Once handed to a programmer, the input data flows to a process (to be implemented as a program) must be sufficient to produce the output data flows.

Process Logic Decomposition diagrams and data flow diagrams will prove very effective tools for identifying processes, but they are not good at showing the logic inside those processes. Eventually, we will need to specify detailed *instructions* for the elementary processes on a data flow diagram. Consider, for example, an elementary process named CHECK CUSTOMER CREDIT. By itself, the named process is

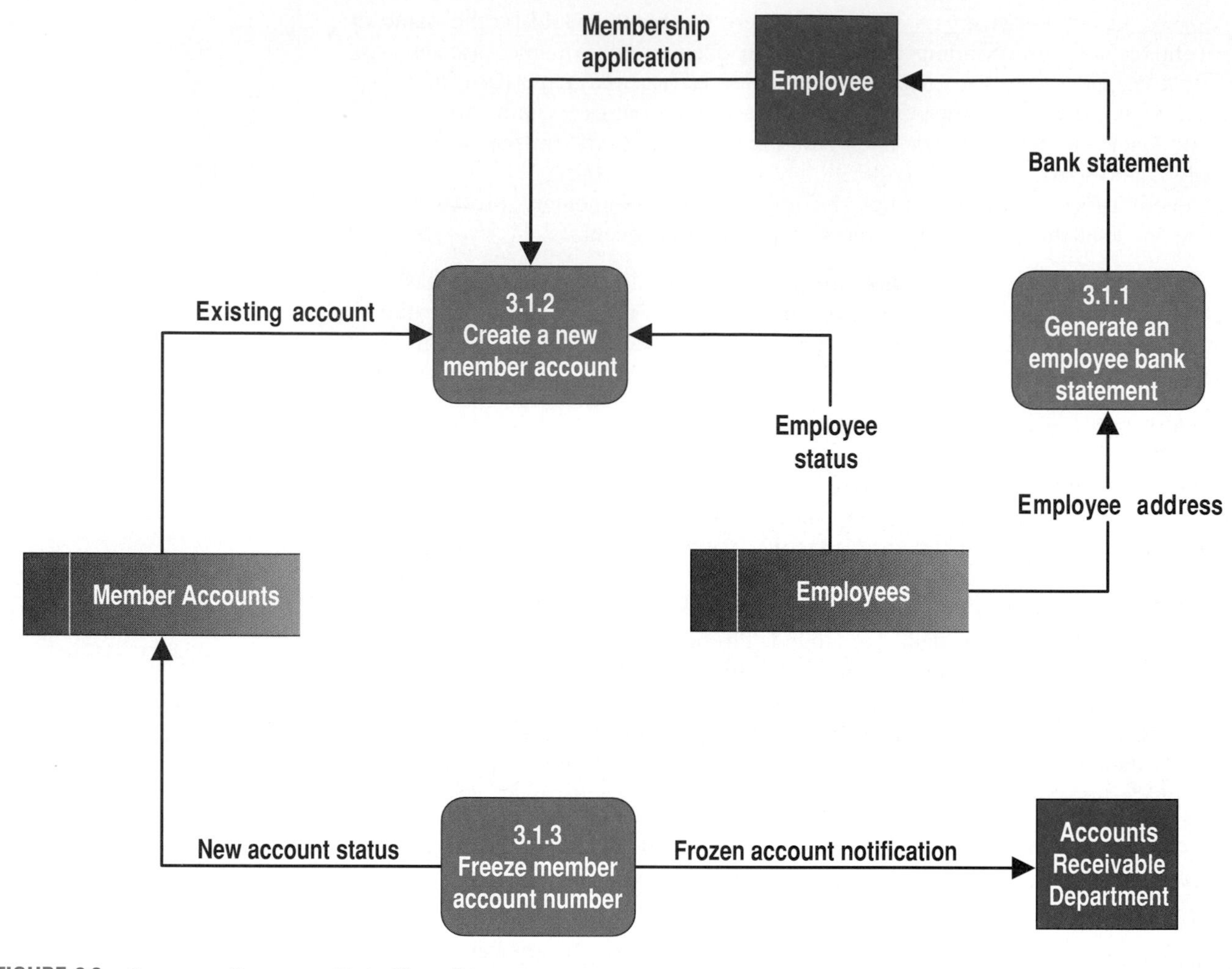

FIGURE 6.6 *Common Errors on Data Flow Diagrams*

insufficient to explain the logic behind CHECK CUSTOMER CREDIT. We need an effective way to model the logic of an elementary process. Ideally, our logic model should be equally effective for communicating with users (who must verify the business accuracy of the logic) and programmers (who may have to implement the business logic in a programming language).

We can rule out flowcharts. While they do model process logic, most end-users tend to be extremely inhibited by them. The same would be true of pseudocode and other popular programming logic tools. We can also rule out natural English. It is all too often imprecise and frequently subject to interpretation (and misinterpretation). Figure 6.7 summarizes some common problems encountered by those who attempt to use natural English as a procedural language.

To address this problem, we require a tool that marries some of the advantages of natural English with the rigor of programming logic tools.

> **Structured English** is a language and syntax, based on the relative strengths of structured programming and natural English, for specifying the underlying logic of elementary processes on process models (such as *data flow diagrams*).

An example of Structured English is shown in Figure 6.8. (The numbers and letters at the beginning of each statement are optional. Some end-users like them because they further remove the programming "look and feel" from the specification.)

FIGURE 6.7 *Problems with Natural English as a Procedure Specification Language*

- Many of us do not write well, and we also tend not to question our writing abilities.
- Many of us are too educated! It's often difficult for a highly educated person to communicate with an audience that may not have had the same educational opportunities. For example, the average college graduate (including most analysts) has a working vocabulary of 10,000 to 20,000 words; on the other hand, the average noncollege graduate has a working vocabulary of around 5,000 words.
- Some of us write everything like it was a program. If business procedures required such precision, we'd write everything in a programming language.
- Too often, we allow the jargon and acronyms of computing to dominate our language.
- English statements frequently have an excessive or confusing scope. How would you carry out this procedure: "If customers walk in the door and they do not want to withdraw money from their account or deposit money to their account or make a loan payment, send them to the trust department." Does this mean that the only time you should not send the customer to the trust department is when he or she wishes to do all three of the transactions? Or does it mean that if a customer does not wish to perform at least one of the three transactions, that customer should not be sent to the trust department?
- We overuse compound sentences. Consider the following procedure: "Remove the screws that hold the outlet cover to the wall. Remove the outlet cover. Disconnect each wire from the plug, but first make sure the power to the outlet has been turned off." An unwary person might try to disconnect the wires before turning off the power!
- Too many words have multiple definitions.
- Too many statements use imprecise adjectives. For example, a loan officer asks a teacher to certify that a student is in good academic standing. What is good?
- Conditional instructions can be imprecise. For example, if we state that "all applicants under the age of 19 must secure parental permission," do we mean less than 19, or less than or equal to 19?
- Compound conditions tend to show up in natural English. For example, if credit approval is a function of several conditions—credit rating, credit ceiling, annual dollar sales for the customer in question—then different combinations of these factors can result in different decisions. As the number of conditions and possible combinations increases, the procedure becomes more and more tedious and difficult to write.

Source: Adapted from Leslie Matthies, *The New Playscript Procedure* (Stamford, CT: Office Publications, Inc., 1977).

For you programmers out there, Structured English is not pseudocode. It does not concern itself with declarations, initialization, linking, and such technical issues. It does, however, borrow some of the logical constructs of *structured programming* to overcome the lack of structure and precision in the English language. Think of it as the marriage of natural English language with the syntax of structured programming.

The overall structure of a Structured English specification is built using the fundamental constructs that have governed structured programming for nearly three decades. These constructs (summarized in Figure 6.9) are:

- A *sequence* of simple, declarative sentences—one after another. Compound sentences are discouraged because they frequently create ambiguity. Each

1. For each CUSTOMER NUMBER in the data store CUSTOMERS:
   a. For each LOAN in the data store LOANS that matches the above CUSTOMER NUMBER:
      1) Keep a running total of NUMBER OF LOANS for the CUSTOMER NUMBER.
      2) Keep a running total of ORIGINAL LOAN PRINCIPAL for the CUSTOMER NUMBER.
      3) Keep a running total of CURRENT LOAN BALANCE for the CUSTOMER NUMBER.
      4) Keep a running total of AMOUNTS PAST DUE for the CUSTOMER NUMBER.
   b. If the TOTAL AMOUNTS PAST DUE for the CUSTOMER NUMBER is greater than 100.00 then
         1) Write the CUSTOMER NUMBER and data in the data flow LOANS AT RISK.
      Else
         1) Exclude the CUSTOMER NUMBER and data from the data flow LOANS AT RISK.

FIGURE 6.8 *Using Structured English to Document an Elementary Process*

Structured English Procedural Structures

<table>
<tr><th>Construct</th><th>Sample Template</th></tr>
<tr><td>Sequence of actions – unconditionally perform a sequence of actions.</td><td>[Action 1]
[Action 2]
. . .
[Action n]</td></tr>
<tr><td>Simple condition actions – if the specified condition is true, then perform the first set of actions. Otherwise, perform the second set of actions.

Use this construct if the condition has only two possible values.

(Note: The second set of conditions is optional.)</td><td>If [truth condition]
  then
    [sequence of actions or other conditional actions]
  else
    [sequence of actions or other conditional actions]
End If</td></tr>
<tr><td>Complex condition actions – test the value of the condition and perform the appropriate set of actions.

Use this construct if the condition has more than two values.</td><td>Do the following based on [condition]:
  Case 1: If [condition] = [value] then
    [sequence of actions or other conditional actions]
  Case 2: If [condition] = [value] then
    [sequence of actions or other conditional actions]
  . . .
  Case n: If [condition] = [value] then
    [sequence of actions or other conditional actions]
End Case</td></tr>
<tr><td>Multiple conditions – test the value of multiple conditions to determine the correct set of actions.

Use a decision table instead of nested if-then-else Structured English constructs to simplify the presentation of complex logic that involves

A decision table is a tabular presentation of complex logic in which rows represent conditions and possible actions, and columns indicate which combinations of conditions result in specific actions.</td><td><table>
<tr><th>DECISION TABLE</th><th>Rule</th><th>Rule</th><th>Rule</th><th>Rule</th></tr>
<tr><td>[Condition]</td><td>value</td><td>value</td><td>value</td><td>value</td></tr>
<tr><td>[Condition]</td><td>value</td><td>value</td><td>value</td><td>value</td></tr>
<tr><td>[Condition]</td><td>value</td><td>value</td><td>value</td><td>value</td></tr>
<tr><td>[Sequence of actions or conditional actions]</td><td>X</td><td></td><td></td><td></td></tr>
<tr><td>[Sequence of actions or conditional actions]</td><td></td><td>X</td><td>X</td><td></td></tr>
<tr><td>[Sequence of actions or conditional actions]</td><td></td><td></td><td></td><td>X</td></tr>
</table>
Although it isn't a Structured English construct, a decision table can be named, and referenced within a Structured English procedure.</td></tr>
<tr><td>One-to-many iteration – repeat the set of actions until the condition is false.

Use this construct if the set of actions must be performed at least once, regardless of the condition's initial value.</td><td>Repeat the following until [truth condition]:
  [sequence of actions or conditional actions]
End Repeat</td></tr>
<tr><td>Zero-to-many iteration – repeat the set of actions until the condition is false.

Use this construct if the set of actions is conditional based on the condition's initial value.</td><td>Do While [truth condition]:
  [sequence of actions or conditional actions]
End Do
- OR -
For [truth condition]:
  [sequence of actions or conditional actions]
End For</td></tr>
</table>

FIGURE 6.9 *Structured English Constructs*

sentence uses strong, action verbs such as GET, FIND, RECORD, CREATE, READ, UPDATE, DELETE, CALCULATE, WRITE, SORT, MERGE, or anything else recognizable or understandable to users. A formula may be included as part of a sentence (e.g., CALCULATE GROSS PAY = HOURS WORKED × HOURLY WAGE).

- A *conditional* or *decision structure* indicates that a process must perform different actions under well-specified conditions. There are two variations (and a departure) on this construct.
  - The IF-THEN-ELSE construct specifies that one set of actions should be taken if a specified condition is true, but a different set of actions should be specified if the specified condition is false. The actions to be taken are typically a *sequence* of one or more sentences as described above.
  - The CASE construct is used when there are more than two sets of actions to choose from. Once again, these actions usually consist of the aforementioned *sequential* statements. The CASE construct is an elegant substitute for an *IF-THEN-ELSE IF-THEN-ELSE IF-THEN* . . . construct (which is very convoluted to the average user).
  - For logic based on multiple conditions and combinations of conditions (which programmers call a *nested IF*), *decision tables* are a far more elegant logic modeling tool. Decision tables will be introduced shortly.
- An *iteration* or *repetition* structure specifies that a set of actions should be repeated based on some stated condition. There are two variations on this construct.
  - The DO-WHILE construct indicates that certain actions (usually expressed as one or more *sequential* and/or *conditional* statements) are repeated zero, one, or more times based on the value of the stated condition. Note that this construct may not execute at all if the condition is not true when the condition is first tested.
  - The REPEAT-UNTIL construct indicates that certain actions (again, usually expressed as one or more *sequential* and/or *conditional* statements) are repeated one or more times based on the value of the stated condition. Note that a REPEAT-UNTIL set of actions must execute at least once, unlike the DO-WHILE set of actions.

Note that some procedure writers suggest that IF, CASE, REPEAT, DO, and FOR constructs be terminated with an END statement. We show these in Figure 6.9; however, we don't recommend them because they make the procedure specification look too much like a computer program.

Additionally, Structured English places the following restrictions on process logic:

- Only strong, imperative verbs may be used.
- Only names that have been defined in the project repository may be used. These names may include those of data flows, data stores, entities (from data models; see Chapter 5), attributes (the specified data fields or properties contained in a data flow, data store, or entity), and domains (the specified legal values for attributes).
- State formulas clearly using appropriate mathematical notations. In short, you can use whatever notation is recognizable to the users. Make sure each operand in a formula is either input to the process in a data flow or a defined constant.
- Undefined adjectives and adverbs (the word *good*, for instance) are not permitted unless clearly defined in the project repository as legal values for data attributes.
- Blocking and indentation are used to set off the beginning and ending of constructs and to enhance readability. (Some authors and models encourage the use of special verbs such as ENDIF, ENDCASE, ENDDO, and ENDREPEAT to terminate constructs. We dislike this practice because it gives the Structured English too much of a pseudocode or programming look and feel.)
- When in doubt, user readability should always take precedence over programmer preferences.

Structured English should be precise enough to clearly specify the required business procedure to a programmer or user. But it should not be so inflexible that you spend hours arguing over syntax.

Many processes are governed by complex combinations of conditions that are not easily expressed with Structured English. This is most commonly encountered in business policies.

A **policy** is a set of rules that governs some process in the business.

In most firms, policies are the basis for decision making. For instance, a credit card company must bill cardholders according to various policies that adhere to restrictions imposed by state and federal governments (maximum interest rates and minimum payments, for instance). Policies consist of *rules* that can often be translated into computer programs if the users and systems analysts can accurately convey those rules to the computer programmer.

Fortunately, there are ways to formalize the specification of policies and other complex combinations of conditions. One such logic modeling tool is a decision table.

A **decision table** is a tabular form of presentation that specifies a set of conditions and their corresponding actions.

Decision tables, unfortunately, don't get enough respect! People who are unfamiliar with them tend to avoid them. Even the CASE tools ignore them. But decision tables are very useful for specifying complex policies and decision-making rules. Figure 6.10 illustrates the three components of a simple decision table.

- Condition stubs (the upper rows) describe the conditions or factors that will affect the decision or policy.
- Action stubs (the lower rows) describe, in the form of statements, the possible policy actions or decisions.

FIGURE 6.10 *A Sample Decision Table*

A SIMPLE POLICY STATEMENT

CHECK CASHING IDENTIFICATION CARD

A customer with check cashing privileges is entitled to cash personal checks of up to $75.00 and payroll checks from companies pre-approved by *LMART*. This card is issued in accordance with the terms and conditions of the application and is subject to change without notice. This card is the property of *LMART* and shall be forfeited upon request of *LMART*.

SIGNATURE *Charles C. Parker, Jr.*
EXPIRES **May 31, 1998**

THE EQUIVALENT POLICY DECISION TABLE

Conditions and Actions	Rule 1	Rule 2	Rule 3	Rule 4
C1: Type of check	personal	payroll	personal	payroll
C2: Check amount less than or equal to $75.00	yes	doesn't matter	no	doesn't matter
C3: Company accredited by *LMART*	doesn't matter	yes	doesn't matter	no
A1: Cash the check	X	X		
A2: Don't cash the check			X	X

- Rules (the columns) describe which actions are to be taken under a specific combination of conditions.

The figure depicts a check-cashing policy that appears on the back of a check-cashing card for a grocery store. This same policy has been documented with a decision table. Three conditions affect the check-cashing decision: the type of check, whether the amount of the check exceeds the maximum limit, and whether the company that issued the check is accredited by the store. The actions (decisions) are either to cash the check or to refuse to cash the check. Notice that each combination of conditions defines a rule that results in an action, denoted by an x.

One final logic modeling comment is in order. Both decision tables and Structured English can describe a single elementary process. For example, a legitimate statement in a Structured English specification might read DETERMINE WHETHER OR NOT TO CASH THE CHECK USING THE DECISION TABLE, LMART CHECK CASHING POLICY.

Data Flows

Processes respond to inputs and generate outputs. Thus, at a minimum, all processes have at least one input and one output *data flow*. Data flows are the communications between processes and the system's environment. Let's examine some of the basic concepts and conventions of data flows.

Data in Motion A data flow is *data in motion*. The flow of data between a system and its environment, or between two processes inside a system is *communication*. Let's study this form of communication.

> A **data flow** represents an input of data to a process, or the output of data (or information) from a process. A data flow is also used to represent the creation, deletion, or updating of data in a file or database (called a *data store* on the DFD).

Think of a data flow as a highway down which packets of known composition travel. The name implies what type of data may travel down that highway. This highway is depicted as a solid-line with arrow (see margin).

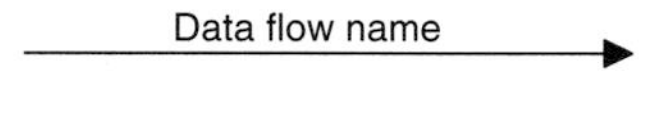

Data Flow Symbol

The *packet* concept is critical. Data that should travel together should be shown as a single data flow, no matter how many *physical* documents might be included. The packet concept is illustrated in Figure 6.11, which shows the correct and incorrect ways to show a logical data flow packet.

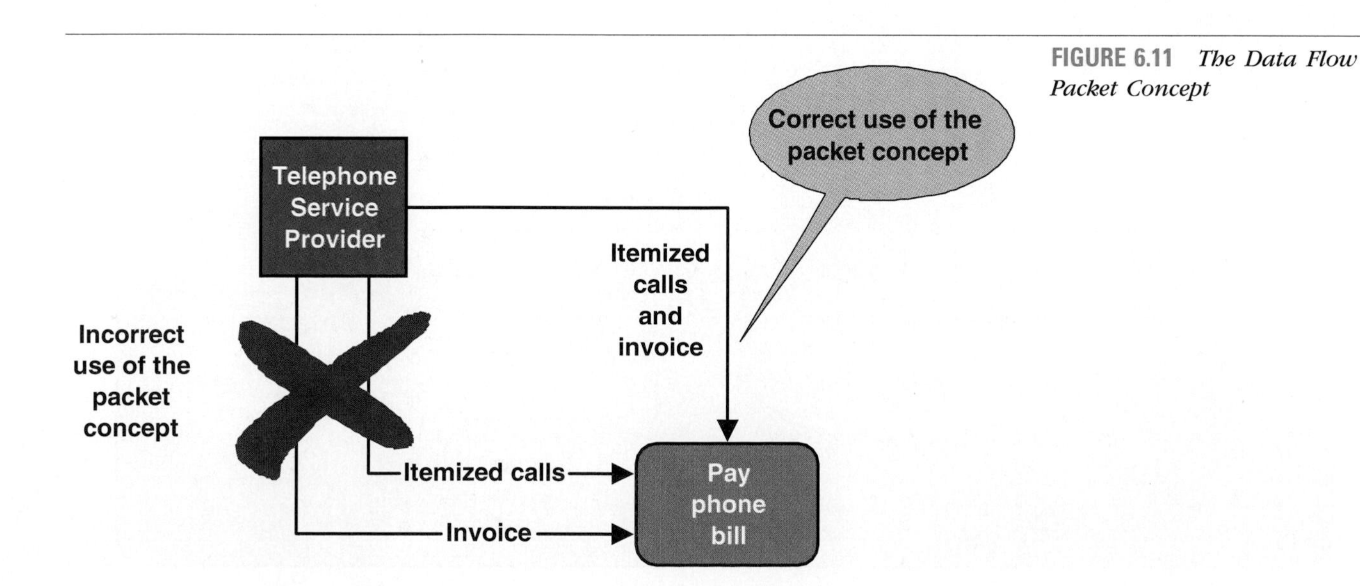

FIGURE 6.11 *The Data Flow Packet Concept*

The *known composition* concept is equally important. A data flow is composed of either actual data attributes (also called *data structures*—more about them later) or other data flows.

> A **composite data flow** is a data flow that consists of other data flows. They are used to combine similar data flows on general-level data flow diagrams to make those diagrams easier to read.

For example, in Figure 6.12(a), a general-level DFD consolidates all types of orders into a composite data flow called ORDER. In Figure 6.12(b), a more detailed data flow diagram shows specific types of orders: STANDARD ORDER, RECURRING ORDER, RUSH ORDER, and EMPLOYEE ORDER. These different orders require somewhat different processing. (The small, black circle is called a **junction.** It indicates that any given ORDER is an instance of only one of the order types.)

Another common use of composite data flows is to consolidate all reports and inquiry responses into one or two composite flows. There are two reasons for this. First, these outputs can be quite numerous. Second, many modern systems pro-

FIGURE 6.12 *Composite and Elementary Data Flows*

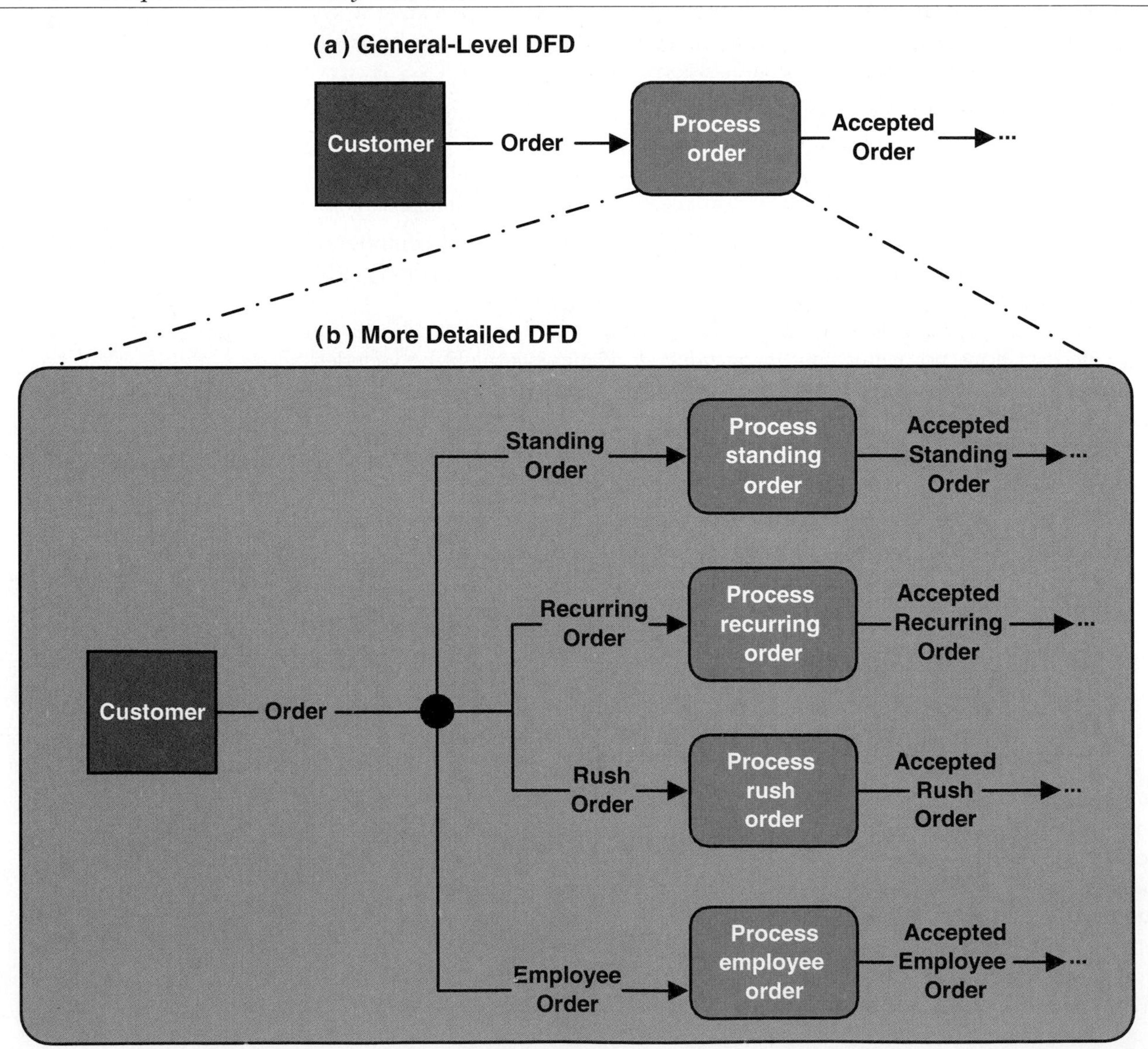

vide extensive user-defined reports and inquiries that cannot be predicted before the system's implementation and use.

Before we exit this introduction to *data* flows, we should acknowledge that some data flow diagramming methods also recognize *nondata* flows called control flows.

A **control flow** represents a condition or nondata event that triggers a process. Think of it as a condition to be monitored while the system works. When the system realizes that the condition meets some predetermined state, the process to which it is input is started.

Control flow name

Control Flow Symbol

The classic information system example is *time*. For example, a report generation process may be triggered by the temporal event END-OF-MONTH. In real-time systems, control flows often represent real-time conditions such as TEMPERATURE and ALTITUDE. In most methodologies that distinguish between data and control flows, the control flow is depicted as a dashed line with arrow (see margin).

Typically, information systems analysts have dealt mostly with data flows; however, as information systems become more integrated with real-time systems (such as manufacturing processes and computer-integrated manufacturing), the need to distinguish the concept of control flows becomes necessary.

***Logical* Data Flows and Conventions** While we recognize that data flows can be implemented a number of ways (e.g., telephone calls, business forms, bar codes, memos, reports, computer screens, and computer-to-computer communications), we are interested only in *logical* data flows. Thus, we are only interested that the flow is needed (not how we will implement that flow). Data flow names should discourage premature commitment to any possible implementation.

Data flow names should be descriptive nouns and noun phrases that are singular, as opposed to plural (ORDER—not ~~ORDERS~~). We do not want to imply that occurrences of the flow must be implemented as a *physical* batch.

Data flow names also should be unique. Use adjectives and adverbs to help to describe how processing has changed a data flow. For example, if an input to a process is named ORDER, the output should not be named ORDER. It might be named VALID ORDER, APPROVED ORDER, ORDER WITH VALID PRODUCTS, ORDER WITH APPROVED CREDIT, or any other more descriptive name that reflects what the process did to the original order.

Logical data flows to and from data stores require special naming considerations (see Figure 6.13). (Data store names are plural, and the numbered bullets match the note to the figure.)

- Only the *net* data flow is shown. Intuitively, you may realize that you have to get a record to update it or delete it. But unless data are needed for some other purpose (e.g., a calculation or decision), the "read" action is not shown. This keeps the diagram uncluttered.

(1) A data flow from a data store to a process indicates that data are to be "read" for some specific purpose. The data flow name should clearly indicate what data are to be read. This is shown in Figure 6.13.

(2) A data flow from a process to a data store indicates that data are to be created, deleted, or updated in/from that data store. Again, as shown in Figure 6.13 these data flows should be clearly named to reflect the specific action performed (such as NEW CUSTOMER, CUSTOMER TO BE DELETED, or UPDATED ORDER ADDRESS).

No data flow should ever go unnamed. Unnamed data flows are frequently the result of flowchart thinking (e.g., step 1, step 2, etc.). If you can't give the data flow a reasonable name, it probably does not exist!

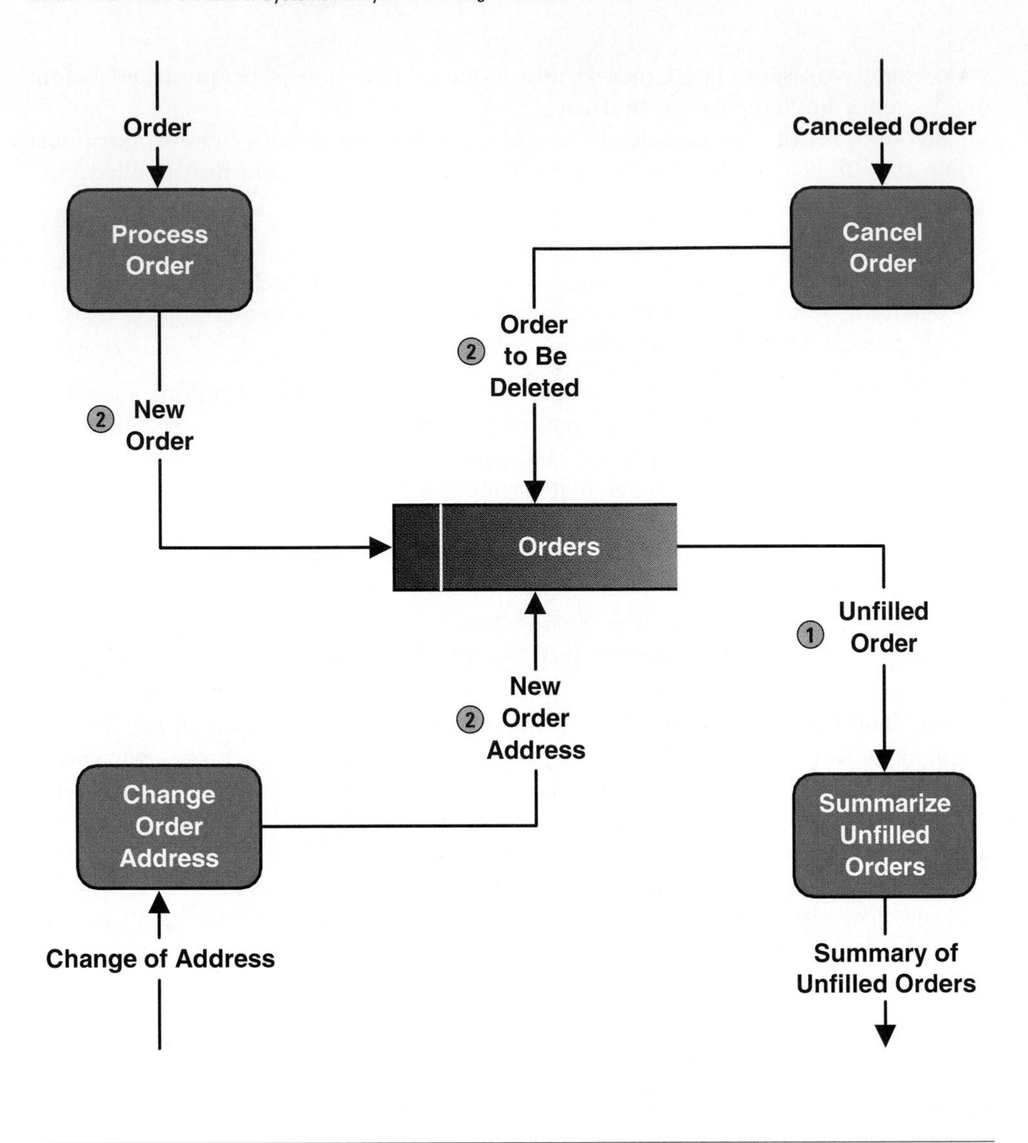

FIGURE 6.13 *Data Flows to and from Data Stores*

Consistent with our goal of *logical* modeling, data flow names should describe the data flow without describing how the flow is or could be implemented. Suppose, for example, that end-users explain their system as follows: *"We fill out Form 23 in triplicate and send it to . . ."* The logical name for the "Form 23" data flow might be COURSE REQUEST. This logical name eliminates physical, implementation biases—the idea that we must use a *paper form,* and the notion that we must use carbon copies. Ultimately, this will free us to consider other physical alternatives such as Touch-Tone phone responses, on-line registration screens, or even long-distance Internet pages!

Finally, all data flows must begin or end at a process because data flows are the inputs and outputs of a process. Consequently, all the data flows on the left side of Figure 6.14 are illegal. The corrected diagrams are shown on the right side.

Data Flow Conservation For many years we have tried to improve business processes by automating them. It hasn't always worked or worked well because the business processes were designed to process data flows in a precomputing era. Consider the average business form. It is common to see the form divided into sections that are designed for different audiences. The first recipient completes his part of the form; the next recipient completes her part, and so forth. At certain points in this processing sequence, a copy of the form might even be detached

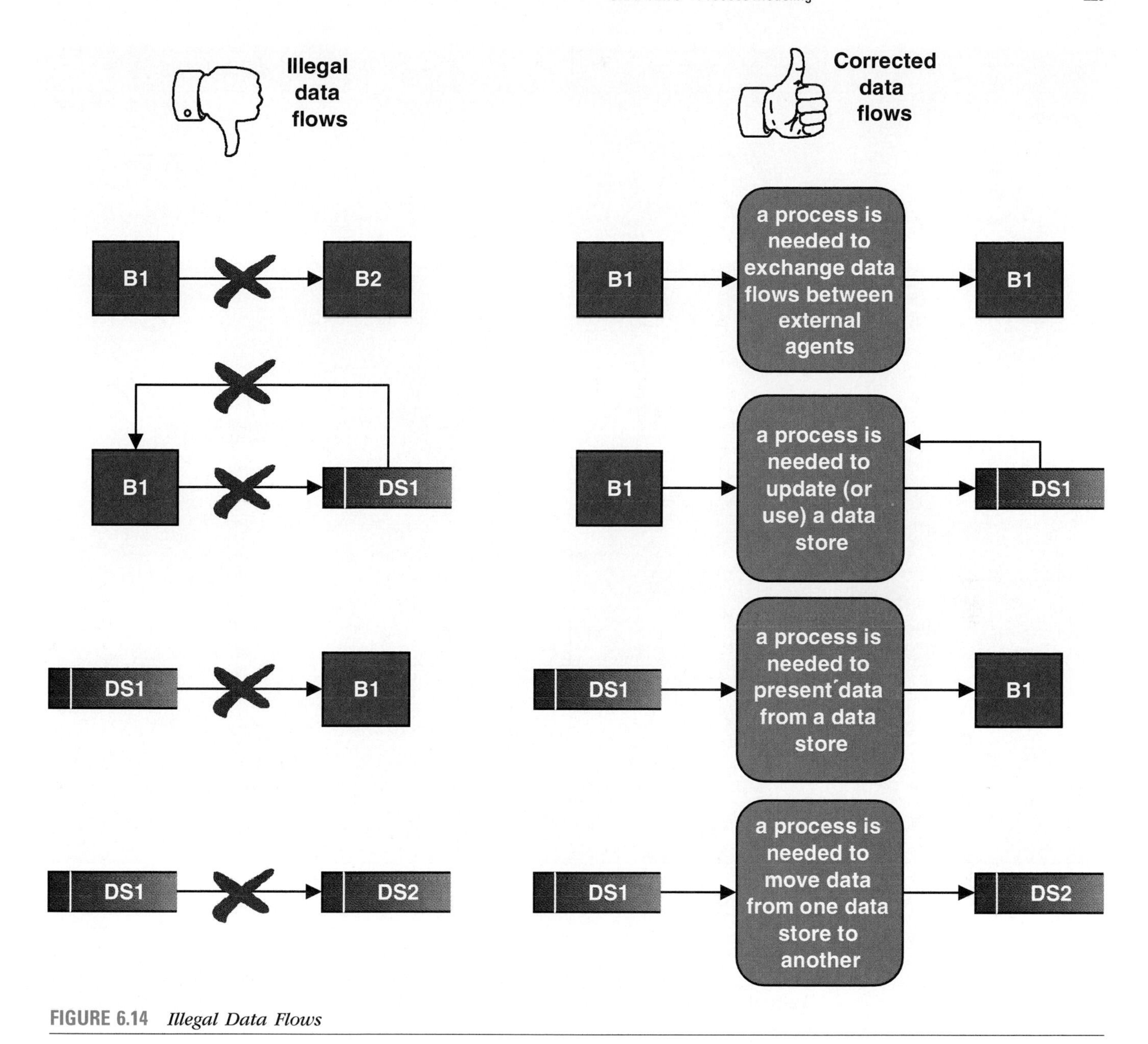

FIGURE 6.14 *Illegal Data Flows*

and sent to another recipient who initializes a new multiple-part form that requires transcribing much of the same data from the initial form. In our own university, we've seen examples where poor form design requires the same data to be typed a dozen times!

Now, if the flow of current data is computerized based on the current business forms and processes, the resulting computer programs will merely automate these inefficiencies. This is precisely what has happened in most businesses! Today, a new emphasis on **business process redesign** encourages management, users, and systems analysts to identify and eliminate these inefficiencies before designing any new information system. We can support this trend in *logical* data flow diagrams by practicing data conservation.

Data conservation, sometimes called "starving the processes," requires that a data flow only contain those data that are truly needed by the receiving process.

DATA STRUCTURE	ENGLISH INTERPRETATION
ORDER = ORDER NUMBER + ORDER DATE + [PERSONAL CUSTOMER NUMBER, CORPORATE ACCOUNT NUMBER] + SHIPPING ADDRESS = ADDRESS + (BILLING ADDRESS = ADDRESS) + 1 { PRODUCT NUMBER + PRODUCT DESCRIPTION + QUANTITY ORDERED + PRODUCT PRICE + PRODUCT PRICE SOURCE + EXTENDED PRICE } N + SUM OF EXTENDED PRICES + PREPAID AMOUNT + (CREDIT CARD NUMBER + EXPIRATION DATE) (QUOTE NUMBER)	An instance of ORDER consists of: ORDER NUMBER and ORDER DATE and Either PERSONAL CUSTOMER NUMBER or CORPORATE ACCOUNT NUMBER and SHIPPING ADDRESS (which is equivalent to ADDRESS) and optionally: BILLING ADDRESS (which is equivalent to ADDRESS) and one or more instances of: PRODUCT NUMBER and PRODUCT DESCRIPTION and QUANTITY ORDERED and PRODUCT PRICE and PRODUCT PRICE SOURCE and EXTENDED PRICE and SUM OF EXTENDED PRICES and PREPAID AMOUNT and optionally: both CREDIT CARD NUMBER and EXPIRATION DATE and optionally: QUOTE NUMBER
ADDRESS = (POST OFFICE BOX NUMBER) + STREET ADDRESS + CITY + [STATE, MUNICIPALITY] + (COUNTRY) + POSTAL CODE	An instance of ADDRESS consists of: optionally: POST OFFICE BOX NUMBER and STREET ADDRESS and CITY and Either STATE or MUNICIPALITY and optionally: COUNTRY and POSTAL CODE

FIGURE 6.15 *A Data Structure for a Data Flow*

By ensuring that processes receive only as much data as they really need, we simplify the interface between those processes. To practice data conservation, we must precisely define the data composition of each (noncomposite) data flow. Data composition is expressed in the form of *data structures.*

Data Structures Ultimately, a data flow contains data items called attributes.

> A **data attribute** is the smallest piece of data that has meaning to the end-users and the business. (This definition also applies to *attributes* as they were presented in Chapter 5.)

Sample attributes for the data flow ORDER might include ORDER NUMBER, ORDER DATE, CUSTOMER NUMBER, SHIPPING ADDRESS (which consists of attributes such as STREET ADDRESS, CITY, and ZIP CODE), ORDERED PRODUCT NUMBERS, QUANTITY(ies) ORDERED, and so on. Notice that some attributes occur once for each instance of ORDER, while others may occur several times for a single instance of ORDER.

The data attributes that comprise a data flow are organized into data structures.

> **Data structures** are specific arrangements of data attributes that define the organization of a single instance of a data flow.

Data flows can be described in terms of the following types of data structures:

- A *sequence* or group of data attributes that occur one after another.
- The *selection* of one or more attributes from a set of attributes.
- The *repetition* of one or more attributes.

The most common data structure notation is a Boolean algebraic notation that is required by many CASE tools. Other CASE tools and methodologies support proprietary, but essentially equivalent notations. A sample data structure for the data flow ORDER is presented in Figure 6.15. This algebraic notation uses the following symbols:

=	Means "consists of" or "is composed of."
•	Means "and" and designates *sequence.*
[. . .]	Means "only one of the attributes within the brackets may be present"—designates *selection.*
{. . .}	Means that the attributes in the braces may occur many times for one instance of the data flow—designates *repetition.*
(. . .)	Means the attribute(s) in the parentheses are optional—no value—for some instances of the data flow.

In our experience, all data flows can be described in terms of these fundamental constructs. Figure 6.16 demonstrates each of the fundamental constructs using examples. Returning to Figure 6.15, notice that the constructs are combined to describe the data content of the data flow.

The importance of defining the data structures for every data flow should be apparent—you are defining the business data requirements for each input and output! These requirements must be determined before any process could be implemented as a computer program. This standard notation provides a simple but effective means for communicating between end-users and programmers.

Before we leave this topic, we should acknowledge an alternative technique for modeling data structures. In Chapter 5, we discussed *data modeling,* a graphical technique for representing data structures. In some methodologies, data models are drawn for each data flow. Each entity in the data model is then described by attributes. This approach is used in CASE tools such as Sterling's *Key:Model* (formerly known as Knowledgeware *ADW*). To learn to draw data models, see Chapter 5.

Domains An attribute is a piece of data. When analyzing a system, it makes sense that we should define those values for an attribute that are legitimate, or that make sense. The values for each attribute are defined in terms of two properties: data type and domain.

> The **data type** for an attribute defines what class of data can be stored in that attribute.

> The **domain** of an attribute defines what values an attribute can legitimately take on.

The concepts of data type and domain were introduced in Chapter 5. See that discussion and Tables 5.1 and 5.2 (pp. 177–178) for a more complete description of data type and domain.

Divergent and Convergent Flows It is sometimes useful to depict diverging or converging data flows on a data flow diagram.

> A **diverging data flow** is one that splits into multiple data flows.

Diverging data flows indicate that all or parts of a single data flow are routed to different destinations.

FIGURE 6.16 *Data Structure Constructs*

Data Structure	Format by Example (relevant portion is **boldfaced**)	English Interpretation (relevant portion is **boldfaced**)
Sequence of Attributes – The sequence data structure indicates one or more attributes that may (or must) be included in a data flow.	WAGE AND TAX STATEMENT = **TAXPAYER IDENTIFICATION NUMBER +** **TAXPAYER NAME +** **TAXPAYER ADDRESS +** **WAGES, TIPS, AND COMPENSATION +** **FEDERAL TAX WITHHELD + ...**	An instance of WAGE AND TAX STATEMENT consists of: **TAXPAYER IDENTIFICATION NUMBER and** **TAXPAYER NAME and** **TAXPAYER ADDRESS and** **WAGES, TIPS, AND COMPENSATION and** **FEDERAL TAX WITHHELD and ...**
Selection of Attributes – The selection data structure allows you to show situations where different sets of attributes describe different instances of the data flow.	ORDER = **(PERSONAL CUSTOMER NUMBER,** **CORPORATE ACCOUNT NUMBER) +** ORDER DATE + ...	An instance of ORDER consists of: **Either PERSONAL CUSTOMER NUMBER or** **CORPORATE ACCOUNT NUMBER;** and ORDER DATE and ...
Repetition of Attributes – The repetition data structure is used to set off a data attribute or group of data attributes that may (or must) repeat themselves a specified number of times for a single instance of the data flow. The minimum number of repetitions is usually *zero* or *one*. The maximum number of repetitions may be specified as "n" meaning "many" where the actual number of instances varies for each instance of the data flow.	CLAIM = POLICY NUMBER + POLICYHOLDER NAME + POLICYHOLDER ADDRESS + **0 { DEPENDENT NAME +** **DEPENDENT'S RELATIONSHIP } N +** **1 { EXPENSE DESCRIPTION +** **SERVICE PROVIDER +** **EXPENSE AMOUNT } N**	An instance of CLAIM consists of: POLICY NUMBER and POLICYHOLDER NAME and POLICYHOLDER ADDRESS and **zero or more instances of:** **DEPENDENT NAME and** **DEPENDENT'S RELATIONSHIP** and **one or more instances of:** **EXPENSE DESCRIPTION and** **SERVICE PROVIDER and** **EXPENSE ACCOUNT**
Optional Attributes – The optional notation indicates that an attribute, or group of attributes in a sequence or selection data structure may not be included in all instances of a data flow. *Note: For the repetition data structure, a minimum of 'zero' is the same as making the entire repeating group 'optional'.*	CLAIM = POLICY NUMBER + POLICYHOLDER NAME + POLICYHOLDER ADDRESS + **(SPOUSE NAME +** **DATE OF BIRTH) + ...**	An instance of CLAIM consists of: POLICY NUMBER and POLICYHOLDER NAME and POLICYHOLDER ADDRESS and **optionally, SPOUSE NAME and** **DATE OF BIRTH** and ...
Reusable Attributes – For groups of attributes that are contained in many data flows, it is desirable to create a separate data structure that can be reused in other data structures.	DATE = MONTH + DAY + YEAR	Then, the reusable structures can be included in other data flow structures as follows: ORDER = ORDER NUMBER ... + DATE INVOICE = INVOICE NUMBER ... + DATE PAYMENT = CUSTOMER NUMBER ... + DATE

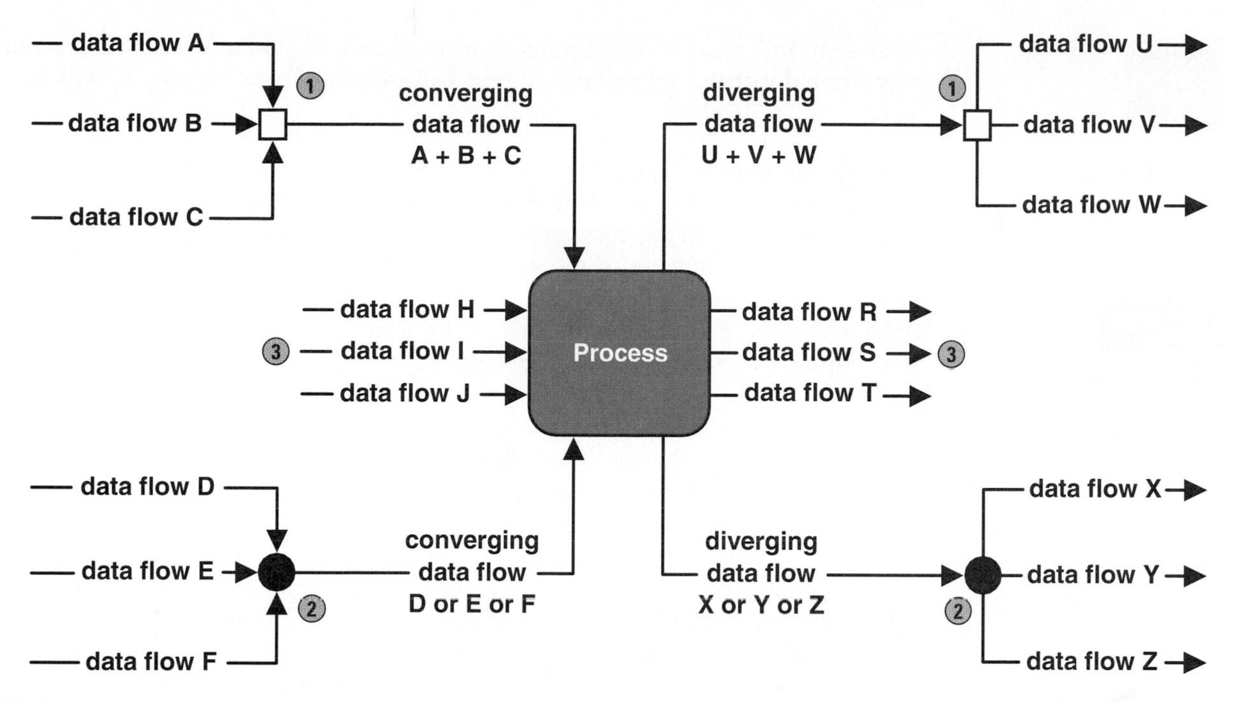

FIGURE 6.17 *Diverging and Converging Data Flows*

A **converging data flow** is the merger of multiple data flows into a single data flow.

Converging data flows indicate that data flows from different sources can (must) come together as a single packet for subsequent processing.

Diverging and converging data flows are depicted as shown in Figure 6.17. Notice that we do not include a process to "route" the flows. The flows simply diverge from or converge to a common flow. The following notations, not supported by all CASE tools, are used in this book.

① The small square *junction* means "and." This means that each time the process is performed, it must input (or output) all the diverging or converging data flows. *(Some DFD notations simply place a + between the data flows.)*

② The small black *junction* means "exclusive or." This means that each time the process is performed, it must input (or output) only one of the diverging or converging data flows. *(Some DFD notations simply place an * between the data flows.)*

③ In the absence of one diverging or converging data flows, the reader should assume an "inclusive or." This means that each time the process is performed, it may input any or all of the depicted data flows.

With the above rules, the most complex of business process and data flow combinations can be depicted.

External Agents

All information systems respond to events and conditions in the environment. The environment of an information system includes *external agents* that form the boundary of the system and define places where the system interfaces with its environment.

An **external agent** defines a person, organization unit, other system, or other organization that lies outside the scope of the project but that interacts with the system being studied. External agents provide the net inputs into a

Gane and Sarson shape

DeMarco/Yourdon shape

External Agent Symbols

system and receive net outputs from a system. Common synonyms include **external entity** (not to be confused with *data entity* as introduced in Chapter 5).

The term *external* means "external to the system being analyzed or designed." In practice, an external agent may actually be outside of the business (such as government agencies, customers, suppliers, and contractors), or it may be inside the business but outside of the project and system scope (such as other departments, other business functions, and other internal information systems). An external agent is represented by a square on the data flow diagram. The DeMarco/Yourdon equivalent is a rectangle (see margin).

It is important to recognize that work and activities are occurring inside the external agent, but that work and those activities are said to be "out of scope" and not subject to change. Thus, the data flows between your system and these boundaries should not cause substantive change to the work or activities performed by the external agents.

To be sure, the external agents of an information system are rarely fixed. As project scope and goals change, the scope of an information system can either grow or shrink. If the system scope grows, it can consume some of the original external agents—in other words, what was once considered outside of the system is now considered inside the system (*as new processes*). Of course, there will now be new external agents.

Similarly, if the system scope shrinks (because of budget or schedule constraints), processes that were once considered to be inside the system may become external agents.

External agents on a logical data flow diagram may include people, business units, other internal systems with which your system must interact, and external organizations. Their inclusion on the logical DFD means that your system interacts with these agents. They are almost always one of the following:

- An office, department, division, or individual within your company that provides net inputs to that system, receives net outputs from that system, or both.
- An organization, agency, or individual that is outside your company but that provides net inputs to, or receives net outputs from, your system. Examples include CUSTOMERS, SUPPLIERS, CONTRACTORS, BANKS, and GOVERNMENT AGENCY(ies).
- Another business or information system—possibly, though not necessarily, computer-based—that is separate from your system but with which your system must interface. Very few information systems do not interface with other information systems. It is becoming common to interface information systems with those of other businesses.
- One of your system's end-users or managers. In this case, the user or manager is either a net source of data to be input to your system and/or a net destination of outputs to be produced by your system.

External agents should be named with descriptive, singular nouns, such as REGISTRAR, SUPPLIER, MANUFACTURING SYSTEM, or FINANCIAL INFORMATION SYSTEM. External agents represent fixed, *physical* systems; therefore, they can have very physical names or acronyms. For example, an external agent representing our school's financial management information system would be called FMIS. If an external agent describes an individual, we recommend job titles or role names instead of proper names (for example, use ACCOUNT CLERK, not *Mary Jacobs*).

To avoid crossing data flow lines on a DFD, it is permissible to duplicate external agents on DFDs. But as a general rule, external agents should be located on the perimeters of the page, consistent with their definition as a system boundary.

Data Stores

Most information systems capture data for later use. The data are kept in the data store, the last symbol on a data flow diagram. It is represented by the open-end box (see margin).

Gane and Sarson shape

DeMarco/Yourdon shape

Data Store Symbols

> A **data store** is an "inventory" of data. Synonyms include *file* and *database* (although those terms are too implementation-oriented for essential process modeling).

If data flows are *data in motion*, think of data stores as *data at rest.*

Ideally, essential data stores should describe "things" about which the business wants to store data. These things include:

- Persons: AGENCY, CONTRACTOR, CUSTOMER, DEPARTMENT, DIVISION, EMPLOYEE, INSTRUCTOR, OFFICE, STUDENT, SUPPLIER. Notice that a person entity can represent either individuals, groups, or organizations.
- Places: SALES REGION, BUILDING, ROOM, BRANCH OFFICE, CAMPUS.
- Objects: BOOK, MACHINE, PART, PRODUCT, RAW MATERIAL, SOFTWARE LICENSE, SOFTWARE PACKAGE, TOOL, VEHICLE MODEL, VEHICLE. An object entity can represent actual objects (such as SOFTWARE LICENSE) or specifications for a type of object (such as SOFTWARE PACKAGE).
- Events: APPLICATION, AWARD, CANCELLATION, CLASS, FLIGHT, INVOICE, ORDER, REGISTRATION, RENEWAL, REQUISITION, RESERVATION, SALE, TRIP.
- Concepts: ACCOUNT, BLOCK OF TIME, BOND, COURSE, FUND, QUALIFICATION, STOCK.

NOTE If the above list looks familiar, it should! A data store represents *all occurrences* of a data entity—defined in Chapter 5 as something about which we want to store data. As such, the data store represents the synchronization of a system's process model with its data model.

If you do data modeling before process modeling, identification of most data stores is simplified by the following rule:

> There should be one data store for each data entity on your entity relationship diagram. (We even include associative entity data stores on our models.)

If, on the other hand, you do process modeling before data modeling, data store discovery tends to be more arbitrary. In that case, our best recommendation is to identify existing implementations of files or data stores (e.g., computer files and databases, file cabinets, record books, catalogs, etc.) and then rename them to reflect the logical "things" about which they store data. Consistent with information engineering strategies, we recommend that data models precede the process models.

Generally, data stores should be named as the plural of the corresponding data model entity. Thus, if the data model includes an entity named CUSTOMER, the process models will include a data store named CUSTOMERS. This makes sense since the data store, by definition, stores all instances of the entity. Avoid physical terms such as *file, database, file cabinet, file folder*, and the like.

As was the case with boundaries, it is permissible to duplicate data stores on a DFD to avoid crossing data flow lines. Duplication should be minimized.

THE PROCESS OF LOGICAL PROCESS MODELING

Now that you understand the basic concepts of process models, we can examine the process of building a process model. When do you do it? How many process models may be drawn? What technology exists to support the development of process models?

Strategic Systems Planning

Many organizations select application development projects based on strategic information system plans. Strategic planning is a separate project that produces an information systems strategy plan that defines an overall vision and architecture for information systems. This architecture frequently includes an **enterprise process model.** (There are other architectural components that are not important to this discussion.)

An enterprise process model typically identifies only business areas and functions. Events and detailed processes are rarely examined. Business areas and functions are identified and mapped to other enterprise models such as the enterprise data model (Chapter 5) and the enterprise geographic model (Chapter 7). Business areas and functions are subsequently prioritized into application development projects. Priorities are usually based on which business areas, functions, and supporting applications will return the most value to the business as a whole.

An enterprise process model usually takes the form of a decomposition diagram and/or very high-level data flow diagram. Relationships to other enterprise models are typically documented with matrices. For example, a process-to-data matrix would show which data entities are used or updated by which functions. Similarly, a process-to-location matrix would show which functions are performed at which geographic locations. By studying these relationships, the planning team can determine logical groupings of processes that can become applications.

An enterprise process model is stored in a corporate repository. Subsequently, as application development projects are started, subsets of the enterprise process model are exported to the project teams to serve as a starting point for building more detailed process models (including data flow diagrams). Once the project team completes systems analysis and design, the expanded and refined process models are imported back into the corporate repository.

Process Modeling for Business Process Redesign

Business process redesign (BPR) has been discussed several times in this book and chapter. Recall that BPR projects analyze business processes and then redesign them to eliminate inefficiencies and bureaucracies before any (re)application of information technology. To redesign business processes, we must first study the existing processes. Process models play an integral role in BPR.

Each BPR methodology recommends its own process model notations and documentation. Most of the models are a cross between data flow diagrams and flowcharts. The diagrams tend to be very *physical* because the BPR team is trying to isolate the implementation idiosyncrasies that cause inefficiency and reduce value. BPR data flow diagrams/flowcharts may include new symbols and information to illustrate timing, throughput, delays, costs, and value. Given this additional data, the BPR team then attempts to simplify the processes and data flows in an effort to maximize efficiency and return the most value to the organization.

Opportunities for the efficient use of information technology may also be recorded on the physical diagrams. If so, the BPR diagram becomes an input to systems analysis (described next).

Process Modeling during Systems Analysis

In systems analysis and in this chapter, we focus exclusively on *logical* process modeling as a part of business requirements analysis. The process model for a system or application is an **application process model.** In your information system framework (review Figure 6.1), logical process models have a process focus and a SYSTEM OWNER and/or SYSTEM USER perspective. Also as shown, they are typically constructed as deliverables of the study and definition phases of a project. Finally, notice that while logical process models are not concerned with implementation details or technology, they may be constructed (through *reverse engineering*) from existing application software, but this technology is much less mature and reliable than the corresponding reverse *data* engineering technology.

In the heyday of the original structured analysis methodologies, process modeling was also performed in the study phase of systems analysis. Analysts would build a *physical process model of the current system*, a *logical model of the current system*, and a *logical model of the target system.* Each model would be built top-down—from very general models to very detailed models. While conceptually sound, this approach led to modeling overkill and significant project delays, so much so that structured techniques guru Ed Yourdon called it "analysis paralysis."

Today, most modern structured analysis strategies focus exclusively on the *logical model of the target system* being developed. Instead of being built either top-down or bottom-up, they are organized according to a commonsense strategy called event partitioning.

Event partitioning factors a system into subsystems based on business events and responses to those events.

This strategy for event-driven process modeling is illustrated in Figure 6.18 and described as follows:

1. A system **context diagram** is constructed to establish *initial* project scope. This simple, one-page data flow diagram shows only the system's interfaces with its environment.
2. A **functional decomposition diagram** is drawn to partition the system into logical subsystems and/or functions. (This step is omitted for very small systems.)
3. An **event-response list** is compiled to identify and confirm the business events to which the system must provide a response. The list will also describe the required or possible responses to each event.
4. One process, called an **event handler,** is added to the decomposition diagram for each event. The decomposition diagram serves as the outline for the system.
5. An **event diagram** is constructed and validated for each event. This simple data flow diagram shows only the event handler and the inputs and outputs for each event.
6. A **system diagram** is constructed by merging the event diagrams. This data flow diagram shows the "big picture" of the system.
7. A **primitive diagram** is constructed for each event process. These data flow diagrams show all the elementary processes, data stores, and data flows for single events. The logic of each elementary process, and the data structure of each elementary data flow, is described using the tools described earlier in the chapter.

The above process models collectively document all the business processing requirements for a system. We'll demonstrate the technique in our SoundStage case study.

Looking Ahead to Systems Configuration and Design

The logical process model from systems analysis describes business processing requirements of the system, not technical solutions. The purpose of the configuration phase is to determine the best way to implement those requirements with technology. In practice, this decision may have already been standardized as part of an *application architecture.* For example, the SoundStage application architecture requires that the development team first determine if an acceptable system can be purchased. If not, the current application architecture specifies that software built in-house be written in either Powersoft's *Powerbuilder* or Microsoft's *Visual C++* (although a new language, Borland's *Delphi,* has been approved for the Member Services Project).

During system design, the logical process model will be transformed into a physical process model (called an application schema) for the chosen technical architecture. This model will reflect the technical capabilities and limitations of the chosen technology. Any further discussion of physical process/application design is deferred until Chapter 12.

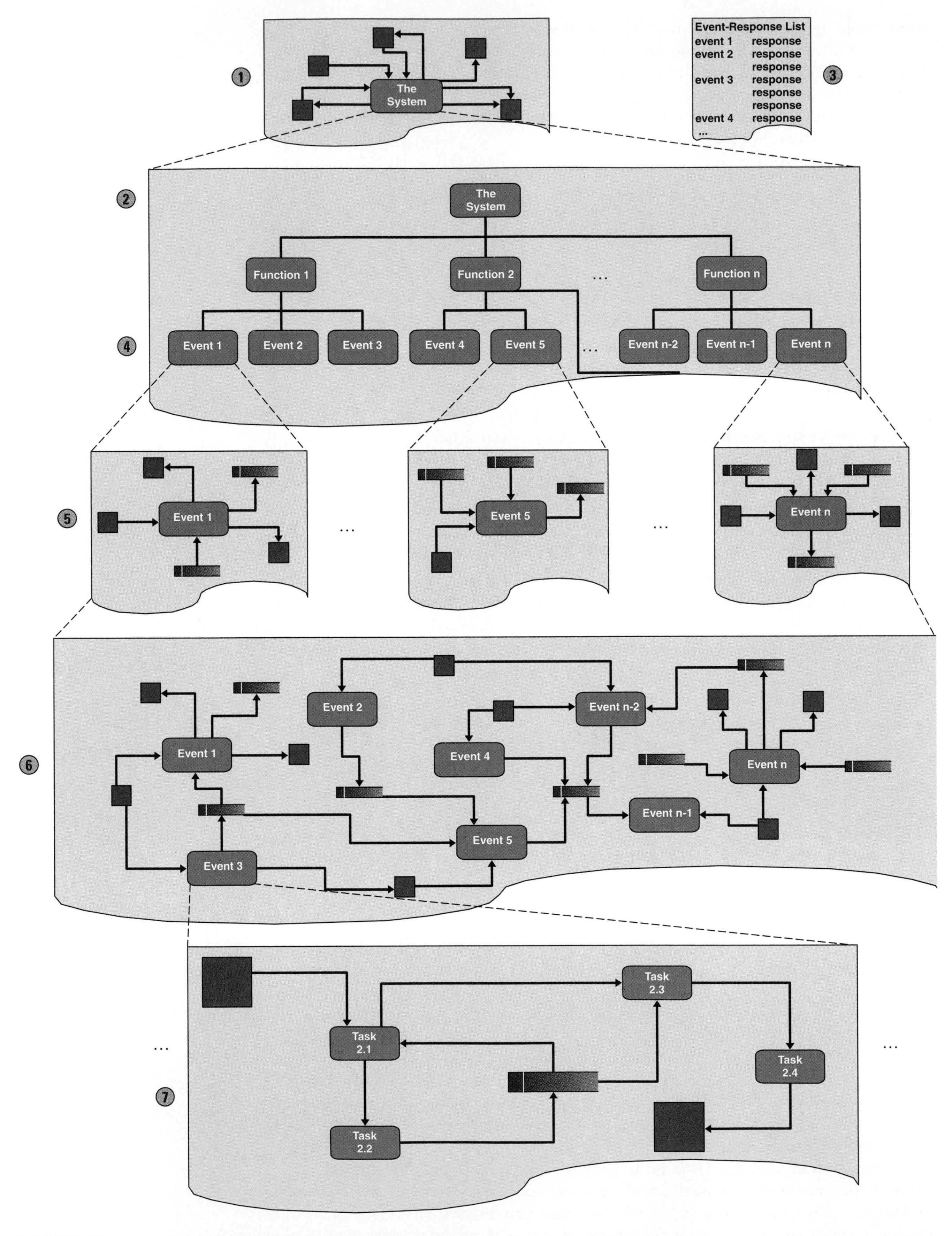

FIGURE 6.18 *Event-Driven Process Modeling Strategy*

Fact-Finding and Information Gathering for Process Modeling

Process models cannot be constructed without appropriate facts and information as supplied by the user community. These facts can be collected by a number of techniques such as sampling of existing forms and files, research of similar systems, surveys of users and management, and interviews of users and management. The fastest method of collecting facts and information and simultaneously constructing and verifying the process models is *joint application development (JAD).* JAD uses a carefully facilitated group meeting to collect the facts, build the models, and verify the models—usually in one or two full-day sessions.

Fact-finding and information gathering techniques are more fully explored in Part Five, Module B. JAD techniques are presented in Part Five, Module D.

Computer-Aided Systems Engineering (CASE) for Process Modeling

Like all system models, process models are stored in the repository. Computer-aided systems engineering (CASE) technology, introduced in Chapter 3, provides the repository for storing the process model and its detailed descriptions. Most CASE products support computer-assisted process modeling. Most support decomposition diagrams and data flow diagrams. Some support extensions for business process analysis and redesign.

Using a CASE product, you can easily create professional, readable process models without the use of paper, pencil, eraser, and templates. The models can be easily modified to reflect corrections and changes suggested by end-users. Also, most CASE products provide powerful analytical tools that can check your models for mechanical errors, completeness, and consistency. Some CASE products can even help you analyze the data model for consistency, completeness, and flexibility. The potential time savings and quality are substantial.

CASE tools do have their limitations. Not all process model conventions are supported by all CASE products. Therefore, it is very likely that any given CASE product may force the company to adapt its methodology's process modeling symbols or approach so that it is workable within the limitations of its CASE tool.

All the SoundStage process models in the next section of this chapter were created with Popkin's CASE tool, *System Architect.* For the case study, we provide you the printouts exactly as they came off our printers. We did not add color. The only modifications by the artist were the bullets that call your attention to specific items of interest on the printouts. All the processes, data flows, data stores, and boundaries on the SoundStage process models were automatically cataloged into *System Architect*'s project repository (which it calls an encyclopedia). Figure 6.19 illustrates some of *System Architect*'s screens as used for data modeling.

HOW TO CONSTRUCT PROCESS MODELS

As a systems analyst or knowledgeable end-user, you must learn how to draw decomposition and data flow diagrams to model business process requirements. We will use the SoundStage Entertainment Club project to teach you how to draw these process models.

The survey and study phases of the *FAST* methodology have been completed and the project team understands the current system's strengths, weaknesses, limitations, problems, opportunities, and constraints. The team has also already built the data model (in Chapter 5) to document business data requirements for the new system. Team members will now build the corresponding process models.

The Context Diagram

Before we construct the actual process model, we need to establish the initial project scope. All projects have scope. A project's scope defines what aspect of the business a system or application is supposed to support. A project's scope also defines how the system or application being modeled must interact with other systems and the business as a whole. In your information system framework (Figure 6.1), scope is defined as the INTERFACE focus from the system owner's perspective. It is documented with a context diagram.

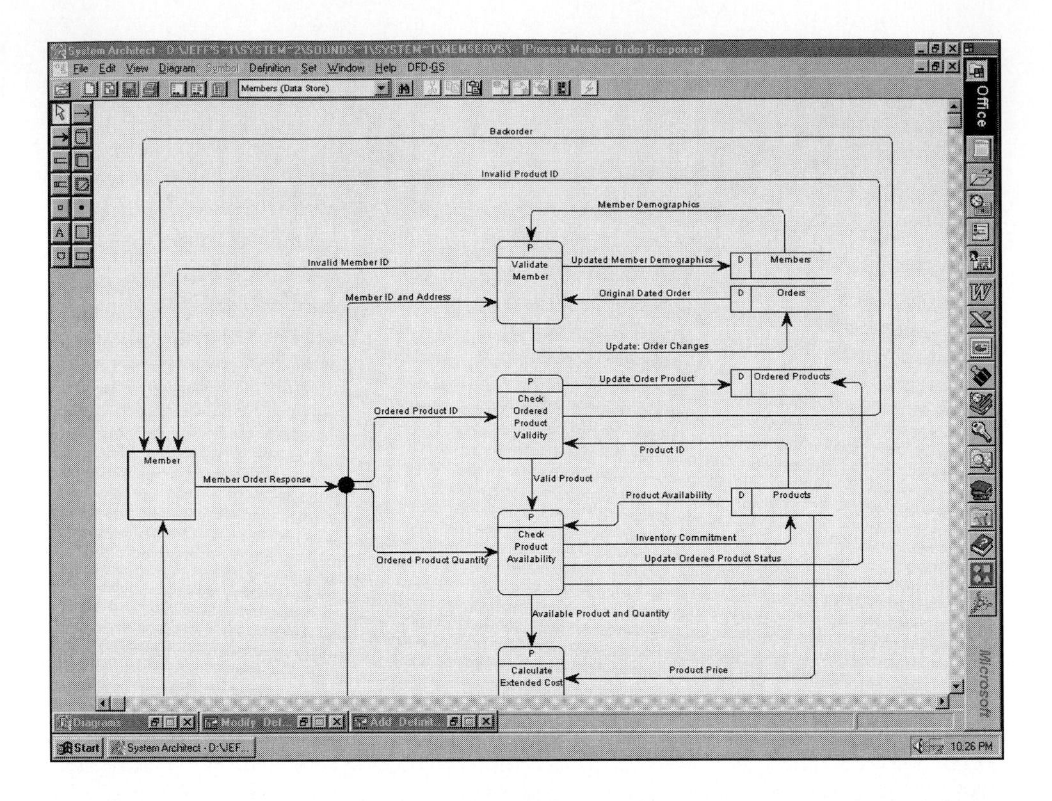

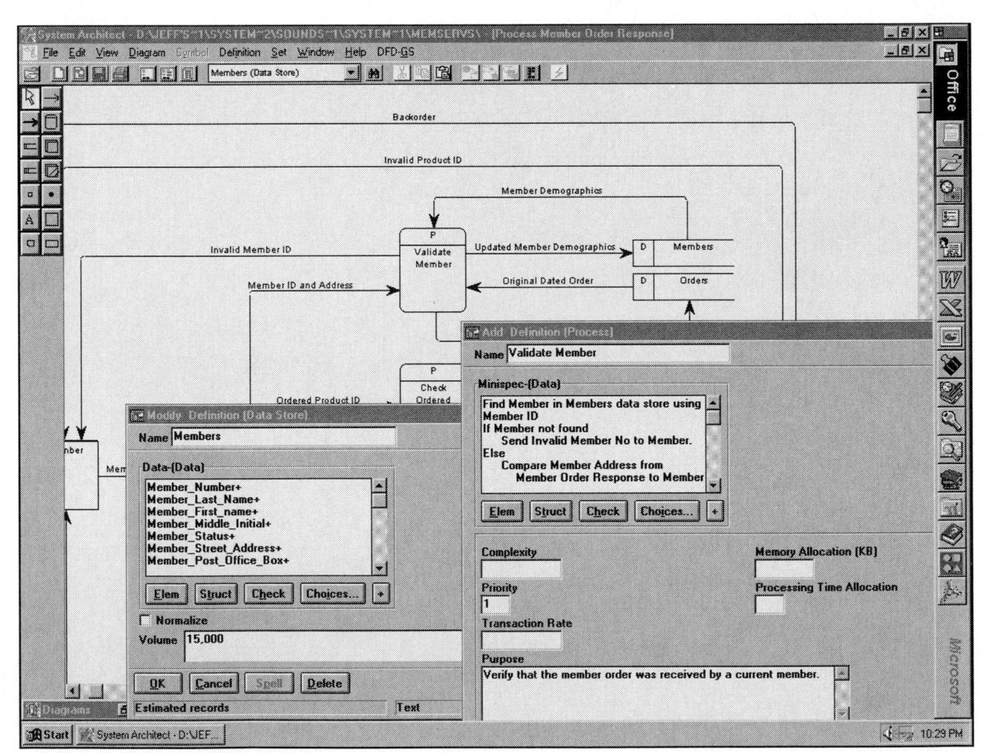

FIGURE 6.19 *CASE for Process Modeling (using* System Architect *by Popkin Software & Systems)*

A **context diagram** [DeMarco, 1978] defines the scope and boundary for the system and project. Because the scope of any project is always subject to change, the context diagram is also subject to constant change. A synonym is *environmental model* [Yourdon, 1990].

We suggest the following strategy for determining the system's boundary and scope:

1. Think of the system as a container in order to distinguish the inside from the outside. Ignore the inner workings of the container. This is the classic black box concept of systems theory.
2. Ask your end-users what business events or transactions a system must respond to. These are the *net inputs* to the system. For each net input, determine its source. Sources will become *external agents* on the context diagram.
3. Ask your end-users what responses must be produced by the system. These are the net outputs to the system. For each net output, determine its destination. Destinations may be *external agents*. Requirements for reports and queries can quickly clutter the diagram. Consider consolidating them into one or two composite data flows.
4. Identify any *external* data stores. Many systems require access to the files or databases of other systems. They may use the data in those files or databases. Sometimes they may update certain data in those files and databases. But generally, they are not permitted to change the structure of those files and databases—therefore, they are outside of the project scope.
5. Draw your context diagram from all of the preceding information.

First-generation context diagramming did not permit data stores on the context diagram. Those data stores were drawn as boundaries. This practice prevented the analyst from defining that data store's "fixed" structure in the CASE tool. Consequently, best current practice now permits external data stores to be included on the context diagram.

If you try to include all the inputs and outputs between a system and the rest of the business and outside world, a typical context diagram might show as many as 50 or more data flows. Such a diagram would have little, if any, communication value. Therefore, we suggest you show only those data flows that represent the main objective or most important inputs and outputs of the system. Defer less common data flows (such as error processing) to more detailed DFDs to be drawn later.

The context diagram contains one and only one process (see Figure 6.20). External agents are drawn around the perimeter. External data stores are also added to the perimeter. Data flows define the interactions of your system with the boundaries and with the external data stores.

The main purpose of our system is to respond to NEW MEMBER SUBSCRIPTIONS (an initial order and request for membership), CLUB PROMOTIONS, and MEMBER ORDER RESPONSES and OTHER ORDERS. (Notice the singular names!) Management has also placed great emphasis on the need for various SALES AND PROMOTION and MEMBER REPORTS (composite data flows; hence the plural names).

Finally, notice that the ACCOUNTS RECEIVABLE DATA BASE data store appears on our context diagram. The Accounts Receivable (A/R) Department has agreed to provide read-only access to its database to facilitate checking MEMBER CREDIT STATUS for member orders. The data store is external because we cannot change its structure or contents without adversely affecting the accounting information system, which is outside our project scope.

The Functional Decomposition Diagram

Recall that a decomposition diagram shows the top-down functional decomposition or structure of a system. It also provides us with the beginnings of an outline for drawing our data flow diagrams.

Figure 6.21 is the functional decomposition diagram for the SoundStage project. Let's study this diagram. First, notice that the processes are depicted as rectangles, not rounded rectangles. This is merely a limitation of our CASE tool's implementation of decomposition diagrams—you also may have to adapt to your CASE tool.

In many decomposition and data flow diagrams, the processes are not only named, but they also are numbered as part of an identification scheme that uses the following guidelines:

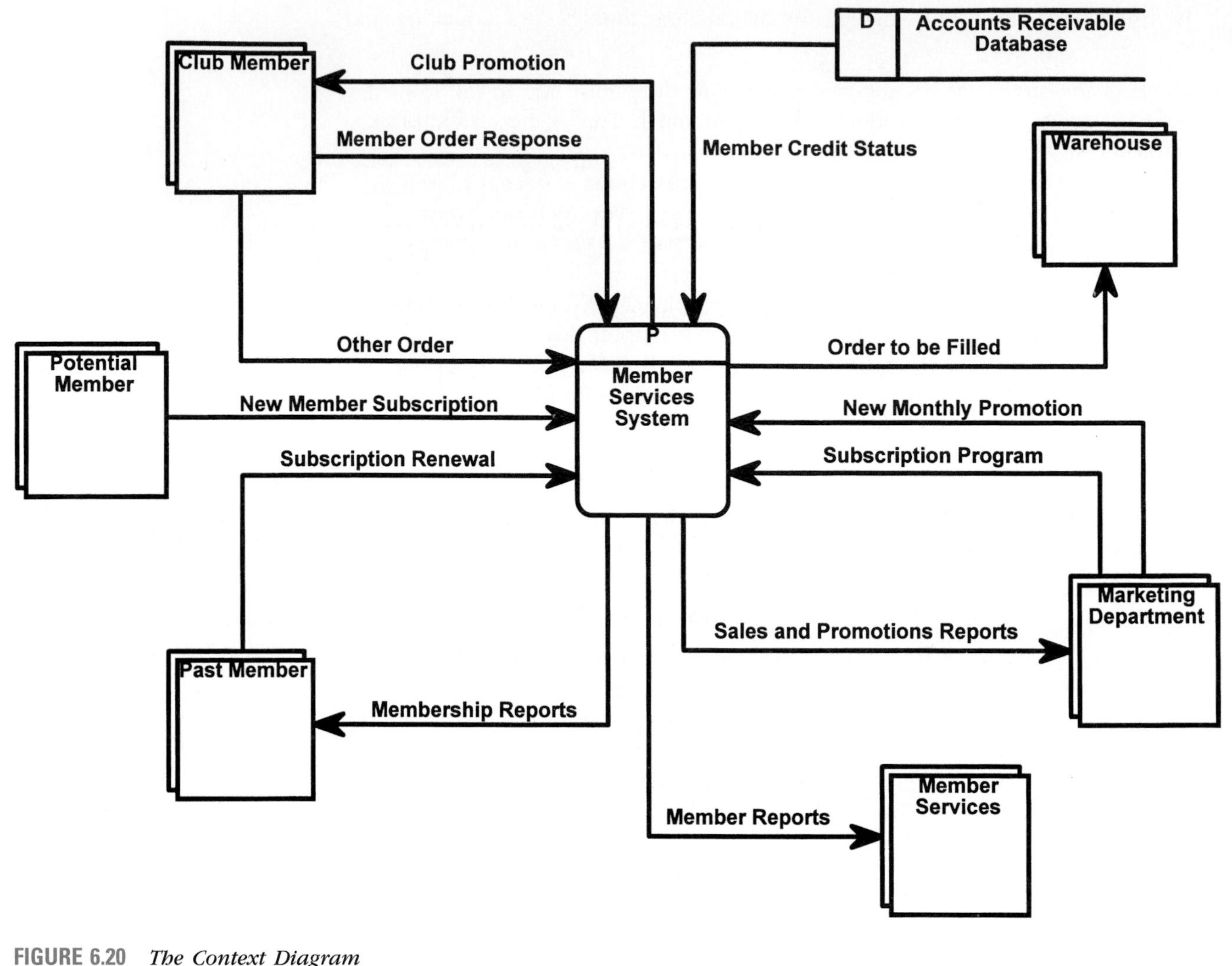

FIGURE 6.20 *The Context Diagram*

1. The root process MEMBER SERVICES SYSTEM would be numbered 0.
2. The three subsystems would be numbered 1, 2, and 3.
3. The subfunctions of the MEMBER SERVICES SUBSYSTEM would be numbered 1.1 and 1.2.
4. If 1.2 is factored into specific transactions, they would be numbered 1.1.1, 1.1.2, 1.1.3, and so forth. Similarly, specific reports would be numbered 1.2.1, 1.2.2, 1.2.3, and so forth.

Although this scheme is well documented in books, we elected *not* to use the numbers. Readers repeatedly misinterpreted them as sequential processes and the numbers were never meant to imply sequence. Furthermore, a numbering scheme discouraged us from later reorganizing the system and diagram; no one wants to change all those ID numbers.

The following is an item-by-item discussion of the decomposition diagram. The circled numbers correspond to specific points of interest on the diagram.

① The root process corresponds to the entire system.

② The system is initially factored into subsystems and/or functions. These subsystems and functions do not necessarily correspond to organization units on an organization chart. Increasingly, analysts and users are being asked to

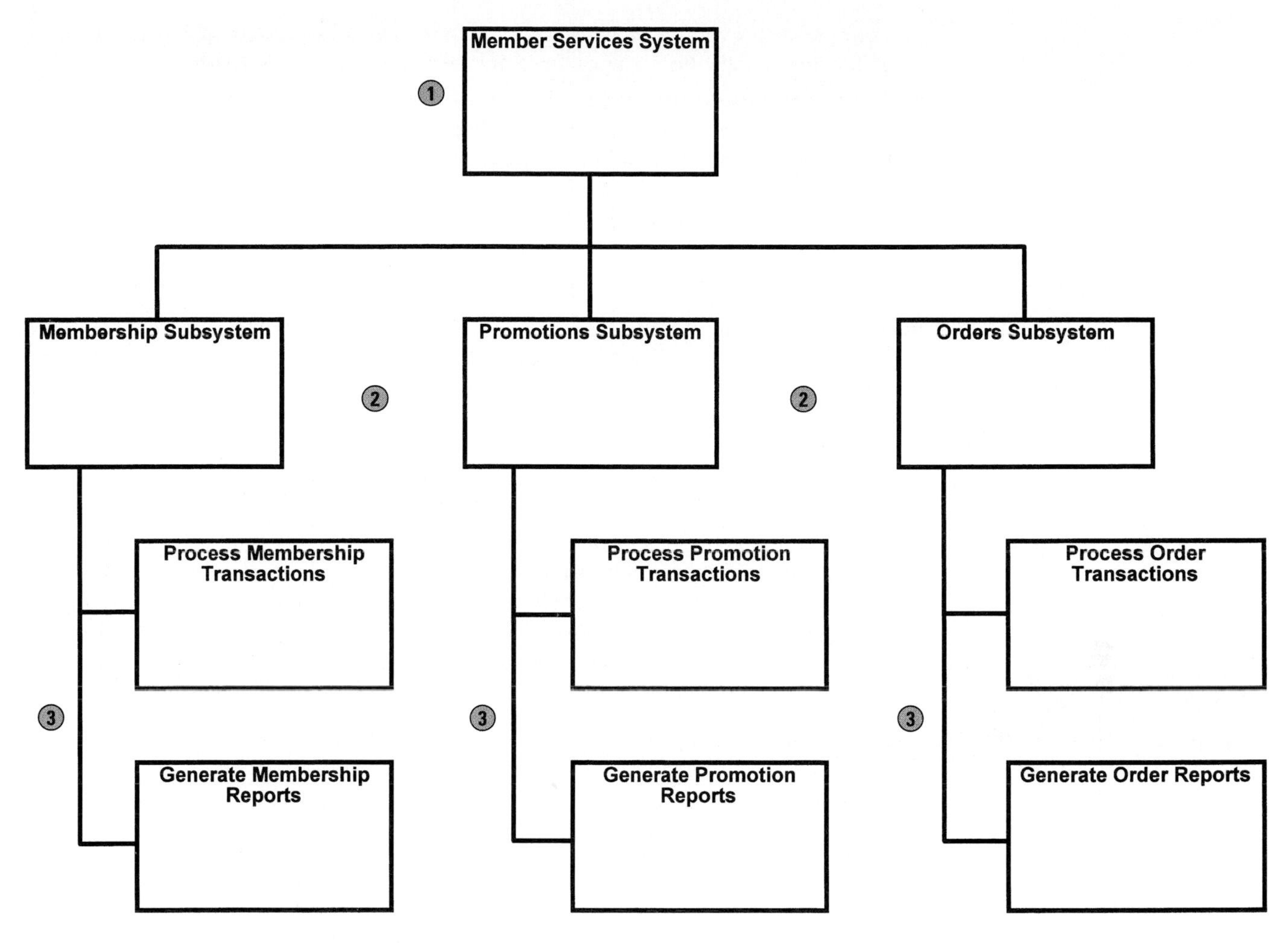

FIGURE 6.21 *A Functional Decomposition Diagram*

ignore organizational boundaries and to build cross-functional systems that streamline processing and data sharing.

③ We like to separate the operational and reporting aspects of a system. Thus, we factored each subsystem accordingly. Later, if this structure doesn't make sense, we can change it.

Larger systems might have first been factored into subsystems *and* functions. There is no limit to the number of child processes for a parent process. Many authors used to recommend a maximum of five to nine processes per parent, but any such limit is too artificial. Instead, structure the system such that it makes sense for the business!

Factoring a parent process into a single child process doesn't make sense. It would provide no additional detail. Therefore, if you plan to factor a process, it should be factored into at least two child processes.

The Event-Response List

The next step is to determine what business events the system must respond to and what responses are appropriate. Events are not hard to find. Some of the inputs on the context diagram are associated with events. But the context diagram rarely shows all the events. Essentially, there are three types of events.

- **External events** are so named because they are initiated by external agents. When these events happen, an input data flow occurs for the system. For example, the event CUSTOMER PLACES A NEW ORDER is recognized in the form of the input data flow ORDER from the external agent CUSTOMER.

- **Temporal events** trigger processes on the basis of time, or something that merely happens. When these events happen, an input *control flow* occurs. Examples of temporal events might include TIME TO REMIND CUSTOMERS TO PAY PAST INVOICES or END OF MONTH.
- **State events** trigger processes based on a system's change from one state or condition to another. Like temporal events, state events will be illustrated as an input *control flow.*

Information systems usually respond mostly to external and temporal events. State events are usually associated with real-time systems such as elevator or robot control.

Events can be discovered in a number of ways. The SoundStage case at the start of the chapter demonstrated one popular technique—study the life histories of the data entities on a data model. Most events update one or more of those entities. But what if you haven't done a data model? It still is not terribly difficult. After giving most users a few examples, they will quickly build a list. After all, events happen! The users live those events!

Each event should be named. The name should reveal the system nature of the event—that is, provide some insight as to at least one appropriate response. We suggest complete, simple sentences. The Robertsons (see the suggested readings) recommend the following guidelines for external and temporal events:

External event	External agent name + reason for the data flow. Example: CUSTOMER REQUESTS ACCOUNT BALANCE.
Temporal event	Time to + action that must be taken. Example: TIME TO BILL CUSTOMER ACCOUNTS.

The number of business events for a typical system can be quite large. But the system must respond to each event in order to support the business operation. Try to be complete in your identification of events, but don't spend too much time on the inputs and outputs; omissions will eventually be identified in the data flow diagrams.

A simple list format illustrated in Figure 6.22 is typical. For each event, we have recorded the input data or control flow. We have also recorded all the associated system responses. Possible responses include:

- One or more output data flows to one or more external agents. The external agent that initiated the event may or may not receive a response. Don't forget to include responses for things that might go wrong. For example, the input ORDER may produce several possible responses including an ORDER CONFIRMATION, an UNKNOWN PRODUCT NOTIFICATION, an ORDER CREDIT PROBLEM, a BACKORDER, and a SHIP ORDER.
- One or more changes to data stores (create, update, or delete). Ideally, data stores correspond to entities in the data model (Chapter 5). For example, the input order may create new ORDER and ORDERED PRODUCT records and update the CUSTOMER and various PRODUCTS records. If a data model doesn't exist, then data stores from appropriate files will need to be brainstormed.

The Event Decomposition Diagram

Now we can further partition our functions in the decomposition diagram. We simply add event handling processes (one per event) to the decomposition (see Figure 6.23). If the entire decomposition diagram will not fit on a single page, add separate pages for subsystems or functions. The root process on a subsequent page should be duplicated from an earlier page to provide a cross-reference. Figure 6.23 shows only the event processes for the MEMBERSHIPS subsystem. Events for the PROMOTIONS and ORDERS functions would be on separate pages.

EVENT LIST

Event Description	Trigger (Inputs)	Responses (Outputs)
Marketing department establishes a new membership plan and offer.	SUBSCRIPTION PROGRAM	SUBSCRIPTION PLAN CONFIRMATION CREATE AGREEMENT
Marketing department terminates a membership offer.	SUBSCRIPTION PROGRAM TERMINATION	SUBSCRIPTION PLAN TERMINATION NOTICE DELETE AGREEMENT UPDATE MEMBERS
Potential member responds to a subscription offer.	NEW MEMBER SUBSCRIPTION	SUBSCRIPTION CONFIRMATION SUBSCRIPTION REJECTION CREATE MEMBER
Potential member is referred to membership by a current member.	REFERRAL SUBSCRIPTION REFERRAL BONUS ORDER	SUBSCRIPTION CONFIRMATION SUBSCRIPTION REJECTION CREATE MEMBER
Potential member exercises 10-day cancellation option.	SUBSCRIPTION CANCELLATION	SUBSCRIPTION CANCELLATION NOTICE DELETE MEMBER
Club member changes name or address.	MEMBER CHANGE OF NAME OR ADDRESS	UPDATE MEMBER
Time to cancel those inactive members.	INACTIVITY CHECK	CANCELLED MEMBERS REPORT UPDATE MEMBER
Marketing department establishes a new monthly or seasonal promotion.	MONTHLY PROMOTION SEASONAL PROMOTION	DATED ORDER CREATE ORDER
Member responds to dated promotional order.	MEMBER ORDER RESPONSE	ORDER TO BE FILLED CREDIT PROBLEM NOTIFICATION UPDATE MEMBER UPDATE ORDER UPDATE PRODUCT
Time to automatically fill order for which member has not replied to dated promotion.	DATED ORDER DEADLINE	ORDER TO BE FILLED DELETE ORDER CREATE ORDER UPDATE ORDER UPDATE MEMBER UPDATE PRODUCT
Time to produce promotion analyses.	END OF PROMOTION	PROMOTION ANALYSIS REPORT
Time to analyze sales.	END OF MONTH END OF QUARTER END OF FISCAL YEAR	SALES ANALYSIS REPORT

FIGURE 6.22 *A Partial Event-Response List*

There is no need to factor the decomposition diagram beyond the events and reports. That would be like outlining down to the final paragraphs or sentences in a paper. The decomposition diagram, as constructed, will serve as a good outline for the later data flow diagrams.

Event Diagram

Using our decomposition diagram as an outline, we can draw one event diagram for each event process.

An **event diagram** is a context diagram for a single event. It shows the inputs, outputs, and data store interactions for the event.

Event diagrams are easy to draw. More importantly, by drawing an event diagram for each process, users do not become overwhelmed by the overall size of the system.

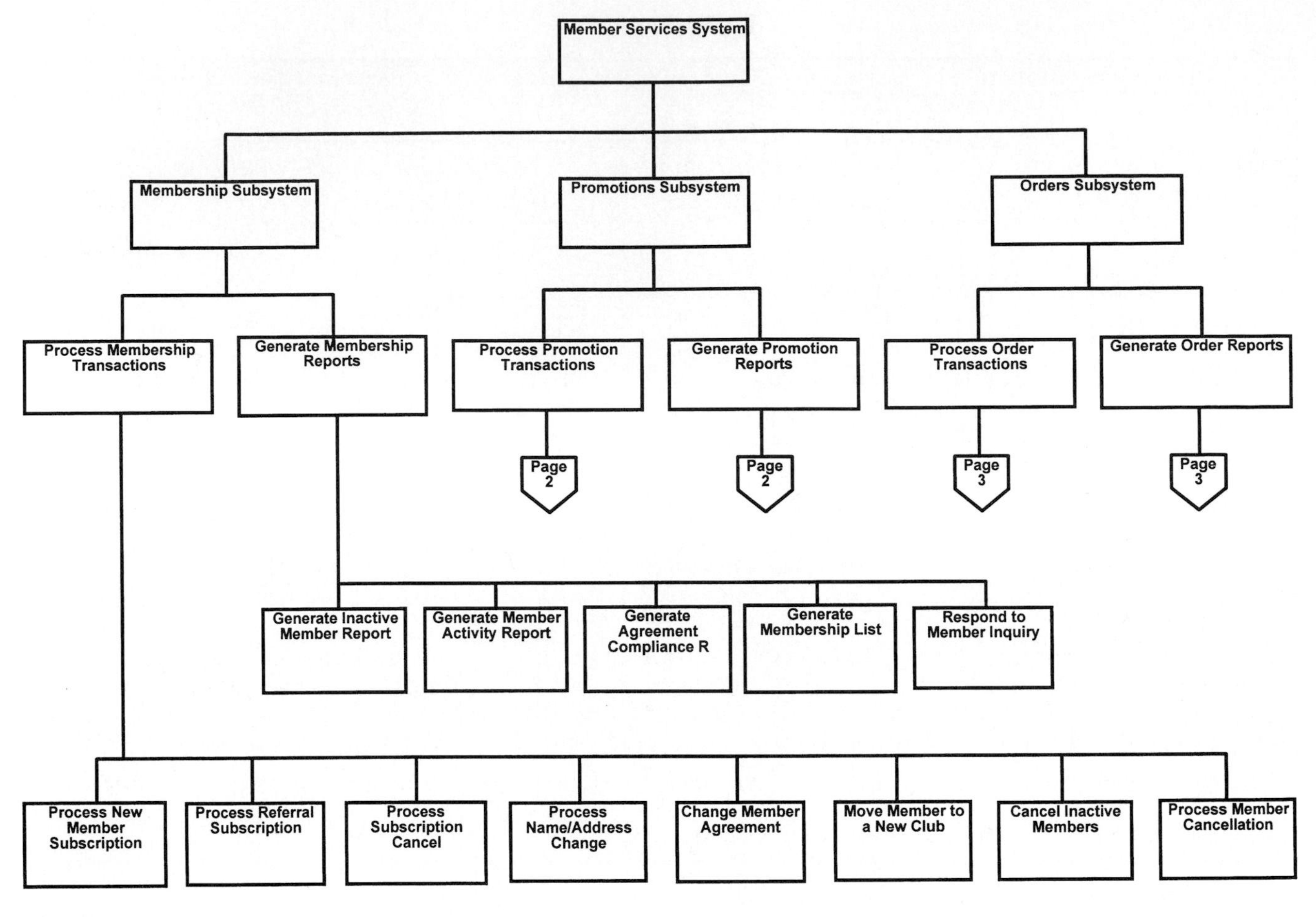

FIGURE 6.23 *A Partial Event Decomposition Diagram*

Before drawing any event diagrams, it is helpful to have a list of all the data stores available. Because SoundStage already completed the data model for this project (Chapter 5), team members simply created a list of the plural for each entity name on that data model. It is useful to review the definition and attributes for each entity/data store on the list.

Most event diagrams contain a single process—the same process that was named to handle the event on the decomposition diagram. For each event, illustrate the following:

- The inputs and their sources. Sources are depicted as external agents. The data structure for each input should be recorded in the repository.
- The outputs and their destinations. Destinations are depicted as external agents. The data structure for each output should be recorded in the repository.
- Any data stores from which records must be "read" should be added to the event diagram. Data flows should be added and named to reflect what data are read by the process.
- Any data stores in which records must be created, deleted, or updated should be included in the event diagram. Data flows to the data stores should be named to reflect the nature of the update.

Very simple, no? The sensibility and simplicity of event diagramming makes the technique a powerful communication tool between users and technical professionals!

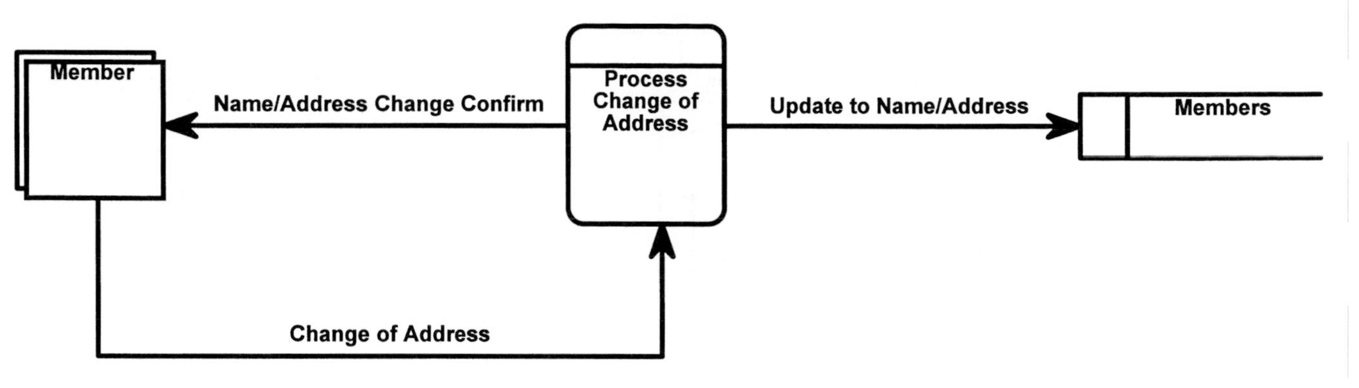

FIGURE 6.24 *A Simple External Event Diagram*

A complete set of event diagrams for the SoundStage case study would double the length of this chapter without adding substantive educational value. Thus, we will demonstrate the model with three simple examples.

Figure 6.24 illustrates a simple event diagram for an external event. Most systems have many such simple event diagrams because all systems must provide for routine maintenance of data stores.

Figure 6.25 depicts a somewhat more complex external event, one for the business transaction MEMBER ORDER RESPONSE. Notice that business transactions tend to use and update more data stores and have more interactions with external agents.

A common question is, Can an event diagram have more than one process on it? The answer is maybe. Some event processes may trigger other event processes. In this case, the combination of events should be shown on a single event diagram. In our experience, most event diagrams have one process. An occasional event diagram may have two or perhaps three processes. If the number of processes exceeds three, you are probably drawing an activity diagram (prematurely), not an event diagram—in other words, you're getting too involved with details. Most event processes do not directly communicate with one another. Instead, they communicate across shared data stores. This allows each event process to do its job without worrying about other processes keeping up.

Figure 6.26 shows an event diagram for a temporal event. Notice that the control flow appears to be coming from nowhere. That is because it merely represents timing. Some modelers will add an external entity CALENDAR or TIME to serve as a source for this control flow.

Each event process should be described to the CASE repository with the following properties:

- Event sentence—for business perspective.
- Throughput requirements—the volume of inputs per some time period.
- Response time requirements—how fast the typical event must be handled.
- Security, audit, and control requirements.
- Archival requirements (from a business perspective).

All the above properties can be added to the descriptions associated with the appropriate processes, data flows, and data stores on the model.

The System Diagram

The event diagrams serve as a meaningful context for users to validate the accuracy of each event to which the system must provide a response. But these events

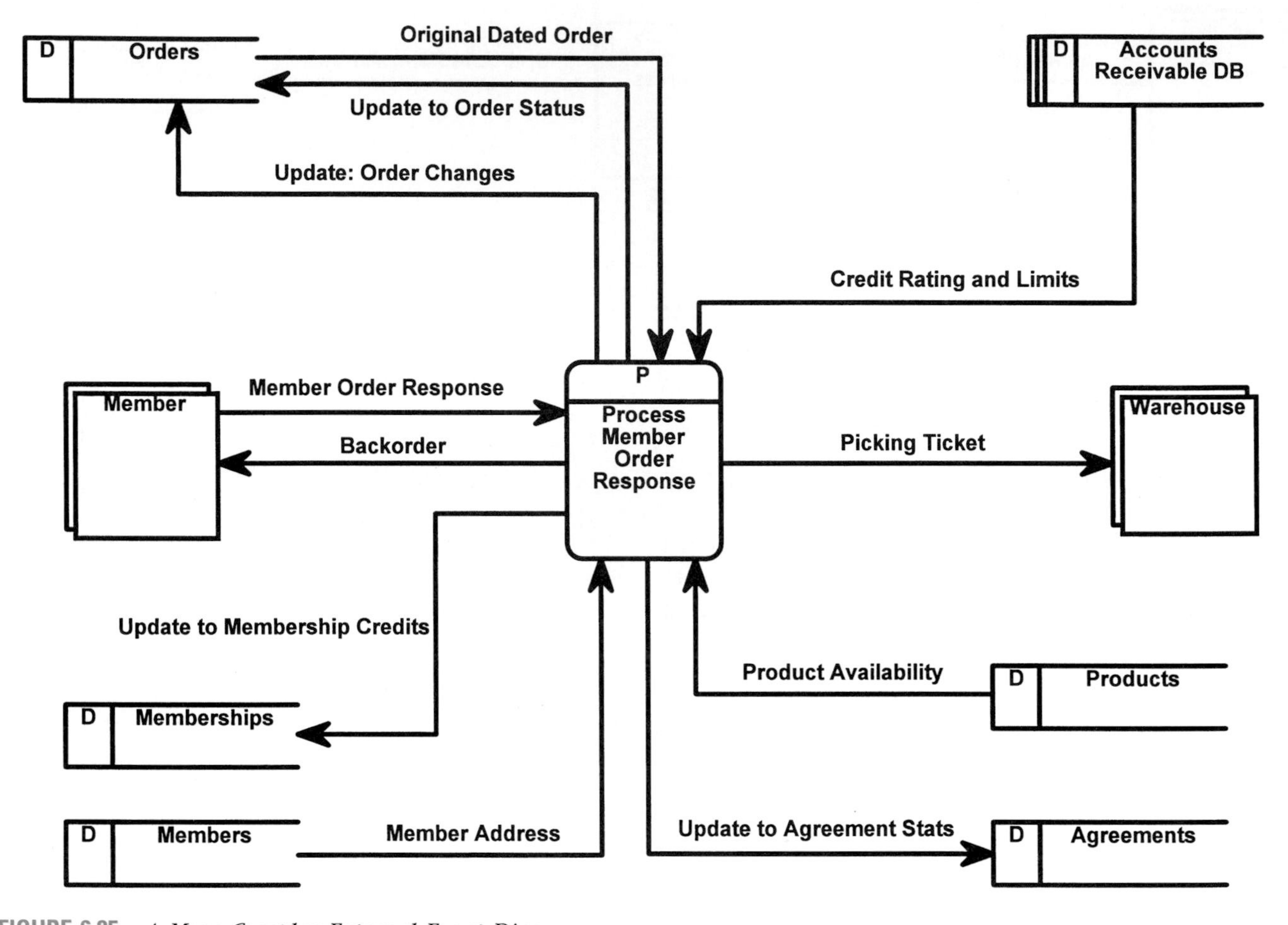

FIGURE 6.25 *A More Complex External Event Diagram*

do not exist in isolation. They collectively define systems and subsystems. It is, therefore, useful to construct one or more system diagrams that show all the events in the system or a subsystem.

The system diagram is said to be "exploded" from the single process that we created on the context diagram (Figure 6.20). The system diagram shows either (1) all the events for the system on a single diagram, or (2) all the events for a single subsystem on a single diagram. Depending on the size of the system, a single diagram may be too large.

Our SoundStage project is moderate in size, but it still responds to too many events to squeeze all those processes onto a single diagram. Instead, Bob Martinez elected to draw a subsystem diagram for each of the major subsystems. Figure 6.27 shows the subsystem diagram for the ORDERS SUBSYSTEM. It consolidates all the transaction and report-writing events onto a single diagram. (The reporting events may be omitted or consolidated into composites if the diagram is too cluttered.) Notice that the system diagram demonstrates how event processes truly communicate—using shared data stores.

If necessary, and after drawing the three subsystem diagrams for this project, Bob could have drawn a system diagram that illustrates only the interactions between the three subsystems. This is a relic of the original top-down data flow

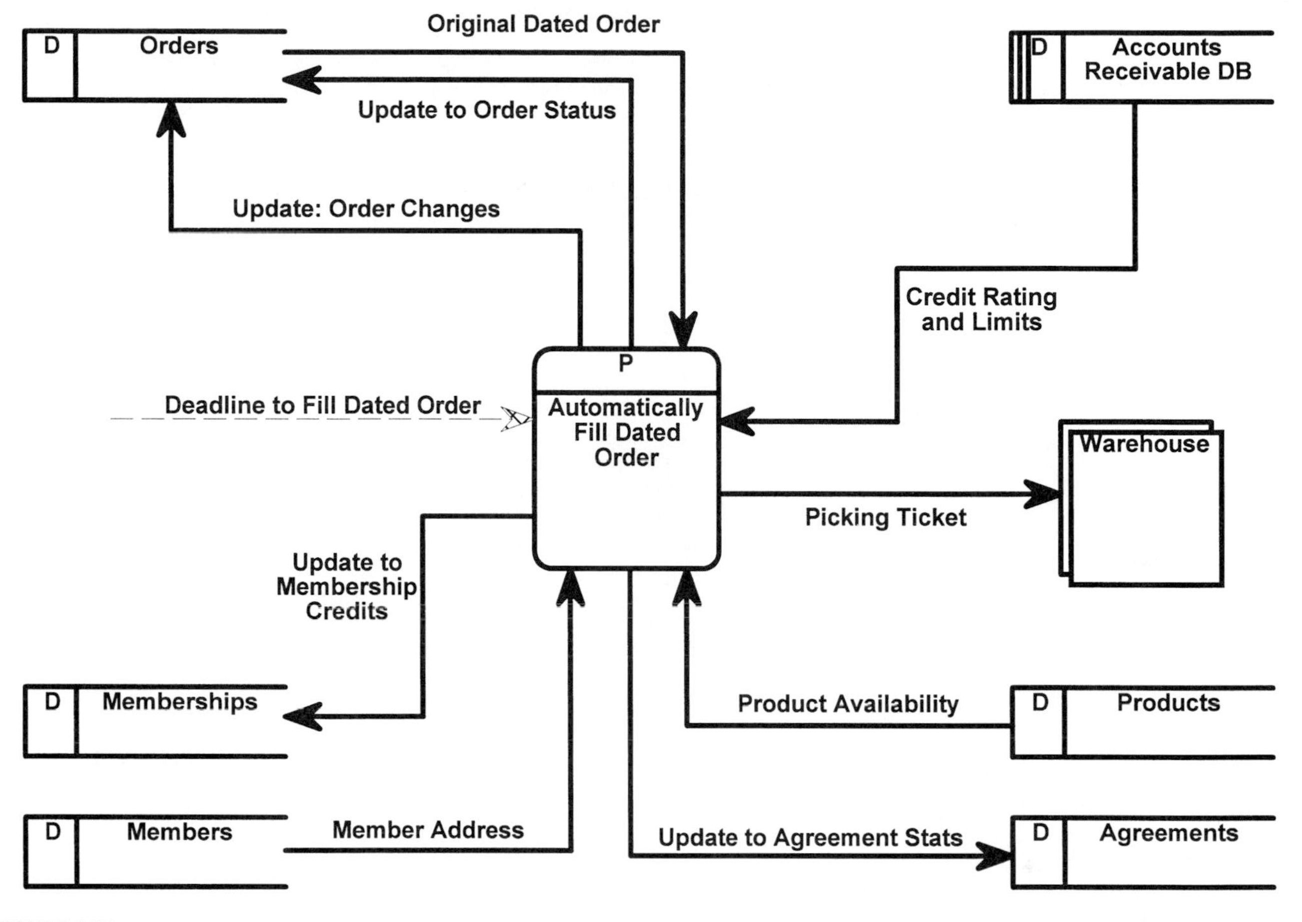

FIGURE 6.26 *A Temporal Event Diagram*

diagramming strategy of the original structured analysis methodology. In practice, this higher-level diagram requires so much consolidation of data flows and data stores that its communication value is questionable. To us (and to Bob) it was busywork, and his time was better spent on the next set of data flow diagrams. Before we leave this topic, we should introduce the concept of *synchronization*.

> **Synchronization** is the balancing of data flow diagrams at different levels of detail to preserve consistency and completeness of the models. Synchronization is a quality assurance technique.

We now have a set of event diagrams (one per business event) and one or more system/subsystem diagrams. The event diagram processes are merged into the system diagrams. It is very important that each of the data flows, data stores, and external agents that were illustrated on the event diagrams be represented on the system diagrams. This is called *balancing*. Most CASE tools include facilities to check for balancing errors.

When creating a system diagram, do not consolidate data stores—otherwise, you will create balancing errors between the system and event diagrams. On the other hand, you may elect to consolidate some data flows (from event diagrams) into composite data flows on the system diagram. If you do, be sure to

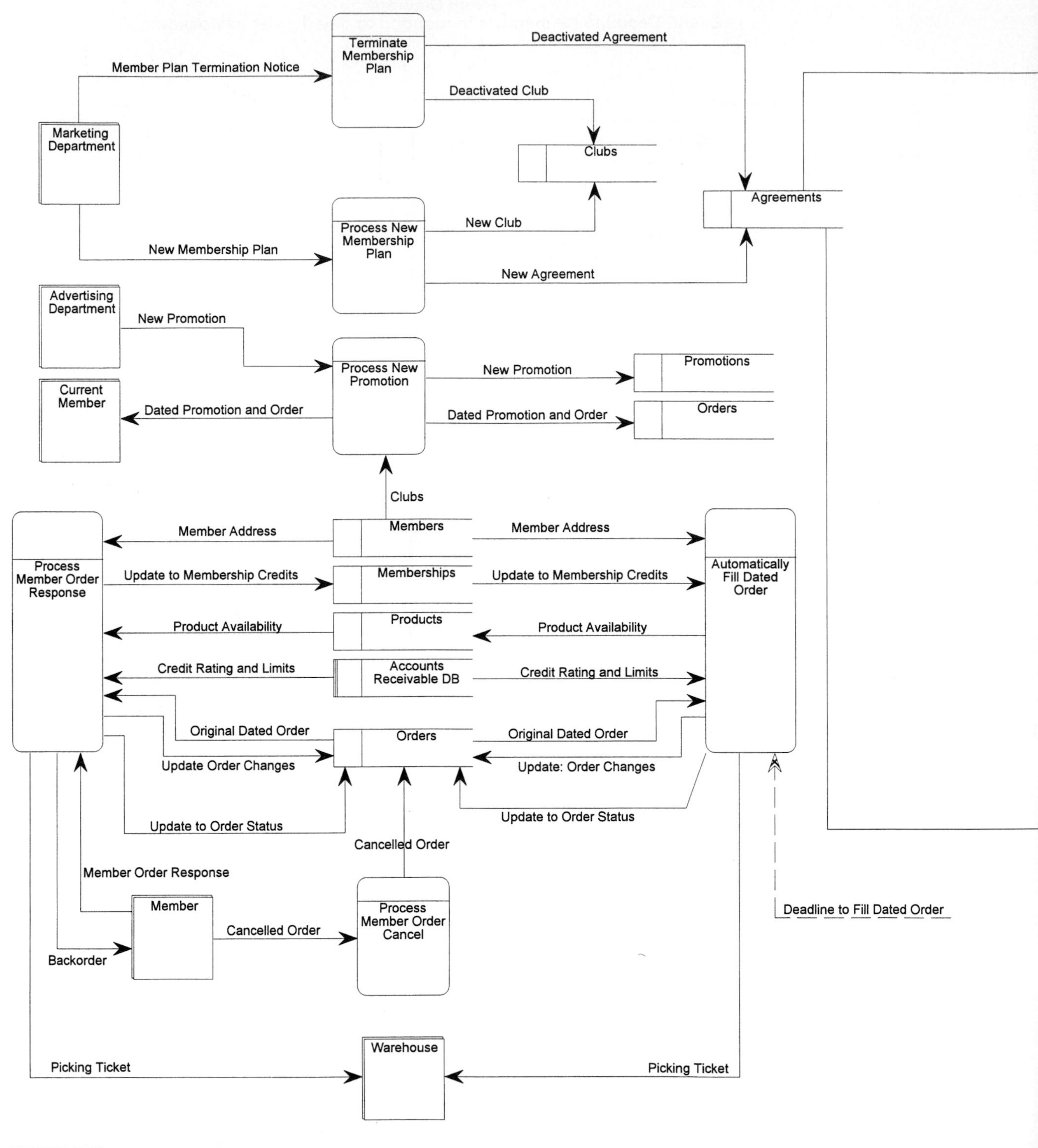

FIGURE 6.27 *A System Diagram*

use junctions on the event diagrams to demonstrate how the elementary data flows are derived from the composite data flows.

Primitive Diagrams

We're almost done! Each event process on the system diagram must be exploded into either (1) a procedural description or (2) a primitive data flow diagram.

Some event processes are not very complex—in other words, they are both an

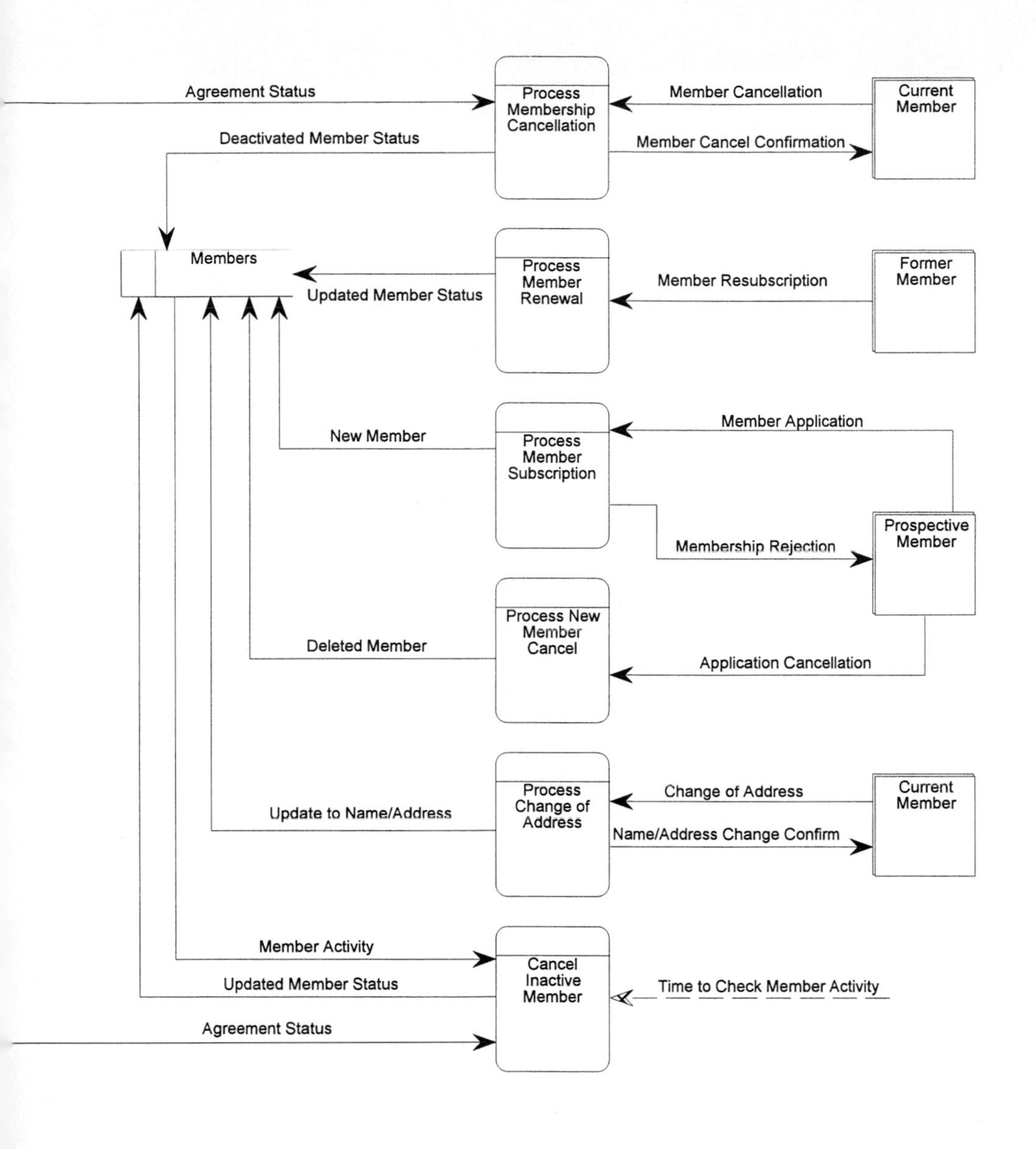

event and an elementary process. They should be described in one page (usually much less) of Structured English. This is true of those event processes for which the event diagrams had relatively few inputs, outputs, and data stores. You simply use the CASE tool to enter the Structured English for the process. (Recall that the data structures for the inputs and outputs were described to the repository as part of the event diagramming step.)

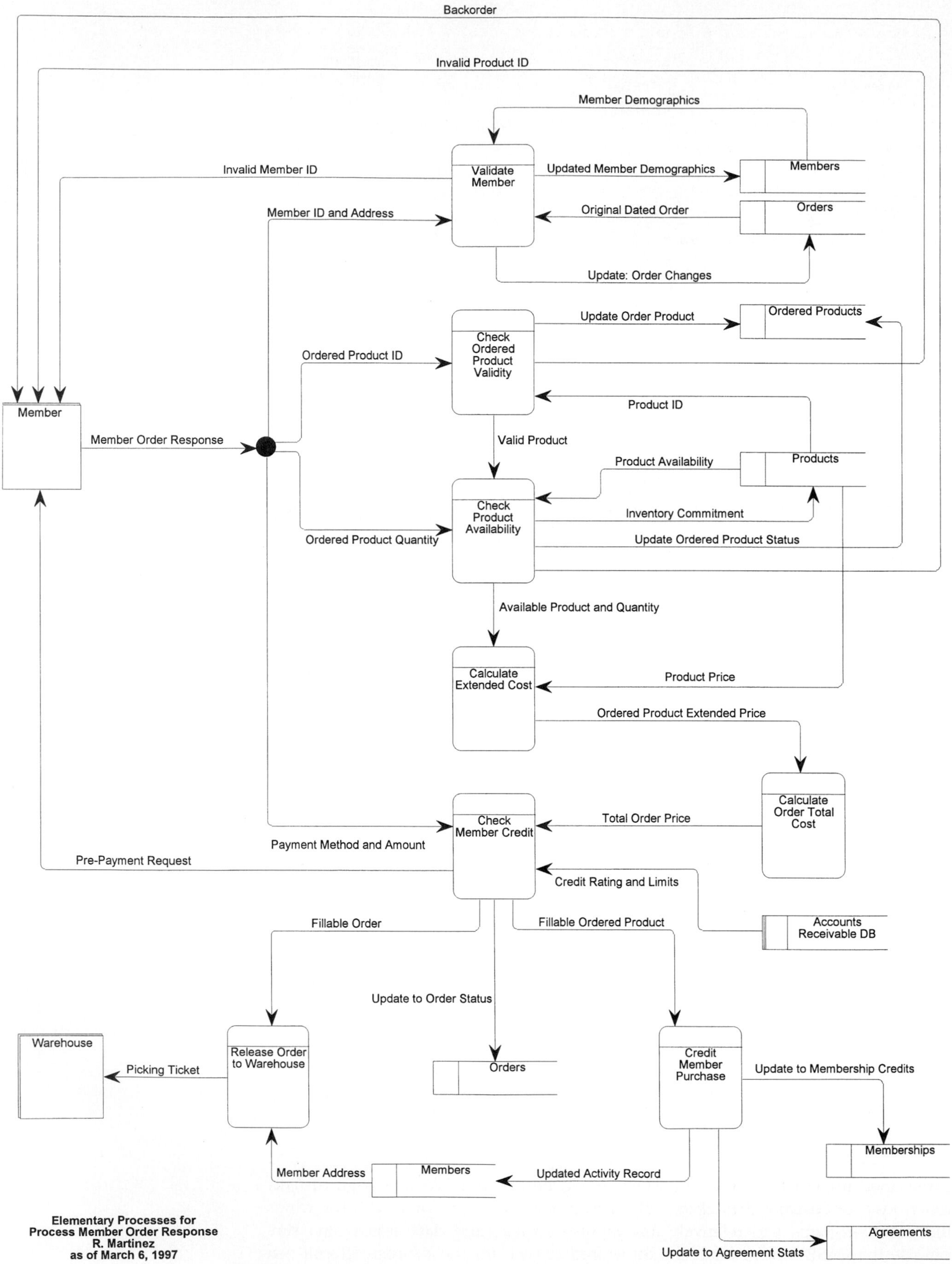

Backorder

Invalid Product ID

Member Demographics

Invalid Member ID

Validate Member

Updated Member Demographics

Members

Member ID and Address

Original Dated Order

Orders

Update: Order Changes

Update Order Product

Ordered Products

Ordered Product ID

Check Ordered Product Validity

Member

Product ID

Member Order Response

Valid Product

Product Availability

Products

Check Product Availability

Inventory Commitment

Ordered Product Quantity

Update Ordered Product Status

Available Product and Quantity

Calculate Extended Cost

Product Price

Ordered Product Extended Price

Calculate Order Total Cost

Check Member Credit

Total Order Price

Payment Method and Amount

Pre-Payment Request

Credit Rating and Limits

Fillable Order

Fillable Ordered Product

Accounts Receivable DB

Update to Order Status

Warehouse

Release Order to Warehouse

Picking Ticket

Orders

Credit Member Purchase

Update to Membership Credits

Memberships

Member Address

Members

Updated Activity Record

Agreements

Update to Agreement Stats

Elementary Processes for
Process Member Order Response
R. Martinez
as of March 6, 1997

FIGURE 6.28 *A Primitive Diagram*

Event processes with more complex event diagrams should be exploded into a more detailed, primitive data flow diagram such as that illustrated in Figure 6.28. This primitive DFD shows detailed processing requirements for the event. This DFD shows several elementary processes *for* the event process. Each elementary process is cohesive—that is, it does only one thing. Each of these elementary processes can now be described with procedural Structured English specifications and, where appropriate, decision tables.

When Bob drew this primitive data flow diagram, he had to add new data flows between the processes. In doing so, he tried to practice good data conservation, making sure each process got only the data it truly needed. The data structure for each data flow had to be described in his CASE tool's repository. Also notice that he used data flow junctions to split and merge appropriate data flows on the diagram.

The combination of the context diagram, system diagram, event diagrams, primitive diagrams, data structures, and procedural logic completes our process requirements package. Collectively, this *is* the process model. A well-crafted and complete process model can effectively communicate business requirements between end-users and computer programmers, eliminating much of the confusion that often occurs in system design, programming, and implementation!

THE NEXT GENERATION

The demand for process modeling skills remains strong. Even though object modeling (Chapter 8) is becoming a new standard in the industry, process modeling skills remain valuable for two reasons:

- The current interest in **business process redesign** requires process models.
- Process models are included in many object modeling strategies such as the *object modeling technique* (OMT).

With respect to the former, business process design emphasizes **physical process modeling.** In a nutshell, physical process models include those processes that reflect the current implementation. This may include sequential processes that merely edit, route, copy, or approve a data flow. Physical data flow diagrams also include additional details such as who or what performs each process, the cost of each process, and a critical evaluation of the value returned by each process. Logical data flow diagrams may also be used to optimize redesigned physical business processes.

Almost without question, leading-edge methodologies are embracing object modeling and *object-oriented analysis* (Chapter 8) as the heir apparent for current data and process models. (Actually, and at the risk of oversimplification, object modeling integrates elements of both data and process modeling.) But object modeling and object-oriented analysis techniques are still young and evolving. Also, many structured analysis practitioners are only beginning to practice the more contemporary, event-driven approach you just learned. Finally, data flow diagrams have become a component of some object-oriented analysis techniques (such as OMT), much as they became an integral component of data-driven analysis techniques such as information engineering. Thus, process modeling will remain a viable skill for many years.

We look for CASE tools to continue to improve. There is still some interest in CASE tools capable of generating program code directly from process models. Early code generators have experienced mixed results, but if process modeling languages continue to improve in their precision and completeness, we may yet realize the vision of generating "software from pictures." Another aspect of CASE to keep your eyes on is **reverse engineering.** Through reverse engineering, CASE tools read the source code of old programs (such as COBOL) to automatically generate physical process models that can then be *forward engineered* into better systems and programs, possibly generated in other languages.

WHERE DO YOU GO FROM HERE?

Most of you will proceed directly to Chapter 7, "Network Modeling." Today's information systems and computer applications are increasingly distributed across a network. As with DATA and PROCESSES, there is a logical (meaning business-oriented) aspect of GEOGRAPHY that should be considered before distributing the data and processes across a *computer* network.

At your instructor's discretion, some of you may jump to Chapter 8, "Object Modeling," especially if you've already completed Chapters 5 and 6 or if your instructor is electing to emphasize object-oriented analysis in your course. Object modeling has many parallels with process modeling. An object includes services that may be modeled with data flow diagrams.

If you want to immediately learn how to implement process models as systems and programs, you should skim or read Chapter 11, "Application Architecture and Process Design," and Chapter 15, "Structured Design." In the former, you'll learn how to transform the logical process models into physical process schemas. In the latter chapter, you'll learn how to transform logical data flow diagrams into program *structure charts*, another type of physical process model.

Finally, process models can be created only through effective interaction with users and managers who know the business. Part Five, Modules B and D teach you how to collect the facts needed to construct real-world process models. Module B teaches fact-finding and information gathering, and Module D teaches joint application development.

SUMMARY

1. A **model** is a representation of reality. We construct **logical models** to better understand business problem domains and business requirements. Eventually, logical models will be transformed into **physical models** to reflect design and implementation decisions.
2. **Process modeling** is a technique for organizing and documenting the process requirements and design for a system. This chapter focused on a popular process model called a **data flow diagram,** which depicts the flow of data through a system's processes.
3. Data flow diagramming is a technique for **systems thinking.**
4. In reality, a system *is* a process. A **process** is work performed on, or in response to, inputs and conditions.
5. Just as systems can be recursively decomposed into subsystems, processes can be recursively decomposed into subprocesses. A **decomposition diagram** shows the functional decomposition of a system into processes and subprocesses. It is a planning tool for subsequent data flow diagrams.
6. **Logical** processes show essential work to be performed by a system without showing how the processes will be implemented. There are three types of logical processes: **functions** (very high level), **events** (middle level of detail), and **elementary processes** (very detailed).
7. Elementary processes are further described by procedural logic. **Structured English** is a tool for expressing this procedural logic. Structured English is a derivative of structured programming logic constructs married to natural English.
8. More complex elementary processes may be described by **policies** that are expressed in decision tables. **Decision tables** show complex combinations of conditions that result in specific actions.
9. **Data flows** are the inputs to and the outputs from processes. They also illustrate data store accesses and updates.
10. All data flows are comprised of either other data flows or discrete data structures that include descriptive **attributes.** A data flow should contain only the amount of data needed by a process; this is called **data conservation.**
11. **External agents** are entities outside the scope of a system and project but that provide net inputs to or net outputs from a system. As such, they form the **boundary** of the system.
12. **Data stores** present files of data to be used and maintained by the system. A data store on a process model corresponds to all instances of an **entity** on a data model.
13. Process modeling may be used in different types of projects including **strategic systems planning, business process redesign,** and **application development.** For application development projects, this chapter taught an event-driven data flow diagramming strategy as follows:
    a. Draw a **context diagram** that shows how the system interfaces to other systems, the business, and external organizations.
    b. Draw a **functional decomposition diagram** that shows the key subsystems and/or functions that comprise the system.
    c. Create an **event list** that identifies the **external** and **temporal events** to which the system must provide a response. External events are triggered by the external agents of a system. Temporal events are triggered by the passing of time.

d. Update the decomposition diagram to include processes to handle the events (one process per event).
e. For each event, draw an **event diagram** that shows its interactions with external entities, data stores, and, on occasion, other triggers to other events.
f. Combine the event diagrams into one or more **system diagrams.**
g. For each event on the system diagram, either describe it as an elementary process using Structured English or *explode* it into a *primitive data flow diagram* that includes elementary process that must be subsequently described by either Structured English, decision tables, or both. When exploding processes on data flow diagrams to reveal greater detail, it is important to maintain consistency between the different types of diagrams; this is called *synchronization.*

14. Most computer-aided software engineering tools support both decomposition diagramming and data flow diagramming.

KEY TERMS

application process model, p. 236
black hole, p. 219
business process redesign, pp. 214, 229, 253
composite data flow, p. 226
context diagram, pp. 237, 240
control flow, p. 227
converging data flow, p. 233
data attribute, p. 230
data conservation, p. 229
data flow, p. 225
data flow diagram, p. 211
data store, p. 235
data structure, p. 230
data type, p. 231
decision table, p. 224
decomposition, p. 216
decomposition diagram, p. 217
diverging data flow, p. 231
domain, p. 231
elementary process, p. 219
enterprise process model, p. 236
essential model, p. 210
event, p. 218
event diagram, p. 245
event handler, p. 237
event partitioning, p. 237
event-response list, p. 237
external agent, p. 233
external entity, p. 234
external event, p. 243
function, p. 218
functional decomposition diagram, p. 237
gray hole, p. 219
junction, p. 226
logical model, p. 210
miracle, p. 219
model, p. 210
physical model, p. 210
physical process modeling, p. 253
policy, p. 224
primitive diagram, p. 237
primitive process, p. 219
process, p. 216
process model, p. 211
process modeling, p. 211
reverse engineering, p. 253
state event, p. 244
Structured English, p. 220
synchronization, p. 249
system diagram, p. 237
systems thinking, p. 214
temporal event, p. 244

REVIEW QUESTIONS

1. What is the difference between logical and physical modeling? Why is logical modeling more important in systems analysis?
2. What is systems thinking, and how do process models represent systems thinking?
3. What are the four symbols on a data flow diagram?
4. What is a process? Name three types of logical processes.
5. What is process decomposition and what role does it play in process modeling?
6. What purpose does Structured English serve in process modeling?
7. What are the basic constructs of Structured English?
8. What is the relationship between a policy and a decision table?
9. What are the components of a decision table?
10. What is the difference between data flows and data stores? What is the difference between data stores and data entities? What is the difference between data entities and external entities?
11. Differentiate between a data flow and a control flow.
12. What is data conservation?
13. What are the basic constructs of a data structure?
14. Differentiate between an enterprise and application process model.
15. During the survey and study phases, an analyst collected numerous samples, including documents, forms, and reports. Explain how these samples will prove useful for process modeling.
16. What is event partitioning?
17. Name and describe three types of business events.
18. Explain the tasks used to construct an application process model.

PROBLEMS AND EXERCISES

1. Compare and contrast process models with data models. What does each model show? Should you choose between the two modeling strategies? Why or why not?
2. A manager who has noted your use of logical data flow diagrams to document a proposed system's requirements has expressed some concern because of the lack of details that demonstrate the computer's role in the system. Defend your use of logical DFDs. Concisely explain the symbols and how to read a DFD. (Note: The answer to this exercise should be a standard component in any report that will include DFDs. You cannot be certain that the person who reads this report will be familiar with the tool.)
3. Explain why you should exclude implementation details when drawing a logical DFD. Can you think of any circumstances in which implementation details might be useful?
4. Explain why a systems analyst might want to draw logical models of an automated portion of an existing information system rather than simply accepting the existing technical information systems documentation, such as systems flowcharts and program flowcharts.
5. Draw a logical DFD to document the flow of data in your school's course registration and scheduling system.
6. Draw a logical DFD for some day-to-day "system" that you use or observe in use—for instance, your morning routine; making your favorite meal, including appetizer, entrée, side dishes, and dessert; constructing something from scratch.
7. Why is a project repository a valuable systems analysis tool? What are the possible consequences of not creating a project repository during systems analysis?
8. Can you think of any specific times that a project repository might have been helpful when you were writing a computer program? Can you think of a situation in which you misinterpreted a computer program requirement because you didn't know something that could have been recorded in a project repository?
9. Dig out your last computer program. Document the data structures for the following:
   a. Inputs (data flows).
   b. Outputs (data flows).
   c. Files or database (data store).
10. You have compiled a complete project repository for your new inventory control system. It is time to verify the contents of three summary reports specified for your end-users. Each data flow (report), record (data structure), and data attribute should be reviewed. Unfortunately, the entire repository is 373 pages. You can't mark the relevant pages and thumb back and forth between pages during your review. How should the report specifications be presented?
11. Why shouldn't a project repository be organized alphabetically independent of type—for example, data flow, data store, data attribute, and so on—like a traditional repository such as a dictionary?
12. Using the algebraic data structure notation in this chapter, create a project repository entry for the following:
    a. Your driver's license.
    b. Your course registration form.
    c. Your class schedule.
    d. IRS Form 1040 (any version).
    e. An account statement and invoice for a credit card.
    f. Your telephone, electric, or gas bill.
    g. An order form in a catalog.
    h. An application for anything (e.g., insurance, housing).
    i. A retail store catalog.
    j. A typical real estate listing.
    k. A computer printout from a business office or computer course.
    l. A catalog that describes the classes to be offered next semester.
    m. Your checkbook.
    n. Your bank statement.
13. During the study phase of systems analysis, the analyst must gather facts concerning both the manual and automated portions of the system. Why would it be desirable for a systems analyst to obtain samples of the existing computer files and computer-generated outputs? What value would project repository entries for computer files and computer-generated outputs have during *logical* modeling?
14. Visit a local business or school office. Ask for samples of five business forms or regular reports. Specify algebraic data structures for each sample.
15. Obtain a formal statement of a policy and procedure, such as a policy for a credit card. Evaluate the policy and procedure statement in terms of the common natural English specification problems identified earlier in this chapter.
16. Reconstruct the policy and procedure used in exercise 15 with the tools you learned in this chapter. Try to specify the decision table with a minimum number of rules.
17. Write a Structured English procedure for balancing your checkbook.
18. Produce a mini spec to describe how to prepare your favorite recipe, tune a car, or perform some other familiar task. Ask a novice to perform the task, working from your specification.

PROJECTS AND RESEARCH

1. Through information systems trade journals, research a commercial CASE product. Evaluate that package's project repository. Can you define both information and process models? Can you describe data structures to the repository? Can you describe individual data attributes to the repository? What types of analytical reports can be generated from the repository?
2. Enter into a contract with a programming instructor at your school. Convert one or more of his or her programming assignments to include data flow diagrams, data structures, and Structured English as described in this chapter. Have students currently enrolled in the course compare and contrast this form of specification with the original assignment's content and style.
3. Research a process modeling tool for business process redesign. How does it differ in symbology and constructs from logical data flow diagrams?

MINICASES

1. Given the following narrative description, draw a context DFD for the portion of the activities described.

   The purpose of the TEXTBOOK INVENTORY SYSTEM at a campus bookstore is to supply textbooks to students for classes at a local university. The university's academic departments submit initial data about courses, instructors, textbooks, and projected enrollments to the bookstore on a TEXTBOOK MASTER LIST. The bookstore generates a PURCHASE ORDER, which is sent to publishing companies supplying textbooks. Book orders arrive at the bookstore accompanied by a PACKING SLIP, which is checked and verified by the receiving department. Students fill out a BOOK REQUEST that includes course information. When they pay for their books, the students are given a SALES RECEIPT.
2. Given the following narrative description, draw a context DFD for the portion of the activities described.

   The purpose of the PLANT SCIENCE INFORMATION SYSTEM is to document the study results from a wide variety of experiments performed on selected plants. A study is initiated by a researcher who submits a RESEARCH PROPOSAL. After a panel review by a group of scientists, the researcher is required to submit a RESEARCH PLAN AND SCHEDULE. An FDA RESEARCH PERMIT REQUEST is sent to the Food and Drug Administration, which sends back a RESEARCH PERMIT. As the experiment progresses, the researcher fills out and submits EXPERIMENT NOTES. At the conclusion of the project, the researcher's results are reported on an EXPERIMENT HISTOGRAM.
3. Given the following narrative description of a system, draw a context diagram, functional decomposition diagram, and an event. Try to brainstorm events that might not be explicitly described. State any assumptions.

   The purpose of the production scheduling system is to respond to a PRODUCTION ORDER (submitted by the SALES DEPARTMENT) by generating a daily PRODUCTION SCHEDULE, generating RAW MATERIAL REQUISITIONS (sent to the MATERIALS MANAGEMENT DEPARTMENT) for all production orders scheduled for the next day, and generating JOB TICKETS for the work to be completed at each workstation during the next day (sent to the SHOP FLOOR SHIFT SUPERVISOR). The work is described in the following paragraphs.

   The production scheduling problem can be conveniently broken down into three functions: routing, loading, and releasing. For each product on a PRODUCTION ORDER, we must determine which workstations are needed, in what sequence the work must be done, and how much time should be necessary at each workstation to complete the work. This data is available from the PRODUCTION ROUTE SHEETS. This process, which is referred to as ROUTING THE ORDER, results in a ROUTE TICKET.

   Given a ROUTE TICKET (for a single product on the original PRODUCTION ORDER), we then LOAD THE REQUEST. Loading is nothing more than reserving dates and times at specific workstations. The reservations that have already been made are recorded in the WORKSTATION LOAD SHEETS. Loading requires us to look for the earliest available time slot for each task, being careful to preserve the required sequence of tasks (determined from the ROUTE TICKET).

   At the end of each day, the WORKSTATION LOAD SHEETS for each workstation are used to produce a PRODUCTION SCHEDULE. JOB TICKETS are prepared for each task at each workstation. The materials needed are determined from the BILL OF MATERIALS data store, and MATERIAL REQUESTS are generated for appropriate quantities.

4. Health Care Plus is a supplemental health insurance company that pays claims after its policyholders' primary insurance benefits through their employer or another policy have been exhausted. The following narrative partially describes its claims processing system. Draw a logical data flow diagram for the following physical narrative. State any assumptions.

   Policyholders must submit an EXPLANATION OF HEALTH CARE BENEFITS (EOHCB) along with proof that their primary health policy claim has been paid. All CLAIMS are mailed to the claims processing department.

   CLAIMS are initially sorted by the claims screening clerk. This clerk returns all requests that do not include the EOHCB. For those requests returned, a PENDING CLAIM is created, dated, and stored by date. Once each week, the clerk deletes all tickets that are more than 45 days old and sends a letter to the policyholders notifying them that their case has been closed. Requests that include the EOHCB are then sorted according to type of claim. Requests that include an EOHCB REFERENCE NUMBER are matched up with an EOHCB form, which is pulled from the OPEN CLAIMS file. At the end of each day, all these claims are forwarded to the preprocessing department.

   In the preprocessing department, clerks screen the EOHCB for missing data. They complete the form if possible. Otherwise, a copy of the claim is returned to the policyholder with a letter requesting the missing data. The original EOHCB is placed in the OPEN CLAIMS file, and a PENDING CLAIM is sent to the claims screening clerk. Completed claims are assigned a claim number, and the claim is microfilmed and filed for archival purposes.

   A different clerk checks to see if the PROOF OF PRIMARY HEALTH CARE POLICY PAYMENT was included or is on file in the PRIMARY PAYMENT file. If it is not available, the policyholder is sent a letter requesting the proof. The EOHCB is placed in a PENDING PROOF file. Claims are automatically purged if they remain in this file for more than 14 days (a letter is sent to policyholders whose claims have been purged).

   If proof is available, another clerk pulls the policyholder's policy record from the POLICY file, records policy and action codes on the EOHCB, and refiles the policy. At the end of the day, all preprocessed claims are forwarded to Information Systems.

5. Given the following narrative description, compile an event-response list and draw a context diagram. State any assumptions.

   The purpose of the GREEN ACRES REAL ESTATE SYSTEM is to assist agents as they sell houses. Sellers contact the agency, and an agent is assigned to help the seller complete a LISTING REQUEST. Information about the house and lot taken from that request is stored in a file. Personal information about the sellers is copied by the agent into a sellers file.

   When a buyer contacts the agency, he or she fills out a BUYER REQUEST. Every two weeks, the agency sends prospective buyers AREA REAL ESTATE LISTINGS and an ADDRESS CROSS REFERENCE LISTING containing actual street addresses. Periodically, the agent will find a particular house that satisfies most or all of a specific buyer's requirements, as indicated in the BUYER'S REQUIREMENTS STATEMENT distributed weekly to all agents. The agent will occasionally photocopy a picture of the house along with vital data and send the MULTIPLE LISTING STATEMENT (MLS) to the potential buyer.

   When the buyer selects a house, he or she fills out an OFFER that is forwarded through the real estate agency to the seller, who responds with either an OFFER ACCEPTANCE or a COUNTEROFFER. After an offer is accepted, a PURCHASE AGREEMENT is signed by all parties. After a PURCHASE AGREEMENT is notarized, the agency sends an APPRAISAL REQUEST to an appraiser, who appraises the value of the house and lot. The agency also notifies its finance company with a FINANCING APPLICATION.

6. Given the following narrative description, draw a context diagram and system-level DFD for the portion of the activities described. State any assumptions.

   The purpose of the OPEN ROAD INSURANCE SYSTEM is to provide automotive insurance to car owners. Initially, customers are required to fill out an INSURANCE APPLICATION. A DRIVER'S TRAFFIC RECORD REQUEST is requested from the local police department. Also, a VEHICLE TITLE AND REGISTRATION is requested from the Bureau of Motor Vehicles. PROPOSED POLICIES are sent in by various insurance companies who will underwrite those policies based on a quoted fee. The agent determines the best policy for the type and level of coverage desired and gives the customer a copy of the INSURANCE POLICY PROPOSAL AND QUOTE. If the customer accepts, he or she pays the INITIAL PREMIUM and is issued both the policy and a state-required INSURANCE COVERAGE STATEMENT (a card to be carried at all times when driving a vehicle). The customer information is now stored. Periodically, a PREMIUM NOTICE is generated, which—along with POLICY COVERAGE CHANGES—is sent to the customer, who responds by sending in a PREMIUM after which new INSURANCE COVERAGE STATEMENTS are issued.

   Both a vehicle owner and the insurance company are required to provide annual PROOF OF LIABILITY INSURANCE to the Bureau of Motor Vehicles.

7. The following case describes how the typical IRS regional center processes your tax return.*

   Initially, postal trucks bring tax returns to the regional center. The envelopes are then sorted by type of return—for example, long form versus short form and whether or not the envelope contains a payment. The sorted envelopes are sent to Receipt and Control, where they are further separated into 27 types falling into three general categories: short forms requesting refunds, long forms requesting refunds, and returns containing tax payments.

   The documents are sorted twice because of the sheer volume of the returns. It's not unusual for the IRS to receive more than 200,000 returns in one day. The first sort divides that total to make the job more manageable.

   Why so many types? Some returns are requests for extensions for filing. Others are quarterly estimated tax

*"The IRS: How Your Return Is Processed," *USA Today,* January 8, 1986, p. 7A.

payments. There are over 500 official government forms for filing tax returns!

For example, to process short forms requesting refunds, operators submit forms to a machine that scans the returns and stores the data for later processing. The data is read by the main computer. It determines the correct tax, decides whether a refund should be sent, updates taxpayers' files, and prints letters, notices, liens, etc.

The refund information is sent to the National Computing Center, which subsequently triggers the Treasury Department to issue the actual refund checks. Letters, notices, and other communications are sent to local IRS sites around the country, from which appropriate information is sent to taxpayers.

The processing of long forms requesting refunds is similar, but not identical, to the processing of the short forms because the long forms usually include multiple schedules of information, such as itemized deductions. First, returns are sorted into blocks of batches to be processed as single units. Batches are numbered to ensure that no returns are lost or excessively delayed. The batches are then forwarded to examiners. The examiners check for and correct errors and code the returns for processing.

The examiners send back to the taxpayers any returns with incomplete or uncorrectable data. Also, clerks stamp a document locator number on each return for additional tracking capability as the return moves through the system. From this point, the processing is similar to the short form. Returns are input to the computer system. Data is stored for subsequent processing. The data is read by the main computer. It determines the correct tax, decides whether a refund should be sent, updates taxpayers' files, selects returns for possible tax audits, and prints letters, notices, liens, and so on. Refund information is sent to the National Computing Center, which subsequently triggers the Treasury Department to issue the actual refund checks. Notices and information regarding audits are sent to local IRS sites around the country, from which appropriate information is sent to taxpayers.

For returns containing tax payments, examiners check for and correct errors, code the returns for processing, and send back to taxpayers any returns with incomplete or uncorrectable data. Returns are entered into the computer. The computer checks taxpayer calculations and amounts, assigns document locator numbers, and stores the data. Then, the preceding steps are repeated using different operators.

The data from the second operators is checked against the first set for accuracy. Error reports are sent to examiners. Accurate data is stored for subsequent processing. Checks are collected for daily deposit into the Federal Reserve Bank.

Examiners check for errors, correcting any errors they can, and write the taxpayers for any missing information. At this point, the returns follow identical processing as described for the long forms requesting refunds.

Draw the logical data flow diagram for the physical description.

8. Prepare a decision table that accurately reflects the following course grading policy:

   A student may receive a final course grade of A, B, C, D, or F. In deriving the student's final course grade, the instructor first determines an initial or tentative grade for the student, which is determined in the following manner:

   A student who has scored a total of no lower than 90 percent on the first three assignments and exams and received a score no lower than 70 percent on the fourth assignment will receive an initial grade of A for the course. A student who has scored a total lower than 90 percent but no lower than 80 percent on the first three assignments and exams and received a score no lower than 70 percent on the fourth assignment will receive an initial grade of B for the course. A student who has scored a total lower than 80 percent but no lower than 70 percent on the first three assignments and exams and received a score no lower than 70 percent on the fourth assignment will receive an initial grade of C for the course. A student who has scored a total lower than 70 percent but no lower than 60 percent on the first three assignments and exams and received a score no lower than 70 percent on the fourth assignment will receive an initial grade of D for the course. A student who has scored a total lower than 60 percent on the first three assignments and exams or received a score lower than 70 percent on the fourth assignment will receive an initial and final grade of F for the course. Once the instructor has determined the initial course grade for the student, the final course grade will be determined. The student's final course grade will be the same as his or her initial course grade if no more than three class periods during the semester were missed. Otherwise, the student's final course grade will be one letter grade lower than his or her initial course grade (for example, an A will become a B).

   Are there any conditions for which there was no action specified for the instructor to take? If so, what would you do to correct the problem? Can your decision table be simplified by eliminating impossible rules or consolidating rules?

9. The Poker Chip Challenge. Joe, Gordon, and Susan own Granger's Restaurant Supply. They are in dire financial straits. They need $250,000 to meet their debts and cannot get a bank loan because of their poor credit rating. Among them, they can collect only $50,000.

   They have decided on a drastic and risky solution to their problem. They will go to Atlantic City and try to gamble their $50,000 into enough money to cover their debts. There is one problem, however. They are lousy gamblers! Within one short hour, they lose the entire $50,000. As they leave the casino, they run into the president of Premier Restaurant & Supply, Inc., their fiercest competitor. He has been trying, unsuccessfully, to buy Granger's for some time. The unlucky trio offers Granger's to the greedy competitor for a bargain basement price. However, their rival, sensing an opportunity to get the business for absolutely nothing, offers the following proposition:

"I have five poker chips in my pocket—three black and two white. I propose to blindfold each of you and then give you each a chip. One by one, I will remove your blindfolds. You will be permitted to see the chip in your colleagues' hands; however, you must keep your own chip concealed in your closed palm. If any one of you can tell me the color of your own chip, then I will give you $1 million, more than enough to ensure the financial future of your business. Each of you has the option of guessing or not guessing. However, if any one of you guesses wrong, you must give me your company, free and clear: Is it a deal?"

The partners have little choice and no other reasonable hope, so they accept the challenge. The competitor then shows them the five chips—three black and two white—and chuckles as he places the blindfolds in place and gives each person one chip. He returns the two unused chips to his pocket.

The blindfold is removed from Joe, the eldest businessman and a world-class chess master. He looks at his partners' chips but, despite his logical mind, cannot determine the color of his own chip. He responds, "I just cannot give an answer: It's too risky. I'm better off giving my partners the opportunity for a better guess."

The blindfold is removed from Gordon, a graduate of a prestigious business school. After looking at the chips of his two partners, he too is unable to guess the color of his own chip. He passes the opportunity to Susan.

The competitor grins as he starts to remove the blindfold from Susan. He doesn't give her any more of a chance than he gave Joe or Gordon.

Susan interrupts confidently, "You can leave my blindfold on. How about double or nothing!" The competitor laughs aloud, "It's your funeral!"

Susan replies, "I'll take that $2 million in cash! I know from the answers of my colleagues that my chip is_______." She is correct, and the winnings save Granger's from financial ruin.

Construct a decision table that shows how Susan knew the color of her chip.

SUGGESTED READINGS

Copi, I. R. *Introduction to Logic*. New York: Macmillan, 1972. Copi provides a number of problem-solving illustrations and exercises that aid in the study of logic. The poker chip problem in our exercises was adapted from one of Copi's reasoning exercises.

DeMarco, Tom. *Structured Analysis and System Specification*. Englewood Cliffs, NJ: Prentice Hall, 1978. This is the classic book on the structured systems analysis methodology, which is built heavily around the use of data flow diagrams. The progression through (1) *current physical system DFDs*, (2) *current logical system DFDs*, (3) *target logical system DFDs*, and (4) *target physical system DFDs* is rarely practiced anymore, but the essence of Tom's pioneering work lives on in event-driven structured analysis. Tom created the data structure and logic notations used in this book.

Gildersleeve, T. R. *Successful Data Processing Systems Analysis*. Englewood Cliffs, NJ: Prentice Hall, 1978. The first edition of this book includes an entire chapter on the construction of decision tables. Gildersleeve does an excellent job of demonstrating how narrative process descriptions can be translated into condition and action entries in decision tables. Unfortunately, the chapter was deleted from the second edition.

Martin, James, and Carma McClure. *Action Diagrams: Towards Clearly Specified Programs*. Englewood Cliffs, NJ: Prentice Hall, 1986. This book describes a formal grammar of Structured English that encourages the natural progression of a process (program) from Structured English to code. Action diagrams are supported directly in some CASE tools.

Matthies, Leslie H. *The New Playscript Procedure*. Stamford, CT: Office Publications, 1977. This book provides a thorough explanation and examples of the weaknesses of the English language as a tool for specifying business procedures.

McMenamin, Stephen M., and John F. Palmer. *Essential Systems Analysis*. New York: Yourdon Press, 1984. This was the first book to suggest event partitioning as a formal strategy to improve structured analysis. The book also strengthened the distinction between logical and physical process models and the increased importance of the logical models (which they called *essential* models).

Robertson, James, and Suzanne Robertson. *Complete Systems Analysis* (Volumes 1 and 2). New York: Dorset House Publishing, 1994. This is the most up-to-date and comprehensive book on the event-driven approach to structured analysis, even though we feel it still overemphasizes the current system and physical models more than the Yourdon book described below.

Seminar notes for *Process Modeling Techniques*. Atlanta, GA: Structured Solutions, Inc., 1991. You probably can't get a copy of these notes, but we wanted to acknowledge the instructors of the *AD/Method* methodology course that stimulated our thinking and motivated our departure from classical structured analysis techniques to the event-driven structured analysis techniques taught in this chapter. By the way, Structured Solutions was acquired by Protelicess, Inc.

Wetherbe, James and Nicholas P. Vatarli. *Systems Analysis and Design: Best Practices.* 4th ed. St. Paul, MN: West Publishing, 1994. Jim Wetherbe has always been one of the strongest advocates of system concepts and system thinking as part of the discipline of systems analysis and design. Jim has shaped many minds, including our own. The authors provide a nice chapter on system concepts in this book—and the rest of the book is must reading for those of you who truly want to learn to "systems think."

Whitten, Jeffrey L., Lonnie D. Bentley, and Victor M. Barlow. *Systems Analysis and Design Methods.* 3rd ed. Burr Ridge, IL: Richard D. Irwin, 1994. This is the previous edition of this textbook. For the current edition, we somewhat reluctantly downsized coverage of data structures, Structured English, and decision tables. For more extensive coverage and examples, see any prior edition.

Yourdon, Edward. *Modern Structured Analysis.* Englewood Cliffs, NJ: Yourdon Press, 1989. This was the first mainstream book to abandon classic structured analysis' overemphasis on the current physical system models and to formalize McMenamin and Palmer's event-driven approach.

7

NETWORK MODELING

Chapter Preview and Objectives

This is the third of four graphic systems modeling chapters. In this chapter you will learn how to use a unique **network modeling** tool, **location connectivity diagrams**, to document a business system's *logical* network locations, independent of a physical computer system's network. A logical network model will be used to determine how actual data and processes will be distributed to those locations. You will know network modeling as a systems analysis tool when you can:

- Describe why network modeling may become an important skill for applications developers in the next several years.
- Define network modeling and explain why it is important.
- Describe a system in terms of locations, location types, and clusters.
- Factor a system's or application's locations into component locations using a special location decomposition diagram.
- Document the connections and essential data flows between locations using location connectivity diagrams (LCDs).
- Explain the complementary relationship among network, process, and data models.
- Synchronize data, process, interface, and network models to provide a complete and consistent logical system specification.
- Explain how network modeling is useful in different types of projects and phases.

SCENE

The project team is busy completing its data and process models. Meanwhile, Sandra, as project manager, has become concerned about the network architecture of the project. The project vision calls for a client/server network architecture that will distribute data and processing across a network that is far more sophisticated than anything SoundStage has ever implemented. Recalling that Bill Ironsides, a fellow senior systems analyst, recently returned from a Client/Server World conference, Sandra has decided to visit him to discuss his perspective on this new application architecture.

SANDRA

Hi, Bill. Do you have a few minutes?

BILL

Sure. I haven't seen much of you since the member services project started. How's it going.

SANDRA

We're on schedule and within budget. It's been interesting, so far, but that's why I'm here. *[brief pause]* Didn't you go to a client/server conference a few months ago?

BILL

Yes I did. Based on the current projects, perhaps you should have gone.

SANDRA

Well, the timing wasn't right. But you are right, I think I could have benefited. I'd like to get your take on this client/server thing. Right now, my team is busy with data flow diagrams and entity relationship diagrams. But I've got this nagging feeling that we should be doing something different—or something extra—to prepare for some of the client/server design issues that we'll surely be facing.

BILL

Well, I'm not experienced, but based on the conference I would tend to agree. Have you read the information strategy plan's network architecture?

SANDRA

Yes, but it's been a while. I don't remember too much.

BILL

Well I read it just before the conference. Let me amplify on it. The plan has some unique elements, most notably its call for application rightsizing through client/server distribution. I don't know how familiar you are with this client/server trend, but let me tell you how I see it.

To date, we've focused on decentralizing data and processes—but keeping all data and processes on a few computers, usually mainframes or minis. In the future, we will further distribute processes. The processes for a single application might be split among multiple computers of different sizes. We'll use a client/server architecture to split these processes.

SANDRA

Like a lot of us, I'm not sure what to make of this client/server thing. I've been reading a lot about this client/server phenomenon. I've seen so many different definitions, I'm not sure what to believe. Some say it's the end of the mainframe. Others say it may include the mainframe. Most recently, I've been reading that some people are already labeling client/server a failure or, at best, a fad.

BILL

Sandra, we've both been around long enough to know these "trends" go through stages. In the beginning, it gets a lot of press, mostly good. Then, publications start to over-glamorize the new technologies. Then, the industry hypes the term to describe everything from soup to nuts. After a while, the press turns bad as early users discover that the technology has problems. In reality, the technology is usually good, but we haven't yet learned how to best apply it. Then, over time, the application techniques catch up and the technology is taken for granted—we wonder why we didn't always do it that way. You've seen that in CASE, right?

SANDRA

Yes, CASE is no longer hyped, but it has proven very useful in virtually all of our most recent projects.

BILL

OK, here's my take on client/server. The technology marketplace has really oversold it. And a lot of companies have experienced client/server project failures. But let's ask ourselves why. I think that these failures are another classic example of technology being better than our ability to apply it.

SANDRA

Again, just like CASE. The CASE industry sold it as a replacement for failed methodologies, but now we know that CASE is dependent on good methodologies.

BILL

Absolutely! Now that the market hype has been redirected to "objects" and the Internet, client/server can begin to move into the comfort zone. We can learn from others' mistakes and properly analyze and design our applications for the world of distributed computing.

SANDRA

So does that mean you think the mainframe/minicomputer is dead, and that all of our future applications are destined for servers?

BILL

Not necessarily, Sandra. Client/server is cooperative processing between computers. The client is usually a personal computer. But who's to say the server can't be a mainframe, minicomputer, or supercomputer. It's a matter of determining where to best locate various pieces of the application. Obviously, the personal computers are viewed as the clients. But there could be several servers, of different sizes, and different purposes. At least, that's what I heard at the conference, and it makes sense to me.

SANDRA

So we might have separate servers for database and business logic?

BILL

That's right. Clients will send requests and data to those servers as needed and get responses as appropriate. This will all be transparent to the users of the client workstations. To the users,

SOUNDSTAGE

SOUNDSTAGE ENTERTAINMENT CLUB

each of them will operate under the perception that their own PC is doing all the work.

SANDRA

Rightsizing! The right sized computer for each function—now I see where the term came from. But how do you make intelligent decisions on assigning the right pieces to the right computer.

BILL

Now therein lies the reason for some of the early failures. Some of the best conference sessions focused on a technique called clean layering. I'm sure you can find information on clean layering in the Internet.

SANDRA

Can you give me a head start?

BILL

Not much, since I haven't actually done this for real. Essentially, clean layering suggests that the database, business logic, and user interface should be cleanly layered as separate elements.

SANDRA

That actually confuses me more.

BILL

Why's that?

SANDRA

Just consider the database layer. It doesn't make sense to me that the entire database necessarily reside on one server. It may make more sense to distribute the data across several servers or even duplicate some of the data on multiple servers.

BILL

I agree. So, on what criteria would you base such a decision.

SANDRA

It's got to come back to the business somehow? We model the business data requirements with entity relationship diagrams. We model the processing requirements with data flow diagrams. It seems that we need some type of model that helps us distribute data and processes—and, of course, interfaces, to the network locations. But users don't necessarily even know where those network nodes will be located.

BILL

But they do know their business. They know the locations where they perform business. Doesn't it make sense that we model those locations so that we can distribute or duplicate logical data and processes to those locations? Sure, we may decide later to insert some network node locations, but those decisions should be predicated on the business locations, no?

SANDRA

That's it! That's what's been bugging me! I've been thinking about this distribution problem, and I don't think our methodology has any provisions to handle distribution, rightsizing, and client/server architecture. Don't we need to extend our modeling to consider distribution issues?

BILL

That sounds like the ticket to me.

SANDRA

This discussion has been most helpful. Now I have some things to think about. But before I go, I have one last question—What's the possibility that client/server is just a fad?

BILL

I really don't think so. The personal computer is here to stay. At the very least, it makes sense to harness the PC as a client in the overall computing landscape. More importantly, these *Windows NT* and *UNIX* servers are packing some very significant processing power into every cost-effective packages. When the system management tools catch up—and I think they will—they will become formidable alternatives to the larger systems of the past. Someday, we may not be able to easily differentiate between types of computers based on physical size.

Sandra, if you're wondering if this client/server project is worthwhile, I have a simple question for you—Do you want to trade projects?

SANDRA

No way! But you knew the answer to that question before you asked!

DISCUSSION QUESTIONS

1. Think of a typical mainframe computer application. How might it be different in the world of client/server technology? How would you expect application development (e.g., programming) to differ?
2. What are some problems or issues that you expect Sandra to encounter in determining the best architecture of clients and servers?
3. How might the concepts of *logical* versus *physical* modeling come into play in network modeling? In other words, what would a logical network model depict? What would a physical network model depict?

NETWORK MODELING—NOT JUST FOR COMPUTER NETWORKS

Computer networks have become the nervous system of today's information systems. It is frequently argued that "the network *is* the computer." The phenomenal growth of the Internet signals that this trend will continue. But the computer network is a *physical* component of an information system. That physical network must be created to support the *logical* distribution of data, processes, and interfaces of an information system. In this chapter we will focus exclusively on *logical* network modeling.

Network modeling is a technique for documenting the geographic structure of a system. Synonyms include *distribution modeling* and geographic modeling.

In the context of your information system building blocks (Figure 7.1), *logical* network models are used to document an information system's GEOGRAPHY focus from the perspective of the system owners and system users (the intersection of the GEOGRAPHY column with the system owner and system user rows). The horizontal arrow suggests the need to synchronize the DATA, PROCESS, and INTERFACE building blocks with those of the network models, especially for the system user perspective.

FIGURE 7.1 *Network Modeling Relationship with the IS Building Blocks*

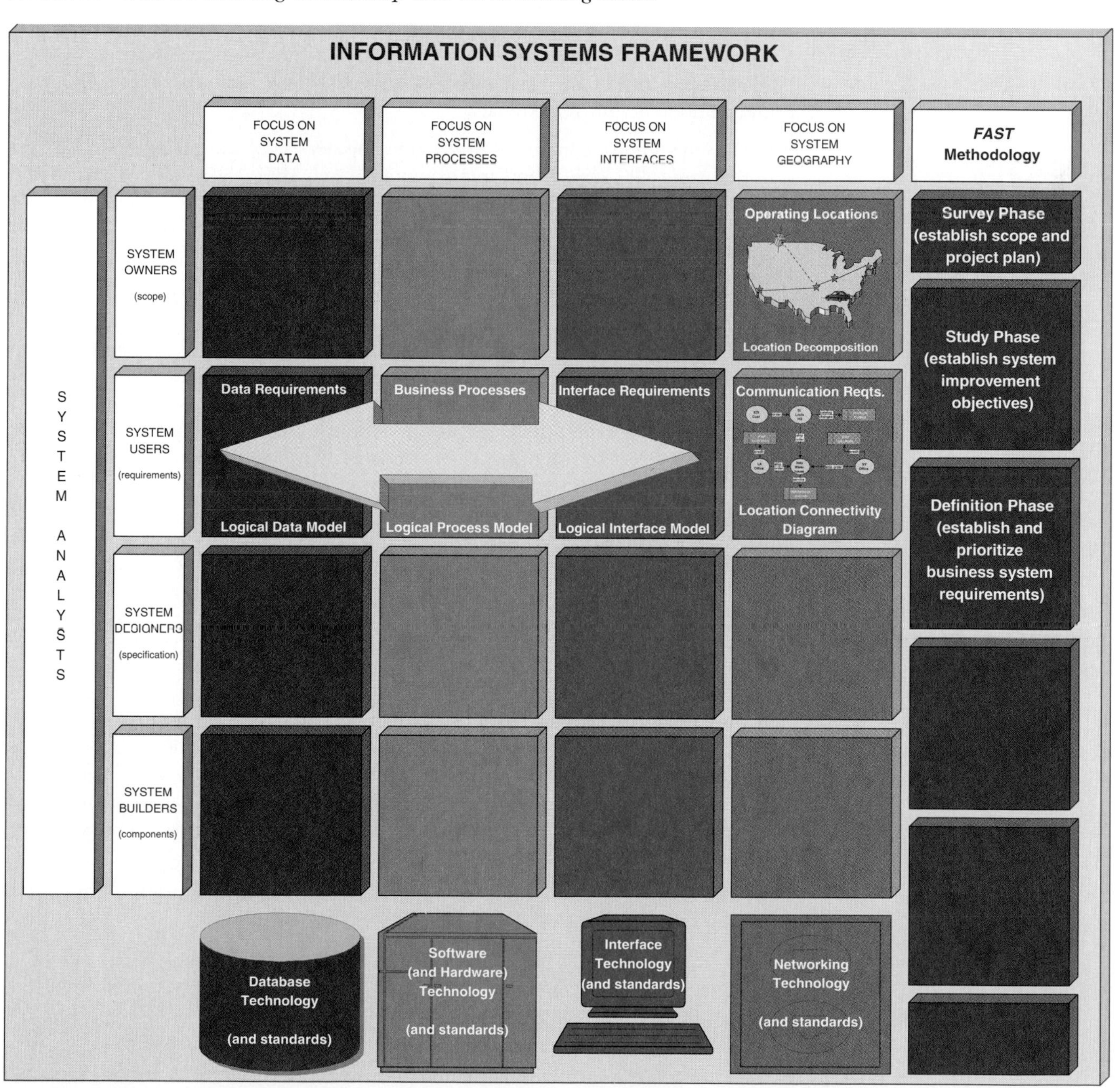

Network modeling is relatively new to the practice of systems analysis. The need for network modeling is being driven by a technical trend—distributed computing.

Distributed computing is the assignment of specific information system elements to different computers that cooperate and interoperate across a computer network. A synonym is **client/server computing;** however, client/server is actually one style of distributed computing.

The distributed computers include:

- Desktop and laptop computers, sometimes called clients. The dominant desktop operating systems are Microsoft's *Windows*, *Windows 95*, and *Windows NT Workstation* with smaller market shares for Apple's *System 7* and *8*, and IBM's *OS/2*.
- Shared network computers, called servers. This market includes Intel- and RISC-based processors that run network operating systems such as *UNIX*, *Novell*, and *Windows/NT Server*. With each passing day, these servers are encroaching on territory that was once the exclusive domain of minicomputers and mainframes.
- Legacy mainframe computers and minicomputers. No, the mainframe (e.g., IBM System 370 series) and minicomputer (e.g., IBM AS/400 series) are not dead! But they are no longer the central focus in distributed computing. Instead, they might best be thought of as a *superserver* in a distributed computing network.

These elements of a distributed computing architecture are interconnected across a computer network.

Any further discussion of the technical aspects of distributed computing would be premature and is deferred until the physical system design unit of this book. Still, the growth of these physical networks has created a need to better understand the logical business networks to be supported by the technology. This need is further amplified by such business trends as (1) globalization of the economy, (2) the vision of an information superhighway, (3) increasing numbers of corporate mergers and acquisitions, and (4) the growth of strategic partnerships with customers, suppliers, contractors, and even competitors.

Let's explore the concepts and constructs for logical network modeling.

SYSTEM CONCEPTS FOR NETWORK MODELING

We begin this section with a famous saying—"Necessity is the mother of invention." Nowhere is that old saying more applicable than in this chapter. To what "necessity" are we referring? Distributed computing technology is evolving faster than our ability to properly apply it. System designers need to make intelligent decisions about the distribution of data, processes, and interfaces when designing today's applications. But how do the system designers make those decisions? The answer is old and proven—"Develop business savvy. Talk to your management and users before you talk to the technical networking specialists!" Today's systems analyst must seek answers to new questions:

- What locations are applicable to this information system or application?
- How many users are at each location?
- Do any users travel while using (or potentially using) the system?
- Are any of our suppliers, customers, contractors, or other external agents to be considered locations for using the system?
- What are the user's data and processing requirements at each location?
- How much of a location's data must be available to other locations? What data is unique to a location?

- How might data and processes be distributed between locations?
- How might data and processes be distributed within a location?

We need a network modeling tool to document what we learn about a business system's geography and requirements.

Network modeling is a diagrammatic technique used to document the shape of a business or information system in terms of its business locations.

Unlike process modeling (with data flow diagrams) and data modeling (with entity relationship diagrams), there are no generally accepted network modeling standards. Thus, we had to invent a tool, *location connectivity diagrams* (LCDs). Why study a topic or technique that is not in the mainstream of current practice? Simple—We are trying to build network-based, distributed systems and applications today!

Business Geography

In this chapter we will focus exclusively on *logical* network modeling—that is, the modeling of business network requirements independent of their implementation. Consistent with data and process modeling, logical models will eventually be followed by physical models that describe the system design of networks and the distributed solution.

Recall from Chapter 2 that one of the fundamental building blocks of information systems is GEOGRAPHY. All information systems have geography—some more complex than others! The need to understand the geography of each information system has driven us to develop a new tool, the location connectivity diagram. This tool models system geography independent of any possible implementation.

A **location connectivity diagram (LCD)** is a logical network modeling tool that depicts the shape of a system in terms of its user, process, data, and interface locations and the necessary interconnections between those locations.

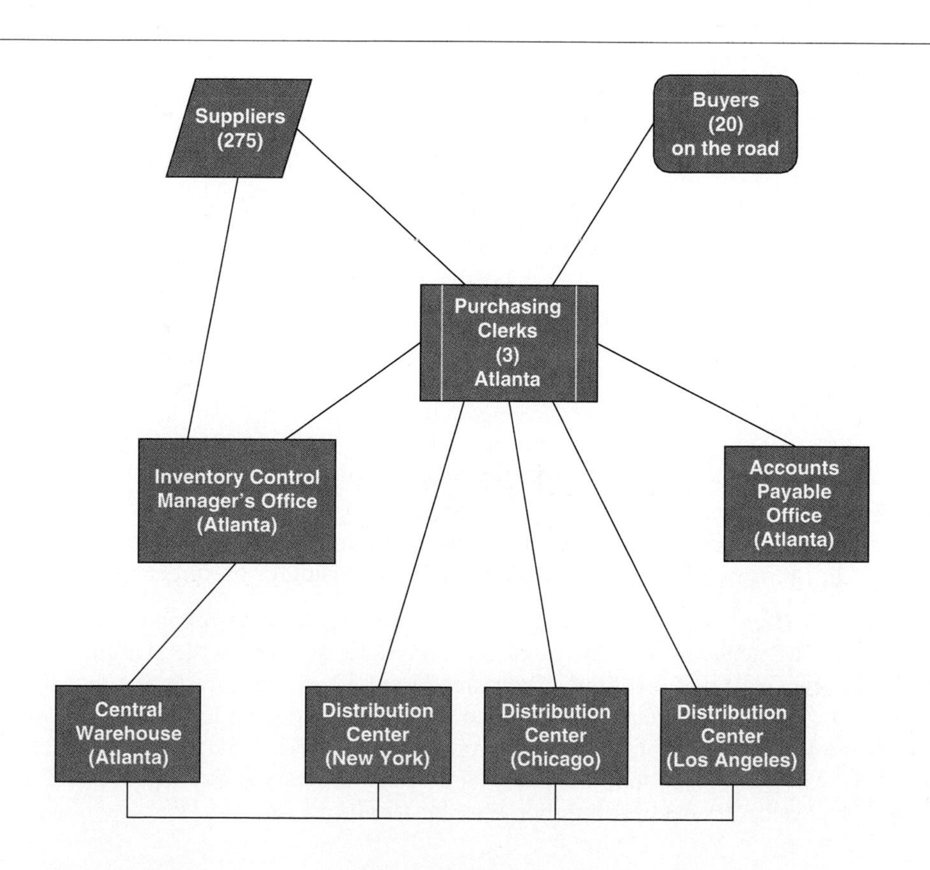

FIGURE 7.2
Logical Network Model

Figure 7.2 (on page 267) illustrates a simple and incomplete location activity diagram. The shapes indicate different types of *locations* and the connections indicate the need for business *connectivity* and *interoperability* between the locations. Basically, the diagram illustrates two concepts—locations and connectivity.

Locations The concept of geography is based on locations.

> A **location** is any place at which users exist to use or interact with the information system or application. It is also any place where business can be transacted or work performed.

There is no universal symbol or symbols for locations. The choice of symbols may be more a function of your CASE tool or drawing package as much as anything. In this edition, we will use a rectangle to depict a location (see margin).

Location

This definition can mean different things to different audiences. Business management and users will tend to identify *logical* locations where people do work or business. Information technologists will tend to discuss *physical* locations where computer and networking technology is located. Examples include:

Logical Locations
(Places where data are collected, work is performed, or information is needed)

- City
- Campus
- Building
- Office
- Work area (e.g., warehouse)
- Subsidiary
- Home office
- Customer, supplier, or contractor

Implementation Locations
(Places where computers, peripherals, and other information technology is located)

- Computer center
- Network server
- PC or terminal location
- Local area network
- Wide area network hub/gateway

Specific Location

In some cases, a logical location might become a physical location. But for the remainder of this chapter, we are concerned exclusively with *logical* locations of a business network. Logical locations can be:

- Scattered throughout the business for any given information system.
- On the move (e.g., traveling sales representatives).
- External to the enterprise for which the system is being built. For instance, customers can become users of an information system via the telephone or the Internet.

Cluster of "Like" Locations

Additionally, logical locations can represent

- Clusters of similar locations—we may elect to show a group of individuals who perform the same duties in the same location as a cluster of "like" locations.
- Organizations and agents outside of the company but that interact with or use the information system, possibly (and increasingly) as direct users

Moving or Mobile Location(s)

We need a more precise symbolic notation for locations to recognize the above variations. We will use derivatives of the rectangle, shown in the margin, to illustrate different types of locations. They are described in the paragraphs that follow.

External Location

Our standard rectangle will be used to represent a specific location. That doesn't mean the location cannot be decomposed into other locations. The West Lafayette, Indiana, campus of Purdue University is a specific location that consists of many buildings, each of which is another specific location.

The rectangle with the double, vertical lines will be used to represent a cluster of locations. Each location within the cluster is actually a simple location, but for the sake of simplicity, we represent "like" locations as a single location. For example, any building at Purdue University may house one or more academic counselors. Clearly, it would not be practical to show every counseling office at Purdue on a logical network model. It is sufficient to recognize that counseling offices perform "like" activities that can be represented as a single shape on the model.

One has to use common sense in deciding when to cluster locations. A group of locations or users should be represented as a single cluster if it is expected that they will likely share the same data and processes (to be assigned from the data and process models). For example, most order clerks share the same data and processes. A single cluster location labeled order clerks is appropriate. In some systems, you might even toss the sales manager in that same symbol. On the other hand, if the sales manager will be assigned unique data or processes, it might be best to show separate locations, even if the locations are in proximity. (This doesn't prevent the network designer from putting the clerks and manager on the same computer network, but it may suggest the need to assign different security levels to clerks and managers.)

Some locations are not stationary. Sales representatives and purchasing agents may be on the road, but they use your information system and must still be considered part of the system or application that you are modeling. Physically, they might interact via dial-up access through modems or cellular modems. Logically, we will represent their mobility with a unique shape, a rounded rectangle.

Finally, some locations represent external organizations and agents (such as customers, suppliers, taxpayers, contractors, and the like). If these external organizations or agents are to directly interact with the system you are designing, then they should be shown on the model. Through the Internet and privately held networks, many organizations are trying to link directly with their customers and with the information systems of external organizations to reduce response time and improve throughput of transactions. For example, Sears, Roebuck mandates electronic data interchange with all its suppliers to improve inventory control at a regional and store level. We'll use a parallelogram to illustrate these external locations.

Location names should describe the location and/or its users. Use proper nouns for locations, but use titles for users. Use singular and plural nouns where appropriate. Plural names are appropriate for clusters. Examples of naming conventions are provided below.

Location Names

Paris, France	Rooms 230–250	Order clerk
Indianapolis, Indiana	Warehouse	Order clerks
Grissom Hall	Shipping dock	Customers
Building 105	User names (as locations)	Suppliers
Grant Street building	Order Entry Department	Students
Room 222		

Decomposition Some locations consist of other locations and clusters. For example, a university may have many campuses and extension sites. A campus may include many buildings. A building may include many types of offices, classrooms, laboratories, and other dedicated space—and many instances of each type. It can be quite helpful to understand the relative decomposition of locations and types of location.

Decomposition should not be a new system concept to most of you.

Decomposition is the act of breaking a system into its component subsystems. Each level of *abstraction* reveals more or less detail (as desired) about the overall system or a subset of that system.

You may recall that we used decomposition in Chapter 6 to factor a system into its component processes. In this chapter, we will use decomposition to factor a system into its component locations.

In systems analysis, decomposition allows you to partition a system into logical subsets of locations for improved communication, analysis, and design. Figure 7.3 demonstrates a location decomposition called a location decomposition diagram.

A **location decomposition diagram** shows the top-down geographic decomposition of the business locations to be included in a system.

A location decomposition diagram is one view of system geography. The location connectivity diagram (next page) is the other.

Connectivity Ultimately, the purpose of network modeling is to help system designers distribute the technical data, processes, and interfaces across the computer network. To that end, the systems analyst needs to specify the technology-independent communications that must occur between business locations. The communication between business locations requires connectivity.

Connectivity defines the need for and provides the means for transporting essential data, voice, and images from one location to another.

Connections between locations represent the possibility of data flows between locations. Connectivity requirements might be expressed as follows:

FIGURE 7.3
Location Decomposition Diagram

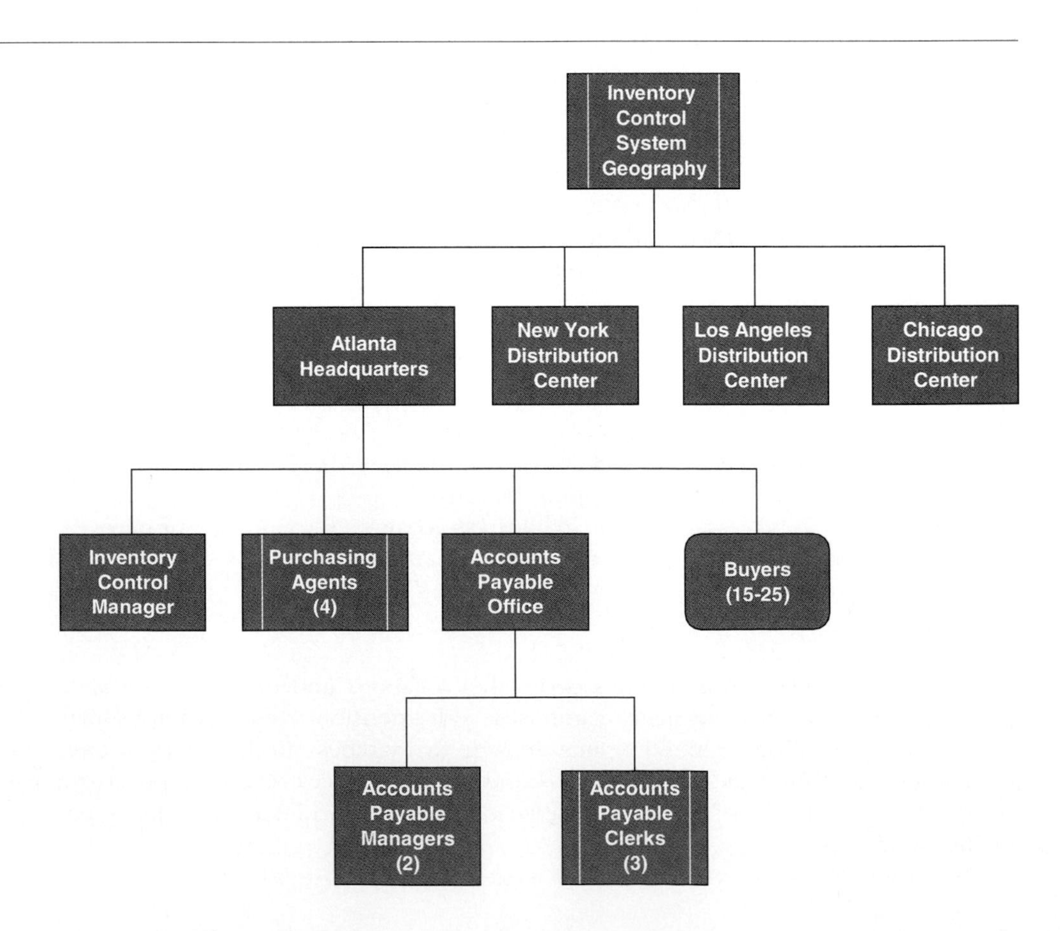

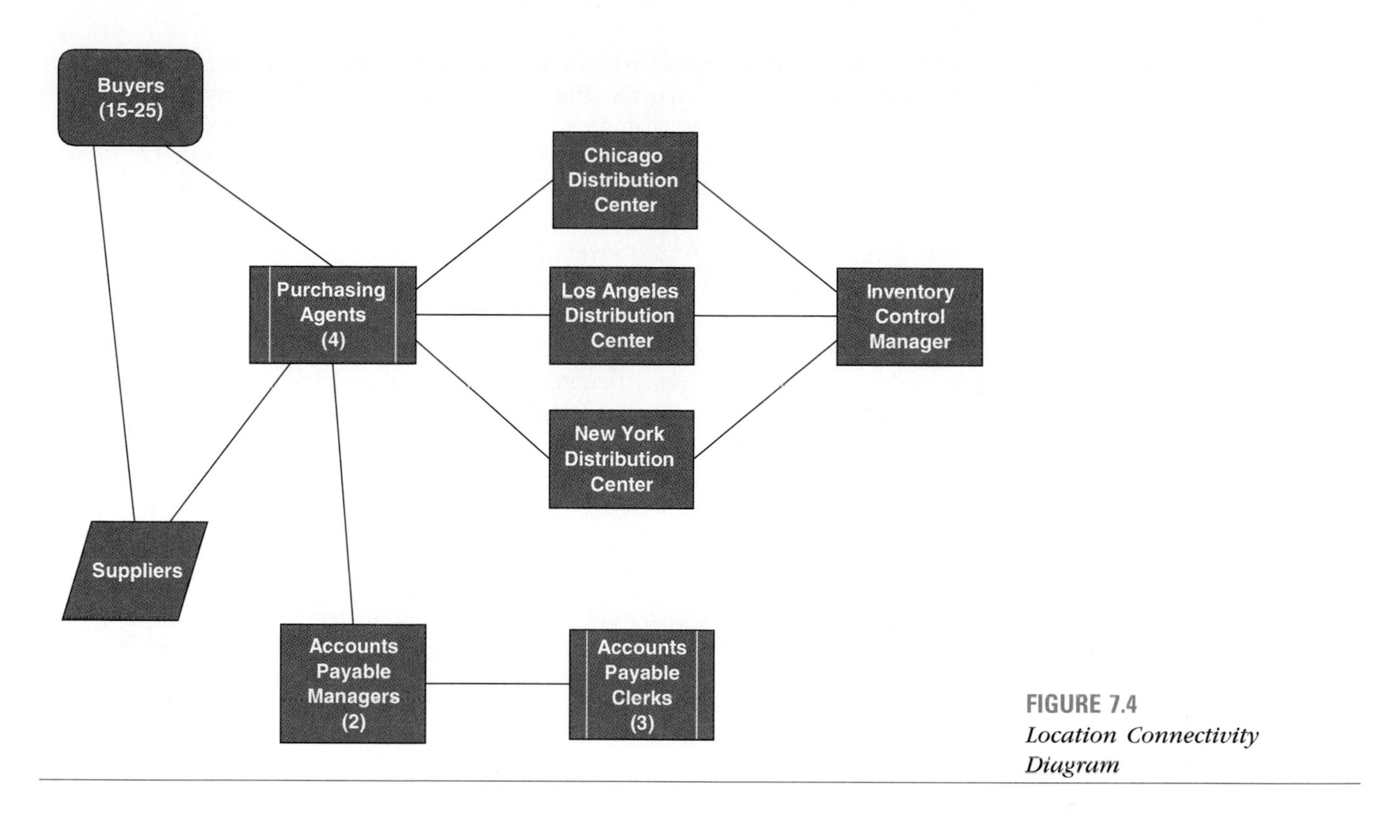

FIGURE 7.4
Location Connectivity Diagram

For this application, the NEW YORK DISTRIBUTION CENTER (a cluster) needs to communicate with or interact with the PURCHASING AGENT OFFICES.

Figure 7.4 illustrates this requirement as a location connectivity diagram (LCD). Connections are drawn without arrows because each connection is a conceptual two-way highway that may support numerous business data flows that must pass between locations. Also, until we know how data and processes will be distributed to locations (a system design decision), we can't possibly know which business data flows will travel each connection or direction.

As an LCD progresses from logical requirements to physical design, specific data flows will need to be associated with connections, and the volume of data traffic for each connection will have to be summed. This will help the network designer or manager determine network capacity requirements and technical options. Data flows and their individual volumes would normally be recorded in a project repository entry for each connection (described in Chapter 10).

Connections are not named on the LCD. It is useful, however, to label each connection by noting the distance between locations. A range of distances should be indicated for mobile locations. A range of distances is also appropriate for geographically scattered locations (e.g., CUSTOMERS or SUPPLIERS).

Do not let current or proposed computer network thinking guide your choice of connections between locations. Like logical data and process models, logical network models are supposed to stimulate your creativity when you eventually make computer networking decisions. Always ask the users if "this" location-to-location connectivity will be useful? If so, include it. If it doesn't make business sense, exclude it. Also, don't get caught up in routing. In other words, don't eliminate a possible connection just because you can get there by following a route of other connections. If a direct connection makes business sense, put the direct connection on the diagram. The network designer will determine the ultimate route later.

Clearly, the best way to identify possible connections is to discuss the business possibilities directly with system owners and users.

Miscellaneous Constructs

There are no universal standards for location connectivity diagrams; therefore, in appropriate situations it is permissible to annotate LCDs with symbols from other models, such as data flow diagrams. For example, external agent symbols might be included to represent external connections that absolutely will not be *directly* connected to your system. Suppose we want to directly place orders for stock with our largest suppliers only. We'll use mail for all other suppliers. This can be represented on our LCD with two symbols: an external location circle labeled DIRECT SUPPLIERS and a DFD external agent labeled INDIRECT SUPPLIERS. In this case both symbols should be drawn in the same general area.

Be very careful to not constrain your creativity by overusing external agents. With evolving technology, it is becoming increasingly possible to at least consider connecting any two locations. For example, the telephone has become a modern terminal and keyboard through creative use of Touch-Tone technology.

Another useful DFD symbol might be the data store. When used on an LCD it distributes or attaches specific data storage to that location. Once again, be careful not to constrain your thinking. Are there other ways to distribute the data and achieve the same, if not better, results?

Synchronizing of System Models

Network, data, interface, and process models represent different views of the same system. But these views are interrelated. Modelers need to synchronize the different views to ensure consistency and completeness of the total system specification. In this section, we'll review the basic synchronization concepts for data, process, and network models.

Data and Process Model Synchronization The linkage between data and process models is almost universally accepted by all major methodologies. In short, there should be one data store in the process models (Chapter 6, data flow diagrams) for each entity in the data model (Chapter 5, entity relationship diagrams). Some methodologies exempt associative entities from this requirement, but we believe it is simpler (and more consistent) to apply the rule to all entities on the data model.

Figure 7.5 illustrates a typical *data-to-process-CRUD matrix*. The decision to include or not include attributes is based on whether processes need to be restricted as to which attributes they can access.

The synchronization quality check is stated as follows:

> Every entity should have *at least* one C, one R, one U, and one D entry for system completeness. If not, one or more event processes were probably omitted from the process models. More importantly, users and management should validate that all possible creates, reads, updates, and deletes have been included.

The matrix provides a simple quality check that is simpler to read than either the data or process models. Of course, any errors and omissions should be recorded both on the matrix and in the corresponding data and process models to ensure proper synchronization.

Data and Network Model Synchronization A data model (Chapter 5, entity relationship diagram) describes the stored data requirements for a system as a whole. The network model (this chapter) describes the business operating locations. Intuitively, it should make sense to identify what data is at which locations. Specifically, we might ask the following business questions:

- Which subset of the entities and attributes are needed to perform the work to be performed at each location?
- What level of access is required?
- Can the location *create* instances of the entity?
- Can the location *read* instances of the entity?
- Can the location *delete* instances of the entity?

Data-to-Process-CRUD Matrix

Entity . Attribute	Process Customer Application	Process Customer Credit Application	Process Customer Change of Address	Process Internal Customer Credit Change	Process New Customer Order	Process Customer Order Cancellation	Process Customer Change to Outstanding Order	Process Internal Change to Customer Order	Process New Product Addition	Process Product Withdrawal from Market	Process Product Price Change	Process Change to Product Specification	Process Product Inventory Adjustment
Customer	C	C			R	R	R	R					
.Customer Number	C	C			R	R	R	R					
.Customer Name	C	C	U		R		R	R					
.Customer Address	C	C	U		RU		RU	RU					
.Customer Credit Rating		C		U	R		R	R					
.Customer Balance Due					RU	U	R	R					
Order					C	D	RU	RU					
.Order Number					C		R	R					
.Order Date					C		U	U					
.Order Amount					C		U	U					
Ordered Product					C	D	CRUD	CRUD		RU			
.Quantity Ordered					C		CRUD	CRUD					
.Ordered Item Unit Price					C		CRUD	CRUD					
Product					R	R	R	R	C	D	RU	RU	RU
.Product Number					R	R	R	R	C			R	
.Product Name					R		R	R	C			RU	
.Product Description					R		R	R	C			RU	
.Product Unit of Measure					R		R	R	C		RU	RU	
.Product Current Unit Price					R		R	R			U		
.Product Quantity on Hand					RU	U	RU	RU					RU

C = create R = read U = update D = delete

FIGURE 7.5 *Data-to-Process-CRUD Matrix*

- Can the location *update* existing instances of the entity?

System analysts have found it useful to define these logical requirements in the form of a *data-to-location-CRUD matrix.*

> A **data-to-location-CRUD matrix** is a table in which the rows indicate entities (and possibly attributes); the columns indicate locations; and the cells (the intersection of rows and columns) document level of access where C = create, R = read or use, U = update or modify, and D = delete or deactivate.

Figure 7.6 illustrates a typical *data-to-location-CRUD matrix*. The decision to include or not include attributes is based on whether locations need to be restricted as to which attributes they can access. Figure 7.6 also demonstrates the ability to document that a location requires access only to a subset (designated SS) of entity instances. For example, each sales office might need access only to those CUSTOMERS in their own region.

In some methodologies and CASE tools, you can define *views* of the data model for each location. A view includes only the entities and attributes to be accessible for a single location. If views are defined, they must also be kept in sync with the master data model. (In practice, most CASE tools do this automatically.)

Data-to-Location-CRUD Matrix

Entity . Attribute	Customers	Kansas City	. Marketing	. Advertising	. Warehouse	. Sales	. Accounts Receivable	Boston	. Sales	. Warehouse	San Francisco	. Sales	San Diego	. Warehouse
Customer	**INDV**					**ALL**	**ALL**		**SS**	**SS**		**SS**		**SS**
.Customer Number	R				R	CRUD	R		CRUD	R		CRUD		R
.Customer Name	RU				R	CRUD	R		CRUD	R		CRUD		R
.Customer Address	RU				R	CRUD	R		CRUD	R		CRUD		R
.Customer Credit Rating	X					R	RU		R			R		
.Customer Balance Due	R					R	RU		R			R		
Order	**INDV**		**ALL**		**SS**	**ALL**			**SS**	**SS**		**SS**		**SS**
.Order Number	SRD		R	CRUD	R	CRUD	R		CRUD	R		CRUD		R
.Order Date	SRD		R	CRUD	R	CRUD	R		CRUD	R		CRUD		R
.Order Amount	SRD		R	CRUD		CRUD	R		CRUD	R		CRUD		R
Ordered Product	**INDV**		**ALL**		**SS**	**ALL**			**SS**	**SS**		**SS**		**SS**
.Quantity Ordered	SUD		R	CRUD	R	CRUD	R		CRUD			CRUD		
.Ordered Item Unit Price	SUD		R	CRUD		CRUD	R		CRUD			CRUD		
Product	**ALL**		**ALL**	**ALL**	**ALL**	**ALL**			**ALL**	**ALL**		**ALL**		**ALL**
.Product Number	R		CRUD	R	R	R			R	R		R		R
.Product Name	R		CRUD	R	R	R			R	R		R		R
.Product Description	R		CRUD	RU	R	R			R	R		R		R
.Product Unit of Measure	R		CRUD	R	R	R			R	R		R		R
.Product Current Unit Price	R		CRUD	R		R			R	R		R		R
.Product Quantity on Hand	X				RU	R			R	RU		R		RU

INDV = individual **ALL** = ALL **SS** = subset **X** = no access

S = submit C = create R = read U = update D = delete

FIGURE 7.6 *Data-to-Location-CRUD Matrix*

Process and Interface Model Synchronization In Chapter 6, we introduced the *context diagram* as an interface model that documents how the system you are developing interfaces to business, other systems, and other organizations. Also in Chapter 6, you learned how to draw *data flow diagrams* that document the system's process response to various business and temporal events. These models should be synchronized.

Purists argue that every business event's trigger (a data or control flow) and the system response (additional data and control flows) should appear on the context diagram as well as in the data flow diagrams. Some methodologies and CASE tools strictly enforce this rule. Because most systems must respond to dozens of events, the net result of this purist approach is a very complex context diagram with large numbers of data flows to and from the single process. (Recall that a context diagram represents the entire system as a single process.)

Pragmatists suggest that the above context diagram loses its communication value. We tend to agree and suggest that the context diagram illustrate the big picture and include only the key data flows that illustrate the main purpose of the system. At the same time, we would argue that all the data flows on the context diagram should be included or represented in the subsequent data flow diagrams. In other words, you can add additional events and responses in the DFDs, but you *must* include or represent the events and responses from the context diagram into the DFDs.

Process and Network Model Synchronization Process models (Chapter 6, data flow diagrams) illustrate the essential work to be performed by the system as a whole.

Process-to-Location-Association Matrix

Process	Customers	Kansas City	. Marketing	. Advertising	. Warehouse	. Sales	. Accounts Receivable	Boston	. Sales	. Warehouse	San Francisco	. Sales	San Diego	. Warehouse
Process Customer Application	X					X			X			X		
Process Customer Credit Application	X						X							
Process Customer Change of Address	X					X			X			X		
Process Internal Customer Credit Change							X							
Process New Customer Order	X					X			X			X		
Process Customer Order Cancellation	X					X			X			X		
Process Customer Change to Outstanding Order	X					X			X			X		
Process Internal Change to Customer Order						X			X			X		
Process New Product Addition			X											
Process Product Withdrawal from Market			X											
Process Product Price Change			X											
Process Change to Product Specification			X	X										
Process Product Inventory Adjustment					X					X				X

FIGURE 7.7 *Process-to-Location-Association Matrix*

Network models (this chapter) identify the locations where work is to be performed. Some work may be unique to one location. Other work may be performed at multiple locations. Before we design the information system, we should identify and document what processes must be performed at which locations.

Synchronization of the models can be accomplished through a *process-to-location-association matrix.*

> A **process-to-location-association matrix** is a table in which the rows indicate processes (event or elementary processes); the columns indicate locations; and the cells (the intersection of rows and columns) document which processes must be performed at which locations.

Figure 7.7 illustrates a typical *process-to-location-association matrix.* Once validated for accuracy, the system designer will use this matrix to determine which processes should be implemented centrally or locally.

Some methodologies and CASE tools may support views of the process model that are appropriate to a location. If so, these views (subsets of the process models) must be kept in sync with the master process models of the system as a whole.

THE PROCESS OF LOGICAL NETWORK MODELING

Now that you understand the basic concepts of network models, we can examine the process of building a network model. When do you do it? What technology exists to support the development of process models?

There are many opportunities to perform network modeling. Some methodologies formally recommend network modeling techniques. For example, *STRADIS* by Structured Solutions includes network modeling as part of application development. (Structured Solutions calls its equivalent of an LCD a *generalized architecture schematic* or *GAS.*) There are also many approaches to performing network modeling. In this section we'll examine when network modeling might be performed during systems development.

Network Modeling during Strategic Systems Planning Projects

Many systems planning methodologies and techniques result in a network architecture to guide the design of all future computer networks and applications that use those networks. Consequently, network modeling is an appropriate technique for systems planning. This usually takes the form of a traditional map or a top-down decomposition diagram that logically groups locations. Association matrices are also typically used to provide an initial mapping of data entities to locations, and processes to locations.

Network Modeling during Systems Analysis

Application development begins with systems analysis. During the study phase of systems analysis, a project team should review any existing network models, logical or physical. It is probably not worthwhile to draw a network model for an existing system since the project vision may radically change the model.

As we move into the definition phase of systems analysis, network modeling becomes more important. If a network model already exists, it is expanded or refined to reflect new application requirements. Otherwise, a network model should be built from scratch.

Looking Ahead to Systems Design

The logical application network model from systems analysis describes *business* networking requirements, not technical solutions. As we proceed to systems design, network models must become more technical—they must become physical network models that will guide the technical distribution and duplication of the other physical system components, namely, DATA, PROCESSES, and INTERFACES. Physical network models will be introduced in Chapter 10.

Fact-Finding and Information Gathering for Network Modeling

Like all system models, network models are dependent on appropriate facts and information as supplied by the user community. These facts can be collected by a number of techniques such as sampling of existing forms and files; research of similar systems; surveys of users and management; and interviews of users and management. The fastest method of collecting facts and information and simultaneously constructing and verifying the process models is joint application development (JAD). JAD uses a carefully facilitated group meeting to collect the facts, build the models, and verify the models—usually in one or two full-day sessions.

Fact-finding and information gathering techniques are more fully explored in Part Five, Module B. JAD techniques are presented in Part Five, Module D.

Computer-Aided Systems Engineering (CASE) for Network Modeling

Like all system models, network models should be stored in the repository. Computer-aided systems engineering (CASE) technology, introduced in Chapter 3, provides the repository for storing various models and their detailed descriptions. Curiously, although CASE tool vendors have embraced distributed computing solutions, they have been slow to provide computer-assisted network modeling. There is, however, a workaround in most CASE tools.

Most CASE tools provide some sort of open-ended modeling tool to support generic presentation graphics or system flowcharting. Such tools can be adapted for use as network models. Obviously, you may have to select different shapes for the different location symbols.

All the SoundStage network models in the next section of this chapter were created with Popkin's CASE tool, *System Architect*. The case study used *System Architect*'s system flowcharting tool. That tool provided the same shapes suggested earlier in this chapter. Our printouts are shown exactly as they came off our printers. We did not add color. The only modifications by the artist were the bullets that call your attention to specific items of interest on the printouts.

Many CASE tools can automatically generate CRUD matrices of various types. Some can even automatically sync the matrices with the diagrams.

HOW TO CONSTRUCT LOGICAL NETWORK MODELS

As a systems analyst or knowledgeable end-user, you should learn how to draw network models of some type to deal with today's distributed business and computing environment. We will use the SoundStage Entertainment Club project to teach you how to draw logical network models. Let's assume we have completed the survey and study phases of the systems development life cycle. We fully understand the current system's strengths, weaknesses, limitations, problems, opportunities, and constraints. We will now model the logical business network requirements. The process, as you will soon see, is fairly straightforward by comparison to data and process models.

Location Decomposition Diagram

Chapter 6 taught you how to draw process decomposition diagrams. In this chapter we use decomposition diagrams to logically decompose and group locations.

First, brainstorm your locations. Think of all the places where direct and indirect users of your system will be located. If you've completed your DFDs already, study the external agents to identify possible external locations. Don't forget to add moving or mobile locations to the list. Be creative in considering how your network may be extended to the outside world. The Internet and telecommunications have opened the doors of opportunity to break down organizational walls and barriers.

Figure 7.8 is the location decomposition diagram for our SoundStage project. Only one symbol is used on the location decomposition diagram—the location—and it is the same location symbol used in LCDs. The locations are connected to

FIGURE 7.8 *SoundStage Location Decomposition Diagram*

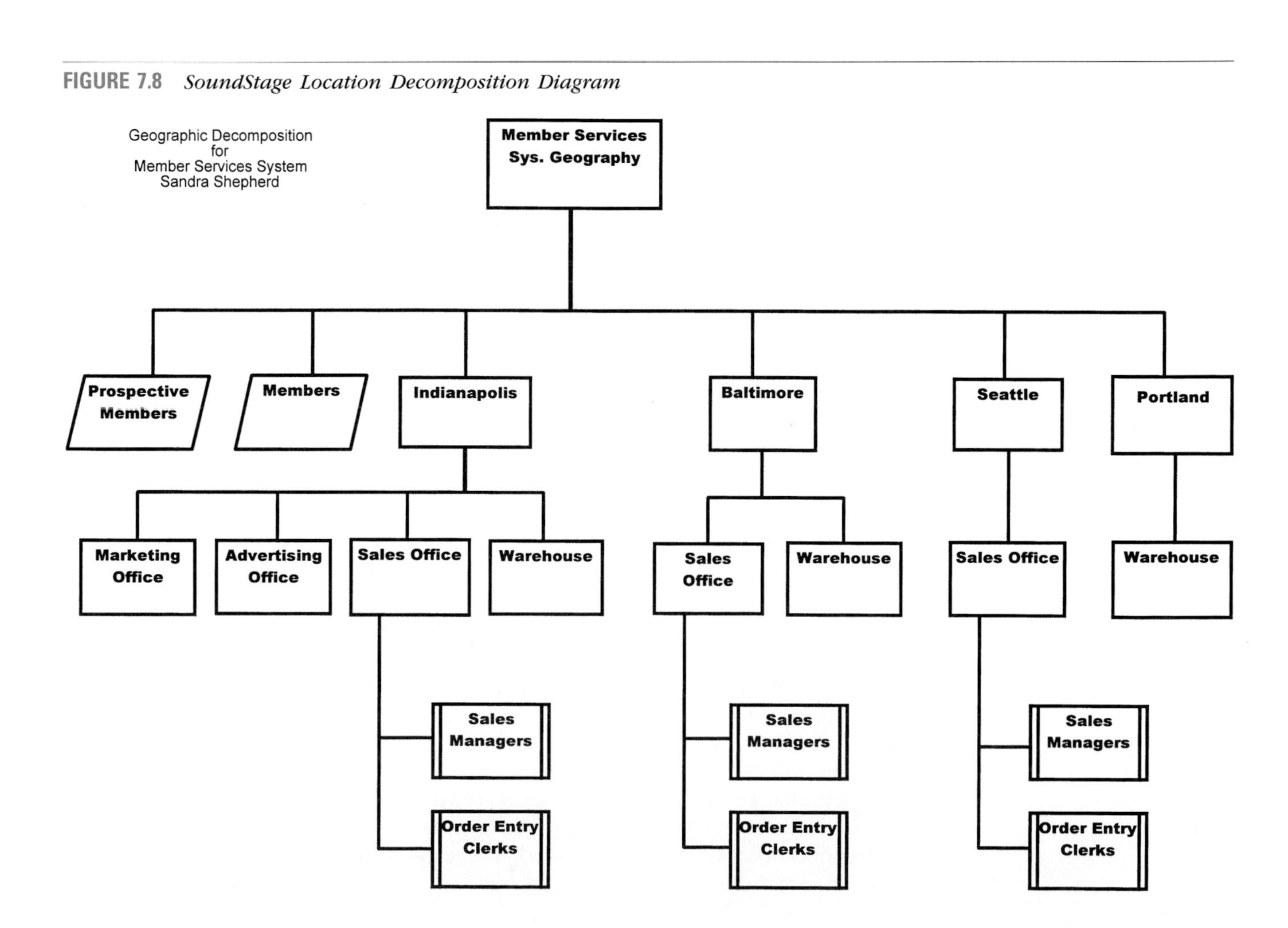

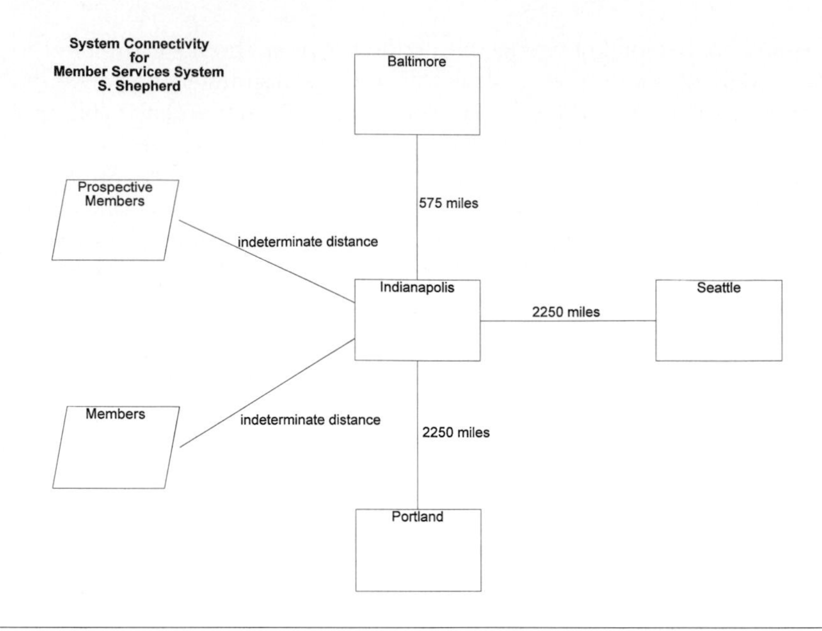

FIGURE 7.9 *High-Level SoundStage Location Connectivity Diagram*

form a top-down, treelike structure. A parent location may consist of those child locations beneath it.

To group locations in the decomposition diagram, keep similar locations on the same level or within the same branch of the tree. For example, don't combine cities with buildings or buildings with rooms. Instead, keep cities with cities, buildings with buildings, and rooms with rooms. It makes the diagrams easier to read. More importantly, it makes it possible to produce a leveled set of LCDs (much in the manner that functional decomposition diagrams made it possible to produce a sensibly partitioned set of data flow diagrams).

Clustering reduces clutter through simplification; however, there is a danger of oversimplifying the model. Once again, common sense should guide the decision to cluster. Cluster a location or its users if the data and processing requirements for all users are expected to be the same. For example, rather than show each ORDER ENTRY CLERK, it probably makes more sense to show one location labeled ORDER ENTRY CLERKS (plural). On the other hand, if these same clerks are in significantly different locations, or they are grouped into classifications that have different authorities or responsibilities, you might want to factor ORDER ENTRY CLERKS into more refined clusters.

Location Connectivity Diagram

The first location connectivity diagram we draw will be a systemwide model. It will include any external locations and locations that have sublocations. The SoundStage system model is shown in Figure 7.9.

Notice that we included external locations MEMBERS and PROSPECTIVE MEMBERS. These external locations were selected to fulfill a system goal to permit customers to directly execute transactions and inquiries, just like SoundStage's own staff. We might implement such a requirement as a Touch-Tone telephone response system or an Internet World Wide Web or Gopher page for customers with their own PCs. The implementation is not yet relevant, but the external location is.

Notice that we also included sublocation symbols for each city. These can now be exploded to reveal the sublocations and their interactions. Finally, notice that each connection's distance is recorded. The explosion diagram is shown in Figure 7.10.

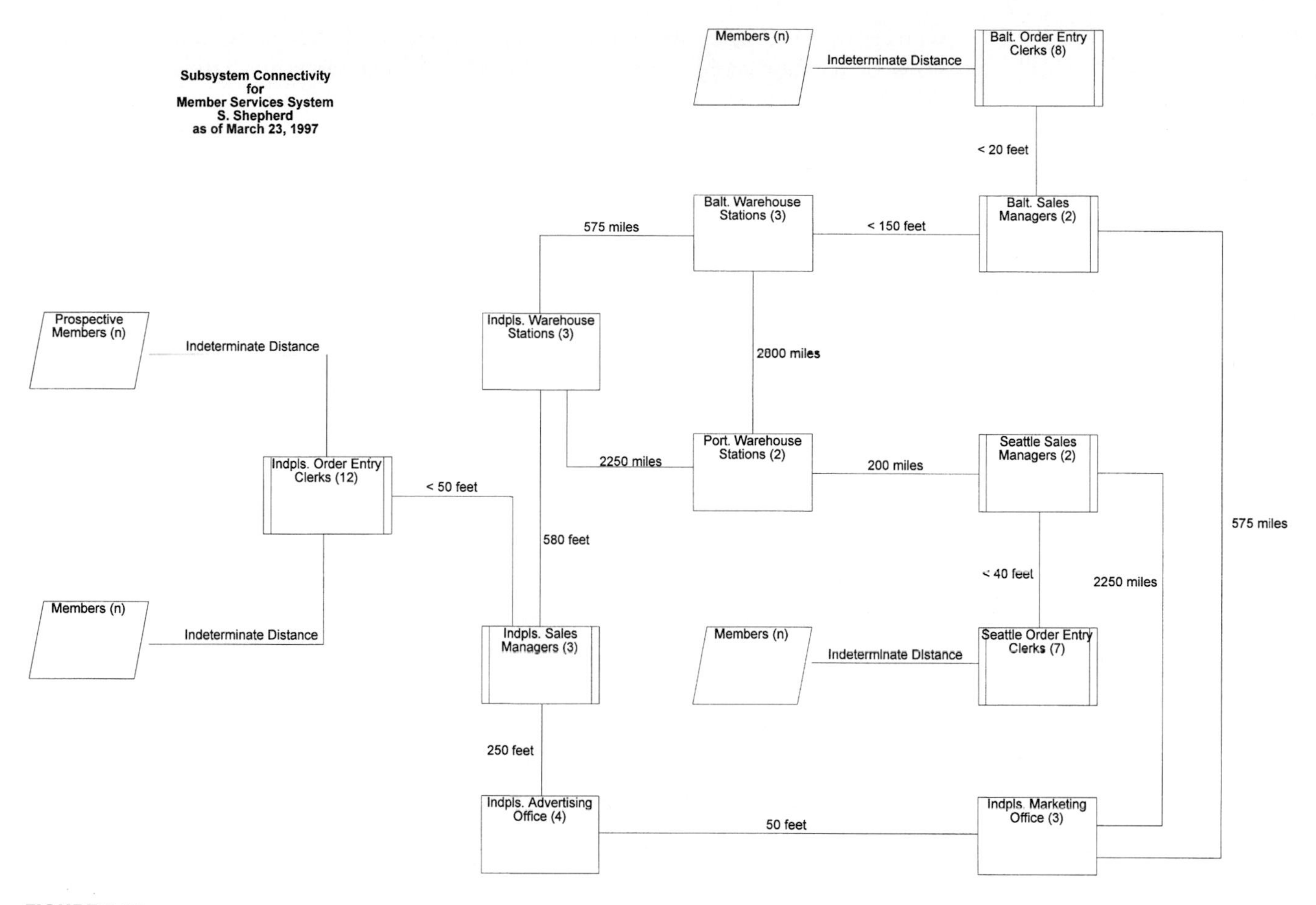

FIGURE 7.10 *Detailed SoundStage Location Connectivity Diagram*

Notice that the connections from the parent location were brought down from the system diagram. This maintains consistency between the diagrams. The new nodes correspond to the parent's child nodes on the decomposition diagram. It's all very straightforward. Once again the connections are labeled to reflect distances.

Although our example doesn't show it, this diagram could have contained additional sublocation nodes. If so, those nodes would have to be exploded to their own diagram. Once again, the parent's connections would be carried down to the more detailed level to preserve balancing.

That's all there is to essential network modeling. Later, processor nodes will be defined, and essential data and processes will be distributed to the processors. But for systems analysis, the network model is somewhat easier to build than the data and process models.

THE NEXT GENERATION

The demand for logical network modeling skills will remain strong as long as the trend toward distributed computing remains strong. We think that is highly likely. All viable methodologies will incorporate some type of logical network modeling paradigm to deal with distributed computing. A variation on data flow diagramming is the most likely candidate to emerge as a de facto standard. (Our system flowcharting approach is loosely based on data flow diagramming although we elected to incorporate unique symbols so as to not confuse data flow diagrams with network models.) As important as the model itself are the underlying descriptions of locations and connections. We also expect de facto standards for such descriptive requirements to eventually emerge.

CASE tool support will evolve after methodologies, since CASE tool engineers are reluctant to invest time and effort before some semblance of a widely accepted methodological standard exists. In the meantime, many CASE tools are providing the technology to grow your own network modeling tool by attaching custom-built descriptions and properties to shapes on a system flowchart or presentation tool.

WHERE DO YOU GO FROM HERE?

If you completed Chapters 5 to 7, then you have completed the modeling chapters associated with the so-called *structured* methods for systems analysts. Many of you will now proceed to Chapter 8, "Object Modeling." Object modeling is a new and exciting approach to system modeling, one that many believe will eventually render structured methods and models obsolete. To say the least, object models combine the best of data and process models into a single model. To say the most, much research and development is being devoted to distributed object technology, that could later render network models obsolete. Think of it! A single model that describes the business data, processing, and geographic requirements!

If you want to immediately learn how to translate logical network models into physical network models, you should skim or read Chapter 11.

Finally, network models can be created only through affective interaction with users and managers who know the business. Part Five, Modules B and D teach you how to collect the facts needed to construct real-world process models. Module B teaches fact-finding and information gathering, and Module D teaches joint application development.

SUMMARY

1. Network modeling is a technique for documenting the geographic structure of a system. While it was created in response to computer networks, it is equally applicable and important for describing business networks.
2. Distributed or client/server computing assigns specific information system building blocks to different computers in a network. This creates the distributed computing architecture that must be mated to a business network.
3. Network modeling is a diagrammatic technique used to document the shape of a business or information system in terms of its locations. This chapter emphasized logical network diagrams that show the business network, independent of any physical, computer network implementation.
4. A location connectivity diagram (LCD) is a logical network modeling tool that depicts the geography of a business network in terms of its user, data, process, and interface locations and the necessary communications lines that must exist between those locations.
5. A location is any place at which the users exist to use or interact with the information system or any place where business is transacted or work is performed. Locations can be (*a*) clustered, (*b*) mobile, or (*c*) external.
6. Locations and clusters of locations can be decomposed from abstract to specific. A location decomposition diagram shows the decomposition of a system into abstract and specific locations and clusters.
7. Connectivity defines the need for and provides the means for transporting essential data, voice, and images between locations.
8. System models must be synchronized for consistency and completeness. With respect to network models, they must be synchronized with both data and process models to determine which data and processes are essential to each geographic location. CRUD matrices and association matrices are tables that conveniently document these requirements and synchronize the data, process, and network models.

9. Logical network models are most applicable to system development projects that will or may be implemented with distributed computing technology. Logical network models are most frequently drawn as part of the definition phase of systems analysis. Later, in system design, these models will be transformed into physical network models of the distributed computing network for the system.
10. Few CASE tools specifically support logical network modeling; however, many CASE tools include system flowchart or presentation graphics capabilities that can be adapted to logical network modeling.

KEY TERMS

client/server computing, p. 266
connectivity, p. 270
data-to-location-CRUD matrix, p. 273
decomposition, p. 270
distributed computing, p. 266
distribution modeling, p. 265
location, p. 268
location connectivity diagram, p. 267
location decomposition diagram, p. 270
logical location, p. 268
network modeling, pp. 265, 267
process-to-location-association matrix, p. 275

REVIEW QUESTIONS

1. What is distributed computing?
2. Explain how distributed computing has created the need for network modeling.
3. Differentiate between logical and physical network modeling. Explain to a manager why we shouldn't restrict network modeling to computer network design only.
4. What is a location connectivity diagram? What are the two primary constructs of a location connectivity diagram.
5. Differentiate between logical and physical locations.
6. What is a cluster? Why is it convenient to use clusters in a logical network model?
7. Why are mobile locations significant?
8. What is a location decomposition diagram? What does it show that a location connectivity diagram cannot show?
9. What is synchronization and why is it important?
10. Name four types of system model synchronization?
11. Describe how network models should be synchronized with data and process models?
12. How is network modeling used in systems analysis and design?
13. Describe the process of constructing logical network models.

PROBLEMS AND EXERCISES

1. Consider a typical mainframe application that you've either used or programmed. In layperson's terms, explain how that application might be redesigned as a client/server application. (Your local course registration system might be a good example.)
2. Compare network, data, and process models. What does each model show? Should you choose between the three modeling strategies? Why or why not?
3. A manager has noted your use of logical location connectivity diagrams to document a proposed system's business network requirements, and he has expressed some concern because of the lack of nodes that depict computer processor locations. Defend your use of logical locations and connections. Concisely explain the symbolism and how to read a location connectivity diagram.
4. Draw a logical location connectivity diagram to document the locations and connections in either your school's admissions system or course registration and scheduling system.
5. Draw a location connectivity diagram for some day-to-day "system" that you use or observe in use—for instance, your morning routine; making your favorite meal, including appetizer, entrée, side dishes, and dessert; constructing something from scratch.

PROJECTS AND RESEARCH

1. Formally research the subject of network modeling, both logical and physical. Compare and contrast techniques uncovered and make recommendations for expanding or complementing the guidelines provided in this chapter.
2. Make an appointment to interview a network manager in your local area. Find out how he or she documents computer networks, then teach him or her to use logical location connectivity diagrams as an applications development tool. Together, analyze the usefulness of the LCD as a means of communicating business network requirements to the network designer. How could it be improved without making the diagram implementation-dependent?
3. Make an appointment with an application development manager in your local community. Find out to what degree the shop is doing downsizing, rightsizing, cooperative processing, and client/server computing. If they aren't moving in these directions, find out why. If they are moving in these directions, how have they adapted their systems development methodology? Explain the purpose of LCDs to the manager, and teach the manager how to use them. Together, analyze the tool's applicability to the shop. Make suggestions for improvement.
4. Make an appointment with a systems analyst in your local community. Discuss an application that he or she is working on or is familiar with. Interview the analyst to discover the locations and potential business connections. Draw the LCD and present it to the analyst. Together, analyze the tool's value for modeling network requirements. Make suggestions for improvements.
5. Make a recommendation on how to adapt your CASE tools to support location connectivity diagrams.

MINICASES

1. Minnesota State University (MSU) consists of a main campus in St. Paul, and two regional campuses in Rochester and Duluth. Additionally, the School of Applied Technology at MSU offers a *GopherTech* program that duplicates selected associate degree programs at non-MSU academic institutions in Camwell, West Grenada, and Southfork. Through *GopherTech*, MSU rents the facilities of the host institution and provides faculty and equipment for the technical courses. The host institution provides the nontechnical courses for the degree. Tuition is shared between the institutions.

   At the St. Paul campus, MSU consists of 19 separate buildings, 14 of which include classrooms, laboratories, offices, conference rooms, and lecture halls. The other three buildings are the administration building (which includes only offices, conference rooms, and storerooms), the library (which includes offices, study rooms, storerooms, and book rooms), and the physical facilities building (which includes storerooms, offices, and work space). The other two facilities are Fox Stadium (for football, track, and soccer), and the Screaming Eagles Fieldhouse (for basketball, volleyball, hockey, minor sports, and recreation—as well as offices and training facilities).

   The Rochester and Duluth regional campuses consist of three and two buildings, respectively, that consolidate small-scale implementations of the main campus. The *GoperTech* sites are single buildings (or parts of buildings) that house offices, storage, classrooms, and labs.

   MSU is also considered a pioneer in distance learning, utilizing the Minnesota *FiberNet* and the *Internet*. Through these technologies, MSU hopes to offer on-line registration and delivery of selected courses to any state citizen.

   Draw a location decomposition diagram for the described geography. Feel free to expand the case study, but state any assumptions. Which elements of your location decomposition diagram would be involved in a systemwide registration information system?
2. American Automotive Parts Association (AAPA) is a national distributor of automotive parts via their chain of independently owned franchises. AAPA is developing a completely reengineered parts sales and distribution systems to improve their ability to supply both franchises and their customers with parts.

   The supply chain for AAPA begins with their suppliers who manufacture the parts that they sell. Some are owned by AAPA, but most are not. Currently, AAPA's 67 suppliers are distributed across the United States, Canada, and Mexico; however, they also deal with non-NAFTA distributors in Germany, Great Britain, and Japan. The ambitious project goal is to link suppliers to a parts data warehouse (a separate, replicated subset of AAPA's primary parts inventory database). Suppliers would use well-defined, contracted business rules to automatically keep AAPA's warehouses stocked with appropriate levels of inventory.

   The corporate headquarters is in Indianapolis. All inventory control and accounts payable functions for all warehouses are managed here.

   AAPA operates nine regional warehouses in Kansas City, Indianapolis, Dallas, Las Vegas, Orlando, Baltimore, Boston, Minneapolis, and Los Angeles. Indianapolis manages inventory for all warehouses; however, manufacturers ship new inventory directly to each warehouse. As the current system limits inter-warehouse inventory transfers, the new system should permit any warehouse to transfer inventory to any other warehouse.

Each warehouse services 100–300 franchises. The current system requires each franchise to manage their own inventory and generate stock orders to the warehouse for new inventory. Most orders are currently generated either via phone or mail or through traveling sales representatives. The new system should additionally and voluntarily permit franchises to directly connect to the new system to order new inventory. Any given franchise could be within a 500-mile radius of the warehouse.

A new service will be provided to service stations located within a 50-mile radius of a warehouse. These service stations, with established credit, can directly order small volumes of parts for overnight delivery. These orders will be called WOGs (Wheels-Off-Ground) because they are typically associated with specific vehicle repairs in progress.

Draw a location connectivity diagram for the new system. State any assumptions you make.

SUGGESTED READINGS

Martin, James. *Information Engineering: Books I–III.* Englewood Cliffs, NJ: Prentice Hall, 1989. This classic set of books about information engineering methods provides numerous references to and examples of geographic considerations in systems analysis and design. The information engineering methodology (and associated CASE tools) also make extensive use of CRUD and association matrices for synchronization of various system models.

Transitioning STRADIS to AD/Method (class handouts and notes). Atlanta, GA: Structured Solutions, Inc., 1993. STRADIS and AD/Method are commercial information system development methodologies that support geographic modeling (although not under that name). *STRADIS* used its *generalized architecture schematic* as a network model. *AD/Method* called its network model a technology model (but it correctly included logical, nontechnical concerns). Our exposure to and experiences with these methodologies played a significant role in our professional appreciation of the need for network modeling. Both methodologies are owned by Structured Solutions. These class notes are probably unavailable to the readers of this textbook; however, they are referenced for the sake of accuracy and with the likelihood that key words may lead to other references (e.g., on the Internet).

Zachman, John. "A Framework for Information Systems Architecture." *IBM Systems Journal* 26, no. 3 (March 1987). This paper is the basis for creating network modeling as a complementary strategy with data and process modeling. You may recall that Zachman's framework is also the basis for the information system framework used throughout this book.

8

OBJECT MODELING

CHAPTER PREVIEW AND OBJECTIVES

This is the first of two chapters on object-oriented tools and techniques for systems development. This chapter focuses on object modeling during systems analysis. You will know object modeling as a systems analysis technique when you can:

- Define object modeling and explain its benefits.
- Recognize and understand the basic concepts and constructs of an object model.
- Read and interpret an object model.
- Describe object modeling in the context of systems analysis.
- Discover objects and classes and their relationships.
- Construct an object model.

Note: The authors gratefully acknowledge the contributions of Kevin C. Dittman, Assistant Professor of Computer Technology, as a partner in the writing of this chapter.

SCENE

Sandra Shepherd, Bob Martinez, and Lonnie Bentley are meeting over lunch to discuss the progress of the special object-oriented version of the Member Services project that Lonnie has been simultaneously working on.

SANDRA

Hello, gentlemen, sorry I'm late.

LONNIE

No problem. Bob was just getting me caught up on the status of your project. Sounds like everything is going well.

SANDRA

Well, I thought so. But after reviewing the system data model I'm not so sure.

BOB

Why? What's wrong? Did I mess the model up with those last minute revisions?

SANDRA

No, the revisions were necessary. It's just that now our process models are no longer in sync. All the process models were drawn based on processes to support the entities and relationships on the original model.

BOB

That's right! I forgot that we would need to make appropriate changes to data flow diagrams (process models).

SANDRA

Oh well, I'm sure that Lonnie doesn't want to hear about our problems. We will talk about what we need to do after lunch. So how goes it, Lonnie?

LONNIE

Actually, I find your discussion quite interesting. I recall that this is precisely why you folks wanted to consider object-oriented techniques to systems development—because of the difficulty of working with two different structured techniques.

BOB

How does the object-oriented approach that you are using on the Member Services system prevent you from encountering synchronization problems with data and processes?

LONNIE

It's simple. First, let's talk about the approach that the two of you are using. The *FAST* methodology that you and Sandra are using on your version of the Member Services development project uses two different models—one to model data and the other to model processes. You're bound to have troubles synchronizing the two models! To make matters worse, you may eventually make the transition over to some other type of model (e.g., structure chart) when you finally get to the design phase. In an object-oriented approach, things are made much simpler because it uses the same model to describe all aspects of the systems, data and processes together. That same model is used throughout analysis, design, and implementation.

SANDRA

Let me interject, if I may. I'm not an expert by any means on object-oriented approaches. Bob, object-oriented models are based on the concept of encapsulation, wherein the object consists of data that describe that object and all the processes that act on that data.

BOB

Why couldn't we just learn the notation for this type of model and use it instead of our separate process and data models?

LONNIE

It's not that simple. You see, there are a number of new concepts to learn, such as the one that Sandra mentioned. Some of these concepts require a totally different way of thinking about systems.

SANDRA

I know that this meeting was intended for you to give us a progress report on your object-oriented development project for the Member Services system. But I think to fully understand the status of your project, it might help if you gave us a brief overview of these new concepts, the object modeling tool, and the overall approach that you are following.

LONNIE

No problem. I understand that part of the agreement for this pilot project was to educate you in object-oriented techniques so that you could incorporate them into your *FAST* methodology. I have an article in my briefcase that provides a good overview of object-oriented concepts, let me . . .

DISCUSSION QUESTIONS

1. What are some difficulties encountered in synchronizing data and process models?
2. What are the advantages of working with a single model to communicate data and process requirements? What are the advantages of working with a single model throughout the project development phases?
3. What are the benefits of simultaneously developing the Member Services system using the *FAST* methodology's structured approach and the object-oriented approach?

AN INTRODUCTION TO OBJECT MODELING

In earlier chapters you were introduced to activities that called for drawing system models. You learned that system models play an important role in systems development by providing a means for dealing with unstructured problems. This chapter will present object modeling during systems analysis as a technique for defining business requirements for a new system. The approach of using object modeling during systems analysis and design is called object-oriented analysis.

Object-oriented analysis (OOA) techniques are used to (1) study existing objects to see if they can be reused or adapted for new uses and (2) define new or modified objects that will be combined with existing objects into a useful business computing application.

Object-oriented analysis techniques are best suited to projects that will implement systems using emerging object technologies to construct, manage, and assemble those objects into useful computer applications. The object-oriented approach is centered around a technique referred to as object modeling.

Object modeling is a technique for identifying objects within the systems environment and the relationships between those objects.

There are many underlying concepts for object modeling. In the next section you will learn about those concepts. Afterward, you will learn how to apply those concepts while developing object models during systems analysis.

SYSTEM CONCEPTS FOR OBJECT MODELING

Object-oriented analysis is based on several concepts. Some of these concepts require a new way of thinking about systems and the development process. These concepts have presented a formidable challenge to veteran developers who must relearn how they have traditionally viewed systems. But, as you will soon see, these concepts are not foreign to how you have already come to view your own environment.

Objects, Attributes, Methods, and Encapsulation

The object-oriented approach to system development is based on the concept of objects that exist within a system's environment. Objects are everywhere. Let's consider your environment. Look around. What objects are present within your environment? Perhaps you see a door, a window, or the room itself. What about this book? It's an object, and so is the page you are reading. And don't forget you are an object, too. Perhaps you also have a student workbook with you and there are other individuals in the room. You may also see a phone, a chair, and perhaps a table. All these are objects that may be clearly visible within your immediate environment.

But let's stop for a moment and consider the *Webster's Dictionary* definition of an object: "Something that is or is capable of being seen, touched, or otherwise sensed."

The objects mentioned earlier were those that one could see or touch. But what about objects that you might sense? Perhaps you are waiting for a phone call. That phone call is something that you are sensing. You may be waiting for a meeting. Once again, that meeting is something that you can identify, relate to, and anticipate even though you can't actually see that meeting. Thus, according to *Webster's Dictionary,* an anticipated phone call or meeting may be considered an object.

The previous examples pertained to objects that may exist within your immediate environment. Similarly, in the object-oriented approach to systems development, it is important to identify those objects that exist within a system's environment. But in object-oriented approaches, objects are considered as being much more than simply "something that is or is capable of being seen, touched, or otherwise sensed." In object-oriented approaches to systems development, the definition of an object is as follows:

An **object** is something that is or is capable of being seen, touched, or otherwise sensed, and about which users store data and associate behavior.

Three portions of this definition need to be examined. First, let's consider the term *something*. That something can be characterized as a type of object much like the objects that we identified within your current environment. The types of objects may include a *person, place, thing,* or *event*. An employee, customer, vendor, and student are examples of person objects. A particular warehouse, regional office, building, and room are examples of place objects. Examples of thing objects include a product, vehicle, equipment, videotape, or a window appearing on a user's display monitor. Finally, examples of event objects include an order, payment, invoice, application, registration, and reservation.

Now let's consider the "data" portion of our definition. In the object-oriented circles, this part of our definition refers to what are called attributes.

Attributes are the data that represent characteristics of interest about an object.

For example, we might be interested in the following attributes for the person object customer: CUSTOMER NUMBER, FIRST NAME, LAST NAME, HOME ADDRESS, WORK ADDRESS, TYPE OF CUSTOMER, HOME PHONE, WORK PHONE, CREDIT LIMIT, AVAILABLE CREDIT, ACCOUNT BALANCE, and ACCOUNT STATUS. In reality, there may be many customer objects for which we would be interested in these attributes. Each individual customer is referred to as an object instance. An **instance** (or *object instance*) of an object consists of the values for the attributes that describe a specific person, place, thing, or event. For example, for each customer the attributes would assume values specific to that customer—such as 123456, Lonnie, Bentley, 2626 Darwin Drive, West Lafayette, Indiana, 47906, and so forth. Let's consider your current environment. Perhaps there's another person in the room. Each of you represents an instance of a person object. Each of you can be described according to some common attributes such as LAST NAME, SOCIAL SECURITY NUMBER, PHONE NUMBER, and ADDRESS. But each of you can be described in terms of your own last name, Social Security number, phone number, and address.

Thus, object-oriented approaches to systems development are concerned with identifying attributes that are of interest regarding an object. It is important to note that with advances in technology, attributes have evolved to include more than simple data characteristics as represented in the previous example. Today, objects may include newer attribute types, such as a bitmap or a picture sound or even video.

Let's now consider the last portion of our definition for an object the "behavior" of an object.

Behavior refers to those things that the object can do and that correspond to functions that act on the object's data (or attributes). In object-oriented circles, an object's behavior is commonly referred to as a *method* or *service* (we may use the terms interchangeably throughout our discussion).

This represents a substantially different way of viewing objects! When you look at the "door" object within your environment, you may simply see a motionless object that is incapable of thinking—much less carrying out some action. In object-oriented approaches to systems development, that door can be associated with behavior that it is assumed can be performed. For example, the door can *open,* it can *shut,* it can *lock,* or it can *unlock*. All these behaviors are associated with the door and are accomplished by the door and no other object.

Let's consider the phone object. What behaviors could be associated with a phone? With advances in technology we actually have phones that are voice activated and can *answer, dial, hang up,* and carry out other behaviors that can be associated with a phone. Thus, object-oriented approaches to systems development simply require an adjustment to how we commonly perceive objects.

Another important object-oriented principle is that an object is solely responsible for carrying out any functions or behaviors that act on its own data (or attributes). For example, only YOU (an object) may CHANGE (behavior) your LAST NAME and HOME ADDRESS (attributes about you). This leads us to an important concept for understanding objects, called encapsulation.

Encapsulation is the packaging of several items together into one unit.

Applied to objects, both attributes and behavior of the object are packaged together. They are considered part of that object. The only way to access or change an object's attributes is through that object's specified behaviors.

In object-oriented development, models depicting objects are often drawn. Let's examine the modeling notation used to represent an object in these object models. Figure 8.1 (a) depicts the symbol for representing an object using the OMT modeling notation. An object is represented using a rounded rectangle. The attribute values for an object instance are optionally recorded within the symbol. Near the top of the rounded rectangle, the name of class in which the object has been categorized appears within parentheses. Let's learn about classifying objects.

Classes, Generalization, and Specialization

Another important concept of object modeling is the concept of categorizing objects into classes.

A **class** is a set of objects that share common attributes and behavior. A class is sometimes referred to as an *object class.*

Let's consider some objects within your current environment. It would be natural for you to classify the textbook and workbook as BOOKS (see Figure 8.1(b)). The textbook and workbook objects represent thing-objects that have some similar attributes and behavior. For example, similar attributes might be ISBN NUMBER, TYPE OF BOOK, TITLE, COPYRIGHT DATE, etc. Likewise, they have similar behavior, such as being able to OPEN and CLOSE. There may be several other objects within your environment that could be classified because of their similarities. For example, you and other individuals in the room might be classified as PERSONS.

We can also recognize subclasses of objects (see Figure 8.1 (c)). For example, some of the individuals in the room might be classified as STUDENTS and others as TEACHERS. Thus, STUDENT and TEACHER object classes are members of the class PERSON. When levels of classes are identified, the concept of inheritance is applied.

Inheritance means that methods and/or attributes defined in an object class can be inherited or reused by another object class.

The approach that seeks to discover and exploit the commonalities between objects/classes is referred to as generalization/specialization.

Generalization/specialization is a technique wherein the attributes and behaviors that are common to several types of object classes are grouped into their own class, called a *supertype.* The attributes and methods of the supertype object class are then inherited by those object classes.

In our example, the object class PERSON is referred to as a *supertype* (or generalization class) whereas STUDENT and TEACHER are referred to as *subtypes* (or specialization class).

A class **supertype** is an object whose instances store attributes that are common to one or more class subtypes of the object.

The class supertype will have one or more *one-to-one* relationships to object class *subtypes*. These relationships are sometimes called "is a" relationships (or "was a", or "could be a") because each instance of the supertype "is also an" instance of one or more subtypes.

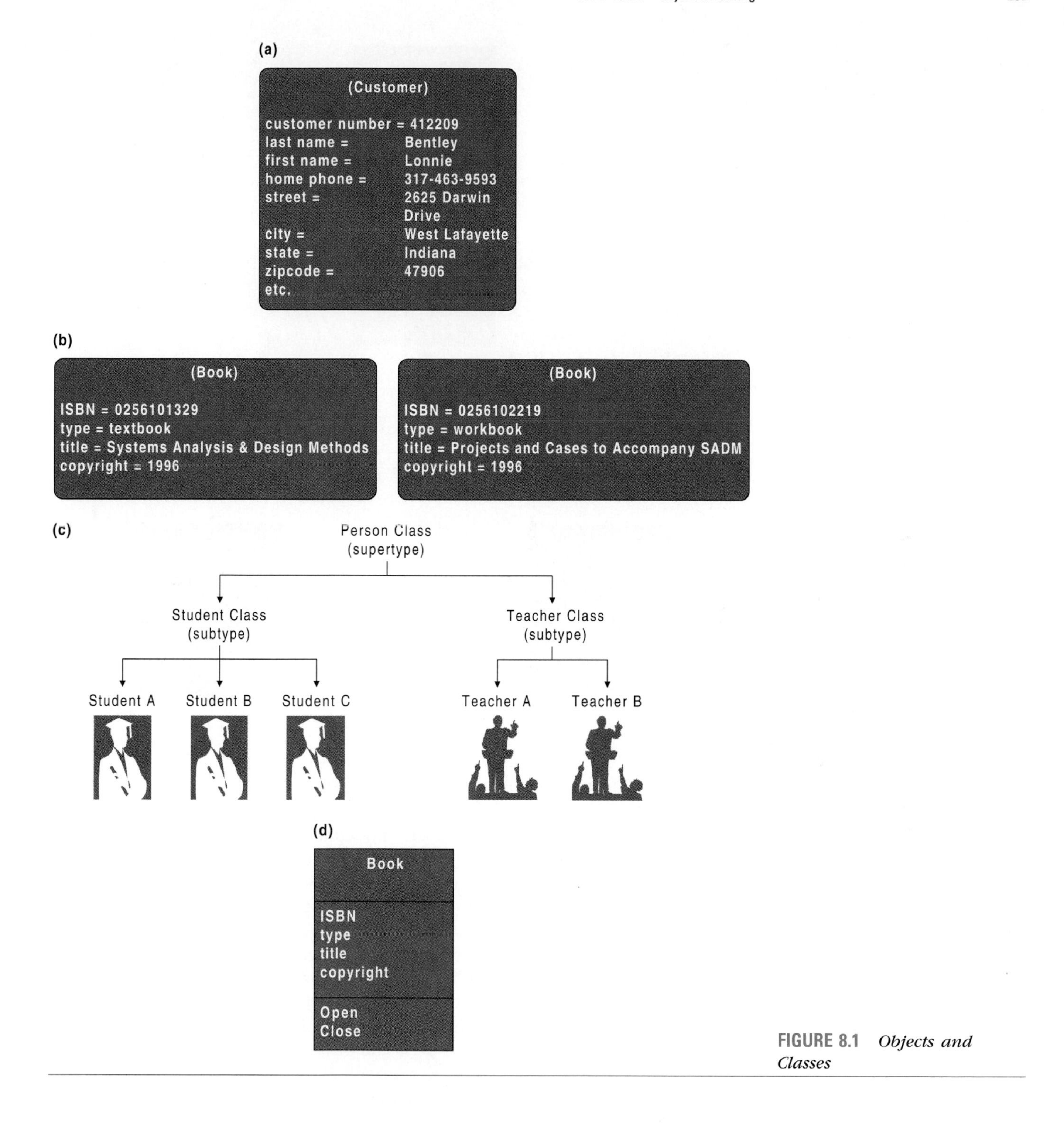

FIGURE 8.1 *Objects and Classes*

A class **subtype** is an object class whose instances inherit some common attributes from a class supertype and then add other attributes that are unique to an instance of the subtype.

In object-oriented systems development, objects are categorized according to classes and subclasses. Identifying classes realizes numerous benefits. For example, consider the fact that a new attribute of interest, called gender, needs to be added to teacher and student objects. Since the attribute is common to both, the attribute could be added once, with the class person—implying both objects within its class share that attribute. Looking down the road toward program maintenance,

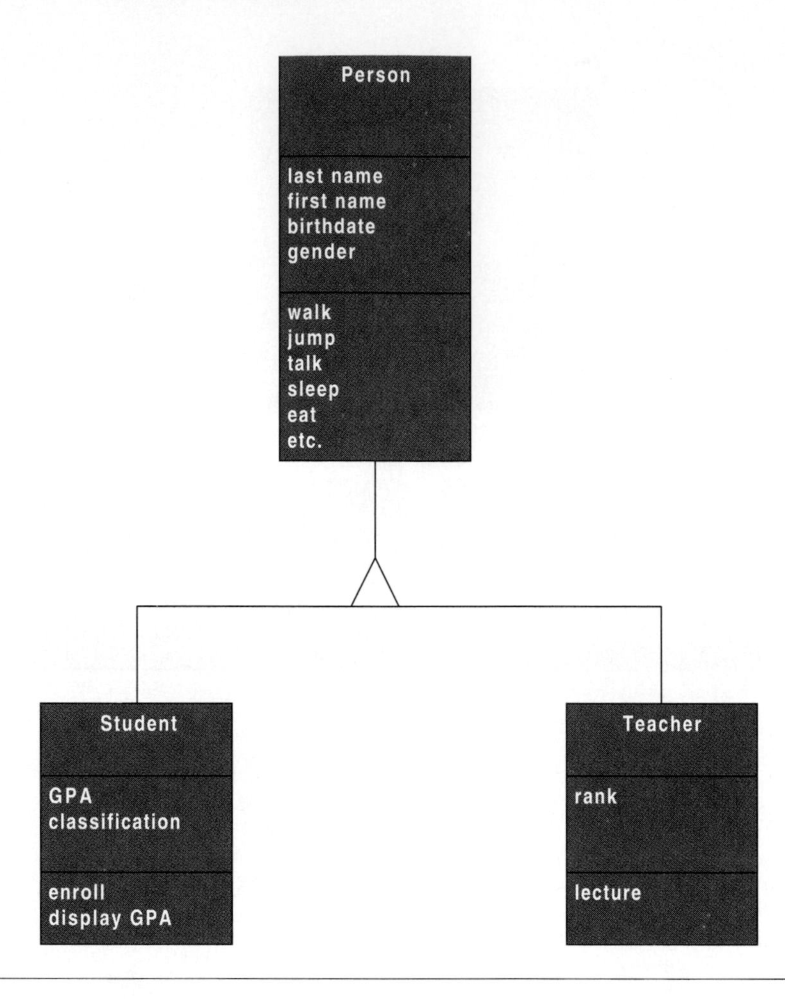

FIGURE 8.2 *Supertype and Subtype Relationships Between Object Classes*

the implication is substantial. Program maintenance is enhanced by the need to simply make modifications in one place.

How are classes represented in object modeling using the OMT approach? As depicted in Figure 8.1 (d), a class is represented using a rectangle. The rectangle is divided into three portions. The top portion contains the name of the class. The middle portion contains the name of the common attributes of interest. The lower portion contains the common behavior (or methods). To simplify the appearance of diagrams containing numerous class symbols, sometimes the classes are drawn without including the list of behaviors and attributes.

How is generalization/specialization (supertype, subtype classes) depicted using the OMT approach? Figure 8.2 illustrates how to depict the supertype, subtype relationship between the PERSON, STUDENT, and TEACHER object classes. All the attributes and behaviors of the PERSON object are inherited by the STUDENT and TEACHER objects. Those attributes and behaviors that uniquely apply to a STUDENT or TEACHER are recorded directly in the subtype class symbol.

Object/Class Relationships

Conceptually, objects and classes do not exist in isolation. The things they represent interact with and impact one another to support the business mission. Thus, we introduce the concept of an object/class relationship.

An **object/class relationship** is a natural business association that exists between one or more objects/classes.

You, for example, interact with this textbook, the phone, the door, and perhaps other individuals in the room. Similarly, objects and classes of objects within a systems environment interact. Consider, for example, the object classes CUSTOMER and

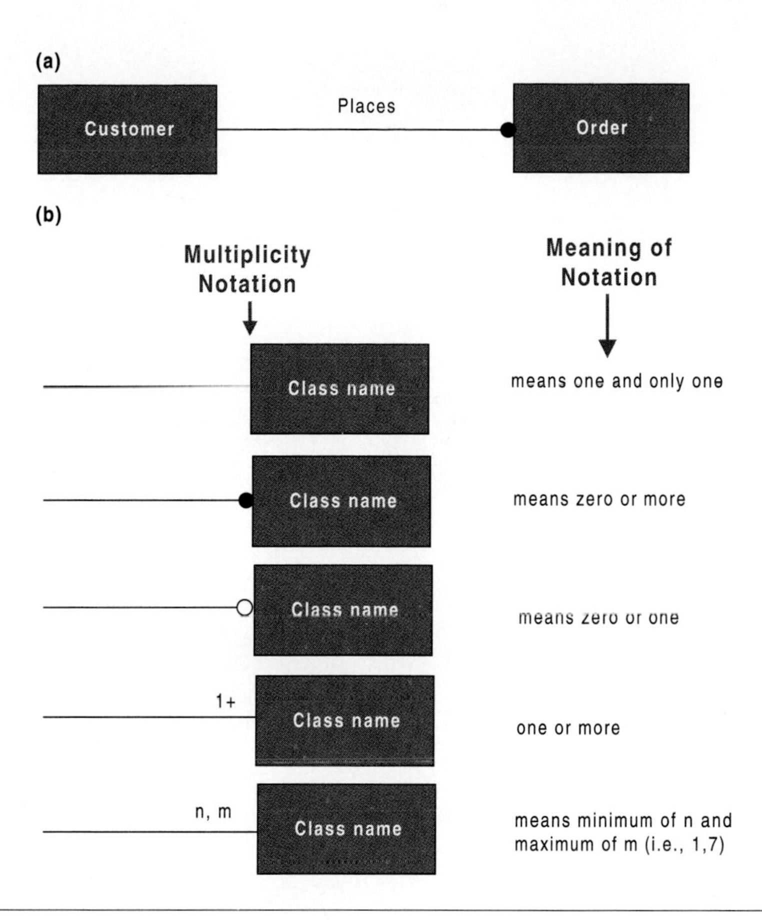

FIGURE 8.3 *Object/Class Relationships and Multiplicity Notations (behaviors omitted for clarity)*

ORDER that may exist in a typical information system. We can make the following business assertions about how customers and orders are associated (or interact):

- A CUSTOMER PLACES zero or more ORDERS.
- An order IS PLACED BY one and only one CUSTOMER.

We can graphically illustrate this association between CUSTOMER and ORDER as shown in Figure 8.3 (a). The connecting line represents a relationship between the classes. The verb phrase describes the relationship. Notice that all relationships are implicitly bidirectional, meaning they can be interpreted in both directions (as suggested by the above business assertions).

Figure 8.3 (a) also shows the complexity or degree of each relationship. For example, in the above business assertions, we must also answer the following questions:

- Must there exist an instance of CUSTOMER for each instance of ORDER? Yes!
- Must there exist an instance of ORDER for each instance of CUSTOMER? No!
- How many instances of ORDER can exist for each instance of CUSTOMER? Many!
- How many instances of CUSTOMER can exist for each instance of ORDER? One!

We call this concept multiplicity.

Multiplicity defines the minimum and maximum number of occurrences of one object/class for a single occurrence of the related object/class.

Because all relationships are bi-directional, multiplicity must be defined in both directions for every relationship. The possible OMT graphical notations for multiplicity between classes is shown in Figure 8.3 (b).

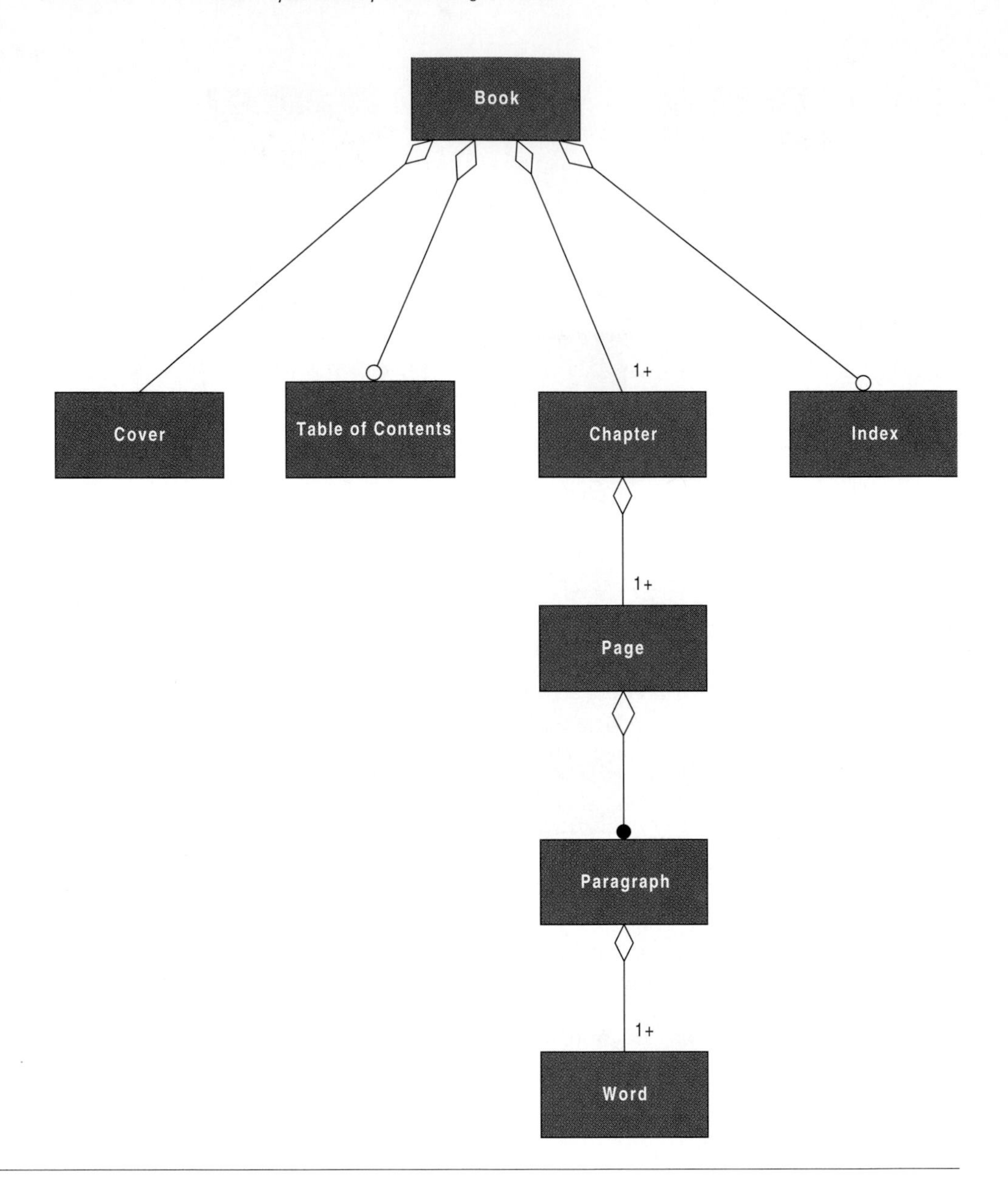

FIGURE 8.4 *Aggregate Relationships (behaviors omitted)*

Let's now consider a special kind of relationship that may exist between objects/classes. Sometimes objects/classes are made up of other objects/classes. This special type of relationship is called aggregation. It is also sometimes referred to as "whole-part" or "part-of" relationships. For example, consider this TEXTBOOK object. This TEXTBOOK contains several objects, including: COVER, TABLE OF CONTENTS, CHAPTER, and INDEX objects. Furthermore, the CHAPTER object contains PAGE objects, which in turn contain PARAGRAPH objects, which in turn contain WORD objects, and so forth.

By identifying aggregation relationships, we can partition a very complex object and assign behaviors and attributes to the individual objects within it. Figure 8.4 depicts the OMT graphical notation for specifying aggregate relationships among object classes. Notice that multiplicity is also specified for aggregate relationships. For example, a BOOK is composed of one and only one COVER, zero or one TABLE OF CONTENTS, one or more CHAPTERS, and zero or one INDEX. Figure 8.4 also depicts multilevel aggregation wherein CHAPTER consists of one or more PAGES, PAGES consists of zero or more PARAGRAPHS, and so forth.

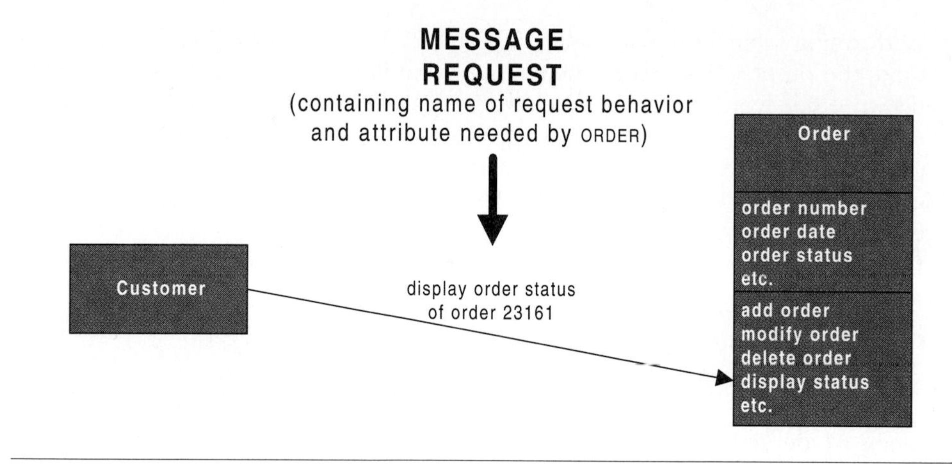

FIGURE 8.5 *Messaging*

Messages and Message Sending

We just learned that objects/classes interact. But how? Objects/classes interact or "communicate" with one another by passing messages.

> A **message** is passed when one object invokes one or more of another object's methods (behaviors) to request information or some action.

Recall the concept of encapsulation wherein an object is a package of attributes and behavior. Only that object can perform its behavior and act on its data. That is, if you want to secure the room that you are sitting in, the door object must carry out the following behaviors: *close* and *lock*. Thus, if YOU (an object) want the ROOM to become secure, YOU must send a message to the DOOR (an object) requesting it to execute the *close* and *lock* behaviors.

Let's consider our CUSTOMER and ORDER objects mentioned earlier. A CUSTOMER object checking the current status of an ORDER sends a message to an ORDER object by invoking the ORDER object's *display status* behavior (a behavior that accesses and displays the ORDER STATUS attribute).

It is important to note that the object sending a message does not need to know how the receiving object is organized internally or how the behavior is to be accomplished, only that it responds to the request in a well-defined way. This concept of messaging is illustrated in Figure 8.5. While some object-oriented approaches show messaging on their object models, the OMT chooses not to depict messaging. Rather, the OMT approach uses a variation on the data flow diagram (covered in Chapter 6) to document messaging.

Polymorphism

An important concept that is closely related to messaging is polymorphism.

> **Polymorphism** means "many forms." Applied to object-oriented techniques, it means that the same named behavior may be completed differently for different objects/classes.

Let's consider the WINDOW and DOOR objects within your environment. Both objects have a common behavior that they may perform—close. But how a DOOR object carries out that behavior may differ substantially from the way in which a WINDOW carries out that behavior. A door "swings shut" and windows "slide downward." Thus, the behavior close may take on two different forms. Once again, let's consider the WINDOW object. Not all windows would actually accomplish the close behavior in the same way. Some window objects, like door objects, "swing shut"! Thus, the close behavior may take on different forms even within a given object class.

So how is polymorphism related to message sending? Once again, the requesting object knows what service (or behavior) to request and from which object. However, the requesting object does not need to worry about how a behavior is accomplished.

THE PROCESS OF OBJECT MODELING

In performing object-oriented analysis (OOA), like any other systems analysis method, the purpose is to gain a better understanding of the system and its requirements. In other words, OOA requires that we identify the objects, their data attributes, associated behavior, and relationships that support the required business system requirements. We perform object modeling to document the identified objects, the data and behavior they encapsulate, plus their relationships with other objects.

There are two general activities when performing object-oriented analysis and they are as follows:

1. Finding and identifying the business objects.
2. Organizing the objects and identifying their relationships.

Finding and Identifying the Business Objects

In trying to identify objects, many methodology experts recommend searching the requirements specifications or other documentation and underlining the nouns that may represent potential objects. This could be a monumental task! There are just too many nouns.

One of the more popular and successful approaches for finding and identifying objects is a technique called use case modeling, developed by Dr. Ivar Jacobson.

> **Use case modeling** is the process of identifying and modeling business events, who initiated them, and how the system responds to them.

Use case modeling provides a solution to this problem by breaking down the entire scope of system functionality into many smaller statements of system functionality called use cases. This smaller format simplifies and makes more efficient the technique of underlining the nouns. One advantage of use case modeling is that it identifies and describes the system functions from the perspective of external users. This is done by identifying and documenting events called use cases, which are initiated by users or systems called actors.

> A **use case** is a behaviorally related sequence of steps (a scenario), both automated and manual, for the purpose of completing a single business task.

> An **actor** represents anything that needs to interact with the system to exchange information. An actor is a user or a role that could be an external system or person.

Actor symbol

An actor initiates system activity, a use case, to complete some business task. An actor represents a role fulfilled by a user interacting with the system and is not meant to portray a single individual or job title. Let's use the example of a college student enrolling for the fall semester. The actor would be the *student* and the business event, or use case, would be *enrolling in course*. What about events that are triggered by time called **temporal events?** Who would be the actor? In the case of temporal events, the actor is the system itself. For example, on a nightly basis a report is automatically generated listing which courses have been closed to enrollment (no open seats available) and which courses are still open. Notice that the report automatically gets generated every night; no one has to request that it be generated. This is a temporal event. The actor for this temporal event is the system itself.

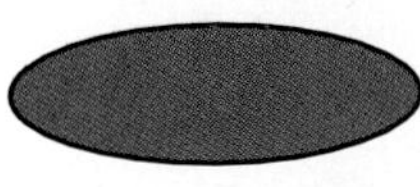

Use case symbol

Use cases are used during the entire system development process. During analysis, the use cases are used to model functionality of the proposed system and are the starting point for identifying the objects of the system. During the whole development process, use cases are continually refined in parallel with the process designing the objects. Because use cases contain an enormous amount of system functionality detail, they will be a constant resource for validating and testing development of the system design.

Use cases provide the following benefits:

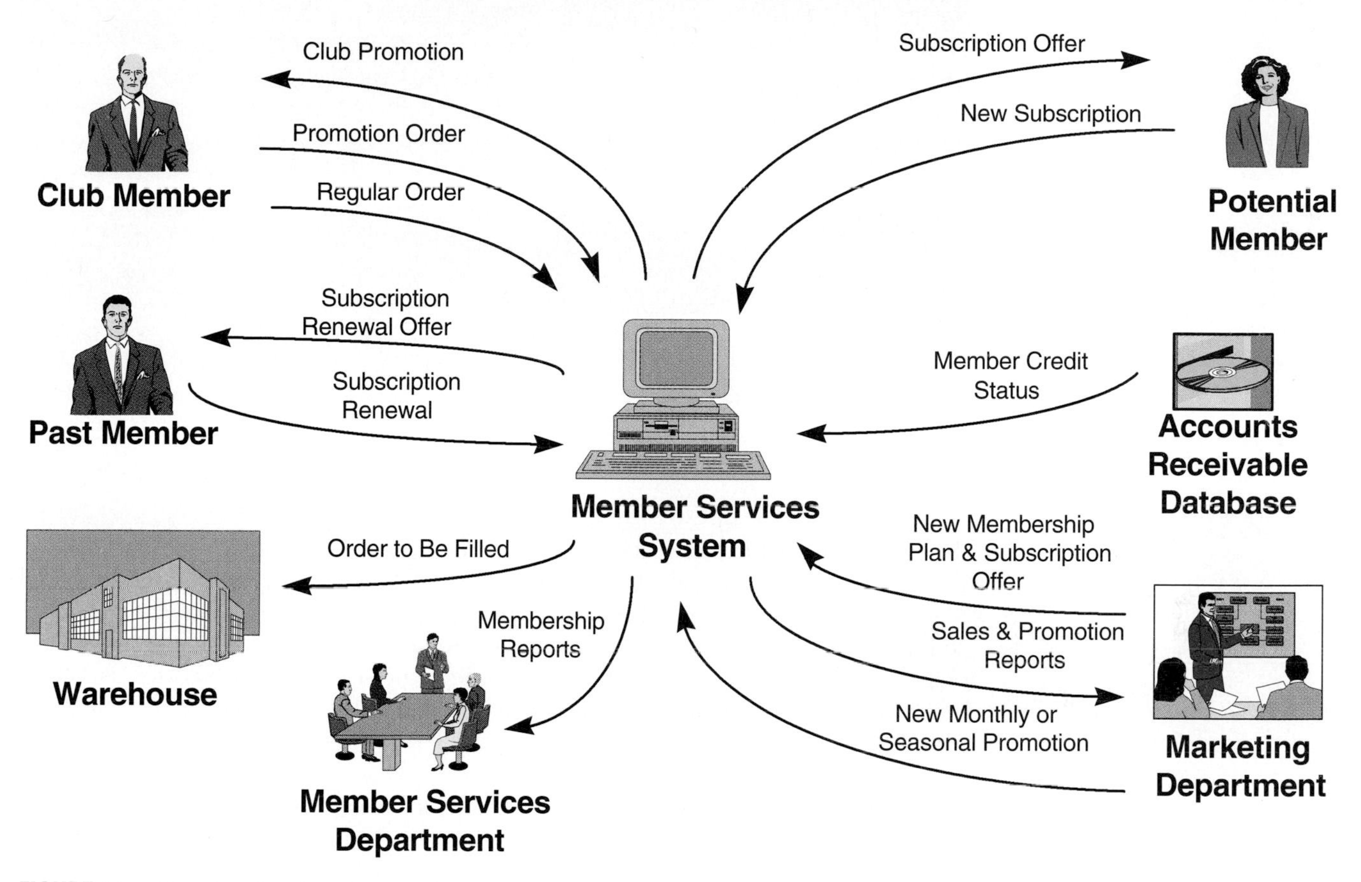

FIGURE 8.6 *Member Services System Context Model Diagram*

- As a basis to help identify objects and their high-level relationships and responsibilities.
- A view of system behavior from an external person's viewpoint.
- An effective tool for validating requirements.
- An effective communication tool.
- As a basis for a test plan.
- As a basis for a user's manual.

There are many approaches to begin use case modeling to help find and identify potential objects. They include prototyping, user and business analyst interviews, plus countless other fact-finding techniques. Usually it involves a combination of several fact-finding techniques. Let's now examine the steps involved in use case modeling to identify and find business objects for object modeling during systems analysis.

Step 1: Identifying Actors and Use Cases A good place to find potential actors and use cases is by analyzing the context model diagram of the system. Recall that a system context model illustrates the external parties that interact with the system to provide inputs and receive outputs. In doing so, it identifies the system's scope and boundaries. Figure 8.6 is a context model for the SoundStage Member Services System. Find the external parties that provide inputs to the system. Does the external party initiate the input or is the input a response to a request from the system? If the external party initiates the input, it is considered an actor. Some of

ACTOR		USE CASE
Club Member	*initiates*	Submit Promotion Order Submit Regular Order
Potential Member	*initiates*	Submit New Subscription
Past Member	*initiates*	Submit Subscription Renewal
Member Services Department	*initiates*	Request Membership Reports
Marketing Department	*initiates*	Create New Monthly Promotion Create New Seasonal Promotion Create New Subscription Program Request Promotion Reports Request Sales Reports
Member Services System	*initiates*	Send New Subscription Offer Send Club Promotion Send Subscription Renewal Offer

FIGURE 8.7 *Listing of Actors and Use Cases for Member Services System*

the inputs are self-explanatory, but others may be misleading. It is always wise to confirm your findings with the system's business analyst. In Figure 8.6, Club Member is an actor that initiates and provides an input called Promotion Order. The sequence of events in which the Member Services System would respond to the club members Submitting a Promotion Order would be considered the use case.

Figure 8.7 presents our findings of analyzing the context diagram. It lists the actors and the use cases they initiate.

Step 2: Constructing a Use Case Model Once all the use cases and actors have been identified, we have defined the total functionality of the entire system. A **use case model diagram** can be used to graphically depict the system scope and boundaries in terms of use cases and actors. The use case model diagram for the Member Services System is shown in Figure 8.8. It was created using Popkin Software's *System Architect* and represents the relationships between the actors and use cases defined for each business subsystem. The subsystems represent logical functional areas of business processes. The partitioning of system behavior into subsystems is very important in understanding the system architecture and is key to defining your development strategy—which use cases will be developed first and by whom.

Step 3: Documenting the Use Case Course of Events For each use case identified, we must now document the use case's normal course of events. A use case's normal course of events is a step-by-step description starting with the actor initiating the use case and ending with the business event. At this point we include only the major steps that happen the majority of the time (its normal course). Exception conditions or conditional branching logic will be documented later.

Figure 8.9 is a use case description for the Member Services System's SUBMIT PROMOTION ORDER use case. Notice it includes the following items:

- (A) The name of the actor who initiated the use case.
- (B) A high-level description of the use case.
- (C) A normal event course describing the use case's major steps, from beginning to end of this interaction with the actor.
- (D) Precondition describing the state the system is in before the use case is executed.
- (E) Post-condition describing the state the system is in after the use case is executed.
- (F) An assumptions section, which includes any nonbehavioral issues, such as

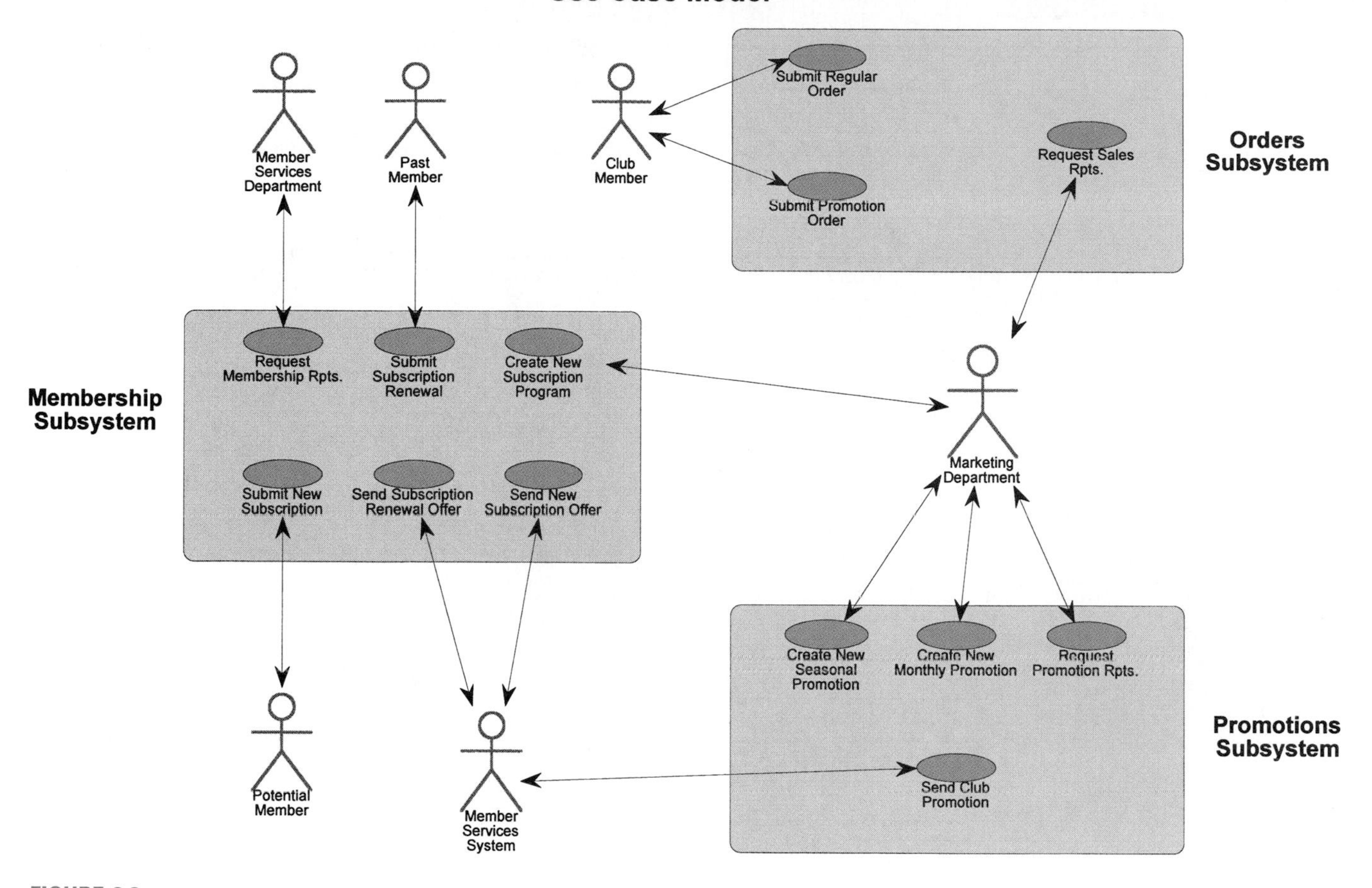

FIGURE 8.8 *Member Services System Use Case Model Diagram*

performance or security, that are associated with the use case but are difficult to model within the use case's course of events.

Step 4: Identifying Use Case Dependencies Some use cases may be dependent on other use cases, with one use case leaving the system in a state that is a precondition for another use case. For example, a precondition of sending a club promotion is that the promotion must first be created. We use a diagram called the **use case dependency diagram** to model such dependencies. The use case dependency diagram provides the following benefits:

- A graphical depiction of the system's events and their states enhances the understanding of system functionality.
- It helps to identify missing use cases. A use case with a precondition that is not satisfied by the execution of any other use case may indicate a missing use case.
- It facilitates project management by depicting which use cases are more critical (have the most dependencies) and thus need to have a higher priority.

Figure 8.10 represents the use case dependency diagram for the Member Services System. The use cases that are dependent on each other are connected with a dashed line labeled *depends on*. In Figure 8.10, the use case SEND CLUB PROMOTION has a dependency (precondition) on either the use case CREATE NEW SEASONAL PROMOTION or CREATE NEW MONTHLY PROMOTION.

USE CASE

Author: L. Bentley

USE CASE NAME:	Submit Promotion Order
ACTOR:	Club Member (A)
DESCRIPTION:	Describes the process when a club member submits a club promotion order to either indicate the products they are interested in ordering or declining to order during this promotion. (B)
NORMAL COURSE: (C)	1. This use case is initiated when the club member submits the promotion order to be processed. 2. The club member's personal information such as address is validated against what is currently recorded in member services. 3. The promotion order is verified to see if product is being ordered. 4. The club member's credit status is checked with Accounts Receivable to make sure no payments are outstanding. 5. For each product being ordered, validate the product number. 6. For each product being ordered, check the availability in inventory and record the ordered product information which includes "quantity being ordered" and give each ordered product a status of "open." 7. Create a Picking Ticket for the promotion order containing all ordered products which have a status "open." 8. Route the Picking Ticket to the Warehouse.
PRECONDITION: (D)	Use case ***Send Club Promotion*** has been processed.
POST-CONDITION: (E)	Promotion order has been recorded and the Picking Ticket has been routed to the Warehouse.
ASSUMPTIONS: (F)	

FIGURE 8.9 *Sample Use Case Description*

Step 5: Documenting the Use Case Alternate Course of Events During the previous steps of use case modeling, we focused on the normal courses of the use cases. This allowed us to concentrate on the system concept without getting bogged down in too many details. At this point the alternate courses and exception conditions in the use cases need to be defined. A use case has one normal event course that was previously defined, and possibly many alternate courses. These alternate courses are deviations, or branches, from the normal event course. Alternate courses are documented in a separate use case course. In Figure 8.11, the use case SUBMIT PROMOTION ORDER has several alternate courses. The first alternate course is for step 2 of the normal course. The club member has indicated an address or telephone change on the promotion order. Normally this does not occur, that's why it is documented as an alternate course. The analyst must refine each use case to include such alternate courses.

After you have defined the primary use cases with their normal and alternate courses, it is now time to start to identify the objects involved in the use cases. These objects represent things or entities in the business domain—things we are interested in and would like to capture information about. At this point we will concentrate on describing these objects with a sentence or two. Later we will

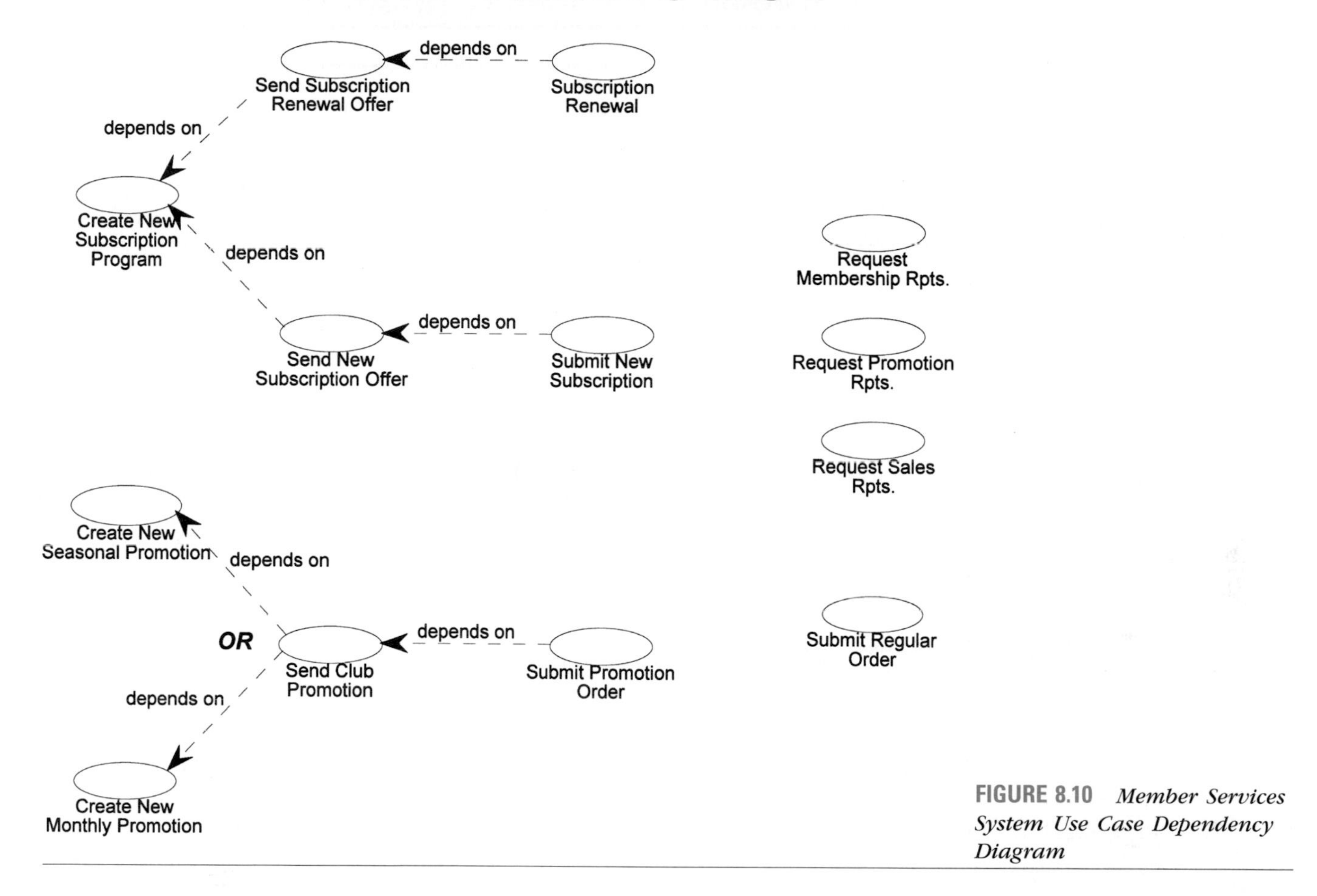

FIGURE 8.10 *Member Services System Use Case Dependency Diagram*

expand our definitions to contain more detailed facts that we learn about each object.

Step 6: Finding the Potential Objects This step is accomplished by reviewing each use case to find nouns that correspond to business entities or events. For example, Figure 8.12 depicts the use case SUBMIT PROMOTION ORDER with all the nouns highlighted. Each noun that is found in reviewing the use case is added to a list of potential objects that will be analyzed further (see Figure 8.13).

Step 7: Selecting the Proposed Objects Not all the candidates on our list represent good business objects. At this time we need to clean up our list by removing the nouns that represent:

- Synonyms.
- Nouns outside the scope of the system.
- Nouns that are roles without unique behavior or are external roles.
- Unclear nouns that need focus.
- Nouns that are really actions or attributes.

Figure 8.14 shows the process of "cleaning up" our list of candidate objects. An ✖ marks the candidates we are discarding and a ✓ marks the candidates we are keeping as objects. Also listed is the explanation of why we are keeping or discarding the candidate. Finally, Figure 8.15 presents the results of our cleaning up

USE CASE

Author: L. Bentley

USE CASE NAME:	Submit Promotion Order
ACTOR:	Club Member
DESCRIPTION:	Describes the process when a club member submits a club promotion order to either indicate the products they are interested in ordering or declining to order during this promotion.
NORMAL COURSE:	1. This use case is initiated when the club member submits the promotion order to be processed.
	2. The club member's personal information such as address is validated against what is currently recorded in member services.
	3. The promotion order is verified to see if product is being ordered.
	4. The club member's credit status is checked with Accounts Receivable to make sure no payments are outstanding.
	5. For each product being ordered, validate the product number.
	6. For each product being ordered, check the availability in inventory and record the ordered product information which includes "quantity being ordered" and give each ordered product a status of "open."
	7. Create a Picking Ticket for the promotion order containing all ordered products which have a status "open."
	8. Route the Picking Ticket to the Warehouse.
ALTERNATE COURSE:	2. If the club member has indicated an address or telephone number change on the promotion order, update the club member's record with the new information.
	3. If the club member is not ordering product at this time, modify the promotion order's status to be "closed" and modify the selection of the month ordered product's record to have a status of "rejected," then cancel the transaction.
	4. If Accounts Receivable returns a credit status that the customer is in arrears, invoke abstract use case ***Send Order Rejection Notice***. Modify the promotion order's status to be "on hold pending payment."
	5a. If the product number is not valid, create an Order Error Report containing the club member's information, the promotion order information, and the product number in error. Each completed report will be routed to a Member Services clerk for resolution.
	5b. If the club member is not ordering the selection of the month, modify the ordered product's record to have a status of "rejected."
	6. If the product being ordered is not available, record the ordered product information which includes "quantity being ordered" and give a status of "backordered."
	7. If there are no ordered product records with a status of "open," cancel the transaction.
PRECONDITION:	Use case ***Send Club Promotion*** has been processed.
POST-CONDITION:	Promotion order has been recorded and the Picking Ticket has been routed to the Warehouse.
ASSUMPTIONS:	

FIGURE 8.11 *Sample Use Case Description with Alternate Courses*

USE CASE

Author: L. Bentley

USE CASE NAME:	Submit **Promotion Order**
ACTOR:	**Club Member**
DESCRIPTION:	Describes the process when a club member submits a promotion order to either indicate the **products** they are interested in ordering or declining to order during this **promotion.**
NORMAL COURSE:	1. This use case is initiated when the club member submits the promotion order to be processed.
	2. The club member's **personal information** such as **address** is validated against what is currently recorded in **member services.**
	3. The promotion order is verified to see if product is being ordered.
	4. The club member's **credit status** is checked with **Accounts Receivable** to make sure no **payments** are outstanding.
	5. For each product being ordered, validate the **product number.**
	6. For each product being ordered, check the **availability** in **inventory** and record the **ordered product information** which includes **"quantity being ordered,"** and give each **ordered product a status** of "open."
	7. Create a **Picking Ticket** for the promotion order containing all ordered products which have a status "open."
	8. Route the Picking Ticket to the **Warehouse.**
ALTERNATE COURSE:	2. If the club member has indicated an address or **telephone number** change on the promotion order, update the **club member's record** with the new information.
	3. If the club member is not ordering product at this time, modify the promotion order's status to be "closed" and modify the **selection of the month** ordered product's record to have a status of "rejected," then cancel the **transaction.**
	4. If **Accounts Receivable** returns a credit status that the customer is in arrears, invoke abstract use case ***Send Order Rejection Notice.*** Modify the **promotion order's status** to be "on hold pending payment."
	5a. If the product number is not valid, create an **Order Error Report** containing the club member's information, the promotion order information, and the product number in error. Each completed report will be routed to a **Member Services clerk** for resolution.
	5b. If the club member is not ordering the selection of the month, modify the ordered product's record to have a status of "rejected."
	6. If the product being ordered is not available, record the ordered product information which includes "quantity being ordered" and give a status of "backordered."
	7. If there are no ordered product records with a status of "open," cancel the transaction.
PRECONDITION:	Use case ***Send Club Promotion*** has been processed.
POST-CONDITION:	Promotion order has been recorded and the Picking Ticket has been routed to the Warehouse.
ASSUMPTIONS:	

FIGURE 8.12 *Sample Use Case Description with Nouns Highlighted*

POTENTIAL OBJECT LIST
Club Member
Potential Member
Past Member
Member Services Department
Marketing Department
Member Services System
Member Address
Promotion Order
Product
Product Inventory
Order Quantity
Ordered Product
Credit Status
Payments
Ordered Product Status
Picking Ticket
Warehouse
Member Telephone Number
Selection Of Month
Transaction
Accounts Receivable
Promotion Order Status
Order Error Report
Member Services Clerk

FIGURE 8.13 *Member Services System Potential Object List*

POTENTIAL OBJECT LIST		REASON
Club Member	✓	Type of "MEMBER"
Potential Member	✓	Type of "MEMBER"
Past Member	✓	Type of "MEMBER"
Member Services Department	✖	Not relevant for current project
Marketing Department	✖	Not relevant for current project
Member Services System	✖	Not relevant for current project
Member Address	✖	Attribute of "MEMBER"
Promotion Order	✓	Result of an event named "PROMOTION"
	✓	Type of "MEMBER ORDER"
Product	✓	"PRODUCT"
Product Inventory	✖	Attribute of "PRODUCT"
Order Quantity	✖	Attribute of "MEMBER ORDER"
Ordered Product	✓	"PRODUCT ON ORDER"
Credit Status	✖	Attribute of "MEMBER"
Payments	✖	Out of Scope
Ordered Product Status	✖	Attribute of "PRODUCT ON ORDER"
Picking Ticket	✖	Potential interface item
Warehouse	✖	Not relevant for current project
Member Telephone Number	✖	Attribute of "MEMBER"
Selection Of Month	✓	Type of "TITLE"
Transaction	✖	Not relevant for current project
Accounts Receivable	✖	Not relevant for current project
Promotion Order Status	✖	Attribute of "MEMBER ORDER"
Order Error Report	✖	Potential interface item
Member Services Clerk	✖	Not relevant for current project

FIGURE 8.14 *Analyzing the Potential Object List*

process, plus we have included other objects we have found on the other use cases.

Organizing the Objects and Identifying Their Relationships

Once we have identified the business objects of the system, it is time to organize those objects and document any major conceptual relationships between the objects. An **object association model** is used to graphically depict the objects and their relationships. On this diagram we will also include multiplicity, generalization/specialization relationships, and aggregation relationships.

When constructing the diagram, we will use James Rumbaugh's object modeling technique (OMT), object model notation. Our examples are created with Popkin Software's *System Architect* CASE tool.

Step 1: Identifying Associations and Multiplicity In this step we need to identify relationships or associations that exist between objects/classes. Recall that a relationship between two objects/classes is what one object/class "needs to know" about the other. This allows for one object/class to cross-reference another object. Once the relationship has been identified, the multiplicity that governs the relationship must be defined.

It is very important that the analyst not just identify relationships that are obvious or recognized by the users. One way to help ensure that possible relationships are identified is to use an object/class matrix. This matrix lists the object/class as column headings as well as row headings. The matrix can then be used as a checklist to ensure that each object/class appearing on a row is checked against *each* object/class appearing in a column for possible relationships. The name of the relationship and the multiplicity can be recorded directly in the intersection cell of the matrix.

Step 2: Identifying Generalization/Specialization Relationships Once we have identified the basic associations and their multiplicity, we must determine if any **generalization/specialization** relationships exist. Recall that generalization/specialization relationships, also known as classification hierarchies or "is a" relationships, consist of superobjects and subobjects. The superobject is general in that it contains the common attributes and behaviors of the hierarchy. The subobjects are specialized in that they contain attributes and behaviors unique to that object, but they inherit the superobject's attributes and behaviors also.

Generalization/specialization relationships may be discovered by looking at the object model association diagram. Do any associations exist between two objects that have a one-to-one multiplicity? If so, can you say the sentence "object X *is a* object Y" and it is true? If it is true, you may have a generalization/specialization relationship. Also look for objects that have common attributes and behaviors. It may be possible to combine the common attributes and behaviors into a new superobject. Generalization/specialization relationships allow us to take advantage of inheritance, which facilitates the reuse of objects and programming code.

PROPOSED OBJECT LIST
MEMBER
MEMBER ORDER
PRODUCT
PRODUCT ON ORDER
TITLE
PROMOTION
-PLUS-
MERCHANDISE
AUDIO TITLE
VIDEO TITLE
GAME TITLE
CLUB
CLUB MEMBERSHIP
AGREEMENT

FIGURE 8.15 *Member Services System Proposed Object List*

Step 3: Identifying Aggregation Relationships In this step we must determine if any aggregation or composition relationships exist. Recall that aggregation is a unique type of association in which one object "is part of" another object. It is often referred to as a **whole/part relationship** and can be read as object A *contains* object B and object B *is part of* object A. Aggregation relationships are asymmetric, in that object B is part of object A, but object A is not part of object B. Aggregation relationships do not imply inheritance, in that object B does not inherit behavior or attributes from object A. Aggregation relationships propagate behavior in that behavior applied to the whole is automatically applied to the parts. For example, if I want to send object A to a customer, object B would be sent also.

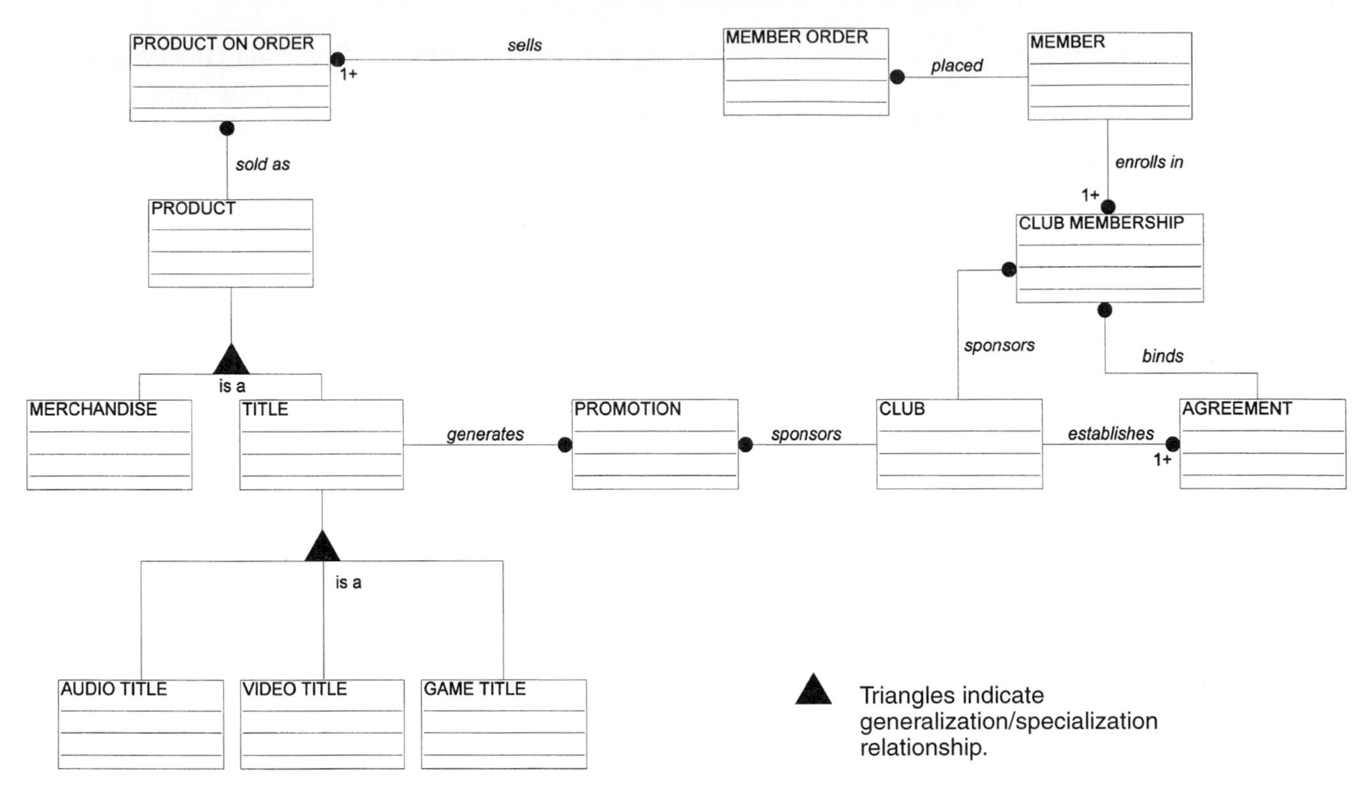

FIGURE 8.16 *Member Services System Object Association Model*

Step 4: Preparing the Object Association Model Figure 8.16 is an OMT object/class association diagram for the Member Services System. The diagram was constructed using Popkin Software's *System Architect*. Notice that the model depicts business objects/classes within the domain of the SoundStage Member Services System. The object class notation on the model does not depict the actual data attributes and behaviors (methods). This is commonly done to prevent the diagram from becoming needlessly cluttered and overwhelming.

The model also reflects the object/class associations and multiplicity that were identified in step 1, and a couple generalization/specialization relationships that were discovered in step 2. Step 3 did not reveal aggregate relationships among our objects/classes. It would be quite easy to imagine a few aggregate relationships. For example, a GAME TITLE may actually consist of a cartridge, a protective case, and an instruction booklet. However, this aggregate relationship (and other possible aggregate structures) were determined to be irrelevant to the users—they have no need to recognize this aggregate relationship since the game title is acquired from suppliers who provide the entire packet of components.

WHERE DO YOU GO FROM HERE?

This chapter introduced the newer object-oriented approach to systems development. Specifically, this chapter focused on object modeling tools and techniques for systems analysis. You are now ready to learn about the object-oriented approach as it applies to systems design. Object-oriented design is covered in Chapter 16. In Chapter 16 you will learn how the object models developed in this chapter are expanded to include design decisions for a new system.

SUMMARY

1. The approach of using object modeling during systems analysis and design is called object-oriented analysis. **Object-oriented analysis (OOA)** techniques are used to (1) study existing objects to see if they can be reused or adapted for new uses, and (2) define new or modified objects that will be combined with existing objects into a useful business computing application.
2. The object-oriented approach is centered around a technique referred to as object modeling. Object modeling is a technique for identifying objects within the systems environment and the relationships between those objects.
3. There are many underlying concepts for object modeling, including:
   a. Systems consist of objects wherein an object is something that is or is capable of being seen, touched, or otherwise sensed, and about which users store data and associate behavior. The data, or attributes, represent characteristics of interest about an object. The behavior of an object refers to those things that the object can do and correspond to functions that act on the object's data (or attributes). Each object encapsulates the attributes and behavior together as a single unit.
   b. Objects can be categorized into classes. A class is a set of objects that share common attributes and behavior. Objects may be grouped into multiple levels of classes. The most general class in the grouping is the supertype (or generalization of the class). The more refined class is referred to as the subtype class (or specialization class). All subtype classes "inherit" the attributes and behavior of the supertype class.
   c. Objects and classes have relationships. A relationship is a natural business association that exists between one or more objects/classes. The degree, or multiplicity, of a relationship specifies the business rules governing the relationship. Some relationships are more "structural"—meaning that a class may be related to another class in that one class may represent an assembly of one or more other class types. This type of relationship is referred to as an aggregation structure.
   d. Objects communicate by passing messages. A message is passed when one object invokes another object's behavior to request information or some action.
   e. A type of behavior may be completed differently for different objects/classes. This concept is referred to as polymorphism.
4. One of the most critical aspects of performing object-oriented development is correctly identifying the objects and their relationships early in the development process. Use case modeling is a popular approach to assist in object identification.
5. In trying to identify objects, many methodology experts recommend searching the requirements document or other associated documentation and underlining the nouns that may represent potential objects. This could be a monumental task! There are just too many nouns. Use case modeling solves this problem by breaking down the entire scope of system functionality into many smaller statements of system functionality called use cases. This smaller format simplifies and makes more efficient the technique of underlining the nouns.
6. Use case modeling utilizes two constructs: actors and use cases. An actor represents anything that needs to interact with the system to exchange information. An actor is a user, a role that could be an external system as well as a person. A use case is a behaviorally related sequence of steps (a scenario), both automated and manual, for the purpose of completing a single business task.
7. An object association model diagram is used to organize the objects found as a result of use case modeling and to document the relationships between the objects.

KEY TERMS

actor, p. 294
attribute, p. 287
behavior, p. 287
class, p. 288
encapsulation, p. 288
generalization/specialization, p. 288
inheritance, p. 288
instance, p. 287
message, p. 293
method, p. 287
multiplicity, p. 291
object, p. 287
object association model, p. 303
object/class relationship, p. 290
object modeling, p. 286
object-oriented analysis, p. 286
polymorphism, p. 293
service, p. 287
subtype, p. 289
supertype, p. 288
temporal event, p. 294
use case, p. 294
use case dependency diagram, p. 297
use case model diagram, p. 296
use case modeling, p. 294
whole/part relationship, p. 303

REVIEW QUESTIONS

1. What is the approach of using object modeling during systems analysis called?
2. Define object modeling.
3. What is an object? Give several examples.
4. What is an attribute? Give several examples of attributes for an object.
5. What is a behavior? Give examples of behaviors for an object.
6. The packaging of an object's attributes and behaviors into a single unit is referred to as what?
7. What is a class? Give an example.
8. Give an example of a supertype and corresponding subtype class. Explain how the concept of inheritance can be applied to the examples.
9. What is the technique called wherein the attributes and behaviors that are common to several types of objects are grouped into their own class, called a supertype?
10. Define object/class relationship and give an example.
11. What purpose does specifying multiplicity play in defining relationships?
12. Give an example of an aggregation structure.
13. Explain the concept of message sending for objects/classes.
14. Define the term *polymorphism*. Give an example.
15. What are the two general activities when performing object-oriented analysis?
16. Define the term *use case*. Give an example.
17. Define the term *actor*. Give an example.
18. Define the term *temporal event*. Who is the actor that initiates a temporal event?
19. What are the different text sections of a use case? Define each one.
20. Describe the process of using use cases to find potential objects.

PROBLEMS AND EXERCISES

1. Why is it desirable to categorize objects into classes?
2. What are some objects within your current environment? What classes can be formed from those objects?
3. Draw an object model to depict relationships between the objects you identified in problem 2. Be sure to specify the multiplicity. Were you able to identify any aggregate relationships?
4. What are some benefits that can be realized through the object-oriented concepts of inheritance and encapsulation?
5. What are the OMT notations for multiplicity? Give examples of object/class relationships that demonstrate each type of multiplicity.
6. Draw an object model to communicate the following: ABC Corporation has several employees. For each employee, the company keeps track of that employee's LAST NAME, FIRST NAME, MIDDLE NAME, GENDER, HOME ADDRESS, DATE HIRED, and BIRTHDATE. Some employees are salaried employees. For those employees, the company is interested in their ANNUAL SALARY. Other employees are hourly employees. For hourly employees, the company is interested in knowing their HOURLY PAY RATE.
7. Why is it important to correctly identify the objects and their relationships early in the development process?
8. Describe why it is important to do use case modeling to identify objects.

9. What are some benefits of using use cases?
10. Compare and contrast a use case normal course versus a use case alternate course.
11. Explain how the precondition of a use case can be used to construct the use case dependency diagram.
12. Prepare a use case complete with normal and alternate courses documenting the event of you buying a can of soda from a soda machine.
13. Based on the results of question 12, identify potential objects and construct an object association model diagram.

PROJECTS AND RESEARCH

1. Research publications to obtain a copy of an object model. Redraw that object model using the OMT notation. Be aware that some object models depict information that is not shown on an OMT object model.
2. Surf the Internet and make a list of links to sites containing information about object-oriented analysis.
3. Research the new unified modeling language (UML) by Rational Software Corporation. Who are the authors of the language? What methodologies is the new language based on? Visit the Rational Software Corporation web site. Get a copy of the UML document.

MINICASES

1. Schedule an interview with the business analyst responsible for your school's registration system. Based on the information provided, perform the following:
   a. Draw a context model diagram.
   b. Identify the actors and the use cases they initiate.
   c. Complete a use case description for the event of a student enrolling for a course. For a student dropping a course.
   d. Use the use cases you have previously prepared and construct an object association model diagram.

SUGGESTED READINGS

Booch, G. *Object-Oriented Design with Applications.* Menlo Park, CA: Benjamin Cummings, 1994.

Coad, P., and E. Yourdon. *Object-Oriented Analysis.* 2nd ed. Englewood Cliffs: NJ: Prentice Hall, 1991. This book provides a very good overview of object-oriented concepts. However, the object model techniques are somewhat limited by comparison to OMT and other object-oriented modeling approaches.

Jacobson, Ivar; Magnus Christerson; Patrik Jonsson; and Gunnar Overgaard. *Object-Oriented Software Engineering—A Use Case Driven Approach.* Wokingham, England: Addison Wesley, 1992. This book presents detailed coverage of the process of use case modeling and how it's used to identify objects.

Martin, J., and J. Odell. *Object-Oriented Analysis and Design.* Englewood Cliffs, NJ: Prentice Hall, 1992.

Rumbaugh, James; Michael Blaha; William Premerlani; Frederick Eddy; and William Lorensen. *Object-Oriented Modeling and Design.* Englewood Cliffs, NJ: Prentice Hall, 1991. This book presents detailed coverage of the object modeling technique (OMT) and its application throughout the entire systems development life cycle.

Taylor, David A. *Object-Oriented Information Systems—Planning and Implementation.* New York: John Wiley & Sons, Inc., 1992. This book is a very good entry-level resource for learning the concepts of object-oriented technology and techniques.

PART THREE

SYSTEMS DESIGN AND CONSTRUCTION METHODS

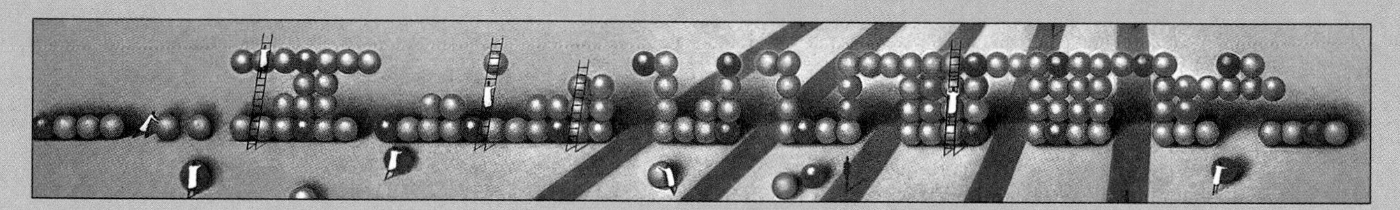

The eight chapters in Part Three introduce you to systems design and construction activities and methods. Chapter 9, "Systems Design and Construction," provides the context for all the subsequent chapters by introducing the activities of systems design. Systems design includes the evaluation of alternative solutions, preparation of detailed computer-based specifications that will fulfill the requirements specified during systems analysis, and construction of system prototypes. With respect to information systems development, systems design consists of the configuration, procurement, and design and integration phases.

Chapter 10, "Application Architecture and Process Design," introduces physical process and data design. It specifically addresses design decisions regarding distribution issues for shared data and processes. This results in an application architecture that consists of design units that can be assigned to different team members for detailed design, construction, and unit testing. The chapter also includes coverage of the new client/server approach.

Chapter 11, "Database Design," introduces the design of physical data stores from the data model developed in Chapter 5. It teaches a procedure that prepares the data model, developed during systems planning and systems analysis, for implementation by ensuring that the model is simple, flexible, and nonredundant. This procedure, called data analysis, results in data entities that are normalized into a third normal form. Database management systems are introduced.

Chapter 12, "Input Design and Prototyping," teaches input design and prototyping. Formats, methods, media, human factors, and internal controls for inputs are stressed. The proper usage of screen-based controls for data input on graphical user interface (GUI) screen designs is discussed. The chapter also emphasizes prototyping as a way of finding, documenting, and communicating input design requirements.

Chapter 13, "Output Design and Prototyping," teaches output design and prototyping. Different types, formats, and media for outputs are presented. The use of the most common types of graphs are discussed. The chapter demonstrates how to design and prototype printed and display outputs.

Chapter 14, "User Interface Design and Prototyping," teaches user interface design and prototyping. You will learn how to develop a friendly and effective interface for an application. The design of the user interface is crucial because user acceptance of the system is frequently dependent on a friendly, easy-to-use interface. A GUI-based interface for obtaining the inputs and outputs designed in Chapters 12 and 13 is demonstrated.

Chapter 15, "Software Design," covers structured design. Tools and techniques for modular design are presented. Also, packaging is presented to ensure that complete design specifications are passed on to computer programmers.

Finally, Chapter 16, "Object-Oriented Design," provides an overview of the increasingly popular object-oriented approach to system design. This chapter is intended to complement Chapter 8's introduction to object modeling.

9

SYSTEMS DESIGN AND CONSTRUCTION

CHAPTER PREVIEW AND OBJECTIVES

In this chapter you will learn more about the design phases in the *FAST* systems development methodology: configuration, procurement, and design and integration phases. You will know that you understand the process of systems design when you can:

- Define the systems design process in terms of the configuration, procurement, and design and integration phases of the life cycle.
- Describe configuration, procurement, and design and integration phases in terms of your information building blocks.
- Describe the configuration, procurement, and design and integration phases in terms of purpose, activities, roles, inputs, outputs, techniques, and steps.
- Identify those chapters and modules in this textbook that can help you perform the activities of systems design.
- Describe traditional and prototyping approaches to systems design.

Although some techniques of systems design are introduced in this chapter, it is *not* the intent of this chapter to teach the *techniques* of systems design. This chapter teaches only the *process* of systems design and introduces you to some of the techniques that will be taught in later chapters.

SOUNDSTAGE

SOUNDSTAGE ENTERTAINMENT CLUB

SCENE

When we last visited SoundStage, Sandra and Bob were meeting with Sarah Hartman to plot the strategy for completing the study phase for the project. Both the study phase and the subsequent definition phase were completed. This episode begins shortly after the project's executive sponsor (the person who pays for the system to be built) has approved the business requirements that came out of the definition phase. Sandra and Bob are in the process of planning the design phases for the project. We join Sandra and Bob in a small conference room.

SANDRA

OK, Bob, here's what I think we should do. Our *FAST* methodology calls for the completion of three design phases. But they are not necessarily sequential. I think we overlap some of the activities. Let's agree on what needs to be done and then determine how we're going to get the job done.

BOB

Sounds good to me.

SARAH

The first phase in *FAST* is the configuration phase. This is where you and I earn our money. Our goal is to come up with hardware and software recommendations for the new system. I thought we should both work together to initially brainstorm some candidate solutions for the new system. Then, I'll take over and try to do further research on each candidate.

BOB

That's fair to me. We already know some possibilities that we have to consider. Remember that Dick Krieger (director of Warehouse Operations) told us the warehousing operations are using bar coding technology. We know we need to consider coming up with solutions that will interface well with their operations.

SANDRA

That's right, I forgot that piece of detail. When we do brainstorm possible solutions, we need to make sure that we consider only hardware and software solutions that meet the company's technology architecture standards. For example, our standards dictate that we use SQL Server or ACCESS on our project. Obviously, we will be using SQL Server.

BOB

We did get an exemption from the standard to use Delphi as part of a pilot project.

SARAH

We are getting ahead of ourselves here. So we will work together to brainstorm candidate solutions. Once we have the solutions, we can evaluate the candidates. I'm sure that we will want to involve other experts in that effort. Then we will select the one solution we want to recommend to our users.

BOB

That brings us to the procurement phase . . . assuming they approve our recommendations.

SARAH

I'll take care of the procurement phase. Our company policy states that all new hardware and software be purchased through Kramer Computers, Inc. I've ordered new equipment and software from them many times.

BOB

Great. I'll start on the last phase, design and integration.

SARAH

I hope you agree that we should use a prototyping approach to completing that phase.

BOB

Do you think the users will be willing to participate in reviewing the screens that I design?

SANDRA

Don't you remember, we already discussed this with the people at our launch meeting? They assured us that they will encourage the users to participate.

BOB

Well, I guess I had better brush up some more on Delphi. I'd like to get started as soon as we finish identifying candidate solutions.

SANDRA

I think we're in agreement on how to attack the design phases. How about we go get some lunch and get started with the configuration phase when we get back?

BOB

Sounds good to me; I'm hungry.

DISCUSSION QUESTIONS

1. How would you characterize the focus of the design phase of this project?
2. What were some advantages that Sandra was seeking by overlapping activities?
3. How could Sandra and Bob present their hardware and software to management without overwhelming them with technical jargon?
4. How does SoundStage's technology architecture impact Sandra and Bob's effort to complete the configuration phase? Why would Sandra and Bob call on other technical experts to be involved in this process?
5. Why would Sandra want to commit to prototyping (building models) as the approach to use in completing the design and integration phase for their project?

WHAT IS SYSTEMS DESIGN?

Let's begin with a definition of systems design.

Systems design is the evaluation of alternative solutions and the specification of a detailed computer-based solution. It is also called **physical design.**

The key term here is design. Whereas systems analysis primarily focused on the logical, implementation-independent aspects of a system (the requirements), systems design deals with the physical or implementation-dependent aspects of a system (the system's technical specifications).

Relative to the information system building blocks, systems design addresses DATA, PROCESSES, INTERFACES, and GEOGRAPHY from the system designer's perspective. What about *technology*? Often technology is in place or specified by a predefined technology architecture. In other cases, the analyst must select or supplement the technology. In all cases, systems design builds on the knowledge derived from systems analysis.

Most of us place too restrictive a definition on the process of design. We envision ourselves drawing blueprints of the computer-based systems to be programmed and developed by ourselves or our own programmers. Thus, we design inputs, outputs, files, databases, and other computer components. Recruiters of computer-educated graduates refer to this restrictive definition as the "not-invented-here syndrome." In reality, many companies purchase more software than they write in-house. That shouldn't surprise you. Why reinvent the wheel? Many systems are sufficiently generic that computer vendors have written adequate—but not perfect—software packages that can be bought and possibly modified to fulfill end-user requirements.

As mentioned in Chapter 4, "configure a feasible solution" would be considered part of systems analysis by some experts. We prefer to think of it as an analysis-to-design transition phase. Since this phase represents a change in the developer's focus toward technology-dependent issues, we have chosen to present systems design as encompassing the configuration, procurement, and design and integration phases introduced in Chapter 3 (see Figure 9.1).

This chapter examines each of the above phases in greater detail. But first, let's examine some overall strategies for systems design.

STRATEGIES FOR SYSTEMS DESIGN

There are many strategies or techniques for performing systems design. They include modern structured analysis, information engineering, prototyping, JAD, RAD, and object-oriented design. These strategies are often viewed as competing alternative approaches to systems design. In reality, certain combinations complement one another. Let's briefly examine these strategies and the scope or goals of the projects to which they are suited. The intent is to develop a high-level understanding only. The subsequent chapters in this unit will actually teach you the techniques.

Modern Structured Design

Structured design techniques help developers deal with the size and complexity of programs.

Modern structured design is a process-oriented technique for breaking up a large program into a hierarchy of modules that result in a computer program that is easier to implement and maintain (change). Synonyms (although technically inaccurate) are *top-down program design* and *structured programming*.

The concept is simple. Design a program as a top-down hierarchy of modules. A module is a group of instructions—a paragraph, block, subprogram, or sub-

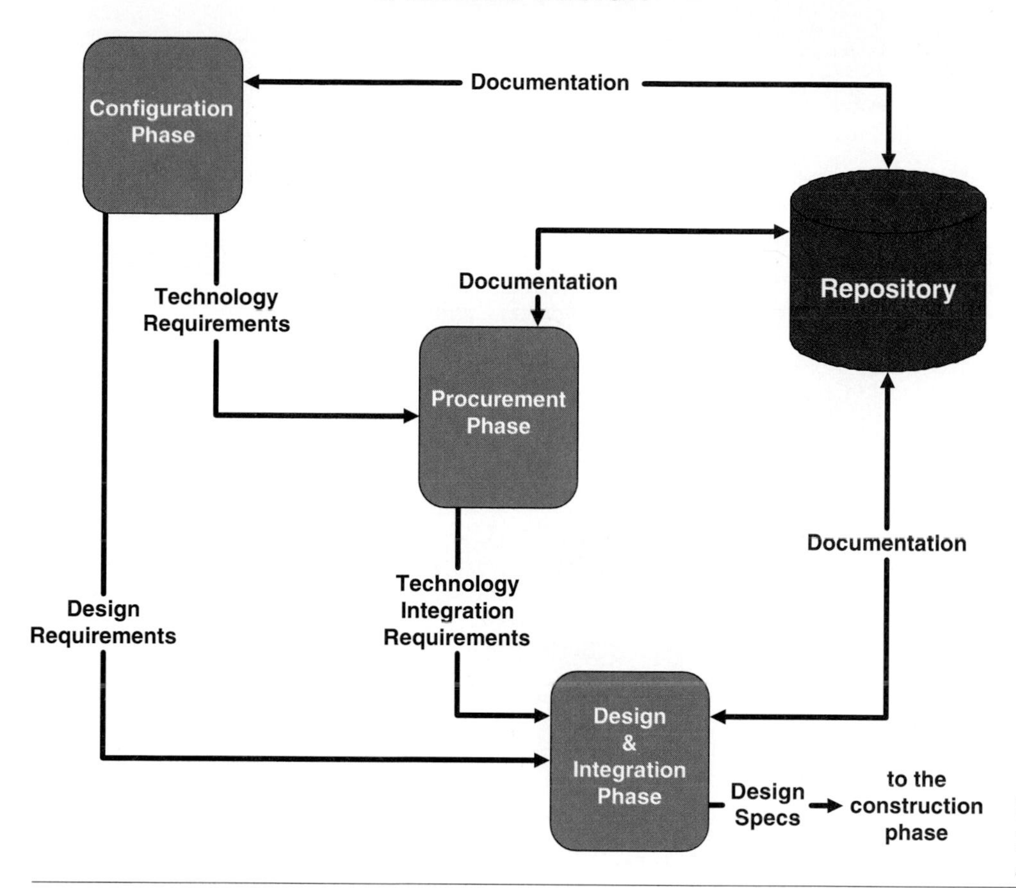

FIGURE 9.1
The Systems Design Phases of a Project

routine. The top-down structure of these modules is developed according to various design rules and guidelines. (Thus, merely drawing a hierarchy or structure chart for a program is *not* structured design.)

Structured design is considered a process technique because its emphasis is on the PROCESS building blocks in our information system—specifically, software processes. Structured design seeks to factor a program into the top-down hierarchy of modules that have the following properties:

- Modules should be highly **cohesive;** that is, each module should accomplish one and only one function. Theoretically this makes the modules reusable in future programs.
- Modules should be loosely **coupled;** in other words, modules should be minimally dependent on one another. This minimizes the effect that future changes in one module will have on other modules.

The software model derived from structured design is called a **structure chart** (Figure 9.2). The structure chart is derived by studying the flow of data through the program. Structured design is performed during systems design. It does not address all aspects of design; for instance, structured design will not help you design inputs, databases, or files.

Structured design has lost some of its popularity with many of today's applications that call for newer techniques that focus on event-driven and object-oriented programming techniques. However, it is still a popular technique for the design of mainframe-based application software and to address coupling and cohesion issues at the system level.

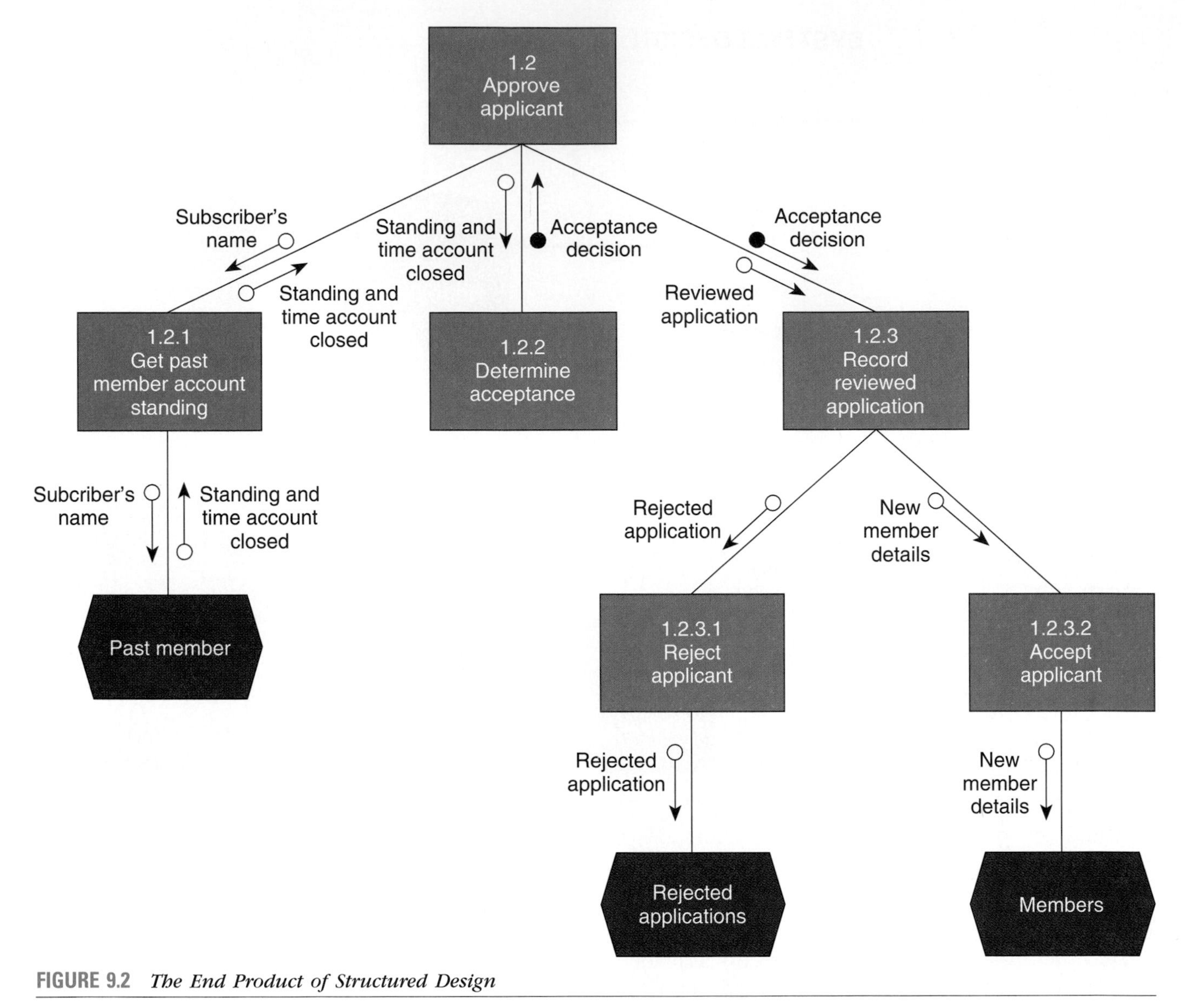

FIGURE 9.2 *The End Product of Structured Design*

Information Engineering (IE)

Recall from Chapter 4, you learned that **information engineering** is a data-centered technique. IE involves conducting a business area requirements analysis from which information system applications are carved out and prioritized. The applications identified in IE become projects to which other systems analysis *and* design methods are intended to be applied to develop the production systems. These methods may include some combination of modern structured analysis (discussed in Chapter 4), modern structured design, prototyping, and object-oriented analysis and design.

Prototyping

Traditionally, physical design has been a paper-and-pencil process. Analysts drew pictures that depicted the layout or structure of outputs, inputs, and files and the flow of dialogue and procedures. This is a time-consuming process that is prone to considerable error and omissions. Frequently, the resulting paper specifications did not prove themselves inadequate, incomplete, or inaccurate until programming started.

Today many analysts are turning to prototyping, a modern engineering-based approach to design. A prototype, according to Webster's dictionary, is "an original or model on which something is patterned" and/or "a first full-scale and usually functional form of a new type or design of a construction (as an airplane)." Engineers build prototypes of engines, machines, automobiles, and the like, before building the actual products. Prototyping allows engineers to isolate problems in both requirements and designs.

The prototyping approach is an iterative process involving a close working relationship between the designer and the users. This approach has several advantages.

- Prototyping encourages and requires active end-user participation. This increases end-user morale and support for the project. End-user morale is enhanced because the system appears real to them.
- Iteration and change are natural consequences of systems development—that is, end-users tend to change their minds. Prototyping better fits this natural situation since it assumes that a prototype evolves, through iteration, into the required system.
- It has often been said that end-users don't fully know their requirements until they see them implemented. If so, prototyping endorses this philosophy.
- Prototypes are an active, not passive, model that end-users can see, touch, feel, and experience. Indeed, if a picture such as a DFD is worth a thousand words, then a working model of a system is worth a thousand pictures.
- An approved prototype is a working equivalent to a paper design specification, with one exception—errors can be detected much earlier.
- Prototyping can increase creativity because it allows for quicker user feedback, which can lead to better solutions. (See the list of disadvantages for ways creativity can be stifled by prototyping.)
- Prototyping accelerates several phases of the life cycle, possibly bypassing the programmer. In fact, prototyping consolidates parts of phases that normally occur one after the other.

There are also disadvantages or pitfalls to using the prototyping approach. Prototyping is not without disadvantages. Most of these can be summed up in one statement: Prototyping encourages ill-advised shortcuts through the life cycle. Fortunately, the following pitfalls can all be avoided through proper discipline.

- Prototyping encourages a return to the "code, implement, and repair" life cycle that used to dominate information systems. As many companies have learned, systems developed in prototyping languages can present the same maintenance problems that have plagued systems developed in languages such as COBOL.
- Prototyping does not negate the need for the survey and study phases. A prototype can just as easily solve the wrong problems and opportunities as a conventionally developed system.
- You cannot completely substitute any prototype for a paper specification. No engineer would prototype an engine without some paper design. Yet many information systems professionals try to prototype without a specification. Prototyping should be used to complement, not replace, other methodologies. The level of detail required of the paper design may be reduced, but it is not eliminated. (In the next section, we'll discuss just how much paper design is needed.)
- Numerous design issues are not addressed by prototyping. These issues can inadvertently be forgotten if you are not careful.
- Prototyping often leads to premature commitment to a design. In other words, the configuration and procurement phases get shortchanged.
- When prototyping, the scope and complexity of the system can quickly expand beyond original plans. This can easily get out of control.

- Prototyping can reduce creativity in designs. The very nature of any implementation—for instance, a prototype of a report—can prevent analysts and end-users from looking for better solutions.
- Prototypes often suffer from slower performance than their third-generation language counterparts.

Building prototypes makes so much sense that you may wonder why we didn't always do it. The reason is simple: The technology wasn't available. Traditional languages such as COBOL, FORTRAN, BASIC, Pascal, and C (often called *third-generation languages*) don't lend themselves to prototyping. Prototypes must be developed and modified quickly, neither of which is possible with third-generation languages. Consider the prospects of continually modifying the DATA and PROCEDURE divisions of a COBOL program as end-users try to make up their mind what they want and how it should look.

Fourth-generation languages (4GLs), *applications generators* (AGs), and some *object-oriented programming languages* (OOPLs) are software tools that make building systems a simpler task. They are less procedural than traditional languages. This means the tools specify more of what the system is or what it should do, and less of how to do it. In other words, they are not as dependent on specification of logic. Finally, it should also be noted that many computer-assisted systems engineering (CASE) products also contain limited prototyping tools for designing screens and reports.

Prototypes can be quickly developed using many of the 4GLs and object-oriented programming languages available today. Prototypes can be built for simple outputs, computer dialogues, key functions, entire subsystems, or even the entire system. Each prototype system is reviewed by end-users and management, who make recommendations about requirements, methods, and formats. The prototype is then corrected, enhanced, or refined to reflect the new requirements. Prototyping technology makes such revisions in a relatively straightforward manner. The revision and review process continues until the prototype is accepted. At that point, the end-users are accepting both the requirements and the design that fulfills those requirements.

Design by prototyping doesn't necessarily fulfill all design requirements. For instance, prototypes don't always address important performance issues and storage constraints. Prototypes rarely incorporate internal controls. These must still be specified by the analyst.

Joint Application Development (JAD)

Joint application development (JAD) was introduced in Chapter 4 as a technique that complements other systems analysis and design techniques by emphasizing *participative development* among system owners, users, designers, and builders. Thus, JAD is frequently used in conjunction with the above design techniques. During the JAD sessions for systems design, the systems designer will take on the role of facilitator for possibly several full-day workshops intended to address different design issues and deliverables.

Rapid Application Development (RAD)

Another popular design strategy used today is rapid application development.

> **Rapid application development (RAD)** is the merger of various structured techniques (especially the data-driven information engineering) with prototyping techniques and joint application development techniques to accelerate systems development.

RAD calls for the interactive use of structured techniques and prototyping to define the users' requirements and design the final system. Using structured techniques, the developer first builds preliminary data and process models of the business requirements. Prototypes then help the analyst and users to verify those requirements and to formally refine the data and process models. The cycle of models, then prototypes, then models, then prototypes, and so forth ultimately results in a combined business requirements and technical design statement to be used for constructing the new system.

Object-Oriented Design (OOD)

Object-oriented design is the newest up-and-coming design strategy. The concepts behind this strategy (and technology) are covered extensively in Chapter 16, "Object-Oriented Design," but a simplified introduction is appropriate here.

This technique is an extension of the object-oriented analysis strategy presented in Chapter 4. Recall that object technologies and techniques are an attempt to eliminate the separation of concerns about DATA and PROCESS.

> **Object-oriented design (OOD)** techniques are used to refine the object requirements definitions identified earlier during analysis and to define design-specific objects.

For example, based on a design implementation decision, during OOD the designer may need to revise the data or process characteristics for an object that was defined during systems analysis. Likewise, a design implementation decision may necessitate that the designer define a new set of objects that will make up an interface screen that the users may interact with in the new system.

FAST SYSTEMS DESIGN METHODS

The *FAST* methodology used by SoundStage Entertainment Club does not impose a single design technique on system developers. Instead, it integrates all the popular design strategies we've discussed: structured design (via process modeling), information engineering (via data modeling), prototyping (via rapid application development), joint application development (for all methods), and rapid application development. Progressive *FAST* developers can use object-oriented design in conjunction with object technology for prototyping to fully exploit the object paradigm.[1]

Consistent with Chapter 4 and its discussion of systems analysis, we will present systems design using the following *FAST* methodology elements:

- *Purpose* (self-explanatory).
- *Roles.* All *FAST* activities are completed by individuals who are assigned to **roles.** Roles are not the same as job titles. One person can play many roles in a project. Conversely, one role may require many people to adequately fulfill that role. For example, a system user role may require several users to adequately represent the interests of an entire system. *FAST* roles are assigned to the following role groups:
  - System owner roles.
  - System user roles.
  - System analyst roles.
  - System designer roles.
  - System builder roles.

Notice that these groups correspond with the perspectives in your information system framework (from Chapters 2 and 3).

Every activity is described with respect to:

- *Prerequisites and inputs* (to the activity).
- *Deliverables and outputs* (produced by the activity).
- *Applicable techniques*—which techniques (from the previous section) are applicable to this phase.
- *Steps*—a brief description of the steps required to complete this activity. In the spirit of continuous improvement, all *FAST* steps are fully customizable for each organization.

[1]Recall from the Chapter 3 opening case that SoundStage has elected to test the use of object techniques and technology in the Member Services Information System project.

Furthermore, all roles, inputs, outputs, techniques, and steps are presented with the following designations:

- (REQ) indicates the role, input, output, technique, or step is REQuired.
- (REC) indicates the role, input, output, technique, or step is RECommended, but not required.
- (OPT) indicates the role, input, output, technique, or step is OPTional, but not required.

FIGURE 9.3 *Building Blocks for the Configuration Phase*

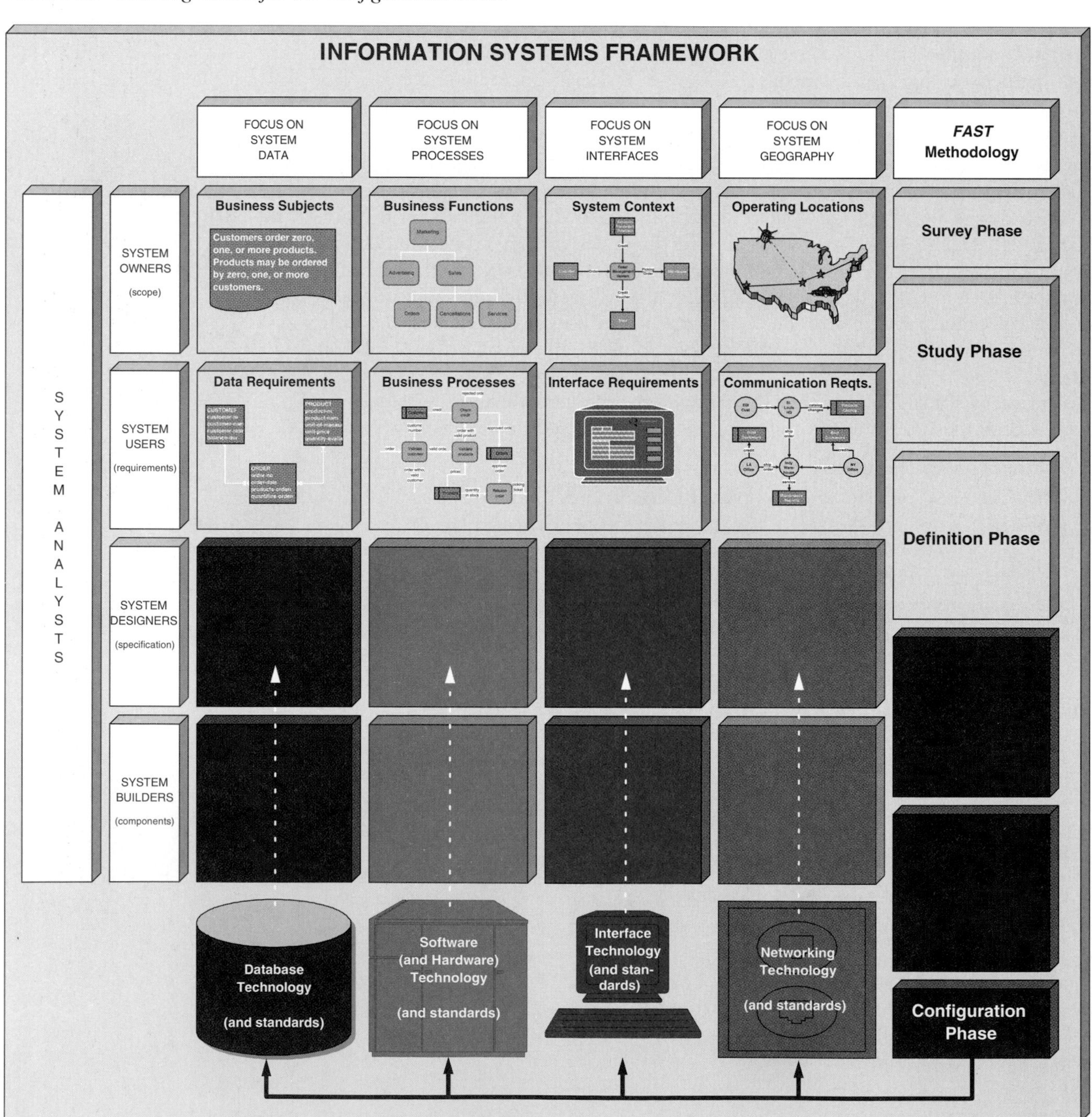

THE CONFIGURATION PHASE OF SYSTEMS DESIGN

Recall from Chapter 3, the purpose of the configuration phase is to identify candidate solutions, analyze those candidate solutions, and recommend a target system that will be designed and implemented. Alternative solutions to be considered should be those that address the business requirements of the information system. Given the business requirements for an improved information system, we can finally address how the new system, including computer-based alternatives, might operate. You should never automatically go with your first hunch. During the configuration phase, it is imperative that you identify options, analyze options, and then sell feasible solutions based on the analysis.

The fundamental objectives of the configuration phase are:

- To identify and research alternative manual and computer-based solutions to support our target information system.
- To evaluate the feasibility of alternative solutions and recommend the best overall alternative solution.

Your information systems framework (Figure 9.3) provides the context for the focus of the configuration phase. As shown in the model, the configuration phase is primarily concerned with the technology requirements for the target system. The configuration phase marks the first point in the systems development process that we have emphasized how the new system might be implemented. Thus, we will address how technology may be used to support the target system. Notice that the technology decisions made during the configuration phase will impact the DATA, PROCESS, INTERFACE, and GEOGRAPHY decisions made in later phases (this is suggested by the upward-pointing dashed lines).

Figure 9.4 is an activity diagram for the configuration phase. The figure illustrates the typical activities of the configuration phase. Let's now examine each activity in greater depth.

Activity: Define Candidate Solutions

Given the business requirements established in the definition phase of systems analysis, we must identify alternative candidate solutions. Some candidate solutions will be posed by design ideas and opinions from system owners and users. Others may come from various sources including: systems analysts, system designers, technical consultants, and other IS professionals. Some technical choices may be limited by a predefined, approved technology architecture provided by system managers. It is not the intent of this activity to evaluate the candidates, rather to simply define possible candidate solutions to be considered.

Purpose The purpose of this activity is to identify alternative candidate solutions to the business requirements defined during systems analysis.

Roles The activity is facilitated by the *project manager*. Other roles are defined as follows:

- System owner roles (REC)—System owners are not normally directly involved in this activity. However, during this activity the system designer will consider any proposed ideas and solutions the owner may have communicated as possibilities. For example, the owner may have read an article or perhaps learned how some competitor or acquaintance's similar system was implemented. In either case, it is politically sound to consider an owner's ideas.
- System user roles (REC)—Users are typically not involved in this activity at this time. Once again, they may be the source of "ideas" for possible solutions, especially manual-based solutions.
- Systems analyst roles (OPT)—As mentioned earlier, the configuration phase is frequently considered an activity of systems analysis. Therefore, this activity

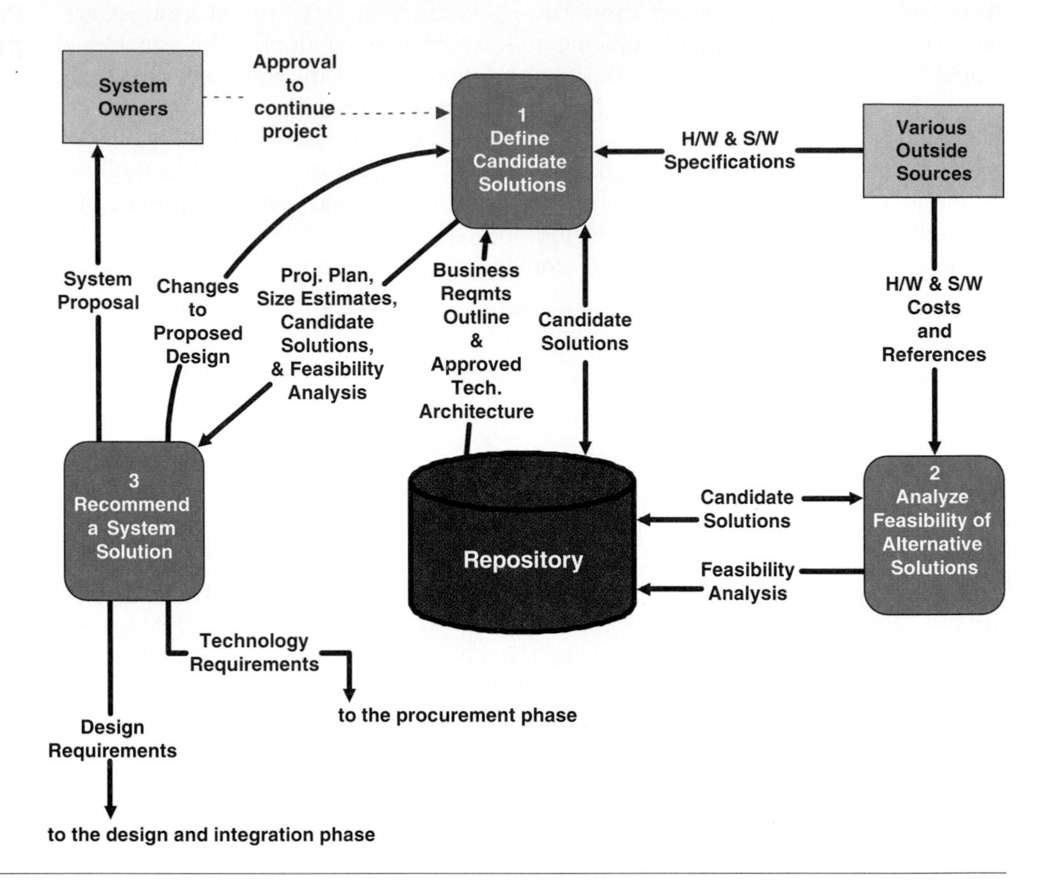

FIGURE 9.4
Activity Diagram for the Configuration Phase

may be completed by the systems analyst. Regardless, the systems analyst is most knowledgeable about the business requirements and therefore should be involved in brainstorming solutions that might fulfill those requirements.

- System designer roles (REQ)—The system designer assumes the major role in this activity. The designer will usually seek input and advice from the following expertises:
  - Database administrator (REC)—This person will be a source of expertise regarding available database technology. He or she will also be familiar with database technology standards.
  - Network administrator (REC)—This person can provide expertise regarding existing network technology. He or she will also be familiar with network architecture standards that must be adhered to for any possible new networking technology.
  - Applications administrator (REC)—This person provides knowledge regarding new and existing applications development tools and standards.
- System builder roles are not typically involved in this activity.

Prerequisites (Inputs) This activity is triggered by the approval from the system owners to continue the project into systems design. The key inputs are the *business requirements outline* (REQ) defined during systems analysis, *hardware and software specifications* (REQ) from various sources such as vendors and customer referrals, and *approved technology architecture* (OPT, but REQ if it exists).

Often during systems analysis, when the primary focus is supposed to be logical business requirements, the analyst learns of a technical solution that he or she anticipates might be a viable alternative to consider when reaching systems design. Or perhaps earlier in the project the system owner encouraged or insisted that a particular technical solution be considered. These facts are frequently documented as future *candidate solutions* (REQ) to be considered when defining candidate solutions.

Deliverables (Outputs) The principal deliverables of this activity are the candidate solutions (REQ) for a new system. The amount of information describing the characteristics of any one candidate solution may become overwhelming. A matrix is a useful tool for effectively capturing, organizing, and communicating the characteristics for candidate solutions.

A partially completed candidate solutions matrix is depicted in Figure 9.5. This allows for a side-by-side comparison of the different characteristics for a number of candidates. In Module C, "Feasibility and Cost–Benefit Analysis," you will learn how to develop a complete candidate matrix.

Applicable Techniques The following technique is applicable to this activity. For each activity, we indicate which chapters or modules of the book teach that technique.

- *Fact finding.* Fact-finding methods (REQ) are used to interact with outside sources such as hardware and software vendors and stores to gather product specifications for each candidate. Fact-finding is discussed in Part Five, Module B, "Fact-Finding and Information Gathering."

Steps The following steps are suggested to complete this activity.

1. (REQ) Review the business requirements outlined in the definition phase of systems analysis.
2. (REQ) If it exists, review the technology architecture to determine any hardware or software standards required for any candidate solution.
3. (REQ) Brainstorm alternative solutions that fulfill the business requirements. Also, identify solutions that were suggested before the design phase.
4. (REQ) Research technical specifications detailing the characteristics of each candidate solution.

Activity: Analyze Feasibility of Alternative Solutions

Once alternative candidate design solutions have been identified, each candidate must be analyzed for feasibility. Feasibility analysis should not be limited to costs and benefits. Most analysts evaluate solutions against four sets of criteria:

- *Technical feasibility.* Is the solution technically practical? Does our staff have the technical expertise to design and build this solution?
- *Operational feasibility.* Will the solution fulfill the user's requirements? To what degree? How will the solution change the user's work environment? How do users feel about such a solution?
- *Economic feasibility.* Is the solution cost-effective?
- *Schedule feasibility.* Can the solution be designed and implemented within an acceptable time period?

When completing this activity, the analyst and users are careful not to compare the candidates. The feasibility analysis is performed on each *individual* candidate without regard to the feasibility of other candidates. This approach discourages the analyst and users from prematurely deciding which candidate is the best.

Purpose The purpose of this activity is to evaluate the alternative candidate solutions according to their economic, operational, technical, and schedule feasibility.

Characteristics	Candidate 1	Candidate 2	Candidate 3	Candidate ...
Portion of System Computerized Brief description of that portion of the system that would be computerized in this candidate.	COTS package Platinum Plus from Entertainment Software Solutions would be purchased and customized to satisfy Member Services required functionality.	Member Services and warehouse operations in relation to order fulfillment.	Same as candidate 2.	
Benefits Brief description of the business benefits that would be realized for this candidate.	This solution can be implemented quickly because it's a purchased solution.	Fully supports user required business processes for SoundStage Inc. Plus more efficient interaction with member accounts.	Same as candidate 2.	
Servers and Workstations A description of the servers and workstations needed to support this candidate.	Technically architecture dictates Pentium Pro, MS Windows NT class servers and Pentium, MS Windows NT 4.0 workstations (clients).	Same as candidate 1.	Same as candidate 1.	
Software Tools Needed Software tools needed to design and build the candidate (e.g., database management system, emulators, operating systems, languages, etc.). Not generally applicable if applications software packages are to be purchased.	MS Visual C++ and MS Access for customization of package to provide report writing and integration.	MS Visual Basic 5.0 System Architect 3.1 Internet Explorer	MS Visual Basic 5.0 System Architect 3.1 Internet Explorer	
Application Software A description of the software to be purchased, built, accessed, or some combination of these techniques.	Package Solution	Custom Solution	Same as candidate 2.	
Method of Data Processing Generally some combination of: on-line, batch, deferred batch, remote batch, and real-time.	Client/Server	Same as candidate 1.	Same as candidate 1.	
Output Devices and Implications A description of output devices that would be used, special output requirements (e.g., network, preprinted forms, etc.), and output considerations (e.g., timing constraints).	(2) HP4MV department laser printers (2) HP5SI LAN laser printers	(2) HP4MV department laser printers (2) HP5SI LAN laser printers (1) PRINTRONIX bar-code printer (includes software & drivers) Web pages must be designed to VGA resolution. All internal screens will be designed for SVGA resolution.	Same as candidate 2.	
Input Devices and Implications A description of input methods to be used, input devices (e.g., keyboard, mouse, etc.), special input requirements (e.g., new or revised forms from which data would be input), and input considerations (e.g., timing of actual inputs).	Keyboard & mouse	Apple "Quick Take" digital camera and software (15) PSC Quickscan laser bar-code scanners (1) HP Scanjet 4C Flatbed Scanner Keyboard & mouse	Same as candidate 2.	
Storage Devices and Implications Brief description of what data would be stored, what data would be accessed from existing stores, what storage media would be used, how much storage capacity would be needed, and how data would be organized.	MS SQL Server DBMS with 100GB arrayed capability.	Same as candidate 1.	Same as candidate 1.	

FIGURE 9.5 *Partially Completed Candidate Matrix*

Roles The activity is facilitated by the *project manager.* Other roles are defined as follows:

- System owner roles (REC)—The opinions of the following individuals may be sought when assessing the operational feasibility of a candidate solution: executive sponsor (REC), user managers (REC), system managers (REC), and project manager (REC).
- System user roles (REC)—Several users may be involved to assess their feelings toward a candidate solution. The financial or business analyst (REC) may be a source for determining the financial techniques to be used when analyzing the economic feasibility of an investment (a new system).
- Systems analyst roles (OPT)—Once again, this activity may be performed by the systems analyst.
- System designers (REQ) are responsible for the completion of this activity. The designer may seek input from the following people regarding the technical feasibility of the candidate solution: database administer (REC), network administrator (REC), and applications administer (REC).
- System builder roles are not typically involved in this activity unless deemed appropriate by a system owner.

Prerequisites (Inputs) This activity is triggered by the definition of one or more *candidate solutions* (REQ). To conduct the feasibility analysis, *hardware and software costs* as well as feedback from customer *references* (REC) are needed.

Deliverables (Outputs) The principal deliverable of this activity is the completed *feasibility analysis* (REQ) for each candidate. Once again, a matrix can be used to communicate the large volume of information about candidate solutions. The matrix in Figure 9.6 allows for a side-by-side unveiling of the different feasibility analyses for a number of candidates.

Applicable Techniques The following techniques are applicable to this activity.

- *Fact-finding.* Fact-finding methods (REQ) are used to obtain costs, opinions, and other facts about candidates from a variety of sources. Fact-finding methods are discussed in Part Five, Module B, "Fact-Finding and Information Gathering."
- *Feasibility analysis.* The ability to perform a feasibility assessment (REQ) is an extremely important skill requirement. Feasibility assessment techniques and skills are more fully covered in Module C, "Feasibility and Cost–Benefit Analysis."

Steps The following steps are suggested to complete this activity.

1. (REQ) Obtain all cost information for each product.
2. (REC) Discuss candidate solutions with system owners and users to obtain a feel for how well received the solution would be from their perspectives.
3. (REC) If possible, obtain feedback from customers who own or have used the hardware and software products. Feedback can also be obtained indirectly from product reviews appearing in various periodicals.
4. (REQ) Determine what economic measures to use to conduct the cost–benefit feasibility analysis. Frequently, a company may have set policies regarding how investments will be measured. Numerous measures such as payback analysis, return on investment, and net present value may be required for this decision.

<table>
<tr><th>Feasibility Criteria</th><th>Weight</th><th>Candidate 1</th><th>Candidate 2</th><th>Candidate 3</th><th>Candidate . . .</th></tr>
<tr><td>Operational Feasibility

Functionality. A description of to what degree the candidate would benefit the organization and how well the system would work.

Political. A description of how well received this solution would be from both user management, user, and organization perspective.</td><td>30%</td><td>Only supports Member Services requirements and current business processes would have to be modified to take advantage of software functionality

Score: 60</td><td>Fully supports user required functionality.

Score: 100</td><td>Same as candidate 2.

Score: 100</td><td></td></tr>
<tr><td>Technical Feasibility

Technology. An assessment of the maturity, availability (or ability to acquire), and desirability of the computer technology needed to support this candidate.

Expertise. An assessment of the technical expertise needed to develop, operate, and maintain the candidate system.</td><td>30%</td><td>Current production release of Platinum Plus package is version 1.0 and has only been on the market for 6 weeks. Maturity of product is a risk and company charges an additional monthly fee for technical support.

Required to hire or train C++ expertise to perform modifications for integration requirements.

Score: 50</td><td>Although current technical staff has only Powerbuilder experience, the senior analysts who saw the MS Visual Basic demonstration and presentation have agreed the transition will be simple and finding experienced VB programmers will be easier than finding Powerbuilder programmers and at a much cheaper cost.

MS Visual Basic 5.0 is a mature technology based on version number.

Score: 95</td><td>Although current technical staff is comfortable with Powerbuilder, management is concerned with recent acquisition of Powerbuilder by Sybase Inc.
MS SQL Server is a current company standard and competes with SYBASE in the Client/Server DBMS market. Because of this we have no guarantee future versions of Powerbuilder will "play well" with our current version SQL Server.

Score: 60</td><td></td></tr>
<tr><td>Economic Feasibility

Cost to develop:

Payback period (discounted):

Net present value:

Detailed calculations:</td><td>30%</td><td>

Approximately $350,000.

Approximately 4.5 years.

Approximately $210,000.

See Attachment A.

Score: 60</td><td>

Approximately $418,040.

Approximately 3.5 years.

Approximately $306,748.

See Attachment A.

Score: 85</td><td>

Approximately $400,000.

Approximately 3.3 years.

Approximately $325,500.

See Attachment A.

Score: 90</td><td></td></tr>
<tr><td>Schedule Feasibility

An assessment of how long the solution will take to design and implement.</td><td>10%</td><td>Less than 3 months.

Score: 95</td><td>9–12 months

Score: 80</td><td>9 months

Score: 85</td><td></td></tr>
<tr><td>Ranking</td><td>100%</td><td>60.5</td><td>92</td><td>83.5</td><td></td></tr>
</table>

FIGURE 9.6 *Partially Completed Feasibility Matrix*

5. (REQ) Evaluate each candidate solution independently for operational, technical, economic, and schedule feasibility. Document your analysis of each candidate solution.

Activity: Recommend a System Solution

Once the feasibility analysis has been completed for each candidate solution, we can select a candidate solution to recommend. First, any infeasible candidates are usually eliminated from further consideration. Since we are looking for the most feasible solution of those remaining, we will identify and recommend the candidate that offers the "best overall" combination of technical, operational, economic, and schedule feasibilities. Rarely is one candidate found to be the most operational, technical, economic, and schedule feasible.

Purpose The purpose of this activity is to select a candidate solution to recommend. The candidate having the "best overall" operational, technical, economic, and schedule feasibility should be selected.

Roles The activity is facilitated by the *project manager*. Other roles are defined as follows:

FIGURE 9.7 *Outline of a Typical System Proposal*

I. Introduction
   A. Purpose of the report
   B. Background of the project leading to this report
   C. Scope of the project
   D. Structure of the report
II. Tools and techniques used
   A. Solution generated
   B. Feasibility analysis (cost–benefit)
III. Information systems requirements
IV. Alternative solutions and feasibility analysis
V. Recommendations
VI. Appendices

- System owner roles (defined in the previous activity).
  - Executive sponsor (REQ)—As the final spending authority, the sponsor must approve recommendations and project continuation.
  - User managers (REC)—The system belongs to these managers; therefore, their input is crucial.
  - System managers (REC)—System managers commit information services resources to projects; therefore, they need to be made aware of any scope, schedule, or budget changes for the project.
  - Project manager (REQ).
  - Steering body (OPT)—Many organizations require that all system proposals be formally presented to a steering body (sometimes called a steering committee) for final approval.
- System users (OPT) are not normally involved in this process.
- Systems analysts (OPT) may assume responsibility for this activity.
- Systems designer (REC) must make and defend the recommendation.
- System builder roles are not typically involved in this activity unless deemed necessary by the project manager.

Prerequisites (Inputs) This activity is triggered by the completion of the feasibility analysis of all candidate solutions. The key inputs to this activity include the *project plan* (REC), *size estimates* (REC), *candidate solutions* (REQ), and completed *feasibility analysis* (REQ).

Deliverables (Outputs) The principal deliverable of this activity is a formal written or verbal *system proposal* (REQ). This proposal is usually intended for the system owners who will normally make the final decision. The proposal will contain the project plans, size estimates, candidate solutions, and feasibility analysis. A typical outline for a system proposal is depicted in Figure 9.7. Based on the outcome of the proposal, *changes to proposed design* (REC) requirements are established for the new systems components we will "buy" or "make."

Applicable Techniques Finally, the techniques and skills needed to complete this activity are all cross life cycle skills:

- *Feasibility assessment.* Feasibility assessment (REQ) is discussed in Module C, "Feasibility and Cost–Benefit Analysis."
- *Report writing.* Report writing (REQ) is discussed in Module E, "Interpersonal Skills and Communications."
- *Verbal presentations.* Verbal presentations (REQ) are discussed in Module E, "Interpersonal Skills and Communications."

Steps The following steps are suggested to complete this activity.

1. (REC) Not all feasibility criteria are necessarily viewed as having equal importance in deciding which candidate is the best overall candidate. If appropriate, establish the weighting to be given to each feasibility criterion.
2. (REQ) Rank the candidates and determine the candidate with the best overall feasibility criteria ranking.
3. (REQ) Prepare a formal written systems proposal containing your analysis and recommendations.
4. (REC) Prepare and present an oral recommendation to management.

This concludes the configuration phase. With approval of the system proposal by the system owners, the project can now proceed to the next phase.

THE PROCUREMENT PHASE OF SYSTEMS DESIGN

The procurement of software and hardware is not necessary for all new systems. On the other hand, when new software or hardware is needed, the selection of appropriate products is often difficult. Decisions are complicated by technical, economic, and political considerations. A poor decision can ruin an otherwise successful analysis and design. The systems analyst is becoming increasingly involved in the procurement of software packages, peripherals, and computers to support specific applications being developed by that analyst.

There are four fundamental objectives of the configuration phase:

- To identify and research specific products that could support our recommended solution for the target information system.
- To solicit, evaluate, and rank vendor proposals.
- To select and recommend the best vendor proposal.
- To establish requirements for integrating the awarded vendor's products.

Your information systems framework (Figure 9.8) provides the context for the focus of the procurement phase. Alternative vendor hardware or software products to be considered should be those that provide the best overall support for the target information system. As shown in the model, the procurement phase is concerned with addressing specific build/buy technology products that support the DATA, PROCESS, INTERFACE, and GEOGRAPHY requirements for the target system.

Figure 9.9 is an activity diagram that illustrates the typical activities of the procurement phase. Let's now examine each activity in greater depth.

Activity: Research Technical Criteria and Options

The first activity is to research technical alternatives. This activity identifies specifications that are important to the hardware and/or software that is to be selected. The activity involves focusing on the hardware and/or software requirements established in the configuration phase. These requirements specify the functionality, features, and critical performance parameters for our new software/hardware.

Most analysts read appropriate magazines and journals to help them identify those technical and business issues and specifications that will become important to the selection decision. Other sources of information for conducting research include the following:

- *Internal standards* may exist for hardware and software selection. Some companies insist that certain technology will be bought from specific vendors if those vendors offer it. For instance, some companies have standardized on specific brands of microcomputers, terminals, printers, database management systems, network managers, data communications software, spreadsheets, and programming languages. A little homework here can save you a lot of unnecessary research.

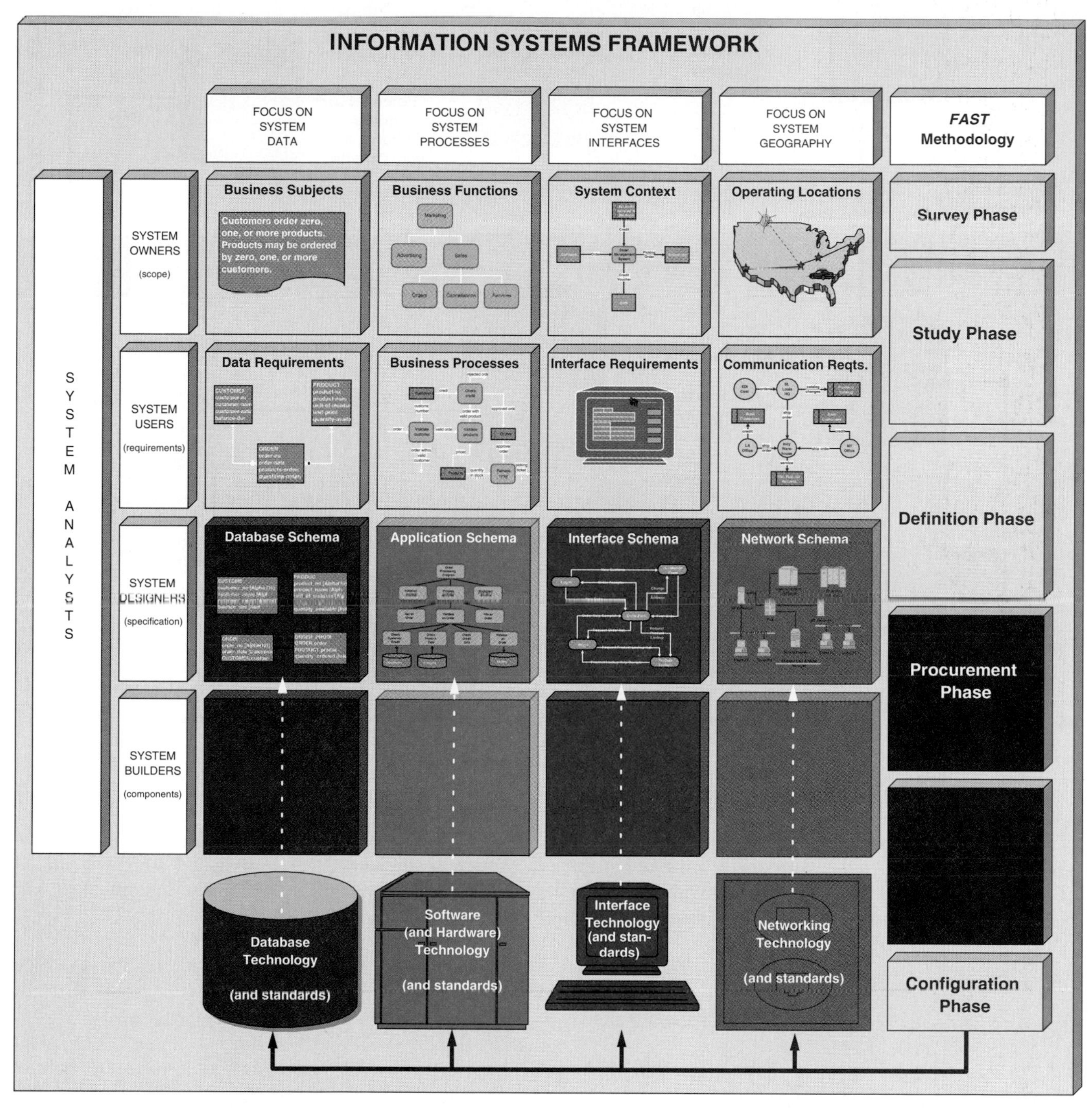

FIGURE 9.8 *Building Blocks for the Procurement Phase*

- *Information services* are primarily intended to constantly survey the marketplace for new products and advise prospective buyers on what specifications to consider. They also provide information such as the number of installations and general customer satisfaction with the products.
- *Trade newspapers and periodicals* offer articles and experiences on various types of hardware and software that you may be considering. Many can be found in school and company libraries. Subscriptions (sometimes free) are also available.

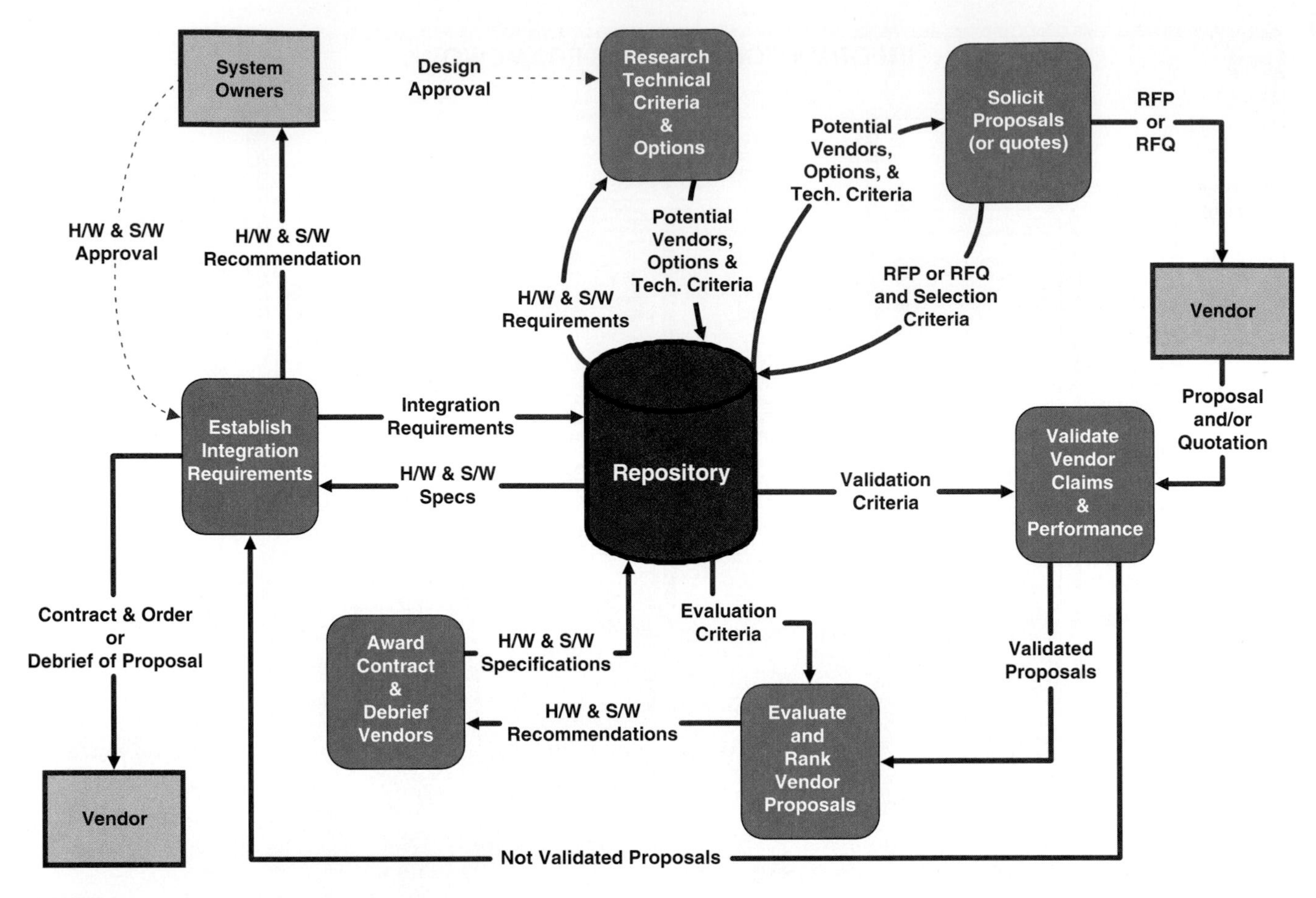

FIGURE 9.9 *Activity Diagram for the Procurement Phase*

The research should also identify potential vendors that supply the products to be considered. After the analysts have completed their homework, they will initiate contact with these vendors. Thus, the analysts will be better equipped to deal with vendor sales pitches after doing their research!

Purpose The purpose of this activity is to research technical alternatives to specify important criteria and options that will be important for the new hardware and/or software that is to be selected.

Roles The activity is facilitated by the *project manager.* Other roles are defined as follows:

- System owners are not involved in this activity.
- System users are usually not involved in this activity.
- Systems analyst roles are not normally involved in this activity.
- System designers (REQ) are responsible for the completion of this activity. The designer may seek input from the following people regarding the technical criteria: database administrator (REC), network administrator (REC), and applications administrator (REC).
- System builder roles are not typically involved in this activity unless deemed appropriate by a system owner.

Prerequisites (Inputs) This activity is triggered by the system owners' approval of a system proposal requiring new software or hardware. As is illustrated in Figure 9.9, a key input to this activity is the *hardware and/or software requirements* (REQ) established in the configuration phase. The analysts will also obtain additional *product and vendor facts* (REQ) from various sources. They are careful not to get their information solely from a salesperson because the number one rule of salesmanship is to emphasize the product's strengths and deemphasize its weaknesses.

Deliverables (Outputs) The principal deliverables of this activity include lists of *potential vendors* (OPT), *product options* (REQ), and *technical criteria* (REQ).

Applicable Techniques The following technique is applicable to this activity. For each activity, we indicate which chapters or modules of the book teach that technique.

- *Fact-finding.* Fact-finding methods (REQ) are used to obtain additional facts about products from various sources. Fact-finding is discussed in Part Five, Module B, "Fact-Finding and Information Gathering."

Steps The following steps are suggested to complete this activity.

1. (REQ) Conduct research to gain important facts concerning the hardware/software product and vendor. Carefully screen the various sources that may be utilized.
2. (REC) Identify potential vendors from which the products might be obtained. This step may be optional if your company has a commitment or contract to acquire certain products from a particular source.
3. (REC) Review the product, vendor, and supplier findings.

Activity: Solicit Proposals (or Quotes) from Vendors

The next activity is to solicit proposals from vendors. If your company is committed to buying from a single source (IBM, for example), the task is quite informal. You simply contact the supplier and request price quotations and terms. On the other hand, most decisions offer numerous alternatives. In this situation, good business sense dictates that you use the competitive marketplace to your advantage.

The solicitation activity requires the preparation of one of two documents: a **request for quotations (RFQ)** or a **request for proposals (RFP).** The request for quotations is used when you have already decided on the specific product, but that product can be acquired from several distributors. Its primary intent is to solicit specific configurations, prices, maintenance agreements, conditions regarding changes made by buyers, and servicing. The request for proposals is used when several different vendors and/or products are candidates and you want to solicit competitive proposals and quotes. RFPs can be thought of as a superset of RFQs. Both define selection criteria that will be used in a later validation activity.

The primary purpose of the RFP is to communicate requirements and desired features to prospective vendors. Requirements and desired features must be categorized as mandatory (must be provided by the vendor), extremely important (desired from the vendor but can be obtained in-house or from a third-party vendor), or desirable (can be done without). Requirements might also be classified by two alternate criteria: those that satisfy the needs of the systems and those that satisfy our needs from the vendor (for example, service).

Purpose The purpose is to solicit product proposals or quotes from candidate vendors.

Roles The activity is facilitated by the *project manager.* Other roles are defined as follows:

- System owners are not involved in this activity.
- System users are not involved in this activity.

- Systems analyst roles are not involved in this activity.
- System designers (REQ) are responsible for the completion of this activity. The designer may seek input from the following people in writing the RFP or RFQ: database administrator (REC), network administrator (REC), and applications administrator (REC).
- System builder roles are not typically involved in this activity unless deemed appropriate by a system owner.

Prerequisites (Inputs) The key inputs to this activity are the *potential vendors* (REQ), *options* (OPT), and *technical criteria* (REQ) that resulted from the previous research activity.

Deliverables (Outputs) The principle deliverable of this activity is the *RFP* (REQ) or *RFQ* (REQ) that is to be received by candidate vendors. The quality of an RFP has a significant impact on the quality and completeness of the resulting proposals. A suggested outline for an RFP is presented in Figure 9.10, since an actual RFP is too lengthy to include in this book.

Applicable Techniques Many of the skills you developed in Part Two, such as process and data modeling (REQ), can be very useful for communicating requirements in the RFP. We have found that vendors are very receptive to these tools because they find it easier to match products and options and package a proposal that is directed toward your needs. Other important skills include:

- *Report writing.* Report writing (REQ) is discussed in Part Five, Module E, "Interpersonal Skills and Communications."
- *Developing questionnaires.* Questionnaires (REC) are covered in Part Five, Module B, "Fact-Finding and Information Gathering."

Steps The following steps are suggested to complete this activity.

1. (REC) Collect and review the facts pertaining to potential vendors, options, and technical criteria.
2. (REQ) If your company buys from a single source, or if the desired product can only be obtained from a single source, contact that source and request a price quotation and terms.
3. (REQ) Prepare a request for quotation (RFQ) and send to all distributors from which the products can be obtained.
4. (REQ) Prepare a request for proposals (RFP) for those products for which you want to solicit competitive proposals and quotes. It is recommended that the RFP categorize product features as mandatory, extremely important, or desirable—or in some fashion aimed at clearly communicating their importance to the vendors.
5. (REC) If deemed necessary or helpful, hold a vendors bidding meeting to address important issues and questions.

Activity: Validate Vendor Claims and Performance

Soon after the RFPs or RFQs are sent to prospective vendors, you will begin receiving proposals and/or quotations. Because proposals cannot and should not be taken at face value, claims and performance must be validated. This activity is performed independently for each proposal; proposals are not compared with one another.

Purpose The purpose of this activity is to validate requests for proposals and/or quotations received from vendors.

Roles The activity is facilitated by the *project manager*. Other roles are defined as follows:

FIGURE 9.10 *Request for Proposals (RFP)*

I. Introduction
  A. Background
  B. Brief summary of needs
  C. Explanation of RFP document
  D. Call for action on part of vendor
II. Standards and instructions
  A. Schedule of events leading to contract
  B. Ground rules that will govern selection decision
    1. Who may talk with whom and when
    2. Who pays for what
    3. Required format for a proposal
    4. Demonstration expectations
    5. Contractual expectations
    6. References expected
    7. Documentation expectations
III. Requirements and features
  A. Hardware
    1. Mandatory requirements, features, and criteria
    2. Essential requirements, features, and criteria
    3. Desirable requirements, features, and criteria
  B. Software
    1. Mandatory requirements, features, and criteria
    2. Essential requirements, features, and criteria
    3. Desirable requirements, features, and criteria
  C. Service
    1. Mandatory requirements
    2. Essential requirements
    3. Desirable requirements
IV. Technical questionnaires
V. Conclusion

- System owners are not involved in this activity.
- System users are not involved in this activity.
- Systems analyst roles are not involved in this activity.
- System designers (REQ) are responsible for the completion of this activity. The designer may involve the following individuals in validating the proposals: database administrator (REC), network administrator (REC), and applications administrator (REC).
- System builder roles are not typically involved in this activity unless deemed appropriate by a system owner.

Prerequisites (Inputs) This activity is triggered by the receipt of *proposals and/or quotations* (REQ) from prospective vendors.

Deliverables (Outputs) The key outputs of this activity are those vendor proposals that proved to be *validated proposals* (REQ) or claims, and others whose claims were *not validated* (OPT).

Applicable Techniques The following technique is applicable to this activity.

- *Interpersonal skills.* Interpersonal skills (REQ) impact the way we communicate and negotiate with one another. Clearly, good interpersonal relations are essential to this activity. Interpersonal skills are taught in Part Five, Module E, "Interpersonal Skills and Communications."

Steps The following steps are suggested to complete this activity.

1. (REC) Collect and review all facts pertaining to product requirements and features.
2. (REQ) Review vendor proposals and eliminate any proposal that does not meet all your mandatory requirements. If you clearly specified your requirements, no vendor should have submitted such a proposal. For proposals that cannot meet one or more extremely important requirements, verify that the requirements or features can be fulfilled by some other means.
3. (REQ) For each vendor proposal not eliminated, validate the vendor claims and promises against validation criteria. Claims about mandatory, extremely important, and desirable requirements and features can be validated by completed questionnaires and checklists (included in the RFP) with appropriate vendor-supplied references to user and technical manuals. Promises can be validated only by ensuring that they are written into the contract. Performance is best validated by a demonstration, which is particularly important when you are evaluating software packages. Demonstrations allow you to obtain test results and findings that confirm capabilities, features, and ease of use.

Activity: Evaluate and Rank Vendor Proposals

The validated proposals can now be evaluated and ranked. The evaluation and ranking task is, in reality, another cost–benefit analysis performed during systems development. The evaluation criteria and scoring system should be established before the actual evaluation occurs so as not to bias the criteria and scoring to subconsciously favor any one proposal.

Purpose The purpose of this activity is to evaluate and rank all *validated* vendor proposals.

Roles Ideally, the activity should be facilitated by the *executive sponsor*. Other roles are defined as follows:

- System owners are not involved in this activity.
- System users are not involved in this activity.
- Systems analyst roles are not involved in this activity.
- System designers (REQ) are responsible for the completion of this activity. The designer may involve the following individuals in evaluating and ranking the proposals: database administrator (REC), network administrator (REC), and applications administrator (REC).
- System builder roles are not typically involved in this activity unless deemed appropriate by a system owner.

Prerequisites (Inputs) The inputs include *validated proposals* (REQ) and the *evaluation criteria* (REQ) to be used to rank the proposals.

Deliverables (Outputs) The key deliverable of this activity is the *hardware and/or software recommendation* (REQ).

Applicable Techniques The following technique is applicable to this activity.

- *Feasibility assessment.* Once again the ability to perform a feasibility assessment (REQ) is an extremely important skill requirement. Feasibility assessment techniques and skills are covered more fully in Part Five, Module C, "Feasibility and Cost–Benefit Analysis."

Steps The following steps are suggested to complete this activity.

1. (REC) Collect and review all details concerning the validated proposals.
2. (REQ) Establish an evaluation criteria and scoring system. Some methods suggest that requirements be weighted on a point scale. Better approaches use dollars and cents! Monetary systems are easier to defend to management than points. One such technique is to evaluate the proposals on the basis of hard and soft dollars. Hard-dollar costs are the costs you will have to pay to the selected vendor for the equipment or software. Soft-dollar costs are additional costs you will incur if you select a particular vendor. (For instance, if you select vendor A, you may incur an additional expense to vendor B to overcome a shortcoming of vendor A's proposed system.) This approach awards the contract to the vendor who fulfills all essential requirements while offering the lowest total hard-dollar cost plus soft-dollar penalties for desired features not provided (for a detailed explanation of this method see Isshiki, 1982, or Joslin, 1977).
3. (REQ) Evaluate and rank the vendor proposals.

Activity: Award (or Let) Contract and Debrief Vendors

Having ranked the vendor proposals, the next activity usually includes presenting a recommendation to management for final approval. Once again, communication skills, especially salesmanship, are important if the analyst is to persuade management to follow the recommendations. Given management's approval of the recommendation, a contract must then be drawn up and awarded to the winning vendor. This activity often also includes debriefing losing vendors, being careful not to burn bridges.

Purpose The purpose of this activity is to negotiate a contract with the vendor who supplied the winning proposal and to debrief those vendors that submitted losing proposals.

Roles Ideally, the activity should be facilitated by the *executive sponsor.* Other roles are defined as follows:

- System owner roles (defined in the previous activity).
  - Executive sponsor (REQ)—As the final spending authority, the sponsor must approve recommendations and project continuation.
  - User managers (REC)—The system belongs to these managers; therefore, their input is crucial.
- System users (OPT) are not normally involved in this process.
- Systems analyst (OPT) may assume responsibility for this activity.
- System designer (REC) must make and defend the recommendation and award the contract. The system designer may involve a company lawyer in drafting the contract.
- System builder roles are not typically involved in this activity unless deemed necessary by the project manager.

Prerequisites (Inputs) The inputs include *validated proposals* (REQ) and the *evaluation criteria* (REQ) to be used to rank the proposals.

Deliverables (Outputs) This activity results in a *hardware and software recommendation* (REQ) that must receive final approval from the system owners. Pending the approval, a *contract order* (REQ) would subsequently be produced for the "winning" vendor. A *debriefing of proposals* (OPT) would be provided for the losing vendors.

Applicable Techniques The following techniques are applicable to this activity.

- *Report writing.* Many of the report writing (REQ) skills discussed in Module E, "Interpersonal Skills and Communications," can be transferred to the contract writing effort.
- *Verbal presentations.* Verbal presentations (REQ) are discussed in Module E, "Interpersonal Skills and Communications."

Steps The following steps are suggested to complete this activity.

1. (REC) Having ranked the proposals (see previous activity), the analyst usually presents a hardware and software recommendation for final approval.
2. (REQ) Once the final hardware and software approval decision is made, a contract must be negotiated with the winning vendor. Certain special conditions and terms may have to be written into the standard contract and order. Ideally, no computer contract should be signed without the advice of a lawyer. For microcomputers and software, legal advice can be prohibitively expensive (compared to the cost of the products). In this case the analyst must be careful to read and clarify all licensing agreements. No final decision should be approved without the consent of a qualified accountant or management. Purchasing, leasing, and leasing with a purchase option involve complex tax considerations.
3. (REC) Out of common courtesy and to maintain good relationships, provide a debriefing of proposals for losing vendors. The purpose of this meeting is not to allow the vendors a second-chance opportunity to be awarded the contract; rather, the briefing is strictly intended to inform the losing vendors of precise weaknesses in their proposal and/or products.

Activity: Establish Integration Requirements

Finally, given the hardware and software specifications of the awarded vendor's products, the next activity is to determine how they can be integrated with other existing information systems. It is not merely enough to purchase or build systems that fulfill the target system requirements. The analyst must integrate or interface the new system to the myriad of other existing systems that are essential to the business. Many of these systems may use dramatically different technology, techniques, and file structures.

The analyst must consider how the target system fits into the federation of systems of which it is a part. The integration requirements that are specified are vital to ensuring that the target system will work in harmony with those systems.

Purpose The purpose of this activity is to establish requirements necessary for integrating the awarded vendor's products into the company's existing federation of information systems.

Roles The activity is facilitated by the *project manager*. Other roles are defined as follows:

- System owners are not involved in this activity.
- System users are not involved in this activity.
- Systems analyst roles are not normally involved in this activity.
- System designers (REQ) are responsible for the completion of this activity. The designer may seek input from the following people regarding the integration of new technology into existing applications: database administrator (REC), network administrator (REC), and applications administrator (REC).
- System builder roles are not typically involved in this activity unless deemed appropriate by a system owner.

Prerequisites (Inputs) The input to this activity is the *hardware and/or software specifications* (REQ) of the awarded vendor's products.

Deliverables (Outputs) The principal deliverable of this activity is a set of *integration requirements* (REQ) for ensuring the systems will work in harmony with other production systems.

Applicable Techniques The following techniques are applicable to this activity. For each activity, we indicate which chapters or modules of the book teach that technique.

- *Data and process modeling.* Data and process models (REC) are frequently used to document systems. These "blueprints" can depict "integration" or interfacing points for different systems and business processes. Data modeling is covered in Chapter 5. Process modeling is covered in Chapter 6.

Steps The following steps are suggested to complete this activity.

1. (REQ) Collect and review the hardware and software specifications of the awarded vendor's products.
2. (REC) Review data and process models for the new system to discover how the vendor products will fit into the overall scheme of the new system.
3. (REQ) Revise data and process models to reflect integration or impact of new products.

One thing will *not* change! We will continue to need systems analysts and business analysts who understand how to fundamentally analyze business problems and define logical business requirements as a preface to system design. But we will all have to do that with increased speed and accuracy to meet the accelerated systems development schedules required in tomorrow's faster-paced economy.

THE DESIGN AND INTEGRATION PHASE OF SYSTEMS DESIGN

Now we come to a more traditional phase of the systems design, the design and integration phase. Given design and integration requirements for the target system, this phase involves developing technical design specifications.

The goal of the design and integration phase is twofold. First and foremost, the analyst seeks to design a system that both fulfills requirements and will be friendly to its end-users. Human engineering will play a pivotal role during design. Second, and still very important, the analyst seeks to present clear and complete specifications to the computer programmers and technicians.

Your information system framework (Figure 9.11) provides the context for the focus of the design and integration phase.

GEOGRAPHY. During the systems analysis phase, we established the network requirements for the target system. Now we need to analyze and distribute the systems data and processes.

DATA. We specified the content of each data and information flow during the definition phase. We specified the media during the selection phase. Now we need to design the style, organization, and format of all inputs and outputs. We also must specify format, organization, and access methods for all files and databases to be used in the computer-based system.

PROCESSES. During design, the sequence of steps and flow of control through the new system must be specified. The processing methods and intermediate manual procedures must also be clearly documented.

INTERFACE. During design, the designer must specify the interfaces that will exist between both systems and people.

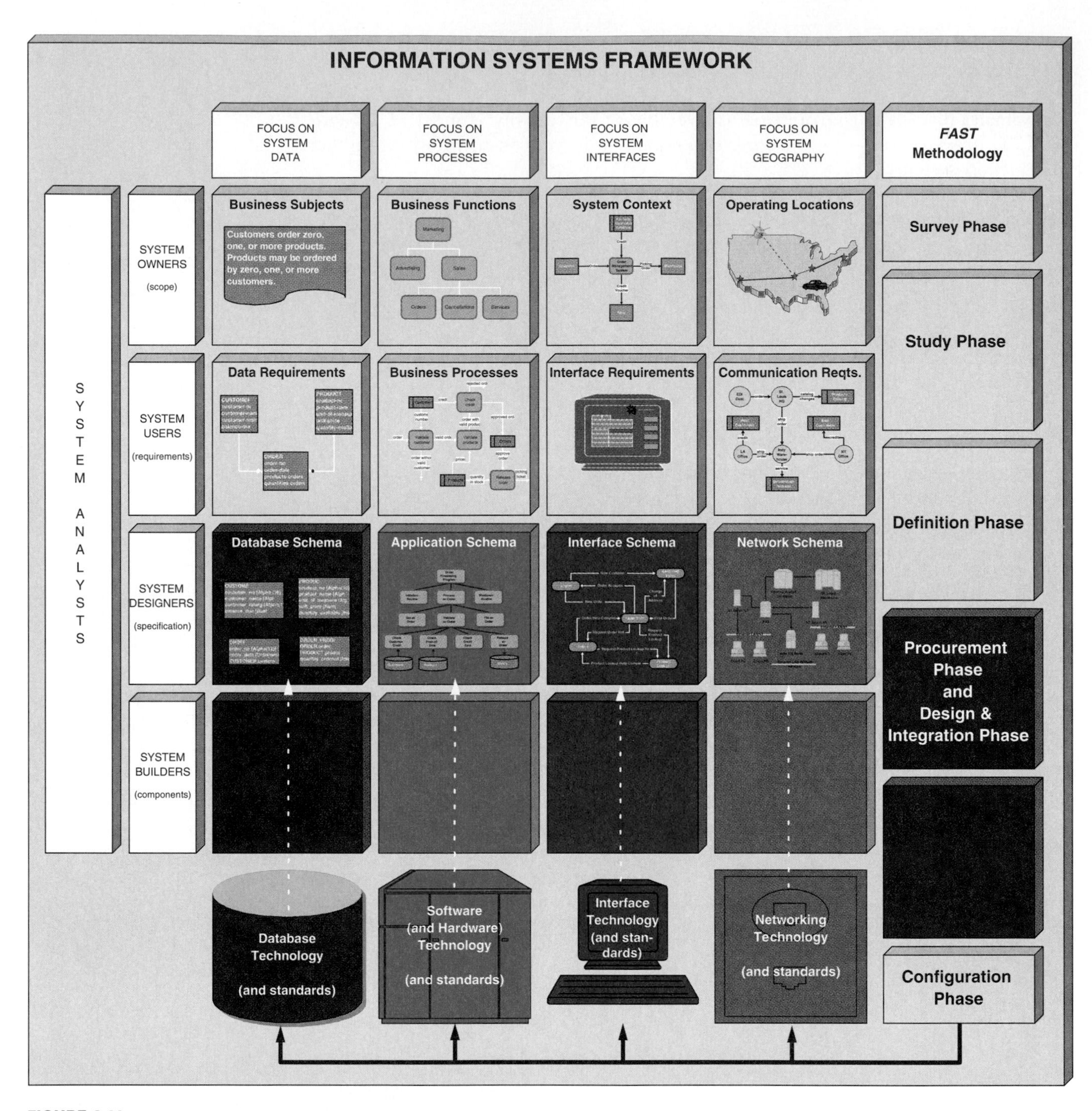

FIGURE 9.11 *Building Blocks for the Design and Integration Phase*

Clearly, the physical design phase gets into considerably greater detail than any of the previous phases of the life cycle.

Figure 9.12 is an activity diagram that illustrates the typical activities of the design and integration phase.

Activity: Analyze and Distribute Data

Before the analyst can design computer files and/or databases for the target system, the analyst must perform some additional analysis and address distribution

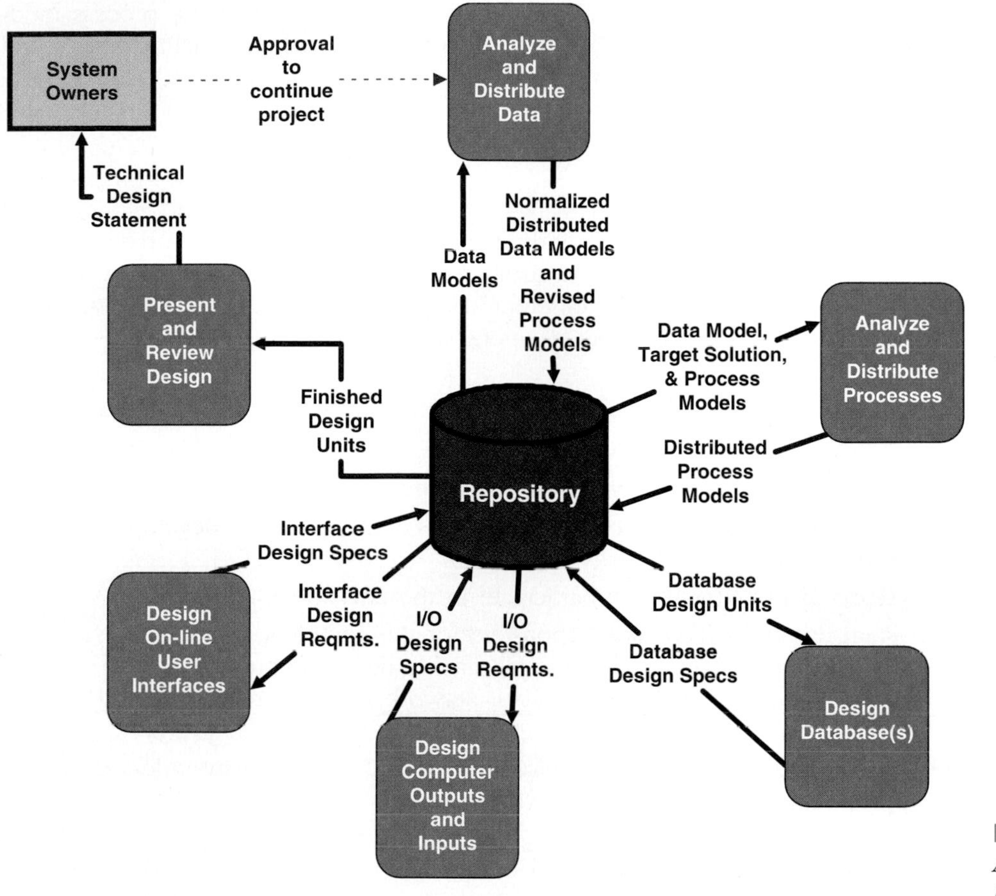

FIGURE 9.12
Activity Diagram for the Design and Integration Phase

issues of the data. If you recall, a data model of the target system already exists. That model was created during systems analysis as a tool for communicating and documenting the data requirements for the target system. However, that data model does not usually represent a good file or database design. In fact, it may contain structural characteristics that may lead to numerous problems.

During this activity, the analyst will work closely with users to develop a good data model—that is, a data model that will allow development of ideal file and database solutions. Data analysis is the technique used to derive a good data model.

Data analysis is a procedure that prepares a data model for implementation as a nonredundant, flexible, and adaptable file/database.

Normalization is the procedure that is used to simplify entities, eliminate redundancy, and build flexibility and adaptability into the data model.

Normalization of data refers to the way data attributes are grouped to form stable, flexible, and adaptive entities.

Once data analysis has been completed, event analysis will be performed to address the analyst's obligations to ensure that the end-users' data will be kept accurate and up to date.

Event analysis is a technique that studies the entities of a fully normalized data model to identify business events and conditions that cause data to be created, deleted, or modified.

Since data and event analysis will likely have an impact on the process models for the target system, the target system data flow diagrams (DFDs) may need to be revised. Data analysis, event analysis, and their impact on DFDs were covered in Chapter 6. The end products of this first activity are the normalized distributed data models and revised process models.

Purpose The purpose of this activity is to develop a *good* data model—one that is simple, nonredundant, flexible and adaptable to future needs, and that will allow the development of ideal file and database solutions. In addition, when the systems data are to be shared by different sites, this activity will make decisions on how the data are to be distributed among the locations.

Roles The activity is facilitated by the *project manager*. Other roles are defined as follows:

- System owners are not involved in this activity.
- System users (REC) may be involved in this activity to help develop the data model.
- Systems analysts (REC) may participate in the data modeling effort.
- System designers (REQ) are responsible for the completion of this activity. The following individuals may play a role in the data distribution decision making: database administrator (REC), network administrator (REC), and applications administrator (REC).
- System builder roles are not typically involved in this activity unless deemed appropriate by a system owner.

Prerequisites (Inputs) As is illustrated in Figure 9.12, a key input to this activity is the existing *data model* from systems analysis. In addition, this activity may have an impact on existing *process models,* which would then have to be revised.

Deliverables (Outputs) The principal deliverables of this activity are the *normalized distributed data models* (REQ) and revised *process models* (OPT). You will learn how to develop these models in Chapters 11 and 10, respectively.

Applicable Techniques The following techniques are applicable to this activity. For each activity, we indicate which chapters or modules of the book teach that technique.

- *Data modeling*. Data modeling (REQ) is covered in Chapter 5.
- *Process modeling*. Process modeling (REQ) is covered in Chapter 6.
- *Data analysis and normalization*. Data analysis and normalization (REQ) are covered in Chapter 11, "Database Design."

Steps The following steps are suggested to complete this activity.

1. (REQ) Collect existing data and process models constructed during systems analysis.
2. (REQ) Perform data analysis and normalization on the data models. Be sure to involve the users.
3. (REQ) If the system has different locations, determine how the data will be distributed across the locations.
4. (REQ) Perform event analysis on each data item on the data model.
5. (REQ) If process models were previously completed, revise any impacted models to reflect new business events and conditions.

Activity: Analyze and Distribute Processes

We can now shift our focus away from data and toward processes. The revised data and process models can now be used to address distribution issues for the target system processes. In Chapter 10, we detail network requirements for the new system, and you'll learn how those requirements can be addressed and reflected within the design of the target system.

Given the data model diagram, target solution, and process models, the analyst will develop distributed process models. To complete this activity, the analyst may involve a number of system designers and users.

Purpose The purpose of this activity is to analyze and distribute system processes to fulfill network requirements for the new system.

Roles The activity is facilitated by the *project manager.* Other roles are defined as follows:

- System owners are not involved in this activity.
- System users (REC) may be involved in this activity to help address business process issues.
- Systems analysts (REC) may participate in the data modeling effort.
- System designers (REQ) are responsible for the completion of this activity. The following individuals may play a role in the process distribution decision making: database administrator (REC), network administrator (REC), and applications administrator (REC).
- System builder roles are not typically involved in this activity unless deemed appropriate by a system owner.

Prerequisites (Inputs) As is illustrated in Figure 9.12, key inputs to this activity include the existing *entity relationship diagrams* (REQ), details about the *application architecture* (REQ), and the *process models* (REQ).

Deliverables (Outputs) The principal deliverable of this activity is the *distributed process models* (REQ) and *design units* (REQ).

Applicable Techniques The following techniques are applicable to this activity. For each activity, we indicate which chapters or modules of the book teach that technique.

- *Physical database design.* Database design (REQ) is covered in Chapter 11.
- *Physical process modeling.* Process modeling (REQ) is covered in Chapter 10.

Steps The following steps are suggested to complete this activity.

1. (REQ) Collect and review existing data and process models.
2. (REQ) Determine which essential processes will be implemented as computer processes and which as manual.
3. (REQ) Based on response time requirements, establish batch versus on-line computer processes.
4. (REQ) Factor the new systems into separate design units. Group processes that are related because they are involved in the processing of a particular business transaction or because they are triggered by common business process cycles, or events (daily, weekly, monthly, etc.).

5. (REQ) Develop network topology diagrams to document the locations or geography of the system.
6. (REQ) Distribute data and processes to these locations. Document these decisions in design unit data flow diagrams.
7. (REQ) Assign technology to design units. Using the technology approved in the earlier design phases, assign appropriate technology to the different design units.

This concludes those activities that are considered part of general design. Let's now discuss those activities that make up what is considered the more traditional design activities, or detailed design.

Activity: Design Databases

Typically the first activity of detailed design is to develop the corresponding database design specifications. The design of data goes far beyond the simple layout of records. Databases are a shared resource. Many programs will use them. Future programs may use databases in ways not originally envisioned. Consequently, the designer must be especially attentive to designing databases that are adaptable to future requirements and expansion.

The designer must also analyze how programs will access the data in order to improve performance. You may already be somewhat familiar with various programming data structures and their impact on performance and flexibility. These issues affect database organization decisions. Other issues to be addressed during database design include record size and storage volume requirements. Finally, because databases are shared resources, the designer must also design internal controls to ensure proper security and disaster recovery techniques, in case data are lost or destroyed.

Purpose The purpose of this activity is to prepare technical design specifications for a database that will be adaptable to future requirements and expansion.

Roles The activity is facilitated by the *project manager*. Other roles are defined as follows:

- System owners are not involved in this activity.
- System users are not involved in this activity.
- Systems analysts (REC) may participate in the data modeling effort.
- System designers (REQ) are responsible for the completion of this activity. The data administrator (REC) may participate (or complete) the database design. Most likely the new system uses some portion of an existing database. This is where the knowledge of the database administrator is crucial.
- System builders (OPT) may become involved at this stage of design. They may be asked to build a prototype database for the project.

Prerequisites (Inputs) As is illustrated in Figure 9.12, a key input to this activity is the *database design units* (REQ).

Deliverables (Outputs) The principle deliverable of this is the *database design specifications* (REQ).

Applicable Techniques The following technique is applicable to this activity. For each activity, we indicate which chapters or modules of the book teach that technique.

- *Database design*. Database design (REQ) is covered in Chapter 11.

Steps The following steps are suggested to complete this activity.

1. (REQ) Collect and review requirements for database design units.
2. (REQ) Design the logical schema for the database. A **schema** is the structural model for a database. It is a picture or map of the records and relationships to be implemented by the database.
3. (REC) Prototype the database (if necessary). Prototype databases should be quickly created, loaded with test data, and tested.

Activity: Design Computer Outputs and Inputs

Once the database has been designed and possibly a prototype built, the system designer can work closely with system users to develop input and output specifications. Because end-users and managers will have to work with inputs and outputs, the designer must be careful to solicit their ideas and suggestions, especially regarding format.

Transaction outputs will frequently be designed as preprinted forms onto which transaction details will be printed. Reports and other outputs are usually printed directly onto paper or displayed on a terminal screen. In any event, the precise format and layout of the outputs must be specified. Finally, internal controls must be specified to ensure the outputs are not lost, misrouted, misused, or incomplete.

For inputs, it is crucial to design the data capture method to be used. For instance, you may design a form on which data to be input will be initially recorded. You want to make it easy for the data to be recorded on the form, but you also want to simplify the entry of the data from the form into the computer or onto a computer-readable medium. This is particularly true if the data are to be input by people who are not familiar with the business application.

Finally, any time you input data to the system, you can make mistakes. We need to define editing controls to ensure the accuracy of input data.

Purpose The purpose of this activity is to prepare technical design specifications for user inputs and outputs.

Roles The activity is facilitated by the *project manager*. Other roles are defined as follows:

- System owners are not involved in this activity.
- System users (REQ) should be involved in this activity! The inputs and outputs are what they will see and work with. The degree to which they are involved is emphasized in design efforts that involve prototyping. They will be asked to provide feedback regarding each input/output prototype.
- System analysts (REC) may participate in the data modeling effort.
- System designers (REQ) are responsible for the completion of this activity. They may draw on the expertise of system designers that specialize in graphical user interface design.
- System builders (OPT) may prototype the inputs and outputs for the system.

Prerequisites (Inputs) As is illustrated in Figure 9.12, the key inputs to this activity are the *input and output design requirements* (REQ) specified during systems analysis.

Deliverables (Outputs) The principal deliverables of this activity are the *input and output design specifications* (REQ).

Applicable Techniques The following techniques are applicable to this activity. For each activity, we indicate which chapters or modules of the book teach that technique.

- *Input design and prototyping.* Input design and prototyping (REQ) are covered in Chapter 12.
- *Output and prototyping.* Output design and prototyping (REQ) are covered in Chapter 13.

Steps The following steps are suggested to complete this activity.

1. (REQ) Collect and review input and output design requirements.
2. (REQ) Determine methods and medium for each input and output.
3. (REC) Prototype inputs and outputs. Optionally, and although not common, traditional paper documentation could substitute or complement prototypes.

Activity: Design On-line User Interface

This activity is omitted from many designs. However, for on-line systems, the development of interface (the dialogue between the end-user and the computer) design specifications from interface design requirements may be the most critical design activity. Too many on-line systems are difficult to learn and use because they exhibit poor human engineering.

The idea behind user interface design is to build an easy-to-learn and easy-to-use dialogue for the user's new system. This dialogue must consider such factors as terminal familiarity, possible errors and misunderstandings that the end-user may have or may encounter, the need for additional instructions or help at certain points, and screen content and layout. Essentially, you are trying to anticipate every little error or keystroke that an end-user might make—no matter how improbable. Furthermore, you are trying to make it easy for the end-user to understand what the screen is displaying at any given time.

Purpose The purpose of this activity is to prepare technical design specifications for an on-line user interface.

Roles The activity is facilitated by the *project manager.* Other roles are defined as follows:

- System owners are not involved in this activity.
- System users (REQ) should be involved in this activity! The on-line interface is what they will see and work with. The degree to which they are involved is emphasized in design efforts that involve prototyping.
- System analysts (REC) may participate in the data modeling effort.
- System designers (REQ) are responsible for the completion of this activity. They may draw on the expertise of system designers that specialize in graphical user interface design.
- System builder (OPT) roles are not typically involved in this activity unless deemed appropriate by a system owner.

Prerequisites (Inputs) As is illustrated in Figure 9.12, the key inputs to this activity are *interface design requirements* (REQ) specified during systems analysis.

Deliverables (Outputs) The principal deliverable of this activity is the *interface design specification* (REQ).

Applicable Techniques The following techniques are applicable to this activity. For each activity, we indicate which chapters or modules of the book teach that technique.

- *User interface design and prototyping.* User interface design (REQ) and prototyping (REC) are covered in Chapter 14.

Steps The following steps are suggested to complete this activity.

1. (REQ) Collect and review input and output design specifications.
2. (REC) Study the users' behavioral characteristics. Are they computer literate? Do they have any handicaps? What type of computer interfaces do they work with? Get to know the users. Your knowledge should be reflected in the overall interface design.
3. (REQ) If they exist, review interface design standards.
4. (REQ) Prototype the user interface; be sure to involve the users. This should be an iterative process of building the model, getting user feedback, and making revisions!

Activity: Present and Review Design

This final detailed design activity packages all the specifications from the previous tasks into computer program specifications that will guide the computer programmer's activities during the construction phase of the systems development life cycle.

There is more to this task than packaging, however. How much more depends on where you draw the line between the system designer's and computer programmer's responsibilities (this issue is moot if the analyst and programmer are the same person). In addition to packaging, this activity may require that you determine the overall program structure. There are numerous strategies for top-down, modular decomposition, which will be surveyed in Chapter 15.

Purpose The purpose of this activity is to prepare technical design specifications for an on-line user interface.

Roles The activity is facilitated by the *project manager.* The systems design should be reviewed with all appropriate audiences, which may include the following:

- System users—End-users have already seen and approved the outputs, inputs, and terminal dialogue. The overall work and data flow for the new system should get a final walk-through and approval.
- Technical support staff—Computer center operations management and staff should get a final chance to review the technical specifications to be sure nothing has been forgotten and so they can commit computer time to the construction and delivery phases of the project.
- Audit staff—Many firms have full-time audit staffs whose job it is to pass judgment on the internal controls in a new system.
- System owner roles
  - Executive sponsor (REQ)—the highest-level manager who will pay for and support the project. Management should get a final chance to question the project's feasibility, given the latest cost-benefit estimates.
  - User managers (REQ)—the manager(s) of the organizational units most likely to be supported by the system developed in this project.
  - System managers (OPT)—the information systems unit manager(s) to whom the project manager report (e.g., the manager of systems development).
  - Project manager (REQ)—the information systems unit manager who will directly manage the construction project team.
- Systems analysts are not normally involved in this activity.
  - System modelers (REC)—systems analysts who are skilled with the system modeling techniques and CASE tools that will be used in the project.

- System designers normally complete this activity and may involve a walk-through with other design specialists to confirm the design.
- System builder roles are not typically involved in this activity.

Prerequisites (Inputs) As is illustrated in Figure 9.12, the key inputs to this activity are *finished design units* (REQ).

Deliverables (Outputs) The principal deliverable of this activity is the *technical design statement* (REQ).

Applicable Techniques The following techniques are applicable to this activity. For each activity, we indicate which chapters or modules of the book teach that technique.

- *Feasibility assessment.* Feasibility assessment (REQ) is discussed in Module C, "Feasibility and Cost–Benefit Analysis."
- *Report writing.* Report writing (REQ) is discussed in Module E, "Interpersonal Skills and Communications."
- *Verbal presentations.* Verbal presentations (REQ) are discussed in Module E, "Interpersonal Skills and Communications."
- *Project management.* Project management (REQ) skills are discussed in Module A, "Project and Process Management."

Steps The following steps are suggested to complete this activity.

1. (REQ) Prepare an implementation plan that presents a proposed schedule for the construction and delivery phases (detailed in Chapter 18).
2. (REQ) Prepare a final cost–benefit analysis that determines if the design is still feasible.
3. (REQ) Prepare a written technical design statement. The final technical design statement specifications are typically organized into a workbook or technical report. Technical design specifications evolve from the essential requirements specifications that were prepared during systems analysis. Thus, the project repository that was started during systems analysis will eventually become the design specifications document.

WHERE DO YOU GO FROM HERE?

This chapter provided a detailed overview of the systems design phases of a project. You are now ready to learn some of the systems design skills introduced in this chapter. Because systems design is dependent on requirements specified during systems analysis, we recommend that you first complete Chapters 4 to 8. Chapter 4 gives you an overview of systems analysis. Chapters 5 to 8 teach different system modeling techniques that provide for basic inputs to the systems design activities presented in Part Three.

The order of the systems design chapters that follow is flexible; however, the authors did present the subsequent techniques in the sequence they are commonly completed for systems design projects.

SUMMARY

1. Formally, systems design is the evaluation of alternative solutions and the specification of a detailed computer-based solution. With respect to information system development, systems design consists of the configuration, procurement, and design and integration phases.
2. There are many popular strategies or techniques for performing systems design. These techniques can be used in combination with one another.

   Modern structured design, a technique that focuses on processes.

   Information engineering (IE), a technique that focuses on data and strategic planning to produce application projects.

   Prototyping, a technique that is an iterative process involving a close working relationship between designers and users to produce a model of the new system.

   Joint application development (JAD), a technique that emphasizes participative development among system owners, users, designers, and builders. During JAD sessions for systems design, the system designer takes on the role of the facilitator.

   Rapid application development (RAD), a technique that represents a merger of various structured techniques with prototyping and JAD to accelerate systems development.

   Object-oriented design (OOD), a new design strategy that follows up object-oriented analysis to refine object requirement definitions and to define new design specific objects.

3. Each phase of systems analysis (survey, study, and definition) can be understood in the context of the information system building blocks: PEOPLE, DATA, PROCESSES, INTERFACES, GEOGRAPHY, and TECHNOLOGY.
4. The purpose of the configuration phase is to identify candidate solutions, analyze those candidate solutions, and recommend a target system that will be designed and implemented. To accomplish the configuration phase objectives, the system designer will work with people to: (*a*) define candidate solutions, (*b*) analyze the feasibility of alternative solutions, and (*c*) recommend a system solution. The output of the configuration phase is a system proposal intended for system owners or a steering committee who will make the final decision.
5. The procurement phase, if necessary, is aimed at selecting appropriate hardware and/or software products for the new system. The fundamental objectives of the procurement phase are to:
   a. Identify and research specific products that could support our recommended solution for the target information system.
   b. Solicit, evaluate, and rank vendor proposals.
   c. Select and recommend the best vendor proposal.
   d. Establish requirements for integrating the awarded vendor's products.

   The deliverable for the procurement phase is the contract order that would be sent to the winning vendor. In addition, a set of integration requirements is created for ensuring that the vendor's products will work in harmony with other product systems.
6. The design and integration phase represents the more traditional phase of systems design. This phase involves developing technical design specifications that will guide the construction of the new system. To complete the design and integration phase, the system designer must complete the following activities:
   a. Analyze and distribute data.
   b. Analyze and distribute processes.
   c. Design databases.
   d. Design computer inputs and outputs.
   e. Design on-line user interfaces.
   f. Present and review the design.

   The output of the design and integration phase is the technical design statement. This output will guide the system builders as the project moves on to construction.

KEY TERMS

cohesive modules, p. 313
coupled modules, p. 313
data analysis, p. 337
economic feasibility, p. 321
event analysis, p. 337
information engineering (IE), p. 314
joint application development (JAD), p. 316
modern structured design, p. 312
normalization, p. 337
object-oriented design, p. 317
operational feasibility, p. 321
physical design, p. 321
rapid application development (RAD), p. 316
request for proposal (RFP), p. 329
request for quotation (RFQ), p. 329
roles, p. 317
schedule feasibility, p. 321
schema, p. 341
structure chart, p. 313
systems design, p. 312
technical feasibility, p. 321

REVIEW QUESTIONS

1. What is the difference in *focus* between systems analysis and systems design?
2. Explain the relationship between prototyping, JAD, and RAD.
3. What three phases make up systems design?
4. What are the objectives of the configuration phase? How might the information system building blocks be used to identify the general level of understanding required for fulfilling these objectives?
5. How might an existing technology architecture impact the configuration phase?
6. Define the terms *economic feasibility, technical feasibility, operational feasibility,* and *schedule feasibility.*
7. What are the objectives of the procurement phase? How might the information system building blocks be used to identify the general level of understanding required for fulfilling these objectives?
8. Is the procurement activity of the configuration phase always required? Why or why not?
9. Explain the difference between a request for proposal (RFP) and a request for quotation (RFQ).
10. What are the objectives of the design and integration phase? How might the information system building blocks be used to identify the general level of understanding required for fulfilling these objectives?
11. What are the end products of the configuration, procurement, and design and integration phases?

PROBLEMS AND EXERCISES

1. How can a successful and thorough systems analysis be ruined by a poor systems design? Answer the question relative to these factors:
   a. The impact on the subsequent implementation (in other words, the systems implementation phases, which you studied in Chapter 3).
   b. The lifetime of the system after it is placed into operation.
   c. The impact on future projects.
2. What skills are important during systems design? Create an itemized list of these skills. Identify other computer, business, and general education courses that would help you develop or improve your skills. Prepare a plan and schedule for taking the courses. (If you are not in school, prepare a plan for using available corporate training resources, reading appropriate books, enrolling in seminars or continuing education courses, etc.) Review your plan with your counselor, advisor, or instructor.
3. What by-products of the systems analysis phases are used in the systems design phases? Why are they important? How are they used? What would happen if they were incomplete or inaccurate?
4. What are the end products of the configuration, procurement, and design and integration phases? What is the content of each end product?
5. United Films Cinemas has asked you to help it acquire microcomputer systems for its theaters and main office. Write a letter that proposes a disciplined approach to acquiring an appropriate system. Assume your end-user is inclined to ignore a disciplined approach and would prefer to go to the local computer store and just buy something. In other words, defend your approach.
6. Distinguish between the terms *validation* and *evaluation* as they apply to the configuration phase of computer equipment and software.
7. Explain what you would do if a vendor said the following in response to an RFP.

   > This thing is not useful to you or me. It rarely tells me what you really want or need. I can do a better job by visiting your business and configuring a system to meet your needs. Also, it takes too long for me to answer all the questions in the RFP. And even if I do, you may not fully understand or appreciate the answers and their implications.

8. A programming assignment in the classroom is a subset of a systems design. Obtain a copy of a programming assignment from a current course. Evaluate the design from the perspective of the systems design phase tasks and the completeness of the design specifications.
9. Obtain a copy of a computer programming assignment. Assume the assignment is to be implemented on a microcomputer that has not been acquired. Estimate the costs necessary to complete the project (hardware, programming, etc.). State your assumptions about salaries, supplies, and other relevant factors.

PROJECTS AND RESEARCH

1. Make an appointment with or write to a hardware and software vendor. Tell the vendor you would like to see and discuss a typical RFP. Ask the vendor how he or she feels about RFPs. If the vendor doesn't like them, find out why. How could RFPs be improved from the vendor's point of view? Do the vendor's attitudes about RFPs help the vendor, the end-user, or both?
2. Make an appointment to discuss physical design standards of a local information systems operation. Does it have standards? Does it follow them? Why or why not? Does the company use prototyping during systems design? Why or why not? If it does prototype systems, what products are used? Has the approach been successful?
3. The city of Granada's art museum recently purchased a Hewlett-Packard Vectra XU 5/90 microcomputer. Museum officials read an article about an art collection inventory system software package that they want to put on that computer. You, having experienced end-users who too hastily purchased software that didn't fulfill promises and expectations, are concerned that they are jumping the gun and should approach the software selection decision with great care. Write a letter to the museum's board of trustees that expresses your concerns and proposes a better approach.
4. Write a letter to your last (or favorite) programming instructor. Suggest a disciplined approach to developing a systems specification to guide the programming assignments for the next term. Your goal should be a system (of programming) specification that will eliminate or drastically reduce the need for students to request clarification from the systems analyst, played by the instructor. Defend your approach.

MINICASE

1. Keith Stallard is a relatively new programmer/analyst at Schuster and Petrie, Inc. Having spent two years as a programmer, he was promoted one year ago. The information services division of S & P requires job performance reviews twice a year. Tim Hayes, associate director of financial systems, has scheduled a job performance review with Keith.

"Well, Keith, do you still want this programmer/analyst job?" asks Tim. "You've had about six months to get used to your new responsibilities."

"More than ever!" responds Keith. "Now that I've had a taste of systems work, I know it's right for me. I assume this meeting will determine if I'm making progress. How am I doing?"

Tim responds, "You're right. I've discussed your performance on the job cost accounting system project with both your supervisor and your key user contact. Your technical design statement was quite impressive. But I have to ask you, where did you learn to complete such thorough design specifications? The implementation appears to be moving along more smoothly than expected, largely because of your specifications."

"I did a lot of reading in systems analysis and design textbooks at the college library," Keith answers. "I also queried both users and programmers about problems with typical specifications. My own experience as a programmer has influenced my specifications. But to be honest, I was really embarrassed by my performance on the account aging project. That's why I did all those things!"

"I don't understand," states a puzzled Tim. "We gave you acceptable ratings. The account aging project was a little off schedule, but that's the only problem I recall."

Keith explains, "It was a little more complicated than that. Bill had done the systems analysis and was supervising me since it was my first experience with systems design. But he had to be called off the project to fix a major flaw in another system. I kept working on the design and passed the design document to Rita [a programmer]. Then lightning struck for a second time. Rita had to go into the hospital, and I had to assume her programming responsibilities. It was the first time I ever had to cut code from my own specifications. There were so many details, and I hadn't documented all of them. Surprisingly, I couldn't even remember all of the thought processes that went into my own specifications. Now I know how the maintenance programmers feel!

"Eventually I got the system up and running—only to find out that some of the reports were not acceptable to my users. The content was there, but the format was wrong. I had to take certain liberties with the format. In my school days, that was what we did with all programming assignments. I just didn't appreciate the importance of user involvement in the design process. I assumed that systems analysis took care of all the user issues. And then, the internal audit department got ahold of my design specifications. They didn't like them at all! There weren't enough internal controls to satisfy their standards. By that time, I had half the programs written and tested. I had to redesign many system components and rewrite several affected programs. To make a long story short, I never want to go through that kind of design experience again. So I learned about systems design."

"So . . . and you still want to be an analyst? After all that?" asks Tim with a big smile.

"Yes," answers Keith with a smile and a nod. "Despite the problems, I found the work to be so much more satisfying than programming. I knew it wasn't going to be easy. But it was enjoyable."

Tim takes over the conversation, "Well, we've discussed your strengths. But we do need to work on a few things. First, as you know, most of our older systems are being converted to on-line systems using databases. I'm sending you to a one-week intensive training course on this new ORACLE database management system. I also want you to go through our user interface course the next time it's offered. I don't know if you've heard that the on-line interface on your accounts aging system hasn't lived up to expectations. You also need to work on your writing and speaking skills. The report you did to sell the new job costing system wasn't well organized, was too wordy, and contained numerous grammatical errors and typos. You were lucky that Bill got it before your users. You might have lost the sale. And your presentation of that system to management could have gone a little smoother. Public speaking is tough, I realize that. But you did not seem confident. You made a good recommendation. But if you don't seem confident and comfortable with your own recommendation, how will management feel about it? We did get it through, though. But your communications skills need improvement, especially if you want my job in the future. You have that potential, Keith. Don't waste it!"

"I understand," replies an accepting Keith. "And I appreciate your honesty. I've suspected the problem. I guess I never took those English and communications courses seriously. I'll get enrolled in some evening continuing education courses for the next term. I'm not going to let poor communications skills get in the way of my future."

"Let's get to the bottom line, Keith," says Tim. "You've shown better than average progress in your new assignment. That's why, effective next month, you'll see a little increase in your paycheck. If you keep up the good work and improve in the areas we've outlined, I'm certain you'll be promoted to systems analyst within two years. Now, let's talk about design specification some more. Do you think we could teach our other analysts to do that?"

a. Think back to your programming courses (or experiences). What are some problems you've had responding to programming assignments?
b. What did Keith learn about working from his own specifications?
c. As a systems analyst working on the design phase of a project, what types of people did Keith have to communicate with? Why does communication become tougher during systems design than during systems analysis?

SUGGESTED READINGS

Application Development Strategies (monthly periodical). Arlington, MA: Cutter Information Corporation. This is our favorite theme-oriented periodical that follows system development strategies, methodologies, CASE, and other relevant trends. Each issue focuses on a single theme.

Boar, Benard. *Application Prototyping: A Requirements Definition Strategy for the 80s*. New York: John Wiley & Sons, 1984. This is one of the first books to appear on the subject of systems prototyping. It provides a good discussion of when and how to do prototyping, as well as thorough coverage of the benefits that may be realized through this approach.

Coad, Peter, and Edward Yourdon. *Object-Oriented Design*. 2nd ed. Englewood Cliffs, NJ: Yourdon Press, 1991, chap. 1. This chapter is a great way to expose yourself to objects and the relationship of object methods to everything that preceded them.

Connor, Denis. *Information System Specification and Design Road Map*. Englewood Cliffs, NJ: Prentice Hall, 1985. This book compares prototyping with other popular analysis and design methodologies. It makes a good case for not prototyping without a specification.

Gane, Chris. *Rapid Systems Development*. Englewood Cliffs, NJ: Prentice Hall, 1989. This book presents a nice overview of RAD that combines model-driven development and prototyping in the correct balance.

Isshiki, Koichiro R. *Small Business Computers: A Guide to Evaluation and Selection*. Englewood Cliffs, NJ: Prentice Hall, 1982. Although it is oriented toward small computers, this book surveys most of the better-known strategies for evaluating vendor proposals. It also surveys most of the steps of the selection process, although they are not put in the perspective of the entire systems development life cycle.

Joslin, Edward O. *Computer Selection*. Rev. ed. Fairfax Station, VA: Technology Press, 1977. Although somewhat dated, the concepts and selection methodology originally suggested in this classic book are still applicable. The book provides keen insights into vendor, customer, and end-user relations.

Lantz, Kenneth E. *The Prototyping Methodology*. Englewood Cliffs, NJ: Prentice Hall, 1986. This book provides excellent coverage of the prototyping methodology.

Wood, Jane, and Denise Silver. *Joint Application Design: How to Design Quality Systems in 40% Less Time*. New York: John Wiley & Sons, 1989. This book provides an excellent in-depth presentation of joint application development.

Yourdon, Edward. *Modern Structured Analysis*. Englewood Cliffs, NJ: Yourdon Press, 1989. Chapter 4, "Moving into Design," shows how modern structured design picks up from modern structured analysis.

Zachman, John A. "A Framework for Information System Architecture." *IBM Systems Journal* 26, no. 3 (1987). This article presents a popular conceptual framework for information systems design.

10

APPLICATION ARCHITECTURE AND PROCESS DESIGN

CHAPTER PREVIEW AND OBJECTIVES

This chapter teaches you techniques for designing the overall information **application architecture** with a focus on **process design**. Information application architecture and process design include techniques for distributing data, processes, and interfaces to network locations in a distributed computing environment. Physical data flow diagrams are used to document the architecture and design in terms of design units—cohesive collections of data and processes at specific locations—that can be designed, prototyped, or constructed in greater detail and subsequently implemented as stand-alone subsystems. You will know that you understand application architecture and process design when you can:

- Define an information system's architecture in terms of DATA, PROCESSES, INTERFACES, and NETWORKS—the building blocks of all information systems.
- Describe both centralized and distributed computing alternatives for information system design, including various client/server and Internet/intranet options.
- Describe various networking topologies and their importance in information system design.
- Describe database and data distribution alternatives for information system design.
- Describe user and system interface alternatives for information system design.
- Describe various software development environments for information system design.
- Describe strategies for developing or determining the architecture of an information system.
- Differentiate between logical and physical data flow diagrams, and explain how physical data flow diagrams are used to model an information system's architecture.
- Draw physical data flow diagrams for an information system's architecture and processes.

SCENE

Robert is putting the finishing touches on the business requirements statement, the final deliverable of systems analysis. Meanwhile, Sandra is attending a meeting of the application architecture variance committee. She has requested a variance from that committee. Attendees include Patricia Wilcox, chair; Gary Bond, representing the Technology Architecture Committee; Helen Grissom, representing the data architecture team; and Brock Peterson, representing the development center.

PATRICIA

Thank you for all coming! I hope that you heard that Gary has been appointed to this committee to replace Jonathan. I think it appropriate to review our charter.

As part of the strategic information planning project that was completed last year, we established an application architecture that standardized on the use of various technologies for all future information systems. But that project team also realized that there may be a need, from time to time, to consider variances from that technology. That is the charge of this committee.

Today, we consider our first such variance. I see that all of you have the written request. I'll ask Sandra Shepherd, the project manager for the Member Services Information System, to present the request now.

SANDRA

Thanks, Patricia. As you know, the current application architecture standards were largely based on our internal expertise at the time of the plan. I am managing the Member Services project as the first strategic project to result from the information systems plan. You may not be aware that we have been underwriting a parallel research project in conjunction with that development project.

Two professors at State University have been tagging along with my team to apply a true object-oriented development methodology on our project. My team is using our information engineering-based methodology, *FAST*. At the end of the project, the professors will share their experiences with the development team and compare our two approaches. Since we know that the next release of *FAST* will incorporate object-oriented methods, and since a shift to object technologies and methods is in our strategic vision, we think this experience is a valuable one.

BROCK

The Development Center made the first contact with the professors and suggested this project as a learning experience.

PATRICIA

So, Sandra, what variance are you requesting?

SANDRA

As you are well aware, our software development environments of choice have been *COBOL* and *C*, with some use of *Powerbuilder* for rapid application development projects. With the possible exception of *Powerbuilder*, these languages are not well suited to object-oriented programming. If we are to get a true test of the object paradigm, we should implement the system in an object-oriented programming language.

GARY

OK, why not have your team use *COBOL*, *C*, or *Powerbuilder*, and let the professors use their object-oriented language of choice?

SANDRA

We had planned to use *Powerbuilder*, but we concluded that it would make more sense for both teams to use the same language. First, we feel that the comparison of the two methodologies would be more valid if both teams use the same language. Second, we feel that an alternate language could be a transferable skill.

PATRICIA

What language are you proposing?

SANDRA

We are proposing *Visual Basic*. The professors and their students have experience in that language. They are also willing to train our team. And we do have a few analysts and programmers who are at least somewhat familiar with the language.

GARY

As I understand it, *VB* is not a pure object-oriented technology.

SANDRA

It depends on how you define object-oriented. Our research, and that of our academic friends, tells us that there are degrees of object-orientation. Everything we've read says that *VB* is somewhere in the continuum, but not pure object oriented.

GARY

So why didn't you choose *C++*, *Smalltalk*, or *Delphi*?

SANDRA

We considered those. We can't afford the premium charged for *Smalltalk* programmers. *C++* doesn't fit our rapid application development needs—some even say that it would slow us down. *Delphi* is an option that is similar to *VB*, but supports pure OO. But again, *Delphi* programmers are not readily available in our market, and while its market share is impressive for its age, it still isn't anywhere close to the size of *VB*.

GARY

Isn't *VB* essentially in the same class as *Powerbuilder*?

SANDRA

Correct, but *VB* is considered a general-purpose programming language like *C++* and *COBOL*. That better suits our needs.

GARY

But *Powerbuilder* is a client/server programming language, and Member Services is a client/server application, no?

SANDRA

That's correct. But *VB* is a client/server language *and* a general-purpose programming language. And it is being taught at State University, our largest supplier of interns and new hires. That's a big plus.

BROCK

Gary, I hope we're not getting hung up on the academic definition of client/

SOUNDSTAGE

SOUNDSTAGE ENTERTAINMENT CLUB

server. In the development center, we've been studying object methods. You may not be aware that the three leading object method gurus, Booch, Rumbaugh, and Jacobson, have merged to form a unified object modeling language and methodology at a company called Rationale. They are designing the UML to be compatible with various OO programming languages, and they include *VB* among those languages.

GARY
I didn't know that, and I am comforted to know that. I'm not trying to be negative. I just believe that we should always play devil's advocate and thoroughly discuss any proposed variances to the application architecture—we should never be a rubber stamp.

PATRICIA
I concur, and I think Gary's questions have been most appropriate.

HELEN
Can I assume that you are not proposing a variance for the database architecture? You still plan to use our Oracle standard?

SANDRA
Absolutely! Because we are designing a two-tiered architecture, we may even decide to move some of the business logic to the Oracle server as stored procedures.

HELEN
I'd like to see us take advantage of that technology! Sandra, what will we have to do to make *VB* work against an Oracle database server? Will you use middleware such as ODBC? From what I understand, that is never as efficient as having native SQL calls in the programming language as is possible with *Powerbuilder.*

SANDRA
Actually, the version of *VB* we'll use includes native SQL for Oracle!

HELEN
That's great!

PATRICIA
Gary, do you have any network concerns?

GARY
No. *VB* is compatible with our network.

BROCK
I have one last question. *VB* is one of the fastest success stories around. But it is interpreted as opposed to compiled. Do you have any concerns about that?

SANDRA
No, the new version is compiled. They had to move that direction to compete with languages like *Powerbuilder* and *Delphi.* I should acknowledge that it is a risk, but not a great one. By the way, we expect that future versions will become pure OO languages.

By the way, I'm surprised that no one asked about Java. We did consider Java, but it just isn't mature at this time. We encourage Brock's group to continue to study that language for future consideration.

PATRICIA
Any other questions? Do I hear a motion to grant this variance?

DISCUSSION QUESTIONS

1. Why would an organization establish a standard set of technologies for all projects?
2. Why might the committee want to grant variances to its technology standards? Why might they not want to grant such variances?
3. Visit the Internet site of Rationale software. How does it define object-orientation?
4. What decision should the committee make? Obviously, there is no right or wrong answer here.

GENERAL SYSTEM DESIGN

Chapter 9 presented a high-level overview of systems design. You learned that **general design** provides the blueprint for subsequent **detailed design** and implementation. This chapter focuses exclusively on general design. (The subsequent chapters then focus on detailed design.) It is during general system design that basic technical decisions are made. These decisions include:

- Will the system use centralized or distributed computing?
- Will the system's data stores be centralized or distributed? If distributed, how so? What data storage technology will be used?
- Will software be purchased, built in-house, or both? For programs to be written, what technology will be used?
- How will users interface with the system? How will data be input? How will outputs be generated?
- How will the system interface to other existing systems?

The answers to these questions (and others) constitute the general design or *application architecture* of the system.

An **application architecture** defines the technologies to be used by (and to build) one, more, or all information systems in terms of its data, process, interface, and network components. It serves as a framework for general design.

In the next section, we'll introduce some of the architectural choices in contemporary system design.

INFORMATION TECHNOLOGY ARCHITECTURE

Information technology (IT) architecture can be a complex subject worthy of its own course and textbook. In this section, we will attempt to *summarize* contemporary IT alternatives and trends available as we go to press. It should be noted that new alternatives are continuously evolving. The best systems analysts will not only know *about* these technologies, but will also understand how they work and their limitations. Such a level of detail is beyond the scope of this book, but you should find various technology, database, and networking courses that will provide greater insight. And of course, systems analysts must continuously read popular trade journals to stay abreast of the latest technologies and techniques that will keep their customers and their information systems competitive.

Your information system framework (Figure 10.1) provides one suitable framework for understanding IT architecture. A few explanations are in order:

- The grayed blocks represent building blocks and phases completed during systems analysis.
- Our goal in general design is to specify the technical solution (at a high level) for various system components—DATA, PROCESSES, INTERFACES, and NETWORKS (the third row in the framework).
- Architectural standards and/or technology constraints are represented in the bottom row of the framework. Notice that these standards or decisions are determined either as part of a separate architecture project (preferred and increasingly common) or as part of each system development project (e.g., the *FAST* methodology's configuration phase). If decisions are made as part of the configuration phase, that phase must be completed before the design phase.
- The upward-pointing arrows are technology decisions that will constrain the design models.

Today, the NETWORK building blocks drive the decision process because the DATA, PROCESS, and INTERFACE components must be distributed across the NETWORK. For this reason, we'll discuss network alternatives first.

NETWORK Architectures for Client/Server Computing

The prevailing computing model of the current era is called *client/server computing*. There are nearly as many definitions of that term as there are books and products. At least one technology corporation CEO has sarcastically defined the term as "anything that sells new IT products." For our purposes, let's define the term and its subparts as follows:

A **client** is single-user computer that provides (1) user interface services and appropriate database and processing services, and (2) connectivity services to servers (and possibly other clients).

A **server** is a multiple-user computer that provides (1) shared database, processing, and interface services, and (2) connectivity to clients and other servers.

In **client/server computing** an information system's database, software, and interfaces are distributed across a network of clients and servers that communicate and cooperate to achieve system objectives. Despite the distribution of computing resources, each system user perceives that a single computer

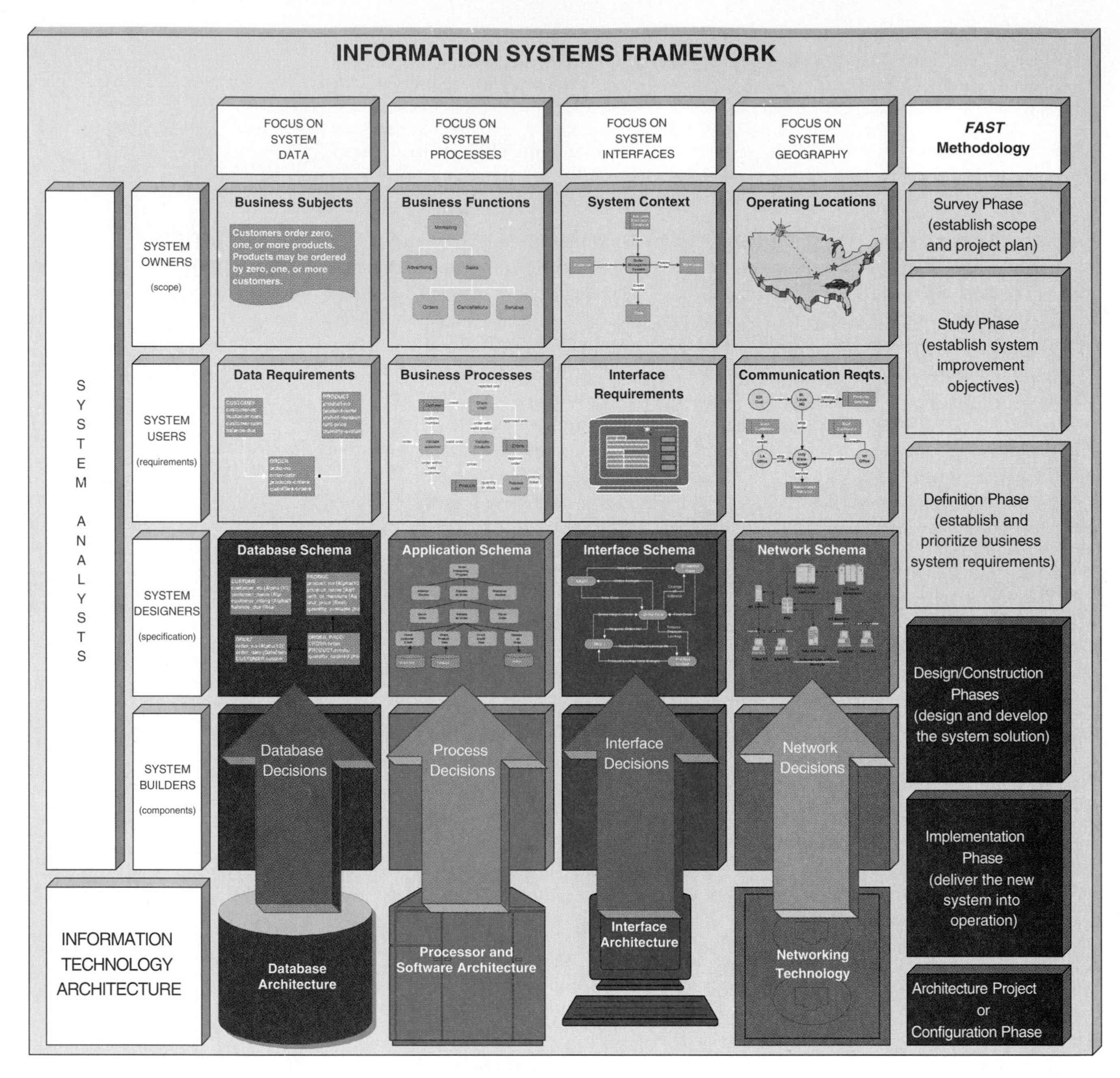

FIGURE 10.1 *Application Architecture and General System Design in the Information Systems Framework*

(the user's own client PC) is doing all the work. Synonyms include **distributed computing** and *cooperative computing.*

Contrary to popular belief, client/server computing is not new, although it is frequently defended as the alternative to traditional centralized computing,

In **centralized computing**, a multi-user computer (usually a mainframe or minicomputer) hosts all the information system components including (1) the data storage (files and databases), (2) the business logic (software and programs), (3) the user interfaces (input and output), and (4) any system interfaces (networking to other computers and systems). The user may interact

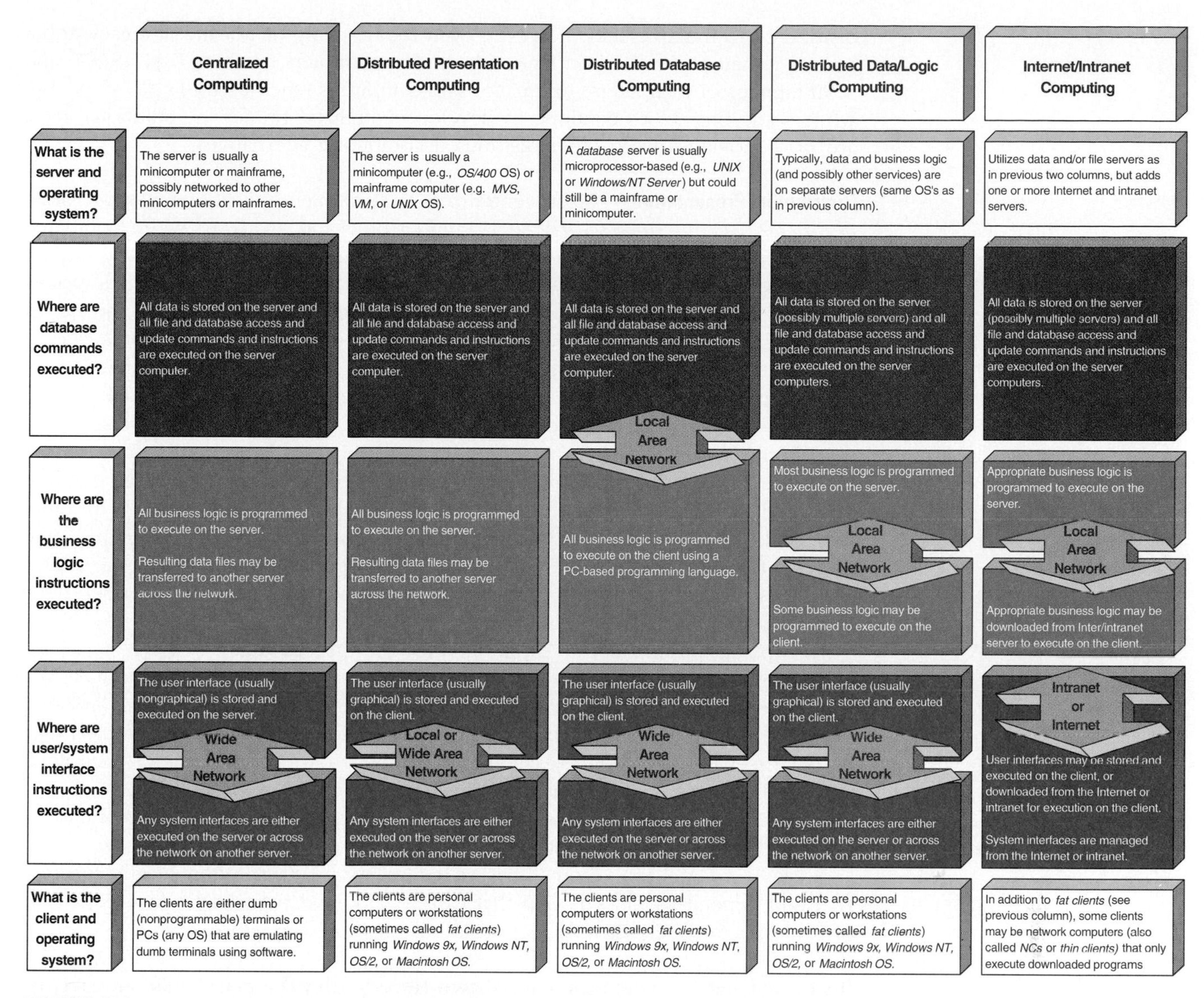

Centralized Computing
Distributed Presentation Computing
Distributed Database Computing
Distributed Data/Logic Computing
Internet/Intranet Computing
What is the server and operating system?
The server is usually a minicomputer or mainframe, possibly networked to other minicomputers or mainframes.
The server is usually a minicomputer (e.g., OS/400 OS) or mainframe computer (e.g. MVS, VM, or UNIX OS).
A database server is usually microprocessor-based (e.g., UNIX or Windows/NT Server) but could still be a mainframe or minicomputer.
Typically, data and business logic (and possibly other services) are on separate servers (same OS's as in previous column).
Utilizes data and/or file servers as in previous two columns, but adds one or more Internet and intranet servers.
Where are database commands executed?
All data is stored on the server and all file and database access and update commands and instructions are executed on the server computer.
All data is stored on the server and all file and database access and update commands and instructions are executed on the server computer.
All data is stored on the server and all file and database access and update commands and instructions are executed on the server computer.
All data is stored on the server (possibly multiple servers) and all file and database access and update commands and instructions are executed on the server computers.
All data is stored on the server (possibly multiple servers) and all file and database access and update commands and instructions are executed on the server computers.
Local Area Network
Where are the business logic instructions executed?
All business logic is programmed to execute on the server.
Resulting data files may be transferred to another server across the network.
All business logic is programmed to execute on the server.
Resulting data files may be transferred to another server across the network.
All business logic is programmed to execute on the client using a PC-based programming language.
Most business logic is programmed to execute on the server.
Local Area Network
Some business logic may be programmed to execute on the client.
Appropriate business logic is programmed to execute on the server.
Local Area Network
Appropriate business logic may be downloaded from Inter/intranet server to execute on the client.
Where are user/system interface instructions executed?
The user interface (usually nongraphical) is stored and executed on the server.
Wide Area Network
Any system interfaces are either executed on the server or across the network on another server.
The user interface (usually graphical) is stored and executed on the client.
Local or Wide Area Network
Any system interfaces are either executed on the server or across the network on another server.
The user interface (usually graphical) is stored and executed on the client.
Wide Area Network
Any system interfaces are either executed on the server or across the network on another server.
The user interface (usually graphical) is stored and executed on the client.
Wide Area Network
Any system interfaces are either executed on the server or across the network on another server.
Intranet or Internet
User interfaces may be stored and executed on the client, or downloaded from the Internet or intranet for execution on the client.
System interfaces are managed from the Internet or intranet.
What is the client and operating system?
The clients are either dumb (nonprogrammable) terminals or PCs (any OS) that are emulating dumb terminals using software.
The clients are personal computers or workstations (sometimes called fat clients) running Windows 9x, Windows NT, OS/2, or Macintosh OS.
The clients are personal computers or workstations (sometimes called fat clients) running Windows 9x, Windows NT, OS/2, or Macintosh OS.
The clients are personal computers or workstations (sometimes called fat clients) running Windows 9x, Windows NT, OS/2, or Macintosh OS.
In addition to fat clients (see previous column), some clients may be network computers (also called NCs or thin clients) that only execute downloaded programs

FIGURE 10.2 *Flavors of Client/Server Computing*

with this host computer via a terminal (or, today, a PC emulating a terminal), but all the work is done on the host computer.

The central computer is a server, and the terminals connected to that server are clients (although those clients are doing little work—they merely display formatted data and information that are presented by the host).

If we can agree that centralized computing is actually a form of client/server computing, then Figure 10.2 provides a useful comparison of the many forms of client/server computing available today. The columns should not be considered mutually exclusive; that is, most businesses will utilize multiple client/server architectures to support legacy, modern, and planned information systems. Let's discuss each column in somewhat greater detail.

Centralized Computing Not too long ago, centralized process architectures were dominant because the cost of placing computers closer to the end-user was prohibitive. Over time, the cost of computers went down, and processors were distributed to multiple sites; however, even today, many (if not most) legacy applications remain centralized on large mainframe computers (such as IBM's S/370 and 3090 families of computers) or smaller minicomputers (such as IBM's AS/400).[1] But while some businesses continue to develop centralized processing solutions, most are now moving toward other distributed computing alternatives.

Distributed Presentation Most centralized computing applications use an older *character user interface (CUI)* that is cumbersome and awkward when compared to today's *graphical user interfaces* (GUIs such as Microsoft's *Windows*, Apple's *Macintosh*, and IBM's *OS/2 Presentation Manager*). As personal computers rapidly replaced dumb terminals, users became increasingly comfortable with the newer technology. And as they developed experience with word processors, spreadsheets, and other personal computing applications, they wanted their centralized computing applications to have a similar look and feel using the GUI model.

Enter distributed presentation! Sometimes called the poor person's client/server, this alternative builds on and enhances centralized computing applications. Essentially, the old character user interfaces are stripped from the centralized applications and regenerated as graphical user interfaces that will run on the PC. In other words, the user interface (or presentation) is distributed off the server and onto the client. All other elements of the centralized application remain on the server, but the system users get a friendlier graphical user interface (usually based on Microsoft *Windows*) to the system.

Distributed presentation computing offers several advantages. First, it can be implemented relatively quickly since most aspects of the legacy application remain unchanged. Second, users get a friendly and familiar interface to existing systems. Finally, the useful lifetime of legacy applications can be extended until such a time as resources warrant a wholesale redevelopment of the application. The disadvantage is that the application's functionality cannot be significantly improved, and the solution does not maximize the potential of the client's desktop computer by dealing only with the user interface.

A class of CASE tools, sometimes called *screen scrapers*, automatically read the CUI and generate a first-cut GUI that can be modified by a GUI editor. Examples of screen scraper CASE tools include *Easal*, *Mozart*, and *Flashpoint*.

Distributed Data Sometimes called **two-tiered client/server**, this architecture places the information system's stored data on a server and the business logic and user interfaces on the clients. A local or wide area network usually connects the clients to the server.

> A **local area network** (or **LAN**) is a set of client computers (usually PCs) connected to one or more server computers (usually microprocessor-based, but could also include mainframes or minicomputers) through cable over relatively short distances—for instance, in a single department or in a single building.
>
> A **wide area network** (or **WAN**) is an interconnected set of LANs, or the connection of PCs over a longer distance—such as between buildings, cities, states, or countries.

It should be noted that a distributed data architecture may include more than one database server.

---

[1]The distinctions between what we used to call small *mainframes* and large *minicomputers* are constantly blurring. Similarly, the distinctions between small minicomputers and larger, microprocessor-based servers (those using Intel, Motorola, or RISC processors; and running network operating systems such as Windows/NT Server, OS/2 Server, and UNIX) are also eroding. Ultimately we expect only two classes of computers—singe-user clients and multiple-user servers.

The *database server* is fundamental to this architecture. You may already be familiar with *file servers* that can store shared PC databases, but file servers and *database servers* are different technologies.

- **File servers** *store* the database, but the client computers must execute all database instructions. This means that entire databases and tables may have to be transported to and from the client across the network. This approach, used by database software such as Microsoft *Access* and Borland *dBASE*, typically generates excessive network traffic.
- **Database servers** also store the database, but the database commands are also executed on those servers. The clients merely send their database commands to the server. The server returns only the result of the database command processing—not entire databases or tables. Thus, database servers generate much less network traffic. This approach is used by high-end database software such as *Oracle* and Microsoft *SQL Server*.

The clients in the distributed database solution typically run the business logic of the information system application. This logic is usually written in a client/server programming language such as Sybase Corporation's *Powerbuilder*, Microsoft's *Visual Basic*, or *C/C++*. Those programs will execute on the client. To improve application efficiency and reduce network traffic, some business logic may be distributed to the database server in the form of *stored procedures* (discussed in the next chapter).

The main advantage of distributed database computing is to separate data and business logic to (1) isolate each from changes to the other, (2) make the data more available to users, and (3) retain the data integrity of centralized computing through centrally managed servers. The key potential disadvantage is that the application logic must be maintained on all the clients, possibly hundreds. The designer must plan for version upgrades and provide controls to ensure that each client is running the most current release of the business logic, as well as ensure that other software on the PC (purchased or developed in-house) does not interfere with the business logic.

Distributed Data and Logic Sometimes called **three-tiered** or **n-tiered client/server computing**, this approach distributes databases *and* business logic to separate servers. When the number of clients grows, two-tiered systems frequently suffer performance problems associated with the inefficiency of executing all the business logic on the clients. Also, in multiple-user transaction processing systems (also called *on-line application processing* or *OLAP*), transactions must be managed by software to ensure that all the data associated with the transaction is processed as a single unit. (In mainframe systems, this task was performed by a transaction monitor such as IBM's *CICS*).

The three-tiered client/server solution uses the same database servers as in the two-tiered approach. Additionally, the three-tiered system introduces an *application server*. Typically, the application server supplies two services. First, it provides a transaction monitor such as *Tuxedo* or *CICS/6000* to manage transactions. Second, some or all of the business logic of the application can be moved from the client to the server. This offers the advantage of not having to maintain business logic on hundreds of clients (at least not all of it).

Three-tiered client/server development tools are only beginning to appear. While code could always be written in languages such as *C* and *C++*, many client development tools (such as Microsoft *Visual Basic* and Borland *Delphi*) now allow programmers to distribute some business logic between clients and servers. High-end tools such as *Forté* provide even greater opportunity to distribute business logic across a complex network. As with the database server solution, some business logic could be distributed to the database server in the form of stored procedures.

In a three-tiered system, the clients execute a minimum of the overall system's components. Only the user interface and some relatively stable or personal business logic need be executed on the clients. This simplifies client configuration and management.

The biggest drawback of three-tiered client/server is its complexity in so far as design and development. The most difficult aspect of a three-tiered client/server application design is partitioning.

> **Partitioning** is the act of determining how to best distribute or duplicate application components (data, process, and interfaces) across the network.

Fortunately, CASE tools are constantly improving to provide greater assistance with partitioning.

The Internet and Intranets The latest extension of the client/server model incorporates Internet and intranet technologies into the client/server picture. At the time of this writing, this is quite immature; however, the technical possibilities are being fueled by the explosive growth of the Internet.

> The **Internet** is an (but not necessarily *the*) information superhighway that permits computers of all types and sizes all over the world to exchange data and information using standard languages and protocols.

The Internet extends the reach of information systems to include customers, potential customers, partners, remotely located employees, suppliers, the government, and even competitors. Today, the Internet is largely being used to market products, information, and services and to provide customer support. As the technology to secure Internet data traffic evolves, the commerce potential of the Internet will grow. On-line transaction processing capabilities will be extended to the Internet. Every computer on the Internet becomes a potential client with access to a world of servers.

But the greatest potential of the technology may be its application to the development of intranets.

> An **intranet** is a secure network, usually corporate, that uses Internet technology to integrate desktop, work group, and enterprise computing into a single cohesive framework.

Many, if not most, companies are investing heavily in developing a corporate intranet. The goal is simple. The intranet could provide management and users with a common interface to applications and information. It could become the new mechanism by which users start all enterprise computing applications, process transactions, initiate data inquiries, and distribute information. It is conceivable that Internet and intranet *browsers* (such as *Netscape* and *Internet Explorer*) will effectively merge with the client and server operating systems (such as *Windows 97*) to make desktop applications virtually indistinguishable from all other server and Internet applications. In a sense, users will be able to surf their intranet for all computing applications. And, of course, the corporate intranet will gain access to the outside world via an Internet gateway.

The most intriguing system development advances to come may be enabled by an emerging Internet/intranet technology called *Java*. Essentially, *Java* is a cross-platform programming language designed specifically to exploit the Internet standards. As with other programming languages, *Java* developers would still be able to write programs to execute on a server. But they could also code modular software components (such as the user interface and some business logic) as *Java applets*. *Java* applets are stored on an Internet or intranet server and downloaded to the client when the client accesses the application. The *Java* standard results in applets that can execute on any client computing platform (meaning *Windows*, Macintosh, *OS/2*, or UNIX). The client downloads (updated) applets automatically to ensure that the latest versions are always used.

Because *Java* applets will be theoretically simpler than most client software, the industry is developing a new type of client computer called the *network computer* to run them. A network computer (or NC) is designed to run only Internet-based applications (such as Web browsers and *Java* applets). The NC (also called a *thin client*) is simpler and much cheaper than personal computers (increasingly called *fat clients*). NCs are intended to substantially reduce the cost of PC ownership and maintenance for businesses—only time will tell. Theoretically, traditional PC applications such as word processors and spreadsheets could be written in *Java* and downloaded from the Internet or intranet server. But whether or not such applications can be as robust (or even sufficiently robust) as their PC counterparts is hotly debated. Most experts predict that PCs and NCs will coexist into the foreseeable future.

All of these Internet and intranet solutions involve leading-edge technologies and standards that have no doubt evolved significantly since these words were written. Technology vendors will undoubtedly play a significant role in the evolution of the technology. This is a newsworthy subject for your continuous education. However the specific technologies play out, we expect the Internet and intranets to become integral models in the future of client/server computing.

The Role of Network Technologies Why the popularity of client/server computing? Initially, the argument was *economics*! Companies speculated that they could downsize applications to much cheaper platforms that, through cooperative processing, could achieve equal or better throughput and response time. Is it true? Perhaps not! Some data now suggest that the cost of maintaining a distributed environment and its applications is equal to that of a centralized solution. Regardless, the client/server model proliferates—perhaps because it harnesses the investment in PCs and because it empowers management and users to a greater degree than did the mainframe computer.

The different client/sever computing architectures are illustrated in Figure 10.3. Regardless of the choice of mainframes and minicomputers, personal computers and workstations, LAN servers and WANs, the key to modern information technology is networking. The well-designed network provides connectivity and interoperability.

> **Connectivity** defines how computers are connected to "talk" to one another.
>
> **Interoperability** is an ideal state in which connected computers cooperate with one another in a manner that is transparent to their users (the clients).

Let's briefly survey some of the underlying connectivity options. These options are covered more extensively in data communications, telecommunications, and networking courses. The following discussion is intended only as a survey.

> **Network topology** describes how a network provides connectivity between the computers on that network.

The simplest network topology is to provide a **bus**—that is, a direct point-to-point link between any two computer systems. This point-to-point networking concept is illustrated in Figure 10.4. Notice that the network can contain mainframes, minicomputers (or midrange computers), personal computers, and dumb and intelligent terminals. To completely connect all points between n computers, you would need $n \times (n-1)/2$ direct paths. Unless each path is heavily utilized, the cost will prove prohibitive.

Only one computer can send data through the bus at any given time. *Ethernet*, a product developed jointly by Xerox, Intel, and Digital Equipment Corporation, is an example of the most common bus network topology used for client/server LANs. *Ethernet*'s bus topology, in cooperation with a LAN operating system (such as Novell *Netware*, Microsoft *Windows NT Server*, and IBM's *OS/2*

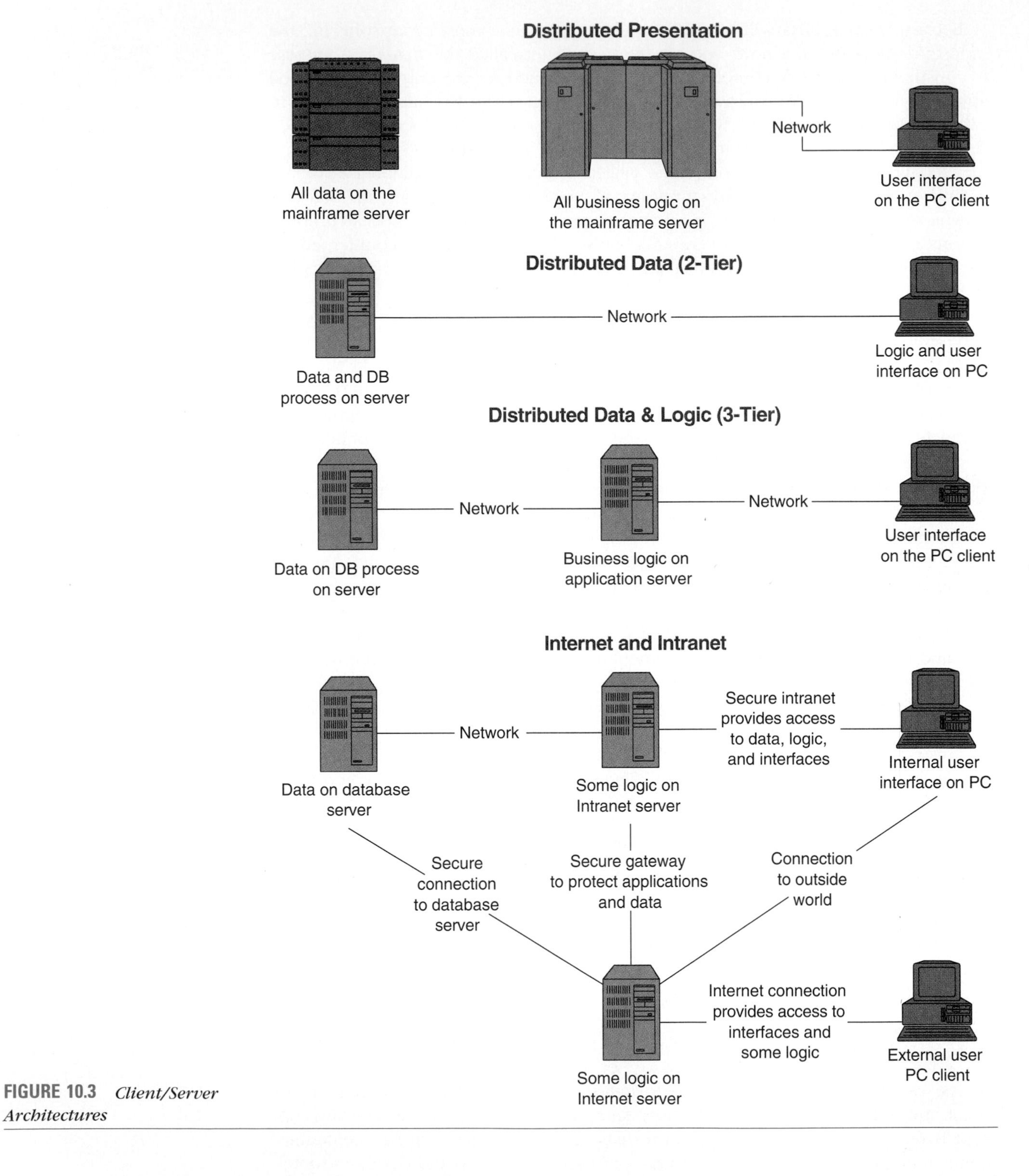

FIGURE 10.3 *Client/Server Architectures*

Server), manages the point-to-point communication between computers and devices on the bus and resolves contention that occurs when more than one computer or device attempts to send a message, instruction, or data across the bus at the same time.

Another network topology, the **ring network**, connects multiple computers and some peripherals into a ring-like structure (see Figure 10.5). Each computer can transmit messages, instructions, and data (called packets) to only one other computer (or node on the network). Every transmission includes an address, similar

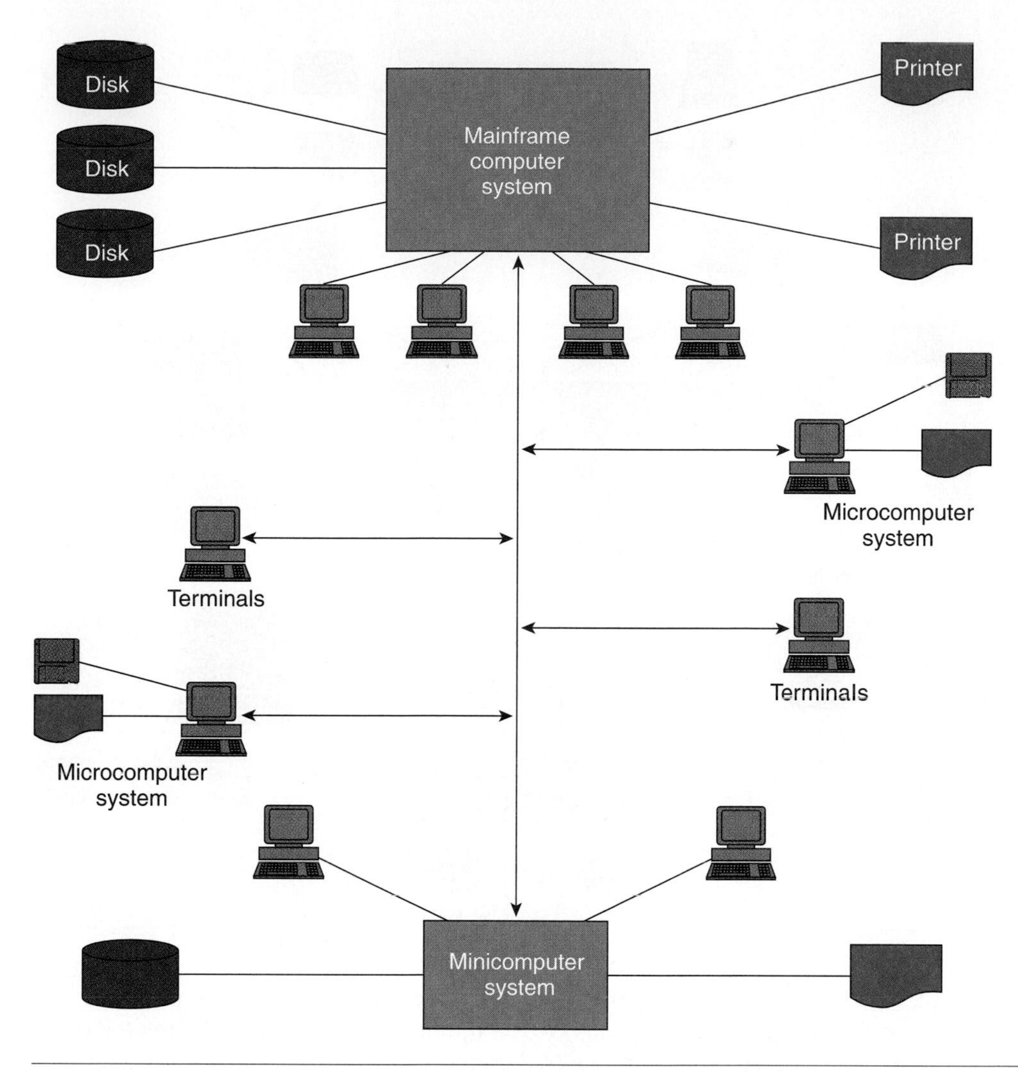

FIGURE 10.4 *Bus Networks*

to an address you print on an envelope. When a computer receives a packet, it checks the address. If the packet's address is different from the computer's address, it passes it on to the next computer or node. Eventually, the packet arrives at its destination. Ring networks generally transmit packets in one direction; therefore, many computers can transmit at the same time to increase network throughput. IBM's *Token Ring Network* is an example of a ring network that competes with bus networks such as *Ethernet.*

Larger mainframe and minicomputer systems may be connected by a **star network** topology that links multiple computer systems through a central computer (see Figure 10.6). Some would argue that this is a throwback to centralized computing. However, the central computer does not have to be a mainframe or minicomputer. It could be an application server that manages the transmission of data and messages between the other clients and servers (as in the n-tier model).

A **hierarchical network** topology can be thought of as a multiple star network, where the communications processors are arranged in a hierarchy (see Figure 10.7). The top computer system (usually a mainframe) controls the entire network. Its satellite processors (in this case, midrange computers) have their own satellites (in this case, personal computers and terminals). Notice that each satellite may have its own complement of peripherals. IBM's *Systems Network Architecture (SNA)* is essentially a hierarchical network for IBM-compatible computers of all sizes.

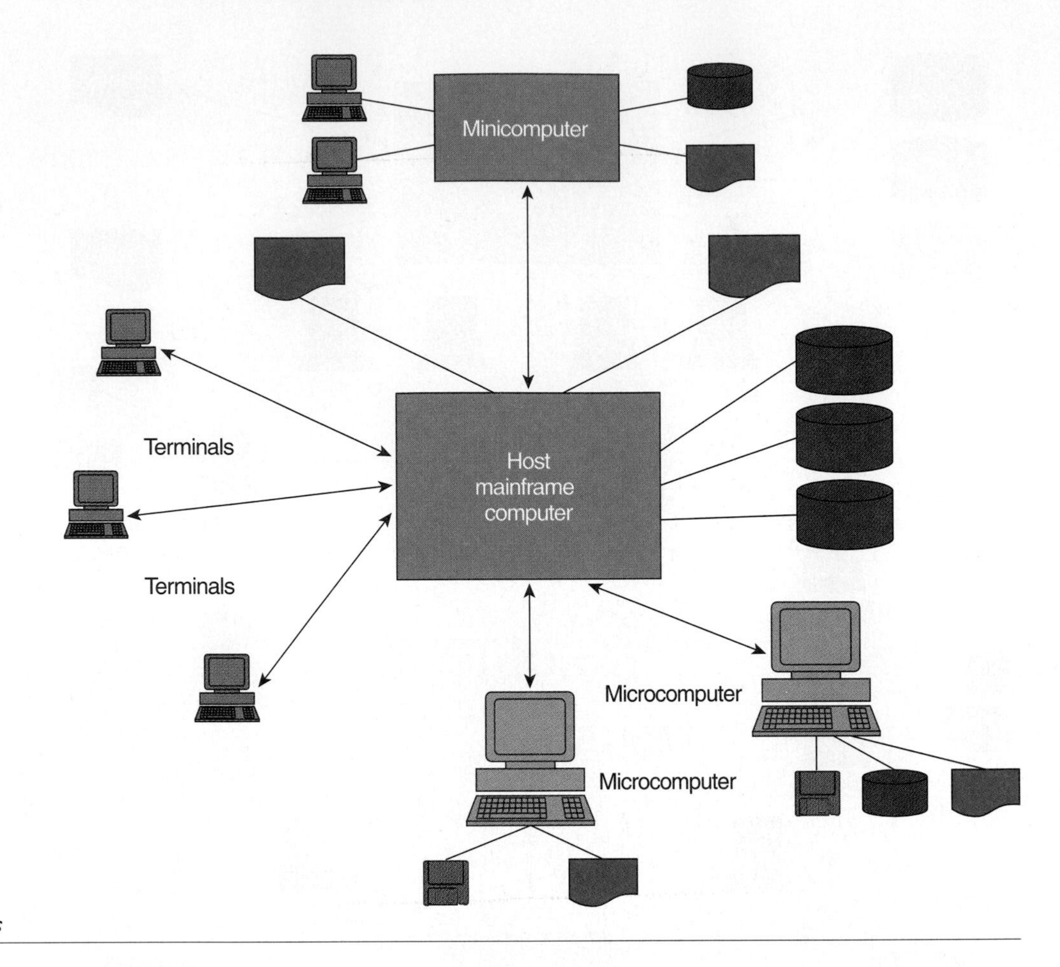

FIGURE 10.5 *Ring Networks*

All network topologies operate according to established network protocols that permit different types of computers to communicate and interoperate. Examples include IBM's *SNA* (commonly used to connect IBM mainframes, minicomputers, and servers), and *TCP/IP* (the current de facto standard for *Ethernet* and networks that contain computers of various architecture).

An application's network architecture is selected based on the client/server model and network topologies needed to support that model. Generally, a qualified network engineer should be included in any discussions about the application's network architecture and design. Network engineers are uniquely qualified to determine the application's impact on network traffic, performance, and security. If the application involves connection to the Internet or an intranet, the appropriate network webmaster should also be involved.

DATA Architectures for Distributed Relational Databases

The network provides the map for distributing data to optimal locations. Historically, the need to control data centrally has been considered essential. As a "shared" resource, centralized data is easier to manage. Until recently, the only practical way to accomplish this goal was to store all data on a central computer, with absolute control by a central data administration group.

The underlying technology of client/server computing has made it possible to distribute data without loss of centralized control. This control is being accomplished through distributed relational databases.

> A **relational database** stores data in a tabular form. Each file is implemented as a *table*. Each field is a *column* in the table. Each record in the file is a *row* in the table. Related records between two tables (e.g., CUSTOMERS and ORDERS) are implemented by intentionally duplicating columns in the two tables (in this example, CUSTOMER NUMBER).

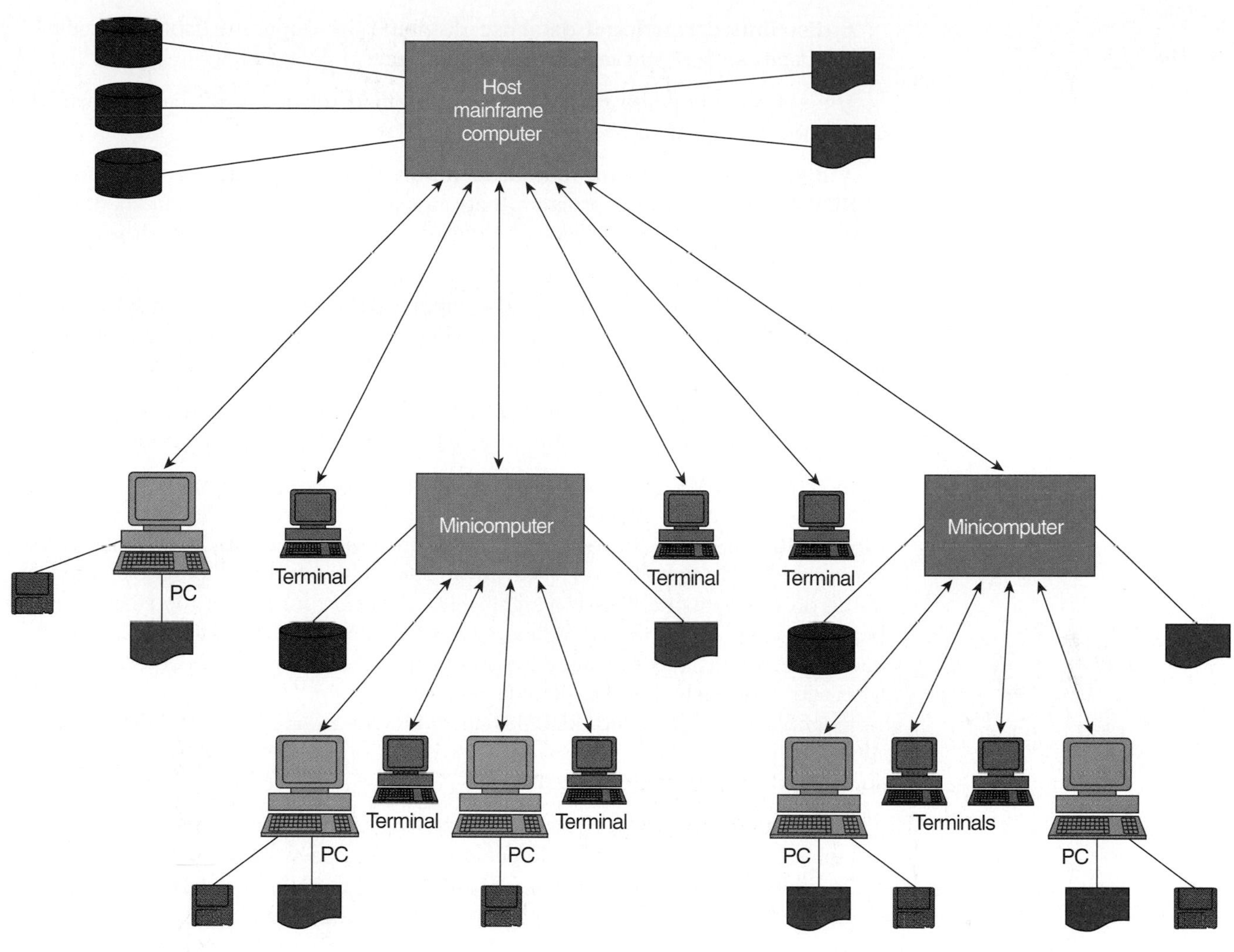

FIGURE 10.6 *Star Networks*

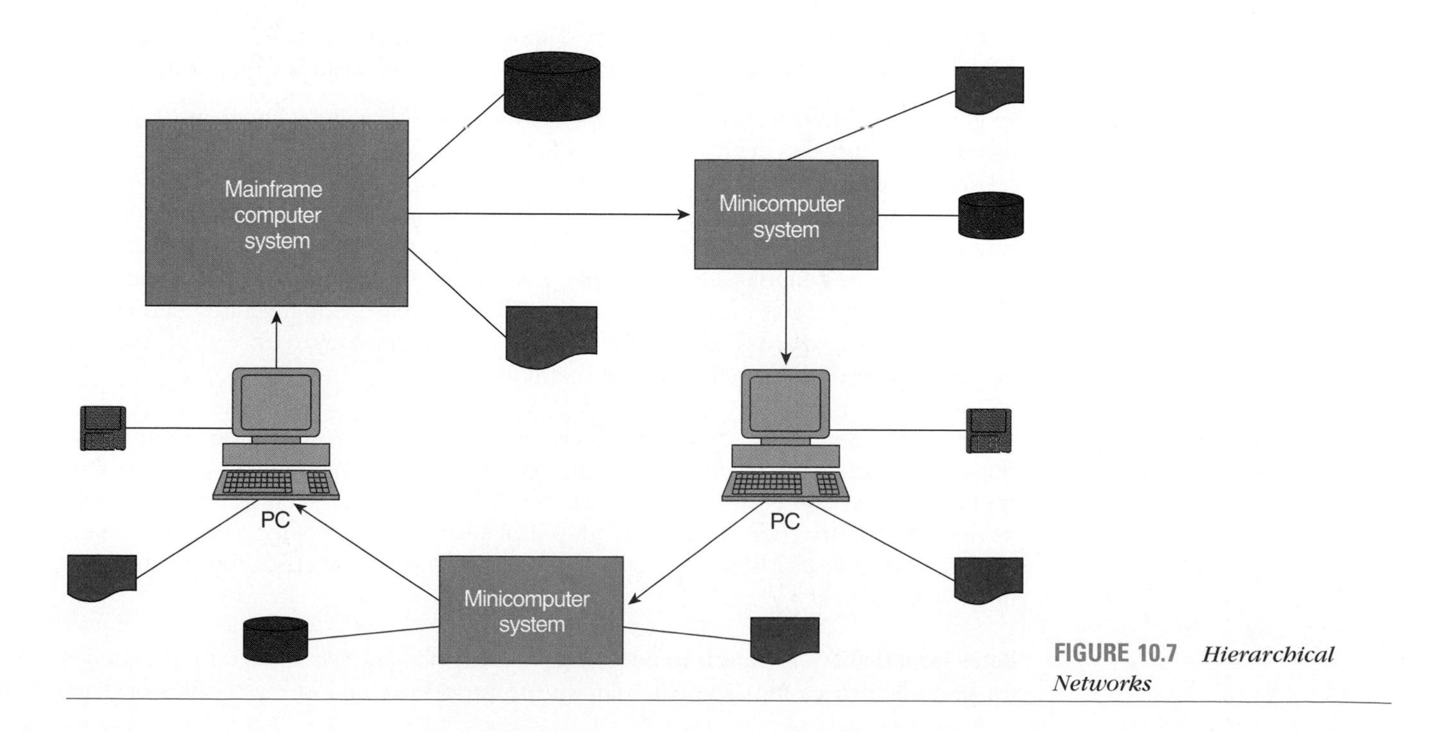

FIGURE 10.7 *Hierarchical Networks*

A **distributed relational database** distributes or duplicates tables to multiple database servers (and in rare cases clients).

The software required to implement distributed relational databases is called a distributed relational database management system.

A **distributed relational database management system** (or **distributed RDBMS**) is a software program that controls access to and maintenance of the stored data. It also provides for backup, recovery, and security. It is sometimes called a client/server database management system.

You may be familiar with personal computer database management systems such as Microsoft *Access* or Borland *dBASE*. They are less complex, but otherwise similar. What sets a distributed RDBMS apart from a PC RDBMS is the database engine.

The **database engine** is that part of the DBMS that executes database commands to create, read, update, and delete records (rows) in the tables.

In a PC RDBMS, the database engine that processes all database commands must execute on the client PC, even if the data is stored on the server. In a distributed RDBMS, the database engine that processes all database commands executes on the database server. The advantage of the latter is to reduce the data traffic on the network. This is a significant advantage for all but the smallest number of users. A distributed relational DBMS also provides more sophisticated backup, recovery, security, integrity, and processing (although the differences seem to erode with each new PC RDBMS release).

Examples of distributed RDBMSs include Oracle Corporation's *Oracle*, IBM's *DB2* family, Microsoft's *SQL Server*, and Sybase Corporation's *Sybase*. Most RDBMSs support two types of distributed data.

True **data distribution** partitions data to one or more database servers. Entire tables can be allocated to different servers, or subsets of rows in a table can be allocated to different servers. An RDBMS controls access to and manages each server.

Data replication duplicates data on one or more database servers. Entire tables can be duplicated on different servers, or subsets of rows in a table can be duplicated to different servers. The RDBMS not only controls access to and management of each server database, but it also ensures that updates on one server are updated on any server where the data are duplicated.

For a given information system application, the DATA architecture must specify the RDBMS and the degree to which data will be distributed or replicated.

An application's DATA architecture is selected based on the desired client/server model and the database technology needed to support that model. Many organizations have standardized on both their PC RDBMS of choice, as well as their preferred distributed RDBMS. For example, SoundStage has standardized on Microsoft *Access* and *SQL Server*. Generally, a qualified database administrator should be included in any discussions about the database technology to be used and the design of any databases to use that technology.

INTERFACE Architectures—Inputs, Outputs, and Middleware

Another fundamental architectural decision must be made regarding inputs, outputs, and intersystem connections. The decision used to be simple—batch versus on-line. Today, we must also consider modern alternatives such as remote batch, keyless data entry, pen data entry, graphical user interfaces, electronic data interchange, imaging and document interchange, among others. Let's briefly examine these alternatives.

Batch Input/Output In **batch processing**, transactions are accumulated into batches for periodic processing. The batch inputs are processed against master files or databases. Transaction files or databases may also be created or updated by the transactions. Most outputs tend to be generated to paper or microfiche on a scheduled basis. Others might be produced on demand or within a specified time period (e.g., 24 hours).

Contrary to popular belief, batch-based applications are not obsolete. Some application requirements lend themselves nicely to batch processing. Perhaps the inputs arrive in natural batches (e.g., mail), or perhaps outputs are generated in natural batches (e.g., invoices). There is, however, the definite trend away from batch input and output to on-line approaches. As the cost of computer processing decreases, and network technology enables paperless document flow, we expect batch inputs and outputs to decline. In any case, as older batch-based systems become candidates for replacement, other alternatives should at least be explored.

On-Line Processing The majority of systems have slowly evolved from batch processing to **on-line processing**. On-line systems provide for a conversational dialogue between the user and computer. Business transactions and inquiries are often best processed when they occur. Errors are identified and corrected more quickly. Transactions tend to be processed earlier since on-line systems eliminate the need for batch data file preparation. Furthermore, on-line methods permit greater human interaction in decision making, even if the data arrive in natural batches. Inquiries and reports can, for the most part, be processed immediately. And as PCs have replaced dumb terminals, PC software can both support more creative formatting of I/O and print to local (and network) printers.

The lower response time requirements of most applications and the desire for human interaction during processing have driven systems development to on-line alternatives. Other contributing factors include faster computers, increased capacity to handle more simultaneous users, and better on-line development and control software.

Client/server applications are simply a new form of on-line processing. Input editing and output formatting occur on client computers in an on-line mode. Input transactions and information requests are transmitted on-line to several computers for processing.

Remote Batch **Remote batch** combines the best aspects of batch and on-line I/O. Distributed on-line computers handle data input and editing. Edited transactions are collected into a batch file for later transmission to host computers that process the file as a batch. Results are usually transmitted as a batch back to the original computers.

Remote batch is hardly a new alternative, but personal computers have made the option increasingly more attractive. PCs provide low-cost, on-line data capture and editing power with adequate storage for resultant batches. PCs with graphical user interfaces (e.g., *Windows*) provide for simpler data entry and user assistance. Finally, PC-to-host communications technology makes it easy to transmit the batch to another computer for processing and to receive the results of that processing.

Remote batch using PCs should get another boost with the advances in laptop and palmtop computer technology. These three- to seven-pound computers can be used to collect batches of everything from mortgage applications to university and housing applications. Don't forget this technology of remote batch I/O as you redesign old batch and on-line systems.

Keyless Data Entry Keying errors have always been a major source of errors in computer inputs (and inquiries). Any technology that reduces or eliminates the possibility of keying errors should be considered for system design. In batch systems, keying errors can be eliminated through **optical character reading (OCR)**

and **optical mark reading (OMR)** technology. Both are still viable options for input design.

The real advances in keyless data entry are coming for on-line systems in the form of **auto-identification** systems. For example, **bar coding** schemes (such as the universal product codes that are common in the retail industry) are widely available for many modern applications. For example, Federal Express creates a bar code-based label for all packages when you take the package to a center for delivery. The bar codes can be read and traced as the package moves across the country to its final destination. Bar code technology is being constantly improved to compress greater amounts of data into smaller labels.

Keyless data entry should be considered for appropriate high-volume transaction-based systems as they become candidates for redesign.

Pen Input Pen-based computing is starting to evolve. As pen-based operating systems (e.g., Microsoft's *Windows CE*) become more widely available and used, and the tools for building pen-based applications become available and standardized, we expect to see more system designs that exploit this technology.

Some businesses already use this technology for remote data collection. For example, UPS uses pen-based notebook systems to communicate deliveries to drivers and to collect delivery confirmation signatures and data from customers and drivers. When a driver returns to the distribution center, the data is transmitted from the pen-based notebook computer to host computers.

A promising technology is emerging in the form of **handheld PCs** (HPCs). Similar to personal organizers (such as HP's *Wizard*) and personal data assistants (such as Apple *Newton*), these HPCs offer greater compatibility with desktop and laptop PCs. Based on Microsoft's *Windows CE* operating system, they can be programmed to become disconnected clients in a client/server application. For example, college admission representatives could take them on the road to high schools and create remote batches of prospective students. Then, from the hotel room, the HPC can be connected by modem to the server systems that accept the batch for processing.

Graphical User Interfaces **Graphical user interfaces** (GUIs) were popularized by the success of Apple's *Macintosh* and Microsoft's *Windows*. While the commercial success has been driven by applications such as word processing and spreadsheets, the popularity of the interface is driving *all* applications to the interface. GUI technology has become the user interface of choice for client/server applications.

GUIs do not automatically make an application better. Poorly designed GUIs can negate the alleged advantages of consistent user interfaces. Fortunately, de facto GUI standards are evolving to guide system designers to create consistent interfaces. For better or for worse, Microsoft's *Windows 97* and *Windows NT* have established the de facto standard for a majority of GUI applications.

The Internet, however, is slowly changing user expectations for interfaces. Most users interface with the Internet via a client software tool called a **browser**. The two dominant browsers are Netscape's *Navigator* and Microsoft's *Internet Explorer*. These browsers currently run under *Windows*, *OS/2*, and Macintosh, but they also take on their own look and feel. The browser paradigm is based on hypertext and hyperlinks. The former are keywords that are clearly highlighted as a link to a new page of information. The latter are links from graphics, buttons, and areas that link to a different page of information. These links make it easy to navigate from page-to-page and application-to-application.

This has not gone unnoticed by operating system makers. Microsoft's next version of *Windows* expects to include an option of a hypertext- and hyperlink-like interface to your entire PC and network. In other words, the browser and the operating system would become one and the same. If this approach were applied to a client/server network, you would have a consistent interface to your PC applications, information system applications, and the outside world. Meanwhile, Netscape envisions an alternative operating system based on its *Navigator* browser and *Java* (described earlier in the chapter).

Electronic Messaging and Work Group Technology Electronic mail has grown up! No longer merely a way to communicate more effectively, information systems are being designed to directly incorporate the technology. For example, Microsoft *Outlook and Exchange Server* and IBM/Lotus *Notes* allow for the construction of intelligent electronic forms that can be integrated into an application.

For example, purchase requisitions could be initiated by any employee via an e-mail-based form. Based on the data submitted on the form, predefined rules automatically route the requisition to the appropriate decision makers. For example, requisitions for inexpensive items might be routed directly to Purchasing. Requisitions for more expensive items might be routed automatically to a department head. Still others might be routed to an investment committee. Eventually forms interface directly to the operational purchasing systems for normal processing. And at each step, the messaging system automatically informs the initiator of progress via e-mail.

Electronic Data Interchange Businesses that operate in many locations and businesses that seek more efficient exchange of transactions with their suppliers and/or customers often utilize electronic data interchange.

> **Electronic data interchange (EDI)** is the electronic flow of business transactions between customers and suppliers.

With EDI, a business can eliminate its dependence on paper documents and mail. For example, Sears uses EDI to directly submit purchase orders from its own computers to suppliers' computers. The competitive advantages of reduced response time should be obvious. It is expected that larger businesses will convert most of their customer and supplier transactions to EDI by the end of the decade.

Various EDI standards exist for the standardized exchange of data between organizations within the same industry. For example, your college may already use academic industry EDI standards to exchange transcript data between the college and its feeder high schools. This eliminates the need to manually transcribe your grades from the high school system to the college record.

The Internet may have the potential to provide a new backbone for EDI, although many security and capacity issues remain to be solved.

Imaging and Document Interchange Another emerging I/O technology is based on image and document interchange. This is similar to EDI except that the actual images of forms and data are transmitted and received. It is particularly useful in applications in which the form images or graphics are required. For example, the insurance industry has made great strides in electronically transmitting, storing, and using claims images. Other imaging applications combine data with pictures or graphs. For example, a law enforcement application can store, transmit, and receive photographic images and fingerprints.

Middleware Most of the above subsections focused on input and output, the so-called user interface. But information systems must also interface to other information systems. These existing systems (called legacy systems) may have been built with various information technologies that are no longer used to build new systems.

> **System integration** is the process of making heterogeneous information systems (and computer systems) interoperate.

A key technology used to interface and integrate systems is middleware.

Middleware is utility software that interfaces systems built with incompatible technologies. Middleware serves as a consistent bridge between two or more technologies. It may be built into operating systems, but it is also frequently sold as a separate product.

A common example of middleware is ODBC, the *open database connectivity* standard. Developed for and bundled into Microsoft operating systems (and improved on by independent software vendors in add-on or replacement products), ODBC allows for simpler sharing of data between incompatible database management systems. Information systems that use ODBC can easily access incompatible databases using ODBC software.

Selecting User and System Interface Technologies The preferred or approved user and system interface technologies may be specified as part of the interface architecture. Alternatively, an organization may leave interface technologies as a decision to be made on a project-by-project basis. Or an organization may establish macro guidelines for interfaces and leave the micro decisions to individual projects. For example, a macro guideline may specify that keyless data entry alternatives, if feasible to the application, should always be pursued. If not feasible, GUI-based keyboard forms should be used. Internal forms should be initiated by e-mail.

PROCESS Architecture—The Software Development Environment and System Management

The PROCESS architecture of an application is defined in terms of the software languages and tools that will be used to develop the business logic and application programs. Typically, this is expressed as a menu of choices since different software development environments (SDEs) are suited to different applications.

A **software development environment** is a language and tool kit for constructing information system applications. They are usually built around one or more programming languages such as *COBOL, Basic, C* or *C++, Pascal, Smalltalk,* or *Java.*

One way to classify SDEs is according to the type of client/server model that they support.

SDEs for Centralized Computing and Distributed Presentation Not that long ago, the software development environment for centralized computing was very simple. It consisted of the following:

- An editor and compiler, usually COBOL, to write programs.
- A transaction monitor, usually CICS, to manage on-line transactions and terminal screens.
- A file management system, such as VSAM, or a database management system, such as DB2.

That was it! Because all these tools executed on the mainframe, only that computer's operating system (more often than not, MVS) was critical.

The personal computer brought many new COBOL development tools down to the mainframe. A PC-based COBOL SDE such as the Micro Focus *COBOL Workbench* usually provided the programmer with more powerful editors and testing and debugging tools at the workstation level. A programmer could do much of the development work at that level and then upload the code to the central computer for system testing, performance tuning, and production. Frequently, the SDE could be interfaced with a CASE tool and code generator to take advantage of process models developed during systems analysis.

Eventually, SDEs provided tools to develop distributed presentation client/server. For example, the Micro Focus *Dialog Manager* provided *COBOL Workbench* users with tools to build *Windows*-based user interfaces that could cooperate with the CICS transaction monitors and the mainframe COBOL programs.

SDEs for Two-Tier Client/Server The typical SDE for two-tiered client/server applications (also called distributed data) consists of a client-based programming language with built-in SQL connectivity to one or more server database engines. Examples of two-tiered client/server SDEs include Powersoft's *Powerbuilder*, Microsoft's *Visual Basic (Client/Server Edition)*, Gupta's *SQL Windows*, and Borland's *Delphi (Client/Server Edition)*. Typically, these SDEs provide the following:

- Rapid application development (RAD) for quickly building the graphical user interface that will be replicated and executed on all of the client PCs.
- Automatic generation of the template code for the above GUI and associated system events (such as mouse-clicks, keystrokes, etc.) that use the GUI. The programmer only has to add the code for the business logic.
- A programming language that is compiled for replication and execution on the client PCs.
- Connectivity (in the above language) for various relational database engines and interoperability with those engines. Interoperability is achieved by including SQL database commands (to, for example, create, read, update, delete, and sort records) that will be sent to the database engine for execution on the server.
- A sophisticated code testing and debugging environment for the client.
- A system testing environment that helps the programmer develop, maintain, and run a reusable test script of user data, actions, and events against the compiled programs to ensure that code changes do not introduce new or unforeseen problems.
- A report writing environment to simplify the creation of new end-user reports off a remote database.
- A *help* authoring system for the client PCs.

Today, most of these tools come in the bundled SDE, but independent software tool vendors have emerged to produce replacement tools that often provide greater functionality and/or productivity than those provided in the basic SDE. To learn more about such add-on tools, search the Internet for Programmers Paradise, a software tool mail-order house.

Some of the process logic of any two-tiered client/server application can be offloaded to the database server in the form of stored procedures. In this case, stored procedures are written in a superset of the SQL language. These procedures are then "called" from the client for execution on the server.

SDEs for Multitier Client/Server The current state of the art in enterprise application development is occurring in SDEs for three-tiered (and beyond) client/server architectures. Unlike two-tiered applications, n-tiered applications must support more than 100 users with mainframe-like transaction response time and throughput, with 100 gigabyte or larger databases. While the two-tiered SDEs described earlier are trying to expand in this market, a different class of SDEs currently dominates the market. Typically, the SDEs in this class must provide all the capabilities typically associated with two-tiered SDEs plus the following:

- Support for heterogeneous computing platforms, both client and server, including *Windows*, *OS/2*, UNIX, Macintosh, and legacy mainframes and minicomputers.
- Code generation and programming for both clients and servers. Most tools in this genre support pure object-oriented languages such as *C++* and *Smalltalk*.

- A strong emphasis on reusability using software application frameworks, templates, components, and objects.
- Bundled minicase tools for analysis and design that interoperate with code generators and editors.
- Tools to help analysts and programmers partition application components between the clients and servers.
- Tools to help developers deploy and manage the finished application to clients and servers. This generally includes security management tools.
- The ability to automatically scale the application to larger and different platforms, client and server. This issue of *scalability* was always assumed in the mainframe computing era but is relatively new to the client/server computing era.
- Sophisticated software version control and application management.

Examples of n-tiered client/server SDEs include Intersolv's *Allegris*, Dynasty's *Dynasty*, Forté Software's *Forté*, Texas Instruments' *Performer* and *Composer*, Compuware's *Uniface*, and IBM's *VisualAge* (a family of products). Again, a large number of independent software tool vendors are building add-on and replacement tools for these SDEs.

SDEs for Internet and Intranet Client/Server Rapid application development tools are emerging to enable client/server Internet and intranet applications. Most of these languages are built around three core standard technologies:

- **HTML (hypertext markup language)**—the language used to construct World Wide Web pages and links.
- **CGI (Computer Gateway Interface)**—a standard for publishing graphical World Wide Web components, constructs, and links.
- ***Java***—a general-purpose programming language for creating platform-independent programs and applets that can execute across the World Wide Web.

Examples of *Java*-specific SDEs include Microsoft's *J++*, Symantec's *Visual Café*, and Sun's *Java Workbench*. These SDEs can create both Internet, intranet, and non-Internet/intranet applications. It should also be acknowledged that virtually all existing two-tiered and n-tiered SDEs are also evolving to support HTML, CGI, and *Java*.

System Management Regardless of the chosen SDE and add-on tools, client/server computing applications usually require one or more of the following common process development and management tools:

Transaction processing (TP) monitors—software that ensures that all the data associated with a single business transaction are processed as a single transaction among all the parallel business transactions that may be in the system at the same time. Examples include IBM's *CICS* and NCR's *Tuxedo*.

Version control and configuration managers—software that tracks ongoing changes to software that is usually developed by teams of programmers. The software also allows management to roll back to a prior version of an application if the current version encounters unanticipated problems. Examples include IBM's *SCM* and Intersolv's *PVCS*.

APPLICATION ARCHITECTURE STRATEGIES AND DESIGN IMPLICATIONS

Regardless of what it is called, all information systems have an application architecture. Different organizations apply different strategies to determining application architecture. Let's briefly classify the two most common approaches.

The Enterprise Application Architecture Strategy

In the enterprise application architecture strategy, the organization develops an enterprisewide information technology architecture to be followed in all subsequent information systems development projects. This IT architecture defines the following:

- The approved network, data, interface, and processing technologies and development tools (inclusive of hardware and software; and clients and servers).
- A strategy for integrating legacy systems and technologies into the application architecture.
- An ongoing process for continuously reviewing the application architecture for currency and appropriateness.
- An ongoing process for researching emerging technologies and making recommendations for their inclusion in the application architecture.
- A process for analyzing requests for variances from the approved application architecture. (You may recall that SoundStage received such a variance to prototype object technology in the member services system project.)

The initial application architecture is usually developed as a separate project, or as part of a strategic information systems planning project. The ongoing maintenance of the application architecture is usually assigned to a permanent information technology research group or to an enterprise application architecture committee.

Subsequent to the approval of the application architecture, every information systems development project is expected to use or choose technologies based on that architecture. In most cases, this greatly simplifies the architecture phase of a system development life cycle (such as *FAST*). You simply select from the approved technologies according to the architecture's rules or guidelines.

Of course, even if a technology is approved in the application architecture, it is subject to a feasibility analysis as described in the next subsection.

The Tactical Application Architecture Strategy

In the absence of an enterprisewide application architecture, each project must define its own architecture for the information system being developed. There still may exist some sort of information technology research and deployment group.

While the proposed application architecture for any new information system may be influenced by existing technologies, the developers usually have somewhat greater latitude in requesting new technologies. Of course, the final decision must be defended and approved as feasible. IT feasibility usually includes the following aspects:

- *Technical feasibility*—This can either be a measure of a technology's maturity, or a measure of the technology's suitability to the application being designed, or a measure of the technology's ability to work with other technologies.
- *Operational feasibility*—This is a measure of how comfortable the business management and users are with the technology and how comfortable the technology managers and support personnel are with the technology.
- *Economic feasibility*—This a measure of both whether or not the technology can be afforded and whether it is cost-effective, meaning the benefits outweigh the costs.

To learn more about analyzing feasibility, see Part Five, Module C.

Build versus Buy Implications

One of the most fundamental software decisions in any project is **build versus buy**. Should we design and build the software in-house, or should we purchase the software as a package? Many organizations have made an enterprise IT decision to always purchase those applications that do not significantly add competitive advantage to the business. Typically, these include human resources (and payroll), financial systems, and systems subject to frequent regulatory change (such as college financial aid).

The purchase of application software does not invalidate systems analysis and design. It does, however, change the methodology and project in the following ways:

- Alternative application software packages must be analyzed against the user requirements, and any unfilled business requirements must be identified.
- Options and preferences within the chosen software package must be analyzed and selected. (Most packages allow various onetime and ongoing customization.)
- Business processes and documents must be analyzed and redesigned to interoperate with the selected software.
- Transition processes must be analyzed and designed to import data from legacy systems into the new software's files and databases.
- The application software's interfaces to other information systems must be analyzed and designed.
- Any unmet business requirements are subject to analysis and design as extensions to the chosen software package

Purchased application software may also be subject to any IT architectural standards adopted by a business. For example, if a business has standardized on *Oracle* as its database management system, the business may also restrict purchased applications to those that use *Oracle* as the underlying database management system. Similarly, an application software package may have to conform to the business's approved network and user interface architectures.

MODELING APPLICATION ARCHITECTURE AND INFORMATION SYSTEM PROCESSES

Just as we modeled business requirements during systems analysis, we should model technology architecture and requirements during system design. The models serve as blueprints for system design, prototyping, and construction. We'll cover two popular tools for modeling the design decisions discussed in the first section of this chapter: *physical data flow diagrams* and *system flowcharts*. The former is the modern tool of choice. The latter used to be the tool of choice and is still encountered when studying documentation of older, legacy systems.

Physical Data Flow Diagrams

Data flow diagrams (DFDs) were introduced in Chapter 6 as a tool for modeling the *essential* or *logical* (meaning nontechnical) business requirements of an information system. With just a few extensions of the graphical language, DFDs can also be used to model the *physical* (meaning technical) design of the system.

Physical data flow diagrams model the technical and human design decisions to be implemented as part of an information system. They communicate technical and other design constraints to those who will actually implement the system—in other words, they serve as a technical blueprint for the implementation.

Let's examine the graphical conventions for physical DFDs. Physical DFDs use the same shapes and connections as logical DFDs (Chapter 6): *processes*, *external agents*, *data stores*, and *data flows*. Only the naming standards (and a few new rules) are changed to extend the language to document technology and design decisions. In this section, we'll focus only on the *new* naming standards and rules.

A sample physical data flow diagram is shown in Figure 10.8. For now, just notice that the PDFD shows more technical and implementation detail than its logical DFD (LDFD) equivalent.

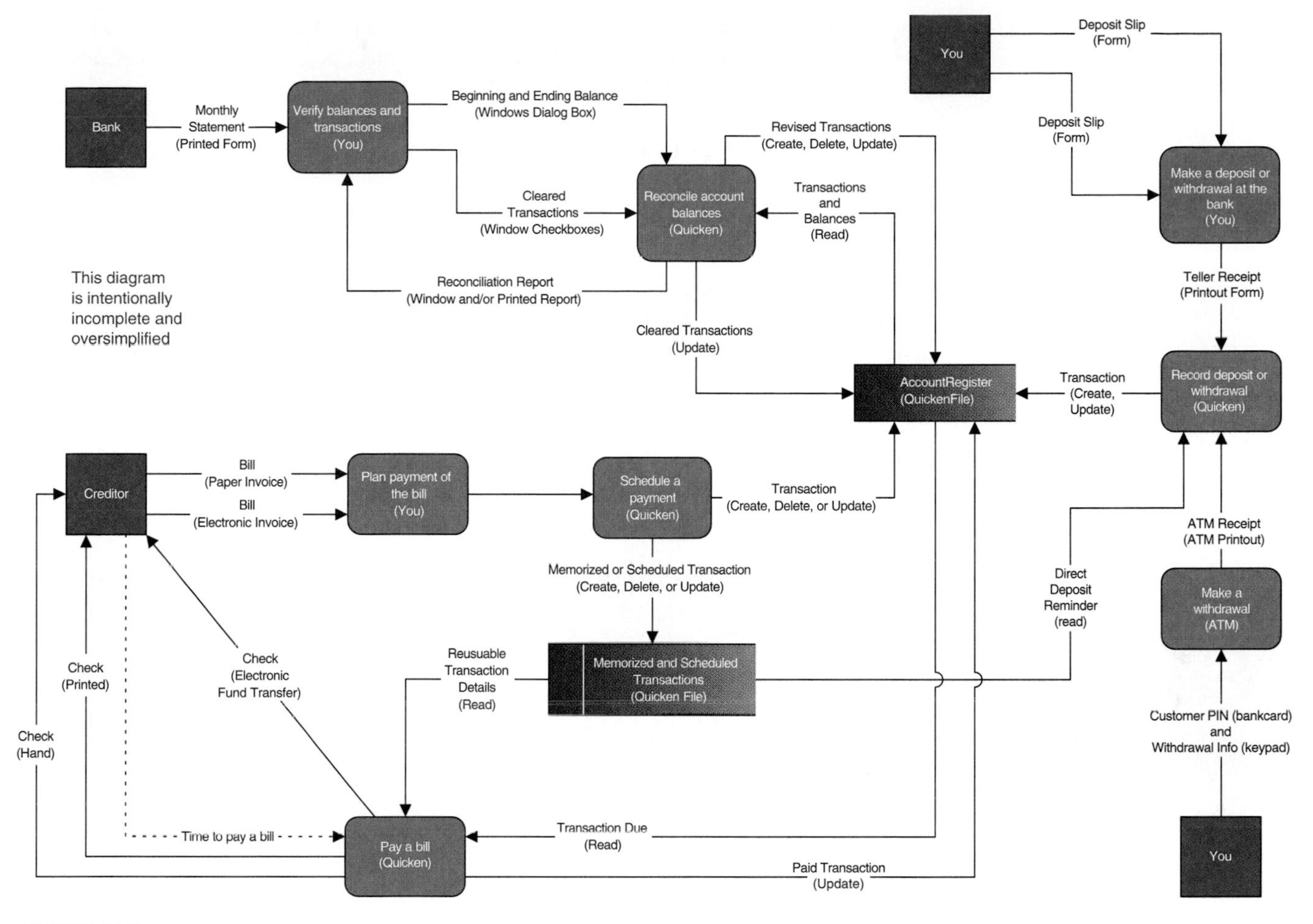

FIGURE 10.8 *A Sample Physical Data Flow Diagram*

Physical Processes Recall that *processes* are the key shapes on any DFD. That's why they are called *process models.*

A **physical process** is either (1) a processor, such as a client PC, network server, or robot, or (2) specific work or actions to be performed on incoming data flows to produce outgoing data flows. In the latter case, the physical process must clearly designate which person or what technology will be assigned to do the work.

Recall that during systems analysis, *logical processes* representing essential business process requirements for the information system were identified and modeled in *logical data flow diagrams.* During system design, we must specify how these logical processes will be physically implemented. As implied in the above definition for *physical processes*, there are two elements to *physical data flow diagrams*:

- Logical processes are frequently assigned to specific physical processors such as clients, servers, or other devices in a computer network. To this end, we might draw a physical DFD called a **network topology data flow diagram** for the information system.
- Subsequently, logical processes are usually implemented as one or more physical processes. Some logical processes must be split into multiple physical processes for the following reasons:
  - To split the process into that portion performed by a person and that portion performed by the computer.

- To split the process into that portion to be implemented with one technology and that portion to be implemented with a different technology.
- To show multiple, different implementations of the same logical process (such as processing a paper order versus processing a phone order).
- To add processes that are necessary to implement audit and control requirements or to handle exceptions.

In all cases, if you split a logical process into multiple physical processes, or add new physical processes, you have to add any necessary physical data flows to preserve the essence of the original logical process. In other words, the physical processes must still meet the logical process requirements.

Process names use the same *action verb + object clause* convention as in Chapter 6. However, the name is preceded or followed by an implementation method. The format, which may be constrained by your choice of CASE tool, is:

Implementation method : Action verb + Object clause
or
Action verb + Object clause : Implementation method

The following names demonstrate three possible implementations of the same logical process:

CICS/COBOL : CHECK CUSTOMER CREDIT
VISUAL BASIC : CHECK CUSTOMER CREDIT
JAVA APPLET : CHECK CUSTOMER CREDIT

If your CASE tool limits the size of names, you may have to develop and use a set of abbreviations for the technology (and possibly abbreviate your action verbs and object clauses). In the margin, we show three different ways to show a physical process (again, usually constrained by the CASE tool of choice).

ID# (opt)
action verb + object clause
implementation method

ID# (opt)
implementation method : action verb + object clause

ID# (opt)
action verb + object clause
(implementation method)

Physical process templates

If a logical process is to be implemented partially by people and partially by software, it must be split into separate physical processes and appropriate data flows must be added between the physical processes. The name of a physical process to be performed by people, not software, should indicate who will perform that process. We recommend titles, not proper names. The following are examples:

CREDIT MANAGER : REVERSE OR LET STAND CREDIT RECOMMENDATION
REVERSE OR LET STAND CREDIT RECOMMENDATION (CREDIT MANAGER)

Why not just change the manual processes to external agents? Because, as developers, a system design is not complete until all new or changed business processes for the system are also described.

For computerized processes, the implementation method is, in part, chosen from one of the following methods:

- A purchased software package, possibly to be selected (e.g., SAP: PROCESS ORDER or FMIS: ENCUMBER FUNDS, where SAP and FMIS are names of purchased software applications).
- A productivity or utility program (e.g., EXCHANGE: MAIL SCHEDULE TO TEAM or QUICKSORT: SORT INVOICES where EXCHANGE is an e-mail productivity application and QUICKSORT is a utility program).
- An existing application program from a program library (e.g., LIB: MOD EDIT_CUST.COB where EDIT_CUST.COB is a COBOL program in the existing library to be MODified).
- A program to be written (e.g., PB : COMMIT INVENTORY is a *Powerbuilder* program to be written, or ACCESS: COMMIT INVENTORY is an equivalent

Microsoft *Access* module to be written, or EXCEL: IMPORT CLASS SCHEDULE is a Microsoft *Excel* spreadsheet application to be developed.

Again, the number of processes on a physical DFD will usually be greater than the number of processes on its equivalent logical DFD. For one thing, processes may be added to reflect data flow collection, filtering, forwarding, preparation, business controls—all in response to the implementation target that has been selected. Also, some logical processes may be split into multiple physical processes to reflect portions of a process to be done manually versus by a computer, to be implemented with different technology, or to be distributed to clients, servers, or different host computers. It is important that the final physical DFDs reflect all manual and computer processes required for the chosen implementation strategy.

Physical Data Flows Recall that all processes have at least one input and one output data flow.

A **physical data flow** represents the planned implementation of an input to or output from a physical process. It can also indicate database action such as create, delete, read, or update a record. It can also represent the import of data from or the export of data to another information system across a network. Finally, it can represent the data flows between two modules or subroutines within the same program.

Physical data flow names use one of the following general formats:

implementation medium : data flow name

or

data flow name : implementation method

Template examples are depicted in the margin.

implementation method :
data flow name

data flow name :
(implementation method)

Physical data flow templates

Representative input methods include GUI (graphical user interface—or, more specifically, WIN95, NT, MAC, OS2 etc.), BARcode, OMR (optical mark reader), OCR (optical character reader), EDI (electronic data interchange), TONE (phone Touch-Tone response), PEN (pen-based interface), and FILE (remote batch file).

Representative output media include PRINT, FICHE (microfiche), GUI (graphical user interface, which might be more specific), EDI, and IMAGE.

For data transmitted across a network, the implementation media should indicate the file transfer protocol to be used. Examples include SNA or TCP/IP. This notation will be especially important for communication between cooperative processes that are to be implemented on clients and servers.

For data transmitted between processes that are to be part of the same program, you could specify parameters and variables to be passed between the program modules.

Physical DFDs must also indicate any data flows to be implemented as *business forms*. For instance, FORM 23: COURSE REQUEST might be a one-part business form used by students to register for classes. Business forms frequently use a multiple (carbon or carbonless) copy implementation. At some point in processing, the different copies are split and travel to different manual processes. This is shown on a physical DFD as a diverging data flow (review Chapter 6). Each copy should be uniquely named. For example, at a restaurant, the customer receives FORM: CREDIT CARD VOUCHER (CUSTOMER COPY).

Most logical data flows are carried forward to the physical DFDs. Some may be consolidated into single physical data flows that represent business forms. Others may be split into multiple flows as a result of having split logical processes into multiple physical processes. Still others may be duplicated as multiple flows with different technical implementations. For example, the logical data flow ORDER might be implemented as all of the following: OMR : ORDER FORM, PHONE : ORDER (verbal order taken over the phone), TONE : ORDER (Touch-Tone order submitted over the phone), WWW : ORDER (electronic form on the World Wide Web), FAX : ORDER (order received by fax), and MODEM : ORDER (an order submitted by a direct on-line modem connection).

Physical External Agents External agents are carried over from the logical DFD to the physical DFD unchanged. Why? By definition, external agents were classified during systems analysis as outside the scope of the systems and, therefore, not subject to change. Only a change in requirements can initiate a change in external agents.

Physical Data Stores From Chapter 6 you know that each data store on the logical DFD now represents a data entity on an entity relationship diagram (Chapter 5). Accordingly,

> Most **physical data stores** represent a single file or a single database or table in the database. Additional physical data stores may be added to represent temporary files or batches necessitated by physical processes.

The name of a physical data store uses the following format (see the margin samples):

> file or database implementation method : file, database, or table name
> file, database, or table name : file or database implementation method

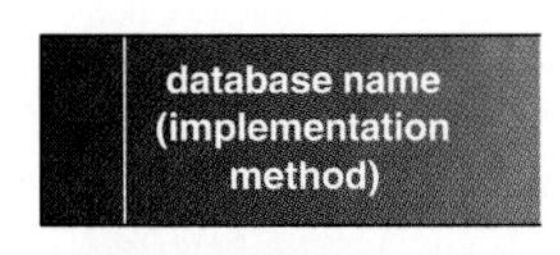

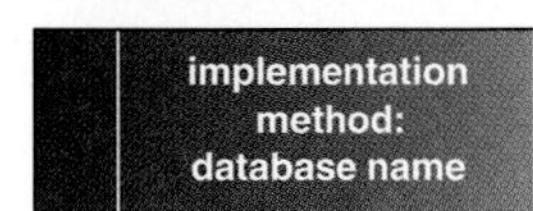

Physical data store templates

Examples include VSAM : CUSTOMERS (a conventional vsam file), ORACLE : ORDER ENTRY DB (a database consisting of many tables), and SQL SERVER : CUSTOMERS TABLE (a table in a SQL SERVER database).

Some designs require that temporary files be created to act as a queue or buffer between processes. Such files are documented in the same manner except their name should include some indication of their temporary status.

A design may also include noncomputerized files. If this is the case, the storage mechanism name replaces the file organization. For example, FILE CABINET: RADIOACTIVE ISOTOPE PROJECT RECORDS indicates that records describing research involving radioactive isotopes are stored in file cabinets. Despite futurist predictions about the demise of paper files, they will remain a part of many systems well into the foreseeable future—if for no other reasons than (1) there is psychological comfort in paper and (2) government frequently requires it!

System Flowcharts

Before we teach you how to draw implementation (and design unit) DFDs, we should introduce an alternative. Before the popularity of DFDs, system flowcharts were a popular tool for modeling design decisions.

> **System flowcharts** are diagrams that show the flow of control through a system while specifying all programs, inputs, outputs, and file/database accesses and retrievals.

While the popularity of system flowcharts is clearly on the decline, you may encounter them (especially as existing system documentation for older information systems). You need to learn how to read them.

The American National Standards Institute (ANSI) has established certain symbols that have been widely used in the computer industry to describe the flow of process control in systems. Although the symbols have been standardized, their use has not. Thus, many system flowcharts look incomprehensible to those who would like to use them.

System flowcharts are supposed to be the basis for communication among systems analysts, end-users, applications programmers, and computer operators. Think

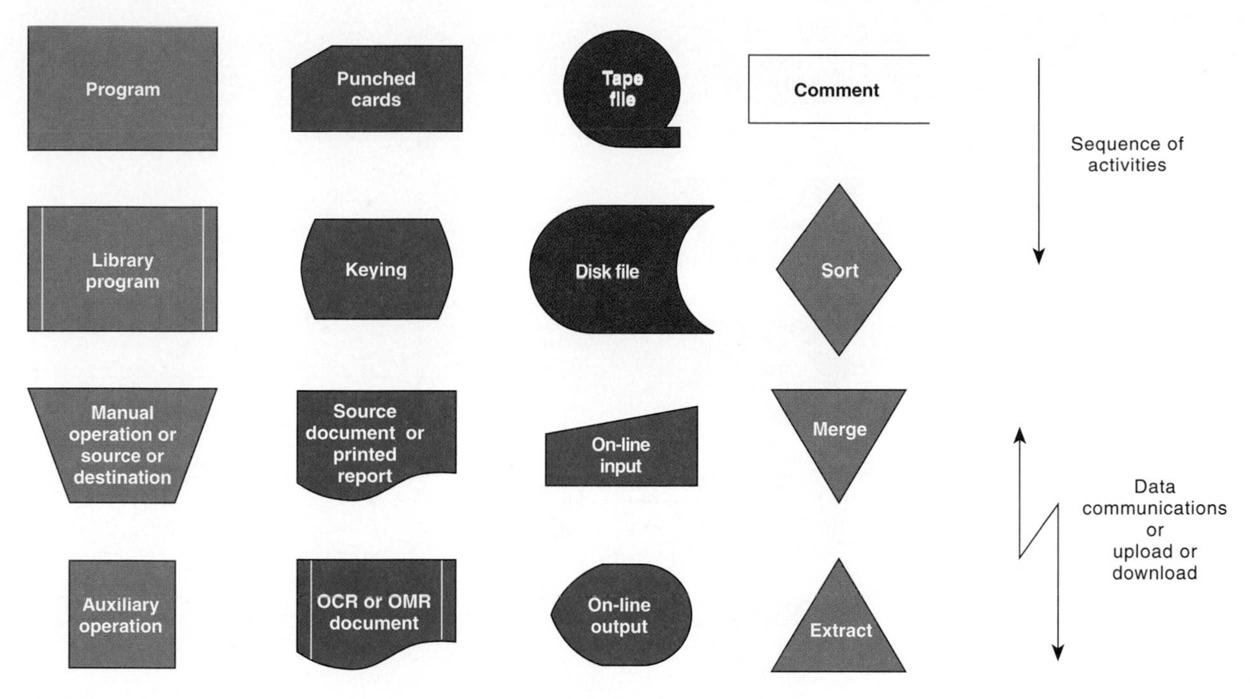

FIGURE 10.9 *System Flowchart Shapes and Connections*

of system flowcharts as a chance to prove or disprove that a specific technical solution to the end-users' requirements will work. In this section, you'll learn how to read system flowcharts.

System Flowchart Symbols Figure 10.9 shows the system flowchart symbols and their meanings. The symbols can be conveniently classified into six subsets.

There are four primary symbols for processing: the *computer program* (to be written), the *library program* (that already exists—possibly a utility program), the *manual operation* (indicating who and/or what—the symbol is also used as a start or finish symbol in a flowchart), and the *auxiliary operation* (used to indicate operations performed by other office equipment). The name (or identification) of the process is recorded in all the symbols.

There are four symbols for batch input: the *source document* or *form* (from which data will be keyed into the system), the *key-to-punched-card* (a dated and increasingly rare batch input medium), the *key-to-disk (KTD)* or *key-to-tape (KTT)*, and the *optical character* (or *mark*) *document* (which could also be used for EDI and imaged documents). The name of the document or input file is recorded in the symbol.

Batch output forms or files are represented by a single *printed output* symbol. The symbol could also be used for microfiche output. The name or identification of the output is recorded in the symbol.

System flowcharts show only those files and databases stored on computers. There are only two symbols: the *tape file* and the *disk file* or database. Because tape files are always sequential, updates always occur in pairs—an original file is input to a program and a new file is produced by the program. Disk files need not be duplicated since reads and writes are processed against the same copy of the file or database. On the other hand, applications sometimes create distinct subsets or reorganizations of files for a program's exclusive use. These versions are shown as separate symbols.

Databases serve to integrate many files. They can be depicted as one integrated

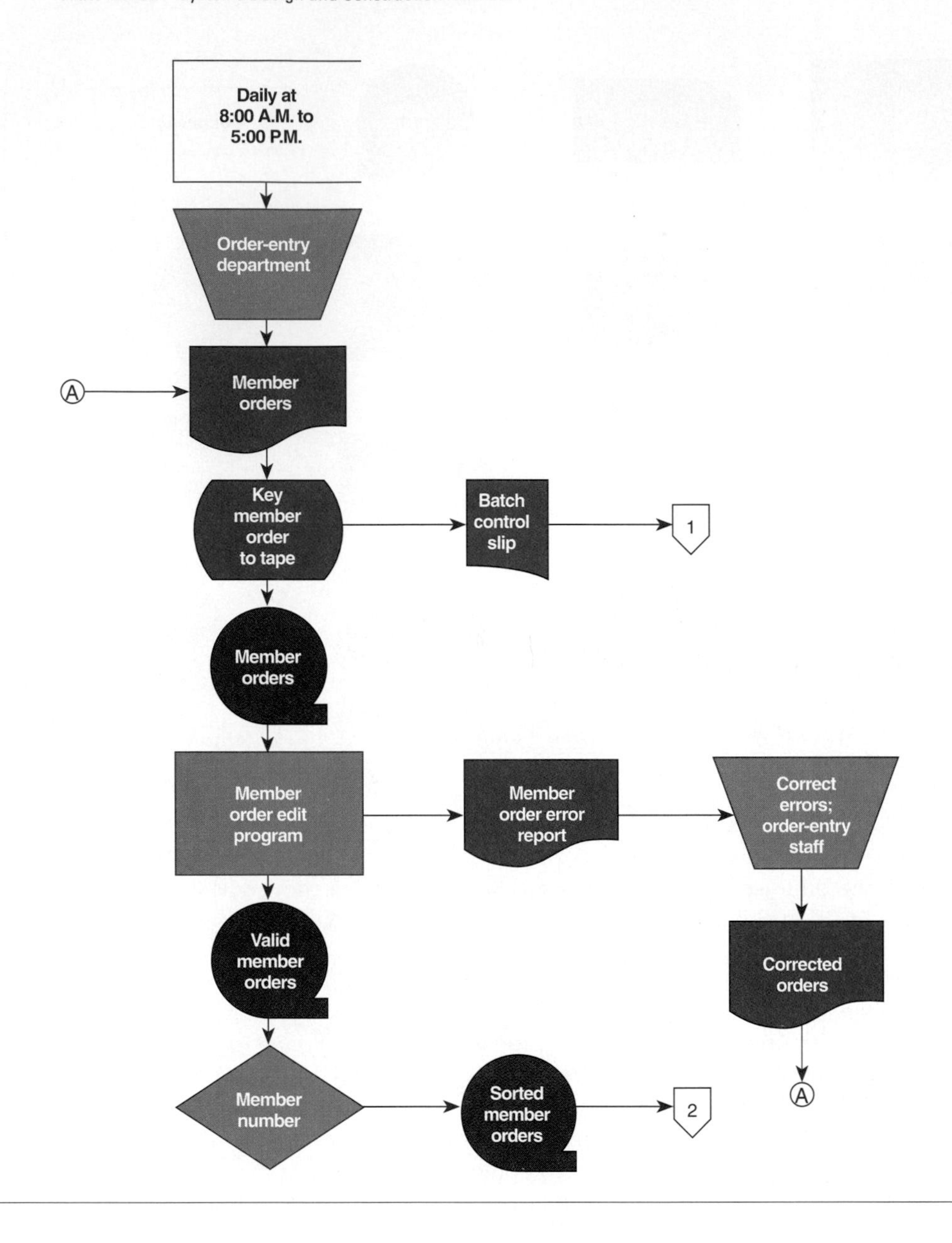

FIGURE 10.10 *A Sample System Flowchart (Page 1)*

symbol or as one symbol per file, record type, or table—depending on what level of detail you are trying to depict. For all tape and disk symbols, the label indicates the name or identification of the file, database, record type, or table.

The symbolism for on-line inputs and outputs can be somewhat tricky. If we had to show every possible screen that could occur, the systems flowchart would get very cluttered. Therefore, most designers show only the net inputs and outputs and exclude the on-line dialogue that gets you to those inputs and outputs. There are two on-line symbols: the *on-line input* and the *on-line output.* As usual, the name of the input or output is recorded in the symbol.

There are a number of miscellaneous symbols in the ANSI standard. They are used to document aspects of methods not covered by the other symbols. The most important of these symbols is the *comment.* It may be connected via a dashed line to any other symbol to add information about such things as timing, security, or instructions. Other miscellaneous symbols include *sort, merge,* and *extract.*

Reading System Flowcharts The symbols are connected by one of three lines: a single-ended arrow (indicating either the sequence of activities or processing in the

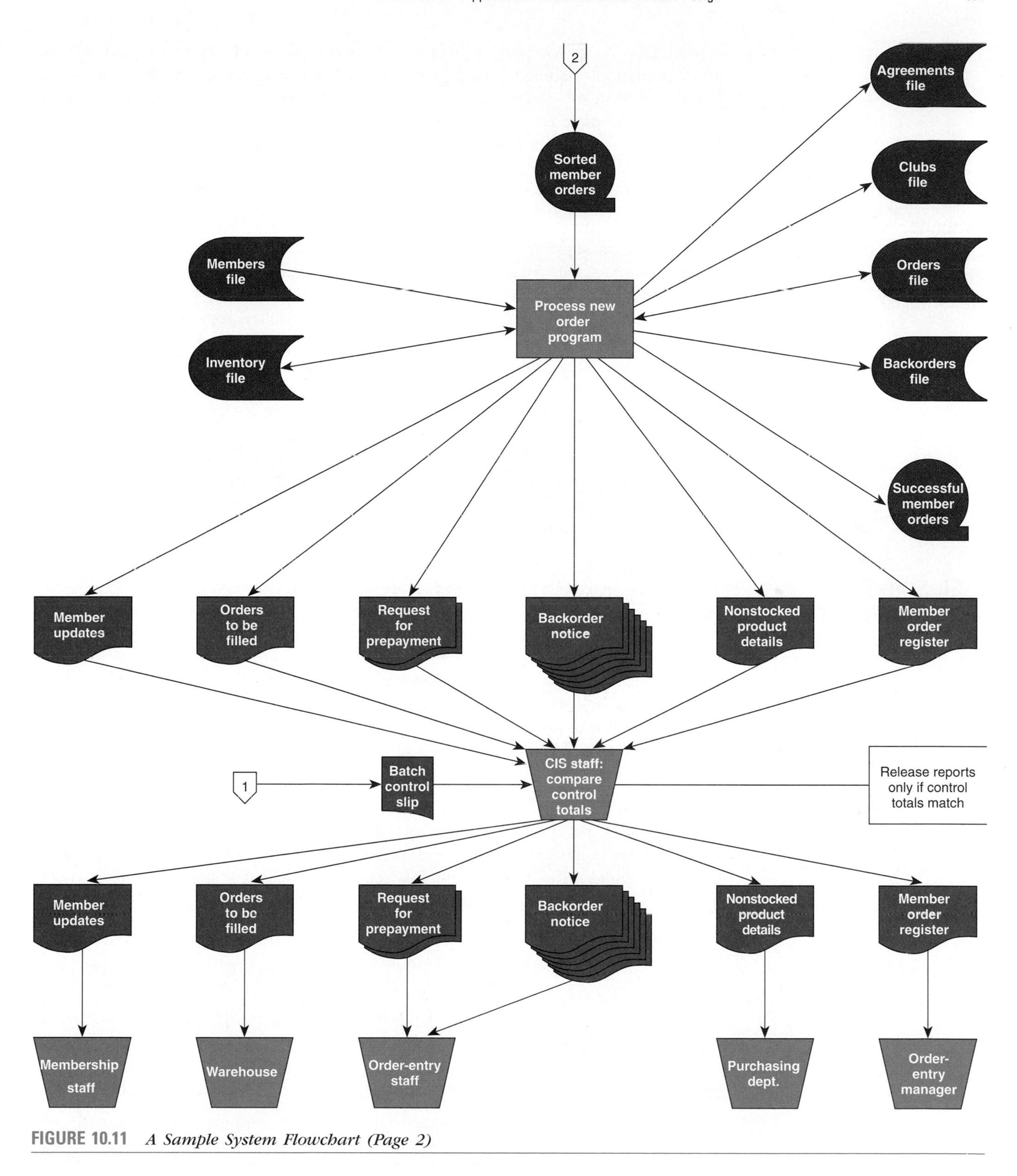

FIGURE 10.11 *A Sample System Flowchart (Page 2)*

system or a read-only or write-only access to a file or database), a double-ended arrow (indicating read-write operations against files and databases), or a jagged-double-ended arrow (indicating on-line dialogue or data flow). Unlike DFDs, connections on system flowcharts are not labeled or named.

Symbols and connections are combined in classic input-process-output patterns to document the design of a system. Figures 10.10 and 10.11 demonstrate sample system flowcharts for part of a system. Notice the small circle (labeled A) and small pentagons (labeled 1 and 2). The circles are on-page connectors that show transfer of flow of control on the same page (to avoid crossed lines). The pentagons are off-page connectors that show transfer of flow of control to the matching symbols on another page. The rest of the diagram should be fairly self-explanatory.

Computer-Aided Systems Engineering (CASE) for Physical DFDs and Flowcharts

CASE was introduced in Chapter 3 as a technology for systems development. Most CASE products support DFDs, but don't distinguish between logical and physical models. Some CASE users simply modify the logical DFDs that they created during systems analysis. The problem with that approach is that you are overwriting the logical DFDs—the requirements themselves! A better approach is to *copy* the logical models and then transform the copies into physical DFDs.

Many CASE tools cannot completely support the graphical language extensions described in this chapter. Limitation on name length is the most common problem. You can overcome the limitations by agreeing on a standard set of technical abbreviations. For example, BC can mean bar code. Alternatively, implementation details can be recorded in the repository description of objects on the DFDs.

The ultimate goal of CASE technology is to be able to automatically generate application programs (code) from these models and supporting detailed specifications (all stored in the CASE tools' repositories).

DESIGNING THE APPLICATION ARCHITECTURE AND THE INFORMATION SYSTEM PROCESSES

The use of *logical* DFDs to model process requirements is a fairly accepted practice. However, the transition from analysis-oriented *logical* DFDs to design-oriented *physical* DFDs has historically been somewhat mysterious and elusive. What we desire is a high-level general design that can serve as an application architecture for the system and a general design for the processes that make up the system. From this model, we can perform detailed design and/or prototyping of the data stores, data flows, and processes.

In this section, we teach the logical-to-physical transformation.

Drawing Physical DFDs

The mechanics for drawing physical DFDs are virtually identical to those of logical DFDs. The rules of correctness are also identical. We'll not repeat those rules here. (Review Chapter 6 for those rules.) Physical DFDs document the high-level, general design of the new system. An acceptable design results in

- A system that works.
- A system that fulfills user requirements (specified in the logical DFDs).
- A system that provides adequate performance (throughput and response time).
- A system that includes sufficient internal controls (to eliminate human and computer errors, ensure data integrity and security, and satisfy auditing constraints).
- A system that is adaptable to ever-changing requirements and enhancements.

We could develop a single physical DFD or a set of physical DFDs for the target system. The *FAST* methodology used by SoundStage calls for the following:

- A physical data flow diagram should be developed for the network topology. Each process on this diagram is a processor in the system. Each server is its own processor; however, it is usually impractical to show each client. Instead, each *class* of clients (e.g., an order entry clerk) is represented by a single processor.
- For each processor on the above model, a physical data flow diagram should be developed to show those event processes (see Chapter 6) that will be

assigned to that processor. It is possible that you would choose to duplicate event processes on multiple processors. For instance, orders may be processed on regional servers or clients.

- For all but the simplest event processes, they should be factored into *design units* and modeled as a single physical data flow diagram.

A **design unit** is a self-contained collection of processes, data stores, and data flows that share similar design attributes. A design unit serves as a subset of the total system whose inputs, outputs, files and databases, and programs can be designed, constructed, and unit tested as a single subsystem. (The concept of design units was first proposed by McDonnell Douglas in its *STRADIS* methodology.)

An example would be a set of processes (one or more) to be designed as a single program. The design unit could then be assigned to a single programmer (or team) who (which) can work independently of other programmers and teams without adversely affecting the work of the other programmers. The implemented units would then be assembled into the final application system. Design units can also be prioritized for implementing versions of a system.

Prerequisites

Let's set the table by describing the prerequisites to creating physical DFDs. They include:

- A logical data model (*entity relationship diagram* created in Chapter 5).
- Logical process models (*data flow diagrams* created in Chapter 6).
- A logical network model (optional—*location connectivity diagrams* created in Chapter 7).
- Repository details for all of the above.

Given these models and details, we can distribute data and processes to create a general design. Your general design will normally be constrained by one or more of the following:

- Architectural standards that predetermined the choice of database management systems, network topology and technology, user interface(s), and/or processing methods.
- Project objectives that were defined at the beginning of systems analysis and refined throughout systems analysis.
- Feasibility of chosen or desired technology and methods. Feasibility analysis methods are covered in Part Four, Module C.

Within any restrictions of those constraints, the ensuing techniques can be applied.

The Network Topology DFD

The first physical DFD to be drawn is the network topology DFD.

A **network topology DFD** is a physical data flow diagram that allocates processors (clients and servers) and devices (e.g., machines and robots) to a network and establishes (1) the connectivity between the clients and servers, and (2) where users will interact with the processors (usually only the clients).

To identify the processors and their locations, the developer utilizes three resources:

- If available, the location connectivity diagram from systems analysis (Chapter 7) models the network from a technology-independent perspective.
- If an enterprise information technology architecture exists, that architecture likely specifies the client/server vision that should be targeted.
- The advice of competent network managers and/or specialists should be

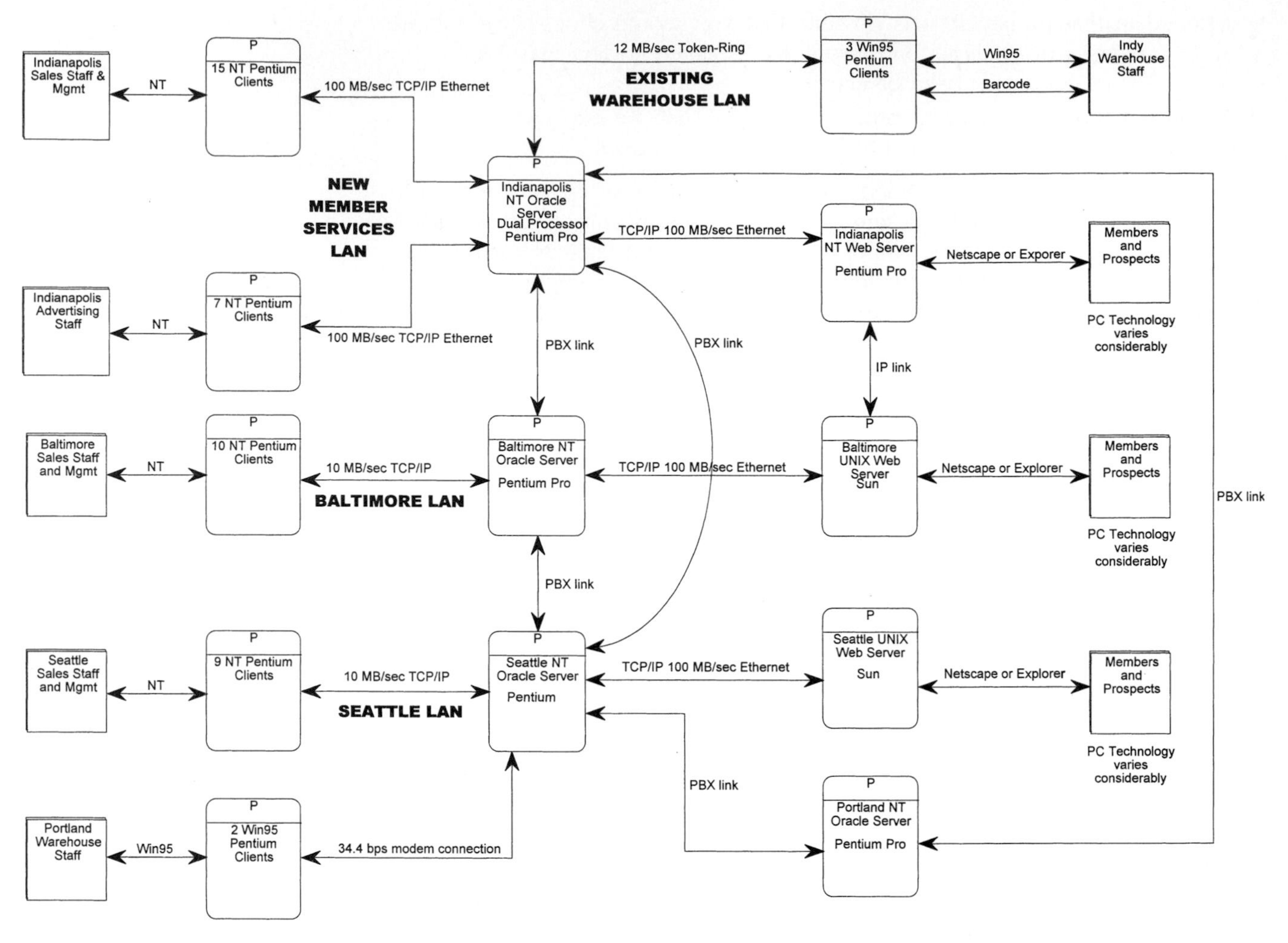

FIGURE 10.12 *Network Topology Data Flow Diagram*

solicited to determine what's in place, what's possible, and what impact the system may have on the computer network.

Network topology DFDs (see Figure 10.12) need to be labeled to show somewhat different information than normal DFDs. They don't show specific data flows per se. Instead, they show highways over which data flows may travel in either direction. Also, network topology DFDs indicate the following:

- *Servers and their physical locations.* Servers are not always located at the sites indicated on a location connectivity diagram. Network staff access to servers is usually an issue. Some network management tasks can be accomplished remotely, and some tasks also require hands-on access.
- *Clients and their physical locations.* In this case, the location connectivity diagram is quite useful in identifying "classes" of like users (e.g., ORDER CLERKS, SALES REPPRESENTATIVES, etc) who will be serviced by similar clients. A single processor should represent the entire class at a single location. The same class may be replicated in multiple locations. For example, you would expect each SALES REGION to have similar types of employees.
- *Processor specifications.* The repository descriptions of processors can be used to define processor specifications such as RAM, hard disk capacity, and display.
- *Transport protocols.* Connections are labeled with transport protocols (e.g., TCP/IP) and other relevant physical parameters.

The network topology DFD can be used to either design a computer network or to document the design of an existing computer network.

Data Distribution and Technology Assignments

The next step is to distribute data stores to the network processors. The required logical data stores are already known from systems analysis—as data stores on the logical DFDs (Chapter 6) or as entities on the logical ERDs (Chapter 5). We need only determine where each will be physically stored and how they will be implemented.

To distribute the data and assign their implementation methods, the developers utilize three resources:

- If available, the data distribution matrices from systems analysis (Chapter 7) model the data needs at business locations from a technology-independent perspective.
- If an enterprise information technology architecture exists, that architecture likely specifies the database vision and technologies that should be targeted.
- The advice of data and database administrators should be solicited to determine what's in place, what's possible, and what impact the database may have on the overall system.

The distribution options were described earlier in the chapter and are summarized as follows:

- Store all data on a single server. Until recently, this was the most common solution. The database (consisting of multiple tables) should be named, and that named database and its implementation method (e.g., ORACLE : MEMBER_SERVICES) should be added to the physical DFD and connected to the appropriate processor.
- Store specific tables on different servers. In this case, and for clarity's sake, we should record each table as a data store on the physical DFD and connect each to the appropriate server.
- Store subsets of specific tables on different servers. In this case we record the tables exactly as above except that we indicate which tables are subsets of the total set of records. For example, the label DB2 : ORDERS TABLE (REG SUBSET) indicates a subset of all orders for a region are stored in a DB2 database table.
- Replicate (duplicate) specific tables or subsets on different servers. In this case, replicated data stores are shown on the physical DFD. One copy of any replicated table is designated as the MASTER, and all other copies are designated as COPY.

Why distribute data storage? There are many possible reasons. First, some data instances are of local interest only. Second, performance can often be improved by subsetting data to multiple locations. Finally, some data needs to be localized to assign custodianship of that data. The data distribution and technology assignments for the SoundStage case study are shown in Figure 10.13.

Data distribution decisions can be very complex—normally the decisions are guided by data and database professionals and taught in data management courses and textbooks. In this book we want to consider only how to document the partition and duplication decisions.

Process Distribution and Technology Assignments

Information system processes can now be assigned to processors as follows:

- For single-tiered client/server systems, all the *logical event diagrams* (Chapter 6) are assigned to the server.
- For two-tiered client/server systems, all the *logical event diagrams* (Chapter 6) are assigned to the client.

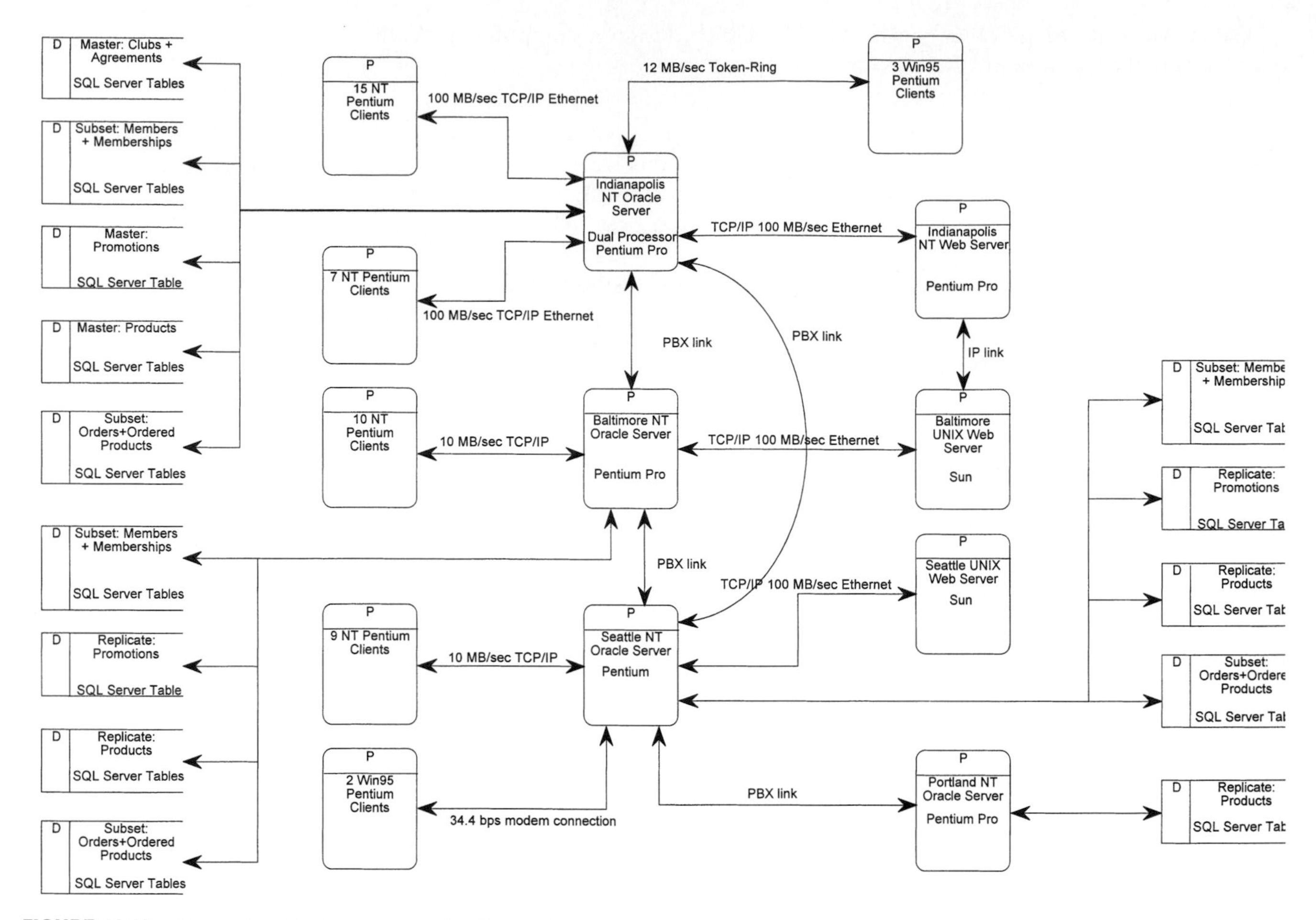

FIGURE 10.13 *Data Distribution across the Network*

- For three-tiered client/server systems, you must closely examine each event's primitive (detailed) data flow diagram. You need to determine which primitive processes should be assigned to the client and which should be assigned to an application server. In general, data capture and editing is assigned to clients and other business logic is assigned to servers. If you partition different aspects of a logical DFD to different clients and servers, you should draw separate physical DFDs for the portions on each client and server.

After partitioning, each physical DFD corresponds to a design unit for a given business event. (Business events were discussed in Chapter 6.) For each of these design units, you must assign an implementation method, the SDE that will be used to implement that process. You must also assign implementation methods to the data flows.

SoundStage's Member Services System will be implemented with a two-tiered client/server architecture. A sample DFD for one event to be assigned to a client is shown in Figure 10.14. Notice that the data stores are shown even though we know they have been partitioned to a database server. This is for the benefit of the programmers who must implement the DFD. Why virtual?

The Person/Machine Boundaries

The last step of process design is to factor out any portion of the physical DFDs that represent manual, not computerized processes. This is sometimes called "establishing a person/machine boundary." Establishing a person/machine boundary is not difficult, but it is not as simple as you might first think. The difficulty arises

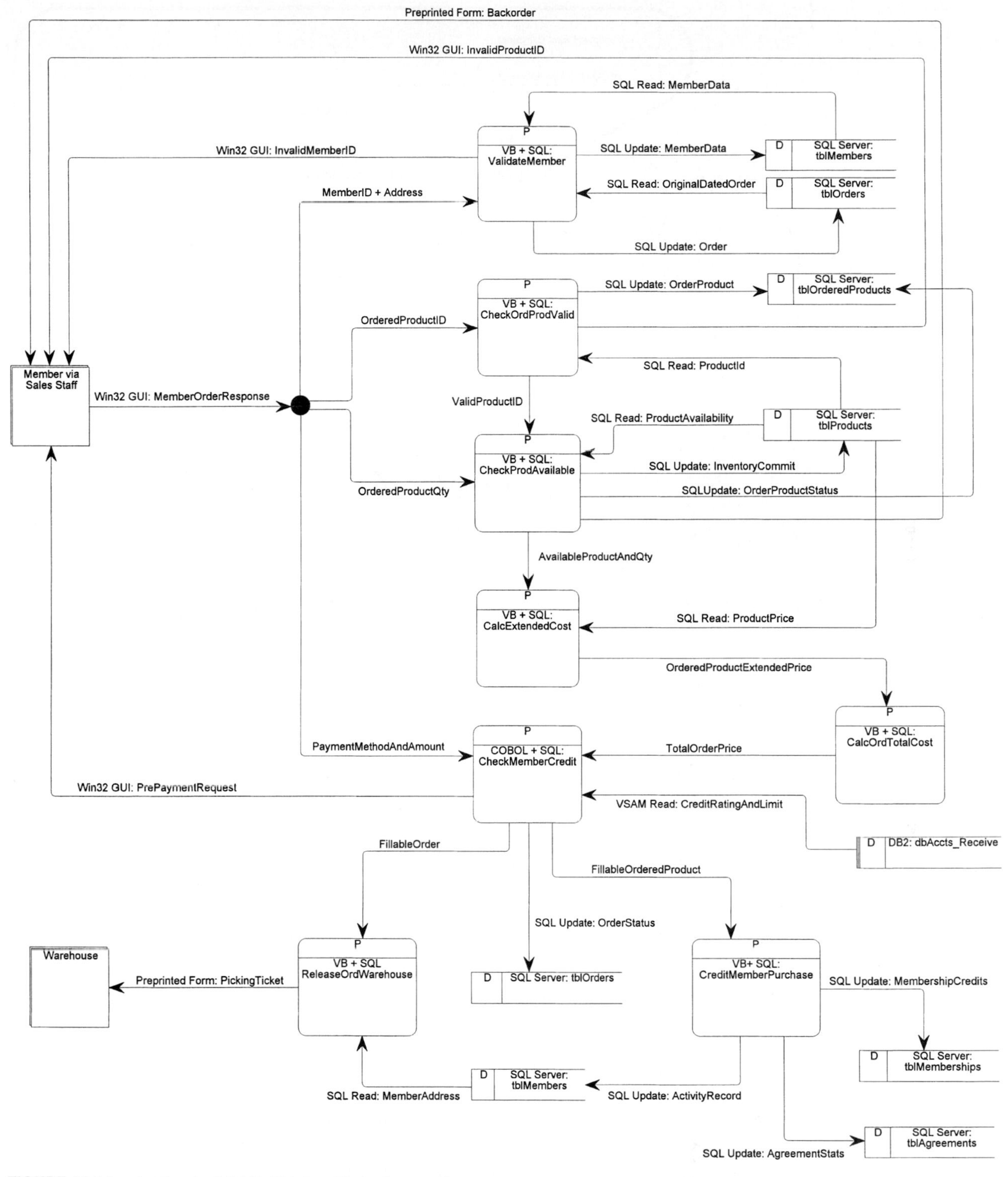

FIGURE 10.14 *A Physical DFD Design Unit for an Event*

when the person/machine boundary cuts *through* a logical process—in other words, part of the process is to be manual and part is to be computerized. This situation is common on logical DFDs since they are drawn without regard to implementation alternatives.

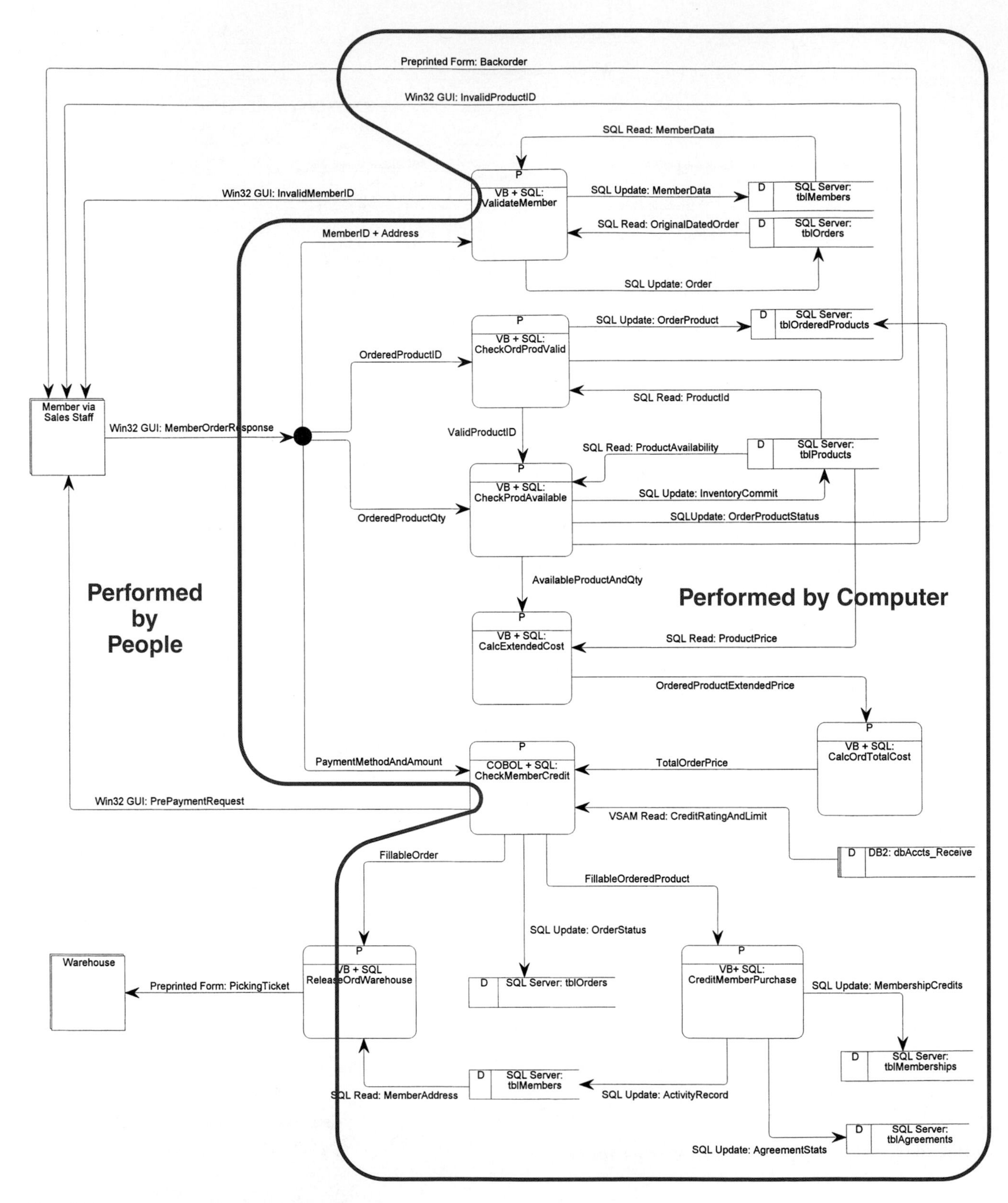

FIGURE 10.15 *A Person/Machine Boundary*

Figure 10.15 adds the person/machine boundary to a physical DFD. Notice that our boundary cuts through several processes, including the CHECK MEMBER CREDIT process. The solution to this process requires two steps.

1. The manual process portions are pulled out as a separate design unit (Figure

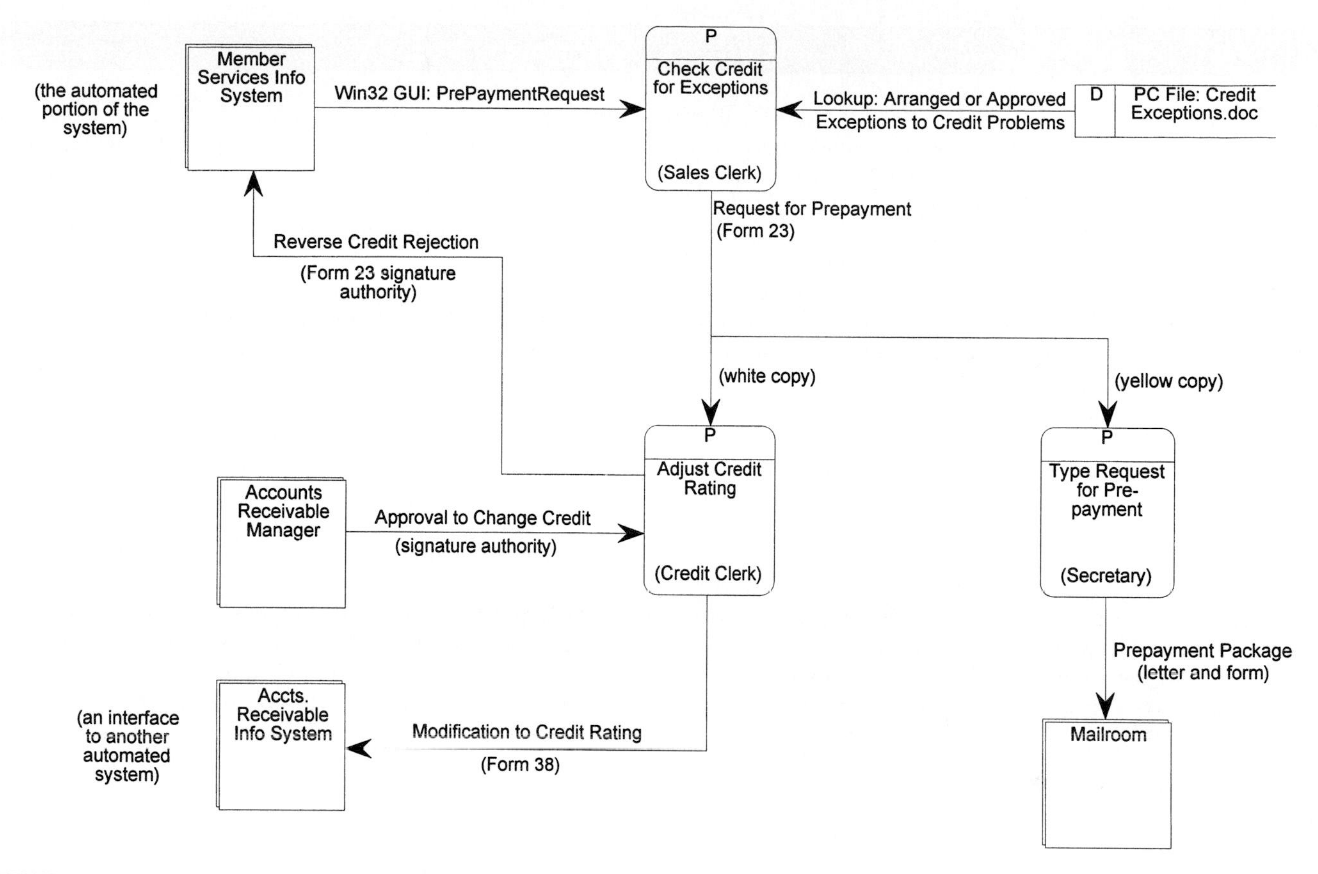

FIGURE 10.16 *A Manual Design Unit*

10.16). All of these processes are completely manual. The interfaces of the manual design units to the computerized processes (on Figure 10.15) are depicted as external agents. Ultimately, the manual processes in the design unit must be clearly described to those people who will have to perform them.

2. If necessary, the processes on the original diagram should be renamed to reflect only the computerized portion. (In practice, the processes were already named that way.)

WHERE DO YOU GO FROM HERE?

In this chapter, you have learned how to outline the design of a new information system to fulfill the requirements identified and modeled during systems analysis. This **general design** for the new system will guide the detailed design and construction of that system.

Because the general design of the system will usually require a commitment of significant resources to detailed design, prototyping, and construction of the final system, management may require a detailed feasibility analysis. If you haven't already done so, this may be a good time to read Module C, "Feasibility and Cost-Benefit Analysis," which covers various aspects of feasibility analysis with a focus on economic analysis and justification.

Most readers will now progress to the detailed design chapters that build upon the general design for the new system. For most of you, we recommend you start with Chapter 11, "Database Design." Most design-by-prototyping and rapid application development techniques are absolutely dependent on the existence of the planned information system's database. Databases must be carefully designed to ensure adaptability and flexibility during the system's lifetime. Thus, Chapter 11 is the best place to begin your study of detailed design. Subsequently, you can move on to chapters that cover other aspects of detailed design including inputs, outputs, and programs.

SUMMARY

1. An application architecture defines the technologies to be used by one, more, or all information systems. There are four categories of technology: NETWORK, DATA, INTERFACE, and PROCESS.
2. The prevailing computing model is currently client/server wherein a network of clients, single-user computers, are connected to and interoperate with servers, multiple-user computers that share their services. This is also called distributed computing.
3. Centralized computing, distributed presentation, distributed data, distributed data and logic, and Internet/intranet computing are flavors of client/server computing.
4. Client/server computing can be based on different network topologies including bus, ring, star, and hierarchical networks.
5. Data storage is typically implemented using distributed relational database technology that either partitions data to different servers or replicates data on multiple servers.
6. User interface options include batch, on-line, remote batch, keyless data entry (including optical character/mark and bar coding methods), pen input, graphical user interfaces, electronic messaging, electronic data interchange, and imaging.
7. System interfacing is typically implemented using middleware.
8. Processes are implemented using highly integrated tool kits called software development environments.
9. Application architectures may be developed and enforced strategically, or they may tactically evolve on a project-by-project basis.
10. Physical data flow diagrams model an information system's application architectures and processes. Because they show the planned implementation of all processes, data stores, and data flows, they serve as a general system design or blueprint for subsequent detailed design, prototyping, and construction.
11. System flowcharts are a lesser used diagram to show the same implementation features as physical data flow diagrams. Although they are rarely drawn today, many older, legacy systems use them for documentation.

KEY TERMS

application architecture, p. 353
auto-identification, p. 366
bar coding, p. 366
batch processing, p. 364
browser, p. 366
build versus buy, p. 371
bus network, p. 359
centralized computing, p. 354
client, p. 353
client/server computing, p. 353
Computer Graphics Interface (CGI), p. 370
connectivity, p. 359
data distribution, p. 364
data replication, p. 364
database engine, p. 364
database server, p. 357
design unit, p. 381
detailed design, p. 352
distributed computing, p. 354
distributed data, p. 356
distributed data and logic, p. 357
distributed presentation, p. 356
distributed RDBMS, p. 364
distributed relational database, p. 364
distributed relational database management system, p. 364
electronic data interchange (EDI), p. 367
file server, p. 357
general design, p. 352
graphical user interface (GUI), p. 366
handheld PC, p. 366
hierarchical network, p. 361
hypertext markup language (HTML), p. 370
Internet, p. 358
interoperability, p. 359
intranet, p. 358
Java, p. 370
local area network (LAN), p. 356
middleware, p. 368
network topology, p. 359
network topology data flow diagram, p. 381
n-tiered client/server computing, p. 357
on-line processing, p. 365
optical character reading (OCR), p. 365
optical mark reading (OMR), p. 366
partitioning, p. 358
physical data flow, p. 375
physical data flow diagram, p. 372
physical data store, p. 376
physical process, p. 373
process design, p. 350
relational database, p. 362
remote batch, p. 365
ring network, p. 360
server, p. 353
software development environment, p. 368
star network, p. 361
system flowchart, p. 376
system integration, p. 367
three-tiered client/server computing, p. 357
transaction processing (TP) monitor, p. 370
two-tiered client/server computing, p. 356
version control and configuration manager, p. 370
wide area network (WAN), p. 356

REVIEW QUESTIONS

1. What is an application architecture? What role does it play in information systems development?
2. What are clients and servers? What role does each play in client/server architecture?
3. Differentiate between centralized and distributed processing alternatives. Why is new application development seriously considering the distributed processing alternative?
4. Differentiate between two-tiered and three-tiered client/server computing.
5. Differentiate between LANs and WANs.
6. What is partitioning and why is it applicable to client/server computing?
7. How do intranets fit into the client/server computing approach?
8. Differentiate between connectivity and interoperability. Why can you have connectivity and lack interoperability? Which is more important?
9. Differentiate between bus and ring networks.
10. Differentiate between star and hierarchical networks.
11. What is a distributed relational database? How does it differ from a PC database?
12. Differentiate between distributed and replicated data.
13. Explain the difference between batch, on-line, and remote batch input methods. Define an input and conceive a situation that might call for each of the three methods to be used.
14. Why are keyless data entry alternatives attractive to business? Describe three keyless data entry technologies.
15. Why have graphical user interfaces become an issue for systems analysts designing custom applications (such as order entry and inventory control) for a business?
16. What is middleware?
17. What is a software development environment?
18. Differentiate between logical and physical data flow diagrams. When is each relevant?
19. Differentiate between a processor and a process as it relates to physical data flow diagrams.
20. What is a network topology DFD?
21. What is a design unit?

PROBLEMS AND EXERCISES

1. Your boss proclaims, "We haven't invested in client/server architecture yet. We are a mainframe shop." Tactfully explain to your boss why she or he is wrong. Explain how your boss might evolve the client/server architecture while continuing to use the mainframe.
2. Respond to the following editorial: "Client/server computing is too big a step for us at this time. I realize that PCs and networks may—and I emphasize *may*—be cheaper to acquire, operate, and maintain, but we can't just get rid of our mainframe and all its legacy applications just like that. We can't rush into this mainframe versus server issue just yet."
3. You have been hired as a consultant to a small copy shop. The owner proudly proclaims, "We are ahead of our time. We are already doing two-tiered client/server. We use Microsoft *Access* as our database server." Explain to the owner why simply placing the *Access* data on a file server does not implement two-tiered client/server. Explain the advantage of true two-tiered client/server computing. What would it take?
4. What are the options for partitioning business logic across a three-tiered client/server network?
5. Explain how a corporate intranet might serve as a framework for a company's client/server information systems.
6. Why are networking and data communications essential subjects for future systems designers who will build client/server applications? How do you plan to acquire this knowledge?
7. What is the topology of your school's local area network? If you are employed, what is the topology of your employer's network?
8. Develop a set of business guidelines to determine whether data should be centralized, distributed, or replicated (including combination approaches).
9. Describe the software development environment in your last programming course or project. What elements of an SDE as described in this chapter were missing? How might they have helped?
10. Prepare a physical data flow diagram (or, if your instructor prefers, a systems flowchart) to describe the backup and recovery procedures for a lost or damaged master file or database.
11. Your organization uses a tactical, project-by-project approach to determining application architecture. Write a two-page proposal to management that defends a more strategic approach, and adapt the *FAST* methodology to propose a project approach to develop an initial enterprisewide application architecture. How do you propose to deal with legacy applications?
12. What criteria can be used to measure information system design quality?
13. Your business has many old-style systems flowcharts that you want converted to physical data flow diagrams since all of your staff are familiar with PDFDs. Create a translation legend that suggests how symbols on a flowchart correspond to their equivalent symbols on data flow diagrams. A small team should be able to use this as a conversion guide for the project.

PROJECTS AND RESEARCH

1. Recent business reports and editorials have concluded that client/server computing is not the economic panacea that most initially believed. Why might cost *not* be a factor in the decision to go client/server for all new applications?
2. Research the evolving subject of client/server applications development. Study the costs and benefits and the issues and problems to be addressed. Be sure to find pro and con examples—both exist! Present your findings in a recommendations report to information systems management.
3. Client/server technology has yet to stabilize. Research and prepare a technology update report on one of the following subjects:
   a. Database management systems for servers in a client/server architecture.
   b. Network topology and operating systems for a client/server environment.
   c. Distributed computing environment (DCE), an evolving standard for open systems interoperability for client/server computing. (Note: Open systems is a goal whereby dissimilar computers and software can communicate and cooperate in a manner that is transparent to users.)
   d. Client programming languages versus downsized COBOL/CICS. Examples of the former include *Powerbuilder*, *Object View*, and *Visual BASIC*.
   e. Middleware for easy, transparent access to existing data stored using dissimilar structure and different servers and hosts.
4. Make an appointment to visit a local business computing facility. How has the growth in numbers of microcomputers affected the networking strategy of the business? What is the networking strategy? What topology was chosen and why?
5. Acquire a network topology model for your school (or a local business). Convert it to a network topology DFD as described in this chapter.
6. Visit a systems analyst or programmer in a local business. Describe the software development environments used in that shop. Were any elements missing?
7. Research the SDEs available for a single programming language such as *Visual Basic*, *C++*, *COBOL*, *Delphi*, or *Java*. Report on third-party extensions available for that environment.
8. At the time of this writing, *Java* was beginning to emerge as a candidate programming language (with SDEs) for mainstream information systems (as opposed to just Internet and intranet applets). Research the current viability and penetration of this new object-oriented programming language.
9. Acquire a systems flowchart from a local business application, an older systems analysis and design textbook, or your instructor (who probably has access to many in his or her course archives). Convert that systems flowchart to a physical data flow diagram.

MINICASES

1. Chapter 5 provided a data model. Using your school or business's technical environment as a constraint (or one provided by your instructor), use the process analysis and design technique in this chapter to derive a design. The design should be documented using physical data flow diagrams. (Note: Your technical environment should at least specify processors, operating systems, languages, network topology, network operating systems and protocols, and database management systems.)
2. Pacific Imports, a wholesale distributor of a wide variety of imported products, is located in Los Angeles, California. Pacific has just completed the automation of its mail-order system. This minicase reviews the design and implementation.
   a. At approximately 8:30 each morning, the Order Entry department receives all new sales order forms. These forms include orders received both by mail and over the phone. Order Entry enters and edits all orders on personal computers, creating a batch of orders on a network file server.
   b. The batch of orders is transmitted from the server's disk to an AS/400 midrange computer via a network gateway over a TCP/IP point-to-point network connection. There, the order-processing program checks the inventory master file (actually, a database table) to validate products ordered and to determine the availability and price of the products ordered. For any products that are out of stock, a backorder is created in the database and printed as a notice for the customer. Fillable orders (and partial orders) are then processed as follows.
   c. Another program checks the customer accounts master file to check credit. For customers who have a poor credit rating, a prepayment request is printed. The order is placed on hold in the order master file. Also, a report of these notices is prepared for management. The program prints an order confirmation letter for those orders that will be filled.
   d. Another program produces a four-part warehouse order form that includes a picking copy, a packing copy, a shipping copy, and an invoicing copy. The processed order is moved to a sales master file, a copy of which is archived on magnetic tape for backup and later use.

The system also includes an on-line program that allows the sales manager both to query the inventory master file to obtain prices and to query the customer accounts master file to obtain credit information needed to manually override a credit rejection. This program is available to the sales manager from 8 A.M. until noon each working day.

e. Customer order cancellations are processed immediately upon receipt (by mail or phone). When the request is received by a clerk, he or she enters the order cancellation using the PC as a terminal into the AS/400. An on-line program reads the sales master file to determine the order's status. If the order has not been filled, the clerk phones the warehouse to have the order terminated. At the end of the day, the program generates cancellation notices for customers and a cancellation report for the sales manager.

i. What are the limitations of the English language for describing a design to programmers who must implement that design?
ii. Did you spot any errors of omission in the narrative description—that is, important questions or processes that were not covered?
iii. Draw a physical data flow diagram for this case.
iv. Draw a system flowchart for this case.

SUGGESTED READINGS

Gane, Chris, and Trish Sarson. *Structured Systems Analysis: Tools and Techniques.* Englewood Cliffs, NJ: Prentice Hall, 1979. This classic on process modeling became the basis of physical data flow diagrams.

Goldman, James. *Applied Data Communications: A Business-Oriented Approach.* New York: John Wiley & Sons, Inc., 1995. Our colleague at Purdue has written an excellent textbook for those seeking to learn about data communications and networking from a business perspective.

Kara, Daniel A., et al. *Enterprise Application Development: Seminar Notes.* Chicago: Software Productivity Group, November 12, 1996. This seminar and the writings of the SPG have strengthened our understanding of two-tiered and n-tiered software application development techniques and technologies.

Lucas, Henry C. *Information Technology for Management.* 6th ed. New York: McGraw-Hill Companies, Inc., 1997. The management focus of this textbook provides a useful business-oriented perspective on how different information technologies can be used in information systems. The author is highly respected as both an author, teacher, and practitioner.

Smith, Patrick, and Steve Guengerich. *Client/Server Computing.* 2nd ed. Indianapolis, IN: SAMS Publishing, 1994. This professional book has been used to teach the basics of client/server technology and architecture to our students at Purdue. Given the rapid evolution of this technology, there may now exist a third edition. Check out the technology case studies in the appendices.

Theby, Stephen E. "Derived Design: Bridging Analysis and Design." McDonnell Douglas Professional Services: Improved System Technologies, 1987. The techniques described in this paper are the basis for a phase in STRADIS (Structured Analysis, Design, and Implementation of Information Systems), a systems development methodology. The technique was altered and simplified to make it suitable to the level of this textbook. As authors, we were quite impressed with the full derived design technique as advocated in the STRADIS methodology.

Whitten, Jeffrey L.; Lonnie D. Bentley; and Victor M. Barlow. *Systems Analysis and Design Methods.* 2nd ed. Burr Ridge, IL: Richard D. Irwin, 1989. If you are interested in a more comprehensive tutorial on systems flowcharts, this older edition of our book covered that material in Chapter 18.

11

DATABASE DESIGN

CHAPTER PREVIEW AND OBJECTIVES

Data storage is a critical component of most information systems. Some people consider it to be the critical component. This chapter teaches the design and construction of physical databases. You will know that you have mastered the tools and techniques of database design when you can:

- Compare and contrast conventional files and modern, relational databases.
- Define and give examples of fields, records, files, and databases.
- Describe a modern data architecture that includes files, operational databases, data warehouses, personal databases, and work group databases.
- Compare the roles of systems analyst, data administrator, and database administrators as they relate to databases.
- Describe the architecture of a database management system.
- Describe how a relational database implements entities, attributes, and relationships from a logical data model.
- Normalize a logical data model to remove impurities that can make a database unstable, inflexible, and nonscalable.
- Transform a logical data model into a physical, relational database schema.
- Generate SQL code to create the database structures in a schema.

SCENE

The application architecture and general system design for the new Member Services System has been completed and approved. Sandra and Bob are now working with Steve Watson, the database administrator, to design the database for the new system.

BOB

I have finished the first draft of the database schema for the new system. As specified in the application architecture, the entire database will be replicated at each distribution center. Each database will be stored on a Compaq quad-processor Pentium Pro with RAID disks. The network operating system will be *Windows NT*, and the database engine will be *SQL Server*.

STEVE

How much disk, Bob?

BOB

We need to calculate that, Steve. I want to get the database schema completed before I calculate the storage space requirements.

SANDRA

Besides, it should not be difficult to calculate the storage. In the logical data model we recorded the number of instances of each entity and expected growth. We also recorded the type and size of each data attribute. As soon as we complete the schema, we can calculate each table's storage requirement from those numbers.

STEVE

Thanks, you're doing me a big favor. I will add an appropriate buffer for data management overhead and growth. We need to get those servers ordered.

SANDRA

I just put that on my action item list. Bob, why don't you share the preliminary database design with us.

BOB

Right! This was really quite straightforward. After confirming that each entity on the data model was normalized, I used our CASE tool, *System Architect*, to generate the first cut database design. I had to tweak the design based on *SA's* generation algorithm, but it turned out to be pretty complete.

[Bob hands out copies of a diagram — Figure 11.A.]

This is the current schema.

SANDRA

This will sure come in handy in my screen design meeting Monday. Do you mind if I generate a prototype of these database tables?

BOB

Don't bother, I already generated an *Access* prototype in anticipation that you would need it. It's in the project directory in a subdirectory called "Test Data"—but I haven't entered any test data yet.

SANDRA

Thanks, I'll take care of that.

STEVE

You know, if the prototype is solid, we might test that new Microsoft wizard that converts an *Access* database into an *SQL Server* database.

BOB

Wouldn't you rather have the standard SQL code instead?

STEVE

Yes, but that takes a while to write.

BOB

Unnecessary! *System Architect* can automatically generate the SQL code to construct the database as soon as you sign off on this database schema.

STEVE

Righteous! This is getting easier all the time.

BOB

Essentially, there is one table for each entity in the approved data model. All primary and secondary keys will be indexes into the tables. Notice that all attributes have been set to the correct data types and null options . . .

STEVE

I don't mean to interrupt, but these foreign keys won't work.

BOB

Excuse me?

STEVE

This is not a complaint. Many books and methodologies allow identical names for the primary keys and their corresponding foreign keys, especially in the logical model. But that becomes a problem in the physical model. Ideally, every attribute is stored in one and only one table. That way, whenever anyone refers to the attribute, you know what they are talking about. Looking at your model, if I refer to MEMBER_NUMBER, what table am I talking about? The MEMBERS table? The MEMBER ORDERS table? The MEMBERSHIPS table? You see? You and I know about foreign keys, but users do not.

SANDRA

I see your point, but we can't delete foreign keys from the tables that are related to members—that's how the database establishes links between the tables.

STEVE

Yes, and I'm sure that using the same names for the primary and foreign keys was useful during logical systems analysis. It helped the users see how you intended to use the foreign keys. But *SQL Server* does not require identical names for the primary and foreign keys. We've found it useful to change the physical name of the foreign key to describe the *role* that the foreign key plays in that table.

BOB

You know, I saw that property, "role," in the *System Architect* screens but did not bother to look it up.

STEVE

Actually, that's great news. It means that your CASE tool supports roles. Sandra, you look puzzled. Maybe an example would help. Consider the table MEMBERS. What is the role of MEMBER_NUMBER in that table?

SANDRA

It's the primary key; therefore, the role is to uniquely identify a member.

STEVE

Correct. Now what is the role of MEMBER_NUMBER in the MEMBER ORDERS table?

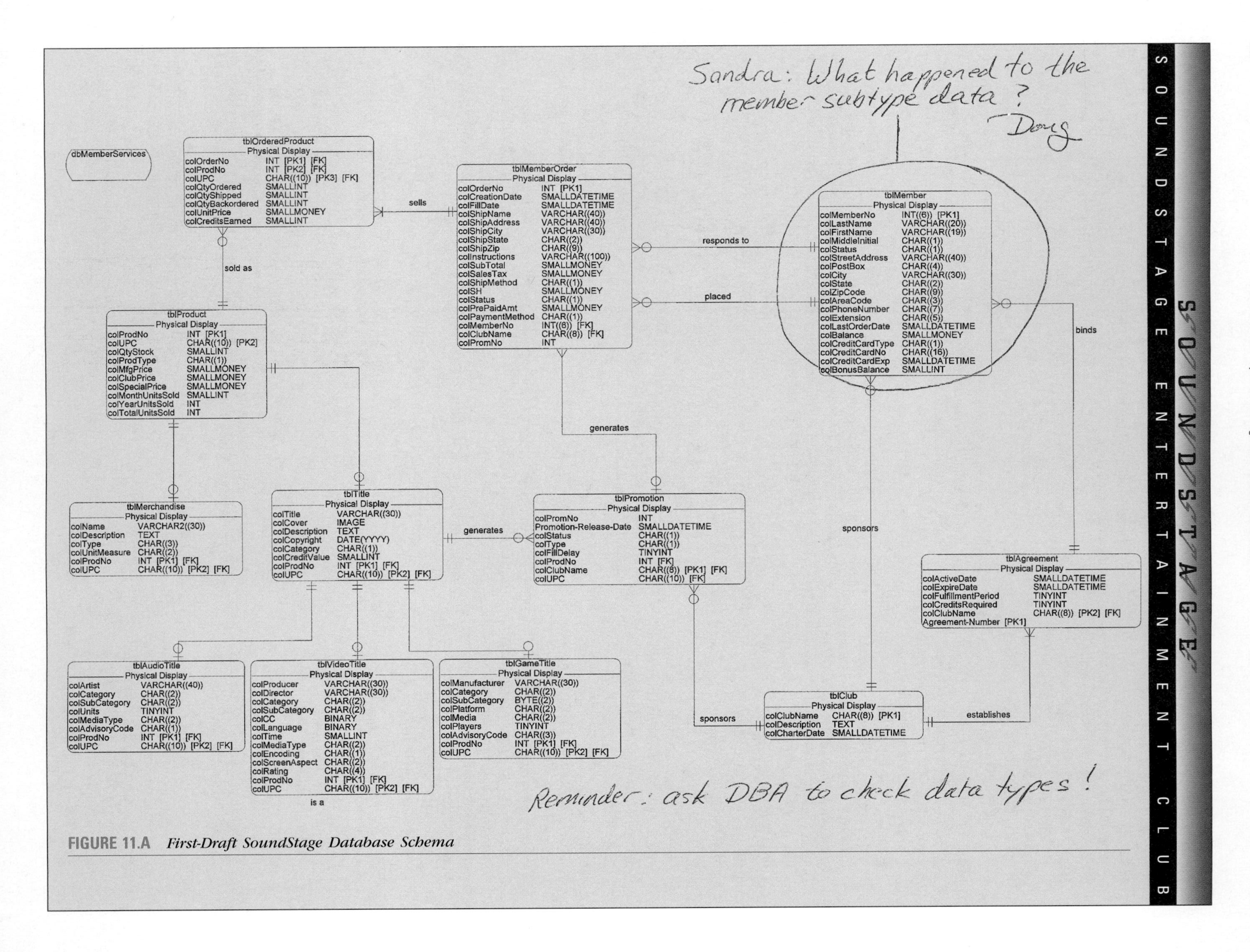

FIGURE 11.A *First-Draft SoundStage Database Schema*

SOUNDSTAGE
SOUNDSTAGE ENTERTAINMENT CLUB

SANDRA
Obviously, you are looking for a different answer . . . Let's see . . . OK, I would say its role is to identify the member *that placed an order.*

STEVE
Absolutely correct! So why not change the foreign key name to ORDERED_BY_MEMBER_NUMBER?

BOB
I get it! And I'm on it. I'll have all these fixed as soon as I figure out how *System Architect* handles it.

STEVE
Great! And I'll bet you credits to navy beans that *System Architect* can automatically generate the correct SQL code for these roles . . .

DISCUSSION QUESTIONS

1. What role does the logical data model (Chapter 5) play in database design?
2. How do the entities, attributes, and relationships from the data model impact disk storage capacity in database design?
3. What value would a PC database prototype (e.g., Microsoft *Access*) return to a project that has targeted to a high-end database system (e.g., *Oracle* or *SQL Server*)?
4. What is a database schema and how is it different from a logical data model?

CONVENTIONAL FILES VERSUS THE DATABASE

All information systems create, read, update, and delete data. This data is stored in files and databases.

Files are collections of similar records.

Examples include a CUSTOMER FILE, ORDER FILE, and PRODUCT FILE.

Databases are collections of *interrelated* files.

The key word is *interrelated*. A database is *not* merely a collection of files. The records in each file must allow for relationships (think of them as "pointers") to the records in other files. For example, a SALES database might contain ORDER records that are somehow "linked" to their corresponding CUSTOMER and PRODUCT records.

Let's compare the file and database alternatives. Figure 11.1 illustrates the fundamental difference between the file and database environments. In the file environment, data storage is built around the applications that will use the files. In the database environment, applications will be built around the integrated database. Accordingly, the database is not necessarily dependent on the applications that will use it. In other words, given a database, new applications can be built to share that database. Each environment has its advantages and disadvantages.

The Pros and Cons of Conventional Files

In most organizations, many or most existing information systems and applications are built around conventional files. You may already be familiar with various conventional file organizations (e.g., indexed, hashed, relative, and sequential) and their access methods (e.g., sequential and direct) from a COBOL course. These conventional files will likely be in service for quite some time.

Conventional files are relatively easy to design and implement because they are normally based on a single application or information system, such as ACCOUNTS RECEIVABLE or PAYROLL. If you understand the end-user's output needs for that system, you can easily determine the data that will have to be captured and stored to fulfill those needs and define the best file organization for those requirements.

Historically, another advantage of conventional files has been processing speed. They can be optimized for the access of a single application. At the same time, they can rarely be optimized for shared use by different tasks in an application, or different applications. Still, files have generally outperformed their database

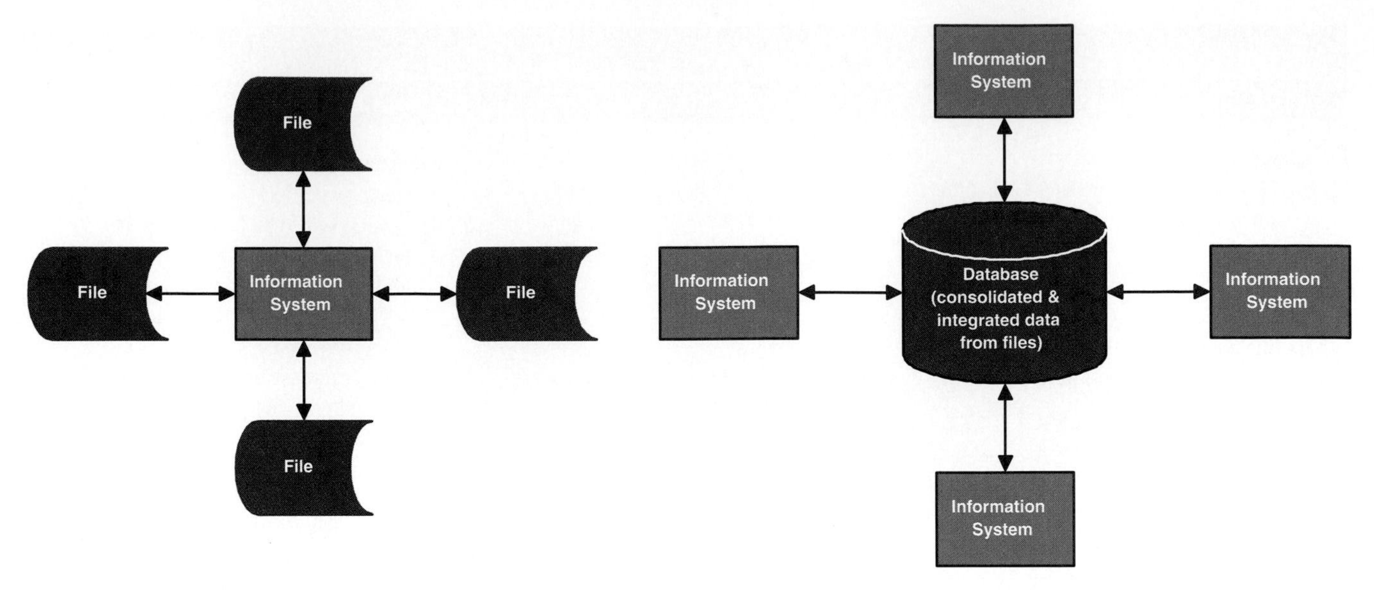

FIGURE 11.1 *Conventional Files versus the Database*

counterparts; however, this limitation of database technology is rapidly disappearing thanks to cheaper and more powerful computers and more efficient database technology.

Conventional files also have numerous disadvantages. Duplication of data items in multiple files is normally cited as the principal disadvantage of file-based systems. Files tend to be built around single applications without regard to other (future) applications. Over time, because many applications have common data needs, the common data elements get stored redundantly in many different systems and files. This duplicate data result in duplicate input, duplicate maintenance, duplicate storage, and possibly data integrity problems (different files showing different values for the same data item).

And what happens if the data format needs to change? Consider the problem faced by many firms if all systems must support a nine-digit ZIP code or four-digit years (to accommodate the year 2000). How many redundant files would have to be located and changed in a typical organization? Add to this the enormous volume of programs that use these ZIP code and date fields and you have some sense of the nightmare that a file structure change can become.

A significant disadvantage of files is their inflexibility and nonscalability. Because files are typically designed to support a single application's *current* requirements and programs, future needs—such as new reports and queries—often require files to be restructured because the original file structure cannot support the new requirements. But if we elect to restructure those files, all programs using those files would also have to be rewritten. In other words, the current programs have become dependent on the files, and vice versa. This usually makes reorganization impractical; therefore, we elect to create new, redundant files to meet the new requirements. But that exasperates the aforementioned redundancy problem. Thus, the inflexibility and redundancy problems tend to escalate one another!

As legacy file-based systems and applications become candidates for reengineering, the trend is overwhelmingly in favor of replacing file-based systems and applications with database systems and applications. For that reason, we have elected to focus this chapter on database design. (Those who encounter the need to learn and apply older file design techniques will find an abundance of reference material in the form of both COBOL file processing and systems analysis and design textbook in their local college library. Prior editions of this textbook covered file design.)

The Pros and Cons of Database

We've already stated the principal advantage of database—the ability to share the same data across multiple applications and systems. A common misconception about the database approach is that you can build a single superdatabase that contains all data items of interest to an organization. This notion, however desirable, is not currently practical. The reality of such a solution is that it would take forever to build such a complex database. Realistically, most organizations build several databases, each one sharing data with several information systems. Thus, there will be *some* redundancy between databases. However, this redundancy is both greatly reduced and, ultimately, controlled.

Database technology offers the advantage of storing data in flexible formats. This is made possible because databases are defined separately from the information systems and application programs that will use them. Theoretically, this allows us to use the data in ways not originally specified by the end-users. Care must be taken to truly achieve this *data independence.* If the database is well designed, different combinations of the same data can be easily accessed to fulfill future report and query needs. The database scope can even be extended without impacting existing programs that use it. In other words, new fields and record types can be added to the database without affecting current programs.

On the other hand, database technology is more complex than file technology. Special software, called a *database management system (DBMS)*, is required. While a DBMS is still somewhat slower than file technology, these performance limitations are rapidly disappearing. Considering the long-term benefits described earlier, most new information systems development is using database technology.

But the advantages of data independence, greatly reduced data redundancy, and increased flexibility come at a cost. Database technology requires a significant investment. The cost of developing databases is higher because analysts and programmers must learn how to use the DBMS. Finally, to achieve the benefits of database technology, analysts and database specialists must adhere to rigorous design principles.

Another potential problem with the database approach is the increased vulnerability inherent in the use of shared data. You are placing all your eggs in one basket. Therefore, backup and recovery and security and privacy become important issues in the world of databases.

Despite the problems discussed, database usage is growing by leaps and bounds. The technology will continue to improve, and performance limitations will all but disappear. Design methods and tools will also improve. For these reasons, this chapter will focus on database design as an important skill for tomorrow's systems analysts.

Database Design in Perspective

To fully exploit the advantages of database technology, a database must be carefully designed. In your information systems framework (Figure 11.2), database design is concerned with the DATA focus from the perspective of the system designer. The end product is called a *database schema,* a technical blueprint of the database. Database design translates the data models that were developed for the system users during the definition phase into data structures supported by the chosen database technology. Subsequent to database design, system builders will construct those data structures using the language and tools of the chosen database technology.

DATABASE CONCEPTS FOR THE SYSTEMS ANALYST

We should begin with a disclaimer. Many of the concepts and issues that are important to database design are also taught in database and data management courses. Most information systems curricula include at least one such course. It is not our intent in this chapter to replace that course. Students of information systems should actively seek out courses that *focus* on data management and database techniques; those courses will cover many more relevant technologies and techniques than we can cover in this chapter.

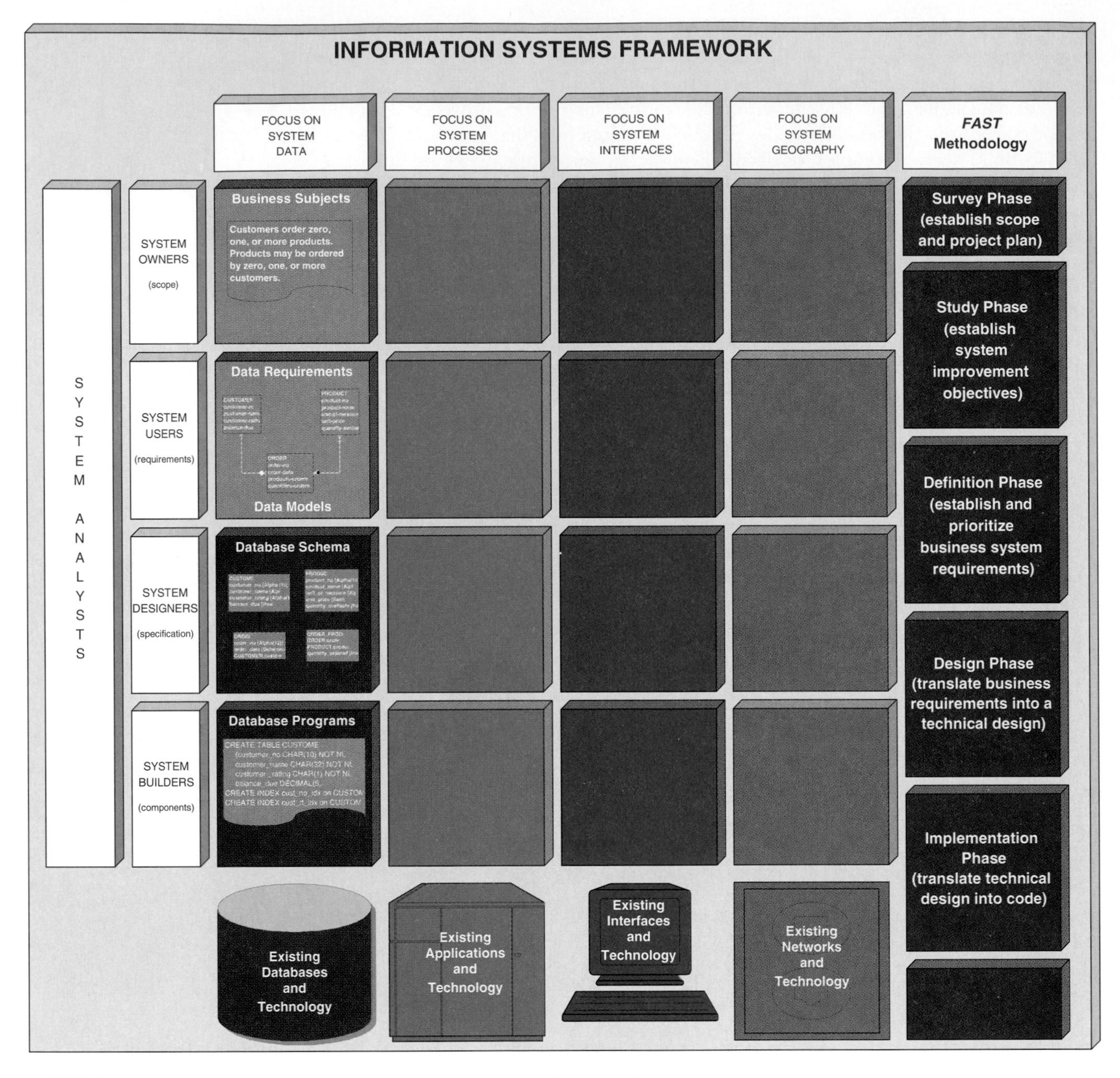

FIGURE 11.2 *Database Design in the Information Systems Framework*

That said, we will first introduce (or, for some of you, review) those database concepts and issues that are pertinent to the systems analyst's responsibilities in information system design. Although the chapter focus is on database design, experienced readers will immediately notice that many of the concepts transcend the differences between files and databases.

Fields

Fields are common to both files and databases.

> A **field** is the implementation of a data attribute (introduced in Chapter 5). Fields are the smallest unit of *meaningful* data to be stored in a file or database.

There are four types of fields that can be stored: *primary keys*, *secondary keys*, *foreign keys*, and *descriptive fields*.

Primary keys are fields whose values identify one and only one record in a file. (This concept was introduced previously in Chapter 5.)

For example, CUSTOMER NUMBER uniquely identifies a single CUSTOMER record in a database, and ORDER NUMBER uniquely identifies a single ORDER record in a database.

Secondary keys are *alternate* identifiers for a database. A secondary key's value may identify either a single record (as with a primary key) or a subset of all records (such as all ORDERS that have the ORDER STATUS of "backordered").

A single file in a database[1] may only have one primary key, but it may have several secondary keys.

Foreign keys (also introduced in Chapter 5) are pointers to the records of a different file in a database. Foreign keys are how the database links the records of one type to those of another type.

For example, an ORDER record contains the foreign key CUSTOMER NUMBER to "identify" or "point to" the CUSTOMER record that is associated with the ORDER. Notice that a foreign key in one file requires the existence of the corresponding primary key in another table—otherwise, it does not "point" to anything! Thus, the CUSTOMER NUMBER in an ORDERS file requires the existence of a CUSTOMER NUMBER in the CUSTOMERS file in order to link those files.

Descriptive fields are any other fields that store business data.

For example, given an EMPLOYEES file, some descriptive fields include EMPLOYEE NAME, DATE HIRED, PAY RATE, and YEAR-TO-DATE WAGES.

The business requirements for both keys and descriptors were defined when you performed data modeling in systems analysis (Chapter 5).

Records

Fields are organized into records. Like fields, records are common to both files and databases.

A **record** is a collection of fields arranged in a predefined format.

For example, a CUSTOMER record may be described by the following fields (notice the common notation):

CUSTOMER (NUMBER, LAST_NAME, FIRST_NAME, MIDDLE_INITIAL, POST_OFFICE_BOX_NUMBER, STREET_ADDRESS, CITY, STATE, COUNTRY, POSTAL_CODE, DATE_CREATED, DATE_OF_LAST_ORDER, CREDIT_RATING, CREDIT_LIMIT, BALANCE, BALANCE_PAST_DUE . . .)

During systems design, records will be classified as either fixed-length or variable-length records. Most database systems impose a **fixed-length record structure,** meaning that each record instance has the same fields, same number of fields, and same logical size. Some database systems will, however, compress unused fields and values to conserve disk storage space. The database designer must generally understand and specify this compression in the database design.

In your prior programming courses (especially COBOL), you may have encountered **variable-length record structures** that allow different records in the same file to have different lengths. For example, a variable-length order record might contain certain common fields that occur once for every order (e.g., ORDER NUMBER, ORDER DATE, and CUSTOMER NUMBER) and other fields that repeat some number of times based on order size (e.g., PRODUCT NUMBER and QUANTITY ORDERED, which depend on the number of items ordered). Database systems typically

[1] For those familiar with relational database technology, a file in a database is typically referred to as a *table*. That term will be introduced shortly.

disallow (or at least discourage) variable-length records. This is not a problem, as we'll show later in the chapter.

Finally, it should be noted that when a computer program "reads" a record from a database, it actually retrieves a group or *block* of records at a time. This approach minimizes the number of actual disk accesses.

A **blocking factor** is the number of *logical records* included in a single read or write operation (from the computer's perspective). A block is sometimes called a *physical record.*

Today, the blocking factor is usually determined and optimized by the chosen database technology, but a qualified database expert may be allowed to fine-tune that blocking factor for performance. Database tuning considerations are best deferred to a real database course or textbook.

Files and Tables

Similar records are organized into groups called files. In database systems, a file corresponds to a set of similar records, usually called a *table.*

A **file** is the set of all occurrences of a given record structure.

A **table** is the relational database equivalent of a file. Relational database technology will be introduced shortly.

Some types of files and table include:

- **Master files** or tables contain records that are relatively permanent. Thus, once a record has been added to a master file, it remains in the system indefinitely. The values of fields for the record will change over its lifetime, but the individual records are retained indefinitely. Examples of master files and tables include CUSTOMERS, PRODUCTS, and SUPPLIERS.
- **Transaction files** or tables contain records that describe business events. The data describing these events normally have a limited useful lifetime. For instance, an INVOICE record is ordinarily useful until the invoice has been paid or written off as uncollectible. In information systems, transaction records are frequently retained *on-line* for some period of time. Subsequent to their useful lifetime, they are *archived* off-line. Examples of transaction files include ORDERS, INVOICES, REQUISITIONS, and REGISTRATIONS.
- **Document files** and tables contain stored copies of historical data for easy retrieval and review without the overhead of regenerating the document.
- **Archival files** and tables contain master and transaction file records that have been deleted from on-line storage. Thus, records are rarely deleted; they are merely moved from on-line storage to off-line storage. Archival requirements are dictated by government regulation and the need for subsequent audit or analysis.
- **Table look-up files** contain relatively static data that can be shared by applications to maintain consistency and improve performance. Examples include sales tax tables, ZIP code tables, and income tax tables.
- **Audit files** are special records of updates to other files, especially master and transaction files. They are used in conjunction with archive files to recover "lost" data. Audit trails are typically built into better database technologies.

In the not too distant past, file design methods required the analyst to specify precisely how the records in a database should be sequenced (called **file organization**) and accessed (called **file access**). In today's database environment, the database technology itself usually predetermines and/or limits the file organization for all tables contained in the database. Once again, a trained database technology expert may be given some control over organization and storage location for performance tuning.

Databases

As described earlier, stand-alone, application-specific files were once the lifeblood of most information systems; however, they are being slowly but surely replaced with databases. Recall that a database may loosely be thought of as a set of interrelated files. By interrelated, we mean that records in one file may be associated with the records in a different file.

For example, a STUDENT record may be linked to all of that student's COURSE records. In turn, a COURSE record may be linked to the STUDENT records that indicate completion of that course. This two-way linking and flexibility allow us to eliminate *most* of the need to redundantly store the same fields in the different record types. Thus, in a very real sense, multiple files are consolidated into a single file—the database.

The idea of relationships between different collections of data was introduced in Chapter 5. In that chapter, you learned to discover an application's data requirements and model those requirements as *entities* and *relationships*. The database now provides for the technical implementation of those entities and relationships.

So many applications are now being built around database technology that database design has become an important skill for the analyst. Indeed, database technology, once considered important only to the largest corporations with the largest computers, is now common for applications developed on microcomputers and departmental networks.

The history of information systems has led to one inescapable conclusion:

Data is a resource that must be controlled and managed!

Few, if any, information systems staffs have avoided the frustration of uncontrolled growth and duplication of data stored in their systems. As systems were developed, implemented, and maintained, the common data needed by the different systems were duplicated in multiple conventional files. This duplication carried with it a number of costs: extra storage space required, duplicated input to maintain redundantly stored data and files, and data integrity problems (e.g., the ADDRESS for a specific customer not matching in the various files that contain that customer's ADDRESS).

Out of necessity, database technology was created so an organization could maintain and use its data as an integrated whole instead of as separate data files. We can now develop a shared data resource that can be used by several information systems.

Data Architecture Data becomes a business resource in a database environment. Information systems are built around this resource to give both computer programmers and end-users flexible access to data.

> A business's **data architecture** is comprised of the files and databases that store all of the organization's data, the file and database technology used to store the data, and the organization structure set up to manage the data resource.

Figure 11.3 illustrates the data architecture into which many companies have evolved. As shown in the figure, most companies still have numerous conventional file-based information system applications, most of which were developed before the emergence of high-performance database technology. In many cases, the processing efficiency of these files or the projected cost to redesign these files has slowed conversion of the systems to database.

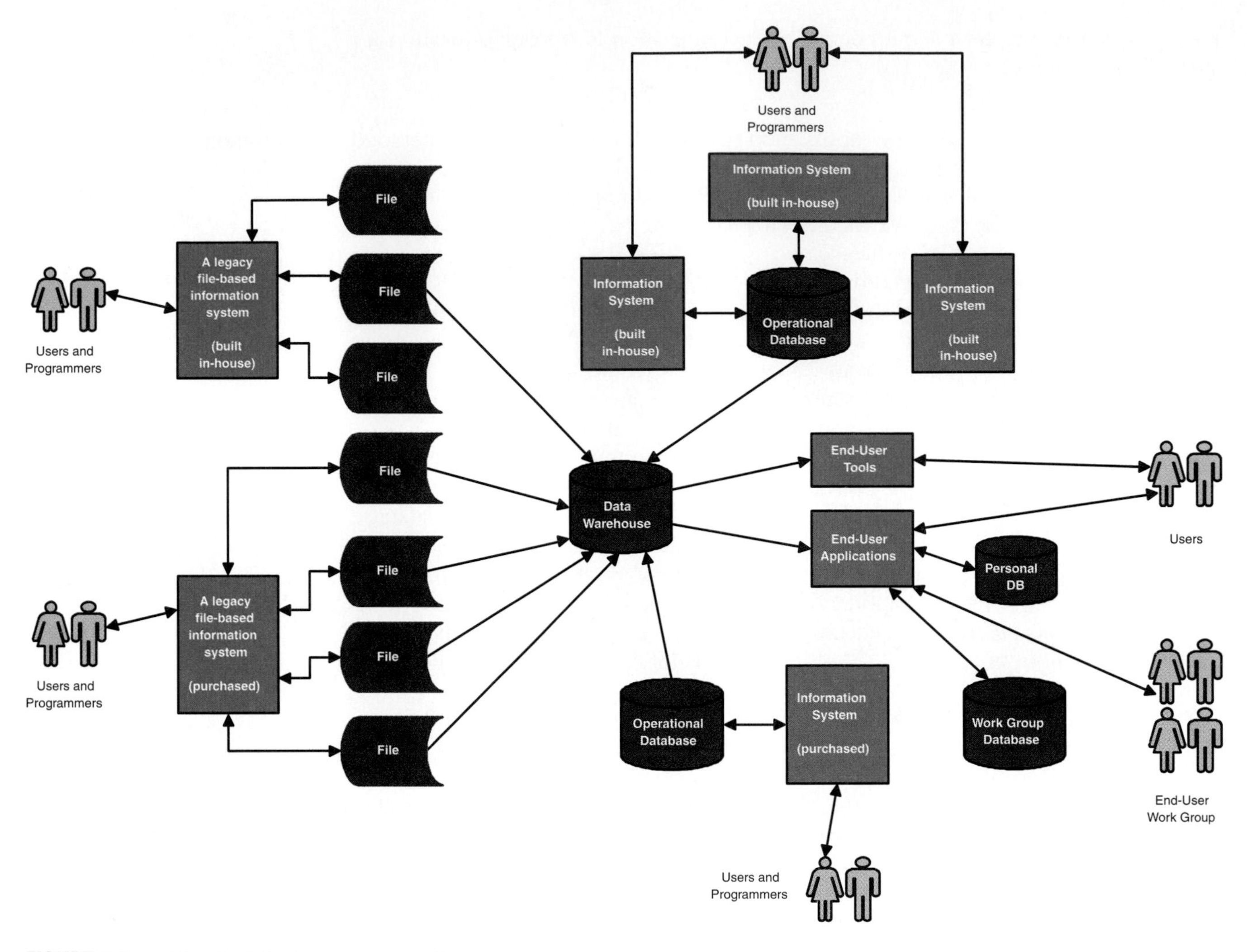

FIGURE 11.3 *A Typical, Modern Data Architecture*

Also shown in Figure 11.3 are the **operational databases** that have been developed to support day-to-day operations and business transaction processing for major information systems. These systems were (and are) developed over time to replace the conventional files that formerly supported the applications. Access to these databases is limited to computer programs that use the DBMS to process transactions, maintain the data, and generate regularly scheduled management reports. Some query access may also be provided.

Many information systems shops hesitate to give end-users access to operational databases for queries and reports. The volume of unscheduled reports and queries could overload the computers and hamper business operations that the databases were intended to support. Instead, data warehouses are developed, possibly on separate computers.

Data warehouses store data extracted from the production databases and conventional files. Fourth-generation programming languages, query tools, and decision support tools are then used to generate reports and analyses off these data warehouses.

Figure 11.3 also shows the emergence of **personal** and **work group** (or departmental) **databases.** Personal computer and local network database technology has rapidly matured to allow end-users to develop personal and departmental databases. These databases may contain unique data, or they may import data from conventional files, operational databases, and/or data warehouses. Personal databases are built using PC database technology such as *Access, dBASE, Paradox,* and *FoxPro.*

Admittedly, this overall scenario is advanced, but many firms are currently using variations of it. To manage the enterprisewide data resource, a staff of database specialists may be organized around the following administrators:

A **data administrator** is responsible for the data planning, definition, architecture, and management.

One or more **database administrators** are responsible for the database technology, database design and construction, security, backup and recovery, and performance tuning.

In smaller businesses, these roles may be combined, or perhaps assigned to a systems analyst.

Database Architecture So far, we have made several references to the *database technology* that makes the above data architecture possible.

Database architecture refers to the database technology including the database engine, database management utilities, database CASE tools for analysis and design, and database application development tools.

The control center of a database architecture is its database management system.

A **database management system (DBMS)** is specialized computer software available from computer vendors that is used to create, access, control, and manage the database. The core of the DBMS is often called its **database engine.** The engine responds to specific commands to create database structures and then to create, read, update, and delete records in the database.

The database management system is purchased from a database technology vendor such as Oracle, IBM, Microsoft, or Sybase.

Figure 11.4 depicts a typical database architecture. A systems analyst, or database analyst, designs the structure of the data in terms of record types, fields contained in those record types, and relationships that exist between record types. These structures are defined to the database management system using its data definition language.

Data definition language (DDL) is used by the DBMS to physically establish those record types, fields, and structural relationships. Additionally, the DDL defines views of the database. Views restrict the portion of a database that may be used or accessed by different users and programs. DDLs record the definitions in a permanent data repository.

Some data dictionaries include formal, elaborate software that helps database specialists track **metadata**—the data about the data—such as record and field definitions, synonyms, data relationships, validation rules, help messages, and so forth. The metadata is stored in a data dictionary or repository (which may or may not be provided by the DBMS vendor).

To help design databases, CASE tools may be provided either by the database technology vendor (e.g., Oracle) or from a third-party CASE tool vendor (e.g., Popkin, Logic Works, etc.).

The database management system also provides a data manipulation language to access and use the database in applications.

A **data manipulation language (DML)** is used to create, read, update, and delete records in the database and to navigate between different records and types of records—for example, from a CUSTOMER record to the ORDER records for that customer. The DBMS and DML hide the details concerning how records are organized and allocated to the disk.

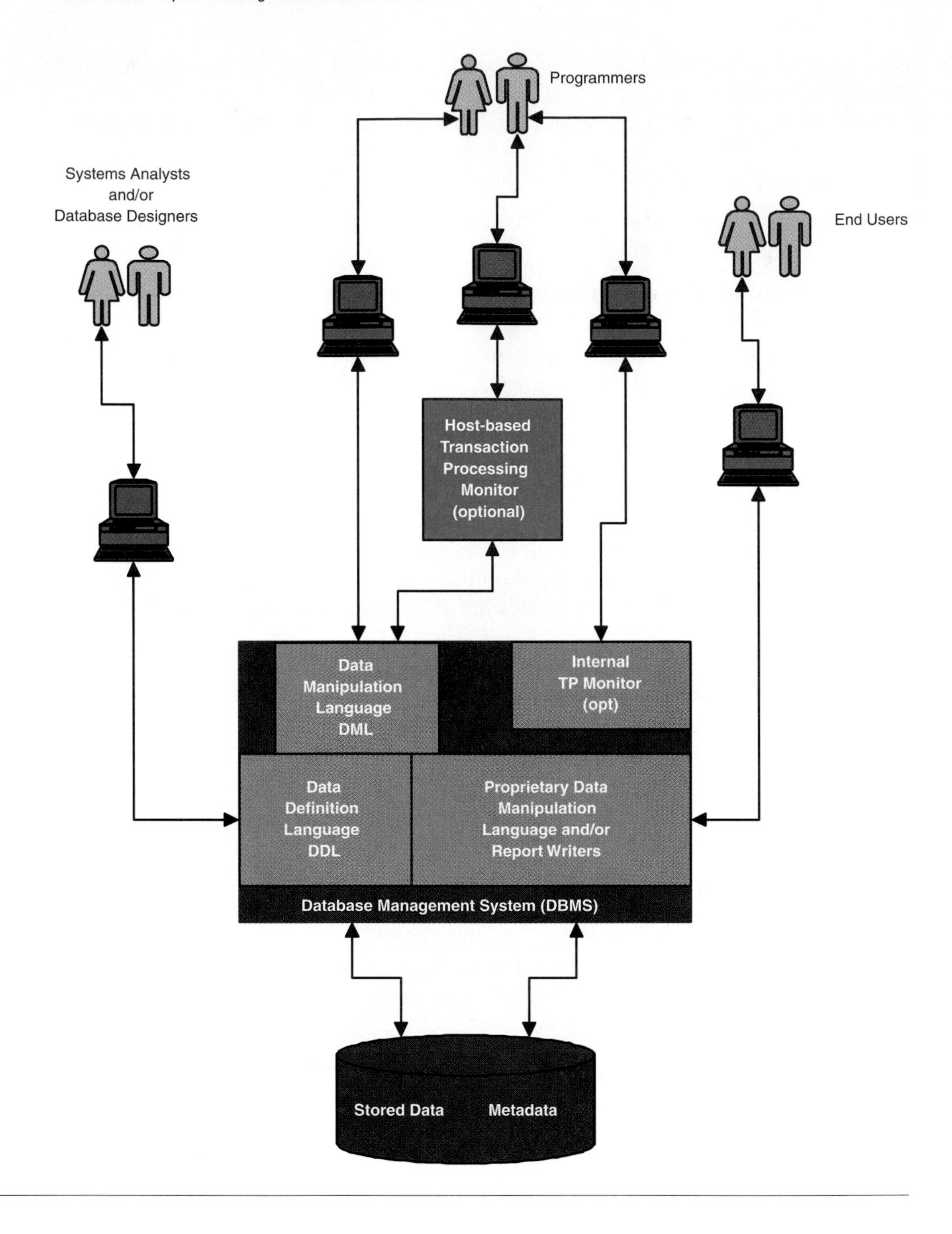

FIGURE 11.4 *A Typical Database Architecture*

In general, the DML is very flexible in that it may be used by itself to create, read, update, and delete records; or its commands may be "called" from a separate host programming language such as COBOL, *Visual Basic,* or *Powerbuilder.*

Many DBMSs don't require the use of a DDL to construct the database or a DML to access the database. Instead (or in addition), they provide their own tools and commands to perform those tasks. This is especially true of PC-based DBMSs such as Microsoft *Access. Access* provides a simple graphical user interface to create the tables and a form-based environment to access, browse, and maintain the tables.

Many DBMSs also include proprietary report writing and inquiry tools to allow users to access and format data without directly using the DML.

Some DBMSs include a **transaction processing monitor** (or *TP monitor*) that manages on-line accesses to the database and ensures that transactions that impact multiple tables are fully processed as a single unit. Alternatively, most high-end

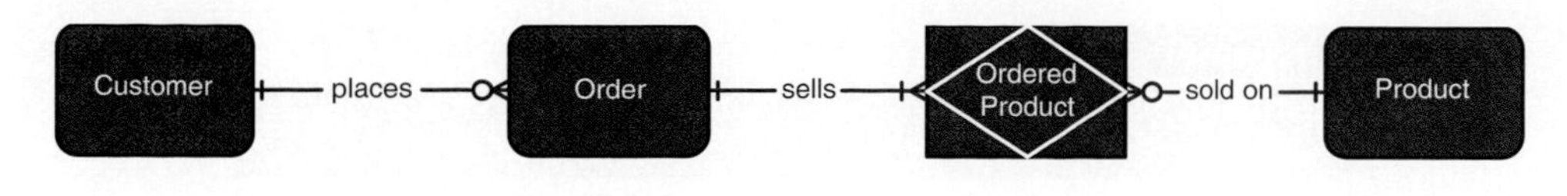

FIGURE 11.5 *A Simple, Logical Data Model*

DBMSs are designed to interact with popular third-party transaction processing monitors such as *CICS* and *Tuxedo.*

All of the above technology is illustrated in Figure 11.4. Today, almost all new database development is using relational database management systems.

Relational Database Management Systems There are several types of database management systems. They can be classified according to the way they structure records. Early database management systems organized records in hierarchies or networks implemented with indexes and linked lists. You may study these further in a database course. But today, most successful database management systems are based on relational technology.

> **Relational databases** implement data in a series of tables that are "related" to one another via foreign keys.

Figure 11.5 illustrates a logical data model. Figure 11.6 is the physical, relational database implementation of that data model (called a schema). In a relational database, files are seen as simple two-dimensional tables, also known as relations. The rows are records. The columns correspond to fields.

Both the DDL and DML of most relational databases is called **SQL** (which stands for Structured Query Language). Despite the name, SQL supports not only queries but also complete database creation and maintenance. To access tables and records, SQL provides the following basic commands:

- SELECT specific records from a table based on specific criteria (e.g., SELECT CUSTOMER WHERE BALANCE > 500.00).
- PROJECT out specific fields from a table (e.g., PROJECT CUSTOMER TO INCLUDE ONLY CUSTOMER_NUMBER, CUSTOMER_NAME, BALANCE).
- JOIN two or more tables across a common field—a primary and foreign key (JOIN CUSTOMER AND ORDER USING CUSTOMER_NUMBER).

When used in combination, these basic commands can address most database access requirements. A fundamental characteristic of relational SQL is that commands return a set of records, not necessarily just a single record (as in nonrelational database and file technology). Most SQL databases also provide commands for creating, updating, and deleting records, as well as sorting records.

High-end relational databases also extend the SQL language to support triggers and stored procedures.

> **Triggers** are programs embedded within a table that are automatically invoked by updates to another table. For example, if a record is deleted from a PASSENGER AIRCRAFT table, a trigger can force the automatic deletion of all corresponding records in a SEATS table for that aircraft.

> **Stored procedures** are programs embedded within a table that can be called from an application program. For example, a complex data validation algorithm might be embedded in a table to ensure that new and updated records contain valid data *before* they are stored.

Both triggers and stored procedures are reusable because they are stored with the tables themselves. This eliminates the need for application programmers to create the equivalent logic within each application that uses the tables.

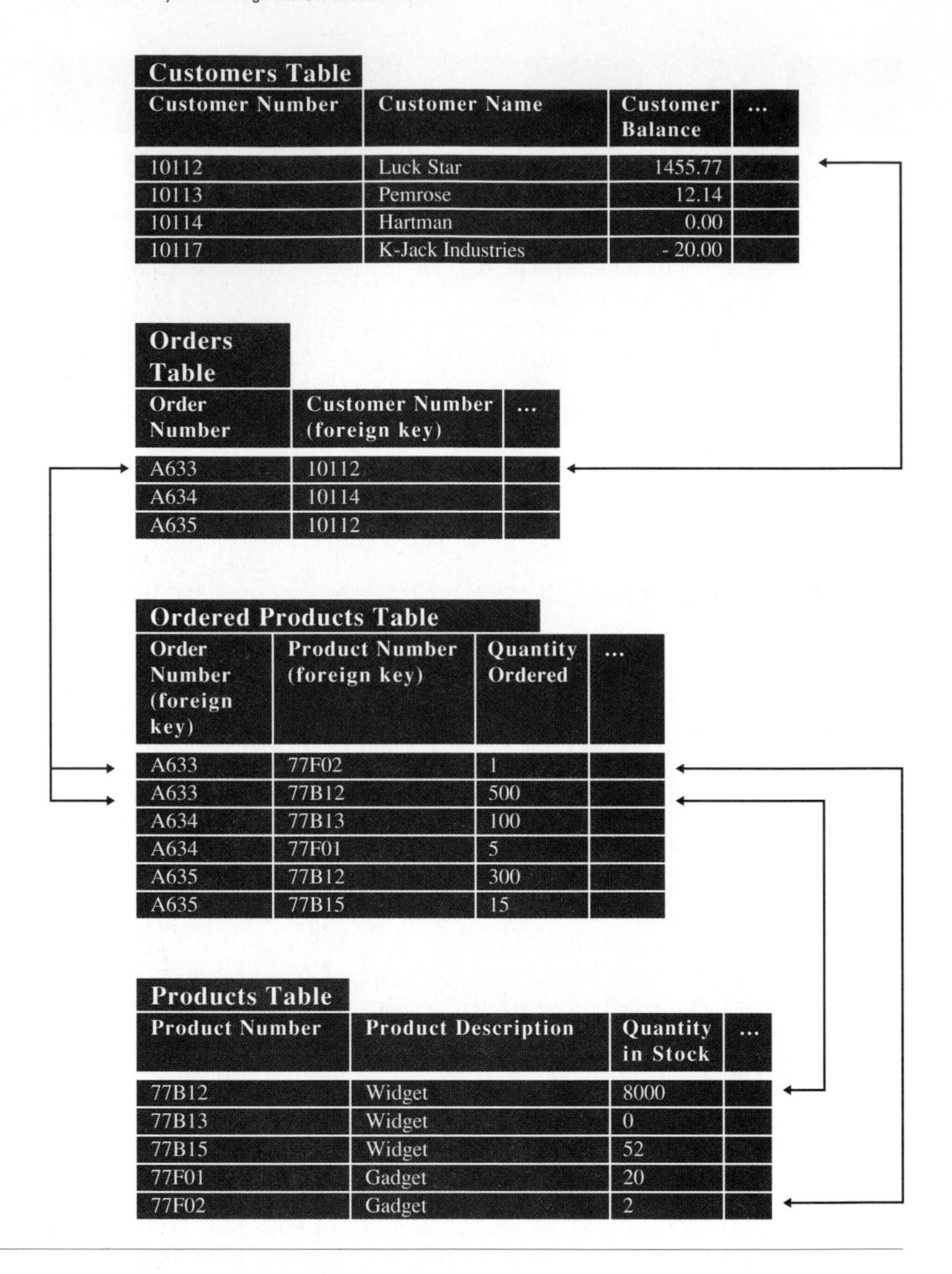

Customers Table

Customer Number	Customer Name	Customer Balance	...
10112	Luck Star	1455.77	
10113	Pemrose	12.14	
10114	Hartman	0.00	
10117	K-Jack Industries	- 20.00	

Orders Table

Order Number	Customer Number (foreign key)	...
A633	10112	
A634	10114	
A635	10112	

Ordered Products Table

Order Number (foreign key)	Product Number (foreign key)	Quantity Ordered	...
A633	77F02	1	
A633	77B12	500	
A634	77B13	100	
A634	77F01	5	
A635	77B12	300	
A635	77B15	15	

Products Table

Product Number	Product Description	Quantity in Stock	...
77B12	Widget	8000	
77B13	Widget	0	
77B15	Widget	52	
77F01	Gadget	20	
77F02	Gadget	2	

FIGURE 11.6 *A Simple, Physical Database Schema*

The SQL language for using and manipulating a relational database is covered extensively in most database courses and textbooks. All high-end relational database management systems (e.g., *Oracle, DB2,* and *SQL Server*) and many personal computer relational database management systems (such as Microsoft *Access*) support the SQL language standards.

Examples of high-performance relational DBMSs include Oracle Corporation's *Oracle*, IBM's *Database Manager*, Microsoft's *SQL Server* (being used in the SoundStage project), and Sybase Corporation's *Sybase*. Many of these databases run on mainframes, minicomputers, and network database servers. Additionally, most personal computer DBMSs are relational (or partially so). Examples include Microsoft's

FIGURE 11.7 *User/Designer Interface for a Relational PC DBMS (Microsoft* Access*)*

Access and *Foxpro* and Borland's *Paradox* and *dBASE*. These systems can run on both stand-alone personal computers and local area network file servers. Figure 11.7 illustrates a relational database management system's user interface.

Database textbooks and courses offer entire chapters and units on relational databases. This is only an introduction to database. We encourage you to learn more!

DATA ANALYSIS FOR DATABASE DESIGN

In Chapter 5 you learned how to model data requirements for an information system. That model took the form of a fully attributed entity relationship diagram and a repository of metadata. These data modeling specifications provided a vehicle for communication with end-users during systems analysis. However, in this chapter our emphasis shifts away from using the data model as a vehicle for communicating business requirements (to system users). Instead, we move toward preparing that data model to communicate database design requirements (to system designers and builders).

While a data model effectively communicates database requirements, it does not necessarily represent a *good* database design. It may contain structural characteristics that reduce flexibility and expansion or create unnecessary redundancy. Therefore, we must prepare the data model for database design and implementation.

This section will discuss the characteristics of a *quality* data model—one that will allow us to develop an ideal database structure. We'll also present the process used to analyze data model quality and make necessary modifications before database design.

What Is a Good Data Model?

What makes a data model "good"? We suggest the following criteria:

- *A good data model is simple.* As a general rule, the data attributes that describe an entity should describe only that entity. Consider, for example, the following entity definition:

COURSE REGISTRATION = COURSE REGISTRATION NUMBER (primary key) +
COURSE REGISTRATION DATE +
STUDENT ID NUMBER (a foreign key) +
STUDENT NAME +
STUDENT MAJOR +
1 { COURSE NUMBER } N

Do STUDENT NAME and STUDENT MAJOR really describe an instance of course registration? Or do they describe a different entity, say STUDENT? The same argument could be applied to STUDENT ID NUMBER, but on further inspection, that attribute is needed to "point" to the corresponding instance of the STUDENT entity. Another aspect of simplicity is stated as follows: Each attribute of an entity instance can have only one value. Looking again at the previous example, COURSE NUMBER can have as many values for one COURSE REGISTRATION as the student elects.

- *A good data model is essentially nonredundant.* This means that each data attribute, other than foreign keys, describes at most one entity. In the prior example, it is not difficult to imagine that STUDENT NAME and STUDENT MAJOR might also describe a STUDENT entity. We should choose. Based on the previous bullet, the logical choice would be the STUDENT entity. There may also exist subtle redundancies in a data model. For example, the same attribute might be recorded more than once under different names (synonyms).
- *A good data model should be flexible and adaptable to future needs.* In the absence of this criteria, we would tend to design databases to fulfill only *today's* business requirements. Then, when a new requirement becomes known, we can't easily change the databases without rewriting many or all of the programs that used those databases. While we can't change the reality that most projects are application-driven, we can make our data models as application-independent as possible to encourage database structures that can be extended or modified without impact to current programs.

So how do we achieve the above goals? How can you design a database that can adapt to future requirements that you cannot predict? The answer lies in data analysis.

Data Analysis

The technique used to improve a data model in preparation for database design is called data analysis.

Data analysis is a process that prepares a data model for implementation as a simple, nonredundant, flexible, and adaptable database. The specific technique is called normalization.

Normalization is a technique that organizes data attributes such that they are grouped to form stable, flexible, and adaptive entities.

Normalization is a three-step technique that places the data model into first normal form, second normal form, and third normal form. Don't get hung up on the terminology—it's easier than it sounds. For now, let's establish an initial understanding of these three formats.

- Simply stated, an entity is in **first normal form (1NF)** if there are no attributes that can have more than one value for a single instance of the entity (frequently called repeating groups). Any attributes that can have multiple values actually describe a separate entity, possibly an entity (and relationship) that we haven't yet included in our data model.
- An entity is in **second normal form (2NF)** if it is already in 1NF, and if the

values of all nonprimary key attributes are dependent on the full primary key—not just part of it. Any nonkey attributes that are dependent on only part of the primary key should be moved to any entity where that partial key becomes the full key. Again, this may require creating a new entity and relationship on the model.

- An entity is in **third normal form (3NF)** if it is already in 2NF, and if the values of its nonprimary key attributes are not dependent on any other non-primary key attributes. Any nonkey attributes that are dependent on other nonkey attributes must be moved or deleted. Again, new entities and relationships may have to be added to the data model.

Normalization Example

There are numerous approaches to normalization. We have chosen to present a nontheoretical approach. We'll leave the theory and detailed implications to the database courses and textbooks.

As usual, we'll use the SoundStage case study to demonstrate the steps. Unfortunately, Sandra and Robert, like most experienced analysts, normalize their entities as they assign attributes to those entities. To them, normalization is a quality check, and they usually identify few errors. To demonstrate the technique, we'll begin with a less than perfect version of the SoundStage data model, and we'll normalize it. The initial model is illustrated in Figure 11.8.

First Normal Form The first step in data analysis is to place each entity into 1NF. In Figure 11.8, which entities are not in 1NF?

HINT: We have used the algebraic notation that was introduced in Chapter 5 to show groups of attributes that can have multiple values for a single entity instance.

You should find three—MEMBER, MEMBER ORDER, and CLUB. Each contains a repeating group, that is, a group of attributes that have multiple values for a single instance of the entity (denoted by the brackets). Consider, for example, the entity MEMBER. A single MEMBER can belong to multiple CLUBs and, therefore, have multiple values for CLUB NAME and AGREEMENT NUMBER—one for each club to which he or she belongs. For a single instance of MEMBER, the number of clubs and agreements may vary.

Similarly, a MEMBER ORDER can contain data about more than one ORDERED PRODUCT. And a CLUB can sponsor more than one AGREEMENT. How do we fix these anomalies in our model?

Figures 11.9 through 11.11 demonstrate how to place these three entities into 1NF. The original entity is depicted on the left side of the page. The 1NF entities are on the right side of the page. Each figure shows how normalization changed the data model and attribute assignments.

Let's examine the MEMBER ORDER entity (Figure 11.9). First, we remove the attributes that can have more than one value for an instance of the entity. That alone places MEMBER ORDER in 1NF. But what do we do with the removed attributes? These attributes repeat many times "as a group." Therefore, we moved the entire group of attributes to a new entity, MEMBER ORDERED PRODUCT. Each instance of these attributes describes one PRODUCT on a single MEMBER ORDER. Thus, if a specific order contains five PRODUCTS, there will be five instances of the new MEMBER ORDERED PRODUCT entity. Each entity instance has only one value for each attribute; therefore, the new entity is also in first normal form.

Notice how the primary key of the new entity was created—by combining the primary key of the original entity, ORDER NUMBER, with the implicit key attribute of the group, PRODUCT NUMBER. Thus, we have what was described in Chapter 5 as a concatenated key. Since we know from Chapter 5 that each part of a concatenated key is a foreign key back to another entity, we added relationships (and

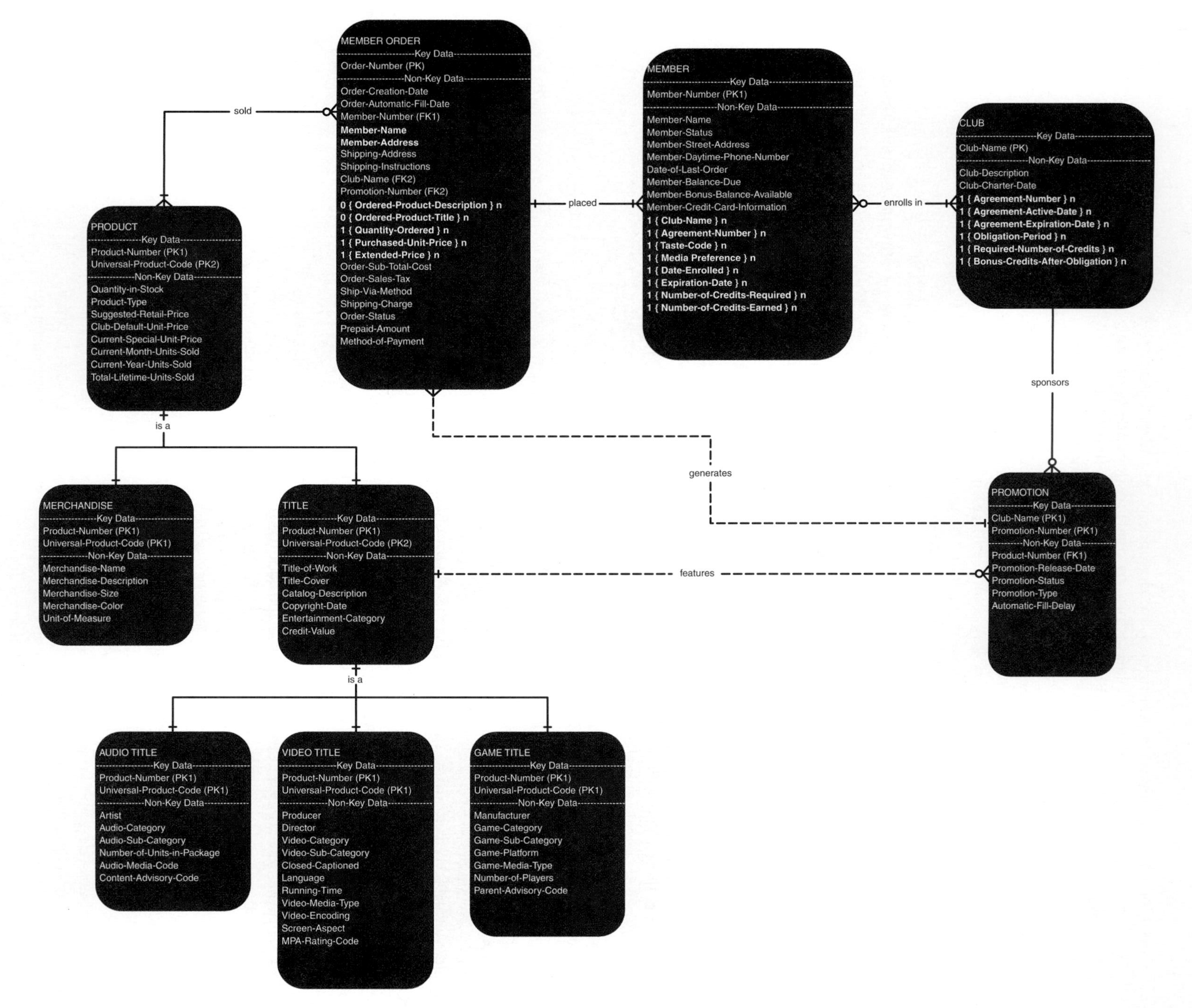

FIGURE 11.8 *An Unnormalized SoundStage Data Model*

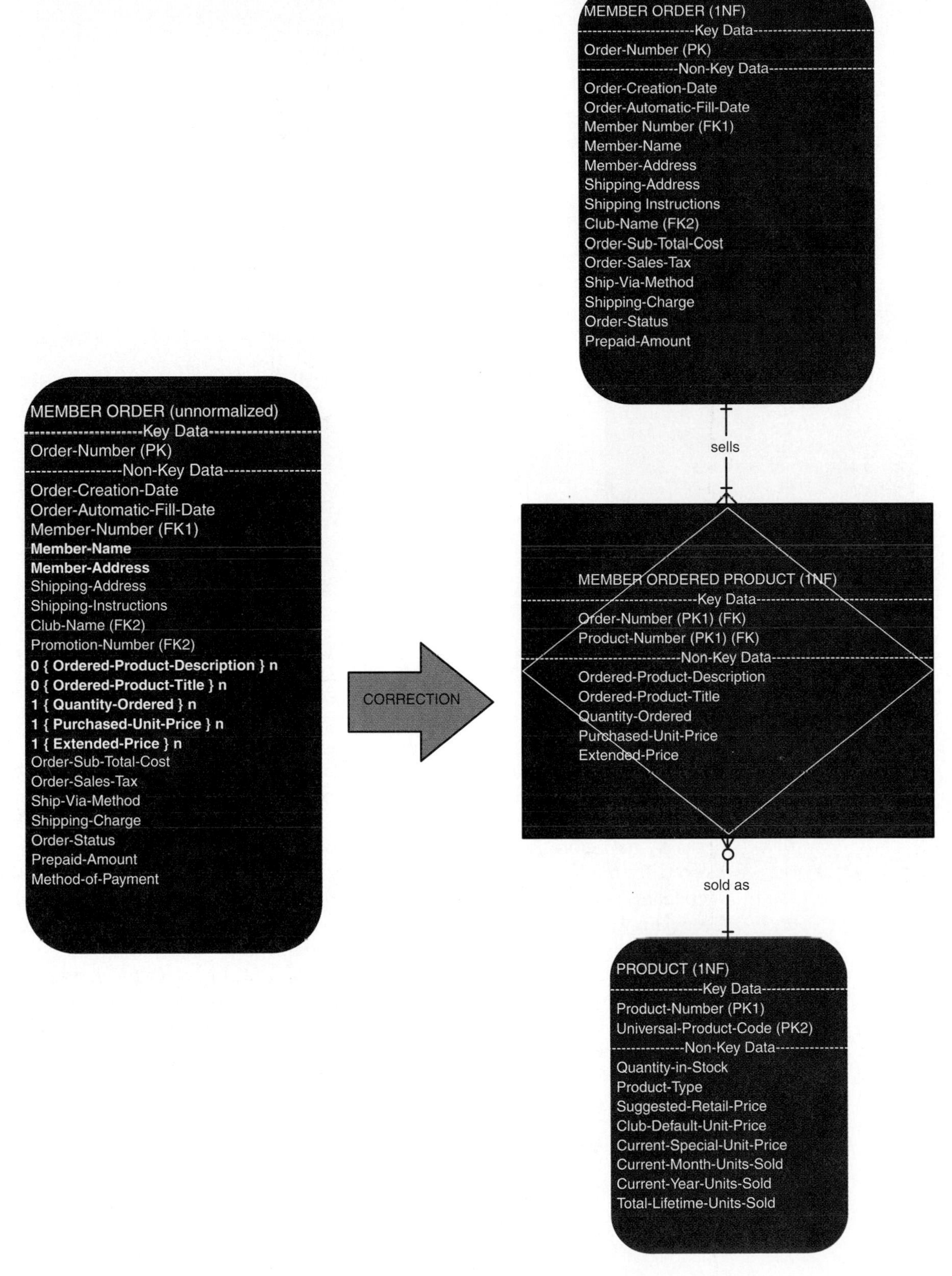

FIGURE 11.9 *First Normal Form*

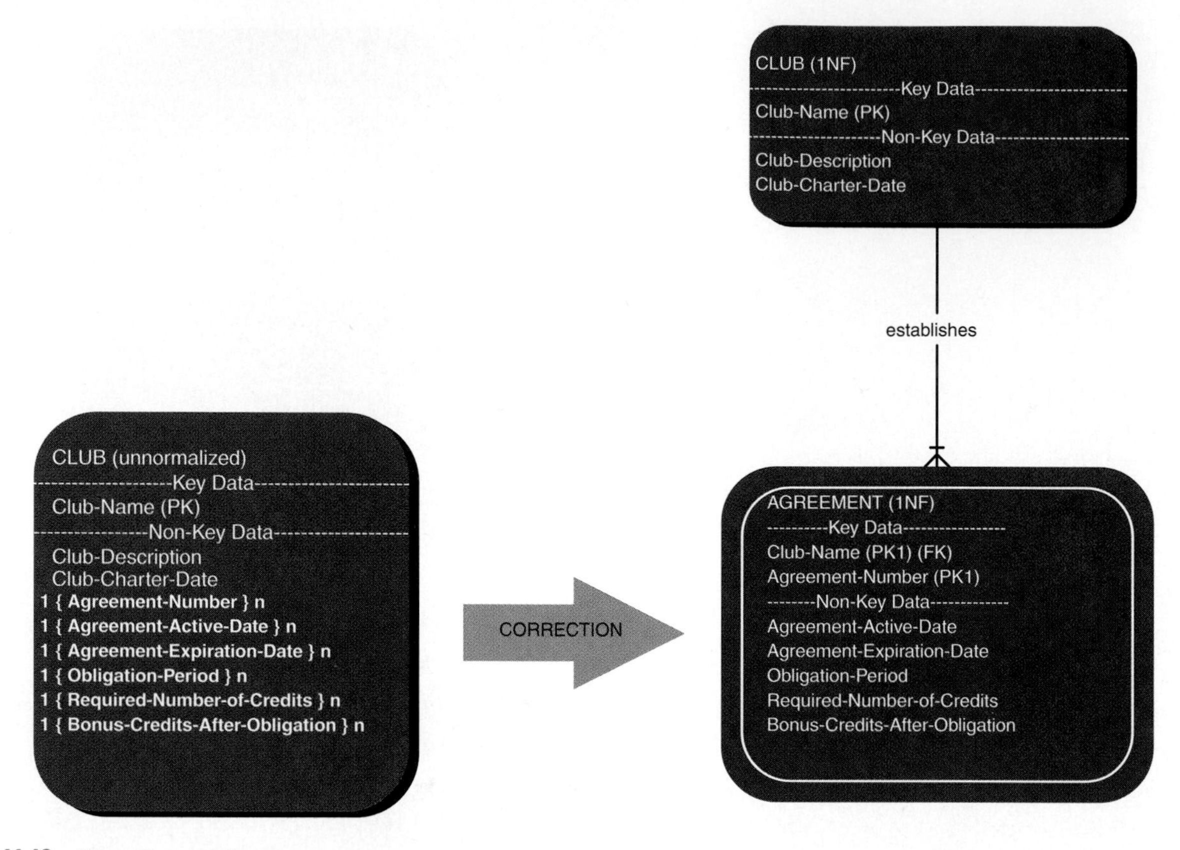

FIGURE 11.10 *First Normal Form*

cardinality) from the new MEMBER ORDERED PRODUCT entity to both the MEMBER and PRODUCT entities.[2]

Another example of 1NF is shown in Figure 11.10 for the CLUB entity. The attributes that can have many values (commonly called repeating attributes) are easy to spot. They include attributes such as AGREEMENT ACTIVE DATE and OBLIGATION PERIOD. As before, we created a new entity, AGREEMENT (as named by the users), keyed by the concatenation of CLUB NAME and AGREEMENT NUMBER. We moved the repeating attributes to that new entity. Once again, we also created a relationship between AGREEMENT and CLUB.

To place the MEMBER entity in 1NF, we removed the repeating attributes. Those attributes seemed dependent on a combination of CLUB NAME and AGREEMENT NUMBER, so we created a new entity called CLUB MEMBERSHIP with that key (see Figure 11.11). The repeating attributes were then moved to that entity.

It was then that we noticed that the club membership entity was, in fact, a ternary associative entity (review Chapter 5). Each part of the concatenated key (MEMBER NUMBER, CLUB NAME, and AGREEMENT NUMBER) was a foreign key back to different entities. Thus, we completed our model by adding relationships (with cardinality) from that associative entity back to the MEMBER, CLUB, and AGREEMENT entities.

All other entities are already in 1NF because they do not contain any repeating groups.

---

[2] The astute reader will recognize that we could have accomplished this same normalization result by merely "resolving" the many-to-many relationship between MEMBER ORDER and PRODUCT entities as was described and demonstrated in Chapter 5. Many 1NF problems can be prevented by using that technique.

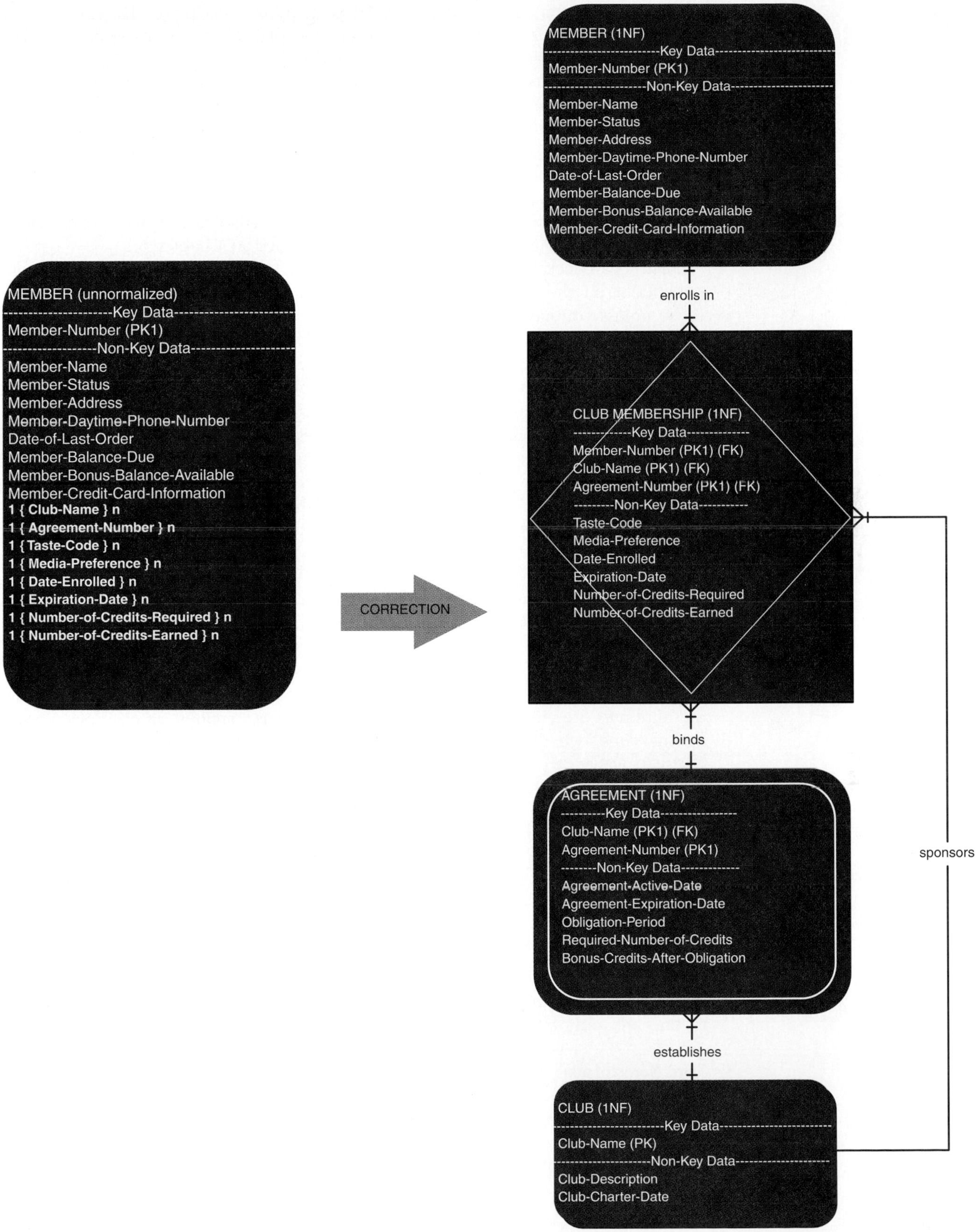

MEMBER (1NF)
----Key Data----
Member-Number (PK1)
----Non-Key Data----
Member-Name
Member-Status
Member-Address
Member-Daytime-Phone-Number
Date-of-Last-Order
Member-Balance-Due
Member-Bonus-Balance-Available
Member-Credit-Card-Information

enrolls in

MEMBER (unnormalized)
----Key Data----
Member-Number (PK1)
----Non-Key Data----
Member-Name
Member-Status
Member-Address
Member-Daytime-Phone-Number
Date-of-Last-Order
Member-Balance-Due
Member-Bonus-Balance-Available
Member-Credit-Card-Information
1 { Club-Name } n
1 { Agreement-Number } n
1 { Taste-Code } n
1 { Media-Preference } n
1 { Date-Enrolled } n
1 { Expiration-Date } n
1 { Number-of-Credits-Required } n
1 { Number-of-Credits-Earned } n

CORRECTION

CLUB MEMBERSHIP (1NF)
----Key Data----
Member-Number (PK1) (FK)
Club-Name (PK1) (FK)
Agreement-Number (PK1) (FK)
----Non-Key Data----
Taste-Code
Media-Preference
Date-Enrolled
Expiration-Date
Number-of-Credits-Required
Number-of-Credits-Earned

binds

AGREEMENT (1NF)
----Key Data----
Club-Name (PK1) (FK)
Agreement-Number (PK1)
----Non-Key Data----
Agreement-Active-Date
Agreement-Expiration-Date
Obligation-Period
Required-Number-of-Credits
Bonus-Credits-After-Obligation

sponsors

establishes

CLUB (1NF)
----Key Data----
Club-Name (PK)
----Non-Key Data----
Club-Description
Club-Charter-Date

FIGURE 11.11 *First Normal Form*

Second Normal Form The next step of data analysis is to place the entities into 2NF. Recall that it is assumed you have already placed all entities into 1NF. Also recall that 2NF looks for an anomaly called a *partial dependency*, meaning an attribute whose value is determined by only part of the primary key.

Entities that have a single attribute primary key are already in 2NF. That takes care of PRODUCT (and its subtypes), MEMBER ORDER, MEMBER, and CLUB. Thus, we need to check only those entities that have a concatenated key—MEMBER ORDERED PRODUCT, PROMOTION, CLUB MEMBERSHIP, and AGREEMENT. (Notice that the list includes the new entities that we created as a result of placing the model into 1NF, which explains why the model had to already be in 1NF. Also notice that we primarily check only associative and weak entities, as defined in Chapter 5.)

First, let's check the MEMBER ORDERED PRODUCT entity. Most of the attributes are dependent on the full primary key. For example, QUANTITY ORDERED makes no sense unless you have *both* an ORDER NUMBER and a PRODUCT NUMBER. Think about it! By itself, ORDER NUMBER is inadequate since the order could have as many quantities ordered as there are products on the order. Similarly, by itself, PRODUCT NUMBER is inadequate since the same product could appear on many orders. Thus, QUANTITY ORDERED requires both parts of the key and is fully dependent on the key. The same could be said of QUANTITY SHIPPED and PURCHASE UNIT PRICE.

But what about ORDERED PRODUCT DESCRIPTION and ORDERED PRODUCT TITLE? Do we really need ORDER NUMBER to determine a value for either? No! Instead, the values of these attributes are dependent only on the value of PRODUCT NUMBER. Thus, the attributes are *not* dependent on the full key; we have uncovered a *partial dependency* error that must be fixed. How do we fix this type of normalization error?

To fix the problem, we simply move the nonkey attributes, ORDERED PRODUCT DESCRIPTION and ORDERED PRODUCT TITLE, to an entity that only has PRODUCT NUMBER as its key. If necessary, we would have to create this entity, but the PRODUCT entity with that key already exists. But we have to be careful because PRODUCT is a supertype. Upon inspection of the subtypes (see Figure 11.12), we discover that the attributes are already in the MERCHANDISE and TITLE entities, albeit under a synonym. Thus, we didn't actually have to move the attributes from the MEMBER ORDERED PRODUCT entity; we just deleted them as redundant data.

Next, let's examine the PROMOTION entity. The concatenated key is the combination of CLUB NAME and PROMOTION NUMBER. At first, you may be tempted to label the foreign key PRODUCT NUMBER as a partial dependency. It is not! PRODUCT NUMBER is dependent on *both* CLUB NAME and PROMOTION NUMBER to fulfill its role of indicating which product is being featured on the promotion. The other attributes in PROMOTION are more clearly dependent on the full concatenated key. Thus, PROMOTION is already in 2NF.

Convince yourself that CLUB MEMBERSHIP and AGREEMENT are, similarly, already in 2NF. In the final analysis, 2NF errors will be rare if you follow a simple rule when initially assigning attributes to all entities—all nonkey attributes should be assigned to one and only one entity; the entity that the attribute *best* describes.

Third Normal Form We can further simplify our entities by placing them into 3NF. Entities are assumed to be in 2NF before beginning 3NF analysis. Third normal form analysis looks for two types of problems, *derived data* and *transitive dependencies.* In both cases, the fundamental error is that nonkey attributes are dependent on other nonkey attributes.

The first type of 3NF analysis is easy—examine each entity for derived attributes.

> **Derived attributes** are those whose values can either be calculated from other attributes or derived through logic from the values of other attributes.

If you think about it, storing a derived attribute makes little sense. First, it wastes

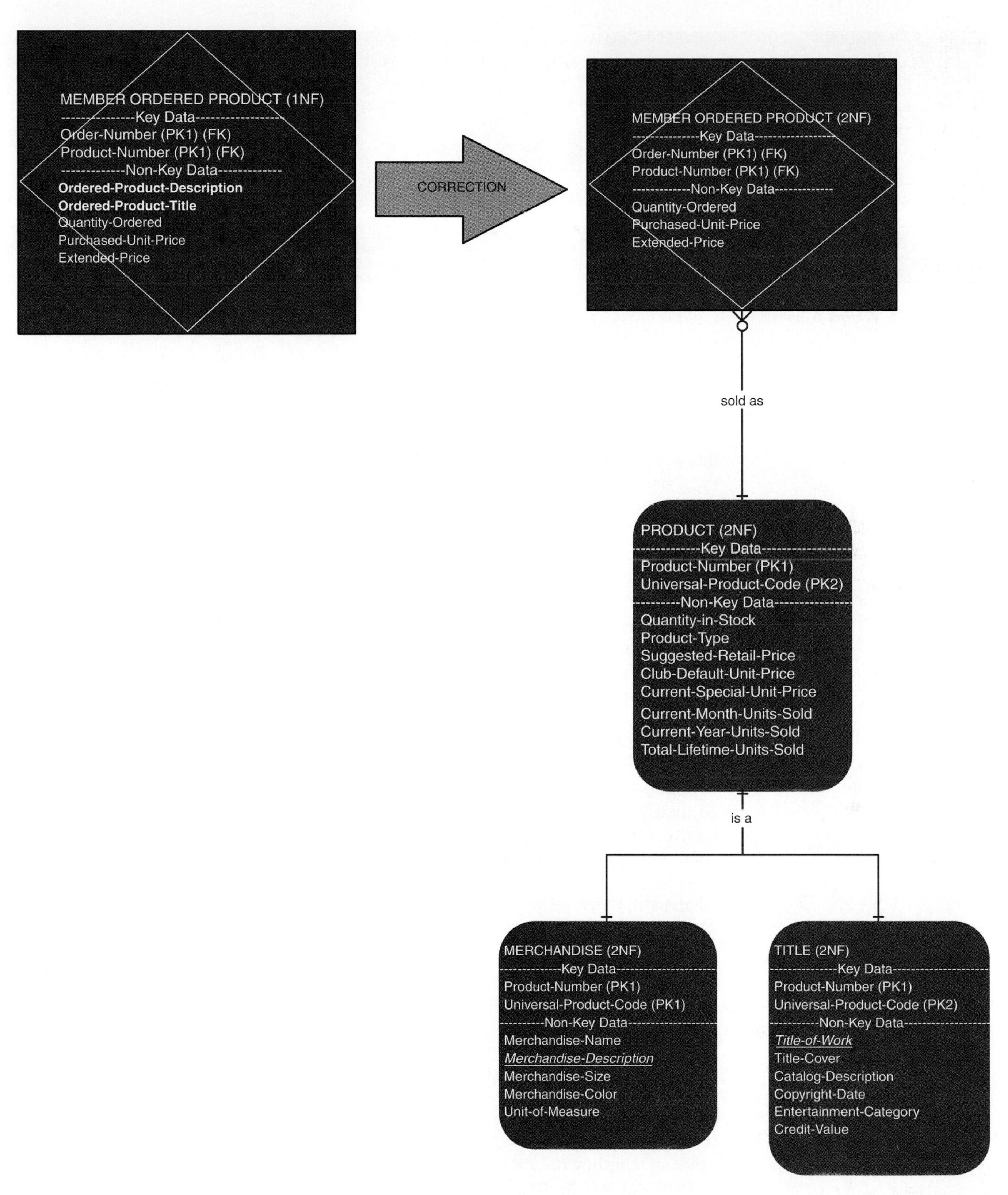

FIGURE 11.12 *Second Normal Form*

disk storage space. Second, it complicates simple updates. Why? Everytime you change the base attributes, you must remember to reperform the calculation and also change its result.

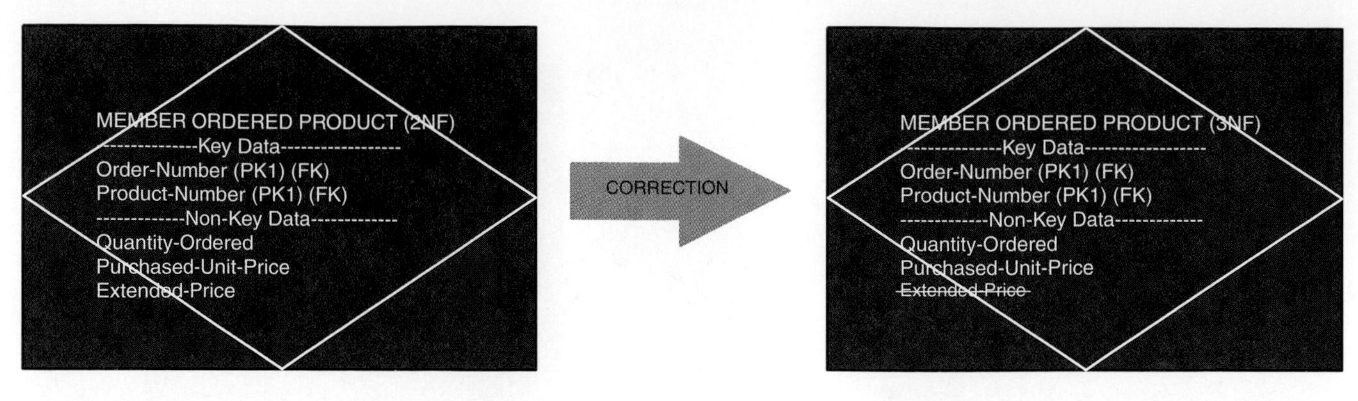

FIGURE 11.13 *Third Normal Form*

For example, look at the MEMBER ORDERED PRODUCT entity in Figure 11.13. The attribute EXTENDED PRICE is calculated by multiplying QUANTITY ORDERED by PURCHASE UNIT PRICE. Thus, EXTENDED PRICE (a nonkey attribute) is not dependent on the primary key as much as it is dependent on QUANTITY ORDERED and PURCHASE UNIT PRICE. Thus, we simplify the entity by deleting EXTENDED PRICE.

Sounds simple, right? Well, not always! There is disagreement on how far you take this rule. Some experts argue that the rule should be applied only within a single entity. Thus, these experts would not delete a derived attribute if the attributes required for the derivation are assigned to *different* entities. Other experts argue that the rule should be required regardless of where the base attributes are stored. We tend to agree based on the argument that a derived attribute that involves multiple entities presents a greater danger for data inconsistency caused by updating an attribute in one entity and forgetting to subsequently update the derived attribute in another entity. (The exception to this rule would be those databases that support *triggers*, described earlier in this chapter, that could automatically update the derived attributes.)

Another form of 3NF analysis checks for transitive dependencies. A transitive dependency exists when a nonkey attribute is dependent on another nonkey attribute (other than by derivation). This error usually indicates that an undiscovered entity is still embedded within the problem entity. Such a condition, if not corrected, can cause future flexibility and adaptability problems if a new requirement eventually requires us to implement that undiscovered entity as a separate database table.

Transitive analysis is performed only on those entities that do not have a concatenated key. In our example, this includes PRODUCT, MEMBER ORDER, MEMBER, and CLUB. For the entity PRODUCT, all the nonkey attributes are dependent on the primary key, and only the primary key. Thus, PRODUCT is already in third normal form.

But look at the entity MEMBER ORDER in Figure 11.14. In particular, examine the attributes MEMBER NAME and MEMBER ADDRESS. Are these attributes dependent on the primary key, MEMBER ORDER NUMBER? No! The primary key MEMBER ORDER NUMBER in no way determines the value of MEMBER NAME and MEMBER ADDRESS. On the other hand, the values of MEMBER NAME and MEMBER ADDRESS are dependent on the value of another non*primary* key in the entity, MEMBER NUMBER.

How do we fix this problem? MEMBER NAME and MEMBER ADDRESS need to be moved from the MEMBER ORDER entity to an entity whose key is just MEMBER NUMBER. If necessary, we would create that entity, but in our case we already have a MEMBER entity with the required primary key. And as it turns out, we don't need to really move the problem attributes since they are already assigned to the MEMBER entity. We did, however, have to notice that MEMBER ADDRESS was a synonym for MEMBER STREET ADDRESS. We elected to keep the latter term in MEMBER.

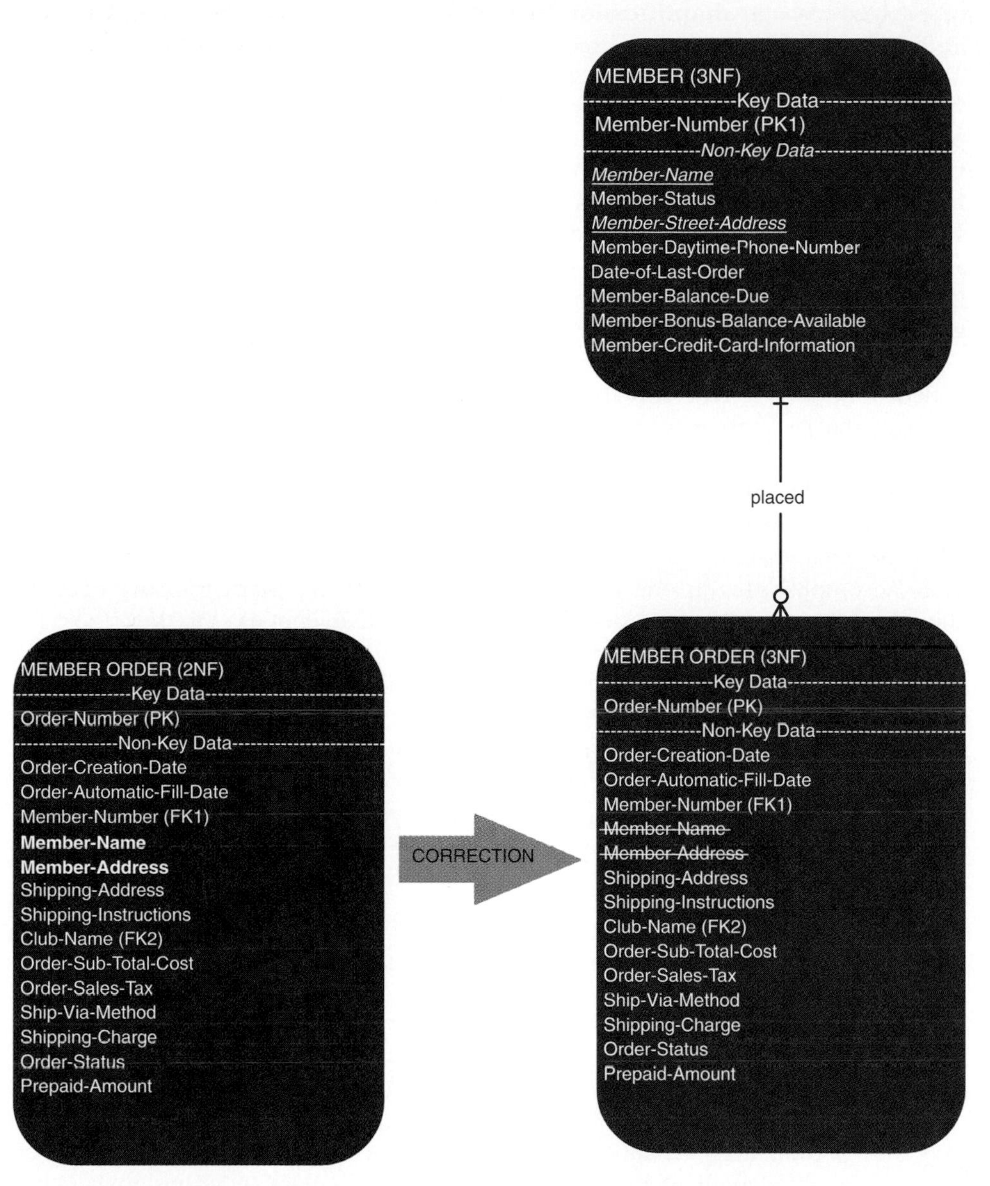

FIGURE 11.14 *Third Normal Form*

Before we leave the subject of normalization, we should acknowledge that several normal forms beyond 3NF exist. Each successive normal form makes the data model simpler, less redundant, and more flexible. However, systems analysts (and most database experts) rarely take data models beyond 3NF. Consequently, we will leave further discussion of normalization to database textbooks.

The first few times you normalize a data model, the process will appear slow and tedious. However, with time and practice, it becomes quick and routine. It may help to always remember the following ditty (source unknown), which nicely summarizes first, second, and third normal forms:

An entity is said to be in third normal form if every nonprimary key attribute is dependent on the primary key, the whole primary key, and nothing but the primary key.

Simplification by Inspection Normalization is a fairly mechanical process. But it is dependent on naming consistencies in the original data model (before normalization). When several analysts work on a common application, it is not unusual to create problems that won't be taken care of by normalization. These problems are best solved through **simplification by inspection,** a process wherein a data entity in 3NF is further simplified by such efforts as addressing subtle data redundancy.

Also through inspection, we realized that the CLUB MEMBERSHIP attributes for "taste" and "media" preferences were different depending on the club. For example, "media" has a different set of possible values based on club. In an AUDIO CLUB, the value set is CASSETTE, COMPACT DISC, MINI-DISC, and DIGITAL VERSATILE DISC. In the VIDEO CLUB, the value set is VHS TAPE, LASER DISC, 8MM TAPE, and DIGITAL VERSATILE DISC. In the GAME CLUB, media values include CD-ROM, DIGITAL VERSATILE DISC, and various CARTRIDGE formats. Thus, what we thought was one attribute, MEDIA PREFERENCE, was three attributes, AUDIO MEDIA PREFERENCE, VIDEO MEDIA PREFERENCE, and GAME MEDIA PREFERENCE.

The final, normalized data model is presented in Figure 11.15. Notice that we added a new supertype/subtype hierarchy to house the attributes created as a result of the last paragraph.

CASE Support for Normalization Many CASE tools claim to support normalization concepts. They read the data model and attempt to isolate normalization errors. On close examination, most CASE tools can normalize only to first normal form. They accomplish this in one of two ways. They look for many-to-many relationships and resolve those relationships into associative entities. Or they look for attributes specifically described as having multiple values for a single entity instance. (Of course, one could argue that the analyst should have recognized that as a 1NF error and not described the attributes as such.)

It is exceedingly difficult for a CASE tool to identify second and third normal form errors. That would require the CASE tool to have the intelligence to recognize partial and transitive dependencies. In reality, such dependencies can be discovered only through cooperation between users and analysts.

FILE DESIGN

The focus of this chapter is on database design; however, we would be remiss to not say a few words about conventional file design. First, file design is simplified because of its orientation to a single application. Typically, the output and input designs (Chapters 12 and 13) would be completed first since the file design is dependent on supporting those application requirements.

Most fundamental entities from the data model would be designed as master or transaction records. The master files are typically fixed-length records. Associative entities from the data model are typically joined into the transaction records to form variable-length records (based on the one-to-many relationships). Other types of files (not represented in the data model) are added as necessary.

Two important considerations of file design are *file access* and *organization.* The systems analyst usually studies how each program (from Chapter 10) will access the records in the file (sequentially or randomly) and then select an appropriate file organization (e.g., sequential, indexed, hashed, etc.). In practice, many systems analysts select an indexed sequential (or ISAM/VSAM) organization to support the likelihood that different programs will require different access methods into the records.

DATABASE DESIGN

We are now ready to design our database. The design of any database will usually involve the DBA and database staff. They will handle the technical details and cross-application issues. Still, it is useful for the systems analyst to understand the basic design principles for relational databases.

The design rules presented here are, in fact, guidelines. We cannot cover every idiosyncrasy. Also, because SoundStage has elected to use Microsoft's *SQL Server* as its database management system, our design will be constrained by that technology. Each relational DBMS presents its own capabilities and constraints. Fortunately, the guidelines presented here are fairly generic and applicable to most

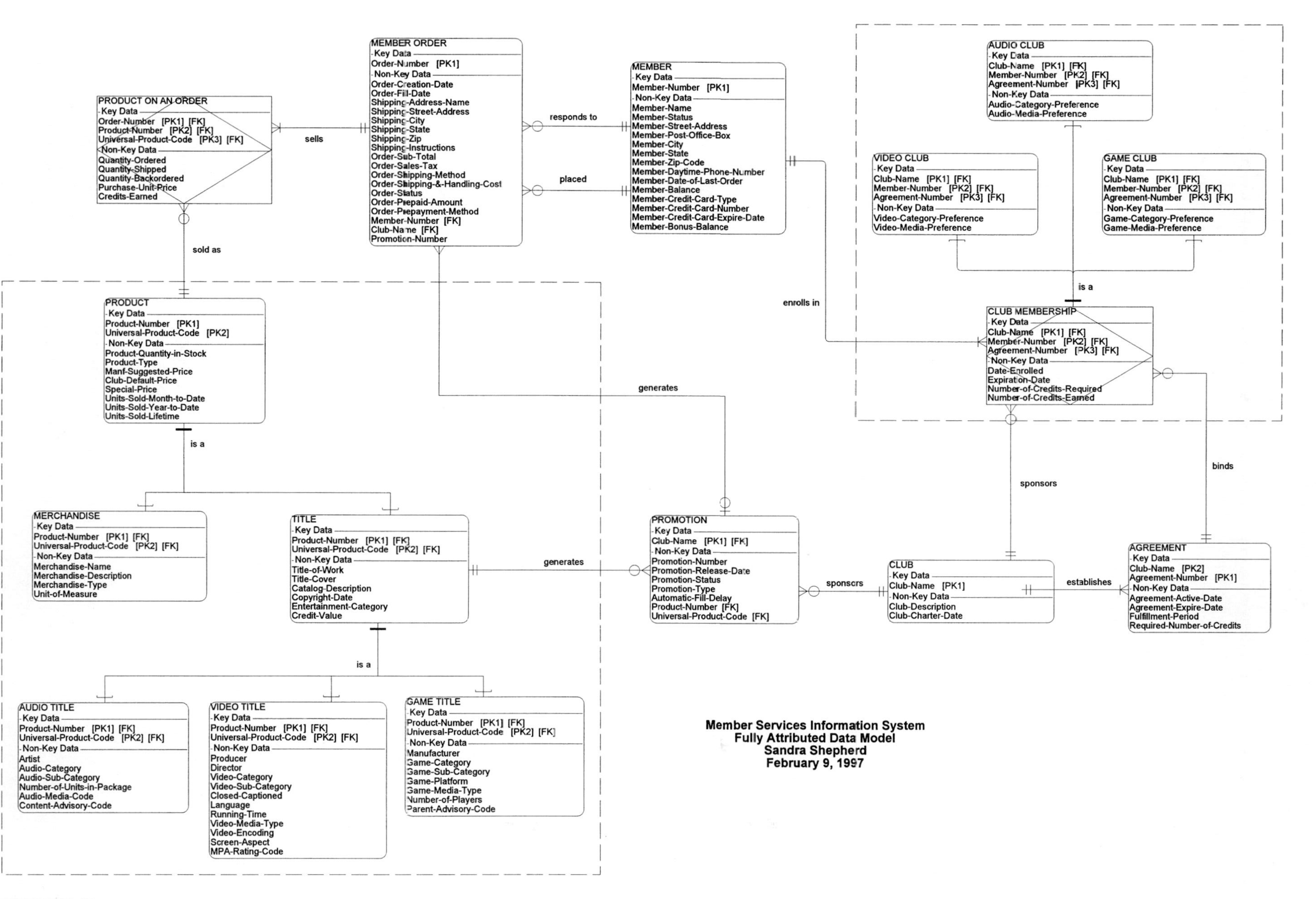

FIGURE 11.15 *SoundStage Logical Data Model in Third Normal Form*

DBMS environments. Database courses and textbooks tend to cover a wider variety of technology and issues.

Computer-assisted systems engineering (CASE) has been a continuing theme throughout this book. There are specific CASE products that address database analysis and design (e.g., Logic Works's *ERwin*). Also, most general-purpose CASE tools now include database design tools. In this example, we continued to use Popkin's *System Architect* CASE product for the SoundStage case study. Finally, most CASE tools (including *System Architect*) can automatically generate SQL code to construct the database structures for the most popular database management systems. This code generation capability is an enormous time-saver.

Goals and Prerequisites to Database Design

The goals of database design are as follows:

- A database should provide for the efficient storage, update, and retrieval of data.
- A database should be reliable—the stored data should have high integrity to promote user trust in that data.
- A database should be adaptable and scalable to new and unforeseen requirements and applications.

The system's data model—in our case, a fully attributed and normalized entity relationship diagram (ERD)—serves as the starting point. This model was previously introduced as Figure 11.15.

The data model may have to be divided into multiple data models to reflect database distribution and database replication decisions made in Chapter 10.

> **Data distribution** refers to the distribution of either specific tables, records, and/or fields to different physical databases.

> **Data replication** refers to the duplication of specific tables, records, and/or fields to multiple physical databases.

Each submodel or view should reflect the data to be stored on a single server. Recall that distributed and replicated databases may share some (or all) entities or entity instances. The SoundStage system is being replicated to each distribution center; therefore, our single data model is sufficient.

The Database Schema

The design of a database is depicted as a special model called a database schema.

> A **database schema** is the *physical* model or blueprint for a database. It represents the technical implementation of the logical data model.

A relational database schema defines the database structure in terms of tables, keys, indexes, and integrity rules. A database schema specifies details based on the capabilities, terminology, and constraints of the chosen database management system. Each DBMS supports different data types, integrity rules, and so forth.

The transformation of the logical data model into a physical relational database schema is governed by some fairly generic rules and options. These rules and guidelines are summarized as follows:

1. Each fundamental, associative, and weak entity is implemented as a separate table. Table names may have to be formatted according to the naming rules and size limitations of the DBMS. For example, a logical entity named MEMBER ORDERED PRODUCT might be changed to a physical table named MEMB_ORD_PROD. Naming conventions might also be governed by internal standards.
   a. The primary key is identified as such and implemented as an index into the table.
   b. Each secondary key is implemented as its own index into the table.

c. Each foreign key will be implemented as such. The inclusion of these foreign keys implements the relationships on the data model and allows tables to be JOINED in SQL and application programs.
d. Attributes will be implemented with fields. These fields correspond to columns in the table. The following technical details must usually be specified for each attribute. (These details may be automatically inferred by the CASE tool from the logical descriptions in the data model.)

Field names may have to be shortened and reformatted according to DBMS constraints and internal rules. For example, in the logical data model, most attributes might be prefaced with the entity name (e.g., *MEMBER* NAME). In the physical database, we might simply use NAME.

i. *Data type*. Each DBMS supports different data types and terms for those data types. For example, different systems may designate a large alphanumeric field differently (e.g., MEMO in *Access* and LONG VARCHAR in *Oracle*). Also, some databases allow the choice of no compression versus compression of unused space (e.g., CHAR versus VARCHAR in *Oracle*).
ii. *Size of the field*. Different DBMSs express precision of real numbers differently. For example, in *Oracle*, a size specification of NUMBER (3,2) supports a range from −9.99 to 9.99.
iii. *NULL or NOT NULL*. Must the field have a value before the record can be committed to storage? Again, different DBMSs may require different reserved words to express this property. Primary keys can never be allowed to have NULL values.
iv. *Domains*. Many database management systems can automatically edit data to ensure that fields contain legal data. This can be a great benefit to ensuring data integrity independent from the application programs. If the programmer makes a mistake, the DBMS catches the mistake. But for DBMSs that support data integrity, the rules must be precisely specified in a language that is understood by the DBMS.
v. *Default*. Many database management systems allow a default value to be automatically set in the event that a user or programmer submits a record without a value.
vi. Again, many of the above specifications were documented as part of a complete data model. If that data model was developed with a CASE tool, the CASE tool may be capable of automatically translating the data model into the language of the chosen database technology.

2. Supertype/subtype entities present additional options as follows:
   a. Most CASE tools do not currently support object-like constructs such as supertypes and subtypes. Consequently, most CASE tools default to creating a separate table for each entity supertype and subtype.
   b. Alternatively, if the subtypes are of *similar* size and data content, a database administrator may elect to collapse the subtypes into the supertype to create a single table. This presents certain problems for setting defaults and checking domains. In a high-end DBMS, these problems can be overcome by embedding the default and domain logic into *stored procedures* for the table.
3. Evaluate and specify referential integrity constraints (described in the next section).

The SoundStage database schema was automatically generated from the logical data model by our CASE tool *System Architect*. It is illustrated in Figure 11.16.

Would you ever want to compromise the third normal form entities when designing the database? For example, would you ever want to combine two third normal form entities into a single table (that would, by default, no longer be in third normal form)? Usually not! Although a database administrator may create such

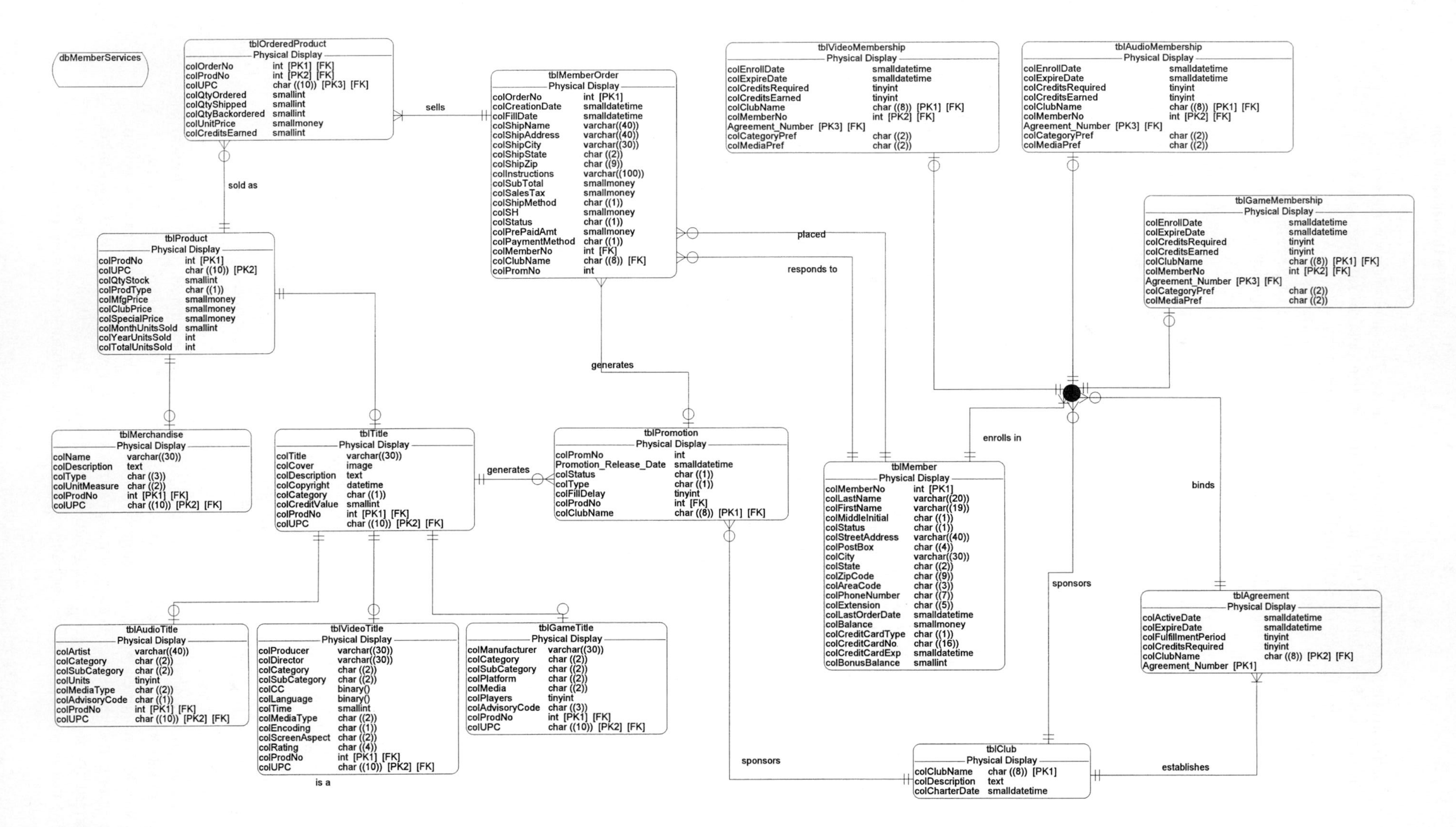

FIGURE 11.16 *Initial SoundStage Physical Database Schema*

a compromise to improve database performance, he or she should carefully weigh the advantages and disadvantages. Although such compromises may mean greater convenience through fewer tables or better overall performance, such combinations may also lead to the possible loss of data independence—should future new fields necessitate resplitting the table into two tables, programs will have to be rewritten. As a general rule, combining entities into tables is not recommended.

Data and Referential Integrity

Database integrity is about trust. Can the business and its users trust the data stored in the database? Data integrity provides necessary internal controls for the database. There are at least three types of data integrity that must be designed into any database.

Key Integrity Every table should have a primary key (which may be concatenated). The primary key must be controlled such that no two records in the table have the same primary key value. (Note that for a concatenated key, the concatenated value must be unique—not the individual values that make up the concatenation.)

Also, the primary key for a record must never be allowed to have a NULL value. That would defeat the purpose of the primary key, to uniquely identify the record.

If the database management system does not enforce these rules, other steps must be taken to ensure them.

Domain Integrity Appropriate controls must be designed to ensure that no field takes on a value that is outside of the range of legal values. For example, if GRADE POINT AVERAGE is defined to be a number between 0.00 and 4.00, then controls must be implemented to prevent negative numbers and numbers greater than 4.00.

Not long ago, application programs were expected to perform all data editing. Today, most database management systems are capable of data editing. For the foreseeable future, the responsibility for data editing will continue to be shared between the application programs and the DBMS.

Referential Integrity The architecture of relational databases implements relationships between the records in tables via *foreign keys.* The use of foreign keys increases the flexibility and scalability of any database, but it also increases the risk of referential integrity errors.

A referential integrity error exists when a foreign key value in one table has no matching primary key value in the related table. For example, an INVOICES table usually includes a foreign key, CUSTOMER NUMBER, to "reference back to" the matching CUSTOMER NUMBER primary key in the CUSTOMERS table. What happens if we delete a CUSTOMER record? There is the *potential* that we may have INVOICE records whose CUSTOMER NUMBER has no matching record in the CUSTOMERS table. Essentially, we have compromised the referential integrity between the two tables.

How do we prevent referential integrity errors? One of two things should happen. When considering the deletion of CUSTOMER records, either we should automatically delete all INVOICE records that have a matching CUSTOMER NUMBER (which doesn't make much business sense), or we should disallow the deletion of the CUSTOMER record until we have deleted all INVOICE records.

Referential integrity is specified in the form of deletion rules as follows:[3]

- *No restriction.* Any record in the table may be deleted without regard to any records in any other tables.

  In looking at the final SoundStage data model, we could not apply this rule to any table.

[3] Knowledgeable database students know that there are also insertion and update rules for referential integrity. A full discussion of these rules is deferred to database courses and textbooks.

- *Delete:Cascade.* A deletion of a record in the table must be automatically followed by the deletion of matching records in a related table. Many relational DBMSs can automatically enforce delete:cascade rules using triggers.
  In the SoundStage data model, an example of a valid delete:cascade rule would be from MEMBER ORDER to MEMBER ORDERED PRODUCT. In other words, if we delete a specific MEMBER ORDER, we should automatically delete all matching MEMBER ORDERED PRODUCTS for that order.
- *Delete:Restrict.* A deletion of a record in the table must be disallowed until any matching records are deleted from a related table. Again, many relational DBMSs can automatically enforce delete:restrict rules.
  For example, in the SoundStage data model, we might specify that we should disallow the deletion of any PRODUCT so long as there exists MEMBER ORDERED PRODUCTS for that product.
- *Delete:Set null.* A deletion of a record in the table must be automatically followed by setting any matching keys in a related table to the value NULL. Again, many relational DBMSs can enforce such a rule through triggers.
  The Delete:Set null option was not used in the SoundStage data model. It is used only when you are willing to delete a master table record, but you don't want to delete corresponding transaction table records for historical reasons. By setting the foreign key to NULL, you are acknowledging that the record does not point back to a corresponding master record, but at least you don't have it pointing to a nonexisting master record.

The final database schema, complete with referential integrity rules, is illustrated in Figure 11.17. This is the blueprint for writing the SQL code (or equivalent) to create the tables and data structures.

Roles

Some database shops insist that no two fields have exactly the same name. This constraint simplifies documentation, help systems, and metadata definitions. This presents an obvious problem with foreign keys. By definition, a foreign key must have a corresponding primary key. During *logical* data modeling, using the same name suited our purpose of helping the users understand that these foreign keys allow us to match related records in different entities. But in a *physical* database, it is not always necessary or even desirable to have these redundant field names in the database.

To fix this problem, foreign keys can be given role names.

> A **role name** is an alternate name for a foreign key that clearly distinguishes the purpose that the foreign key serves in the table.

For example, in the SoundStage database schema, PRODUCT_NUMBER is a primary key for the PRODUCTS table and a foreign key in the MEMBER ORDERED PRODUCTS table. The name should not be changed in the PRODUCTS table. But it may make sense to rename the foreign key to ORDERED_PRODUCT_NUMBER to more accurately reflect its role in the MEMBER ORDERED PRODUCTS table.

The decision to require role names or not is usually established by the data or database administrator.

Database Prototypes

Prototyping is not an alternative to carefully thought-out database schemas. On the other hand, once the schema is completed, a prototype database can usually be generated very quickly. Most modern DBMSs include powerful, menu-driven database generators that automatically create a DDL and generate a prototype database from that DDL. A database can then be loaded with test data that will prove useful for prototyping and testing outputs, inputs, screens, and other systems components.

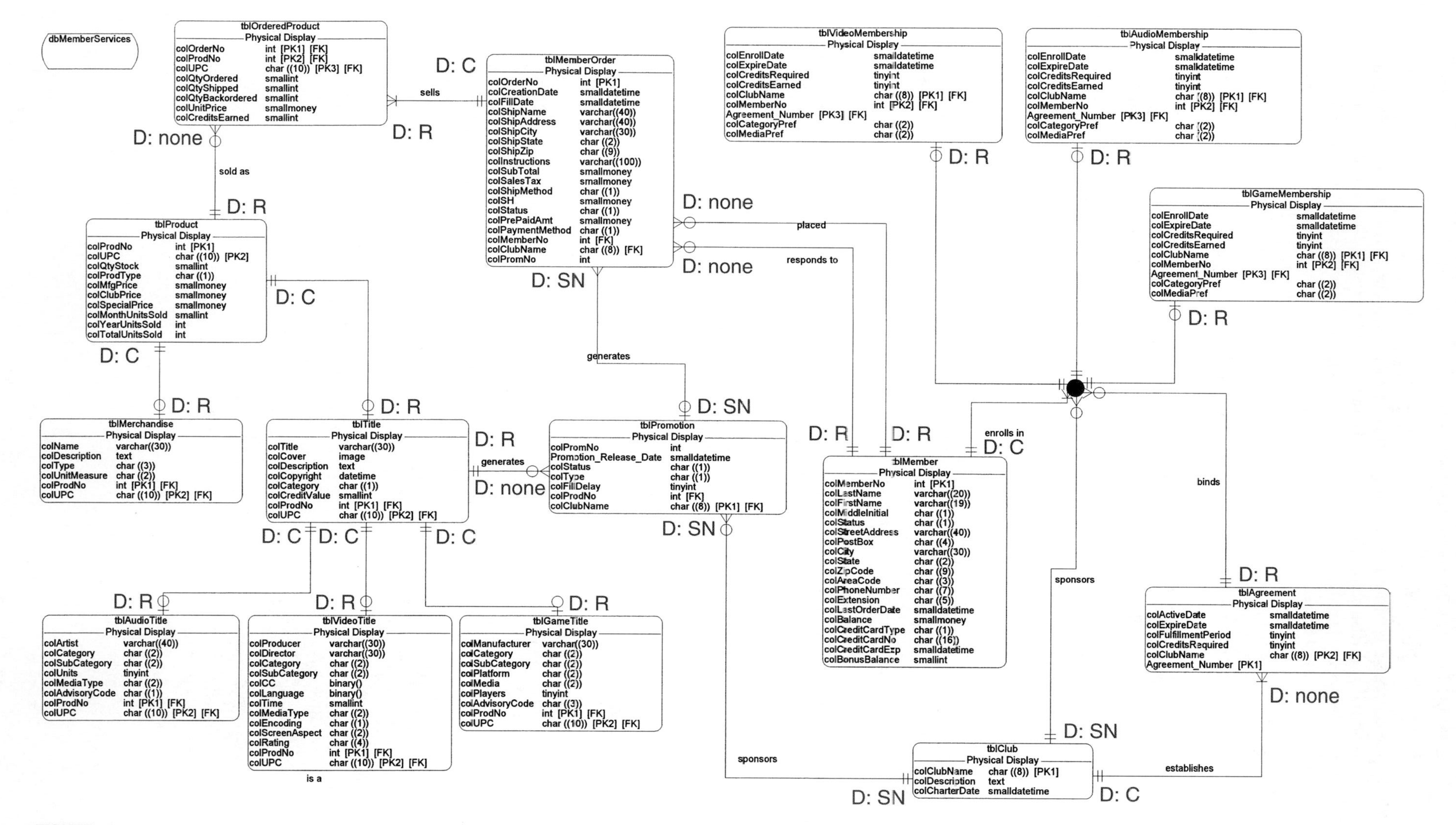

FIGURE 11.17 *Final SoundStage Physical Database Schema*

Database Capacity Planning

A database is stored on disk. Ultimately, the database administrator will want an estimate of disk capacity for the new database to ensure that sufficient disk space is available. Database capacity planning can be calculated with simple arithmetic as follows. This simple formula ignores factors such as packing, coding, and compression, but by leaving out those possibilities, you are adding slack capacity.

1. For each table, sum the *field sizes*. This is the *record size* for the table. Avoid the implications of compression, coding, and packing—in other words, assume that each stored character and digit will consume one byte of storage. Note that formatting characters (e.g., commas, hyphens, slashes, etc.) are almost never stored in a database. Those formatting characters are added by the application programs that will access the database and present the output to the users.
2. For each table, multiply the *record size* times the number of entity instances to be included in the table. It is recommended that growth be considered over a reasonable time period (e.g., three years). This is the *table size*.
3. Sum the *table sizes*. This is the *database size*.
4. Optionally, add a slack capacity buffer (e.g., 10 percent) to account for unanticipated factors or inaccurate estimates above. This is the *anticipated database capacity*.

Database Structure Generation

CASE tools are frequently capable of generating SQL code for the database directly from a CASE-based database schema. This code can be exported to the DBMS for compilation. Even a small database such as the SoundStage model can require 50 pages or more of SQL data definition language code to create the tables, indexes, keys, fields, and triggers. Clearly, a CASE tool's ability to automatically generate syntactically correct code is an enormous productivity advantage. Furthermore, it almost always proves easier to modify the database schema and regenerate the code than to maintain the code directly. Figure 11.18 is a sample page of code generated by *System Architect* from the SoundStage database schema.

THE NEXT GENERATION OF DATABASE DESIGN

Relational database technology is widely deployed and used in contemporary information systems shops. The skills taught in this chapter will remain viable well into the foreseeable future. But one new technology is slowly emerging that could ultimately change the landscape dramatically—*object* database management systems.

The heir apparent to relational DBMSs, object database management systems store true objects, that is, encapsulated data and all of the processes that can act on that data. Because relational database management systems are so widely used, we don't expect this change to happen quickly. Furthermore, the relational DBMS vendors are not likely to give up their market share without a fight. It is expected that these vendors will either build object technology into their existing relational DBMSs or they will create new, object DBMSs and provide for the transition between relational and object models. Regardless, this is one technology to keep an eye on.

WHERE DO YOU GO FROM HERE?

Let's begin with the obvious! If you have information systems career aspirations, you had better plan to take one or more true database courses. The topics presented in this chapter represent only the tip of the iceberg as it relates to database technology, development, and management. Most IS curricula include at least one good database or data management course to add value to your education. Take it!

You have only begun your journey through system design. The database is the *brain* of a new system or application. The subsequent chapters focus on the design of other crucial body parts. Chapters 12 through 14 teach input, output, and user interface design, respectively. Think of inputs, outputs, and interfaces as the *soul* of the system. Finally, Chapters 15 and 16 teach program design. Programs are the *heart* of the new system, that which gives the system life.

```
                    Member Services Database DDL/SQL
go
CREATE TABLE tblVideoMembership(
         colCategoryPref                        char ((2)) NOT NULL,
         colMediaPref                           char ((2)) NOT NULL,
         colClubName                            char ((8)) NOT NULL
                UNIQUE NONCLUSTERED ,
         colMemberNo                            int NOT NULL
                UNIQUE NONCLUSTERED ,
         Agreement_Number                       MISSING_SQL_DATATYPE)
go

sp_bindrule domVideoCategory, 'tblVideoMembership.colCategoryPref'
go

sp_bindrule domVideoMedia, 'tblVideoMembership.colMediaPref'
go

sp_bindrule domMemberNumber, 'tblVideoMembership.colMemberNo'
go

ALTER TABLE tblVideoMembership ADD
    PRIMARY KEY CLUSTERED  (colClubName,colMemberNo,Agreement_Number)
go
CREATE TABLE tblVideoTitle(
         colProducer                            varchar((30)) NULL,
         colDirector                            varchar((30)) NULL,
         colCategory                            char ((2)) NOT NULL,
         colSubCategory                         char ((2)) NULL,
         colCC                                  binary NOT NULL,
         colLanguage                            binary NOT NULL,
         colTime                                smallint NOT NULL,
         colMediaType                           char ((2)) NOT NULL,
         colEncoding                            char ((1)) NOT NULL,
         colScreenAspect                        char ((2)) NOT NULL,
         colRating                              char ((4)) NOT NULL,
         colProdNo                              int NOT NULL
                UNIQUE NONCLUSTERED ,
         colUPC                                 char ((10)) NOT NULL
                UNIQUE NONCLUSTERED )
go

sp_bindrule domVideoCategory, 'tblVideoTitle.colCategory'
go

sp_bindrule domVideoCategory, 'tblVideoTitle.colSubCategory'
go

sp_bindefault defNo, 'tblVideoTitle.colCC'
                              Page 10
```

FIGURE 11.18 *Partial SQL Code to Construct the SoundStage Database*

```
                    Member Services Database DDL/SQL
go

sp_bindrule domYesNo, 'tblVideoTitle.colCC'
go

sp_bindefault defEnglish, 'tblVideoTitle.colLanguage'
go

sp_bindrule domLanguage, 'tblVideoTitle.colLanguage'
go

sp_bindrule domVideoMedia, 'tblVideoTitle.colMediaType'
go

sp_bindrule domScreenAspect, 'tblVideoTitle.colScreenAspect'
go

sp_bindrule domVideoRatingCodes, 'tblVideoTitle.colRating'
go

ALTER TABLE tblVideoTitle ADD CONSTRAINT idxVideoTitle
    PRIMARY KEY CLUSTERED  (colProdNo,colUPC)
go
/* FK for reference to tblClub through relation establishes */
ALTER TABLE tblAgreement ADD CONSTRAINT establishes
    FOREIGN KEY (colClubName)
    REFERENCES tblClub (colClubName)
go


/* FK for reference to tblMembership through relation is a */
ALTER TABLE tblAudioMembership ADD CONSTRAINT is_a
    FOREIGN KEY (colClubName,colMemberNo,Agreement_Number)
    REFERENCES tblMembership (colClubName,colMemberNo,Agreement_Number)
go


/* FK for reference to tblTitle through unnamed relation */
ALTER TABLE tblAudioTitle ADD
    FOREIGN KEY (colProdNo,colUPC)
    REFERENCES tblTitle (colProdNo,colUPC)
go


/* FK for reference to tblMembership through unnamed relation */
ALTER TABLE tblGameMembership ADD
    FOREIGN KEY (colClubName,colMemberNo,Agreement_Number)
    REFERENCES tblMembership (colClubName,colMemberNo,Agreement_Number)
                                   Page 11
```

FIGURE 11.18 *(Concluded)*

SUMMARY

1. The data captured by an information system is stored in files and databases. A file is a collection of similar records. A database is a collection of interrelated files.
2. Many legacy systems were built with file technology. Because files were built for specific applications, their design was optimized for those applications. This close relationship between the files and their applications made it difficult to restructure the files to meet future requirements. And because many applications use the same data, it is not uncommon to find redundant files with data values that do not always match.
3. As the above legacy systems are slowly reengineered, they are usually converted to database technology. Well-designed databases share nonredundant data and overcome all the limitations of conventional files.
4. Database design is the process of translating logical data models (Chapter 5) into physical database schemas.
5. The smallest unit of meaningful data that can be stored is called a field. There are four types of fields.
   a. A primary key is a field that uniquely identifies one and only one record in a file or table.
   b. A secondary key is a field that may either uniquely identify one and only one record in a file or table or identify a set of records with some common, meaningful characteristic.
   c. A foreign key is a field that points to a related record in a different table.
   d. All other fields are called descriptive fields.
6. Fields are organized into records, and similar records are organized into files or tables.
7. A database is a collection of tables (files) with logical pointers that relate records in one table to records in a different table.
8. The data architecture that has evolved in most organizations includes conventional files, operational databases, data warehouses, and personal and work group databases. To coordinate this complex infrastructure, many organizations assign a data administrator to plan and manage the overall data resource and database administrators to implement and manage specific databases and database technologies.
9. A database architecture is built around a database management system (DBMS) that provides the technology to define the database structure and then to create, read, update, and delete records in the tables that make up that structure. A DBMS provides a data language to accomplish this. That language provides at least two components:
   a. A data definition language to create and maintain the database structure and rules.
   b. A data manipulation language to create, read, use, update, and delete records in the database.
10. Today, relational database management systems are used to support the development and reengineering of the overwhelming number of information systems. Relational databases store data in a collection of tables that are related via foreign keys.
    a. The data definition and manipulation languages of most relational DBMSs are consolidated into a standard language known as SQL.
    b. High-end relational database management systems support triggers and stored procedures, programs that are stored with the tables and callable from other SQL-based programs.
11. Data analysis and normalization are techniques for removing impurities from a data model as a preface to designing the database. These impurities can make a database unreliable, inflexible, and nonscalable.
12. Normalization involves checking each entity (table) for first, second, and third normal form impurities.
    a. An entity is in first normal form if it contains no repeating attributes (that is, attributes that can have multiple values for a single instance of the entity).
    b. An entity is in second normal form if it contains no partial dependencies (that is, a nonkey attribute whose value is dependent only on part of the entity's primary key).
    c. An entity is in third normal form if it contains no derived attributes (that is, calculated or logic-based attributes) or no transitive dependencies (that is, a nonkey attribute whose value is dependent on another nonkey attribute).
13. Distribution and replication decisions should be made before database design. Each unique database should be represented by its own logical data submodel.
14. A database schema is the physical model for a database based on the chosen database technology. The rules for transforming a logical data model into a physical database schema are generalized as follows:
    a. Each entity becomes a table.
    b. Each attribute becomes a field (column in the table).
    c. Each primary and secondary key becomes an index into the table.
    d. Each foreign key implements a possible relationship between instances of the table.
15. Database integrity should be checked and, if necessary, improved to ensure that the business and its users can trust the stored data.
    a. Key integrity ensures that every record will have a unique, non-NULL primary key value.
    b. Domain integrity ensures that appropriate fields will store only legitimate values from the set of all possible values.
    c. Referential integrity ensures that no foreign key value points to a nonexistent primary key value. A deletion rule should be specified for every relationship with another table. The deletion rules either cascade the deletion to related records in other tables, disallow the deletion until related records in other tables are first deleted, or allow the deletion but set any foreign keys in related tables to NULL.

KEY TERMS

archival file, p. 400
audit file, p. 400
blocking factor, p. 400
data administrator, p. 403
data analysis, p. 408
data architecture, p. 401
data definition language (DDL), p. 403
data distribution, p. 420
data manipulation language (DML), p. 403
data replication, p. 420
data warehouse, p. 402
database, p. 395
database administrator, p. 403
database architecture, p. 403
database engine, p. 403
database management system (DBMS), p. 403
database schema, p. 420
descriptive field, p. 399
domain integrity, p. 423
field, p. 398
file, p. 395, 400
file access, p. 400
file organization, p. 400
first normal form (1NF), p. 408
fixed-length record structure, p. 399
foreign key, p. 399
key integrity, p. 423
master file, p. 400
metadata, p. 403
normalization, p. 408
operational database, p. 402
personal database, p. 402
primary key, p. 399
record, p. 399
referential integrity, p. 423
relational database, p. 405
role name, p. 424
second normal form (2NF), p. 408
secondary key, p. 399
simplification by inspection, p. 418
SQL, p. 405
stored procedure, p. 405
table, p. 400
third normal form (3NF), p. 409
transaction file, p. 400
transaction processing monitor (TP monitor), p. 404
trigger, p. 405
variable-length record structure, p. 399
work group database, p. 402

REVIEW QUESTIONS

1. Differentiate between conventional files and databases.
2. What is a database? What is the difference between a production database and an end-user database?
3. Explain the advantages and disadvantages of conventional files versus databases.
4. Define the terms *field, record,* and *file.*
5. Differentiate between primary, secondary, and foreign keys.
6. Identify six types of files, and give several examples of each.
7. Differentiate between fixed- and variable-length records. What impact does a record storage format have on a file design?
8. Differentiate between file access and file organization.
9. Differentiate between an operational database and a data warehouse. What types of applications does each serve?
10. Differentiate between a data and database administrator. What is the relationship between these positions and the systems analyst?
11. Briefly explain the differences between a data definition language, a host programming language, and a data manipulation language.
12. List and briefly describe the three table operations used to manipulate relational tables.
13. Differentiate between triggers and stored procedures.
14. Give three characteristics of a good data model.
15. List and briefly describe the three steps of normalization.

PROBLEMS AND EXERCISES

1. Eudrup University's current student registration system consists of the following ISAM files: STUDENTS, COURSES, INSTRUCTORS, REGISTRATIONS, SPACE, and SCHEDULES. The first three files are fixed-length master files. The last three are variable-length transaction files. The administration would like to integrate all this data into a single database. Write a step-by-step project plan to accomplish this. (Hint: You may need to review Chapter 5.)
2. Kevin, an inventory manager, is considering a DBMS for his microcomputer. He's not certain that he really understands what a database is. In college, he took an introductory computer course and learned about files. He assumed database is the current buzzword for a collection of files. Write him a memo explaining the difference between a file and database environment. What are the advantages and disadvantages of each environment?
3. Draw a data model for your school's course registration system and then convert the data model into a relational database schema.
4. If databases were created with the ability to solve many of the problems characteristic of conventional file-based systems, why aren't all information systems shops operating in a database environment?
5. Transform the following entities into 3NF entities. Draw

the original and final data models, and state any reasonable assumptions. (Note: Primary keys are underlined.)

AIRCRAFT (AIRCRAFT ID NUMBER, AIRCRAFT CODE, AIRCRAFT DESCRIPTION, NUMBER OF SEATS).

FLIGHT (FLIGHT NUMBER, DEPARTURE CITY, 1 { ARRIVAL CITY } n, MEAL CODE, 1 { FLIGHT DATE, CURRENT SEAT PRICE } m).

PASSENGER (PASSENGER NUMBER, PASSENGER NAME, FREQUENT FLYER NUMBER, 1 { FLIGHT NUMBER, SEAT NUMBER, QUOTED SEAT PRICE, AMOUNT PAID, BALANCE DUE }).

6. How are relationships appearing on an entity relationship diagram implemented by a relational database management system?
7. How would you implement a recursive relationship in a relational database system?
8. Explain the role of data analysis in database design. Why not just go straight to database design?
9. What is the difference between a data entity in first normal form (1NF) and second normal form (2NF)? Give an example of an entity in 1NF and show its conversion to 2NF.
10. What is the difference between a data entity in second normal form (2NF) and third normal form (3NF)? Give an example of an entity in 2NF and show its conversion to 3NF.

PROJECTS AND RESEARCH

1. Visit a local information systems shop that uses a DBMS. Describe the existing database environment. Does it have production oriented databases or end-user databases? What host programming language(s) is utilized to load, maintain, and use the data? Ask the systems analyst or database administrator to give you a brief orientation on the physical and logical structures supported by the DBMS. To what extent are conventional files used?
2. Visit a local information systems shop that operates in a strictly conventional file environment. Ask the systems analyst for information describing several of the master and transaction files. Do some of the files contain duplicate data? Is this data input several times? What impact has the duplicated data had on maintenance? Do they experience problems with data integrity? Have the analyst explain the impact of changing the format of one of the files.

MINICASES

1. Sunset Valley Distributors recently completed a major conversion project. Several months ago, Sunset decided to move into the database era. Many of its computer-based files had become unreliable, difficult to maintain, and too inflexible to be used to fulfill many end-user reporting and inquiry requests. A DBMS seemed to be the obvious solution. Two systems analysts were primarily responsible for the conversion project, which took several months to complete. The systems analysts had decided to simply implement each of the computer-based files as a separate table in their relational database. Once the conversion was completed, the same problems that existed with the file-based system reappeared in the database system. Reports contained inaccurate data, report and inquiry requests could not easily be obtained, and data maintenance was still difficult. A consultant was hired to investigate the problems. The consultant acknowledged that many of the problems resulted because the analysts failed to do data modeling. Explain the importance of doing data modeling ahead of time when designing databases.
2. Design the logical schema for a relational database using the entity relationship diagram that follows. The primary keys of the entities are as follows:

Data Structure		**Field Size**
EMPLOYEE	= SOCIAL SECURITY NUMBER (PK)	9
	+ EMPLOYEE NAME	32
	+ EMPLOYEE STREET ADDRESS	32
	+ EMPLOYEE CITY	12
	+ EMPLOYEE STATE	2
	+ EMPLOYEE ZIP CODE	9
	+ EMPLOYEE HOME PHONE NUMBER	10
	+ EMPLOYEE EMAIL ADDRESS	15
	+ 1 { DEPARTMENT CODE +	2
	OFFICE LOCATION +	3
	OFFICE PHONE NUMBER } 4	5
	+ DATE EMPLOYED	8
	+ DATE OF BIRTH	8
	+ (SPOUSE NAME +	
	SPOUSE DATE OF BIRTH)	20 + 8
	+ 0 { DEPENDENT NAME +	20
	DEPENDENT RELATIONSHIP +	1
	DEPENDENT DATE OF BIRTH }n	8
	+ [MONTHLY SALARY |	5 or 3.2
	HOURLY PAY RATE]	
	+ VACATION DAYS DUE	2
	+ SICK DAYS DUE	2

Data Structure		Field Size
	+ GROSS PAY YEAR-TO-DATE	6.2
	+ FEDERAL TAX WITHHELD YEAR-TO-DATE	5.2
	+ STATE TAX WITHHELD YEAR-TO-DATE	4.2
	+ FICA TAX WITHHELD YEAR-TO-DATE	5.2

x.y *indicates number of digits to the left and right of the decimal point, respectively*

3. Design the 3NF logical schema for a relational database using the entity relationship diagram that follows:

Data Structure		Field Size
SUPPLIER	= SUPPLIER IDENTIFICATION NUMBER	12
	+ SUPPLIER NAME	30
	+ SUPPLIER STREET ADDRESS	30
	+ SUPPLIER CITY	15
	+ SUPPLIER STATE	2
	+ SUPPLIER PHONE NUMBER	10
	+ 1 { PAYMENT METHOD +	1
	EARLY DISCOUNT PERIOD +	2
	EARLY DISCOUNT RATE +	0.2
	PAYMENT DEADLINE } 3	2
	+ 1 { MATERIAL NUMBER +	9
	MATERIAL DESCRIPTION +	30
	UNIT PRICE +	5.2
	QUANTITY DISCOUNT THRESHOLD +	3
	QUANTITY DISCOUNT PERCENTAGE } n	0.2

x.y *indicates number of digits to the left and right of the decimal point, respectively*

4. Given the sample form that appears at the top of p. 433, prepare a list of entities and their associated data attributes as determined from the document. Then, completely normalize the entities to 3NF and draw a hypothetical ERD. Your instructor should be the final interpreter for the form.
5. Design the relational database schema for the following entity relationship diagram at the bottom of p. 433.
6. Precious Jewels Diamond Centers is a franchised jewelry store that specializes in diamonds and other gems, custom selected by and for customers. Gems are custom set into rings, pendants, and other pieces. Precious Jewels also serves as a diamond broker, providing gems to other franchises and jewelry stores. These gems are sent out on approval. The stores have the option of purchasing the gems or returning them.

   Jeff Kassels, vice president, is looking for a consultant to improve information systems for its IBM and Compaq microcomputers. About two years ago, executives decided to purchase two microcomputers. On the recommendation of the computer superstore, they also bought Microsoft *Excel* (a spreadsheet), Borland *Paradox* (a PC-DBMS), and Lotus *Word Pro* (a word processor). Initially and unfortunately, they didn't invest in the training to exploit these packages, especially the *Paradox* database package.

   Eventually, they hired some young students who were into PCs. They wrote some *Paradox* programs and *Excel* macros for inventory control and sales. The programs seemed to work. Precious Jewels entered lots of data into the system, and it generated several reports.

   Later, Precious Jewels staff realized the need for new reports and inquiries. They tried to generate them themselves, but they just didn't understand the report writer in *Paradox*. The original students were unavailable, so they hired a woman who does *Paradox* programming on the side. She couldn't seem to generate the reports from the stored data—even though the data is in there.

   To compound matters, there are growing problems with the data already in the system. As Precious Jewels staff regenerated the original reports, they noticed that records existed that should have been deleted a long time ago. And to make matters worse, they found records of gems for which there was no associated purchase order. This caused insurance problems.

   When they asked their new consultant to add some new fields to existing programs, she informed Precious Jewels management that many of the existing programs would have to be rewritten because fields would need to be moved to different or new files for efficiency.

   a. What went wrong and why?
   b. What benefits do you think can be derived from studying data before you study output needs and processing requirements?
   c. Why do you think that consultants—and experienced analysts—so frequently ignore or do not adequately consider the future implications of the databases they design?

For Minicase 4

PURCHASING REQUISITION Form 12 Rev.1988

INSTRUCTIONS — INCLUDE IN EACH REQUISITION ONLY SUCH ARTICLES AS MAY BE PURCHASED FROM ONE FIRM. IF SPECIAL HANDLING IS DESIRED, NOTE. SEE REVERSE SIDE FOR SPECIAL COMMENTS BY REQUESTOR.

DEPARTMENT COMPLETES UNSHADED AREA | PURCHASING COMPLETES SHADED AREA | ORDER NO.

DEPT. OR FUNCTION: Computer Information Systems | COMMITMENT NO. | COMMODITY CODE | ORDER TYPE

M F C	RES CODE.	ACCOUNT NUMBER FUND	CENTER DEPT.— PROJ.	OBJECT	DEPT. REFERENCE	AMOUNT	FUND EXPIRATION DATE
1				5-6207		8,736.00	
2				5 6106		399.00	
3				5-6107		84.00	

SHIP TO STAFF MEMBER: Jonathan Doe
DEPT.: 242
BUILDING & ROOM: Administration

ORDER DATE
FOLLOW UP
PRICING METHOD
☐ RQ #
☐ 1 Phone/Verbal Quote
☐ 2 Agreement/Contract
☐ 3 Price List on File
☐ 4 Repair Negotiation
☐ 5 None of the above
BUYER

REQUISITIONER'S PHONE NO.: 555-4545
MATERIAL WILL BE USED FOR
VENDOR SUGGESTED: IBM, Main Street, Somewhere, IN 47906
VENDOR NAME
VENDOR NUMBER

FOB: ☐ 1 DESTINATION ☐ 2 DESTINATION PREPAY & ADD ☐ 3 SHIPPING POINT ☐ 4 SHIPPING POINT FREIGHT ALLOWED ☐ 5 SEE BELOW | VIA | TERMS

ITEM #	ITEM DESCRIPTION	MFC	QUANTITY	UNIT	UNIT PRICE	EXTENDED PRICE	DELIVER ON	EST.	COMM.
	IBM PS/2 Model 70 86 8570-121	1	1		7,995.00	4,797.00			
	IBM PS/2 2-8 MB Memory Module Expansion Option #5211	1	1		1,695.00	1,017.00			
	IBM PS/2 2MB Memory Module Kit #5213	1	3		1,395.00	2,511.00			
	IBM 8513 PS/2 Color Display	1	1		685.00	411.00			
	IBM 8770 PS/2 Mouse	2	1		95.00	57.00			
	IBM PC Network Adapter II/A #150122	2	1		570.00	342.00			
	IBM DOS 3.3	3	1		120.00	84.00			

REQUESTED — HEAD OF DEPT.: *Thomas J. Mathien* | DATE: 6-3-89 | APPROVED — FOR THE COMPTROLLER | DATE | BYPASS APPROVAL REQUESTED ☐ | APPROVAL SIGNATURE/DATE | PURCHASING APPROVALS: PA, AD, DIR

RECOMMENDED — DEAN OR ADMINISTRATOR | DATE | APPROVED — FOR THE EXECUTIVE VICE PRESIDENT AND TREASURER | DATE | OCGBA PREAUDIT BY: DATE:

For Minicase 5

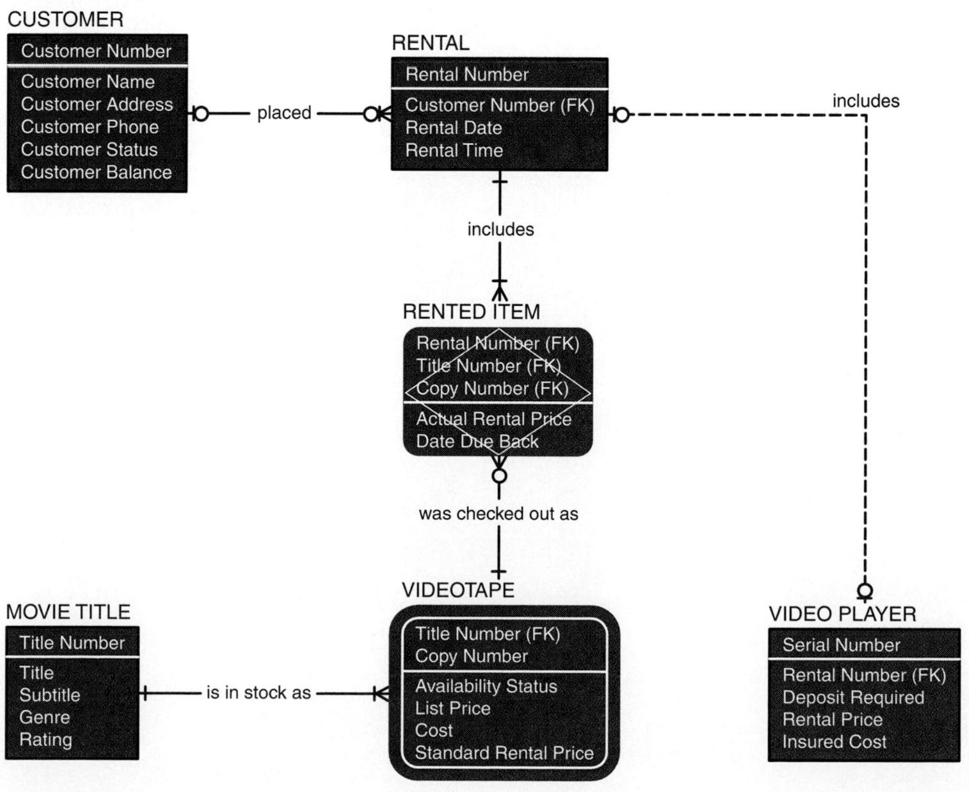

SUGGESTED READINGS

Bruce, Thomas. *Designing Quality Databases with IDEF1X Information Models.* New York: Dorset House Publishing, 1992. This has rapidly become our favorite practical database design book. Incidentally, the foreword was written by John Zachman whose *Framework for Information Systems Architecture* inspired our own information system building blocks framework.

Martin, James. *Managing the Database Environment.* Englewood Cliffs, NJ: Prentice Hall, 1983. Martin is one of the most noted authorities, writers, and lecturers in the database field. No database list would be complete without one of his many titles. We chose this title because of its management orientation and readability.

Martin, James, and Clive Finkelstein. *Information Engineering.* 2 vols. New York: Savant Institute, 1981. Information engineering is a formal, database-, and fourth-generation-language-oriented methodology. The method is logically equivalent; however, the authors use entity diagrams instead of entity relationship diagrams. ERDs could easily be substituted.

McFadden, Fred, and Hoffer, Jeffrey. *Modern Database Management.* 4th ed. Redwood City, CA: Benjamin Cummings Publishing Co., 1994. For those seeking to expand their overall data management and database education, this is one of the most popular introductory textbooks on the market and our own favorite.

Teorey, Toby. *Database Modeling & Design: The Fundamental Principles.* 2nd ed. San Francisco: Morgan Kaufman Publishers, Inc., 1990. This is our favorite database design conceptual book. Appendix A provides a concise review of the SQL language.

12

INPUT DESIGN AND PROTOTYPING

CHAPTER PREVIEW AND OBJECTIVES

In this chapter you will learn how to complete a preliminary design for computer inputs. It is the first of three chapters that address the design of on-line systems using a graphical user interface. You will know how to design inputs when you can:

- Define the appropriate format and media for a computer input.
- Explain the difference between data capture, data entry, and data input.
- Identify and describe several automatic data collection technologies.
- Apply human factors to the design of computer inputs.
- Design internal controls for computer inputs.
- Design a good source document for capturing transaction data.
- Select proper screen-based controls for input attributes that are to appear on a GUI input screen.

We begin this episode in the conference room where Sandra and Bob have scheduled a Saturday morning meeting to review input design screens with the order processing staff. Sally Hoover suggested a Saturday morning overtime meeting because she wanted her entire staff to become familiar with the new input and on-line methods that would be used in the new system.

SANDRA

OK, let's call this meeting to order. I think we should get started so we don't take away too much of everyone's weekend. I know all of you are eager to enjoy your weekend, so we'll make this as brief as possible. First, Bob and I would like to thank you for your cooperation so far. We realize that we have spent a great deal of time defining what you need in order to do your jobs. We have been deliberately avoiding specific details of how to do things because we wanted to build a system that would make your jobs easier. It is your system, and it is shaping up very well. Today, we want to review system inputs. Let's begin with a member order. (*Sandra adjusts the computer project pad so the screen image is in clear focus on the pull-down projection screen.*) This is the proposed member order response form that you will be receiving from our customers. Bob has begun to build a screen that you folks will use to enter data from these forms. We want your feedback. It is not the intent of this meeting to focus on the overall appearance of this screen. Rather, we simply want to make sure that we are capturing all the data that we need to capture about our customer orders. We are also concerned with the manner in which that data is captured. So, Bob, why don't you take it from here.

BOB

Thanks, Sandra. First, notice that you will not have to enter the order date anymore; the system will do that for you by using the current date. In fact, you'll notice that the background color of that field is grayed to suggest that you can't change the date.

SALLY

Can't change the date? What if the order date is different from the date when the order is entered?

SANDRA

Why would they be different?

SALLY

Sometimes orders are sent to the wrong department, or they don't get entered on the day they are received due to a high volume of orders. Since we generate automatic orders to members based on the date, it's very important that the date on a member's order be correct.

BOB

That's my fault. Sally and I discussed this, but I forgot to make the change to the specifications. If we have the system fill the date field with the current date but allow the order processing staff to override that value we can solve that problem and save them a lot of typing at the same time.

SALLY

I have another question, Sandra. Do we have to enter all those fields in order to begin processing an order?

SANDRA

The order number is automatically generated by the system. Also, when you enter the member number, the member's name will appear for you to verify. In addition to the order date we just talked about, the selection of the month accepted field is represented by a check box that will have a check mark to indicate a default answer of yes. And to save you some more typing, the product number will be automatically inserted.

CLERK A

But 70 percent of our members reject the selection of the month. That means we will have to change those fields most of the time.

SANDRA

That's interesting. We will change the default value to no and leave the product number blank. Are there any more questions or concerns?

BOB

If not, what we need to do now is to verify the size of each field and get an idea of the range of values each field can assume. This will help us write programs to check the accuracy of the data before they're processed.

CLERK B

Do I have to use a computer? They really make me nervous. I am afraid I'm going to blow something up!

SANDRA

Don't worry. I remember my first experience with a computer. I thought if I did something wrong, I'd break the machine or something, but that can't happen. First, we are going to develop an overall interface that will be very intuitive. As much as possible, we will try to make you forget that you are talking with a machine. Also, you will be asked to give us feedback as we attempt to develop such screens. What would you think if there wasn't any reference guide and everything you needed to know was right on the screen?'

CLERK B

What happens if I enter something several times and the computer won't take it, but I don't understand what I'm doing wrong?

BOB

That's a good question. As you can see with this screen example, we will try to provide you with the best mechanism for ensuring that correct data are selected and entered. For example, notice the field with the label "Closed Caption?" That field contains two possible values, yes or no. I provided a check box where all you do is click to toggle its value from yes to no. If a box contains a check mark, that means its value is yes. Over here you see what we call radio buttons. Each button corresponds with a possible value of this field. And here, we used a box with a list of possible values. So in either case, you could not possibly enter any other value than one that is provided. I should also point out that anytime you need more information about a particular field, you will be able to get immediate help.

SOUNDSTAGE

SOUNDSTAGE ENTERTAINMENT CLUB

CLERK B
Well, if all the screens work like that I think there's hope for me.

DISCUSSION QUESTIONS

1. Why did Sandra and Bob want to focus on the content of the input screen and its individual fields? What would be the advantages of this emphasis?
2. Do users of applications that have a graphical interface such as those provided by *Windows*-based products really need help features? Explain.
3. Why would Sandra and Bob be concerned with having the computer provide values for fields where possible, rather than the users.
4. What benefits are derived by Sandra and Bob involving the users during input design?

METHODS AND ISSUES FOR DATA CAPTURE AND INPUT

"Garbage in! Garbage out!" This overworked expression is no less true today than it was when we first studied computer programming. Management and users make important decisions based on system outputs. These outputs are produced from data that are either input or retrieved from databases. And data in databases must have been input first. In this chapter, you are going to learn how to design computer inputs. Input design serves an important goal—capture and get the data into a format suitable for the computer. And data constitute one of the fundamental building blocks for information systems.

One of the first things you must learn is the difference between data capture and data input. Alternative input media and methods must also be understood before designing the inputs. And because accurate data input is so critical to successful processing, file maintenance, and output, you should also learn about human factors and internal controls for input design. After learning these fundamental concepts, we will study the tools and techniques of input design and prototyping.

Data Capture, Data Entry, and Data Input

When you think of "input," you usually think of input devices, such as keyboards and mice. But input begins long before the data arrive at the device. To actually input business data into a computer, the analyst may have to design source documents, input screens, and methods and procedures for getting the data into the computer (from *customer* to *form* to *data entry clerk* to *disk* to *computer*).

This brings us to our fundamental question. What is the difference among data capture, data entry, and data input? *Data happens!* It accompanies business events called **transactions.** Examples include orders, time cards, reservations, and the like. We must determine *when* and *how* to capture the data.

Data capture is the identification of new data to be input.

When is easy! It's always best to capture the data as soon as possible after it is originated. *How* is another story! Traditionally, special paper forms called source documents were used.

A **source document** is a paper form used to record data that will eventually be input to a computer.

With advances in video display technology, screen display forms can duplicate the appearance of almost any paper-based form. Most applications' data capture involves the use of source documents and screen display forms. Their design is not easy. Screen display forms and source documents must be designed to be easy for the system user to complete and should facilitate rapid data entry.

Data entry is not the same as data capture.

Data entry is the process of translating the source document into a machine-

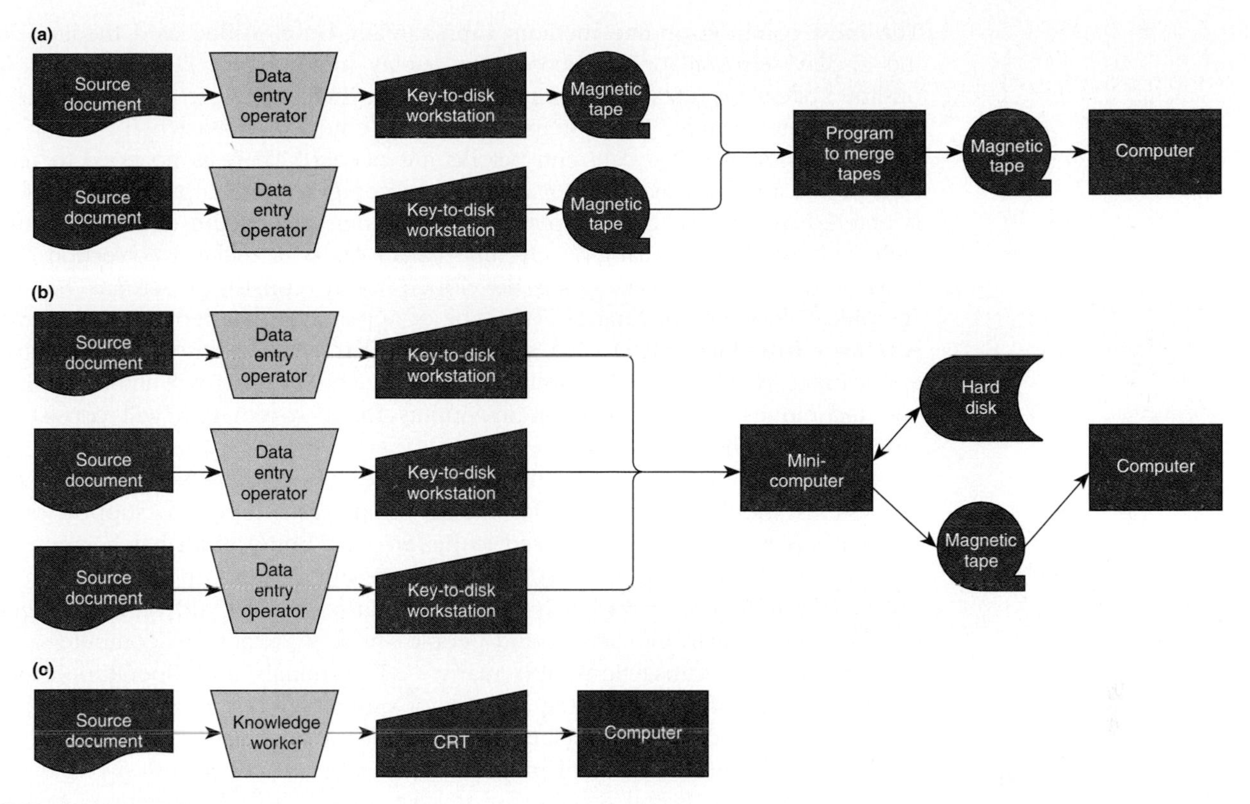

FIGURE 12.1 *Input Methods and Media*

readable format. That format may be a magnetic disk, an optical-mark form, a magnetic tape, or a floppy diskette, to name a few.

Once data entry has been performed, we are ready for data input.

Data input is the actual entry of data in a machine-readable format into the computer.

Let's examine some data capture and data entry issues you should consider during systems design.

Modern Input Methods: Batch versus On-Line Inputs

The systems analyst usually selects the method and medium for all inputs. Input methods can be broadly classified as either batch or on-line.

Batch input is the oldest and most traditional input method. Source documents or forms are collected and then periodically forwarded to data entry operators, who key the data using a data entry device that translates the data into a machine-readable format.

Traditional media for batch input data included **key-to-disk (KTD)** and **key-to-tape (KTT)** workstations that transcribe data to magnetic disks and magnetic tape, respectively. The data can be corrected, because they are initially placed into a buffer.

Figures 12.1(a) and 12.1(b) illustrate the key-to-tape and key-to-disk input procedures, respectively. We have distinguished the data capture activities, the data entry activities, and the data input activities discussed in the previous section.

Today, most, but not all, systems have been converted or are being converted to on-line methods.

On-line input is the capture of data at its point of origin in the business and the direct inputting of that data to the computer, preferably as soon as possible after the data originates.

The most common on-line medium cannot really be classified as a medium; it is the display terminal, or microcomputer display monitor [see Figure 12.1(c)]. The on-line system includes a monitor screen and keyboard that are directly connected to a computer system. The system user directly enters the data when or soon after that data originates. No data entry clerks are needed! There is no need to record data onto a medium that is later input to the computer; this input is direct! If data is entered incorrectly, the computer's edit program detects the error and immediately requests that the cathode ray tube (CRT) operator make a correction.

Most new applications being developed today consist of screens having a "graphical" looking appearance. This type of appearance is referred to as a **graphical user interface (GUI).** You are likely familiar with Microsoft *Windows*-based applications, which have a graphical interface. This chapter will introduce issues and techniques for designing on-line inputs for a system that will consist of a graphical user interface.

Now that you understand batch versus on-line, let's address the issue of whether all systems should be designed for on-line input? Technology to support on-line applications is cheaper than it used to be. So why bother with batch input?

No matter how cheap and fast on-line processing gets, an on-line program cannot be nearly as fast as its batch equivalent. Many (but not all) on-line programs require some human interaction, and people are slow, relative to computers. Also, for large-volume transactions, too many CRT terminals and operators may be needed to meet demand. As the number of on-line CRTs grows, the overall performance of the computer declines. Furthermore, many inputs naturally occur in batches. For instance, our mail may include a large batch of customer payments on any given day. Postal delivery is, at least today, a batch operation. Additionally, some input data may not require immediate attention. Finally, batch processing may be preferable because internal controls (discussed shortly) are simpler. So you see, batch inputs can still be justified.

But there is a compromise solution, the remote batch.

> **Remote batch** offers on-line advantages for data that is best processed in batches. The data is input on-line with on-line editing. Microcomputers or minicomputer systems can be used to handle this on-line input and editing. The data is not immediately processed. Instead, it is batched, usually to some type of magnetic media. At an appropriate time, the data is uploaded to the main computer, merged, and subsequently processed as a batch. Remote batch is also called *deferred batch* or *deferred processing.*

Trends in Automatic Data Collection Technology

With the advancement in today's technology, data input has become more sophisticated. We can eliminate much (and sometimes all) human intervention associated with the input methods discussed in the previous section. By eliminating human intervention we can decrease the time delay and errors associated with human interaction. This opportunity is especially important to businesses operating in today's globally competitive environment!

A number of alternative **automatic data collection (ADC)** technologies are available today and finding their way into batch and on-line applications. Some of these are presented in the sections that follow (Dunlap, 1995).

Biometric **Biometric** ADC technology is based on unique human characteristics or traits. For example, individuals can be identified by their own unique fingerprint, voice pattern, or pattern of certain veins (retina or wrist). Biometric ADC systems consist of sensors that capture an individual's characteristic or trait, digitize the image pattern, and then compare the image to stored patterns for identification. Biometric ADC is popular because it offers the most accurate and reliable means for identification. This technology is particularly popular for systems that require security access.

Electromagnetic **Electromagnetic** ADC technology is based on the use of radio frequency to identify physical objects. This technology involves attaching a tag and antenna to the physical object that is to be tracked. The tag contains memory that is used to identify the object being tracked. The tag can be read by a reader whenever the object resides within the electromagnetic field generated by the reader. This identification technology is becoming very popular in applications that involve tracking physical objects that are out of sight and on the move. For example, electromagnetic ADC is being used for public transportation tracking and control, tracking manufactured products, and tracking animals, to name a few.

Magnetic **Magnetic** ADC technology is one you will likely recognize. It usually involves using magnetic stripe cards, but it also may include the use of magnetic ink character recognition (MICR). Over 1 billion magnetic stripe cards are in use today! They have found their way into a number of business applications, such as credit card transactions, building security access control, and employee attendance tracking. MICR is most widely used in the banking industry.

Optical You have likely encountered an example of **optical** technology most every day, **bar coding.** Point-of-sale terminals in retail and grocery stores frequently include bar code and optical-character readers. Everyone has seen the bar codes recorded on today's grocery products. These **bar codes** eliminate the need for keying data, either by data entry clerks or end-users. Instead, sophisticated laser readers read the bar code and send the data represented by that code directly to the computer for processing. Frequently items are encountered in which a bar code can't physically be attached. This is typically overcome by providing the data entry clerk with a poster sheet containing a picture and accompanying bar code of those items. The clerk simply scans the bar code of the appropriate picture appearing on the sheet.

Another optical ADC alternative is the optical-mark form. You may have encountered this medium in machine-scored tests. Optical-mark forms eliminate most or all of the need for data entry. Essentially, the source document becomes the input medium and is directly read by an **optical-mark reader (OMR)** or **optical-character reader (OCR).** The computer records the data to magnetic tape, which is then input to the computer. OCR and OMR input are generally suitable only for high-volume input activities. By having data directly recorded on a machine-readable document, the cost of data entry is eliminated. This technology is commonly used for applications involving surveys, questionnaires, or testing.

Smart Cards Smart card technology has the ability to store a massive amount of information. **Smart cards** are similar to, albeit slightly thicker than, credit cards. They also differ in that they contain a microprocessor, memory circuits, and a battery. Think of it as a credit card with a computer on board. They represent a portable storage medium from which input data can be obtained. While this technology is only beginning to make inroads in the United States, smart cards are used on a daily basis by over 60 percent of the French population. Smart card applications are particularly promising in the area of health records where a person's blood type, vaccinations, and other past medical history can be made readily available. Other uses may include such applications as passports, financial information for point-of-sale transactions, and pay television, to name a few.

Touch Touch-based ADC systems include **touch** screens, buttons, and pen-based computing technology. In particular, touch screen technology has been very popular in restaurant or point-of-sale business applications. Recently, manufacturing companies have begun to use touch screens throughout the manufacturing shop floor as a means to capture data pertaining to such things as work orders, machine setup, material requisitions, employee attendance, and scheduling. Pen-based computing is popular for applications that require handwriting recognition. You may have experienced this technology for capturing data when you were asked to sign for a special delivery package.

Technology will continue to evolve. It is the systems analyst's responsibility to be aware of trends in new technology to enhance data capture and input. With an ever-increasing emphasis on helping companies gain a competitive advantage, those analysts that continue to grow professionally by keeping abreast of technological advances in the area of data capture and input will certainly enhance their careers.

System User Issues for Input Design

Because inputs originate with system users, human factors play a significant role in input design. Inputs should be as simple as possible and designed to reduce the possibility of incorrect data being entered. Furthermore, the needs of data entry clerks must also be considered. With this in mind, several human factors should be evaluated.

The volume of data to be input should be minimized. The more data that are input, the greater the potential number of input errors and the longer it takes to input that data. Thus, numerous considerations should be given to the data that are captured for input. These general principles should be followed for input design:

- *Capture only variable data.* Do not enter constant data. For instance, when deciding what elements to include in a SALES ORDER input, we need PART NUMBERS for all parts ordered. However, do we need to input PART DESCRIPTIONS for those parts? PART DESCRIPTION is probably stored in a database table. If we input PART NUMBER, we can look up PART DESCRIPTION. Permanent (or semipermanent) data should be stored in the database. Of course, inputs must be designed for maintaining those database tables.
- *Do not capture data that can be calculated or stored in computer programs.* For example, if you input QUANTITY ORDERED and PRICE, you don't need to input EXTENDED PRICE, which is equal to QUANTITY ORDERED × PRICE. Another example is incorporating FEDERAL TAX WITHHOLDING data in tables (arrays) instead of keying in that data every time.
- *Use codes for appropriate attributes.* Codes were introduced earlier. Codes can be translated in computer programs by using tables.

Second, if source documents are used to capture data they should be easy for system users to complete and subsequently enter into the system. The following suggestions may help:

- *Include instructions for completing the form.* Also, remember that people don't like to have to read instructions printed on the back side of a form.
- *Minimize the amount of handwriting.* Many people suffer from poor penmanship. The data entry clerk or CRT operator may misread the data and input incorrect data. Use check boxes wherever possible so the system user only needs to check the appropriate values.
- *Data to be entered (keyed) should be sequenced so it can be read like this book, top to bottom and left to right* [see Figure 12.2(a)]. The data entry clerk should not have to move from right to left on a line or jump around on the form [see Figure 12.2(b)] to find data items to be entered.
- *Ideally, portions of the form that are not to be input are placed in or about the lower right portion of the source document* (the last portion encountered when reading top to bottom and left to right). Alternatively, this information can be placed on the back of the form.

(a)

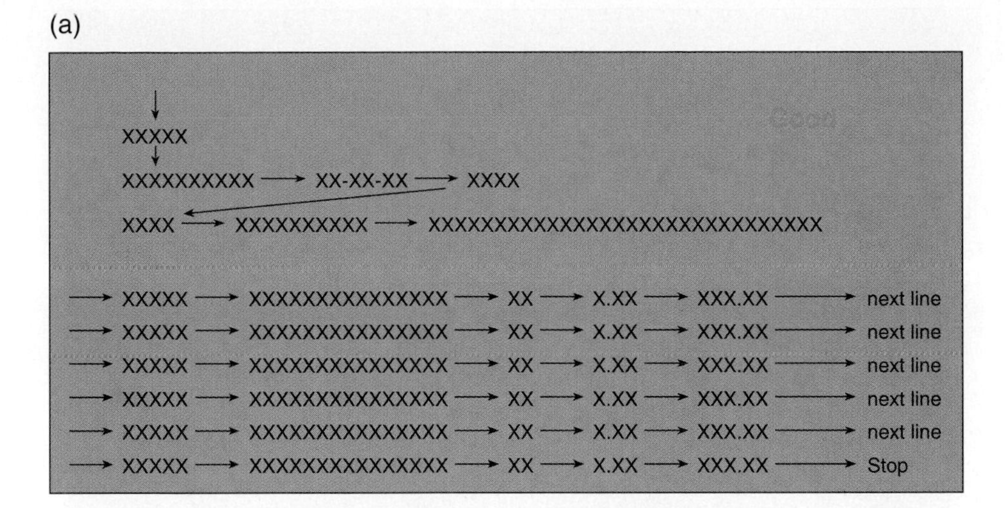

(b)

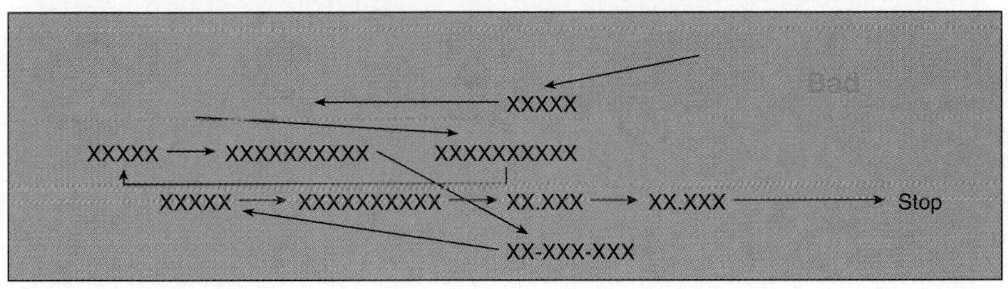

FIGURE 12.2 *Keying from Source Documents*

There are several other guidelines and issues specific to data input for GUI screen designs. We'll introduce these guidelines as appropriate when we discuss GUI controls for input design later in this chapter, as well as in the subsequent chapters on output design and user interface design.

Internal Controls for Inputs

Internal controls, a continuing theme throughout the design chapters of this book, are a requirement in all computer-based systems. Input controls ensure that the data input to the computer is accurate and that the system is protected against accidental and intentional errors and abuse, including fraud. The following internal control guidelines are offered:

1. *The number of inputs should be monitored.* This is especially true with the batch method, because source documents may be misplaced, lost, or skipped.
   - In batch systems, data about each batch should be recorded on a batch control slip. Data include BATCH NUMBER, NUMBER OF DOCUMENTS, and CONTROL TOTALS (e.g., total number of line items on the documents). These totals can be compared with the output totals on a report after processing has been completed. If the totals are not equal, the cause of the discrepancy must be determined.
   - In batch systems, an alternative control would be one-for-one checks. Each source document would be matched against the corresponding historical report detail line that confirms the document has been processed. This control check may be necessary only when the batch control totals don't match.
   - In on-line systems, each input transaction should be logged to a separate audit file so it can be recovered and reprocessed in the event of a processing error or if data is lost.

MODULUS 11

The following procedure is used to assign a check digit to a key field:

STEP 1: Determine the size of the key field in digits.
2 4 1 3 5 = 5 digits

STEP 2: Number each digit location from *right* or *left* beginning with the number "2."
2 4 1 3 5
6 5 4 3 2

STEP 3: Multiply each digit in the key field by its assigned location number.
2 × 6 = 12
4 × 5 = 20
1 × 4 = 4
3 × 3 = 9
5 × 2 = 10

STEP 4: Sum the products from step 3.
12 + 20 + 4 + 9 + 10 = 55

STEP 5: Divide the sum from step 4 by 11.
55/11 = 5 Remainder 0

STEP 6: If the remainder is less than 10, append the remainder digit to the key field. If the remainder is equal to 10, append the character "X" to the key field.
2 4 1 3 5 0

FIGURE 12.3 *Modulus 11 Self-Checking-Digit Technique*

2. *Care must also be taken to ensure that the data is valid.* Two types of errors can infiltrate the data: data entry errors and invalid data recorded by system users. Data entry errors include copying errors, transpositions (typing 132 as 123), and slides (keying 345.36 as 3453.6). The following techniques are widely used to validate data:

   - **Completeness checks** determine whether all required fields on the input have actually been entered.
   - **Limit and range checks** determine whether the input data for each field falls within the legitimate set or range of values defined for that field. For instance, an upper-limit range may be put on PAY RATE to ensure that no employee is paid at a higher rate.
   - **Combination checks** determine whether a known relationship between two fields is valid. For instance, if the VEHICLE MAKE is Pontiac, then the VEHICLE MODEL must be one of a limited set of values that comprises cars manufactured by Pontiac (Firebird, Grand Prix, and Bonneville to name a few).
   - **Self-checking digits** determine data entry errors on primary keys. A check digit is a number or character that is appended to a primary key field. The check digit is calculated by applying a formula, such as Modulus 11, to the actual key (see Figure 12.3). The check digit verifies correct data entry in one of two ways. Some data entry devices can automatically validate data by applying the same formula to the data as it is entered by the data entry clerk. If the check digit entered doesn't match the check digit calculated, an error is displayed. Alternatively, computer programs can also validate check digits by using readily available subroutines.
   - **Picture checks** compare data entered against the known COBOL picture or other language format defined for that data. For instance, the input field may have a picture clause XX999 AA (where X can be a letter or

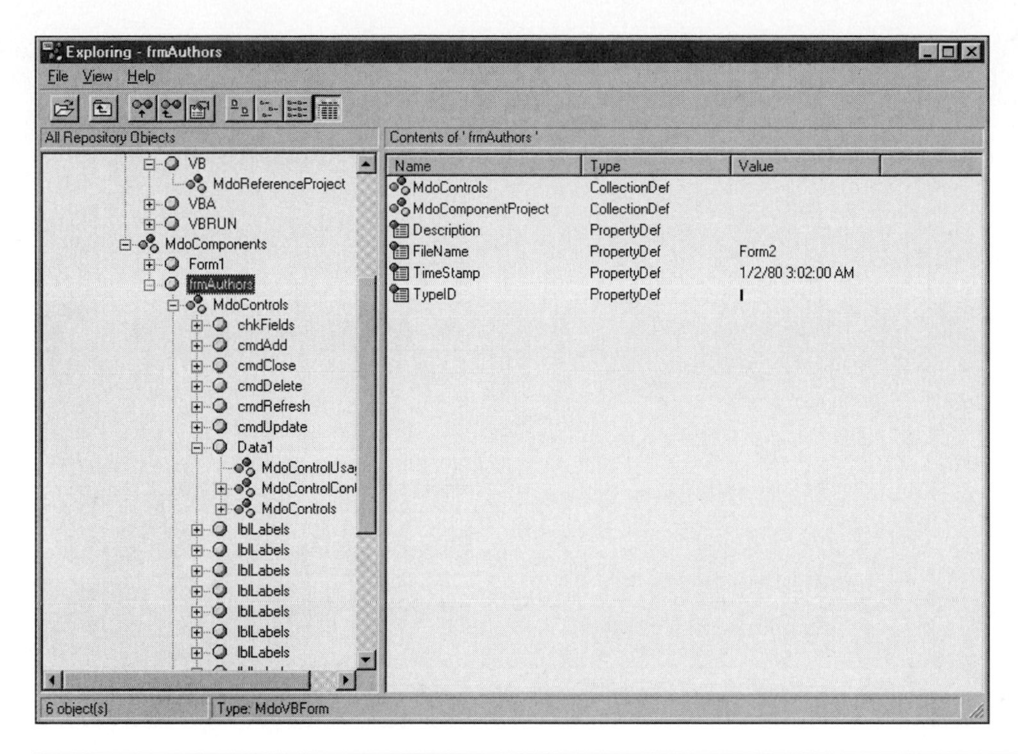

FIGURE 12.4 *Sample Entry Using Repository-Based Programming*

number, 9 must be a number, and A must be a letter). The field "A4898 DH" would pass the picture check, but the field "A489 ID8" would not.

Data validation requires that special edit programs be written to perform checks. However, the input validation requirements should be designed when the inputs themselves are designed.

GUI CONTROLS FOR INPUT DESIGN

As mentioned earlier, most new applications being developed today include a GUI. These types of interfaces are rapidly replacing the more traditional text-based screen designs that characterized mainframe-based applications. While GUI designs provide a more user-friendly interface, they also present many more design issues that must be considered. This chapter will not attempt to address all the GUI design issues; entire books have been written on the subject. Rather, this chapter will focus on selecting the proper screen-based controls for entering data on a GUI screen. This approach is influenced by a new trend in programming, called **repository-based programming.**

Repository-based programming is demonstrated in Figure 12.4. This figure depicts information entered by a *Visual Basic* developer in a repository for the physical data attribute frmAuthors. The developer can in a single location define most characteristics for a particular data element. Once the developer defines this information, it can be used by multiple developers in an organization. This repository-based approach guarantees that every instance of the attribute frmAuthors will be used in a consistent manner. Furthermore, the dictionary entries can be changed if business rules dictate and no additional changes to the applications will be required.

This section takes a similar approach to GUI input screen design. We will first learn about available screen-based controls for inputting data. We address the purpose, advantages, disadvantages, and guidelines for each control. Given this understanding, we are then in a good position to make decisions concerning which controls should be considered for each data attribute that will be input on our

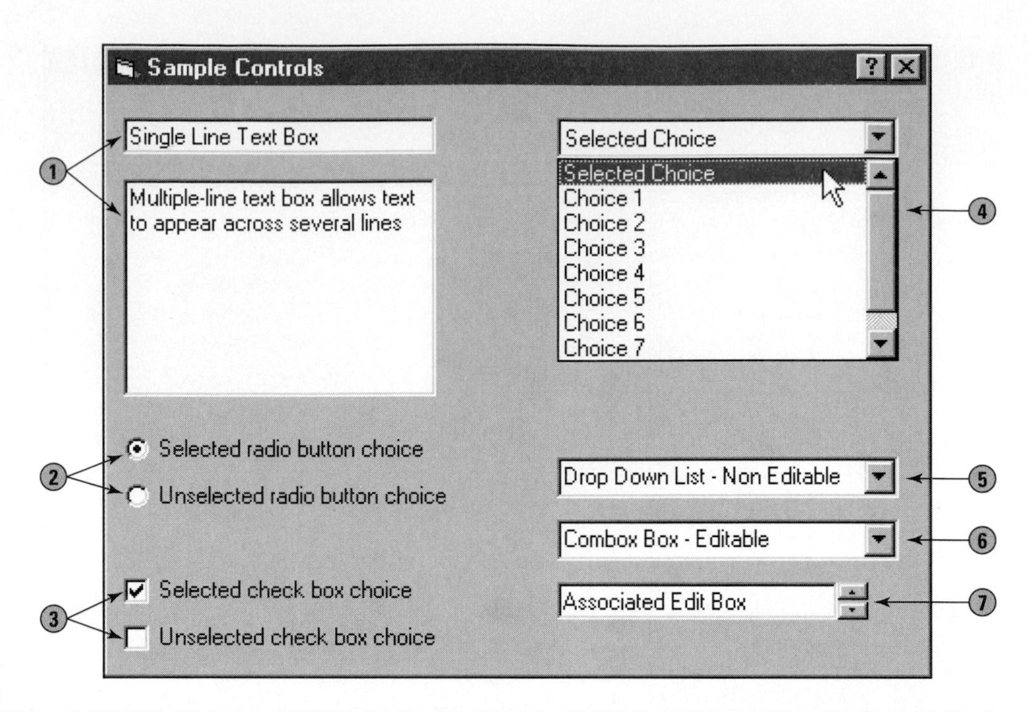

FIGURE 12.5 *Common Screen-Based Controls for Input Data*

screens. We will defer the overall look and feel of the screen designs for our application to Chapter 14, "User Interface Design and Prototyping."

Refer to Figure 12.5 as a study of screen-based controls for input data. Figure 12.5 contains each of the controls to be discussed.

Text Box ①

Perhaps the most common control used for input of data is the **text box.** A text box consists of a rectangular shaped box that is usually accompanied by a caption. This control requires the user to type the data inside the box. A text box can allow for single or multiple lines of data characters to be entered. When a text box contains multiple lines of data, scrolling features are also normally included.

When to Use Text Boxes for Input A text box is most appropriately used in those situations where the input data values are unlimited in scope and the analyst is unable to provide the users with a meaningful list of values from which they can select. For example, a single-line text box would be an appropriate control for capturing a new customer's LAST NAME—since the possibilities for the customer's LAST NAME are virtually impossible to predetermine. A text box would also be appropriate for capturing data about SHIPPING INSTRUCTIONS that describe a particular order that was placed by a customer. Once again, the possible values for SHIPPING INSTRUCTIONS are virtually unlimited. In addition, the multiple-line text box would be appropriate due to the unpredictable length of the SHIPPING INSTRUCTIONS. In those cases where the text box is not large enough to view the entire input data values, the text box may use scrolling and word-wrap features.

Suggested Guidelines for Using Text Boxes Numerous guidelines should be followed when using a text box on an input screen. Let's first address the captions for text boxes. A text box should be accompanied by a descriptive caption. To avoid possible confusion, the user should be provided with a meaningful caption. Avoid using abbreviations for captions. Finally, only the first character of the caption's text should be capitalized.

The location of the caption is also significant. The user should be able to clearly associate the caption with the text box. Therefore, the caption should be located to the left of the actual text box or left-aligned immediately above the text box.

Finally, it is also generally accepted that the caption be followed by a colon to help the user visually distinguish the caption from the box.

There are also several guidelines relating to the text box. Generally, the size of the text box should be large enough for all characters of fixed-length input data to be entered and viewed by the user. When the length of the data to be input is variable and could become quite long, the text box's scrolling and word-wrapping features should be applied.

Radio Button ②

Radio buttons provide the user with an easy way to quickly identify and select a particular value from a value set. A radio button consists of a small circle and an associated textual description that corresponds to the value choice. The circle is located to the left of the textual description of the value choice. Radio buttons normally appear in groups—a radio button per value choice. When a user selects the appropriate choice from the value set, the circle corresponding to that choice is partially filled to indicate it has been selected. When a choice is selected, any default or previously selected choice's circle is deselected. Radio buttons also offer the advantage of allowing the user the flexibility of selecting via the keyboard or mouse.

When to Use Radio Buttons for Input Radio buttons are most appropriately used in those cases where a user may be expected to input data that have a limited predefined set of mutually exclusive values. For example, a user may be asked to input an ORDER TYPE and GENDER. Each of these has a limited, predefined, mutually exclusive set of valid values. For example, when the users are to input an ORDER TYPE, they might be expected to indicate one and only one value from the value set "regular order," "rush order," or "standing order." For GENDER, the user would be expected to indicate one and only one value from the set "female," "male," or "unknown."

Suggested Guidelines for Using Radio Buttons There are several guidelines to consider when using radio buttons as a means for data input. First, radio buttons should present the alternatives vertically aligned and left-justified to aid the user in browsing. If necessary the choices can be presented where they are aligned horizontally, but adequate spacing should be used to help visually distinguish the choices. Also, the group of choices should be visually grouped to set them off from other input controls appearing on the screen. The grouping should also contain an appropriate meaningful caption. For example, radio buttons for male, female, and unknown might be vertically aligned and left-justified with the heading/caption "Gender" left-justified above the set.

The sequencing of the choices should also be given consideration. The larger the number of choices the more thought should be given to the ease of the scanning and identifying the choices. For example, in some cases it may be more natural for the user to locate choices that are presented in alphabetical order. In other cases, the frequency in which a value is selected may be important in regards to where it is located in the set of choices.

Finally, it is not recommended that radio buttons be used to select the value for an input data whose value is simply a Yes/No (or On/Off state). Instead, a check box control should be considered.

Check Box ③

As with text boxes and radio buttons, a **check box** also consists of two parts. It consists of a square box followed by a textual description of the input field for which the user is to provide the Yes/No value. Check boxes provide the user the flexibility of selecting the value via the keyboard or mouse. An input data field whose value is yes is represented by a square that is filled with a "✓." The absence of a "✓" means the input field's value is no. The user simply toggles the input field's value from one value/state to the other as desired.

When to Use Check Boxes for Input Often a user needs to input a data field whose value set consists of a simple yes or no value. For example, a user may be asked for a Yes/No value for such items as the following input data: CREDIT APPROVED? SENIOR CITIZEN? HAVE YOU EVER BEEN CONVICTED OF FRAUD? and MAY WE CONTACT YOUR PREVIOUS EMPLOYER? In each situation a check box control could be used. A check box control offers a visual and intuitive means for the user to input such data.

The previous example represented a simplified scenario for the use of a stand-alone check box. Often on a single input screen it may be desirable to ask a user to enter values for a number of related input fields having a Yes/No value. For example, a receptionist at a health clinic may be entering data from a completed patient form. On a section of that form, the patient may have been asked about a number of illnesses. They may have been asked their past medical history and instructed to "check all that apply" from a list of types of various illnesses. If properly designed, the receptionist's input screen would represent each illness as a separate input field using a check box control. The controls would be physically associated into a group on the screen. The group would also be given an appropriate heading/caption. Recognize that even though the check boxes may be visually grouped on the screen, each check box operates as a separate independent input field.

Suggested Guidelines for Using Check Boxes Here are some recommended guidelines for using check box controls. Once again, make sure the textual description is meaningful to the user. Look for opportunities to group check boxes for related Yes/No input fields and provide a descriptive group heading.

To aid in the user's browsing and selecting from a group of check boxes, arrange the group of check box controls where they are aligned vertically and left-justified. If necessary, align horizontally and be sure to leave adequate space to visually separate the controls from one another. Finally, provide further assistance to the user by appropriately sequencing the input fields according to their textual description. In most cases, where the number of check box controls is large, the sequencing should be alphabetical. In those cases where the text description describes dollar ranges or some other measurement, the sequencing may be according to the numerical order. Still, in other cases such as those where a very limited number of controls are grouped, the basis for sequencing may be according to the frequency that a given input data field's Yes/No value is selected. (All input data fields represented using a check box have a default value.)

List Box ④

A **list box** is a control that requires the user to select a data item's value from a list of possible choices. The list box is rectangular and contains one or more rows of possible data values. The values may appear as either a textual description or graphical representation. List boxes having a large number of possible values may consist of scroll bars to navigate through the row of choices.

It is also common for a list box's row to contain more than one column. For example, a list box could simply contain rows having a single column of permissible values for an input data item called JOB CODE. However, it may be asking too much to expect the user to recognize what each job code actually represented. In this case, to place the values of JOB CODE into a meaningful perspective, the list box could include a second column containing the corresponding JOB TITLE for each job code.

When to Use List Boxes for Input How does one choose between a radio button and a list box control? Both controls are useful in ensuring that the user enters the correct value for a data item. Both are also appropriate when it is desirable to have the value choices constantly visible to the user.

The decision is normally driven by the number of possible values for the data item and the amount of screen space that is available for the control. Scrolling capabilities make list boxes appropriate for use in those cases where there is limited screen space available and the input data item has a *large* number of predefined, mutually exclusive set of values from which to choose.

Suggested Guidelines for Using List Boxes There are several guidelines to consider when using a list box as a means for data input. A list box should be accompanied by a descriptive caption. Avoid using abbreviations for captions and capitalize only the first character of the caption's text. Finally, it is also generally accepted that the caption be followed by a colon to help the user visually distinguish the caption from the box.

The location of the caption is also significant. The user should be able to clearly associate the caption with the list box. Therefore, the caption should appear left-justified immediately above the actual list box.

There are also several guidelines relating to the list box. First, it is recommended that a list box contain a highlighted default value. Second, consider the size of the list box. Generally, the width of the list box should be large enough for most characters of fixed-length input data to be entered and viewed by the user. The length of the box should allow for at least three choices and be limited in size to containing about seven choices. In both cases scrolling features should be used to suggest additional choices are available to the user.

If graphical representations are used for value choices, make sure the graphics are meaningful and truly representative of the choice. If textual descriptions are used, use mixed-case letters and ensure that the descriptions are meaningful. It is important that these decisions or judgments be based on the perspective and opinions of the user!

You should also give careful thought to the ease with which a user can scan and identify the choices appearing in the list box. The list of choices should be left-justified to aid in browsing. Be sure to involve the user when addressing the order in which choices will appear in the list. In some cases it may be natural to the user if the list of choices appeared in alphabetical order. In other cases, the frequency in which a value is selected may be important in regards to where it is located in the list.

Drop-Down List ⑤

A **drop-down list** is another control that requires the user to select a data item's value from a list of possible choices. A drop-down list consists of a rectangular selection field with a small button connected to its side. The small button contains the image of a downward pointing arrow and bar. This button is intended to suggest to the user the existence of a hidden list of possible values for a data item.

When requested, the hidden list appears to "drop or pull down" beneath the selection field to reveal itself to the user. The revealed list has characteristics similar to the list box control mentioned in the previous section. When the user selects a value from the list of choices, the selected value is displayed in the selection field and the list of choices once again becomes hidden from the user.

When to Use Drop-Down Lists for Input A drop-down list should be used in those cases where the data item has a large number of predefined values and screen space availability prohibits the use of a list box to provide the user with a list box. One disadvantage of a drop-down list is that it requires extra steps by the user, in comparison to the previously mentioned controls.

Suggested Guidelines for Drop-Down Lists Many of the guidelines for using list boxes directly apply to drop-down lists. One exception is the placement of the caption. The caption for a drop-down list is generally either left-aligned immediately above the selection field portion of the control or located to the left of the control.

Combination (Combo) Box ⑥

A **combination box,** often simply called a combo box, is a control whose name reflects the fact that it combines the capabilities of a text box and list box. A combo box gives the user the flexibility of entering a data item's value (as with a text box) or selecting its value from a list (as with a list box).

At first glance, a combo box closely resembles a drop-down list control. Unlike the drop-down list control, however, the rectangular box can serve as an entry field for the user to directly enter a data item's value. Once the small button is selected, a hidden list is revealed. The revealed list appears slightly indented beneath the rectangular entry field.

When the user selects a value from the list of choices, the selected value is displayed in the entry field and the list of choices once again becomes hidden from the user.

When to Use Combo Boxes for Input A combo box is most appropriately used in those cases where limited screen space is available and it is desirable to provide the user with the option of selecting a value from a list or typing a value that may or may not appear as an option in the list.

Suggested Guidelines for Combo Boxes The same guidelines for using drop-down lists directly apply to combo boxes.

Spin (Spinner) Box ⑦

A **spin box** is a screen-based control that consists of a single-line text box followed immediately by two small buttons. The two buttons are vertically aligned. The top button has an arrow pointing upward and the bottom button has an arrow pointing down. This control allows the user to enter data directly into the associated text box or to select a value by using the mouse to scroll (or "spin") through a list of values using the buttons. The buttons have a unit of measure associated with them. When the user clicks on one of the arrow buttons, a value will appear in the text box. The value in the text box is manipulated by clicking on the arrow buttons. The upward pointing button will increase the value in the text box by a unit of measure, whereas the downward pointing button will decrease the value in the text box by the same unit of measure.

When to Use Spin Boxes for Input A spin box is most appropriately used to allow the user to make an input selection by using the buttons to navigate through a small set of meaningful choices or by directly keying the data value into the textbox. The data values for a spin box should be capable of being sequenced in a predictable manner.

Suggested Guidelines for Spin Boxes Spin boxes should contain a label or caption that clearly identifies the input data item. This label should be located to the left of the text box or left-aligned immediately above the text box portion of the control. Finally, spin boxes should always contain a default value in the text box portion of the control.

That completes our discussion of screen-based controls for designing GUI input screens. Many more controls are available for designing graphical user interfaces. The above are the most common controls for capturing input data. There are others, and you should make yourself familiar with them and their proper usage for inputting data. In later chapters you will be exposed to several other controls used for other purposes. Keep on top of developments in the area of GUI as new controls are sure to be made available.

HOW TO PROTOTYPE AND DESIGN COMPUTER INPUTS

How do you design on-line? Traditionally, the designer was concerned with the overall content, appearance, and functionality of the input screen—in relative isolation of other screens that needed to be designed. The designers knew they would simply design a subsequent set of menu screens from which the users would select an option that would lead them to the appropriate input screen. Simple enough. However, given today's graphical environments, there is an emphasis on developing an overall system that blends well into the user's overall workplace environment. This emphasis rarely results in a hierarchical, menu-driven application interface that characterized the more traditional text- or command-based applications of old.

Developing graphical user interfaces for new applications involves two stages for input design. In the first stage, the designer focuses on correctly identifying the confirming content of the input and, consistent with the repository-based programming emphasis discussed earlier, identifying properties or characteristics for that input data.

The second stage deals with the overall appearance or look and feel of the input. This stage is typically deferred until the designer has given consideration to the *overall* appearance and working of the entire application's interface.

The following section will demonstrate how the first stage of input design is completed. We will draw on examples from our SoundStage case study. Thus, focus your attention on the content of the sample screens presented. In Chapter 14, "User Interface Design and Prototyping," you will learn about stage two of input design and how decisions were made that influenced the overall look and feel and functionality of the sample input screens provided.

Step 1: Review Input Requirements

Input requirements may have been defined during systems analysis. Thus, a good starting point for input design is the design unit data flow diagrams (DFDs) for the new system. The design unit DFDs depict inputs to be designed. These inputs are represented as data flows that connect external entities to processes.

Given an input to be designed, we should review the required attributes. The basic content of these inputs should have been recorded in the project repository during systems analysis. If the content has not been recorded, we can define input requirements by studying the output and database requirements or designs. An output attribute that can't be retrieved from database tables or calculated from attributes that are retrieved from tables must be input! Additionally, inputs must be designed to maintain the database tables in the system.

There are a number of design considerations for attributes to be included in the input. For each attribute, the designer must identify an appropriate caption or label they will use to clearly identify to the user the attribute appearing on the screen. In addition, the size and edit mask (or format) of the attribute must be predetermined.

This activity is commonly supported by most database, CASE, and programming products. *Windows*-based versions of such products typically provide a graphical facility for viewing an existing data model or database. From that view, the designer can simply select (point and click) appropriate tables and attributes for which the input screen will capture data. In many cases such products will provide default labels, sizes, and edit masks for entries in the database/data model. Thus, the designer will simply modify those entries as appropriate.

After reviewing input requirements specified during systems analysis for our SoundStage case study, it was determined that there were three inputs that pertained to the subject VIDEO TAPE. It was determined that a single input screen could be used to support the three inputs NEW VIDEO TITLE, DISCONTINUED VIDEO TITLE, and VIDEO TITLE UPDATE. The data content for the three inputs should capture or display the following data:

PRODUCT NUMBER
UNIVERSAL PRODUCT CODE
QUANTITY IN STOCK
PRODUCT TYPE
MANUFACTURER'S SUGGESTED RETAIL UNIT PRICE
CLUB DEFAULT UNIT PRICE
CURRENT SPECIAL UNIT PRICE

CURRENT MONTH UNITS SOLD
CURRENT YEAR UNITS SOLD
TOTAL LIFETIME UNITS SOLD
TITLE OF WORK
CATALOG DESCRIPTION
COPYRIGHT DATE
CREDIT VALUE
PRODUCER
DIRECTOR
VIDEO CATEGORY
VIDEO SUBCATEGORY
CLOSED CAPTION?
LANGUAGE
RUNNING TIME
VIDEO MEDIA TYPE
VIDEO ENCODING
SCREEN ASPECT
MOTION PICTURE ASSOCIATION RATING CODE

The attributes PRODUCT NUMBER, MONTHLY UNIT SALES, YEAR UNIT SALES, and TOTAL UNIT SALES are not to be entered by the user. Rather, these attributes are to be automatically generated by the system. Also, for the TITLE COVER, the user will be expected to simply specify a bitmap file that will contain an actual image of the new video title.

Step 2: Select the GUI Controls

Now that we have an idea of the content for our input, we can address the proper screen-based control to use for each attribute to appear on our screen. Using the repository-based programming approach, we would first check to see if such decisions and other attribute characteristics have already been made and recorded as repository entries. If so, we would simply reuse those repository entries that correspond to the attributes we will use on our input screens. In those cases where there is no repository entry, we will have to simply create them.

To choose the correct control for our attributes, we must begin by examining the possible *values* for each attribute. Here are some preliminary decisions regarding our input attributes identified in the previous step:

- PRODUCT NUMBER, CURRENT MONTH UNITS SOLD, CURRENT YEAR UNITS SOLD, TOTAL LIFETIME UNITS SOLD, UNIVERSAL PRODUCT CODE, MANUFACTURER'S SUGGESTED RETAIL UNIT PRICE, CLUB DEFAULT UNIT PRICE, CURRENT SPECIAL UNIT PRICE, PRODUCER, and DIRECTOR attributes all have input data values that are unlimited in scope or noneditable. Since the designer is unable to provide the user with a meaningful list of values from which to choose, a single-line text box was chosen. Since the attribute CATALOG DESCRIPTION also fits this criteria, a multiple-line text box (referred to as a memo box by some products) was selected.
- PRODUCT TYPE, LANGUAGE, VIDEO ENCODING, SCREEN ASPECT, and VIDEO MEDIA TYPE all contain a limited predefined set of values. Therefore, it was determined that radio buttons would be the preferred screen-based control for these input items.
- It was determined CLOSED CAPTION? is an input attribute that contains a yes/no value. Therefore, a check box was selected as the control for this attribute.
- QUANTITY IN STOCK, RUNNING TIME, COPYRIGHT DATE, and CREDIT VALUE contain data values that can be sequenced in a predictable manner. Thus, a spin box with an associated text box would be a good choice for these attributes.
- The attributes VIDEO CATEGORY and VIDEO SUBCATEGORY contain a large number of predefined values. With so many attributes to display on our screen, it was determined that a drop-down list would be the best control choice.
- TITLE COVER presented an interesting challenge. Its value is actually a drive, directory, and name of a file that contains a bitmap image of the cover of the video title. This attribute will make use of an advanced control called an image box to store a picture of the video title cover. When this object is selected by the user, a set of controls and special dialogue (user interaction)

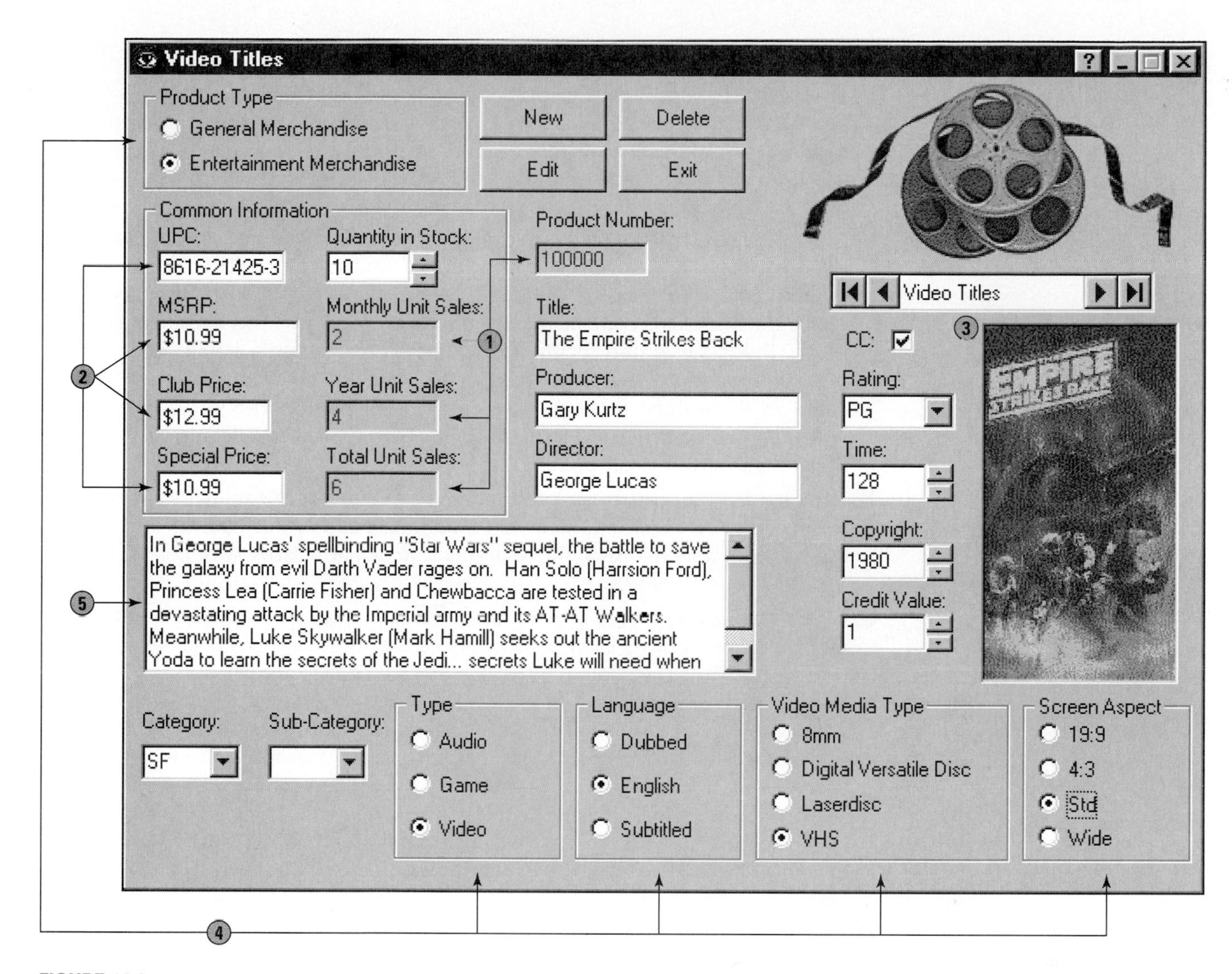

FIGURE 12.6 *SoundStage Prototype for* NEW VIDEO TITLE, DISCONTINUED VIDEO TITLE, *and* VIDEO TITLE UPDATE *inputs*

will be used to capture the input for this item. We'll illustrate this input later in step 3.

Once again, we remind you that there are many other screen-based controls that could be used to input data. Our examples focus on the most commonly used controls. How well you complete this activity will be a function of how knowledgeable you are with these common controls *and* other more advanced controls.

Step 3: Prototype the Input Screen

This step involves developing prototype screens for the user to review and test. Their feedback may result in the need to return to steps 1 and 2 to add new attributes and address their characteristics.

Let's take a look at a SoundStage screen prototype. Figure 12.6 represents a possible prototype screen for handling NEW VIDEO TITLE, DISCONTINUED VIDEO TITLE, and VIDEO TITLE UPDATE. First, the logo appearing in the upper right portion of the screen was included to adhere to a company standard—all screens must display the company logo. The buttons also appearing in the upper center and right portion of the screen were added because of the decision to combine the three inputs into a single screen. They were needed to allow the user the option of selecting the desired type of input and record action. We will discuss these buttons and other command and navigation controls and their use in Chapter 14.

Draw your attention to the following issues addressed during this activity:

1. The PRODUCT NUMBER, MONTHLY UNIT SALES, YEAR UNIT SALES, and TOTAL UNIT SALES are screened in a special color as a visual clue to the user that these fields are locked and they cannot enter data into them. These fields are automatically generated by the system. Other fields appearing on the screen have a white background as a visual clue that they are editable.
2. Notice that edit masks were specified for these input fields. The UNIVERSAL PRODUCT CODE field contains dashes in specified locations. The user does not actually enter these dashes. Rather, the user simply types in the numbers and afterward the entire content is redisplayed according to the specified edit mask. The same is true for the MANUFACTURER'S SUGGESTED RETAIL PRICE, CLUB DEFAULT UNIT PRICE, and CURRENT SPECIAL UNIT PRICE fields. For example, in either of these three fields the user could type the number 9, press enter, and the content would be redisplayed (according to the edit mask) with a dollar sign and decimal point.
3. Each field on a screen has been given a label that is meaningful to the users. Feedback from users indicated "CC" was a commonly recognized abbreviation for "closed caption." Also, the users indicated that a label was not necessary for CATALOG DESCRIPTION.
4. Notice that related radio buttons have been arranged in a group box that contains a descriptive label. Group boxes are frequently used to visually associate a variety of controls that are related. For example, notice the group box labeled "Common Information." The fields located inside this group box were grouped because the user associates these attributes to any type of SoundStage product. Also, realize that each label that corresponds to a radio button option is not what is actually input and stored in the database. Rather, what you see is the meaning of the value. The actual value that is stored is a code. For example, the code value E would actually be stored instead of "English" if the user selects the radio button labeled "English" for the attribute LANGUAGE.
5. Notice that the multiple-line text box has a vertical scroll bar feature. This is a visual clue that there is additional text not appearing inside the CATALOG DESCRIPTION field.

When prototyping input screens, it is important to actually let the user exercise or test the screens. Part of that experience should involve demonstrating how the user may obtain appropriate help or instructions. New versions of Microsoft products use what is called tooltips to provide a brief description of buttons and boxes that appear on a screen. The tooltip description displays when the user positions the mouse over the top of the object. This is demonstrated in Figure 12.7(a). When the user places the mouse over the image box for TITLE COVER, a brief description/instruction appears. When the user carries out the instruction she sees the dialogue box that appears in Figure 12.7(b). The dialogue box contains controls that enable the user to easily select from a set of files containing images of covers for video titles.

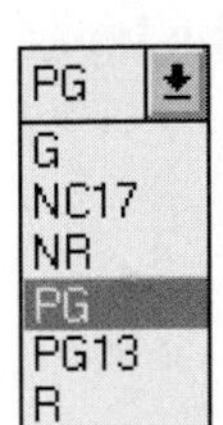

Selected Drop-Down List

Finally, as with tooltips, some controls do not display all details to a user unless they are requested (or triggered by a user action). For example, the drop-down list for Motion Picture Association RATING code displays only a default value. However, the downward pointing arrow is a visual clue that a list box containing possible values exists. The list box may be viewed by simply clicking on the downward pointing arrow. The result of that action is illustrated in the margin.

Earlier you learned the importance of internal controls for inputs. Prototypes should also demonstrate to the user how security will be handled. Also, the prototype should demonstrate how error handling will be accomplished. Figure 12.8 represents a screen prototype that shows users how only authorized persons will

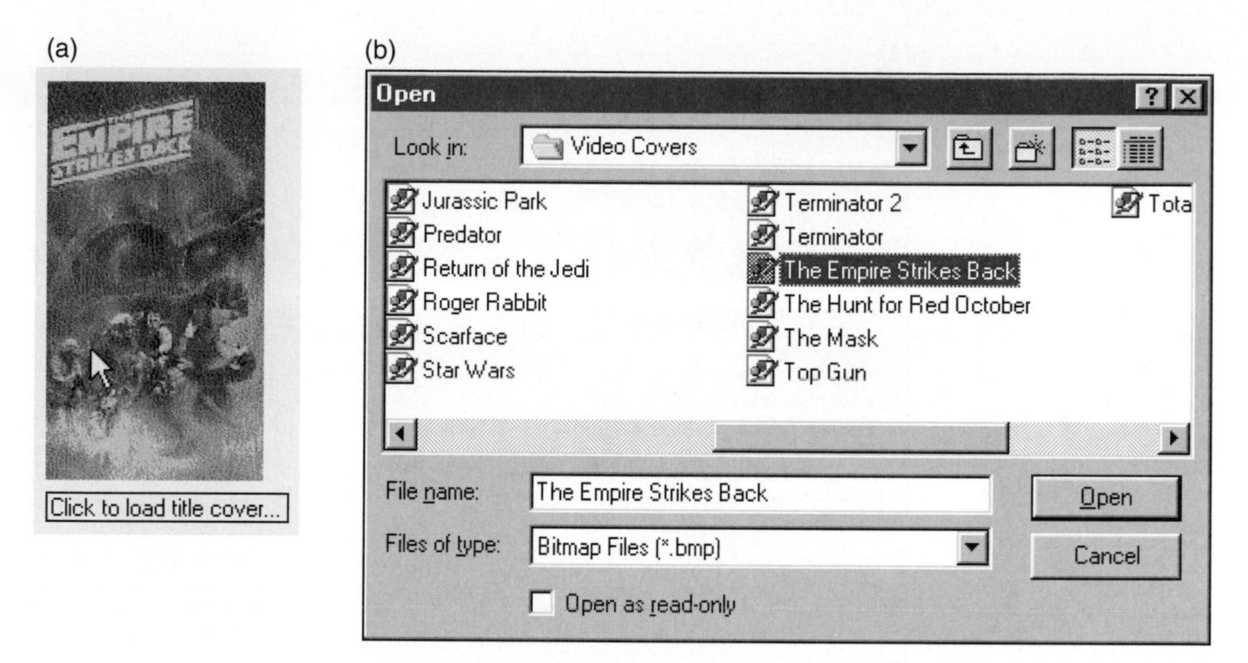

FIGURE 12.7 *SoundStage Prototype Demonstrating Help Instructions*

be able to gain access to their system and its input screens. Notice in Figure 12.8(a), that the text box for entering the user password does not display the actual password itself; rather, a series of asterisks appear in its place. This same screen demonstrates how entry errors will be handled. Figure 12.8(b) shows the resulting error message that gets displayed when a user provides an invalid user ID or password.

Let's see one more example of an input prototype screen. Figure 12.9 represents a prototype screen for entering data for three different inputs: NEW MEMBER, MEMBER CANCELLATION, and MEMBER UPDATE. Notice once again the attribute content and types of controls used to capture the data values. You've now seen two prototype screens that handle six different inputs. Such consolidation is common for applications built using a graphical interface. Traditional text-based screen design of old typically would have resulted in six separate screen designs—one for each input.

Modern prototyping tools have made screen design an infinitely easy task. Most CASE products include facilities for rapid prototyping of input screens. They are especially useful since they can use the project repository data recorded during systems analysis. Throughout this book, we've used the CASE product *System Architect* to capture requirements. *System Architect,* like other CASE products, can pass the repository data to other products for prototyping (such as *Powerbuilder*) or provide some simple prototyping capabilities of its own. A continuing theme in this book has been the use of database management systems and fourth-generation languages (4GLs) to prototype systems. Virtually all such tools include powerful screen design facilities that make it possible to quickly develop screens. These screens can be directly tested by system users and modified to reflect their opinions. Most 4GL-developed prototypes can eventually evolve into finished production systems, although some must be reprogrammed in traditional languages to improve processing efficiency or security. These 4GL prototyping capabilities can be found in both mainframe and microcomputer databases.

Step 4: If Necessary, Design or Prototype the Source Document

If a source document will be used to capture data, we must also design that document. The source document is for the system user. In its simplest form, the prototype may be a simple sketch or an industrial artist's rendition.

(a)

Security Authorization

Welcome to the

Soundstage Entertainment Club

Warning: This computer program is protected by copyright law and international treaties. Unauthorized reproduction or distribution of this program, or any portion of it, may result in civil and criminal penalties, and will be prosecuted to the maximum extent possible under the law.

User ID Information

User ID: Rick

Password: *******

Login Cancel

(b)

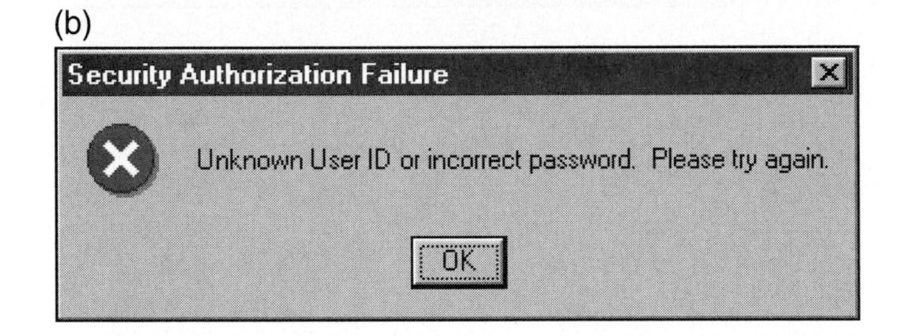

FIGURE 12.8 *SoundStage Prototype Demonstrating Security and Error Handling*

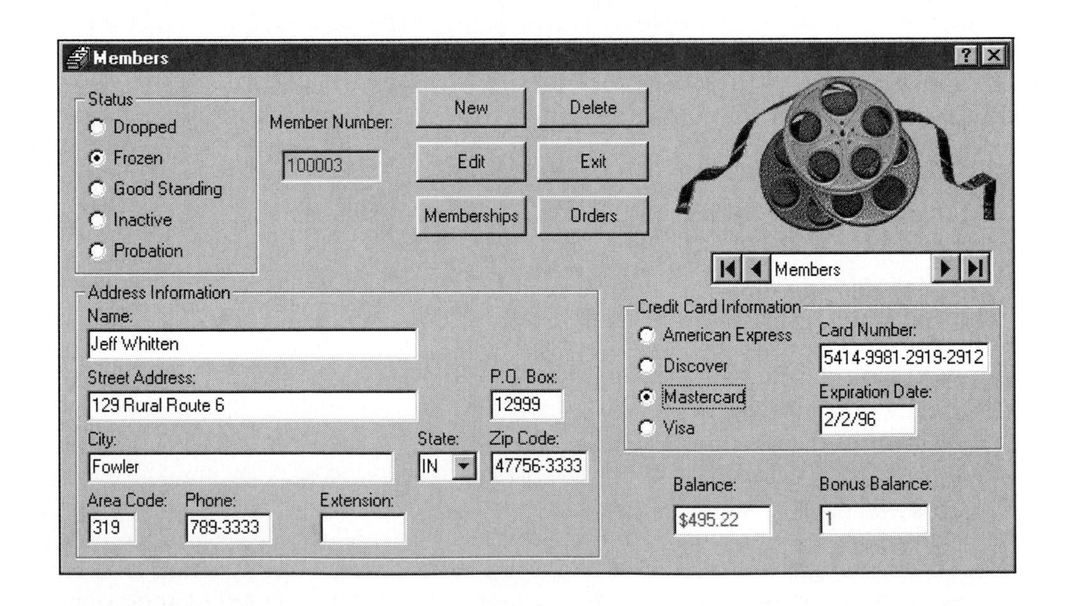

FIGURE 12.9 *SoundStage Prototype for* NEW MEMBER, MEMBER CANCELLATION, *and* MEMBER UPDATE *Inputs*

A well-designed source document will be divided into zones. Some zones are used for identification; these include company name, form name, official form number, date of last revision (an important attribute that is often omitted), and logos. Other zones contain data that identify a specific occurrence of the form, such as form sequence number (possibly preprinted) and date. The largest portion of the document is used to record transaction data. Data that occur once and data that repeat should be logically separated. Totals should be relegated to the lower portion of the form because they are usually calculated and, therefore, not input. Many forms include an authorization zone for signatures. Instructions should be placed in a convenient location, preferably not on the back of the form

Prototyping tools have become more advanced in recent years. Spreadsheet programs such as Microsoft's *Excel* can make very realistic models of forms. These tools give you outstanding control over font styles and sizes, graphics for logos, and the like. Laser printers can produce excellent-quality printouts of the prototypes.

Another way to prototype source documents is to develop a rough model using a word processor. Next, pass the model to one of the growing number of desktop publishing systems that can transform the rough model into impressive looking forms (so impressive, in fact, that some companies now develop forms this way instead of subcontracting their design to a forms manufacturer).

WHERE DO YOU GO FROM HERE?

This chapter provided a detailed overview of the systems design phases of a project. You are now ready to learn some of the systems design skills introduced in this chapter. Because systems design is dependent on requirements specified during systems analysis, we recommend that you first complete Chapters 4 to 8. Chapter 4 gives you an overview of systems analysis. Chapters 5 to 8 teach different system modeling techniques that provide for basic inputs to the systems design activities presented in Part Three.

The order of the system design chapters that follow are flexible; however, the authors did present the subsequent techniques in the sequence they are commonly completed for systems design projects.

SUMMARY

1. Several concepts are important to input design. One of the first things you must learn is the difference between data capture and data input. Alternative input media and methods must also be understood before designing the inputs. And because accurate data input is so critical to successful processing, file maintenance, and output, you should also learn about human factors and internal controls for input design.
2. *Data happens!* It accompanies business events called **transactions.** Examples include orders, time cards, reservations, and the like. This is an important concept because system designers must determine *when* and *how* to capture the data. The designer must understand the difference between the following:
   - **Data capture** is the identification of new data to be input.
   - A **source document** is a paper form used to record data that will eventually be input to a computer.
   - **Data entry** is the process of translating the source document into a machine-readable format. That format may be a magnetic disk, an optical-mark form, a magnetic tape, or a floppy diskette, to name a few.
   - **Data input** is the actual entry of data in a machine-readable format into the computer.
3. The systems analyst usually selects the method and medium for all inputs. Input methods can be broadly classified as either batch or on-line.
   - **Batch input** is the oldest and most traditional input method. Source documents or forms are collected and then periodically forwarded to data entry operators, who key the data using a data entry device that translates the data into a machine-readable format.

- **On-line input** is the capture of data at its point of origin in the business and the direct inputting of that data to the computer, preferably as soon as possible after the data originates.
- **Remote batch** offers on-line advantages for data that is best processed in batches. The data is input on-line with on-line editing. Microcomputers or minicomputer systems can be used to handle this on-line input and editing. The data is not immediately processed. Instead, it is batched, usually to some type of magnetic media. At an appropriate time, the data is uploaded to the main computer, merged, and subsequently processed as a batch. Remote batch is also called *deferred batch* or *deferred processing.*

4. Most new applications being developed today consist of screens having a "graphical" looking appearance. This type of appearance is referred to as a **graphical user interface (GUI)**.
5. A number of alternative **automatic data collection (ADC)** technologies are available today and finding their way into batch and on-line applications. With advances in these technologies, we can eliminate much human intervention associated with most traditional input methods: **biometric, electromagnetic, magnetic, optical—bar coding, bar codes, optical-mark reader (OMR), optical-character reader (OCR)—smart cards, touch.**
6. Inputs should be as simple as possible and designed to reduce the possibility of incorrect data being entered. Furthermore, the needs of data entry clerks must also be considered. With this in mind, system designers should understand human factors that should be evaluated during input design.
7. Input controls ensure that the data input to the computer is accurate and that the system is protected against accidental and intentional errors and abuse, including fraud.
8. When designing input screens for an application that will contain a GUI appearance, the designer must be careful to select the proper control object for each input attribute. Each control serves a specific purpose, has certain advantages and disadvantages, and should be used according to guidelines. Some of the most commonly used screen-based controls for inputting data include: text box, radio button, check box, list box, drop-down list, combination box, and spin box.

KEY TERMS

automatic data collection (ADC), p. 440
bar codes, p. 441
bar coding, p. 441
biometric, p. 440
check box, p. 447
combination box, p. 450
combination check, p. 444
completeness check, p. 444
data capture, p. 438
data entry, p. 438
data input, p. 439
drop-down list, p. 449
electromagnetic, p. 441
graphical user interface (GUI), p. 440
key-to-disk (KTD), p. 439
key-to-tape (KTT), p. 439
limit and range check, p. 444
list box, p. 448
magnetic, p. 441
optical, p. 441
optical-character reader (OCR), p. 441
optical-mark reader (OMR), p. 441
picture check, p. 444
radio button, p. 447
remote batch, p. 440
repository-based programming, p. 445
self-checking digit, p. 444
smart cards, p. 441
source document, p. 438
spin box, p. 450
text box, p. 446
touch, p. 441
transactions, p. 438

REVIEW QUESTIONS

1. What is the difference among data capture, data entry, and data input?
2. Explain the difference between batch and on-line input methods.
3. Should all new applications be built using on-line input methods? Explain your answer.
4. Why are automatic data collection (ADC) technologies finding their way into batch and on-line applications? Identify six types of ADC technology and the types of business applications where they may be utilized.
5. List and describe several input data validation techniques.
6. What is the purpose of a text box control? When should a text box be used to input a data attribute? Identify some guidelines that should be followed when using a text box control on an input screen.
7. What is the purpose of a radio button box control? When should a radio button be used to input a data attribute? Identify some guidelines that should be followed when using a radio button control on an input screen.
8. What is the purpose of a check box control? When should a check box be used to input a data attribute? Identify some guidelines that should be followed when using a check box control on an input screen.
9. What is the purpose of a list box control? When should a

list box be used to input a data attribute? Identify some guidelines that should be followed when using a list box control on an input screen.

10. What is the purpose of a drop-down list control? When should a drop-down list be used to input a data attribute? Identify some guidelines that should be followed when using a drop-down list control on an input screen.
11. What is the purpose of a combo box control? When should a combo box be used to input a data attribute? Identify some guidelines that should be followed when using a combo box control on an input screen.
12. What is the purpose of a spin box control? When should a spin box be used to input a data attribute? Identify some guidelines that should be followed when using a spin box control on an input screen.

PROBLEMS AND EXERCISES

1. Define an appropriate input method and medium for each of the following inputs:
   a. Customer magazine subscriptions.
   b. Hotel reservations.
   c. Bank account transactions.
   d. Customer order cancellations.
   e. Employee weekly time cards.
2. What effects can be caused by the lack of internal controls for inputs?
3. What implications would input design likely have on output design?
4. What are some advantages to graphical screen interfaces over traditional, text-based screens?
5. Why are source documents appropriate for on-line inputs?

PROJECTS AND RESEARCH

1. Review some sample text-based input screens of an application. For each input data attribute appearing on those screens, select a proper screen-based control that would be used if the screens were redeveloped as a GUI.
2. Obtain a copy of an application form—such as a loan, housing, or school form—or any other document used to capture data (such as a course scheduling form, credit card purchase slip, or time card). Do not be concerned whether the application is currently input to a computer system. How do the people who initiate or process the form feel about it? Comment on the human engineering. How well is it divided into zones? Comment on the suitability of the application for data entry. Are data fields that wouldn't be keyed properly located? What changes would you make to the form?
3. The order-filling operation for a local pharmacy is to be automated. Customer prescriptions are to be entered online by pharmacists. The pharmacy expects the new system to consist of user-friendly, graphical screens. The content of the input PRESCRIPTION contains the following attributes. For each input attribute, indicate a proper screen-based control to use on the screens that will be used to enter a PRESCRIPTION.

   A PRESCRIPTION contains the following attributes:

   CUSTOMER NAME—20 characters.
   DOCTOR NAME—20 characters.
   1 to 10 occurrences of the following:
      DRUG NAME—30 characters.
      QUANTITY PRESCRIBED—4 digits.
      MEDICAL INSTRUCTIONS—120 characters.
   RX NUMBER—a federal licensing number of 6 digits.
   1 to 10 occurrences of the following added by pharmacist:
      DRUG NUMBER—a number that uniquely identifies a prescription drug—6 characters.
      LOT NUMBER—a number that uniquely identifies the lot from which a chemical was produced—6 characters
      DOSAGE FORM—the form of the medication issued, such as "pill." P=PILL C=CAPSULE L=LIQUID I=INJECTION R=LOTION.
      UNIT OF MEASURE—G=GRAMS O=OUNCES M=MILLILITERS.
      QUANTITY DISPENSED—4 digits.
      NUMBER OF REFILLS—2 digits.
      and optionally:
      EXPIRATION DATE—date prescription expires.
4. A moving company maintains data concerning fuel-tax liability for its fleet of trucks. When truck drivers return from a trip, they submit a journal describing mileage, fuel purchases, and fuel consumption for each state traveled through. These data are to be input daily to maintain records on trucks and fuel stations. For each TRIP JOURNAL data attribute, indicate the GUI control to be used on an input screen.

   A TRIP JOURNAL consists of the following attributes:
   TRUCK NUMBER—4 characters.
   DRIVER NUMBER—9 characters.
   CODRIVER NUMBER—9 characters.
   TRIP NUMBER—3 characters.

DATE DEPARTED.
DATE RETURNED.
1 to 20 of the following:
  STATE CODE—2 characters.
  MILES DRIVEN—5 digits.
  FUEL RECEIPT NUMBER—9 characters.
  GALLONS PURCHASED—3 digits (1 decimal).
  TAXES PAID—4 digits (2 decimals).
  STATION NAME—10 characters.
  STATION LOCATION—15 characters.

MINICASES

1. Kathleen Smathers, a programmer/analyst for the Wholesale Cost-Plus Club, had an early afternoon appointment with Linda Pratney, the Accounts Payable assistant manager. Wholesale Cost-Plus Club is a large, citywide warehouse outlet that sells virtually any type of merchandise to club members for a cost very close to wholesale price (significantly below the retail prices charged by grocery stores, drugstores, department stores, and other retail stores). Kathleen had been largely responsible for the Accounts Payable system implemented last fall. Accounts Payable pays off invoices from the suppliers of the club's merchandise. The meeting's purpose was a mandatory post-implementation review of the new system. (Post-implementation reviews occur one month after a new system replaces an old system.) It was no company secret that Linda was displeased with some aspects of the new system.

LINDA Kathleen, I guess you heard that I'm having a little trouble with this new payables system. You told me that the computer screens would make my people's job much easier, and that just hasn't happened. I'm also receiving a lot of complaints from our suppliers.

KATHLEEN I'm really sorry it hasn't worked for you. But I'm willing to work overtime to make this system work the way you need it to work. What isn't working? I thought I included all the fields you requested for entering the invoices. Is it that I didn't provide adequate training for your people?

LINDA That's not the problem. Everyone is saying that the system is simply not easy to use, that it takes them longer than it should to enter invoices. They're asking why the interface doesn't have the same type of simple looking and easy-to-use screens as their other PC programs.

KATHLEEN They must be talking about their *Windows*-based products. They are all running the standard Microsoft *Office* products. That would have been difficult to do because I'm not up to speed on these new programming languages that allow you to develop those kinds of graphical screens.

LINDA I don't get it, Kathleen. Why couldn't you learn? Because of that, my people have to put up with this unforgiving system. I think we all would rather have waited for you to learn.

KATHLEEN You are right. But I thought the data entry screens were easy to use, even if they aren't graphical. In fact, I was pretty proud of them. I used whatever text-based functions I could to make the input screens exciting. I used a lot of fancy blinking screens, reverse video, and other things like that to make their job of entering invoices more interesting. And besides, what do supplier complaints have to do with my system?

LINDA According to my people, the new system isn't recording the supplier invoices accurately. They . . .

KATHLEEN That's impossible! That can't be. The program asks the data entry clerks to enter all the information that's on the supplier's invoice forms. If the suppliers are sending us a correct invoice, then there shouldn't be any problems.

LINDA Kathleen, you're denying that there are problems. I'm telling you that there are. And I expect you to resolve them. If you can't or are unwilling, then I'll get someone else.

a. Do you think Linda is overreacting at the end of their meeting?
b. Was Kathleen right in choosing to develop an interface that was not graphical?
c. Why might the supplier invoices be captured incorrectly?
d. Do the use of blinking screens, reverse video, and other fancy functions ensure a good input screen design?

SUGGESTED READINGS

Application Development Strategies (monthly periodical). Arlington, MA: Cutter Information Corporation. This is our favorite theme-oriented periodical that follows system development strategies, methodologies, CASE, and other relevant trends. Each issue focuses on a single theme. This periodical will provide a good foundation for how to develop input prototypes.

Boar, Benard. *Application Prototyping: A Requirements Definition Strategy for the 80s*. New York: John Wiley & Sons, 1984. This is one of the first books to appear on the subject of systems prototyping. It provides a good discussion of when and how to do prototyping, as well as thorough coverage of the benefits that may be realized through this approach.

Connor, Denis. *Information System Specification and Design Road Map*. Englewood Cliffs, NJ: Prentice Hall, 1985. This book compares prototyping with other popular analysis and design methodologies. It makes a good case for not prototyping without a specification.

Dunlap, Duane. *Understanding and Using ADC Technologies.* A White Paper for the ADC Industry. A SCAN TECH 1995 Presentation. October 23, 1995, Chicago. We are indebted to our friend and colleague. Professor Dunlap is a leader in the field of ADC. This paper was the basis for much of our discussion on the trends in ADC technology.

Fitzgerald, Jerry. *Internal Controls for Computerized Information Systems*. Redwood City, CA: Jerry Fitzgerald & Associates, 1978. This is our reference standard on the subject of designing internal controls into systems. Fitzgerald advocates a unique and powerful matrix tool for designing controls. This book goes far beyond any introductory systems textbook; it is must reading.

Gane, Chris. *Rapid Systems Development.* Englewood Cliffs, NJ: Prentice Hall, 1989. This book presents a nice overview of RAD that combines model-driven development and prototyping in the correct balance.

Kozar, Kenneth. *Humanized Information Systems Analysis and Design*. New York: McGraw-Hill, 1989. A good user-oriented treatment of input design.

Lantz, Kenneth E. *The Prototyping Methodology*. Englewood Cliffs, NJ: Prentice Hall, 1986. This book provides excellent coverage of the prototyping methodology.

Wood, Jane, and Denise Silver. *Joint Application Design: How to Design Quality Systems in 40% Less Time*. New York: John Wiley & Sons, 1989. This book provides an excellent in-depth presentation of joint application development. JAD techniques may be applied during input design to obtain consensus agreement on inputs.

13

OUTPUT DESIGN AND PROTOTYPING

CHAPTER PREVIEW AND OBJECTIVES

In this chapter you will learn how to design and prototype computer outputs. You will know how to design and prototype outputs when you can:

- Identify and describe two basic types of computer outputs.
- Explain the difference between medium and format for outputs.
- Differentiate among tabular, zoned, graphic, and narrative formats for presenting information.
- Distinguish among bar, column, pie, line, and scatter charts and their uses.
- Describe several general principles that are important to output design.
- Design and prototype computer outputs.

SCENE

We begin this episode in the conference room where Sandra and Bob scheduled a Tuesday morning meeting to review output designs with the order processing staff. Sally Hoover required her entire staff to be present to become familiar with the new outputs that would be generated by the new system. Dick Krieger, warehouse manager, has also been invited to review an output that Order Processing will send to his area.

SANDRA

OK, let's call this meeting to order. I think we should go ahead and get started so each of you can get back to normal jobs as quickly as possible. First, Bob and I would like to thank you for coming this morning. Today, we want to review some of the system outputs. Let's begin with a picking ticket form that is sent to the Warehouse. We'll discuss that output first so that Dick can then be excused while we discuss other outputs that are specific to the Order Processing staff. [*Sandy adjusts the computer project pad so that the screen image is in clear focus on the pull-down projection screen.*] Based on the requirements statement we created earlier, I've developed a prototype of what the form might look like. If it's OK, I'd like to confirm a few facts before we review this output. [*Hearing no objections, Sandra continues.*] These forms will be printed each day, but I need to know how many could be printed in a single day.

SALLY

Currently, we never process more than 4,000 a day.

ANN

[*A member of the Order Processing staff.*] But what about growth? We expect to continue adding 100 new members a month. Maybe we had better plan on 5,000 orders a day.

SALLY

[*Sally watched as Sandra made some notations on the form.*] Sandra, what's the form you're writing on?

SANDRA

This is a list of output specifications I'm making for the information systems people. It helps them to anticipate the impact the system outputs will have on their facilities. It will also help us to choose the best printer to meet your needs.

SALLY

OK, I can't wait to comment. This looks nothing like our old picking ticket form. It looks like a sheet of mailing labels with bar codes.

SANDRA

That's not too far from the truth. This form will go to Dick's people in the warehouse. Notice that there will be a peel-off bar code for each product on the picking ticket (customer order) and a bar code with the customer and order information. Is this what you had in mind, Dick?

DICK

Absolutely! Our warehouse personnel can run the scanner over the product bar code and it will instantaneously tell them if the item is in stock and precisely where they should go in the warehouse to pick it up. As I understand it from our previous conversation, when they actually retrieve the product they will scan the label again to reflect the fact that the product has been picked, and then remove and attach the bar code sticker. I assume the name and order label would also be scanned and attached to the packaged order.

SANDRA

That's correct. Also, if the product is not available, an automatic back order would be generated. And the partially completed picking ticket would be processed again at a later time. That's why we needed two separate labels, each containing the customer and order information. We may need one to go on the partial order shipment, and the other for subsequent backorder processing.

ANN

What do you mean by automatic?

SANDRA

Well, as the information is scanned, necessary data will be fed into the computer regarding the status of the order, what products were picked, what products are backordered, and other information that the Warehouse has to write in on the picking ticket forms that they currently receive and send back to you folks. This way, you can simply track the status of customer orders by calling them up on the computer. No paper exchange is necessary. Thus, the form serves both to communicate information as well as to subsequently collect information.

SALLY

That sounds great! Between us and the Warehouse, we've lost a fair number of those completed picking tickets in the past.

SANDRA

Dick, what do you think?

DICK

I think this will work. I'm eager to try it.

SANDRA

I guess if this is OK by you, you're free to leave. I know you have to get back to other duties. Thanks for coming.

DICK

Thanks for the invitation. I must say that this should not only streamline my operations, but I think it will do a lot to help the two different groups communicate much more effectively.

SANDRA

Now that Dick is gone, let's take a look at some outputs that the system will generate for you folks. Bob came up with prototypes of a number of printed reports that we understand you will find useful in running your operations.

BOB

Don't forget the screens. I also prototyped several screens that are intended to provide information that you will likely request on an ad hoc basis.

SALLY

Before we get started. Did you ever receive the memo or complaint list that I sent you regarding the reports that we are currently receiving? I don't want to experience the same problems with the new system. I'm tired of receiving reports when they are too outdated to be

SOUNDSTAGE

SOUNDSTAGE ENTERTAINMENT CLUB

of use. I want to receive them on time. I want them to be accurate. And I want them to be relevant. I can't have myself and my people making decisions with poor, or absent, information.

SANDRA

Yes, I did receive it. I sent you a copy, did I not Bob?

BOB

I got it. That's part of what this meeting is about today. As I show you the various reports and screens, give me feedback. I want to ensure the content and style are acceptable. I also need you to indicate who will receive the reports and when. Realize that some of the screens and reports are new. Some you requested, some Sandra and I thought of. So be sure to tell us whether or not they would be of value. We have no desire to overwhelm you with unnecessary information.

SALLY

Great. Let's see those screens and reports.

DISCUSSION QUESTIONS

1. Sandra and Bob designed the picking ticket output such that it could also be used to subsequently aid in data collection. What would be the advantages of this solution?
2. Why was Sandra concerned with knowing the volume of picking tickets that would be generated?
3. Why would Sandra and Bob want Warehouse and Order Processing representatives present to talk about the picking ticket output?
4. Do you think the kinds of concerns expressed by Sally regarding the current outputs they receive are common?

PRINCIPLES AND GUIDELINES FOR OUTPUT DESIGN

Outputs present information to system users. Outputs, the most visible component of a working information system, are the justification for the system. During systems analysis, you defined output needs and requirements, but you didn't design those outputs. In this section, you will learn how to design effective outputs for system users.

Types of Outputs

There are two basic types of computer outputs. The first type is external outputs.

> **External outputs** leave the system to trigger actions on the part of their recipients or confirm actions to their recipients.

Examples of external outputs are invoices, paychecks, course schedules, airline tickets, boarding passes, travel itineraries, telephone bills, and purchase orders. Also, Figure 13.1 illustrates a sample purchase order output for SoundStage Entertainment Club. Most external outputs are created as preprinted forms that are designed and duplicated by forms manufacturers for use on computer printers. Some are designed as turnaround documents.

> **Turnaround outputs** are those that are typically implemented as a form that eventually reenters the system as an input.

The SoundStage Entertainment Club invoice depicted in Figure 13.2 is a typical external turnaround document. Notice that the invoice has a top and lower portion. The top portion is to be detached and returned with the customer payment.

Some outputs do not leave the information system. These outputs are called internal outputs.

> **Internal outputs** stay inside the system to support the system's users and managers.

These outputs fulfill management reporting and decision support requirements. Recall from Chapter 2, management information systems typically produce three types of reports: detailed, summary, and exception. Examples of each are depicted in Figure 13.3.

> **Detailed reports** present information with little or no filtering or restrictions.

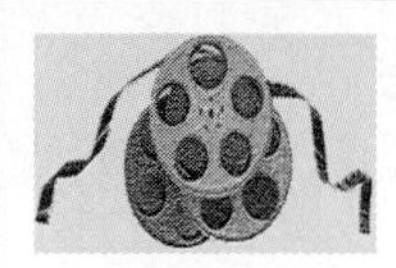

SoundStage Entertainment Club
Fax 317-494-0999

The following number must appear on all related correspondence, shipping papers, and invoices:
P.O. NUMBER: 712812

To:
SoundStage Entertainment Club
2625 Darwin Drive
Indianapolis, IN 45213

Ship To:
SoundStage Entertainment Club
Shipping/Receiving Station
Building A
2630 Darwin Drive
Indianapolis, IN 45213

P.O. DATE	REQUISITIONER	SHIP VIA	F.O.B. POINT	TERMS
5-3-96	ldb	ups		N30

QTY	DESCRIPTION	UNIT PRICE	TOTAL
10000	Powder - VHS	19.99	199,900.00
5000	Now and Then - VHS	15.95	79,750.00
2500	Pulp Fiction Soundtrack - CD	7.99	19,975.00
450	U2 on Tour - T-shirt	3.49	1,570.50
		Subtotal	301,195.50
		Tax	15,059.77
		Total	316,255.27

1. Please send two copies of your invoice.
2. Enter this order in accordance with the prices, terms, delivery method, and specifications listed above.
3. Please notify us immediately if you are unable to ship as specified.

Madge Worthy — Authorized by
5-4-96 — Date

FIGURE 13.1 *Typical External Output*

An example of a detailed report is depicted in Figure 13.3(a). This example is a listing of all purchase orders that were generated on a particular date. Other examples of detail reports would be a detailed listing of all customer accounts, orders, or products in inventory. Some detailed reports are historical. They confirm and document the successful processing of transactions and serve as an audit trail for subsequent management inquiry. These reports assist management planning and control by generating schedules and analysis. Other detailed reports are regulatory, that is, required by government.

Summary reports categorize information for managers who do not want to wade through details.

A sample summary report is depicted in Figure 13.3(b). This report summarizes the month's and year's total sales by product type and category. The data for summary reports are typically categorized and summarized to indicate trends and potential problems. The use of graphics (charts and graphs) on summary reports

SoundStage
Entertainment Club
2630 Darwin Drive - Bldg B
Indianapolis, IN 45213
317-496-0998 fax 317-494-0999

Invoice No. 301231

INVOICE

Customer
Name Carlina Smith
Address 3019 Duroc Drive
City Little Rock State AR ZIP 42653
Phone 502-430-4545

Date 6/5/96
Order No. 346910
Payment Amt

Detach and return top portion with payment

Qty	Description	Unit Price	TOTAL
1	Star Wars - Empire Strikes Back VHS	$19.99	$19.99
1	Eric Clapton Unplugged CD	$13.99	$13.99
1	Aladdin VHS	$17.95	$17.95
		SubTotal	$51.93
		Shipping & Handling	$7.00
		Taxes	$2.95
		TOTAL	$61.88

Payment Details
Cash
Check
Credit Card
Name
CC #
Expires

Office Use Only

Please return top portion of invoice with payment. Make checks payable to: SoundStage Entertainment Club.

RETURN TOP PORTION WITH PAYMENT

FIGURE 13.2 *Typical External Turnaround Document*

is also rapidly gaining acceptance because it more clearly summarizes trends at a glance.

> **Exception reports** filter data before they are presented to the manager as information.

Exception reports only report exceptions to some condition or standard. An example is depicted in Figure 13.3(c) where delinquent member accounts are identified. Another classic example of an exception report is a report that identifies items that are low in stock (soon to run out).

A basic understanding of these types of outputs and report styles is essential to good output design. Such outputs can be designed using the principles and techniques covered in this chapter.

Output Media and Formats

We assume you are familiar with different output devices, such as printers, plotters, computer output on microfilm (COM), and CRT display terminals. These are standard topics in most introductory information systems courses. In this chapter,

(a)

Page 1

SOUNDSTAGE ENTERTAINMENT CLUB
– Products Ordered on 6-31-1996 –

PO Number	Product Number	Product Type	Quantity In Stock	Quantity On Order
112312	102774	Merchandise	273	450
	202653	Title	75	325
	393752	Title	251	125
112313	109833	Merchandise	0	200
	111340	Title	46	150
	231045	Title	225	1,500
	253967	Title	332	850
112314	287904	Title	0	2,000
	699034	Merchandise	0	300
	836785	Merchandise	35	175
	984523	Title	213	250

(b)

Page 1

SOUNDSTAGE ENTERTAINMENT CLUB
– Product Sales Summary as of 7-2-1996 –

Product Type	Product Category	Current Month's Unit Sales	Current Year Unit Sales
Merchandise	Clothing	784	4,312
	Media Accessory	541	2,079
	Total:		
Title	Audio	3,815	20,175
	Game Title	1,247	5,671
	Video Title	2,136	9,032
	Total:		

(c)

Page 1

SOUNDSTAGE ENTERTAINMENT CLUB
– Delinquent Member Accounts as of 7-9-1996 –
– (90 Days Overdue) –

Number	Name	Area Code	Phone	Extension	Balance Due
137842	Joe Dunn	317	490-0012	111	29.43
142314	Bob Fischer	501	282-7996		43.97
157723	Mary Slatter	218	993-9901		56.99
209438	Harold Martin	823	231-8355		33.17
237121	Kevin Ditmano	655	219-0988		99.23
384563	Rick Carlina	501	454-6311		11.23
421134	Barb Kitts	393	789-5412	231	23.66
476688	Kenny Bum	443	234-8845		123.77

FIGURE 13.3 *Sample Detailed, Summary, and Exception Reports*

we are more concerned with the actual output than with the device. A good systems analyst will consider all available options for implementing an output, especially output medium and output format.

A **medium** is what the output information is recorded on, such as paper or video display device.

Now that we have defined what the information will be stored on, we need to determine exactly how the information will appear.

Format is the way the information is displayed on a medium; for instance, columns of numbers.

The selection of an appropriate medium and format for an output depends on how the output will be used and when it is needed.

Alternative Media for Presenting Information A common medium for computer outputs is paper; such outputs are called printed output. Currently, paper is the cheapest medium we will survey. Although the paperless office (and business) has been predicted for several years, it has not yet become a reality. Perhaps there is a psychological dependence on paper as a medium. In any case, paper output will be with us for a long time.

However, paper is bulky and requires considerable storage space. To overcome the storage problem, many businesses have turned to the use of film as an output medium. The first film medium is microfilm.

Microfilm is a roll of photographic film that is used to record information in a reduced size.

Another similar medium is microfiche.

Microfiche is a single sheet of film that is capable of storing many pages of reduced output.

The use of film presents its own problems; microfiche and microfilm can be produced and read only by special equipment. Therefore, other than paper, the most common output medium is video.

Video, the fastest-growing medium for computer outputs, is the on-line display of information on a visual display device, such as a CRT terminal or microcomputer display.

Although this medium provides the system user with convenient access to information, the information is only temporary. When the image leaves the screen, that information is lost unless it is redisplayed. If a permanent copy of the information is required, paper and film are superior media. Often when system designers build systems that include video outputs, they also may provide the user with the ability to obtain that output on paper.

Looking toward the future of output design, we recommend that you keep abreast of changes in the area of output medium. Several trends should be monitored. Companies are beginning to use the Internet as a medium for outputs. You will likely be asked to develop Web pages that allow individuals across the world to share or simply view outputs. Companies are also exploiting electronic data interchange (EDI) as a means of sharing output files. Some companies are developing outputs that are intended to be transmitted via e-mail, faxes, and bulletin board messages. Products such as Delrina Forms provide a number of support functions that support sending and receiving documents. For example, one user can send a document to another user who is able to review and "annotate" the document electronically before sending it back to the originator.

Keep abreast of these trends. Your familiarity will give you an advantage and help you to design system outputs that better serve your employer.

Alternative Formats for Presenting Information There are several different formats you can choose for communicating information on a medium.

Tabular output using columns of text and numbers is the oldest and most common format for computer outputs. This format presents information as columns.

Most of the computer programs you've written probably generated tabular reports.

Another similar output format is zoned output.

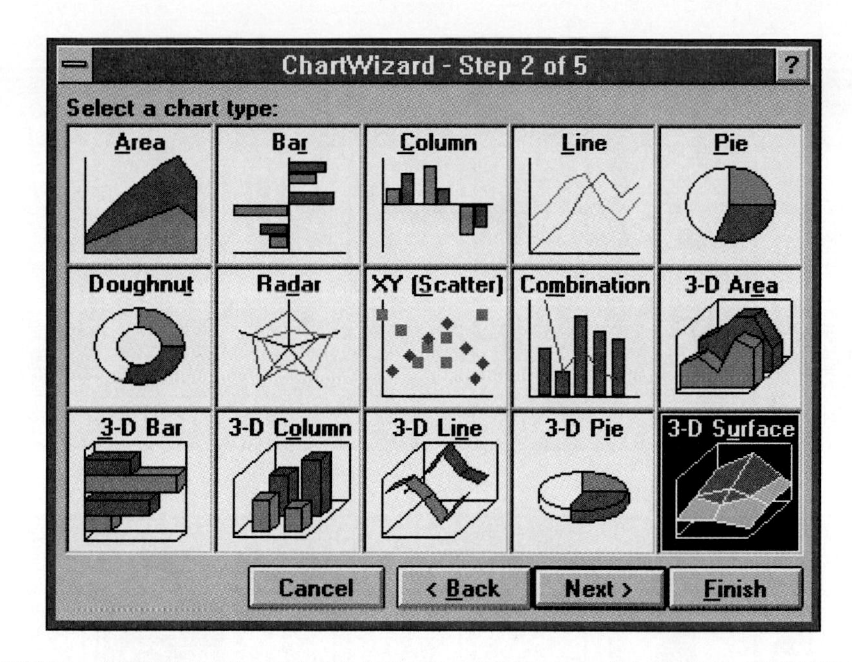

FIGURE 13.4 *Chart Types Available in Microsoft's Excel Spreadsheet*

Zoned output places text and numbers into designated areas of a form or screen.

Zoned output is often used in conjunction with tabular output. For example, an order output contains zones for customer and order data in addition to tables (or rows of columns) for ordered items.

An increasingly popular alternative format for information is graphic output.

Graphic output is the use of a graph or chart to convey information.

To the system user, a picture can be more valuable than words. Graphs can help system users grasp trends and data relationships that cannot be easily seen in columns of numbers. There are numerous types and styles of graphs for presenting information. Figure 13.4 depicts the types of graphs available in Microsoft's *Excel* spreadsheet product. The most commonly used graphs are the bar, column, pie, and line charts and scatter diagrams. Various formats or styles of each of these are available. Some can be seen in Figure 13.5. A system designer must be careful to select a graph type that is most effective for presenting the output information. The five most common can be distinguished as follows:

Bar charts are used to show individual figures or values at a specific time or to depict comparisons among items. The categories to be compared are organized vertically, while the values are organized horizontally—for example, see illustration number 3 in Figure 13.5(a). This layout allows emphasis to be placed on the "comparison" rather than time. As is illustrated in example 5 in Figure 13.5(a), a stacked bar chart style may be used to show the relationship of individual items to the whole.

A **column chart** is a simple variation of the bar chart and is used when there is a desire to show the variation over a period of time or to depict comparisons among items. In a column chart, categories are organized horizontally and values are organized vertically. This appearance emphasizes variations over a period of time.

Pie charts are used to show the relationship or proportions of parts to the whole at a specific period of time. Examples of different pie chart formats are illustrated in Figure 13.5(c). Notice that some styles (example 3 and 4)

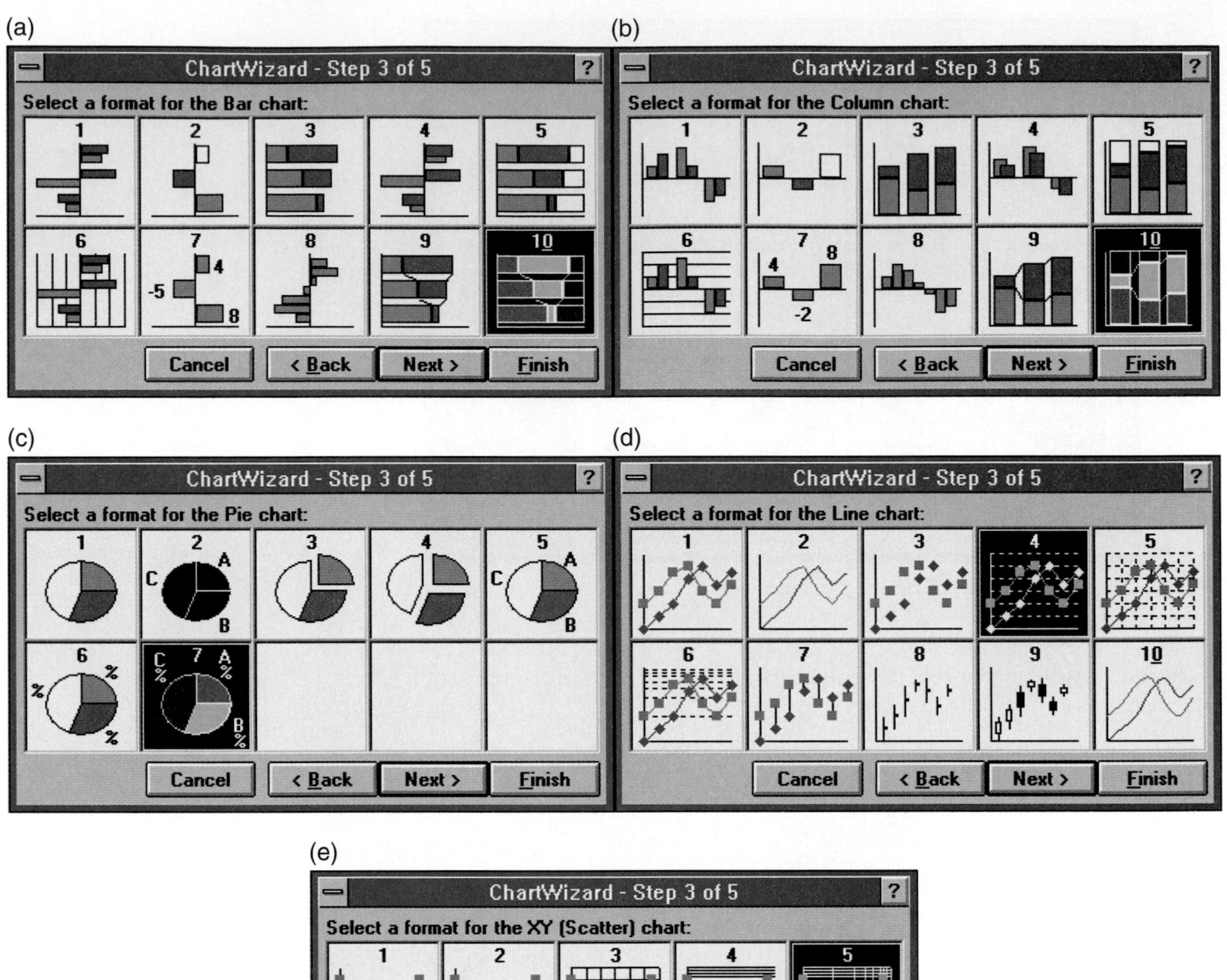

FIGURE 13.5 *Chart Styles in Microsoft's Excel Spreadsheet*

represent explosions and are used to emphasize a particular item. As a general rule, a pie chart should be used to show comparisons that involve 7 or fewer portions.

Line charts are used to show trends over a period of time, at even intervals. It is most common to organize the item being charted on the horizontal axis and the measurement along the vertical axis. Various line chart formats are illustrated in Figure 13.5(d).

Scatter charts or diagrams are used to plot the data values of two items to show uneven intervals or clusters of data. Various standard statistical techniques can then be applied to determine the degree of correlation that exists. Variations of scatter charts are depicted in Figure 13.5(e).

The popularity of graphic output has been stimulated by the availability of low-cost, easy-to-use graphics printers and software, especially in the microcomputer industry. Later in this chapter we will show you an alternate graphic design for a SoundStage output.

Another increasingly popular output format is the narrative format.

Narrative output format uses sentences and paragraphs to replace or supplement standard text, numbers, and pictures.

Word processing technology has exploited the narrative format for reports, business letters, and personalized form letters. For example, an accounts receivable system might interface with a word processor to provide names, addresses, and past due data for personalized credit-reminder letters.

System User Issues for Output Design

There are many system user issues that apply to output design. The following general principles are important for output design:

1. *Computer outputs should be simple to read and interpret.* These guidelines may enhance readability:
   a. Every report or output screen should have a title.
   b. Reports and screens should include section headings to segment large amounts of information.
   c. Information in columns should have column headings.
   d. Because section headings and column headings are sometimes abbreviated to conserve space, reports should include legends to interpret those headings.
   e. Legends should also be used to formally define all fields on a report. You never know whose hands a report might end up in!
   f. Computer jargon and error messages should be omitted from all outputs.

   On many computer outputs, these guidelines are ignored or overlooked; consequently, the outputs appear cluttered and disorganized.
2. *The timing of computer outputs is important.* Outputs must be received by their recipients while the information is pertinent to transactions or decisions. This can affect how the output is designed and implemented.
3. *The distribution of computer outputs must be sufficient to assist all relevant system users.*
4. *The computer outputs must be acceptable to the system users who will receive them.* An output design may contain the required information and still not be acceptable to the system user. To avoid this problem, the systems analyst must understand how the recipient plans to use the output.

HOW TO PROTOTYPE AND DESIGN COMPUTER OUTPUTS

In this section, we'll discuss and demonstrate the process of output design and prototyping. We'll introduce some tools for documenting and prototyping output design, and we'll also apply the concepts you learned in the last section. We will demonstrate how CASE and object-oriented programming languages can be used to prototype outputs and layouts to system users and programmers. As usual, each step of the output design technique will be demonstrated using examples drawn from our SoundStage Entertainment Club case study.

Step 1: Identify System Outputs

Output requirements should have been defined during systems analysis. A good starting point for output design is the design unit DFDs for the new system. The design unit DFDs identify one output requirement that must be designed. These outputs can easily be identified by examining the DFD for data flows that are connected to external entities. Content and other requirements for these outputs may have been documented in the project dictionary. In the absence of such models, as is the case during discovery prototyping, the designer is expected to interview users and to brainstorm outputs of the system.

Brainstorming outputs can easily be done by examining the data model for the new system. By examining the various data entities, content, and their relationships with other entities, the designer tries to think of possible outputs (containing information describing those entities and their relationships) that may be *useful* to the system users. To ensure that the possible report is useful, the designer must be knowledgeable of the overall business side of the system being supported.

Step 2: Select Output Medium and Format

Recall that the selection phase of the system design determined how the output data flows will eventually be implemented. Relative to outputs, the decisions were made by determining the best medium and format for the design and implementation based on:

- The type and purpose of the output.
- The technical and economic feasibility.

Since feasibility is important to more than just outputs, the techniques for evaluating feasibility are covered separately (in Part Five, Module C, "Feasibility and Cost-Benefit Analysis"). The first set of criteria, however, is described in the following paragraphs.

First, you must understand the type and purpose of the output. Is the output an internal or external report? If it's an internal report, is it a historical, detailed, summary, or exception report? If it's an external report, is the form a turnaround document? After assuring yourself that you understand what type of report the output is and how it will be used, you need to address several design issues.

1. *What medium would best serve the output?* Various media were discussed earlier in the chapter. You will have to understand the purpose or use of the output to determine the proper medium. You can select more than one medium, for instance, video with optional paper. All these decisions are best addressed with the system users.
2. *What would be the best format for the report?* Tabular? Zoned? Graphic? Narrative? Some combination of these? After establishing the format, you can determine what type of form or paper will be used. Computer paper comes in three standard sizes: 8½ by 11, 11 by 14, and 8½ by 14 inches. Many printers can now easily compress 132 columns of print into an 8-inch width. You need to determine the capabilities and limitations of the intended printer. Despite the increase in larger 17-inch and 21-inch high-resolution monitors today, it is still recommended that display outputs (thus, the entire application screen) are designed for the lowest common denominator to ensure that all users be able to run the application and see the screens on their computers. Thus, it is still recommended that screen applications be able to run on systems having 640 by 480 screen resolution.

   If a preprinted form is to be used, requirements for that form must be specified. Should the form be designed for mailing? What will be the form's size? Will the form be perforated for bursting into several sections? What legends and instructions need to be printed on the form (both front and back)? What colors will be used and for which copies?

   Incidentally, form images can be stored and printed with modern laser printers, thereby eliminating the need for dealing with forms manufacturers in some businesses.
3. *How frequently is the output generated?* On demand? Hourly? Daily? Monthly? For scheduled outputs, when do system users need the report? Today, reports are more commonly generated by the users themselves. However, in the event that reports are to be printed by the information services department,

they must be worked into the information systems operations schedule. For instance, a report the system user needs by 9:00 A.M. on Thursday may have to be scheduled for 5:30 A.M. Thursday. No other time may be available.

4. *How many pages or sheets of output will be generated for a single copy of a report?* This information is necessary to accurately plan paper and form consumption.
5. *Does the output require multiple copies?* If so, how many? Photocopy (doesn't tie up printer)? Carbons (most printers can make no more than six legible carbons)? Duplicates (requires the most printer time, although laser printers are changing this situation)?

   For external documents, there are also several alternatives. Carbon and chemical carbon are the most common duplicating techniques. Selective carbons are a variation whereby certain fields on the master copy will not be printed on one or more of the remaining copies. The fields to be omitted must be communicated to the forms manufacturer. Two-up printing is a technique whereby two sets of forms, possibly including carbons, are printed side by side on the printer.
6. *For printed outputs, have distribution controls been finalized?* For on-line outputs, access controls should be determined.
7. *For attributes contained on the output, what format should be followed?*

These design decisions should be recorded in the project repository. Let's consider an example from our SoundStage Entertainment Club case.

One output for SoundStage is the MEMBER RESPONSE SUMMARY. This report was requested to provide internal management with information regarding customer responses to the monthly promotional offers. The following decisions regarding the previous list of design issues were made:

1. The manager will request the report from his or her own workstation. It was determined that the output should be presented as a display with the manager having the option of obtaining a laser printer output of the report.
2. Review of the output's data content requirements supports the fact that the report is presenting summary data. It was determined that the user should be given the option of viewing the data in a tabular or graphic format. It was determined that a printed version of the report would be made available on 8½-by-11-inch laser paper.
3. Since the manager will be fully responsible for generating the report, scheduling its display or printing is unnecessary.
4. A single printed copy of the report will require 2 to 10 pages of paper. It is assumed that a paper version of the report will be requested no more than once per week. Therefore, it is estimated that this output will require approximately 500 sheets of 8½-by-11-inch laser paper per year.
5. The report is an internal report that does not require additional copies.
6. The report does not have distribution requirements. It can be accessed by any individual having security access to use the new system.
7. The numerical data appearing in the tabular version of the output should be right justified and zero suppressed.

Step 3: Prototype the Output for System Users

After design decisions and details have been recorded in the project repository, we must create the format of the report. The format or layout of an output directly affects the system user's ability to read and interpret it. The best way to lay out outputs is to sketch or, better still, generate a sample of the report or document. We need to show that sketch or prototype to the system user, get feedback, and modify the sample. It's important to use realistic or reasonable data and demonstrate all control breaks.

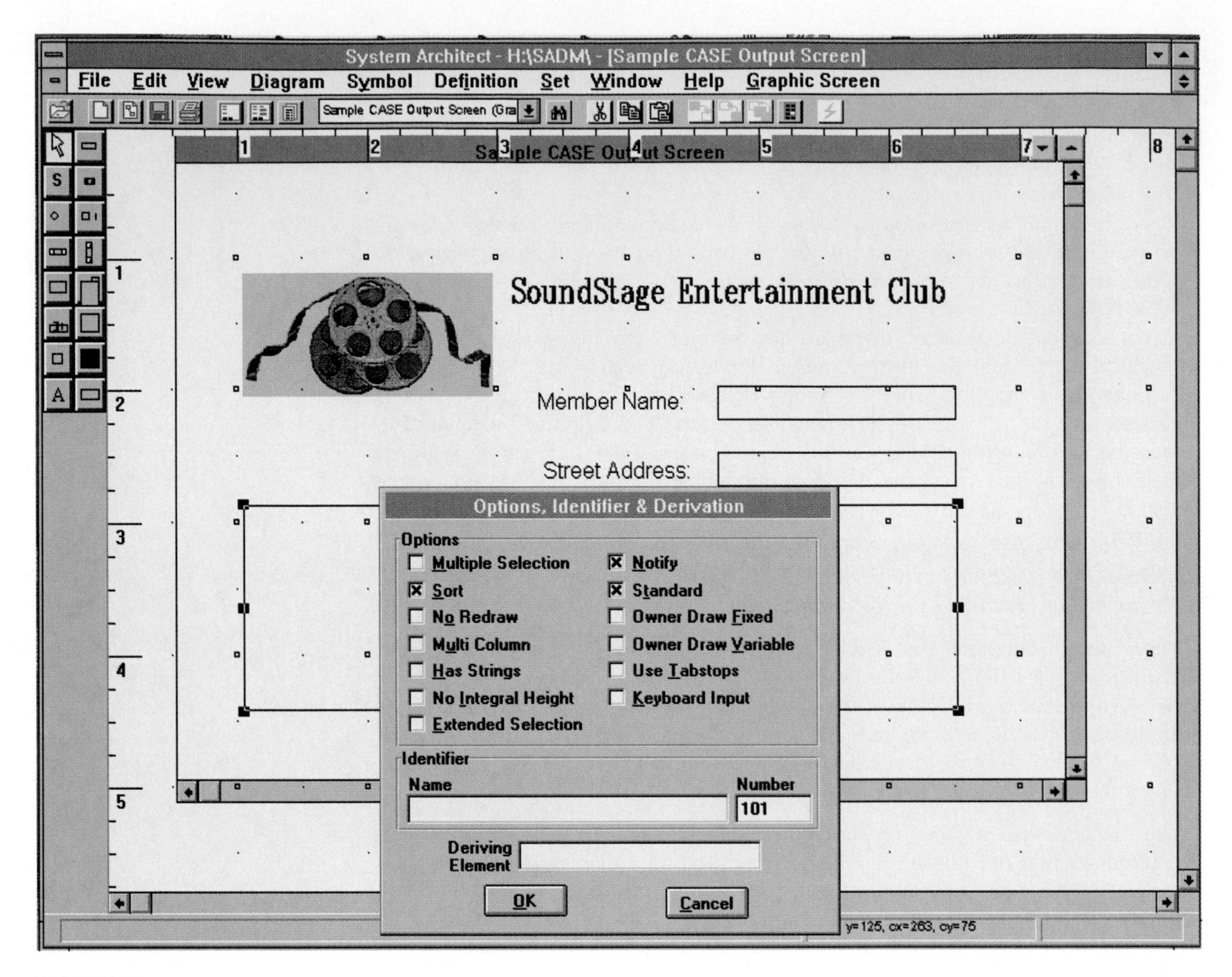

FIGURE 13.6 *Sample of CASE tool for prototyping*

Before the availability of prototyping tools, analysts could sketch only rough drafts of outputs to get a feel for how system users wanted outputs to look. With modern tools, we can develop more realistic prototypes of these outputs. Perhaps the least expensive and most overlooked prototyping tool is the common spreadsheet. Examples include *Lotus 1-2-3,* Microsoft's *Excel,* and Borland's *Quattro.* A spreadsheet's tabular format is ideally suited to the creation of rapid prototypes. Arithmetic and logical formulas and functions can be placed in cells (a cell is the intersection of a row and a column); therefore, spreadsheets can automatically calculate and recalculate some cells to make the information accurate. Finally, most spreadsheets now include facilities to quickly convert tabular data into a variety of popular graphic formats. Consequently, spreadsheets provide an unprecedented way to prototype graphs for system users.

Many CASE products support or include facilities for report and screen design and prototyping via a project repository created during systems analysis. Some CASE products provide links to various programming languages. The link may allow the prototype screens to be fed to the programming language for subsequent enhancement and use or simply allow repository access. A screen design in process is shown in Figure 13.6.

Recall that most database management systems or fourth-generation languages include powerful applications generators for quickly prototyping fully functional

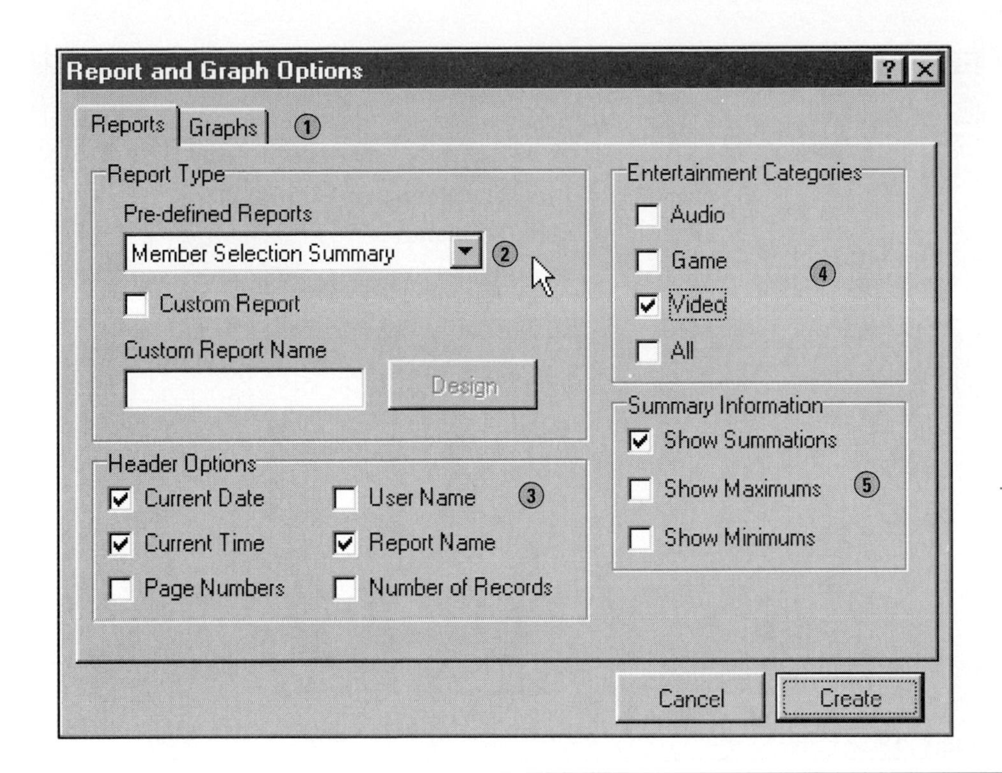

FIGURE 13.7 *Sample Report Customization Prototype Screen*

systems. If a prototype database was created during database design, the test data stored in the database can be used to prototype reports and screens. Most systems prototyping tools include report writers and query languages that allow analysts (or systems users) to quickly design and generate samples of outputs.

In the SoundStage case study, prototypes were built using *Delphi* to access a SQL/Server database. Let's look at some of the sample output screens.

The SoundStage management expressed concern that the MEMBER RESPONSE SUMMARY output could potentially become too lengthy. Often the manager is interested in seeing only information pertaining to member responses for one or a few different product promotions. Thus, it was decided that the manager needed the ability to "customize" the output. The screens used to allow the manager to specify the customization desired should be prototyped as well as the report and graph containing the actual information. Figure 13.7 represents the prototype of the screen the user can use to choose a particular report (or graph) and customize its content. The following points should be noted:

1. A *tab dialogue box* is used to allow the user to select between obtaining a report or graph. A *tab control* is used to present a series of related information. If the user clicks on the tab labeled "Graph", information would be displayed for customizing the output as a graph.
2. A *drop-down list* is used to select the desired report. The user can click on the downward arrow to obtain a list of possible reports to choose from.
3. The user is provided with a series of *check boxes* that correspond to general options for customizing the selected report. The user simply "checks" those options he wishes to appear on the report.
4. A group of check boxes is also used to allow the user to select one or more product categories she wishes to include on the report.
5. Once again, a group of check boxes is used to allow the user to further customize the report. Here the user is allowed to indicate the type of summary information or totals desired for each product category.

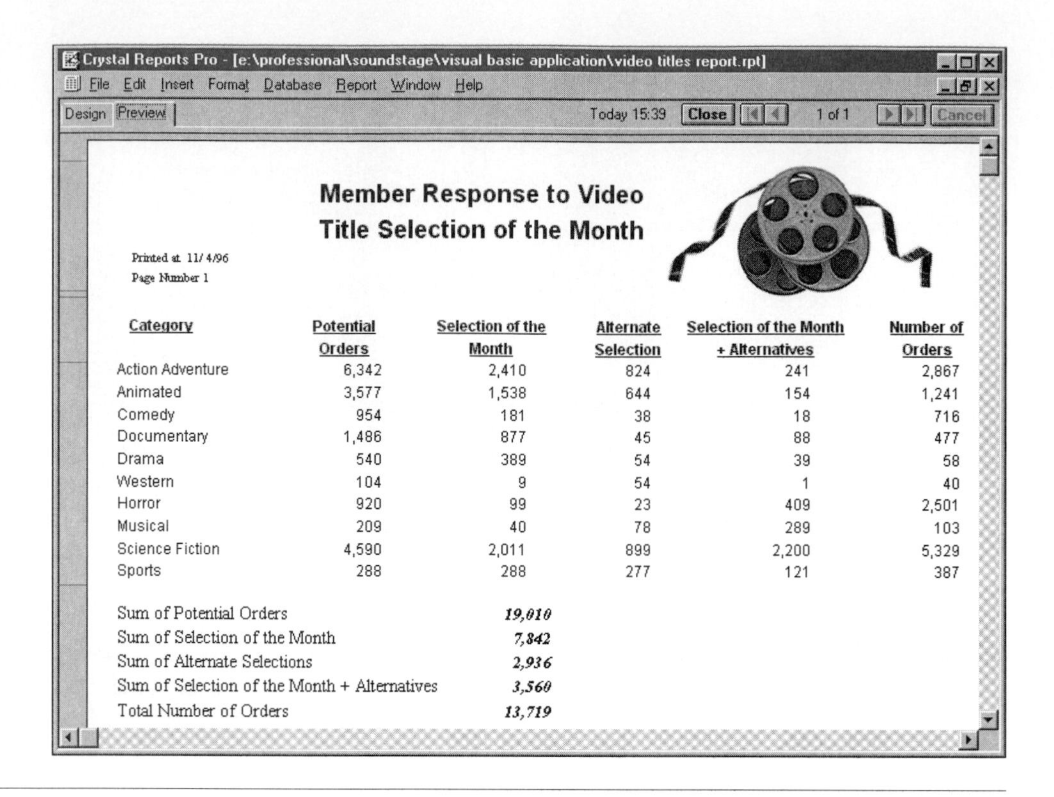

Member Response to Video Title Selection of the Month

Printed at 11/4/96
Page Number 1

Category	Potential Orders	Selection of the Month	Alternate Selection	Selection of the Month + Alternatives	Number of Orders
Action Adventure	6,342	2,410	824	241	2,867
Animated	3,577	1,538	644	154	1,241
Comedy	954	181	38	18	716
Documentary	1,486	877	45	88	477
Drama	540	389	54	39	58
Western	104	9	54	1	40
Horror	920	99	23	409	2,501
Musical	209	40	78	289	103
Science Fiction	4,590	2,011	899	2,200	5,329
Sports	288	288	277	121	387

Sum of Potential Orders	*19,010*
Sum of Selection of the Month	*7,842*
Sum of Alternate Selections	*2,936*
Sum of Selection of the Month + Alternatives	*3,560*
Total Number of Orders	*13,719*

FIGURE 13.8 *Sample Report Prototype Screen*

Let's now look at a prototype of a typical report that may result from the previous customization screen. Figure 13.8 is a prototype of a preview screen containing the report. Examine the content and appearance of the tabular report.

Notice that the user is allowed to scroll vertically and horizontally to view the entire report. In addition, buttons are provided to allow the user to toggle forward and backward to view different report pages.

Finally, let's look at a prototype of a graphic version of the MEMBER RESPONSE SUMMARY output (see Figure 13.9). Note the following:

① Notice that the graph is clearly labeled along the vertical and horizontal axis.

② A legend has been provided to aid in interpreting the graph bars.

③ The designer has implemented drill-down capability. Drill-down is used to allow the user to get additional, more detailed data about a component. Here, the user clicked on a portion of a graph bar and as a result was provided with the *specific* data value.

When prototyping outputs, it is important to involve the user to obtain feedback. The user should be allowed to actually exercise or test the screens. Part of that experience should involve demonstrating how the user may obtain appropriate help or instructions, drill-down to obtain additional information, navigate through pages, request different formats that are available, size the outputs, and perform test customization capabilities. All features should be demonstrated or tested.

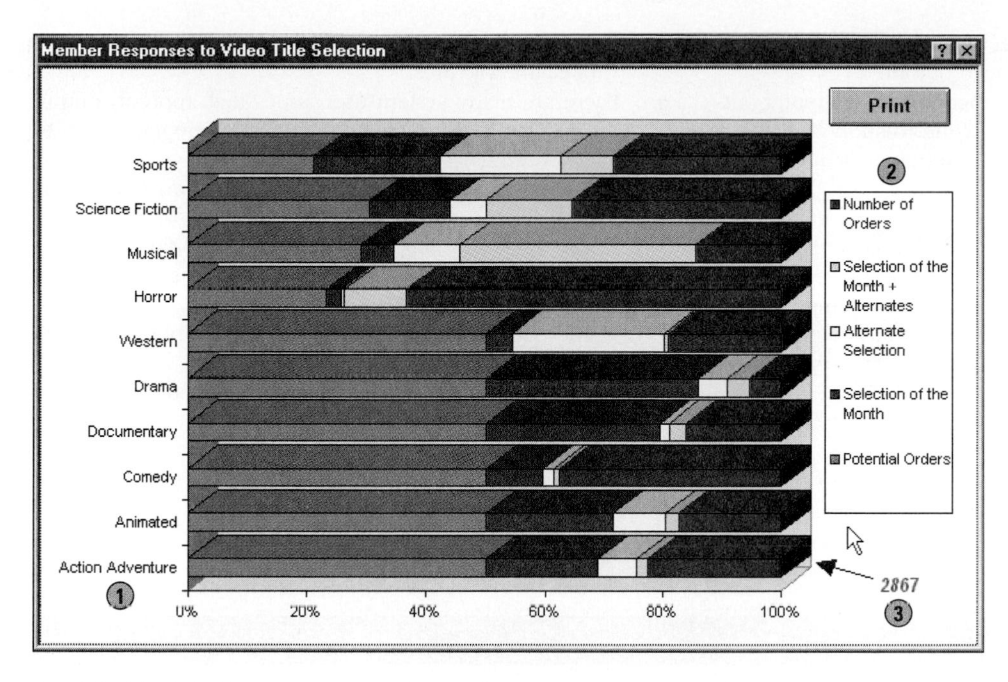

FIGURE 13.9 *Sample Graph Prototype Screen*

WHERE DO YOU GO FROM HERE?

This chapter provided a detailed overview of the design and prototyping of a computer outputs project. If you haven't completed Chapter 12, it is recommended that you do so before covering Chapter 14. Chapter 14 deals with designing an application's overall interface. As such, it involves tying together and presenting the applications functions addressed in Chapters 12 and 13.

SUMMARY

1. Several concepts are important to output design. One of the first things you must learn is the difference between **external** and **internal outputs.** Some external outputs are designed as turnaround outputs that leave and re-enter the system later. Such outputs are usually designed for printing on specially designed manufactured forms.
2. Some outputs are generated to fulfill management information system requirements. These outputs can be classified as one of the following three types of reports:
   - **Detailed reports,** which present information with little or no filtering or restrictions.
   - **Summary reports,** which categorize information for managers who do not want to wade through details.
   - **Exception reports,** which filter data before they are presented to the manager as information. Such reports usually report exceptions to some condition or standard.
3. The system designer usually selects the media and format for all outputs.
   - **Medium** is what the output information is recorded on, such as paper or video display.
   - **Format** is the way the information is displayed on a medium, for instance, columns of numbers or a graph.
4. Common alternative media include the following:
   - **Microfilm** is a roll of photographic film that is used to record information in a reduced size.

   - **Microfiche** is a single sheet of film that is capable of storing many pages of reduced output.
   - Video is the fastest-growing medium for computer outputs; it is the on-line display of information on a visual display device, such as a CRT terminal or microcomputer display.
5. There are several different formats you can choose for communicating information on a medium.
   - **Tabular output** using columns of text and numbers is the oldest and most common format for computer outputs. The format presents information as columns.
   - **Zoned output** places text and numbers into designated areas of a form or screen.
   - **Graphic output** is the use of a graph or chart to convey information. Some types are **bar charts, column charts, pie charts, line charts,** and **scatter charts.**
   - **Narrative output** format uses sentences and paragraphs to replace or supplement standard text, numbers, and pictures.
6. There are many system user issues that apply to output design. The following general principles are important for output design:
   - Computer outputs should be simple to read and interpret.
   - The timing of computer outputs is important.
   - The distribution of computer outputs must be sufficient to assist all relevant system users.
   - The computer outputs must be acceptable to the system users who will receive them.
7. The design and prototyping of computer outputs involve the following steps:
   - Identify system outputs.
   - Select output medium and format.
   - Prototype the output for systems users.

KEY TERMS

bar chart, p. 469
column chart, p. 469
detailed report, p. 464
exception report, p. 466
external output, p. 464
format, p. 468
graphic output, p. 469
internal output, p. 464
line chart, p. 470
medium, p. 467
microfiche, p. 468
microfilm, p. 468
narrative output, p. 471
pie chart, p. 469
scatter chart, p. 470
summary report, p. 465
tabular output, p. 468
turnaround output, p. 464
video, p. 468
zoned output, p. 469

REVIEW QUESTIONS

1. What is the difference between external and internal outputs? Give examples of each.
2. What are turnaround documents? Give several examples.
3. Differentiate between medium and format.
4. What are two film media that are frequently used to overcome paper storage problems?
5. What is the fastest-growing medium for computer outputs?
6. Describe four alternative formats for presenting information.
7. List five common format styles for graphic outputs.
8. Identify several system user issues that apply to output design.
9. What are the steps for prototyping and designing computer outputs?
10. Identify three types of tools that can be used to prototype computer outputs.

PROBLEMS AND EXERCISES

1. To what extent should system users be involved in output design? How would you get system users involved? What would you ask them to do for themselves?
2. What are the three most commonly used media for outputs? What are the advantages and disadvantages of each?
3. Obtain sample outputs of each of the following format types: zoned, tabular, graphic, and narrative. Was the format type of each output the most effective of the four alternatives? If not, why? What format type, or combination of format types, would you have chosen to implement the output? Sketch the layout you would have chosen.
4. Explain how a data flow diagram is useful in identifying outputs that need to be designed?
5. Prepare an expanded data dictionary to describe the following outputs:
   a. Your driver's license.

b. Your course schedule.
c. Your bank statement.
d. Your phone bill.
e. Your W-2 statement.
f. A bank or credit card account statement and invoice.
g. An external document printed on a computer

6. Using a spreadsheet, CASE tool, or 4GL, prototype one of the outputs from exercise 5.
7. What effects may be caused by the lack of well-defined internal controls for outputs?
8. Develop sample outputs that utilize a bar chart, column chart, pie chart, line chart, and scatter chart. Explain why each chart was chosen over the other chart styles to communicate the data relationship.
9. Talk to a user regarding a computer-generated output that he or she receives. Compare the comments to the general principles for output design presented in this chapter.
10. Explain the importance of good input and database design to output design.

PROJECTS AND RESEARCH

1. The sales manager for SoundStage Entertainment Club has requested a daily report. This report should describe the nearly 1,000 customer order responses received for a given day. A response is a member decision on whether to accept the record-of-the-month selection, request an alternate selection, request both, or request that no selection be sent that month. The report is to be sequenced by MEMBERSHIP NUMBER and CATALOG NUMBER. The data repository for the report follows:

   The order response report consists of the following attributes:

   DATE * of the report
   PAGE NUMBER
   1 to 1,000 of the following:
   - MEMBERSHIP NUMBER * 5 digits
   - MEMBER NAME * which consists of the following:
     - MEMBER LAST NAME * 15 characters
     - MEMBER FIRST NAME * 15 characters
     - MEMBER MIDDLE INITIAL * 1 character
   - MUSICAL PREFERENCE * possible values are:
     "EASY LISTENING" "TEEN HITS"
     "CLASSICAL" "COUNTRY" "JAZZ"
   - SELECTION OF MONTH DECISION * possible values are:
     "YES" "NO" "NONE"

   1 to 15 of the following:
   - CATALOG NUMBER * 5 digits
   - MEDIA * possible values are:
     "RECORD" "CASSETTE" "COMPACT DISC"
     "AUDIOPHILE" "8 TRACK" "REEL"
     - NUMBER OF PURCHASE CREDITS NEEDED * 2 digits
     - PERIOD AGREEMENT EXPIRES * date membership expires

   What type of output was being requested by the sales manager? Prepare an expanded data repository for the output. Prototype the requested output. Verify the output with your instructor (serving as the sales manager or system user). Be sure to include appropriate report headings, edit masks, and timing entries.

2. The sales manager has also requested that the sales staff be able to obtain information concerning a particular customer's order response at any time during normal working hours. Prepare an expanded data repository for the output. Prototype the requested output. Verify the output with your instructor (serving as the sales manager or system user).

3. The order filling operation for a local pharmacy is to be automated. Customer prescriptions are to be entered online by pharmacists. The pharmacists insist that they be able to immediately obtain information pertaining to all previously filled prescriptions for a given customer. Using the following data that were captured each time a customer's prescription was filled, prototype the output screen(s).

   A PRESCRIPTION contains the following attributes:
   - CUSTOMER NAME—20 characters
   - DOCTOR NAME—20 characters
   - 1 to 10 occurrences of the following:
     - DRUG NAME—30 characters
     - QUANTITY PRESCRIBED—4 digits
     - MEDICAL INSTRUCTIONS—120 characters
     - RX NUMBER * a federal licensing number of 6 digits
   - 1 to 10 occurrences of the following * added by pharmacist:
     - DRUG NUMBER * a number that uniquely identifies a prescription drug—6 characters
     - LOT NUMBER * a number that uniquely identifies the lot from which a chemical was produced—6 characters
     - DOSAGE FORM * the form of the medication issued, such as "pill"
     - P=PILL C=CAPSULE L=LIQUID I=INJECTION R=LOTION
     - UNIT OF MEASURE * G=GRAMS O=OUNCES M=MILLILITERS
     - QUANTITY DISPENSED—4 digits
     - NUMBER OF REFILLS—2 digits
     - and optionally:
     - EXPIRATION DATE * date prescription expires

4. A moving company maintains data concerning fuel tax liability for its fleet of trucks. When truck drivers return from a trip, they submit a journal describing mileage, fuel purchases, and fuel consumption for each state traveled through. This data is to be input daily to maintain records on trucks and fuel stations. Using the following data, brainstorm and then prototype a detailed, exception, and summary report. Explain the usefulness (value) of each report; do so by describing the types of decisions that are supported by each report.

   A TRIP JOURNAL consists of the following attributes:
   - TRUCK NUMBER—4 characters
   - DRIVER NUMBER—9 characters
   - CO-DRIVER NUMBER—9 characters

TRIP NUMBER—3 characters
DATE DEPARTED
DATE RETURNED
1 to 20 of the following:
STATE CODE—2 characters
MILES DRIVEN—5 digits
FUEL RECEIPT NUMBER—9 characters
GALLONS PURCHASED—3 digits (1 decimal)
TAXES PAID—4 digits (2 decimals)
STATION NAME—10 characters
STATION LOCATION—15 characters

MINICASE

1. Sharon Miller, a programmer/analyst for SaveBig, was meeting with Betty Carlton, the Accounts Payable assistant manager. The meeting's purpose was to discuss problems with a new system that Sharon was largely responsible for developing. It was no company secret that Betty was displeased with some aspects of the new system.

BETTY Sharon, I guess you heard that I'm having a little trouble with this new payables system. You told me that these computer reports would make my job much easier, and that just hasn't happened.

SHARON I'm really sorry it hasn't worked for you. But I'm willing to work overtime to make this system work the way you need it to work. What isn't working? I thought I included all the fields you requested. Is it that you're not getting the reports on time?

BETTY That's not the problem. The information is there. It's just that I find the report difficult to use. For example, this report doesn't tell me what I need to know. I spend most of my time trying to interpret it.

SHARON Why don't you tell me how you use the report?

BETTY Well, first I read down the report, line by line, and attempt to count and classify the number of invoices that are less than 10 days old, between 10 and 30 days old, and between 31 and 60 days old. You see, I have this graph here that shows the totals for each category over the past 12 months. I compare the new totals from this report with the totals on the graph to identify any significant trends in payment activities.

SHARON Why?

BETTY Because we pay some invoices off early to gain a 2 percent discount. Others we defer to the final due date of the supplier, which is usually 30 or 60 days after receipt of the invoice.

SHARON I see. Go on.

BETTY I also use the report to identify the costs of discounts that we did not take, choosing instead to defer payment until close to the final due date. To do that, I locate the invoice on the report and look up that same invoice on the previous copy of the report. Then I log the amount as a lost discount to a particular supplier. At the end of the week, I sum the lost discounts by supplier and send the report to my superiors. It really helps. I can think of several occasions when my superiors authorized a change in payment policy for a particular supplier in order to take better advantage of discounts.

SHARON When I wrote down specifications for this report, I assumed I understood what information you were requesting. It sounds as if there is some additional data that should have been included on the report. You're wasting a lot of time looking for information that could be automatically generated by the system.

BETTY I'm glad that you are sympathetic to my problem. I hope you can do something about it. I'm pretty frustrated with that report.

SHARON I'm sure I can have it fixed in no time. I'm really sorry. I guess I just didn't understand what information you wanted and how you would be using it. Let's see, how should we proceed? I don't want to spend a lot of time designing a new report that isn't precisely what you need. I've got an idea. I just got a new package on my office microcomputer. It's called *Excel*. I'm pretty sure that I can quickly mock up and, if necessary, change a sample of the report you want. I can even simulate this pie chart graph you want. Once we get the reports and graphs looking the way you want, we'll use them as models for the programmers.

BETTY That's a good idea.

SHARON OK, what about this other report?

BETTY I'm getting some flak about this report I asked you to produce for my clerks. It was supposed to list those supplier accounts that we owe payment on. Anyhow, my clerks claim the report is too cluttered and is difficult to use. All they really need to know is the supplier's account number, current balance due, discount date, and the final due date. As you can see, there are a number of unnecessary fields, and the report lists two supplier accounts on the same line. Could you clean this report up too?

SHARON No problem. Again, I'm sorry I was so off target on those reports. I can probably use *Excel* to mock up these reports as well.

a. What type of reports was Betty receiving?
b. What type of reports does Betty really need?
c. What did you think of Sharon's new strategy of using *Excel* for mocking up new reports? Does the approach

sound similar to a strategy that has been frequently described in this book?

d. Why do you suppose the report for Betty's subordinates was inappropriate? Who's to blame, Betty or Sharon? Why?

SUGGESTED READINGS

Application Development Strategies (monthly periodical). Arlington, VA: Cutter Information Corporation. This is our favorite theme-oriented periodical that follows system development strategies, methodologies, CASE, and other relevant trends. Each issue focuses on a single theme. This periodical will provide a good foundation for how to develop prototypes.

14

USER INTERFACE DESIGN AND PROTOTYPING

CHAPTER PREVIEW AND OBJECTIVES

In this chapter you will learn how to design and prototype the user interface for a system. The user interface should provide a friendly means by which the user can interact with the application to process inputs and obtain outputs. Recall that in Chapters 12 and 13, you learned how to design and prototype inputs and outputs. User interface design and prototyping addresses the overall presentation of the application and may require revisions to those preliminary input and output prototypes. Today, there are two commonly encountered interfaces: terminals (or microcomputers behaving as terminals) used in conjunction with mainframes and the more common display monitors connected to microcomputers. There are also several strategy styles for designing the user interface for systems. You will know that you've mastered user interface design when you can:

- Determine which features on available terminal and microcomputer displays can be used for effective user interface design.
- Identify the backgrounds and problems encountered by different types of terminal and microcomputer users.
- Design and evaluate the human engineering in a user interface for a typical information system.
- Apply appropriate user interface strategies to an information system. Use a state transition diagram to plan and coordinate a user interface for an information system.
- Describe how prototyping can be used to design a user interface.

SCENE

We begin this episode in the conference room where Sandra and Bob are reviewing a prototype of the new system interface. The meeting is occurring one week after their last meeting with the users to review input and output design screens.

SANDRA

OK, let's call this meeting to order. I should remind you that when Bob and I previously met with you, we reviewed several rough prototypes of screens that you would be using to enter business transaction data and to view various report information. Since that time, Bob and I made some final revisions to the content of those screens. We will show those revised screens to you. However, we then developed additional screens used to present the overall application and its functions to you. It is the purpose of this meeting to demonstrate the screens that each of you will see when you use the new system. We'd like very much to have your input.

SALLY

I'm not sure we understand what these new screens are?

BOB

Let me address that, Sandra. Sandra and I realize that it would be unreasonable for us to expect you folks to know how to access each of the many individual screens that we reviewed during our last system. Therefore, we developed sort of a menu screen where each of you will simply select the appropriate business function you wish to process. Given that selection, we could have the system then display the appropriate screens to you. It's sort of like your ATM machine. Most ATM machines ask you to enter an access ID and if successful then ask you what you want to do: deposit, withdraw, transfer, or make a payment. Based on your answer, the ATM then knows how to interact with you.

SANDRA

Good analogy, Bob. But if that's unclear, let's get right to the very first screen that each one of you would see. Bob, if you will please activate the new system so that the first screen appears on our projector . . . thanks. As you can see, the first screen you will see is this screen prompting you to enter your ID and password. As with the ATM machine, no one gets past this point without proper authorization!

SALLY

I assume that I will have some say regarding whom those individuals will be.

BOB

Absolutely. I've already drafted a memo asking for those people's names.

SANDRA

Next screen, Bob. OK, on successfully logging onto the new system you will then see this screen. Notice that the screen provides a menu of choices that correspond to the many business functions that are supported by the new member services system, again, much like an ATM that would respond with a menu of transaction choices.

SALLY

OK. Now I understand. That's a cool screen!

CLERK A

That is a nice screen. It looks so simple and intuitive. But what is that button with the picture of the man holding a sheet of paper?

CLERK B

Yeah! And what's the button with the balloons?

SANDRA

Bob? You designed this screen. Do you want to address those questions?

BOB

Well, the man with the paper was intended to represent "reports." You'd press that button to obtain a listing of possible reports. The balloons were used to represent "promotions."

CLERK A

That's funny, I thought the man with the reports might have something to do with promotions. He looks like he is selling or promoting something.

CLERK B

I would never have guessed that the balloons implied promotions. There are a few others that don't make sense to me either.

SANDRA

Bob, it sounds like we may have trouble using pictures to clearly suggest most of the available functions. I think we should use clear meaningful text labels for our buttons rather than pictures. That's the problem with pictures. Unless the picture is universally recognized among the people who will use the new system, now and in the future, it's only going to cause confusion.

SALLY

What about the color? Some earlier screens were one color and now this screen is a different color.

BOB

I did that on purpose. Which did you prefer? We will need to make that decision today. I'll make all screens the same color.

SALLY

Definitely the color used on those earlier screens we saw. This bright purple is nice, but it is already annoying me.

SANDRA

I just noticed something, Bob. You have a button labeled "Quit." On a previous screen you called a similar button "Exit." Which is it? We need to be consistent.

BOB

My mistake. According to our standards, it should be called "Exit" thoughout.

SANDRA

Let's have one of you volunteer to come up front and take over the keyboard and mouse. Let's test the system. We all can learn better by having each of you give the system a test run.

SALLY

I hate to pull rank, but I go first.

BOB

Great! Here's a test ID and password. Let's start the system as if you just showed up for work . . .

SOUNDSTAGE
SOUNDSTAGE ENTERTAINMENT CLUB

DISCUSSION QUESTIONS

1. What would be the benefits that Sandra and Bob (as well as the users) would receive by having the users test the application?
2. Why is it important that the application have consistency regarding colors, labels of buttons, and other features?
3. Why might pictures (graphic representations) not always be preferred on a screen?
4. Why would it be appropriate for Sandra and Bob to have the users view the entire application—since they had already reviewed input and output screens?

STYLES OF USER INTERFACES

User interface design is the specification of a conversation between the system user and the computer. This conversation generally results in either input or output—possibly both. There are several types of user interface styles. Traditionally these styles were viewed as alternatives. However, with recent movements toward designing systems with graphical user interfaces, a blending of all styles can be found. This section presents an overview of several different styles or strategies used for designing user interfaces and how they are being incorporated into today's applications.

Menu Selection

The more traditional user interaction dialogue strategy is menu selection.

> The **menu selection** strategy of dialogue design presents a list of alternatives or options to the user. The system user selects the desired alternative or option by keying in the number or letter that is associated with that option.

More sophisticated technology allows menu selection by touching the screen or selecting menu options with a pen, mouse, cursor keys, or other pointing devices.

A classic hierarchical menu dialogue is illustrated in Figure 14.1. If there are so many menu alternatives that the menu screen is too small or becomes cluttered, menus can be designed hierarchically. Small lists of related menu options can be grouped into a single menu. These menus can then be grouped into a higher-level menu. This approach was applied in Figure 14.1. If the option DISPLAY WARRANTY REPORTS is selected, the submenu WARRANTY SYSTEM REPORT MENU will appear. Then, if the PART WARRANTY SUMMARY option is selected, the next-to-the-bottom screen shown in Figure 14.1 will appear. Specific reports can be selected from that screen. There is no technical limit to how deeply hierarchical menus can be nested. However, the deeper the nesting, the more you should consider providing direct paths to deeply rooted menus for the experienced system user who may find navigating through multiple levels annoying.

The example in Figure 14.1 was typical of traditional character-based screen designs. Today, systems with graphical user interfaces are being developed. Developers are now using newer 4GLs and object-oriented languages to develop applications that take advantage of the look and feel of *Windows*-based applications. Let's examine some menu features, their advantages and disadvantages, being used in these types of newer GUI-based applications.

Menu Bars **Menu bars** are used to display horizontally across the top of the screen/window a series of choices from which the user can select (see Figure 14.2). The choices typically correspond to commands or properties that the user can select or toggle. The choices themselves are typically organized from left to right on the basis of the frequency that a choice is selected. In some cases, the sequencing of the choices is predetermined by standards (for example, *Windows*-based products always have the File command as the first option in the menu).

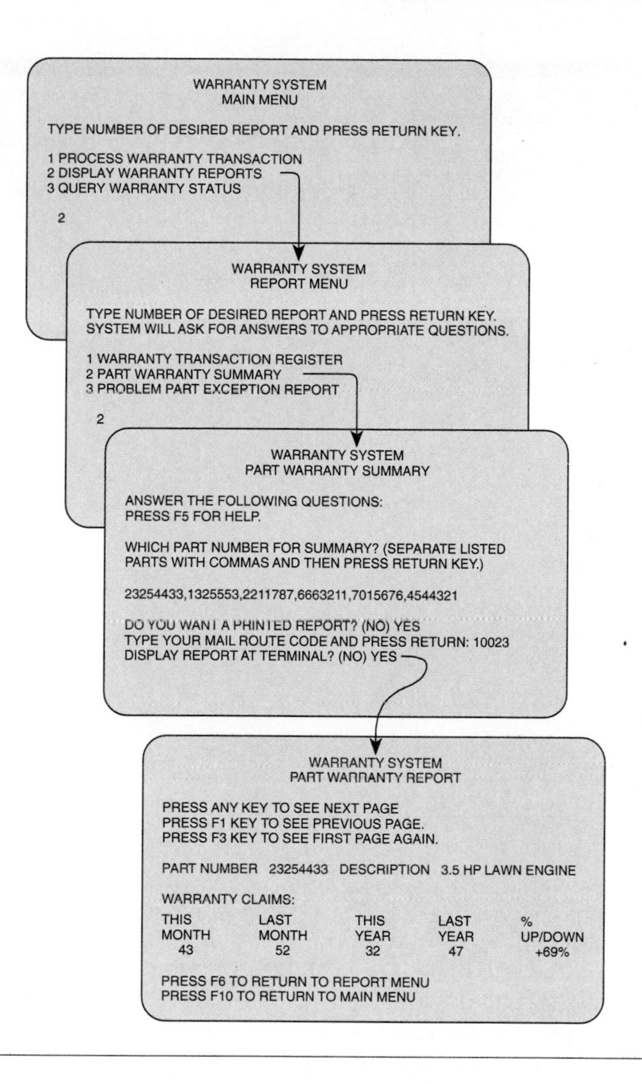

FIGURE 14.1 *A Classical Hierarchical Menu Dialogue*

Menu bars are used to identify common and frequently used actions that occur in a wide variety of different windows that make up the application. They offer the advantage of always being readily visible to the user, consistently located, and easily selected via the keyboard or mouse.

One disadvantage of menu bars is that the menu choices are organized for left-to-right scanning. Studies have shown that users can more easily browse and select from a list that is vertically arranged. To aid in clearly scanning the list, adequate spacing between choices is necessary.

Pull-Down Menus Pull-down menus provide a vertical list of choices to the user. A **pull-down menu** is made available once the user selects a choice from a menu bar. The choices are typically organized from top to bottom according to the frequency in which they are chosen. Some lists also are segmented into groups of related choices as is demonstrated in Figure 14.3 (notice that all save actions for a file have been visually set off from other actions that the user might select).

One special type of pull-down menu is called a **tear-off menu.** Occasionally there is a need to extensively use or refer to the choices available on a pull-down menu. With a tear-off menu, the user can select the menu and drag to relocate it elsewhere on the screen. The tear-off menu is then available for continual referencing. A tear-off menu is a common feature for *Presentation Manager* and on OSF/Motif graphical systems.

Pull-down menus offer the advantage of allowing the designer to simplify a menu bar that may otherwise contain too many choices. They group a related set

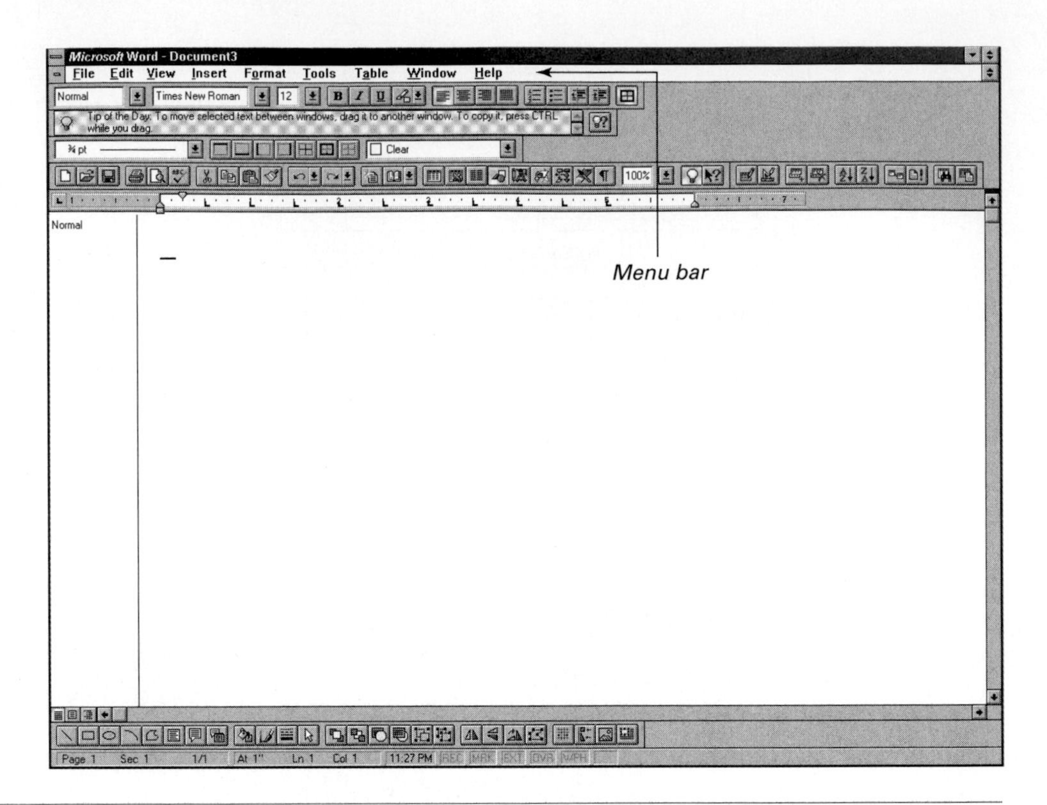

FIGURE 14.2 *Sample Screen with Menu Bar*

of choices into its own separate list. As with choices on the menu bar, pull-down menu items can be selected via the keyboard or mouse.

A disadvantage of pull-down menus is that the user is not provided with a visual clue that suggests the menu exists. This is becoming less of an issue as more and more users are becoming familiar with using graphical applications. Another disadvantage is that when the pull-down menu does appear, it occasionally obstructs the user's view of other areas of interest within the body of the screen/window.

Finally, in some cases a choice appearing on a pull-down menu may result in yet another list or menu of choices. In such cases, a visual clue such as a right-pointing arrow or ellipsis is used. These symbols appear next to the label that describes the menu choice. When an item containing an ellipsis is selected, a dialogue box containing commands and properties for the user to complete a task is displayed. For example, the dialogue box appearing in Figure 14.4 appears when the user selects the option Open . . . from the pull-down menu in Figure 14.3. When an item containing an arrow is selected, a cascading menu appears.

Cascading Menus A **cascading menu** is a menu that must be requested by the user from another higher-level menu. As mentioned in the discussion of pull-down menus, the cascading menu's existence is suggested by the visual clue of a right-pointing arrow appearing next to the higher-level menu choice. When requested, the menu list will appear to the immediate right (in some cases, to the left) of the selected choice from the higher-level menu (see Figure 14.5). Cascading menus are used heavily in Microsoft's new *Windows 95* and *Encarta* (see Figure 14.6) products.

As with pull-down menus, cascading menus offer the advantage of simplifying higher-level menus into a smaller set of related choices. The vertical arrangement of the choices also makes scanning the choices easier.

A disadvantage of cascading menus is that the menu must be requested by the

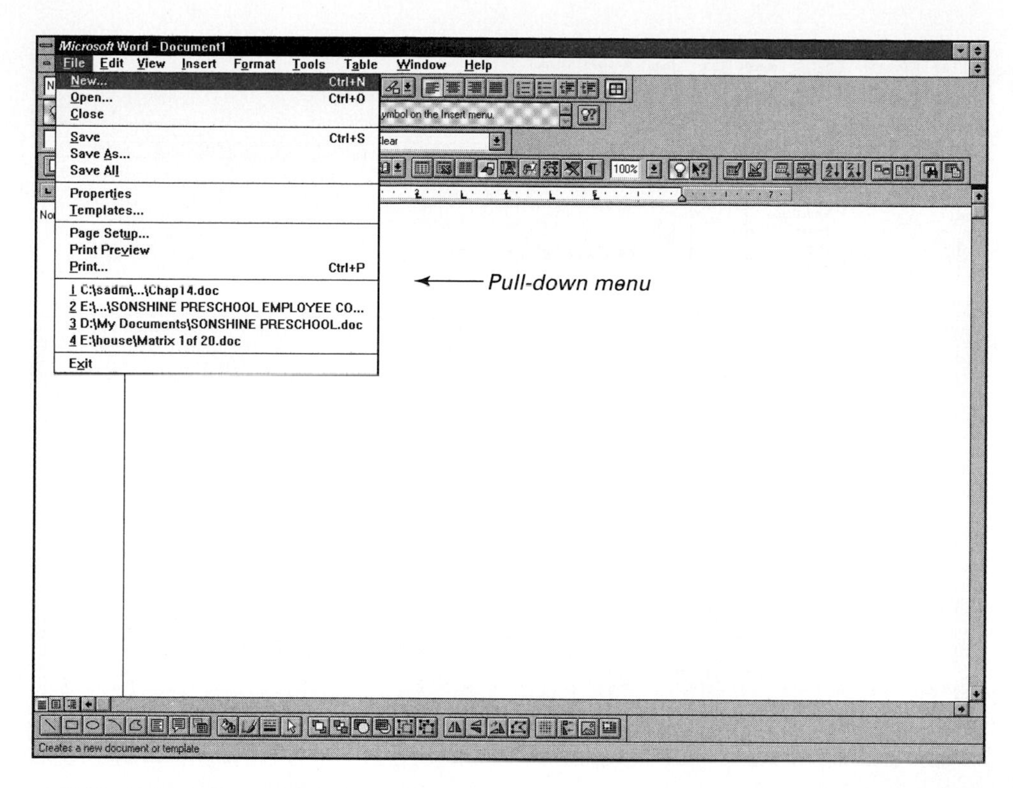

FIGURE 14.3 *Sample Screen with Pull-Down Menu*

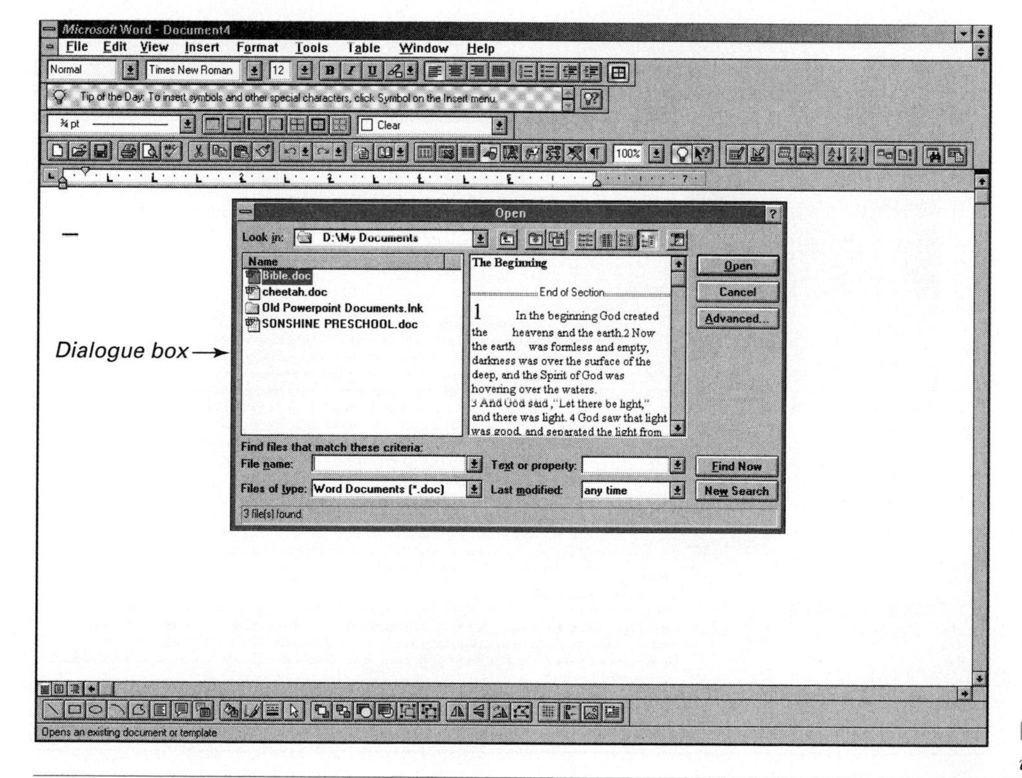

FIGURE 14.4 *Sample Screen with Dialogue Box*

user. Some also believe that applications should not go more than two levels deep with menus. This opinion would prohibit the use of cascading menus—limiting an application to using only a menu bar and associated pull-down menus. Perhaps this thinking is extreme, but some consideration should be given to limiting the effort required for the user to locate and select a desired option.

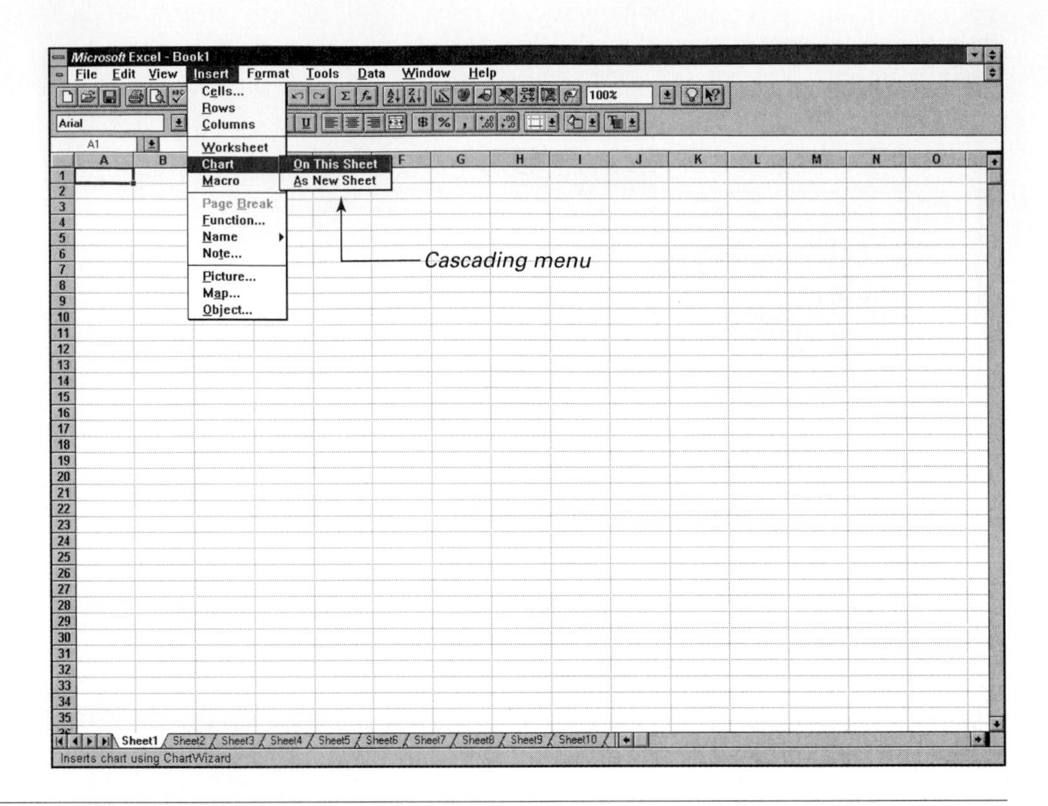

FIGURE 14.5 *Sample Screen with Cascading Menu*

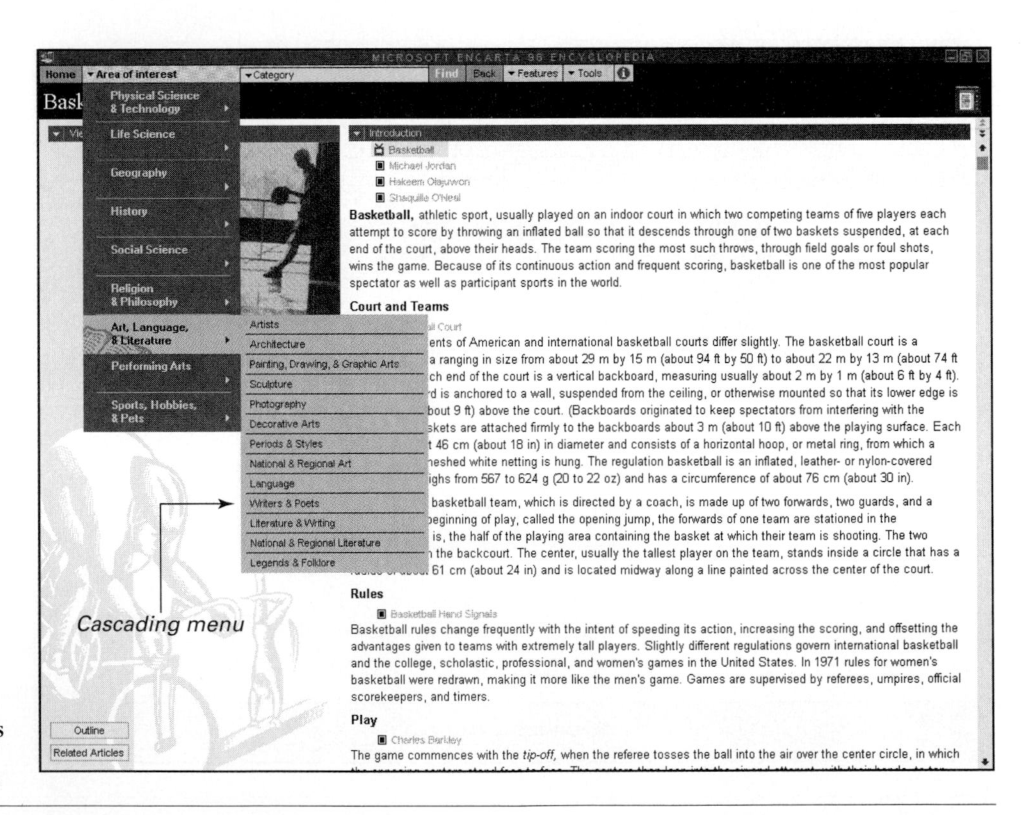

FIGURE 14.6 *Sample* Windows 95 *Product Containing Cascading Menu*

Pop-up menus A **pop-up menu** is another type of vertical listing of choices that must be requested by the user. As shown in the example in Figure 14.7, the user activated the menu by selecting the text "The End" and clicking the right mouse button. Unlike pull-down and cascading menus, the pop-up menu's appearance depends on where the pointer was located when the menu was requested.

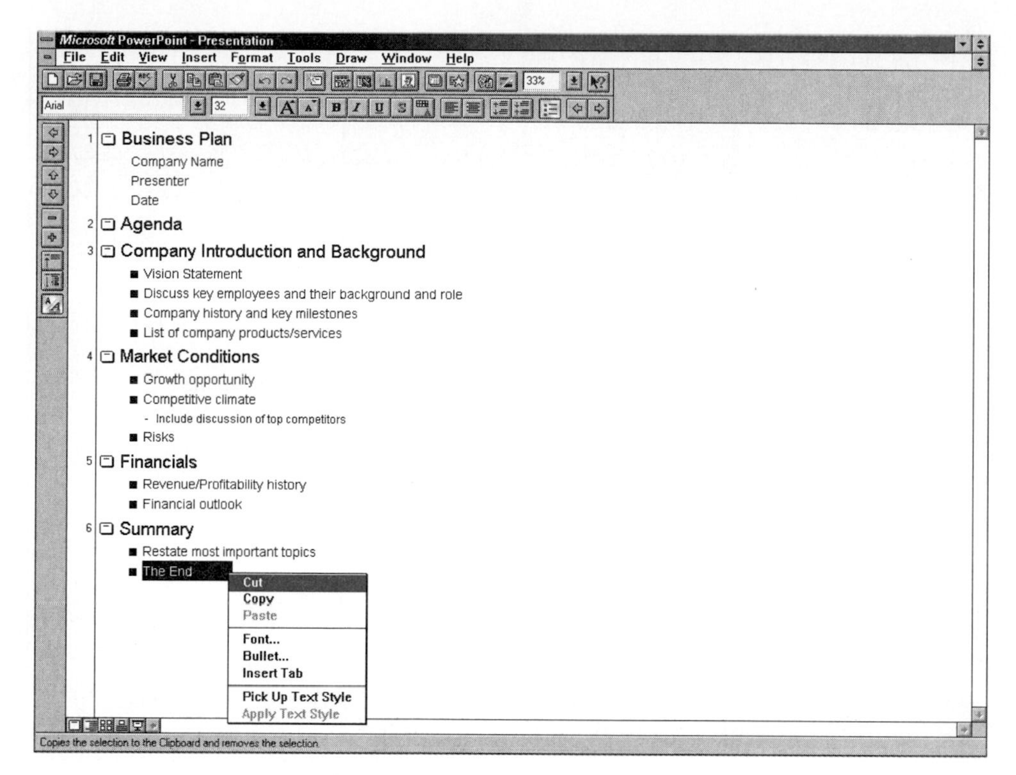

FIGURE 14.7 *Sample Screen with Pop-Up Menu*

Pop-up menus offer the ability to provide a list of options that pertain to a specific object that the user selected. When the user positions the cursor over an object of interest and clicks the right mouse button, a pop-up menu containing commands or properties pertaining to that object appears in the vicinity of the object. This allows users to obtain a list of actions without shifting their focus from the object or work area on the screen.

Unfortunately, no visual clues suggest the existence of pop-up menus. Also, as with pull-down and cascading menus, an activated pop-up menu may obstruct portions of the viewing area of interest to the user.

Iconic menus An **iconic menu** uses graphic representations for menu options (see Figure 14.8). These types of menus are typically used to present the user with options that pertain to special functions that can be performed within the application.

Iconic menus offer the advantage of easy recognition. The use of graphic images helps the user to memorize and recognize the functions available within an application. The choice presented in the form of an icon also provides a relatively larger selection target than the previously discussed menus.

Iconic menus do require special considerations. It is often difficult to find or create meaningful graphic images. While there are numerous sources for image files, not everything can easily be represented as a picture. Moreover, what is a readily identifiable and meaningful picture to one person may not be to the next person.

Instruction Sets

Instead of menus, or in addition to menus, some traditional applications were designed using a dialogue around an **instruction set** (also called a command language interface). Because the user must learn the syntax of the instruction set, this approach is suitable only for dedicated users. There are three types of syntax that can be defined. Determining which type should be used depends on the available technology.

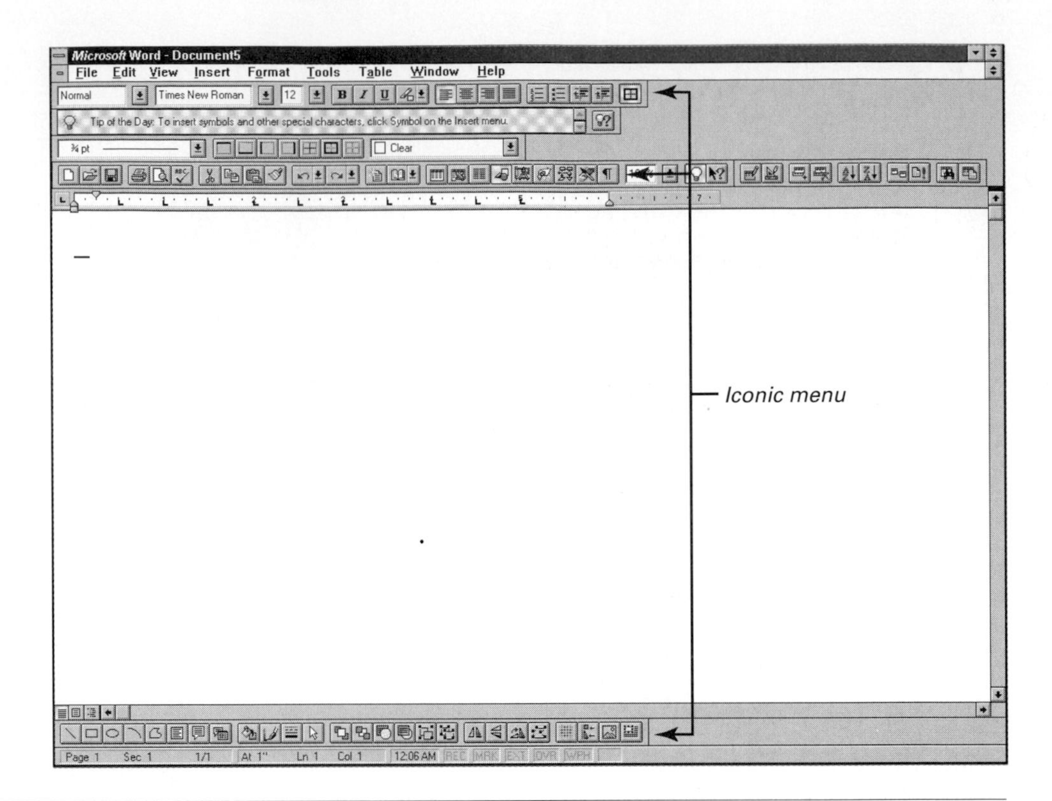

FIGURE 14.8 *Sample Screen with Iconic Menu*

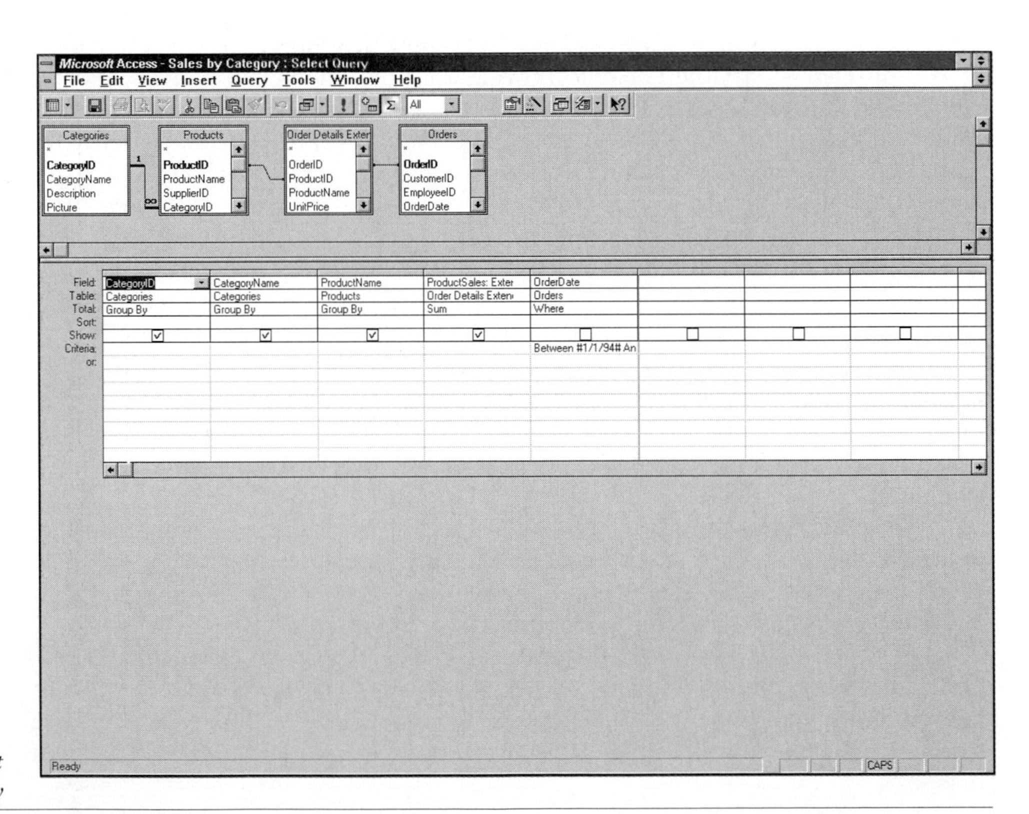

FIGURE 14.9 *Sample Microsoft* Access *Screen of Query Facility*

1. A form of **structured English** can be defined as a set of commands that control the system. In this type of dialogue, an elaborate HELP system should be created so the user who forgets the syntax can get assistance quickly.
2. A **mnemonic syntax** is built around meaningful abbreviations for all commands. Once again, a HELP facility is highly recommended.

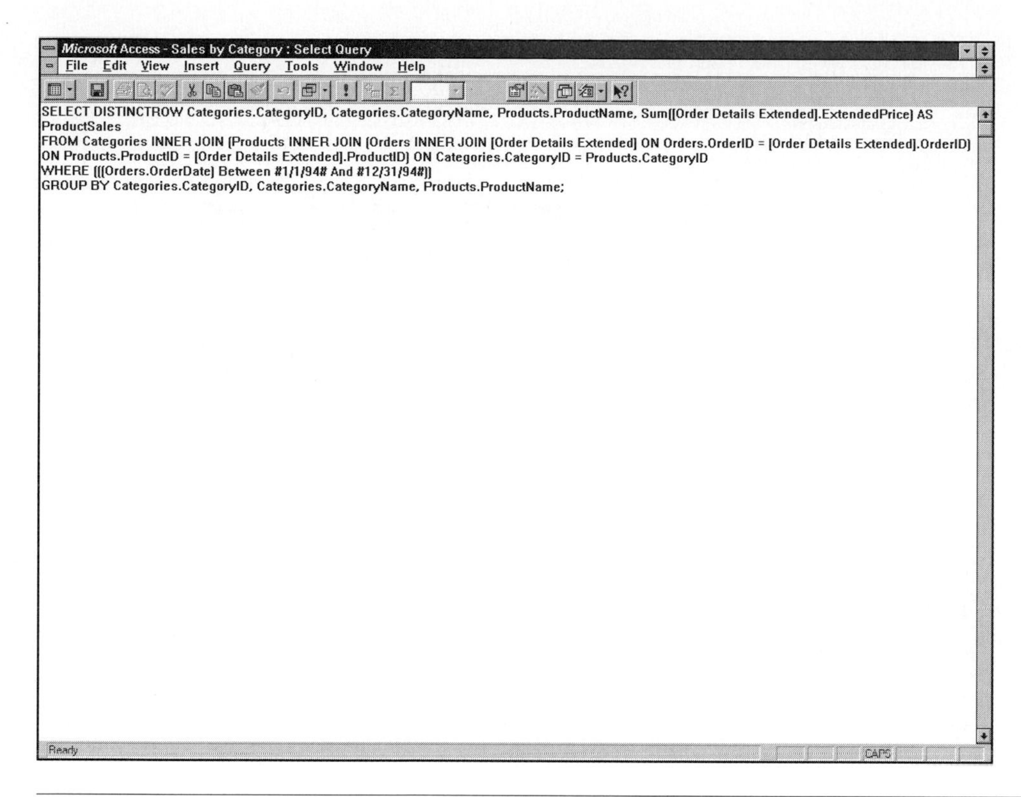

FIGURE 14.10 *Sample Microsoft* Access *Screen of Query Facility with SQL*

3. **Natural language syntax** interpreters are now becoming available. When employing natural language syntax, the system user enters commands using natural English (either conversational or formal, written English). The system interprets these commands against a known syntax and requests clarification if it doesn't understand what the user wants. As new interpretations become known, the system learns the system user's vocabulary by saving it for future reference.

Although this style was primarily used to develop applications for highly skilled computer users of mainframe-based applications, this style of interaction can still be found in graphical applications of today. For example, Microsoft's *Access* database product contains a query facility that allows the developer to visually (point and click) develop a query (see Figure 14.9). The developer simply selects from lists, databases, tables, and columns to include in a query. The developer subsequently can identify selection criteria to be applied to the records. Once complete, the developer can execute the query to see the results. If desired, the developer can view the SQL code that implements the query (see Figure 14.10). This is where the instruction set comes into play. If the developer knows SQL, he or she can examine the SQL instruction set and make any desired modifications. Once again, this approach requires a degree of user experience and know-how.

Question-Answer Dialogues

Question-answer dialogue strategy is a style that was primarily used to supplement either menu-driven or syntax-driven dialogues. The simplest questions involve yes or no answers—for instance, "Do you want to see all records? [NO]." Notice how the user was offered a default answer! Questions can be more elaborate. For example, the system could ask, "Which part number are you interested in? [last part number queried]." This strategy requires that you consider all possible correct answers and deal with the actions to be taken if incorrect answers are entered. Question-answer dialogue is difficult because you must try to consider everything that the system user might do wrong!

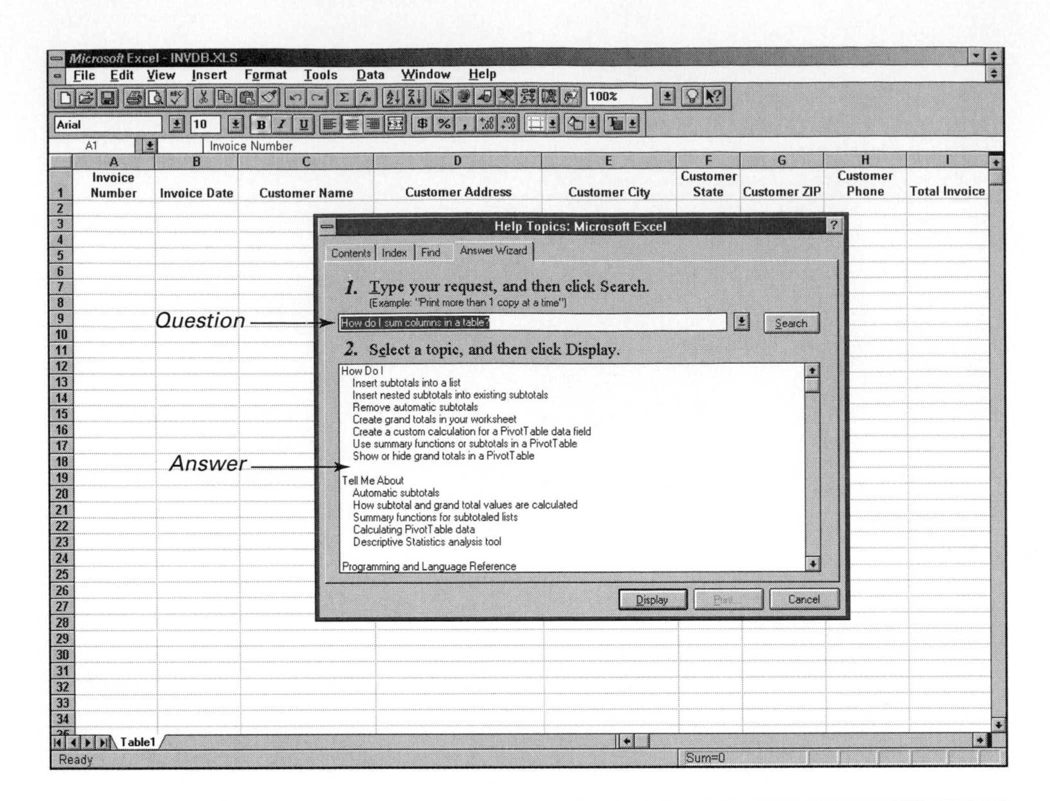

FIGURE 14.11 *Sample Microsoft* Excel *Screen with Question-Answer Dialogue*

Question-answer dialogue strategy was also popular in developing interfaces for character-based screens used in mainframe applications. But they are still commonly found in today's GUI applications. Take for instance the Answer Wizard provided by Microsoft's *Windows* products. Notice in Figure 14.11 that the user is instructed to "Type your request". Users can type a key word or type an entire sentence describing what they are seeking help for. For example, users could formulate a question in their own words such as, "How do I sum columns in a table?" The exact wording is not important. The Answer Wizard simply looks for key words and responds.

Direct Manipulation

The newest and most popular of the four user interface styles allows **direct manipulation** of graphical objects appearing on a screen. In a sense, we've already demonstrated some common aspects of this interface style by showing how the other interface styles are often represented in a graphical environment such as Microsoft's *Windows.*

Essentially, this user interface style focuses on using icons, or small graphic images, to suggest functions to the user. A trash can icon, for instance, might symbolize a delete command. Selecting the icon with a pointing device such as a mouse or light pen executes the function. Also, icons can work in conjunction with one another. For instance, a pointing device can be used to drag the icon of a file folder (representing a named file) to a trash can icon—instructing the system to delete (or throw away) the file.

Entire applications are being developed using this visual approach to more effectively interact with users. Intuit's very successful product, *Quicken for Windows,* is one such application. Figure 14.12 depicts a sample *Quicken* screen. Notice how icons are used to provide a visual, intuitive clue to the user. This visual means of communicating with users has proven to substantially improve user acceptance and productivity.

Several direct manipulation interface products are available. It was the Apple

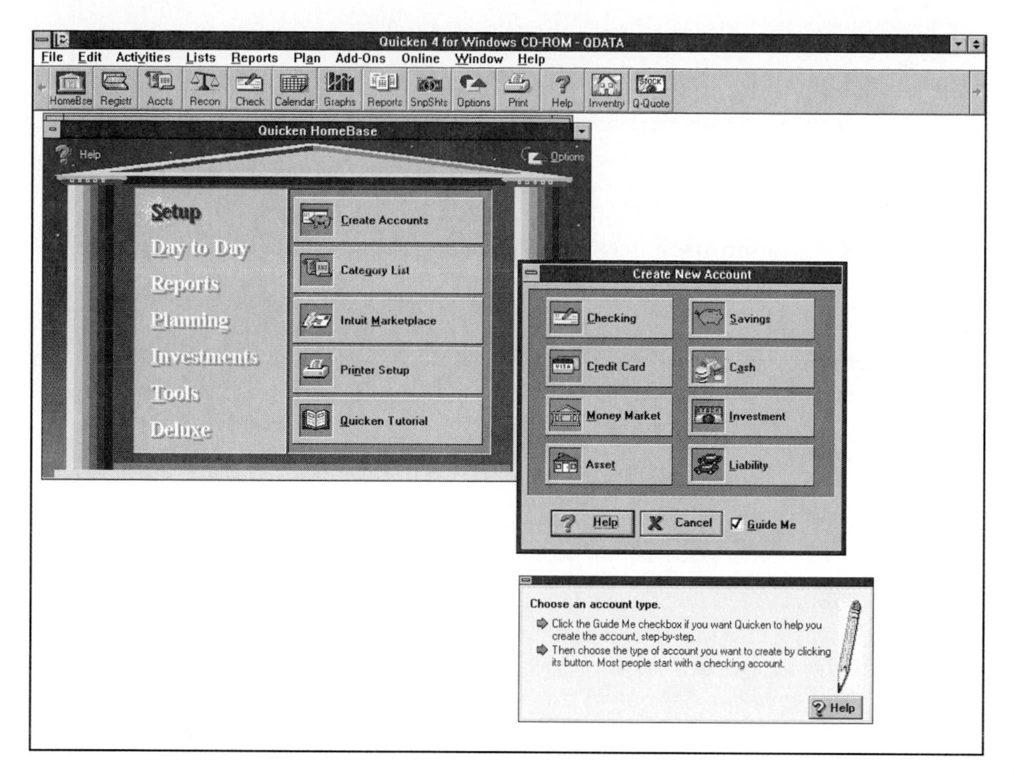

FIGURE 14.12 *Sample Intuit* Quicken *Screen*

Macintosh interface (called *Finder*) that popularized the use of icons. IBM's *Presentation Manager* and Microsoft's *Windows 95* are the equivalent in the IBM/clone marketplace. Development products are available for developing applications that take on the graphical characteristics of these interface products. The development products provide numerous graphic controls and features that simplify the development process—and the number of graphic controls and features is growing each day. Some of the graphic controls and features utilized in direct manipulation applications have been discussed earlier in this chapter as advances in usage of the other alternative user interface styles. Also, some features were demonstrated in Chapters 12 and 13, which focused on the design and prototyping of graphical inputs and outputs, respectively. But there are many other features commonly used in many direct manipulation applications, including **windows, menu bars, menus, tool bars, buttons** (with text and/or picture), **scroll bars** (horizontal and vertical), **dialogue boxes, tabs,** and others. It is essential that today's designer be aware of the graphic controls and features available and their appropriate usage. The Suggested Readings includes some good references for learning more about direct manipulation interface features.

The four strategies we've presented should be integrated with human considerations in mind. If you evaluate your dialogue against fundamental human factors, you may save yourself from that dreaded 2:00 A.M. phone call: "Betty? Did I wake you? Sorry! But we have a problem. The system is asking us for . . ." Let's examine some human factors that should be considered when designing and prototyping the user interface for an application.

HUMAN FACTORS FOR USER INTERFACE DESIGN

Nowhere are human factors as important as they are in user interface design. Just ask the typical systems analyst who spends half the day answering phone calls from system users who are having difficulty using the computer system. That's why we want to discuss **human engineering.**

System users can be broadly classified as either dedicated or casual.

A **dedicated system user** is one who will spend considerable time using specific programs. This user is likely to become comfortable and familiar with the terminal or PC's operation.

However, some system users are only casual users.

The **casual system user** may use a specific program only on an occasional basis. This user may never become truly comfortable with the terminal or the program.

The user who hasn't used a terminal or a microcomputer is becoming less common in this computer-literate age. It is difficult to imagine today's youth being uncomfortable with the computer or display terminal. Still, most of today's systems are designed for the casual system user, with an emphasis on user friendliness.

General Human Engineering Guidelines

Given the type of user for a system, there are a number of important human engineering factors that should be incorporated into the design:

- *The system user should always be aware of what to do next.* The system should always provide instructions on how to proceed, back up, exit, and the like. There are several situations that require some type of feedback:[1]
  - *Tell the user what the system expects right now.* This can take the form of a simple message such as READY, ENTER COMMAND, ENTER CHOICE, or ENTER DATA.
  - *Tell the user that data have been entered correctly.* This can be as simple as moving the cursor to the next field in a form or displaying a message such as INPUT OK.
  - *Tell the user that data have not been entered correctly.* Short, simple messages about the correct format are preferred. Help functions can supplement these messages with more extensive instructions.
  - *Explain to the user the reason for a delay in processing.* Some actions require several seconds or minutes to complete. Examples include sorting, indexing, printing, and updating. Simple messages such as "SORTING—PLEASE STAND BY" or "INDEXING—THIS MAY TAKE A FEW MINUTES. PLEASE WAIT" tell the user that the system has not failed.
  - *Tell the user that a task was completed or was not completed.* This is especially important in the case of delayed processing, but it is also important in other situations. A message such as "PRINTING COMPLETE" or "PRINTER NOT READY—PLEASE CHECK AND TRY AGAIN" will suffice.
- *The screen should be formatted so that the various types of information, instructions, and messages always appear in the same general display area.* This way, the system user knows approximately where to look for specific information. Note: In a windowing environment, the window standards (e.g., *CUA, Windows, Motif,* etc.) provide their own explicit standards or guidelines for where items are to appear.
- *Messages, instructions, or information should be displayed long enough to allow the system user to read them.*
- *Use display attributes sparingly.* Display attributes, such as blinking, highlighting, and reverse video, can be distracting if overused. Judicious use allows you to call attention to something important—for example, the next field to be entered, a message, or an instruction.
- *Default values for fields and answers to be entered by the user should be specified.* In windowing environments, valid values are frequently presented

[1] Adapted from Kenneth Kendall and Julie Kendall, *Systems Analysis and Design* (Englewood Cliffs, NJ: Prentice Hall, 1988).

in a separate window or dialogue box as a scrollable region. The default is typically the first value.

- *Anticipate the errors users might make.* System users will make errors, even when given the most obvious instructions. If it is possible for the user to execute a dangerous action, let it be known (a message or dialogue box such as "ARE YOU SURE YOU WANT TO DELETE THIS FILE?" is appropriate). An ounce of prevention goes a long way!

With respect to errors, a user should not be allowed to proceed without correcting the error. Instructions on how to correct the error should be displayed. In windowing environments, the error can be highlighted using a display property and then explained in a pop-up window or dialogue box. A HELP option can be defined to display additional instructions or give clarification in the body zone or a pop-up zone. In any event, the system user should never get an operating system message or fatal error. If the user does something that could be catastrophic, the keyboard should be locked to prevent any further input. An instruction to call the analyst or computer operator should be displayed in this situation.

Dialogue Tone and Terminology

The overall tone and terminology of a dialogue are also important human engineering considerations. The session should be user friendly (a goal that is frequently not achieved). With respect to the tone of the dialogue, the following guidelines are offered:

- *Use simple, grammatically correct sentences.* It is best to use conversational English rather than formal, written English.
- *Don't be funny or cute!* When someone has to use the system 50 times a day, the intended humor quickly wears off.
- *Don't be condescending; that is, don't insult the intelligence of the system user.* For instance, don't offer rewards or punishment.

With respect to the terminology used in the dialogue, the following suggestions may prove helpful:

1. *Don't use computer jargon.*
2. *Avoid most abbreviations.* Abbreviations assume that the user understands how to translate them. Check first!
3. *Use simple terms.* Use NOT CORRECT instead of INCORRECT. There is less chance of misreading or misinterpretation.
4. *Be consistent in your use of terminology.* For instance, don't use EDIT and MODIFY to mean the same action.
5. *Instructions should be carefully phrased, and appropriate action verbs should be used.* The following recommendations should prove helpful:
   - Try SELECT instead of PICK when referring to a list of options. Be sure to indicate whether the user can select more than one option from the list of available options.
   - Use TYPE, not ENTER, to request the user to input specific data or instructions.
   - Use PRESS, not HIT or DEPRESS, to refer to keyboard actions. Whenever possible, refer to keys by the symbols or identifiers that are actually printed on the keys. For instance, the ↵ is used on some terminals to designate the RETURN or ENTER key.
   - When referring to the cursor, use the term POSITION THE CURSOR, not POINT THE CURSOR.

DISPLAY FEATURES THAT AFFECT USER INTERFACE DESIGN

The design of a user interface can be enhanced or restricted by the available features of your terminal display or monitor/keyboard. Let's examine some of these features.

Display Area

The size of the display area is critical to user interface design. For terminal displays, the two most common display areas are 25 (lines) by 80 (columns) and 25 by 132. Some displays can be easily shifted between these two sizes. Some newer displays are designed to show more lines, for example, 65 lines on one screen. Some terminals can show four complete 25 by 80 character screens simultaneously.

For microcomputer and workstation display monitors, display size is measured in pixels. The greater the number of pixels, the more information can be displayed. Pixel display areas are specified in width by height.

Character Sets and Graphics

Every display uses a predefined character set. Most displays use the common ASCII character set. Some displays allow the programmer to supplement or replace the predefined character set. Additionally, most displays today offer graphics capabilities. Graphics capabilities must be supported by graphics controllers and software that allow the programmer to take advantage of the graphics capabilities. Graphics-based displays may support a virtually unlimited character set.

Paging and Scrolling

The manner in which the display area is shown to the user is controlled by both the technical capabilities of the display and the software capabilities of the computer system. Paging and scrolling are the two most common approaches to showing the display area to the user.

> **Paging** displays a complete screen of characters at a time. The complete display area is known as a page (also called a screen or frame). The page is replaced on demand by the next or previous page; much like turning the pages of a book.

The other common alternative to paging is called scrolling.

> **Scrolling** moves the displayed characters up or down, one line at a time. This is similar to the way movie and television credits scroll up the screen at the end of a movie.

Once again, PC displays offer a wider range of paging and scrolling options.

Display Properties

Most displays in use today provide a wide variety of display properties that may be manipulated to more effectively present data and information. **Display** properties are characteristics that change the way in which a character or group of characters is displayed on a screen. For example, displays allow color to be used to highlight specific messages, data, or areas of the screen. Nondisplay of a selected field (for example, passwords) is another example. PC displays and software offer a wide array of display properties that can simplify and improve user interfaces.

Split-Screen and Windowing Capabilities

Split-screen capability is a variation on the windows concept. The display screen, under software control, can be divided into different areas (called windows). Each window can act independently of the other windows, using features such as paging, scrolling, display attributes, and color. Each window can be defined to serve a different purpose. Windows can be resized, moved, and hidden or recalled on user demand.

Windowing is rapidly becoming accepted as a standard user interface. Most microcomputer products use windowing interfaces such as Microsoft's *Windows,* IBM's *Presentation Manager,* OSF's *Motif,* and Apple's *Finder.*

Keyboards and Function Keys

Although not a display feature, most modern terminals and monitors are integrated with keyboards. The number of keys and their layout may vary, but most keyboards contain special keys called function keys.

> **Function keys** (usually labeled F1, F2, and so on) can be used to implement certain common, repetitive operations in a user interface (for example,

START, HELP, PAGE UP, PAGE DOWN, EXIT). These keys can be programmed to perform common functions.

Function keys should be used consistently. That is, a system's programs should consistently use the same function keys for the same purposes.

Pointer Options

We are no longer restricted to the keyboard as the only input technology for displays and terminals. Today, we are encountering many other selection options, such as touch-sensitive screens, voice recognition, and pointers. The most common pointer is the mouse.

A **mouse** is a small hand-sized device that sits on a flat surface near the terminal. It has a small roller ball on the underside. As you move the mouse on the flat surface, it causes the pointer to move across the screen. Buttons on the mouse allow you to select objects or commands to which the cursor has been moved. Alternatives include trackballs, pens, and trackpoints.

Since GUIs have for the most part replaced the older, more traditional character-based interface methods, this chapter will focus on how to design and prototype GUI applications.

HOW TO DESIGN AND PROTOTYPE A USER INTERFACE

It shouldn't surprise you that a typical user interface may involve many possible screens (which in turn may consist of several windows), perhaps hundreds! Each screen can be designed and prototyped. But what about the coordination of these screens?

Screens typically occur in a specific order. You may also be able to toggle among the screens. Additionally, some screens may appear only under certain conditions. And to make matters more difficult, some screens may occur repetitively until some condition is fulfilled. This sounds almost like a programming problem, doesn't it? We need a tool to coordinate the screens that can occur in a user interface.

Step 1: Chart the Dialogue

A **state transition diagram** is used to depict the sequence and variations of screens that can occur when the system user sits at the terminal. (Note: The authors are using the term *screen* in a general sense. When designing graphical interfaces the term may refer to an entire display screen, a window, or a dialogue box.) You can think of it as a road map. Each screen is analogous to a city. Not all roads go through all cities. Rectangles are used to represent display screens. The arrows represent the flow of control and triggering event causing the screen to become active or receive focus. The rectangles only describe what can appear during the dialogue. The direction of the arrows indicates the order in which these screens occur. Notice that a separate arrow, each with its own label, is drawn for each direction. Why? Because different actions trigger flow of control from and flow of control to a given screen.

Let's examine a dialogue that is under construction for the SoundStage project (see Figure 14.13). The partially completed SoundStage state transition diagram is being developed using a CASE product, Popkin's *System Architect*. Draw your attention to the following:

1. The partial state transition diagram includes all the SoundStage input screens developed in Chapter 12.
2. The diagram also includes all output screens designed in Chapter 13.
3. The MEMBER SERVICES SYSTEM screen will be a new screen that will need to be designed and prototyped. This screen will serve as the application's main window. It will play a major role in providing the user with the ability to get access to the system's input and output screens that were designed earlier. It will also provide the user with the ability to complete a number of

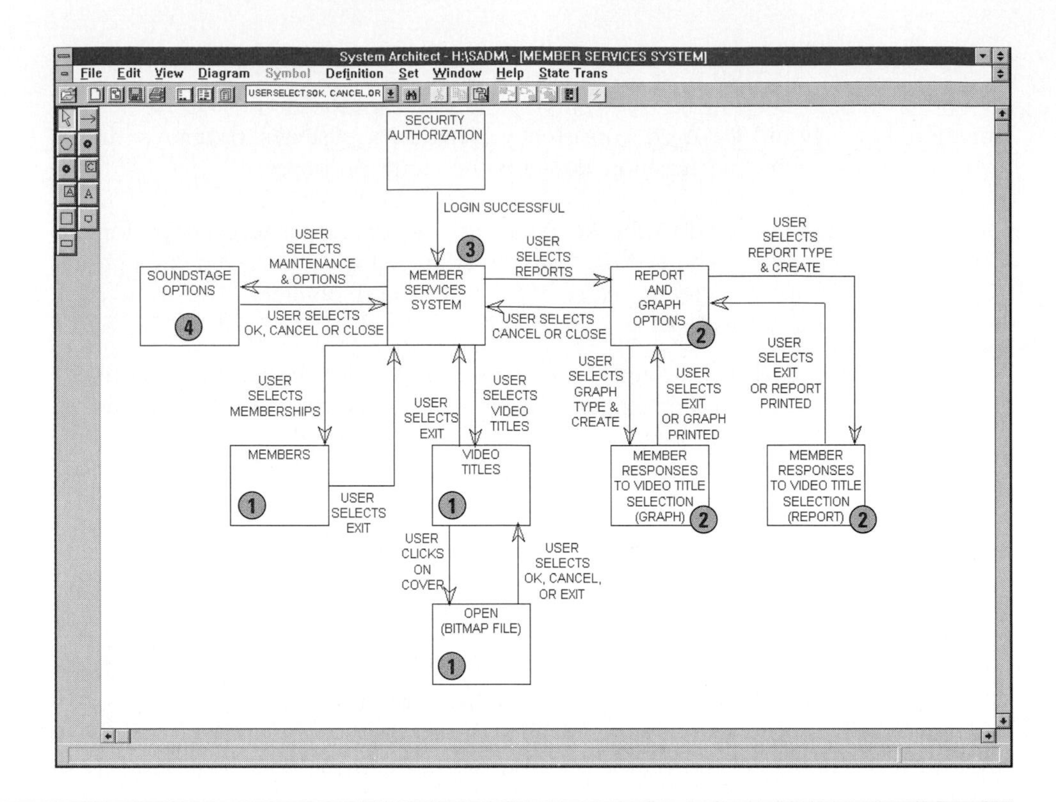

FIGURE 14.13 *SoundStage State Transition Diagram*

FIGURE 14.14 *SoundStage Log-in Screen*

additional functions (beyond input and output processing) that are commonly established during user interface design. Notice that it will be accessible only when the users have first been provided with the SECURITY AUTHORIZATION screen and have successfully logged into the system.

(4) The SOUNDSTAGE OPTIONS screen is another new screen to be created. This screen is not being added to coordinate or bring together the application's screens and functions. Rather, this screen will allow users to set various user options and defaults to be used during their session. For example, selecting a printer, zooming, and many other options.

State transition diagrams such as the one presented in Figure 14.13 can become

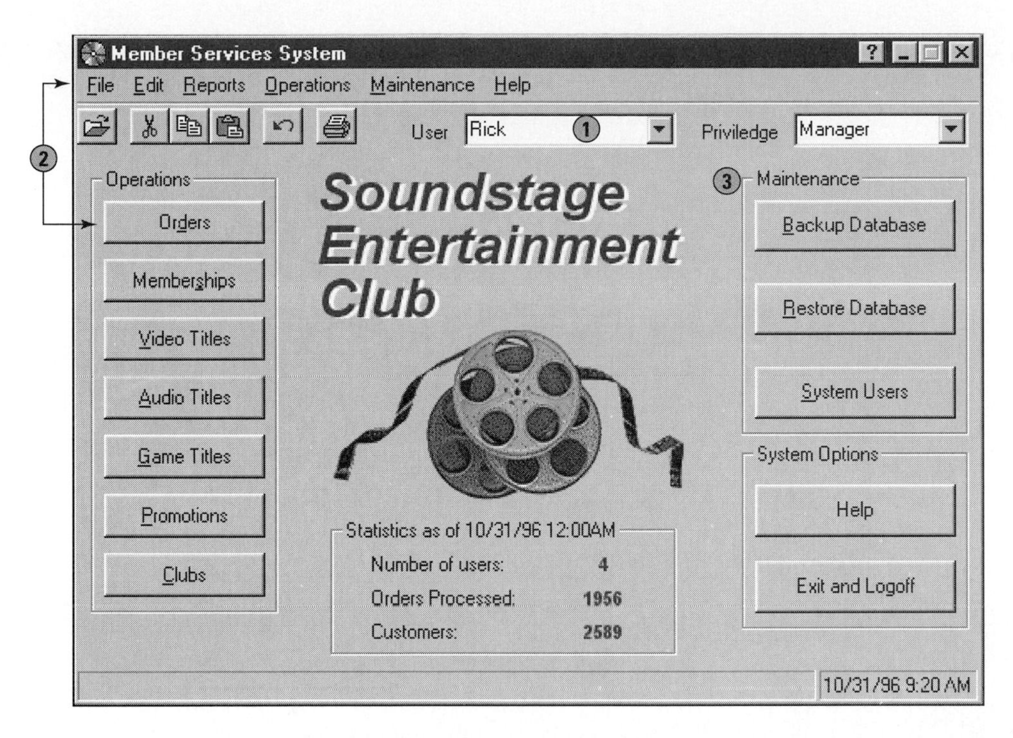

FIGURE 14.15 *SoundStage Main System Screen Prototype*

quite large. Especially when *all* input, output, help, and other screens are added to the diagram. Therefore, it is common to partition the diagram into a set of separate simpler and easier-to-read diagrams.

Step 2: Prototype the Dialogue and User Interface

Recall that we have some new screens to design and prototype. Some of these new screens were identified to bring together the application and its input and output screens that were designed earlier. Some screens were identified to provide the users some flexibility with customizing the application's interaction to suit their own preferences. Still others may have been identified to deal with system controls, such as backup and recovery.

Let's look at some new screens that were to be created for SoundStage Member Services System. We first draw your attention to the screen that the user first sees, Figure 14.14. This screen was created during input design as a means of security. According to our state transition diagram, the successful log-in of a user results in the screen depicted in Figure 14.15. Notice the following:

① The users and their access privileges are confirmed. Based on their access privilege, certain functions will be enabled and disabled.

② Through a menu bar selection or through a vertical menu of buttons, the user is able to complete common member services business operations. These buttons will lead to screens that allow the user to process appropriate transactions via input screens designed and prototyped earlier. Text labels were used for buttons since the analyst was unable to establish pictures that all users could readily identify with as a representation of the operations. Notice that the menu bar and buttons contain hot keys to provide the user with the flexibility of selecting via the keyboard or mouse. A group box was used to visually associate the buttons that represent operations.

③ In addition to common business operations, the user is provided with the ability to complete various routine maintenance operations.

Via the menu bar of the MEMBER SERVICES SYSTEM screen, users can choose to set options for their work session. This new screen is depicted in Figure 14.16.

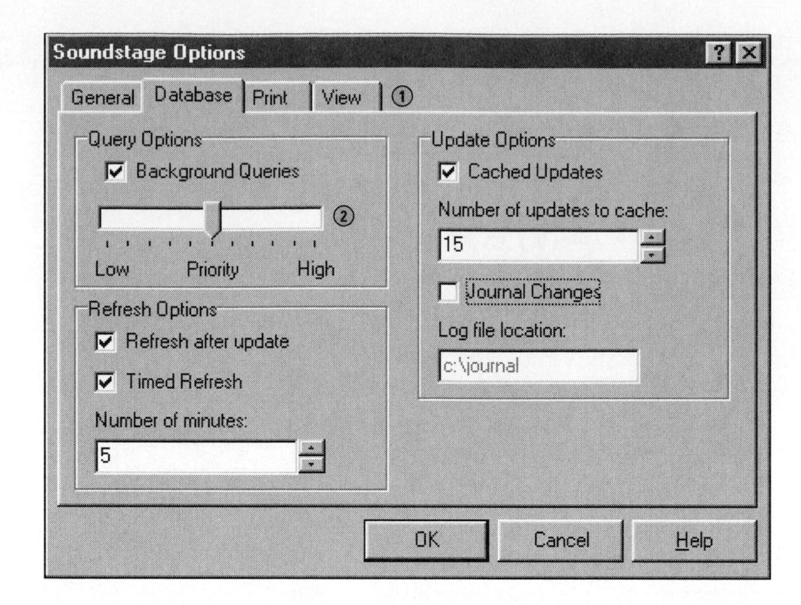

FIGURE 14.16 *SoundStage Prototype of Options Screen*

① This screen utilizes tabs as a means of allowing the user to alter four different options. A **slider** control is used to allow the user to adjust the priority for background queries.

② This control is often used for items whose values are best presented as a spatial representation, and when an approximate rather than precise value is sufficient.

In reality, the analyst would need to prototype the content and appearance of the General, Print, and View tabs as well as the Database tab. According to our state transition diagram, this screen will return control to the parent window, MEMBER SERVICES SYSTEM.

The remaining screens depicted in Figure 14.17 were created during input and output design (see Chapters 12 and 13, respectively). Study the state transition diagram and the screens that we just examined to see how this portion of the overall system dialogue will work together. There is one more point that needs to be stressed relative to the input and output screens that were previously developed. By studying the entire collection of screens you may discover the need to revise some screens. Such issues as color, naming consistencies of common buttons and menu options, and other look-and-feel conflicts may need to be resolved. Once again, adherence to any standards governing GUIs should be confirmed.

Step 3: Obtain User Feedback

There are several ways to prototype a dialogue and user interface. Many computer-aided systems engineering (CASE) products include dialogue and interface prototyping tools. Some CASE products can support both screen design as well as testing (through simulation). Most database management systems and fourth-generation languages include screen and dialogue generators that can be used to prototype the user interface. Exercising (or testing) the user interface is a key advantage of all these prototyping environments.

Exercising (or testing) the user interface means that system users experiment with and test the interface design before extensive programming and actual implementation of the working system. Analysts can observe this testing to improve on the design.

In the absence of prototyping tools, the analyst should at least simulate the dialogue by walking through the screen sketches with system users. Regardless, user feedback is essential. The analyst should encourage the user to participate in testing the application's interface. Finally, the analyst should expect to revisit steps 1 and 2 as needed changes become known.

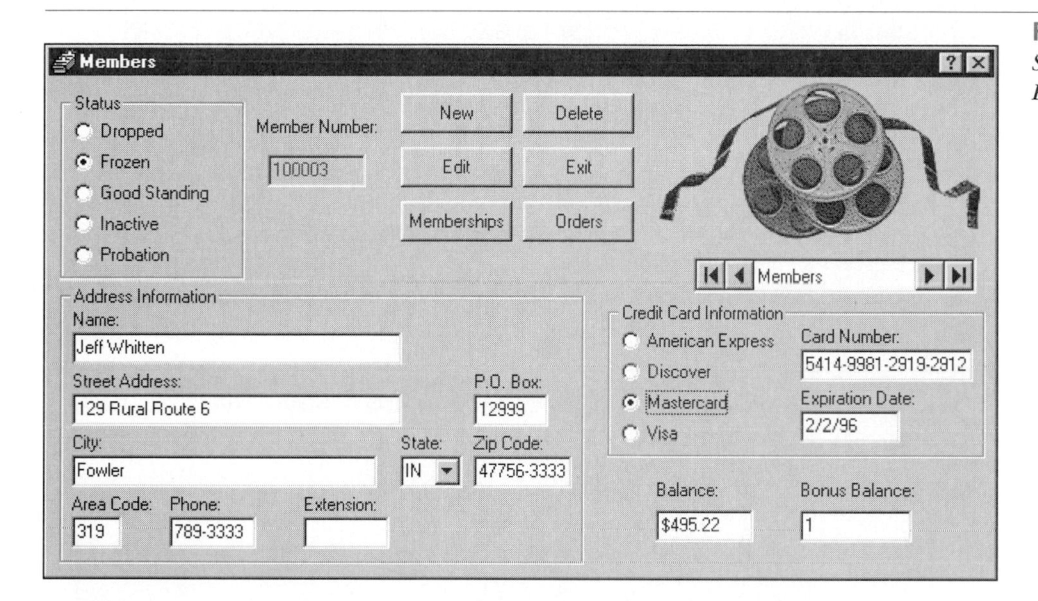

FIGURE 14.17
SoundStage Members Prototype Screen

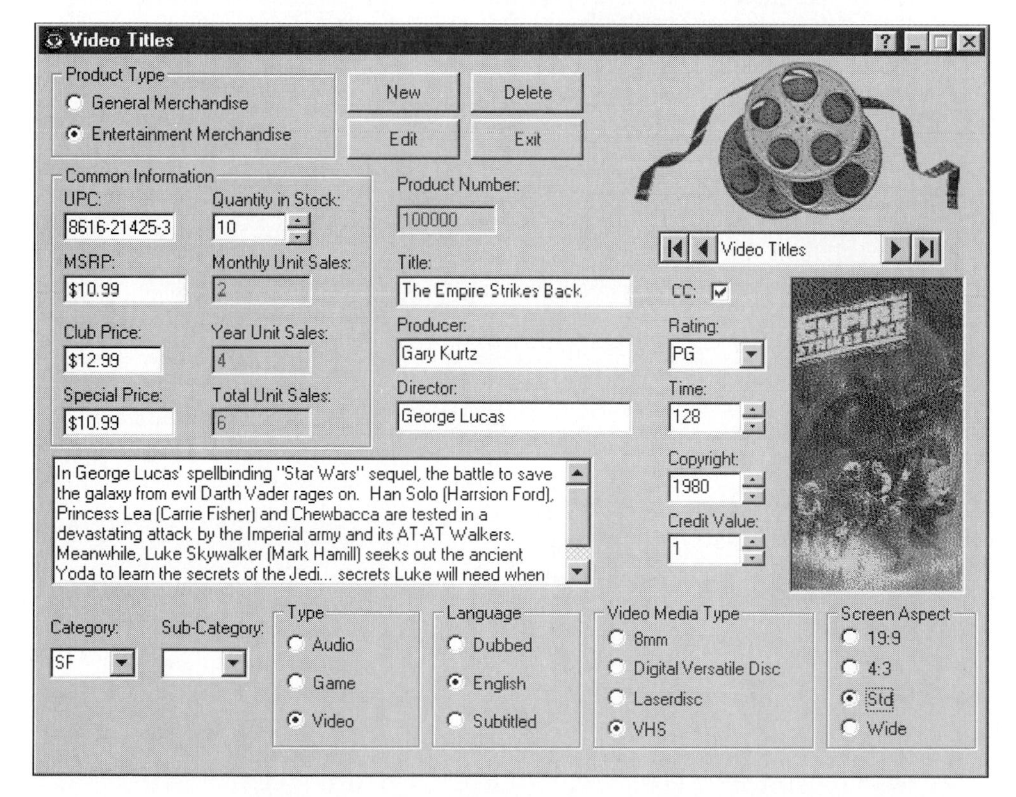

Soundstage Video Titles Prototype Screen

FIGURE 14.17 (*continued*)
Soundstage Open Prototype Screen

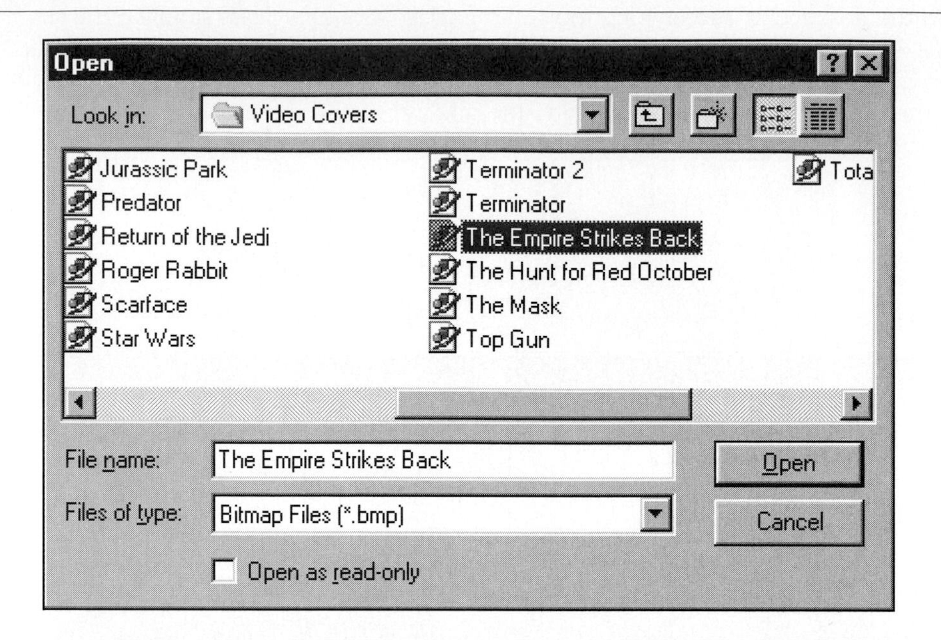

SoundStage Report and Graph Options Prototype Screen

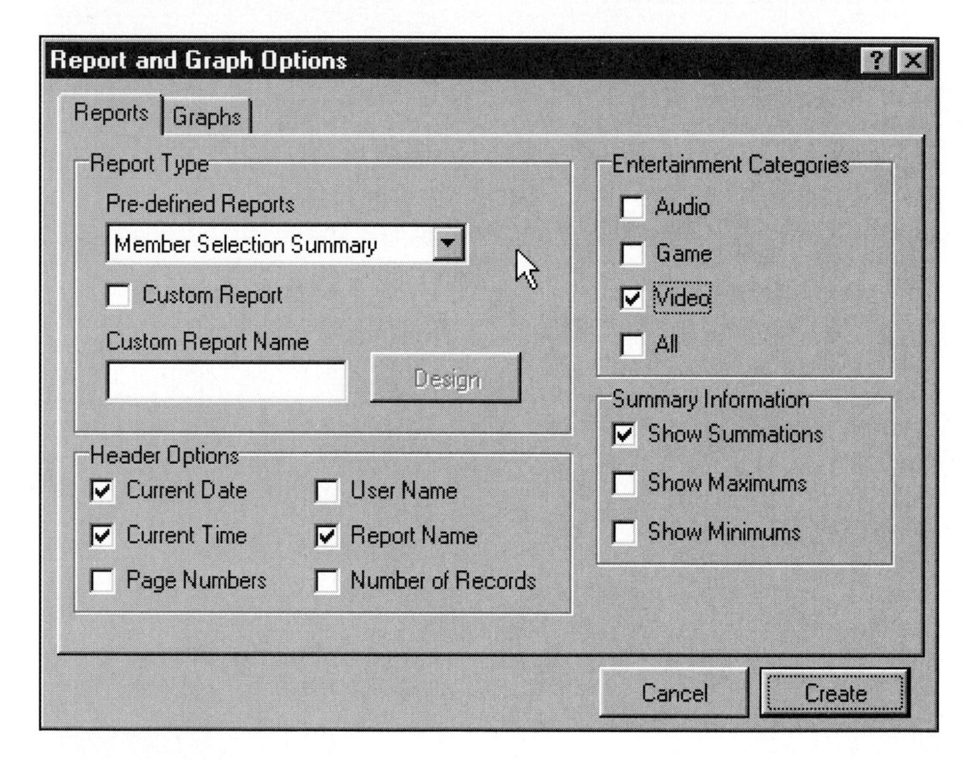

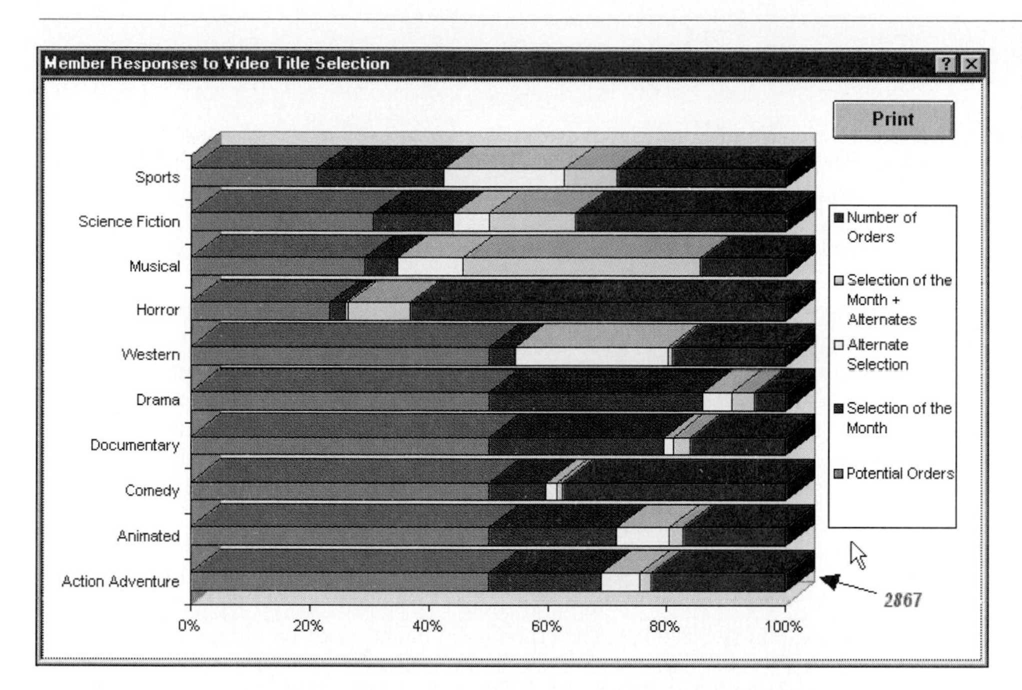

FIGURE 14.17 (*concluded*)
SoundStage Member Response to Video Title Selection Graph Prototype Screen

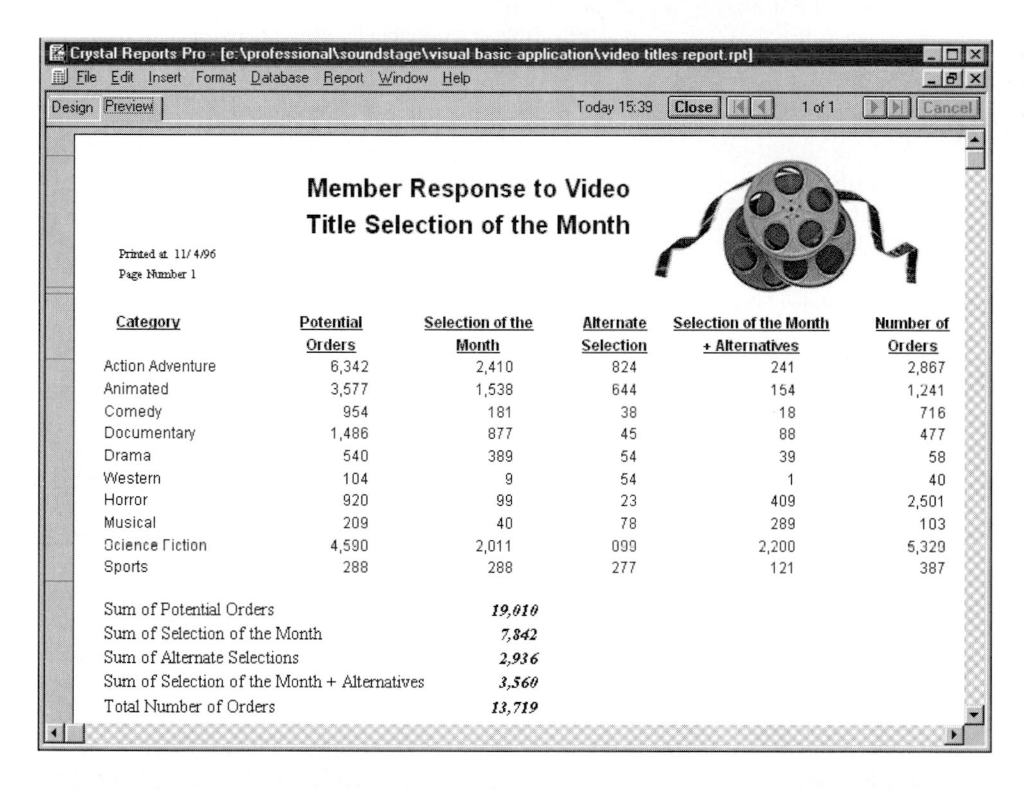

Member Response to Video Title Selection of the Month

Printed at 11/4/96
Page Number 1

Category	Potential Orders	Selection of the Month	Alternate Selection	Selection of the Month + Alternatives	Number of Orders
Action Adventure	6,342	2,410	824	241	2,867
Animated	3,577	1,538	644	154	1,241
Comedy	954	181	38	18	716
Documentary	1,486	877	45	88	477
Drama	540	389	54	39	58
Western	104	9	54	1	40
Horror	920	99	23	409	2,501
Musical	209	40	78	289	103
Science Fiction	4,590	2,011	099	2,200	5,329
Sports	288	288	277	121	387

Sum of Potential Orders	*19,010*
Sum of Selection of the Month	*7,842*
Sum of Alternate Selections	*2,936*
Sum of Selection of the Month + Alternatives	*3,560*
Total Number of Orders	*13,719*

SoundStage Member Response to Video Title Selection Report Prototype Screen

WHERE DO YOU GO FROM HERE?

This chapter provided a detailed overview of user interface design and prototyping. You are now ready to learn how to complete software design. Software design in covered in the next chapter and is concerned with two aspects: the modular design (or decomposition) of a program into manageable pieces for implementation, and the packaging of program specifications. Before proceeding, we recommend that you first review Chapter 9 to see where software design falls into the overall systems development process.

SUMMARY

1. User interface design is the specification of a conversation between the system user and the computer. This conversation generally results in either input or output—possibly both. There are several types of user interface styles, including menu selection, instruction sets, question-answer dialogues, and direct manipulation. Traditionally these styles were viewed as alternatives. However, with recent movements toward designing systems with graphical user interfaces, a blending of all styles can be found.
   - **Menu selection** is a strategy of dialogue design that presents a list of alternatives or options to the user. The system user selects the desired alternative or option by keying in the number or letter that is associated with that option. In today's graphical user interface designs, menu selection is commonly represented using **menu bars, pull-down menus, cascading menus, pop-up menus,** and **iconic menus.**
   - **Instruction sets** is a strategy where the application is designed using a dialogue syntax that the user must learn. There are three types of syntax: structured English, mnemonic syntax, and natural language. Each of these types can be found in today's applications using a graphical user interface.
   - **Question-answer dialogue strategy** is a style that was primarily used to supplement either menu-driven or syntax-driven dialogues. The simplest questions involve yes or no answers. Question-answer dialogue strategy was also popular in developing interfaces for character-based screens used in mainframe applications. But they are commonly found in today's GUI applications.
   - The newest and most popular of the four user interface styles allows **direct manipulation** of graphical objects appearing on a screen. Essentially, this user interface style focuses on using icons, small graphic images, to suggest functions to the user.
2. Nowhere are human factors as important as they are in user interface design. Therefore, user interface design must be completed with human engineering considerations. System users can be broadly classified as either dedicated or casual.
   - A **dedicated system user** is one who will spend considerable time using specific programs. This user is likely to become comfortable and familiar with the terminal or PC's operation.
   - The **casual system user** may use a specific program only occasionally. This user may never become truly comfortable with the terminal or the program.
3. Given the type of user for a system, there are a number of important human engineering factors that should be incorporated into the design:
   - The system user should always be aware of what to do next.
   - The screen should always be formatted so that the various types of information, instructions, and messages always appear in the same general display area.
   - Messages, instructions, or information should be displayed long enough to allow the system user to read them.
   - Display attributes should be used sparingly.
   - Default values for fields and answers to be entered by the user should be specified.
   - The errors users might make should be anticipated.
4. With respect to the tone of the dialogue, the following guidelines are offered:
   - Use simple, grammatically correct sentences.
   - Don't be funny or cute!
   - Don't be condescending; that is, don't insult the intelligence of the system user.

   With respect to the terminology used in the dialogue, the following suggestions may prove helpful:
   - Don't use computer jargon.
   - Avoid most abbreviations.
   - Use simple terms.
   - Be consistent in your use of terminology.
   - Instructions should be carefully phrased, and appropriate action verbs should be used.
5. The design of a user interface can be enhanced or restricted by the available features of your terminal display or monitor/keyboard. Such features that should be considered include display area, character sets and graphics, paging and scrolling, display properties, split-screen and windowing capabilities, keyboards and function keys, and printer options.
6. The steps involved in designing and prototyping a user interface include charting the dialogue, prototyping the dialogue and user interface, and obtaining user feedback.

KEY TERMS

buttons, p. 493
cascading menu, p. 486
casual system user, p. 494
dedicated system user, p. 494
dialogue boxes, p. 493
direct manipulation, p. 492
display, p. 496
exercising user interfaces, p. 500
function keys, p. 496
human engineering, p. 493
iconic menu, p. 489
instruction set, p. 489
menu bar, p. 484
menus, p. 493
menu selection, p. 484

mnemonic syntax, p. 490
mouse, p. 497
natural language syntax, p. 491
paging, p. 496
pop-up menu, p. 488
pull-down menu, p. 485
question-answer dialogue, p. 491
scroll bars, p. 493
scrolling, p. 496
slider, p. 500
state transition diagram, p. 497
structured English, p. 490
tabs, p. 493
tear-off menu, p. 485
tool bars, p. 493
user interface design, p. 484
windows, p. 493

REVIEW QUESTIONS

1. What is user interface design?
2. Identify four styles of user interface design.
3. What five types of menus are commonly used to develop modern, graphical-based screens?
4. What are three types of instruction set syntax?
5. Differentiate between the two types of system users—dedicated and casual.
6. List several important general human engineering guidelines for user interface design.
7. List several important guidelines for dialogue tone used in user interface design.
8. List several display features that may affect user interface design.
9. What are the steps for designing and prototyping an application's interface?
10. How are state transition diagrams used in designing screens?
11. What is meant by "exercising" a user interface?

PROBLEMS AND EXERCISES

1. To what extent should the system user be involved during user interface design? What would you do for the user? What would you ask the user to do for you? Detail a strategy that consists of specific steps you and the system users would follow.
2. What documentation prepared during input design and output design is needed during user interface design? How does that input and output design documentation relate to user interface design?
3. Explain the difference between a dedicated and a casual terminal user. How would your strategy for designing user interfaces for a dedicated user differ from that for designing user interfaces for a casual user?
4. Display properties can be overused and frequently hinder a user's performance during a terminal session. Cite some examples in which display properties are appropriate and in which display properties may hinder a user's performance.
5. Describe four strategies for designing user interfaces. What criteria would you consider when choosing between the strategies?

PROJECTS AND RESEARCH

1. Arrange to study a microcomputer application. It may be either a business system (such as an inventory, accounts receivable, or personnel system) or a productivity tool (such as a word processor, spreadsheet, or database system). Analyze the human engineering of the user interface. Analyze the human engineering of the display screens. If possible, discuss the dialogue and screens with system users. What do they like and dislike about the design?
2. Redesign the application in Project 1 to improve the user interface and screens. If possible, discuss your improved design with users. Do they like your new design better? Did they raise any concerns?
3. Obtain documentation or magazine reviews on an automated screen-design aid. If possible, arrange for a demonstration. How would the product improve your productivity? How would the product decrease your productivity? What features do you dislike or would you prefer to see?

MINICASES

1. Richards & Sons, Inc., is a large investment company located in Tampa, Florida, that buys stocks, bonds, commodities, and various other assets for its clients. The company also manages clients' investment portfolios for a variety of investment objectives. Finding new people with money to invest is crucial to its success, so keeping track of clients and potential clients is very important. Morgan Adamson is the senior analyst in charge of the new sales prospect and contact management system at Richards & Sons, Inc. The new system has just been installed and the project team is working with the system users who are doing acceptance testing. Morgan is talking to Kevin Brock, the junior analyst who was responsible for the design of the system's user interface.

MORGAN I guess you've heard that some of the system users are not very happy with the new contact tracking system. In particular, they are expressing dissatisfaction with the user interface that you designed.

KEVIN I don't understand what the problem could be—I put a lot of time and effort into that design. What are the specific problems?

MORGAN Some of the users are complaining that they don't know what to do next or how to use some of the screens. Are you sure that all of the screens are consistent throughout this system?

KEVIN I didn't think it was important where the information was as long as I clearly labeled it. Besides, the users should be expected to read the screen. I deliberately put lots of highlighting, blinking, and reverse video fields on the screens to draw attention to important information.

MORGAN Yes, I know. Some of the users claim that there is so much highlighted information that it distracts from the purpose of the screen. I've also had complaints that proper default values were not specified for some of the fields.

KEVIN Default values were specified for the most common fields, but the users should expect to have to type some of the information in—that's their job, isn't it?

MORGAN In some cases, maybe several possible default values should have been provided in a pop-up window to eliminate possible keying errors. Remember, we want to reduce the amount of user entry keystrokes as much as possible. Some of the clerks say that they don't understand some of the terminology and abbreviations on some of the screens.

KEVIN Oops! I must not have checked the screens carefully enough to catch some of the computer terminology. That's no big deal. I can fix that right away. Are there any other problems?

MORGAN Other users have indicated that the use of certain menu options is not consistent across all of the screens. Didn't you use the same labels and screen buttons or menus for the same actions throughout the entire interface?

KEVIN No, I didn't. I thought that I could reuse the function keys to mean different things as long as I clearly labeled them on each screen.

MORGAN Why?

KEVIN Some of the keyboards have only 12 function keys, and I was afraid I might run out of keys to assign unless I reused them.

MORGAN You need to realize that the system users don't always read the instructions that you provide. Whether that is right or wrong is not important; you need to be consistent so that the users don't have to learn a different set of function keys for each screen.

KEVIN I guess I really didn't think that through very well; it shouldn't be too difficult to make all the function key assignments consistent.

MORGAN There have also been some complaints about an insufficient amount of help messages for some of the input screens. For instance, one clerk said the contact entry screen consistently refused to accept the date of contact he tried to enter. The system did output an error message indicating that the date was incorrect and should be reentered, but the clerk doesn't understand why the date was invalid. He says the system doesn't provide any information about the correct format of the date or any valid examples. You did design help screens and messages for each of the input screens, didn't you?

KEVIN Uh, well, I'm not exactly sure what you mean by a "help" screen. I did very thorough input error checking so that invalid data could not be entered. I created error messages for each of the edited fields. I thought that would be sufficient for the users to identify the input error and make the necessary correction.

MORGAN I think I'm beginning to understand the problem. We need to talk about some very important human engineering guidelines that you need to follow whenever you are designing a user interface. First . . .

a. What did Kevin do wrong in designing the user interface? What are some of the other mistakes that an analyst might make when designing a user interface?
b. How could these mistakes have been avoided? (What would you have done differently?) What role should the system user play in interface design?

SUGGESTED READINGS

Fitzgerald, Jerry. *Internal Controls for Computerized Information Systems*. Redwood City, CA: Jerry Fitzgerald & Associates, 1978. This is our reference standard on the subject of designing internal controls into systems. Fitzgerald advocates a unique and powerful matrix tool for designing controls. This book goes far beyond any introductory systems textbook—must reading.

Galitz, Wilbert. *It's Time to Clean Your Windows: Designing GUIs That Work*. New York: John Wiley & Sons, 1994. This is an excellent book that provides an unbiased reference on designing graphical interfaces.

Kendall, Kenneth, and Julie Kendall. *Systems Analysis and Design*. Englewood Cliffs, NJ: Prentice Hall, 1988. Chapter 16 provides another look at user interface design.

Kozar, Kenneth. *Humanized Information Systems Analysis and Design*. New York: McGraw-Hill, 1989. A good user-oriented treatment of user interface design.

Martin, Alexander, and David Eastman. *The User Interface Design Book for the Applications Programmer*. New York: John Wiley & Sons, 1996.

Microsoft Corporation. *The Windows Interface Guidelines for Software Design*. Microsoft Press, 1996.

Nielsen, Jakob. *Coordinating User Interfaces for Consistency*. San Diego, CA: Academic Press, 1989

Weinschenk, Susan, and Sarah C. Yeo. *Guidelines for Enterprise-Wide GUI Design*. New York: John Wiley & Sons, 1995.

15

SOFTWARE DESIGN

CHAPTER PREVIEW AND OBJECTIVES

In this chapter you will learn how to design good programs and how to package program design specifications. You will know that you understand how to design programs and package design specifications into a format suitable for programmers when you can:

- Factor a program into manageable program modules that can be easily modified and maintained.
- Recognize a popular structured design tool for depicting the modular design.
- Revise a data flow diagram to reflect necessary program detail before program design.
- Describe two strategies for developing structure charts by examining data flow diagrams.
- Design programs into modules that exhibit loose coupling and high cohesive characteristics.
- Package program design specifications for communicating program requirements for implementation.

SOUNDSTAGE

SOUNDSTAGE ENTERTAINMENT CLUB

SCENE

We begin this episode in the conference room where Sandra and Bob are busy putting together their technical design statement.

SANDRA

I hope you don't mind, Bob, but I decided to get an early start. I just have a few more items to include in our technical design statement.

BOB

Not at all. Sorry I'm late. That workshop went a little over.

SANDRA

We have to make sure this document contains everything. If we don't include all the design specifications, we can't expect our programming team to accurately implement everything. By the way, do you have a diskette copy of our final prototype?

BOB

Yup. Got it right here in my pocket. Here you go.

SANDRA

That's great. But I have a big favor to ask. Could you get a hard-copy printout of prototype screens and supporting documentation? I'd like to provide them with a diskette so that they can run the application and see how it works, but I also know that they will be needing some of the supporting documentation that you can get from *Delphi*.

BOB

No problem. I'll have to go to my office to reload it on my computer. This will take a while. Can I have it to you tomorrow afternoon?

SANDRA

Sure, that would be great. Say . . . what's this workshop that you've been attending?

BOB

Well, I've been assigned a new project. We're going to be working on some legacy systems. There are several COBOL programs that are 10 to 15 years old. They want to build a *Windows* front end to those applications. They don't have time to rebuild the applications as a client/server application. So they feel that a simple *Windows* front-end appearance would be an enhancement, especially since the users are crying out for a better interface for those applications. In addition to the front end, they want us to clean up the program structures and code. The programs have been modified so many times and by so many people that they've gotten to the point that the programs are too difficult to maintain.

SANDRA

So what kind of training are you receiving in the workshop?

BOB

It's a workshop on structured design.

SANDRA

I used to do structured design all the time. Back when I was working on developing mainframe-based applications. They have some cool techniques for deriving a program structure that ensures that it will be easy to read, understand, and maintain.

BOB

I did have some exposure to it in school. The course is a nice refresher. You know, they also teach you how to measure the quality of the program design. I didn't realize structured design gave you a way of evaluating the quality of your program design.

SANDRA

Hah! I bet you skipped that day. Well I hope you learn a lot.

BOB

I'm sure I will. I'm going to get these printouts for you. See you tomorrow.

DISCUSSION QUESTIONS

1. Sandra and Bob are creating a technical design statement that will be used for implementation. Why wouldn't they simply provide the team of people who will implement the system with a diskette containing the system prototype?
2. Why do you think all older systems are not simply redeveloped as client/server applications?
3. What value would a *Windows*-based front end to a mainframe application have to the users?
4. Why are legacy programs difficult to modify and maintain? What kinds of program restructuring and code revisions do you suspect need to be made to the older programs?

WHAT IS SOFTWARE DESIGN?

Our study of systems design is nearly complete. You've designed the databases, inputs, outputs, and on-line user interfaces. You've selected appropriate computer equipment and packaged software (which it is hoped has been delivered and installed during systems design). The final step involves **software design.**

From your programming courses, you may think of software design as algorithm or logic design. That is not the subject of this chapter. We don't intend to reteach you how to draw structured program flowcharts, to prepare pseudocode, or to construct box charts (sometimes called Nassi-Schneidermann charts). That is clearly a subject for a programming textbook. Instead, we are concerned with how the programming specifications are presented to the computer programmer for implementation. To this end, we view software design as consisting of two components—modular design and packaging.

> **Modular design** is the decomposition of a program into modules.

What is a module? It could be a subroutine or subprogram. And it could be a main program. It also could be a unit of measure smaller than any of those. For instance, a module could be a paragraph in a COBOL program. So what is a module?

> A **module** is a group of executable instructions with a single point of entry and a single point of exit.

> **Packaging** is the assembly of DATA, PROCESS, INTERFACE, and GEOGRAPHY design specifications for each module.

Software design received much emphasis throughout the 80s and early 90s. However, as newer software development tools (such as *Powerbuilder* and *Delphi*) and techniques (such as RAD) for developing client/server applications evolved, software design received less emphasis. Why? These tools redirected programmers' attention away from thinking in terms of a program's decomposition into manageable pieces. Rather, the emphasis is placed on event-driven programming wherein the focus is more directed to the components that make up the application's screens (for example, the windows, buttons, menu bars) and events that are triggered by the user's interaction with those components. In addition, the nature of the tools and newer development techniques (such as RAD) emphasized quickly building prototypes of the application for the users to review—with little or no regard for the need to do modular decomposition.

Given this trend, you might ask why software design is covered in this book. The answer is simple. While most newer applications being developed today are client/server applications and are being developed using RAD-like techniques, most organizations are simply reengineering their older applications. Their strategy is to simply build a front-end graphical interface for those older applications. This allows the users the advantage of interacting with a much more effective and intuitive interface that is used in the newer client/server applications. What about the back end or actual code? That is where modular design comes into play. Organizations don't have the time to rewrite or redevelop all their old applications. Thus, organizations are using software products to simply reengineer or improve the code. Those improvements are achieved by applying many of the modular design principles and techniques taught in this chapter.

Finally, some readers are likely to interpret the material covered in this chapter as an invasion of the programmer's turf. It really varies from one computer information systems shop to another. Some shops insist that the analyst prepare detailed modular designs and program specifications (at a level close to pseudocode). Other shops believe the analyst's job ends with general programming specification, leaving modular design to the programmer. Depending on your opinion, you may want to omit this chapter. However, we recommend the chapter for the following reasons:

- Your career may take you to organizations or management that prefer both of the approaches.
- The chapter helps tie the previously developed design specifications to the program specifications that normally initiate systems implementation.
- In the absence of a company standard, you may want to consider a rigorous personal standard for presenting specifications. In Chapter 16 you will learn that the analyst is frequently engaged in a large number of activities during systems implementation. The more thorough and complete your programming specifications are, the less time you'll have to spend clarifying those specifications for the programmer.

Let's begin with a study of one of the most popular modular design techniques used to develop those more traditional mainframe-based applications of the 80s.

STRUCTURED DESIGN

Ed Yourdon and Larry Constantine (Page-Jones, 1980) developed a popular strategy for determining an optimal modular design for programs. Their technique is called **structured design.** This technique deals with the size and complexity of a program by breaking up the program into a hierarchy of modules that result in a computer program that is easier to implement and maintain.

The concept of structured design is simple—design a program as a top-down hierarchy of modules. The top-down structure of these modules is developed according to various design rules and guidelines. The resulting hierarchy of modules can then be evaluated according to certain quality acceptance criteria to ensure the best modular design for the program. Upon completion, the modules are to be implemented using structured programming principles.

Thus, as depicted in Figure 15.1, during structured design the system designer's primary focus is on the process component of the information systems framework. And as you will soon see, the business process model is instrumental in identifying and deriving the hierarchy of modules for a program.

The following sections present the various tools and techniques used in structured design. While our SoundStage case study is a client/server solution and would negate the need for modular design, you are provided with SoundStage examples of how structured design might have been applied.

Structure Charts

The primary tool used in structured design is the structure chart. **Structure charts** are used to graphically depict a modular design of a program. Specifically, they show how the program has been partitioned into smaller more manageable modules, the hierarchy and organization of those modules, and the communication interfaces between modules. Structure charts, however, do not show the internal procedures performed by the module or the internal data used by the module.

Figure 15.2 depicts the symbol set used for structure charts. We draw your attention to the following:

1. Structure chart modules are depicted by named rectangles. Modules are factored, from the top down, into submodules. The highest-level module is referred to as the system or root module. It serves to coordinate or "boss" the modules appearing directly beneath it. In turn, those modules may coordinate those modules appearing immediately below them.
2. Structure chart modules are presumed to execute in a top-to-bottom, left-to-right sequence. The line connecting two modules represents a normal call. For example, SYSTEM MODULE calls MODULE B.
3. An arc-shaped arrow located across a line (representing a module call) means the module makes iterative calls. Thus, SYSTEM MODULE calls MODULE A to be performed *N* number of times, or until some condition is met.
4. A diamond symbol located at the bottom of a module means the module calls one and only one of the other lower modules that are connected to the

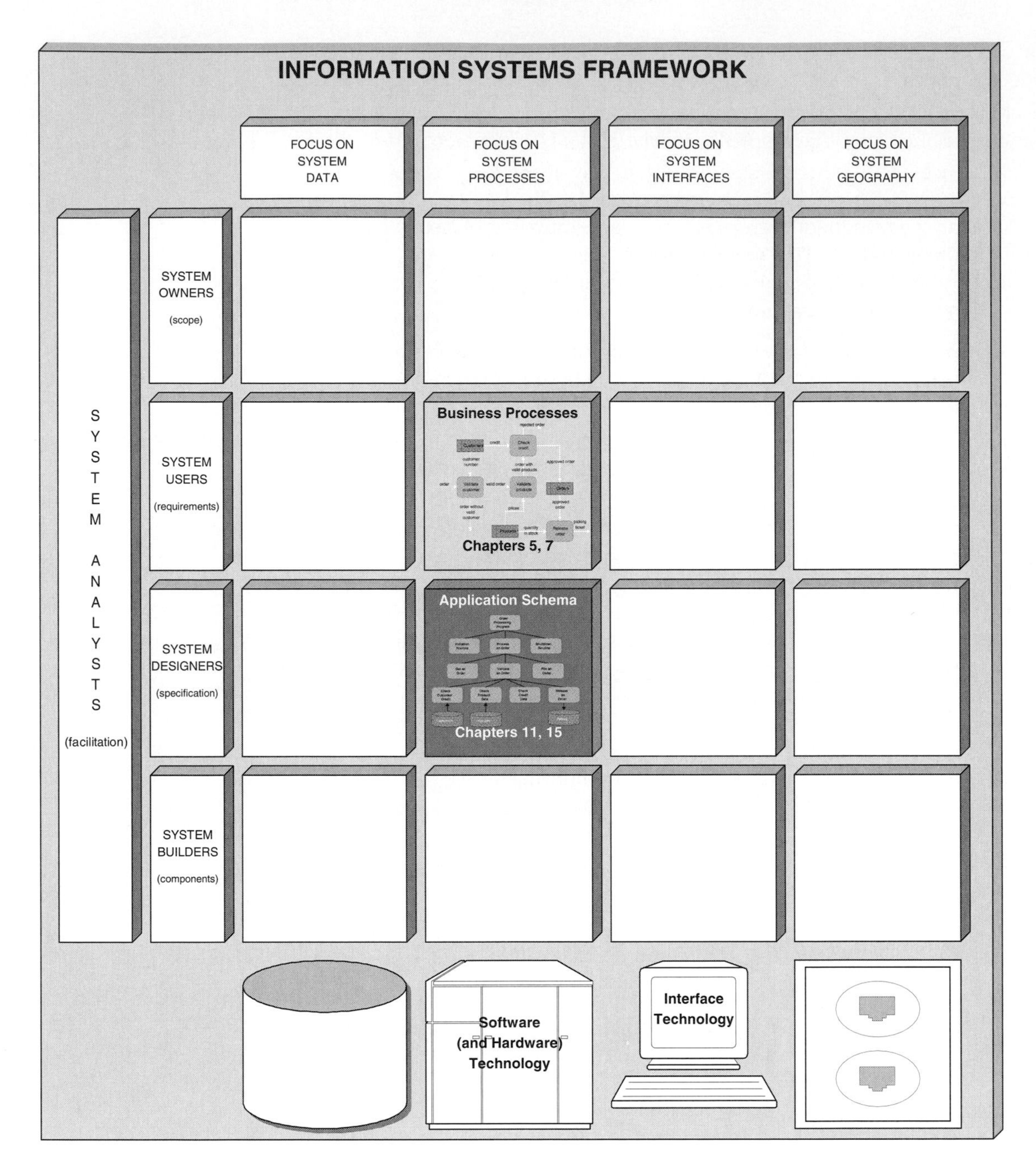

FIGURE 15.1 *Information Systems Framework Focus of Structured Design*

diamond. Thus, MODULE A can call MODULE C or MODULE D. Notice, however, that MODULE B can call MODULE G and it can call *either* MODULE E or MODULE F. This diagram construct is referred to as a *transaction center.*

5. Program modules communicate with each other through passing of data. Data being passed are represented by named arrows with a small circle on one end. The direction of the arrow is significant. Note that DATA A is being passed "up" from MODULE C to its parent, MODULE A. The downward direction of the arrow for DATA B implies that SYSTEM MODULE is passing it to MODULE B.

6. Programs may also communicate with each other through passing of messages or control parameters, called **flags.** Control flags are depicted by an arrow with a darkened circle on one end. As with data, the direction of the arrow implies the source and receiving modules.

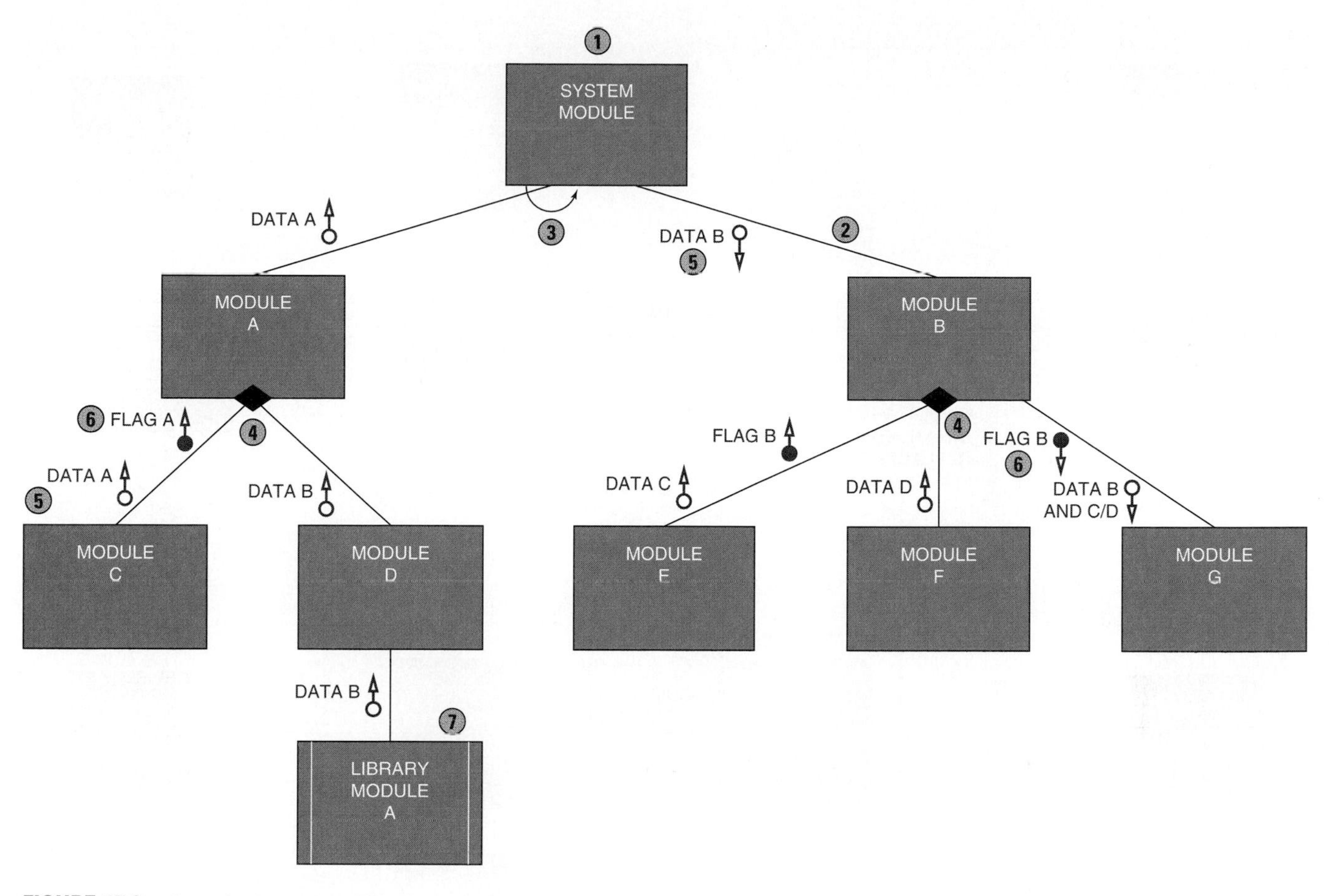

FIGURE 15.2 *Sample Structure Chart*

(7) Often a system designer recognizes the opportunity to use a predefined or library module. For example, perhaps one function of a program is to perform a special sort routine that was previously developed and placed in the program library for use by other developers for applications that might reuse that function. Such modules are depicted on a structure chart as a rectangle containing a vertical line on each side.

Structured design provides strategies for developing structure charts. Each of these strategies will be covered in detail later. But it is important to note that both strategies are based on the use of data flow diagrams to derive the structure chart.

Data Flow Diagrams of Programs

Structured design requires that data flow diagrams (DFDs) first be drawn for the program. You were introduced to data flow diagrams in Chapter 6, "Process Modeling." A DFD depicting the elementary processes for each business event were drawn in Chapter 6. Processes appearing on these DFDs may represent modules on a structure chart. However, because the audience was business users and management, not programmers, it is likely that those DFDs and their processes are not sufficiently detailed enough to be used to generate a structure chart. Thus, those DFDs need to be revised to show more detail. The following revisions may be necessary:

- Processes appearing on the DFD should do one function. Thus, some elementary processes may need to be expanded into two or more smaller processes that each accomplishes a single function. As a general rule, a process should have either one input or one output. Thus, as is demonstrated in Figure 15.3(a), PROCESS WITH MANY INPUTS & OUTPUTS should be expanded. It

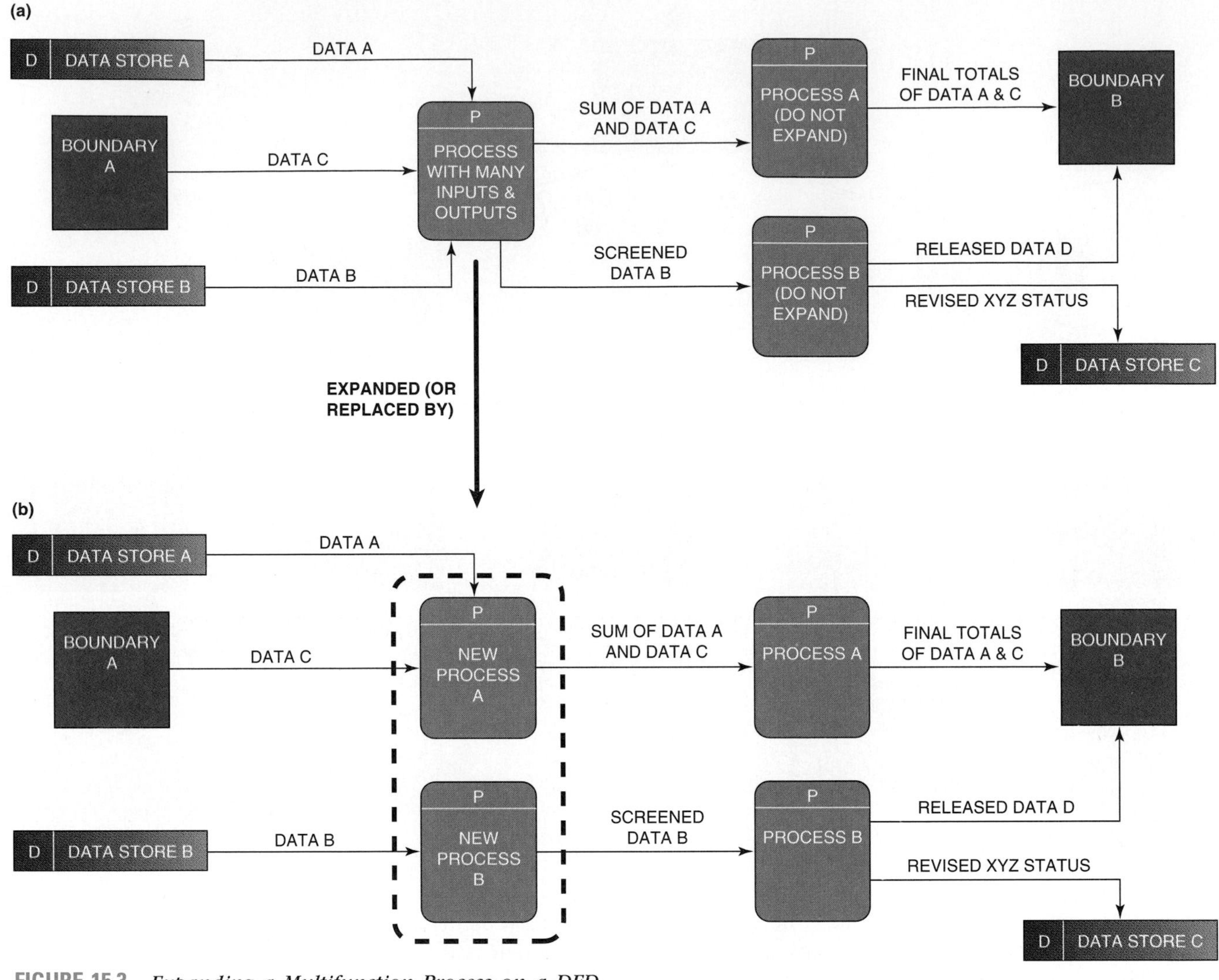

FIGURE 15.3 *Expanding a Multifunction Process on a DFD*

should be expanded into two or more processes—each smaller process should contain either one input or one output as demonstrated in Figure 15.3(b).

- To keep DFDs from becoming overly cluttered and overwhelming the user, processes for reading, modifying, and deleting data in a data store were not included on the elementary DFDs. Thus, processes now need to be added to handle data access and maintenance. Thus, you would need to locate each process that accesses or maintains data in a data store and add an intermediate process that will be responsible for that database action. For example, Figure 15.4(a) contains a process with four data flow connections to data stores—representing a read, add, update, and delete action. Figure 15.4(b) shows the addition of four new processes to accomplish those database functions.
- If you recall, data flow diagrams are often drawn from the perspective of a perfect world—thus, many of the trivial business processing exceptions and internal controls are not shown. Let's consider three typical situations. First, perhaps a DFD may show a process receiving data from a boundary (such as a customer), doing some processing of that data, and then passing the output to another process. In reality, the original input data may need to be edited

(a)

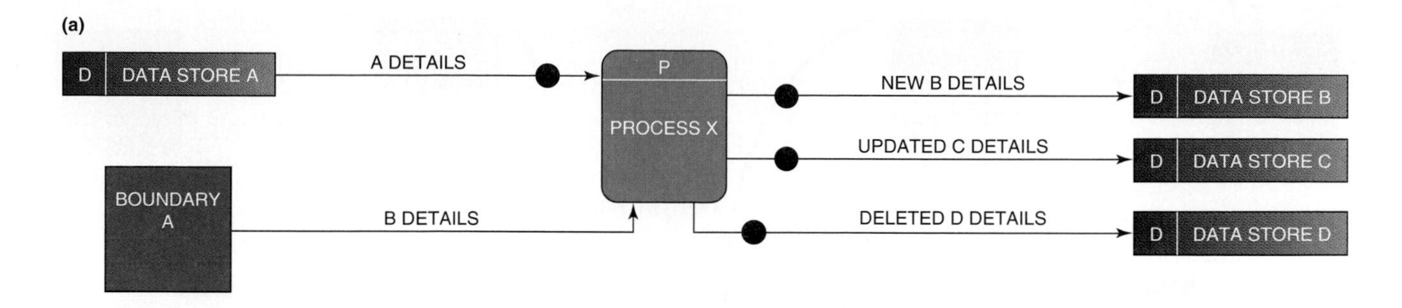

(b)

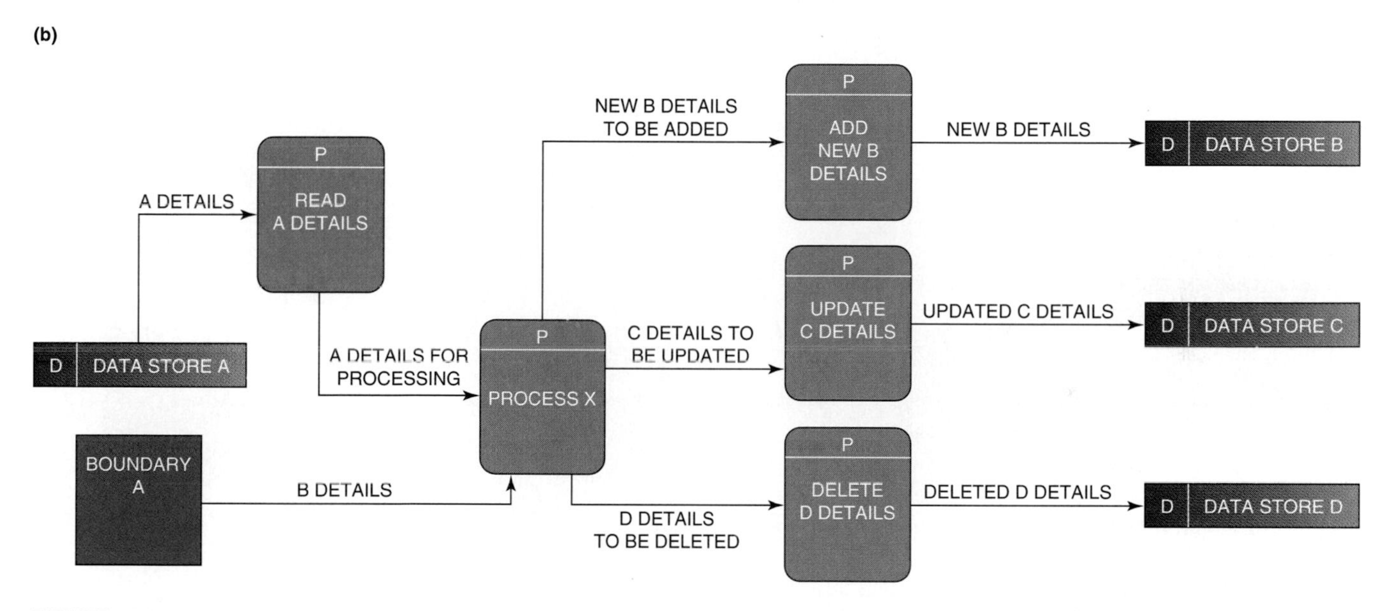

FIGURE 15.4 *Adding Data Access and Maintenance Process to a DFD*

and proper error handling routines performed. Second, perhaps a process accesses data from a data store, yet the data cannot be found. The data flow diagram may depict processes for handling this exception. Third, perhaps internal controls were established requiring that processing data for a particular business event be logged to provide an audit trail. But a programmer must know the details! Therefore, DFDs must be revised to include editing and error handling processes that were purposefully ignored during systems analysis and processes to implement internal controls established during systems design.

Once the data flow diagram has been revised, a structure chart can be derived. There are two strategies for developing the structure charts from data flow diagrams. Let's examine each strategy.

Transform Analysis

One approach used to derive a program structure chart from program DFD is transform analysis.

Transform analysis is an examination of the DFD to divide the processes into those that perform input and editing, those that do processing or data transformation (e.g., calculations), and those that do output.

Essentially, this strategy is based on the input-process-output concept. The data flow diagram is examined and partitioned into three areas—portions containing

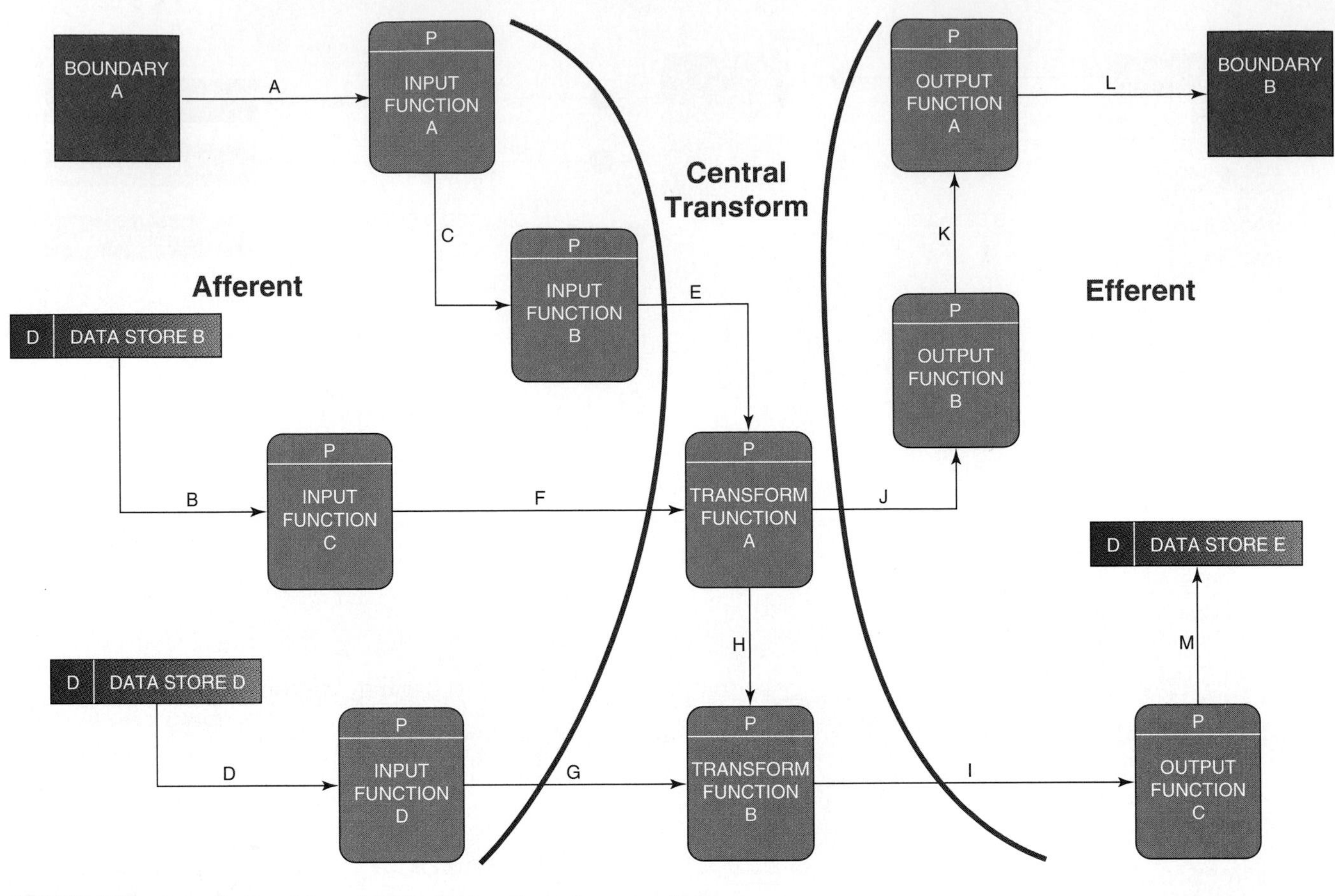

FIGURE 15.5 *Sample Partitioned DFD during Transform Analysis*

processes that appear to be doing input and editing, processes that appear to be doing processing, and those that appear to be doing output activities (see Figure 15.5). Structured design refers to that portion of the DFD consisting of processes that perform input and editing as the **afferent.** That portion of the DFD consisting of processes that do actual processing or transformations of data is referred to as the **central transform.** The portion of the DFD consisting of processes that do output is referred to as the **efferent.**

The strategy for identifying the afferent, central transform, and efferent portions of a DFD begins by first tracing the sequence of processing for each input. There may be several sequences of processing. As is depicted in Figure 15.6, each sequence of processing begins with an input and ends with an output. For example, the sequence of processing followed by input D concludes with the output M. A sequence of processing for a given input may actually split to follow different paths. For example, the sequence path followed by both input A and input B splits at TRANSFORM FUNCTION A into two separate paths, resulting in the output L and the output M.

Once sequence paths have been identified, each sequence path is examined to identify processes along that path that are afferent processes. The steps are as follows:

1. Beginning with the input data flow, the data flow is traced through the sequence until it reaches a process that does processing (transformation of data) or an output function. For example, the path of input data flow A can be traced until it reaches TRANSFORM FUNCTION A. The data flow going to a process that does transformation is then typically marked with an arc as a

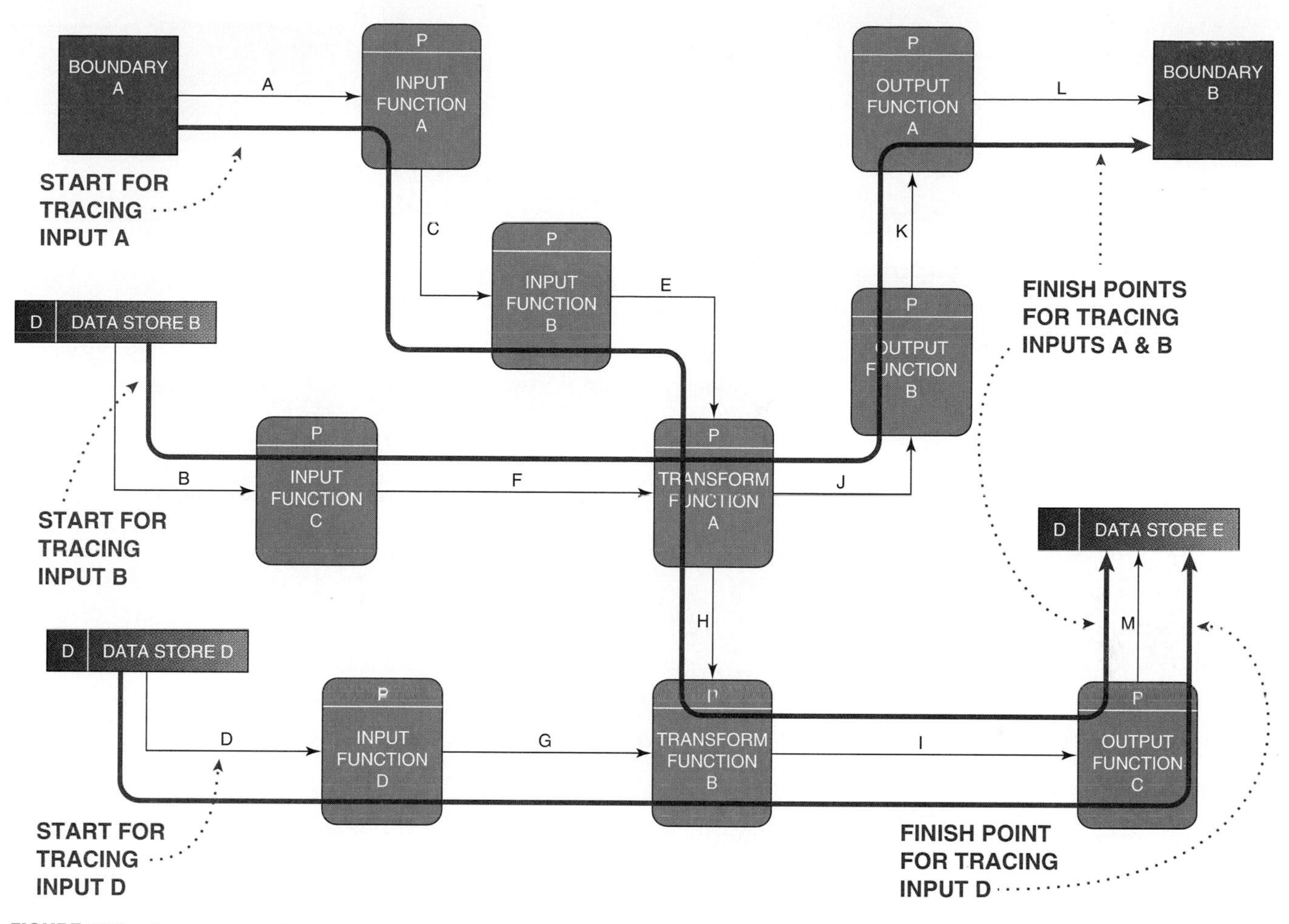

FIGURE 15.6 *Sequences of Processing on a DFD*

visual clue. Thus, the data flow E was marked accordingly in Figure 15.5. This mark suggests that all processes encountered along the path until that arc is reached are said to be afferent processes. They represent processes that perform input-oriented functions (such as initialization of counters, reading the first record, editing incoming data before processing).

2. Beginning with an output data flow from a path, the data flow is traced backward through connected processes until a transformation process is reached (or a data flow is encountered that first represents output). For example, the output data flow L can be traced to the point where data flow J is exiting TRANSFORM FUNCTION A. Likewise, that data flow is marked as a visual clue. All processes encountered during the backward tracing along the path are considered efferent processes. They are said to perform output preparation or presentation functions (such as writing an updated record to a database, formatting a report line, printing report totals, or displaying information to a user).
3. All other processes are then considered to be part of the central transform! These represent the processes that do the real work—making decisions or transforming data (such as checking a customer's credit or calculating an employee's pay). They can be characterized as processes that produce an output that can clearly be distinguished from the input data flows. In other words, the output data flow clearly is different in content or meaning from the incoming data flow.

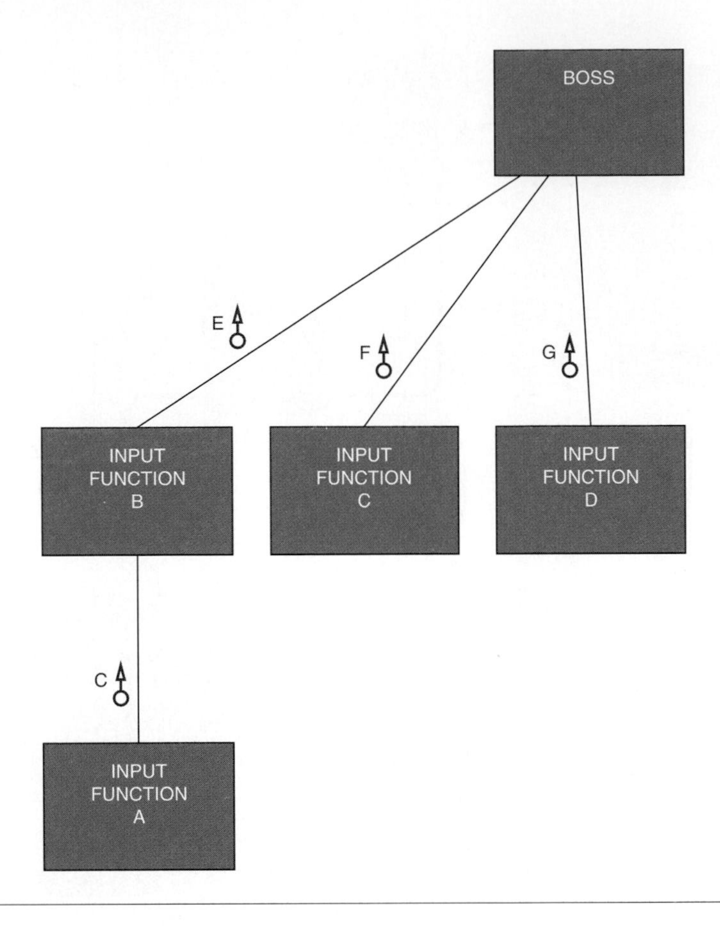

FIGURE 15.7 *Afferent Portion of Structure Chart*

Now that we have partitioned the DFD into groups of processes that correspond to afferent, central transform, or efferent categories, we can now begin to unveil a structure chart that communicates the modular design of our program.

In deriving the structure chart, we first create a process that will serve as the overall boss, sort of the "commander-in-chief" of all other modules (see Figure 15.7). This module manages or coordinates the execution of the other program modules. The last process encountered in a path that identifies afferent processes becomes a second-level module on the structure charts. Beneath that module should be a module that corresponds to its preceding process on the DFD. This would continue until all afferent processes in the sequence path are included on the structure chart. Notice that when a process appearing on the DFD is added to the structure chart, data flows from the DFD are also brought along. For example, notice that the output data flow F from the process INPUT FUNCTION C on our DFD was shown as a data communication between the module INPUT FUNCTION C and the module BOSS.

Next, if there is only one transformation process, it should appear as a single module directly beneath the boss module. Otherwise, a coordinating module for the transformation processes should be created and located directly above the transformation process (see Figure 15.8). Subsequently, a module per transformation process on the DFD should be located directly beneath the controller module. Notice that the structure chart has a notation to communicate that the module CENTRAL TRANSFORM CONTROLLER makes an iterative call to TRANSFORM FUNCITON B. As modules are added to a structure chart, such procedural information needs to be added.

The last process encountered in a path that identifies efferent processes (see step 2 above) becomes a second-level module on the structure charts (see

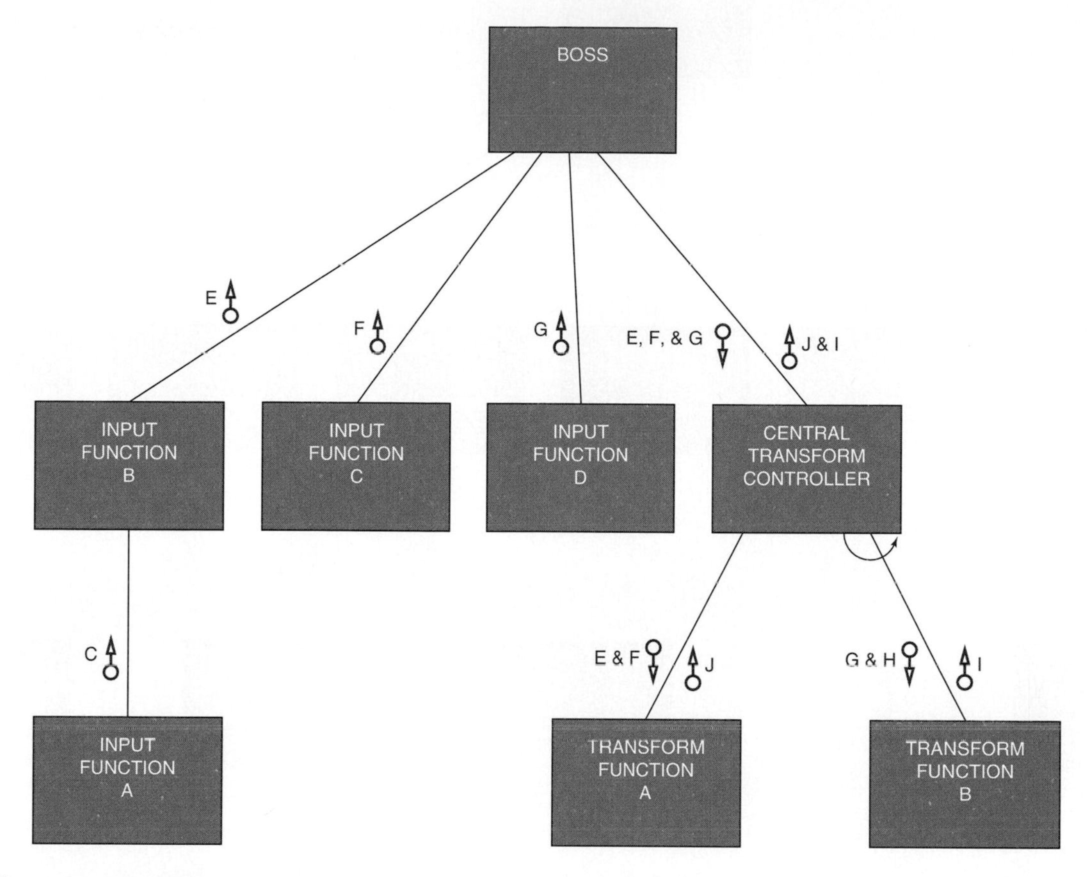

FIGURE 15.8 *Central Transform Portion of Structure Chart*

Figure 15.9). Beneath that module should be a module that corresponds to the succeeding process appearing on the sequence path. Likewise any process immediately following that process would appear as a module beneath it on the structure chart.

The above steps represent a simplified discussion of the transform-centered approach to structured chart development. In reality, a designer is not guaranteed that these steps will automatically result in a good modular design for a program. In reality, the designer would review the quality of the design. They may revisit this approach after evaluating the final design—making numerous revisions to the DFD and structure chart. Later you will learn about many of the issues that would be considered in evaluating the quality of a structure chart. For now, let's examine a SoundStage example where transform analysis is applied.

Recall that the SoundStage project called for a client/server solution using an object-oriented language and prototyping. Thus, structured design would not be used. However, for demonstration purpose, a sample DFD and structure chart demonstrating how transform analysis may have been completed is provided in Figures 15.10 and 15.11.

As you can see in Figure 15.10, the transform center consists of a single process, CALCULATE ORDER VOLUMES. Therefore, there was no need to add a controller module on our structure chart. Notice that the GET ORDER DETAILS receives input from three processes. On our structure chart (see Figure 15.11), the sequence for which GET ORDER DETAILS calls the three input processes is very significant. It is essential that READ MEMBER precedes READ MEMBER ORDER, which in turn must precede READ PRODUCT CONTAINED ON ORDER. How might we have known this sequencing requirement? The designer must consider facts that can be communicated in previous

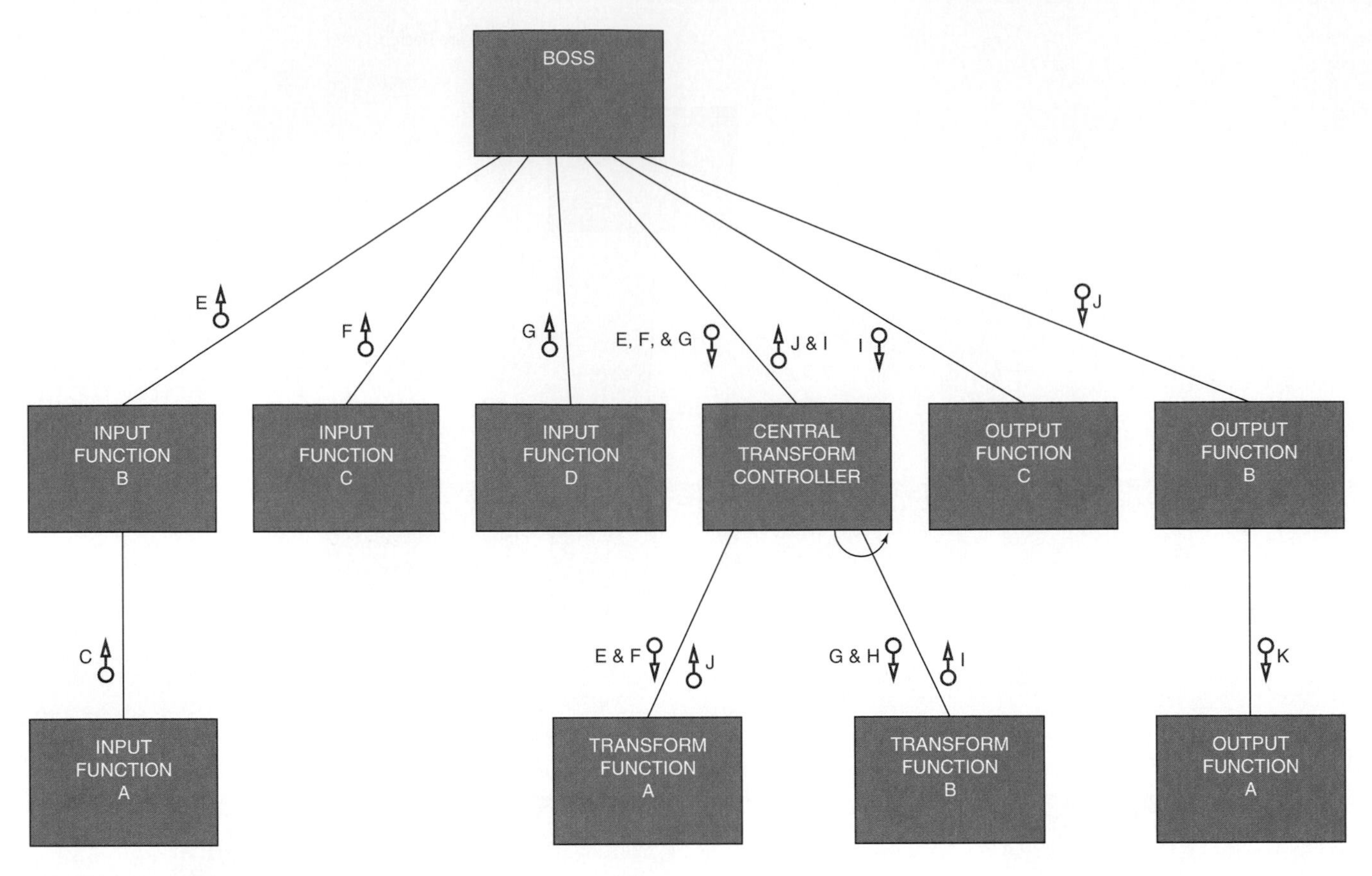

FIGURE 15.9 *Efferent Portion of Structure Chart*

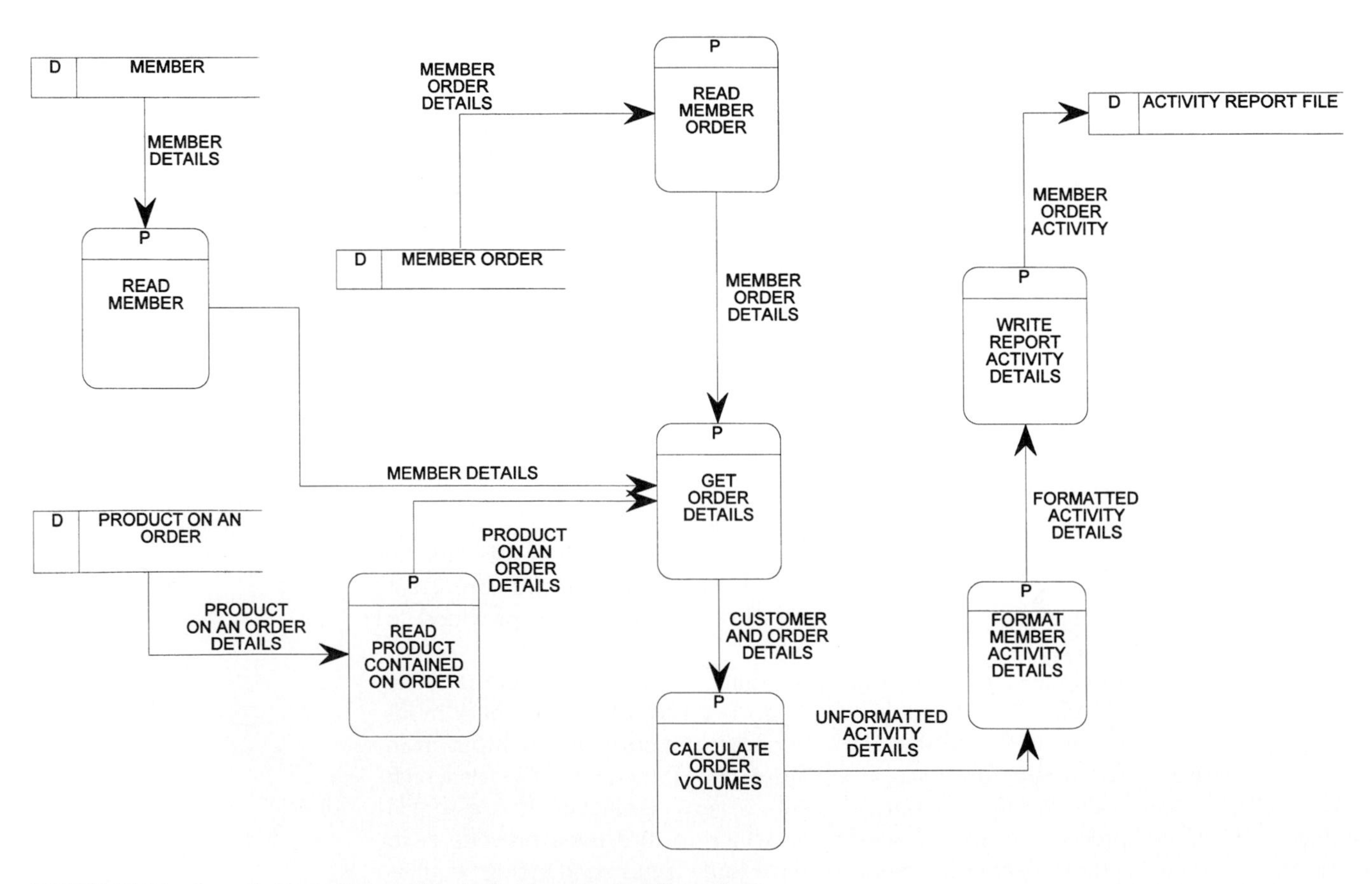

FIGURE 15.10 *Sample SoundStage DFD Reflecting Central Transform*

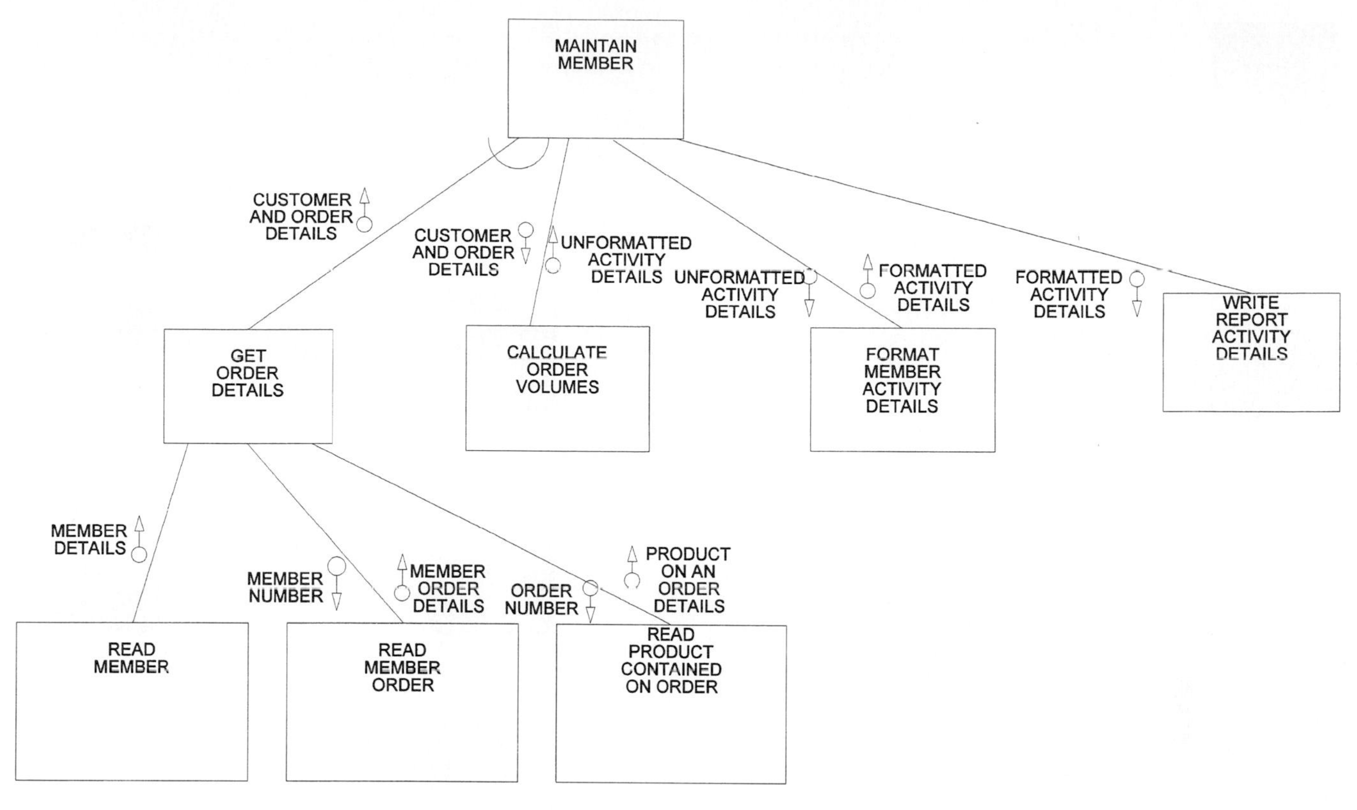

FIGURE 15.11 *Sample SoundStage Structure Chart from Transform Analysis*

design specifications. For example, the purpose of this DFD is to generate a report. The design of this output (recall Chapter 13) would communicate that data on the output are organized by member, then by orders for each member, followed by details regarding products appearing on each order. The fact that READ MEMBER ORDER must precede READ PRODUCT CONTAINED ON ORDER is also reinforced by examining our data model diagram (recall Chapter 5) and noticing the relationship paths (for example, MEMBER is related to MEMBER ORDER, which is related to PRODUCT CONTAINED ON ORDER).

Thus, structured design is not accomplished in isolation using only DFDs and structure charts. Let's now examine our second structured design approach.

Transaction Analysis

An alternative structured design strategy for developing structure charts is called transaction analysis.

> **Transaction analysis** is the examination of the DFD to identify processes that represent transaction centers.

A **transaction center** is a process that does not do actual transformation on the incoming data (data flow); rather, it serves to route the data to two or more processes. You can think of a transaction center as a traffic cop that directs traffic flow. Such processes are usually easy to recognize on a DFD, because they usually appear as a process containing a single incoming data flow to two or more other processes (see Figure 15.12). The primary difference between transaction analysis and transform analysis is that transaction analysis recognizes that modules can be organized around the transaction center rather than a transform center. For example, the structure chart for the DFD depicted in Figure 15.12 would be organized as shown in Figure 15.13. Notice that the transaction center is shown as a high-level module containing a diamond to suggest that it will call one and only one of the subordinate modules.

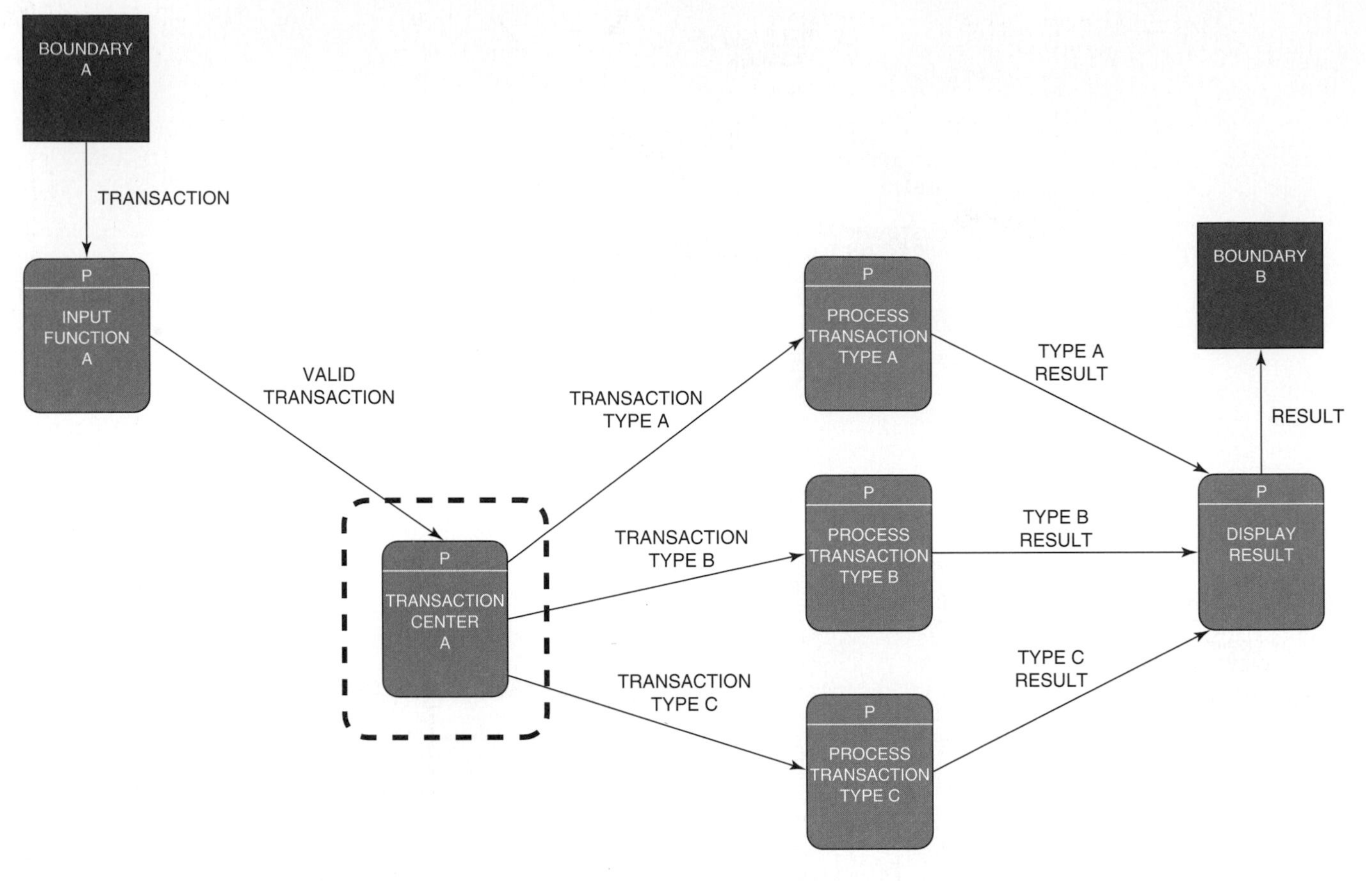

FIGURE 15.12 *Sample DFD with Transaction Center*

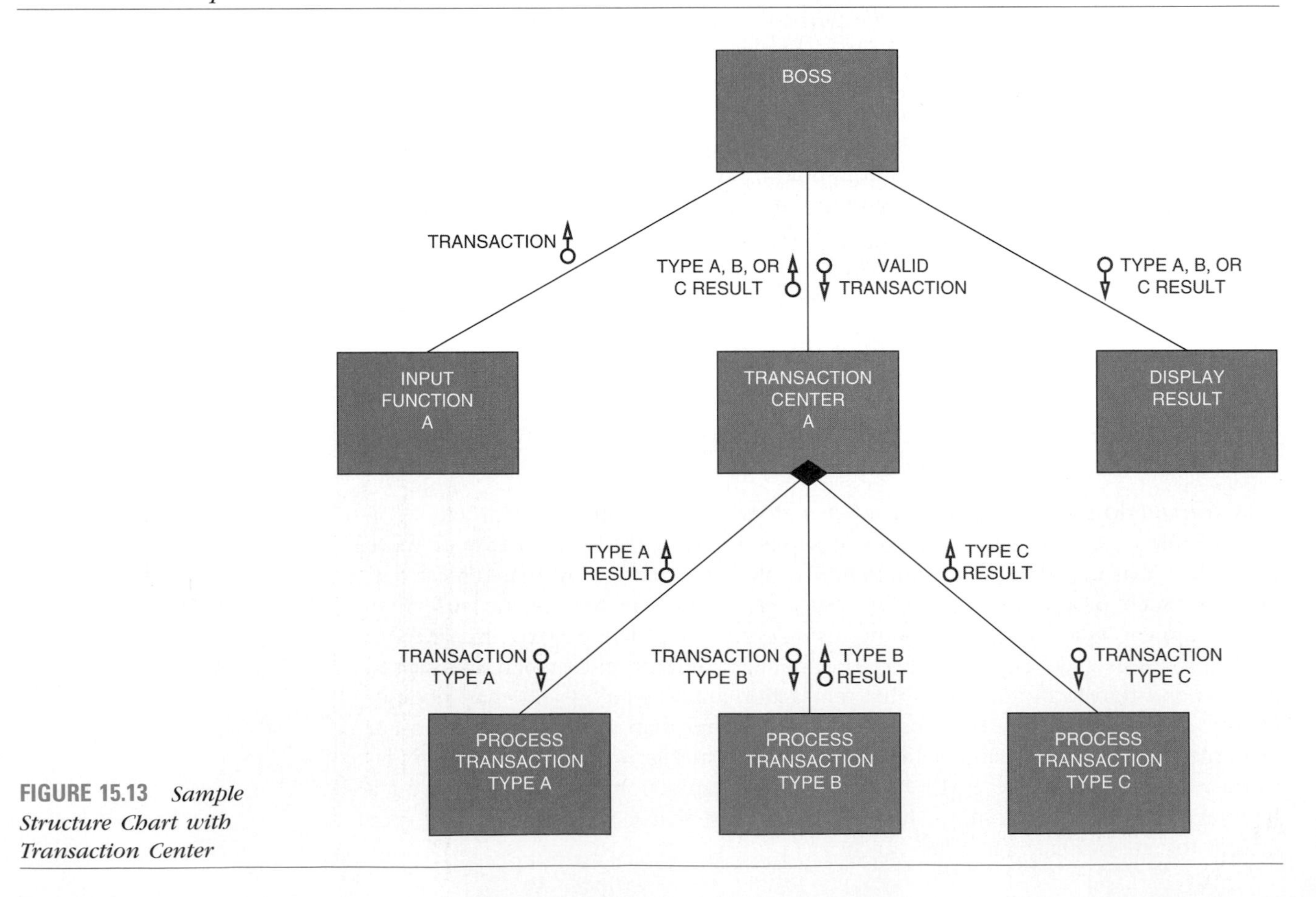

FIGURE 15.13 *Sample Structure Chart with Transaction Center*

There are a few classic examples of applications that would lend themselves to this strategy. A data flow diagram of each of these applications would contain a transaction center that routes the application's input to the appropriate function.

- A file maintenance program typically supports at least three functions that might be performed based on the type of input transaction: adding a new record, deleting an existing record, or modifying an existing record.
- An on-line system typically supports multiple levels of transactions. For instance, the main menu may offer three choices: EMPLOYEE FILE MAINTENANCE, PERSONNEL TRANSACTION, and EMPLOYEE INQUIRY. Each of these subfunctions may consist of multiple transactions. EMPLOYEE FILE MAINTENANCE might be factored as described in the preceding example. PERSONNEL TRANSACTION could be factored into SICK LEAVE PROCESSING, TIME CARD PROCESSING, VACATION PROCESSING, and so on. Thus, it is possible that an application contains multiple transaction centers—in this case, a transaction center may route a transaction to a module that also serves as a transaction center.

Although DFDs may help you identify transaction centers found in such applications, it depends on how detailed the analyst drew those DFDs (for instance, many analysts won't factor the DFD process MAINTAIN EMPLOYEE FILE into three separate processes). As with the transform analysis strategy discussed earlier, DFDs will likely first need to be revised to show missing details.

Let's examine a SoundStage example where transaction analysis resulted in a structure chart. Figure 15.14 is a sample DFD for maintaining member data. Notice that the DFD contains a transaction center, ROUTE MEMBER TRANSACTION. Figure 15.15 depicts the structure chart that might be derived by applying transaction analysis. Notice that the READ MEMBER process appearing on our DFD was added as a library module on the structure chart. The symbol placed immediately above it is called a connector. This symbol is often used to avoid crossing of lines on complex diagrams. The downward pointing direction implies the library module is being called. The number serves as a reference to help find the calling modules, in this case PROCESS UPDATE MEMBER TRANX is the calling module.

Structure Chart Quality Assurance Checks

Structure charts developed through structured design are evaluated for quality. By using the Yourdon/Constantine strategy to divide a program into modules, you can end up with modules that are said to be loosely coupled and highly cohesive. In the following sections we examine these two measures of program design quality. As you study these two measures, recognize that the data and control flow symbols depicted on a structure chart can serve as aids in determining the degree of coupling and cohesion of modules.

Coupling A fundamental principle of structured design is that the program should be decomposed into smaller more manageable modules. But the decomposition should be done in such a way that the modules are as independent as possible from one another. In structured design, programs appearing on a structure chart are evaluated relative to their degree of coupling.

Coupling refers to the level of dependency that exists between modules.

Loosely coupled modules are less likely to be dependent on one another. There are several levels or types of coupling. Let's briefly examine each type. The types will be presented in order of best (loosely coupled) to worst (tightly coupled).

Data coupling—two modules are said to be data coupled if their dependency is based on the fact that they communicate by passing of data. Other than communicating through data, the two modules are independent; that is, each module performs its own function with no regard to what or how the other module completes its functions. In

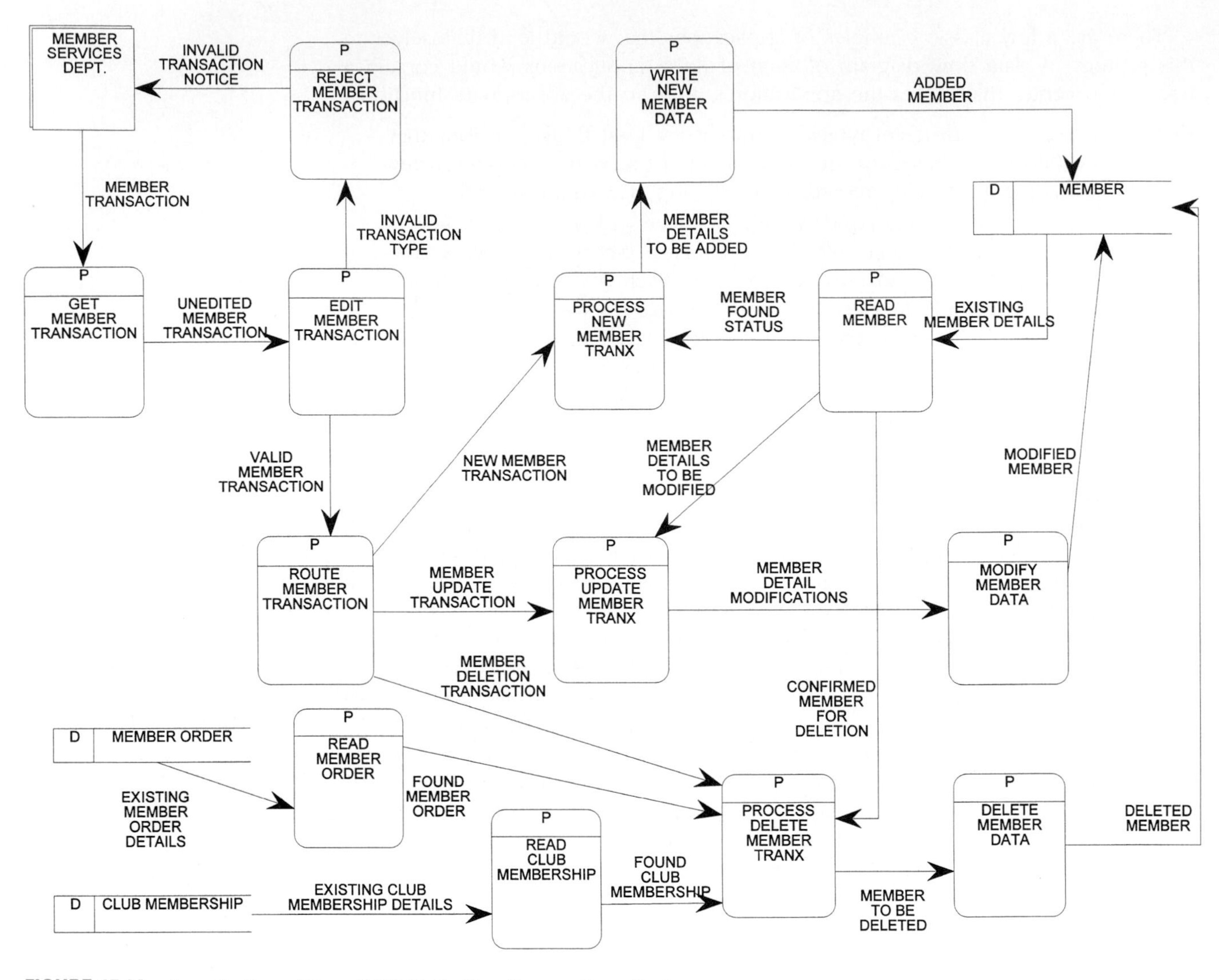

FIGURE 15.14 *Sample SoundStage DFD Reflecting Transaction Center*

examining modules for data coupling, careful attention should be made to ensure that no module communication contains "tramp" data. Tramp data are any unnecessary data communicated between the modules. For example, a module might call a second module to have it calculate an employee's net pay. Only data needed by the second module to complete its task should be passed from the calling module. By ensuring that modules communicate only necessary data, module dependency is minimized. This helps to avoid the "ripple effect" wherein making changes in one module inadvertently affects another module that happens to receive the same data.

Stamp coupling—two modules are said to be stamp coupled if their communication of data is in the form of an entire data structure or record. Since not all data making up the structure are usually necessary in the communication between the modules, stamp coupling typically involves tramp data. The passing of an entire data structure or record is also undesirable because any changes to the data structure or record may adversely affect any module that uses it.

Control coupling—two modules are said to be control coupled if their dependency is based on the fact that they communicate by passing of

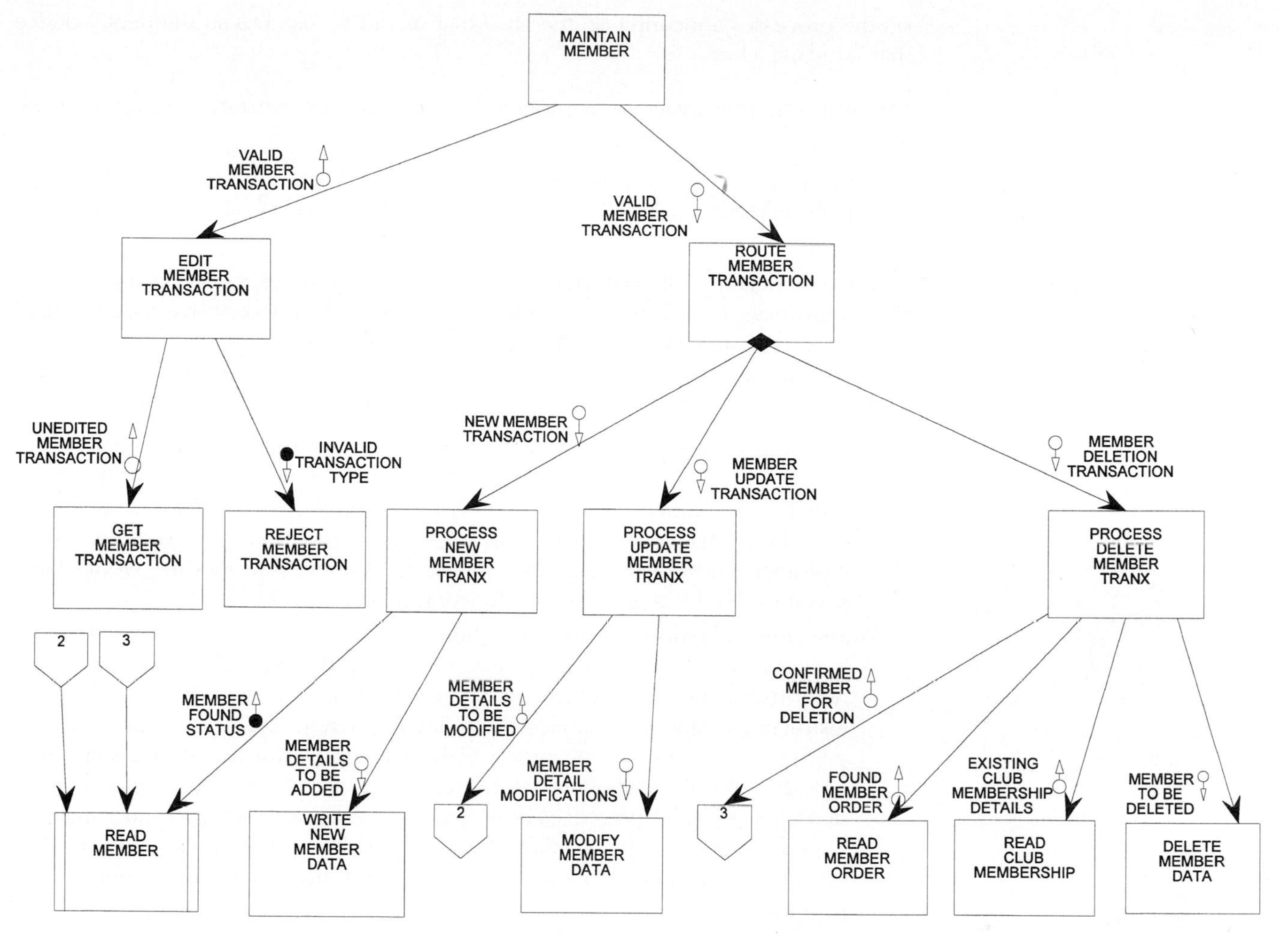

FIGURE 15.15 *Sample SoundStage Structure Chart from Transaction Analysis*

control information or flags. Control coupled modules represent a higher level of dependency. The mere fact that one module passes control information to another module suggests that the first module is involved in changing or coordinating the functions to be accomplished in the receiving module.

Common coupling—modules are said to be common coupled if they refer to the same global data area. Global data areas are commonly found in third-generation programming languages (3GLs) such as COBOL. Common coupling represents an even higher level of module dependency. For example, all modules that reference the global data area could be adversely affected by any changes that any of the other modules made to data in that global data area.

Content coupling—two modules are said to be content coupled (also referred to as *hybrid coupled*) when one module actually modifies the procedural contents of another module. In essence, the connection that exists between the two modules represents control. Content coupling represents the highest degree of module dependency.

Finally, it should be pointed out that quality checks for coupling are typically performed on the modules appearing on a structure chart. However, those experienced in structured design are sometimes capable of performing coupling checks on the processes appearing on the DFD that would be used to subsequently derive that structure chart.

Cohesion Another measure of program quality used in structured design is cohesion.

Cohesion refers to the degree to which a module's instructions are functionally related.

Thus, highly cohesive modules contain instructions that collectively work together to solve a specific task. The goal is to ensure that modules exhibit a high degree of cohesiveness. Programs that are implemented with highly cohesive modules tend to be easier to understand, modify, and maintain.

There are seven types or levels of cohesion. Let's briefly examine each type. The types will be presented in order of most desirable to least desirable.

Functional cohesion occurs in modules whose instructions are related because they collectively work together to accomplish a single well-defined function. Examples of functionally cohesive modules would be modules that check customer account balance, add a new customer, delete a customer, and query a customer. In each example, the module would be accomplishing a single, simple function.

Sequential cohesion occurs in modules whose instructions are related because the output data from one instruction are used as input data to the next instruction. An example of this type of module would be one whose instructions might accomplish the following series of tasks: get an order, edit the order, release the order to the warehouse for filling and shipping, and then bill the customer for the shipped order. This type of module does not typically present serious coupling and cohesion problems that would affect its maintainability. However, since several functions have been included in a single module, the reuse of any given function is not possible.

Communicational cohesion occurs in modules whose instructions accomplish tasks that utilize the same pieces of data. For example, a module may consist of numerous instructions that each accomplishes a task using customer data, such as checking a customer balance, adding a new customer, canceling a customer, updating a customer's record, changing a customer's status, or querying a customer. Such modules are easier to modify and maintain if they are expanded into separate modules that accomplish their own separate task.

Procedural cohesion occurs in modules whose instructions accomplish different tasks yet have been combined because there is a specific order in which the tasks are to be completed. These types of modules are typically the result of first flowcharting the solution to a program and then selecting a sequence of instructions to serve as a module. Since these modules consist of instructions that accomplish several tasks that are virtually unrelated, these types of modules tend to be less maintainable.

Temporal cohesion occurs in modules whose instructions appear to have been grouped together into a module because of "time." For example, a temporal cohesive module may contain instructions that were grouped together because they perform start-up or initialization activities (such as setting program counters or control flags) associated with the program. Or perhaps the instructions were to be performed at the end of the program, such as printing final report totals, closing a file, or displaying an end-of-job message to the user.

Logical cohesion occurs in modules that contain instructions that appear to be related because they fall into the same logical class of functions. For example, the instructions were grouped together as a module perhaps because they all involve editing or arithmetic operations. Unfortunately, logically cohesive modules do not meet our goal of a module containing instructions that belong together because they collectively serve to accomplish a single function or task.

Coincidental cohesion occurs in modules that contain instructions that have little or no relationship to one another. Coincidental cohesive modules appear to have been derived with no attention given to the actual "function" being served by the module. In fact, their existence is typically based on coincidence. For example, a designer may decide to create a module that will consist of a series of program instructions encountered several times elsewhere in the program's logic.

Finally, it should be pointed out that quality checks for cohesion can be performed on the modules appearing on a structure chart or on the processes appearing on DFDs that would be used to subsequently derive that structure chart.

PACKAGING PROGRAM SPECIFICATIONS

Using the design techniques presented in this unit, you have accumulated a good number of design specifications for the new system. Perhaps some or all of the documentation was done using a CASE product and resides in a repository. What do we actually package together? Put yourself in the role of the computer programmer. Are those specifications in a format that will help you write the programs? Not really. As a systems analyst, you are responsible for packaging that set of design documentation into a format suitable for the programmer. You learned in Chapter 9 that this package is typically referred to as a **technical design statement.**

As is suggested by our information systems framework depicted in Figure 15.16, the technical design statement should include all DATA, PROCESS, INTERFACE, and GEOGRAPHY building block specifications developed by the designer. The manner in which these items are organized and presented to the programmer may vary. But the organization must make it easy for the programmer to readily identify the program requirements.

WHERE DO YOU GO FROM HERE?

This chapter provided a detailed overview of software design—modular design and packaging. You are now ready to learn some systems implementation, the process that implements a new system using the technical design statement. Systems implementation and support are covered in Chapters 17 and 18, respectively. Before moving on, you may wish to revisit Chapter 9 and its introduction to systems implementation and how it fits into the overall systems development life cycle.

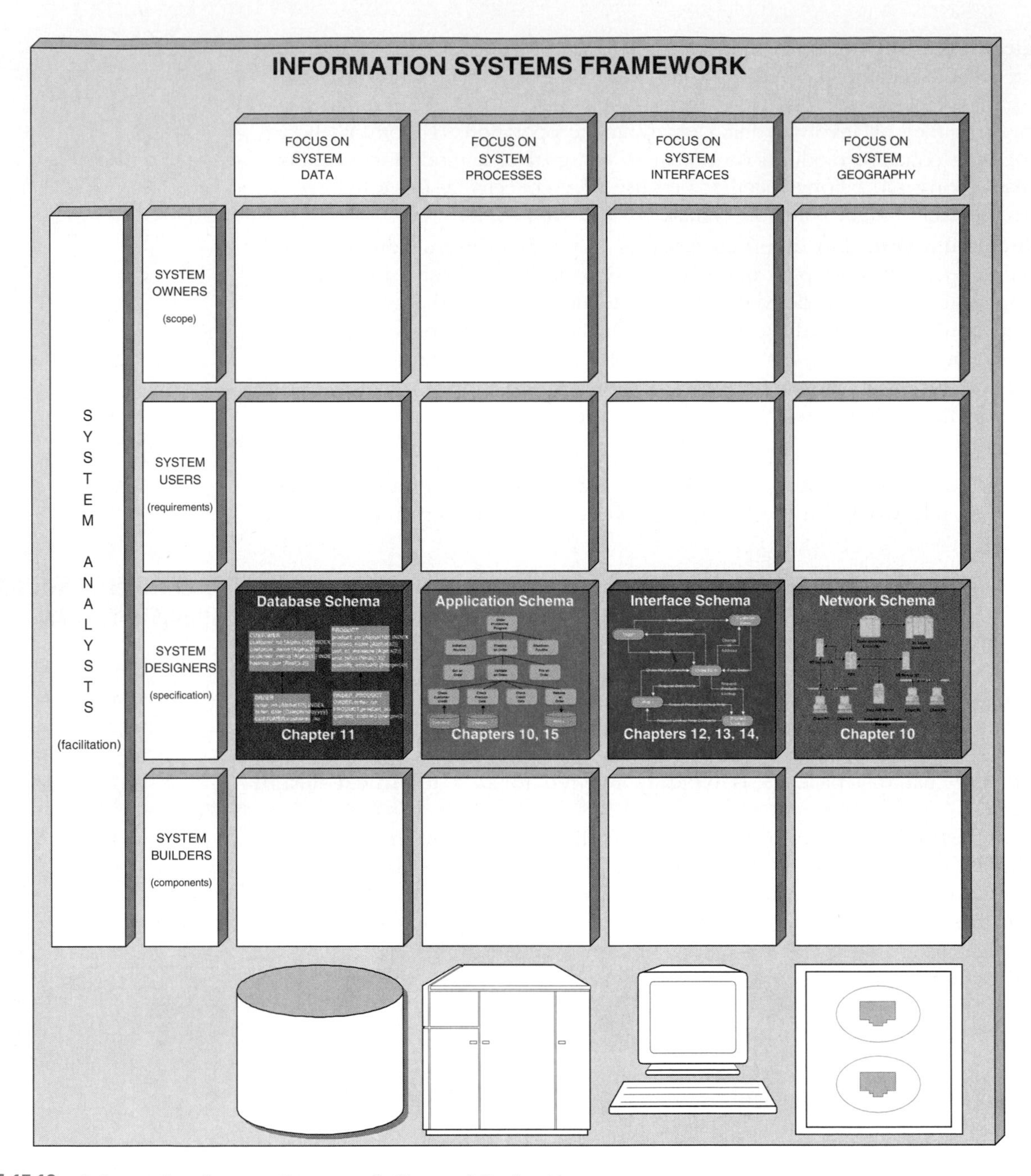

FIGURE 15.16 *Information Systems Framework Focus of Packaging*

SUMMARY

1. Software design consists of two components—modular design and packaging.
   a. Modular design is the decomposition of a program into modules. A module is a group of executable instructions with a single point of entry and a single point of exit.
   b. Packaging is the assembly of DATA, PROCESS, INTERFACE, and GEOGRAPHY design specifications for each module.
2. Structured design is a popular strategy for determining an optimal modular design for programs. This technique deals with the size and complexity of a program by breaking up a program into a hierarchy of modules and results in a computer program that is easier to implement and maintain.
3. The primary tool used in structured design is the structure chart. Structure charts are used to graphically depict a modular design of a program. Specifically, they show how the program has been partitioned into smaller more manageable modules, the hierarchy and organization of those modules, and the communication interfaces between modules.
4. Structured design provides two strategies for developing structure charts, transform analysis and transaction analysis. Both strategies are based on the use of data flow diagrams to derive the structure chart.
   a. Transform analysis is an examination of the DFD to divide the processes into those that perform input and editing, those that do processing or data transformation (e.g., calculations), and those that do output.
   b. Transaction analysis is the examination of the DFD to identify processes that represent transaction centers.
5. Relative to transform analysis, structured design refers to that portion of the DFD consisting of processes that perform input and editing as the afferent. That portion of the DFD consisting of processes that do actual processing or transformations of data is referred to as the central transform. The portion of the DFD consisting of processes that do output is referred to as the efferent.
6. A transaction center is a process that does not do actual transformation on the incoming data (data flow); rather, it serves to route the data to two or more processes.
7. Two measures of the quality of a structure chart are coupling and cohesion.
   a. Coupling refers to the level of dependency that exists between modules.
   b. Cohesion refers to the degree to which a module's instructions are functionally related.
8. There are five types or levels of coupling. In order of most desirable to least desirable, they are data, stamp, control, common, and content coupling.
9. There are seven types or levels of cohesion. In order of most desirable to least desirable, they are functional, sequential, communicational, procedural, temporal, logical, and coincidental cohesion.
10. Packaging results in a technical design statement that will be used during systems implementation.

KEY TERMS

afferent, p. 516
central transform, p. 516
cohesion, p. 526
coincidental cohesion, p. 527
common coupling, p. 525
communicational cohesion, p. 526
content coupling, p. 525
control coupling, p. 524
coupling, p. 523
data coupling, p. 523
efferent, p. 516
flag, p. 512
functional cohesion, p. 526
logical cohesion, p. 527
modular design, p. 510
module, p. 510
packaging, p. 510
procedural cohesion, p. 526
sequential cohesion, p. 526
software design, p. 510
stamp coupling, p. 524
structure chart, p. 511
structured design, p. 511
temporal cohesion, p. 526
transaction analysis, p. 521
transaction center, p. 521
transform analysis, p. 515

REVIEW QUESTIONS

1. What are the two activities of software design?
2. Define modular design. What is a module?
3. Define packaging. What building blocks of our information systems framework are addressed during modular design? Which are addressed during packaging?
4. How does structured design deal with the size and complexity of a program?
5. What are the two primary tools used in structured design?
6. Identify and describe two structured design strategies used to derive structure charts.

7. What two types of module communication are depicted on a structure chart?
8. What is the difference between "data" and "flags" that appear on a structure chart?
9. Distinguish between an afferent, transform, and efferent process/module.
10. What is a transaction center?
11. What are two typical applications that lend themselves to applying transaction analysis?
12. Name two quality measures for structure charts.
13. Identify several types or degrees of coupling. Rank them from best to worst.
14. Identify several types of cohesion. Rank them from best to worst.
15. What is the deliverable of packaging?

PROBLEMS AND EXERCISES

1. What value would an existing structure chart of an existing program be to a systems analyst during the study phase of systems development?
2. What correlations can be drawn between a DFD and a structure chart?
3. Differentiate between coupling and cohesion.
4. Give an example of an afferent, efferent, and transform process.
5. Give several examples of transaction centers.
6. What are the transaction centers in the following program?

   An on-line program allows an end-user to perform inquiries to obtain information concerning customer accounts, orders, invoices, and products. The end-user is allowed to obtain general information concerning an order or information about specific orders that have been placed on backorder. The end-user who wishes to obtain information concerning orders placed on backorder may request information describing orders that have been backordered for less than one week, backordered for more than one week but less than two weeks, or backordered for more than a two-week period. The end-user may also perform inquiries to retrieve general information about a specific part and information concerning backordered parts.

PROJECTS AND RESEARCH

1. Obtain a copy of the documentation for a completed programming assignment. Study the program's source code to identify all referenced modules. Using a structure chart, document the existing modular structure implemented by the program.
2. Develop a DFD for the program in project 1. Which structured design strategy would you use to develop a structure chart? Explain why.
3. Study the processing requirements for the program you used for project 1. Which structured design strategy would you use to develop a new structure chart for the program? Explain why. Develop a new structure chart for the program. Compare the structure chart with the one derived in project 1. Which would you prefer to work from as a programmer? Why?

MINICASES

1. George Amana is a programmer/analyst for Tower Lawn and Garden, Inc. Tower is a distribution center for lawn and garden equipment in northern Louisiana. George just sat down to lunch in the company cafeteria. Pete Wilcox, a senior partner in the firm, joins him.

   PETE Hey, George, you look pretty frustrated. What's the problem?

   GEORGE I had to take over the sales information systems project that Judy left behind when she quit. It's total chaos. I was told that it was all but finished. Come to find out, she didn't finish several of the programs.

   PETE But she did do a good job of specifying all the program requirements. What's so tough about the programs? Judy always preached about the benefits of structured programming. In fact, she taught me how to do it. Don't tell me she doesn't practice what she preaches.

   GEORGE No, her code is very well structured. And her documentation is adequate. It's just that the programs seem so poorly designed. Some of her subroutines are so long and complex that it's difficult to get a grasp on small enough pieces to test them for correctness. It seems like an all-or-nothing proposition. If I encounter a bug, I have to test large sections of code to zero in on the problem. Sometimes the bug turns out to be in an entirely different subroutine!

PETE Why didn't Judy break the system into smaller pieces?

GEORGE Oh, she did! The modules are evidence of that. But it almost seems like she generated the modules on the fly—as if to say, "Well, this piece of code is getting complex. I'd better put in a module to finish it." She left me a rough draft of a structure chart, but I just don't understand the reasons she factored the system the way she did.

PETE That's the way I write programs. I start by trying to draw a flowchart on a single page—sort of the high-level flowchart. Then I factor the more complex processes into more detailed processes that I implement as modules. It sounds like that may be what Judy did.

GEORGE Maybe she did. But that strategy causes the lower-level modules to be very dependent on other routines. I frequently encounter bugs that get traced back to other, seemingly unrelated routines. I'm just getting further behind schedule. I may just have to write the programs from scratch.

PETE Why don't you get some help? Barbara just finished her project. Maybe she can help you. You could divide up the work and get it done faster.

GEORGE Divide up the work? I don't see how. Judy's program specifications are just one big unorganized document. I'm not sure which file and report specifications to match up to which modules. For that matter, I'm not sure the programs themselves are fully documented.

PETE Well, I'm sorry, George. I don't know what to tell you.

a. If design specifications are thorough and complete and program code is well structured, how can the system still be difficult to construct?
b. How should modules in a program be conceived? How does Judy seem to have created them? What is the potential problem with creating modules during coding?
c. What effect does the program and module size have on testing?
d. What would Barbara require in order to take on responsibility for some of the programs that haven't been written? What does the programmer need to be able to write a new program? How would you organize the necessary documentation of program requirements?

SUGGESTED READINGS

Adams, David R., Gerald E. Wagner, and Terrence J. Boyer. *Computer Information Systems: An Introduction*. Cincinnati: South-Western Publishing, 1983. This book suggests how most programs can be factored into initiate, main process, and terminate functions (Chapter 8).

Boehm, Barry. "Software Engineering." IEEE Transactions on Computers C-25, December 1976, pp. 1226–41. This paper, a classic, describes the logarithmic relationship between time and the cost to correct an error in the systems specifications.

Page-Jones, Meiler. *A Practical Guide to Structured Systems Design*. 2nd ed. New York: Yourdon Press, 1980. This book discusses the modular design methodology called Structured Systems Design. The book includes discussion and numerous examples of both transform and transaction analysis. The concepts of coupling and cohesion are also covered in great detail.

Yourdon, Edward, and Larry Constantine. *Structured Design: Fundamentals of a Discipline of Computer Program and Systems Design*. Englewood Cliffs, NJ: Prentice Hall, 1979.

16

OBJECT-ORIENTED DESIGN

CHAPTER PREVIEW AND OBJECTIVES

This is the second of two chapters on object-oriented tools and techniques for systems development. This chapter focuses on **object-oriented modeling** tools and techniques that are used during systems design. You will know **object-oriented systems** design when you can

- Differentiate between entity, interface, and control objects.
- Understand the basic concept of *object responsibility* and how it is related to message sending between object types.
- Explain the importance of considering object reuse during systems design.
- Describe three activities involved in completing object design.
- Construct an ideal object model diagram, CRC card, and object interaction diagram.

Note: The authors gratefully acknowledge the contributions of Kevin C. Dittman, Assistant Professor of Computer Technology, Purdue University, as a partner in the writing of this chapter.

SCENE

We begin this episode with Sandra Shepherd, Bob Martinez, and Lonnie Bentley meeting to discuss the progress of the special object-oriented version of the Member Services project. Lonnie has just completed the object-oriented design phase for the system. As part of his consulting obligation, Lonnie is explaining the approach that he used to Sandra and Bob.

SANDRA

So, I understand that you've completed the design phase for your OO project. You're on pretty much the same pace as Bob and I are with our version of the project.

LONNIE

That's what Bob was telling me during lunch yesterday. Let's all keep our fingers crossed and hope that there are no surprises when we start the implementation effort.

SANDRA

Bob and I are really excited to hear about your experiences thus far. It's going to be interesting to see the transition process that you went through going from OO analysis to OO design.

BOB

I agree. The last time we talked, Sandra and I were having trouble coordinating our data and process models. You said that the object-oriented approach uses a single object model to deal with data and processes throughout the life cycle. I'm curious to see how you were able to carry the model over into the design phase.

LONNIE

OK. Let's start with the object model. Here's the original model that I developed during analysis. Now, here's the model upon completing the design phase.

BOB

They look very similar.

SANDRA

It looks like you have expanded the object symbols to indicate their behaviors.

LONNIE

That's exactly what I did. During object-oriented design, an attempt is made to identify the systems' behaviors and associate or assign those behaviors to the objects that will be responsible for performing those behaviors.

BOB

That's right, I forgot. The object model specifies both data and processes. So you waited until the design phase to identify the object behaviors that correlate to processes for the system. How did you determine the behaviors?

LONNIE

Do you recall the use cases that I developed during analysis?

SANDRA

I remember them. You had one for each business event.

LONNIE

Well, first I refined each of those use cases to include more detail that reflected the physical implementation of the use case. For example, the use case referred to windows, buttons, and other items involved in the processing. Once I had completed the detailed use cases I examined them to identify behaviors or actions that were involved in the use case scenario. I simply highlighted each verb phrase. Each verb phrase suggested an action or behavior that must be accomplished. Those behaviors are subsequently associated with objects.

SANDRA

I like it! That sounds simple. The simpler the better as far as I'm concerned.

BOB

I like it too. But how can you be sure that your use case is accurate and complete?

LONNIE

Good question. There are some other tools I used during design. One tool is the object interaction diagram. This diagram depicts all the objects involved in the use case and how they interact.

SANDRA

So it's a pictorial equivalent to the use case?

LONNIE

That's right. And it's very easy to walk through the diagram to see if the objects and their interactions work toward fulfilling the processing of the business event.

BOB

That makes a lot of sense. What are some of the other tools you used?

LONNIE

There are a few more tools. But as with any development tool, there needs to be a well-disciplined set of procedures that guide the development and use of the tools. So why don't we back up to square one? I'll go through each step that I followed during the design phase and introduce the tools in the proper context.

SANDRA

That's a good idea. Bob and I are really intrigued with the tools. But we really do need to understand the process of object-oriented design.

DISCUSSION QUESTIONS

1. What are some advantages of being able to carry the object model over into the design phase?
2. Why did the use case scenarios need to be refined during the design phase?
3. What important concept introduced in Chapter 8 is reflected in the object interaction diagrams that Lonnie was describing?

AN INTRODUCTION TO OBJECT-ORIENTED DESIGN

This chapter will present object-oriented techniques for designing a new system. The approach of using object-oriented techniques for designing a system is referred to as **object-oriented design.** Recall that object-oriented development approaches are best suited to projects that will implement systems using emerging object technologies to construct, manage, and assemble those objects into useful computer applications. Object-oriented design is the continuation of object-oriented analysis, continuing to center the development focus around object modeling techniques.

There are many underlying concepts for object modeling during systems design. In this section you will learn about those concepts. Afterward, you will learn how to apply those concepts while learning the process of object-oriented design

Design Objects

In object-oriented analysis we concentrated on identifying the objects that represented actual data and methods that act of the data within the business domain. These objects are called **entity objects.** During object-oriented design we continue to refine these entity objects while identifying other types of objects that will be introduced as the result of physical implementation decisions for the new system. Two additional types of objects will be introduced during design. New objects will be introduced to represent a means through which the user will interface with the system. These objects are called **interface objects.** Other types of objects that are introduced are objects that hold application or business rule logic. These objects are called **control objects.**

The structuring of an object-based system into three types of objects was proposed by Dr. Ivar Jacobson and is similar to the mechanism used in *Smalltalk* programming called **model-view-controller (MVC).** The underlying principle for using three types of objects is that responsibilities and behaviors needed to support a system's functionality are distributed across these three types of objects that work together to carry out the service. This practice makes the maintenance, enhancement, and abstraction of those objects simpler and easier to manage than having entity objects encapsulate all required data and behavior, a practice used by some object-oriented design technologies.

The three object types also correlate well with the client/server model. The client is responsible for the application logic (control objects) and presentation method (interface objects), and the server is responsible for the repository (entity objects). Let's further examine each object type.

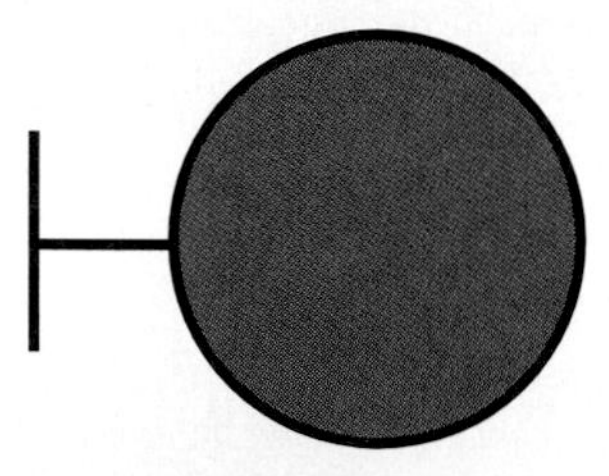

Interface Object

Interface Objects It is through interface objects that the users communicate with the system. The use case functionality that describes the user directly interacting with the system should be placed in interface objects. The symbol appearing in the margin is used to represent an interface object. The responsibility of the interface object is twofold:

1. It translates the user's input into information that the system can understand and use to process the business event.
2. It takes data pertaining to a business event and translates the data for appropriate presentation to the user.

Each actor or user needs its own interface object to communicate with the system. In some cases the user may need multiple interface objects. Take for example the ATM machine. Not only is there a display for presenting information, but there is also a card reader, money dispenser, and receipt printer. All of these would be considered interface objects.

Entity Object

Entity Objects Entity objects usually correspond to items in real life and contain information, known as attributes, that describes the different instances of the entity. They also encapsulate those behaviors that maintain its information or attributes. The symbol appearing in the margin is used to represent an entity object.

An entity object is said to be **persistent**—meaning the object typically "lives on" after the execution of a method. An entity object exists between method executions because the information about that entity object is typically stored in a database (allowing for later retrieval and manipulation).

Control Objects When distributing behavior and responsibilities among objects, there is often behavior that does not naturally reside in either the interface or entity objects. In other words, the behavior is not related to how the user interacts with the system, nor is it related to how the data in the system are handled. Rather, such behavior is related to the management of the interactions of objects to support the functionality of the use case. Control objects serve as the "traffic cop" containing the application logic or business rules of the event for managing or directing the interaction between the objects. Control objects allow the scenario to be more robust and simplify the task of maintaining that process once it is implemented. As a general rule of thumb, within a use case, a control object should be associated with one and only one actor. Control objects are depicted using the symbol appearing in the margin.

Control Object

Object Responsibilities

In object-oriented systems, objects encapsulate both data and behaviors. In design we focus on identifying the behaviors a system must support and, in turn, design the methods to perform those behaviors. Along with behaviors, we determine the responsibilities an object must have.

In Chapter 8 you learned that objects have behaviors, or things that it can do. In object-oriented design it is important to recognize an object has responsibility.

> An **object responsibility** is the obligation that an object has to provide a service when requested, thus collaborating with other objects to satisfy the request if required.

Object responsibility is closely related to the concept of objects being able to send and/or respond to messages. For example, an ORDER object may have the responsibility to display a customer's order, but it may need to collaborate with the CUSTOMER object to get the customer data, the PRODUCT object to get the product data, and the ORDER LINE object to get specific order data about each product being ordered. Thus, CUSTOMER, PRODUCT, and ORDER LINE have an obligation to provide the requested service (provide requested data) to the ORDER object.

Object Reusability

The number one driving force for developing systems using object-oriented technology is the potential for **object reusability.** Developers and managers strive to create quality applications cheaper and in less time. Object technology appears to aid in accomplishing that goal. Several studies have documented the success of object reuse. In fact, an article that appeared in *ComputerWorld* tells how Electronic Data Systems (EDS) initiated two projects to develop the same system using two different programming languages.[1] One project used a traditional 3GL language called *PL/1,* and the other used *Smalltalk,* an object-oriented-based language. The results were impressive as indicated in the following table:

PL/1	19 calendar months	152 person months	265,000 lines of code
Smalltalk	3.5 calendar months	10.4 person months	22,000 lines of code

Similar studies have produced similar results. To maximize the ability to reuse objects, objects have to be correctly designed initially, experts say. This means they have to be defined within a good generalization/specialization hierarchy. The

[1] "White Paper on Object Technology: A Key Software Technology for the 90s," *ComputerWorld,* May 11, 1992.

goal is to make objects general enough to be easily used in other applications. For example, when designing a STUDENT object, make sure it is general enough to use in a Student Registration system as well as in a system that tracks student financial aid or student housing. Any attributes that may be related to a student, but unique to a particular type of student, should be abstracted and placed in a newer, more specialized object.

Many companies achieve their highest level of reuse by exploiting object frameworks.

An **object framework** is a set of related, interacting objects that provide a well-defined set of services for accomplishing a task.

An example of an object framework is a calendar routine, used for calculating or displaying dates. Routines used for charting, printing, or any type of application utility would be good candidates for object frameworks. By using object frameworks, developers can concentrate on developing the logic that is new or unique to the application, thus reducing the overall time required to build the entire system.

THE PROCESS OF OBJECT DESIGN

In performing object-oriented analysis (OOA) we identified objects and use cases based on ideal conditions and independent of any hardware or software solution. During object-oriented design (OOD) we want to refine those objects and use cases to reflect the actual environment of our proposed solution.

Object-oriented design includes the following activities:

1. Refining the use case model to reflect the implementation environment.
2. Modeling object interactions and behavior that support the use case scenario.
3. Updating the object model to reflect the implementation environment.

In the following sections we will review each of these activities to learn what steps, tools, and techniques are used to complete object-oriented design.

Refining the Use Case Model to Reflect the Implementation Environment

In this iteration of use case modeling, the use cases will be refined to include details of how the actor (or user) will actually interface with the system and how the system will respond to that stimulus to process the business event. The manner in which the user accesses the system—via a menu, window, button, bar code reader, printer, and so on—should be explicitly described in detail. The contents of windows, reports, and queries should also be specified within the use case. While refining use cases is often time consuming and tedious, it is essential that it be completed. These use cases will be the basis on which subsequent user manuals and test scripts are developed during systems implementation. In addition, these use cases will be used by programmers to construct application programs during systems implementation.

When refining each use case, you might discover the need for additional use cases. A use case may contain complex functionality consisting of several steps that are difficult to understand. To simplify the use case and make it more easily understood, we can extract the more complex steps into its own use case. This type of use case is called an **extension use case** in that it extends the functionality of the original use case. An extension use case can be invoked only by the use case it is extending. More commonly, you may discover two or more use cases that perform steps of identical functionality. It is best to extract these common steps into their own separate use case called an **abstract use case.** An abstract use case represents a form of reuse. An abstract use case is available for referencing (or use) by any other use case that requires its functionality. Figure 16.1 is an example of graphically depicting a use case with its extensions and abstractions.

In the following steps we will adapt each use case to the implementation environment or "reality" and document the results. It is important that each use case

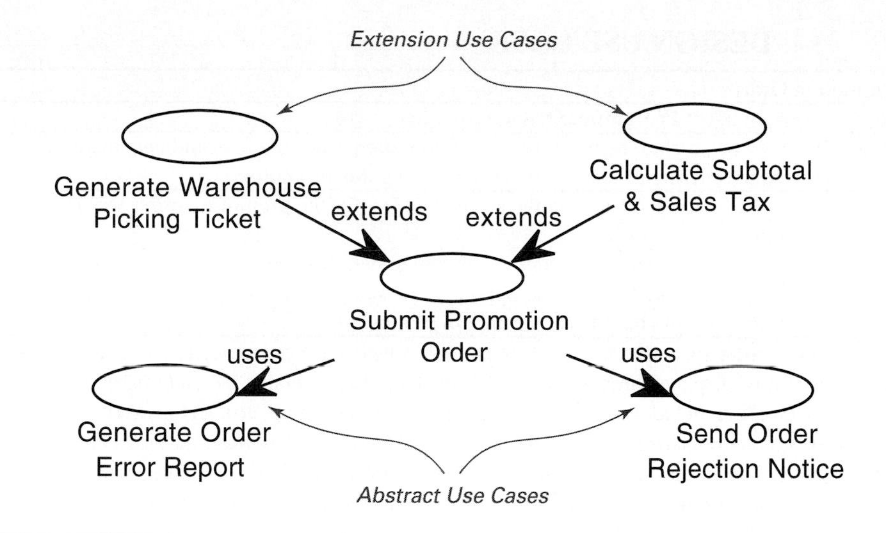

FIGURE 16.1 *Use Case Extensions and Abstractions*

be highly detailed in describing the user interaction with the system. These refined use cases can be used by the user to validate systems design and by the programmer for process and interface specifications.

Step 1—Transforming the "Analysis" Use Cases to "Design" Use Cases In Chapter 8 you learned how to do use case modeling during systems analysis to document user requirements for a given business scenario. In this step, we refine each of those use cases to reflect the physical aspects of the implementation environment for our new system.

Figure 16.2 illustrates the refinement of the "Submitting a Promotion Order" use case that was originally defined during systems analysis. This copy is labeled as a design use case to distinguish it from the analysis version previously completed. We want to keep the original analysis use cases separate from the refined design use cases to allow maximum flexibility in reusing use cases for variations of different physical implementations. We draw your attention to the following refinements to our use case description in Figure 16.2.

① We have included an entry that specifies the system user. In analysis, we concentrated on the actor—the party that initiates the business event. In design, we begin to think in terms of "how" the business event is accomplished and by whom. Thus, we are concerned with identifying the party or "system user" that is involved in processing the business event or interacting with the system. In some cases, the actor and the system user may be the same person.

In this implementation of the system, the actor, "club member," is not the party actually using the system. The club member simply triggers the business event by supplying information to the "promotion order specialist" who then enters it into the system. The system could be designed to allow club members to input orders themselves (via a World Wide Web page or an automated phone system), thus participating as a system user.

② Step 2 now describes the windows that will be displayed to the user and the contents (i.e., field names) of the windows.

③ We have included the physical address of the printer in the warehouse that will be used to print the picking tickets.

④ Descriptions of error messages, special action buttons, possible cursor movements, and other window characteristics should be included in each design use case step.

Author: K. Dittman

DESIGN USE CASE

USE CASE NAME:	Submit Promotion Order
ACTOR:	Club Member - **System user: Promotion Order Specialist** ①
DESCRIPTION:	Describes the process when a club member submits a promotion order to either indicate the products he or she is interested in ordering or declining to order during this promotion.
NORMAL COURSE:	The **main window** is currently displayed on the screen waiting for the promotion order specialist to select a menu option. 1. When a promotion order is received from the club member, this use case is initiated when the promotion order specialist selects the option "Process Promotion Order," displaying a window requesting the club member's NUMBER to be entered.
	2. The promotion order specialist enters the MEMBER NUMBER. The system verifies that the number is valid and if it is, displays three windows. The first window contains the club member's personal information including: ② **NAME, STATUS, STREET ADDRESS, P. O. BOX, CITY, STATE, ZIP, DAYTIME PHONE NUMBER, DATE OF LAST ORDER, BALANCE DUE, and BONUS BALANCE.** The second window contains the promotion order header record information that was created when the promotion was sent to the club member. The second window contains: ORDER NUMBER, ORDER CREATION DATE, ORDER AUTOMATIC FILL DATE, SUBTOTAL COST, SALES TAX, and ORDER STATUS. The third window contains the individual ordered products.This is a **multi-lined window** with the first order line containing the selection of the month information. This record was also created when the promotion was sent to the club member. The third window contains: PRODUCT NUMBER, PRODUCT NAME, QUANTITY AVAILABLE, QUANTITY ORDERED, QUANTITY BACKORDERED, SUGGESTED UNIT PRICE, PURCHASED UNIT PRICE, EXTENDED COST, CREDITS EARNED, and ORDERED PRODUCT STATUS.
	3. The promotion order specialist checks to see if the club member has made any address or phone number changes on the promotion order. If no changes have been made, the promotion order specialist checks the promotion order to see if any products are being ordered. If products are being ordered, the promotion order specialist positions the cursor in the ordered product window.
	4. The promotion order specialist checks the promotion order to see if the club member has accepted or declined the selection of the month. If accepted, the promotion order specialist checks quantity ordered to see if it is greater than one. If so, the promotion order specialist edits the QUANTITY ORDERED field to reflect the new quantity. The system calculates the new EXTENDED COST by multiplying the PURCHASED UNIT PRICE times the QUANTITY ORDERED and then advances the cursor to the next line.
	5. The promotion order specialist checks the promotion order to see if the club member has ordered additional products. If so, the promotion order specialist enters the PRODUCT NUMBER of the next item being ordered. The system displays the PRODUCT NAME, QUANTITY AVAILABLE, and the SUGGESTED UNIT PRICE. The promotion order specialist enters the QUANTITY ORDERED, and the PURCHASE UNIT PRICE. The system calculates the new EXTENDED COST by multiplying the PURCHASE UNIT PRICE times the QUANTITY ORDERED. It also updates the ORDERED PRODUCT STATUS field to be "open," and then advances the cursor to the next line. This step is repeated until all ordered products have been input.
	6. Once all ordered products have been entered into the system, the promotion order specialist commits the order. Invoke extension use case ***Calculate Subtotal and Sales Tax***. The system checks the club member's STATUS. If the club member is in good standing, ⑤ invoke extension use case ***Generate Warehouse Picking Ticket***. The picking ticket will be routed to the warehouse and printed on their printer ③ (**HP.PRT.20195**). For each product ordered, the system reduces the QUANTITY AVAILABLE by the QUANTITY ORDERED and saves the changes.
	7. Once the order has been committed, the system will prompt the user if he or she wants to process a new order or cancel the transaction.

FIGURE 16.2 *Design Use Case*

⑤ The design use case step includes references to extension and abstract use cases. Recall that extension use cases extend the functionality of the original use case by extracting complex or hard to understand logic into its own use

ALTERNATE COURSE:	2. If the MEMBER NUMBER is invalid, the system displays a window with the error message "**Member Number Not on File.**" The promotion order specialist can then re-enter the number or cancel the transaction.
	3a. If the club member indicates an address or telephone number change on the promotion order, the promotion order specialist selects the option to update the club member's personal information. The personal information window is activated for update. The promotion order specialist updates the required fields and saves the changes. 3b. If the club member is not ordering product at this time, the promotion order specialist **clicks the button "Order Declined."** The system modifies the ORDER STATUS field to have a status of "closed" and modifies the ORDERED PRODUCT STATUS field (for the selection of the month) to have a status of "rejected." The changes are saved and then the system prompts the user to process a new order or cancel the transaction.
	4. If the club member is not ordering the selection of the month, the promotion order specialist clicks the button "Selection of the Month Declined." The system modifies the ORDERED PRODUCT STATUS field (for the selection of the month) to have a status of "rejected." **The cursor is then advanced to the next line.**
	5a. If no additional products are being ordered, proceed to step 6 of the normal course. 5b. If the PRODUCT NUMBER or QUANTITY ORDERED is not valid, the **system highlights the fields** indicating those fields have invalid entries. The promotion order specialist either corrects the entries or advances to the next line. In a later step an error report will be generated for any invalid entries.
	6a. If the club member's status is not acceptable, invoke abstract use case ***Send Order Rejection Notice***. The system modifies the ORDER STATUS field to have a status of "on hold pending payment." The changes are saved and the system prompts the user to process a new order or cancel the transaction. 6b. If the QUANTITY ORDERED is greater than the QUANTITY AVAILABLE, the NUMBER TO BE PICKED will be set to the QUANTITY AVAILABLE. The system will reduce the QUANTITY AVAILABLE to zero (if it is not already) and update QUANTITY BACKORDERED to be equal to the QUANTITY ORDERED minus the NUMBER TO BE PICKED. 6c. If there are invalid product numbers or quantity ordered entries, invoke abstract use case ***Generate Order Error Report***.
PRECONDITION:	Use case ***Send Club Promotion*** has been processed.
	Use case ***User Logs In*** has been processed.
POST-CONDITION:	Promotion order has been recorded and the Picking Ticket has been routed to the Warehouse.
ASSUMPTIONS:	

FIGURE 16.2 *(concluded)*

case. Abstract use cases are those that contain steps that are used by more than one design use case.

Step 2—Updating the Use Case Model Diagram and Other Documentation to Reflect Any New Use Cases After all the analysis use cases have been transformed to design use cases, it is possible that new use cases, use case dependencies, or even actors have been discovered. It is very important that we keep our documentation accurate and current. Thus, in this step the use case model diagram, the use case dependency diagram, and the actor and use case glossaries should be updated to reflect any new information introduced in step 1.

Modeling Object Interactions and Behaviors that Support the Use Case Scenario

In the previous section we refined the use cases to reflect the implementation environment. In this activity we want to identify and categorize the design objects required by the functionality that was specified in each use case and identify the object interactions, their responsibilities, and their behaviors.

① INTERFACE OBJECTS	② CONTROL OBJECTS	③ ENTITY OBJECTS
Member Services Main Window	Order Processor	MEMBER
Order Processing Window	Ticket Generator	PRODUCT
Picking Ticket Printer		MEMBER ORDER
		MEMBER ORDERED PRODUCT

FIGURE 16.3 *Interface, Control, and Entity Objects of "Submit Promotion Order" Use Case*

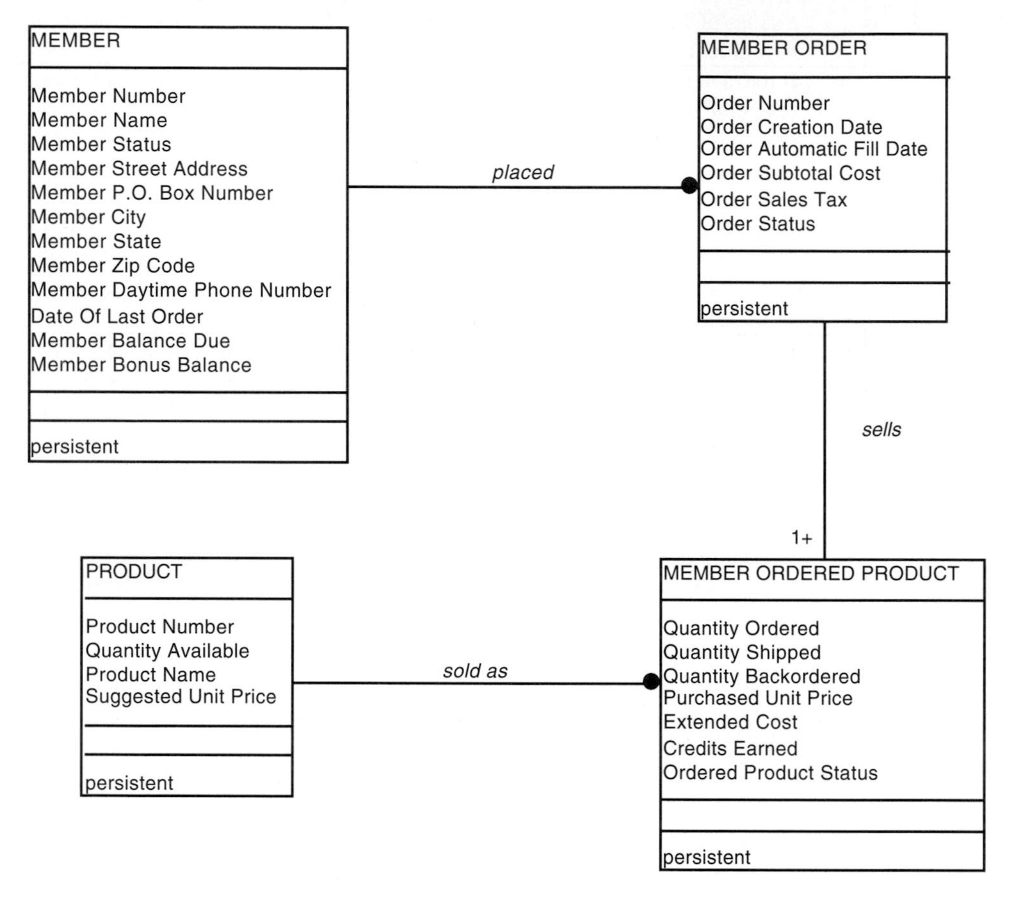

FIGURE 16.4 *Object Association Diagram for "Submit Promotion Order" Use Case*

Step 1—Identify and Classify Use Case Design Objects Earlier we learned there are three categories of design objects: interface, control, and entity. In this step we examine each design use case to identify and classify the types of objects required by the logic of the use case or business scenario. Figure 16.3 depicts the result of analyzing the use case shown in Figure 16.2. We draw your attention to the following:

① The interface object column contains a list of objects mentioned in the use case with which the users directly interface, such as screens, windows, card readers, and printers. The only way an actor or user can interface with a system is via an interface object. Therefore, there should be at least one interface object per actor or user.

② The control object column contains a list of objects that encapsulate application logic or business rules. As a reminder, a use case should reveal one control object per unique user or actor.

③ The entity object column contains a list of objects that correspond to the business domain objects whose attributes were referenced in the use case.

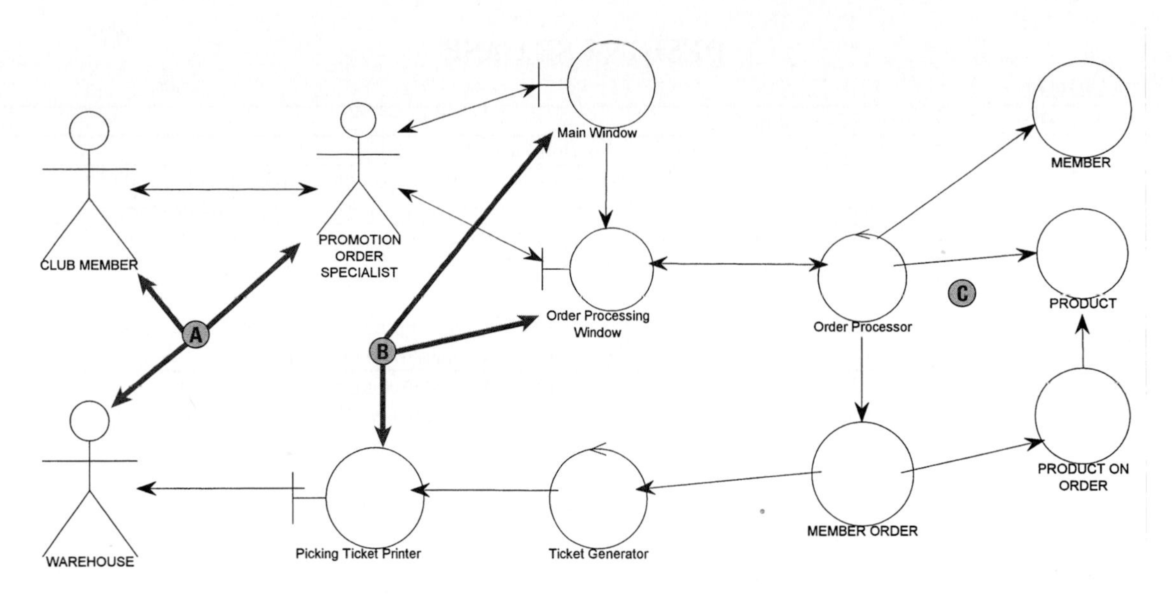

FIGURE 16.5 *Ideal Object Model Diagram for "Submit Promotion Order" Use Case*

Step 2—Identify Object Attributes During both analysis and design, object attributes may be discovered. In efforts to transform analysis use cases into design use cases, we begin referencing the attributes in the use case text. In this step we examine each use case for additional attributes that haven't been previously identified, and we update our object association diagram to include attributes. Figure 16.4 is an object association diagram containing only the entity objects referenced in the "Submit Promotion Order" use case.

Step 3—Model High-Level Object Interactions for a Use Case After identifying and categorizing the design objects involved in a use case, we need to model those objects and their interactions. Such models are called **ideal object model diagrams.** Figure 16.5 depicts an ideal object model diagram for the "Submit Promotion Order" use case. This diagram was developed using Popkin Software's *System Architect* CASE tool. An ideal object model diagram includes symbols to represent actors, interface, control, and entity objects, and arrows that represent messages or communication between the objects. We draw your attention to the following:

1. We have included both the actor that "initiated" the use case (the CLUB MEMBER) and the actor who is the direct "user" of the system (the PROMOTION ORDER SPECIALIST). Another actor represented is the WAREHOUSE. The WAREHOUSE is a "receiver." In this scenario, the warehouse receives the picking ticket information from the system.
2. Actors may interact with the system only via interface objects. In our example above, the PROMOTION ORDER SPECIALIST (direct user of the system) interacted with the system by way of the MAIN WINDOW and ORDER PROCESSING WINDOW interface objects. Notice that the WAREHOUSE (a receiver of the system) interacts with the PICKING TICKET PRINTER interface object.
3. By following the logic of the use case we can construct a high-level, first-cut view of how the objects need to interact with each other to support the functionality of the use case. Notice that the ORDER PROCESSOR control object sends messages (depicted using arrows) to the MEMBER entity object. Based on the text of the use case, we know that MEMBER information needs to be retrieved, displayed, and updated. Therefore the messages between ORDER PROCESSOR and MEMBER objects request the MEMBER object to fulfill its object responsibility by providing those services.

DESIGN USE CASE

Author: K. Dittman

USE CASE NAME:	Submit Promotion Order
ACTOR:	Club Member - System user: Promotion Order Specialist
DESCRIPTION:	Describes the process when a club member submits a promotion order to either indicate the products he or she is interested in ordering or declining to order during this promotion.
NORMAL COURSE:	The main window is currently displayed on the screen waiting for the promotion order specialist to select a menu option. 1. When a **promotion order is received** from the club member, this use case is initiated when the promotion order **specialist selects the option** "Process Promotion Order," **displaying a window** requesting the club member's NUMBER to be entered.
	2. The promotion order **specialist enters the MEMBER NUMBER**. The **system verifies that the number is valid** and if it is, **displays three windows**. The first window contains the club member's personal information including: NAME, STATUS, STREET ADDRESS, P. O. BOX, CITY, STATE, ZIP, DAYTIME PHONE NUMBER, DATE OF LAST ORDER, BALANCE DUE, and BONUS BALANCE. The second window contains the promotion order header record information that was created when the promotion was sent to the club member. The second window contains: ORDER NUMBER, ORDER CREATION DATE, ORDER AUTOMATIC FILL DATE, SUB-TOTAL COST, SALES TAX, and ORDER STATUS. The third window contains the individual ordered products. This is a multi-lined window with the first order line containing the selection of the month information. This record was also created when the promotion was sent to the club member. The third window contains: PRODUCT NUMBER, PRODUCT NAME, QUANTITY AVAILABLE, QUANTITY ORDERED, QUANTITY BACKORDERED, SUGGESTED UNIT PRICE, PURCHASED UNIT PRICE, EXTENDED COST, CREDITS EARNED, and ORDERED PRODUCT STATUS.
	3. The promotion order **specialist checks to see if the club member has made any address or phone number changes** on the promotion order. If no changes have been made, the promotion order **specialist checks the promotion order to see if any products are being ordered**. If products are being ordered, the promotion order **specialist positions the cursor in the ordered product window**.
	4. The promotion order **specialist checks the promotion order to see if the club member has accepted or declined** the selection of the month. If accepted, the promotion order **specialist checks quantity ordered** to see if it is greater than one. If so, the promotion order **specialist edits the QUANTITY ORDERED field** to reflect the new quantity. The **system calculates the new EXTENDED COST** by multiplying the PURCHASED UNIT PRICE times the QUANTITY ORDERED and then **advances the cursor** to the next line.
	5. The promotion order **specialist checks the promotion order to see if the club member has ordered additional products**. If so, the promotion order **specialist enters the PRODUCT NUMBER** of the next item being ordered. The **system displays the PRODUCT NAME, QUANTITY AVAILABLE, and the SUGGESTED UNIT PRICE**. The promotion order **specialist enters the QUANTITY ORDERED, and THE PURCHASE UNIT PRICE**. The **system calculates the new EXTENDED COST** by multiplying the PURCHASE UNIT PRICE times the QUANTITY ORDERED. **It also updates the ORDERED PRODUCT STATUS** field to be "open," and then **advances the cursor** to the next line. This step is repeated until all ordered products have been input.
	6. Once all ordered products have been entered into the system, the promotion order **specialist commits the order**. Invoke extension use case ***Calculate Subtotal and Sales Tax***. The **system checks the club member's STATUS**. If the club member is in good standing, invoke extension use case ***Generate Warehouse Picking Ticket***. The **picking ticket will be routed** to the warehouse and **printed on their printer** (HP.PRT.20195). For each product ordered, the **system reduces the QUANTITY AVAILABLE** by the QUANTITY ORDERED and **saves the changes**.
	7. Once the order has been committed, the **system will prompt the user** if he or she wants to **process a new order or cancel the transaction**.

FIGURE 16.6 *Design Use Case with Verb Phrases Suggesting Behaviors*

Remember at this point the ideal object diagram is our first attempt at determining how the objects should interact with each other to support the business event. Later we will use this diagram to model more detailed object interactions.

ALTERNATE COURSE:	2. If the MEMBER NUMBER is invalid, the **system displays a window** with the error message "Member Number Not on File." The promotion order **specialist can then re-enter the number or cancel the transaction.**
	3a. If the club **member indicates an address or telephone number change** on the promotion order, the promotion order **specialist selects the option** to update the club member's personal information. The personal information **window is activated for update**. The promotion order **specialist updates the required fields and saves** the changes. 3b. If the club member is not ordering product at this time, the promotion order **specialist clicks the button** "Order Declined." The **system modifies the ORDER STATUS field** to have a status of "closed" and **modifies the ORDERED PRODUCT STATUS field** (for the selection of the month) to have a status of "rejected." The **changes are saved** and then the **system prompts the user to process a new order or cancel the transaction.**
	4. If the club member is not ordering the selection of the month, the promotion order **specialist clicks the button** "Selection of the Month Declined." The **system modifies the ORDERED PRODUCT STATUS field** (for the selection of the month) to have a status of "rejected." The **cursor is then advanced** to the next line.
	5a. If no additional products are being ordered, proceed to step 6 of the normal course. 5b. If the PRODUCT NUMBER or QUANTITY ORDERED is not valid, the **system highlights the fields** indicating those fields have invalid entries. The promotion order **specialist either corrects the entries or advances to the next line.** In a later step an error report will be generated for any invalid entries.
	6a. If the club member's status is not acceptable, invoke abstract use case ***Send Order Rejection Notice.*** The **system modifies the ORDER STATUS field** to have a status of "on hold pending payment." The **changes are saved** and the **system prompts the user to process a new order or cancel the transaction.** 6b. If the QUANTITY ORDERED is greater than the QUANTITY AVAILABLE, the **NUMBER TO BE PICKED will be set** to the QUANTITY AVAILABLE. The **system will reduce the QUANTITY AVAILABLE** to zero (if it is not already) and **update QUANTITY BACKORDERED** to be equal to the QUANTITY ORDERED minus the NUMBER TO BE PICKED. 6c. If there are invalid product numbers or quantity ordered entries, invoke abstract use case ***Generate Order Error Report.***
PRECONDITION:	Use case ***Send Club Promotion*** has been processed. Use case ***User Logs In*** has been processed.
POST-CONDITION:	Promotion order has been recorded and the Picking Ticket has been routed to the Warehouse.
ASSUMPTIONS:	

FIGURE 16.6 *(concluded)*

Step 4—Identify Object Behaviors and Responsibilities Once we have identified all the objects needed to support the functionality of the use case, we shift our attention to defining their specific behaviors and responsibilities. This step involves the following tasks:

1. Analyze the use cases to identify required system behaviors.
2. Associate behaviors and responsibilities with objects.
3. Examine object model for additional behaviors.
4. Verify classifications.

In Chapter 8 you learned objects encapsulate data and behavior. Our first task in identifying the object behaviors and responsibilities is accomplished by once again examining our use case. The use case description is examined to identify all *action verb phrases.* Action verb phrases suggest behaviors that are required to complete a use case scenario. Figure 16.6 illustrates the "Submit Promotion Order" use case where all the action verb phrases and their associated nouns have been

Behavior	Automated/Manual	Object Type
Process promotion order	Manual/Automated	Control
Send promotion order	Manual	
Select system option	Manual	
Display window	Automated	Interface
Enter member number	Manual	
Verify member number	Automated	Entity
Report member information	Automated	Entity
Report member order information	Automated	Entity
Check order for changes	Manual	
Check order for products being ordered	Manual	
Position cursor	Manual/Automated	Interface
Check order for selection of the month	Manual	
Check quantity ordered	Manual	
Enter quantity ordered	Manual	
Calculate extended cost	Automated	Entity
Enter product number	Manual	
Report product information	Automated	Entity
Update ordered product status	Automated	Entity
Commit order	Manual	
Verify member status	Automated	Entity
Process picking ticket	Automated	Control
Route picking ticket to warehouse	Automated	Interface
Print picking ticket	Automated	Interface
Prompt User	Automated	Interface
Cancel transaction	Automated	Control
Enter member information	Manual	
Update member information	Automated	Entity
Decline order	Manual	
Update order status	Automated	Entity
Decline selection of month	Manual	
Highlight field	Automated	Interface
Calculate number to be picked	Automated	Entity
Calculate quantity available	Automated	Entity
Calculate quantity backordered	Automated	Entity

FIGURE 16.7 *"Submit Promotion Order" Use Case Behaviors*

highlighted. These verb phrases correlate to the system behaviors required to respond to a business event of a club member submitting a promotion order. Each use case should be separately examined to identify behaviors associated with the use case.

Once the behaviors have been identified, we must determine if the behaviors are manual or they will be automated. If they are to be automated, they must be associated with the appropriate object type that will have the responsibility of carrying out that behavior. Figure 16.7 lists each verb phrase or behavior in the "Submit Promotion Order" use case, whether it's manual or to be automated, and the design object type it is to be associated with.

In Figure 16.8 we have condensed the behavior list to show only the behaviors that need to be automated. Recall that the object types were defined earlier in step 1. We will use this list as the source of behaviors to be allocated in the next task.

Our third task is aimed toward identifying all behaviors that can be associated with an object type and identifying collaborations among those objects. A popular tool for documenting the behaviors and collaborations for an object is the **class responsibility collaboration (CRC) card.** A CRC card for the object type MEMBER ORDER is depicted in Figure 16.9. The CRC card contains all use case behaviors and responsibilities that have been associated with the object type MEMBER

Behavior	Object Type
Process picking ticket	Control
Cancel transaction	Control
Process promotion order	Control
Verify member number	Entity
Report member information	Entity
Report member order information	Entity
Calculate extended cost	Entity
Report product information	Entity
Update ordered product status	Entity
Verify member status	Entity
Update member information	Entity
Update order status	Entity
Calculate number to be picked	Entity
Calculate quantity available	Entity
Calculate quantity backordered	Entity
Display window	Interface
Route picking ticket to warehouse	Interface
Print picking ticket	Interface
Prompt user	Interface
Highlight field	Interface
Position cursor	Interface

FIGURE 16.8 *Condensed Behavior List for "Submit Promotion Order" Use Case*

Object Name: Member Order	
Sub Object:	
Super Object	
Behaviors and Responsibilities	**Collaborators**
Report order information	Member Ordered Product
Calculate subtotal cost	
Calculate sales tax	
Update order status	
Create ordered product	
Delete ordered product	

FIGURE 16.9 *CRC Card for "Member Order" Object*

ORDER. Analysis of the use case scenarios may not reveal all behaviors for any given object type. On the other hand, by examining the object model, you may find additional behaviors (not mentioned in the use case scenarios) that need to be assigned to an object type. For example, analyze the relationships between the objects in Figure 16.4. How are those relationships created or deleted? Which object should be assigned that responsibility? As a general rule, the object that controls the relationship should be responsible for creating or deleting the relationship. Reference Figure 16.4 again and draw your attention to the relationship between MEMBER ORDER and MEMBER ORDERED PRODUCT. By designing the system to have the MEMBER ORDER object have a behavior to "add ordered product," we have effectively given the MEMBER ORDER object control to create this relationship. Also, recall from Chapter 8 that there are four "implicit" behaviors that can be associated with any object class. Those are the ability to create new instances, change its data or attributes, delete instances, and display information about the object class. While examining the use cases to identify and associate behaviors with object types, we also focus on identifying the **collaboration** or cooperation that is necessary between object types. In Figure 16.9 the MEMBER ORDER object needs collaboration from the MEMBER ORDERED PRODUCT object to retrieve information about each of

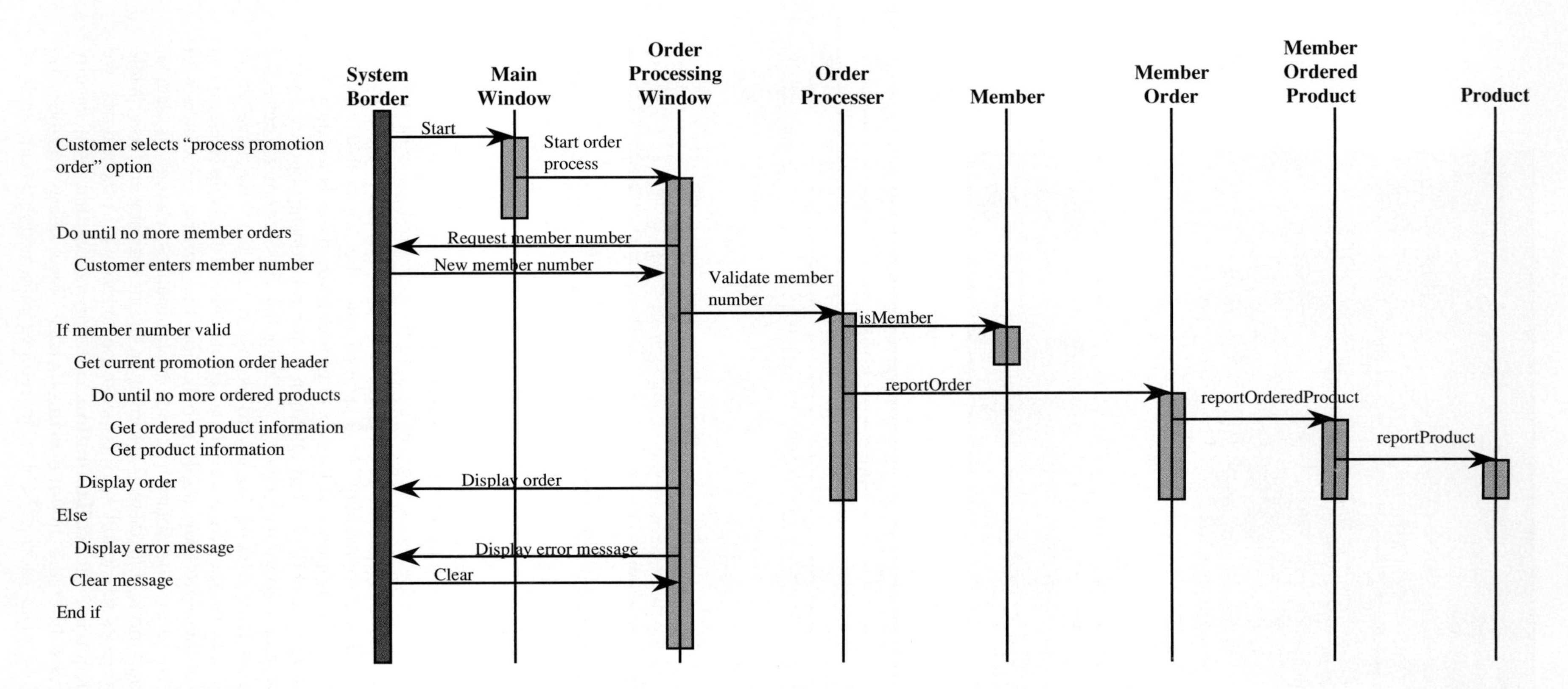

FIGURE 16.10 *Partial Interaction Diagram for the "Submit Promotion Order" Use Case*

the products being ordered. Remember if an object needs another object's attribute to accomplish a behavior, the collaborating object needs to have a behavior or method to provide that attribute.

Identifying the collaboration of object types is necessary to ensure that all use case objects work in harmony to complete the processing required for the business event that triggers the use case scenario.

Finally, our last task is to verify the results from the previous tasks. This consists of conducting walk-throughs with the appropriate users. One verification approach that is commonly used is **role playing.** In role playing, the use case scenarios are acted out by the participants. The participants may assume the role of actors or object types that collaborate to process a hypothetical business event. Message sending is simulated by using an item such as a ball that is passed (or sometimes thrown) between the participants. Role playing is quite effective in discovering missing objects and behaviors, as well as verifying the collaboration among objects.

Step 5—Model Detailed Object Interactions for a Use Case Once we have determined the objects' behaviors and responsibilities, we can create a detailed model of how the objects will interact with each other to provide the functionality specified in each design use case. We use a model called an **object interaction diagram** to graphically depict these interactions. Interaction diagrams show us in great detail how the objects interact with each other over time

Figure 16.10 is a partial interaction diagram for the "Submit Promotion Order' use case. It is read from top to bottom, following the logic of the use case, which is written in pseudo code at the left of the diagram. Each object referenced in the use case is symbolized by a vertical line. The behaviors or operations that each object needs to fulfill an obligation is represented by a gray box. These boxes represent program code. The arrows between the lines represent interactions or messages being sent to a particular object to invoke one of its operations to satisfy a request.

Updating the Object Model to Reflect the Implementation Environment

Once we have designed the objects and their required interactions, we can refine our object model to include the behaviors or implementation methods it needs to possess. Figure 16.11 is a partial view of our object model that correlates to the objects used in the "Submit Promotion Order" use case. We have given each behavior or method a name. Normally these names reflect the programming language used to develop the system.

WHERE DO YOU GO FROM HERE?

This chapter provided an introduction to the newer object-oriented approach to systems design. Since prototyping is an integral part of object-oriented design, it is recommended that you learn about prototyping during systems design. Thus, if you haven't already covered these chapters, it is recommended that you now read Chapters 11, 12, and 13 to learn how to prototype inputs, outputs, and user interfaces (respectively).

To provide a better understanding of object-oriented design and its impact on the subsequent construction and implementation of a new system, it is recommended that you consider learning about object-oriented programming (OOP). Consider taking a course dealing with object-oriented programming. Numerous books have recently been written on object-oriented programming. Many of these books demonstrate how the object-oriented concepts explained in Chapters 8 and 16 are implemented in an object-oriented programming language environment.

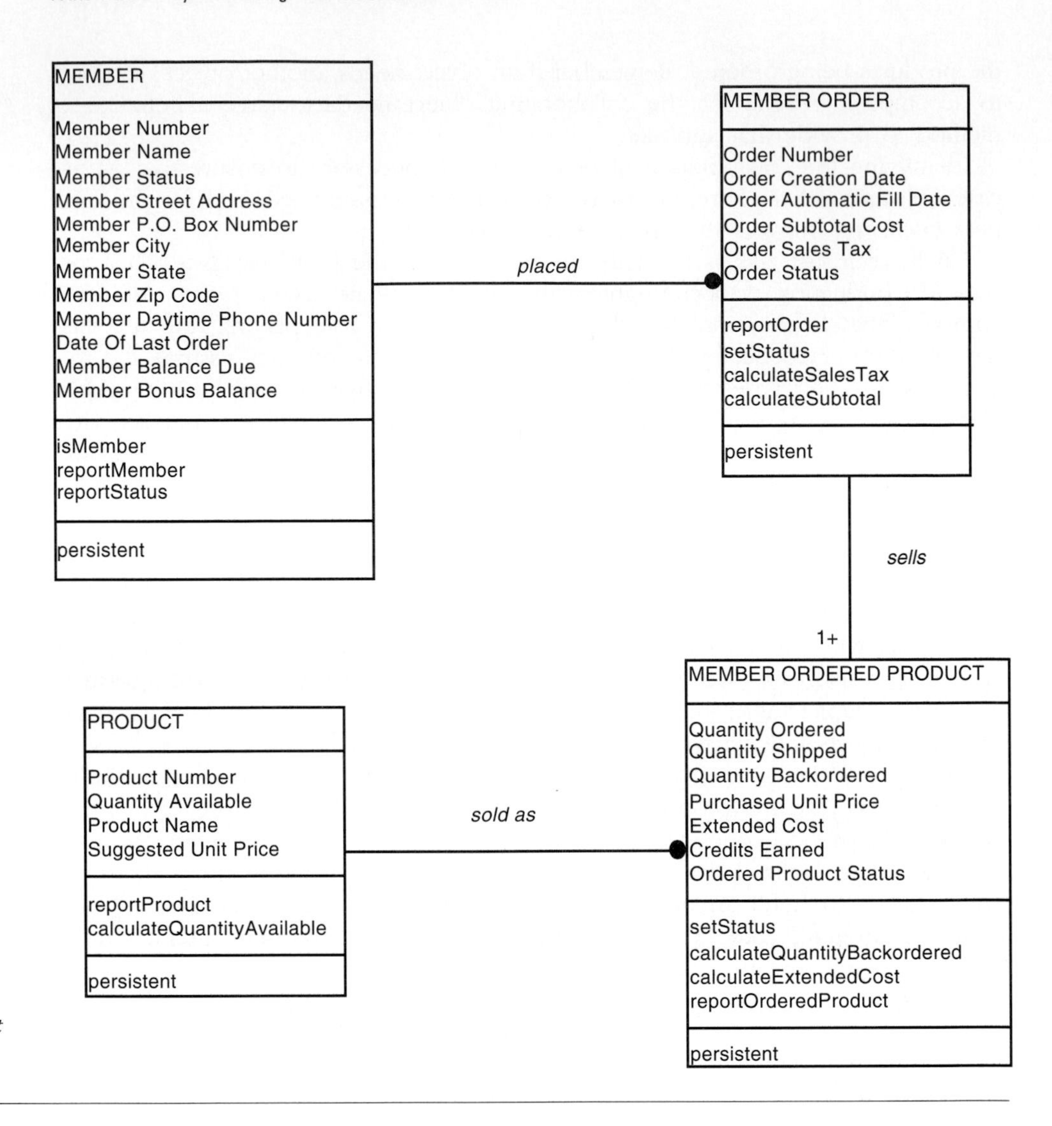

FIGURE 16.11 *Partial Object Model Correlating to "Submit Promotion Order" Use Case Objects*

SUMMARY

1. Using object-oriented techniques for designing a system is referred to as object-oriented design.
2. Object-oriented design is concerned with identifying and classifying three object types—interface, entity, and control. Interface and control object types are objects that are introduced as a result of implementation decisions that were made during systems design. Entity objects are identified during systems analysis and usually correspond to items in real life and contain information, known as attributes, that describes the different instances of the entity. Interface objects are objects that are introduced to represent a means through which the user will interface with the system. The responsibility of the interface object is twofold: (1) it translates the user's input into information that the system can understand and use to process the business event, and (2) it takes data pertaining to a business event and translates the data for appropriate presentation to the user. Control objects are those that hold application or business rule logic. Controller objects serve as the "traffic cop" containing the application logic or business rules of the event for managing or directing the interaction between the objects.
3. Object responsibility is the obligation that an object has to provide a service when requested, thus collaborating with other objects to satisfy the request if required. Object responsibility is closely related to the concept of objects being able to send and/or respond to messages.
4. Object-oriented design includes the following activities: refining the use case model to reflect the implementation environment, modeling object interactions and behavior that support the use case scenario, and updating the object model to reflect the implementation environment.
5. A use case may contain complex functionality that merits those steps being their own use case in order for it to be

simplified and more easily understood. This type of use case is called an extension use case in that it extends the functionality of the original use case.

6. It is common for two or more use cases to perform steps of identical functionality. The common steps should be extracted into their own separate use case called an abstract use case. An abstract use case represents a form of "reuse." An abstract use case is available for referencing (or use) by any other use case that requires its functionality.
7. An ideal object model diagram is used to document objects and their interactions. An ideal object model diagram includes symbols to represent actors, interface, control, and entity objects, and arrows that represent messages or communication between the objects.
8. During systems design, use case descriptions are examined to identify all action verb phrases. Action verb phrases suggest behaviors required to complete a use case scenario. These behaviors must be associated with a system object.
9. A popular tool for documenting the behaviors and collaborations for an object is the class responsibility collaboration (CRC) card.
10. An interaction diagram is used to document the detailed interactions between objects.

KEY TERMS

abstract use case, p. 536
class responsibility collaboration (CRC) card, p. 544
collaboration, p. 545
control object, p. 534
entity object, p. 534
extension use case, p. 536
ideal object model diagram, p. 541
interface object, p. 534
model-view-controller (MVC), p. 534
object framework, p. 536
object interaction diagram, p. 547
object-oriented design, p. 534
object responsibility, p. 535
object reuse ability, p. 535
persistent, p. 535
role playing, p. 547

REVIEW QUESTIONS

1. What is the approach of using object modeling during systems design called?
2. Differentiate between entity, interface, and control objects.
3. Explain the concept of model-view-controller.
4. What does it mean when an entity object is said to be persistent?
5. Define the term *object responsibility*. Give an example.
6. Explain the importance of object reuse.
7. What is an object framework? Give an example.
8. Identify the three object-oriented design activities.
9. What is an abstract use case?
10. What is an extension use case?
11. What is an ideal object model diagram used for in object-oriented design?
12. Explain the difference between an "initiator" actor and a "receiver" actor. Give an example of each actor.
13. Explain how CRC cards are used during object-oriented design.
14. What do verb phrases in a use case suggest?
15. What are four implicit behaviors for any entity object?
16. What is the purpose of an object interaction diagram?

PROBLEMS AND EXERCISES

1. What are the benefits of structuring an object-based system into entity, interface, and control objects?
2. What correlation can be drawn between the three object types in problem 1 and the client/server model?
3. What correlation can be drawn between object responsibility and message sending?
4. What correlation can be drawn between an abstract use case and object reusability?
5. In transforming the analysis use cases to design use cases, what types of refinement might be done? Can such refinements result in new objects?

6. Analyze the following use case to identify different object types.

USE CASE NAME:	Purchasing a soda
ACTOR:	Customer
DESCRIPTION:	This use case describes the scenario of a customer purchasing a can of soda from a soda machine. The soda machine has six selections of soda, is able to accept coins and $1 bills, and has a digital display.
BASIC COURSE:	**Step 1**. This use case is initiated when the customer wants to purchase a can of soda. **Step 2**. The customer inserts a coin into the machine's coin receptacle. **Step 3**. The machine verifies whether the coin is legitimate, records the amount deposited, then updates the display with the total amount deposited. Go to **Step 2** if another coin is deposited. **Step 4**. The customer pushes the selection button of the type of soda desired. **Step 5**. The machine calculates whether change is due, then releases the can of soda from its holding bin, and it drops to the can retrieval receptacle. Machine records transaction, including date, time, type of purchase, and amount of purchase. **Step 6**. The customer retrieves the can of soda.
ALTERNATE COURSES	**Step 2**. The customer inserts a $1 bill into the machine's dollar bill receptacle. **Step 3A**. If the coin is not legitimate, release the coin to the change receptacle. **Step 3B**. If the dollar bill is not legitimate, reject bill out of receptacle. **Step 4A**. If customer depresses change return lever, release held coin(s) to change receptacle or reject dollar bill out of dollar bill receptacle. **Step 5A**. If customer has not deposited enough money for the soda selected, display message "Not enough money deposited" for five seconds, then display total amount deposited. Return to **Step 2**. **Step 5B**. If the soda the customer has selected is out of stock, display message "Out of Stock, Please Make Another Selection" for five seconds, then display total amount deposited. Return to **Step 4**. **Step 5C**. If change is due, release appropriate coins to the change receptacle.
PRECONDITION:	None
POSTCONDITION:	Customer has purchased soda.
ASSUMPTIONS:	The soda machine is operating normally and is able to give change.

7. Analyze the use case in problem 6 to identify behaviors.
8. Using the results of problems 6 and 7, develop a CRC card for one of the use case objects.
9. Using the results of problems 6, 7, and 8, construct an ideal object diagram.
10. Using the results of problems 9, construct an interaction diagram.

PROJECTS AND RESEARCH

1. Surf the Internet and make a list of links to sites containing information about object-oriented design.
2. Research the new Unified Modeling Language (UML) sponsored by the Rational Software Corporation. How are the diagrams that are used in the UML similar to the diagrams that we used in this chapter? How are they different?
3. Perform role playing using the use case in problem 6 above as the script.

MINICASES

1. Schedule an interview with the business analyst responsible for your school's registration system. Based on the information provided perform the following:
   a. Draw a context model diagram.
   b. Identify the actors and the use cases they initiate.
   c. Complete a use case description for the event of a student enrolling for a course. For a student dropping a course.
   d. Use the use cases you have previously prepared and construct an object association model diagram.
   e. Transform the use cases you previously prepared into design use cases.
   f. Identify the design objects in each of the design use cases.
   g. Construct ideal object model diagrams for each of the design use cases.
   h. Using the previously prepared design use cases, identify the required responsibilities and behaviors, and document them on CRC cards.
   i. Construct interaction diagrams for each of the design use cases.
   j. Refine the object association diagram of step d to include attributes and methods.

SUGGESTED READINGS

Booch, G. *Object-Oriented Design with Applications*. Redwood City, CA: Benjamin/Cummings, 1994.

Booch, G. *Object Solutions—Managing the Object-Oriented Project*. Reading, MA: Addison-Wesley, 1996.

Coad, P., and E. Yourdon. *Object-Oriented Analysis*. 2nd ed. Englewood Cliffs, NJ: Prentice Hall, 1991. This book provides a very good overview of object-oriented concepts. However, the object model techniques are somewhat limited by comparison to OMT and other object-oriented modeling approaches.

Goldberg, Adele, and Kenneth S. Rubin. *Succeeding with Objects*. Reading, MA: Addison-Wesley, 1995.

Jacobson, Ivar, Magnus Christerson, Patrik Jonsson, and Gunnar Overgaard. *Object-Oriented Software Engineering—A Use Case Driven Approach*. Wokingham, England: Addison-Wesley, 1992. This book presents detailed coverage of the process of use case modeling and how it's used to identify objects.

Martin, J., and J. Odell. *Object-Oriented Analysis and Design*. Englewood Cliffs, NJ: Prentice Hall, 1992.

Rumbaugh, James, Michael Blaha, William Premerlani, Frederick Eddy, and William Lorensen. *Object-Oriented Modeling and Design*. Englewood Cliffs, NJ: Prentice Hall, 1991. This book presents detailed coverage of the object modeling technique (OMT) and its application throughout the entire systems development life cycle.

Taylor, David A. *Object-Oriented Information Systems—Planning and Implementation*. New York: John Wiley & Sons, 1992. This book is a very good entry-level resource for learning the concepts of object-oriented technology and techniques.

Tkach, Daniel, and Richard Puttick. *Object Technology in Application Development*. Redwood City, CA: Benjamin/Cummings, 1994.

Wilkinson, Nancy M. *Using CRC Cards*. New York: SIGS Books, 1995.

P A R T   F O U R

BEYOND SYSTEMS ANALYSIS AND DESIGN

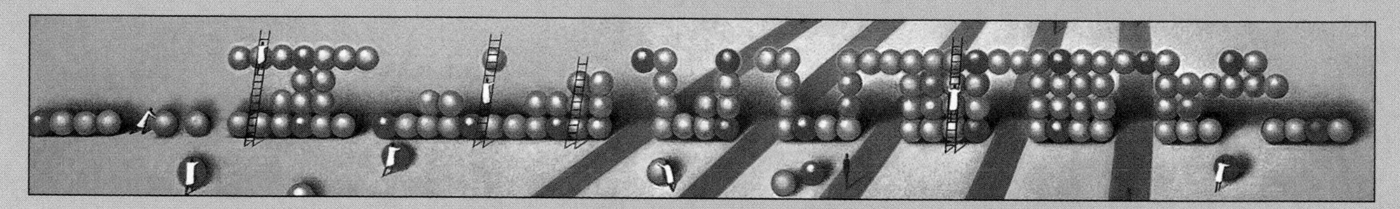

Part Four introduces you to the systems implementation and systems support phases of the systems development life cycle. Two chapters make up this unit. First, Chapter 17, Systems Implementation, presents systems implementation, the process of putting the design specifications for the new information system in actual operation. Two implementation phases are discussed: construction and delivery. Each of these phases is discussed in terms of purpose, activities, roles, inputs and outputs, techniques, and steps.

Chapter 18, Systems Support, discusses four types of systems support for an application. This ongoing maintenance of a system after it has been placed into production consists of correcting errors, recovering the system, assisting users, and adapting the system. Systems support is very important because it is likely that young systems analysts will be responsible for maintaining legacy systems. This chapter concludes our exploration of the systems development life cycle.

17

SYSTEMS IMPLEMENTATION

CHAPTER PREVIEW AND OBJECTIVES

In this chapter you will learn more about the systems implementation phases in the *FAST* systems development methodology. The systems implementation phases serve to construct and deliver the final system into operation. You will know that you understand the process of systems implementation when you can:

- Define the systems implementation process in terms of the construction and delivery phases of the life cycle.
- Describe systems implementation phases in terms of your information building blocks.
- Describe the systems implementation phases in terms of purpose, activities, roles, inputs and outputs, techniques, and steps.
- Identify those chapters and modules in this textbook that can help you actually perform the activities of systems implementation.

Although some of the techniques of systems implementation are introduced in this chapter, it is *not* the intent of this chapter to teach the *techniques* of systems implementation. This chapter teaches only the process of systems implementation.

SCENE

When we last visited SoundStage, Sandra and Bob were busy finishing the technical design statement for their project. Sandra and Bob have since progressed to implementation of the new system. They are meeting to plot the strategy for completing this final phase to deliver the new system into production.

SANDRA

Well, Bob, I think we are finally starting to see the light at the end of the tunnel. I'm getting really excited, but at the same time a little nervous. We're going to be cutting it close on our deadline.

BOB

You're telling me. I worked all weekend to try to get caught up. I'm tired, but I feel a lot better about where I'm at with construction of the final pieces for the new system.

SANDRA

I thought that I was the only one working this weekend. I spent most of the weekend planning how we are going to complete the delivery of the new system once we finished the construction of all the final pieces. You've been working on their construction. So where do we stand in that respect?

BOB

I think we're in pretty good shape! Fortunately the new system is to use one of our existing LANs. So we gained some time by not having to oversee the installation and testing of a LAN. Since we used a prototyping approach to development, the database was already built and tested before we built any screens. The new system didn't involve purchasing a software package, so we didn't need to install and test any vendor packages. All that left us with was a few new programs that needed to be written and tested. So . . . now you know what I did this weekend.

SANDRA

Sounds like we're in better shape than I thought.

BOB

Yes. I just have a few more new programs to write and test. Then I'll have to run a system test to ensure that they all work together properly.

SANDRA

Great. I've been generating test data for when we do a system test. I've also been coming up with a conversion plan. We'll do a staged conversion. We will install it at our pilot warehouse location and then transport it to others later. In talking with the managers, we concluded that the conversion should be a parallel conversion. They really wanted to continue using the existing system at the same time as the new system.

BOB

That's probably a very good idea since this system represents a critical component of the business.

SANDRA

Since you were working on the new database, I thought I'd ask you to proceed with installing that database.

BOB

No problem. I can quickly crank out some programs to extract data from the production databases to load into the new one. There may be some data that need to be keyed in from other sources though.

SANDRA

I was afraid of that, so I asked permission to hire a few people from a local temporary employment company to help us out. I've studied the training needs of the users.

BOB

Since we used a prototyping approach to involve the users throughout the development process, they should be relatively familiar with the new system. Don't forget, I provided content sensitive help inside the application. So they should be relatively easy to take care of when they use the application.

SANDRA

That's true. That strategy should pay off. However, what about any future employee who will be a user of the system?

BOB

Good point. I guess we will have to schedule a few training sessions as needed.

SANDRA

I agree. We should provide training sessions now and for any new employee. But also, I have been outlining a user manual that highlights the application for the user.

BOB

What else do we have left to do?

SANDRA

That's it . . . except the actual conversion itself. Here, take a look at my draft of the conversion plan and tell me what you think.

DISCUSSION QUESTIONS

1. How would you characterize the focus of the implementation phase of this project?
2. What were some of the benefits that Sandra and Bob realized during systems implementation by following the prototyping approach to systems development.
3. What impact would the need to build a network or install a purchased software package have on Sandra and Bob's implementation of the new system.
4. How does the system test that Bob and Sandra mentioned differ from the testing that Bob had done over the weekend?
5. What type of concerns would the managers have with discarding the existing system and immediately placing the final system into production?

WHAT IS SYSTEMS IMPLEMENTATION?

Let's begin with a definition of systems implementation.

> **Systems implementation** is the construction of the new system and the delivery of that system into production (meaning day-to-day operation). Unfortunately, *systems development* is a common synonym. (We dislike that synonym since it is more frequently used to describe the *entire* life cycle.)

Relative to the information systems building blocks, systems implementation addresses DATA, PROCESSES, INTERFACES, and GEOGRAPHY, primarily from the system builders' perspective.

Figure 17.1 illustrates the phases of a typical systems implementation—construction and delivery. The trigger for systems implementation is the approval of the technical design statement and prototypes. This chapter examines each of these phases in detail.

FAST SYSTEMS IMPLEMENTATION METHODS

Consistent with the discussion of systems analysis and design in Chapters 4 and 9, this chapter will present systems implementation using the following *FAST* methodology elements:

- *Purpose* (self-explanatory).
- *Roles.* All *FAST* activities are completed by individuals who are assigned to **roles.** Roles are not the same as job titles. One person can play many roles in a project. Conversely, one role may require many people to adequately fulfill that role. For example, a system user role may require several users in order to adequately represent the interests of an entire system. *FAST* roles are assigned to the following role groups: system owner roles, system user roles, systems analyst roles, system designer roles, and system builder roles.

These groups correspond with the perspectives in your information systems framework (from Chapters 2 and 3).

Every activity is described with respect to:

- *Prerequisites and inputs* (to the activity).

FIGURE 17.1
The Systems Implementation Phases of a Project

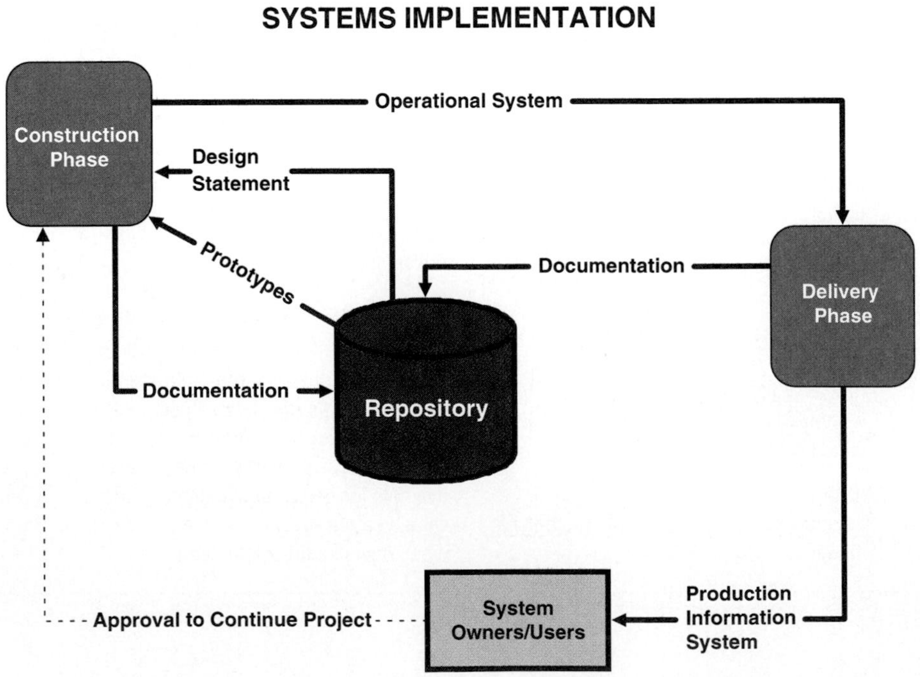

SYSTEMS IMPLEMENTATION

Construction Phase
Operational System
Design Statement
Prototypes
Documentation
Delivery Phase
Documentation
Repository
Approval to Continue Project
System Owners/Users
Production Information System

- *Deliverables and outputs* (produced by the activity).
- *Applicable techniques*—which techniques (from the previous section) are applicable to this phase.
- *Steps*—a brief description of the steps required to complete this activity. In the spirit of continuous improvement, all *FAST* steps are fully customizable for each organization.

Furthermore, all roles, inputs, outputs, techniques, and steps are presented with the following designations:

- (REQ) indicates the role, input, output, technique, or step is REQuired.
- (REC) indicates the role, input, output, technique, or step is RECommended, but not required.
- (OPT) indicates the role, input, output, technique, or step is OPTional, but not required.

THE CONSTRUCTION PHASE OF SYSTEMS IMPLEMENTATION

In Chapter 3, you learned that in the *FAST* methodology, the construction phase is actually part of a design/construction loop that implements *rapid application development*. Given some aspect of the system design, we construct and test the system components in that design. After several iterations of the design/construction loop, we will have built the functional system to be implemented.

The purpose of the construction phase is twofold:

- To build and test a functional system that fulfills business and design requirements.
- To implement the interfaces between the new system and existing production systems.

Your information system framework (Figure 17.2) identifies the relevant building blocks and activities for the construction phase—the project team must construct the database, application programs, user and system interfaces, and networks. Some of these elements may already exist as prototypes from design, or as existing system components (subject to enhancement).

Figure 17.3 is an activity diagram for the construction phase. The figure illustrates the typical activities of the construction phase. Let's now examine each activity in greater detail.

Activity: Build and Test Networks (if Necessary)

Recall that in the definition phase of systems analysis, we established network requirements. Subsequently, during the design and integration phase we developed distributed data and process models. Using these specifications to implement the network blocks for an information system is a prerequisite for the remaining construction and delivery activities.

In many cases, new or enhanced applications are built around existing networks. If so, skip this activity. However, if the new application calls for new or modified networks, they must normally be implemented before building and testing databases and writing or installing computer programs that will use those networks. Thus, the first activity of the construction phase may be to build and test networks.

Purpose The purpose of this activity is to build and test new computer networks or modify existing networks for use by the new system.

Roles This activity will normally be completed by the same telecommunications specialists that designed the network. Other roles are defined as follows:

- System owners and system users are rarely involved in this activity.

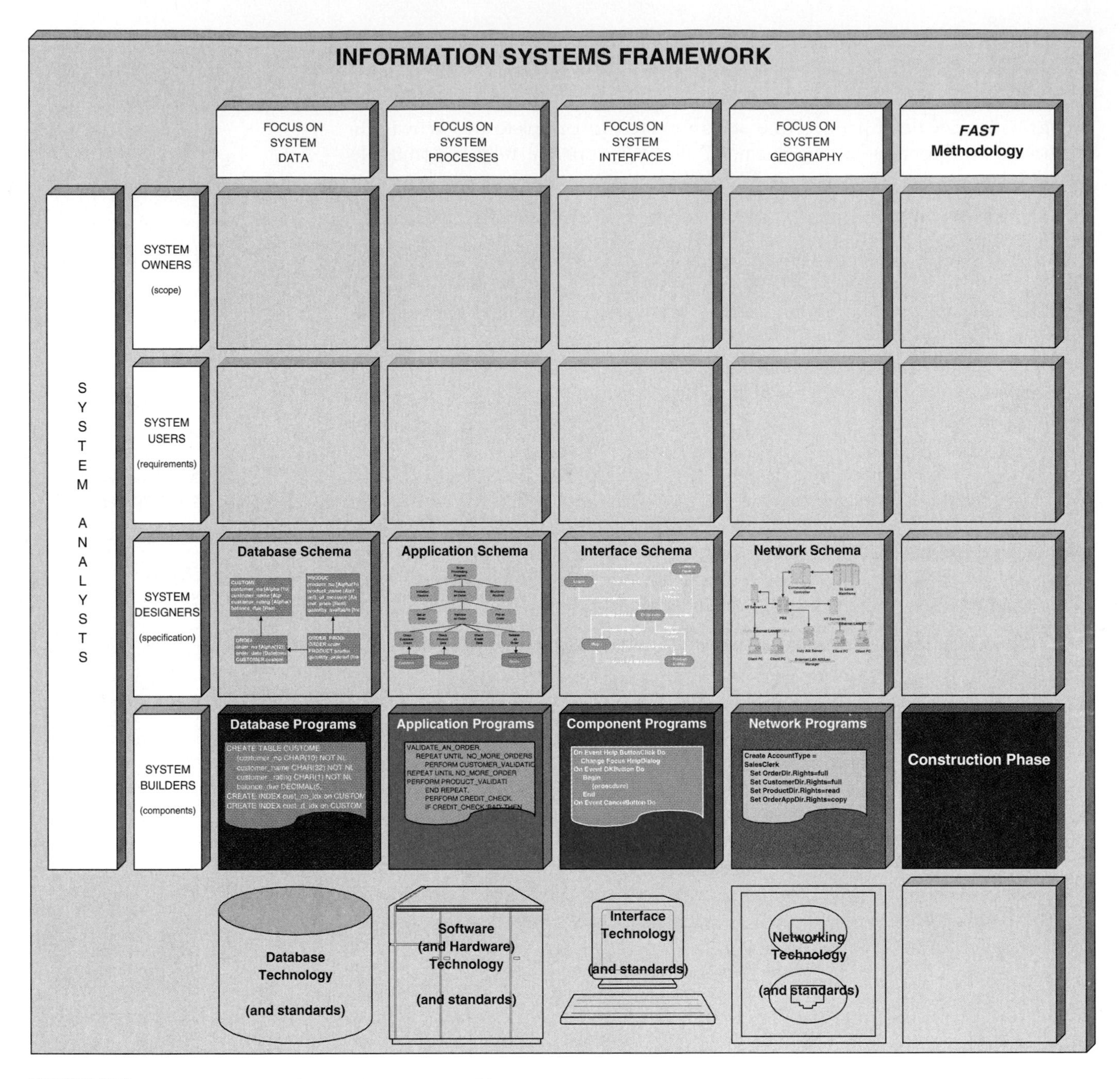

FIGURE 17.2 *Building Blocks for the Construction Phase*

- Systems analyst (REQ)—The systems analyst's role is more in terms of a facilitator, and ensures that business requirements are not compromised by the network solution.
- Network designer (REQ)—The network designer is a specialist in the design of local and wide area networks and their connectivity.
- System builders.
  - Network administrator (REQ)—This person has the expertise for building and testing network technology for the new system. He or she will also be familiar with network architecture standards that must be adhered to for any possible new networking technology. This person is also responsible for security. (Note: the network designer and network administrator may be one and the same person.)

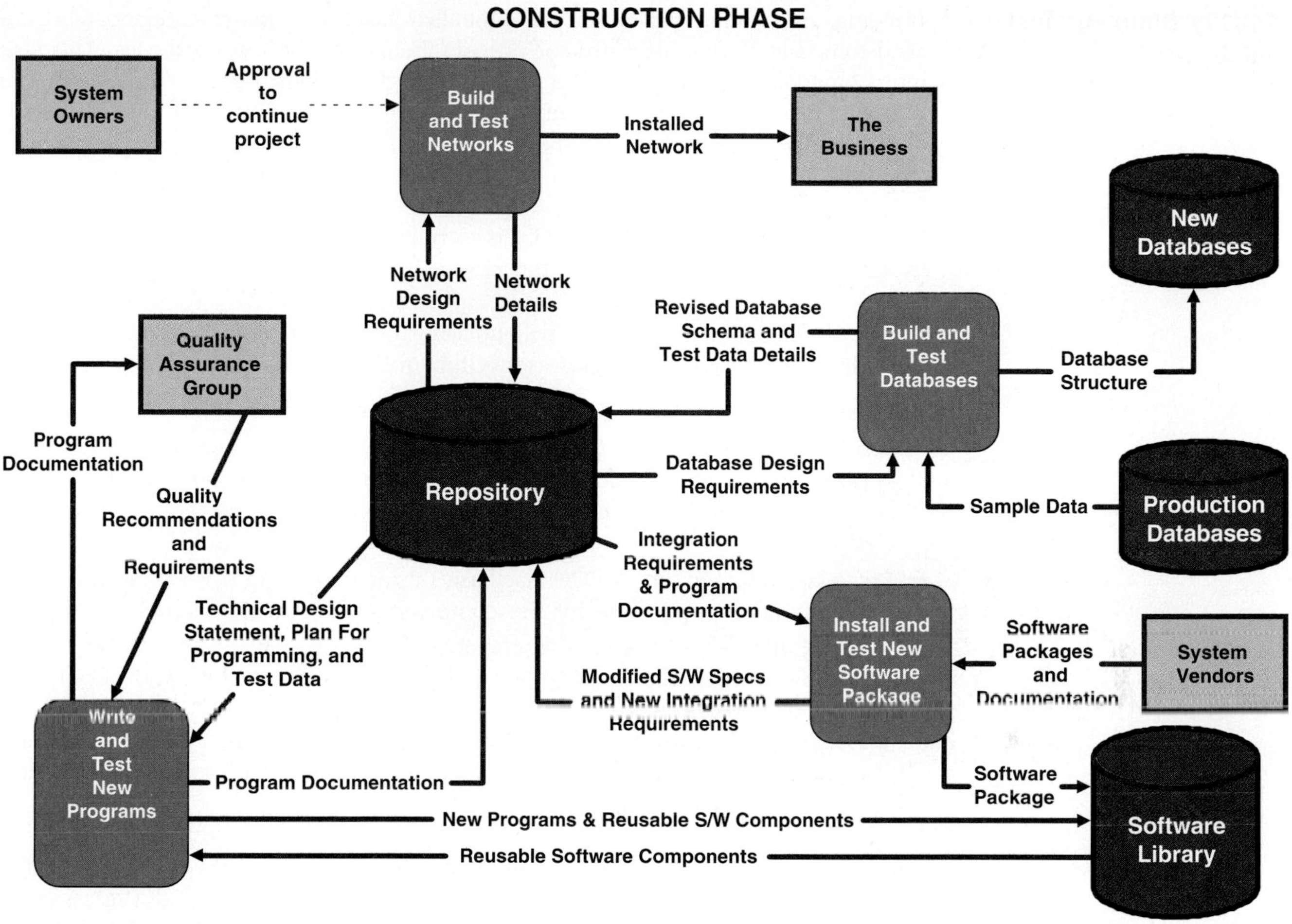

FIGURE 17.3 *Activity Diagram for the Construction Phase*

Prerequisites (Inputs) This activity is triggered by the approval from the system owners to continue the project into systems design. The key input is the *network design requirements* (REQ) defined during systems design

Deliverables (Outputs) The principal deliverable of this activity is an *installed network* (REQ) that is placed into operation. *Network details* (REQ) will be recorded in the project repository for future reference.

Applicable Techniques Knowledge of networks is becoming increasingly important for systems analysts, especially given the industry trend toward network-based computing. It is not the intent of this book to teach network implementation skills. You should, however, consider taking one or more courses on networking. However, Chapters 7 and 10 in this book provide a good overview of networking.

Steps The following steps are suggested to complete this activity.

1. (REQ) Review the network design requirements outlined in the technical design statement developed during systems design.
2. (REQ) Make any appropriate modifications to existing networks and/or construct and test new networks.
3. (REC) Revise network specifications for future reference.

Activity: Build and Test Databases

Building and testing databases are unfamiliar tasks for many students, who are accustomed to having an instructor provide them with the test databases. This task must immediately precede other programming activities because databases are the resources shared by the computer programs to be written. If new or modified databases are required for the new system, we can now build and test those databases.

Purpose The purpose of this activity is to build and test new databases and modify existing databases for use by the new system.

Roles Once again, this activity will typically be completed by the same system specialist that designed the databases. Other roles are defined as follows:

- System users (REC)—should provide or approve test data.
- Systems analyst (REC)—When the database to be built is a noncorporate, applications-oriented database, the systems analyst often completes this activity. Otherwise, systems analysts mostly ensure business requirements compliance.
- Database designer (REQ)—The database designer will often become the system builder responsible for the completion of this activity.
- System builders are primarily responsible for this activity.
  - Database programmers (REQ)—Database programmers build and populate the initial database.
  - Database administrator (REC)—When the database to be built is a corporate database, the database administrator tunes performance, adds security, and provides for backup and recovery.

Prerequisites (Inputs) The primary inputs to this activity are the *database design requirements* (REQ) specified in the technical design statement during systems design. *Sample data* (REC) from production databases may be loaded into tables for testing the databases.

Deliverables (Outputs) The end-product of this activity is an unpopulated *database structure* (REQ) for the new database. The term *unpopulated* means the database structure is implemented but data have not been loaded into the database structure. As you'll soon see, programmers will eventually write programs to populate and maintain those new databases. *Revised database schema and test data details* (REC) are also produced during this activity and placed in the project repository for future reference.

Applicable Techniques The following techniques are applicable to this activity.

- *Sampling*. Sampling methods (REC) are used to obtain representative data for testing database tables. Sampling methods are discussed in Part Five, Module B, "Fact-Finding and Information Gathering."
- *Data modeling*. This activity requires a good understanding of data modeling. Data modeling is covered in Chapter 5.
- *Database design*. Understanding database design requirements is essential for completing this phase. Database design is introduced in Chapter 11.

Steps The following steps are suggested to complete this activity.

1. (REQ) Review the technical design statement for database design requirements.
2. (REC) Locate production databases that may contain representative data for testing database tables. Otherwise, generate test data for database tables.

3. (REQ) Build/modify databases per design specifications.
4. (REQ) Load tables with sample data.
5. (REC) Revise database schema and store as necessary for future reference.

Activity: Install and Test New Software Package (if Necessary)

Some systems solutions may have required the purchase or lease of software packages. If so, once networks and databases for the new system have been built, we can direct our attention toward installing and testing the new software.

Purpose The purpose of this activity is to install any new software packages and make them available in the software library.

Roles This is the first activity in the life cycle that is specific to the applications programmer. Other roles are defined as follows:

- System owners and system users are not involved in this activity.
- Systems analyst (OPT)—The systems analyst typically participates in the testing of the software package by clarifying requirements.
- System designer (REC)—The system designer may be involved in clarifying of integration requirements and program documentation that is used in testing the software.
- System builders are primarily responsible for this activity.
  - Applications programmer (REQ)—The applications programmer (or programming team) is responsible for the installation and testing of new software packages.
  - Network administrators (OPT)—Network administrators may be involved in actually installing the software package on the network server.

Prerequisites (Inputs) The main input to this activity is the new *software packages and documentation* (REQ) that is received from the system vendors. The applications programmer will complete the installation and testing of the package according to *integration requirements and program documentation* (REQ) that was developed during system design.

Deliverables (Outputs) The principal deliverable of this activity is the installed and tested *software package* (REQ) that is made available in the software library. Any modified software specifications and new integration requirements that were necessary are documented and made available in the project repository to provide a history and serve as future reference.

Applicable Techniques There are no specific techniques applicable to this activity. Successful completion of this activity is primarily dependent on programming experience and knowledge of testing.

Steps The following steps are suggested to complete this activity.

1. (REQ) Obtain the software package and review documentation.
2. (REQ) Install the software package.
3. (REQ) Conduct tests on the software package to ensure that it works properly, making revisions as necessary.
4. (REC) For future reference, revise software specifications to reflect modifications.
5. (REC) Add the software to the information systems shop's software library.

Activity: Write and Test New Programs

We are now ready to develop (or complete) any in-house programs for the new system. Recall that prototype programs are frequently constructed in the design phase. These prototypes are included in the technical design statement that specifies a schedule for completing systems implementation. However, these prototypes are rarely not fully functional or complete. Therefore, this activity may involve developing or refining those programs.

Purpose The purpose of this activity is to write and test all programs to be developed in-house.

Roles This activity is specific to the applications programmer. Other roles are defined as follows:

- System owners and system users are usually not involved in this activity.
- Systems analyst (REC)—The systems analyst typically serves merely in clarifying business requirements to be implemented by the programs.
- System designer (REC)—The system designer may be involved in clarification of the program design, integration requirements, and program documentation (developed during systems design) that is used in writing and testing the programs.
- System builders are primarily responsible for this activity.
  - Applications programmer or programming team (REQ)—The applications programmer is responsible for writing and testing in-house software. Most large programming projects require a team effort. One popular organization strategy is the use of *chief programmer teams.* The team is managed by the *chief programmer,* a highly proficient and experienced programmer who assumes overall responsibility for the program design strategy, standards, and construction. The chief programmer oversees all coding and testing activities and helps out with the most difficult aspects of the programs. Other team members include a *backup chief programmer, program librarian, programmers,* and *specialists.*
  - Application tester (REC)—an application or software tester specializes in building and running *test scripts* that are consistently applied to programs to test all possible events and responses.

Prerequisites (Inputs) The primary inputs to this activity are the *technical design statement, plan for programming,* and *test data* (REQ) developed during systems design. Since any new programs or program components may have already been written and be in use by other existing systems, the experienced applications programmer will know to first check for possible *reusable software components* (REC) available in the software library.

Some information systems shops have a quality assurance group staffed by specialists who review the final program documentation for conformity to standards. This group will provide appropriate feedback regarding *quality recommendations and requirements* (OPT).

Deliverables (Outputs) The principal deliverables of this activity are the *new programs* and *reusable software components* (REC) that are placed in the software library. This activity also results in *program documentation* that may need to be approved by a quality assurance group. The program documentation is placed in the project repository for future reference.

Applicable Techniques Testing is an important skill that is often overlooked in academic courses on computer programming. If modules are coded top-down, they should be tested and debugged top-down and as they are written. Testing should not be deferred until after the entire program has been written! There are three levels of testing to be performed: stub testing, unit or program testing, and systems testing.

> **Stub testing** is the test performed on individual modules, whether they be main program, subroutine, subprogram, block, or paragraph.

How can you test a higher-level module before coding its lower-level modules? Easy! You simulate the lower-level modules. These lower-level modules are often called stubs. Stub modules are subroutines, paragraphs, and the like that contain no logic. Perhaps all they do is print that they have been correctly called, and then control goes back to the parent module.

> **Unit or program testing** is a test whereby all the modules that have been coded and stub tested are tested as an integrated unit.

Eventually, all modules will have been implemented, and that unit equals the program itself. Unit testing uses the test data created during the design phase.

> **Systems testing** is a test that ensures that application programs written in isolation work properly when they are integrated into the total system.

Just because a single program works properly doesn't mean that it works properly with other programs. The integrated set of programs should be run through a systems test to make sure one program properly accepts, as input, the output of other programs.

Steps The following steps are suggested to complete this activity.

1. (REQ) Review the design specifications. One major controversy that should be addressed is the freezing of the design specification. By freezing, we mean that changes to the design specifications are discouraged or prohibited. Advocates of freezing argue that with no discouragement against changes, users will continually be permitted to identify something they forgot or some new need or idea, and the system may never be constructed and delivered. On the other hand, some experts dispute the idea of freezing the specifications. They argue that such an action is artificial and is not consistent with our goal to serve the end-user.

   Both sides are right! We suggest that you tentatively freeze the document. If changes are proposed, ask yourself a simple question: Is this a critical change that will make or break the system, or is it an enhancement that could be added later? Critical changes require modifying the specifications document. If the change isn't critical, log the change as future enhancement requirements.
2. (REC) Develop a detailed programming plan. You don't just start programming. Most design specifications include numerous programs. Which programs should be written first? Many systems are built in versions. The first version implements the most critical aspects of the system, so that a version can be placed into operation before the system has been completely constructed.

   Another appropriate approach is to construct event-processing programs first. Implement these programs in the same sequence as that in which they would have to be run (customer order processing, then order cancellation processing, then customer billing processing. Then implement management reporting and decision support programs according to their relative importance. General file maintenance and backup and recovery programs are written last.
   a. (OPT) Formulate the project team and assign responsibilities.
   b. (REQ) Write and document programs and perform unit testing.

c. (REC) Review program documentation for quality standards. Note that this step may be required in some organizations.
d. (REQ) Conduct system testing to ensure all programs work properly together.
e. (OPT) Update the project repository with revised program documentation for future referencing.
f. (OPT) Place new programs and reusable components in the software library.

THE DELIVERY PHASE OF SYSTEMS IMPLEMENTATION

Now we come to the last systems implementation phase in our life cycle—deliver the new system into operation. The analyst is the principal figure in the delivery phase, regardless of his or her role in the construction effort.

The purpose of the delivery phase is to smoothly convert from the old system to the new system. To achieve this, we must accomplish the following objectives:

- Conduct a system test to ensure that the new system works properly.
- Prepare a conversion plan to provide a smooth transition to the new system.
- Install databases to be used by the new system.
- Provide training and documentation for individuals who will be using the new system.
- Convert from the old system to the new system and evaluate the project and final system.

Your information systems framework (Figure 17.4) provides the context for the focus of the delivery phase. All building blocks are pertinent to this phase!

Figure17.5 is an activity diagram that illustrates the typical activities of this final phase of systems implementation. Let's examine each of the activities.

Activity: Conduct System Test

Now that the software packages and in-house programs have been installed and tested, we need to conduct a final system test. All software packages, custom-built programs, and any existing programs that comprise the new system must be tested to ensure that they all work together.

Purpose The purpose of this activity is to test all software packages, custom-built programs, and any existing programs that comprise the new system to ensure they all work together.

Roles The systems analyst facilitates this activity. Other roles are as follows:

- System owners and system users are the ultimate authority on whether or not a system is operating correctly.
- Systems analyst (REQ)—The systems analyst typically facilitates this activity by communicating testing problems and issues with the project team members.
- System builders (REQ)—System builders, of various specialties, are involved in the systems testing. Applications programmers, database programmers, and networking specialists may need to resolve problems revealed during systems testing.

Prerequisites (Inputs) The primary inputs to this activity include the *software packages, custom-built programs,* and *any existing programs* (REQ) comprising the new system. The system test is done using the *system test data* (REQ) that were developed earlier by the systems analyst.

Deliverables (Outputs) As with previous tests that were performed, our system test may result in required *modifications to programs* (OPT), thus, once again prompt-

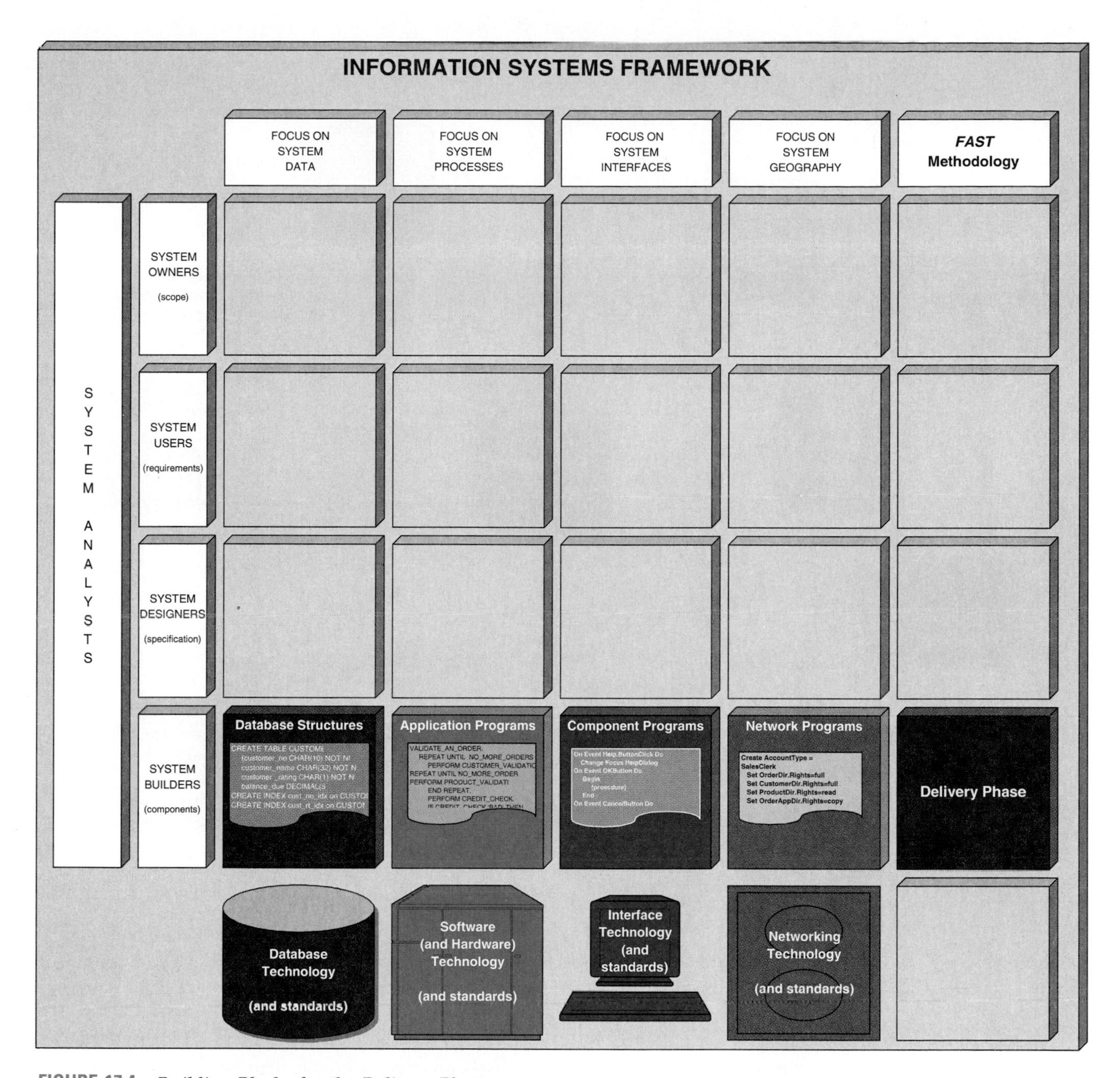

FIGURE 17.4 *Building Blocks for the Delivery Phase*

ing the return to a previous activity in the implementation phase. This iteration would continue until a *successful system test* was experienced (REQ).

Applicable Techniques Once again, a good understanding of testing is essential. Whereas this activity is concerned with systems testing, the identification of program-specific problems may necessitate a return to previous activities and subsequent stub and unit level testing.

Steps The following steps are suggested to complete this activity.

1. (REQ) Obtain system test data.
2. (REC) Ensure that all software packages, custom-built programs, and existing programs have been installed and that unit testing has been completed.

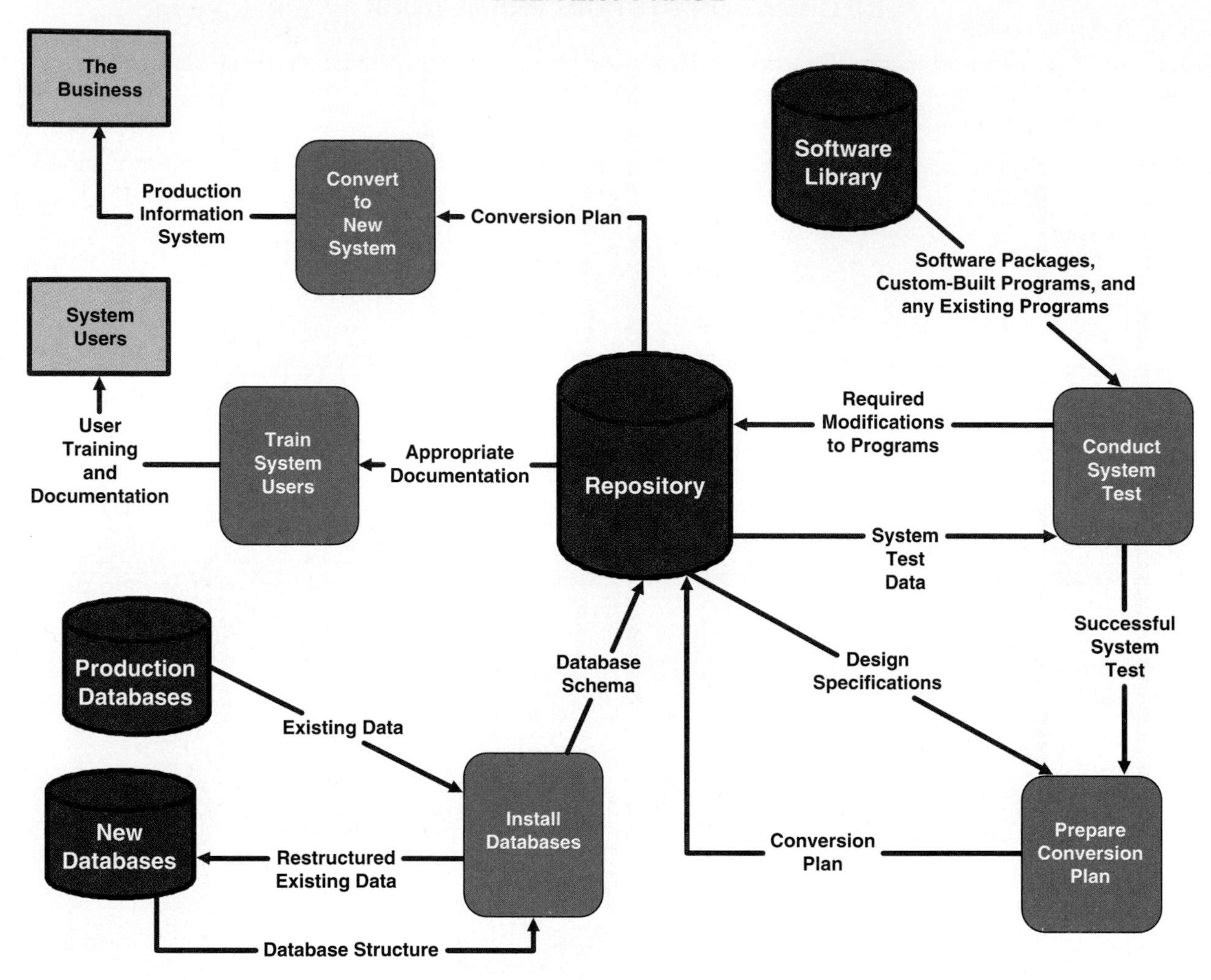

FIGURE 17.5 *Activity Diagram for the Delivery Phase*

3. (REQ) Perform tests to check that all programs work properly together, making appropriate revisions as needed and testing again.
4. (REC) Record any required modifications to programs in the project repository.

Activity: Prepare Conversion Plan

Once a successful system test has been completed, we can begin preparations to place the new system into operation. Using the design specifications for the new system, the systems analyst will develop a detailed conversion plan. This plan will identify databases to be installed, end-user training and documentation that need to be developed, and a strategy for converting from the old system to the new system.

Purpose Prepare a detailed conversion plan to provide a smooth transition from the old system to the new system.

Roles The activity is facilitated by the *project manager*. Other roles are defined as follows:

- System owner roles (defined in the previous activity).
  —Executive sponsor (REQ).
  —User managers (REC).

—System managers (REC).

—Project manager (REQ).

—Steering body (OPT)—Many organizations require that all project plans be formally presented to a steering body (sometimes called a steering committee) for final approval.

- System user roles.

  —Business analysts (OPT).

- Systems analyst, system designer, and system builder roles are not typically involved in this activity unless deemed necessary by the project manager.

Prerequisites (Inputs) This activity is triggered by the completion of a *successful system test* (REQ). Using the *design specifications* (REC) for the new system, a detailed conversion plan can be assembled.

Deliverables (Outputs) The principal deliverable of this activity is the *conversion plan* (REQ). This plan will identify databases to be installed, end-user training and documentation that need to be developed, and a strategy for converting from the old system to the new system.

Applicable Techniques The following technique is applicable to this activity:

- *Project management* (REC) techniques are taught in Part Five, Module A, "Project and Process Management."

Steps The following steps are suggested to complete this activity.

1. (REQ) Collect and review design specifications for the new system to identify databases to be installed and user training needs.
2. (REQ) Establish a schedule for installation of databases.
3. (REQ) Identify a training program and schedule for the system users.
4. (REQ) Develop a detailed installation strategy to follow for converting from the existing to the new production information system. Some commonly used strategies include:
   - **Abrupt cut-over.** On a specific date (usually a date that coincides with an official business period such as month, quarter, or fiscal year), the old system is terminated and the new system is placed into operation. This is a high-risk approach because there may still be major problems that won't be uncovered until the system has been in operation for at least one business period. On the other hand, there are no transition costs. Abrupt cut-over may be necessary if, for instance, a government mandate or business policy becomes effective on a specific date and the system couldn't be implemented before that date.
   - **Parallel conversion.** Under this approach, both the old and new systems are operated for some time period. This is done to ensure that all major problems in the new system have been solved before the old system is discarded. The final cut-over may be either abrupt (usually at the end of one business period) or gradual, as portions of the new system are deemed adequate. This strategy minimizes the risk of major flaws in the new system causing irreparable harm to the business; however, it also means the cost of running two systems over some period must be incurred. Because running two editions of the same system on the computer could place an unreasonable demand on computing resources, this may be possible only if the old system is largely manual.
   - **Location conversion.** When the same system will be used in numerous geographical locations, it is usually converted at one location (using either

abrupt or parallel conversion). As soon as that site has approved the system, it can be farmed to the other sites. Other sites can be cut over abruptly because major errors have been fixed. Furthermore, other sites benefit from the learning experiences of the first test site. Incidentally, the first production test site is often called a beta test site.
 - **Staged conversion.** Like location conversion, staged conversion is a variation on the abrupt and parallel conversions. A staged conversion is based on the version concept introduced earlier. Each successive version of the new system is converted as it is developed. Each version may be converted using the abrupt, parallel, or location strategies.

5. (REQ) Develop a systems acceptance test plan. The systems acceptance test is the final opportunity for end-users, management, and information systems operations management to accept or reject the system. A **systems acceptance test** is a final system test performed by end-users using real data over an extended time period. It is an extensive test that addresses three levels of acceptance testing: verification testing, validation testing, and audit testing.
   - **Verification testing** runs the system in a simulated environment using simulated data. This simulated test is sometimes called alpha testing. The simulated test is primarily looking for errors and omissions regarding end-user and design specifications that were specified in the earlier phases but not fulfilled during construction.
   - **Validation testing** runs the system in a live environment using real data. This is sometimes called beta testing. During this validation, we are testing a number of items.
     a. *Systems performance.* Is the throughput and response time for processing adequate to meet a normal processing workload? If not, some programs may have to be rewritten to improve efficiency or processing hardware may have to be replaced or upgraded to handle the additional workload.
     b. *Peak workload processing performance.* Can the system handle the workload during peak processing periods? If not, we may have to improve hardware and/or software to increase efficiency or rethink our scheduling of processing—that is, consider doing some of the less critical processing during nonpeak periods.
     c. *Human engineering test.* Is the system as easy to learn and use as anticipated? If not, is it adequate? Can enhancements to human engineering be deferred until after the system has been placed into operation?
     d. *Methods and procedures test.* During conversion, the methods and procedures for the new system will be put to their first real test. Methods and procedures may have to be modified if they prove to be awkward and inefficient from the end-users' standpoint.
     e. *Backup and recovery testing.* Now that we have full-sized computer files and databases with real data, we should test all backup and recovery procedures. We should simulate a data loss disaster and test the time required to recover from that disaster. Also, we should perform a before-and-after comparison of the data to ensure that data were properly recovered. It is crucial to test these procedures. Don't wait until the first disaster to find an error in the recovery procedures.
   - **Audit testing** certifies that the system is free of errors and is ready to be placed into operation. Not all organizations require an audit. But many firms have an independent audit or quality assurance staff that must certify a system's acceptability and documentation before that system is placed into final operation. There are independent companies that perform systems and software certification for end-users' organizations.

Activity: Install Databases

In a previous phase you built and tested databases. To place the system into operation, you need fully loaded (or "populated") databases. Therefore, the next activity we'll survey is installation of databases.

At first, this activity may seem trivial. But consider the implications of loading a typical table, say, MEMBER. Tens or hundreds of thousands of records may have to be loaded. Each must be input, edited, and confirmed before the database table is ready to be placed into operation.

Purpose The purpose of this activity is to populate the new systems databases with existing data from the old system.

Roles This activity will normally be completed by application programmers. Other roles are defined as follows:

- System owners and system users are not involved in this activity.
- System analysts and designers may play a small role in completing this activity. Their primary involvement will be the calculation of database sizes and estimating of time required to perform the task of installing them.
- System builders play a primary role in this activity.
- Application programmers (REQ) will write the special programs to extract data from existing databases and programs to populate the new databases.
- Date entry personnel or hired help (OPT) may be assigned to do data entry for data that needs to be keyed.

Prerequisites (Inputs) Special programs will have to be written to populate the new databases. *Existing data* (REQ) from the production databases, coupled with the database schema(s) models and *database structures* (REQ) for the new databases will be used to write computer programs to populate the new databases with restructured existing data.

Deliverables (Outputs) The principal deliverables of this activity are the *restructured existing data* (REQ) that have been populated in the databases for the new system.

Applicable Techniques Database and application programming skills are essential to the successful completion of this activity.

Steps The following steps are suggested to complete this activity.

1. (REQ) Review the database structures for new databases.
2. (REQ) Identify existing data currently in production databases (and other sources) to be used to populate the databases for the new system.
3. (OPT) Obtain additional manual resources to do on-time keying of data not obtained from existing production databases.
4. (REQ) Write programs to extract data from production databases.
5. (REQ) Write programs to load new databases.
6. (REC) Conduct another system test to ensure new system is unaffected (see "Activity: Conduct System Test"). This preventative step will ensure that no task accomplished in this activity adversely affects the new system.
7. (REC) If necessary, revise the database schema and update the project repository.

Activity: Train System Users

Change may be good, but it's not always easy. Converting to a new system necessitates that system users be trained and provided with documentation (user manuals) that guides them through using the new system.

Training can be performed one on one; however, group training is generally preferred. It is a better use of your time, and it encourages group learning possibilities. Think about your education for a moment. You really learn more from your fellow students and colleagues than from your instructors. Instructors facilitate learning and instruction, but you master specific skills through practice with large groups where common problems and issues can be addressed more effectively. Take advantage of the ripple effect of education. The first group of trainees can then train several other groups.

Purpose Provide training and documentation to system users to prepare them for a smooth transition to the new system.

Roles The activity is facilitated by the systems analyst. Other roles are defined as follows:

- System owner (REQ)—System owners must support this activity. They must be willing to approve the release time necessary for their people to obtain the training needed to become successful users of the new system.
- System user (REQ)—Remember, the system is for the user! User involvement is important in this activity because the end-users will inherit your successes and failures from this effort. Fortunately, user involvement during this activity is rarely overlooked. The most important aspect of their involvement is training and advising the users. They must be trained to use equipment and to follow the procedures required of the new system. But no matter how good the training is, users will become confused at times. Or perhaps they will find mistakes or limitations (an inevitable product, despite the best of planning, analysis, design, and implementation techniques). The analyst will help the users through the learning period until they become more familiar and comfortable with the new system.
- Systems analyst (REQ)—This activity typically performed by systems analysts is to train system users to use the new system. Given appropriate documentation for the new system, the systems analysts will provide end-user documentation (typically in the form of manuals) and end-user training for the system users.
- System designers and builders are not normally involved in this activity.

Prerequisites (Inputs) Given *appropriate documentation* (REQ) for the new system, the systems analyst will provide the system users with documentation and training needed to properly use the new system.

Deliverables (Outputs) The principal deliverable of this activity is *user training and documentation* (REQ). Many organizations hire special systems analysts who do nothing but write user documentation and training guides. If you have a skill for writing clearly, the demand for your services is out there! Figure 17.6 is a typical outline for a training manual. The Golden Rule should apply to user manual writing: "Write unto others as you would have them write unto you." You are not a business expert. Don't expect the reader to be a technical expert. Every possible situation and its proper procedure must be documented.

Applicable Techniques Written and oral communications skills are critical. (These skills are more fully covered in Part Five, Module E, "Interpersonal Skills and Communications.") Familiarity with organizational behavior and psychology may also prove helpful. Converting to a new system represents change, and people tend to resist change or to look for fault in change. There is comfort in the status quo—even if the current system is fraught with problems.

Training Manual End-Users Guide Outline

I. Introduction.
II. Manual.
  A. The manual system (a detailed explanation of people's jobs and standard operating procedures for the new system).
  B. The computer system (how it fits into the overall workflow).
    1. Terminal/keyboard familiarization.
    2. First-time end-users.
      a. Getting started.
      b. Lessons.
  C. Reference manual (for nonbeginners).
III. Appendixes.
  A. Error messages.

FIGURE 17.6 *An Outline for a Training Manual*

Steps The following steps are suggested to complete this activity.

1. (REQ) Collect documentation that may prove useful in developing user documentation and training guides.
2. (REQ) Write user documentation manuals.
3. (REQ) Referring to the conversion plan, review the training needs of the system users.
4. (REQ) Schedule training sessions (individual and group sessions).
5. (REQ) Conduct training sessions and distribute user documentation.

Activity: Convert to New System

Conversion to the new system from the old system is a significant milestone. After conversion, the ownership of the system officially transfers from the analysts and programmers to the end-users. The analyst completes this activity by carefully carrying out the conversion plan. Recall that the conversion plan includes detailed installation strategies to follow for converting from the existing to the new production information system. This activity also involves completing what is called a systems audit.

Purpose Convert to the new system from the old system and evaluate the project experience and final system.

Roles The activity is facilitated by the *project manager* (REQ) who will oversee the conversion process. Other roles are defined as follows:

- System owners (REC) provide feedback regarding their experiences with the overall project. They may also provide feedback regarding the new system that has been placed into operation.
- System users (REC) will provide valuable feedback pertaining to the actual use of the new system. They will be the source of the majority of the feedback used to measure the system's acceptance.
- Systems analysts, designers, and builders will be involved in assessing the feedback received from the system owners and users once the system is in operation. In many cases, that feedback may stimulate actions on their part to correct shortcomings that have been identified. Regardless, the feedback will be used to help benchmark new systems projects down the road.

Prerequisites (Inputs) The input to this activity is the *conversion plan* (REQ) that was created in an earlier delivery phase activity.

Deliverables (Outputs) The principal deliverable of this is the *product information system* (REQ) that is placed into operation in the business.

Applicable Techniques The following techniques are applicable to this activity.

- *Project and process management.* Project and process management (REC) techniques are taught in Part Five, Module A, "Project and Process Management."

Steps The following steps are suggested to complete this activity.

1. (REC) Review the conversion plan.
2. (REQ) Complete the detailed steps outlined in the conversion plan.
3. (REC) Schedule meeting with project team to evaluate the development project and the production system.
4. Conduct review meeting (REC); record enhancement/fix requirements that are identified.

WHERE DO YOU GO FROM HERE?

This chapter provided a detailed overview of the systems implementation phases of systems development. Specifically, you learned about the construction and delivery of a new system. You are now ready to learn systems support. Systems support is covered in Chapter 18.

Before proceeding to learn about systems support, you may wish to revisit Chapter 3 and its introduction to the systems development life cycle. This review will help you to first understand how systems support fits into the overall systems development life cycle.

SUMMARY

1. **Systems implementation** is the construction of the new system and the delivery of that system into production (meaning day-to-day operation). Unfortunately, *systems development* is a common synonym.
2. The purpose of the construction phase of systems implementation is twofold: To build and test a functional system that fulfills business and design requirements and to implement the interfaces between the new system and existing production systems.
3. The construction phase consists of four activities: Build and test networks, build and test databases, install and test new software packages, and write and test new programs.
4. There are three levels of testing performed on new programs:
   - **Stub testing** is the test performed on individual modules, whether they be main program, subroutine, subprogram, block, or paragraph.
   - **Unit or program testing** is a test whereby all the modules that have been coded and stub tested are tested as an integrated unit.
   - **Systems testing** ensures that application programs written in isolation work properly when they are integrated into the total system.
5. The purpose of the delivery phase is to smoothly convert from the old system to the new system.
6. The delivery phase of systems implementation consists of the following activities: Conducting a system test, preparing a systems conversion plan, installing databases, training system users, and converting from the old system to the new system.
7. There are several commonly used strategies for converting from an existing to a new production information system, including:
   - **Abrupt cut-over.** On a specific date, the old system is terminated and the new system is placed into operation.

- **Parallel conversion.** Both the old and new system are operated for some time period to ensure that all major problems in the new system are solved before the old system is discarded.
- **Location conversion.** When the same system will be used in numerous geographical locations, it is usually converted at one location and, following approval, farmed to the other sites.
- **Staged conversion.** Each successive version of the new system is converted as it is developed. Each version may be converted using the abrupt, parallel, or location strategies.

8. The systems acceptance test is the final opportunity for end-users, management, and information systems operations management to accept or reject the system. A systems acceptance test is a final system test performed by end-users using real data over an extended period. It is an extensive test that addresses three levels of acceptance testing: verification testing, validation testing, and audit testing.
   - **Verification testing** runs the system in a simulated environment using simulated data.
   - **Validation testing** runs the system in a live environment using real data. This is sometimes called beta testing.
   - **Audit testing** certifies that the system is free of errors and is ready to be placed into operation.

KEY TERMS

abrupt cut-over, p. 567
audit testing, p. 569
location conversion, p. 568
parallel conversion, p. 567
staged conversion, p. 568
stub testing, p. 563
systems acceptance test, p. 568
systems implementation, p. 556
systems testing, p. 563
unit or program testing, p. 563
validation testing, p. 568
verification testing, p. 568

REVIEW QUESTIONS

1. Define systems implementation.
2. What are the two phases in systems implementation? Define the purpose of each phase.
3. What are the four major activities in constructing the new system?
4. What are the five major activities in delivering a new system?
5. List and briefly describe three levels of testing that may be performed on new programs.
6. List and briefly describe four strategies commonly used to convert from an old system to a new production system.
7. What is a systems acceptance test? When is this test performed?
8. What are three levels of acceptance testing?

PROBLEMS AND EXERCISES

1. How can successful and thorough systems planning, analysis, and design be ruined by a poor systems implementation? How can poor systems analysis or design ruin a smooth implementation? For both questions, list some implementation consequences.
2. What skills are important during systems implementation? Create an itemized list. Identify computer, business, and general education courses that would help you develop or improve those skills.
3. How do your information systems architecture framework blocks aid in systems implementation? Examine each building block and address issues and relevance to the systems implementation phases of construction and delivery.
4. What products of the systems design phases are used in the systems implementation phases? Why are they important? How are they used? What would happen if they were incomplete or inaccurate?
5. Differentiate between the types of testing done on application programs and the types of tests conducted for an overall system?
6. Why should a systems analyst perform a postimplementation review? What types of benefits can be derived from it?

PROJECTS AND RESEARCH

1. You are preparing to meet with your end-users to discuss converting from their old system to a new system. In this meeting you wish to discuss alternative strategies that could be used. Prepare a brief description of the alternative strategies along with a description of situations for which each approach would be preferred and required.
2. Visit a local information systems shop. Ask to see samples of the conversion plan for a recent implementation of a new system. What items are detailed in that plan? Can you think of any items that should have been included? What type of conversion strategy did they use?
3. Visit a local company and obtain a copy of a user manual for a system. Evaluate the manual. Is it written clearly using nontechnical terminology? How do the users feel about the manual? Did the manual help with their transition to the new system? Why or why not?

MINICASES

1. Tim Stallard is a systems analyst at Beck Electronic Supply. He has been a systems analyst for only six months. Unusual personnel turnover thrust him into the position after only 18 months as a programmer. Now it is time for his semiannual job performance review. *(Tim enters the office of Ken Delphi. Ken is the assistant director of MIS at Beck.)*

KEN Another six months! It hardly seems that long since your last job performance review.

TIM I personally feel very good about my progress over the last six months. This new position has been an eye-opener. I didn't realize that analysts do so much writing. I enrolled in some continuing education writing classes at the local junior college. The courses are helping . . . I think.

KEN I wondered what you did. It shows in everything from your memos to your reports. More than any technical skills, your ability to communicate will determine your long-term career growth here at Beck. Now, let's look at your progress in other areas. Yes, you've been supervising the materials requirements planning project implementation for the last few months. This is your first real experience with the entire implementation process, right?

TIM Yes. You know, I was a programmer for 18 months. I thought I knew everything there was to know about systems implementation. But this project has taught me otherwise.

KEN How's that?

TIM The computer programming tasks have gone smoothly. In fact, we finished the entire system of programs six weeks ahead of schedule.

KEN I don't mean to interrupt, but I just want to reaffirm the role your design specifications played in accelerating the computer programming tasks. Bob has told me repeatedly that he had never seen such thorough and complete design specifications. The programmers seem to know exactly what to do.

TIM Thanks! That really makes me feel good. It takes a lot of time to prepare design specifications like that, but I think that it really pays off during implementation. Now, what was I going to say? Oh yes. Even though the programming and testing were completed ahead of schedule, the system still hasn't been placed into operation; it's two weeks late.

KEN That means you lost the six-week buffer plus another two weeks. What happened?

TIM Well, I'm to blame. I just didn't know enough about the nonprogramming activities of systems implementation. First, I underestimated the difficulties of training. My first-draft training manual made too many assumptions about computer familiarity. My end-users didn't understand the instructions, and I had to rewrite the manual. I also decided to conduct some training classes for the end-users. My instructional delivery was terrible, to put it mildly. I guess I never really considered the possibility that, as a systems analyst, I'd have to be a teacher. I think I owe a few apologies to some of my former instructors. I can't believe how much time needs to go into preparing for a class.

KEN Yes, especially when you're technically oriented and your audience is not.

TIM Anyway, that cost me more time than I had anticipated. But there are still other implementation problems that have to be solved. And I didn't budget time for them!

KEN Like what?

TIM Like getting data into the new databases. We have entered several thousand new records. And to top it off, management is insisting that we operate the new system in parallel with the old system for at least two months. Then, and only then, will they be willing to allow the old system to be discarded.

KEN Well, Tim, I think you're learning a lot. Obviously, we threw you to the wolves on this project. But I needed Bob's experience and attention elsewhere. I knew when I pulled Bob off the project that it could introduce delays—I call it the rookie factor. Under normal circumstances, I would never have let you work on this alone. But you're doing a good job and you're learning. We have to take the circumstances into consideration. You'll obviously feel some heat from your end-users because the implementation is behind schedule, and I want you to deal with that on your own. I think you can handle it. But don't hesitate to call on Bob or me for advice. Now, let's talk about some training and job goals for the next six months.

a. Above and beyond programming, what activities do you think make up systems implementation? Can you think of any activities that weren't described in this minicase?
b. Why is training so difficult? How do you feel about the prospects of becoming a "teacher"? How long do you think it takes to prepare for one hour of classroom instruction? What activities do you think would be involved in preparing for a lesson plan?
c. A 3,000-record (row) database table must be created for a new system. Each record consists of 15 fields/attributes. The record length is 200 bytes. How long do you suppose it would take to create that database table? If necessary, use your own typing speed as a performance gauge. What factors would affect how long it may take to get the table up and running?
d. What assumption did Tim make about transition from the old system to the new system? Why was it wrong? Can you think of any circumstances under which it would be correct?

SUGGESTED READINGS

Bell, P., and C. Evans. *Mastering Documentation.* New York: John Wiley & Sons, 1989.

Boehm, Barry. "Software Engineering." *IEEE Transactions on Computers, C-25,* December 1976. This classic paper demonstrated the importance of catching errors and omissions before programming begins.

Metzger, Philip W. *Managing a Programming Project.* 2nd ed. Englewood Cliffs, NJ: Prentice Hall, 1981. This is one of the few books to place emphasis solely on systems implementation.

Mosely, D. J. *The Handbook of MIS Application Software Testing.* Englewood Cliffs, NJ: Yourdon Press, 1993.

18

SYSTEMS SUPPORT

CHAPTER PREVIEW AND OBJECTIVES

Recall from Chapter 3 that systems support involves refining a production system by iterating through the previous four phases on a smaller scale to refine and improve that system. In this chapter, you will learn more about systems support. It is very likely that a young systems analyst or user will become directly involved in a systems support project. Most analysts carry responsibility for one or more legacy systems. Therefore, it is useful to understand the different types of systems support provided for a production system. You will know that you understand the process of systems support when you can:

- Define systems support.
- Describe the role of a repository in systems support.
- Differentiate between maintenance, enhancement, reengineering, and design recovery.
- Understand the maintenance challenge presented by the year 2000.

Although some of the techniques of systems support are introduced in this chapter, it is *not* the intent of this chapter to teach the *techniques* of systems support. This chapter teaches only the process of systems support.

SOUNDSTAGE

SOUNDSTAGE ENTERTAINMENT CLUB

SCENE

When we last visited SoundStage, Sandra and Bob were implementing the new system. Since then, both have been providing ongoing support. Bob has been simultaneously working on his new project assignment involving many of the company's legacy COBOL applications.

SANDRA

I'm glad you could get free from your new project, Bob. The Member Services System has been working really well since we put it into production—with the exception of a few client systems crashing every once in a while. But I think we've gotten them straightened out.

BOB

Crashes? When did that happen?

SANDRA

I got a phone call, a couple actually, this past Friday. You weren't around so I handled it by myself. That's all taken care of, but we have a few problems that need to be corrected pronto. There are a few errors that will require some program corrections. Since you did most of the programming, I thought you should take care of them. Here, I wrote up the problems for you.

BOB

[*Bob reviews the problem descriptions*]

I can't believe these! We conducted all sorts of program and system tests before and after the system was installed.

SANDRA

Don't get too upset. The problems are relatively minor. But I do want to get them corrected as soon as possible. I want the users to know that we intend to provide top-notch support for their new system. Besides, in all my years of working on projects, I've never seen a new system placed into operation that did not have a few errors that slip by. It is nearly impossible to test a system for all possible errors.

BOB

That's comforting to know. I'll get right on these corrections. Anything else?

SANDRA

Yes, but I will take care of it. We do have a user returning from maternity leave who still isn't up to speed on the new system. She has been calling occasionally for assistance. I gave her the user manual and other users have been trying to help her get up to speed. I will be meeting with her today to do some further training.

BOB

What about enhancements to the new system? If you recall, we froze the specifications. But there were a few new requirements that the users were asking us to implement.

SANDRA

I thought we should give the system a little more time to work out the bugs and allow time for the users to become proficient with the new system. But you're right. We will need to begin planning for those enhancements.

BOB

Sounds good. If you get any more feedback concerning errors, let me know.

DISCUSSION QUESTIONS

1. How would you characterize the focus of the support phase of this project?
2. What are some activities that Bob will need to perform to make corrections to the system?
3. What other types of user assistance might be expected of Bob and Sandra?
4. What types of activities are likely to be included in Bob and Sandra's plan for making enhancements to the new system?

WHAT IS SYSTEMS SUPPORT?

Systems support was first defined in Chapter 3:

> **Systems support** is the ongoing maintenance of a system after it has been placed into operation. This includes program maintenance and system improvements.

Systems support is often ignored in systems analysis and design textbooks. Young analysts are often surprised to learn that half of their duties (or more) are associated with supporting existing systems. Systems support often requires developers to revisit activities typically performed in systems analysis, design, and implementation.

Let's first set the stage for systems support. In Figure 18.1, applications maintenance projects are contrasted with systems planning and application development projects. Notice that once applications are implemented, they are said to be in *production*. Production is the day-to-day, week-to-week, month-to-month, and year-to-year execution of the application programs to process business data (inputs) and generate useful information (outputs).

In Figure 18.1, we also see three distinct types of system-level data storage. First, we see our **central repository**. This repository stores all system models and detailed specifications. Subsets of the central repository are checked out to sup-

FIGURE 18.1 *Context for Systems Support Projects and Activities*

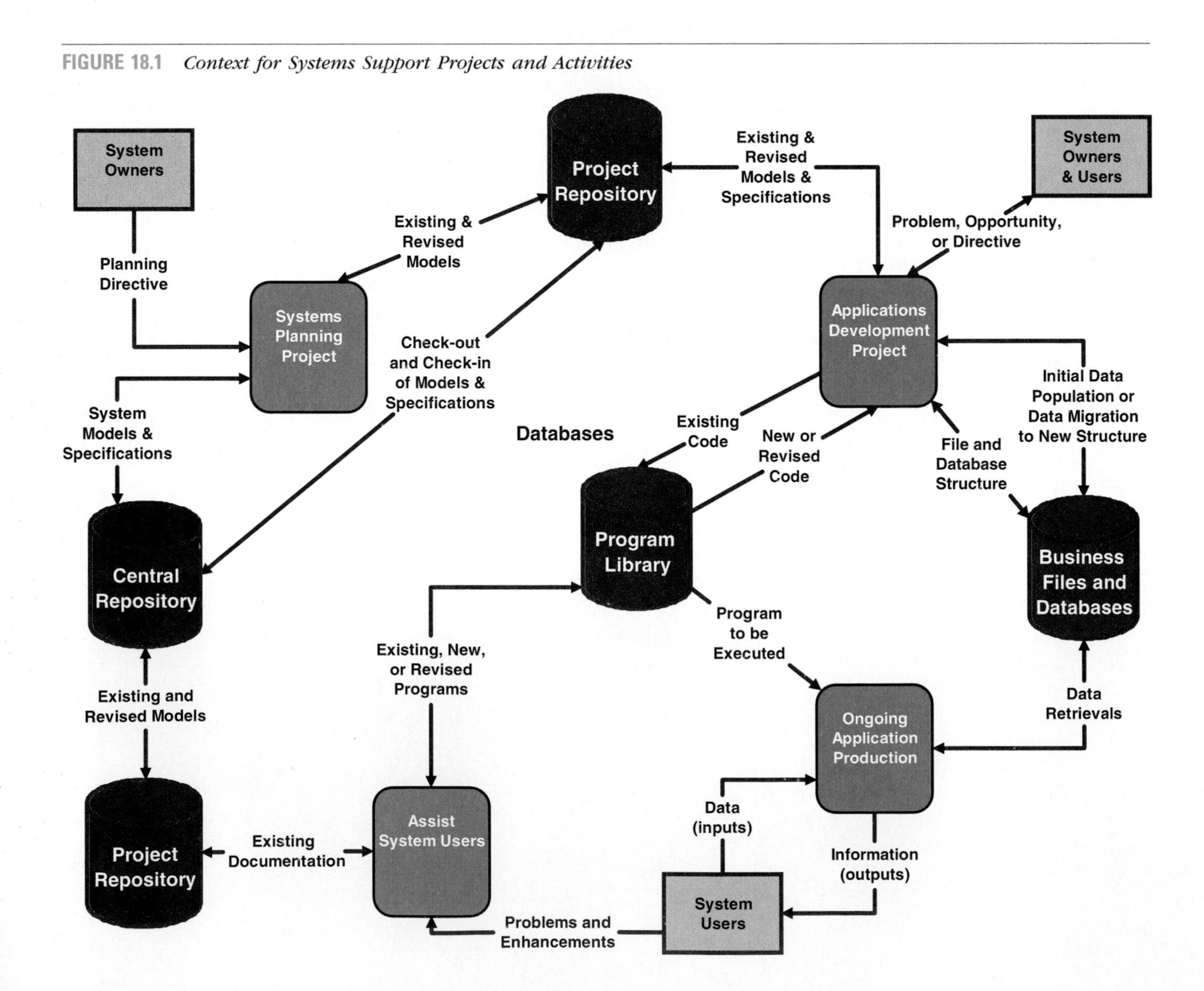

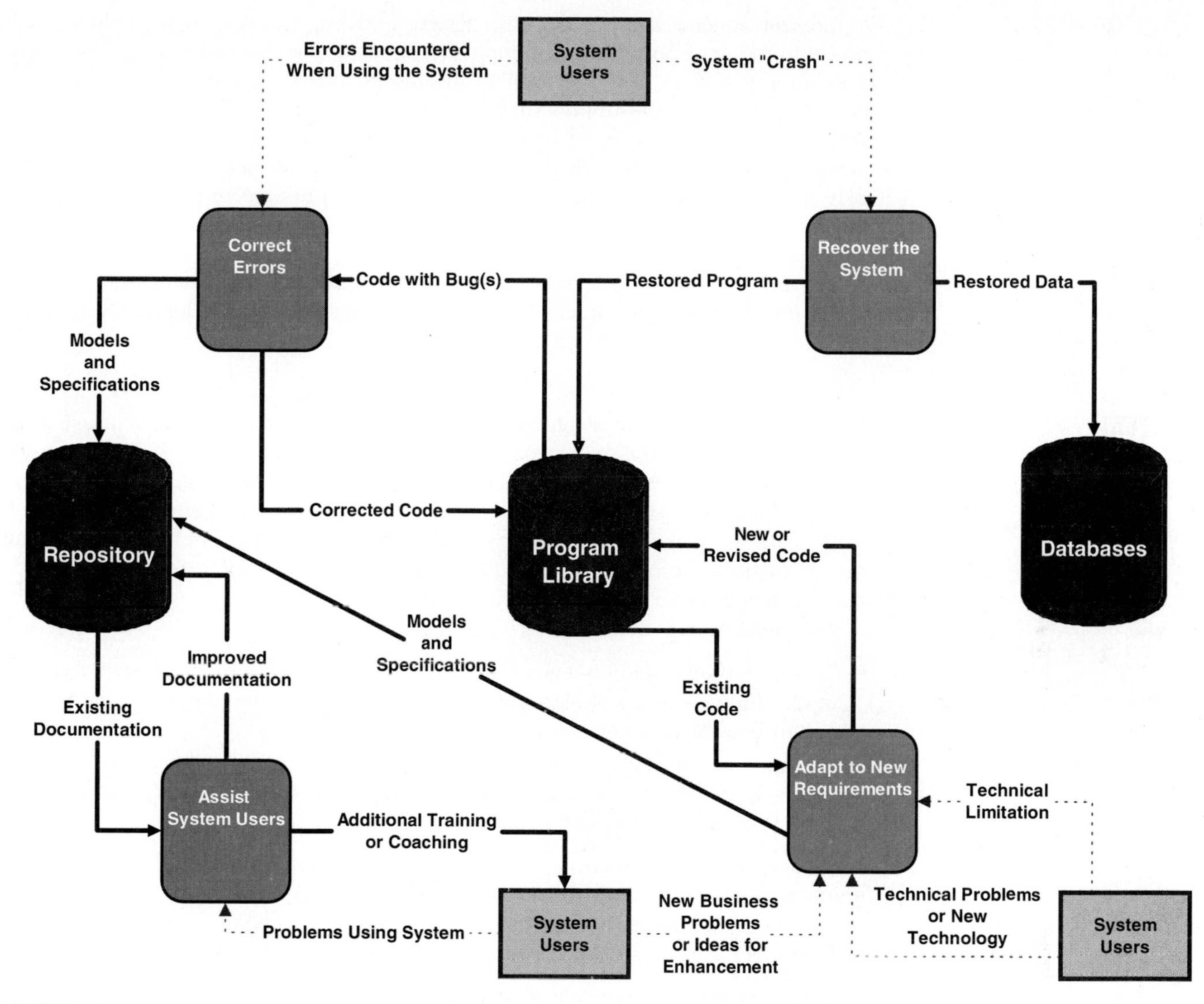

FIGURE 18.2 *Types of System Support*

port various planning and development projects. These subsets are stored as project repositories, usually implemented through various CASE tools. Second, we see **program libraries** that store the actual application programs that have been placed into production. In most shops, a software-based librarian will track changes and maintain a few previous versions of the software in case a problem arises with a new version. Third, we see actual business databases that store the operational data created and maintained by the production application programs.

Systems support is primarily driven by system designers and system builders in support of system users. Relative to your information systems framework, systems support maintains all of the building blocks (and their documentation) for a production information system. Unlike systems analysis, systems design, and systems implementation, systems support cannot be broken down into actual phases that the system goes through. Rather, systems support consists of four ongoing activities—each activity is triggered by a particular type of problem encountered with the implemented system. Figure 18.2 is a life cycle diagram that illustrates the four types of support.

This chapter examines each type of support in detail. Before we begin, we want to point out that this chapter abandons the approach used in previous chapters to present systems survey, analysis, design, and implementation to address each phase in terms of purpose, roles, inputs, outputs, applicable techniques, and steps. This approach does not lend itself to the discussion of systems support. As mentioned earlier, systems support cannot be viewed in terms of phases. Also, as you will soon learn, the types of activities performed during systems support are triggered by problems. The actual activities that need to be performed, what people play a role in those activities, what inputs and outputs are required, which techniques are applicable, and what specific steps are taken all depend on the problem that needs to be resolved. Finally, since systems support may in some cases simply involve returning to survey, analysis, design, and implementation activities that were discussed in earlier chapters, we wanted to avoid redundantly presenting the details of those activities.

SYSTEM MAINTENANCE—CORRECTING ERRORS

Regardless of how well designed, constructed, and tested a system or application may be, errors or bugs will inevitably occur. Some **bugs** will be caused by miscommunication of requirements. Others will be caused by design flaws. Others will be caused by situations that were not anticipated and, therefore, not tested. And finally bugs may be caused by unanticipated misuse of the programs. In all these situations, corrective action must be taken. We call this corrective action *system maintenance* or *program maintenance.*

The fundamental objectives of system maintenance are:

- To make predictable changes to existing programs to correct errors that were made during systems design and implementation. Consequently, we exclude enhancements and new requirements from this activity.
- To preserve those aspects of the programs that were already correct. Inversely, we try to avoid the possibility that "fixes" to programs cause other aspects of those programs to behave differently.

To achieve these objectives, you need an appropriate understanding of the programs you are fixing and of the applications in which those programs participate. This prerequisite understanding is often the downfall of systems maintenance!

How does system maintenance map to your information systems building blocks?

DATA—System maintenance rarely impacts data, except for the possibility of improving data editing.

PROCESSES—Business and information systems processes are implemented as application programs. System maintenance is all about fixing errors made when those programs were implemented.

INTERFACES—System maintenance may involve correcting problems related to how the application interfaces with the users or another system.

GEOGRAPHY—System maintenance rarely involves computer networks, although on occasion, computer networks can be the root cause of bugs.

TECHNOLOGY—System maintenance, as defined in this activity, does not deal with changing technology.

Figure 18.3 illustrates some typical activities of system maintenance. Let's examine the activities that must be completed during system maintenance.

Define and Validate the Problems

The first task of the assigned team is to define and validate problems. Ideally, this activity will be facilitated by the analyst and/or programmer, but it should clearly involve the users. The *problem programs* are retrieved from the program library.

Working with the users, the team should attempt to validate the problem by reproducing it. If the problem cannot be reproduced, the project should be suspended until the problem reoccurs and the user can explain the circumstances under which it occurred. The inputs are the *errors encountered when using the*

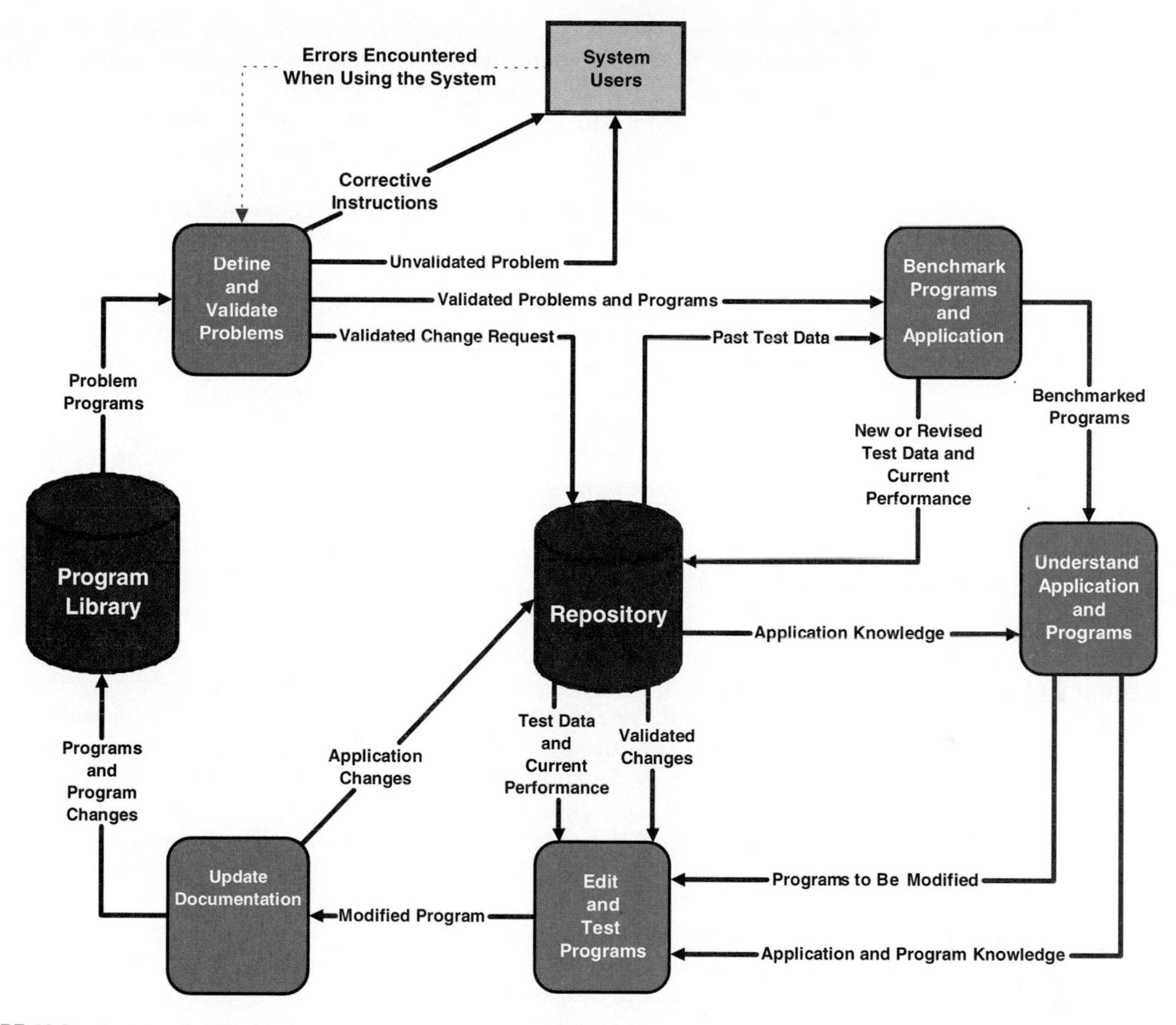

FIGURE 18.3 *Activity Diagram for System Maintenance*

system (usually called bugs). One possible output is *validated change requests.* These change requests should define expectations of the solution.

Another possible output is an *unvalidated problem.* In the event that the bug reoccurs, users should be instructed as to how to better document circumstances that led to the bug and symptoms of the problem.

In some cases the bug arises from simple misunderstandings or misuse, and *corrective instructions* can bring the project to closure. On the other hand, if the bug has been validated, the *validated problems and programs* are passed to the next task.

NOTE All subsequent maintenance will be performed on a copy of the program. The original program remains in the program library and can be used in production systems until it is fixed.

The analyst or programmer requires appropriate fact-finding skills (Part Five, Module B) and interpersonal skills (Part Five, Module E). The analyst or programmer also needs to be familiar with the application.

Benchmark the Programs and Application

The program isn't all bad, or it would have never been placed into production in the first place. The team should next benchmark the programs and application. System maintenance can result in unpredictable and undesirable side effects that impact the programs' or application's overall functionality and performance. For this reason, we highly recommend that, before making any changes to programs, the programs be executed and tested to establish a baseline against which the modified programs and applications can be measured.

This step is performed by the systems analyst and/or programmer. The users may also participate. The primary inputs are the *validated problems and programs.*

Test cases can be defined in either of two ways. First, you may find *past test data* in the repository. If so, that data should be reexecuted to provide the benchmark. It should also be analyzed for completeness and, if necessary, revised. The correct responses, including error handling, should be recorded in the repository.

Alternatively, test data can be automatically captured using a test tool. The advantage is that the test cases and responses are recorded in the repository for later playback. Test tools also measure response time and throughput using the test cases. *SQA Suite* by SQA, Inc., consists of a set of tools for performing a variety of tests on client/server applications.

For either alternative, the outputs are *new or revised test data* and *current performance* and *benchmarked programs.*

The analyst or programmer needs to have good testing skills (usually taught in programming courses) and may require training in test tools. Neither is explicitly taught in this textbook.

Understand the Application and Its Programs

Frequently, system maintenance is not performed by the same people who wrote the program. In fact, several people may have written parts of any program or application, and those people may no longer be available for clarification. For this reason, we need to gain an understanding of the application and the problematic programs. You may be surprised to learn that most analysts and programmers spend more time in this task than any other!

Ideally, *application knowledge* comes from the repository. This assumes, of course, that application knowledge has been maintained throughout the application's lifetime. Too often this is not true, especially for older systems. In non-repository-based shops, application knowledge may be available from prior programmers and analysts, but it is usually not up-to-date. Still, it may be useful for gaining a sufficient understanding of the application and where the problematic programs fit into that application.

Application and program knowledge usually comes from studying the source code from the *benchmarked programs.* Unfortunately, program understanding can take considerable time. This activity is slowed by some combination of the following limitations:

- Poor modular structure.
- Unstructured logic (from prestructured era code).
- Prior maintenance (quick fixes and poorly designed extensions).
- Dead code (instructions that cannot be reached or executed—often leftovers from prior testing and debugging).
- Poor or inadequate documentation.

The purpose of application understanding is to see the big picture—that is, how the programs fit into the total application and how they interact with other programs. The purpose of program understanding is to gain insight into how the program works and doesn't work. You need to understand the fields (variables) and where and how they are used, and you need to determine the potential impact of changes throughout the program. Program understanding can also lead to better estimates of the time and resources that will be required to fix the errors.

There used to be no shortcuts for program understanding. Today, maintenance CASE technology can help. For example, VIASOFT's VIA/Insight provides studies and analyzes programs to provide the programmer with considerable insight and information about unfamiliar code. It reveals program structure, marks or traces related code even if it is in different paragraphs and subroutines, isolates dead code (code that cannot be executed), traces field usage (and nonusage) and relationships between different field names and structures, and provides numerous cross-references. This information can reduce the time it takes to understand programs by hours, days, and even weeks. Poorly structured programs can then be modified to conform to accepted structured programming practices. VIASOFT's *VIA/Smart Doc* can then generate a wealth of information, graphical and textual, for the program.

Edit and Test the Programs

Given *application and program knowledge* and *validated changes*, you can now make changes to the *programs to be modified*. This task, performed by the programmer, is not dissimilar from that described in the last chapter on systems implementation. The result is a *modified program*.

There is a big difference between editing a new program and editing an existing program. As the designer and creator of a new program, you are probably intimately familiar with the structure and logic of the program. By contrast, as the editor of the existing program, you are not nearly as familiar (or current) about that program. Changes that you make may have an undesirable ripple effect through other parts of the program or, worse still, other programs in the application.

It is hoped your code changes will benefit from your understanding (or review) of the application and program. But testing takes on even greater importance in system maintenance. The following tests are essential and recommended:

- **Unit testing** (essential) ensures that the stand-alone program fixes the bug without side effects. The *test data and current performance* that you recovered, created, edited, or generated when the programs were benchmarked are used here.
- **System testing** (essential) ensures that the entire application, of which the modified program was a part, still works. Again, the *test data and current performance* are used here.
- **Regression testing** (recommended) extrapolates the impact of the changes on program and application throughput and response time from the before-and-after results using the *test data and current performance*.

Tested programs will be returned to production. Generally, when the programs are returned to the program library, they are subject to version control.

Version control is a process whereby a librarian (usually software-based) keeps track of changes made to programs. This allows recovery of prior versions of the programs in the event that new versions cause unexpected problems. In other words, version control allows users to return to a previously accepted version of the system.

One example of version control software is INTERSOLV's *PVCS* (for local area networks).

Update Documentation

The high cost of system maintenance is due, in large part, to failure to update application and program documentation. If application documentation has changed in the slightest, it should be modified in the repository and program library. Application documentation is usually the responsibility of the systems analyst who supports that application. Program documentation is usually the responsibility of the programmer who made the program changes.

The programmer is responsible for this activity. The input is the *modified program. Application changes* (changes in models and specifications) are recorded in the repository. The new *programs and program changes* are stored in the program library. Once returned to the library, they are available for production.

Recording application and program changes in the repository and program library will help future programmers and analysts (including yourself) reduce application understanding time during future maintenance. You will forget changes, however small, unless they are properly recorded. The long-term benefit comes when the application is due for major redevelopment. The study phase of systems analysis will pass quickly if existing documentation is up-to-date.

SYSTEM RECOVERY—OVERCOMING THE "CRASH"

From time to time a system failure is inevitable. It generally results in an aborted or "hung" program (also called an ABEND or crash) and possible loss of data. The systems analyst often fixes the system or acts as intermediary between the users and those who can fix the system. The purpose of this section is to quickly summarize the analyst's role in system recovery.

System recovery activities can be summarized as follows:

1. In many cases the analyst can sit at the user's terminal and recover the system. It may be something as simple as pressing a specific key or rebooting the user's personal computer. Corrective instruction may be required to prevent the crash from recurring. In some cases the analyst may arrange to observe the user during the next use of the program or application.
2. In some cases the analyst must contact systems operations personnel to correct the problem. Operations can usually terminate an on-line session and reinitialize the application and its programs.
3. In some cases the analyst may have to call data administration to recover lost or corrupted data files or databases. Data backup and recovery is beyond the scope of this book. It is covered extensively in most data and database management courses and textbooks.
4. In some cases the analyst may have to call network administration to fix a local, wide, or internetworking problem. Network professionals can usually log out an account and reinitialize programs.
5. In some cases the analyst may have to call technicians or vendor service representatives to fix a hardware problem.
6. In some cases the analyst will discover a bug caused the crash. The analyst attempts to quickly isolate the bug and trap it (automatically or by coaching users to manually avoid it) so that it can't cause another crash. Bugs are then handled as described in the previous section of this chapter.

END-USER ASSISTANCE

Another relatively routine ongoing activity of systems support is routine end-user assistance. No matter how well users have been trained or how well documentation has been written, users will require additional assistance. The systems analyst is generally on call to assist users with the day-to-day use of specific applications. In mission critical applications, the analyst must be on call day and night.

The most typical activities include: routinely observing the use of the system, conducting user-satisfaction surveys and meetings, changing business procedures for clarification (written and in the repository), providing additional training, and logging enhancement ideas and requests in the repository.

SYSTEMS ENHANCEMENT AND REENGINEERING

Adapting an existing system to new requirements is an expectation for all newly implemented systems. Adaptive maintenance forces an analyst to analyze the new requirement and return to the appropriate phases of systems analysis, design, and implementation. In this section we will examine two types of adaptive maintenance—systems enhancement and systems reengineering.

Most adaptive maintenance is in response to new business problems, new information requirements, or new ideas for enhancement. It is reactionary in nature—fix it when it breaks or when users make a request. We call this **system enhancement**. The objective of system enhancement is to modify or expand the application system in response to constantly changing requirements. This objective can be linked to your information system building blocks as follows:

- DATA—Many system enhancements are requests for new information that can be derived from existing stored data. Some data enhancements call for expansion of data storage.
- PROCESSES—Most system enhancements require the modification of existing programs or the creation of new programs to extend the overall application system.
- INTERFACES—Many enhancements require modifications to how the users interface with the system, and how the system interfaces with other systems.
- GEOGRAPHY—Most system enhancements are not driven by networks (see reengineering).

Another type of reactionary maintenance deals with changing technology. Information system staffs have become increasingly reluctant to wait until systems break. Instead, they analyze their program libraries to determine which applications and programs are costing the most to maintain or which ones are the most difficult to maintain. These systems might be adapted to reduce the costs of maintenance. The preceding examples of adaptive maintenance are classified as *reengineering*. The objectives of reengineering are to either adapt the system to a major change in technology, fix the system before it breaks, or make the system easier to fix when it breaks or needs to be adapted. These objectives can be linked to your information system building blocks as follows:

- DATA—Many reengineering projects are driven by the need to restructure stored data, either to make it more flexible and adaptable or to convert it to a new technology.
- PROCESSES—Many reengineering projects attempt to restructure or reorganize application programs to make them more maintainable or to convert them to a new technology (e.g., language). Many others change the input or output methods for programs (e.g., from batch to on-line or from on-line to graphical user interfaces).
- INTERFACES—Many reengineering projects are driven by the desire to replace old system interfaces with newer easier-to-use graphical user interfaces.
- GEOGRAPHY—Some application projects seek to change applications to new network technology.

Before we move on, we should acknowledge another trend in systems support. As a system's useful lifetime approaches its end, systems analysts and designers will turn to design and analysis recovery technology to automatically discover the models and specifications hidden in old COBOL programs.

Figure 18.4 expands on the activities of systems enhancement and reengineering. In this section, we briefly describe each activity, participants and roles, inputs and outputs, and techniques.

Analyze Enhancement Request

The purpose of this activity is to determine the appropriate course of action to either a *new business problem or idea for enhancement, technical limitation or problem*, or *enhancement idea* (from other system support activities). Recall that the support, in general, does not actually enhance the system. Instead, it studies

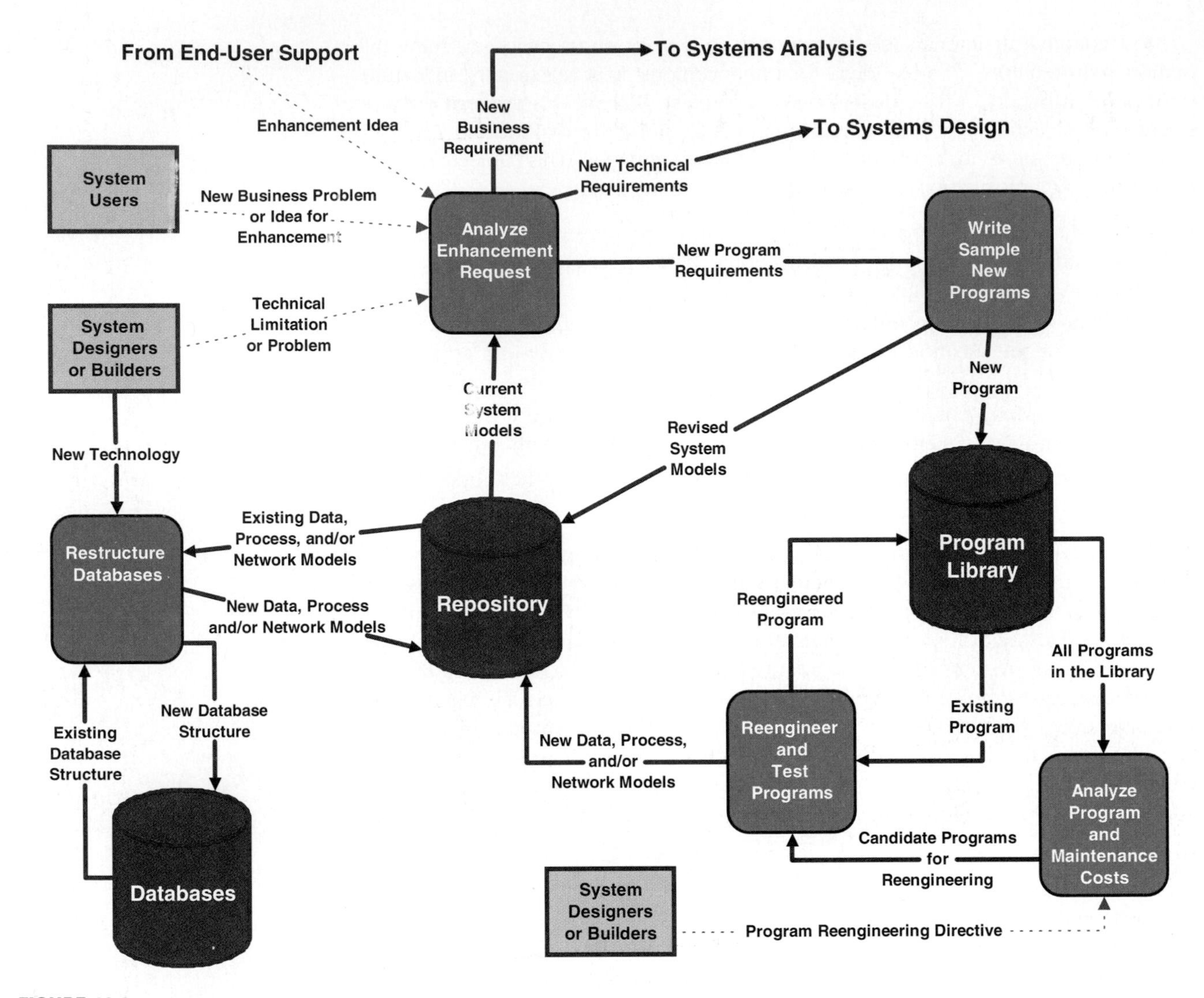

FIGURE 18.4 *Systems Enhancement and Reengineering Activities*

existing documentation to determine the appropriate action. Based on analysis of *current system models*, that action may include:

- Define **new business requirements** and return to systems analysis.
- Define **new technical requirements** and return to systems design.
- Define **new program requirements** and proceed to the next activity, Write Sample New Programs.

In the latter case, new programs are generally restricted to those that generate new information from existing data stores. Anything more complex should go through systems analysis and design.

The systems analyst should be skilled in project planning (Part Five, Module A), fact-finding (Part Five, Module B), and cost-benefit analysis (Part Five, Module C). The latter may be necessary to justify the enhancement project.

Write Simple, New Programs

Many enhancements can be accomplished quickly by writing simple, new programs. Simple programs are those that use existing data, do not update existing data, and do not input new data (for purposes of storing that data). In other words, these programs generate new reports and answer new inquiries. *New program requirements* represent the majority of today's enhancements.

NOTE It is our belief that any new program requirements that exceed our definition of *simple* should be treated as new business requirements and subjected to a systems analysis and design to more fully consider implications within the complete application system's structure.

Most such programs can be easily written by end-users with a minimal knowledge of a fourth-generation language or a PC-to-host database retrieval language (such as Q&E), but also becoming available in most PC database packages (such as *Access, Approach,* and *Paradox*). Programmers and analysts are also capable of writing such programs, but some shops question whether this is a valuable use of their time.

With today's fourth-generation and database languages/tools, these programs can be completed within hours. Since they generally do not enter or update data stores, testing requirements are not nearly as stringent. Once implemented, a *new program* may be stored locally (on a PC or LAN server), or it may be added to the program library (if many people could benefit from its use). Since the "local" programs are created for specific users, they generally do not qualify for IS support; however, in most cases the area systems analyst or local end-user computing guru provides some support.

Optionally, *revised system models* may be updated in the repository to reflect the existence of the new processes (programs) added to the system.

Restructure Files or Databases

From time to time, systems analysts help in the reengineering of files and databases. Many of today's data stores are implemented with traditional file structures (e.g., ISAM and VSAM) or early database structures (e.g., hierarchical IMS structures and network structures). Today's database technology of choice is SQL-based relational databases (which store data in tables that are integrated through redundant fields that act as pointers). Tomorrow, object database technology may present yet another shift in popularity.

Migrating data structures from one data storage technology to another is a major endeavor, wrought with opportunities to corrupt essential business data and programs. Thus, reengineering file and database structures has become an important task.

Database reengineering is usually covered more extensively in data and database management courses and textbooks; however, a brief explanation is in order here. The key player in database restructuring is the database analyst (or database administrator). The systems analyst plays a role because of the potential impact on existing applications. Network analysts may also be involved if databases are (to be) distributed across computer networks.

The key inputs are the *existing database structure* (which can be read from the file or database management system's dictionary that is included in most data stores) and *existing data, process, and network models* as stored in the repository. The outputs are a *new database structure* and a *new data, process, and network model.*

Data restructuring can and has been done by hand in many businesses. However, database reengineering CASE tools are increasingly used to read the data structure, produce the implementation model, perform data analysis to improve the model, and regenerate the new database structure. For the most part, database reengineering tools convert only data stores. Processes must also be converted to execute the new database retrieval and update commands against the new data structures. Technology for such conversions is available, but generally it is very expensive.

We've said it before—today's systems analyst must be both database literate and data management literate. Courses exist in most IS programs to provide such knowledge.

Analyze Program Library and Maintenance Costs

As mentioned earlier, many businesses are questioning the return on investment in corrective and adaptive maintenance. They realize that if complex and high-cost software can be identified, it might be reengineered to reduce complexity and maintenance costs. The first activity required to achieve this goal is to analyze program library and maintenance costs. This activity almost always requires software capable of performing the analysis. Systems analysts usually interpret the results.

Software tools such as VIASOFT's *VIA/Recap* measures your software library using a variety of widely accepted software metrics.

Software metrics are mathematically proven measurements of software quality and productivity.

Examples of software metrics applicable to maintenance include:

- **Control flow knots**—the number of times logic paths cross one another. Ideally, a program should have zero control flow knots. (We have seen knot counts in the thousands on some older, poorly structured programs.)
- **Cycle complexity**—the number of unique paths through a program. Ideally, the fewer, the better.

Software metrics, in combination with cost accounting (on maintenance efforts) can help identify those programs that would benefit from restructuring.

The input to this task is *all (or most) programs in the library*. The output is a *candidate program for reengineering*.

Reengineer and Test Programs

Given a *candidate program for reengineering*, there are three types of reengineering that can be performed on that program: code reorganization, code conversion, and code slicing.

- **Code reorganization** restructures the modular organization and/or logic of the program. For example, modules may be combined or separated to reduce coupling or increase cohesion (see Chapter 15). Logic may be restructured to eliminate control flow knots and reduce cycle complexity.
- **Code conversion** translates the code from one language to another. Typically, this translation is from one language version to another. There is a debate on the usefulness of translators between different languages. If the languages are sufficiently different, the translation may be very difficult. If the translation is easy, the question is "why change?"
- **Code slicing** is the most intriguing program-reengineering option. Many programs contain components that could be factored out as subprograms. If factored out, they would be easier to maintain. More importantly, if factored out, they would be reusable. Code slicing cuts out a piece of a program to create a separate program or subprogram. This may sound easy, but it is not! Consider your average COBOL program. The code you want to slice out may be located in many paragraphs and have dependent logic in many other paragraphs. Futhermore, you would have to simultaneously slice out a subset of the data division for the new program or subprogram.

  Code slicing is greatly simplified with reengineering software. VIASOFT's *VIA/Renaissance* is a code-slicing tool that automatically traces code back to prerequisite code and data structures. It can quickly create a stand-alone executable program from the sliced code. Furthermore, it can modify the original program to call the new program (deleting the sliced code from the original), or it can keep the original program unchanged.

The *candidate program for reengineering* is copied from the program library. It is reengineered using one or more of the preceding methods, it is thoroughly tested (as described earlier in the chapter), and the *reengineered program* is returned to the program library where it is available for production. Any *new data, process, and/or network models* are updated in the repository.

THE YEAR 2000 AND SYSTEMS SUPPORT

We've talked about systems support and how it addresses different types of problems. But there is one enormous problem that is just waiting to explode on the scene—the arrival of the year 2000. This event has the potential of triggering widespread computer application disasters across many corporations.

In the early 1960s and 1970s storage space was precious. Thus, millions of applications were built with efforts to utilize as little storage space as possible. To save two bytes of storage space, dates for this century were stored without the first two digits "19." For example, the date January 1, 1996, might be stored in a YYMMDD format as 960101. Now consider the date January 1, 2000. That date would be represented in the YYMMDD format as 000101. Many applications developed during the 60s and 70s stored dates in this format and used dates in arithmetic operations within applications. A quick comparison reveals that the numbers used to store the January 1, 2000, date yield a smaller number (meaning that it occurred earlier in time), implying that the date occurs before the January 1, 1996, date. If the dates were stored in a YYYYMMDD format, a comparison would have been accurate.

The bottom line is that corporations are racing against time to locate and correctly revise the millions of programs, library utilities, files, and databases that use/store the abbreviated dates. This is a monumental feat in most cases. Simply locating the programs to be modified is made difficult because programmers likely used different names for the data fields. This is even more complicated in those cases where a program passes the data field over to a utility that uses the date, which in turn may refer to the date field by a different name. Even when the programs are identified, often you may be dealing with a program that has been modified and modified until it is now extremely difficult to understand what impact the change will have.

Needless to say, this challenge has spurred a number of new software tools and consulting companies specifically targeted to helping companies revise their applications to deal with year 2000. Despite millions and millions of dollars spent, it is predicted that many corporations will have applications that fail to handle the turn of the century. To gain an interesting insight into possible fallouts that companies may experience on the arrival of the year 2000, read Warren Keuffel's article titled "Coping with the Year 2000, Rollover" (see Suggested Readings).

WHERE DO YOU GO FROM HERE?

This chapter provided a detailed overview of the systems support phase of systems development. You learned about the different types of systems support provided for a system: maintenance, enhancement, reengineering, and design recovery.

If you have been covering the chapters in order, you are now prepared to do systems development. Otherwise, you may wish to return to previous chapters to learn more about the tools and techniques used in systems development. Regardless, completion of this book does not guarantee your future success in systems development. Systems development approaches, tools, and techniques continue to evolve. Thus, your learning will be an ongoing process.

SUMMARY

1. **Systems support** is the ongoing maintenance of a system after it has been placed into operation. This includes program maintenance and system improvements.
2. Systems support involves solving different types of problems with the system. There are several different types of systems support: systems maintenance, systems recovery, end-user assistance, systems enhancement, and reengineering.
3. Regardless of how well designed, constructed, and tested a system or application may be, errors or bugs will inevitably occur. The corrective action that must be taken is called systems maintenance.
4. From time to time a system failure is inevitable. It generally results in an aborted or "hung" program (also called an ABEND or crash) and possible loss of data. The systems analyst often fixes the system or acts as intermediary between the users and those who can fix the system; this is referred to as systems recovery.
5. Another relatively routine ongoing activity of systems support is routine end-user assistance. No matter how well users have been trained or how well documentation has been written, users will require additional assistance. The systems analyst is generally on call to assist users with the day-to-day use of specific applications. In mission critical applications, the analyst must be on call day and night.
6. Most adaptive maintenance is in response to new business problems, new information requirements, or new ideas for enhancement. It is reactionary in nature—fix it when it breaks or when users make a request. We call this **system enhancement**. The objective of system enhancement is to modify or expand the application system in response to constantly changing requirements. Another type of reactionary maintenance deals with changing technology. Information system staffs have become increasingly reluctant to wait until systems break. Instead, they choose to analyze their program libraries to determine which applications and programs are costing the most to maintain or which ones are the most difficult to maintain. These systems might be adapted to reduce the costs of maintenance. The preceding examples of adaptive maintenance is classified as *reengineering*.

KEY TERMS

bugs, p. 580
central repository, p. 578
code conversion, p. 588
code reorganization, p. 588
code slicing, p. 588
control flow knots, p. 588
cycle complexity, p. 588
program libraries, p. 579
regression testing, p. 583
software metrics, p. 588
system enhancement, p. 585
system recovery, p. 584
system testing, p. 583
systems support, p. 578
unit testing, p. 583
version control, p. 583

REVIEW QUESTIONS

1. Define systems support.
2. What are the four different systems support activities? Define the purpose of each phase.
3. What is a repository? How does it differ from a program library and a business database?
4. How do application development projects and application maintenance projects differ in their use of a project repository and program library?
5. What are the objectives of systems maintenance? How do the information systems building blocks apply to systems maintenance?
6. How does a systems analyst or programmer validate a bug? Why do they validate a bug?
7. What is the purpose of program benchmarking as it pertains to systems maintenance?
8. What are the inhibitors to program understanding?
9. What are three types of program tests?
10. Define software metrics. Give two examples of software metrics. How are software metrics used in systems reengineering?
11. What are three types of reengineering that can be performed on a program?
12. What is version control? Why is version control a necessity in systems maintenance?
13. Briefly describe anticipated problems that the year 2000 will have on many existing systems.

PROBLEMS AND EXERCISES

1. How is a systems support request handled differently from a project request for a new information system?
2. Obtain a copy of a computer program. Analyze the program for control flow knots and cycle complexity.

PROJECTS AND RESEARCH

1. There are a number of CASE tools specifically oriented to system maintenance, redevelopment, and reengineering. Through research, do a market survey to identify the characteristics and capabilities of these tools. Briefly compare several products. Select one product and write the vendor. Request product literature, company information, success stories, and the like. Complete your report with an in-depth discussion of the CASE tool.
2. A number of commercial methodologies include system maintenance or redevelopment within their approach. Through research, do a market survey to identify the characteristics and capabilities of the planning capabilities of these methodologies. Select one methodology and write the vendor. Request product literature, company information, success stories, and the like. Compare and contrast the methodology with the generic methodology presented in this chapter. Force yourself to identify at least three major features or capabilities that you like better and three that you have some concerns about.
3. Make an appointment with a systems analyst or programmer. Conduct an interview on either the problems and issues encountered in system maintenance or the person's standard approach or methodology for system maintenance. Compare and contrast this with the information and approach presented in this chapter. Write a report of your findings.

MINICASES

1. The Minnesota State University is a large, public, metropolitan university located within 20 miles of four cities in Minnesota. *(Scene: Kurt Wilson, Director of Administrative Information Management, is meeting with Paula Teague, assistant director of Applications Development.)*

KURT Good morning, Paula. How's the cold?

PAULA Much better, thank you. It's good to be back. I assume this is the meeting I had to cancel when I got sick?

KURT Right. As you know, the administrative information systems master plan will be complete within the next three months. Assuming the executive committee approves the plan, the real work begins—delivering the new business processes and applications outlined in the plan.

PAULA I've been wondering when you were going to address that issue. We can't keep up with new systems development requests as it is. Am I going to get additional staff?

KURT I'm afraid not. In this era of staff downsizing, I suspect that we'll be lucky to hold on to what we have.

PAULA Well, I know we can increase productivity using CASE tools driven off the planning models your staff has recorded in the new repository. But there is a learning curve with CASE technology, as well as the new methodology. Also, these new applications call for a greater degree of adaptability and integration than we have historically expected. I just don't see how we can deliver more systems with the same or fewer people.

KURT I've got an idea. I've been running some numbers against the time accounting system. According to our own records, we are using 19.3 FTE (full-time equivalent staff) to simply support existing system maintenance.

PAULA That wouldn't surprise me. Legacy code is the anchor that inhibits new systems development in all shops. Don't tell me you are going to eliminate existing systems support? I think there would be an immediate and fatal backlash from the user community.

KURT True, but that's not exactly what I had in mind. Don't have a cardiac, but what would you say if I told you that I wanted you to reduce your maintenance effort to 8.5 FTE? [*Paula does not respond.*] Your silence indicates that you are concerned.

PAULA And rightfully so, don't you think? The user community will scream for my head on a platter! You are talking about cutting support by more than 50 percent.

KURT Actually, Paula, I'm asking you to cut support by less than 25 percent, but to use 50 percent fewer people!

PAULA And how am I supposed to do that?

KURT I have a couple of ideas for you to consider. First, according to my analysis of change request forms, almost two-thirds of all requests fall into the category of enhancements. Half of those enhancements can be characterized as desirable, not essential. It seems to me that we could declare a temporary moratorium on such maintenance projects.

PAULA I can't confirm your numbers, but I'll agree that we are going to have to be pickier about what we choose to do and not do. I'd like to see your data after this meeting.

KURT No problem! Second, I'd like you to consider a SWAT team approach to maintenance.

PAULA SWAT? Like the police?

KURT SWAT stands for "Specialists With Automated Tools." With only 8.5 FTE for maintenance, it seems to me that it would be a mistake to assign one or two persons per existing development team. Instead, I see a maintenance SWAT team that takes over all maintenance.

PAULA It might work. I'd want to make sure the SWAT team had at least one member from each current development team to preserve application knowledge. But what's the "automated tools" angle?

KURT Glad you asked that. Our luncheon meeting will be with a sales representative from a company called VIASOFT. They sell what amounts to CASE tools for maintaining, enhancing, and reengineering existing COBOL programs. They promise to increase productivity of maintenance programmers who specialize in the tools.

PAULA Now we're talking. There would be the usual learning curve, but once the team is comfortable with the technology, we might just be able to do 75 percent of our existing maintenance with less than 50 percent of the existing staff.

KURT I think so! In fact, I'll make a bolder prediction. In time, I think you'll eventually be able to increase support over today's levels with less than 50 percent staff. Anyway, you're a pretty creative person. Give some more thought to ways we can make this work. We need to go to our luncheon date.

a. Why does system support consume up to 80 percent of some systems development budgets?
b. Can you think of other ways to provide adequate systems support while reducing systems support staff?
c. Can you think of any aspects of systems support that have not been addressed by Kurt's SWAT and technology proposals?

SUGGESTED READINGS

Hammer, M. And J. Champy. *Reengineering the Corporation*. New York: Harper Business, 1993.

Keuffel, Warren. "Coping with the Year 2000 Rollover." *Software Development*. 4, no. 8 (August 1996), pp. 23–24, 26, 28.

Martin, E. W.; D. W. DeHayes; J. A. Hoffer; and W. C. Perkins. *Managing Information Technology: What Managers Need to Know*. New York: Macmillan, 1994.

PART FIVE

CROSS LIFE CYCLE ACTIVITIES AND SKILLS

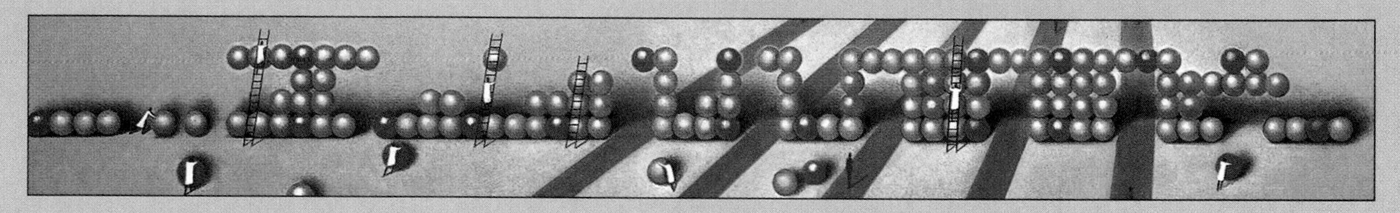

The modules in Part Five are not appendixes! A number of skills, tools, and techniques are important to multiple phases of systems development—analysis, design, and implementation. We feel strongly that these cross life cycle skills are as important as any tool and technique taught in this book. In fact, they may be the ultimate critical success factor for all systems work. So why are they at the end of the book?

Placing these modules (a name chosen to distinguish them from the chapters you've been reading) in the analysis, design, or implementation units would have understated their value to the other phases. On the other hand, these modules do require some prerequisite knowledge—in most cases, Chapters 1 through 3. By placing these modules at the end of the book, you and your instructor have the flexibility to introduce them when you prefer. Each module begins by describing not only objectives, but also prerequisite chapters.

Module A, Project and Process Management, introduces *project and process management techniques.* All projects are dependent on the planning, scheduling, control, and leadership principles that are surveyed. The module also presents two popular modeling techniques for project management: Gantt and PERT. These tools help you schedule activities, evaluate progress, and modify schedules.

Module B, Fact-Finding and Information Gathering, surveys *fact-finding and information gathering techniques.* These techniques are used to solicit factual information, opinions, and requirements from end-users. They are a crucial prerequisite to all modeling techniques—for example, entity relationship diagrams, data flow diagrams, object models, and so forth. The techniques surveyed include sampling, research, observation, questionnaires, and interviews.

Module C, Feasibility and Cost-Benefit Analysis, presents *feasibility* and *cost-benefit analysis techniques.* For any potential solution that you evaluate and recommend to management, you must be prepared to defend its operational, technical, schedule, and economic feasibility. Economic feasibility is especially important, since most organizations are either profit-oriented, cost-reduction-oriented, or both. Your ability to estimate costs and benefits and then analyze those numbers for cost-effectiveness is critical.

Module D, Joint Application Development, introduces *joint application development (JAD),* a technique used to accelerate systems analysis and design phases. Developers, users, and management come together in highly structured and facilitated workshops to reduce fact-finding, problem solving, requirements analysis, system modeling, and prototyping activities to a few intensive daylong workshops instead of the usual weeks or months required for those activities.

Finally, Module E, Interpersonal Skills and Communications, introduces *interpersonal skills and communications* for the systems analyst. After you've collected facts and modeled systems, you must be able to present your findings and recommendations. In this module you will learn how to plan and run meetings, conduct brainstorming sessions, conduct walkthroughs of documentation, make oral presentations, and write reports.

PROJECT AND PROCESS MANAGEMENT TECHNIQUES

MODULE PREVIEW, PREREQUISITES, AND OBJECTIVES

Most of you are familiar with Murphy's Law, which suggests that "if anything can go wrong, it will." Murphy has motivated numerous pearls of wisdom about projects, machines, people, and why things go wrong. The purpose of this module is to teach you strategies, tools, and techniques for project management.

Prerequisites—It is assumed that you have read Chapter 3. For maximum impact, this module should be sequenced to follow either Chapter 3, 4, or 9.

You will know project and process management techniques when you can:

- Define a project and the need for project management.
- Define project management and the consequences of mismanagement.
- Compare and contrast project and process management.
- Develop or modify a work breakdown structure for a project.
- Read Gantt charts as a model of project activities, schedules, and progress.
- Read PERT charts as a model of project activities, schedules, and progress.
- Describe a typical software approach to project modeling and management.

WHAT IS PROJECT MANAGEMENT?

Before we can define *project management,* we should first define *project.* There are as many definitions as there are authors, but we like the definition put forth by Wysocki, Beck, and Crane:

> "A **project** is a sequence of unique, complex, and connected activities having one goal or purpose and that must be completed by specific time, within budget, and according to specification." [1]

The keywords are underlined. As applied to information system development, we note the following:

- A system development methodology, such as *FAST,* defines a sequence of activities, mandatory and optional.
- Every system development project is unique; that is, it is different from every system development project that preceded it.
- The activities that comprise systems development are relatively complex. They require the skills that you are learning in this book, and they require that you be able to adapt concepts and skills to changing conditions and unanticipated events.
- By now, you've already learned that the activities that make up a system development methodology are generally sequential. While some tasks may overlap, many tasks depend on the completion of other tasks.
- The development of an information system represents a goal. Of course, there may be several objectives to be met in order to achieve that goal.
- Although many information system development projects do not have absolute deadlines or specified times (there are exceptions), they are notoriously completed later than originally projected. This is becoming less acceptable to upper management given the organizationwide pressures to reduce cycle times for products and business processes.
- Few information systems are completed within budget. Again, this tendency is being increasingly rejected by upper management.
- Information systems must satisfy the business, user, and management expectations and specifications (which we call *requirements* throughout this book).

For any systems development project, effective *project management* is necessary to ensure that the project meets the deadline, is developed within an acceptable budget, and fulfills expectations and specifications.

> **Project management** is the process of defining, planning, directing, monitoring, and controlling the development of an acceptable system at a minimum cost within a specified time frame.

Corporate rightsizing has changed the structure and culture of most organizations. Rigid hierarchical command structures and permanent teams have been replaced by more flexible and temporary interdepartmental teams that are given greater responsibility and authority for the success of organizations. This is definitely true in information systems development. Contemporary system development methodologies depend on building teams to include both technical and nontechnical people, users and managers, and information specialists all directed to the project goal. These dynamic teams require leadership and project management.

Different organizations take different approaches to project management. One approach is to appoint a project manager from the ranks of the team (once it has been formed). This approach is a result of the *self-directed* team paradigm. But

[1] Robert K. Wysocki, Robert Beck, Jr., and David B. Green, *Effective Project Management: How to Plan, Manage, and Deliver Projects on Time and within Budget* (New York: John Wiley & Sons, 1995), p. 38.

many organizations have found that successful project managers apply a unique body of knowledge and skills that must be learned. These organizations tend to hire and/or develop professional project managers who are assigned to one or more projects at any given time.

This module introduces project management tools and techniques; however, the methods are not sufficient on their own. Some schools offer project management courses that would be a valuable addition to your curriculum (regardless of whether or not they are oriented to information systems professionals).

Project Management Causes of Failed Projects

We can develop an appreciation for the importance of project management by studying the mistakes of other project managers. Failures and limited successes far outnumber successful information systems. Why is that? True, many systems analysts and information technologists are unfamiliar with or undisciplined in the tools and techniques of *systems analysis and design.* But that only partially explains the shortcomings of many information system projects. Many projects suffer from poor leadership and management. Project mismanagement can sabotage the best application of the systems analysis and design methods taught in this book. Let's examine some of the project mismanagement problems and consequences.

One of the most common causes of project failure is taking shortcuts through or around the methodology. Throughout this book, we apply a hypothetical system development methodology called *FAST.* Many companies have developed or purchased a methodology, but project teams often take shortcuts for one or more of the following reasons:

- The project gets behind schedule and the team wants to catch up.
- The project is over budget and the team wants to make up costs by skipping methodology steps.
- The team members are not trained or skilled in some of the methodology's activities and requirements, so they skip them.

Some analysts and systems development managers might argue: *The methodology is the problem! It is an excessively long, drawn-out process that leads to schedule and cost overruns.*

We could not disagree more! In fact, we view that argument as an excuse to justify not managing projects. Methodologies define best practices and processes for developing information systems. Methodologies apply engineering rigor to the project. Methodologies ensure that important activities that could (and frequently do) compromise expectations and quality are not omitted. In a nutshell, project and process management are each dependent on the other!

Another common cause of project failures is poor *expectations management.* All users and managers have expectations of the project. Over time, these expectations change. Unfortunately, this change takes the form of scope creep.

Scope creep is the unexpected growth of user expectations and business requirements for an information system as the project progresses.

Unfortunately, the schedule and budget are rarely modified at the same time. This is a mistake, and the project manager is ultimately held accountable for the inevitable and unavoidable schedule and budget overruns. In other words, the users' expectations of schedule and budget did not change as the scope changed. But there are ways to manage expectations. We'll discuss a simple tool and technique later in this module.

A similar phenomenon and problem is caused by feature creep.

Feature creep is the uncontrolled addition of technical features to a system under development without regard to schedule and budget.

Each unplanned feature, however impressive, adds time and costs to the overall schedule.

One major problem with cost overruns is that many methodologies or project plans call for an unreasonably precise estimate of costs before the project begins. These estimates are made after a quick preliminary study or feasibility study. Think about it! Can you accurately estimate project costs before making a detailed study of the current system or defining end-user requirements? Can you estimate the costs of computer programming before a detailed systems design has been completed? It's not very likely. The cost estimates of a project will change as you get further into the systems development process.

Poor estimating techniques are another cause of cost overruns. We suspect that many systems analysts estimate by making the best calculated estimate (guesstimate?) and then doubling that number. This is hardly a scientific approach. There are better approaches available; some useful techniques are discussed in Module C.

And finally, cost overruns are often caused by schedule delays. Once again, we can point to premature estimates as a problem. These early estimates are based on the initial scope of the project. Because systems analysts (and information systems professionals in general) are eternal optimists, they often quote optimistic schedules and fail to modify those schedules as the true scope of the project becomes apparent.

Because many managers and analysts are often poor time managers, project schedules slip slowly but steadily. "So we've lost a day or two! It's no big deal. We can make it up later." This may be true, but then again, it might not. They fail to recognize that in the systems development life cycle, certain tasks are dependent on other tasks. Because of these dependencies, a one-day slip can set the whole schedule back. And when those one-day delays pile up, we inevitably find ourselves working 15-hour days at the end of the project.

Another cause of missed schedules is what Brooks has described as the *mythical man-month.*[2] As the project gets behind schedule, the project leaders frequently try to solve the problem by assigning more people to the project team. It just doesn't work! There is no linear relationship between time and number of personnel. The addition of personnel creates more communications and political interfaces. The result? The project gets even further behind schedule.

Poor people management can also cause projects to fail. Managers tend to be thrust into management, not prepared for management. These projects are easy to identify—no one seems to be in charge; customers don't know the status of the project; teams don't meet regularly to discuss and monitor progress; team members aren't communicating with one another; and the project (or phase) is always said to be "95 percent complete."

Yet another cause of project failure is that the business is in a constant state of change. If the project's importance changes, or if the management and business reorganize, all projects should be reassessed for compatibility with changes and importance to the business.

You've probably noticed that the causes of failed projects are related. For instance, missed requirements may cause schedule slippages that, in turn, cause cost overruns. You might ask why somebody isn't able to recognize these problems and correct them. Somebody should. And that person is supposed to be the project manager or leader. This brings us to a major cause of project failure: lack of project management and leadership. Good computer programmers don't always go on to become good analysts. Similarly, good analysts don't automatically perform well as project managers and leaders. To be a good project manager, the analyst must possess or develop skills in the basic functions of management.

The Basic Functions of the Project Manager

The project manager is not just a senior analyst who happens to be in charge. A project manager must apply a set of skills different from those applied by the analyst. What skills must the project manager possess or learn? The basic functions

[2] Fred Brooks, *The Mythical Man-Month* (Reading, MA: Addison-Wesley, 1975).

of a manager or leader have been studied and refined by management theorists for many years. These functions include planning, staffing, organizing, scheduling, directing, and controlling.

Scoping the Project If you cannot scope project expectations and constraints, you can't manage expectations. At a minimum, a complete project definition should include the following:

- *A project champion and executive sponsor.* The champion is the person who sells the project to management. The executive sponsor underwrites the cost of the project.
- *A brief statement of the problem or opportunity to be addressed by the project.* Ideally, the problem or opportunity statement should be worded such that *everybody* agrees with it.
- *The project goal.* Recall that every project has one overriding goal. Obviously, the goal should be directed to the stated problem or opportunity.
- *The project objectives.* As described in Chapter 4, objectives are dated and measurable actions, results, or conditions that are directed toward achievement of the project goal. Think of them as a "grading key" against which we can ultimately measure the success or failure of the project.
- *Project assumptions and constraints.* Assumptions and constraints define those factors that cannot change. Examples might include absolute deadlines, available resources, the current technology, expected technological advances, and the like.

Failure to achieve consensus on the above dooms a project before it starts. In fact, we would argue that the written *project definition* should be a visible element in every project deliverable. Then, if any of these parameters change, that change would result in a reevaluation of the entire project management plan.

Planning Project Tasks and Staffing the Project Team A good manager always has a plan. The manager estimates resource requirements and formulates a plan to deliver the target system. This is based on the manager's understanding of the requirements of the target system at that point in its development. A basic plan for developing an information system is provided by the systems development life cycle. Many firms have their own standard life cycles, and some firms have standards for the methods and tools to be used.

Each task required to complete the project must be planned. How much time will be required? How many people will be needed? How much will the task cost? What tasks must be completed before other tasks are started? Can some of the tasks overlap? These are all planning issues. Some of these issues can be resolved with the project modeling tools that will be discussed later in this module.

Project managers frequently build the project team. Which users and managers should be assigned to the team? How much release time will they need to be granted to participate in project tasks? The project manager should carefully consider the business and technical expertise that may be needed to successfully finish the project. Which systems analysts and programmers should be assigned? The key is to match the personnel to the required tasks that have been identified as part of project planning. Most system development methodologies, including SoundStage's *FAST,* recommend mandatory and optional project roles to be staffed by the project manager.

Organizing and Scheduling the Project Effort Given the project plan and the project team, the project manager is responsible for organizing and scheduling the project. Members of the project team should understand their own individual roles and responsibilities as well as their reporting relationship to the project manager.

The project schedule should be developed with an understanding of task time requirements, personnel assignments, and intertask dependencies. Many projects present a deadline or requested delivery date. The project manager must determine whether a workable schedule can be built around such deadlines. If not, the deadlines must be delayed or the project scope must be trimmed. We will soon introduce project modeling tools that can be used for project scheduling.

Directing and Controlling the Project Once the project has begun, the project manager becomes a supervisor. As a supervisor, the project manager directs the team's activities and evaluates progress. Therefore, every project manager must demonstrate such people management skills as motivating, rewarding, advising, coordinating, delegating, and appraising team members. Additionally, the manager must frequently report progress to superiors.

Perhaps the manager's most difficult and important function is controlling the project. Few plans will be executed without problems and delays. We've already discussed the causes and effects of unsuccessful projects. The project manager's job is to monitor tasks, schedules, costs, and expectations in order to control those elements. If the project scope is increasing, the project manager is faced with a decision: Should the scope be reduced so the original schedule and budget will be met, or should the schedule and budget be revised? The project manager must be able to present the alternatives and their implications for the budget and schedule in order to manage expectations.

Project Management Software

Today, project management software is routinely used to help project managers plan projects, develop schedules, develop budgets, monitor progress and costs, generate management reports, and effect change. Examples include Microsoft's *Project* and Applied Business Technology's *Project Manager Workbench*. These packages greatly simplify the preparation of the project management models presented in the next section. In fact, the models and techniques would be difficult to apply without software assistance.

For this reason, we will teach you project modeling and management techniques in the context of project management software. We used Microsoft *Project* because that tool is frequently available to students at special academic prices through their college bookstore. Microsoft *Project*, like most project management software tools, defaults to a project model called a *Gantt chart*. And like most project management software tools, it supports *PERT charts*, another useful type of project model. Figures A.1(a) and (b) show Gantt and PERT charts, respectively, as displayed by Microsoft *Project*.

PROCESS MANAGEMENT

Today, many organizations are paying a lot of attention to their *business processes*. These are the fundamental processes that the business uses to fill orders, maintain inventory, handle purchases, and so forth. Organizations are trying to streamline those business processes to improve efficiency, quality, and value to the business.

Information system development and maintenance are business processes—very complex business processes. Like most business processes, information system development processes must be managed. In fact, process management is a prerequisite to systems development project management.

> **Process management** is the planning, selection, deployment, and consistent application of standard system development methods, tools, techniques, and technologies to all information system projects.

For most information system organizations, process management is built around a system development methodology. A methodology provides an organization with a consistent, repeatable process to develop information systems. While

(a)

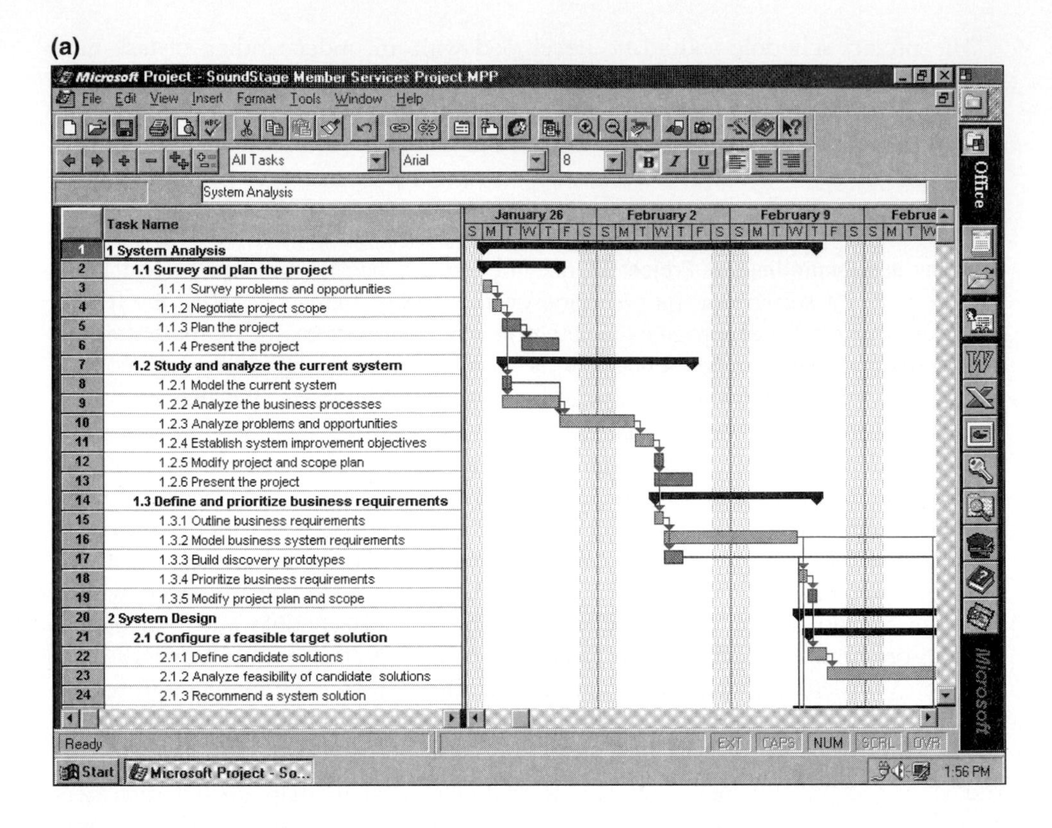

(b)

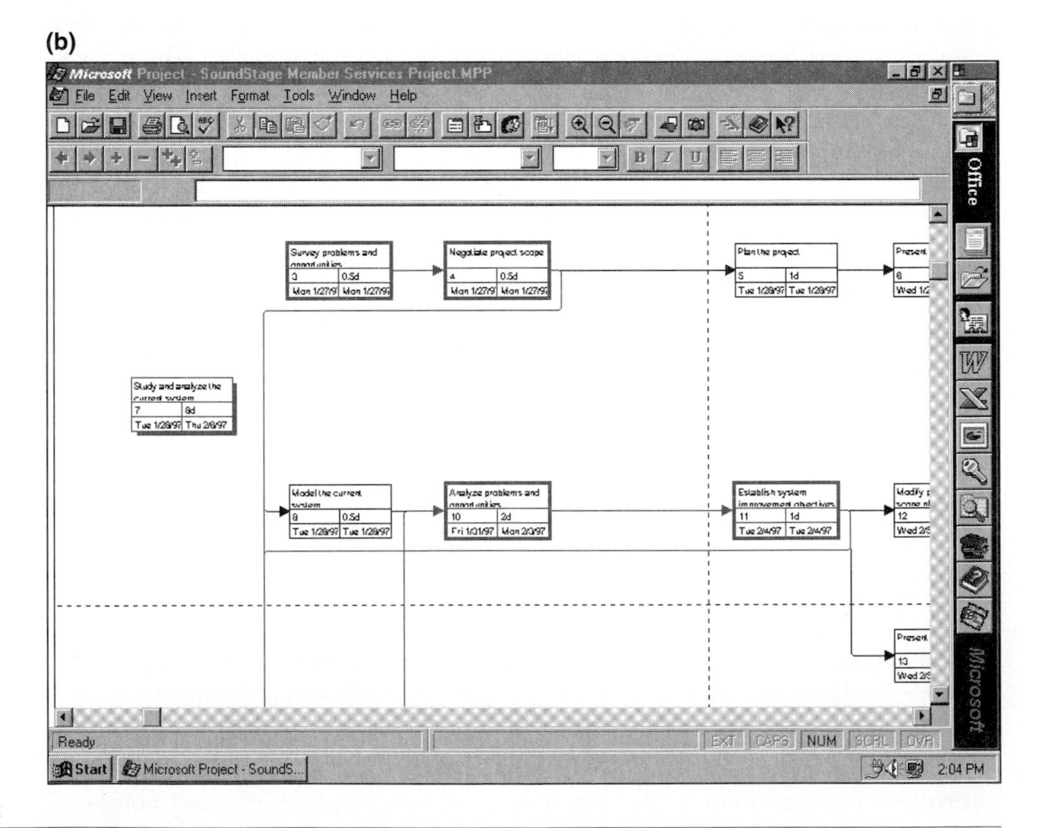

FIGURE A.1 *Microsoft* Project

methodologies can be developed in-house, most companies purchase their methodology. (Recall that SoundStage purchased its methodology, *FAST.*)

Process management includes a number of basic functions described in the following subsections.

Management of the Methodology

Once a methodology has been selected (or developed), it must be implemented. This can be an enormous undertaking that requires the following:

- Establishing visibility for the methodology by educating all developers, managers, and technical support staff in the basic development process, tools, and techniques to be used.
- Providing just-in-time detailed training to development teams as each team begins its first project. This training usually expands on many of the tools and techniques taught in this book.
- Providing consultation to project teams as they apply the methodology. Initially, this consultation may be subcontracted to the methodology vendor, but ultimately, most organizations want to build the expertise internally.
- Improving the methodology. Invariably, some aspects of the methodology will work better than others. While methodology vendors usually release improved versions of their methodologies, organizations should implement their own improvements as needed.

Methodologies are notorious for becoming shelfware, that is, not used at all. Methodology management is the key to avoiding the following common consequences.

- No consistency in the processes used to develop systems—even successes cannot be reliably repeated.
- No flexibility in the process used to develop systems—project teams are unable to adapt the methodology to new or unique situations.
- Failure to follow the methodology—either because management doesn't expect it, or project managers inappropriately skip or accelerate activities in response to schedule or budget problems.

Despite these common problems, the establishment and consistent application of a methodology is widely recognized as an important measure of an organization's system development maturity. In fact, the Software Engineering Institute's (SEI) *Capability Maturity Model* considers methodology implementation, application, and management as critical to its measure of system and software development sophistication.

Management of System Development Technology

The selection and use of computer-aided systems (software) engineering CASE tools, software development environments (SDEs), and project management software are a consistent theme in this book. Not surprisingly, the planning, selection, and deployment of this development technology is an important part of process management.

Development technology must be carefully evaluated and selected based on the technology architecture and vision of the business and compatibility with (or adaptability to) the chosen methodology. Once a technology has been chosen, developers must be trained in its correct use. Again, the most effective training occurs just-in-time (JIT) for the project team.

Total Quality Management

A development process (methodology) does not ensure quality. Quality must be managed. Quality management begins with establishing quality standards. While better methodologies provide quality guidelines, only an individual organization can set the detailed standards. Internal standards applicable to system development may include the following:

- Standards for project deliverables such as reports and documentation.
- Modeling techniques and standards.
- Naming standards for models, objects, programs, databases, and so on.
- Quality checkpoints, deliverables, and sign-offs at various stages of the projects.
- Technology standards such as approved graphical user interface components and placement.
- Testing procedures and tolerances.
- Acceptance criteria for system implementation.

Metrics and Measurement

This is a relatively new dimension of process management. According to the SEI *Capability Maturity Model,* sophisticated development organizations measure their productivity and quality with formal metrics and adjust the development process to effect continuous improvement. *System and software metrics* is a relatively new and rapidly changing discipline. The interest in that discipline is being increasingly driven by upper management's desire to improve the accountability of developers and the entire information system unit to the overall organization.

The Development Center

The ultimate process management infrastructure is a development center.

> A **development center** is a central group of information system development consultants and managers who plan, implement, and support all aspects of process management, including methodology, technology, quality, and measurement.

Development center staff do not develop information systems. Rather, staff members provide consulting services to those who do develop information systems (including systems analysts and programmers).

As shown in Figure A.2, a development center typically reports to that manager who oversees all information systems development. A fully staffed and mature development center might include the following specialists:

- A methodology coordinator.
- Methodology experts (sometimes called methodologists) who train developers and consult to developers.
- Tool experts (sometimes called CASE tool analysts) who train developers, consult to developers, and install and support development technologies.
- Quality analysts who establish and document development standards and implement processes to evaluate quality of appropriate project deliverables.
- Measurement analysts who establish metrics and measure productivity and quality.
- Repository managers who manage the consolidation of system models, specifications, and components.

PROJECT MANAGEMENT TOOLS AND TECHNIQUES

There are many project management tools and techniques—enough for an entire book. In this section we will introduce two project planning and control tools and a simple tool for managing expectations.

Gantt Charts

The Gantt chart, conceived by Henry L. Gantt in 1917, is the most commonly used project scheduling and progress evaluation tool in use.

> A **Gantt chart** is a simple horizontal bar chart that depicts project tasks against a calendar. Each bar represents a named project task. The tasks are listed vertically in the left-hand column. On a Gantt chart, the horizontal axis is a calendar timeline.

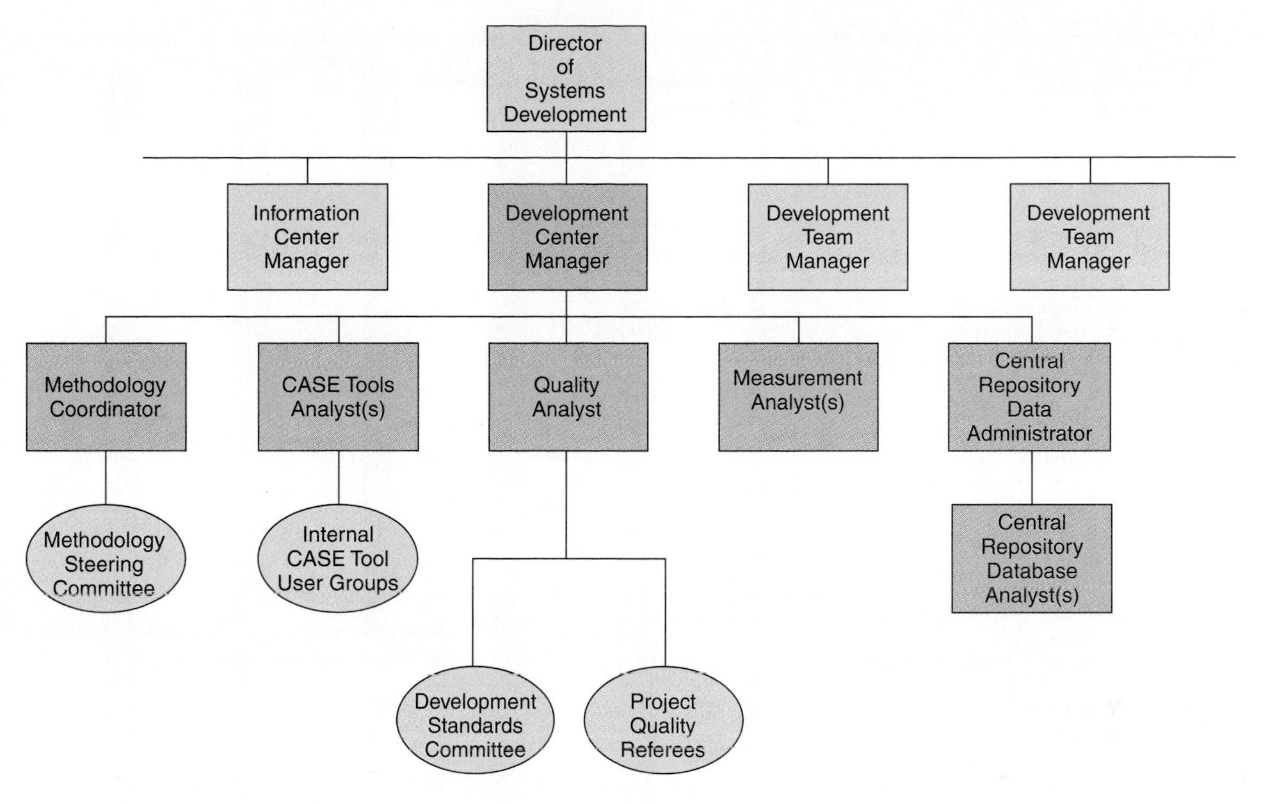

FIGURE A.2 *A Development Center Organization*

The popularity of Gantt charts stems from their simplicity—they are easy to learn, read, prepare, and use.

Figure A.3 illustrates a partial Microsoft *Project* Gantt chart for the SoundStage Member Services project (at the start of the project). You should note the following:

1. The solid black bars are summary tasks that represent project *phases* that are further decomposed into other tasks. In the *FAST* methodology, phases are broken down into activities. These activities could be broken down into steps (but the figure purposefully omits that level of detail).
2. The red and blue bars represent detailed tasks. The length represents the duration of the task. The bar is positioned according to its planned start and finish dates. The color in Microsoft *Project* is significant as follows.
   - Red bars indicate tasks have been determined to be critical to the schedule, meaning any extension to the duration of those tasks will delay other tasks and the project as a whole. We'll talk more about critical tasks later.
   - Blue bars indicate tasks that are not critical to the schedule, meaning they have some slack time during which delays will not impact other tasks and the project as a whole.
3. The red and blue arrows indicate prerequisites between detailed tasks. The red arrows define the project's critical path—the sequence of tasks that determine the project's final completion date.
4. The diamonds indicate milestones. Milestones are events that have no duration. They signify the end of some significant phase or deliverable.

NOTE Gantt charts clearly depict the overlap of scheduled tasks. Because systems development tasks frequently overlap, this is a major advantage.

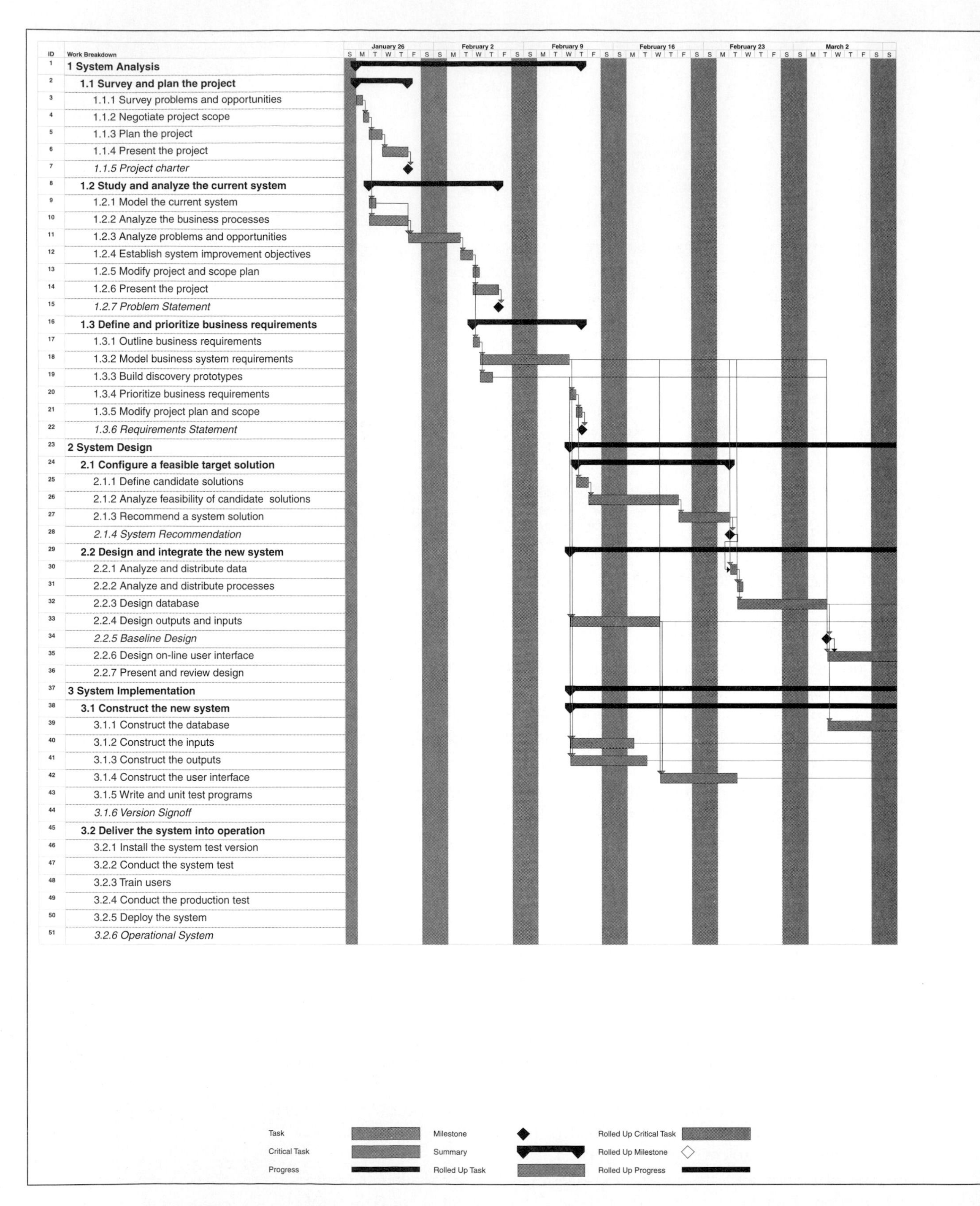

FIGURE A.3 *A Gantt Chart*

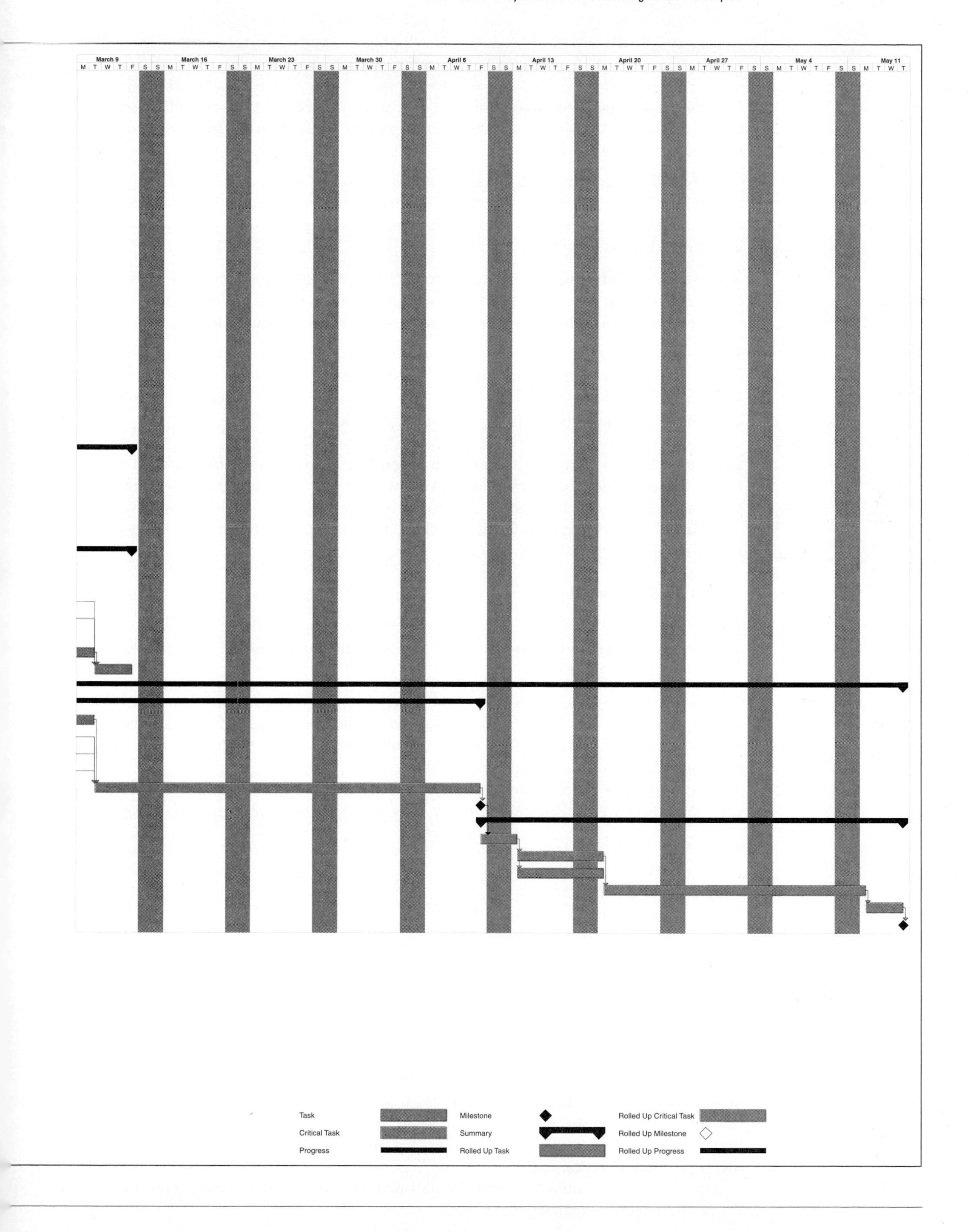
March 9
March 16
March 23
March 30
April 6
April 13
April 20
April 27
May 4
May 11
Task
Critical Task
Progress
Milestone
Summary
Rolled Up Task
Rolled Up Critical Task
Rolled Up Milestone
Rolled Up Progress

Let's briefly discuss the process of developing a Gantt chart using the Microsoft *Project* approach. Refer back to Figure A.3 as we progress through the discussion.

Forward and Reverse Scheduling Initially, you must determine the scheduling strategy to be used. Two basic scheduling approaches are supported by most project management software tools.

> **Forward scheduling** establishes a project start date and then schedules forward from that date. Based on the planned duration of required tasks, and the allocation of resources to complete those tasks, a projected project completion date is calculated.
>
> **Reverse scheduling** establishes a project deadline and then schedules backward from that date. Essentially, tasks, their duration, and resources must be chosen to ensure that the project can be completed by the deadline.

The former strategy is used in most information system projects. If the completion date is unacceptable, then project scope must be reduced to make the project feasible. The latter strategy is obviously used in those projects where an absolute deadline (such as a government mandate) is applicable.

Calendars Every business operates on a **calendar.** In Microsoft *Project* a base calendar is established to identify the work week (e.g., Monday through Friday), workday (e.g., 8 A.M. to noon; 1 P.M. to 5 P.M.), and holidays and conflict days.

After the project team has been identified, personal calendars can also be established to block out individual vacations and commitments.

Work Breakdown Structures Most projects can be defined by a hierarchical breakdown of the required work.

> A **work breakdown structure** is a hierarchical decomposition of the project into phases, activities, and tasks.

A template of the work breakdown structures for most information system development projects exists in the form of the chosen methodology. In fact, many commercial methodologies sell electronic work breakdown structures for use with the most popular project management software.

Work breakdown structures may be depicted with decomposition diagrams (Chapter 6); however, Microsoft *Project* uses a simpler *outline* format as follows:

1. Phase 1 of the project
    - 1.1 Activity 1 of Phase 1
        - 1.1.1 Task 1 of Activity 1.1
            - 1.1.1.1 Step 1 of Task 1.1.1
            - 1.1.1.2 Step 2 of Task 1.1.1
            - 1.1.1.3 . . .
        - 1.1.2. Task 2 of Activity 1.1
        - 1.1.3 . . .
    - 1.2 Activity 2 of Phase 1 . . .
2. Phase 2 of the project . . .

Those work units that are broken down into more detailed work units are called **summary tasks** in Microsoft *Project.* They are not scheduled, per se. The duration of summary tasks will be automatically calculated based on the duration of those tasks that will not be broken down into more granular work units. We'll call those **primitive tasks.**

Another type of entry in a work breakdown structure is a milestone. Milestones do not represent actual work, per se.

Milestones are events that signify major accomplishments or events during a project.

Milestones do not require work. They just happen. An example might be completion of a phase (e.g., definition phase) or deliverable (e.g., requirements statement).

Effort and Duration For each primitive task, we need to estimate its duration. This will determine the length of the bars in the Gantt chart. There is no foolproof technique for estimating work duration. The methodology may provide a baseline duration (for example, for a one-year project). But for the sake of demonstration, we offer the following technique. Because people resources have yet to be assigned, we will base our initial estimate on a single individual doing the work.

1. Estimate the minimum amount of time it would take to perform the task. We'll call this the optimistic time (OT). The optimistic time estimate assumes that even the most likely interruptions or delays, such as occasional employee illnesses, will not happen.
2. Estimate the maximum amount of time it would take to perform the task. We'll call this the pessimistic time (PT). The pessimistic time estimate assumes that anything that can go wrong will go wrong. All possible interruptions or delays—such as labor strikes, illnesses, training, inaccurate specification of requirements, equipment delivery delays, and underestimation of the system's complexity—are assumed to be inevitable.
3. Calculate the most likely time (MLT) that will be needed to perform the task. Don't just take the median of the optimistic and pessimistic times. Attempt to identify interruptions or delays that are likely to occur, such as occasional employee illnesses, inexperienced personnel, and occasional training. Calculate the expected duration (ED) as follows:

$$ED = \frac{OT + (4 \times MLT) + PT}{6}$$

This formula provides a weighted average of the various estimates. The formula is based on experience and may be modified to reflect project history in any firm.

Milestones are a special case. Because milestones signify accomplishments or events, they have no duration. In fact, specifying a task as having a duration of zero is how you tell Microsoft *Project* that the task is actually a milestone (automatically marked by a diamond).

Predecessors and Constraints If you have been following the above steps in Microsoft *Project*, you will have noticed that, so far, all the bars in the Gantt chart start on the same date. Obviously, this won't happen. The start of any given task may depend on the start or completion of another previous task. Additionally, the completion of a task frequently depends on the completion of a prior task.

In Microsoft *Project*, these relationships are expressed as predecessors and constraints. Each task has an ID number in the far left column of the Gantt chart. (Note that the ID number is not the same as the work breakdown structure outline number.) A predecessor column is provided to record those tasks that must be completed before the start of the task in question. A task may have zero, one, or more than one predecessor.

If a task is dependent on the start of another task, as opposed to its completion, you can open a task dialog box that allows you to specify this less common constraint. Other constraints such as dependence on specific dates can also be specified.

Milestones almost always have several predecessors that signify those tasks that must be completed before you can say that the milestone has been achieved.

Once again, a methodology template may include predefined predecessors and constraints (and other useful information such as descriptions and exceptions). Once predecessors have been recorded, Microsoft *Project* automatically schedules the task by shifting it to the right of the start date (or left of the deadline) based on prerequisite tasks and durations. Arrows depict the precedents between the bars.

Before we leave the subject of predecessors, we should acknowledge that there is a tool that naturally lends itself to modeling predecessors—the PERT chart. Because PERT charts will be covered later in the module, we'll defer that discussion until then.

Critical Path and Slack Resources Like most project management software packages, Microsoft *Project* automatically determines a critical path and marks those bars and arrows in a predetermined color (the default is red).

> The **critical path** is a sequence of dependent project tasks that have the largest sum of estimated durations.

It is the path that has no slack time built in. If any of these tasks fall behind schedule, the project's completion date will be delayed.

> The **slack time** available for any task is equal to the difference between the earliest and latest completion times.

Tasks that have slack time can get behind schedule by an amount less than or equal to that slack time without having any impact on the project's final completion date. Tasks with slack time are illustrated in blue.

Understanding the critical path and slack resources in a project are indispensable to the project manager. Knowledge of such project parameters influences the people management decisions to be made by the project manager. Emphasis can be placed on the critical path tasks, and if necessary, resources might be temporarily diverted from tasks with slack time to help get one or more critical tasks back on schedule.

Resource Assignments and Management Resources complete the tasks that you have included in a Gantt chart.

> **Resources** are people, material, and tools that you assign to the completion of a task.

In system development methodologies, most resources are expressed in terms of *roles* that must be assigned to individuals. Resources may be constrained by the following: resources available to the project manager; competition with other managers and projects for a resource's time; and calendars of resources (discussed earlier).

In Microsoft *Project*, you must specify how much of a resource is available to your project and any other pertinent constraints on the resource. (*Project* provides forms and tables to enter this data.) Then, based on this information, the schedule bars are adjusted to reflect the available resources and constraints. For example, if a task normally takes a full-time person one day to complete, but you can only assign a specific person half-time to that task, *Project* will allocate two days to complete the task. This approach is called **resource-driven scheduling.** A wealth of options for balancing resources, delaying resources, and overloading resources are provided.

Costs can be assigned to resources to assist in budgeting the project. And if actual time spent on tasks is also recorded, budgets can be compared to actual expenses.

Using Gantt Charts to Evaluate Progress One of the project manager's frequent responsibilities is to report project progress to superiors. Gantt charts frequently find their way into progress reports because they can conveniently compare the original schedule with actual performance. To report progress we must expand our Gantt charting conventions. If a task has been completed, completely shade in the bar corresponding to that task. If a task is partially completed, partially shade in the bar. The percentage of the bar that is shaded should correspond to the percentage of the task completed. Unshaded bars represent tasks that have not begun. Next, draw a bold vertical line that is perpendicular to the horizontal axis and that intersects the current date. You can now evaluate project progress.

As tasks are completed, in part or in whole, that progress can be added to the task information in Microsoft *Project*. Subsequently, *Project* can generate progress views of the Gantt chart that use color or shading to clearly highlight those tasks that have been completed, those tasks in progress (visually showing percentage complete), and those tasks not yet started. These progress-oriented charts can also use color to show those tasks that are on schedule, ahead of schedule, and behind schedule. Finally, color can be set to compare the original schedule against any revisions to that schedule. All of these views are useful to project managers who may need to determine the impact of missed deadlines or unexpected delays.

PERT Charts

PERT, which stands for *project evaluation and review technique,* was developed in the late 1950s to plan and control large weapons development projects for the U.S. Navy. It was developed to make clear the interdependence of project tasks when projects are being scheduled. Essentially, PERT is a graphic networking technique.

In Microsoft *Project* and other project management software packages, PERT charts represent another view of the project. Although this view does not communicate schedule as effectively as a calendar or Gantt chart, it does show intertask relationships much more effectively. Let's take a closer look at PERT charts—what they are, how to draw them, and how to use them.

PERT Definitions and Symbols Like Gantt charts, PERT chart projects are organized in terms of tasks and milestones. A variety of symbols—circles, squares, and the like—have been used to depict tasks and milestones on PERT charts. Microsoft *Project* uses rectangles (see Figure A.4).

Each rectangle is divided into sections. The task name is at the top. The task ID and the estimated duration are in the middle section of the rectangle. The start and end dates, which are automatically calculated by *Project,* are in the lower section of the rectangle.

Primitive tasks (only) are connected by arrows that indicate predecessors and successors. This information, if recorded on the Gantt chart, is carried over to the PERT chart.

The Critical Path in a PERT Network Some project managers like to use the PERT chart to record predecessors and constraints. For example, in *Project,* you can use the mouse to drag a line from a predecessor task to a successor task. This can be much faster than looking up task IDs and recording them in tables (as we suggested in the discussion of Gantt charts).

If durations have also been recorded and resources have been assigned, Microsoft *Project* schedules the tasks as it did for Gantt charts. In PERT charts, the red rectangles are critical tasks, and the arrows that connect them make up the project's critical path.

As a reminder, the **critical path** is a sequence of dependent project tasks that have the largest sum of estimated durations.

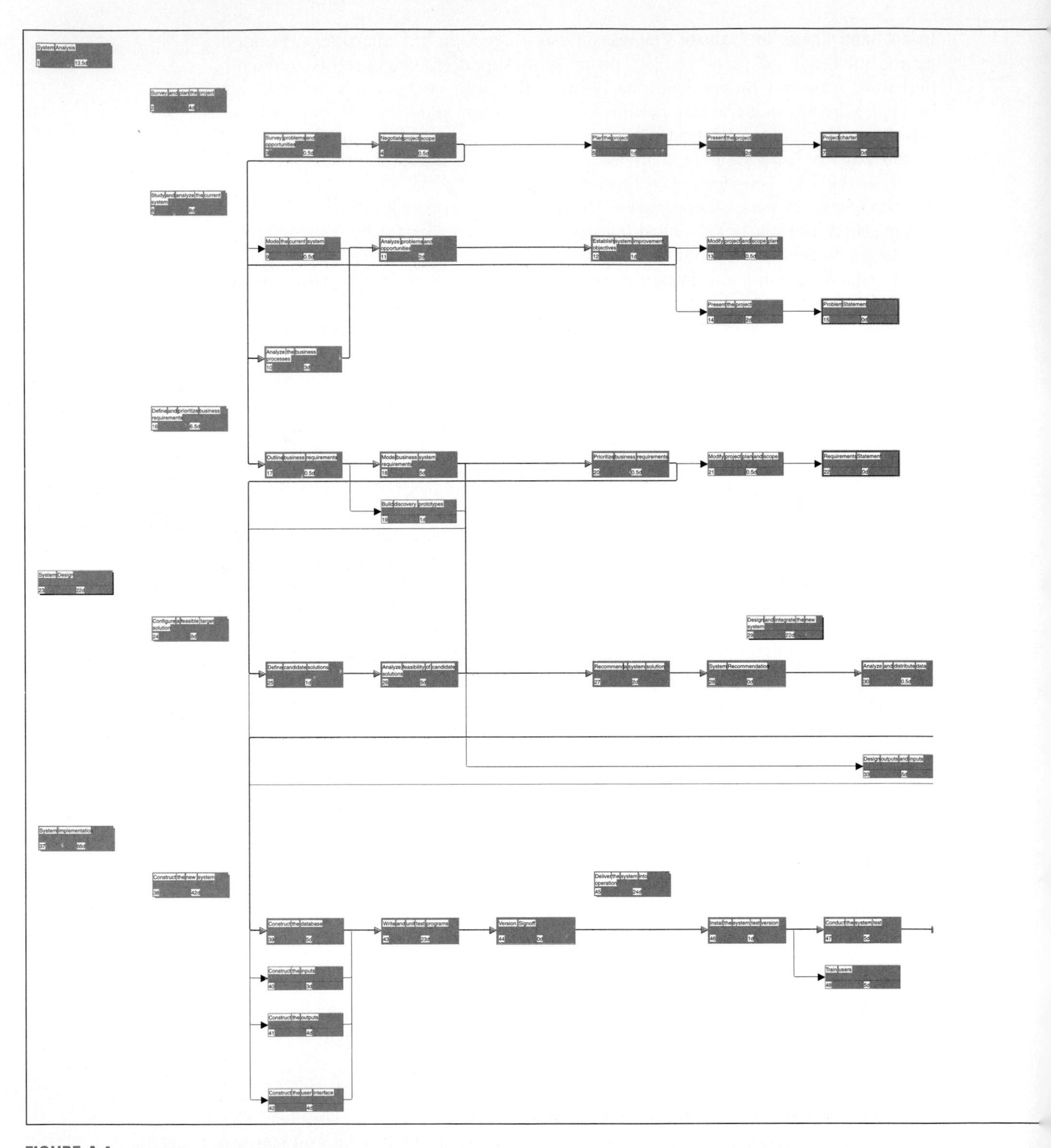

FIGURE A.4 *A PERT Chart*

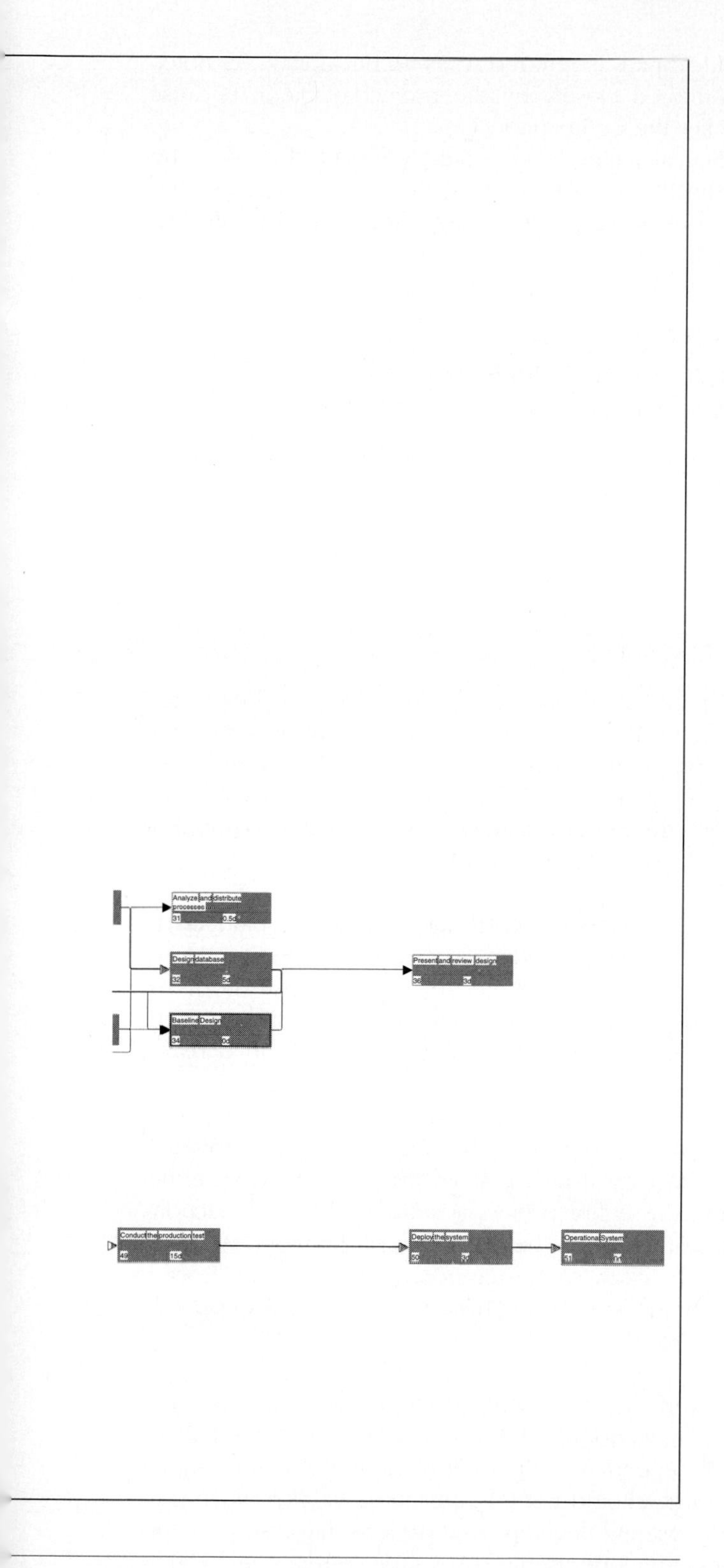
Analyze and distribute processes
31
0.5d
Design database
32
5d
Present and review design
36
3d
Baseline Design
34
0d
Conduct the production test
49
15d
Deploy the system
50
3d
Operational System
51

Each task appearing on the critical path is referred to as a critical task. As noted earlier, critical tasks must be monitored closely by the project manager because any delays in those tasks will delay the entire project.

Consider the following hypothetical example. A project consists of eight primitive tasks that have been given the IDs A, B, C, D, E, F, G, and H. Each task has an estimated duration recorded below in parentheses. There are four distinct sequences of tasks in a project. They are:

Path 1: A(3) → B(2) → C(2) → D(7) → H(5)
Path 2: A(3) → B(2) → C(2) → E(6) → H(5)
Path 3: A(3) → B(2) → C(2) → F(3) → H(5)
Path 4: A(3) → B(2) → C(2) → G(2) → H(5)

The total expected duration time for a path is equivalent to the sum of the expected duration times for each task in the path. For example,

Path 1: $3 + 2 + 2 + 7 + 0 + 5 = 19$
Path 2: $3 + 2 + 2 + 6 + 0 + 5 = 18$
Path 3: $3 + 2 + 2 + 3 + 0 + 5 = 15$
Path 4: $3 + 2 + 2 + 2 + 5 = 14$

You can now identify the critical path as the one having the largest total expected duration time. In our example, path 1 is the critical path. It indicates that the expected time for completing the programming project is 19 days. But what if task G in path 4 had an expected duration time of 7 days? We would then have two critical paths containing tasks that the project manager would have to monitor closely! Tasks that have slack capacity are shown in black.

Using PERT for Planning and Control Project managers find PERT charts particularly useful for communicating schedules of large systems projects to superiors. However, the primary uses and advantages of the PERT chart lie in its ability to assist in the planning and controlling of projects. In planning, the PERT chart aids in determining the estimated time required to complete a given project, in deriving actual project dates, and in allocating resources.

As a control tool, the PERT chart helps the manager identify current and potential problems. Particular attention should be paid to the critical path of a project. When a project manager identifies a critical task that is running behind schedule and that is in danger of upsetting the entire project schedule, alternative actions are examined. Corrective measures, such as the shuffling of human resources, might be taken. These resources are likely to be temporarily taken away from a noncritical task that is currently running smoothly. These noncritical tasks normally offer some slack time for the project.

PERT versus Gantt Charting PERT and Gantt charting are frequently presented as mutually exclusive project management tools. PERT is usually recommended for larger projects with high intertask dependency. Gantt is recommended for simpler projects. But PERT and Gantt charting should not be considered as alternative project management approaches. All systems development projects have some intertask dependency, and all projects offer opportunities for task overlapping. Therefore, PERT and Gantt charts can be used in a complementary manner to plan, schedule, evaluate, and control systems development projects.

Still, most information systems project managers seem to prefer Gantt charts because of their simplicity and ability to show the schedule of a project. Fortunately, project management software allows the best feature of PERT—the critical path analysis—to be incorporated into Gantt charts. As activities are entered, their duration and dependencies are entered. Gantt bars are scheduled to take into consideration the dependencies. Usually, the critical path is highlighted with boldfacing or color. Additionally, the amount of slack time in noncritical path activities is also highlighted. This can prove useful when deciding which activities to delay to get off-schedule activities back on track.

EXPECTATIONS MANAGEMENT

Experienced project managers often complain that managing management's expectations of a project is more difficult than managing cost, schedule, people, or quality. In this section we introduce a simple tool that we'll call an expectations management matrix that can help project managers deal with the problem. We first learned about this tool from Dr. Phil Friedlander, a consultant and trainer then with McDonnell Douglas in the Improved Systems Technologies business unit. He attributes the matrix to "folklore" but also credits Jerry Gordon of Majer LTD and Ron Leflour, a large project management educator/trainer. Dr. Friedlander's paper is listed in the suggested readings for this module. We have slightly adapted the tool for this presentation.

The Expectations Management Matrix

Every project has goals and constraints when it comes to cost, schedule, scope, and quality. In an ideal world, you could optimize each of these parameters. Management often has that expectation. Reality, however, suggests that you can't optimize them all—you must strike a balance that is both feasible and acceptable to management. That is the purpose of the expectations management matrix.

An **expectations management matrix** is a rule-driven tool for helping management appreciate the dynamics of changing project parameters. The parameters include cost, schedule, scope, and quality.

Who is management? In the case of systems development, management is defined as the system owner(s)—the individual(s) who sponsors and pays for a system to be developed or modified. To achieve true project success, it is extremely important that the project manager and system owner come to an understanding about assumptions and expectations of the project and that they review the implications of changing budgets, schedules, and system requirements on a regular basis. The matrix is a tool for doing just that.

The basic matrix, shown in Figure A.5, consists of three rows and three columns (plus headings). The rows correspond to the measures of success in any project: cost, schedule, and scope and/or quality. The columns correspond to priorities: first, second, and third. To establish expectations, we assign names to the priorities as follows:

- Maximize or minimize—The most important of the three measures in a given project.
- Constrain—The second most important of the three measures in a project.
- Accept—The least important of the three measures in a project.

Again, while most managers would ideally like to give equal priority to all three measures, experience suggests that the three measures tend to balance themselves

FIGURE A.5 *A Management Expectations Matrix*

Priorities / Measures of Success	Max or Min	Constrain	Accept
Cost			
Schedule			
Scope and/or Quality			

Priorities / Measures of Success	Max or Min	Constrain	Accept
Cost Estimated at $20 billion			X
Schedule Deadline = December 31, 1969		X	
Scope and/or Quality 1. Land a man on the moon. 2. Get him back safely.	X		

FIGURE A.6 *Managing Expectations in the Lunar Landing Initiative*

naturally. For example, if you increase scope or quality requirements it will take more time and/or money. On the contrary, if you try to get any job done faster, you generally have to reduce scope or quality requirements or pay more money to compensate. The management expectations matrix helps (forces) management to understand this through three simple rules:

- For any project, you must record three Xs within the nine available cells.
- No row may contain more than one X. In other words, a single measure of success must have one and only one priority.
- No column may contain more than one X. In other words, there must be a first, second, and third priority.

Let's illustrate the tool using Dr. Friedlander's own example. In 1961 President John F. Kennedy established a major project—before the end of the decade, land a man on the moon and return him. Figure A.6 shows the realistic expectations of the project. Let's walk through the example. John Kennedy (and the public who elected him) is the system owner.

1. The system owner had both scope and quality expectations. The scope (or requirement) was to successfully land a man on the moon. The quality measure was to return the man (or men) safely. Because the public would expect no less from the new space program, this had to be made the first priority. In other words, we had to maximize safety and minimize risk as a first priority. Hence, we record the X in column 1, row 3.
2. At the time of the project's inception, the Soviet Union was ahead in the race to space. This was a matter of national pride; therefore, the second priority was to get the job done by the end of the decade. We call this the project constraint—there is no need to rush the deadline, but we don't want to miss the deadline. Thus, we record the second X in column 2, row 2.
3. By default, the third priority had to be cost (estimated at $20 billion in 1961). By making cost the third priority, we are not stating that cost will not be controlled. We are merely stating that we may have to "accept" cost overruns to achieve the scope and quality requirement by the constrained deadline.

History records that we achieved the scope and quality requirement, and did so in 1969. The project actually cost well in excess of $30 billion, a 50 percent cost overrun. Did that make the project a failure? On the contrary, most people perceived the project a grand success. The government managed the public's expectations of the project in realizing that maximum safety and minimum risk,

Priorities / Measures of Success	Max or Min	Constrain	Accept
Cost		X	
Schedule			X
Scope and/or Quality	X		

FIGURE A.7 *An Initial Expectations Matrix*

plus meeting the deadline (beating the Soviets!), was an acceptable trade-off for the cost overrun! The government brilliantly managed public opinion. Systems development project managers can learn a valuable lesson from this balancing act.

At the beginning of any project, the project manager should consider introducing the system owner to the matrix concept and should work with the system owner to complete the matrix. For most projects, it would be difficult to record all the scope and quality requirements in the matrix. Instead, they would be listed on an attached document. The matrix heading for scope and/or quality would be stated as "See Attachment A." The estimated costs and deadlines could be recorded directly on the matrix.

The project manager must never establish the priorities or even suggest those priorities. The project manager merely enforces the rules: 3 Xs and no more than one X in any row or column. This sounds easy. It rarely is. Many managers are unwilling to be pinned down on the priorities—"Shouldn't we be able to maximize everything?" These managers need to be educated about the reason for the priorities. Of course we always try to maximize all three measures of success because it makes us (the project managers) look that much better. But we need to know priorities in the event that we cannot maximize all three measures. This helps us make intelligent compromises instead of merely guessing right or wrong.

What if your system owner refuses to prioritize? The tool is less useful then, except as a mechanism for documenting your concerns before they become disasters. A system owner who refuses to set priorities is a manager who may be setting the project manager up for a no-win performance review. And as Dr. Friedlander points out, "Those who do not 'believe' the principles [of the matrix] will eventually 'know' the truth. You do not have to believe in gravity, but you will hit the ground just as hard as the person who does."

Using the Expectations Management Matrix

Let's assume you have a management expectations matrix that conforms to the aforementioned rules. How does this help you manage expectations?

During the course of the average systems development project, priorities are not stable. Various factors such as the economy, government, and politics can change the priorities. Budgets may become more or less constrained. Deadlines may become more or less important. Quality can become more (rarely less) important. And, most frequently, requirements increase (rarely decrease). As already noted, these changing factors affect all the measures in some way. The trick is to manage expectations despite changing project parameters.

The technique is relatively straightforward. Whenever the max/min measure or the constrain measure begins to slip, you have a potential expectations management problem. For example, suppose you are faced with the following common priorities (see Figure A.7):

Priorities / Measures of Success	Max or Min	Constrain	Accept
Cost (Record New Budget)		X+ Increase budget	
Schedule (Record New Deadlines)			X− Extend deadline
Scope and/or Quality (Revise attachment that describes scope and quality. Be sure to date all new requirements to distinguish them from original requirements.)	X+ Accept expanded requirements		

FIGURE A.8 *Adjusting Expectations*

1. Explicit requirements and quality expectations were established at the start of a project and given the highest priority.
2. An absolute maximum budget was established for the project.
3. You agreed to shoot for the desired deadline, but the system owner(s) accepted the reality that if something must slip, it should be schedule.

Now suppose that during systems analysis, significant and unanticipated business problems arise. The analysis of these problems has placed the project somewhat behind schedule. Furthermore, solving the new business problems substantially expands the user requirements for the system. How do you, as project manager, react? First, don't overreact to the schedule slippage—schedule slippage was the "accept" priority in the matrix. The scope increase (in the form of several new requirements) is the more significant problem—not because implementing new requirements might further delay the schedule! No, the real problem is that the added requirements will increase the cost of the project. Cost is the constrained measure of success. As it stands, we have an expectations problem. It is time to review the matrix with the system owner.

First, the system owner needs to be made aware of which measure or measures are in jeopardy and why. Then together, the project manager and system owner can discuss courses of action. Several are possible:

- The resources (cost and/or schedule) can be reallocated. Perhaps the system owner can find more money somewhere. All priorities would remain the same (noting, of course, the revised deadline based on schedule slippages already encountered during systems analysis).
- The budget might be increased, but it would be offset by additional planned schedule slippages. For instance, by extending the project into a new fiscal year, additional money might be allocated without taking any money from existing projects or uses. This solution is shown in Figure A.8.
- The user requirements (or quality) might be reduced through prioritizing those requirements and deferring some number of those requirements until Version 2 of the system. This alternative would be appropriate if the budget cannot be increased.
- Finally, measurement priorities can be changed. Dr. Friedlander calls this priority migration.

Only the system owner may initiate priority migration. For example, the system owner may agree that the expanded requirements are worth the additional cost. He or she allocates sufficient funds to cover the requirements but migrates

Priorities / Measures of Success	Max or Min	Constrain	Accept
Cost	← Step	1 X	
Schedule			X
Scope and/or Quality	X Step	2 →	

FIGURE A.9 *Changing Priorities*

priorities such that minimizing cost becomes the highest priority (see Figure A.9, step 1). But now the matrix violates a rule—there are two Xs in column 1. To compensate, we must migrate the scope and/or quality criteria to another column, in this case, the constrain column (see Figure A.9, step 2). Expectations have been adjusted. In effect, the system owner is freezing growth of requirements and still accepting schedule slippage.

There are three final comments about priority migration. First, priorities may migrate more than once during a project. Expectations can be managed through any number of changes as long as the matrix is balanced (meaning it conforms to our rules). Second, expectations management can be achieved through any combination of priority migrations and resource adjustments. Finally, it should be noted that system owners can initiate priority migration even if the project is on schedule. For example, government regulation might force an uncompromising deadline on an existing project. That would suddenly migrate our "accept" schedule slippages to "max" schedule. The other Xs would have to be migrated to rebalance the matrix—such as "accepting" any cost and "constraining" scope.

The expectations management matrix is a simple tool, but sometimes simple tools are the most effective!

PEOPLE MANAGEMENT

The management or supervision of project team members is equally important to planning and controlling the project schedule, budget, and expectations. The topic could easily require an entire module on its own. In the interest of space, we will direct you to a couple of the shortest and most valuable books ever written on the subject. You could easily read both books overnight!

The One Minute Manager

The One Minute Manager by Kenneth Blanchard and Spencer Johnson is a classic, fun, and indispensable aid to anyone managing people for the first time. In just over 100 pages, the authors share the simple secrets of managing people and achieving success through the actions of your subordinates. The book highlights three basic secrets: one-minute goal setting, one-minute praisings, and one-minute reprimands. This book should be a part of every college graduate's personal library!

The Subtle Art of Delegation and Accountability

Most young and many experienced managers have difficulty delegating responsibilities. Worse still, they let subordinates reverse-delegate tasks back to the manager. This leads to poor time management and manager frustration. In *The One Minute Manager Meets the Monkey,* Kenneth Blanchard teams with William Oncken and Hal Burrows to help managers overcome this problem.

The solution is based on Oncken's classic principle of "the care and feeding of monkeys." *Monkeys* refers to problems that managers delegate to their subordinates who, in turn, attempt to reverse-delegate back to the manager. In this 125-

page book the authors teach managers how to keep the monkeys on the subordinates' backs. Doing so increases the manager's available work time, accelerates task accomplishment by subordinates, and teaches subordinates how to solve their own problems.

SUMMARY

1. A project is a sequence of unique, complex, and connected activities having one goal or purpose and that must be completed by a specific time, within budget, and according to specification.
2. Project management is the process of defining, planning, directing, monitoring, and controlling the development of an acceptable system at a minimum cost within a specified time frame.
3. Projects can be mismanaged through shortcuts in the development process, poor management of expectations, premature estimates, poor estimates, cost overruns, schedule delays, poor resource management, people mismanagement, and business change.
4. The project manager is responsible for (1) scoping the project, (2) planning project tasks and assigning staff, (3) organizing and scheduling the project effort, and (4) directing and controlling the project.
5. Process management is the planning, selection, deployment, and consistent application of standard system development methods, tools, techniques, and technologies to all information systems.
6. A development center is a central group of information system development consultants and managers who plan, implement, and support all aspects of process management including methodology, technology, quality, and measurement.
7. Gantt charts are simple horizontal bar charts that depict project tasks against a calendar.
8. Forward scheduling establishes a project start date and schedules forward from that date. Reverse scheduling establishes a project deadline and then schedules backward from that date.
9. A work breakdown structure is a hierarchical decomposition of the project into phases, activities, tasks, and milestones. Milestones are events that signify major accomplishments or events during a project.
10. A critical path is a sequence of dependent project tasks that have the largest sum of estimated durations. The slack time available for any task is equal to the difference between the earliest and latest completion times.
11. Resources are people, material, and tools that you assign to the completion of a task.
12. An expectations management matrix is a rule-driven tool for helping management appreciate the dynamics of changing project parameters.

KEY TERMS

calendar, p. 606
critical path, pp. 608, 609
development center, p. 602
expectations management matrix, p. 613
feature creep, p. 596
forward scheduling, p. 606
Gantt chart, p. 602
milestone, p. 607
PERT, p. 609
primitive task, p. 606
process management, p. 599
project, p. 595
project management, p. 595
resources, p. 608
resource-driven scheduling, p. 608
reverse scheduling, p. 606
scope creep, p. 596
slack time, p. 608
summary task, p. 606
work breakdown structure, p. 606

REVIEW QUESTIONS

1. What are the characteristics that define a project?
2. What is project management?
3. What is process management?
4. Compare and contrast Gantt and PERT charts.
5. What is the critical path through a project? What is the slack resource in a critical task?
6. Differentiate between forward and reverse scheduling.
7. What is a work breakdown structure? What purpose does it serve?
8. Differentiate between summary and primitive tasks.
9. What is a resource?
10. Why don't milestones have duration?
11. What is resource-driven scheduling?
12. What is a management expectations matrix?

PROBLEMS AND EXERCISES

1. Using a project (school or work) that you have mismanaged (It happens to all of us!), what are some causes of mismanaged projects that result in missed requirements and needs, cost overruns, and late delivery? Explain how these problems that result from mismanaged projects are related.
2. Give some examples that differentiate between scope and feature creep.
3. Write a job advertisement for a professional project manager. Prepare a set of interview questions for the applicants.
4. Systems analysts tend to assign additional people to a project that is running behind schedule. What are some potential problems with such an action?
5. For a programming course or other IT project in a past or current course, write a complete project definition.
6. Differentiate between project and process management.
7. Your financial aid office must respond annually to changing regulations. Financial aid applications must be sent out annually by February 15 in order to allow for sufficient processing time for applications. Therefore, each year the information systems must be ready to process applications by March 1. What type of project scheduling strategy must be used and why?
8. For a project that will last this entire academic term (at your school), define basic project calendars for faculty and staff.
9. Prepare a work breakdown structure for freshmen in a computer programming course.
10. Using system development, give examples of summary and primitive tasks.
11. How can expectations mismanagement lead to perceived project failure?
12. Why shouldn't estimated project time requirements be stated in terms of person-days?
13. Calculate the expected duration for the following tasks:

Task ID	Optimistic Time	Pessimistic Time	Most Likely Time
A	3	6	4
B	1	3	2
C	4	7	6
D	2	5	3
E	3	9	6
F	3	3	4

14. Derive the earliest completion time and latest completion time for each of the following:

Task ID	Event ID	Predecessors	Successors	Duration
A	2	1	3	2
B	3	2	4	3
C	4	3	5,6	4
D	4	4	7	5
E	6	4	7	4
F	7	5	8,9	3
G	7	6	8,9	0
H	8	7	10	6
I	9	7	10	5
J	10	8	none	0
K	10	9	none	6

15. Draw the PERT chart described in problem 13. Be sure to include sequencing and identification for all tasks and events along with their time estimates. What is the critical path? What is the total expected duration time represented by the critical path?
16. Make a list of the tasks that you performed on your last programming assignment. Alternatively, list the tasks required to complete your next programming assignment. Develop a PERT chart to depict the tasks and events and the dependency of tasks on one another. What is the critical path? How can the PERT chart aid in planning and scheduling the programming assignment?
17. Derive a Gantt chart to graphically depict the project schedule and overlapping of tasks for the programming assignment chosen in problem 16. How can the Gantt chart be used to evaluate the progress that is being or has been made?
18. Draw a PERT chart for the curriculum in which you are enrolled. Be sure to consider the prerequisites for all courses.
19. Draw a Gantt chart for your plan of study to get your degree. Annotate the graph to indicate your progress toward your degree or job objectives.
20. At the beginning of a project, management decided that highest priority should be placed on meeting an absolute deadline for a project. Second highest priority is to be placed on living within the allocated project budget. Draw the expectations management matrix for this project.
21. During the project initiated in problem 20, things started going wrong. Creeping requirements set in. Both the deadline and budget are in jeopardy. Using the expectations management matrix, identify alternatives for adjusting the project. Who should make the decision?

PROJECTS AND RESEARCH

1. Research project management software in your local library. Present a report to management that identifies important selection criteria for selecting a project management software package.
2. Complete the tutorial for Microsoft *Project* (or its equivalent). Evaluate the package and analyze its strengths and weaknesses.
3. Make an appointment to visit an information systems project manager (or systems analyst with project management experience). What techniques does he or she use to plan and control projects? Why? Is project management software used? If so, what does the project manager like and dislike about that software?
4. If your systems course requires you to complete a real or simulated development project, prepare a management expectations matrix with your client (or your instructor acting as your client). Over the course of the semester, review the matrix with your client (or instructor) and make appropriate adjustments. At the end of the project, be prepared to defend your management of expectations as well as your progress. (Note: Your instructor may assign a subjective grade to your management of expectations. That may not seem fair; however, it is realistic. Project success is as much perceived as it is real.)

MINICASE

1. Fun & Games, Inc., is a successful developer and manufacturer of board, electronic, and computer games. The company is headquartered in Cleveland, Ohio. Jan Lampert, applications development manager, has requested a meeting with Steven Beltman, systems analyst and project manager for a new distribution project recently placed into production.

"Steven, I want to discuss the distribution project your team completed last month. Now that the system has been operational for a few weeks, we need to evaluate the performance of you and your team. Frankly, Steven, I'm a little disappointed."

"Me too! I don't know what happened! We used the standard methodology and tools, but we still had problems."

"You still have some, Steven. The production system isn't exactly getting rave reviews from either users or managers."

Steven replies, " I know."

Jan continues, "Well, I've talked to several of the analysts, programmers, and end-users on the project, and I've drawn a few conclusions. Obviously, the end-users are less than satisfied with the system. You took some shortcuts in the methodology, didn't you?"

"We had to, Jan! We got behind schedule. We didn't have time to follow the methodology to the letter."

Jan explains, "But now we have to do major parts of the system over. If you didn't have time to do it right, where will you find time to do it over? You see, Steven, systems development is more than tools, techniques, and methodologies. It's also a management process. In addition to your missing the boat on end-user requirements, I note two other problems. And both of them are management problems. The system was over budget and late. The projected budget of $35,000 was exceeded by 42 percent. The project was delivered 13 weeks behind schedule. Most of the delays and cost overruns occurred during programming. The programmers tell me that the delays were caused by rework of analysis and design specifications. Is this true?"

Steven answers, "Yes, for the most part."

Jan continues, "Once again, those delays were probably caused by the shortcuts taken earlier. The shortcuts you took during analysis and design were intended to get you back on schedule. Instead, they got you further behind schedule when you got into the programming phase."

"Not all the problems were due to shortcuts," says Steven. "The users' expectations of the system changed over the course of the project."

"What do you mean?" asks Jan.

Steven answers, "The initial list of general requirements was one page long. Many of those requirements were expanded and supplemented by the users during the analysis and design phases."

Jan interrupts, "The old 'creeping requirements syndrome.' How did you manage that problem?"

Steven replies, "Manage it? Aren't we supposed to simply give in? If they want it, you give it to them."

"Yes," answers Jan, "but were the implications of the creeping requirements discussed with the project's management sponsor?"

Steven answers, "Not really! I don't recall any schedule or budget adjustments. We should explain that to them now."

"An excuse?" inquires Jan.

Steven replies, "I guess that's not such a good idea. But the project grew. How would you have dealt with the schedule slippage during analysis?"

Jan answers, "If I were you, I would have reevaluated the scope of the project when I first saw it changing. In this case, either project scope should have been reduced or project resources—schedule and budget—should have been increased. *[pause]* Don't be so glum! We all make mistakes. I had this very conversation with my boss seven years ago. You're going to be a good project manager. That's why I've decided to send you to this project management course and workshop."

a. What did Steven do wrong? How would you have done it differently?
b. Should Jan share any fault for the problems encountered in this project?
c. Why would it be a mistake to use creeping requirements as an excuse for the project mismanagement?

SUGGESTED READINGS

Blanchard, Kenneth, and Spencer Johnson. *The One Minute Manager.* New York: Berkley Publishing Group, 1981, 1982. Arguably, this is one of the best people management books ever written. Available in most bookstores, it can be read overnight and used for discussion material for the lighter side of project management (or any kind of management). This is must reading for all college students with management aspirations.

Blanchard, Kenneth; William Oncken, Jr.; and Hal Burrows. *The One Minute Manager Meets the Monkey.* New York: Simon & Schuster, 1988. A sequel to *The One Minute Manager,* this book effectively looks at the topic of delegation and time management. The monkey refers to Oncken's classic article, "Managing Management Time: Who's Got the Monkey?" as printed in the *Harvard Business Review* in 1974. The book teaches managers how to achieve results by helping their staff (their monkeys) solve their own problems.

Brooks, Fred. *The Mythical Man-Month.* Reading, MA: Addison-Wesley, 1975. A classic set of essays on software engineering, also known as systems analysis, design, and implementation. Emphasis is on managing complex projects.

Bucki, Lisa A. *Managing with Microsoft Project.* Rocklin, CA: Prima Publishing. This professional market book provides a good tutorial for Microsoft *Project.*

Friedlander, Phillip. "Ensuring Software Project Success with Project Buyers." *Software Engineering Tools, Techniques, and Practices* 2, no. 6 (March/April 1992), pp. 26–29. We adapted our expectations management matrix from Dr. Friedlander's work.

Gildersleeve, Thomas. *Data Processing Project Management.* New York: Van Nostrand Reinhold, 1974. This book offers no explicit PERT or Gantt coverage, but it does provide excellent coverage of the people side of project management. A classic, hypothetical series of memos that documents a failed project precedes the topical coverage.

Kernzer, Harold. *Project Management: A Systems Approach to Planning, Scheduling, and Controlling.* 4th ed. New York: Van Nostrand Reinhold, 1989. Many experts consider this book to be the definitive work in the field of project management. Dr. Kernzer's seminars and courses on the subject are renowned.

London, Keith. *The People Side of Systems.* New York: McGraw-Hill, 1976. Chapter 8, "Handling a Project Team," does an excellent job of teaching the people and leadership aspects of project management.

Wiest, Jerome D., and Ferdinand K. Levy. *A Management Guide to PERT-CPM: With PERT-PDM, DCPM, and Other Networks.* 2nd ed. Englewood Cliffs, NJ: Prentice Hall, 1977. A good source for more on PERT/CPM and other project planning and control networks.

Wysocki, Robert K.; Robert Beck, Jr.; and David B. Crane, *Effective Project Management: How to Plan, Manage, and Deliver Projects on Time and within Budget.* New York: John Wiley & Sons, 1995. Buy this book! This is our new benchmark for introducing project management. It is easy to read and worth its weight in gold. We were surprised how compatible the book is with past editions of our book, and our project management directions continue to be influenced by this work.

B

FACT-FINDING AND INFORMATION GATHERING

MODULE PREVIEW AND OBJECTIVES

Effective fact-finding techniques are crucial to the development of systems projects. Fact-finding is performed during all phases of the systems development life cycle. To support systems development, the analyst must collect facts about DATA, PROCESSES, INTERFACES, and GEOGRAPHY. This module introduces seven popular fact-finding techniques and suggests a strategy for conducting fact-finding efforts and the ethics involved. You will know that you understand the process of fact-finding when you can:

- Identify the seven fact-finding techniques and characterize the advantages and disadvantages of each.
- Identify the types of facts a systems analyst must collect.
- Develop a questionnaire and interview agenda.
- Describe a fact-finding strategy that will make the most of your time with end-users.
- Describe the role of ethics in the process of fact-finding.

Note: The authors gratefully acknowledge the contributions of Professor Kevin C. Dittman to the development of this module.

WHAT IS FACT-FINDING?

Applying the tools and techniques for systems development in the classroom is easy. Applying those same tools and techniques in the real world may not work—if they are not complemented by effective methods for fact-finding.

> **Fact-finding** is the formal process of using research, interviews, questionnaires, sampling, and other techniques to collect information about systems, requirements, and preferences. It is also called information gathering or data collection.

Tools, such as data and process models, document facts, and conclusions are drawn from facts. If you can't collect the facts, you can't use the tools. Fact-finding skills must be learned and practiced.

Systems analysts need an organized method of collecting facts. They especially need to develop a detective mentality to be able to discern relevant facts! This module presents popular alternative fact-finding techniques. Although an entire textbook could be devoted to fact-finding techniques and strategies, no introductory systems course would be complete without the survey provided in this module.

Before we leap headfirst into specific fact-finding techniques, let's make sure we understand what we are trying to accomplish. The tools of systems analysis and design are used to document facts about an existing or proposed information system. These facts are in the domain of the business application and its end-users. Therefore, the analyst must collect those facts in order to effectively apply the documentation tools and techniques. When might the analyst use fact-finding techniques? What kinds of facts should be collected? And how are facts collected?

WHAT FACTS DOES THE SYSTEMS ANALYST NEED TO COLLECT AND WHEN?

There are many occasions for fact-finding during the systems development life cycle. However, fact-finding is most crucial to the systems planning and systems analysis phases. It is during these phases that the analyst learns about the vocabulary, problems, opportunities, constraints, requirements, and priorities of a business and a system. Fact-finding is also used during the systems design and support phases, but to a lesser extent. During systems design, fact-finding becomes technical as the analyst attempts to learn more about the technology selected for the new system. During the systems support phase, fact-finding is important in determining that a system has decayed to a point where the system needs to be redeveloped.

Fortunately, we have a framework to help us determine what facts need to be collected, no matter what project we are working on. Throughout the systems development process, we are looking at an existing or target information system. In Chapter 2 we saw that any information system can be examined in terms of four building blocks: DATA, PROCESSES, INTERFACES, and GEOGRAPHY. Those four building blocks were depicted using our matrix model. As it turns out, the facts that describe any information system also correspond nicely with the building blocks of that matrix model.

WHAT FACT-FINDING METHODS ARE AVAILABLE?

Now that we have a framework for our fact-finding activities, we can introduce seven common fact-finding techniques: (1) Sampling of existing documentation, forms, and databases; (2) Research and site visits; (3) Observation of the work environment; (4) Questionnaires; (5) Interviews; (6) Rapid application development (RAD); and (7) Joint application development (JAD).

An understanding of each of these techniques is essential to your success. An analyst usually applies several of these techniques during a single systems project. To select the most suitable technique for use in any given situation, you will have to learn the advantages and disadvantages of each. Techniques 1 through 6 are presented in this module. Technique 7, joint application development, receives extensive coverage in Module D.

SAMPLING OF EXISTING DOCUMENTATION, FORMS, AND FILES

Particularly when you are studying an existing system, you can develop a good feel for the system by studying existing documentation, forms, and files. A good analyst always gets facts first from existing documentation rather than from people.

Collecting Facts from Existing Documentation

The first document the analyst should seek out is the organizational chart. Next, the analyst may want to trace the history that led to the project. To accomplish this, the analyst may want to collect and review documents that describe the problem. These include:

- Interoffice memoranda, studies, minutes, suggestion box notes, customer complaints, and reports that document the problem area.
- Accounting records, performance reviews, work measurement reviews, and other scheduled operating reports.
- Information systems project requests—past and present.

In addition to documents that describe the problem, there are usually documents that describe the business function being studied or designed. These documents may include:

- The company's mission statement and strategic plan.
- Formal objectives for the organization subunits being studied.
- Policy manuals that may place constraints on any proposed system.
- Standard operating procedures (SOPs), job outlines, or task instructions for specific day-to-day operations.
- Completed forms that represent actual transactions at various points in the processing cycle.
- Samples of manual and computerized databases.
- Samples of manual and computerized screens and reports.

Also, don't forget to check for documentation of previous system studies and designs performed by systems analysts and consultants. This documentation may include:

- Various types of flowcharts and diagrams.
- Project dictionaries or repositories.
- Design documentation, such as inputs, outputs, and databases.
- Program documentation.
- Computer operations manuals and training manuals.

All documentation collected should be analyzed to determine currency of the information. Don't discard outdated documentation. Just keep in mind that additional fact-finding will be needed to verify or update the facts collected. As you review existing documents, take notes, draw pictures, and use systems analysis and design tools to model what you are learning or proposing for the system.

Document and File Sampling Techniques

Because it would be impractical to study every occurrence of every form, analysts normally use sampling techniques to get a large enough cross section to determine what can happen in the system.

Sampling is the process of collecting sample documents, forms, and records.

Experienced analysts avoid the pitfalls of sampling blank forms—they tell little about how the form is used, not used, or misused. When studying documents or records from a database table, you should study enough samples to identify all the possible processing conditions and exceptions. You use statistical sampling techniques to determine if the sample size is large enough to be representative.

How to Determine the Sample Size The size of the sample depends on how representative you want the sample to be. There are many sampling issues and factors, which is a good reason to take an introductory statistics course. One simple and reliable formula for determining sample size is

$$\text{Sample size} = 0.25 \times (\text{Certainty factor}/\text{Acceptable error})^2$$

The certainty factor depends on how certain you want to be that the data sampled will not include variations not in the sample. The certainty factor is calculated from tables (available in many industrial engineering texts). A partial example is given here.

Desired Certainty	Certainty Factor
95%	1.960
90	1.645
80	1.281

Suppose you want 90-percent certainty that a sample of invoices will contain no unsampled variations. Your sample size, SS, is calculated as follows:

$$SS = 0.25(1.645/0.10)^2 = 68$$

We need to sample 68 invoices to get the desired accuracy.

Now suppose we know from experience that 1 in every 10 invoices varies from the norm. Based on this knowledge we can alter the above formula by replacing the heuristic .25 with $p(1-p)$.

$$SS = p(1-p)(1.645/0.10)^2$$, where p is the proportion of invoices with variances

By using this formula, we can reduce the number of samples required to get the desired accuracy.

$$SS = .10(1-.10)(1.645/0.10)^2 = 25$$

Selecting the Sample How do we choose our 25 invoices? Two commonly used sampling techniques are randomization and stratification.

Randomization is a sampling technique characterized as having no predetermined pattern or plan for selecting sample data.

Therefore, we just randomly choose 25 invoices.

Stratification is a systematic sampling technique that attempts to reduce the variance of the estimates by spreading out the sampling—for example, choosing documents or records by formula—and by avoiding very high or low estimates.

For computerized files, stratification sampling can be executed by writing a sample program. For instance, suppose our invoices were on a computer file that had a volume of approximately 250,000 invoices. Recall that our sample size needs to include 25 invoices. We will simply write a program that prints every 10,000th record (= 250,000/25). For manual files and documents, we could execute a similar scheme.

We could also use a spreadsheet package such as Microsoft *Excel* to assist us in selecting a random sample. Suppose we had a volume of 1,000 invoices last week and they were sequentially numbered from 2,000 to 2,999. We can create a spreadsheet with 1,000 rows with column A containing the number of each invoice. Using the data analysis sampling tool we can let *Excel* select which invoices we should randomly sample as shown in Figure B.1.

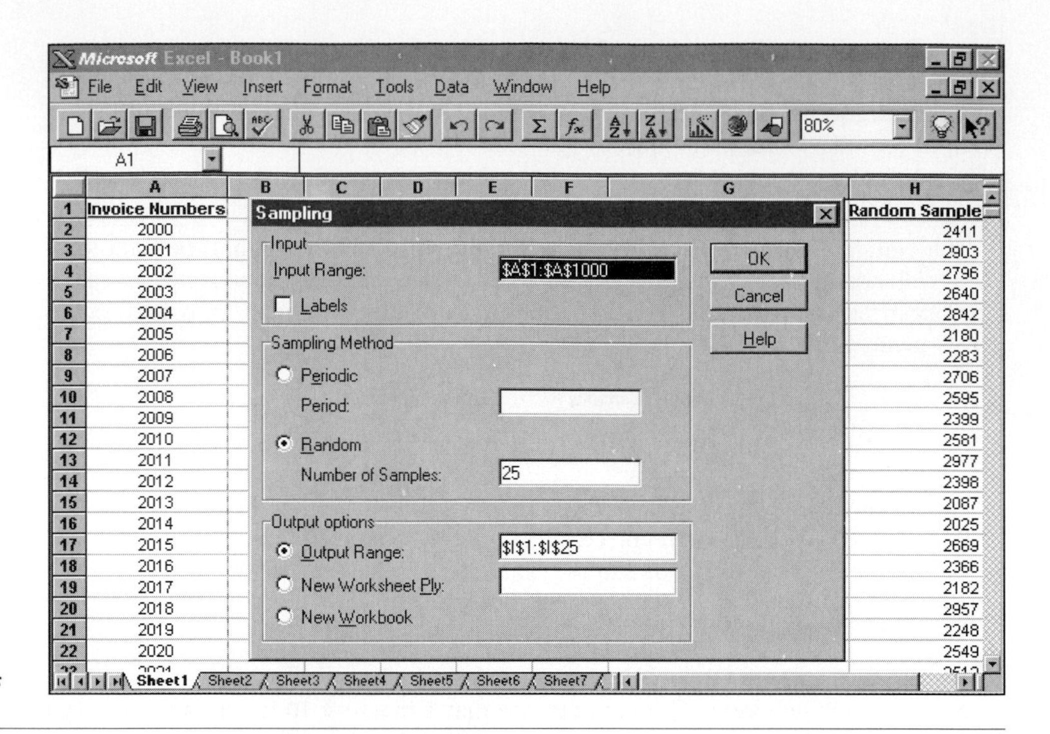

FIGURE B.1 *Microsoft* Excel*'s Data Analysis Sampling Tool*

RESEARCH AND SITE VISITS

A second fact-finding technique is to thoroughly research the application and problem. Computer trade journals and reference books are good sources of information. They can provide you with information on how others have solved similar problems, plus you can learn whether or not software packages exist to solve your problem. And now with recent advances in cyberspace, you don't even have to leave your desk to do it. Exploring the Internet and World Wide Web (WWW) via your personal computer can provide you with immeasurable amounts of information.

Internet is a global network of networks. It was conceived in 1964 by the U.S. Department of Defense to create a national military communications network that would be impervious to attacks. This network concept has exploded to include or link networks from all over the world and is used by all types of organizations and private citizens.

World Wide Web (WWW) was proposed in 1989 by a group of European physics researchers as a means for communicating research and ideas throughout the organization. It now has evolved to become the primary navigational, information management, and information distribution system, which permits users to easily travel the Internet. It provides the capability to transmit different types of information including sound, video, still images, and text.

All kinds of organizations now use the WWW to distribute information, advertise their services, and even deliver their product. Figure B.2 shows the home page of Ziff-Davis Publications, which publishes several popular computer periodicals. Here Ziff-Davis provides links to articles within the periodicals, which can be read on-line, printed, or even downloaded and saved.

Corporations have also recognized the power and benefit of the Internet and now use it as an effective means of communicating with their employees. These corporate networks, called **intranets**, function and provide the same assets as the WWW but can restrict access from anyone outside the corporation. Figure B.3 shows a Web page from Lockheed Martin Corporation. This page provides links to the company's policy and procedures, position on ethics, and organizational structure.

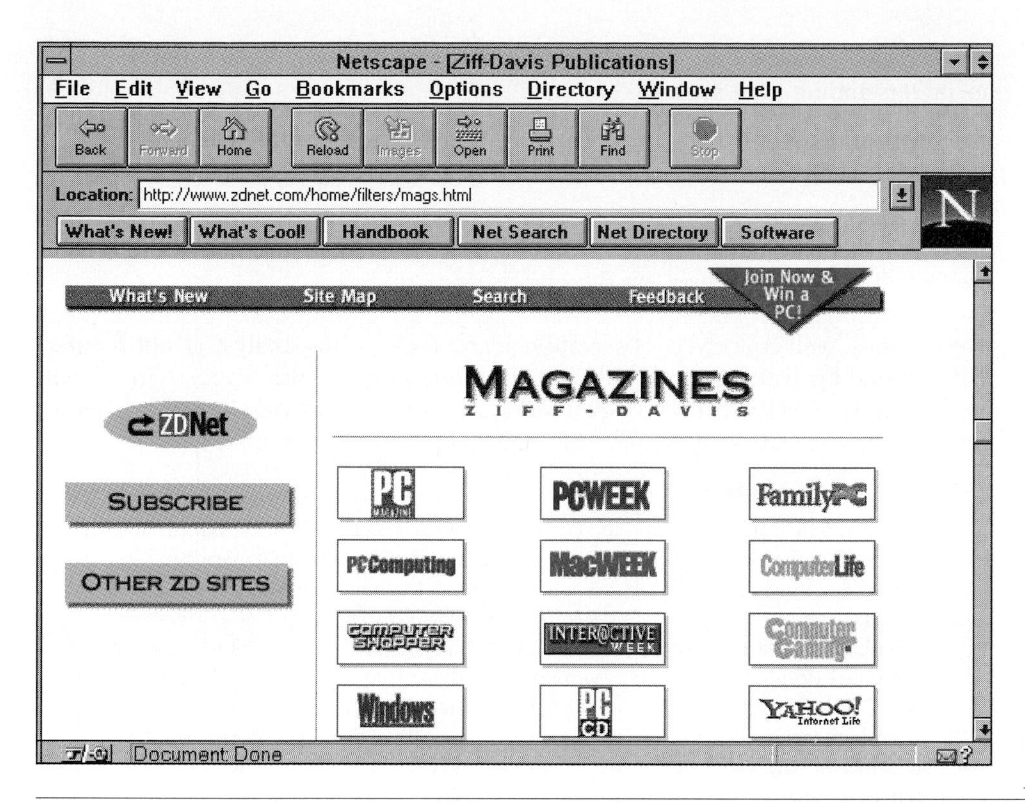

FIGURE B.2 *Ziff-Davis WWW Home Page*

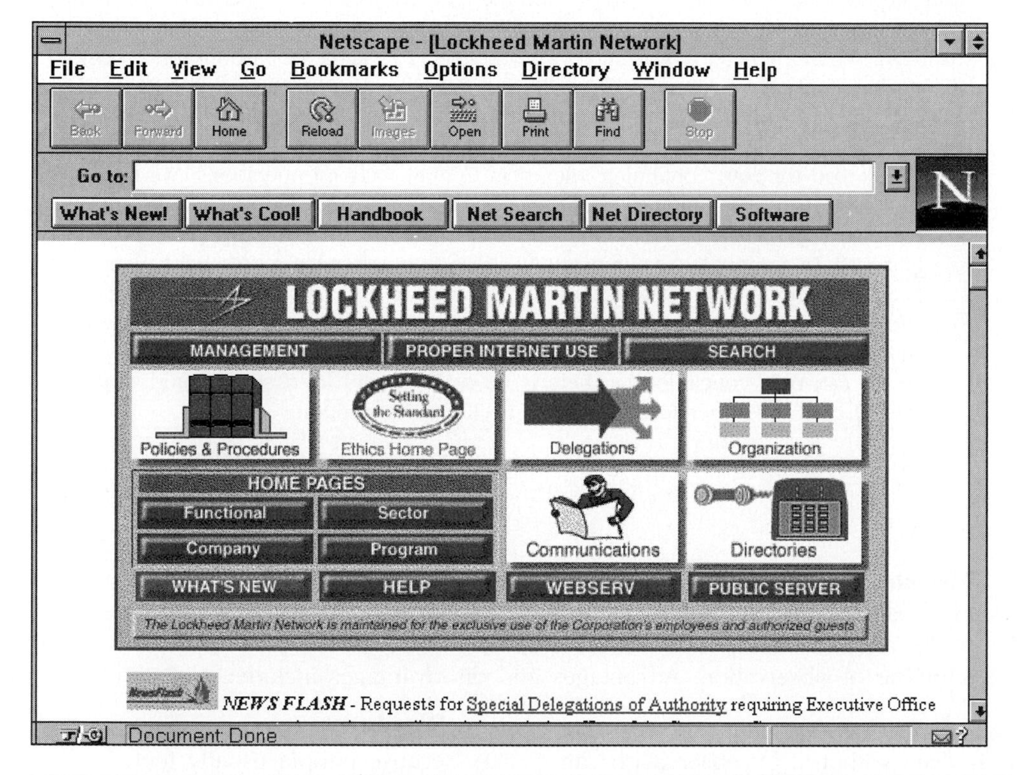

FIGURE B.3 *Lockheed Martin Intranet Web Page*

A similar type of research involves visiting other companies or departments that have addressed similar problems. Memberships in professional societies such as the Data Processing Management Association (DPMA) or the Association for Information Systems (AIS), among others, can provide a network of useful contacts.

OBSERVATION OF THE WORK ENVIRONMENT

Observation is one of the most effective data collection techniques for obtaining an understanding of a system.

Observation is a fact-finding technique where the systems analyst either participates in or watches a person perform activities to learn about the system.

This technique is often used when the validity of data collected through other methods is in question or when the complexity of certain aspects of the system prevents a clear explanation by the end-users.

Collecting Facts by Observing People at Work

Even with a well-conceived observation plan, the systems analyst is not assured that fact-finding will be successful. The following story, which appears in a book by Gerald M. Weinberg called *Rethinking Systems Analysis and Design,* gives an entertaining yet excellent example of some pitfalls of observation.[1]

The Railroad Paradox

About 30 miles from Gotham City lay the commuter community of Suburbantown. Each morning, thousands of Suburbanites took the Central Railroad to work in Gotham City. Each evening, Central Railroad returned them to their waiting spouses, children, and dogs.

Suburbantown was a wealthy suburb, and many of the spouses liked to leave the children and dogs and spend an evening in Gotham City with their mates. They preferred to precede their evening of dinner and theater with browsing among Gotham City's lush markets. But there was a problem. To allow time for proper shopping, a Suburbanite would have to depart for Gotham City at 2:30 or 3:00 in the afternoon. At that hour, no Central Railroad train stopped in Suburbantown.

Some Suburbanites noted that a Central train did pass through their station at 2:30, but did not stop. They decided to petition the railroad, asking that the train be scheduled to stop at Suburbantown. They readily found supporters in their door-to-door canvass. When the petition was mailed, it contained 253 signatures. About three weeks later, the petition committee received the following letter from the Central Railroad:

Dear Committee,

Thank you for your continuing interest in Central Railroad operations. We take seriously our commitment to providing responsive service to all the people living along our routes, and greatly appreciate feedback on all aspects of our business. In response to your petition, our customer service representative visited the Suburbantown station on three separate days, each time at 2:30 in the afternoon. Although he observed with great care, *on none of the three occasions were there any passengers waiting for a southbound train.*

We can only conclude that there is no real demand for a southbound stop at 2:30, and must therefore regretfully decline your petition.

Yours sincerely,
Customer Service Agent
Central Railroad

Observation Advantages and Disadvantages Observation can be a very useful and beneficial fact-finding technique provided you have the ability to observe thoroughly and accurately. You should become aware of the pros and cons of the technique of observation. Advantages and disadvantages include:

Advantages

1. Data gathered by observation can be highly reliable. Sometimes,

Disadvantages

1. Because people usually feel uncomfortable when being watched,

[1] Gerald M. Weinberg, *Rethinking Systems Analysis and Design,* pp. 23–24. Copyright © 1988, 1982 by Gerald M. Weinberg. Reprinted by permission of Dorset House Publishing, 353 W. 12th St., New York, NY 10014 (212-620-4053/1-800-DH-BOOKS/www.dorsethouse.com). All rights reserved.

observations are conducted to check the validity of data obtained directly from individuals.

2. The systems analyst is able to see exactly what is being done. Complex tasks are sometimes difficult to clearly explain in words. Through observation, the systems analyst can identify tasks that have been missed or inaccurately described by other fact-finding techniques. Also, the analyst can obtain data describing the physical environment of the task (e.g., physical layout, traffic, lighting, noise level).
3. Observation is relatively inexpensive compared with other fact-finding techniques. Other techniques usually require substantially more employee release time and copying expenses.
4. Observation allows the systems analyst to do work measurements.

they may unwittingly perform differently when being observed. In fact, the famous Hawthorne Experiment proved that the act of observation can alter behavior.

2. The work being observed may not involve the level of difficulty or volume normally experienced during that time period.
3. Some systems activities may take place at odd times, causing a scheduling inconvenience for the systems analyst.
4. The tasks being observed are subject to various types of interruptions.
5. Some tasks may not always be performed in the manner in which they are observed by the systems analyst. For example, the systems analyst might have observed how a company filled several customer orders. However, the procedures the systems analyst observed may have been those steps used to fill a number of regular customer orders. If any of those orders had been special orders (e.g., an order for goods not normally kept in stock), the systems analyst would have observed a different set of procedures being executed.
6. If people have been performing tasks in a manner that violates standard operating procedures, they may temporarily perform their jobs correctly while you are observing them. In other words, people may let you see what they want you to see.

Guidelines for Observation

How does the systems analyst obtain facts through observation? Does one simply arrive at the observation site and begin recording everything that's viewed? Of course not. Much preparation should occur. The analyst must determine how data will actually be captured. Will it be necessary to have special forms on which to quickly record data? Will the individuals being observed be bothered by having someone watch and record their actions? When are the low, normal, and peak periods of operations for the task to be observed? The systems analyst must identify the ideal time to observe a particular aspect of the system.

Observation should first be conducted when the workload is normal. Afterward, observations can be made during peak periods to gather information for measuring the effects caused by the increased volume. The systems analyst might also obtain samples of documents or forms that will be used by those being observed. As you can see, much planning and preparation must be done.

The sampling techniques discussed earlier are also useful for observation. In this case, the technique is called work sampling.

Work sampling is a fact-finding technique that involves a large number of observations taken at random intervals.

This technique is less threatening to the people being observed because the observation period is not continuous. When using work sampling, you need to predefine the operations of the job to be observed. Then calculate a sample size as you did for document and file sampling. Make that many random observations, being careful to observe activities at different times of the day. By counting the number of occurrences of each operation during the observations, you will get a feel for how employees spend their days.

With proper planning completed, the actual observation can be done. Effective observation is difficult to carry out. Experience is the best teacher; however, the following guidelines may help you develop your observation skills:

1. Determine the who, what, where, when, why, and how of the observation.
2. Obtain permission from appropriate supervisors or managers.
3. Inform those who will be observed of the purpose of the observation.
4. Keep a low profile.
5. Take notes during or immediately following the observation.
6. Review observation notes with appropriate individuals.
7. Don't interrupt the individuals at work.
8. Don't focus on trivial activities.
9. Don't make assumptions.

QUESTIONNAIRES

Another fact-finding technique is to conduct surveys through questionnaires.

Questionnaires are special-purpose documents that allow the analyst to collect information and opinions from respondents.

The document can be mass produced and distributed to respondents, who can then complete the questionnaire on their own time. Questionnaires allow the analyst to collect facts from a large number of people while maintaining uniform responses. When dealing with a large audience, no other fact-finding technique can tabulate the same facts as efficiently.

Collecting Facts by Using Questionnaires

The use of questionnaires has been heavily criticized and is often avoided by systems analysts. Many systems analysts claim that the responses lack reliable and useful information. But questionnaires can be an effective method for fact gathering, and many of these criticisms can be attributed to the inappropriate use of the questionnaires by systems analysts. Before using questionnaires, you should first understand the pros and cons associated with their use.

Advantages

1. Most questionnaires can be answered quickly. People can complete and return questionnaires at their convenience.
2. Questionnaires provide a relatively inexpensive means for gathering data from a large number of individuals.
3. Questionnaires allow individuals to maintain anonymity. Therefore, individuals are more likely to provide the real facts, rather than telling you what they think their boss would want them to.

Disadvantages

1. The number of respondents is often low.
2. There's no guarantee that an individual will answer or expand on all the questions.
3. Questionnaires tend to be inflexible. There's no opportunity for the systems analyst to obtain voluntary information from individuals or to reword questions that may have been misinterpreted.

4. Responses can be tabulated and analyzed quickly.	4. It's not possible for the systems analyst to observe and analyze the respondent's body language.
	5. There is no immediate opportunity to clarify a vague or incomplete answer to any question.
	6. Good questionnaires are difficult to prepare.

Types of Questionnaires

There are two formats for questionnaires, free-format and fixed-format.

Free-format questionnaires offer the respondent greater latitude in the answer. A question is asked, and the respondent records the answer in the space provided after the question.

Here are two examples of free-format questions:

1. What reports do you currently receive and how are they used?
2. Are there any problems with these reports (e.g., are they inaccurate, is there insufficient information, or are they difficult to read and/or use)? If so, please explain.

Such responses may be difficult to tabulate. It is also possible that the respondents' answers may not match the questions asked. To ensure good responses in free-format questionnaires, the analyst should phrase the questions in simple sentences and not use words, such as *good*, that can be interpreted differently by different respondents. The analyst should also ask questions that can be answered with three or fewer sentences. Otherwise, the questionnaire may take up more time than the respondent is willing to sacrifice.

The second type of questionnaire is fixed-format.

Fixed-format questionnaires contain questions that require specific responses from individuals.

Given any question, the respondent must choose from the available answers. This makes the results much easier to tabulate. On the other hand, the respondent cannot provide additional information that might prove valuable. There are three types of fixed-format questions.

1. For **multiple-choice questions**, the respondent is given several answers. The respondent should be told if more than one answer may be selected. Some multiple-choice questions allow for very brief free-format responses when none of the standard answers apply. Examples of multiple-choice, fixed-format questions are:

   Do you feel that backorders occur too frequently?

   ❑ YES ❑ NO

   Is the current accounts receivable report that you receive useful?

   ❑ YES ❑ NO

   If no, please explain.

2. For **rating questions**, the respondent is given a statement and asked to use supplied responses to state an opinion. To prevent built-in bias, there should be an equal number of positive and negative ratings. The following is an example of a rating fixed-format question:

   The implementation of quantity discounts would cause an increase in customer orders.

   ❑ Strongly agree ❑ Agree ❑ No opinion
   ❑ Disagree ❑ Strongly disagree

3. For **ranking questions**, the respondent is given several possible answers, which are to be ranked in order of preference or experience. An example of a ranking fixed-format question is:

   Rank the following transactions according to the amount of time you spend processing them:

________ % new customer orders	________ % order cancellations
________ % order modifications	________ % payments

Developing a Questionnaire

Good questionnaires are designed. If you write your questionnaires without designing them first, your chances of success are limited. The following procedure is effective:

1. Determine what facts and opinions must be collected and from whom you should get them. If the number of people is large, consider using a smaller, randomly selected group of respondents.
2. Based on the needed facts and opinions, determine whether free- or fixed-format questions will produce the best answers. A combination format that permits optional free-format clarification of fixed-format responses is often used.
3. Write the questions. Examine them for construction errors and possible misinterpretations. Make sure the questions don't offer your personal bias or opinions. Edit the questions.
4. Test the questions on a small sample of respondents. If your respondents had problems with them or if the answers were not useful, edit the questions.
5. Duplicate and distribute the questionnaire.

INTERVIEWS

The personal interview is generally recognized as the most important and most often used fact-finding technique.

> **Interviews** are a fact-finding technique whereby the systems analyst collects information from individuals face to face.

Interviewing can be used to achieve any of the following goals: find facts; verify facts; clarify facts; generate enthusiasm; get the end-user involved; identify requirements; and solicit ideas and opinions. There are two roles assumed in an interview. The systems analyst is the **interviewer**, responsible for organizing and conducting the interview. The system user, system owner, or advisor is the **interviewee**, who is asked to respond to a series of questions. Unfortunately, many systems analysts are poor interviewers. In this section you will learn how to conduct proper interviews.

Collecting Facts by Interviewing People

The most important element of an information system is people. More than anything else, people want to be in on things. No other fact-finding technique places as much emphasis on people as interviews. But people have different values, priorities, opinions, motivations, and personalities. Therefore, to use the interviewing technique, you must possess good human relations skills for dealing effectively with different types of people. And like other fact-finding techniques, interviewing isn't the best method for all situations. Interviewing has its advantages and disadvantages, which should be weighed against those of other fact-finding techniques for every fact-finding situation.

Advantages	**Disadvantages**
1. Interviews give the analyst an opportunity to motivate the interviewee to respond freely and	1. Interviewing is a very time-consuming, and therefore costly, fact-finding approach.

openly to questions. By establishing rapport, the systems analyst is able to give the interviewee a feeling of actively contributing to the systems project.

2. Interviews allow the systems analyst to probe for more feedback from the interviewee.
3. Interviews permit the systems analyst to adapt or reword questions for each individual.
4. Interviews give the analyst an opportunity to observe the interviewee's nonverbal communication. A good systems analyst may be able to obtain information by observing the interviewee's body movements and facial expressions as well as by listening to verbal replies to questions.

2. Success of interviews is highly dependent on the systems analyst's human relations skills.
3. Interviewing may be impractical due to the location of interviewees.

Interview Types and Techniques

There are two types of interviews, unstructured and structured.

Unstructured interviews are conducted with only a general goal or subject in mind and with few, if any, specific questions. The interviewer counts on the interviewee to provide a framework and direct the conversation.

This type of interview frequently gets off track, and the analyst must be prepared to redirect the interview back to the main goal or subject. For this reason, unstructured interviews don't usually work well for systems analysis and design.

In **structured interviews** the interviewer has a specific set of questions to ask of the interviewee.

Depending on the interviewee's responses, the interviewer will direct additional questions to obtain clarification or amplification. Some of these questions may be planned and others spontaneous. **Open-ended questions** allow the interviewee to respond in any way that seems appropriate. An example of an open-ended question is "Why are you dissatisfied with the report of uncollectable accounts?" **Closed-ended questions** restrict answers to either specific choices or short, direct responses. An example of such a question might be "Are you receiving the report of uncollectable accounts on time?" or "Does the report of uncollectable accounts contain accurate information?" Realistically, most questions fall between the two extremes.

How to Conduct an Interview

Your success as a systems analyst is at least partially dependent on your ability to interview. A successful interview will involve selecting appropriate individuals to interview, preparing extensively for the interview, conducting the interview properly, and following up on the interview. Here we examine each of these aspects in more detail. Let's assume that you've identified the need for an interview and you have determined exactly what kinds of facts and opinions you need.

Select Interviewees You should interview the end-users of the information system you are studying. A formal organizational chart will help you identify these individuals and their responsibilities. You should attempt to learn as much as possible about each individual before the interview. Attempt to learn what their strengths, fears, biases, and motivations might be. The interview can then be geared to take the characteristics of the individual into account.

Always make an appointment with the interviewee. Never just drop in. Limit the appointment to somewhere between a half hour and an hour. The higher the management level of the interviewee, the less time you should schedule. If the interviewee is a clerical, service, or blue-collar worker, get the permission of the person's supervisor before scheduling the interview. Be certain the location you want for the interview will be available during the time the interview is scheduled. Never conduct an interview in the presence of your officemates or the interviewee's peers.

Prepare for the Interview Preparation is the key to a successful interview. An interviewee can easily detect an unprepared interviewer. In fact, the interviewee may resent the lack of preparation because it is a waste of valuable time. When the appointment is made, the interviewee should be notified about the subject of the interview. To ensure that all pertinent aspects of the subject are covered, the analyst should prepare an interview guide.

An **interview guide** is a checklist of specific questions the interviewer will ask the interviewee.

The interview guide may also contain follow-up questions that will be asked only if the answers to other questions warrant the additional answers. A sample interview guide is presented in Figure B.4. Notice that the agenda is carefully laid out with the specific time allocated to each question. Time should also be reserved for follow-up questions and redirecting the interview. Questions should be carefully chosen and phrased. Most questions begin with the standard who, what, when, where, why, and how much type of wording. Avoid the following types of questions:

- *Loaded questions*, such as "Do we have to have both of these columns on the report?" The question conveys the interviewee's personal opinion on the issue.
- *Leading questions*, such as "You're not going to use this OPERATOR CODE, are you?" The question leads the interviewee to respond, "No, of course not," regardless of actual opinion.
- *Biased questions*, such as "How many codes do we need for FOOD CLASSIFICATION in the INVENTORY FILE? I think 20 ought to cover it." Why bias the interviewee's answer with your own?

Additional guidelines for questions are provided below. You should especially avoid threatening or critical questions. The purpose of the interview is to investigate, not to evaluate or criticize.

Interview Question Guidelines

1. Use clear and concise language.
2. Don't include your opinion as part of a question.
3. Avoid long or complex questions.
4. Avoid threatening questions.
5. Don't use "you" when you mean a gorup of people.

Conduct the Interview The actual interview can be characterized as consisting of three phases: the opening, body, and conclusion. The **interview opening** is intended to influence or motivate the interviewee to participate and communicate by establishing an ideal environment. When establishing an environment of mutual

INTERVIEWEE:

Jeff Bentley, Accounts Receivable Manager

DATE:

Tuesday, March 23, 1993

TIME:

1:30 P.M.

PLACE:

Room 223, Admin. Bldg.

SUBJECT:

Current Credit-Checking Policy

1 to 2 min.

Open the interview.

Introduce ourselves.

Thank Mr. Bentley for his valuable time.

State the purpose of the interview -- to obtain an understanding of the existing credit-checking policies.

5 min.

What conditions determine whether a customer's order is approved for credit?

5 min.

What are the possible decisions or actions that might be taken once these conditions have been evaluated?

3 min.

How are customers notified when credit is not approved for their order?

1 min.

After a new order is approved for credit and placed in the file containing orders that can be filled, a customer might request that a modification be made to the order. Would the order have to go through credit approval again if the new total order cost exceeds the original cost?

1 min.

Who are the individuals that perform the credit checks?

1 to 3 min.

May I have permission to talk to those individuals to learn specifically how they carry out the credit-checking process?

If so:

When would be an appropriate time to meet with each of them?

1 min.

Conclude the interview:

Thank Mr. Bentley for his cooperation and assure him that he will be receiving a copy of what transpired during the interview.

21 minutes

+9 minutes for follow-up questions and redirection

30 minutes allotted for interview (1:30 p.m. - 2:00 p.m.)

FIGURE B.4 *Sample Interview Guide*

trust and respect, you should identify the purpose and length of the interview and explain how the gathered data will be used. Here are three ways to effectively begin an interview:

1. Summarize the apparent problem and explain how the problem was discovered.
2. Offer an incentive or reward for participation.
3. Ask the interviewee for advice or assistance.

The **interview body** represents the most time-consuming phase. During this phase, you obtain the interviewee's responses to your list of questions. Listen closely and observe the interviewee. Take notes concerning both verbal and non-verbal responses from the interviewee. It's very important for you to keep the interview on track. Anticipate the need to adapt the interview to the interviewee. Often questions can be bypassed if they have been answered earlier in part of an answer to another question, or they can be deleted if determined to be irrelevant, based on what you've already learned during the interview. Finally, probe for more facts when necessary.

During the **interview conclusion**, you should express your appreciation and provide answers to any questions posed by the interviewee. The conclusion is very important for maintaining rapport and trust with the interviewee.

The importance of human relations skills in interviewing cannot be overemphasized. These skills must be exercised throughout the interview. Here is a set of rules that should be followed during an interview.

Conducting the Interview

DO

1. Be courteous.
2. Listen carefully.
3. Maintain control.
4. Probe.
5. Observe mannerisms and nonverbal communication.
6. Be patient.
7. Keep interviewee at ease.
8. Maintain self-control.

AVOID

1. Continuing an interview unnecessarily.
2. Assuming an answer is finished or leading nowhere.
3. Revealing verbal and nonverbal clues.
4. Using jargon.
5. Revealing your personal biases.
6. Talking instead of listening.
7. Assuming anything about the topic and the interviewee.
8. Tape recording—a sign of poor listening skills.

Follow Up on the Interview To help maintain good rapport and trust with interviewees, you should send them a memo that summarizes the interview. This memo should remind the interviewees of their contributions to the systems project and allow them the opportunity to clarify any misinterpretations that you may have derived during the interview. In addition, the interviewees should be given the opportunity to offer additional information they may have failed to bring out during the interview.

RAPID APPLICATION DEVELOPMENT (RAD)

Rapid application development is gaining popularity as a fact-finding technique for discovering user requirements. This technique allows analysts to quickly create mock forms and tables to simulate the implemented system. Users can suggest changes to the prototype and in most cases watch as the analyst tweaks the software to produce the desired look and feel. This process may take several iterations to correctly capture the functions necessary to automate the required business processes. Once the prototype is completed, you have the basis for a users manual, a requirements specification, and a template for a test plan.

Many analysts are adopting this **user-centered approach** in developing systems in that the emphasis for the scope and the functionality of the system is based primarily on the user's requirements. Past history and studies have shown us that requirements errors are the hardest and most expensive to fix, and therefore it is in our best interest to identify, define, and validate the user's requirements as early as possible. Using rapid application development tools and techniques can reduce the risk of discovering requirements errors later in the life cycle. RAD has shown to be an excellent technique in discovering and documenting the system data and processes. Several CASE vendors today have tools to rapidly build prototype windows, reports, and databases; record the information in a repository; plus upload the identified objects to a development environment in which to construct the final system.

FACT-FINDING ETHICS

More often than not during your fact-finding exercises you may come across or be analyzing sensitive information. It could be a file of an aerospace company's pricing structure for a contract it wishes to bid on or even employee profile data, which may include salaries, performance history, medical history, and career plans. Analysts must take great care to protect the data they have been entrusted with. Many people and organizations in this highly competitive atmosphere are looking for an "edge" to get ahead. Talking with friends about something you read in the data or leaving sensitive documents unsecured could cause great harm to the organization or the individuals. If the data would fall in the wrong people's hands, a competing aerospace company could submit a lower bid for the contract, steal away expert engineers by offering higher salaries and better benefits, or make public a person's medical history to question his or her character or competence. In some cases you would be responsible for the invasion of a person's privacy and could be liable!

Most computer professional societies such as AITP have a code of conduct and code of ethics their members must adhere to and abide by in conducting business. The following paragraphs are a fragment of AITP's Code of Ethics relating to the protection of information:[2]

> **Code of Ethics**
>
> I acknowledge:
>
> . . . Further, I shall not use knowledge of a confidential nature to further my personal interest, nor shall I violate the privacy and confidentiality of information entrusted to me or to which I may gain access.
>
> That I have an obligation to my employer whose trust I hold, therefore, I shall endeavor to discharge this obligation to the best of my ability, to guard my employer's interest, and to advise him or her wisely and honestly.
>
> That I have an obligation to my country, therefore, in my personal, business, and social contacts, I shall uphold my nation and shall honor the chosen way of life of my fellow citizens.
>
> I accept these obligations as a personal responsibility and as a member of this Association, I shall actively discharge these obligations and I dedicate myself to that end.

Many corporations require their employees to attend annual training seminars on company ethics. Code of conduct and ethics statements are displayed in places highly visible to employees. Ethics manuals may be stored at a convenient location or even ethics WWW pages may exist so all employees have easy access to

[2] Courtesy of AITP. Reprinted by permission.

the policies to know what is required and expected of them. Violations of the ethics statements could lead to disciplinary action or even termination. Most corporations make every effort to ensure they conduct business in an ethical manner. In some cases they are required by law. Many corporations have been levied heavy fines for not conducting business properly.

Washington, D.C., is the home of the Computer Ethics Institute, a nonprofit research, education, and policy study organization. It strives to make people more aware of computer ethics and to use computers more responsibly. A primary goal is to make computer ethics part of the standard school curriculum. To promote more awareness, the institute has published "The 10 Commandments of Computer Ethics," which appears in Figure B.5.

A FACT-FINDING STRATEGY

At the beginning of this module, we suggested that an analyst needs an organized method for collecting facts. An inexperienced analyst will frequently jump right into interviews. "Go to the people. That's where the real facts are!" Wrong! This attitude fails to recognize an important fact of life: People must complete their day-to-day jobs! Your job is not their main responsibility. Your demand on their time is their money lost. Now you may be thinking, "But I thought you've been saying that the system is for people and that direct end-user involvement in systems development is essential! Aren't you contradicting yourselves?"

Not at all! Time is money. To waste your end-users' time is to waste your company's money. To make the most of the time that you spend with end-users, don't jump right into interviews. Instead, first collect all the facts you can by using other methods. Consider the following step-by-step strategy:

1. Learn all you can from existing documents, forms, reports, and files. You'll be surprised how much of the system becomes clear without any people contact.
2. If appropriate, observe the system in action. Agree not to ask questions. Just watch and take notes or draw pictures. Make sure the workers know that you're not evaluating individuals. Otherwise, they may perform in a more efficient manner than normal.
3. Given all the facts that you've already collected, design and distribute questionnaires to clear up things you don't fully understand. This is also a good time to solicit opinions on problems and limitations. Questionnaires do require your end-users to give up some of their time. But they choose when to best make that sacrifice.
4. Conduct your interviews (or group work sessions, such as JAD or RAD). Because you have already collected most of the pertinent facts by low-user-contact methods, you can use the interview to verify and clarify the most difficult issues and problems.
5. Follow up. Use appropriate fact-finding techniques to verify facts (usually interviews or observation).

The strategy is not sacred. Although a fact-finding strategy should be developed for every pertinent phase of systems development, every project is unique. Sometimes observation and questionnaires may be inappropriate. But the idea should always be to collect as many facts as possible before using interviews.

1. **Thou shalt not use a computer to harm other people.**
2. **Thou shalt not interfere with other people's computer work.**
3. **Thou shalt not snoop around in other people's computer files.**
4. **Thou shalt not use a computer to steal.**
5. **Thou shalt not use a computer to bear false witness.**
6. **Thou shalt not copy or use proprietary software for which you have not paid.**
7. **Thou shalt not use other people's computer resources without authorization or proper compensation.**
8. **Thou shalt not appropriate other people's intellectual output.**
9. **Thou shalt think about the social consequences of the program you are writing or the system you are designing.**
10. **Thou shalt always use a computer in ways that insure consideration and respect for your fellow humans.**

Source: Computer Ethics Institute.

FIGURE B.5 *The 10 Commandments of Computer Ethics*

SUMMARY

1. Effective fact-finding techniques are crucial to the application of systems analysis and design methods during systems projects. Fact-finding is performed during all phases of the systems development life cycle. To support development activities, the analyst must collect facts about end-users, the business, data and information resources, and information systems components.
2. There are seven common fact-finding techniques.
   a. The sampling of existing documents and files can provide many facts and details with little or no direct personal communication being necessary. The analyst should collect historical documents, business operations manuals and forms, and information systems documents.
   b. Research is an often overlooked technique based on the study of other similar applications. It now has become more convenient with the Internet and World Wide Web (WWW). Site visits are a special form of research.
   c. Observation is a fact-finding technique in which the analyst studies people doing their jobs.
   d. Questionnaires are used to collect similar facts from a larger number of individuals.
   e. Interviews are the most popular but the most time-consuming fact-finding technique. When interviewing, the analyst meets individually with people to gather information.
   f. Rapid application development is becoming more popular as a fact-finding method. It facilitates the user and analyst working together and allows the user to see requirements on a computer screen.
   g. Many analysts find flaws with interviewing—separate interviews often lead to conflicting facts, opinions, and priorities. The end result is numerous follow-up interviews and/or group meetings. For this reason, many shops are using the group work session known as a joint application development session as a substitute for interviews.
3. Conducting business in an ethical manner is a required practice and analysts need to be more aware of the implications of not being ethical.
4. Because "time is money," it is wise and practical for the systems analyst to use a fact-finding strategy to maximize the value of time spent with end-users.

KEY TERMS

closed-ended questions, p. 633
fact-finding, p. 623
fixed-format questionnaires, p. 631
free-format questionnaires, p. 631
Internet, p. 626
interview body, p. 636
interview conclusion, p. 636
interview guide, p. 634
interview opening, p. 634
interviewee, p. 632
interviewer, p. 632
interviews, p. 632
Intranet, p. 626
multiple-choice questions, p. 631
observation, p. 628
open-ended questions, p. 633
questionnaires, p. 630
randomization, p. 625
ranking questions, p. 632
rating questions, p. 631
sampling, p. 624
stratification, p. 625
structured interviews, p. 633
unstructured interviews, p. 633
user-centered approach, p. 637
work sampling, p. 630
World Wide Web, p. 626

REVIEW QUESTIONS

1. What is fact-finding?
2. What are the seven common fact-finding techniques?
3. When collecting facts from existing documentation, what is the first document the analyst should seek out? What should the analyst do next?
4. What is sampling? What are two commonly used sampling techniques? How are they different?
5. Explain the difference between the Internet and intranet. What is the World Wide Web and how is it related to the Internet?
6. What is work sampling?
7. What are four advantages of using questionnaires? What are four disadvantages?
8. Identify and briefly describe the two types of questionnaires?
9. Which fact-finding technique is generally recognized as the most important and most often used?
10. What is rapid application development? Why is it so popular?
11. What is one of the primary goals of the Computer Ethics Institute? What document did it publish to promote more awareness of computer ethics?
12. Why should a systems analyst use a fact-finding strategy when working with an end-user?

PROBLEMS AND EXERCISES

1. Explain how the information system building blocks can serve as a framework in determining what facts need to be collected during systems development.
2. Explain how an organizational chart can aid in planning for fact-finding. What are some potential drawbacks to using an existing organizational chart?
3. A systems analyst wants to study documents stored in a large metal file cabinet. The cabinet contains several hundred records describing product warranty claims. The analyst wishes to study a sample of the records in the file and to be 95 percent certain (certainty factor = 1.960) that the data from which the sample is taken will not include variations not in the sample. How many sample records should the analyst retrieve to get this desired accuracy?
4. For the sample size in exercise 3, explain two specific strategies for selecting the samples.
5. Describe how you would use form and/or file sampling in each phase of the systems development life cycle. If you think sampling would be inappropriate for any of these phases, explain why.
6. Repeat exercise 5 for the technique of observation.
7. Make a list of things that might affect your work performance when you are being observed performing your job. What could an observer do to eliminate these concerns or problems?
8. Repeat exercise 5 for the questionnaire technique.
9. Give two examples of free-format questions and two examples of each of the following types of fixed-format questions:
   a. Multiple choice.
   b. Rating.
   c. Ranking.
10. Repeat exercise 5 for the interviewing technique.
11. Explain the difference between a structured and an unstructured interview. When is each type of interview appropriately used?
12. Prepare a sample interview guide to use in obtaining from your academic advisor facts describing the course registration policies and procedures.
13. List five examples of violating "The 10 Commandments of Computer Ethics." What measures is the Computer Ethics Institute taking to promote more ethics awareness?

PROJECTS AND MINICASES

1. Mr. Art Pang is the Accounts Receivables manager. You have been assigned to do a study of Mr. Pang's current billing system, and you need to solicit facts from his subordinates. Mr. Pang has expressed his concern that, although he wishes to support you in your fact-finding efforts, his people are extremely busy and must get their jobs done. Write a memo to Mr. Pang describing a fact-finding strategy that you could follow to maximize your fact-finding while minimizing the release time required for his subordinates.

SUGGESTED READINGS

Berdie, Douglas R., and John F. Anderson. *Questionnaires: Design and Use*. Metuchen, NJ: Scarecrow Press, 1974. A practical guide to the construction of questionnaires. Particularly useful because of its short length and illustrative examples.

Davis, William S. *Systems Analysis and Design*. Reading, MA: Addison-Wesley, 1983. Provides useful pointers for preparing and conducting interviews.

Dejoie, Roy; George Fowler; and David Paradice. *Ethical Issues in Information Systems*. Boston: Boyd and Fraser, 1991. Focuses on the impact of computer technology on ethical decision making in today's business organizations.

Eager, Bill. *Using the World Wide Web*. Indianapolis, IN: Que, 1994. A comprehensive guide to navigating the WWW and the Internet.

Fitzgerald, Jerry; Ardra F. Fitzgerald; and Warren D. Stallings, Jr. *Fundamentals of Systems Analysis*. 2nd ed. New York: John Wiley & Sons, 1981. A useful survey text for the systems analyst. Chapter 6, "Understanding the Existing System," does a good job of presenting fact-finding techniques in the study phase.

Gildersleeve, Thomas R. *Successful Data Processing System Analysis*. Englewood Cliffs, NJ: Prentice Hall, 1978. Chapter 4, "Interviewing in Systems Work," provides a comprehensive look at interviewing specifically for the systems analyst. A thorough sample interview is scripted and analyzed in this chapter.

London, Keith R. *The People Side of Systems*. New York: McGraw-Hill, 1976. Chapter 5, "Investigation versus Inquisition," provides a very good people-oriented look at fact-finding, with considerable emphasis on interviewing.

Lord, Kenniston W., Jr., and James B. Steiner. *CDP Review Manual: A Data Processing Handbook*. 2nd ed. New York: Van Nostrand Reinhold, 1978. Chapter 8, "Systems Analysis and Design," provides a more comprehensive comparison of the merits and demerits of each fact-finding technique. This material is intended to prepare data processors for the Certificate in Data Processing examinations, one of which covers systems analysis and design.

Miller, Irwin, and John F. Freund. *Probability and Statistics for Engineers*. Englewood Cliffs, NJ: Prentice Hall, 1965. Introductory college textbook on probability and statistics.

Mitchell, Ian; Norman Parrington; Peter Dunne; and John Moses. "Practical Prototyping, Part One." *Object Currents*, May 1996. First of a three-part series of articles that explores prototyping and how you can benefit from it.

Salvendy, G., ed. *Handbook of Industrial Engineering*. New York: John Wiley & Sons, 1974. A comprehensive handbook for industrial engineers; systems analysts are, in a way, a type of industrial engineer. Excellent coverage on sampling and work measurement.

Stewart, Charles J., and William B. Cash, Jr. *Interviewing: Principles and Practices*. 2nd ed. Dubuque, IA: Wm. C. Brown, 1978. Popular college textbook that provides broad exposure to interviewing techniques, many of which are applicable to systems analysis and design.

Weinberg, Gerald M. *Rethinking Systems Analysis and Design*. Boston: Little, Brown and Company, 1982. A book created to stimulate a new way of thinking.

Wood, Jane, and Denise Silver. *Joint Application Design*. New York: John Wiley & Sons, 1989. This book provides a comprehensive overview of IBM's joint application design technique.

C

FEASIBILITY AND COST-BENEFIT ANALYSIS

MODULE PREVIEW AND OBJECTIVES

This module is about feasibility and cost-benefit analysis, two of the most nontechnical skills that any systems analyst must develop. Systems analysts sell change. Good systems analysts thoroughly evaluate alternative solutions before recommending change. In this module you will learn how to analyze and document those alternatives on the basis of four feasibility criteria: operational, technical, schedule, and economics. You will know that you understand the feasibility and cost-benefit analysis skills needed by the systems analyst when you can:

- Identify feasibility checkpoints in the systems development life cycle.
- Define and describe four types of feasibility and their respective criteria.
- Perform various cost-benefit analyses using time-adjusted costs and benefits.

Note: The authors gratefully acknowledge the contributions of Professor Kevin C. Dittman who developed this module.

In today's business world, it is becoming more and more apparent that analysts must learn to think like business managers. Computer applications are expanding at a record pace. Now more than ever, management expects information systems to pay for themselves. Information is a major capital investment that must be justified, just as marketing must justify a new product and manufacturing must justify a new plant or equipment. Systems analysts are called on more than ever to help answer the following questions: Will the investment pay for itself? Are there other investments that will return even more on their expenditure?

This module deals with cost-benefit analysis and other feasibility issues of interest to the systems analyst and users of information systems. Few topics are more important. Feasibility analysis isn't really systems analysis, and it isn't systems design either. Instead, feasibility analysis is a cross life cycle activity and should be continuously performed throughout a systems project.

FEASIBILITY ANALYSIS—A CREEPING COMMITMENT APPROACH

Feasibility Checkpoints in the Life Cycle

Let's begin with a formal definition of feasibility and feasibility analysis.

Feasibility is the measure of how beneficial or practical the development of an information system will be to an organization.

Feasibility analysis is the process by which feasibility is measured.

Feasibility should be measured throughout the life cycle. In earlier chapters we called this a **creeping commitment** approach to feasibility. The scope and complexity of an apparently feasible project can change after the initial problems and opportunities are fully analyzed or after the system has been designed. Thus, a project that is feasible at one point may become infeasible later. Let's study some checkpoints for our systems development life cycle.

If you study your company's project standards or systems development life cycle (SDLC), you'll probably see a feasibility study phase or deliverable, but not an explicit ongoing process. But look more closely! On deeper examination, you'll probably identify various go/no-go checkpoints or management reviews. These checkpoints and reviews identify specific times during the life cycle when feasibility is reevaluated. A project can be canceled or revised in scope, schedule, or budget at any of these checkpoints. Thus, an explicit feasibility analysis phase in any life cycle should be considered to be only an initial feasibility assessment.

Feasibility checkpoints can be installed into any SDLC that you are using. Figure C.1 shows feasibility checkpoints for a typical life cycle (similar to, but not identical to, the life cycle used in this book). The checkpoints are represented by red diamonds. The diamonds indicate that a feasibility reassessment and management review should be conducted at the end of the prior phase (before the next phase). A project may be canceled or revised at any checkpoint, despite whatever resources have been spent.

This idea may bother you at first. Your natural inclination may be to justify continuing a project based on the time and money you've already spent. Those costs are sunk! A fundamental principle of management is never to throw good money after bad—cut your losses and move on to a more feasible project. That doesn't mean the costs already spent are not important. Costs must eventually be recovered if the investment is ever to be considered a success. Let's briefly examine the checkpoints in Figure C.1.

Systems Analysis—A Survey Phase Checkpoint The first feasibility analysis is conducted during the survey phase. At this early stage of the project, feasibility is rarely more than a measure of the urgency of the problem and the first-cut estimate of development costs. It answers the question: Do the problems (or opportunities) warrant the cost of a detailed study of the current system? Realistically, feasibility can't be accurately measured until the problems (and opportunities) and requirements (definition phase) are better understood.

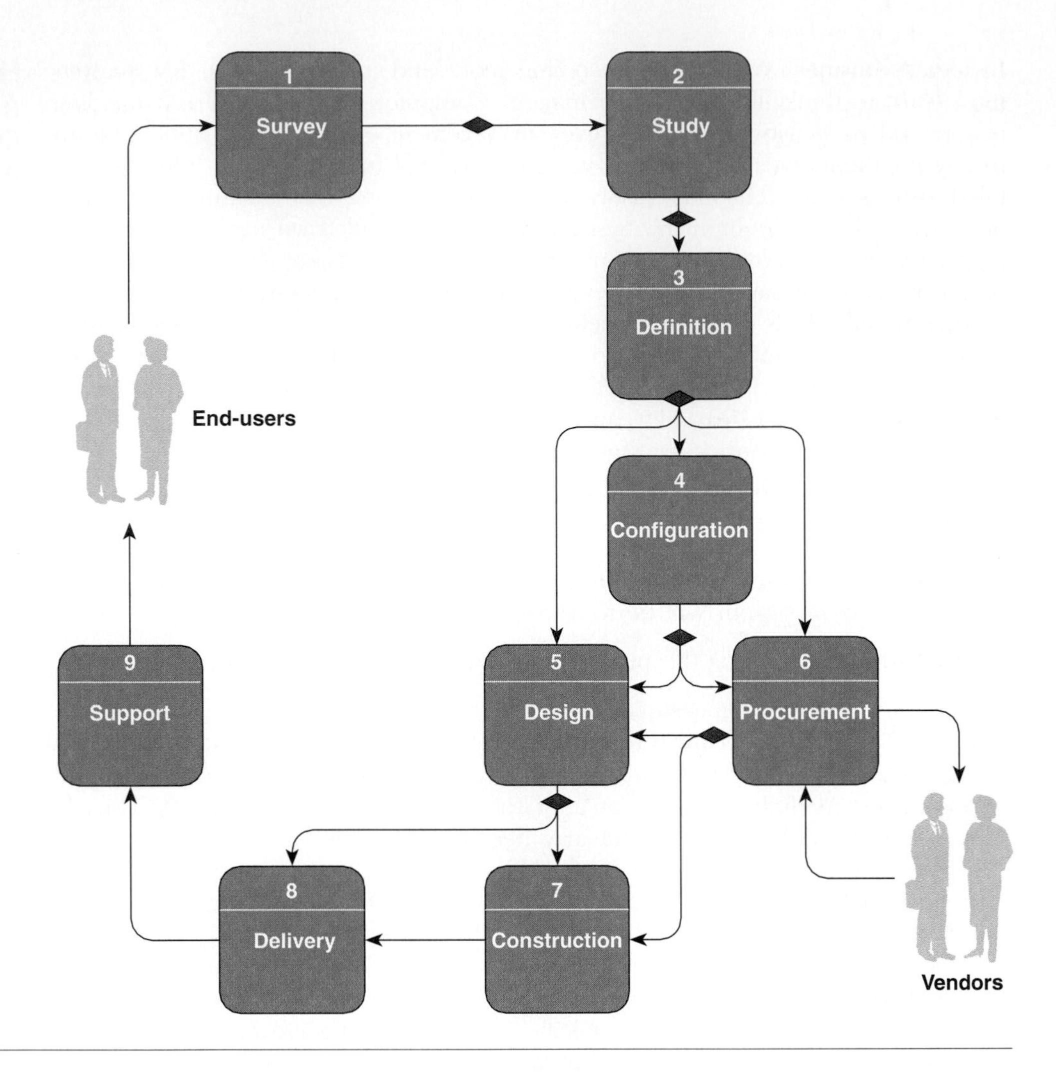

FIGURE C.1 *Feasibility Checkpoints in the Systems Development Life Cycle*

After estimating benefits of solving the problems and opportunities, analysts will estimate costs of developing the expected system. Experienced analysts routinely increase these costs by 50 percent to 100 percent (or more) because experience tells them the problems are rarely well-defined and user requirements are typically understated.

Systems Analysis—A Study Phase Checkpoint The next checkpoint occurs after a more detailed study of the current system. Because the problems are better understood, the analysts can make better estimates of development costs and of the benefits to be obtained from a new system. The minimum value of solving a problem is equal to the cost of that problem. For example, if inventory carrying costs are $35,000 over acceptable limits, then the minimum value of an acceptable information system would be $35,000. It is hoped an improved system will be able to do better than that; however, it must return this minimum value!

Development costs, at this point, are still just guesstimates. We have yet to fully define user requirements or to specify a design solution to those requirements.

If the cost estimates significantly increase from the survey phase to the study phase, the likely culprit is scope. Scope has a tendency to increase in many projects. If increased scope threatens feasibility, then scope might be reduced. See Module A, "Project and Process Management Techniques," for a discussion on expectations management.

Systems Analysis—A Definition Phase Checkpoint The next checkpoint occurs after the definition of user requirements for the new system. These requirements frequently prove more extensive than originally stated. For this reason, the analyst must frequently revise cost estimates for design and implementation. Once again, feasibility is reassessed. If feasibility is in question, scope, schedule, and costs must be rejustified. (Again, Module A offers guidelines for adjusting project expectations.)

If early estimates were adjusted up, you may still be within the range despite an increase in scope. If not, the project need not always be canceled or reduced in scope. If you have kept track of the increase in problems and requirements since the beginning of the project, your system owner may be willing to pay for the increased requirements (and adjust the schedule accordingly).

Systems Design—A Selection Phase Checkpoint The SDLC in Figure C.1 is the design decision-making phase. This SDLC separates design decision making from the actual design phase. In any case, the selection phase represents a major feasibility analysis activity since it charts one of many possible implementations as the target for systems design.

Problems and requirements should be known by now. During the selection phase, alternative solutions are defined in terms of their input/output methods, data storage methods, computer hardware and software requirements, processing methods, and people implications. The following list presents the typical range of options that can be evaluated by the analyst.

- Do nothing! Leave the current system alone. Regardless of management's opinion or your own opinion of this option, it should be considered and analyzed as a baseline option against which all others can and should be evaluated.
- Reengineer the (manual) business processes, not the computer-based processes. This may involve streamlining activities, reducing duplication and unnecessary tasks, reorganizing office layouts, and eliminating redundant and unnecessary forms and processes, among others.
- Enhance existing computer processes.
- Purchase a packaged application.
- Design and construct a new computer-based system. This option presents numerous other options: centralized versus distributed versus cooperative processing; on-line versus batch processing; and files versus database for data storage. Of course, an alternative could be a combination of the preceding options.

After defining these options, each option is analyzed for operational, technical, schedule, and economic feasibility. This module will closely examine these four classes of feasibility criteria. One alternative is recommended to system owners for approval. The approved solution becomes the basis for general and detailed design.

Systems Design—A Procurement Phase Checkpoint Because the procurement of hardware and applications software involves economic decisions that may require sizable outlays of cash, it shouldn't surprise you that feasibility analysis is required before a contract is extended to a vendor. The procurement phase (covered more extensively in Chapter 9) may be consolidated into the selection phase because hardware and software selection may have a significant impact on the feasibility of the solutions being considered.

Systems Design—A Design Phase Checkpoint A final checkpoint is completed after the system is designed. The general and detailed design specifications have been completed. The complexity of the solution should be apparent. Because implementation is often the most time-consuming and costly phase, the checkpoint after design gives us one last chance to cancel or downsize the project. Downsizing is the act of reducing the scope of the initial version of the system. Future versions can address other requirements after the system goes into production.

FOUR TESTS FOR FEASIBILITY

So far, we've defined feasibility and feasibility analysis, and we've identified feasibility checkpoints in the life cycle. Most analysts agree that there are four categories of feasibility tests:

- **Operational feasibility** is a measure of how well the solution will work in the organization. It is also a measure of how people feel about the system/project.
- **Technical feasibility** is a measure of the practicality of a specific technical solution and the availability of technical resources and expertise.
- **Schedule feasibility** is a measure of how reasonable the project timetable is.
- **Economic feasibility** is a measure of the cost-effectiveness of a project or solution. This is often called a *cost-benefit analysis.*

Operational and technical feasibility criteria measure the worthiness of a problem or solution. Operational feasibility is people oriented. Technical feasibility is computer oriented.

Economic feasibility deals with the costs and benefits of the information system. Actually, few systems are infeasible. Instead, different options tend to be more or less feasible than others. Let's take a closer look at the four feasibility criteria.

Operational Feasibility

Operational feasibility criteria measure the urgency of the problem (survey and study phases) or the acceptability of a solution (definition, selection, acquisition, and design phases). How do you measure operational feasibility? There are two aspects of operational feasibility to be considered:

1. Is the problem worth solving, or will the solution to the problem work?
2. How do the end-users and management feel about the problem (solution)?

Is the Problem Worth Solving, or Will the Solution to the Problem Work? Do you recall the PIECES framework for identifying problems (Chapters 3 and 4)? PIECES can be used as the basis for analyzing the urgency of a problem or the effectiveness of a solution. The following is a list of the questions that address these issues:

P *Performance.* Does the system provide adequate throughput and response time?

I *Information.* Does the system provide end-users and managers with timely, pertinent, accurate, and usefully formatted information?

E *Economy.* No, we are not prematurely jumping into economic feasibility! The question here is, Does the system offer adequate service level and capacity to reduce the costs of the business or increase the profits of the business?

C *Control.* Does the system offer adequate controls to protect against fraud and embezzlement and to guarantee the accuracy and security of data and information?

E *Efficiency.* Does the system make maximum use of available resources including people, time, flow of forms, minimum processing delays, and the like?

S *Services.* Does the system provide desirable and reliable service to those who need it? Is the system flexible and expandable?

NOTE The term *system,* used throughout this discussion, may refer either to the existing system or a proposed system solution, depending on which phase you're currently working in.

How Do the End-Users and Managers Feel about the Problem (Solution)? It's important not only to evaluate whether a system *can* work, but we must also evaluate whether a system *will* work. A workable solution might fail because of end-user or management resistance. The following questions address this concern:

- Does management support the system?
- How do the end-users feel about their role in the new system?
- What end-users or managers may resist or not use the system? People tend to resist change. Can this problem be overcome? If so, how?
- How will the working environment of the end-users change? Can or will end-users and management adapt to the change?

Essentially, these questions address the political acceptability of solving the problem or the solution.

Usability Analysis When determining operational feasibility in the later stages of the development life cycle, **usability analysis** is often performed with a working prototype of the proposed system. This is a test of the system's user interfaces and is measured in how easy they are to learn and to use and how they support the desired productivity levels of the users. Many large corporations, software consultant agencies, and software development companies employ user interface specialists for designing and testing system user interfaces. They have special rooms equipped with video cameras, tape recorders, microphones, and two-way mirrors to observe and record a user working with the system. Their goal is to identify the areas of the system where the users are prone to make mistakes and processes that may be confusing or too complicated. They also observe the reactions of the users and assess their productivity.

How do you determine if a system's user interface is usable? There are certain goals or criteria that experts agree help measure the usability of an interface and they are as follows:

- *Ease of learning*—How long it takes to train someone to perform at a desired level.
- *Ease of use*—You are able to perform your activity quickly and accurately. If you are a first-time user or infrequent user, the interface is easy and understandable. If you are a frequent user, your level of productivity and efficiency is increased.
- *Satisfaction*—You, the user, are favorably pleased with the interface and prefer it over types you are familiar with.

Technical Feasibility

Technical feasibility can be evaluated only after those phases during which technical issues are resolved—namely, after the evaluation and design phases of our life cycle have been completed. Today, very little is technically impossible. Consequently, technical feasibility looks at what is practical and reasonable. Technical feasibility addresses three major issues:

1. Is the proposed technology or solution practical?
2. Do we currently possess the necessary technology?
3. Do we possess the necessary technical expertise, and is the schedule reasonable?

Is the Proposed Technology or Solution Practical? The technology for any defined solution is normally available. The question is whether that technology is mature enough to be easily applied to our problems. Some firms like to use state-of-the-art technology, but most firms prefer to use mature and proven technology. A mature technology has a larger customer base for obtaining advice concerning problems and improvements.

Do We Currently Possess the Necessary Technology? Assuming the solution's required technology is practical, we must next ask ourselves, Is the technology available in our information systems shop? If the technology is available, we must ask if we have the capacity. For instance, Will our current printer be able to handle the new reports and forms required of a new system?

If the answer to either of these questions is no, then we must ask ourselves, Can we get this technology? The technology may be practical and available, and, yes, we need it. But we simply may not be able to afford it at this time. Although this argument borders on economic feasibility, it is truly technical feasibility. If we can't afford the technology, then the alternative that requires the technology is not practical and is technically infeasible!

Do We Possess the Necessary Technical Expertise, and Is the Schedule Reasonable? This consideration of technical feasibility is often forgotten during feasibility analysis. We may have the technology, but that doesn't mean we have the skills required to properly apply that technology. For instance, we may have a database management system (DBMS). However, the analysts and programmers available for the project may not know that DBMS well enough to properly apply it. True, all information systems professionals can learn new technologies. However, that learning curve will impact the technical feasibility of the project; specifically, it will impact the schedule.

Schedule Feasibility

Given our technical expertise, are the project deadlines reasonable? Some projects are initiated with specific deadlines. You need to determine whether the deadlines are mandatory or desirable. For instance, a project to develop a system to meet new government reporting regulations may have a deadline that coincides with when the new reports must be initiated. Penalties associated with missing such a deadline may make meeting it mandatory. If the deadlines are desirable rather than mandatory, the analyst can propose alternative schedules.

It is preferable (unless the deadline is absolutely mandatory) to deliver a properly functioning information system two months late than to deliver an error-prone, useless information system on time! Missed schedules are bad. Inadequate systems are worse! It's a choice between the lesser of two evils.

Economic Feasibility

The bottom line in many projects is economic feasibility. During the early phases of the project, economic feasibility analysis amounts to little more than judging whether the possible benefits of solving the problem are worthwhile. Costs are practically impossible to estimate at that stage because the end-user's requirements and alternative technical solutions have not been identified. However, as soon as specific requirements and solutions have been identified, the analyst can weigh the costs and benefits of each alternative. This is called a cost-benefit analysis. Cost-benefit analysis is discussed in the last section of this module.

The Bottom Line

You have learned that any alternative solution can be evaluated according to four criteria: operational, technical, schedule, and economic feasibility. How do you pick the best solution? It's not always easy. Operational and economic issues often conflict. For example, the solution that provides the best operational impact for the end-users may also be the most expensive and, therefore, the least economically feasible. The final decision can be made only by sitting down with end-users, reviewing the data, and choosing the best overall alternative.

COST-BENEFIT ANALYSIS TECHNIQUES

Economic feasibility has been defined as a cost-benefit analysis. How do you estimate costs and benefits? And how do you compare those costs and benefits to determine economic feasibility? Most schools offer complete courses on these subjects—courses on financial management, financial decision analysis, and engineering economics and analysis. Such a course should be included in your plan of study. This section presents an overview of the techniques.

How Much Will the System Cost?

Costs fall into two categories. There are costs associated with developing the system, and there are costs associated with operating a system. The former can be estimated from the outset of a project and should be refined at the end of each phase of the project. The latter can be estimated only once specific computer-based solutions have been defined (during the selection phase or later). Let's take a closer look at the costs of information systems.

The costs of developing an information system can be classified according to the phase in which they occur. Systems development costs are usually onetime costs that will not recur after the project has been completed. Many organizations have standard cost categories that must be evaluated. In the absence of such categories, the following lists should help:

- *Personnel costs*—The salaries of systems analysts, programmers, consultants, data entry personnel, computer operators, secretaries, and the like who work on the project make up the personnel costs. Because many of these individuals spend time on many projects, their salaries should be prorated to reflect the time spent on the projects being estimated.
- *Computer usage*—Computer time will be used for one or more of the following activities: programming, testing, conversion, word processing, maintaining a project dictionary, prototyping, loading new data files, and the like. If a computing center charges for usage of computer resources such as disk storage or report printing, the cost should be estimated.
- *Training*—If computer personnel or end-users have to be trained, the training courses may incur expenses. Packaged training courses may be charged out on a flat fee per site, a student fee (such as $395 per student), or an hourly fee (such as $75 per class hour).
- *Supply, duplication, and equipment costs.*
- *Cost of any new computer equipment and software.*

Sample development costs for a typical solution are displayed in Figure C.2.

Almost nobody forgets systems development budgets when itemizing costs. On the other hand, it is easy to forget that a system will incur costs after it is operating. The lifetime benefits must recover both the developmental and operating costs. Unlike systems development costs, operating costs tend to recur throughout the lifetime of the system. The costs of operating a system over its useful lifetime can be classified as fixed and variable.

Fixed costs occur at regular intervals but at relatively fixed rates. Examples of fixed operating costs include:

- Lease payments and software license payments.
- Prorated salaries of information systems operators and support personnel (although salaries tend to rise, the rise is gradual and tends not to change dramatically from month to month).

Variable costs occur in proportion to some usage factor. Examples include:

- Costs of computer usage (e.g., CPU time used, terminal connect time used, storage used), which vary with the workload.
- Supplies (e.g., preprinted forms, printer paper used, punched cards, floppy disks, magnetic tapes, and other expendables), which vary with the workload.

Estimated Costs for Client-Server System Alternative

DEVELOPMENT COSTS:

Personnel:

2	Systems Analysts (400 hours/ea $35.00/hr)		$28,000
4	Programmer/Analysts (250 hours/ea $25.00/hr)		$25,000
1	GUI Designer (200 hours/ea $35.00/hr)		$7,000
1	Telecommunications Specialist (50 hours/ea $45.00/hr)		$2,250
1	System Architect (100 hours/ea $45.00/hr)		$4,500
1	Database Specialist (15 hours/ea $40.00/hr)		$600
1	System Librarian (250 hours/ea $10.00/hr)		$2,500

Expenses:

4	Smalltalk training registration ($3,500.00/student)		$14,000

New Hardware & Software:

1	Development Server (Pentium Pro class)		$18,700
1	Server Software (operating system, misc.)		$1,500
1	DBMS server software		$7,500
7	DBMS Client software ($950.00 per client)		$6,650

Total Development Costs: $118,200

PROJECTED ANNUAL OPERATING COSTS

Personnel:

2	Programmer/Analysts (125 hours/ea $25.00/hr)		$6,250
1	System Librarian (20 hours/ea $10.00/hr)		$200

Expenses:

1	Maintenance Agreement for Pentium Pro Server		$995
1	Maintenance Agreement for Server DBMS software		$525
	Preprinted forms (15,000/year @ .22/form)		$3,300

Total Projected Annual Costs: **$11,270**

FIGURE C.2 *Costs for a Proposed Systems Solution*

- Prorated overhead costs (e.g., utilities, maintenance, and telephone service), which can be allocated throughout the lifetime of the system using standard techniques of cost accounting.

Sample operating cost estimates for a solution are displayed in Figure C.2. After determining the costs and benefits for a possible solution, you can perform the cost-benefit analysis.

What Benefits Will the System Provide?

Because benefits or potential benefits become known before costs, we'll discuss benefits first. Benefits normally increase profits or decrease costs, both highly desirable characteristics of a new information system.

As much as possible, benefits should be quantified in dollars and cents. Benefits are classified as tangible or intangible.

Tangible benefits are those that can be easily quantified.

Tangible benefits are usually measured in terms of monthly or annual savings or of profit to the firm. For example, consider the following scenario:

> While processing student housing applications, we discover that considerable data is being redundantly typed and filed. An analysis reveals that the same data is typed seven times, requiring an average of 44 additional minutes of clerical work per application. The office processes 1,500 applications per year. That means a total of 66,000 minutes or 1,100 hours of redundant work per year. If the average salary of a secretary is $6 per hour, the cost of this problem and the benefit of solving the problem is $6,600 per year.

Alternatively, tangible benefits might be measured in terms of unit cost savings or profit. For instance, an alternative inventory valuation scheme may reduce inventory carrying cost by $0.32 per unit of inventory. Some examples of tangible benefits are: fewer processing errors; increased throughput; decreased reponse time; elimination of job steps; reduced expenses; increased sales; better credit; and reduced credit losses.

Other benefits are intangible.

> **Intangible benefits** are those benefits believed to be difficult or impossible to quantify.

Unless these benefits are at least identified, it is entirely possible that many projects would not be feasible. Examples of intangible benefits are: improved customer goodwill; improved employee morale; better service to community; and better decision making.

Unfortunately, if a benefit cannot be quantified, it is difficult to accept the validity of an associated cost-benefit analysis that is based on incomplete data. Some analysts dispute the existence of intangible benefits. They argue that all benefits are quantifiable; some are just more difficult than others. Suppose, for example, that improved customer goodwill is listed as a possible intangible benefit. Can we quantify goodwill? You might try the following analysis:

1. What is the result of customer ill will? The customer will submit fewer (or no) orders.
2. To what degree will a customer reduce orders? Your user may find it difficult to specifically quantify this impact. But you could try to have the end-user estimate the possibilities (or invent an estimate to which the end-user can react). For instance,
   a. There is a 50 percent (.50) chance that the regular customer would send a few orders—fewer than 10 percent of all its orders—to competitors to test their performance.
   b. There is a 20 percent (.20) chance that the regular customer would send as many as half its orders (.50) to competitors, particularly those orders we are historically slow to fulfill.
   c. There is a 10 percent (.10) chance that a regular customer would send us an order only as a last resort. That would reduce that customer's normal business with us to 10 percent of its current volume (90 percent or .90 loss).
   d. There is a 5 percent (.05) chance that a regular customer would choose not to do business with us at all (100 percent or 1.00 loss).
3. We can calculate an estimated business loss as follows:

$$\begin{aligned}\text{Loss} &= .50 \times (.10 \text{ loss of business})\\ &\quad +.20 \times (.50 \text{ loss of business})\\ &\quad +.10 \times (.90 \text{ loss of business})\\ &\quad +.50 \times (1.00 \text{ loss of business})\\ &= .29\\ &= 29\% \text{ statistically estimated loss of business}\end{aligned}$$

4. If the average customer does $40,000 per year of business, then we can expect to lose 29 percent or $11,600 of that business. If we have 500 customers, this can be expected to amount to a total of $5,800,000.
5. Present this analysis to management, and use it as a starting point for quantifying the benefit.

Is the Proposed System Cost-Effective?

There are three popular techniques to assess economic feasibility, also called *cost-effectiveness:* payback analysis, return on investment, and net present value.

The choice of techniques should consider the audiences that will use them. Virtually all managers who have come through business schools are familiar with all three techniques. One concept that should be applied to each technique is the adjustment of cost and benefits to reflect the time value of money.

The Time Value of Money A concept shared by all three techniques is the **time value of money**—a dollar today is worth more than a dollar one year from now. You could invest that dollar today and, through accrued interest, have more than one dollar a year from now. Thus, you'd rather have that dollar today than in one year. That's why your creditors want you to pay your bills promptly—they can't invest what they don't have. The same principle can be applied to costs and benefits *before* a cost-benefit analysis is performed.

Some of the costs of a system will be accrued after implementation. Additionally, all benefits of the new system will be accrued in the future. Before cost-benefit analysis, these costs should be brought back to current dollars. An example should clarify the concept.

Suppose we are going to realize a benefit of $20,000 two years from now. What is the current dollar value of that $20,000 benefit? The current value of the benefit is the amount of money we would need to invest today to have $20,000 two years from now. If the current return on investments is running about 10 percent, an investment of $16,528 today would give us our $20,000 in two years (we'll show you how to calculate this later). Therefore, the current value of the estimated benefit is $16,528—that is, we'd rather have $16,528 today than the promise of $20,000 two years from now.

The same adjustment could be made on costs that are projected into the future. For example, suppose we are projecting a cost of $20,000 two years from now. What is the current dollar value of that $20,000 cost? The current value of the cost is the amount of money we would need to invest today to have $20,000 to pay the cost two years from now. Again, if we assume a 10 percent return on current investments, an investment of $16,528 today would give us the needed $20,000 in two years. Therefore, the current value of the estimated cost is $16,528—that is, we can fulfill our cost obligation of $20,000 in two years by investing $16,528 today.

Why go to all this trouble? Because projects are often compared against other projects that have different lifetimes. Time value analysis techniques have become the preferred cost-benefit methods for most managers. By time-adjusting costs and benefits, you can improve the following cost-benefit techniques.

Payback Analysis The **payback analysis** technique is a simple and popular method for determining if and when an investment will pay for itself. Because systems development costs are incurred long before benefits begin to accrue, it will take some time period for the benefits to overtake the costs. After implementation, you will incur additional operating expenses that must be recovered. Payback analysis determines how much time will lapse before accrued benefits overtake accrued and continuing costs. This period of time is called the **payback period**.

In Figure C.3 we see an information system that will be developed at a cost of $418,040. The estimated net operating costs for each of the next six years are also

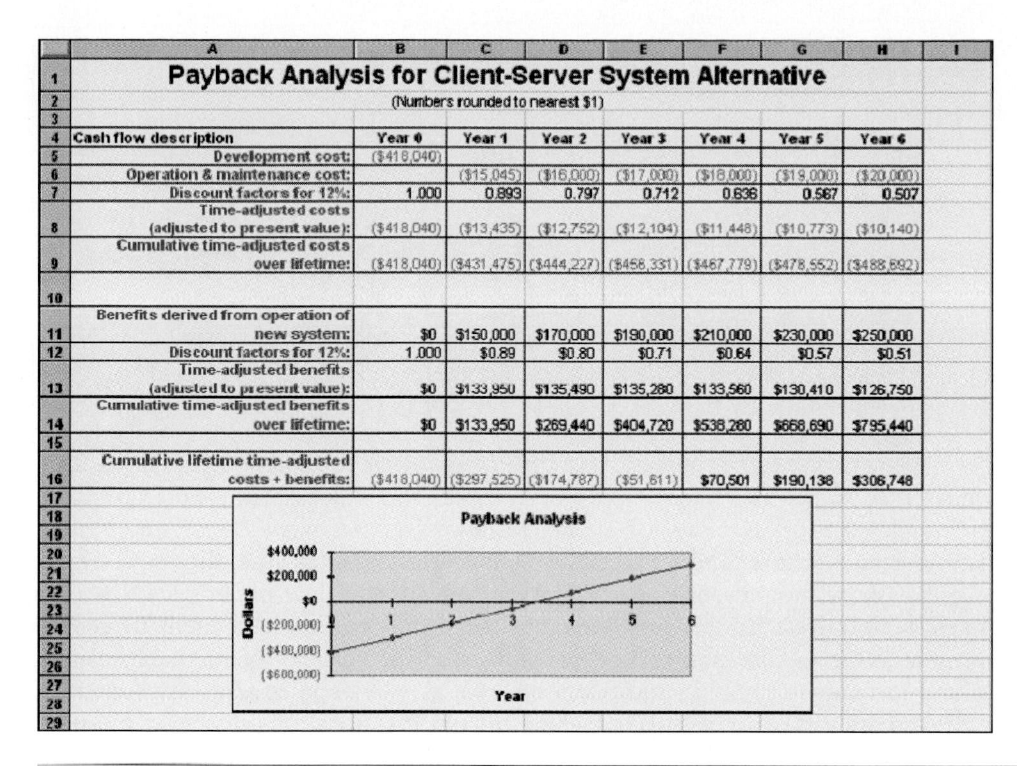

Payback Analysis for Client-Server System Alternative

(Numbers rounded to nearest $1)

Cash flow description	Year 0	Year 1	Year 2	Year 3	Year 4	Year 5	Year 6
Development cost:	($418,040)						
Operation & maintenance cost:		($15,045)	($16,000)	($17,000)	($18,000)	($19,000)	($20,000)
Discount factors for 12%:	1.000	0.893	0.797	0.712	0.636	0.567	0.507
Time-adjusted costs (adjusted to present value):	($418,040)	($13,435)	($12,752)	($12,104)	($11,448)	($10,773)	($10,140)
Cumulative time-adjusted costs over lifetime:	($418,040)	($431,475)	($444,227)	($456,331)	($467,779)	($478,552)	($488,892)
Benefits derived from operation of new system:	$0	$150,000	$170,000	$190,000	$210,000	$230,000	$250,000
Discount factors for 12%:	1.000	$0.89	$0.80	$0.71	$0.64	$0.57	$0.51
Time-adjusted benefits (adjusted to present value):	$0	$133,950	$135,490	$135,280	$133,560	$130,410	$126,750
Cumulative time-adjusted benefits over lifetime:	$0	$133,950	$269,440	$404,720	$538,280	$668,690	$795,440
Cumulative lifetime time-adjusted costs + benefits:	($418,040)	($297,525)	($174,787)	($51,611)	$70,501	$190,138	$306,748

FIGURE C.3 *Payback Analysis for a Project*

recorded in the table. The estimated net benefits over the same six operating years are also shown. What is the payback period?

First, we need to adjust the costs and benefits for the time value of money (that is, adjust them to current dollar values). Here's how! The present value of a dollar in year n depends on something typically called a **discount rate**. The discount rate is a percentage similar to interest rates that you earn on your savings account. In most cases the discount rate for a business is the **opportunity cost** of being able to invest money in other projects, including the possibility of investing in the stock market, money market funds, bonds, and the like. Alternatively, a discount rate could represent what the company considers an acceptable return on its investments. This number can be learned by asking any financial manager, officer, or comptroller.

Let's say the discount rate for our sample company is 12 percent. The current value, actually called the **present value**, of a dollar at any time in the future can be calculated using the following formula:

$$PV_n = 1(1 + i)^n$$

where PV_n is the present value of $1.00 n years from now and i is the discount rate. Therefore, the present value of a dollar two years from now is

$$PV_2 = 1(1 + .12)^2 = 0.797$$

Does that bother you? Earlier we stated that a dollar today is worth more than a dollar a year from now. But it looks as if it is worth less. This is an illusion. The present value is interpreted as follows. If you have 79.7 cents today, it is better than having 79.7 cents two years from now. How much better? Exactly 20.3 cents better since that 79.7 cents would grow into one dollar in two years (assuming our 12 percent discount rate).

To determine the present value of any cost or benefit in year 2, you simply multiply 0.797 times the estimated cost or benefit. For example, the estimated operating expense in year 2 is $16,000. The present value of this expense if $16,000 × 0.797, or $12,752 (rounded up). Fortunately, you don't have to calcu-

FIGURE C.4 *Partial Table for Present Value of a Dollar*

Periods	....	8%	10%	12%	14%	....
1		0.926	0.909	0.893	0.877	
2		0.857	0.826	0.797	0.769	
3		0.794	0.751	0.712	0.675	
4		0.735	0.683	0.636	0.592	
5		0.681	0.621	0.567	0.519	
6		0.630	0.564	0.507	0.456	
7		0.583	0.513	0.452	0.400	
8		0.540	0.467	0.404	0.351	
.						
.						
.						

late discount factors. There are tables similar to the partial one shown in Figure C.4 that show the present value of a dollar for different time periods and discount rates. Simply multiply this number times the estimated cost or benefit to get the present value of that cost or benefit. More detailed versions of this table can be found in many accounting and finance books as well as in spreadsheet functions.

Better still, most spreadsheets include built-in functions for calculating the present value of any cash flow, be it cost or benefit. All the examples in this module were done with Microsoft *Excel*. The same tables can be prepared with *Lotus 1-2-3*. The beauty of a spreadsheet is that once the rows, columns, and functions have been set up, you simply enter the costs and benefits and let the spreadsheet discount the numbers to present value. (In fact, you can also program the spreadsheet to perform the cost-benefit analysis.)

Returning to Figure C.3, we have brought all costs and benefits for our example back to present value. Notice that the discount rate for year 0 is 1.000. Why? The present value of a dollar in year 0 is exactly $1. It makes sense. If you hold a dollar today, it is worth exactly $1!

Now that we've discounted the costs and benefits, we can complete our payback analysis. Look at the cumulative lifetime costs and benefits. The lifetime costs are gradually increasing over the six-year period because operating costs are being incurred. But also notice that the lifetime benefits are accruing at a much faster pace. Lifetime benefits will overtake the lifetime costs between years 3 and 4. By charting the cumulative lifetime time-adjusted Costs + Benefits, we can estimate that the break-even point (when Costs + Benefits = 0) will occur approximately 3.5 years after the system begins operating.

Is this information system a good or bad investment? It depends! Many companies have a payback period guideline for all investments. In the absence of such a guideline, you need to determine a reasonable guideline before you determine the payback period. Suppose that the guideline states that all investments must have a payback period less than or equal to four years. Because our example has a payback period of 3.5 years, it is a good investment. If the payback period for the system were greater than four years, the information system would be a bad investment.

It should be noted that you can perform payback analysis without time-adjusting the costs and benefits. The result, however, would show a 2.8-year payback that looks more attractive than the 3.5-year payback that we calculated. Thus, non-time-adjusted paybacks tend to be overoptimistic and misleading.

Return-on-Investment Analysis The **return-on-investment (ROI) analysis** technique compares the lifetime profitability of alternative solutions or projects. The ROI for a solution or project is a percentage rate that measures the relationship

	A	B	C	D	E	F	G	H	I	J
1	Net Present Value Analysis for Client-Server System Alternative									
2	(Numbers rounded to nearest $1)									
3										
4	Cash flow description	Year 0	Year 1	Year 2	Year 3	Year 4	Year 5	Year 6	Total	
5	Development cost:	($418,040)								
6	Operation & maintenance cost:		($15,045)	($16,000)	($17,000)	($18,000)	($19,000)	($20,000)		
7	Discount factors for 12%:	1.000	0.893	0.797	0.712	0.636	0.567	0.507		
8	Present value of annual costs:	($418,040)	($13,435)	($12,752)	($12,104)	($11,448)	($10,773)	($10,140)		
9	Total present value of lifetime costs:								($488,692)	
10										
11	Benefits derived from operation of new system:	$0	$150,000	$170,000	$190,000	$210,000	$230,000	$250,000		
12	Discount factors for 12%:	1.000	$0.89	$0.80	$0.71	$0.64	$0.57	$0.51		
13	Present value of annual benefits:	$0	$133,950	$135,490	$135,280	$133,560	$130,410	$126,750		
14	Total present value of lifetime benefits:								$795,440	
15										
16	NET PRESENT VALUE OF THIS ALTERNATIVE:								$306,748	
17										

FIGURE C.5 *Net Present Value Analysis for a Project*

between the amount the business gets back from an investment and the amount invested. The ROI for a potential solution or project is calculated as follows:

ROI = (Estimated lifetime benefits − Estimated lifetime costs) / Estimated lifetime costs

Let's calculate the ROI for the same systems solution we used in our discussion of payback analysis. Once again, all costs and benefits should be time-adjusted. The time-adjusted costs and benefits were presented in rows 9 and 16 of Figure C.3. The estimated lifetime benefits minus estimated lifetime costs equal

$795,440 − $488,692 = $306,748

Therefore, the ROI is

ROI = $306,748/$488,692 = .628 = 63%

This is a lifetime ROI, not an annual ROI. Simple division by the lifetime of the system yields an average ROI of 10.5 percent per year. This solution can be compared with alternative solutions. The solution offering the highest ROI is the best alternative. However, as was the case with payback analysis, the business may set a minimum acceptable ROI for all investments. If none of the alternative solutions meets or exceeds that minimum standard, then none of the alternatives is economically feasible. Once again, spreadsheets can greatly simplify ROI analysis through their built-in financial analysis functions.

We could have calculated the ROI without time-adjusting the costs and benefits. This would, however, result in a misleading 129.4 percent lifetime or a 21.6 percent annual ROI. Consequently, we recommend time-adjusting all costs and benefits to current dollars.

Net Present Value The **net present value** of an investment alternative is considered the preferred cost-benefit technique by many managers, especially those who have substantial business schooling. Once again, you initially determine the costs and benefits for each year of the system's lifetime. And once again, we need to adjust all the costs and benefits back to present dollar values.

Figure C.5 illustrates the net present value technique. Costs are represented by negative cash flows while benefits are represented by positive cash flows. We have brought all costs and benefits for our example back to present value. Notice again that the discount rate for year 0 (used to accumulate all development costs) is 1.000 because the present value of a dollar in year 0 is exactly $1.

After discounting all costs and benefits, subtract the sum of the discounted costs from the sum of the discounted benefits to determine the net present value. If it is positive, the investment is good. If negative, the investment is bad. When comparing multiple solutions or projects, the one with the highest positive net present value is the best investment. (This works even if the alternatives have different lifetimes!) In our example the solution being evaluated yields a net present value of $306,748. This means that if we invest $306,748 at 12 percent for six years, we will make the same profit that we'd make by implementing this information systems solution. This is a good investment provided no other alternative has a net present value greater than $306,748.

Once again, spreadsheets can greatly simplify net present value analysis through their built-in financial analysis functions.

FEASIBILITY ANALYSIS OF CANDIDATE SYSTEMS

During the systems selection and procurement phases of system design, the systems analyst identifies candidate system solutions and then analyzes those solutions for feasibility. We discussed the criteria and techniques for analysis in this chapter. In this concluding section, we evaluate a pair of documentation techniques that can greatly enhance the comparison and contrast of candidate system solutions. Both use a matrix format. We have found these matrices useful for presenting candidates and recommendations to management.

Candidate Systems Matrix

The first matrix allows us to compare candidate systems on the basis of several characteristics. The **candidate systems matrix** documents similarities and differences between candidate systems; however, it offers no analysis.

The columns of the matrix represent candidate solutions. Better analysts always consider multiple implementation options. At least one of those options should be the existing system because it serves as our baseline for comparing alternatives.

The rows of the matrix represent characteristics that differentiate the candidates. For purposes of this book, we based our characteristics on the information system building blocks. The breakdown is as follows:

- TECHNOLOGY—describe the technical solution represented by the candidate system.
- INTERFACES—identify how the system will interact with people and other systems.
- DATA—identify how data stores will be implemented (e.g., conventional files, relational databases, other database structures), how inputs will be captured (e.g., on-line, batch, etc.), how outputs will be generated (e.g., on a schedule, on demand, printed, on screen, etc.).
- PROCESSES—identify how (manual) business processes will be modified, how computer processes will be implemented. For the latter, we have numerous options, including on-line versus batch processes and packaged versus built-in-house software.
- GEOGRAPHY—identify how processes and data will be distributed. Once again, we might consider several alternatives—for example, centralized versus decentralized versus distributed (or duplicated) versus cooperative (client/server) solutions.

The cells of the matrix document whatever characteristics help the reader understand the differences between options. Figure C.6 demonstrates the basic structure of the matrix.

Before considering any solutions, we must consider any constraints on solutions. Solution constraints take the form of architectural decisions intended to bring order and consistency to applications. For example, a technology architecture may restrict solutions to relational databases or client/server networks.

	Candidate 1 Name	Candidate 2 Name	Candidate 3 Name
Technology			
Interfaces			
Data			
Processes			
Geography			

FIGURE C.6 *Candidate Systems Matrix Template*

A sample, partially completed candidate systems matrix listing three of the five candidates is shown in Figure C.7. In Figure C.7, the matrix is used to provide overview characteristics concerning the portion of the system to be computerized, the business benefits, and software tools and/or applications needed. Subsequent pages would provide additional details concerning other characteristics such as those mentioned previously. Two columns can be similar except for their entries in one or two cells. Multiple pages would be used if we were considering more than three candidates. A simple word processing "table" template can be duplicated to create a candidate systems matrix.

Feasibility Analysis Matrix

The second matrix complements the candidate systems matrix with an analysis and ranking of the candidate systems. It is called a **feasibility analysis matrix.**

The columns of the matrix correspond to the same candidate solutions as shown in the candidate systems matrix. Some rows correspond to the feasibility criteria presented in this chapter. Rows are added to describe the general solution and a ranking of the candidates. The general format is shown in Figure C.8.

The cells contain the feasibility assessment notes for each candidate. Each row can be assigned a rank or score for each criterion (e.g., for operational feasibility, candidates can be ranked 1, 2, 3, etc.). After ranking or scoring all candidates on each criterion, a final ranking or score is recorded in the last row. Be careful. Not all feasibility criteria are necessarily equal in importance. Before assigning final rankings, you can quickly eliminate any candidates for which any criterion is deemed infeasible. In reality, this doesn't happen very often.

A completed feasibility analysis matrix is presented as Figure C.9. In Figure C.9 the feasibility assessment is provided for each candidate solution. In this example, a score is recorded directly in the cell for each candidate's feasibility criteria assessment. Again, this matrix format can be most useful for defending your recommendations to management.

Characteristics	Candidate 1	Candidate 2	Candidate 3	Candidate ...
Portion of System Computerized Brief description of that portion of the system that would be computerized in this candidate.	COTS package Platinum Plus from Entertainment Software Solutions would be purchased and customized to satisfy Member Services required functionality.	Member Services and warehouse operations in relation to order fulfillment.	Same as candidate 2.	
Benefits Brief description of the business benefits that would be realized for this candidate.	This solution can be implemented quickly because it's a purchased solution.	Fully supports user required business processes for SoundStage Inc. Plus more efficient interaction with member accounts.	Same as candidate 2.	
Servers and Workstations A description of the servers and workstations needed to support this candidate.	Technically architecture dictates Pentium Pro, MS Windows NT class servers and Pentium, MS Windows NT 4.0 workstations (clients).	Same as candidate 1.	Same as candidate 1.	
Software Tools Needed Software tools needed to design and build the candidate (e.g., database management system, emulators, operating systems, languages, etc.). Not generally applicable if applications software packages are to be purchased.	MS Visual C++ and MS Access for customization of package to provide report writing and integration.	MS Visual Basic 5.0 System Architect 3.1 Internet Explorer	MS Visual Basic 5.0 System Architect 3.1 Internet Explorer	
Application Software A description of the software to be purchased, built, accessed, or some combination of these techniques.	Package Solution	Custom Solution	Same as candidate 2.	
Method of Data Processing Generally some combination of: on-line, batch, deferred batch, remote batch, and real-time.	Client/Server	Same as candidate 1.	Same as candidate 1.	
Output Devices and Implications A description of output devices that would be used, special output requirements (e.g., network, preprinted forms, etc.), and output considerations (e.g., timing constraints).	(2) HP4MV department laser printers (2) HP5SI LAN laser printers	(2) HP4MV department laser printers (2) HP5SI LAN laser printers (1) PRINTRONIX bar-code printer (includes software & drivers) Web pages must be designed to VGA resolution. All internal screens will be designed for SVGA resolution.	Same as candidate 2.	
Input Devices and Implications A description of input methods to be used, input devices (e.g., keyboard, mouse, etc.), special input requirements (e.g., new or revised forms from which data would be input), and input considerations (e.g., timing of actual inputs).	Keyboard & mouse	Apple "Quick Take" digital camera and software (15) PSC Quickscan laser bar-code scanners (1) HP Scanjet 4C Flatbed Scanner Keyboard & mouse	Same as candidate 2.	
Storage Devices and Implications Brief description of what data would be stored, what data would be accessed from existing stores, what storage media would be used, how much storage capacity would be needed, and how data would be organized.	MS SQL Server DBMS with 100GB arrayed capability.	Same as candidate 1.	Same as candidate 1.	

FIGURE C.7 *Sample Candidate Systems Matrix*

	Candidate 1 Name	Candidate 2 Name	Candidate 3 Name
Description			
Operational Feasibility			
Technical Feasibility			
Schedule Feasibility			
Economic Feasibility			
Ranking			

FIGURE C.8 *Feasibility Analysis Matrix Template*

Feasibility Criteria	Wt.	Candidate 1	Candidate 2	Candidate 3	Candidate ..
Operational Feasibility **Functionality**. A description of to what degree the candidate would benefit the organization and how well the system would work. **Political**. A description of how well received this solution would be from both user management, user, and organization perspective.	30%	Only supports Member Services requirements and current business processes would have to be modified to take advantage of software functionality **Score: 60**	Fully supports user required functionality. **Score: 100**	Same as candidate 2. **Score: 100**	
Technical Feasibility **Technology**. An assessment of the maturity, availability (or ability to acquire), and desirability of the computer technology needed to support this candidate. **Expertise**. An assessment of the technical expertise needed to develop, operate, and maintain the candidate system.	30%	Current production release of Platinum Plus package is version 1.0 and has only been on the market for 6 weeks. Maturity of product is a risk and company charges an additional monthly fee for technical support. Required to hire or train C++ expertise to perform modifications for integration requirements. **Score: 50**	Although current technical staff has only Powerbuilder experience, the senior analysts who saw the MS Visual Basic demonstration and presentation have agreed the transition will be simple and finding experienced VB programmers will be easier than finding Powerbuilder programmers and at a much cheaper cost. MS Visual Basic 5.0 is a mature technology based on version number. **Score: 95**	Although current technical staff is comfortable with Powerbuilder, management is concerned with recent acquisition of Powerbuilder by Sybase Inc. MS SQL Server is a current company standard and competes with SYBASE in the client/server DBMS market. Because of this we have no guarantee future versions of Powerbuilder will "play well" with our current version of SQL Server. **Score: 60**	
Economic Feasibility **Cost to develop:** **Payback period (discounted):** **Net present value:** **Detailed calculations:**	30%	Approximately $350,000. Approximately 4.5 years. Approximately $210,000. See Attachment A. **Score: 60**	Approximately $418,040. Approximately 3.5 years. Approximately $306,748. See Attachment A. **Score: 85**	Approximately $400,000. Approximately 3.3 years. Approximately $325,500. See Attachment A. **Score: 90**	
Schedule Feasibility An assessment of how long the solution will take to design and implement.	10%	Less than 3 months. **Score: 95**	9-12 months **Score: 80**	9 months **Score: 85**	
Ranking	100%	60.5	92	83.5	

FIGURE C.9 *Sample Feasibility Analysis Matrix*

SUMMARY

1. Feasibility is a measure of how beneficial the development of an information system would be to an organization. Feasibility analysis is the process by which we measure feasibility. It is an ongoing evaluation of feasibility at various checkpoints in the life cycle. At any of these checkpoints, the project may be canceled, revised, or continued. This is called a creeping commitment approach to feasibility.
2. There are four feasibility tests: operational, technical, schedule, and economic.
   a. Operational feasibility is a measure of problem urgency or solution acceptability. It includes a measure of how the end-users and managers feel about the problems or solutions.
   b. Technical feasibility is a measure of how practical solutions are and whether the technology is already available within the organization. If the technology is not available to the firm, technical feasibility also looks at whether it can be acquired.
   c. Schedule feasibility is a measure of how reasonable the project schedule or deadline is.
   d. Economic feasibility is a measure of whether a solution will pay for itself or how profitable a solution will be. For management, economic feasibility is the most important of our four measures.
3. To analyze economic feasibility, you itemize benefits and costs. Benefits are either tangible (easy to measure) or intangible (hard to measure). To properly analyze economic feasibility, try to estimate the value of all benefits. Costs fall into two categories: development and operating.
   a. Development costs are onetime costs associated with analysis, design, and implementation of the system.
   b. Operating costs may be fixed over time or variable with respect to system usage.
4. Given the costs and benefits, economic feasibility is evaluated by the techniques of cost-benefit analysis. Cost-benefit analysis determines if a project or solution will be cost-effective—if lifetime benefits will exceed lifetime costs. There are three popular ways to measure cost-effectiveness: payback analysis, return-on-investment analysis, and net present value analysis.
   a. Payback analysis defines how long it will take for a system to pay for itself.
   b. Return-on-investment and net present value analyses determine the profitability of a system.
   c. Net present value analysis is preferred because it can compare alternatives with different lifetimes.
5. A candidate systems matrix is a useful tool for documenting the similarities and differences between candidate systems being considered.
6. A feasibility analysis matrix is used to evaluate and rank candidate systems. Both the candidate systems matrix and the feasibility analysis matrix are useful for presenting the results of a feasibility analysis as part of a system proposal.

KEY TERMS

candidate systems matrix, p. 656
creeping commitment, p. 643
discount rate, p. 653
economic feasibility, p. 646
feasibility, p. 643
feasibility analysis, p. 643
feasibility analysis matrix, p. 657
intangible benefits, p. 651
net present value, p. 655
operational feasibility, p. 646
opportunity cost, p. 653
payback analysis, p. 652
payback period, p. 652
present value, p. 653
return-on-investment (ROI) analysis, p. 654
schedule feasibility, p. 646
tangible benefits, p. 650
technical feasibility, p. 646
time value of money, p. 652

REVIEW QUESTIONS

1. What is the difference between feasibility and feasibility analysis?
2. What are the four tests for project feasibility? How is each test for feasibility measured?
3. How can the PIECES framework be used in operational feasibility analysis? Explain.
4. When performing usability analysis, list the three goals that help measure the usability of an interface.
5. What are the two categories of costs? Give several examples of each.
6. What is the difference between fixed and variable operating costs? Give several examples of each.
7. What is the difference between a tangible and an intangible benefit? Give several examples of each.
8. What are the three popular techniques to assess economic feasibility?
9. Of the three techniques in question 8, which one is most preferred by managers today?

PROBLEMS AND EXERCISES

1. Explain what is meant by the creeping commitment approach to feasibility. What feasibility checkpoints can be built into a systems development life cycle?
2. Visit a local information systems shop. Try to obtain documentation of their systems development life cycle standards or guidelines. What feasibility checkpoints have been installed? What feasibility checkpoints do you think should be installed? (Don't be misled into believing that only during phases labeled "feasibility" is feasibility analyzed. There may be other points in the life cycle where this also happens.)
3. What feasibility criteria does the information systems shop you visited for Exercise 2 use to evaluate projects? How do the criteria compare to the criteria in this book? Have we omitted any tests that they feel are important? Have they omitted any tests we use?
4. Can you think of any technological trends or products that may be technically infeasible for the small- to medium-sized business at the current time? Defend your reasoning.
5. Whether or not you have information systems experience, you have experience with people who use computers (including friends, relatives, acquaintances, teachers, and fellow employees). Considerating their biases for and against computers, identify issues that may make a proposed system operationally infeasible or unacceptable to those individuals.
6. List several intangible benefits. How would you quantify each intangible benefit in terms of dollars and cents (a measure that management can understand)?
7. What are some of the advantages and disadvantages of the payback analysis, return-on-investment analysis, and present value analysis cost-benefit techniques?
8. A new production scheduling information system for XYZ Corporation could be developed at a cost of $125,000. The estimated net operating costs and estimated net benefits over five years of operation would be:

Year	Estimated Net Operating Costs	Estimated Net Benefits
0	$125,000	$ 0
1	3,500	26,000
2	4,700	34,000
3	5,500	41,000
4	6,300	55,000
5	7,000	66,000

   Assuming a 12 percent discount rate, what would be the payback period for this investment? Would this be a good or bad investment? Why?
9. What is the ROI (return on investment) for the project in exercise 8?
10. What is the net present value of the investment in exercise 8 if the current discount rate is 12 percent?

SUGGESTED READINGS

Gildersleeve, Thomas R. *Successful Data Processing Systems Analysis.* 2nd ed. Englewood Cliffs, NJ: Prentice Hall, 1985. This book provides an excellent chapter on cost-benefit analysis techniques. We are indebted to Gildersleeve for the creeping commitment concept.

Gore, Marvin, and John Stubbe. *Elements of Systems Analysis.* 4th ed. Dubuque, IA: Brown, 1988. The feasibility analysis chapter suggests an interesting matrix approach to identifying, cataloging, and analyzing the feasibility of alternative solutions for a system.

Wetherbe, James. *Systems Analysis and Design: Traditional, Structured, and Advanced Concepts and Techniques.* 2nd ed. St. Paul, MN: West, 1984. Wetherbe pioneered the PIECES framework for problem classification. In this module we extended that framework to analyze operational feasibility of solutions.

D

JOINT APPLICATION DEVELOPMENT

MODULE PREVIEW AND OBJECTIVES

Effective joint application development (JAD) techniques are crucial to the rapid development of systems projects. JAD is performed during many phases of the systems development life cycle to collect business requirements and confirm design decisions regarding DATA, PROCESSES, INTERFACES, and GEOGRAPHY. This module introduces this increasingly popular alternative for fact-finding. You will know that you understand the process of JAD when you can:

- Recognize joint application development as an alternative fact-finding technique used throughout systems development, and how it can expedite the process.
- Characterize the typical participants in a JAD session and describe their roles.
- Complete the planning process for conducting a JAD session, including selecting and equipping the location, selecting the participants, and preparing an agenda to guide the JAD session.
- Describe several benefits of using JAD as a fact-finding technique.

JOINT APPLICATION DEVELOPMENT

Separate interviews of end-users and management have always been the classic fact-finding technique practiced during systems development. However, many analysts and organizations have discovered the great flaw of interviewing—separate interviews often lead to conflicting facts, opinions, and priorities, not to mention significant time and effort being expended. For these reasons, many organizations are using the group work session as a substitute for interviews. One example of the group work session approach is IBM's joint application design. This and similar techniques generally require extensive training to work as intended. However, they can significantly decrease the time spent on fact-finding in one or more phases of the life cycle.

> **Joint application design** (JAD) is a process whereby highly structured group meetings or miniretreats involving system users, system owners, and analysts occur in a single room for an extended time (four to eight hours per day, anywhere from one day to a couple of weeks).

JAD-like techniques are becoming increasingly common in systems planning and systems analysis to obtain group consensus on problems, objectives, and requirements. Therefore, it is more commonly referred to as **joint application development**—to appropriately reflect that it includes more than simply systems design.

In this module, you will learn about the participants of a JAD session and their roles. We will also learn how to go about planning and conducting a JAD session, the tools and techniques that are used during a JAD session, and the benefits to be achieved through JAD.

JAD PARTICIPANTS

Joint application development sessions include a wide variety of participants and roles. Each participant is expected to attend and actively participate for the entire JAD session. Let's examine the different types of individuals involved in a typical JAD session and their roles.

Sponsor

Any successful JAD session requires a single person, called the **sponsor,** to serve as its *champion.* This person is normally an individual who is in top management who has authority that spans the different departments and users who are to be involved in the systems project. The role of the sponsor is to give full support to the systems project by encouraging designated users to willingly and actively participate in the JAD session(s). Recalling the "creeping commitment" approach to systems development, it is the sponsor who usually makes final decisions regarding the go or no-go direction of the project.

The sponsor plays a visible role during a JAD session by "kicking off" the meeting with introductions of the participants. Often, the sponsor will also make closing remarks for the session. The sponsor also works closely with the JAD leader to plan the session by helping identify individuals from the user community who should attend and determining the time and location for the JAD session.

JAD Leader (or Facilitator)

JAD sessions involve a single individual who plays the role of the leader or facilitator. The **JAD leader** is usually responsible for leading all sessions that are held for a systems project. This individual is someone who:

- Has excellent communication skills.
- Possesses the ability to negotiate and resolve group conflicts.
- Has a good knowledge of the business.
- Has strong organizational skills.
- Is impartial to decisions that will be addressed .
- Does not report to any of the JAD session participants.

It often is difficult to find an individual within the company who possesses all these traits. Thus, companies often must provide extensive JAD training or hire an expert from outside the organization to fill this role.

The role of the JAD leader is to plan the JAD session, conduct the session, and follow through on the results. During the session, the leader is to serve as a facilitator, leading the discussion, encouraging the attendees to actively participate, resolving issue conflicts that may arise, and ensuring the goals and objectives of the meeting are fulfilled. It is the JAD leader's responsibility to establish the ground rules that will be followed during the meeting and ensure that the participants abide by these rules.

Users and Managers

Joint application development includes a number of participants from the user and management sectors of an organization who are given release time from their day-to-day job to devote themselves to active involvement in the JAD sessions. These participants are normally chosen by the project sponsor, who must be careful to ensure that each person has the business knowledge to contribute during the fact-finding sessions. The project sponsor must exercise authority and encouragement to ensure that these individuals will be committed to actively participate.

A typical JAD session may involve anywhere from a relatively small number of user/management people to a dozen or more. The role of the users during a JAD session is to effectively communicate business rules and requirements, review design prototypes, and make acceptance decisions. The role of the managers during a JAD session is to approve project objectives, establish project priorities, approve schedules and costs, and approve identified training needs and implementation plans. Overall, both users and managers are depended on to ensure that their critical success factors are being addressed.

Scribe(s)

A JAD session also includes one or more **scribes** who are responsible for keeping records pertaining to everything discussed in the meeting. These records are published and disseminated to the attendees immediately following the meeting in order to maintain the momentum that has been established by the JAD session and its members. This need to quickly publish the records is reflected in more scribes using CASE tools to capture many facts (documented using data and process models) that are communicated during a JAD session. Thus, it is advantageous for the scribe to possess strong knowledge of systems analysis and design and skill with using CASE tools.

IS Staff

A JAD session may also include a number of IS personnel who primarily listen and take notes regarding issues and requirements voiced by the users and managers. Normally, IS personnel do not speak up unless invited to do so. Rather, any questions or concerns they have are usually directed to the JAD leader immediately after or before the JAD session. It is the JAD leader who initiates and facilitates discussion of issues by users and managers.

The IS staff in the JAD session usually consists of members of the project team. These members may work closely with the scribe to develop models and other documentation related to facts communicated during the meeting. Specialists may also be called on to gain information regarding special technical issues and concerns that may arise. When the situation warrants, the JAD leader may prompt the IS professional to address the technical issue.

HOW TO PLAN AND CONDUCT JAD SESSIONS

Most JAD sessions span three to five days and occasionally last up to two weeks. The success of any JAD session depends on proper planning and effectively carrying out that plan. In this section, we will examine how a JAD session is planned and conducted.

Planning the JAD Session

Some preparation is necessary well before the JAD session can be performed. Before planning a JAD session, the analyst must work closely with the executive sponsor to determine the scope of the project that is to be addressed through JAD sessions. It is also important that the high-level requirements and expectations of each JAD session are determined. This normally involves interviewing selected individuals who are responsible for departments or functions that are to be addressed by the systems project. Finally, before planning the JAD session, the analyst must ensure that the executive sponsor is willing to commit people, time, and other resources to the JAD session.

Planning for a JAD session involves three steps: selecting a location for the JAD session, selecting JAD participants, and preparing an agenda to be followed during the JAD session. Let's examine each of these planning steps in detail.

Selecting a Location for JAD Sessions When possible, JAD sessions should be conducted away from the company workplace. Most local hotels or universities have facilities designed to host group meetings. By holding the JAD session at an off-site location, the attendees can concentrate on the issues and activities related to the JAD session and avoid interruptions and distractions that would occur at their regular workplace. Regardless of the location of the JAD session, all attendees should be required to attend and be prohibited from returning to their regular workplace.

A JAD session typically requires several rooms. A conference room is required in which the entire group can meet to address JAD issues. Also, if the JAD session includes many people, several small breakout rooms may be needed for separate groups of people to meet and focus discussion on specific issues.

The conference or main meeting room should comfortably hold all the attendees. The room should be fully equipped with tables, chairs, and other items that meet the needs of all attendees. Figure D.1 depicts a typical room layout for a JAD session. Typical visual aids for a JAD room should include: white board or blackboard; one or more flipcharts; and overhead projector(s).

The room should also include computer equipment needed by scribes to record facts and issues communicated during the session. The computer itself should include software packages to support the various types of records or documentation to be captured and later published by the scribe. Such software may include: CASE tool; word processing; spreadsheet; presentation package; prototyping software (i.e., 4GL); printer; copier (or quick access); and computer projection capability.

Finally, the room should include notepads and pencils for users, managers, and other attendees. Attendees should also be provided with name tags and place cards. Plan on also providing snacks and drinks to make the attendees as comfortable as possible. Creature comforts are very important since JAD sessions are very intensive and often run the entire day.

Selecting JAD Participants As mentioned earlier, the analyst, executive sponsor, and managers establish the needs and expectations for a JAD session. It is also their responsibility to select the JAD leader. Ideally, an experienced JAD leader may be available in-house. If not, an individual may be selected to obtain the extensive training needed to conduct JAD sessions. Many companies opt to hire a qualified person from outside the organization. Hiring an individual outside the organization provides the benefit of having a JAD leader who will not be biased.

One or more scribes must also be selected. Since scribes must possess technical skills (word processing, CASE, data and process modeling, etc.), they are usually selected from among the ranks of the organization's IS professionals. Sometimes the duties of the scribe are shared—one scribe may be responsible for taking notes and meeting presentation materials; whereas another scribe may be focused on documenting technical requirements and issues such as developing data models, process models, and prototypes.

In addition to scribes, other IS professionals must be selected to be involved in the JAD session. Once again, the role of these IS professionals is primarily to

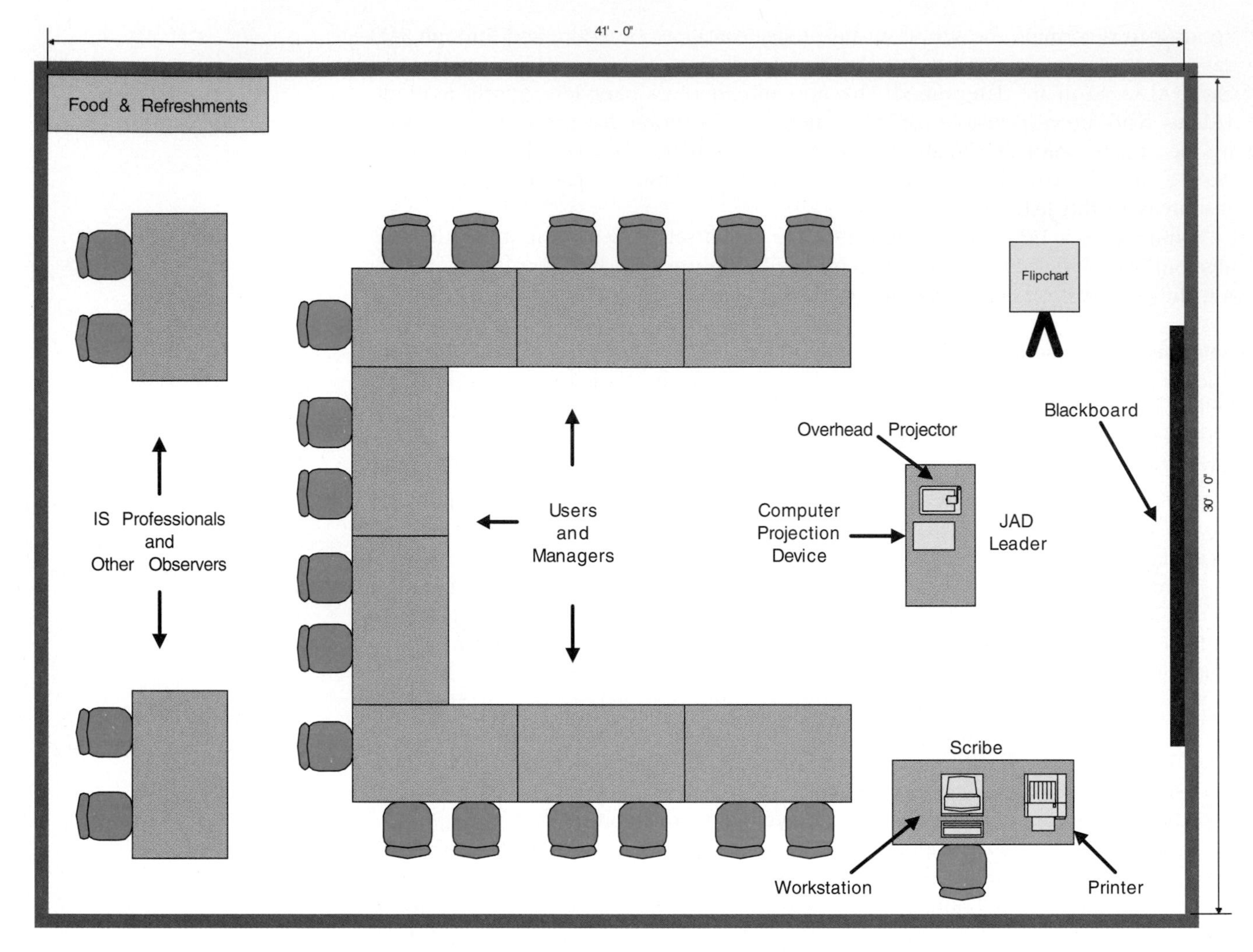

FIGURE D.1 *Typical Room Layout for JAD Session*

listen and learn about the users' and manager' needs. Usually all IS individuals assigned to the project team are involved in the JAD session. Other IS specialists may also be assigned to address specific technical issues pertaining to the project.

Finally, the analyst and managers must select individuals from the user community. While all managers will participate, clearly not all users can attend and participate. The number of attendees must be kept at a level that provides a conducive environment for each attendee to be able to actively participate. Thus, only those users who are able to clearly articulate facts and opinions will be invited. These individuals should be key individuals who are knowledgeable about their business area. Unfortunately, managers are often very dependent on these individuals to run their business area and are therefore often hesitant to release them from their duties. Thus, the analyst must ensure that management is committed to the JAD project and willing to not only permit but also require these key individuals to participate.

Preparing a JAD Session Agenda Preparation is the key to a successful JAD session. The JAD leader must prepare documentation to brief the participants about the scope and objectives of the sessions. In addition, an agenda for each JAD session should be prepared and distributed before each session. The agenda dictates issues to be discussed during the session and the amount of time allotted to each item.

The agenda should contain three parts: the opening, body, and conclusion. The opening is intended to communicate the expectations of the session, to communicate the ground rules, and to influence or motivate the attendees to participate. The body is intended to detail the topics or issues to be addressed in the JAD session. Finally, the conclusion is intended to allow time to summarize the day's session and to remind the attendees of unresolved issues (to be carried forward).

Conducting a JAD Session

The JAD session begins with opening remarks, introductions, and a brief overview of the agenda and objectives for the session. The JAD leader will direct the session by following the prepared script. To successfully conduct the session, the leader should follow these guidelines:

- Do not unreasonably deviate from the agenda.
- Stay on schedule (agenda topics are allotted specific time).
- Ensure that the scribe is able to take notes (this may mean having the users and managers restate their points more slowly or clearly).
- Avoid the use of technical jargon.
- Apply conflict resolution skills.
- Allow for ample breaks.
- Encourage group consensus.
- Encourage user and management participation without allowing individuals to dominate the session.
- Make sure that attendees abide by the established ground rules for the session.

As mentioned earlier, the success of a JAD session is highly dependent on planning and the skills of the JAD leader and scribes. These skills only get better through proper training and experience. Therefore, JAD sessions are usually concluded with an evaluation questionnaire for the participants to complete. The responses will help ensure the likelihood of future JAD successes.

The end product of a JAD session is typically a formal written document. This document is essential in confirming the specifications agreed on during the session to all participants. The content and organization of the specification is obviously dependent on the objectives of the JAD session. The analyst may provide a different set of specifications to different participants based on their role—for example, a manager may receive more of a summary version of the document provided to the user participants (especially in those cases in which the system owners had minimal actual involvement in the JAD session).

BENEFITS OF JAD

Joint application development offers many benefits as an alternative fact-finding and development approach. More and more companies are beginning to realize its advantages and are incorporating JAD into their existing methodologies. An effectively conducted JAD session offers the following benefits:

- JAD actively involves users and management in the development project (encouraging them to take "ownership" in the project).
- JAD reduces the amount of time required to develop systems. This is achieved by replacing traditional, time-consuming one-on-one interviewing of each user and manager with group meetings. The group meetings allow for more easily obtaining consensus among the users and managers, as well as resolving conflicting information and requirements.
- When JAD incorporates prototyping as a means for confirming requirements and obtaining design approvals, the benefits of prototyping are realized.

Once again, achieving a successful JAD session depends on the JAD leader and the leader's ability to plan and facilitate the JAD session.

SUMMARY

1. Separate interviews of end-users and management have always been the classic fact-finding technique practiced during systems development. However, many analysts and organizations have discovered the great flaw of interviewing—separate interviews often lead to conflicting facts, opinions, and priorities, not to mention significant time and effort being expended. For these reasons, many organizations are using the **joint application development (JAD)** alternative to interviewing. JAD is a process whereby highly structured meetings or miniretreats involving users, managers, analysts, and other IS personnel meet collectively for an extended period to define system needs and expectations.
2. Joint application development sessions include a wide variety of participants and roles. Each participant is expected to attend and actively participate for the entire duration of the JAD session. Participants include:
   a. A sponsor who serves as the JAD project's champion. This person is normally an individual who is in top management who has authority that spans the different departments and users who are to be involved in the systems project. The role of the sponsor is to give full support to the systems project by encouraging designated users to willingly and actively participate in the JAD session.
   b. A JAD leader. JAD sessions involve a single individual who plays the role of the leader or facilitator. The JAD session leader is usually responsible for leading all sessions that are held for a systems project.
   c. Several users and managers. Joint application development includes a number of participants from the user and management sectors of an organization who are given release time from their day-to-day jobs to devote themselves to active involvement in the JAD sessions.
   d. One or more scribes. Scribes are primarily responsible for keeping records pertaining to everything discussed in the JAD session.
   e. Several IS personnel. A JAD session may also include a number of IS personnel who are primarily in attendance to listen and take notes regarding issues and requirements voiced by the users and managers. Normally, IS personnel do not speak up unless invited to do so.
3. An effective JAD session requires extensive planning. Planning for a JAD session involves three steps: selecting a location for the JAD session, selecting JAD participants, and preparing an agenda.
4. It is essential the JAD leader be experienced or fully trained on how to conduct JAD sessions. When leading the session, the JAD leader should follow a prepared agenda.
5. JAD offers many benefits to systems development, including:
   a. Actively involving users.
   b. Reducing the overall development time.
   c. Allowing for prototyping and its benefits to also be realized.

KEY TERMS

JAD leader, p. 663
JAD sponsor, p. 663
joint application design, p. 663
joint application development, p. 663
scribe, p. 664

REVIEW QUESTIONS

1. What is joint application design?
2. Why is the term joint application development more commonly accepted than joint application design?
3. Who are the participants in a JAD session?
4. Who is a JAD sponsor and what is that person's role?
5. Who is a JAD leader and what is that person's role?
6. What characteristics are important for a JAD leader?
7. What is a scribe and what is that person's role in JAD?
8. What steps are involved in planning a JAD session?
9. Where should a JAD session be located and why?
10. What resources should be available at the location for a JAD session?
11. What important guidelines should a JAD leader adhere to in running a JAD session?
12. What benefits can be derived through a successful JAD project?

PROBLEMS AND EXERCISES

1. Explain how JAD may significantly decrease the amount of systems development time.
2. Why might an organization choose to use a JAD leader from outside the company?
3. Explain why IS personnel should primarily take a back-seat role in a JAD session?
4. Research the topic of conflict resolution. Describe one or more approaches that the leader may use to resolve conflicts.
5. Write a memo that would defend your recommendations to conduct a JAD session at an off-site location.

PROJECTS AND RESEARCH

1. Group decision support systems (GDSS) are popular for dealing with conflict resolutions. Research this topic and write a brief overview of its principles and how it is applied to JAD-like situations.
2. Research available software GDSS products (see item 1). What would the resource implications be for using a GDSS product in a JAD session?

SUGGESTED READINGS

Andrews, D. C., and N. S. Leventhal. *Fusion Integrating IE, CASE and JAD: A Handbook for Reengineering the Systems Organization*. Englewood Cliffs, NJ: Prentice Hall, 1993.

Gane, C. *Rapid Systems Development*. New York: Rapid Systems Development, Inc., 1987. This book provides a good discussion on how to lead a group meeting/interview.

Mitchell, Ian; Norman Parrington; Peter Dunne; and John Moses. "Practical Prototyping, Part One." *Object Currents*, May 1996. First of a three-part series of articles that explores prototyping and how you can benefit from it. Prototyping is an integral part of JAD.

Wood, Jane, and Denise Silver. *Joint Application Design*. New York: John Wiley & Sons, 1989. This book provides a comprehensive overview of IBM's joint application design technique.

E

INTERPERSONAL SKILLS AND COMMUNICATIONS

MODULE PREVIEW AND OBJECTIVES

Despite the availability of improved tools and methodologies, many information systems projects still fail due to breakdowns in communications. Information systems projects are frequently plagued by communications barriers between the analyst and the system users. The business world has its own language to describe forms, methods, procedures, financial data, and the like. And the information systems industry has its own language of acronyms, terms, buzzwords, and procedures. A communications gap has developed between the system users and the system designers.

The systems analyst is supposed to bridge this communications gap. A typical project requires the participation of a diverse audience, both technical and nontechnical. In this module you will receive a survey of interpersonal skills, the cornerstone of successful systems development. You will know that you understand the interpersonal and communication skills needed by the systems analyst when you can:

- Understand and perform the six guidelines for effective listening.
- Identify the four speaking styles and the situations where you would use each.
- Identify and list examples of both benefit terms and loss terms, and the responses that they would elicit from the audience.
- Understand what body language and proxemics are and why a systems analyst should care about them. Know and understand the procedures to be able to prepare for, conduct, and follow up on meetings, formal presentations, and project walkthroughs.
- Demonstrate and perform the proper methods in writing business and technical reports.

Note: The authors gratefully acknowledge the contributions of Professor Kevin C. Dittman who developed this module.

COMMUNICATING WITH PEOPLE

Organizations within the information technology arena are placing a greater premium on communications skills. No longer are data processing professionals thought of as individuals who spend most of their careers behind closed doors programming away, shielded from outside influences especially users, only to pause while their supervisor cautiously slides a plate of nourishment under the door from time to time. In today's age, not many companies have the luxury of allowing employees to specialize in a single skill. With downsizing, or to use the politically correct term *rightsizing,* the information technology professional, including the systems analyst, must be multiskilled and an excellent problem solver. In doing that, the professional must have good if not impeccable communications skills. Outstanding communication skills will aid the climb up the corporate ladder. Don Walton, a consultant on communications, quotes the CEO and chairman of The Prudential Insurance Company of America in his book, *Are You Communicating?*

> Starting my Prudential career as an agent, I understood quickly that although people may listen, they don't always hear. I had to make sure, therefore, that my presentations were clear, concise, and to the point. In addition, I taught myself to listen and understand others, another crucial point in making sales. Clear communication is an important component of any career foundation. I have seen bright, ambitious people fail simply because they were unable to understand the importance of this. The person who has the ability to make his or her point simply and effectively, while clearly understanding what is being said by others, will have the best chance of success in a society and business environment as complex and multidimensional as ours.[1]

We continue this module with one of the earliest recorded stories of communication problems, which we feel is an excellent lead-in to the material we are going to present. The story is called the Tower of Babel and is as follows:

> Once upon a time all the world spoke a single language and used the same words. As men journeyed in the east, they came upon a plain in the land of Shinar and settled there. They said to one another, "Come, let us make bricks and bake them hard"; they used bricks for stone and bitumen for mortar. "Come," they said, "let us build ourselves a city and a tower with its top in the heavens, and make a name for ourselves; or we shall ever be dispersed all over the earth." Then the Lord came down to see the city and tower which mortal men had built, and he said, "Here they are, one people with a single language, and now they have started to do this; henceforward nothing they have a mind to do will be beyond their reach. Come, let us go down there and confuse their speech, so that they will not understand what they say to one another." So the Lord dispersed them from there all over the earth, and they left off building the city. That is why it is called Babel, because the Lord there made a babble of the language of all the world; from that place the Lord scattered men all over the face of the earth.[2]

The Tower of Babel project, like many information systems projects, failed because of a breakdown in communications. Information systems projects are frequently plagued by communications barriers, usually created intentionally or accidentally by the project participants. The system's owners and users have their own language to describe forms, methods, procedures, and so on. System designers and builders have their own terms, acronyms, and buzzwords for describing the same things. As a result, a communications gap has developed between these groups.

Because systems are built by people for people, understanding people is an appropriate introduction to communications skills. With whom does the systems analyst communicate? How are they different? What words influence these people, and in what ways?

---

[1] Donald Walton, *Are You Communicating? You Can't Manage without It* (New York: McGraw-Hill, 1989), p. 24.

[2] Genesis 11:1–9 from *The New English Bible* (The Delegates of the Oxford University Press and The Syndics of the Cambridge Press, 1961, 1970.) Reprinted by permission.

Four Audiences for Interpersonal Communication during Systems Projects

For years English and communications scholars have told us that the secret of effective oral and written communications is to know the audience. Who is the audience during a systems development project? We can identify at least four distinct groups:

1. System designers, consisting of your colleagues—other analysts and information systems specialists.
2. System builders, the programmers and technical specialists who will actually construct the system.
3. System users, the people whose day-to-day jobs will be affected, directly or indirectly, by the new system.
4. System owners, who, in addition to possibly being system users, sponsor the project and approve systems expenditures.

You should recognize all these audiences as end-users. Each audience has different levels of technical expertise, different perspectives on the system, and different expectations. System users and system owners, in particular, present special problems. These people have day-to-day responsibilities and time constraints. Before communicating with any of them, ask yourself the following questions:

- What are the responsibilities of, and how might the new system affect, this person?
- What is the attitude of this person toward the existing system or the target system?
- What kind of information about the project does this person really need or want?
- How busy is this person? How much of this person's time and attention can I reasonably expect?

Listening

When most people talk about communication skills, they think of speaking and writing. The skill of listening hardly gets mentioned, but it may be the most important. As a systems analyst you will be working with customers or users trying to solve their system problems. To be successful, you must be able to listen to their problems and understand what they are asking you to do. As Thomas Gildersleeve states in his book, *Successful Data Processing Systems Analysis,* you must distinguish between hearing and listening. "To hear is to recognize that someone is speaking. To listen is to understand what the speaker wants to communicate."

We have been conditioned most of our lives to learn how not to listen. For example, take how we can ignore our quarreling brothers and sisters while we enjoy our favorite show on television, or as a student we learn to study by blocking out distractions such as noisy roommates, or our personal favorite of being oblivious to the outside world, including our spouses, while "Monday Night Football" is on. We have learned not to listen, but we can also learn how to listen effectively and productively.

Guidelines in Effective Listening

When a trying to solve a customer's problems, your most difficult task may be getting the customer to communicate to you what you want to know. The following guidelines can open the lines of communication.

Approach the Session with a Positive Attitude No matter what your feelings are for the person you are working with, or the project as whole, approaching it with a negative attitude is fighting a losing battle. You have a job to do! You might as well make the best of it and look at it as a fun, pleasurable experience. Your performance appraisal and career growth may depend on it.

Set the Other Person at Ease It's no secret that a good way to get a person talking is to present a nice, cheerful attitude. A good approach is to start by talking about the person's interest or hobbies. Showing an interest in the individual's personal life sometimes can serve as an icebreaker and put the person more at ease.

Let Them Know You Are Listening Few things are more irritating than having someone slump back in his chair and stare into space while you are speaking to him This gives you the impression that the other person really doesn't care what you have to say. When you are listening, always maintain eye contact and use a response such as a head nod or say "uh-huh" to indicate you acknowledge what the other person is saying. Always maintain good posture and even sit on the edge of your seat and lean forward. This will give the other person the impression that you are really interested in what he or she is saying.

Ask Questions To make sure you clearly understand what the person is saying or to clarify a point, ask a question to help you understand. This will show that you are listening and will also give the other person the opportunity to expand on the answer.

Don't Assume Anything One of the worst things to do is to get in a hurry and be impatient with the speaker. For example, you assume you know what the other person is going to say so you cut in and finish the sentence, possibly missing entirely what the person was going to say, plus irritating the speaker in the process. Or you interrupt or stop the speaker because you may have already heard that information before or you believe it is not applicable to what you are doing, thus risking missing a valuable piece of information. Don't assume anything! Art Linkletter learned this lesson on his popular television show, "Kids Say the Darndest Things":

> On my show I once had a child tell me he wanted to be an airline pilot. I asked him what he'd do if all the engines stopped out over the Pacific Ocean. He said, "First I would tell everyone to fasten their seatbelts, and then I'd find my parachute and jump out."
>
> While the audience rocked with laughter, I kept my attention on the young man to see if he was being a smart alec. The tears that sprang into his eyes alerted me to his chagrin more than anything he could have said, so I asked him why he'd do such a thing. His answer revealed the sound logic of a child: "I'm going for gas . . . I'm coming back!"[3]

Take Notes Taking notes serves two purposes. First, jotting down brief notes while the other person is speaking gives the impression that what the person has to say is important enough that you want to write it down. Second, it helps you remember the major points of the meeting when you reference your notes later.

Speaking

Given a choice most people would rather talk to each other than use any other form of communication. Learning to speak interestingly and effectively is one of the most admired skills in our business. It is an attribute that separates the leaders from the followers.

Systems analysts need to be able to speak effectively in their work to be successful. An effective speaker delivers a clear and concise message that is received and understood for its intended purpose. As a systems analyst you will need to give instruction, conduct interviews, conduct and attend meetings, give presentations, and answer questions. This form of conversation is sometimes called **business or intellectual speaking.** It is a formal method of oral communication in which successful individuals learn to think before they speak by following a logical process. They approach it very similarly to the way they approach business

[3] Donald Walton, *Are You Communicating? You Can't Manage without It* (New York: McGraw-Hill, 1989), p. 31.

writing communication. Before they speak, they organize their thoughts to think about what is the purpose for speaking, what is the main point, who is the intended audience, and what are the desired results. As they speak they obtain feedback via oral response or body language to see if the message is being received and the desired results are being obtained. If not, they can change their approach and try again.

Keep in mind that different situations may call for different speaking styles, just as different business writings call for different writing styles. Courtland L. Bovee and John V. Thill (1989), noted authors in business communications, have identified four speaking styles and the situations when they are used.

- **Expressive style**—spontaneous, conversational, and uninhibited. We use this style when we are expressing our feelings, joking around, complaining, being intimate, or socializing. For example: "Over my dead body will I let that deadbeat be the leader of this project!"
- **Directive style**—authoritative and judgmental. We use this style to give orders, give instruction, exert leadership, pass judgment, or state our opinions. For example: "I want John to be the leader of this project!"
- **Problem-solving style**—rational, objective, unbiased, and bland. This is the style most used in business dealings. For example: "I believe John's past performance qualifies him to be an ideal candidate to be the leader of this project."
- **Meta style**—used to discuss the communications process itself. Meta language enables us to talk about our interactions. For example: "It appears we have some disagreement over who should be the leader of this project."

When you think about the message you are trying to convey, be sure to match the style to its intended purpose.

Use of Words: Turn-ons and Turnoffs

We communicate with words, both oral and written. How important are words? Ask any politician. The wrong words at the wrong time, no matter what the intention, and the next election is history. But that's just politics, right? No. All businesses are political. And choosing the right words is important, especially to the systems analyst who must effectively communicate with a diverse group of system users, owners, and builders. What words affect the feelings, attitudes and decisions of your audience?

First, let's talk about words and phrases that appeal to system users and owners. Leslie Matthies (1976), a noted author and consultant in the systems development field, has identified two categories of terms that influence managers: benefit terms and loss terms. Both can be used to sell ideas.

> **Benefit terms** are words or phrases that evoke positive responses from the audience. Benefit terms can be used very effectively to sell proposed changes. Managers will usually accept ideas that produce benefit terms.

Some examples of benefit terms are: increase productivity; reduce inventory costs; increase profit margin; improve customer relations; increase sales; and reduce risk.

People like to feel they are part of the systems development effort. Avoid using the first-person pronoun *I*. People also like words of appreciation for their time and effort—systems development is your job, not theirs, and they are helping you to do your job. Make their names and department names a vivid part of any presentation. Most of all, people want respect. Words should be carefully chosen to show respect for people's feelings, knowledge, and skills.

> **Loss terms** are words or phrases that evoke negative responses from the audience. Loss terms can also be used very effectively to sell proposed changes. Managers will usually accept ideas that eliminate loss terms.

Some examples of loss terms are: higher costs; increased processing errors; higher credit losses; excessive waste; higher taxes; delays; and increased stockouts.

What about turnoff words or phrases? These can kill projects by changing the attitudes and opinions of management. Let's start with the oldest turnoff, the use of jargon. Jargon is important to the analyst and technician because it helps us to easily communicate with the computing industry and our colleagues. But jargon has no place in the business system user's world. Avoid terms such as JCL, EBCDIC, CPU, ROM, and DOS—leave your acronyms in the CIS offices! This includes the jargon you've learned in this book. For example, instead of saying "This is a DFD of your materials handling system," try saying "This is a picture of the work and data flow in your materials handling operation."

Other red-flag terms include those that attack people's performance or threaten their job. Before you candidly state that the current system is inefficient and cumbersome, consider the possibility that a system user who had a major role in its development and approval may be in your audience. Consider potential threats to job security when you get ready to propose the elimination of job steps. In other words, be diplomatic and tactful when you speak.

Electronic Mail

We are constantly finding newer and more efficient ways of communicating with other people. One of the newer forms of interpersonal communication of particular importance to the systems analyst is **electronic mail (e-mail).**

Electronic mail gives us the ability to create, edit, send, and receive information electronically, usually using some type of computer network. All that's required is a computer and some type of mail software. The advantages of this form of communication are obvious. A person can send messages to and receive messages from someone almost instantaneously practically anywhere in the world (provided both people are linked by some type of computer network). These messages can be read, stored, printed, edited, or deleted. Also, once the mail system software and computer network are in place, the actual cost of sending a message is very small. Many mail packages allow individual users to be grouped together so that one message can be simultaneously sent to many different people (for example, a letter to all programmers in a company with multiple sites).

Unfortunately, electronic mail has some disadvantages. First, the sheer volume of electronic mail an individual receives may be overwhelming. This can be particularly true if the user is automatically receiving mail from special-interest mailing list servers. Also, because it is so quick and easy to create a response to an electronic mail message and because mail users sometimes forget that they are communicating with another person via a machine, not with the machine directly, electronic mail messages are sometimes blunt, tactless, or inflammatory. Personal privacy is another concern. An electronic mail message is only as private as the security built into the mail software and the computer network that carries the message. Finally, electronic mail deprives its users of some of the richness of other forms of communication, such as tone of voice, facial expression, body language, and so on. Even with these drawbacks, electronic mail is growing rapidly and will be a major form of communication for many people for years to come.

Body Language and Proxemics

What is body language, and why should a systems analyst care about it?

> **Body language** is all of the information being communicated by an individual other than spoken words. Body language is a form of nonverbal communication that we all use and are usually unaware of.

Why should the analyst be concerned with body language? Research studies have determined a startling fact: Of a person's total feelings, only 7 percent are communicated verbally (in words), 38 percent are communicated by the tone of voice used, and 55 percent of those feelings are communicated by facial and body expressions. If you only listen to someone's words, you are missing most of what he or she has to say!

For this discussion, we will focus on just three aspects of body language: facial disclosure, eye contact, and posture. Facial disclosure means you can sometimes understand how people feel by watching the expressions on their faces. Many common emotions have easily recognizable facial expressions associated with them. However, you need to be aware that the face is one of the most controlled parts of the body. Some people who are aware that their expressions often reveal what they are thinking are very good at disguising their faces.

Another form of nonverbal communication is eye contact. Eye contact is the least controlled part of the face. Have you ever spoken to someone who will not look directly at you? How did it make you feel? A continual lack of eye contact may indicate uncertainty. A normal glance is usually from three to five seconds in length; however, direct eye contact time should increase with distance. As an analyst, you need to be careful not to use excessive eye contact with a threatened user so that you won't further intimidate that person. Direct eye contact can cause strong feelings, either positive or negative, in other people. If eyes are "the window to the soul," be sure to search for any information they may provide.

Posture is the least controlled aspect of the body, even less than the face or voice. As such, body posture holds a wealth of information for the astute analyst. Members of a group who are in agreement tend to display the same posture. A good analyst will watch the audience for changes in posture that could indicate anxiety, disagreement, or boredom. An analyst should normally maintain an "open" body position signaling approachability, acceptance, and receptiveness. In special circumstances, the analyst may choose to use a confrontation angle of head-on or at a 90-degree angle to another person in order to establish control and dominance.

In addition to the information communicated by body language, individuals also communicate via proxemics.

Proxemics is the relationship between people and the space around them. Proxemics is a factor in communications that can be controlled by the knowledgeable analyst.

People tend to be very territorial about their space. Observe where your classmates sit in a course that does not have assigned seats. Or the next time you are involved in a conversation with someone, deliberately move much closer or farther away from the person and see what happens. A good analyst is aware of four spatial zones:

1. Intimate zone—closer than 1.5 feet.
2. Personal zone—from 1.5 feet to 4 feet.
3. Social zone—from 4 feet to 12 feet.
4. Public zone—beyond 12 feet.

Certain types of communications occur only in some of these zones. For example, an analyst conducts most interviews with system users in the personal zone. But the analyst may need to move back to the social zone if the user displays any signs (body language) of being uncomfortable. Sometimes increasing eye contact can make up for a long distance that can't be changed. Many people use the fringes of the social zone as a "respect" distance.

We have examined some of the informal ways that people communicate their feelings and reactions. A good analyst will use all the information available, not just the written or verbal communications of others. The remainder of this module will survey several common interpersonal communications techniques—specifically meetings, formal presentations, project walkthroughs, and written reports.

MEETINGS

During a systems development project, many meetings are usually held.

A **meeting** is an attempt to accomplish an objective as a result of discussion under leadership. Some possible meeting objectives are: presentation; problem solving; conflict resolution; progress analysis; gathering and merging facts; decision making; training; and planning.

The ability to coordinate or participate in a meeting is critical to the success of any project. In this section we will discuss how to prepare for, conduct, and follow up on a meeting. The following section will focus on two special types of meetings: formal presentations and project walkthroughs.

Preparing for a Meeting

Many people have a very negative image of meetings because many meetings are poorly organized and/or poorly conducted. Meetings are also very expensive because they require several people to dedicate time that could be better spent on other productive work. The more individuals involved in a meeting, the more the meeting costs. But because meetings are an essential form of communication, we must strive to offset the meeting costs by maximizing benefits (in terms of project progress) realized during the meeting. It is not difficult to run a meeting if you are well prepared. Without good organization, however, the meeting may prove chaotic or worthless to the participants. When planning and conducting meetings, use the following steps.

Step 1: Determine the Need for and Purpose of the Meeting Why do you need a meeting? Every meeting should have a well-defined purpose that can be communicated to its participants. Meetings without a well-defined purpose are rarely productive. Some of the possible objectives of a meeting were listed earlier.

The purpose of every meeting should be attainable within 60 to 90 minutes because longer meetings tend to become unproductive. However, when necessary, longer meetings are possible if they are divided into well-defined submeetings that are separated by breaks that allow people to catch up on their normal responsibilities. But it must be remembered that longer meetings are more likely to conflict with the participants' day-to-day responsibilities. The impact on the business can be the same as if everyone took a vacation on the same day.

Step 2: Schedule the Meeting and Arrange for Facilities After deciding the purpose of the meeting, determine who should attend. The proper participants should be chosen to ensure that the purpose of the meeting can be attained. (The larger the number of participants, the less the amount of work likely to be completed.) Some research indicates that the most creative problem solving and decision making is done in small, odd-numbered groups. Given the appropriate participants, the meeting can now be scheduled. The date and time for the meeting will be subject to the availability of the meeting room and the prior commitments of the various participants. Morning meetings are generally better than afternoon meetings because the participants are fresh and not yet caught up in the workday's problems. It is best to avoid scheduling meetings in the late afternoon (when people are eager to go home), before lunch, before holidays, or on the same day as other meetings involving the same participants.

The meeting location is very important. Important factors to consider when selecting a meeting location are: size of room; lighting; outside distractions; seating arrangements; temperature; and audiovisual needs. Seating arrangement is particularly important. If leader-to-group interaction is required, the group should face the leader but not necessarily other members of the group. If group-to-group interaction is needed, the team members, including the leader, should all face one another. Make sure that any necessary visual aids (flip charts, overhead projectors, chalk, and so forth) are also available in the room.

Step 3: Prepare an Agenda A written agenda for the meeting should be distributed well in advance. The agenda confirms the date, time, location, and duration of the meeting. It also states the meeting's purpose and offers a tentative timetable for discussion and questions. If participants should bring specific materials with them or review specific documents before the meeting, specify this in the agenda. Finally, the agenda may include any supplements—for example, reports, documentation, or memoranda—that the participants will need to refer to or study before or during the meeting.

Conducting a Meeting

Try to start on time, but do not start the meeting until everyone is present. If an important participant is more than 15 minutes late, consider canceling the meeting. Once the meeting has started, try to discourage interruptions and delays, such as phone calls. Have enough copies of handouts for all participants. Get off to a good start by listing or reviewing the agenda so that the discussion items become group property. Cover each item on the agenda according to the timetable developed when the meeting was scheduled. The group leader should ensure that no one person or subgroup dominates or is left out of the discussion. Decisions should be made by consensus opinion or majority vote. One rule is always in order: Stay on the agenda and end on time! If you do not finish discussing all items on the agenda, schedule another meeting.

Meetings often offer the analyst a unique opportunity to assess the true attitudes of project participants by observing their nonverbal behavior (or body language). For a variety of reasons, people are sometimes reluctant to verbalize their thoughts and ideas. And while it is relatively easy to refrain from speaking, it is not very easy to disguise your true emotions as displayed by your body language. A really good systems analyst will listen to what users say with their words and (frequently more importantly) what they say with their actions.

Sometimes, the purpose of a meeting is to generate possible ideas to solve a problem. One approach is called brainstorming.

> **Brainstorming** is a technique for generating ideas during group meetings. Participants are encouraged to generate as many ideas as possible in a short period of time without any analysis until all the ideas have been exhausted.

Contrary to what you might believe, brainstorming is a formal technique that requires discipline. These guidelines should be followed to ensure effective brainstorming:

1. Isolate the appropriate people in a place that will be free from distractions and interruptions.
2. Make sure everyone understands the purpose of the meeting (to generate ideas to solve the problem) and focus on the problem.
3. Appoint one person to record ideas. This person should use a flip chart, chalkboard, or overhead projector that can be viewed by the entire group.
4. Remind everyone of the brainstorming rules:
   a. Be spontaneous. Call out ideas as fast as they occur.
   b. Absolutely no criticism, analysis, or evaluation of any kind is permitted while the ideas are being generated. Any idea may be useful, if only to spark another idea.
   c. Emphasize quantity of ideas, not necessarily quality.
5. Within a specified time period, team members call out their ideas as quickly as they can think of them.
6. After the group has run out of ideas and all ideas have been recorded, then and only then should the ideas be analyzed and evaluated.
7. Refine, combine, and improve the ideas that were generated earlier.

With a little practice and attention to these rules, brainstorming can be a very effective technique for generating ideas to solve problems.

Following Up on a Meeting

As soon as possible after the meeting is over, the minutes of the meeting should be published. The minutes are a brief, written summary of what happened during the meeting—items discussed, decisions made, and items for future consideration. The minutes are usually prepared by the **recording secretary,** a team member designated by the group leader.

FORMAL PRESENTATIONS

To communicate information to the many different people involved in a systems development project, a systems analyst is frequently required to make a formal presentation.

> **Formal presentations** are special meetings used to sell new ideas and gain approval for new systems. They may also be used for any of these purposes: sell new system; sell new ideas; sell change; head off criticism; address concerns; verify conclusions; clarify facts; and report progress. In many cases, a formal presentation may set up or supplement a more detailed written report.

Effective and successful presentations require three critical ingredients: preparation, preparation, and preparation. The time allotted to presentations is frequently brief; therefore, organization and format are critical issues. You cannot improvise and expect acceptance.

Presentations offer the advantage of impact through immediate feedback and spontaneous responses. The audience can respond to the presenter, who can use emphasis, timed pauses, and body language to convey messages not possible with the written word. The disadvantage to presentations is that the material presented is easily forgotten because the words are spoken and the visual aids are transient. That's why presentations are often followed by a written report, either summarized or detailed.

Preparing for the Formal Presentation

As mentioned earlier, it is particularly important to know your audience. This is especially true when your presentation is trying to sell new ideas and a new system. The systems analyst is frequently thought of as the dreaded agent of change in an organization. As Machiavelli wrote in his classic book *The Prince,*

> There is nothing more difficult to carry out, nor more dangerous to handle, than to initiate a new order of things. For the reformer has enemies in all who profit by the old order, and only lukewarm defenders in all those who would profit from the new order, this lukewarmness arising partly from fear of their adversaries—and partly from the incredulity of mankind, who do not believe in anything new until they have had actual experience of it.[4]

People tend to be opposed to change. There is comfort in the familiar way things are today. Yet a substantial amount of the analyst's job is to bring about change (in methods, procedures, technology, and the like). A successful analyst must be an effective salesperson. It is entirely appropriate (and strongly recommended) for an analyst to formally study salesmanship. To effectively present and sell change, you must be confident in your ideas and have the facts to back them up. Again, preparation is the key!

First, define your expectations of the presentation—for instance, that you are seeking approval to continue the project, that you are trying to confirm facts, and so forth. A presentation is a summary of your ideas and proposals that is directed toward your expectations.

---

[4] Niccolo Machiavelli, *The Prince and Discourses,* trans. Luigi Ricci (New York: Random House, Inc., 1940, 1950). Reprinted by permission of Oxford University Press.

FIGURE E.1 *Typical Outline and Time Allocation for an Oral Presentation*

I. Introduction (one-sixth of total time available)
   A. Problem statement
   B. Work completed to date
II. Part of the presentation (two-thirds of total time available)
   A. Summary of existing problems and limitations
   B. Summary description of the proposed system
   C. Feasibility analysis
   D. Proposed schedule to complete project
III. Questions and concerns from the audience (time here is not to be included in the time allotted for presentation and conclusion; it is determined by those asking the questions and voicing their concerns)
IV. Conclusion (one-sixth of total time available)
   A. Summary of proposal
   B. Call to action (request for whatever authority you require to continue systems development)

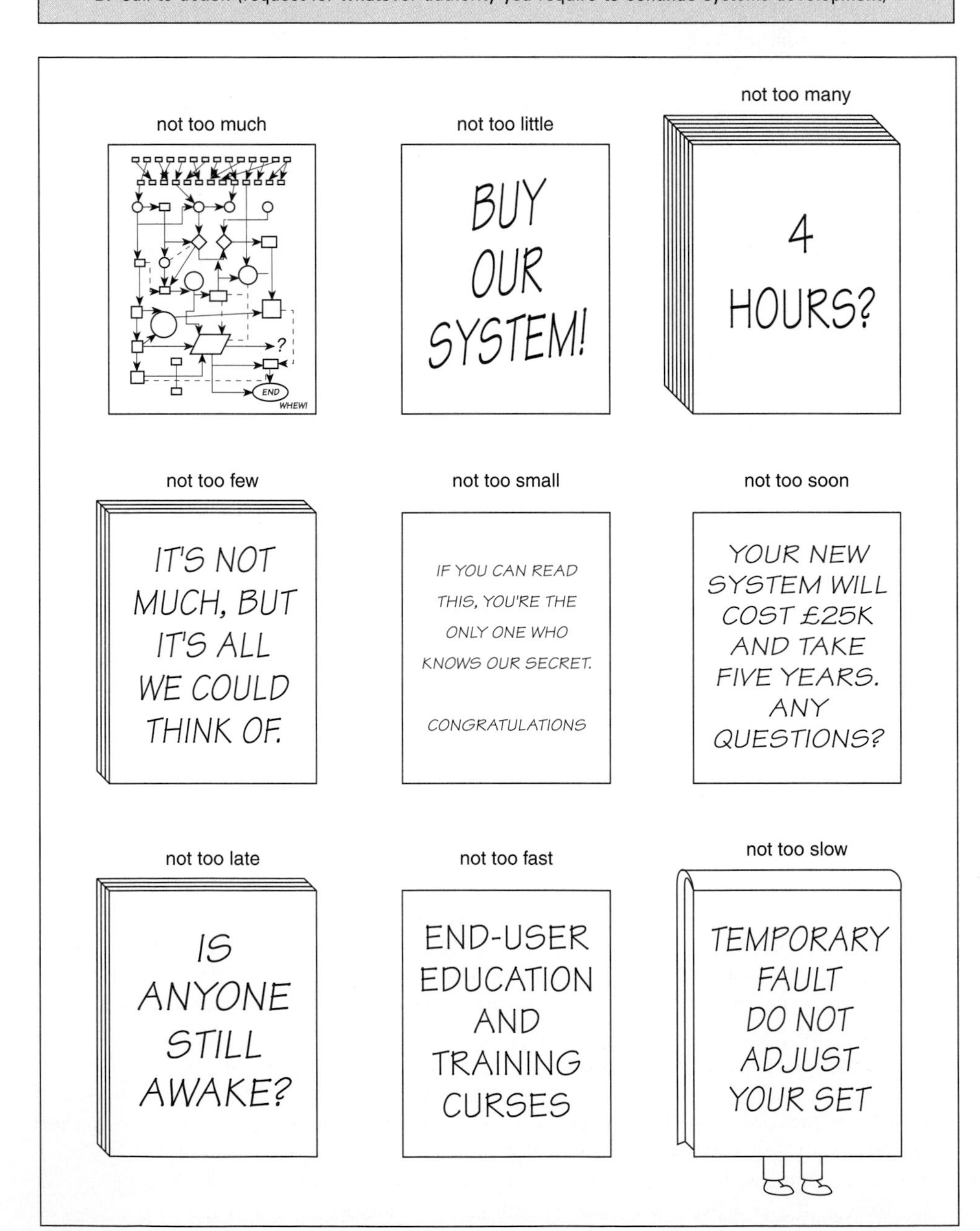

FIGURE E.2 *Guidelines for Visual Aids*

Source: Copyright Keith London.

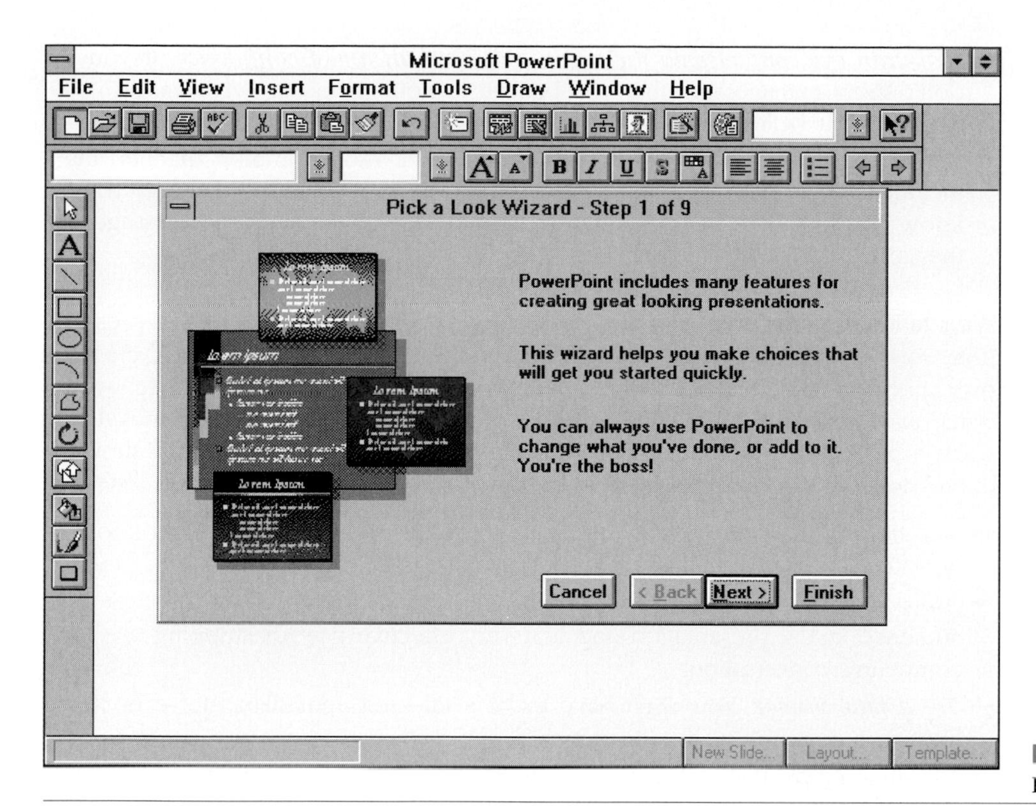

FIGURE E.3 *Microsoft* PowerPoint *Wizard*

Executives are usually put off by excessive detail. To avoid this, your presentation should be carefully organized around the allotted time (usually 30 to 60 minutes). Although each presentation differs, you might try the organization and time allocation suggested in Figure E.1. This figure illustrates some typical topics of an oral presentation and the amount of time to allow for those topics. Note that this particular outline is for a systems analysis presentation. Other types of presentations might be slightly different.

What else can you do to prepare for the presentation? Because of the limited time, use **visual aids**—predrawn flip charts, overhead slides, Microsoft *PowerPoint* slides, and the like—to support your position. Just like a written paragraph, each visual aid should convey a single idea. When preparing pictures or words, use the guidelines shown in Figure E.2.

Microsoft *PowerPoint* contains software guides called **wizards** to assist the most novice users to create professional-looking presentations (see Figure E.3). The wizard steps the user through the development process by asking a series of questions and tailoring the presentation based on responses. To hold your audience's attention, consider distributing photocopies of the visual aids at the start of the presentation. This way, the audience doesn't have to take as many notes.

Finally, practice the presentation in front of the most critical audience you can assemble. Play your own devil's advocate or, better yet, get somebody else to raise criticisms and objections. Practice your responses to these issues.

Conducting the Formal Presentation

If you are well prepared, the presentation is 80 percent complete. There are a few additional guidelines that may improve the actual presentation:

- *Dress professionally.* The way you dress influences people. John T. Malloy's books, *Dress for Success* and *The Woman's Dress for Success Book,* are excellent reading for both wardrobe advice and the results of studies regarding the effects of clothing on management.
- *Avoid using the word "I"* when making the presentation. Use *you* and *we* to assign ownership of the proposed system to management.

- *Maintain eye contact with the group and keep an air of confidence.* If you don't show management that you believe in your proposal, why should management believe in it?
- *Be aware of your own mannerisms.* Some of the most common mannerisms include using too many hand gestures, pacing, and repeatedly saying "you know" or "OK." Although mannerisms alone don't contradict the message, they can distract the audience.

Ways to Keep the Audience Listening Sometimes while you are making a presentation, some members of the audience may not be listening. This lack of attention may take several forms. Some people may be engaged in competing conversations, some may be daydreaming, some may be busy glancing at their watches, some who are listening may have puzzled expressions, and some may show no expression. The following suggestions may prove useful to keep people listening:

- *Stop talking.* The silence can be deafening. The best public speakers know how to use dramatic pauses for special emphasis.
- *Ask a question, and let someone in the audience answer it.* This involves the audience in the presentation and is a very effective way of stopping a competing conversation.
- *Try a little humor.* You don't have to be a talented comedian. But everybody likes to laugh. Tell a joke on yourself.
- *Use some props.* Use some type of visual aid to make your point clearer. Draw on the chalkboard, illustrate on the back of your notes, create a physical model to make the message easier to understand.
- *Change your voice level.* By making your voice louder or softer, you force the audience to listen more closely or make it easier for them to hear. Either way, you've made a change from what the audience was used to, and that is the best way to get and hold attention.
- *Do something totally unexpected.* Drop a book, toss your notes, jingle your keys. Doing the unexpected is almost always an attention grabber.

Answering Questions Usually a formal presentation will include a time for questions from the audience. This time is very important because it allows you to clarify any points that were unclear and draw additional emphasis to important ideas. It also allows the audience to interact with you. However, sometimes answering questions after a presentation may be difficult and frustrating. We suggest the following guidelines when answering questions:

- *Always answer questions seriously, even if you think it is a silly question.* Remember, if you make someone feel stupid for asking a "dumb" question, that person will be offended. Also, other members of the audience won't ask their questions for fear of the same treatment.
- *Answer both the individual who asked the question and the entire audience.* If you direct all your attention to the person who asked the question, the rest of the audience will be bored. If you don't direct enough attention to the person who asked the question, that person won't be satisfied. Try to achieve a balance. If the question is not of general interest to the audience, answer it later with that specific person.
- *Summarize your answers.* Be specific enough to answer the question, but don't get bogged down in details.
- *Limit the amount of time you spend answering any one question.* If additional time is needed, wait until after the presentation is over.
- *Be honest.* If you don't know the answer to a question, admit it. Never try to bluff your way out of a question. The audience will eventually find out, and

you will have destroyed your credibility. Instead, promise to find out and report back. Or ask someone in the audience to do some research and present the findings later.

Following Up the Formal Presentation

As mentioned earlier, it is extremely important to follow up a formal presentation because the spoken word and impressive visual aids used in a presentation do not usually leave a lasting impression. For this reason, most presentations are followed by written reports that provide the audience with a more permanent copy of the information that was communicated. Written reports will be covered in a later section of this module.

PROJECT WALKTHROUGHS

A special type of meeting conducted by the analyst is called a project walkthrough.

The **project walkthrough** is a peer group review of systems development documentation. Walkthroughs may be used to verify almost any type of detailed documentation such as location connectivity diagrams; data flow diagrams; entity relationship diagrams; input designs; output designs; file designs; database designs; policies and procedures; user manauals; and program code.

Why does peer group review tend to identify errors that go unnoticed by the analyst who prepared the documentation? Consider the last paper or report you wrote. You probably gave that report to a colleague or teacher to review. That colleague or teacher caught obvious errors that you didn't, right? You didn't catch them because, like any author, you have mental blocks that prevent you from discovering errors in your own products. We tend to read what we meant to say rather than reading what we actually said.

Who Should Participate in the Walkthrough?

A walkthrough group should consist of seven or fewer participants. All members of the walkthrough must be treated as equals. The analyst who prepared the documentation to be reviewed should present that documentation to the group during the walkthrough. Another analyst or key system user should be appointed as **walkthrough coordinator.** The coordinator schedules the walkthrough and ensures that each participant gets the documentation well before the meeting date. The coordinator also makes sure that the walkthrough is properly conducted and mediates disputes and problems that may arise during the walkthrough. The coordinator has the authority to ask participants to stop a disagreement and move on. Finally, the coordinator designates a **walkthrough recorder** to take notes during the walkthrough.

The remaining participants include system users, analysts, or specialists who evaluate the documentation. These reviewers may also assume roles. For example, some reviewers may evaluate the accuracy of the documentation, while other reviewers comment on quality, standards, and technical issues. Participants must be willing to devote time to details. However, walkthroughs should never last more than 90 minutes. Our experience indicates that system users particularly enjoy walkthroughs because the meetings encourage a sense of personal involvement and importance in the project.

Conducting a Walkthrough

All participants must agree to follow the same set of rules and procedures. Also, the participants must agree to review the documentation; this should not be done by the person who prepared the documentation. The basic purpose of the walkthrough is **error detection,** not error correction. The analyst who is presenting the documentation should seek only whatever clarification is needed to correct the errors. This approach maximizes the use of time! The analysts should never argue with the reviewers' comments. A defensive attitude inhibits constructive criticism. The coordinator is responsible for seeing that these rules are properly explained,

WALKTHROUGH REPORT

Coordinator	Project
Segment for Review	

Coordinator's checklist:

1. Confirm with developer that material is ready and stable ______

2. Issue invitations, assign responsibilities, distribute materials

Date ______ Time ______ Duration ______

Place ______

Responsibilities		Participants	Can attend	Received materials?

Agenda

______ 1. All participants agree to follow the *SAME* set of rules.

______ 2. New segment: walkthrough of material

______ 3. Old segment: item-by-item checkoff of previous action list

______ 4. Group decision

______ 5. Deliver copy of this form to project management.

Decision:

______ Accept product as is

______ Revise (no further walkthrough)

______ Revise and schedule another walkthrough

Signatures		

FIGURE E.4 *Typical Project Walkthrough Form*

understood, and followed. Reviewers should be encouraged to offer at least one positive and one negative comment to guarantee that the walkthrough is not superficial.

After the walkthrough, the coordinator should ask the reviewers for a recommendation. There are three possible alternatives:

1. Accept the documentation in its present form.
2. Accept the documentation with the revisions noted.
3. Request another walkthrough because a large number of errors were found or because criticisms created controversy.

Following Up on the Walkthrough

The walkthrough should be promptly followed by a written report from the coordinator. The report contains a management summary that states what was reviewed, when the walkthrough occurred, who attended, and the final recommendation. A sample form used for walkthroughs in a real company is displayed in Figure E.4.

WALKTHROUGH ACTION LIST—SCRIBE'S REPORT

Coordinator	Scribe	Date
Project	Segment	

= fixed	Issues raised in review

FIGURE E.4
(concluded)

This walkthrough form can be completed by the recorder and distributed to all participants as a record of the walkthrough.

WRITTEN REPORTS

The **business and technical report** is the primary method used by analysts to communicate information about a systems development project. The purpose of the report is to either inform or persuade, possibly both. In a few pages, it is not possible to provide a comprehensive discussion of report writing. But because people make judgments about who we are and what we can accomplish based on our writing ability, we can offer some motivation for further study and some guidelines for writing reports.

Business and Technical Reports

What types of formal reports are written by the systems analyst? Content outlines for several reports can be found in Chapters 6, 7, and 12, which place those reports in the context of the systems development life cycle phases. But an overview (or review) is appropriate here.

Systems Planning Reports The first planning phase of the systems development life cycle is the study phase. While studying the business mission, the analyst will usually prepare a planning project charter for review, correction, and approval by the appropriate managers and staff.

Next, during the definition phase, the analyst must prepare and present the information architecture and plan. This architecture and plan must be approved by both information systems manager and staff and system owners and users.

The third planning phase of the life cycle, the evaluation phase, results in several important reports, including the business area plan, planned database and/or network development projects, and planned application development projects. The last often serves as the trigger for systems analysis.

Systems Analysis Reports The next major phase of the life cycle is the survey phase. After completing this phase, the analyst normally prepares a preliminary feasibility assessment and a statement of project scope, both of which are presented to a steering committee, which makes a decision concerning the continuation or cancellation of the project.

Next, during the study phase, the analyst prepares and presents a business problem statement and new system objectives to verify with system users their understanding of the current system and analyses of problems, limitations, and constraints in that system.

The third analysis phase of the life cycle, the definition phase, results in a business requirements statement. This specification document is often large and complex and is rarely written up as a single report to system users and owners. It is best reviewed in walkthroughs (in small pieces) with users and maintained as a reference for analysts and programmers.

Systems Design Reports The next formal report, the systems proposal, is generated after the selection phase has been completed. This report combines an outline of the system user requirements from the definition phase with the detailed feasibility analysis of alternative solutions that fulfill those requirements. The report concludes with a recommended or proposed solution. This report is normally preceded or followed by a presentation to those managers and executives who will decide on the proposal.

The design phase results in detailed design specifications that are often organized into a technical design report. This report is quite detailed and is primarily intended for information systems professionals. It tends to be quite a large report because it contains numerous forms, charts, and technical specifications.

The acquisition phase of systems development is undertaken only if the new system requires the purchase of new hardware or software. Several reports can be generated during this phase. The most important report—the request for proposals—is used to communicate requirements to prospective vendors who may respond with specific proposals. It was covered in Chapter 9. Especially when the selection decision involves significant expenditures, the analyst may have to write a report that defends the recommended proposal to management.

Systems Implementation Reports In a sense, the most important report is written during the construction and delivery phases. Actually, it isn't a report; it's a manual, a users manual and reference guide. This document explains how to use the computer system (such as what keys to push, how to react to certain messages, and where to get help). How well this manual is written will frequently determine how many phone calls you'll get over the months that follow the conversion to the new system. In addition to computer manuals, the analyst may rewrite the standard operating procedures for the system. A standard operating procedure explains both the noncomputer and computer tasks and policies for the new system.

FIGURE E.5 *Formats for Written Reports*

Factual Format	Administrative Format
I. Introduction	I. Introduction
II. Methods and procedures	II. Conclusions and recommendations
III. Facts and details	III. Summary and discussion of facts and details
IV. Discussion and analysis of facts and details	IV. Methods and procedures
V. Recommendations	V. Final conclusion
VI. Conclusion	VI. Appendices with facts and details

Length of a Written Report

Unfortunately, the written report is the most abused method used by analysts to communicate with system users. We have a tendency to generate large, voluminous reports that look quite impressive. Sometimes such reports are necessary, but often they are not. If you lay a 300-page technical report on a manager's desk, you can expect that manager will skim it but not read it—and you can be certain it won't be studied carefully!

Report size is an interesting issue. After many bad experiences, we have learned to use the following general guidelines to restrict report size:

- To executive-level managers—one or two pages.
- To middle-level managers—three to five pages.
- To supervisory-level managers—less than 10 pages.
- To clerk-level personnel—less than 50 pages.

It is possible to organize a larger report to include subreports for managers who are at different levels. These subreports are usually included as early sections in the report and summarize the report, focusing on the bottom line: What's wrong? What do you suggest? What do you want?

Organizing the Written Report

There is a general pattern to organizing any report. Every report consists of both primary and secondary elements.

Primary elements present the actual information that the report is intended to convey. Examples include the introduction and the conclusion.

While the primary elements present the actual information, all reports also contain secondary elements.

Secondary elements package the report so the reader can easily identify the report and its primary elements. Secondary elements also add a professional polish to the report.

Primary Elements As indicated in Figure E.5, the primary elements can be organized in one of two formats: factual and administrative. The **factual format** is very traditional and best suited to readers who are interested in facts and details as well as conclusions. This is the format we would use to specify detailed requirements and design specifications to system users. On the other hand, the factual format is not appropriate for most managers and executives.

The **administrative format** is a modern, result-oriented format preferred by many managers and executives. This format is designed for readers who are interested in results, not facts. It presents conclusions or recommendations first. Any reader can read the report straight through, until the point at which the level of detail exceeds their interest.

Both formats include some common elements. The **introduction** should include four components: purpose of the report, statement of the problem, scope of the project, and a narrative explanation of the contents of the report. The **methods and procedures section** should briefly explain how the information contained in

FIGURE E.6 *Secondary Elements for a Written Report*

Letter of transmittal
Title page
Table of contents
List of figures, illustrations, and tables
Abstract or executive summary
(The primary elements—the body of the report, in either the factual or administrative format—are presented in this portion of the report.)
Appendices

the report was developed—for example, how the study was performed or how the new system will be designed. The bulk of the report will be in the **facts section.** This section should be named to describe the type of factual data to be presented (e.g., "Existing Systems Description," "Analysis of Alternative Solutions," or "Design Specifications"). The **conclusion** should briefly summarize the report, verifying the problem statement, findings, and recommendations.

Secondary Elements Figure E.6 shows the secondary, or packaging, elements of the report and their relationship to the primary elements. Many of these elements are self-explanatory. We briefly discuss here those that may not be. No report should be distributed without a **letter of transmittal** to the recipient. This letter should be clearly visible, not inside the cover of the report. A letter of transmittal states what type of action is needed on the report. It can also call attention to any features of the project or report that deserve special attention. In addition, it is an appropriate place to acknowledge the help you've received from various people.

The **abstract or executive summary** is a one- or two-page summary of the entire report. It helps readers decide if the report contains information they need to know. It can also serve as the highest-level summary report. Virtually every manager reads these summaries. Most managers will read on, possibly skipping the detailed facts and appendices.

Writing the Business or Technical Report

This is not a writing textbook. You should take advantage of every opportunity to improve your writing skills, through business and technical writing classes, books, audiovisual courses, and seminars. Writing can greatly influence career paths in any profession. Figure E.7 illustrates the proper procedure for writing a formal report. Here are some guidelines to follow:

- *Paragraphs should convey a single idea.* They should flow nicely, one to the next. Poor paragraph structure can almost always be traced to outlining deficiencies.
- *Sentences should not be too complex.* The average sentence length should not exceed 20 words. Studies suggest that sentences longer than 20 words are difficult to read and understand.
- *Write in the active voice.* The passive voice becomes wordy and boring when used consistently.
- *Eliminate jargon, big words, and deadwood.* For example, replace "DBMS" with "database management system," substitute "so" for "accordingly," try "useful" instead of "advantageous," and use "clearly" instead of "it is clear that."

Get yourself a copy of *The Elements of Style* by William S. Strunk, Jr., and E. B. White. This classic paperback may set a record in value-to-cost ratio. Barely bigger than a pocket-sized book, it is a virtual gold mine of information. Anything we might suggest about grammar and style can't be said any more clearly than in *The Elements of Style.*

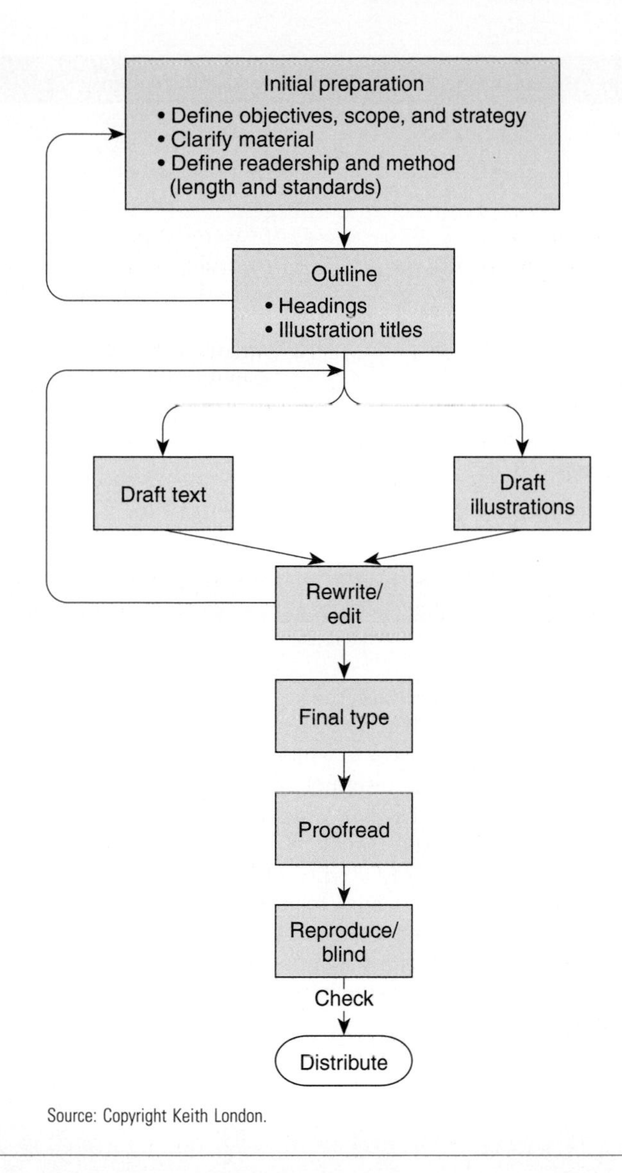

Source: Copyright Keith London.

FIGURE E.7 *Steps in Writing a Report*

SUMMARY

1. Four distinct groups of people are part of the analyst's audience: system designers, system builders, system users, and system owners. Before directing any communication to any of these audiences, the analyst should profile the audience.
2. Two important communication skills are listening and speaking. For systems analysts to be successful, they must be highly skilled in both.
3. Systems analysts can expect to spend a considerable amount of time in meetings. To maximize the use of meeting time, the analyst should follow these steps:
   a. Determine the purpose of the meeting.
   b. Schedule the meeting at an appropriate time.
   c. Arrange for adequate facilities.
   d. Conduct the meeting according to an agenda.
   e. Follow up on meeting results.
4. If the meeting is intended to generate ideas, brainstorming is an effective technique.
5. Formal presentations are a special type of meeting at which a person presents conclusions, ideas, or proposals to an interested audience. Preparation is the key to effective presentations.
6. Walkthroughs are peer group evaluation meetings that seek to identify (but not correct) errors in systems development documentation.
7. Written reports are the most common communications vehicle used by analysts. Reports consist of both primary and secondary elements. Primary elements contain factual information. Secondary elements package the report for ease of use. Reports may be organized in either the factual or administrative format. The factual format presents details before conclusions; the administrative format reverses that order. Managers like the administrative format because it is results-oriented and gets right to the bottom-line question.

KEY TERMS

abstract or executive summary, p. 688
administrative format, p. 687
benefit terms, p. 674
body language, p. 675
brainstorming, p. 678
business and technical report, p. 685
business or intellectual speaking, p. 673
conclusion, p. 688
directive style, p. 674
electronic mail (e-mail), p. 675
error detection, p. 683
expressive style, p. 674
facts section, p. 688
factual format, p. 687
formal presentations, p. 679
introduction, p. 687
letter of transmittal, p. 688
loss terms, p. 674
meeting, p. 677
meta style, p. 674
methods and procedures section, p. 687
primary elements, p. 687
problem-solving style, p. 674
project walkthrough, p. 683
proxemics, p. 676
recording secretary, p. 679
secondary elements, p. 687
visual aids, p. 681
walkthrough coordinator, p. 683
walkthrough recorder, p. 683
wizards, p. 681

REVIEW QUESTIONS

1. List the four styles of speaking.
2. What are benefit terms? What are loss terms? Give some examples of each.
3. What is body language? Give an example of using body language.
4. List the four spatial zones and the characteristics of each.
5. What are the three rules of brainstorming?
6. What are visual aids? List three examples. List three common mistakes people make when preparing overhead slides.
7. List four ways to keep an audience listening. From your experience, give an example of how you kept an audience listening.
8. When you are asked a question that you don't know the answer to while giving a formal presentation, how should you respond?
9. What is a project walkthrough and who should participate?
10. List four types of business and technical reports. List the guidelines for the length of the report based on the intended audience.
11. When writing a business or technical report, why should you write in the active voice?

PROBLEMS AND EXERCISES

1. Identify the four audiences that a systems analyst must effectively communicate with. How are they different? How would you allow for these differences in communicating with each group?
2. Explain the meaning of the statements, "To hear is to recognize that someone is speaking. To listen is to understand what the speaker wants to communicate."
3. List and explain six guidelines for effective listening.
4. List three aspects of body language. Why should a systems analyst care about body language?
5. What are the possible recommendations after completing a walkthrough?
6. Identify two formats for a written report. What are the elements common to both formats? Should the length of a written report vary by audience? Why or why not?
7. The secret of effective oral and written communications is to know your audience. What are some things you would want to know about your audience before making a formal presentation to them? How could this knowledge be used to your advantage in formulating a presentation?
8. Why do formal presentations usually accompany written reports?
9. Systems analysts have a tendency to generate written reports that are much too large for managers to read. How would you handle a size problem with a technical report?
10. What are some ways you might improve your written communications skills? Identify specific courses that help you improve your skills.
11. Take one of the report outlines in Chapters 4, 9, 17, or 18 and prepare formal outlines that include primary and secondary elements for both the factual and administrative formats.

PROJECTS AND MINICASES

1. Get permission to attend a board meeting or subcommittee meeting of a local organization (e.g., the Data Processing Management Association or the Association of Computing Machinery), school committee, or some other business meeting. Observe how the meeting is run by the leader. What was the purpose of the meeting? Did the purpose of the meeting appear to be understood by all the participants? Why or why not? Did the meeting start and end on time? If not, what caused the delay? Was the meeting room reserved ahead of time? Did the room provide a comfortable atmosphere? Were there any problems with the meeting location? Was an agenda distributed before or during the meeting? Did the leader follow the agenda during the meeting? Were arrangements made for the minutes of the meeting to be published and distributed to appropriate individuals?
2. While attending the next lecture in each of your classes, observe the instructor's presentation of class material. Make a note of and learn from the techniques the professor uses to clearly deliver difficult material. If you feel comfortable about discussing the lecture with your professor, discuss your findings.
3. Arrange a formal walkthrough of one of your systems analysis and design assignments. Try to include your instructor and a few students. Prepare a walkthrough report.
4. In a current or future programming class, discuss the possibility of formal walkthroughs on one programming assignment. You'll need to secure permission so that your instructor will not consider the walkthrough cheating. Conduct walkthroughs as soon as you've completed the program design and immediately after coding (but before you compile or interpret the program). Analyze the impact the walkthroughs had on your productivity by comparing the number of compiles that you required to finish the assignment against the number of compiles that programmers who didn't use walkthroughs required.
5. Try to obtain a systems development report outline or table of contents from an information systems shop. Was the report organized using the factual format or the administrative format? Do you think everybody who should have read that report did read it? Why or why not? Reorganize the outline or table of contents into an alternative format. Be sure to include secondary elements, even if they weren't included in the original report.

SUGGESTED READINGS

Bovee, Courtland L., and John V. Thill. *Business Communication Today*. 2nd ed. New York: Random House, 1989. A textbook teaching the skills and concepts people need to communicate effectively in the business world.

Gildersleeve, Thomas R. *Successful Data Processing Systems Analysis*. Englewood Cliffs, NJ: Prentice Hall, 1978. Gildersleeve doesn't talk too much about tools in his books—that's why we like him! Chapter 5 discusses presentations, and Chapter 10 discusses interpersonal relations. They are worthwhile additional readings for any analyst.

Malloy, John T. *Dress for Success*. 2nd ed. New York: Warner, 1987. Based on this best-selling book, John Malloy has been labeled "America's first wardrobe engineer." Like its sequel for women, this book teaches people how to dress for power and prestige. The guidelines are based on research conducted by Malloy.

Malloy, John T. *The Woman's Dress for Success Book*. New York: Warner, 1975. The working woman's version of Malloy's successful book on how to dress for power and respect.

Matthies, Leslie H. *The Management System: Systems Are for People*. New York: John Wiley & Sons, 1976. Chapter 10 explains how to present and sell a new system to management. Some concepts we use were initially presented in this book.

Smith, Randi Sigmund. *Written Communication for Data Processing*. New York: Van Nostrand Reinhold, 1976. An excellent book on written communications for DP professionals—not just reports, but memos and letters too!

Stuart, Ann. *Writing and Analyzing Effective Computer System Documentation*. New York: Holt, Rinehart & Winston, 1984. At last! A book for students about writing in the information systems environment. And a good book at that. Must reading!

Uris, Auren. *The Executive Deskbook*. 3rd ed. New York: Van Nostrand Reinhold, 1988. An excellent executive reference that has entire chapters on effective communication, meetings, decision making, problem solving, and planning.

Walton, Donald. *Are You Communicating? You Can't Manage without It*. New York: McGraw-Hill, 1989. An easy-to-use guidebook on the process of communications and must for anyone who must work with people and influence them.

GLOSSARY/INDEX

A

ABEND, 584

Abrupt cut-over *On a specific date (usually a date that coincides with an official business period such as month, quarter, or fiscal year), the old system is terminated and the new system is placed into operation. This is a high-risk approach because there may still be major problems that won't be uncovered until the system has been in operation for at least one business period. On the other hand, there are no transition costs. Abrupt cut-over may be necessary if, for instance, a government mandate or business policy becomes effective on a specific date and the system couldn't be implemented before that date.,* **567**

Abstract or executive summary, 688

Abstract use case, 536–537

Activity diagram, systems analysis
- definition phase, 153
- study phase, 139
- survey phase, 129

Actor *Represents anything that needs to interact with the system to exchange information. An actor is a user or a role that could be an external system or a person.,* **294,** 295–296

AD Consulting, 15

Adams, David R., 531

Adaptability, systems analysts and, 24

ADC (automatic data collection), systems design and; *see* Inputs

Administrative format, reports, 687

Afferent, 516

Agendas, for JAD sessions, 665

Aggregation relationships, 303

Alpha testing, 568

Alphabetic codes, 195

Alternate key *Any candidate key that is not selected to become the primary key.,* **179**

American Management Systems, 15

American National Standards Institute (ANSI), 376, 378

Analysis
- business analysts, 9
- business area (BAA), 124
- cause-effect, 144
- cost-benefit; *see* Cost-benefit analysis
  - data; *see* **Data analysis**
- event, 337–338
- feasibility; *see* **Feasibility analysis**
- impact; *see* Impact analysis
- management and user involvement in, 17
- modern structured, 122–123, 157

Analysis (*continued*)
- object-oriented; *see* **Object-oriented analysis (OOA)**
- paralysis, 236–237
- payback, 652–654
- in problem solving, 9
- requirements, 163–164
- systems; *see* Systems analysis; Systems analysts
- transaction, 521–523
- transform, 515–521
- usability, 647

Analysis use cases, versus design use cases, 537–539

Andersen Consulting, 14, 15

Anderson, John F., 641

Andrews, D.C., 669

Apple Computer Co., 59, 266, 356, 366, 492–493, 496

Application analysts *Systems analysts that specialize in application design and technology-dependent aspects of development. A synonym is* system *or* application architect., **9**

Application architecture *Defines an approved set of technologies to be used when building any new information system., 56,* **92**

Application architecture *Defines the technologies to be used by (and to build) one, more, or all information systems in terms of its data, process, interface, and network components. It serves as a framework for general design.,* 350–391, **353;** *see also* **Application programs**
- build vs. buy implications, 371–372
- data for relational databases, 362–364, 399n
- design phase, 380–387
  - data distribution and technology assignments, 383
  - design unit, 381
  - general, 352–353
  - network topology DFD, 373, 381–383
  - person/machine boundaries, 384–387
  - physical DFDs, 380–387
  - process distribution and technology assignments, 383–384
- economic feasibility, 371
- enterprise application, 371
- information technology, 353–370
- interface architectures, 364–368
  - batch input/output, 364–365
  - electronic data interchange (EDI), 367
  - electronic messaging, 367
  - graphical user interfaces (GUI), 366–367
  - imaging and document interchange, 367
  - keyless data entry, 365–366

Application architecture *(continued)*
middleware, 367–368
on-line processing, 365
pen input, 366
remote batch processing, 365
user and system interfaces, selecting, 368
modeling for, 372–380
CASE for, 380
physical data flows, 380
diagrams, 372–376
physical data stores, 376
physical external agents, 376
physical processes, 373–375
system flowcharts, 376–380
network architectures for client/server computing; *see* Network architectures
operational feasibility, 371
process architecture, 368–370
centralized computing/distributed presentation, 368–369
the Internet and intranet, 370
multi-tier client/server, 369–370
software development environment (SDE), 368–370
systems management, 370
transaction processing (TP) monitors, 370
two-tier client/server, 369
version control and configuration managers, 370
SoundStage case study, 351–352
strategies for, 370–372
tactical application, 371
technical feasibility, 371
Application Development Strategies, 169
Application programs *Language-based, machine-readable representations of what a computer process is supposed to do, or how a computer process is supposed to accomplish its task.,* 16, **57;** *see also* **Application architecture**
benchmarking, 581–582
cross-functional, 13, 14
data model, 187–188
development of, 12, 14
life cycle for, 72; *see also* **Systems development life cycle (SDLC)**
rapid; *see* **Rapid application development (RAD)**
partitioning, 64
process model, 236–237
understanding, 582–583
Application schema *A model that communicates how selected business processes are, or will be, implemented using the computer and programs.,* **56**
Applied Business Technology, 599
Architecture; *see also* Building blocks of information systems
applications; *see* **Application architecture**
data architecture, 401–403
database, 403–405
information systems, 42
information technology, 13–14
tools, CASE, 105–107
Archival files/tables *Contain master and transaction file records that have been deleted from on-line storage. Thus, records are rarely deleted; they are merely moved from on-line storage to off-line storage. Archival requirements are dictated by government regulation and the need for subsequent audit or analysis.,* **400**
Artificial intelligence technology, 40; *see also* **Expert systems**
Association for Computing Machinery (ACM), 22
Association for Information Systems (AIS), 627
Association for Systems Management (ASM), 22
Associations, object modeling and, 303–304
Associative entity *An entity that inherits its primary key from more that one other entity (parents). Each part of that concatenated key points to one and only one instance of each of the connecting entities.,* **181**
Attribute *A descriptive property or characteristic of an entity. Synonyms include* element, property, and field., **176**
Attribute *The data that represent characteristics of interest about an object., 230,* **287**
derived, 414–416
Audiences
communication skills and, 672
formal presentations and, 682
Audit files *Special records of updates to other files, especially master and transaction files. They are used in conjunction with archive files to recover "lost" data. Audit trails are typically built into better database technologies.,* **400**
Audit testing *Certifies that the system is free of errors and is ready to be placed into operation. Not all organizations require an audit. But many firms have an independent audit or quality assurance staff that must certify a system's acceptability and documentation before that system is placed into final operation. There are independent companies that perform systems and software certification for end-users' organizations.,* **568**
Auto-identification systems, 366
Automatic data collection (ADC), 440
systems design and; *see* Inputs
Auxiliary operation, symbol for, 377

B

Backlogs, 78
Backup and recovery testing, 568
Bar charts *Used to show individual figures or values at a specific time or to depict comparisons among items. The categories to be compared are organized vertically, while the values are organized horizontally—for example,* see *illustration number 3 in Figure 13.5(a). This layout allows emphasis to be placed on the "comparison" rather than time. As is illustrated in example 5 in Figure 13.5(a), a stacked bar chart style may be used to show the relationship of individual items to the whole.,* **469**
Bar coding, 366, 441
Barlow, Victor M., 261, 391
BASIC, 56, 368
Batch input/processing *The oldest and most traditional input method. Source documents or forms are collected and then periodically forwarded to data entry operators, who key the data*

Batch input/processing *(continued)*
using a data entry device that translates the data into a machine-readable format., **439**–440
interface architectures and, 364–365
Beck, Robert, Jr., 595, 595n, 621
Behavior *Refers to those things that the object can do and that correspond to functions that act on the object's data (or attributes). In object-oriented circles, an object's behavior is commonly referred to as a* method *or* service *(we may use the terms interchangeably throughout our discussion).*, **287**
Bell, P., 575
Benchmarking, of programs, 581–582
Benefit terms *Words or phrases that evoke positive responses from the audience. Benefit terms can be used very effectively to sell proposed changes. Managers will usually accept ideas that produce benefit terms.*, **674**
Benefits; *see* Cost-benefit analysis
Benjamin, R.I., 73n, 115
Bentley, Lonnie D., 261, 391
Berdie, Douglas R., 641
Beta testing, 568
The *Bible,* 671n
Biometric ADC, 440
Black holes, 219
Blaha, Michael, 307, 551
Blanchard, Kenneth, 617–618, 621
Block codes, 195
Blocking factor *The number of* logical records *included in a single read or write operation (from the computer's perspective). A block is sometimes called a* physical record., **400**
Boar, Bernard, 348, 461
Body language *All of the information being communicated by an individual other than spoken words. Body language is a form of nonverbal communication that we all use and are usually unaware of.*, **675**–676
Boehm, Barry, 575
Booch, G., 307, 551
Borland Technology, 97, 105, 237, 357, 369, 407, 474, 654
Bouldin, Barbara, 115
Bovee, Courtland L., 674, 691
Boxes
check, 447–448, 475
combination (combo), 450
dialogue, 493
list, 448–449
spin (spinner), 450
tab dialogue, 475
text, 446–447
Boyer, Terrence J., 531
Brainstorming *A technique for generating ideas during group meetings. Participants are encouraged to generate as many ideas as possible in a short period of time without any analysis until all the ideas have been exhausted.*, **678**–679
Brooks, Fred, 597, 597n, 621
Browsers, 358, 366
Bruce, Thomas A., 69, 194n, 206, 434
Bubble chart transformation graph, 381–383; *see also* **Data flow diagram (DFD)**
Bucki, Lisa, 621
Budget constraints, 595–596
Bugs, 580
Build versus buy implications, 371–372
Building blocks of information systems, 32–69; *see also* Architecture
data-focused, 50–53
business resources and, 52
data requirements and, 52–53
system builders' view, 53
system designers' view, 53
system owners' view, 51–52
system users' view, 52–53
focus of systems and, 48–50
fundamentals of, 37–42
architecture for, 42
business process redesign, 39, 55
data, 37
data processing systems, 38
data warehouse, 40
decision support systems (DSS), 39–40
expert systems, 40
information systems/technology, 38
management information systems, 39
office automation systems, 40–41
personal information systems, 41
perspectives on systems; *see* Perspectives (in the framework)
transaction processing systems, 38–39
work group information systems, 41
geography-focused, 60–64
client/server computing and, 61
communications requirements and, 63
distributed computing and, 61
network programs and, 64
system builders' view, 64
system designers' view, 64
system owners' view, 61–63
system users' view, 63–64
interface-focused, 57–60
context model, 58–59
interface schema, 59
middleware and, 60
system builders' view, 60
system designers' view, 59–60
system owners' view, 57–59
system users' view, 59
processes-focused, 54–57
application schema, 56
applications programs, 57
business functions and, 55
business processes defined, 55–56
cross-functional information systems and, 55
policies and, 56
procedures and, 56
prototyping, 57
system builders' view, 56–57
system designers' view, 56
system owners' view, 55
system users' view, 55–56
SoundStage case study, 33–37
Burrows, Hal, 617–618, 621
Bus, 359
Business analysts *Systems analysts specializing in business problem analysis and technology-independent requirements analysis.*, **9**

Business and technical report *The primary method used by analysts to communicate information about a systems development project. The purpose of the report is to either inform or persuade, possibly both. In a few pages, it is not possible to provide a comprehensive discussion of report writing. But because people make judgments about who we are and what we can accomplish based on our writing ability, we can offer some motivation for further study and some guidelines for writing reports.*, **685**–689
Business area analysis (BAA), 124
Business databases, 587; *see also* **Databases**
Business functions *Ongoing activities that support the business. Functions can be decomposed into other functions and eventually into discrete processes that do specific tasks.*, 48, **55**
Business geography, network modeling and, 267–271
Business models, 172–173; *see also* **Logical models; Models/Modeling**
Business object identification; *see* **Object modeling**
Business or intellectual speaking, 673–674
Business process redesign (BPR) *The study, analysis, and redesign of fundamental business processes to reduce costs and improve value added to the business.*, **18**–19
Business process redesign (BPR) *(Also called* business process reengineering*) is the application of systems analysis (and design) methods to the goal of dramatically changing and improving the fundamental business processes of an organization, independent of information technology.*, 39, 55, 78–79, **126**
process modeling for, 236, 253
Business process reengineering; *see* **Business process redesign (BPR)**
Business processes *Discrete activities that have inputs and outputs, as well as starting times and stopping times. Some business processes happen repetitively, while others happen occasionally or even rarely. Business processes may be implemented by people, machines, computers, or a combination of all three.*, 48, **55**–56
analyzing, 142–143
reengineering, 126
Business requirements
new, 586
outlining, 151–154
prioritizing, 159–161
Business resources *Are (1) things that are essential to the system's purpose or mission; or (2) things that must be managed or controlled to achieve business goals and objectives.*, 51–**52**
Business trends, 18–20
Buttons, 493

C

Calendars, 606
Cancellation, of development, 76
Candidate key *A candidate to become the primary identifier of instances of an entity. It is sometimes called a* candidate identifier. *(A candidate key may be a single attribute or a concatenated key.)*, **179**
Candidate programs/systems
feasibility analysis of, 656–659
for reengineering, 588–589
Capacity planning, for database design, 426
Cap/Gemini, 14
Capital investments, justification of, 45, 76
Cardinality *Defines the minimum and maximum number of occurrences of one entity for a a single occurrence of the related entity. Because all relationships are bidirectional, cardinality must be defined in both directions for every relationship.*, **180**–181
Career preparation, for systems analysts, 20–22; *see also* Employment
Cascading menus, 486–488
Case model diagram, 539
CASE repository *A developers' database. It is a place where the developers can store diagrams, descriptions, specifications, and other by-products of systems development. Synonyms include* **dictionary** *and* **encyclopedia., 106**
Case studies; *see* SoundStage case study
CASE tools; *see* **Computer-aided systems engineering (CASE)**
Cash, William B., Jr., 641
Casual system user *May use a specific program only on an occasional basis. This user may never become truly comfortable with the terminal or the program.*, **494**
Cause-effect analysis, 144
Central information services, 13–14
Central repository, 578–579
Central transform, 516
Centralized computing *When a multi-user computer (usually a mainframe or minicomputer) hosts all the information system components including (1) the data storage (files and databases), (2) the business logic (software and programs), (3) the user interfaces (input and output), and (4) any system interfaces (networking to other computers and systems). The user may interact with this host computer via a terminal (or, today, a PC emulating a terminal), but all the work is done on the host computer.*, **354**–356
and distributed presentation, 368–369
network architectures and, 354–356
CGI (Computer Gateway Interface) *A standard for publishing graphical World Wide Web components, constructs, and links.*, **370**
Champy, J., 592
Change, systems development and, 77–78
Character, systems analysts and, 24–25
Character sets, display features and, 496
Character user interface (CUI), 356
Chart transformation graph, 211; *see also* **Data flow diagram (DFD)**
Check boxes, 447–448, 475
Chief Information Officer (CIO), 12
Christerson, Magnus, 307
CICS, 370
Class *A set of objects that share common attributes and behavior. A class is sometimes referred to as an* object class., **288**–290
relationships, 290–292
Class responsibility collaboration (CRC) card, 544–547

Clerical workers, 45

Client *Single-user computer that provides (1) user interface services and appropriate database and processing services, and (2) connectivity services to servers (and possibly other clients).*, 16, **353**

Client/server computing *When an information system's database, software, and interfaces are distributed across a network of clients and servers that communicate and cooperate to achieve systems objectives. Despite the distribution of computing resources, each system user perceives that a single computer (the user's own client PC) is doing all the work. Synonyms include* **distributed computing** *and* cooperative computing., 266, **353–354**

network architectures for; *see* Network architectures

Client/server computing applications *Information system building blocks are distributed between client personal computers and server shared computers. The clients and servers effectively interoperate to share the overall workload.*, 22, **61**

Closed-ended questions, 631

Coad, Peter, 169, 307, 348, 551

Code changes, reengineering and, 588–589

Code conversion *Translates the code from one language to another. Typically, this translation is from one language version to another. There is a debate on the usefulness of translators between different languages. If the languages are sufficiently different, the translation may be very difficult. If the translation is easy, the question is "why change?"*, **588**

Code generator tools, 107

Code reorganization *Restructures the modular organization and/or logic of the program. For example, modules may be combined or separated to reduce coupling or increase cohesion (*see *Chapter 18). Logic may be restructured to eliminate control flow knots and reduce cycle complexity.*, **588**

Code slicing *The most intriguing program-reengineering. Many programs contain components that could be factored out as subprograms. If factored out, they would be easier to maintain. More importantly, if factored out, they would be reusable. Code slicing cuts out a piece of a program to create a separate program or subprogram. This may sound easy, but it is not!*, **588**

Cohesion *Refers to the degree to which a module's instructions are functionally related.*, **526**–527, 588

cohesive modules, 313

Coincidental cohesion *Occurs in modules that contain instructions that have little or no relationship to one another. Coincidental cohesive modules appear to have been derived with no attention given to the actual "function" being served by the module. In fact, their existence is typically based on coincidence. For example, a designer may decide to create a module that will consist of a series of program instructions encountered several times elsewhere in the program's logic.*, **527**

Collaboration card, class responsibility (CRC), 544–547

Column charts *Simple variations of the bar chart; used when there is a desire to show the variation over a period of time or to depict comparisons among items. In a column chart, categories are organized horizontally and values are organized vertically. This appearance emphasizes variations over a period of time.*, **469**

Combination checks *Determine whether a known relationship between two fields is valid. For instance, if the vehicle make is a Pontiac, then the vehicle model must be one of a limited set of values that comprises cars manufactured by Pontiac (Firebird, Grand Prix, and Bonneville to name a few).*, **444**

Combination (combo) boxes, 450

Comment section, symbol for, 378

Common coupling *Modules are said to be common coupled if they refer to the same global data area. Global data areas are commonly found in third-generation programming languages (3GLs) such as COBOL. Common coupling represents an even higher level of module dependency. For example, all modules that reference the global data area could be adversely affected by any changes that any of the other modules made to data in that global data area.*, **525**

Communicational cohesion *Occurs in modules whose instructions accomplish tasks that utilize the same pieces of data. For example, a module may consist of numerous instructions that each accomplish a task using customer data, such as checking a customer balance, adding a new customer, canceling a customer, updating a customer's record, changing a customer's status, or querying a customer. Such modules are easier to modify and maintain if they are expanded into separate modules that accomplish their own separate task.*, **526**

Communications, interpersonal skills and; *see* Interpersonal skills

Communications requirements *Define the information resource requirements for operating locations and how different operating locations need to communicate with one another. These communications requirements are expressed independently of any specific technology.*, **63**

Completeness checks *Determine whether all required fields on the input have actually been entered.*, **444**

Composite attribute; *see* **Compound attribute**

Composite data flow *A data flow that consists of other data flows. They are used to combine similar data flows on general-level data flow diagrams to make those diagrams easier to read.*, **226**–227

Composite key; *see* **Concatenated key**

Compound attribute *One that actually consists of more primitive attributes. Synonyms in different data modeling languages are numerous:* concatenated attribute, composite attribute, *and* data structure., **176**

Compound key; *see* **Concatenated key**

Computer application *Computer-based solution to one or more business problems and needs. One or more computer applications are typically contained within an information system.*, **7**

Computer Ethics Institute, 638, 639

Computer Gateway Interface (CGI) *A standard for publishing graphical World Wide Web components, constructs, and links.*, **370**

Computer operations, 12

Computer programs/programming, 583–584; *see also* Software
- benchmarking and, 581–582
- changes to, 583–584
- code, 48
- design, 48
- editing and testing of, 583
- enhancement and reengineering, 588
- experience and expertise in, 22
- languages, 22, 60, 368
  - fourth-generation (4GLs), 158, 587
- libraries for, 82, 579
- programmers, 48; *see also* **Systems builders**
- requirements, new, 586
- structured, 221–223
- symbols and, 377
- understanding, systems support and, 582–583
- writing and testing, 562–564, 586–587

Computer Sciences Corp., 14

Computer usage costs, 649

Computer-aided systems engineering (CASE) *The application of information technology to systems development activities, techniques, and methodologies.* CASE tools *are programs (software) that automate or support one or more phases of a systems development life cycle. The technology is intended to accelerate the process of developing systems and to improve the quality of the resulting systems.*, **103**–108, 155, 163–164
- benefits of, 107–108
- CASE repository, 106
- data modeling and, 189–190, 202
- database design and, 418, 420
- development centers and, 108
- history and evolution of, 103–104
- lower-CASE, 105
- network modeling and, 276, 280
- normalization of data and, 418
- for physical DFDs and flowcharts, 380
- process modeling and, 239, 241, 253
- prototyping screen design, 455
- tool architecture, 105–107, 356
- tool framework, 104–105
- upper-CASE, 104

ComputerWorld, 21, 535

Compuware, 370

Concatenated attribute; *see* **Compound attribute**

Concatenated key *A group of attributes that uniquely identifies an instance of an entity. Synonyms include composite key and compound key.*, **177,** 178

Conceptual models, 172–173; *see also* **Logical models**

Conclusion section, 688

Conditional structure, 223

Configuration
- network(s), 64
- systems; *see also* **Systems design**
  - data modeling and, 188
  - phase, *FAST* development methodology, 82–83, 91–93
  - process modeling and, 237

Connectivity *Defines the need for and provides the means for transporting essential data, voice, and images from one location to another.*, **270**–271

Connectivity *Defines how computers are connected to "talk" to one another.*, **359**

Connor, Denis, 348, 461

Conservation, of data flows, 228–230

Constantine, Larry, 511

Constraint *Something that will limit your flexibility in defining a solution to your objectives. Essentially, constraints cannot be changed.*, 145–147, **146**

Construction; *see also* **Systems implementation**
- of network models; *see* **Network modeling**
- phase, *FAST* development methodology, 83, 96–97
- of process models; *see* **Process modeling**

Consulting, 15–16

Content coupling *Two modules are said to be content coupled (also referred to* ashybrid coupled*) when one module actually modifies the procedural contents of another module. In essence, the connection that exists between the two modules represents control. Content coupling represents the highest degree of module dependency.*, **525**

Context diagram *Defines the scope and boundary for the system and project. Because the scope of any project is always subject to change, the context diagram is also subject to constant change. A synonym is* environmental model., 156, 237, 239–241, **240,** 242

Context models, 58–59, 188, 193–194

Continuous process improvement (CPI) *Continuous monitoring of business processes to effect small but measurable improvements to cost reduction and value added.*, **19**

Control, in PIECES framework, 79–81

Control coupling *Two modules are said to be control coupled if their dependency is based on the fact that they communicate by the passing of control information or flags. Control coupled modules represent a higher level of dependency. The mere fact that one module passes control information to another module suggests that the first module is involved in changing or coordinating the functions to be accomplished in the receiving module.*, **524–525**

Control flow *Represents a condition or nondata event that triggers a process. Think of it as a condition to be monitored while the system works. When the system realizes that the condition meets some predetermined state, the process to which it is input, is started.*, **227**

Control flow knots *The number of times logic paths cross one another. Ideally, a program should have zero control flow knots. (We have seen knot counts in the thousands on some older, poorly structured programs.)*, **588**

Control objects, 534–535

Converging data flow *The merger of multiple data flows into a single data flow.*, 231–**233**

Conversion, 571–572
- code, 588
- location, 568
- parallel, 567

Conversion (*continued*)
plan preparation, 566–568
staged, 568
Cooperative computing, 266, 354; *see also* **Client/server computing**
Copi, I.R., 260
Cost-benefit analysis, 646, 649–656; *see also* **Economic feasibility**
cost-effectiveness, 652–656
net present value analysis, 655–656
payback analysis, 652–654
return-on-investment (ROI) analysis, 654–655
time value of money, 652
system benefits, 650–652
system cost, 649–650
Cost-effectiveness *The result obtained by striking a balance between the cost of developing and operating a system, and the benefits derived from that system.*, **76**
Coupling *The level of dependency that exists between modules.*, **523**
structured software design and, 313, 523–526, 588
Crane, David B., 621
Crashes, system recovery from, 584
Creeping commitment, 129n, 643
Critical path *A sequence of dependent project tasks that have the largest sum of estimated durations.*, **608**, 609, **612**
Cross-functional informational systems *Systems that support relevant business processes from several business functions without regard to traditional organizational boundaries such as divisions, departments, centers, and offices.*, 13, 14, **55**
Cross life cycle activities *Activities that overlap many or all phases of the methodology. In fact, they are normally performed in conjunction with several phases of the methodology.*, **99**–103
documentation and presentations, 100
estimation and measurement, 100–102
fact-finding process, 100
feasibility analysis, 102
project and process management, 102–103
Customer service, in PIECES framework, 79–81
Cycle, defined, 10
Cycle complexity *The number of unique paths through a program. Ideally, the fewer, the better.*, **588**

D

Data *Raw facts about the organization and its business transactions. Most data items have little meaning and use by themselves.*, 10, **37**, 63; *see also* Building blocks of information systems
administration of, 12
as focus of information systems, 48, 50–53, 335–338
integrity of, 423–424
in motion, 225–227
in PIECES framework, 79–81
sharing tools, 107
systems design and, 336–338
Data administrator *Person responsible for the data planning, definition, architecture, and management.*, **403**
Data analysis *A procedure that prepares a data model for implementation as a nonredundant, flexible, and adaptable file/database.*, **337**–338
Data analysis *Process that prepares a data model for implementation as a simple, nonredundant, flexible, and adaptable database. The specific technique is called normalization.*, **408**–409
Data architecture *The files and databases that store all of the organization's data, the file and database technology used to store the data, and the organization structure set up to manage the data resource.*, **401**–403
Data attribute *The smallest piece of data that has meaning to the end-users and the business. (This definition also applies to* attributes *as they are presented in Chapter 5.)*, **230**
process modeling and, 230–231
Data capture *The identification of new data to be input.*, **438**–439
Data conservation *Sometimes called "starving the processes," requires that a data flow only contain those data that are truly needed by the receiving process.*, 228–230, **229**
Data coupling *Two modules are said to be data coupled if their dependency is based on the fact that they communicate by the passing of data. Other than communicating through data, the two modules are independent; that is, each module performs its own function with no regard to what or how the other module completes its functions. In examining modules for data coupling, careful attention should be made to ensure that no module communication contains "tramp" data. Tramp data are any unnecessary data communicated between the modules. For example, a module might call a second module to have it calculate an employee's net pay. Only data needed by the second module to complete its task should be passed from the calling module. By ensuring that modules communicate only necessary data, module dependency is minimized. This helps avoid the "ripple effect" wherein making changes in one module inadvertently affects another module that happens to receive the same data.*, **523–524**
Data definition language (DDL) *Used by the DBMS to physically establish record types, fields, and structural relationships. Additionally, the DDL defines views of the database. Views restrict the portion of a database that may be used or accessed by different users and programs. DDLs record the definitions in a permanent data repository.*, **403**
Data distribution *Partitions data to one or more database servers. Entire tables can be allocated to different servers, or subsets of rows in a table can be allocated to different servers. An RDBMS controls access to and manages each server.*, **364**
Data distribution *The distribution of either specific tables, records, and/or fields to different physical databases.*, **420**
and technology assignments, 383
Data entry *The process of translating the source document into a machine-readable format. That format may be a magnetic disk , an optical*

Data entry *(continued)*
mark form, a magnetic tape, or a floppy diskette, to name a few., **438–439**
Data flow *Represents an input of data to a process, or the output of data (or information) from a process. A data flow is also used to represent the creation, deletion, or updating of data in a file or database (called* data store *on the DFD).*, **225**
composite, 226–227
converging, 231–233
diverging, 231–233
logical, 227–228
network topology data flow diagram (DFD), 373, 381–383
physical, 372–376, 380–387
system modeling and; *see* **Process modeling**
Data flow diagram (DFD) *A tool that depicts the flow of data through a system and the work or processing performed by that system. Synonyms include* bubble chart transformation graph *and* **process model.**, 122–123, **211**–214, 220
network topology, 373, 381–383
physical, 372–376, 380–387
of programs, 513–515
transaction analysis and, 521–523
transform analysis and, 515–521
synchronization, 249, 272–275
Data independence, 397
Data input *The actual entry of data in a machine-readable format into the computer.*, **439**
Data maintenance, 38
Data manipulation language (DML) *Used to create, read, update, and delete records in the database and to navigate between different records and types of records—for example, from a customer record to the order records for that customer. The DBMS and DML hide the details concerning how records are organized and allocated to the disk.*, **403**–404
Data modeling *A technique for organizing and documenting a system's data. Data modeling is sometimes called database modeling because a data model is usually implemented as a database. It is sometimes called* information modeling., 53, 170–206, **173**
alternate keys, 179
application data model, 187–188
attributes and, 176–179
candidate keys, 179
CASE, 189–190, 202
compound attributes and, 176
concatenated keys, 177–178
construction of models, 190–202
alphabetic codes, 195
block codes, 195
context data model, 193–194
entity discovery, 190–192
fully attributed model, 197–199
fully described model, 199
generalized hierarchies, 185, 197
hierarchical codes, 195, 197
identifying relationships, 196
independent entity, 193
intelligent keys, 194
key-based data model, 194–197
nonidentifying relationships, 197

Data modeling *(continued)*
serial codes, 195
significant position codes, 195
context data model, 188
data types and, 177
database design and, 407–408
defaults, 177
definition phase of systems analysis, 155
domains and, 176–177
enterprise data model, 186
entities and, 175–176
fact-finding and information gathering, 189
fully attributed data model, 188
fully described data model, 188
identification and, 177–179
information gathering, 189
key-based data model, 188
keys, 177–179
logical models, 172–173
logical process for, 186–190
model defined, 172
next generation, 199–202
physical models, 173
relationships and, 179–186
associative entity, 181
cardinality, 180–181
degree, 181
foreign keys, 182–183
generalization, 183–186
nonspecific (many-to-many), 179, 183
recursive, 181
subtype, 185–186
supertype, 184–186
SoundStage case study, 171–172
strategic data modeling, 186
synchronization of
with network modeling, 272–274
with process modeling, 272
system concepts for, 175–186
during systems analysis, 187–188
systems configuration and design, 188
systems modeling, 172–175
systems thinking, 175
Data Processing Management Association (DPMA), 22, 627
fact-finding ethics and, 637–638
Data processing systems, 38; *see also* **Information systems**
Data replication *Duplicates data on one or more database servers. Entire tables can be duplicated on different servers, or subsets of rows in a table can be duplicated to different servers. The RDBMS not only controls access to and management of each server database, but it also ensures that updates on one server are updated on any server where the data are duplicated.*, **364**
Data replication *Duplication of specific tables, records, and/or fields to multiple physical databases.*, **420**
Data requirements *A representation of users' data in terms of entities, attributes, relationships, and rules. They should be expressed in a format that is independent of the technology that can or will be used to implement the data.*, **52**–53
Data store *An "inventory" of data. Synonyms include* file *and* database *(although those terms*

Data store *(continued)*
are too implementation-oriented for essential process modeling)., **235**
physical, 376
Data structures *Specific arrangements of data attributes that define the organization of a single instance of a data flow.*, 176, **230**–231; *see also* **Compound attribute**
Data type *Defines what class of data can be stored in an attribute.*, **177, 231**
Data warehouse *Read-only, informational database that is populated with detailed, summary, and exception information that can be accessed by end-users and managers with DSS tools that generate a virtually limitless variety of information in support of unstructured decisions.*, **40,** 402
Database administrators *Persons responsible for the database technology, database design and construction, security, backup and recovery, and performance tuning.*, **403,** 421–423
Database architecture *Database technology including the database engine, database management utilities, database CASE tools for analysis and design, and database application development tools.*, **403**–405
Database design, 392–434
capacity planning for, 426
CASE support for, 418, 420
concepts, 397–407
blocking factor, 400
data administrator, 403
data architecture, 401–403
data definition language (DDL), 403
data manipulation language (DML), 403–404
database architecture, 403–405
database management system (DBMS), 397, 403
descriptive fields, 399
fields, 398–399
files and tables, 400–401
foreign keys, 399
primary keys, 399
records, 399–400
relational database management systems, 405–407
secondary keys, 399
stored procedures, 405
triggers, 405
conventional files vs. databases, 395–397
data analysis for, 407–418
data models, good, 407–408
defined, 408
normalization, 408–418
attributes, derived, 414–416
CASE support for, 418
example, 409–418
first normal form (1NF), 408, 409–413, 412n, 418
second normal form (2NF), 408–409, 414
simplification by inspection, 417–418
third normal form (3NF), 409, 414–418
data distribution, 420
data replication, 420
database management systems (DBMS), 397, 403
database prototypes, 424
database schema, 53, 188, 341, 397, 420–423
domain integrity, 423
Database design, *(continued)*
file design, 418
goals and prerequisites to, 420
integrity
data and referential, 423n, 423–424
domain, 423
key, 423
referential, 423n, 423–424
next generation of, 426
pros and cons of databases, 397
referential integrity, 423n, 423–424
roles, 424
SoundStage case study, 393–395
structure generation, 426
Database engine *That part of the DBMS that executes database commands to create, read, update, and delete records (rows) in the tables.*, **364,** 403
Database management system (DBMS) *Specialized computer software available from computer vendors that is used to create, access, control, and manage the database. The core of the DBMS is often called its* database engine. *The engine responds to specific commands to create database structures and then to create, read, update, and delete records in the database.*, 397, **403**
Database Programming and Design, 206
Database schema *The* physical *model or blueprint for a database. It represents the technical implementation of the logical data model.*, 53, 188, 341, 397, **420**–423
Database servers *Store the database, but the database commands are also executed on those servers. The clients merely send their database commands to the server. The server returns only the result of the database command processing—not entire databases or tables. Thus, database servers generate much less network traffic. This approach is used by high-end database software such as* Oracle *and Microsoft* SQL Server., **357**
Databases *Collections of* interrelated *files.*, 40, **395**
building and testing, 560–561
data modeling of business requirements and; *see* **Data modeling**
Database Programming and Design, 69
designing; see Database design
FAST development methodology and, 82
installing, 569
prototypes of, 424
relational, 362–364, 399n
restructuring, 587–588
schema for; *see* **Database schema**
systems design and, 340–341
technology, 51
Data-to-location-CRUD matrix *A table in which the rows indicate entities (and possibly attributes); the columns indicate locations; and the cells (the intersection of rows and columns) document level of access where C = create, R = read or use, U = update or modify, and D = delete or deactivate.*, **273**
Data-to-process-CRUD matrix, 272, 273
Davis, Gordon B., 69

Davis, William S., 641
De Marco, Tom, 260
Decision structure, process modeling and, 223
Decision support system (DSS) *Information system application that provides its users with decision-oriented information whenever a decision-making situation arises. When applied to executive managers, these systems are sometimes called* **executive information systems., 39**–40
Decision table *A tabular form of presentation that specifies a set of conditions and their corresponding actions.,* **224**
Decisions/Decision-making
  support tools, 107
  unstructured, 39–40
Decomposition *The act of breaking a system into its component subsystems, processes, and subprocesses. Each level of* abstraction *reveals more or less detail (as desired) about the overall system or a subset of that system.,* **216**–218, **270**
Decomposition diagram *Also called a hierarchy chart, shows the top-down functional decomposition and structure of a system.,* **217**–218, 237, 241–243
  location type; *see* **Location decomposition diagram**
Dedicated system user *One who will spend considerable time using specific programs. This user is likely to become comfortable and familiar with the terminal or PC's operation.,* **494**
Default *Attribute value that will be recorded if not specified by the user.,* **177**
Deferred batch; *see* **Remote batch**
Deferred processing; *see* **Remote batch**
Definition phase
  FAST development methodology, 82–83, 88–91
  logical models, 154
  of systems analysis, 151–163
    activity diagram, 153
    business requirements
      model of, 154–157
      outline of, 151–154
      prioritization of, 159–161
    data models, 155
    discovery prototypes, building, 157–159
    distribution models, 155
    feasibility checkpoint, 643
    interface models, 155
    object models, 155
    process models, 155
    project plan and scope modifications, 161–162
    timeboxing, 160
Degree *The number of entities that participate in the relationship.,* **181**
DeHayes, D.W., 592
Dejoie, Roy, 641
Delivery phase, of systems implementation; *see* **Systems implementation**
  FAST development methodology, 83, 98
DeMarco/Yourdon shape for DFDs, 214, 215, 234, 235, 240
Departmental computing coordination, 14
Dependencies, object modeling and, 297
Derived attributes *Attributes whose values can either be calculated from other attributes or*
Derived attributes (*continued*)
  derived through logic from the values of other attributes., **414**–416
Description tools, 107
Descriptive fields *Any other fields that store business data.,* **399**
Design; *see* **Systems design**
Design objects, of OOD, 534–535
Design recovery, 583, 584
Design unit *A self-contained collection of processes, data stores, and data flows that share similar design attributes. A design unit serves as a subset of the total system whose inputs, outputs, files and databases, and programs can be designed, constructed, and unit tested as a single subsystem. (The concept of design units was first proposed by McDonnell Douglas in its* STRADIS *methodology.),* **381**
Design use cases, analysis use cases versus, 537–539
Detailed design, 352
Detailed object interactions, OOD, 547
Detailed reports *Present information with little or no filtering or restrictions.,* **464**–465
Development center *A central group of information system professionals who plan, implement, and support a systems development environment for other developers. They provide training and support for both the methodology and CASE tools.,* **108**
Development center *A central group of information system development consultants and managers who plan, implement, and support all aspects of process management, including methodology, technology, quality, and measurement.,* 12, **602**
Diagramming tools, 107
Dialogue
  boxes, 493
  charting, 497–499
  user interface design and
    prototyping, 499–500
    tone and terminology, 495
Dictionary, 106; *see also* **CASE repository**
Digital Equipment Corp., 359
Digitalk, 97
Direct manipulation, user interface design and, 492–493
Directing/controlling projects, project managers, 599
Directive *A new requirement that's imposed by management, government, or some external influence. (Note: You could argue that until a directive is fully complied with, it is, in fact, a problem.),* **79**
  FAST development methodology and, 79
  survey phase of systems analysis, 129–132
Directive style (of speaking) *Authoritative and judgmental. We use this style to give orders, give instruction, exert leadership, pass judgment, or state our opinions. For example: "I want John to be the leader of this project!",* **674**
Discount rate, 653
Discovery prototypes *Simple mock-ups of screens and reports that are intended to help systems analysts discover requirements. The discovered requirements would normally be added to system models. A synonym is requirements prototypes.,* **125,** 157–159, **158**

Disk file, symbol for, 377
Displays, features of, 495–497
Distributed computing *The decentralization of applications and databases to multiple computers across a computer network.*, 22, **61**
Distributed computing *Assignment of specific information system elements to different computers that cooperate and interoperate across a computer network. A synonym is* **client/server computing;** *however, client/server is actually one style of distributed computing.*, **266,** 354
Distributed data, networks, 356–358
Distributed presentation
 centralized computing and, 368–369
 network architectures and, 356
Distributed relational database *Distributes or duplicates tables to multiple database servers (and in rare cases clients).*, **364,** 399n
Distributed relational database management system (RDBMS) *A software program that controls access to and maintenance of the stored data. It also provides for backup, recovery, and security. It is sometimes called a client/server database management system.*, **364**
Distribution modeling, 265; *see also* **Network modeling**
 definition phase of systems analysis, 155
Dittman, Kevin C., 206
Diverging data flow *One that splits into multiple data flows.*, **231**–233
Divide and conquer, system development and, 76–77
Document files/tables *Contain stored copies of historical data for easy retrieval and review without the overhead of regenerating the document.*, **400**
Documentation *The activity of recording facts and specifications for a system.*, **100**
 organization tools, 107
 system development standards for, 75–76
 updating, systems support and, 583–584
Domain *Defines what values an attribute can legitimately take on.*, **177, 231**
 integrity of, 423
 process modeling and, 231
Drop-down lists, 449, 475
Dunlap, Duane, 461
Dunne, Peter, 641, 669
Duration, Gantt charts and, 605–606
Dynasty, 370

E

Eager, Bill, 641
Eastman, David, 507
Economic feasibility *A measure of the cost-effectiveness of a project or solution. This is often called a* cost-benefit analysis., 321–324, **646,** 648; *see also* Cost-benefit analysis
 architecture strategies and, 371
Economics, in PIECES framework, 79–81
Economy, globalization of, 19–20
Eddy, Frederick, 307, 551
Efferent, 516
Efficiency, in PIECES framework, 79–81
Effort and duration, Gantt charts and, 605–606
Electromagnetic ADC, 441
Electronic data interchange (EDI) *The electronic flow of business transactions between customers and suppliers.*, 46, **367**
Electronic Data Systems (EDS), 14
Electronic forms, 41
Electronic mail (e-mail), 675
Electronic messaging, 41
 interface architectures and, 367
Element, 176; *see also* **Attribute**
Elementary processes *Discrete, detailed activities or tasks required to complete the response to an event. In other works, they are the lowest level of detail depicted in a process model. A common synonym is* **primitive process, 219**
The Elements of Style, 688
Employment, systems analysts and, 11–16
 application software solution providers, 16
 career preparation, 20–22
 consulting, 15–16
 information services, 12–14
 outsourcing, 14–15
Empowerment *Driving authority to make decisions to nonmanagers and teams.*, **20**
Encapsulation *The packaging of several items together into one unit.*, **288**
Encyclopedia, 106
End-users, 12; *see also* **User(s)**
 systems support assistance for, 584
Enterprise application, architecture strategies, 371
Enterprise data model, 186
Enterprise process model, 236
Entity *Something about which we want to store data. Synonyms include* entity type *and* entity class., **175**
Entity *A class of persons, places, objects, events, or concepts about which we need to capture and store data.*, **176,** 401
 class, 175
 data modeling and, 175–176
 discovery and, 190–192
 objects, 534–535
 relationship(s)
  data model, 174–175
  diagram (ERD), 174–175
 type, 175
Entropy, 77
Ergonomics, 57
Ernst & Young, 15
Errors, correcting; *see also* **Systems support**
 detection versus, 683–684
Essential models, 154, 172–173, 210; *see also* **Logical models**
Entity instance *A single occurrence of an entity.*, **176**
Estimation *The activity of approximating the time, effort, costs, and benefits of developing systems. The term guesstimation (as in "make a guess") is used to describe the same activity in the absence of reliable data.*, **101**–102
Ethernet, 359–360, 362
Ethics *Personal character trait in which an individual understands the difference between "right" and "wrong," and acts accordingly.*, **24**–25
 for computer usage, 637
 fact-finding and, 637–638
Evans, C., 575

Event *A logical unit of work that must be completed as a whole. An event is triggered by a discrete input and is completed when the process has responded with appropriate outputs. Events are sometimes called transactions.*, **218**
decomposition diagram, 244–245
object modeling and, 296–299
Event analysis *A technique that studies the entities of a fully normalized data model to identify business events and conditions that cause data to be created, deleted, or modified.*, **337**–338
Event diagram *A context diagram for a single event. It shows the inputs, outputs, and data store interactions for the event.*, 237, **245**–248
Event handler, 237
Event partitioning *Factors a system into subsystems based on business events and responses to those events.*, **237**
Event-response list, 237, 243–244
Excelerator, 104
Exception reports *Filter data before they are presented to the manager as information.*, **466**
Executive information systems, 39
Executive managers, 46
Executive sponsor, 84
Executive summary, 688
Exercising (or testing) the user interface *System users experiment with and test the interface design before extensive programming and actual implementation of the working system. Analysts can observe this testing to improve on the design.*, **500**–501
Expectations management matrix *A rule-driven tool for helping management appreciate the dynamics of changing project parameters. The parameters include cost, schedule, scope, and quality.*, **613**–617
Expert systems *Information system application that captures the knowledge and expertise of a problem solver or decision maker and then simulates the "thinking" of that expert for those who have less expertise.*, **40**
Expressive style (of speaking) *Spontaneous, conversational, and uninhibited. We use this style when we are expressing our feelings, joking around, complaining, being intimate, or socializing. For example: "Over my dead body will I let that deadbeat be the leader of this project!"*, **674**
Extension use case, 536–537
External agent *A person, organization unit, other system, or other organization that lies outside the scope of the project but that interacts with the system being studied. External agents provide the net inputs into a system and receive net outputs from a system. Common synonyms include* **external entity** *(not to be confused with* data entity *as introduced in Chapter 5).*, **233–234**
physical, 376
External entity; *see* **External agent**
External events, 243–244, 247
External outputs *Leave the system to trigger actions on the part of their recipients or confirm actions to their recipients.*, **464**
External users, 17, 46–47

F

Fact-finding *Also called information gathering or data collection. The formal process of using research, interviews, meetings, questionnaires, sampling, and other techniques to collect information about systems, requirements, and preferences.*, **100,** 622–641, **623**
for data modeling, 189
ethics and, 637–638, 639
interviews, 632–636
types of, 633
methods for, 623
network modeling and, 276
process modeling and, 239
questionnaires, 630–632
rapid application development (RAD) and, 636–637
research and site visits, 626–628
the Internet, 626
World Wide Web (WWW), 626
sampling existing documentation, forms, and files, 624–626
randomization, 625
sample selection, 625
sample size determination, 625
stratification, 625
strategy for, 638
work environment observations, 628–630
work sampling, 630
Facts section, 688
Factual format, reports, 687
Failed projects, project management as cause, 596–597
FAST **activity diagram** *Shows the activities or work that must be completed to accomplish a* FAST *phase.*, **129**
FAST development methodology, 73, 78–99, 121, 122, 317–318
business process redesign and, 78–79
data modeling during systems analysis, 187–188
databases and, 82
directives and, 79
FAST activity diagram, 129
impact analysis; *see* Impact analysis
information strategy plan and, 78
opportunities and, 79
phases of, 82–98
construction, 83, 96–97
definition, 82–83, 88–91
design, 83, 94–96
prototypes and, 95–96
rapid application development (RAD) and, 95–96
implementation, 83, 98
purchasing, 83, 93–94
study, 82–83, 87–88
survey, 82–87
targeting, 82–83, 91–93
PIECES classification framework, 79–81
planned system initiative and, 78–79, 82, 86
problems and, 79
program library and, 82
repository and, 82
starting, 78–79
systems analysis strategies, 127–128
systems implementation and, 556–557
systems support after, 98–99
systems tests, 97

FAST development methodology (*continued*)
unit tests, 97
unplanned system request and, 78, 79–80, 86
Feasibility *A measure of how beneficial or practical the development of an information system would be to an organization.*, **102, 643**
of systems design solutions, 321–324
Feasibility analysis *The activity by which feasibility is measured.*, 86–87, **102,** 642–661
of candidate systems, 656–659
candidate systems matrix, 656–657, 658
feasibility analysis/comparison matrix, 657, 659
checkpoints, life cycle, 643–646
systems analysis
definition phase checkpoint, 645
study phase checkpoint, 644
survey phase checkpoint, 643–644
systems design
design phase checkpoint, 645–646
procurement phase checkpoint, 645
selection phase checkpoint, 645
economic, 371, 646, 648
matrix, 656
operational feasibility, 371, 646–647
schedule feasibility, 646, 648
technical feasibility, 371, 646–648
tests for feasibility, 646–648
usability analysis, 647
Feasibility prototyping *Is used to test the feasibility of specific technology that might be applied to the business problem. For example, we might use Microsoft* Access *to build a quick-but-incomplete prototype of the feasibility of moving a mainframe application to a PC-based environment.*, **125**
Feature creep *The uncontrolled addition of technical features to a system under development without regard to schedule and budget,* **596**
Fields *Implementation of a data attribute (introduced in Chapter 5). Fields are the smallest unit of* meaningful *data to be stored in a file or database.*, 176, **398**–399; *see also* **Attribute**
File access, 400
File organization, 400
File servers Store *the database, but the client computers must execute all database instructions. This means that entire databases and tables may have to be transported to and from the client across the network. The approach, used by database software such as Microsoft* Access *and Borland* dBASE, *typically generates excessive network traffic.*, **357**
Files *Collections of similar records,* **395**–397
Files *The set of all occurrences of a given record structure.*, 399n, **400**–401
pros and cons of, 395–396
Finkelstein, Clive, 206, 434
First normal form (1NF) *Entity such that there are no attributes that can have more than one value for a single instance of the entity (frequently called repeating groups). Any attributes that can have multiple values actually describe a separate entity, possibly an entity (and relationship) that we haven't yet included in our data model.*, **408,** 409–413, 412n, 418
Fitzgerald, Ardra F., 641
Fitzgerald, Jerry, 461, 507, 641
Fixed-format questionnaires *Contain questions that require specific responses from individuals.*, **631**
Fixed-length record structure, 399
Flags, 512
Flat-file technologies, 53
Flexibility, systems analysts and, 24
Flowcharts, system; *see* **Systems flowcharts**
Foci, of information systems, 48–50
Foreign keys *A primary key of one entity that is contributed to (duplicated in) another entity to identify instances of a relationship. A foreign key (always in a child entity) always matches the primary key (in a parent entity).*, 182–**183**
Foreign keys *(also introduced in Chapter 5) Pointers to the records of a different file in a database. Foreign keys are how the database links the records of one type to those of another type.*, **399**
Formal presentations *Special meetings used to sell new ideas and gain approval for new systems. They may also be used for any of these purposes: sell new system; sell new ideas; sell change; head off criticism; address concerns; verify conclusions; clarify facts; and report progress. In many cases, a formal presentation may set up or supplement a more detailed written report.*, **679**–683
Format *The way the information is displayed on a medium; for instance, columns of numbers.*, **468**
output, systems design and; *see* Outputs
Forté Software, 370
Forward scheduling *Establishes a project start date and then schedules forward from that date. Based on the planned duration of required tasks, and the allocation of resources to complete those tasks, a projected project completion date is calculated.*, **606**
Fourth-generation languages (4GLs), 158, 587
Fowler, George, 641
Free-format questionnaires *Offer the respondent greater latitude in the answer. A question is asked, and the respondent records the answer in the space provided after the question.*, **631**
Freund, John F., 641
Friedlander, Phillip, 613–616, 621
Fully attributed data model, 188, 197–199
Fully described data model, 188, 199
Function *A set of related and ongoing activities of the business. A function has no start or end; it just continuously performs its work as needed.*, **218**
Function keys *(usually labeled F1, F2, and so on) can be used to implement certain common repetitive operations in a user interface (for example, START, HELP, PAGE UP, PAGE DOWN, EXIT). These keys can be programmed to perform common functions.*, **496–497**
Functional cohesion *Occurs in modules whose instructions are related because they collectively work together to accomplish a single well-defined function. Examples of functionally cohesive modules would be modules that check customer account balance, add a new customer, delete a*

Functional cohesion *(continued)* *customer, and query a customer. In each example, the module would be accomplishing a single, simple function.,* **526**
Functional decomposition diagram; see **Decomposition diagram**
Fundamentals of information systems; see Building blocks of information systems

G

Galitz, Wilbert, 507
Gane, Chris, 115, 169, 348, 391, 461, 669
Gane/Sarson shape for DFDs, 214, 215, 234, 235
Gantt, Henry L., 602
Gantt chart *A simple horizontal bar chart that depicts project tasks against a calendar. Each bar represents a named project task. The tasks are listed vertically in the left-hand column. On a Gantt chart, the horizontal axis is a calendar timeline.,* 599, 602–609
  critical path and slack resources, 608
  effort and duration, 607
  forward and reverse scheduling, 606
  milestones, 606–607
  PERT chart versus, 609–613
  predecessors and constraints, 607–608
  resource assignments and management, 608
  work breakdown schedules, 606
Gartner Group, 14n, 15, 20n, 31
Gause, Donald C., 169
General business knowledge, 22–23
General system design, application architecture, 352–353, 380–387; *see also* **Systems design**
Generalization/specialization *A technique wherein the attributes that are common to several types of an entity are grouped into their own entity, called a* supertype., **183**–186
Generalization/specialization *A technique wherein the attributes and behaviors that are common to several types of object classes are grouped into their own class, called a* supertype. *The attributes and methods of the supertype object class are then inherited by those object classes.,* **288**–290, 303
Generalized hierarchies, 185, 197
Generations, future, of systems analysts, 25–26
Geographic modeling, 265; *see also* **Network modeling**
Geography
  as focus of information systems, 48–50, 335–336; *see also* Building blocks of information systems
  network modeling concepts for, 267–271
George, Joey F., 189n
Gildersleeve, Thomas R., 76n, 115, 260, 621, 641, 672, 691
Globalization of the economy, systems analysts and, 19–20
Goldberg, Adele, 551
Goldman, James E., 69, 391
Gordon, Jerry, 613
Graphic output *The use of a graph or chart to convey information.,* **469**
Graphical user interfaces (GUI), 59, 356, 440
  interface architectures and, 366–367
  systems design and, input design; *see* Inputs
Gray holes, 219
Green, David B., 595, 595n
Growth, future, information system development and, 77–78
Gupta, 369

H

Hammer, Mike, 115, 169, 592
Handheld PCs (HPCs), 366
Hay, David C., 206
Hewlett-Packard, 366
Hierarchical codes, data modeling and, 195, 197
Hierarchical network, 361–362
Hierarchy chart, 217; *see also* **Decomposition diagram**
High-level object interactions, OOD, 541–543
Hobus, James J., 115
Hoffer, Jeffrey A., 189n, 434, 592
Hopper, Grace, 23
Housekeeping tools, 107
HPCs (handheld PCs), 366
HTML (hypertext markup language) *The language used to construct World Wide Web pages and links.,* **370**
Human engineering, 57, 493–495, 568
Hybrid coupling, 525; *see also* **Content coupling**

I

IBM, 14, 127, 266, 328, 356, 361, 362, 364, 367, 370, 403, 406, 493, 496
ICCS, 14
Iconic menus, 489–490
Ideal object model diagrams, 541–543
Identifier, 177; *see also* **Key**
Identifying relationships, data modeling and, 196
Imaging, 41
  and document interchange, 367
Impact analysis, **FAST** development methodology
  construction phase, 97
  definition phase, 90–91
  design phase, 96
  implementation phase, 98
  purchasing phase, 94
  study phase, 88
  survey phase, 87
  targeting phase, 92
Implementation; *see also* **Systems implementation**
  environment, object-oriented design (OOD), 547–548
  management and user involvement in, 17
  model, 173; *see also* **Physical models**
  network modeling and, 268–269
  in problem solving, 10
Independent entity *One that exists regardless of the existence of any other entity. Its primary key contains no attributes that would make it dependent on the existence of another entity.,* **193**
Index Technology, 104
Information *Data that has been refined and organized by processing and purposeful intelligence. The latter, purposeful intelligence, is crucial to the definition—people provide the purpose and the intelligence that produces true information.,* **38;** *see also* **Data** *in PIECES framework,* 79–81
Information engineering (IE) *A data-centered, but process-sensitive technique that is applied to*

Information engineering (IE) *(continued)*
the organization as a whole (or a significant part, such as a division), rather than on an ad-hoc, project-by-project basis (as in structured analysis)., **123**–124
systems design and, 78, 157, 314
Information gathering; *see also* **Fact-finding**
for data modeling, 189
network modeling and, 276
process modeling and, 239
Information hiding *Principle which suggests that models should hide inappropriate details in an effort to focus attention on what's really important. In other words, "If we want to learn* anything, *we must not try to learn* everything—*at least not all at once."*, **140**
Information modeling; *see* **Data modeling**
Information services, 7
in a business, 12–14
organization of, 11–12
systems design and, 327
Information strategy planning (ISP), 13, 78, 123
Information systems architecture *Unifying framework into which various people with different perspectives can organize and view the fundamental building blocks of information systems.*, **42**
Information systems development, 70–115
cancellation or revision options, 76
CASE; *see* **Computer-aided systems engineering (CASE)**
cost-effectiveness and, 76
cross life cycle activities; *see* **Cross-life cycle activities**
development and documentation standards, 75–76
development centers and, 108
divide and conquer, 76–77
FAST methodology; *see FAST* development methodology
growth and change, design for, 77–78
justify systems as capital investments, 76
owner and user involvement, 73–74
phases and activity establishment, 74–75
principles for, 73–78
problem-solving approach, 74
software and systems metrics, 102
SoundStage case study, 71–72
system development life cycles and methodologies, 72–73
systems support after development, 98–99
Information systems framework; *see also* Building blocks of information systems
systems design and integration, 335–336
Information systems *Arrangement of people, data, processes, interfaces, networks and technology that interact to support and improve both day-to-day operations in a business (sometimes called data processing), as well as support the problem-solving and decision-making needs of management (sometimes called information services).*, **7**
Information systems *An arrangement of people, data, processes, interfaces, and geography that are integrated for the pur-poses of supporting and improving the day-to-day operations in a business, as well as fulfilling the problem-solving*
Information systems *(continued)*
and decision-making information needs of business managers., **38**
building blocks of; *see* Building blocks of information systems
development; *see* Information systems development
fundamentals of; *see* Building blocks of information systems
justification of, 45
Information technologists, 20
Information technology *Combination of computer technology (hardware and software) with telecommunications technology (data, image, and voice networks).*, **8, 38**
architecture, 13–14; *see also* **Application architecture**
competency centers, 14
working knowledge of, 21–22
Information Week, 21
Information worker *(Also called knowledge worker.) Those people whose jobs involve the creation, collection, processing, distribution, and use of information.*, 42–44; *see also* **Knowledge worker**
Inheritance *Means that methods and/or attributes defined in an object class can be inherited or reused by another object class.*, **288**
Initial study report, 86
Inputs, systems design and, 341–342, 436–461
automatic data collection (ADC), 440–442
biometric, 440
electromagnetic, 441
magnetic, 441
optical, 441
smart cards, 441
touch-based, 441–442
batch versus online, 439–440
data capture, entry, and input, 438–439
graphical user interface (GUI) controls, 440, 445–450
check boxes, 447–448
combination (combo) boxes, 450
drop-down lists, 449
list boxes, 448–449
radio buttons, 447
repository-based programming, 445
spin (spinner) boxes, 450
text boxes, 446–447
internal controls and, 443–445
prototyping for, 450–457
GUI controls selection, 452–453
input requirements review, 451–452
input screen prototyping, 453–455
source document design, 455–457
remote batch, 440
SoundStage case study, 437–438
source document, 438
user issues, 442–443
Inquiry and reporting tools, 107
Instances, 288–289
Instruction sets, 489–491
Intangible benefits *Those benefits believed to be difficult or impossible to quantify.*, **651**
Integration; *see* **Systems integration**
Intel, 59, 266, 356n, 359
Intellectual speaking, 673–674
Intelligence, purposeful, 38

Intelligent keys, data modeling and, 194
Interaction diagram, 547
Interface(s); *see also* **User interface**
  building blocks; *see* Building blocks of information systems
  as focus of information systems, 48–50, 335–336
  models of, definition phase of systems analysis, 155, 156
  objects, 534
  on-line, systems design and, 342–343
  schema, 59
  systems/applications and, 10, 63
Internal controls, for inputs and outputs, 443–445
Internal outputs *Stay inside the system to support the system's users and managers.*, **464**
Internal standards, 326–327
Internal users, 17
The Internet *An (but not necessarily* the*) information superhighway that permits computers of all types and sizes all over the world to exchange data and information using standard languages and protocols.*, 47, **358,** 370
The Internet *A global network of networks. It was conceived in 1964 by the U.S. Department of Defense to create a national military communications network that would be impervious to attacks. This network concept has exploded to include or link networks from all over the world and is used by all types of organizations and private citizens.*, **626**
Interoperability *An ideal state in which connected computers cooperate with one another in a manner that is transparent to their users (the clients).*, **359**
Interpersonal skills
  communications and, 670–691
    audiences, 672
    benefit terms, 674
    body language and proxemics, 675–676
    brainstorming, 678–679
    communicating with people, 671–676
    electronic mail, 675
    formal presentations, 679–683
      audience attention span, 682
    listening, 672–673
    loss terms, 674–675
    meetings, 677–679
    project walkthroughs, 683–685
    speaking, 673–674
    speaking styles, 674
    word usage turn-ons and turn-offs, 674–675
    written reports, 685–689
      primary elements, 687–688
      secondary elements, 687, 688
  systems analysis, 23–24
Intersolv, 104, 370, 583
Interview guide *A checklist of specific questions the interviewer will ask the interviewee.*, **634**
Interviews *Fact-finding technique whereby the systems analyst collects information from individuals face to face.*, **632**–636
  body of, 636
  conclusion of, 636
  interviewee, 632
  interviewer, 632
  opening, 634–635
  types of, 633
Intranet *A secure network, usually corporate, that uses Internet technology to integrate desktop, work group, and enterprise computing into a single cohesive framework.*, **358,** 370, 626
Introduction section, reports, 687
Intuit, 492
Inversion entry; *see* **Subsetting criteria**
Isshiki, Koichiro, 333, 348
Iteration structure, process modeling and, 223

J

Jacobson, Ivar, 307, 534, 551
JAD leader, 663–664
JAD sponsor, 663
Java *A general-purpose programming language for creating platform-independent programs and applets that can execute across the World Wide Web.*, 358–359, 368, **370**
Johnson, Spencer, 614, 618
Joint application design *A process whereby highly structured group meetings or miniretreats involving system users, system owners, and analysts occur in a single room for an extended time (four to eight hours per day, anywhere from one day to a couple of weeks).*, *125,* **663;** *see also* **Joint application development (JAD)**
Joint application development (JAD) *Uses highly organized and intensive workshops to bring together owners, users, analysts, designers, and builders to jointly define and design systems. Synonyms include* joint application design *and* joint requirements planning., **125**–126, 316, 662–669
  benefits of, 667
  leader, 663–664
  participants in, 663–664, 665–666
  planning and conducting, 664–667
Joint requirements planning; *see* **Joint application development (JAD)**
Jonsson, Patrik, 307
Joslin, Edward O., 333, 348
Justification, of information systems, 45
  system development and, 76

K

Kara, Daniel A., 391
Kendall, Julie, 494n, 507
Kendall, Kenneth, 494n, 507
Kennedy, John F., 614
Kernzer, Harold, 621
Keuffel, Warren, 589, 592
Key *Attribute, or group of attributes, that assumes a unique value for each entity instance. It is sometimes called an* identifier., **177**
  integrity of, 423
Key-based data model, 188, 194–197
Keyboards, 496–497
Keyless interfaces, 59–60
  data entry interface architectures and, 365–366
Key-to-disk (KTD)
  symbol for, 377
  workstations, 439
Key-to-punched-card, symbol for, 377
Key-to-tape (KTT)
  symbol for, 377
  workstations, 439

Knots, control flow, 588

Knowledge worker *Subset of information workers whose responsibilities are based on a specialized body of knowledge.,* **46;** *see also* **Information worker**

Kozar, Kenneth, 461, 507

L

Lantz, Kenneth E., 349, 461
Laudon, Jane P., 69
Laudon, Kenneth C., 69
Leflour, Ron, 613
Legacy systems, 77
Letter of transmittal, 688
Leventhal, N.S., 669
Levy, Ferdinand K., 621
Library program, 377

Limit and range checks *Determine whether the input data for each field falls within the legitimate set or range of values defined for that field. For instance, an upper-limit range may be put on pay rate to ensure that no employee is paid at a higher rate.,* **444**

Line charts *Used to show trends over a period of time, at even intervals. It is most common to organize the item being charted on the horizontal axis and the measurement along the vertical axis. Various line chart formats are illustrated in Figure 13.5(d).,* **470**

Line of business, 11n
Linkletter, Art, 673
List boxes, 448–449
Listening, communication skills and, 672–673

Local area network (or **LAN**) *A set of client computers (usually PCs) connected to one or more server computers (usually microprocessor-based, but could also include mainframes or minicomputers) through cable over relatively short distances—for instance, in a single department or in a single building.,* **356**

Location *Any place at which users exist to use or interact with the information system or application. It is also any place where business can be transacted or work performed.,* **268**–271
  for JAD sessions, 665
  logical, network modeling and, 268–269
  naming of, network modeling and, 269

Location connectivity diagram (LCD) *A logical network modeling tool that depicts the shape of a system in terms of its user, process, data, and interface locations and the necessary interconnections between those locations.,* **267**–268
  network modeling and, 278–279

Location conversion *When the same system will be used in numerous geographical locations, it is usually converted at one location (using either abrupt or parallel conversion). As soon as that site has approved the system, it can be farmed to the other sites. Other sites can be cut over abruptly because major errors have been fixed. Furthermore, other sites benefit from the learning experiences of the first test site. Incidentally, the first production test site is often called a beta test site.,* **568**

Location decomposition diagram *Shows the top-down geographic decomposition of the business locations to be included in a system.,* **270**
  network modeling and, 277–278

Lockheed Martin Corp., 626, 627
Logic, process modeling and, 219–225
Logic Works, 403, 420

Logical cohesion *Occurs in modules that contain instructions that appear to be related because they fall into the same logical class of functions. For example, the instructions were grouped together as a module perhaps because they all involve editing or arithmetic operations. Unfortunately, logically cohesive modules do not meet our goal of a module containing instructions that belong together because they collectively serve to accomplish a single function or task.,* **527**

Logical data flows and conventions, 227–228
Logical design, 121; *see also* **Systems analysis**
Logical DFDs, 380
Logical locations, network modeling and, 268–269

Logical models *Depict what a system is or what a system must do*—not *how the system will be implemented. Because logical models depict the* essence *of the system, they are sometimes called essential models.,* **154**

Logical models *Show* what *a system is or does. They are implementation* independent*; that is, they depict the system independent of any technical implementation. As such, logical models illustrate the* essence *of the system. Popular synonyms include* **essential model,** conceptual model, *and* business model., **172, 210**–211

Logical processes
  for data modeling; *see* **Data modeling**
  for network modeling; *see* **Network modeling**
  for process modeling; *see* **Process modeling**

Logical records, 400
London, Keith R., 37n, 621, 641
Lord, Kenniston W., Jr., 23n, 641
Lorensen, William, 307, 551

Loss terms *Words or phrases that evoke negative responses from the audience. Loss terms can also be used very effectively to sell proposed changes. Managers will usually accept ideas that eliminate loss terms. Some examples of loss terms are: higher costs; increased processing errors; higher credit losses;, excessive waste; higher taxes; delays; and increased stockouts.,* **674**–675

Lotus Development, 41, 64, 474, 654

Lower-CASE *Describes tools that automate or support the lower or later phases of systems development—detailed design, construction, and delivery (and also support).,* **105**

Lucas, Henry C., 391

M

McClure, Carma, 260
MacDonald, Robert D., 31
McDonnell Douglas, 381, 613
McFadden, Fred, 434
Machiavelli, Niccolo, 679, 679n
McMenamin, Stephen M., 260, 261
Magnetic ADC, 441
Mainframes, 356, 356n

Maintenance of systems—correcting errors; *see* **Systems support**
Majer LTD, 613
Malloy, John T., 691
Management, 73n
analysis and, 17
cross-life cycle activities and, 102–103
expectations management matrix, 613–617
failed projects and, 596–597
information systems, 39
of people, 617–618
of processes; *see* **Process management**
of projects; *see* **Project management**
reporting oriented towards, 39
of resources, 608
systems, 370
Management consultants, 15–16
Management information systems (MIS) *Information system application that provides for management-oriented reporting, usually in a predetermined, fixed format.*, **39**
Man-month, mythical, 597
Manual operation, symbol for, 377
Many-to-many (nonspecific) relationship *One in which many instances of one entity are associated with many instances of another entity. Such relationships are suitable only for preliminary data models and should be resolved as quickly as possible.*, 179, **183**
Martin, Alexander, 507
Martin, E.W., 592
Martin, James, 31, 115, 123n, 206, 260, 283, 307, 434, 551
Master files *Contain records that are relatively permanent. Thus, once a record has been added to a master file, it remains in the system indefinitely. The values of fields for the record will change over its lifetime, but the individual records are retained indefinitely. Examples of master files and tables include customers, products, and suppliers.*, **400**
Materials requirements planning (MRP), 39
Matthies, Leslie H., 260, 674, 691
Measurement *The activity of measuring and analyzing developer productivity and quality (and sometimes costs).*, 100, **101**–102
development centers and, 602
Medium *What the output information is recorded on, such as paper or a video display device.*, **467**
output formats, system design and; *see* Outputs
Meeting *Attempt to accomplish an objective as a result of discussion under leadership. Some possible meeting objectives are: presentation; problem solving; conflict resolution; progress analysis; gathering and merging facts; decision making; training; and planning.*, **677**–679
Mellor, Stephen J., 206
Menu selection *Strategy of dialogue design that presents a list of alternatives or options to the user. The system user selects the desired alternative or option by keying in the number or letter that is associated with that option.*, **484**–*489*
types of menus, 493
cascading, 486–488
iconic, 489
menu bars, 484–485, 493
pop-up, 488–489
Menu selection (*continued*)
pull-down, 485–486
tear-off, 485
Message *Item passed when one object invokes one or more of another object's methods (behaviors) to request information or some action.*, **293**
Meta style (of speaking) *Used to discuss the communications process itself. Meta language enables us to talk about our interactions. For example: "It appears we have some disagreement over who should be the leader of this project.",* **674**
Metadata, 403
Methodology *The physical implementation of the logical life cycle that incorporates (1) step-by-step activities for each phase, (2) individual and group roles to be played in each activity, (3) deliverables and quality standards for each activity, and (4) tools and techniques to be used for each activity.*, **73;** *see also* Methods
development centers and, 12, 108, 602
management of, for process management, 601
Methods, 287; *see also* **Behavior; Methodology**
object modeling and, 286–288
and procedures section, reports, 687–688
and procedures test, 568
Metrics, software and systems, 102
Metzger, Philip, 575
Microfiche *A single sheet of film that is capable of storing many pages of reduced output.*, **468**
Microfilm *A roll of photographic film that is used to record information in a reduced size.*, **468**
Microsoft Corp., 41, 59, 97, 127, 136, 144, 148, 149, 153, 160, 162, 188, 237, 266, 356, 357, 364, 366, 367, 368, 369, 370, 375, 403, 404, 406–407, 418, 440, 457, 474, 486, 491, 492, 493, 496, 507, 599, 600, 603, 606, 608, 654, 681
Middle managers, 46
Middleware *A layer of utility software that sits in between applications software and systems software to transparently integrate differing technologies so that they can operate.*, **60**
Middleware *Utility software that interfaces systems built with incompatible technologies. Middleware serves as a consistent bridge between two or more technologies. It may be built into operating systems, but it is also frequently sold as a separate product.*, 367–**368**
Milestones *Events that signify major accomplishments or events during a project.*, **607**
Miller, Irwin, 641
Minicomputers, 356, 356n
Mitchell, Ian, 641, 669
Mnemonic syntax, 490
Mobile users, 46
Model-driven development *Techniques that emphasize the drawing of models to define business requirements and information system designs. The model becomes the design blueprint for constructing the final system.*, **122,** 125
Modeling *The act of drawing one or more graphical (meaning* picture-oriented*) representations of a system. The resulting picture represents the users' DATA, PROCESS, INTERFACE, or GEOGRAPHY requirements from a business point of view.*, **89**–90

Modeling object interactions and behaviors, object-oriented design (OOD) process; *see* **Object-oriented design (OOD)**

Models/Modeling *A representation of reality. Just as a picture is worth a thousand words, most models use pictures to represent reality.*, **122, 172, 210**
for application architecture and process design; see **Application architecture**
of current system, 139–142

Model-view-controller (MVC), 534

Modern structured analysis *A process-centered technique that is used to model business requirements for a system. The models are structured pictures that illustrate the processes, inputs, outputs, and files required to respond to business events (such as orders).*, **122**–123, 157

Modern structured design *A process-oriented technique for breaking up a large program into a hierarchy of modules that result in a computer program that is easier to implement and maintain (change). Synonyms (although technically inaccurate) are top-down program design and structured programming.*, **312**

Modular design *The decomposition of a program into modules.*, **510**

Module *A group of executable instructions with a single point of entry and a single point of exit.*, **510**

Mosely, D.J., 575

Moses, John, 641, 669

Motorola, 59, 356n

Mouse *A small hand-sized device that sits on a flat surface near the terminal. It has a small roller ball on the underside. As you move the mouse on the flat surface, it causes the pointer to move across the screen. Buttons on the mouse allow you to select objects or commands to which the cursor has been moved. Alternatives include trackballs, pens, and trackpoints.*, **497**

Multiple-choice questions, 631

Multiplicity *Defines the minimum and maximum number of occurrences of on object/class for a single occurrence of the related object/class.*, **291**, 303

Multi-tier client/server, SDEs for, 369–370

Mythical man-month, 597

N

Narrative output *Format that uses sentences and paragraphs to replace or supplement standard text, numbers, and pictures.*, **471**

Nassi-Schneidermann charts, 510

Natural language syntax, 491

Navigator, 366–367

NCR, 370

Negotiation, of project scope, 132–134

Net present value, 655–656

Netscape, Inc., 366–367

Network architectures, 353–362
centralized computing, 354–356
clients defined, 353
client/server computing defined, 353–354
database servers, 357
distributed data, 356–358
distributed presentation, 356

Network architectures (*continued*)
file servers, 357
the Internet and intranets, 358–359
local area network (LAN), 356
network technologies, role of, 359–362
connectivity, 359
hierarchical network, 361–362
interoperability, 359
network topology, 359–362
ring network, 360–361
star network, 361
partitioning, 358
servers defined, 353
wide area network (WAN), 356

Network computers (NC), 359

Network modeling *A technique for documenting the geographic structure of a system. Synonyms include* **distribution modeling** *and geographic modeling.*, **265**

Network modeling *A diagrammatic technique used to document the shape of a business or information system in terms of its business locations.*, 262–283, **267**
computer networks, nonexclusive to, 264–266
construction of models, 277–279
location connectivity diagram, 278–279
location decomposition diagram, 277–278
distributed computing and, 266
logical process for, 275–276
CASE, 276, 280
fact-finding/information gathering, 276
strategic systems planning projects, 276
systems analysis, 276
systems design, 276
next generation of, 279–280
SoundStage case study, 263–264
system concepts for, 266–275
business geography, 267–271
data-to-location-CRUD matrix, 273–274
implementation locations and, 268–269
location connectivity diagram (LCD), 267–268
decomposition of, 269–270
locations and, 267–271
connectivity and, 270–271
logical, 268–269
naming of, 269
miscellaneous constructs, 272
model synchronization
data and network, 272–274
data and process, 272
process and interface, 274
process and network, 274–275
system, 272–275
process-to-location-association matrix, 275

Network programs *Machine-readable specifications of computer communications parameters such as node addresses, protocols, line speeds, flow controls, security, privileges, and other complex, networking parameters.*, **64**

Network schema *A technical model that identifies all the computing centers, computers, and networking hardware that will be involved in a computer application.*, **64**

Network topology *Describes how a network provides connectivity between the computers on that network.*, 64, **359**, 373, 381–383

Network topology data flow diagram (DFD) *A physical data flow diagram that allocates processors (clients and servers) and devices (e.g., machines and robots) to a network and establishes (1) the connectivity between the clients and servers, and (2) where users will interact with the processors (usually only the clients).*, 373, **381**–383
Network(s)
architecture; *see* Network architectures
building and testing, 557–559
configuration, 64
modeling; *see* **Network modeling**
systems/applications and, 10
technologies, 359–362
topology, 64, 359, 373, 381–383
Next generation
of data modeling, 199–202
of database design, 426
in process modeling, 250
of systems analysts, 25–26
Nielsen, Jakob, 507
Nonidentifying relationships, data modeling and, 197
Nonspecific (many-to-many) relationship *One in which many instances of one entity are associated with many instances of another entity. Such relationships are suitable only for preliminary data models and should be resolved as quickly as possible.*, 179, **183**
Normalization (of data) *The way data attributes are grouped to form stable, flexible, and adaptive entities.*, **337**
Normalization (of data) *A technique that organizes data attributes such that they are grouped to form stable, flexible, and adaptive entities.*, **408**–418
Novell, 359
N-tiered client/server computing, 357–358

O

Object *Something that is or is capable of being seen, touched, or otherwise sensed, and about which users store data and associate behavior.*, **287**
Object association model, 304
Object attribute identification, OOD, 541
Object behavior and responsibilities identification, OOD, 543–547
Object framework *A set of related, interacting objects that provide a well-defined set of services for accomplishing a task.*, **536**
Object instances, 288–289
Object interaction diagram, 547
Object modeling *A technique for identifying objects within the systems environment and the relationships between those objects.*, 284–307, **286**
in definition phase of systems analysis, 155
Object Modeling Technique (OMT), 253
object-oriented analysis, 286
process of, 294–304
business object identification, 294–303
actors and use cases, 294, 295–296
case modeling, 294, 296
potential objects, 299
proposed objects, 299–303
use case; *see* **Use case**
Object modeling *(continued)*
object organization and relationship, 303–304
SoundStage case study, 285
system concepts for, 286–293
classes, generalization, and specialization, 288–290
inheritance, 288
subtype, 289
supertype, 288–289
messages and message sending, 293
object/class relationships, 290–292
multiplicity, 291
objects, attributes, methods and encapsulation, 286–288
polymorphism, 293
Object responsibility *The obligation that an object has to provide a service when requested, thus collaborating with other objects to satisfy the request if required.*, **535**
Object reusability, OOD, 535–536
Object/class relationship *A natural business association that exists between one or more objects/classes.*, 288–292, **290**
Objective *A measure of success. It is something that you expect to achieve, if given sufficient resources.*, **145**–147
Object-oriented analysis (OOA) *Techniques that are used to (1) study existing objects to see if they can be reused or adapted for new uses, and to (2) define new or modified objects that will be combined with existing objects into a useful business computing application.*, **126**–127, 157, 164, 253, **286**
Object-oriented design (OOD) *Techniques used to refine the object requirements definitions identified earlier during analysis and to define design-specific objects.*, **317**, 532–551
control objects, 535
design objects, 534–535
entity objects, 534–535
interface objects, 534
object framework, 536
object responsibilities, 535
object reusability, 535–536
process of, 536–548
detailed object interactions for use case, 547
high-level object interactions for use case, 541–543
modeling object interactions and behaviors, 539–547
object attribute identification, 541
object behavior and responsibilities identification, 543–547
updating object model for implementation environment, 547–548
use case design object identification, 540
use case model refining, 536–539
analysis use case vs. design use case, 537–539
case model diagram updating, 539
SoundStage case study, 533
Object-oriented programming languages, 22
O'Brien, James, 69
Observation *A fact-finding technique where the systems analyst either participates in or watches a person perform activities to learn about the system.*, **628**–630
Odell, J., 307, 551
Office automation suite, 41

Office information systems *Provide support for the wide range of business office activities that provide for improved work flow and communications between workers, regardless of whether or not those workers are physically located in an office.*, **40**–41
Oncken, William, 614–615, 618
The One Minute Manager, 617
The One Minute Manager Meets the Monkey, 617–618
On-line application processing (OLAP), 357
On-line input *The capture of data at its point of origin in the business and the direct inputting of that data to the computer, preferably as soon as possible after the data originates.*, **439**–440
  symbol for, 378
On-line output, symbol for, 378
On-line processing, interface architectures and, 365
On-line user interfaces, systems design and, 342–343
Open Database Connectivity (ODBC), 60, 368
Open-ended questions, 633
Operating locations, 61
Operational databases, 402
Operational feasibility *A measure of how well the solution will work in the organization. It is also a measure of how people feel about the system/project.*, 321–324, **646–647**
  architecture strategies and, 371
Opportunity *A chance to improve the organization even in the absence of specific problems. (Note: You could argue that any unexploited opportunity is, in fact, a problem.)*, **79**
  FAST development methodology and, 79
  in study phase of systems analysis, 143–145
  in survey phase of systems analysis, 129–132
Opportunity cost, 653
Optical ADC, 366, 441
Optical character, symbol for, 377
Optical character reading (OCR), 365–366, 441
Optical mark reading (OMR), 366, 441
Oracle Corp., 364, 403, 406
Organization charts, 11
Organizing and scheduling, by project managers, 598–599
Orr, Ken, 115
OSF, 496
Outputs, systems design and, 341–342, 462–481
  internal controls and, 443–445
  media and formats, 466–471
    alternative, 468–471
    bar charts, 469
    column charts, 469
    format defined, 468
    graphic output, 469
    line charts, 470
    medium defined, 467
    microfiche, 468
    microfilm, 468
    narrative output, 471
    pie charts, 469–470
    scatter charts, 470–471
    selecting, 472–473
    tabular output, 468
    video, 468
    zoned output, 469
  principles and guidelines, 464–471
  prototyping for, 471–477
Outputs (*continued*)
    check boxes, 475
    drop-down list, 475
    medium and format, selecting, 472–473
    system outputs, identifying, 471–472
    system users and, 473–477
    tab dialogue box, 475
  SoundStage case study, 463–464
  system user issues, 471
    types of outputs, 464–466
    detailed, 464–465
    exception reports, 465–466
    external, 464
    internal, 464
    summary reports, 465–466
    turnaround, 464
Outsourcing *Contracting a service or function to an external third party.*, 11, **14**–15
Overgaard, Gunnar, 307
Owners; *see* **Systems owners**

P

Packaging *The assembly of DATA, PROCESS, INTERFACE, and GEOGRAPHY design specifications for each module.*, **510**
Paging *Displays a complete screen of characters at a time. The complete display area is known as a page (also called a screen or frame). The page is replaced on demand by the next or previous page; much like turning the pages of a book.*, **496**
Palmer, John F., 260, 261
Paradice, David, 641
Parallel conversion *Under this approach, both the old and new systems are operated for some time period. This is done to ensure that all major problems in the new system have been solved before the old system is discarded. The final cut-over may be either abrupt (usually at the end of one business period) or gradual, as portions of the new system are deemed adequate. This strategy minimizes the risk of major flaws in the new system causing irreparable harm to the business; however, it also means the cost of running two systems over some period must be incurred. Because running two editions of the same system on the computer could place an unreasonable demand on computing resources, this may be possible only if the old system is largely manual.*, **567–568**
Parameters, expectations management matrix for, 613–617
Parrington, Norman, 641, 669
Participative development, 125
Partitioning *The act of determining how to best distribute or duplicate application components (data, process, and interfaces) across the network.*, **358**
Payback analysis, 652–654
Payback period, 652
Peak workload processing performance test, 568
Pen input, 366
People
  communicating with; *see* Interpersonal skills
  information systems framework and, 10, 63, 335–336
  management of, 617–618

Performance
in PIECES framework, 79–81
vendor validation, 330–332
Periodicals, systems design and, 327
Perkins, W.C., 592
Persistent objects, 535
Personal databases, 402
Personal information systems *Those designed to meet the needs of a single user. They are designed to boost an individual's productivity.*, **41**
Person/machine boundaries, 384–387
Personnel costs, 649
Perspectives (in the framework), information systems, 42–48
executive managers, 46
external users, 46–47
information worker, 42–44
knowledge workers, 46
middle managers, 46
mobile users, 46
remote users, 46
supervisors, 46
systems builders, 47
systems designers, 47
systems owners, 44–45
systems users, 45–47
PERT charts, 599
Gantt charts versus, 612–613
as project management tool, 609–613
Phases, of system development, 74
FAST development methodology; *see FAST* development methodology
Physical data flow *Represents the planned implementation of an input to or output from a physical process. It can also indicate database action such as create, delete, read, or update a record. It can also represent the import of data from or the export of data to another information system across a network. Finally, it can represent the data flows between two modules or subroutines within the same program.*, **375–376**
Physical data flow diagrams (DFDs) *Model the technical and human design decisions to be implemented as part of an information system. They communicate technical and other design constraints to those who will actually implement the system—in other words, they serve as a technical blueprint for the implementation.*, **372**–376, 380–387
Physical data stores *Represent a single file or a single database or table in the database. Additional physical data stores may be added to represent temporary files or batches necessitated by physical processes.*, **376**
Physical design; *see* **Systems design**
Physical external agents, 376
Physical locations, 268–271
Physical models *Show not only* what *a system is or does, but also* how *the system is physically and technically implemented. They are implementation dependent because they reflect technology choices and the limitations of those technology choices. Synonyms include* implementation model *and* technical model., **173, 210**–211
Physical process *Either (1) a processor, such as a client PC, network server, or robot, or (2) specific work or actions to be performed on incoming data flows to produce outgoing data flows. In the latter case, the physical process must clearly designate which person or what technology will be assigned to do the work.*, **373**–375
modeling, 253
Physical records, 400
Picture checks *Compare data entered against the known COBOL picture or other language format defined for that data. For instance, the input field may have a picture clause XX999 AA (where X can be a letter or number, 9 must be a number, and A must be a letter). The field "A4898 DH" would pass the picture check, but the field "A489 ID8" would not.*, **444**
Pie charts *Used to show the relationship or proportions of parts to the whole at a specific period of time. Examples of different pie chart formats are illustrated in Figure 13.5(c). Notice that some styles (example 3 and 4) represent explosions and are used to emphasize a particular item. As a general rule, a pie chart should be used to show comparisons that involve 7 or fewer portions.*, **469–470**
PIECES classification framework, 79–81
operational feasibility and, 646–647
Planning
management and user involvement in, 17
in problem solving, 9
of the project
survey phase of systems analysis, 134–135
tasks, 598
of system initiatives, 78–79, 82, 86
Pointer options, user interface design and, 497
Policies *Rules that govern some process in the business.*, 55–**56, 224**
Polymorphism *Means "many forms." Applied to object-oriented techniques, it means that the same named behavior may be completed differently for different objects/classes.*, **293**
Popkin Software, 105, 190, 239, 240, 276, 296, 303, 304, 403, 420
Pop-up menus, 488–489
Powersoft, 97, 105, 237, 369
Predecessors and constraints, Gantt charts and, 606–607
Premerlani, William, 307, 551
Present value, 653–654
Presentation *The activity of formally packaging documentation for review by interested users and managers. Presentations may be either written or verbal.*, **100**
in study phase of systems analysis, 149–150
in survey phase of systems analysis, 135–137
Price Waterhouse, 15
Primary elements *Present the actual information that the report is intended to convey. Examples include the introduction and the conclusion.*, **687**–688
Primary keys *The candidate keys that will most commonly be used to uniquely identify a single entity instance.*, **179**
Primary keys *Fields whose values identify one and only one record in a file.*, **399**
Primitive diagrams, 237, 250–253

Primitive process; *see* **Elementary processes**
Primitive tasks, 606
The Prince, 679
Printed output, symbol for, 377
Problem Statement Analyzer (PSA), 103
Problem Statement Language (PSL), 103
Problem statements, 131–132
Problems *Undesirable situations that prevent the organization from fully achieving its purpose, goals, and objectives.,* **79,** 580
Problem-solving
  skills, systems analysts, 23
  study phase of systems analysis, 143–145
  style, 672
  survey phase of systems analysis, 129–132
  as system development approach, 74
Problem-solving style (of speaking) *Rational, objective, unbiased, and bland. This is the style most used in business dealings. For example: "I believe John's past performance qualifies him to be an ideal candidate to be the leader of this project.",* **674**
Procedural cohesion *Occurs in modules whose instructions accomplish different tasks yet have been combined because there is a specific order in which the tasks are to be completed. These types of modules are typically the result of first flow-charting the solution to a program and then selecting a sequence of instructions to serve as a module. Since these modules consist of instructions that accomplish several tasks that are virtually unrelated, these types of modules tend to be less maintainable.,* **526**
Procedures *Step-by-step instructions and logic for accomplishing a business process., 56*
Process *Work performed on, or in response to, incoming data flows or conditions. A synonym is* transform., **216;** *see also* Process(es)
Process concepts, modeling and; *see* **Process modeling**
Process design, application architecture and; *see* **Application architecture**
Process distribution and technology assignments, 383–384
Process management *An ongoing activity that establishes standards for activities, methods, tools, and deliverables of the life cycle.,* 102–**103**
Process management *The planning, selection, deployment, and consistent application of standard system development methods, tools, techniques, and technologies to all information system projects.,* **599**–602
  development centers and, 602
  methodology management, 601
  system development technology management, 601
  total quality management and, 601–602
Process modeling *A technique for organizing and documenting the structure and flow of data through a system's process and/or the logic, policies, and procedures to be implemented by a system's processes.,* 123, 208–261, **211;** *see also* **Data flow diagram (DFD)**
  construction of models, 239–253
    context diagram, 237, 239–241, 242
    event decomposition diagram, 244–245
    event diagram, 245–247
    event-response list, 237, 243–244
Process modeling (*continued*)
    functional decomposition diagram, 237, 241–243
    primitive diagrams, 237, 250–253
    synchronization and, 249
    system diagram, 247–250
  definition phase of systems analysis, 155
  logical process for, 235–239
    for business process redesign (BPR), 236, 253
    CASE for, 239, 241, 253
    event partitioning, 237
    fact-finding and information gathering, 239
    strategic systems planning, 235–236
    during systems analysis, 236–237
    systems configuration and design, 237
  next generation, 250
  SoundStage case study, 209–210
  synchronization, 249
    with interface model, 274
    with network model, 274–275
  system modeling concepts for, 210–214, 214–235
    data flow diagram (DFD), 211–214
    data flows and, 225–233
      composite, 226
      conservation of, 228–230
      control flow, 227
      data attributes, 230–231
      data in motion, 225–227
      data structures, 230–231
      data type, 231
      divergent and convergent, 231–233
      domains, 231
      logical flows and conventions, 227–228
    data stores, 234
    decision tables and, 224–225
    decomposition diagrams, 217–218
    elementary processes, 219
    events, 218–219
    external agents and, 233–234
    functions, 218
    logic, process, 219–225
    logical conventions, 218–219
    logical models, 210–211
    models defined, 210–211
    physical models, 210–211
    policy and, 224
    process concepts, 215–225
    process decomposition, 216–218
    process logic, 219–225
    structured English, 220–224
    system as process, 215–216
    systems thinking, 214
Process(es)
  building blocks of, 335–336; *see also* Building blocks of information systems
  as focus of information systems, 48–50, 335–336
  systems design and, 339–340
  systems/applications and, 10, 63
Process-to-data-to-CRUD matrix, 273
Process-to-location-association matrix *A table in which the rows indicate processes (event or elementary processes); the columns indicate locations; and the cells (the intersection of rows and columns) document which processes must be performed at which locations.,* **275**

Procurement phase, of systems design; *see* **Systems design**
FAST development methodology, 83, 93–94
Production system, 82
Professional associations, 22
Professional staff, 45–46
Program(s); *see* Computer programs/programming
Project *A sequence of unique, complex, and connected activities having one goal or purpose and that must be completed by a specific time, within budget, and according to specification.*, **595**
Project management *The ongoing activity by which an analyst plans, delegates, directs, and controls progress to develop an acceptable system within the allotted time and budget.*, 85, **102**–103
Project management *The process of defining, planning, directing, monitoring, and controlling the development of an acceptable system at a minimum cost within a specified time frame.*, **595**–599
as cause of failed projects, 596–597
directing and controlling the project, 599
expectations management, 613–617
feature creep, 596
functions of project managers, 597–599
Gantt charts; *see* **Gantt chart**
organizing and scheduling project effort, 598–599
people management, 617–618
PERT charts, 609–613
critical path in, 609–612
planning project tasks, 598
scope creep, 596
scoping the project, 598
software for, 599
staffing project teams, 598
tools and techniques, 602–613
Project plan, 86–87
Project walkthrough *A peer group review of systems development documentation. Walkthroughs may be used to verify almost any type of detailed documentation such as location connectivity diagrams; data flow diagrams; entity relationship diagrams; input designs; output designs; file designs; database designs; policies and procedures; user manuals; and program code.*, **683**–685
Property, 176; *see also* **Attribute**
Proposals, request for, 93–94, 328–330
Prototyping *A technique for quickly building a functioning model of the information system using rapid application development tools (provided with most popular programming languages).*, **57**
Prototyping *The act of building a small-scale, representative or working model of the users' requirements to discover or verify those requirements.*, **90, 157**–159
of databases, 424
discovery prototypes, 125, 157–159
FAST development methodology and, 95–96
feasibility, 125
of inputs, systems design and; *see* Inputs
of outputs, systems design and; *see* Outputs
process focused, 57
requirements, 158–159
Prototyping *(continued)*
systems design and, 314–316
third-generation languages and, 316
tools, 107
user interface design and; *see* **User interface**
Proxemics *The relationship between people and the space around them. Proxemics is a factor in communications that can be controlled by the knowledgeable analyst.*, 675–**676**
The Prudential Insurance Company of America, 671
PSL/PSA, 103
Pull-down menus, 485–486
Purposeful intelligence, 38
Puttick, Richard, 551
PVCS, 370

Q

Quality, development centers and, 602
Quality assurance, 107, 249; *see also* **Synchronization**
for structure charts, 523–527
Question-answer dialogues, 491–492
Questionnaires *Special-purpose documents that allow the analyst to collect information and opinions from respondents.*, **630**–632
Quotes (proposals), solicitation from vendors, 93–94, 328–330

R

Radio buttons, 447
Randomization *Sampling technique characterized as having no predetermined pattern or plan for selecting sample data.*, **625**
Ranking questions, 632
Rapid application development (RAD) *The merger of various structured techniques (especially the data-driven information engineering) with prototyping techniques and joint application development techniques to accelerate systems development.*, 125, 158, **316**–317
construction phase of systems implementation, 557
fact-finding and, 636–637
FAST design phase, 95–96, 556–557
Rating questions, 631
Recording secretary, 679
Records *Collection of fields arranged in a predefined format.*, 399
Recursive relationships, 181
Reengineering
business process, 126; *see also* **Business process redesign (BPR)**
program, 588–589
systems enhancement and; *see* **Systems support**
Referential integrity, database design and, 423n, 423–424
Regression testing *Extrapolates the impact of the changes on program and application throughput and response time from the before-and-after results using the* test data and current performance., **583**
Relational databases *Store data in a tabular form. Each file is implemented as a* table. *Each field is a* column *in the table. Each record in the file is a* row *in the table.*, **362**–364, 399n
Relational databases *Implement data in a series of tables that are "related" to one another via foreign keys.*, **405**–407

Relationship *A natural business association that exists between one or more entities. The relationship may represent an event that links the entities or merely a logical affinity that exists between the entities.*, **179**–186, 401
data modeling systems concepts and; *see* **Data modeling**
Remote batch *Offers on-line advantages for data that is best processed in batches. The data is input on-line with on-line editing. Microcomputers or minicomputer systems can be used to handle this on-line input and editing. The data is not immediately processed. Instead, it is batched, usually to some type of magnetic media. At an appropriate time, the data is uploaded to the main computer, merged, and subsequently processed as a batch. Remote batch is also called* deferred batch *or* deferred processing, **440**
interface architectures and, 365
Remote users, 46
Repetition structure, 223
Reporting, management-oriented, 39
Repository *A collection of those places where we keep all documentation associated with the application and project.*, **121**–122
FAST development methodology and, 82
Repository-based programming, 445
Request for proposals (RFP), 93–94, 329
Request for quotations (RFQ), 329
Requirements, system, 586
analysis of, 163–164
prototypes, 158–159
statement model, 155
Research and site visits, 626–628
Resources *People, material, and tools that you assign to the completion of a task.*, **608**
Rethinking Systems Analysis and Design, 628
Return-on-investment (ROI) analysis, 654–655
Reverse engineering, 253
Reverse scheduling *Establishes a project deadline and then schedules backward from that date. Essentially, tasks, their duration, and resources must be chosen to ensure that the project can be completed by the deadline.*, **606**
Revisions, development and, 76
Ring network, 360–361
Ripple effect, 523–524
RISC processors, 356n
Robertson, James, 260
Robertson, Suzanne, 260
Role name *An alternate name for a foreign key that clearly distinguishes the purpose that the foreign key serves in the table.*, **424**
Role playing, 547
Roles, 127, 317, 424, 556
Rubin, Kenneth S., 551
Rumbaugh, James, 307, 551

S

Sales engineers, 16
Salvendy, G., 641
Sampling *The process of collecting sample documents, forms, and records.*, **624**–626
Sarson, Trish, 391
Scatter charts *Used to plot the data values of two items to show uneven intervals or clusters of data. Various standard statistical techniques can then be applied to determine the degree of correlation that exists. Variations of scatter charts are depicted in Figure 13.5(e).*, **470–471**
Schedule feasibility *A measure of how reasonable the project timetable is.*, 321–324, **646,** 648
Schlaer, Sally, 206
SCM, **370**
Scope creep *The unexpected growth of user expectations and business requirements for an information system as the project progresses.*, **596**
Scoping the project/system, 132–134
definition phase of systems analysis, 161–162
project manager functions and, 598
scope creep, 596
statement of, 86
study phase of systems analysis, 147–149
Scribes, 662
Scroll bars, 493
Scrolling *Moves the displayed characters up or down, one line at a time. This is similar to the way movie and television credits scroll up the screen at the end of a movie.*, **496**
Sears, 46, 367
Second normal form (2NF) *Entity that is already in 1NF, and the values of all nonprimary key attributes are dependent on the full primary key—not just part of it. Any nonkey attributes that are dependent on only part of the primary key should be moved to any entity where that partial key becomes the full key. Again, this may require creating a new entity and relationship on the model.*, **408–409**, 414
Secondary elements *Package the report so the reader can easily identify the report and its primary elements. Secondary elements also add a professional polish to the report,* **687**, 688
Secondary keys Alternate *identifiers for a database. A secondary key's value may identify either a single record (as with a primary key) or a subset of all records (such as all orders that have the order status of "backordered").*, **399**
Security, in PIECES framework, 79–81
Self-checking digits *Determine data entry errors on primary keys. A check digit is a number or character that is appended to a primary key field. The check digit is calculated by applying a formula, such as Modulus 11, to the actual key (*see *Figure 12.3). The check digit verifies correct data entry in one of two ways. Some data entry devices can automatically validate data by applying the same formula to the data as it is entered by the data entry clerk. If the check digit entered doesn't match the check digit calculated, an error is displayed. Alternatively, computer programs can also validate check digits by using readily available subroutines.*, **444**
Sequencing, process modeling and, 221–223
Sequential cohesion *Occurs in modules whose instructions are related because the output data from one instruction are used as input data to the next instruction. An example of this type of*

Sequential cohesion (*continued*)
module would be one whose instructions might accomplish the following series of tasks: get an order, edit the order, release the order to the warehouse for filling and shipping, and then bill the customer for the shipped order. This type of module does not typically present serious coupling and cohesion problems that would affect its maintainability. However, since several functions have been included in a single module, the reuse of any given function is not possible., **526**
Serial codes, 195
Server *A multiple-user computer that provides (1) shared database, processing, and interface services and (2) connectivity to clients and other servers.,* **353;** *see also* **Client/server computing; Client/server computing applications**
Service, 287; *see also* **Behavior**
in PIECES framework, 79–81
Service workers, 45
Significant position codes, 195
Silver, Denise, 169, 349, 461, 641, 669
Simplification by inspection, 417–418
Site visits, 626–628
Slack resources, Gantt charts and, 608
Slack time *Time available for any task equal to the difference between the earliest and latest completion times.,* **608**
Smalltalk, 368
Smart cards, 441
Smith, Randi Sigmund, 691
Software; *see also* Program(s)
application; *see* **Application architecture; Application programs**
designing; *see* Software design
new, install and test, 561
for project management, 599
technology, 54
Software and systems metrics *Provides an encyclopedia of techniques and tools that can both simplify the estimation process and provide a statistical database of estimates versus performance.,* **102**
Software design, 508–531
defined, 510–511
modular, 510
packaging and, 510, 527
SoundStage case study, 509
structured design, 511–527
data flow diagrams of programs, 513–515
transaction analysis and, 521–523
transform analysis and, 515–521
structure charts, 313–314, 511–513
cohesion and, 526–527
common coupling, 525
content coupling, 525
control coupling, 524–525
coupling, 523–526
data coupling, 523–524
quality assurance checks for, 523–527
stamp coupling, 524
transaction analysis, 521–523
transaction center, 521–523
transform analysis, 515–521
Software development environment (SDE) *A language and tool kit for constructing information system applications. They are usually built around one or more programming languages such as* COBOL, BASIC, C *or C++,* Pascal, Smalltalk, *or* Java., **368**
process architecture and, 368–370
Software Engineering Institute (SEI), 601
Software engineers, 16, 48
Software metrics *Mathematically proven measurements of software quality and productivity.,* **588**
SoundStage case study
application architecture and process design, 351–352
building blocks of information systems, 33–37
data modeling, 171–172
database design, 393–395
information system development, 71–72
input design, 437–438
modern systems analysts, 3–7
network modeling, 263–264
object modeling, 285
object-oriented design (OOD), 533
output design, 463–464
process modeling, 209–210
software design, 509
systems analysis, 119–120
systems design, 311
systems implementation, 555
systems support, 577
user interface design, 483–484
Source document *A paper form used to record data that will eventually be input to a computer.,* 377, **438**
Speaking, communication skills and, 673–674
Specialization, 288–290
Specifications, 595–596
packaging, software design and, 527
Spin (spinner) boxes, GUI controls for input design, 450
Split screen capabilities, display features and, 496
Sponsors, 84–85
of JAD, 661
SQL (Structured Query Language), 405–407
SSADM/IDEFO shape for DFDs, 215–216
Staffing of project team, 598
Staged conversion *Like location conversion, staged conversion is a variation on the abrupt and parallel conversions. A staged conversion is based on the version concept introduced earlier. Each successive version of the new system is converted as it is developed. Each version may be converted using the abrupt, parallel, or location strategies.,* **568**
Stakeholders, 42
Stallings, Warren D., 641
Stamp coupling *Two modules are said to be stamp coupled if their communication of data is in the form of an entire data structure or record. Since not all data making up the structure are usually necessary in the communication between the modules, stamp coupling typically involves tramp data. The passing of an entire data structure or record is also undesirable because any changes to the data structure or record may adversely affect any module that uses it.,* **524**

Standards, for system development and documentation, 75–76
Star network, 361
State events, 244
State transition diagrams, 59, 497
Steering body *A committee of executive business and system managers that studies and prioritizes competing project proposals to determine which projects will return the most value to the organization and thus should be approved for continued systems development.*, 78, **135**–136
Steiner, James B., 23n, 641
Stewart, Charles J., 641
Stored procedures *Programs embedded within a table that can be called from an application program. For example, a complex data validation algorithm might be embedded in a table to ensure that new and updated records contain valid data* before *they are stored.*, **405**
STRADIS, 381
Strategic data modeling, 186
Strategic systems planning
  network modeling and, 276
  process modeling and, 235–236
Stratification *A systematic sampling technique that attempts to reduce the variance of the estimates by spreading out the sampling—for example, choosing documents or records by formula—and by avoiding very high or low estimates.*, **625**
Structure charts, 313–314, 511–513
  quality assurance checks for, 523–527
Structure generation, database design and, 426
Structured analysis, modern, 122–123, 157
Structured design; *see* Software design
Structured English *A language and syntax, based on the relative strengths of structured programming and natural English, for specifying the underlying logic of elementary processes on process models (such as* data flow diagrams*).*, **220**–224, 480
Structured interviews *Interview in which the interviewer has a specific set of questions to ask of the interviewee.*, **633**
Structured programming, 221–223, 312; *see also* **Modern structured design**
Structured Solutions, 275
Strunk, William S., Jr., 688
Stuart, Ann, 691
Stub testing *The test performed on individual modules, whether they be main program, subroutine, subprogram, block, or paragraph.*, **563**
Study phase
  FAST development methodology, 82–83, 87–88
  of systems analysis, 137–150
    activity diagram, 139
    business processes analysis, 142–143
    feasibility checkpoint, 644
    findings/recommendations presentation, 149–150
    information hiding, 140
    model current system, 139–142
    problems/opportunities analysis, 143–145
    scope and plan of project, 147–149
    system improvement objectives and constraints, 145–147
Subprograms, 589
Subsetting criteria *An attribute (or concatenated attribute) whose finite values divide all entity instances into useful subsets. Some methods call this an* inversion entry., **179**
Subtype *An entity/object class whose instances inherit some common attributes from an entity/class supertype and then add other attributes that are unique to an instance of the subtype.*, **185**–186, **289**
Successful Data Processing Systems Analysis, 670
Summary reports *Categorize information for managers who do not want to wade through details.*, **465**–466
Summary tasks, 606
Sun Microsystems, 97, 370
Supertype *An object whose instances store attributes that are common to one or more entity/class subtypes of the object.*, **184**–186, **288**–289
Supervisors, 46
Support
  management and user involvement in, 17
  in problem solving, 10
  systems; *see* **Systems support**
Survey phase
  activity diagram, 129
  FAST development methodology, 128–129, 129n
  of systems analysis
    feasibility checkpoint, 643–644
    plan the project, 134–135
    present the project, 135–137
    problems, opportunities, and directives, 129–132
    project scope, 132–134
    steering body, 78, 135–136
Sybase Corp., 357, 364, 403, 406
Symantec, 370
Synchronization *Balancing of data flow diagrams at different levels of detail to preserve consistency and completeness of the models. Synchronization is a quality assurance technique.*, **249**
  network modeling and, 272–275
System Architect, 155
Systems acceptance test, 568–569
Systems analysis reports, 686
Systems analysis *The study of a business problem domain to recommend improvements and specify the business requirements for the solution.*, **7**
Systems analysis *The dissection of a system into its component pieces to study how those component pieces interact and work.*, **120**
Systems analysis *(1) The survey and planning of the system and project, (2) the study and analysis of the existing business and information system, and (3) the definition of business requirements and priorities for a new or improved system. A popular synonym is* **logical design.**, 118–169, **121**
  business area analysis (BAA), 124
  business process redesign (BPR), 126
  data modeling during, 187–188
  defined, 120–122
  definition phase of; *see* Definition phase
  discovery prototyping, 125
  FAST methodology; *see FAST* development methodology
  feasibility prototyping, 125

Systems analysis *(continued)*
information engineering (IE), 123–124
information strategy planning (ISP), 123
joint application development (JAD), 125–126, 316
model-driven development, 122
modern structured analysis, 122–123
network modeling and, 276
object-oriented analysis (OOA), 126–127
phases of a project, 121
process modeling during, 236–237
prototyping, 124–125
repository, 121–122
requirements analysis, 163–164
skills, 25
SoundStage case study, 119–120
strategies for, 122–128
study phase of; *see* Study phase
survey phase of; *see* Survey phase
systems synthesis, 120

Systems analysts *Facilitators of the development of information systems and computer applications.*, 7

Systems analysts *Facilitators of the study of the problems and needs of a business to determine how the business systems and information technology can best solve the problem and accomplish improvements for the business. The* product *of this activity may be improved business processes, improved information systems, or new or improved computer applications—frequently all three.*, 2–31, **8**
adaptability and, 24
application analysts, 9
business analysts, 9
business trend implications for, 18–20
career preparation, 20–22
character and, 24–25
computer programming experience and expertise, 22
customers of, 16–17
database design concepts for; *see* Database design
definition of, expanded, 8–9
employment and; *see* Employment
ethics and, 24–25
flexibility and, 24
functions of, 9–11
general business knowledge, 22–23
globalization of the economy and, 19–20
information technology and, 8, 21–22
interpersonal skills, 23–24
next generation of, 25–26
partners for, 20
as problem solvers, 7–17, 23
SoundStage case study, 3–7
system analysis and design skills, 25
users and, 16–17

Systems builders *Construct the information system components based on the design specifications from the system designers. In many cases, the systems designer and builder for a component are one and the same.*, 42, **47**
data building blocks and, 53
geography building blocks and, 64
information systems focus and framework, 49–50
interface building blocks and, 60
process building blocks and, 56–57

Systems concepts
for data modeling; *see* **Data modeling**
for network modeling; *see* **Network modeling**
for object modeling; *see* **Object modeling**
for process modeling; *see* **Object modeling**

Systems consultants, 15–16

Systems design *Specification or construction of a technical, computer-based solution for the business requirements identified in a systems analysis.*, 7

Systems design *The evaluation of alternative solutions and the specification of a detailed computer-based solution. It is also called physical design.*, 310–349, **312**
application architecture and, 352–353, 380–387
configuration phase, 319–326
candidate solutions, 319–321
feasibility of alternative solutions, 321–324
systems solution recommendation, 324–326
design and integration phase, 335–344
data analysis and distribution, 336–338
databases, 340–341
design presentation and review, 343–344
event analysis, 337–338
feasibility checkpoint, 643–644
normalization, 337–338
on-line user interfaces, 342–343
outputs and inputs, 341–342
process analysis and distribution, 339–340
FAST development methodology phase, 83, 94–96; *see also FAST* development methodology
generation tools, 107
implementation phase; *see* **Systems implementation**
information engineering, 314
joint application development (JAD), 125–126, 316
management and user involvement in, 17
modern structured design, 312–314
network modeling and, 276
object-oriented design (OOD); *see* **Object-oriented design (OOD)**
in problem solving, 9–10
process modeling and, 237
procurement phase, 326–335
contracts, awarding (or letting), 333–334
feasibility checkpoint, 643
integration requirements, 334–335
proposals (or quotes) from vendors, 328–330
technical criteria and options, 326–328
vendor claims and performance, 330–332
vendor debriefing, 333–334
vendor proposals, 332–333
prototyping, 314–316; *see also* **Prototyping**
rapid application development; *see* **Rapid application development (RAD)**
selection phase feasibility checkpoint, 645
SoundStage case study, 311
strategies for, 312–317

Systems design reports, 686

Systems designers *Translate users' business requirements and constraints into technical solutions. They design the computer files, databases, inputs, outputs, screens, networks, and programs that will meet the systems users' requirements. They also integrate the technical solution back into the day-to-day business environment.*, 42, **47,** 48

Systems designers *(continued)*
data building blocks and, 53
geography building blocks and, 64
information systems focus and framework, 49–50
interface building blocks and, 59–60
process building blocks and, 56
Systems development, 12, 556
Systems development life cycle (SDLC) *Systematic and orderly approach to solving system problems.*, **9**
Systems development life cycle (SDLC) *A logical process by which systems analysts, software engineers, programmers, and end-users build information systems and computer applications to solve business problems and needs. It is sometimes called* **application development life cycle.**, **72**–73
Systems development technology management, 601
Systems diagram, 237
process modeling and, 247–250
Systems engineering, 103
Systems engineers, 16
Systems enhancement, 584–585
and reengineering; *see* **Systems support**
Systems flowcharts *Diagrams that show the flow of control through a system while specifying all programs, inputs, outputs, and file/database accesses and retrievals.*, **376**–380
CASE for, 380
reading, 378–380
symbols for, 377–378
Systems implementation *The construction of the new system and the delivery of that system into production (meaning day-to-day operation). Unfortunately,* systems development *is a common synonym. (We dislike that synonym since it is more frequently used to describe the entire life cycle).*, 554–575, **556;** *see also* Construction; Implementation
construction phase, 557–564
build and test databases, 560–561
build and test networks, 557–559
install and test new software package, 561
stub testing, 563
systems testing, 563
unit or program testing, 563
write and test new programs, 562–564
delivery phase, 564–572
abrupt cut-over, 567
audit testing, 569
conversion, 571–572
plan preparation, 566–569
database installation, 569–570
location conversion, 568
parallel conversion, 567–568
staged conversion, 568
system tests, conducting, 564–566
system user training, 570–571
systems acceptance test, 568–569
validation testing, 568–569
verification testing, 568
FAST systems methods, 556–557
SoundStage case study, 555
Systems implementation reports, 686
Systems integration *The process of making heterogeneous information systems (and computer systems) interoperate. A key technology used to interface and integrate systems is middleware.*, 16, 335–344, **367**–**368**
requirements for, 334–335
Systems integrators, 16
Systems maintenance—correcting errors; *see* **Systems support**
Systems management, application architecture and, 370
Systems modeling; *see also* **Process modeling**
data modeling and, 172–175
Systems Network Architecture (SNA), 361–362
Systems owners *Provide sponsorship of information systems and computer applications. In other words, they pay to have the systems and applications developed and maintained. They may also approve technology, and most certainly approve significant business changes caused by using technology.*, **17**
Systems owners *Information system sponsors and chief advocates. They are usually responsible for budgeting the money and time to develop, operate, and maintain the information system. They are also ultimately responsible for the system's justification and acceptance.*, 42, 44–**45,** 73n
data building blocks and, 51–52
geography building blocks and, 61–63
information systems focus and framework, 49–50
interface building blocks and, 57–59
process building blocks and, 55
system development participation, 73–74
Systems performance test, 568
Systems planning reports, 686
Systems problem solving *Studying a problem environment to implement corrective solutions as new or improved systems.*, **9**–10, 17
Systems recovery, from crashes, 584
Systems support *The ongoing maintenance of a system after it has been placed into operation. This includes program maintenance and system improvements.*, **98**–99, 576–592, **578**
end-user assistance, 584
SoundStage case study, 577
system maintenance—correcting errors, 580–584
benchmarking programs and applications, 581–582
define and validate problems, 580–581
edit and test the programs, 583
regression testing, 583
systems testing, 583
understanding programs and applications, 582–583
unit testing, 583
update documentation, 583–584
version control, 107, 370, 583
system recovery, from crashes, 584
systems enhancement and reengineering, 584–589
analyze enhancement request, 585–586
analyze program library and maintenance costs, 588
code conversion, 588
code reorganization, 588
code slicing, 588
control flow knots, 588
cycle complexity, 588

Systems support *(continued)*
reengineer and test programs, 588–589
restructure databases, 587–588
software metrics, 588
write simple, new programs, 586–587
Year 2000 and, 589
Systems synthesis *The re-assembly of a system's component pieces back into a whole system—it is hoped an improved system.*, **120**
Systems testing *Ensures that applications programs written in isolation work properly when they are integrated into the total system.*, **97**
Systems testing *A test that ensures that application programs written in isolation work properly when they are integrated into the total system.*, **563**, 583
Systems testing *Ensures that the entire application, of which the modified program was a part, still works. Again the* test data and current performance *from task 2 are used here.*, **583**
Systems thinking *The application of formal systems theory and concepts to systems problem solving.*, **175**, **214**
Systems users *Individuals who either have direct contact with an information system or application (e.g., they use a terminal or PC to enter, store, or retrieve data) or use information (reports) generated by a system.*, **16**
Systems users *people who use (and directly benefit from) the information systems on a regular basis—capturing, validating, entering, responding to, storing, and exchanging data and information.*, 42, **45**–47; *see also* **User(s)**
casual, 494
data building blocks and, 52–53
dedicated, 494
external, 17, 46–47
geography building blocks and, 63–64
information systems focus and framework, 49–50
input design and, 442–443
interface building blocks and, 59
internal, 17
output design and, 471, 473–477
process building blocks and, 55–56
training of, 570–571

T

Tab dialogue box, 475
Table *The relational database equivalent of a file.*, 399n, **400**–401
Table look-up files *Contain relatively static data that can be shared by applications to maintain consistency and improve performance. Examples include sales tax tables,ZIP code tables, and income tax tables.*, **400**
Tabs, 493
Tabular output *Uses columns of text and numbers. The oldest and most common format for computer outputs. This format presents information as columns.*, **468**
Tactical application, architecture strategies, 371
Tangible benefits *Those that can be easily quantified. Tangible benefits are usually measured in terms of monthly or annual savings or of profit to the firm.*, **650**–651
Tape file, symbol for, 377
Tasks, Gantt charts and; *see* **Gantt chart**
Taylor, David A., 307, 551
Tear-off menu, 485
Technical design statement, 527
Technical feasibility *A measure of the practicality of a specific technical solution and the availability of technical resources and expertise.*, 321–324, **646**, 647–648
architecture strategies and, 371
Technical limitation, 585–586
Technical model, 173; *see also* **Physical models**
Technical report; *see* **Business and technical report**
Technical requirements, new, 586
Technical sponsor, 85
Technical staff, 45–46
Technology
development centers and, 602
information systems framework and, 335–336
Teichrowe, Daniel, 103
Telecommunications, 12
Telecommuting, 46
Temporal cohesion *Occurs in modules whose instructions appear to have been grouped together into a module because of "time." For example, a temporal cohesive module may contain instructions that were grouped together because they perform start-up or initialization activities (such as setting program counters or control flags) associated with the program. Or perhaps the instructions were to be performed at the end of the program, such as printing final report totals, closing a file, or displaying an end-of-job message to the user.*, **526**
Temporal events, 244, 294
Teorey, Toby J., 206, 434
Tests/Testing, 107
alpha version, 568
audit, 569
backup and recovery, 568
beta version, 568–569
databases, 560–561
exercising user interfaces, 500–501
methods and procedures, 568
networks, 557–559
regression, 583
stub, 563
systems tests; *see* **Systems testing**
unit tests; *see* **Unit testing**
validation, 568–569
verification, 568
Texas Instruments, 370
Text boxes, 446–447
Theby, Stephen E., 391
Thill, John V., 674, 691
Thin clients, 359
Third normal form (3NF) *Entity that is already in 2NF, and the values of its nonprimary key attributes are not dependent on any other non-primary key attributes. Any nonkey attributes that are dependent on other nonkey attributes must be moved or deleted. Again, new entities and relationships may have to be added to the data model.*, **409**, 414–418
Third-generation languages (3GLs), 158
prototyping and, 316
Three-tiered client/server computing, 357–358

Time boxing *A technique that divides the set of all business requirements for a system into subsets, each of which will be implemented as a version of the system. Essentially, the project team guarantees that new versions will be implemented on a regular and timely basis.,* **90, 160**
Time constraints, 595–596
Time value of money, 652
Tkach, Daniel, 551
Token Ring Network, 361
Tool bars, 493
Top-down programming, 312; *see also* **Modern structured design**
Topology, network, 64, 359, 373, 381–383
Total quality management (TQM) Comprehensive approach to facilitating quality improvements and management within a business., **18,** 55, 126
  process management and, 601–602
Touch-based ADC, 441–442
Tower of Babel, 671
Trade newspapers, 327
Training
  costs of, 649
  of system users, 569–571
Tramp data, 523–524
Transaction analysis *The examination of the DFD to identify processes that represent transaction centers.,* **521**–523
Transaction center, 521
Transaction files/tables *Contain records that describe business events. The data describing these events normally have a limited useful lifetime. For instance, an invoice record is ordinarily useful until the invoice has been paid or written off as uncollectible. In information systems, transaction records are frequently retained* on-line *for some period of time. Subsequent to their useful lifetime, they are* archived *off-line. Examples of transaction files include orders, invoices, requisitions, and registrations.,* **400**
Transaction processing (TP) monitors *Software that ensures that all the data associated with a single business transaction are processed as a single transaction among all the parallel business transactions that may be in the system at the same time. Examples include IBM's* CICS, *and NCR's* Tuxedo., **370,** 404–405
Transaction processing (TP) systems *Information system applications that capture and process data about (or for) business transactions. They are sometimes called* **data processing systems., 38**–39
Transactions, 218, 438; *see also* **Event**
Transform; *see* **Process**
Transform analysis *An examination of the DFD to divide the processes into those that perform input and editing, those that do processing or data transformation (e.g., calculations), and those that do output.,* **515**–521
Triggers *Programs embedded within a table that are automatically invoked by updates to another table. For example, if a record is deleted from a passenger aircraft table, a trigger can force the automatic deletion of all corresponding records in a seats table for that aircraft.,* **405**
Turnaround outputs *Those that are typically implemented as a form that eventually reenter the system as an input.,* **464**
Turn-key system, 94
Tuxedo, 370
Two-tiered client/server computing, 356–357
  SDEs for, 369

U

Unit testing *Ensure that the applications programs work properly when tested in isolation from other applications programs.,* **97**
Unit testing *A test whereby all the modules that have been coded and stub tested are tested as an integrated unit.,* **563**
Unit testing *Ensures that the stand-alone program fixes the bug without side effects. The* test data and current performance *that you recovered, created, edited, or generated in task 2 are used here.,* **583**
Unplanned system request, 78, 79–82, 86
Unstructured decisions, 39–40
Unstructured interviews *Conducted with only a general goal or subject in mind and with few, if any, specific questions. The interviewer counts on the interviewee to provide a framework and direct the conversation.,* **633**
Upper-CASE *Describes tools that automate or support the upper or earliest phases of systems development—the survey, study, definition, and design phases.,* **104**
UPS, 366
Uris, Auren, 691
Usability analysis, 647
Use case *A behaviorally related sequence of steps (a scenario), both automated and manual, for the purpose of completing a single business task.,* **294,** 295–299
  dependency diagram, 297
  design object identification, 540
  model diagram, 296, 297
  model refining, 536–539
  object modeling
    alternative course of events, 298–299
    business object identification, 294–299
    course of events, 296–297
    dependencies, 297
    models, 296
Use case modeling *The process of identifying and modeling business events, who initiated them, and how the system responds to them.,* **294,** 296, 297, 536–539
User dialogues *Describe how the user moves from screen to screen, interacting with the application programs to perform useful work.,* **59**
User interface *Defines how the system users directly interact with the information system to provide inputs and queries and receive outputs and help.,* **59;** *see also* Interface(s)
  designing, 482–507
  display features, 495–497
  human factors for, 493–495
    casual user, 494
    dedicated user, 494
    dialogue tone and terminology, 495
  keyboards, 496–497

User interface *(continued)*
mouse, 497
pointer options, 497
prototyping and, 497–503
dialogue and user interface, 499–500
dialogue charting, 497–499
exercising (or testing) the user interface, 500–501
state transition diagram, 497
user feedback during, 500–501
SoundStage case study, 483–484
styles for, 484–493
cascading menus, 486–488
direct manipulation, 492–493
iconic menus, 489
instruction sets, 489–491
menu bars, 484–485
menu selection, 484–489
mnemonic syntax, 490
natural language syntax, 491
pop-up menus, 488–489
pull-down menus, 485–486
question-answer dialogues, 491–492
structured English, 490
User(s) *Persons, or groups of persons, for whom the systems analyst builds and maintains business information systems and computer applications. A common synonym is* **client.**, **16**–17; *see also* **Systems users ; User interface**
feedback from, 500–501
JAD and, 664
system development participation and, 73–74
user-centered approach to systems development, 637

V

Valacich, Joseph S., 189n
Validation testing *Runs the system in a live environment using real data. This is sometimes called beta testing.*, **568–569**
Variable-length record structures, 399
Vatarli, Nicholas P., 261
Vendors, 20
awarding contract and debriefing, 333–334
claims and performance validation, 330–332
evaluation and ranking of, 332–333
proposal solicitation and, 328–330
Verification testing *Runs the system in a simulated environment using simulated data. This simulated test is sometimes called alpha testing. The simulated test is primarily looking for errors and omissions regarding end-user and design specifications that were specified in the earlier phases but not fulfilled during construction.*, **568**
Version control *A process whereby a librarian (usually software-based) keeps track of changes made to programs. This allows recovery of prior versions of the programs in the event that new versions cause unexpected problems. In other words, version control allows users to return to a previously accepted version of the system.*, 107, 370, **583**
Version control and configuration managers *Software that tracks ongoing changes to software that is usually developed by teams of programmers. The software also allows management to roll back to a prior version of an application if the current version encounters unanticipated problems. Examples include IBM's* SCM *and Intersolv's* PVCS., 107, **370**
VIASOFT, 588
Video *The fastest-growing medium for computer outputs. The on-line display of information on a visual display device, such as a CRT terminal or microcomputer display.*, **468**
Visual aids, 681
Visual programming languages, 22
Vitalari, Nicholas P., 79n

W

Wagner, Gerald E., 531
Walkthrough
coordinator, 683
of projects, 683–685
recorder, 683
Walton, Donald, 671, 671n, 673n, 691
Weinberg, Gerald M., 30n, 169, 628, 641
Weinschenk, Susan, 507
Wetherbe, James, 79, 79n, 115, 169, 261
White, E.B., 688
Whitten, Jeffrey L., 261, 391
Whole/part relationship, 303
Wide area network (or **WAN**) *An interconnected set of LANs, or the connection of PCs over a longer distance—such as between buildings, cities, states, or countries.*, **356**
Wiest, Jerome D., 620
Wilkinson, Nancy M., 551
Windows 3.1 or 95, 57, 59, 60
Windows/Windowing, 493, 496
Wizards, 681
Wood, Jane, 169, 349, 461, 641, 669
Wood, Michael, 20n
Word usage turn-ons and turn-offs, 674–675
Work breakdown structure *A hierarchical decomposition of the project into phases, activities, and tasks.*, **606**–607
Work environment observations, 628–630
Work group information systems *Those designed to meet the needs of a work group. They are designed to boost the group's productivity.*, **41**
databases, 402
Work sampling *A fact-finding technique that involves a large number of observations taken at random intervals.*, **630**
Workplaces, of systems analysts, 11–16
World Wide Web (WWW) *Proposed in 1989 by a group of European physics researchers as a means for communicating research and ideas throughout the organization. It now has evolved to become the primary navigational, information management, and information distribution system, which permits users to easily travel the Internet. It provides the capability to transmit different types of information including sound, video, still images, and text.*, 47, 358–359, 370, **626**
Written reports, 685–689
Wysocki, Robert K., 595, 595n, 620

X

Xerox Corp., 359

Y

Year 2000, systems support and, 589
Yeo, Sarah C., 507
Yourdon, Edward, 16, 16n, 31, 122n, 169, 237, 261, 307, 348, 349, 511, 551

Z

Zachman, John A., 42n, 69, 115, 169, 283, 349, 434
Ziff-Davis Publications, 626, 627
Zoned output *Places text and numbers into designated areas of a form or screen.*, **469**